International Exhibition & Conference for Power Electronics, Intelligent Motion, Renewable Energy and Energy Management (PCIM Europe 2024)

Nuremberg, Germany
11 – 13 June 2024

Volume 3 of 5

ISBN: 978-1-7138-9966-2

Printed from e-media with permission by:

Curran Associates, Inc.
57 Morehouse Lane
Red Hook, NY 12571

Some format issues inherent in the e-media version may also appear in this print version.

Copyright© (2024) by Mesago Messe Frankfurt GmbH
All rights reserved.

Printed with permission by Curran Associates, Inc. (2024)

For permission requests, please contact VDE VERLAG GMBH
at the address below.

VDE VERLAG GMBH
Bismarckstr. 33
P.O.B. 12 01 43
10625 Berlin, Germany

Phone: +49 30 34 80 01 - 0
Fax: +49 30 34 80 01 - 9088

kundenservice@vde-verlag.de

Additional copies of this publication are available from:

Curran Associates, Inc.
57 Morehouse Lane
Red Hook, NY 12571 USA
Phone: 845-758-0400
Fax: 845-758-2634
Email: curran@proceedings.com
Web: www.proceedings.com

TABLE OF CONTENTS

VOLUME 1

KEYNOTE

K01   AI BETWEEN HYPE AND INDUSTRIAL-GRADE - THE IMPACT OF AI ON THE ENTIRE POWER ELECTRONICS LIFECYCLE.. 1
Rolf Hellinger

K02   INFRASTRUCTURE REQUIREMENTS FOR ELECTRIFIED HEAVY GOODS TRANSPORT IN GERMANY AND THE EU.. 7
Martin Wietschel

K03   CHALLENGES AND SOLUTIONS TO POWER LATEST PROCESSOR GENERATIONS FOR HYPER SCALE DATACENTERS 15
Gerald Deboy

GAN RUGGEDNESS

OP001   AN IMPROVED ULTRAFAST DESATURATION-BASED PROTECTION SCHEME FOR GAN HEMT 19
Juncheng Lu

OP002   THE PERFORMANCE OF A GAN EMODE HEMT IN SURGE CURRENT SCENARIOS SUCH AS THE ACTIVE SHORT CIRCUIT.. 24
Dominik Nehmer

OP003   GATE RESISTANCE EFFECT ON SHORT-CIRCUIT ROBUSTNESS OF P-GAN HEMTS 34
Mohamed Lemine Dedew

ADVANCED PACKAGING TECHNOLOGIES

OP004   NEURAL NETWORK ASSISTED NUMERICAL SIMULATION BENCHMARKING FOR ELECTRIC VEHICLE THERMAL MANAGEMENT SYSTEM 40
Ekin Alp Bicer

OP005   RELATIONSHIP BETWEEN POROSITY IN CU SINTERED BONDING AND BONDING RELIABILITY 49
Hideo Nakako

OP006   HIGH THERMAL DURABILITY OF THIN COPPER DIE-ATTACH LAYERS AND FINITE ELEMENT MODEL SIMULATION.. 56
Takaaki Eyama

THERMAL CYCLING RELIABILITY

OP007   THERMAL SHOCK TEST LIFETIME IMPROVEMENT WITH OPTIMIZED ADHESIVE STRENGTH BETWEEN EPOXY RESIN AND COPPER 62
He Kangjia

OP008   POWER CYCLING RELIABILITY AND FAILURE MODE ANALYSIS OF POL 67
Kenichi Koi

OP009   ACCELERATED POWER CYCLING OF GAN HEMTS USING SWITCHING LOSS
AND FAST TEMPERATURE MEASUREMENT 74
Wing Tai Leung

HIGH POWER CONVERTERS

OP010   CONTROL OF AN MMC-BASED HYBRID TRANSFORMER WITH STAR-POINT
VOLTAGE INJECTION 84
Rui Wang

OP011   PROTECTION AND CONTROL OF A DUAL MMC MEDIUM VOLTAGE SUPPLY 93
Max Dupont

OP012   STATION POWER ELECTRONICS CONVERTER WITH HIGH THERMAL
ENDURANCE TO POLE-TO-POLE SHORT CIRCUITS FOR LVDC DISTRIBUTION GRID 103
Frédéric Reymond-Laruina

GATE DRIVERS

OP013   SUPPRESSION OF OSCILLATIONS IN A SIC BRIDGE-LEG USING A CUSTOM
SINGLE-CHIP DIGITAL ACTIVE GATE DRIVER WITH 2×255 STRENGTH LEVELS 113
Qilei Wang

OP014   SIC MOSFET SHORT-CIRCUIT PROTECTION: A FASTER SOFT SHUT DOWN
METHOD FOR GATE DRIVERS 121
Julien Weckbrodt

OP015   PARAMETER IDENTIFICATION: GATE SENSOR FOR POWER TRANSISTOR
TOLERANCE COMPENSATION IN ADVANCED GATE DRIVER ICS 128
Christopher Wille

ADVANCED CONTROL TECHNIQUES ON ELECTRICAL DRIVES I

OP016   AN INNOVATIVE HIGH-SPEED TRACK RANGE RESTART STRATEGY FOR
PERMANENT MAGNET SYNCHRONOUS MOTOR 135
Anna Corbitt

OP017   STEADY-STATE ERROR REDUCTION OF REINFORCEMENT LEARNING BASED
INDIRECT CURRENT CONTROL OF PERMANENT MAGNET SYNCHRONOUS
MACHINES 140
Tobias Schindler

OP018   PERFORMANCE COMPARISON OF USING SHUNT-BASED AND INTEGRATED
CURRENT SENSING FOR SENSORLESS FIELD-ORIENTED CONTROL 150
John Emmanuel Tan

GAN CONVERTERS

OP019   DESIGN OF HIGH-POWER INVERTER WITH 12 PARALLEL GAN DEVICES 161
Takashi Sawada

OP020　OVER 99.7% EFFICIENT GAN-BASED 6-LEVEL CAPACITIVE-LOAD POWER
CONVERTER 167
    Stefan Mönch

OP021　CASCADED PRIMARY-SIDE-ONLY CONTROL OF A COMPACT 2 MHZ 500 W
WIRELESS POWER TRANSFER SYSTEM... 174
    Tim Krigar

ADVANCED MATERIALS AND TECHNOLOGIES

OP022　POWER MODULE EVALUATION USING ULTRA HIGH HEAT DISSIPATION AND
HIGH HEAT RESISTANCE RESIN SHEET CONTAINING CARD HOUSE TYPE BORON
NITRIDE FILLER 180
    Ayano Imai

OP023　INVESTIGATING TEMPERATURE DEPENDENT WARPAGE IN METAL CERAMIC
SUBSTRATES FOR POWER ELECTRONICS DEVICES 190
    Benjamin Fabian

OP024　DEGRADATION MODE ANALYSIS OF DIFFERENT BONDING TECHNOLOGIES
OF SIC POWER SEMICONDUCTORS STRESSED BY ACTIVE POWER CYCLING 197
    Rasched Sankari

CHARGING STATION TECHNOLOGY

OP025　IMPLEMENTATION AND VERIFICATION OF A 50KW OPPORTUNITY WIRELESS
CHARGER DESIGN 205
    Carlos Costas Sos

OP026　PERFORMANCE EVALUATION OF SILICON-BASED 3-LEVEL VIENNA
RECTIFIER IN ISOPLUS SMPD PACKAGE 214
    Karsten Haehre

OP027　PERFORMANCE ANALYSIS OF A 25-KW SIC-BASED DUAL ACTIVE BRIDGE
CONVERTER BASED ON PARALLEL-CONNECTED DEVICES 222
    Francesco Porpora

MODELLING AND MONITORING

OP028　SEMICONDUCTOR CHIP MODELS ARE THE KEY FOR ENABLING VIRTUAL
DESIGN AND OPTIMIZATION WORKFLOWS OF POWER ELECTRONIC SYSTEMS 230
    Stefan Haensel

OP029　IMPROVED RESONANT FREQUENCY-BASED PARASITIC INDUCTANCE
ESTIMATION METHOD FOR SIC MOSFET HALF-BRIDGE CIRCUIT 238
    Hongpeng Zhang

OP030　FAST SIMULATOR WITH INVERTER TEMPERATURE ESTIMATION FOR
TRACTION EDRIVES IN VEHICLES SUBJECTED TO DRIVING CYCLES......................... 248
    Simone Giuffrida

SOLID STATE TRANSFORMERS

OP031   A NEW FAMILY OF THREE-PHASE-UNFOLDER-BASED MVAC-LVDC SOLID-STATE TRANSFORMERS 254
Jonas Huber

OP032   VOLTAGE BALANCING OF A SPLIT-CAPACITOR IGCT 3L-NPC LEG FOR THE RESONANT DC TRANSFORMER 264
Renan Pillon Barcelos

OP033   COMPARATIVE ANALYSIS OF UNIDIRECTIONAL HIGH STEP-UP CONVERTERS FOR MEDIUM VOLTAGE APPLICATIONS 274
Stefan Subotic

ADVANCED CONTROL TECHNIQUES ON ELECTRICAL DRIVES II

OP034   STARTUP BEHAVIOR OF HARMONIC SUPPRESSION IN ELECTRICAL MACHINES USING ITERATIVE LEARNING CONTROL AND NEURAL NETWORKS......................... 284
Annette Mai

OP035   ANALYTICAL APPROACH OF THE VECTOR CURRENT CONTROL FLUX-WEAKENING STRATEGY FOR PERMANENT MAGNET SYNCHRONOUS MACHINES 290
Oriol Subirats Rillo

POWER ELECTRONICS FOR E-MOBILITY

OP036   INVESTIGATION ON DIRECT LIQUID COOLING DESIGN OF POWER MODULES WITH FLAT BASEPLATE FOR AUTOMOTIVE APPLICATION.. 298
Nobuhide Arai

OP037   A NOVEL APPROACH FOR AFFORDABLE ELECTRIC VEHICLES BASED ON DUAL 48V BATTERY SYSTEM WITH MULTI-FUNCTIONAL 3-LEVEL CONVERTER......................... 305
Radovan Vuletic

OP038   AN INNOVATIVE 3-LEVEL SOLUTION FOR AUTOMOTIVE APPLICATIONS: EMPACK 315
Pranav Panchal

OP039   GATED RECURRENT UNITS-ASSISTED STATE-SPACE MODELING FOR ELECTRIC VEHICLE TEMPERATURE PREDICTION 322
Xinyuan Liao

OP040   NOVEL BIDIRECTIONAL SINGLE-STAGE ISOLATED 600-V GAN M-BDSBASED SINGLE/THREE-PHASE-OPERABLE EV ON-BOARD CHARGER 330
Sven Weihe

ENCAPSULATION MATERIALS

OP041   APPLICATION-SPECIFIC INVESTIGATION OF INORGANIC POTTING MATERIAL IN DRIVE TRAINS 338
Soenke Fleck

OP042 THE INFLUENCE OF THE GLASS TRANSITION TEMPERATURE OF EPOXY
MOLD COMPOUNDS ON THE RELIABILITY OF A SEMICONDUCTOR DEVICE 343
  Stefan Schwab

OP043 CORROSION RESISTANT PACKAGING FOR POWER SEMICONDUCTOR
MODULES - MODIFIED INSULATION MATERIALS FOR CONTAMINATED
ENVIRONMENTS 351
  Michael Hanf

OP044 INVESTIGATION OF INORGANIC ENCAPSULATION MATERIALS IN POWER
ELECTRONIC SYSTEMS FOR HIGH POWER DENSITY APPLICATIONS 361
  Stefan Behrendt

OP045 CHARACTERIZATION OF THERMALLY AGED SILICONE GELS FOR POWER
SEMICONDUCTOR MODULES... 369
  Sonja Madloch

POWER QUALITY

OP046 A COORDINATED CONTROL OF HYBRID SINGLE-PHASE AC/DC MICROGRIDS
BASED ON THE NATURAL HARMONIC INJECTION CONCEPT 378
  Mehdi Baharizadeh

OP047 A HIGH-POWER DENSITY SIC BASED TP PFC WITH HIGH-FREQUENCY RIPPLE
CANCELLATION LEG 383
  Serkan Dusmez

OP048 HIGH FREQUENCY ACTIVE FILTER FOR AC-DC HIGH POWER CONVERTERS 390
  Sarah Sifoune

OP049 LABORATORY SETUP FOR ACCURACY INVESTIGATION OF ELECTRICITY
METERS AND MONITORS UNDER INDUSTRY-TYPICAL OPERATING CONDITIONS 397
  Matthias Schmidt

GRID CONNECTED CONVERTERS

OP050 REAL-TIME EVALUATION OF WEIGHTING FACTORLESS PREDICTIVE
CONTROL OF LCL FILTER EQUIPPED GRID-SIDE CONVERTERS USING SORTING
NETWORKS.. 403
  Kristóf Bándy

OP051 RELAXED ROBUST CONTROL WITH PRAGMATIC SHORTAGE OF PASSIVITY
FOR WIND, STORAGE AND PV POWER CONVERTERS 411
  Sergio De Lopez Diz

OP052 AN EFFECTIVE DC VOLTAGE REGULATION OF ACTIVE FRONT-END
RECTIFIER THROUGH MODEL PREDICTIVE CONTROL.. 419
  Mobina Pouresmaeil

OP053 BI-DIRECTIONAL 11KW MULTI-LEVEL ACTIVE-NEUTRAL-POINT-CLAMPED
AC-DC CONVERTER USING 600V/750V SI SUPER-JUNCTION AND SIC MOSFETS FOR
HIGH-EFFICIENCY AND HIGH-DENSITY APPLICATIONS.. 424
  Mengxing Chen

OP054   A STUDY OF GRID-FORMING INVERTER CONTROL STRATEGY FOR FAULT-RIDE-THROUGH CAPABILITY.. 433
Hirofumi Uemura

PASSIVE COMPONENTS

OP055   FILM CAPACITORS FOR HIGH TEMPERATURE AC-DC INVERTER APPLICATIONS.. 440
Adel Bastawros

OP056   LOSS REDUCTION IN HF-TRANSFORMERS USING LAMINATED FERRITE E-CORES... 447
Lukas Reißenweber

OP057   MULTIGAP TOROIDAL TRANSFORMER AND INDUCTORS FOR OVERCOMING FRINGING LOSSES IN HIGH FREQUENCY CONVERTERS......................... 456
Pau Colomer

OP058   STUDY ON SAMPLE GEOMETRIES FOR FERRITE CHARACTERISATION IN THE MHZ RANGE.. 463
Till Piepenbrock

OP059   FEM-SUPPORTED AND NON-DESTRUCTIVE MAGNETIC CHARACTERIZATION METHOD FOR NON-LAMINATED STEEL.. 472
Stefan Tobler

DRIVES FOR HIGH DEMANDING APPLICATIONS

OP060   HIGHLY-COMPACT BEARINGLESS AXIAL-FLUX MOTOR FOR A PEDIATRIC IMPLANTABLE FONTAN BLOOD PUMP   480
Andreas Horat

OP061   A NOVEL PERMANENT MAGNET SYNCHRONOUS MOTOR DRIVE FOR REACTION WHEELS IN SATELLITES.. 490
Baris Colak

OP062   EXPLORING HIGH FREQUENCY OPERATION OF MOTOR DRIVES: PRACTICAL INSIGHTS ON EFFICIENCY AND LOSS.. 497
Asantha Kempitiya

OP063   HIGH POWER DENSITY SYSTEM DESIGN FOR GAN-BASED LV MOTOR DRIVES.. 502
Marco Cannone

OP064   DESIGN OF GAN TRANSISTOR BASED VARIABLE SPEED DRIVE INVERTER WITH OUTPUT VOLTAGE FILTERING.. 510
Kaspars Kroics

IGBT

OP065   THE 8TH GENERATION LV100 IGBT MODULE WITH HIGHER CURRENT RATING.. 518
Daichi Otori

LOSS REDUCTION BY LAMINATIGN FERRITE E CORES.. 525
Lukas Reißenweber

OP066   NEW PLANAR 4.5 KV SPLIT-GATE (SG) SI-IGBT DEVICE FOR IMPROVED
SWITCHING CHARACTERISTICS AND HIGH FREQUENCY OPERATION 534
Gaurav Gupta

OP067   4.5 KV DOUBLE-GATE REVERSE-CONDUCTING PRESS-PACK IEGT 543
Satoshi Yoshida

DEVICE CONCEPTS

OP068   EVALUATION OF A 3 KV POLARIZATION SUPERJUNCTION GAN HEMT........................... 549
Alireza Sheikhan

OP069   MORE THAN 1200 V BREAKDOWN AND LOW AREA-SPECIFIC ON STATE
RESISTANCES BY PROGRESS IN LATERAL GAN-ON-SI AND GAN-ON-INSULATOR
TECHNOLOGIES.. 557
Richard Reiner

OP070   NOVEL 200 V MOSFET TECHNOLOGY PUSHES MOTOR DRIVE INVERTER
EFFICIENCY TO AN UNPRECEDENTED LEVEL.. 564
Mark Thomas

DEGRADATION MECHANISMS

OP071   MOISTURE ROBUST CHIP DESIGN - IMPROVED EDGE-TERMINATIONS FOR
HIGH LIFETIME UNDER HIGH HUMID CONDITIONS.. 571
Michael Hanf

OP072   METHOD FOR MEASURING THE INITIAL STATE OF A SOLDER JOINT
DELAMINATION IN A 3D PCB INTEGRATION ASSEMBLY OF SIC MOSFETS................................. 581
Souhila Bouzerd

OP073   GENERIC LIFETIME MODEL FOR WIRE BONDS DEGRADATION IN IGBT
MODULES BASED ON A FRACTURE MECHANICS PARAMETER... 589
Merouane Ouhab

ADVANCED CONVERSION CONCEPTS

OP074   MODULAR COAXIAL POWER CONVERTER FOR HIGH-DENSITY INTEGRATION
INTO MEDIUM-VOLTAGE CABLES 599
Mark Cairnie

OP075   CONTROLLED INDUCTOR BASED BCM BUCK CONVERTERS 608
Ziv Gellman

OP076   INFLUENCE OF VARYING COMMON MODE CHOKE SIZES ON THE
PERFORMANCE AND STABILITY OF AN ACTIVE EMI FILTER.. 615
Patrick Körner

PHOTOVOLTAIC SYSTEMS

OP077   A HIGH EFFICIENCY BATTERY CHARGER WITH MAXIMUM POWER POINT TRACKING FOR MAGNETIC ENERGY HARVESTERS 625
Antonio Miguel Munoz Gomez

OP078   SYMMETRIC FLYING-CAPACITOR BOOST CONVERTER FOR MEDIUM-VOLTAGE PHOTOVOLTAIC APPLICATIONS 635
Luis Alves Rodrigues

OP079   COMPARISON OF SI IGBT, SIC MOSFET AND ADJUSTABLE HYBRID SWITCH PV INVERTERS FOR DIFFERENT GEOGRAPHICAL LOCATIONS 645
Tanya Thekemuriyil

MODEL BASED SYSTEM ANALYSIS

OP080   OPTIMISING A POWER MODULE FOR ELECTRICAL AND THERMAL PERFORMANCE AND SYMMETRY USING EDA TOOLS 655
Wilfried Wessel

OP081   CONDUCTOR-BASED MODELING OF VOLTAGE DISTRIBUTION ALONG A SINGLE-TOOTH WINDING OF ELECTRICAL MACHINES 665
Hujun Peng

OP082   REDUCTION OF PWM HARMONICS WITH CARRIER PHASE SHIFTING IN A DUAL-STATOR PMSM WITH MAGNETIC COUPLED WINDINGS 672
Bünyamin Tekir

VOLUME 2

SIC DEVICES

OP083   THE NEW COOLSIC MOSFET 1200 V G2: ELECTRICAL PERFORMANCE AND COMPACT MODELLING 681
Andreas Huerner

OP084   PARALLELING SIC-POWER-MOSFET BODY DIODES UNDER HARSH SWITCHING CONDITIONS 690
Michael Rauh

OP085   3.3KV SBD-EMBEDDED SIC-MOSFET MODULE FOR TRACTION USE 699
Yoichi Hironaka

OP086   DEAD TIME OPTIMIZATION FOR HIGH POWER SIC MOSFET MODULE IN CONSIDERATION OF PARASITIC COMPONENTS 707
Pham Ha Trieu To

WBG RELIABILITY

OP087   PERFORMANCE INSTABILITY OF 650 V P-GAN GATE HEMT DEVICE UNDER TEMPERATURE-RELATED POSITIVE GATE BIAS STRESSES 717
Renze Yu

OP088   GATE OXIDE RELIABILITY OF CURRENT GENERATION 1.2 KV SIC MOSFETS
UNDER STEP-WISE INCREASED GATE VOLTAGE... 723
Roman Boldyrjew-Mast

OP089   AN ACCELERATED DYNAMIC GATE SWITCHING STRESS TEST CONCEPT OF
SIC MOSFETS AT HIGH DRAIN-SOURCE VOLTAGE (HV-GSS) 731
Clemens Herrmann

OP090   SILICON CARBIDE POWER DEVICE USE IN SPACECRAFT AND AIRCRAFT 739
Akin Akturk

POWER ELECTRONICS FOR E-MOBILITY/ CONTROL

OP091   CURRENT RIPPLE REDUCTION BY COMBINATION OF SI IGBT AND SIC
MOSFETS IN HEAVY DUTY FUEL CELL TRUCKS... 745
Yavuz Gürlek

OP092   EVALUATION OF ACTIVE GATE DRIVERS WITH SWITCHABLE GATE
RESISTORS AND INTERMEDIATE VOLTAGE LEVELS FOR SIC MOSFETS IN WLTC 754
Michael Frank

OP093   PERFORMANCE EVALUATION OF TCM-BASED, ZERO-VOLTAGE SWITCHING
(ZVS) THREE-PHASE INVERTER FOR ELECTRIC VEHICLE DRIVE SYSTEMS 764
Khizra Abbas

OP094   A PARTIAL LOAD THREE-PHASE TRIANGULAR CURRENT MODE
MODULATION CONCEPT WITH AN OPTIMIZED FILTER INDUCTOR FOR HIGH
EFFICIENCY TRACTION DRIVES   774
Bhaskar Chatterjee

DC-DC CONVERTERS I

OP095   GAN VS SI SYNCHRONOUS RECTIFIER FOR LLC CONVERTER 784
Gokhan Sen

OP096   CO-SIMULATION DESIGN OF A GAN-BASED THREE-PHASE LLC CONVERTER
WITH INTEGRATED THREE-PHASE MAGNETICS 791
Jhih-Cheng Hu

OP097   SWITCHING ASSISTING CIRCUIT IMPROVING THE EFFICIENCY OF DC-DC
CONVERTERS BASED ON PIEZOELECTRIC RESONATORS.. 797
Ghislain Despesse

OP098   TRANSFORMER-BASED FIXED-RATIO RESONANT DC-DC CONVERTERS FOR
48V DATA CENTERS 803
Xufu Ren

PFC CONVERTERS

OP099   HIGH-DENSITY 3.3 KW GAN RECTIFIER FOR SERVER APPLICATIONS
COMPRISING A 130 KHZ TOTEM-POLE PFC AND A 500 KHZ LLC... 812
Manuel Escudero Rodriguez

OP100 ADDRESSING POWER SWITCH TECHNOLOGY SELECTION SI/SIC/GAN IN HIGH EFFICIENCY ZVS-PFC RESONANT CONVERTERS 822
Marco Torrisi

OP101 BUCK-TYPE CURRENT UNFOLDING CONVERTER WITH DISCONTINUOUS CONDUCTION MODE IN ULTRA-LOW POWER-FACTOR OPERATION 831
Tomoyuki Mannen

OP102 GAN BASED BI-DIRECTIONAL 6.6KW INTERLEAVED TOTEM-POLE PFC WITH 13KW/L POWER DENSITY AND HIGH EFFICIENCY 837
Juncheng Lu

SIC MODULES

OP103 THE DESIGN OF A 2KV 1700A SIC MOSFET DUAL MODULE 843
Jorge Mari

OP104 TECHNOLOGICAL APPROACHES TO HIGH-POWER DENSITY SIC POWER MODULE FOR AUTOMOTIVE 849
Takeshi Tokorozuki

OP105 EXTREMELY COMPACT SIC POWER MODULE FOR EV TRACTION INVERTERS IN THE 250 KW CLASS 855
Raffael Schnell

OP106 BENEFITS OF .XT INTERCONNECTION TECHNOLOGY FOR 3.3 KV XHP 2 MODULE WITH 3.3 KV COOLSIC MOSFET 863
Matthias Bürger

ADVANCED COOLING

OP107 LARGE-AREA BONDING WITH LMEE: SUPPRESSION OF THE DEGRADATION OF THE JUNCTION-TO-WATER THERMAL RESISTANCE IN POWER MODULES 870
Yo Mochizuki

OP108 ACTIVE THERMAL CONTROL OF SIC MOSFETS UTILIZING TRANSIENT THERMAL CHARACTERIZATION 875
Varaha Satya Bharath Kurukuru

OP109 THERMAL MANAGEMENT SOLUTIONS BY ADDITIVE MANUFACTURING – POWDER BED FUSION AND DIFFUSION BONDING 883
Simon Jahn

OP110 ADVANCED PUMPED TWO-PHASE COLD PLATE FOR COOLING POWER ELECTRONICS 888
Elizabeth Seber

DC-DC CONVERTERS II

OP111 FEASIBILITY STUDY OF HIGH-POWER DENSITY ISOLATED CLLC DC-DC INTERFACE WITH WIDE RANGE OF VOLTAGE/CURRENT REGULATION 893
Oleksandr Husev

OP112   DC-BIAS REDUCTION IN HIGH-FREQUENCY DUAL ACTIVE BRIDGE DC-DC CONVERTERS THROUGH SLOW DC MEASUREMENTS 903
Patrick Lenzen

OP113   OPTIMIZED CURRENT SHARING TECHNIQUE FOR INTERLEAVED CLLC CONVERTERS FOR MINIMAL OUTPUT CURRENT DISTORTION 909
Martin Gendrin

OP114   PRIMARY-SIDE OUTPUT REGULATION PRINCIPLES IN DYNAMIC MULTI-MHZ INDUCTIVE POWER TRANSFER SYSTEMS AND ISOLATED DC/DC CONVERTERS 916
Ioannis Nikiforidis

SMART GRID

OP115   LOW VOLTAGE DC-GRIDS WITH GALVANIC ISOLATION: SYSTEM DISCUSSION, EFFICIENCY AND PERFORMANCE COMPARISON TO AC-FEEDING 926
Lukas Fräger

OP116   IMPLEMENTATION AND EXPERIMENTAL EVALUATION OF AN ADAPTIVE DC GRID CONTROLLER FOR DECENTRALISED GRID CONTROL 933
Steffen Menzel

OP117   DEMONSTRATING THE EFFECTIVENESS OF A DC SOLID-STATE CIRCUIT BREAKER'S FAST RESPONSE TIME 942
Ehab Tarmoom

OP118   MODELLING AND SIZING SENSITIVITY ANALYSIS OF A FULLY RENEWABLE ENERGY-BASED ELECTRIC VEHICLE CHARGING STATION MICROGRID 949
David A. Stone

MEASUREMENT TECHNIQUES AND METHODS

OP119   LED POWERED ROTOR TELEMETRY SYSTEM 958
Raphael Beyerle

OP120   'INFINITY GATE SENSOR': A DIFFERENTIAL MAGNETIC FIELD SENSOR FOR MEASURING GATE CURRENT OF SIC POWER TRANSISTORS 966
Yushi Wang

OP121   CHARACTERISING WIDE BANDGAP POWER MODULES: VALIDATING THE M-SHUNT CONCEPT FOR HIGH-POWER APPLICATIONS IN THE KILOAMPERE RANGE 976
Hauke Lutzen

OP122   CHARACTERIZATION OF POWER-MODULE PARASITICS: SUB-NANOSECOND LARGE SIGNAL PULSING VS. DOUBLE-PULSE TESTING 986
Gerhard Groos

STATISTICAL VARIATIONS IN THE PARASITIC CAPACITANCE OF A COIL 997
Kevin Talits

HIGH VOLTAGE SWITCHES

PP001   A 4.5 KV FAST RECOVERY DIODE PLATFORM FOR HIGH-CURRENT IGBTS 1002
Jan Vobecky

PP002   6.5 KV INNOVATIVE SILICON POWER DEVICE (I-SI) MODULE WITH HIGH POWER DENSITY AND LOW LOSS BY STORED CARRIER CONTROL 1007
Takashi Hirao

PP003   HIGH CURRENT DENSITY 4.5KV PRESSPACK IGBTS PUSH SOA LIMITS 1013
Hossein Davoodi

PP004   2.5KV IGBT MODULE WITH HIGH RELIABILITY FOR RENEWABLE APPLICATIONS 1018
Akiyoshi Masuda

PP005   NEW GENERATION 4.5KV IGCT AND FAST RECOVERY DIODE FOR RAILWAY POWER SUPPLY APPLICATIONS 1025
Umamaheswara Reddy Vemulapati

PP006   NEXT GENERATION 4.5 KV IGBT-ONLY STAKPAK MODULE WITH REDUCED LOSSES AND HIGH TEMPERATURE CAPABILITY 1031
Jeremy Jones

THERMAL MODELLING AND SIMULATIONS

PP007   FINITE ELEMENT ANALYSIS OF THE UPSCALING OF WARPAGE AND BIFURCATION HYSTERESIS LOOPS: FROM CU/SI DIE TO LARGE WAFERS 1039
Vincenzo Vinciguerra

PP009   MAXIMUM JUNCTION TEMPERATURE SIMULATION AND VALIDATION FOR THE HOT SPOT IN MULTI-CHIP SIC POWER MODULE 1046
Wonjin Dylan Cho

PP010   INTEGRATION OF CFD-SIMULATION RESULTS IN PLECS USING LOOKUP TABLES 1051
Simon Cepin

PP011   PCB ONLY THERMAL MANAGEMENT TECHNIQUES FOR EGAN FETS IN A HALF-BRIDGE CONFIGURATION 1057
Adolfo Herrera

HIGH POWER DENSITY DESIGNS

PP013   FROM 4X TO 3X STPAK – OPTIMIZATION FOR A MORE COMPACT EV TRACTION INVERTER SOLUTION 1065
Vittorio Giuffrida

PP014   A MULTI-OBJECTIVE STRUCTURAL OPTIMIZATION METHOD BASED ON MULTI-PHYSICS SIMULATIONS FOR POWER MODULE 1072
Baihan Liu

PP015   HOLISTIC APPROACH TO MAXIMIZE LIFETIME AND POWER DENSITY IN HIGH POWER SEMICONDUCTOR MODULES 1077
Martin Schulz

PP016   REGULATED HIGH DENSITY SWITCH CAPACITOR TOPOLOGY 1082
Pierrick Ausseresse

PP017 SILICON INTERPOSER AS A SUBSTRATE FOR POWER MODULES WITH HIGH POWER DENSITY AND SUPERIOR THERMAL PERFORMANCE 1087
Ahmed Ammar

SPECIAL CONVERTER APPLICATIONS

PP018 ANALYTICAL MODELING AND STABILITY CHARACTERIZATION OF A DAMPED VSCC CM ACTIVE EMI FILTER FOR SINGLE- AND THREE-PHASE AC-DC APPLICATIONS 1092
Timothy Hegarty

PP020 A REPETITIVE HIGH VOLTAGE NANOSECOND PULSE GENERATOR: FIRST PROTOTYPE DESIGN AND TEST RESULTS 1101
Serge Gavin

PP021 FREQUENCY SHIFT KEYED DUAL SIDE CONTROL OF INDUCTIVE POWER TRANSFER: AN APPLICATION OF TALKATIVE POWER CONVERSION 1105
Hamzeh Beiranvand

PP022 STUDY OF A MULTI-ACTIVE BRIDGE CONVERTER FOR A DOMESTIC ELECTRICAL GRID 1113
Abdennour Merrouche

INTEGRATION TECHNOLOGIES AND RELIABILITY DESIGN

PP023 FABRICATION DEVELOPMENT FOR GATE DRIVER EMBEDDED DOUBLE-SIDED COOLING SIC POWER MODULE FOR ELECTRIC VEHICLE APPLICATION 1123
Anna Corbitt

PP024 PRINTED CIRCUIT EMBEDDING OF PREPACKAGED 150V POWER MOSFETS IN A PORTABLE WELDING APPLICATION 1128
Thomas Gebhard

PP025 PROCESS CHALLENGES AND PROGRESS TOWARDS DIRECT CONNECTION OF AUTOMOTIVE POWER MODULES (TMM) TO HEATSINK 1133
Indrajit Paul

PP026 OPTIMIZING PCB STACKUPS FOR ENHANCED GAN TRANSISTOR PERFORMANCE IN HIGH-POWER APPLICATIONS 1139
Philipp Czerwenka

PP027 NEW GENERATION CERAMIC SUBSTRATES – KEY COMPONENTS FOR POWER ELECTRONIC APPLICATIONS: PROCESSING AND CHARACTERIZATION 1147
Stefanie Schindler

PP028 AI-ENHANCED VACUUM REFLOW OVEN: PRECISION CONTROL FOR RELIABLE LARGE-AREA SOLDERING 1152
Chih Hui Lee

PP030 CORROSION-COMPATIBLE DRIVE ELECTRONICS FOR ELECTRIC VEHICLES AND INDUSTRIAL POWER MODULES 1158
Tom Petzold

PP031 EVALUATING THE SAFETY ISOLATION OF THE PACKAGE IN AN INTEGRATED POWER DEVICE 1168
Thomas Anthony Capobianco

CONTROL METHODS I

PP032 FLEXIBLE CONTROL SYSTEM FOR MODULAR ONE-PHASE INTERLEAVED GAN-BASED TOTEM POLE PFC USING REAL-TIME HARDWARE 1174
Oleksandr Solomakha

PP033 A PEAK CURRENT MODE CONTROL METHOD FOR PFC 1180
Sean Yu

PP034 ADAPTIVE RESONANT CONTROLLER FOR A THREE-PHASE PFC CONVERTER FOR AN ON-BOARD CHARGE APPLICATION 1185
Rami Troudi

PP035 SYNTHESIS OF A FIELD ORIENTED CONTROL ALGORITHM BY USING TWO DIFFERENT POLE-ZERO COMPENSATION APPROACHES.................................1192
Marco Denk

PP037 AVERAGE CURRENT MODE CONTROL AND ITS LOOP DESIGN 1200
Niklas Schwarz

PP038 NOVEL POWER FEED-FORWARD REGULATION FOR DUAL STAGE PFC+DCDC CONVERTERS 1207
Alfredo Medina-Garcia

HIGH POWER AC-DC AND DC-AC CONVERTER

PP039 22 KW BI-DIRECTIONAL WALL-BOX CHARGER WITH 1200 V SIC MOSFET....................1212
Sanbao Shi

PP040 DYNAMIC SWITCHING FREQUENCY SELECTION FOR EFFICIENCY OPTIMIZATION IN ON-BOARD CHARGER PFC STAGE BASED ON NOVEL SIC MOSFET POWER MODULE...1217
Giuseppe Aiello

PP041 DESIGN AND OPTIMIZATION OF SIC-BASED 11KW MOTOR DRIVE WITH HIGH EFFICIENCY...1222
Iris Liu

PP042 MODEL DESIGN DEVELOPMENT FOR FALSE TURN-ON CHARACTERIZATION IN SIC-BASED ACTIVE T-TYPE CONVERTER CONSIDERING ALL PARASITICS 1227
Amir Babaki

PP043 EFFICIENCY INVESTIGATIONS OF AN AUXILIARY RESONANT COMMUTATED POLE INVERTER...1233
Markus Zocher

PP044 A NOVEL HYBRID TWO-STAGE AC-DC CONVERTER WITH SOFT-SWITCHED CCM PFC STAGE FOR EVS CHARGING APPLICATIONS.................................1242
Lei Wang

PP045   A METHOD FOR TUNING LEAKAGE INDUCTANCE IN TRANSFORMERS 1249
Rosemary O'Keeffe

PP046   LOW COST HIGH DENSITY 300W/20V AC-DC CONVERTER ENABLED BY GAN
POWER ICS.. 1254
Tom Ribarich

PP047   25KVA GRID-TIED BI-DIRECTIONAL T-TYPE INVERTER WITH HIGH-
EFFICIENCY AND HIGH-POWER DENSITY USING SIC MOSFETS.................................. 1259
Tamanna Bhatia

PP048   COST-EFFECTIVE EFFICIENCY ENHANCEMENT IN AC-DC CONVERTERS: A
STUDY ACROSS THE FULL LOAD CYCLE 1264
Sebastian Gick

E-MOBILITY TRACTION I

PP049   NEXT GENERATION POWER MODULE WITH PARALLEL CONNECTED SIC
MOSFETS FOR BEV TRACTION INVERTERS.. 1272
Kohei Tanikawa

PP051   INVESTIGATION OF COMMON SOURCE FEEDBACK IN SIC POWER MODULES
REGARDING PERFORMANCE AND SHORT CIRCUIT ROBUSTNESS 1277
Dominik Ruoff

PP052   HYBRIDPACK DRIVE POWER MODULES WITH SIC-MOSFET'S AND
MONOLITHIC RC- SNUBBER CHIPS FOR OPTIMIZED POWER DENSITY......................... 1283
Andre Uhlemann

PP053   ROBUST AUXILIARY POWER SUPPLY FOR EVS BASED ON INNOVATIVE
STI2GAN 650V IC.. 1289
Federica Cammarata

PP054   IMPACT OF VARIOUS SILICON DIODES ON THE HYBRID SWITCH INVERTER 1297
Michael Walter

PP055   ADVANCED PULSE SEQUENCE FOR SALIENCY-BASED HIGH-ACCURATE
ROTOR POSITION ESTIMATION OF RAILWAY TRACTION LOCOMOTIVE MOTORS 1307
Markus Vogelsberger

CONTROL TECHNIQUES

PP056   OPTIMIZED HALF-BRIDGE GATE-DRIVE WITH LOW TIME-SKEW FOR RC-
IGBTS AND SIC-MOSFET DEAD-TIME CONTROL 1315
Jan Fuhrmann

PP057   DESIGN OF A TRACTION INVERTER BASED ON PCB-EMBEDDED GAN
DEVICES 1322
Maurizio Tranchero

PP058   OPTIMIZING ELECTRIC VEHICLE PERFORMANCE WITH GAN DESIGN........................ 1330
Andrew Patterson

PP059   FAST ANALYTICAL CALCULATION OF THE MAGNETIC FIELD IN PERMANENT MAGNET SYNCHRONOUS MACHINES WITH FLUX BARRIERS INCLUDING SATURATION 1336
    Martin Ackermann

PP060   MODELING AND CONTROL OF LCL FILTERED 3L-VSCS IN INTERLEAVED TOPOLOGY 1346
    Adeel Jamal

PP062   ENHANCING SAFETY AND EFFICIENCY FOR ISOLATED PLC I/O DESIGNS WITH SPI DAISY CHAIN 1352
    Travis Lenz

VOLUME 3

PP063   COST-EFFECTIVE METHOD TO DISCHARGE DC LINK CAPACITORS WITH SIC POWER MODULES 1361
    Paul Kanatzar

POWER QUALITY

PP064   A STUDY ON CIRCULATION CURRENT IN PARALLEL OPERATION OF TRANSFORMER LESS UPS 1368
    Koji Kato

PP065   DESIGN CHALLENGES AND CONSIDERATIONS FOR GATE DRIVERS OF SIC MOSFETS AND THEIR TESTING 1374
    Niranjan Hegde

PP066   A PORTABLE EFFICIENCY CHARACTERIZATION SETUP FOR TECHNOLOGY DEMONSTRATION OF POWER MODULES 1380
    Sebastian Tengvall

PP067   FAST EME CHARACTERIZATION OF BARE-DIE SIC MOSFETS 1385
    Robert Kragl

PP068   THEORETICAL COMPARISON OF COMPONENT-RELATED MEASUREMENT METHODS OF PHOTOVOLTAIC INVERTERS FOR LONG-TERM TESTING 1393
    Niclas Reitz

DYNAMIC TRANSIENTS AND RELIABILITY OF HIGH-VOLTAGE SILICON & 4H-SIC BIPOLAR JUNCTION TRANSISTORS UNDER AVALANCE AND SHORT-CIRCUITS 1402
    Mana Hosseinzadehlish

PP069   POWER CYCLING TEST OPTIMIZATION TOWARD RELIABILITY ASSESSMENT OF SINTERED POWER MODULES 1410
    Robert Graham

PP070   REAL-TIME ESTIMATION AND SENSITIVITY ANALYSIS OF PARASITIC CAPACITANCES IN ELECTRIC DRIVE SYSTEMS 1418
    Mohammadreza Bagheribavaryani

MODELLING AND TESTING

PP071 PARASITIC COMPONENT EFFECTS OF INTERNAL AND EXTERNAL PACKAGE
LEVEL ON SWITCHING PERFORMANCE OF SIC POWER MODULE 1428
Nguyen Nghia Do

PP072 A MULTI-PHYSICS ITERATIVE APPROACH FOR TEMPERATURE ESTIMATION
IN SIC POWER MODULE FOR ELECTRIC VEHICLE 1434
Stefano Orlando

PP073 VOLTAGE BALANCING METHOD FOR SERIES CONNECTION OF 50 SIC
MOSFETS 1441
Antoine Philippe

VOLTAGE BALANCING METHOD FOR SERIES-CONNECTION OF 50 SIC MOSFETS 1449
Antoine Philippe

PP074 A LABORATORY-SCALE MMC-BASED DC SYSTEM WITH RCP AND PHIL
SIMULATION CAPABILITIES 1457
Marc René Lotz

PP075 FILM CAPACITOR STANDARD SERIES DIGITALIZATION: ELECTROMAGNETIC
& THERMAL MODELLING IMPLEMENTATION IN CLARA WEB TOOL 1467
Fernando Aunon

PP076 ACCURACY EVALUATION AND PROPOSED DYNAMIC TUNING PROCEDURE
OF A COMPACT SIC SPICE MODEL 1475
Austin Curbow

PP077 INVESTIGATION OF USE-CASE-DEPENDENT MODELING APPROACH FOR
SWITCHED-MODE POWER CONVERTER FOR LVDC GRID EVALUATION 1485
Melanie Lavery

PP078 AVERAGED MODEL WITH BLOCKING CAPABILITY FOR SOLID-STATE
TRANSFORMERS 1495
Ahmed Meligy

ADVANCED COMPONENTS

PP080 SURFACANT-MODIFIED NANOCOMPOSITE THIN-FILM CAPACITORS 1504
Bartosz Gackowski

PP081 INCREASING ENERGY STORAGE CAPABILITIES OF POWDER CORES BY
ADAPTING THE WINDING AND THE USE OF FRINGING FLUX 1511
Paul Winkler

PP082 PEEC-BASED THERMAL MODELING OF PASSIVE COMPONENTS 1516
Sascha Langfermann

PP083 GALVANICALLY ISOLATED POWER SUPPLY FOR GATE DRIVERS IN HIGH
VOLTAGE APPLICATIONS 1523
Priyanka Ghosh

PP084　FABRICATION TECHNIQUE FOR NOVEL NANOCRYSTALLINE CORES WITH
HIGH SATURATION POLARIZATION AND LOW LOSSES 1532
　　　　Merlin Thamm

PP085　EXCITATION-DEPENDENT TEMPERATURE BEHAVIOR OF THE QUASI-STATIC
HYSTERESIS LOSS ENERGY DENSITY OF N87 FERRITE MATERIAL.................................. 1538
　　　　Jeremias Kaiser

PP087　PASSIVE METHODS LIMITING LEAKAGE CURRENT IN METAL-OXIDE
VARISTOR AS VOLTAGE CLAMPING DEVICE USED DC LOW VOLTAGE POWER
ELECTRONICS-BASED CIRCUIT BREAKERS 1545
　　　　Kenan Askan

GAN DEVICES AND APPLICATIONS

PP088　ESD SOLUTIONS FOR 650V NORMALLY-OFF ALGAN/GAN HEMTS 1555
　　　　Thanh Hai Phung

PP089　A SIMULATIVE STUDY OF MEASUREMENT ERRORS DURING DOUBLE PULSE
TESTING OF GAN DEVICES.. 1561
　　　　Severin Klever

PP090　PARALLEL CONNECTION OF GAN FETS: AN EXPERIMENTAL INVESTIGATION
APPROACH.. 1568
　　　　Marco Palma

PP091　REPETITIVE SHORT CIRCUITS ON 650 V GAN 1574
　　　　Adrien Lambert

PP092　COMPARISON OF SWITCHING LOSSES AND DYNAMIC ON RESISTANCE OF
600 V-CLASS GAN HEMTS... 1584
　　　　André Thönnessen

PP093　PERFORMANCE EVALUATION OF DEADTIME AND GATE RESISTANCE FOR
PARALLEL CONNECTED GAN HEMTS 1590
　　　　Junhyeok Jegal

PP094　REACHING BEYOND 1200V: LATERAL GAN HEMTS FOR HIGH-RELIABILITY
EV AND INDUSTRIAL APPLICATIONS 1598
　　　　Kamal Varadarajan

SIC DEVICES AND TECHNOLOGIES

PP095　SMARTSIC 150 & 200MM ENGINEERED SUBSTRATE: INCREASING SIC POWER
DEVICE CURRENT DENSITY UP TO 30%.. 1604
　　　　Eric Guiot

PP096　DYNAMIC TRANSIENTS IN HIGH-VOLTAGE SILICON AND 4H-SIC NPN
BIPOLAR JUNCTION TRANSISTORS 1610
　　　　Mana Hosseinzadehlish

PP097　AN ADVANCED MULTI-ASPECT PERFORMANCE ANALYSIS OF PLANAR-GATE
1.2 KV SIC POWER MOSFETS 1613
　　　　Anja Katerina Brandl

PP098   SIC MOSFET DIE SORTING AND PARALLEL FOR OPTIMAL MODULE DESIGN 1621
Zhong Ye

PP099   SIMULATION APPROACH FOR RADIATED ELECTRO-MAGNETIC FIELDS
ESTIMATION ON ACEPACK DRIVE SIC POWER MODULE 1627
Andrea Cusumano

CONTROL METHODS II

PP100   EXACT ANALYSIS OF CONTROL-TO-OUTPUT TRANSFER FUNCTIONS OF
PWM-CONVERTERS - A COMPARISON OF TWO METHODS.. 1634
Daniel Breidenstein

PP101   3-LEVEL FLYING CAPACITOR MULTILEVEL TOPOLOGY WITH DELTA-SIGMA
MODULATION 1642
Jannik Maier

PP102   MODEL BASED CONTROLLED POWER CONVERTER TEST PLATFORM.......................... 1651
Dawid Koczy

PP103   EDUCATIONAL HARDWARE TRAINER FOR TEACHING THE DUAL ACTIVE
BRIDGE IN A DC GRID 1658
Peter Van Duijsen

PP104   STUDY OF THE OPERATING PERFORMANCE OF A FCS-MPC-CONTROLLED
MATRIX-CONVERTER FOR PMSM AT DIFFERENT FREQUENCY RATIOS 1664
Robert Zipprich

PP105   ENHANCING REACTIVE POWER CAPACITY IN BATTERY-FED POWER
CONDITIONING SYSTEMS.. 1673
Lucas Araujo

PP106   PULSE SHARING: ACHIEVING HIGH EFFICIENCY AND EXCELLENT REGULA-
TION IN MULTI-OUTPUT FLYBACK POWER SUPPLIES 1680
Xingda Yan

PP107   RELIABILITY-OPTIMIZED SPACE VECTOR MODULATION (RO-SVM) FOR
SEMICONDUCTORS LIFETIME ENHANCEMENT 1686
Amin Rezaeizadeh

INTELLIGENT POWER MODULES

PP108   ANALYSIS AND OPTIMIZATION OF INTERNAL COUPLING INTERFERENCE IN
INTEGRATED SIC POWER MODULE BASED ON DBC 1693
Chenhang Zeng

PP109   MULTISPECTRAL ELECTROLUMINESCENCE SENSING OF SIC MOSFETS FOR
JUNCTION TEMPERATURE AND CURRENT EXTRACTION.. 1703
Lukas Ruppert

PP110   SIC-IPM FOR COMPACT AND ENERGY EFFICIENT LOW-POWER MOTOR
DRIVES 1712
Jongmu Lee

PP111　CONCEPT FOR A GAN-BASED INTELLIGENT MOTOR CONTROLLER WITH INTEGRATED FAILURE PREDICTION FOR THE INVERTER AND THE DRIVE 1717
Christoph Blechinger

PP112　INTRODUCING THE NEW 1200 V CIPOS MAXI IM817 INTELLIGENT POWER MODULE FOR MOTOR DRIVE APPLICATIONS 1724
Kihyun Lee

PP113　THERMAL PERFORMANCE OF INFINEON'S NEW 600 V CIPOSTM MICRO IM241 IPM FOR LOW POWER MOTOR DRIVE SYSTEMS WITHOUT HEATSINK 1732
David Jo

INTRODUCING THE NEEW 1200 V CIPOSTM MAXI IM12BXXXC1 INTELLIGENT POWER MODULE FOR MOTOR DRIVE APPLICATIONS 1737
Kihyun Lee

INTELLIGENT GATE DRIVE UNITS

PP114　AN ADAPTIVE DEAD TIME CONTROL BASED ON SWITCH NODE VOLTAGE DERIVATIVE.. 1745
Lukas Knappstein

PP115　COUPLING COIL DESIGN AND POSITIONING OPTIMIZATION ON NEW HIGH POWER SEMICONDUCTOR MODULE FOR FAST SHORT CIRCUIT DETECTION 1751
Yannick Dumollard

PP116　ENABLING ACTIVE THERMAL CONTROL VIA AN ADAPTIVE MULTI-VOLTAGE GATE DRIVER.. 1759
Tianlong Albert

PP117　INNOVATIVE GATE DRIVE METHOD TRIC3 FOR MOTOR................................... 1765
Hisashi Sugie

PP118　A NEW CLASS OF SOLID STATE ISOLATORS ENHANCES THE RELIABILITY OF SOLID STATE RELAYS 1770
Wolfgang Frank

PP119　A SELF-DRIVING 3-LEVEL ACTIVE GATE DRIVER NETWORK TO CONTROL THE SWITCHING SLEW RATE FOR SIC MOSFETS 1775
Vin Loong Choo

E-MOBILITY TRACTION II

PP121　ANALYSIS OF LONG-TERM RELIABILITY OF SIC IN TRACTION INVERTER CONSIDERING VTH INSTABILITY 1781
Chi Zhang

PP122　EFFICIENT MAPPING OF ON-DEMAND DRIVE LOAD PROFILES ON INVERTER STRESS.. 1788
Zlatko Bosnjic

PP123　EV TRACTION INVERTER OPTIMAL DESIGN IS DOMINATED BY 3-LEVEL ANPC 1797
Timothé Delaforge

PP124   INTRODUCTION OF POWER SEMICONDUCTOR OPTIONS FOR AN EXCITER OF
ELECTRICALLY EXCITED SYNCHRONOUS MOTOR 1804
   Yeriel Bai

PP125   A NOVEL HIGH POWER DENSITY THREE PHASE TRACTION INVERTER
ARCHITECTURE FOR ELECTRIC VEHICLE (EV) APPLICATIONS...................................... 1809
   Yiyang Yan

PP126   A MODULAR DC-LINK CAPACITOR SOLUTION FOR THE MAIN POWERTRAIN
INVERTER OF XEV 1814
   David Olalla

PP127   FAULT IDENTIFICATION TESTING METHODS FOR A COMMERCIAL TRACTION
INVERTER 1821
   Anna Corbitt

PP128   SHORT CIRCUIT ROBUSTNESS FOR TRACTION INVERTERS FROM AN
APPLICATION POINT OF VIEW 1828
   Karl Oberdieck

INVESTIGATIONS OF PARTICULAR SIC DEVICE PHENOMENON

PP129   THE IMPACT OF THE DEADTIME ON THE STABILITY OF 1.2KV SIC MOSFET
BODY DIODE UNDER HARD SWITCHING WITH SYNCHRONOUS RECTIFICATION...................... 1835
   Mohammed Amer Karout

PP130   RC-DC SNUBBER IMPLEMENTATION FOR SUPPRESSION OF DIODE VOLTAGE
PEAK AND RINGING IN A FULL SIC HALF-BRIDGE POWER MODULE 1844
   Emanuela Alfonzetti

PP131   SUB-5 SECOND WIDE-BANDGAP POWER DEVICE CALORIMETRIC
MEASUREMENTS UTILZIING OPTICAL SENSORS AND PELTIER ELEMENTS 1851
   Ruben Schnitzler

PP132   SIC TRENCH MOSFETS IN AVALANCHE MODE WITH RC SNUBBER CIRCUIT............... 1858
   Sebnem Tuncay

PP133   HIGH-FREQUENCY OSCILLATIONS IN SIC MOSFET POWER MODULES
DURING TURN-ON SWITCHING TRANSIENT – ANALYSIS BASED ON SIMULATIONS
AND MITIGATION METHODS... 1865
   Rajani Kumar Thirukoluri

PP134   A DYNAMIC CURRENT BALANCING METHOD USING FULL-COUPLED
INDUCTORS IN PARALLELED GATE BRANCHES.. 1872
   Jianwei Lv

PP135   QUANTITATIVE PERFORMANCE COMPARISON OF LARGE-FORMAT SIC
MOSFET AND SI IGBT MODULES 1878
   Arthur Boutry

THERMAL MANAGEMENT AND ADVANCED COOLING

PP136   SOLDER PREFORM TECHNOLOGY FOR IMPROVED THERMOMECHANICAL
PERFORMANCE IN MOLDED POWER MODULE PACKAGE-ATTACH 1886
   Joseph Hertline

PP138 EFFECT OF FLIP-CHIP DIE-ATTACH ON THE THERMAL BEHAVIOR OF POWER GAAS DIODES 1891
Felix Steiner

PP139 INFLUENCES OF SOLDER DELAMINATION ON THE THERMAL PERFORMANCE IN AUTOMOTIVE TRACTION MODULE 1896
Hansol Seo

PP141 DEVELOPMENT OF A PASSIVE CAPILLARY-PUMPED COOLING SYSTEM FOR HIGH-PERFORMANCE ELECTRONICS 1902
Justin Fey

PP143 ADVANCED COOLING OF POWER ELECTRONICS WITH COPPER COLD SPRAYED ALUMINIUM HEATSINKS & BUSBARS 1907
Michael Dasch

PP144 COLD PLATE DESIGN FOR COOLING LV100 SILICON CARBIDE POWER MODULE PACKAGING 1910
Wahid Cherief

PP145 AN IMPROVED DOUBLE-LAYER SPACER IN DOUBLE-SIDED COOLING POWER MODULE 1917
Linhao Ren

RELIABILITY TESTING

PP146 POWER CYCLING OF 1.7KV MULTI-CHIP POWER MODULES – SIC MOSFETS VS SILICON IGBTS 1923
Nick Baker

PP147 POWER CYCLING CAPABILITY OF DISCRETE SIC MOSFET DEVICES WITH DIFFERENT DESIGNS 1930
Luhong Xie

PP148 MODEL-BASED PARAMETER TUNING OF SEMICONDUCTOR DEVICES IN DC POWER CYCLING TEST 1936
Yi Zhang

PP149 INFLUENCE OF TRANSFER MOLDING ON THE RELIABILITY OF DCM SIC POW-ER MODULES 1942
Jacek Rudzki

PP150 DAMP HEAT BEHAVIOR OF HIGH HEAT CAPACITORS FOR APPLICATIONS IN ELECTRIC VEHICLES 1951
Adel Bastawros

PP151 INFLUENCE OF THE GATE VOLTAGE DURING ON-TIME ON THE POWER CYCLING CAPABILITY OF SIC MOSFETS 1955
Patrick Heimler

PP152 INVESTIGATION OF THE TEMPERATURE MEASUREMENT VIA VSD(T)-METHOD APPLIED TO PARALLELED SIC MOSFET CHIPS DURING POWER CYCLING 1964
Kevin Ladentin

PP153 APPROACHES OF TSEP MEASUREMENTS FOR POWER SEMICONDUCTORS 1969
Philipp Hauenschild

PP154   REALTIME JUNCTION TEMPERATURE ESTIMATION IN SIC POWER MODULES
BASED ON MULTIPLE TSEP ACQUISITION 1978
    Kevin Muñoz Barón

HIGH VOLTAGE WBG DEVICES

PP155   ENHANCED CURRENT MEASUREMENT APPROACH FOR NON-ISOLATED 6.5
KV SILICON CARBIDE MOSFETS 1987
    Xinyuan Du

PP156   NEW 2KV SIC-MOS TECHNOLOGY FOR APPLICATION FIELDS IN THE
INDUSTRIAL LANDSCAPE 1991
    Igor Kasko

PP157   HIGH TEMPERATURE EXPERIMENTAL CHARACTERIZATIONS OF COSS OF 3.3
KV SIC MOSFET FOR MEDIUM VOLTAGE PV APPLICATIONS 1999
    Paul Schmidt

PP158   IMPACT OF GATE CONTROL ON THE SWITCHING PERFORMANCE OF 3.3KV
SBD-EMBEDDED SIC-MOSFET 2006
    Junya Sakai

PP159   COMPARATIVE ASSESSMENT OF OVERLOADABILITY POTENTIAL OF 3.3 KV
SI-IGBTS AND SIC-MOSFET POWER MODULES 2013
    Muhammad Nawaz

PP160   IMPROVED RELIABILITY OF A 2200 V SIC MOSFET MODULE WITH AN EPOXY-
ENCAPSULATED INSULATED METAL SUBSTRATE 2022
    Hiroshi Kono

PP161   PARALLELING 3.3-KV/800-A RATED SIC-MOSFET MODULES – AN
OPTIMIZATION METHOD 2028
    Hiroyuki Irifune

PP162   PERFORMANCE ASSESSMENT OF 10 KV SIC MOSFET AND PIN DIODE IN 3L-
NPC CONVERTER TOPOLOGY 2036
    Renato Amaral Minamisawa

VOLUME 4

PP163   PERFORMANCE EVALUATION OF COOLSIC 2 KV SIC MOSFET DISCRETE IN
1500 V DC LINK SYSTEMS 2041
    Ajith Kumar Sekar

PP164   A NEW 2.3 KV RATED SIC MOSFET MODULE WITH LOW-INDUCTANCE HIGH-
POWER PACKAGE HPNC FOR 1500 VDC APPLICATIONS 2049
    Junya Kawabata

PACKAGING AND INTERCONNECTION MATERIALS

PP166   MECHANISM FOR IMPROVING THE HEAT-RESISTANCE OF ADHESIVE
INTERFACE IN FLEXIBLE PRINTED CIRCUITS 2053
    Keita Suzuki

PP167 A SYSTEMATIC COMPARISON STUDY OF DIFFERENT BONDING
TECHNOLOGIES FOR SUBSTRATE ATTACHMENT OF POWER ELECTRONICS............................ 2060
  Lisheng Wang

PP168 STABILITY OF PRESSURE SINTERED INTERCONNECTS AS A FUNCTION OF
TEMPERATURE AND ENVIRONMENTAL CONDITIONS 2067
  Kentaro Yoshioka

PP169 THE EFFECT OF NANO-CU INTERCONNECTION MATERIALS ON THE
THERMOMECHANICAL PROPERTIES OF SIC DOUBLE-SIDED POWER MODULES 2074
  Suhang Wei

PP170 ALL-IN-ONE-SINTERING: DIE-ATTACH AND SUBSTRATE-ATTACH ON BARE
COPPER IN A PRESSURE ASSISTED SINTERING ONE-STEP PROCESS.. 2082
  Battist Rabay

PP171 SEQUENTIAL MANUFACTURING OF HIGHLY FUNCTIONALIZED THREE-
DIMENSIONAL CERAMIC COMPONENTS FOR POWER ELECTRONICS.. 2088
  Lars Rebenklau

PP173 PARAMETRIC STUDY OF DAMAGE EVOLUTION IN SILVER SINTERED
LAYERS OF DOUBLE SIDED POWER ELECTRONICS MODULES OF ELECTRICAL
VEHICLES.. 2094
  Saeed Akbari

DC-DC CONVERTER I

PP174 TRISTATE MODIFIED BOOST CONVERTER.. 2104
  Johannes Gragger

PP175 COMPARATIVE EVALUATION OF THE CENTER TAPPED BOOST CONVERTER
TOPOLOGY 2112
  Bryan Radix

PP176 COMPARISON OF MULTI-LEVEL TOPOLOGIES TO REDUCE THE
COMPONENTS VOLTAGE STRESSES WHEN POWERED FROM INDUSTRIAL DC GRIDS..............2119
  Katharina Machtinger

PP177 HARD-SWITCHING HIGH-FREQUENCY GAN-BASED DC-DC CONVERTERS
WITH CONCOMITANT DATA TRANSMISSION FUNCTIONALITY 2128
  Abdelmoumin Allioua

PP178 EFFICIENT DESIGN OF HIGH-CURRENT, LOW-OUTPUT VOLTAGE DC-DC
CONVERTERS USING ARTIFICIAL INTELLIGENCE-BASED TOPOLOGY SELECTION
AND OPTIMIZATION 2138
  Thomas Harmand

HIGH POWER DC-DC CONVERTER I

PP180 A SIC BASED 60KW LLC CONVERTER WITH NOVEL TRANSFORMER DESIGN
FOR IMPROVING VOLTAGE BALANCE.. 2146
  Frank Wei

PP181 ANALYSIS OF INVERTER OPERATION MODES OF AN IGBT-BASED ZCS LLC CONVERTER FOR A 2 KW AUTOMOTIVE ON-BOARD DC-DC 2152
Daniel Urbaneck

PP182 DUAL OUTPUT HYBRID CONVERTER FOR 48 V DATA CENTERS: M-HSC....................... 2162
Simone Mazzer

PP183 3.6KW HIGH EFFICIENCY SIC-BASED HV/LV DC-DC CONVERTER FOR EVS 2167
Veera Bharath Chandra Reddy Gandluru

PP184 BIDIRECTIONAL DC-DC TOPOLOGIES COMPARISON FOR 800 V AUTOMOTIVE APPLICATIONS INTEGRATING 650 V GAN-ON-SI DEVICES.. 2175
Ilias Chorfi

PP185 ANALYSIS OF PHASE SHIELDING METHOD BASED ON ?-CR-Y THREE-PHASE INTERLEAVED LLC CONVERTER.. 2182
Jin Wen

PP186 22KW IMS-BASED BIDIRECTIONAL DC-DC CONVERTER USING SURFACE MOUNT SIC MOSFETS FOR OBCS 2185
Hamlin Wang

PP187 COMPARATIVE ANALYSIS OF DC-DC CONVERTERS FOR ELECTROLYZERS USING GEOMETRIC PROGRAMMING 2190
Tim McRae

PP188 DESIGN CONSIDERATION OF BI-DIRECTIONAL CLLLC RESONANT CONVERTER IN ENERGY STORAGE SYSTEMS 2200
Sheng-Yang Yu

SMART-GRID TECHNOLOGIES

PP189 ADAPTIVE FAST CHARGING SYSTEM WITH SECOND LIFE BATTERIES - AN OVERVIEW OF A RESEARCH PROJECT 2208
Lukas Böhning

PP190 PARALLEL OPERATION AND SYNCHRONIZATION OF MICROGRIDS BY USING THE THEVENIN THEOREM 2217
Marius Block

PP192 21 KA SOLID STATE DC BREAKER FOR SUPERGRID INSTITUTE'S HIGH POWER TEST FACILITY... 2227
Christophe Conilh

PP193 DESIGN AND ANALYSIS OF A 50KW SIC-BASED ACTIVE FRONT END WITH A VERY SMALL LINE CHOKE FOR DC-GRIDS.. 2234
Raphael Otte

PP194 INVESTIGATION OF LOAD TRANSITIONS BETWEEN LOADED AND LOAD FREE CONDUCTOR SEGMENTS IN INDUSTRIAL CONDUCTOR SYSTEMS 2240
Jan-Niklas Koch

PP195 A METHOD TO CONTROL VOLTAGE AND POWER FLOW IN A DC GRID 2248
Peter Van Duijsen

ENERGY STORAGE SYSTEMS

PP196   CONSIDERATIONS ON A HIGH-CELL-COUNT CONVERTER-BASED BATTERY STORAGE SYSTEM WITH REDUCED COMMUNICATION EFFORT 2258
Paul Aspalter

PP197   STUDYING CONVERTORS FOR VOLTAGE EQUALIZATION IN ENERGY STORAGE SYSTEM WITH ACTIVE BMS 2268
Dimitar Arnaudov

PP198   CHALLENGES OF HIGH SIDE GATE DRIVER AND DISCONNECT MOSFET FOR BATTERY PROTECTION UNIT DURING START-UP, TURN-OFF AND OVER CURRENT EVENTS 2273
Niranjan Suravarapu Reddy

PP199   ELECTRIC INSULATION COORDINATION TO PREVENT ELECTRIC ARCS IN LITHIUMION BATTERIES 2278
Daniel Chatroux

PP201   BATTERY CHARGER WITH IMPEDANCE SPECTROSCOPY CAPABILITY FOR LI-ION CELLS 2286
Christian Branas

EMC

PP202   EFFICIENCY, VOLUME AND CO2 EMISSIONS IMPACT IN A PFC CONVERTER WITH AN ACTIVE FILTER SOLUTION FOR OBC APPLICATION 2294
Kelly Ribeiro

PP203   ANALYTICAL AND EXPERIMENTAL VALIDATION COMMON MODE FEEDBACK LOOP FOR A THREE-PHASE_LEVEL VIENNA RECTIFIER 2303
Daniel San Laureano Igartuburu

PP204   ROBUSTNESS OF FREQUENCY-DOMAIN TERMINAL MODELING OF ELECTROMAGNETIC INTERFERENCES IN STATIC CONVERTERS 2309
Mehyeddine Singer

PP205   STUDY OF EMI BEHAVIOR OF A 2-LEVEL GAN-INVERTER – SIMULATION AND MEASUREMENT 2316
Benedikt Kohlhepp

COMMON MODE CURRENTS IN RESONANT CIRCUITS GENERATED WITH A DELTA-SIGMA MODULATED VOLTAGE SOURCE INVERTER 2326
Tobias Haas

PP206   ANALYSIS OF COMMON-MODE NOISE GENERATED DUE TO FAST-SWITCHING GAN DEVICES IN TOTEM-POLE PFCS 2334
Serkan Dusmez

PP207   CONDUCTED EMI FROM GAN-BASED 48V TO 12V DC-DC-CONVERTERS FOR AUTOMOTIVE APPLICATIONS 2342
Erik Kampert

ADVANCED DESIGN

PP208   APPLIED DESIGN AUTOMATION FOR FINDING FEASIBLE DESIGNS FOR HIGH-FREQUENCY PLANAR TRANSFORMERS 2350
Rando Raßmann

PP209   FREQUENCY DEPENDENT AREA PRODUCT METHOD 2359
Alfonso Martínez

HIGH RESOLUTION MIXED-SIGNAL PULSE WIDTH MODULATOR FOR HIGH-FREQUENCY DC-DC CONVERTERS 2364
Tim McRae

PP210   DESIGNING A CONTROL LIBRARY FOR GRID-FOLLOWING AND GRID-FORMING POWER INVERTERS 2370
Lars Lindner

PP211   INTELLIGENT OPTIMISATION OF A WIND TURBINE DIGITAL TWIN MODEL 2377
René Reimann

PP212   THERMAL TRANSIENT DIGITAL TWIN MODELLING FOR POWER CONVERTERS 2386
Xianghao Mo

PP213   A DIGITAL TWIN APPROACH TOWARD LIFETIME ANALYSIS AND PREDICTIVE MAINTENANCE OF POWER SEMICONDUCTORS FOR RAILWAY APPLICATION 2394
Emmanuel Batista

INDUCTORS

PP214   SATURABLE FERRITE CORE INDUCTORS IN LCL FILTERS OF THREE-PHASE VOLTAGE SOURCE INVERTERS 2400
Marius Kaufmann-Bühler

PP215   2D COPPER LOSS ANALYTICAL MODEL FOR PLANAR INDUCTOR COMBINING HIGH AND LOW PERMEABILITY MATERIALS 2408
Idriss Nachete

PP216   CNC-MANUFACTURED POWER INDUCTORS WITH EXCELLENT BANDWIDTH FOR MULTI-MEGAWATT CONVERTERS 2416
Thomas Kreppel

PP217   ANALYTICAL EVALUATION OF DIFFERENTIAL MODEL DC EMI FILTER INDUCTORS USING MATERIAL SATURATION COEFFICIENT 2425
Lukas Mueller

PP218   DESIGN AND PERFORMANCE EVALUATION OF AIR CORE INDUCTORS FOR VERY HIGH FREQUENCY POWER CONVERSION 2431
Florentin Salomez

PP220   IMPROVING MULTI-PHASE FERRITE MAGNETICS BY COUPLING FOR MV AND UPS CONVERTERS 2438
Michael Schmidhuber

E-MOBILITY CHARGING

PP221   22-KW BIDIRECTIONAL SINGLE-STAGE DIRECT-AC-AC POWER CONVERSION ON-BOARD CHARGER WITH HIGH-POWER-DENSITY IMPLEMENTATION.................................. 2448
Oscar Lucia

PP222   BENCHMARKING DC FAST CHARGERS: A COMPARATIVE ANALYSIS OF POWER CONVERTER STRUCTURES FOR WIDE VOLTAGE RANGE 2453
Sadik Cinik

PP223   PERFORMANCE OPTIMIZATION OF SINGLE-PHASE ON-BOARD CHARGERS WITH RIPPLE PORT 2461
Davide Gottardo

PP224   A REDUCED-SENSOR MODULAR DUAL ACTIVE BRIDGE-BASED BATTERY CHARGING SYSTEM FOR ELECTRIC VEHICLES USING AN IMPROVED LINEAR EXTENDED STATE OBSERVER... 2469
Armel Asongu Nkembi

PP225   BIDIRECTIONAL NON-ISOLATED THREE-PHASE ONBOARD CHARGER WITH A LOW-VOLTAGE LOWER-PHASE OPERATION MODE 2478
Steffen Frei

PP226   CONTROL OF A THREE-PHASE INDUCTIVE POWER TRANSFER SYSTEM BASED ON DD²Q COIL TOPOLOGY 2488
Nikola Mirkovic

PP227   COMPARISON OF TWO BIDIRECTIONAL 11KW 400V CLLC AND CLLLC RESONANT CONVERTERS FOR EV APPLICATIONS 2494
Hasan Mousavi Somarin

PP228   DYNAMIC WIRELESS CHARGING SYSTEM DESIGN FOR EXTRA-URBAN AREAS BASED ON RESONANT INDUCTIVE POWER TRANSFER 2503
Irene Maria Torres Alfonso

PP229   BIDIRECTIONAL ISOLATED 400-12V DC-DC CONVERTER WITH IMPROVED POWER DENSITY AND FULL-RANGE OPERAION FOR EV APPLICATIONS 2513
Oscar Lucia

HIGH POWER DC-DC CONVERTER II

PP230   GAIN OPTIMIZATION CONTROL METHOD FOR CLLLC RESONANT CONVERTERS UNDER PHASE SHIFT MODE 2518
Sean Yu

PP231   ANALYSIS OF COMMON AND SPLIT DC-BUS INTERLEAVED H-BRIDGE CONVERTERS FOR HIGH-CURRENT LOW-RIPPLE APPLICATIONS.................................. 2524
Bhavana Gudala

PP232   OPTIMAL FREQUENCY OPERATING POINTS FOR HYBRID SWITCHED CAPACITOR CONVERTERS AND LOSSLESS CURRENT SENSE METHOD 2532
Simone Mazzer

PP233   DESIGN AND TESTING OF A 250 KW 50 KHZ SIC-BASED HALF-BRIDGE-SERIES-RESONANT-CONVERTER 2538
Daniel Haake

PP234   30KW - 97% EFFICIENCY ISOLATED DC-DC CONVERTER WITH LARGE INPUT VOLTAGE RANGE BASED ON A BOOST DAB ASSOCIATION 2547
Jean-Jacques Huselstein

PP235   ANALYSIS OF A FULL-BRIDGE PUSH-PULL FORWARD DUAL ACTIVE BRIDGE DC-DC CONVERTER 2557
Gean Sousa

DC-DC CONVERTER II

PP236   SYMMETRICAL OPERATION OF FOUR CHANNEL RESONANT BOOST DC-DC CONVERTERS IN CONTINUOUS CONDUCTION MODE 2566
Kristóf Bándy

PP237   IMPACT OF MAGNETICS TOLERANCE ON THE POWER SHARING OF PARALLEL DUAL-OUTPUT PHASE-SHIFT FULL-BRIDGE CONVERTERS 2576
Riccardo Mandrioli

PP238   A BALANCING CONVERTER WITH SERIES CONNECTED MOSFETS FOR +/-700V BIPOLAR DC GRIDS 2583
Sachin Yadav

PP239   OPTIMIZATION AND DESIGN OF LOW-VOLTAGE AND HIGH-CURRENT POINT-OF-LOAD CONVERTER UNDER 48V BUS ARCHITECTURE 2591
Jiajia Guan

PP240   INTERLEAVED BOOST CONVERTER EFFICIENCY AND POWER DENSITY MODEL FOR ACTIVE AND PASSIVE COMPONENT DESIGN 2596
Damien Lemaitre

NOVEL AND ADVANCED SEMICONDUCTOR DEVICES

PP241   EVALUATION OF A HYBRID POWER SWITCH BASED ON TRENCH CLUSTERED IGBT AND SIC MOSFET 2606
Alireza Sheikhan

PP242   CONTRIBUTIONS FOR BUILDING BLOCKS FOR NORMALLY-OFF 650V GAN-ON-SI POWER INTEGRATED CIRCUITS 2612
Thanh Hai Phung

PP243   NEW BIDIRECTIONAL ASYMMETRIC HIGH VOLTAGE TVS (TRANSIENT VOLTAGE SUPPRESSOR) DIODE 2620
Boris Rosensaft

PP244   ISO247: HIGH PERFORMANCE CERAMIC BASED ADVANCED ISOLATED DISCRETE PACKAGE TO FULLY EXPLOIT THE ADVANTAGES OF SIC MOSFET 2627
Sachin Shridhar Paradkar

PP245   IMPACT OF CURRENT RIPPLE REDUCTION USING HIGH SWITCHING FREQUENCIES ON PMSM EFFICIENCY 2632
Jannik Fuchs-Gade

PP246   MAXIMIZING COST-EFFICIENCY IN ELECTRIC DRIVETRAINS: A SIC/SI
FUSION SWITCH APPROACH 2638
    Matthias Ippisch

ADVANCED CONTROL

PP247   CONCISE AND RELIABLE SIC MOSFET DRIVER CIRCUITS 2646
    Zhong Ye

PP248   ARTIFICIAL INTELLIGENCE ENHANCED RESOLVER SYSTEM FOR
AUTOMOTIVE TRACTION INVERTER APPLICATIONS BASED ON AURIX TC4X.......................... 2651
    David Zipperstein

PP250   MULTIFUNCTIONAL GRID MANAGER TOPOLOGY WITH CONFIGURABLE
OUTPUT 2657
    Peter Van Duijsen

PP252   CO2 FOOTPRINT OF MEDIUM VOLTAGE DC SOLID STATE
TRANSFORMER   2663
    Adriana Campos

SIC MOSFET

PP253   THERMO-ELECTRICAL ANALYSIS AND PERFORMANCE: A COMPARATIVE
STUDY BETWEEN MODULAR AND DISCRETE APPROACHES.. 2673
    Stefano Orlando

PP254   IMPACT OF PARAMETER SPREAD IN PARALLEL-OPERATED SIC MOSFETS
FOR HARD-SWITCHING CONVERSION.. 2680
    Andrea Piccioni

PP255   ASSESSMENT OF THE RDS,ON OF SIC MOSFET DIES THROUGH KELVIN WIRE
CONNECTION 2686
    Philipp Rehlaender

PP256   CHALLENGES IN SCALING SIC SINGLE-CHIP MEASUREMENTS TO
CORRESPONDING POWER MODULES 2693
    Hao Wang

PP257   SWITCHING PERFORMANCE EVALUATION OF HIGH-POWER 1.7 KV SIC
MOSFET MODULES USING A COMMON BUSBAR DESIGN.. 2700
    Sebastian Neira

PP258   CHARACTERIZING THE SWITCHING BEHAVIOR OF A 1.2 KV MIXED SIC JFET
AND MOSFET HALF BRIDGE.. 2708
    Tim Ringelmann

VOLUME 5

WBG HIGH FREQUENCY APPLICATION

PP259   PERFORMANCE EVALUATION OF THE PACKAGING OF SIC DIODES IN A 6.78
MHZ WIRELESS POWER TRANSFER SYSTEM 2718
    Ioannis Nikiforidis

PP260 VOLTAGE WAVEFORM GENERATION FOR SAWYER-TOWER COSS LOSS
MEASUREMENTS USING A HYBRID POWER CONVERTER 2724
Malachi Hornbuckle

PP261 EVALUATION OF SIC DEVICES FOR OVER 500KHZ APPLICATION BASED ON
BUCK CIRCUIT 2730
Minli Jia

PP262 LINEARIZATION OF DRAIN-SOURCE CAPACITANCES FOR ANTISERIAL
CONFIGURATED SIC MOSFETS IN HIGH FREQUENCY SOLID STATE SWITCHES 2737
Lars Dresel

SIC RUGGEDNESS

PP263 EFFECTS OF NON-KILLER DEFECTS ON SIC MOSFET SHORT-CIRCUIT
RUGGEDNESS AND RELIABILITY 2745
Sara Kuzmanoska

PP264 DYNAMIC REVERSE BIAS TEST: ELECTRO-THERMAL CHARACTERIZATION
OF SIC MOSFETS 2751
Giuseppe Mauromicale

PP266 RADIATION HARDNESS OF SIC BASED INVERTERS BASED ON AN EV
MISSION PROFILE 2758
Hadiuzzaman Syed

PP267 RAPID SHORT CIRCUIT PROTECTION USING DIDT DETECTION FOR SIC
POWER MODULES 2764
Koki Samura

PP268 COMPARISON OF DYNAMIC GATE STRESS TEST RESULTS OF SIC MOSFETS 2769
Mathias Gebhardt

PP279 EXTENDING SIC MOSFET SHORT-CIRCUIT WITHSTANDING TIME BY TWO-
LEVEL TURN-OFF GATE DRIVING 2778
Kwokwai Ma

PP270 EXPERIMENTAL INVESTIGATIONS ON PARASITIC TURN-ON OF 1.2KV SIC
MOSFET DISCRETE DEVICES 2786
Thanh-Toan Pham

PP271 BEHAVIOR MODELLING THE SHORT CIRCUIT CHARACTERISTICS OF SIC
MOSFETS USING COMPACT MODELS 2791
Qing Sun

THERMAL CHARACTERIZATION

PP273 THERMAL ANALYSIS AND MODELLING OF CHARGING STATIONS FOR
ELECTRIC VEHICLES 2796
Ruben Kopischke

PP274 JUNCTION TEMPERATURE MEASUREMENT OF A 3.3 KV SILICON CARBIDE
MOSFET POWER MODULE 2803
Michael Gleissner

PP275 INNOVATIVE 3D POWER MODULE DEFAULTS DETECTION VIA THERMAL IMPEDANCE ANALYSIS AND SIMULATIONS 2811
Louis Alauzet

PP276 THERMAL CHARACTERIZATION OF AN AIR-COOLED PEBB BASED ON SIC MOSFET POWER MODULES 2819
Alexandre Marie

PP277 THERMAL BEHAVIOUR OF SIC MOSFET WITH PLANAR PACKAGING TECHNOLOGY 2826
Yijun Ye

RELIABILITY AND AVAILABILITY

PP279 IMPLEMENTING MODULE HEALTH MONITORING IN EV TRACTION INVERTERS 2831
Karol Rendek

PP280 RELIABILITY TESTS OF COPPER THICK-FILM SUBSTRATES FOR POWER ELECTRONIC APPLICATIONS.. 2838
Henry Barth

PP281 POWER MODULE SOLUTIONS WITH IMPROVED RELIABILITY FOR ELEVATOR DRIVE APPLICATIONS 2843
Tiago Jappe

PP282 FAIL-OPERATIONAL LLC TOPOLOGIES WITH FAULT-TOLERANCE INTEGRATED REDUNDANT CAPABILITIES 2850
Aswathy M. Prince

PP283 THERMAL AND RELIABILITY OPTIMIZATION OF CLIPS IN SIC MOSFET POWER MODULES... 2860
Zexiang Zheng

PP284 CONDITION MONITORING OF A GAN FULL-BRIDGE BY MEANS OF FORWARD VOLTAGE IN CONTINUOUS OPERATION.. 2866
Michael Vogt

PP285 A SIMPLE AND LOW COST OVERCURRENT PROTECTION SYSTEM BASED ON COMMERCIAL SHUNT FOR WIDE-BANDGAP DEVICES... 2874
Emanuele Martano

PP286 SVM-BASED FAULT-TOLERANT CONTROL FOR A CASCADED H-BRIDGE MULTILEVEL CONVERTER UNDER MULTIPLE OPEN-CIRCUIT SWITCH FAULTS........................ 2880
Dong Xie

PP287 REVOLUTIONIZING MOBILITY: THE SECOND LIFE OF ONBOARD CHARGING SYSTEMS IN COMMERCIAL VEHICLES 2886
Ajay Krishna Voppu Muralikrishna

LOW VOLTAGE SWITCHES

PP288 A BEHAVIORAL TRANSIENT MODEL FOR IGBT DEVICE WITH ANTI PARALLEL FREEWHEELING DIODE... 2893
Shiwu Zhu

PP289   PARAMETER EXTRACTION FOR AN ANN-ASSISTED IGBT MODEL IN TRANSIENT SIMULATIONS 2901
Huaiyuan Zhang

PP290   FABRICATION OF 600V RC-IGBT USING 300MM WAFER 2909
Masaki Ueno

PP291   NEXT LEVEL OF POWER MODULE SOLUTION FOR PV C&I STRING INVERTER WITH 1200V H7 TECHNOLOGY IN EASY3B PACKAGE 2914
Tilo Poller

PP292   ANALYSIS OF MOSFET SWITCHING LOSSES IN RESONANT CONVERTERS USING ELECTRICAL AND THERMAL MEASUREMENTS AND LOSS TRENDS WITH MOSFET SIZE VARIATION 2921
Alfio Scuto

PP293   OPTIMOS 6 135V FOR HIGH POWER MOTOR DRIVES 2930
Kunal Jha

PP294   AUTO POWER-SOI: SHAPING THE FUTURE OF BATTERY MONITORING TECHNOLOGY 2937
Alex Lim

LIFETIME MODELLING AND CONDITION MONITORING

PP295   UNDERSTANDING THE IMPACT OF IEC60747-17 ON CAPACITIVE AND MAGNETIC COUPLERS 2942
Shu Ee Ong

PP296   PARIS LAW APPLIED TO WIRE BONDS DEGRADATION USING CRACK GROWTH MEASUREMENT 2948
Merouane Ouhab

PP297   CONDITION MONITORING TECHNIQUE OF POWER ELECTRONIC MODULES VIA SQUARE-WAVE GATE SIGNAL EXCITATION 2956
Isabel Austrup

PP298   STATISTICS-BASED LIFETIME SIMULATION ENVIRONMENT FOR POWER MODULES INCORPORATING DEGRADATION MODELS 2963
Karthik Debbadi

PP299   POWER CYCLING RESULTS FOR RELIABILITY STUDIES OF SIC-INVERTERS 2972
Robert Keilmann

PP300   GAN CASCODE IN HIGH SPEED DRIVEN AIR COMPRESSORS FOR AUTOMOTIVE FUEL CELLS 2981
Florian Lippold

PP301   PROGNOSTIC ANALYSIS OF IGBT HEALTH: REAL-TIME ON-STATE VOLTAGE PREDICTION THROUGH MACHINE LEARNING 2986
Tanya Thekemuriyil

PP302   ROBUSTNESS ANALYSIS OF TEMPERATURE-SENSITIVE ELECTRICAL PARAMETERS OF IGBTS 2995
Laurids Schmitz

PP303 OBSERVATION OF THERMAL-RESISTANCE INCREASE OF DEGRADED IGBT MODULES BY VCE (SAT) MEASUREMENT IN A CHOPPER CIRCUIT 3002
Kazunori Hasegawa

PULSE WITH MODULATION METHODS

PP304 MODULATION TECHNIQUE FOR REDUCED AC CONTENT OF THE DC LINK CURRENT IN THREE-PHASE TWO-LEVEL INVERTERS 3007
Steffen Frei

PP305 COMMON MODE CURRENTS IN RESONANT CIRCUITS GENERATED WITH A DELTA-SIGMA MODULATED VOLTAGE SOURCE INVERTER 3017
Tobias Haas

PP306 EVALUATION OF NEW MODULATION SCHEME FOR 3L-ANPC USING BOTH CURRENT PATHS IN ZERO STATE 3020
Felix Eichler

PP307 AN INNOVATIVE SYNCHRONOUS RECTIFICATION METHOD FOR 11KW CLLC CONVERTER 3029
Sanbao Shi

PP308 INTERLEAVED ASYNCHRONOUS DELTA-SIGMA MODULATION CONCEPT FOR DYNAMIC POWER CONVERTERS 3034
Philipp Czerwenka

PP309 HIGH RESOLUTION MIXED-SIGNAL PULSE WIDTH MODULATOR FOR HIGH-FREQUENCY DC-DC CONVERTERS 3042
Tim McRae

PP310 IMPLEMENTATION AND CONTROL OF OPTIMIZED PULSE PATTERNS FOR SALIENT PERMANENT MAGNET SYNCHRONOUS MACHINES IN ELECTRIC VEHICLES 3045
Maximilian Hepp

PP311 A 3-LEG INTERLEAVED TP PFC WITH A 90° PHASE-SHIFTED ASYMMETRIC LEG FOR REDUCED MAGNETICS 3060
Serkan Dusmez

PP312 FAULT-TOLERANT OPERATION ANALYSIS OF A FIVE-PHASE THREE-LEVEL TNPC INVERTER FOR ELECTRIC AIRCRAFT PROPULSION SYSTEMS 3067
Chanuch Chaisakdanugull

AC-DC AND DC-AC CONVERTER

PP313 CCM TOTEM-POLE PFC FOR ULTRA-HIGH POWER DENSITY USB-PD CHARGERS 3077
Manuel Escudero Rodruigez

PP314 COMPARISON OF HYBRID SI/SIC AND SIC TWO-LEVEL AND THREE-LEVEL CONVERTERS FOR LOW-VOLTAGE LOW-POWER APPLICATIONS 3086
Tim Augustin

PP315 ANALYSIS OF ANALOGUE CURRENT AND FLUX BALANCING FOR THE DUAL-ACTIVE-BRIDGE CONVERTER 3096
Christophe Basso

PP316 DESIGN AND OPTIMIZATION OF A SINGLE-STAGE PHOTOVOLTAIC MICROINVERTER WITH INTEGRATED MAGNETICS 3103
Jin Wen

PP317 EXPERIMENTAL INVESTIGATION OF CLASS F INVERTER UNDER VARIOUS LOAD CONDITIONS........... 3110
Baptiste Daire

PP318 ANALYSIS, MODELING, DESIGN, AND LIMITATIONS OF CURRENT INJECTION BASED UPF RECTIFIER WITH SMALL DC-LINK CAPACITOR........... 3118
Ramkrishan Maheshwari

PP319 HIGH-EFFICIENT ISOLATED AC-DC CONVERTER WITH CIRCULATING CURRENT REDUCTION FOR AC ADAPTERS 3125
Hiroki Watanabe

PP320 A PHASE-LOCKED LOOP (PLL) BASED STRATEGY FOR ACCURATE BLANKING TIMES IN BRIDGELESS TOTEM-POLE PFCS 3130
Sandu Tigira Tigira

PP321 CIRCULATING CURRENTS IN COUPLED MULTI-TERMINAL HYBRID AC-DC GRIDS 3136
Fabian Herzog

ADVANCED CONVERTER TOPOLOGIES

PP322 COMPARISON OF 4500V STATE-OF-THE-ART XHP3 IGBT AND CONVENTIONAL IHV IGBT FOR 3300V 3-LEVEL ANPC MEDIUM VOLTAGE DRIVES 3142
Martin Knecht

PP323 GENERALIZED SWITCHING SEQUENCE FOR VOLTAGE BALANCING IN A FLYING CAPACITOR DC-DC CONVERTER WITH QUASI-2-LEVEL MODULATION 3150
Jose Andres Aguilar Croston

PP324 OPTIMIZATION-BASED SIZING OF A MODULAR MULTILEVEL CONVERTER BASED ON 650 V GAN MODULES FOR NEW LVDC/MVDC GRIDS........... 3160
Gregoire Le Goff

PP325 A NOVEL THREE-PHASE LOW-SWITCH-COUNT AC-DC GRID CONVERTER TOPOLOGY WITH GALVANIC ISOLATION........... 3169
Liska Steenbock

PP326 SINGLE-STAGE LED DRIVER BASED ON COUPLED INDUCTOR POWER FACTOR CORRECTION AND LLC CONVERTER........... 3175
Alireza Ramezan Ghanbari

PP327 A INVERSE COUPLED DC-DC BOOST INDUCTOR WITH 2-KV SIC MOSFET MODULE FOR 1500V SOLAR INVERTER MPPT........... 3181
Yusi Liu

PP328 ENVIRONMENTAL IMPACT OF MODULAR POWER ELECTRONICS SYSTEMS CONSIDERING DIAGNOSTIC-DRIVEN UNIT REPLACEMENT 3187
Briac Baudais

POWER ELECTRONICS FOR RAILWAY APPLICATIONS

PP329   SWITCHING PERFORMANCE COMPARISON OF 3.3 KV SIC MOSFET AND SI IGBT POWER MODULES FOR RAILWAY TRACTION SYSTEMS 3197
Yue Zhao

PP330   COMPARISON OF THREE-LEVEL INVERTER TOPOLOGIES FOR MVDC REVERSIBLE RAILWAY SUBSTATIONS 3206
Luc Bimmel

PP331   CONTROL OF BIDIRECTIONAL POWER FLOW IN RAILWAY CATENARY OVERHEAD LINES.. 3213
Peter Van Duijsen

PP332   A RAIL TRACTION CONVERTER PLATFORM BASED ON POWER MODULE IMPLEMENTATIONS WITH 450 A, 600 A AND 800 A 3.3 KV IGBT MODULES.................................. 3221
Ekrem R. Gunes

PP333   COMPARISON OF SELECTED MEGAWATT-LEVEL TRACTION CONVERTER POWER MODULE IMPLEMENTATIONS IN TERMS OF COMMUTATION INDUCTANCE AND PRACTICALITY.. 3229
Abdulkerim Ugur

CURRENT RELATED TESTING

PP334   PITFALLS AND THEIR AVOIDABILITY IN THE DOUBLE-PULSE TEST 3237
Nikolas Förster

PP335   MODELING AND SIMULATION OF FLUXGATE BASED CURRENT SENSOR 3247
Yunus Çay

PP336   SIGMA-DELTA BASED CURRENT ACQUISITION WITH REDUCED SETTLING TIME 3256
Joschka Randerath

PP337   CHARACTERISATION OF WIDE-BANDGAP SEMICONDUCTORS IN DOUBLE PULSE TESTING USING OPTICALLY ISOLATED PROBES.. 3264
Lennart Hoffmann

PP338   NON-INVASIVE BATTERY CONDITION TESTING USING ELECTRICAL SIGNALS AND OSCILLOSCOPES.. 3269
Srikrishna N. H

PP339   INSTRUMENTATION REQUIREMENTS FOR FAST 130 V/NS SWITCHING OF 1700 V, 35 M? SIC MOSFETS 3276
Matthew Appleby

POWER ELECTRONICS FOR AEROSPACE APPLICATIONS

PP340   CONCEPTUALIZATION AND EXPERIMENTAL ASSESSMENT OF DESIGN ASPECTS FOR 3-LEVEL ANPC INVERTERS 3286
Lukas Radomsky

PP341 DESIGN OF A HIGH POWER DENSITY INVERTER AND FOC IMPLEMENTATION FOR UAVS 3296
Matthias Neuner

PP342 HIGHLY-INTEGRATED, FLEXIBLE POWER SOLUTION FOR AEROSPACE 5KVA – 20 KVA MOTOR DRIVE APPLICATIONS 3305
Alain Calmels

PP343 DATABASE-SUPPORTED PRELIMINARY DESIGN, SIMULATION AND EVALUATION OF POWER CONVERTERS IN ELECTRIC AIRCRAFT PROPULSION SYSTEMS 3315
Jeff Kugener

PP344 DESIGN AND ANALYSIS OF GATE-DRIVER FOR SIC-BASED INVERTER FOR MEGAWATT SCALE ALL ELECTRIC AIRCRAFT 3318
Jeff Kugener

MEASUREMENT TECHNIQUES AND METHODS

PP345 ADDRESSING TESTING CHALLENGES FOR POWER MODULES AND THREE-LEVEL INVERTERS 3328
Oleg Fotteler

PP346 CHARACTERIZATION OF THE BONDING QUALITY OF SILVER SINTERED COMPOUNDS BY MEANS OF LASER-INDUCED BREAKDOWN SPECTROSCOPY 3334
Yannick Bockholt

PP347 INVERTER-INTEGRATED MEASUREMENT OF THE FREQUENCY-DEPENDENT WINDING IMPEDANCE OF ELECTRIC MACHINES 3340
Christian Mühlfeld

PP348 COMPENSATION TECHNIQUES FOR BANDWIDTH-DISTORTED MEASUREMENTS OF FAST TRANSIENTS IN DOUBLE PULSE TESTS 3347
Christian Lottis

PP349 AN AERODYNAMIC LOAD MEASUREMENT TECHNIQUE FOR AUTONOMOUS AERIAL VEHICLES 3353
Mehmet Oguz Girgin

COMPENSATION TECHNIQUES FOR BANDWIDTH-DISTORTED MEASUREMENTS OF FAST TRANSIENTS IN DOUBLE PULSE TESTS 3358
Christian Lottis

PP350 A HIGH-BANDWIDTH MULTILEVEL COUNTER CIRCUIT FOR BEARING CURRENT EVALUATION 3364
Felix Schulte

TRANSFORMERS

PP351 CORE LOSS MODEL FOR CONSIDERING ANISOTROPY AND TEMPERATURE EFFECTS ON ELECTRICAL STEEL UNDER POWER ELECTRONIC CONDITIONS 3371
Michael Owzareck

PP353   CIRCULAR ECONOMY ORIENTED AND RECONFIGURABLE PLANAR
TRANSFORMER DESIGN FOR ISOLATED DC-DC CONVERTERS 3380
    Fabian Groon

PP354   CONTROLLABLE MAGNETICS: VARIABLE TRANSFORMERS AND VARIABLE
INDUCTORS, THEORY – PRODUCTION – APPLICATION.. 3390
    Florian Fenske

PP355   A THREE-PHASE INTERLEAVED LLC INTEGRATED TRANSFORMER USING
PCB WINDINGS FOR FUEL CELL DCDC CONVERTERS 3395
    Jiajia Guan

PP356   TESTING THE PRIMARY-SECONDARY COIL COUPLING OF HIGH-FREQUENCY
TRANSFORMER IMPLEMENTED ON ETD AND TOROIDAL CORES 3400
    Alexis Gioda

Author Index

PCIM Europe 2024, 11– 13 June 2024, Nuremberg   DOI: 10.30420/566262186

Cost-Effective Method to Discharge DC Link Capacitors with SiC Power Modules

Paul T, Kanatzar[1], Brian T. DeBoi[1] , Austin Curbow[1] , Stephanie J. Vinueza[1]
[1] Wolfspeed Inc., USA

Corresponding author: Paul Kanatzar, Paul.Kanatzar@Wolfspeed.com
Speaker: Paul Kanatzar, Paul.Kanatzar@Wolfspeed.com

Abstract

The high bus voltages commonly used in electric vehicles (EVs) present a safety hazard to vehicle operators and technicians. As the number of EVs on the road increases, it is essential to mitigate these risks by implementing safety features to prevent contact with high-voltage conductors within the vehicle. The primary source of shock risk within the vehicle is from the DC link capacitors that source energy from the EV batteries to the inverter. These capacitors must be discharged when the vehicle is not in operation, such as when the vehicle is turned off or in the event of a crash. This paper presents an active discharge technique that dissipates the DC link energy through the switching semiconductors already present in the system, thus not requiring additional components or control circuitry. The proposed method works by commanding the inverter to alternate between zero sequence states at a high switching frequency. The output capacitance (C_{OSS}) of the power devices are charged and discharged each switching period to dissipate the energy stored in the DC link capacitors across the power module. This method is advantageous because the discharge rate is easily controlled and limits thermal heating within the MOSFETs. This paper demonstrates the operation of this method using a SiC power module and presents an analytical model for predicting the discharge time of the DC link capacitor. The analytical model is shown to accurately predict the discharge time for a given system. The power dissipated in the modules during the discharge event is also evaluated and shown to be well within the allowable operating conditions of the power module.

1   Introduction

As EV manufacturers strive to maximize range and efficiency in electric vehicles, 800 V vehicle power-train designs are becoming increasingly common [1], [2]. The higher voltages and large DC link capacitances required for the inverter subsystem can present life-threatening shock, as stated by IEC 60479-2 [3]. Therefore, methods to quickly and reliably discharge the DC link capacitors when the vehicle is not in operation are necessary. Currently, two primary methods are used to discharge the DC link capacitors. The first is a simple method that uses a switch to connect a resistive element across the DC bus to dissipate the energy, as shown in Fig. 1. While this method is effective, it is expensive and adds at least two failure points to the inverter system.

A second method commonly proposed in literature uses advanced control schemes to discharge the DC link capacitors energy in the motor windings [4], [5], [6], [7]. However, these methods require that the load be connected, functional, and ener-

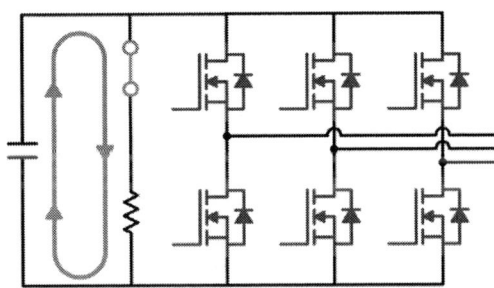

Fig. 1   Simple Resistive Discharge Method

gized. This paper purposes a novel method for discharging the DC link capacitors that does not require additional components or a functional load, and only requires use of the power switching devices. The discharge is achieved by alternating between zero sequence states and using the device output capacitance to discharge the DC link capacitors.

PCIM Europe 2024, 11– 13 June 2024, Nuremberg   DOI: 10.30420/566262186

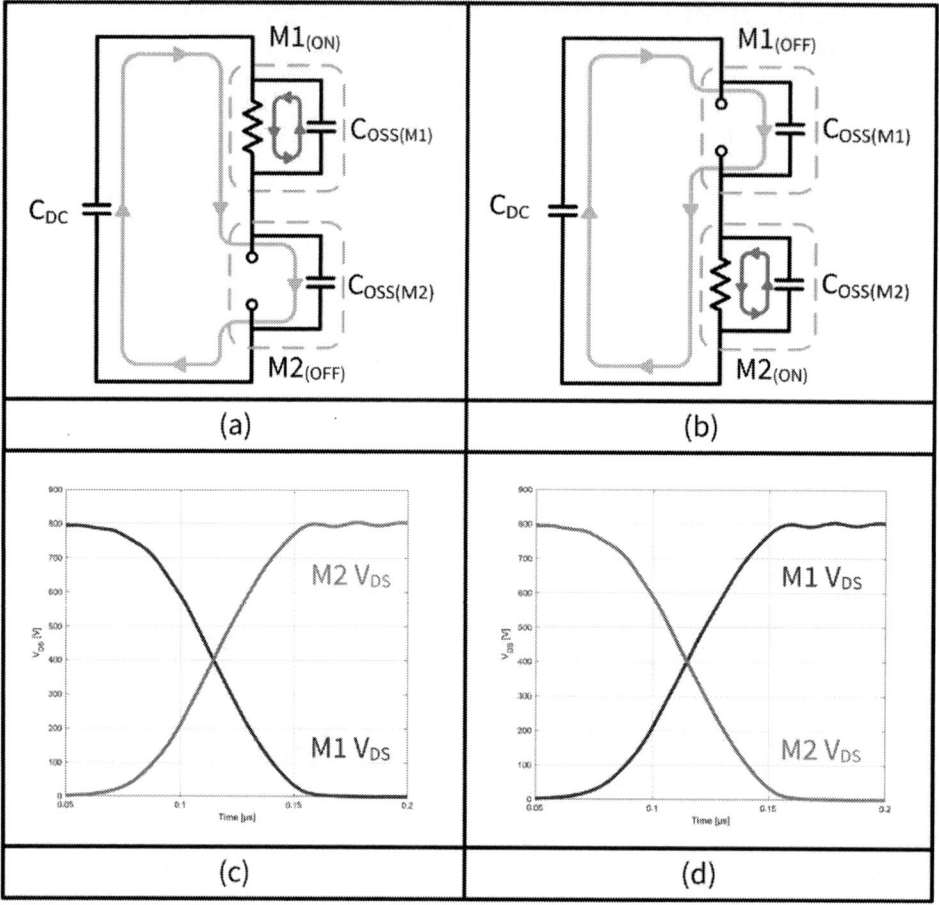

Fig. 2   Half-bridge power module model architecture [3]

2   Theory of Operation

The proposed discharge method involves commanding each phase in the inverter to alternate between zero sequence states each period. At no point during the sequence can DC current flow through the switches; at least one switch position is blocking at all times. For the first half of the period, all high-side switches are on and all low-side switches are off. For the second half of the period, all low-side switches are on and all high-side switches are off. This process sequentially charges and discharges the output capacitance of the power MOSFETs. During the first half of the period shown in Fig. 2 (a), the high-side switch is turned on by the gate driver. In this state, $C_{OSS(M2)}$ charges up to the bus voltage. The energy stored in $C_{OSS(M1)}$ along with the additional current required to charge $C_{OSS(M2)}$ is dissipated as heat across device M1. During the second half of the period shown in Fig. 2 (b), $C_{OSS(M1)}$ is charging, and

the energy stored in $C_{OSS(M2)}$ along with the additional current required to charge $C_{OSS(M1)}$ is dissipated as heat across device M2. This leads to two instances of the MOSFET's stored energy due to the output capacitance, E_{OSS}, being removed from the DC link capacitor for each half-period. Therefore, four times E_{OSS} is removed from the DC link capacitor for each switching period. Note that the currents shown in Fig. 2 (a) and Fig. 2 (b) are only flowing during the dv/dt events shown in Fig. 2 (c) and Fig. 2 (d), respectively. V_{DC} is the voltage across the module defined in (1).

$$V_{DC} = V_{M1} + V_{M2} \qquad (1)$$

1362

2.1 Energy Loss Per Switching Period

For each switching period the voltage transitions shown in Fig. 2 (c) and Fig. 2 (d) occur corresponding to the switch states in Fig. 2 (a) and Fig. 2 (b). Once the voltage has transitioned, there is no current flowing from C_{DC}. Fig. 2 (a) shows the time period where M1 is on and M2 is off. Prior to the switching event, $C_{OSS(M1)}$ is charged and has potential energy stored equal to (2). Switch M1 is turned on and the output capacitance of M2 charges to V_{DC}. Once charged it will have taken energy from the bus C_{DC} and stored it as potential energy in accordance with (3).

$$E_{COSS(M1)} = \frac{1}{2} \cdot Q_{OSS(M1)} \cdot V_{M1,t0} \qquad (2)$$

$$E_{COSS(M2)} = \frac{1}{2} \cdot Q_{OSS(M2)} \cdot V_{M2} \qquad (3)$$

The change in voltage across the module (V_{DC}), is very small due to the capacitance of C_{DC} being much larger than C_{OSS} (4).

$$C_{DC} \ggg C_{OSS} \qquad (4)$$

From (4), we make the simplifying assumption that the change in voltage across C_{DC} (V_{DC}) is zero during the charge time of $C_{OSS(M2)}$. With this simplifying assumption, the total energy loss from C_{DC} can be calculated with (5).

$$\Delta E_{DC} = Q_{OSS} \cdot V_{DC} \qquad (5)$$

Knowing that half of the energy lost from C_{DC} is stored in $C_{OSS(M2)}$ as potential energy (3), we know the other half of the energy lost from C_{DC} is dissipated as heat across M1. However, the total amount of energy dissipated across M1 must include the energy stored in $C_{OSS(M1)}$. The total amount of energy dissipated as heat in M1 is given in (7).

$$E_{M1(Heat)} = 2 \cdot \frac{1}{2} \cdot Q_{OSS} \cdot V_{DC} \qquad (6)$$

$$E_{M1(Heat)} = Q_{OSS} \cdot V_{DC} \qquad (7)$$

During each switching event, the energy removed from the DC link capacitor is given in (5) and is dissipated across the activated switch position. For Fig. 2 (a), heat is dissipated across M1, and for Fig. 2 (b), heat is dissipated across M2.

V_{DC} (V)	C_{DC} (µF)	F_{SW} (kHz)	$F_{SW,Step}$ (kHz)
1000	181	10-100	10
800	181	10-100	10
600	181	10-100	10

Table 1 Test Matrix

2.2 Temperature Rise per Discharge

The total power dissipated in the switch for each discharge cycle is given by (8). From that we can estimate the average die temperature increase from the R_{TH} of the system with (9).

$$P_{Loss} = \frac{\frac{1}{2} \cdot C_{DC} \cdot V_{DC}^2}{Discharge\ Time\ [s]} \qquad (8)$$

$$\Delta T_j = R_{TH} \cdot P_{Loss} \qquad (9)$$

For a 2 mF DC discharged in one second across three half bridge modules with an R_{TH} of 0.15 (K/W) per position the expected temperature rise is 16 °C.

3 Results

Testing was performed using a modified version of the Wolfspeed KIT-CRD-CIL12N-XM3 evaluation board with a single DC link capacitor (shown in Fig. 3) and an EAB450M12XM3 (shown in Fig. 4). Key parameters of the test setup were measured including the capacitance of the DC link capacitor and output capacitance of the EAB450M12XM3.

Fig. 3 KIT-CRD-CIL12N-XM Clamped Inductive Load (CIL) evaluation kit for XM3 power modules

Fig. 4 EAB450M12XM3 Half Bridge Silicon Carbide (SiC) Power Module

The test matrix consisted of three DC link voltages 1000 V, 800 V, and 600 V. For each voltage, the discharge time is defined from the starting voltage to 50 V. Each test was performed for switching frequencies from 10 kHz to 100 kHz with a 10 kHz increment.

3.1 Test Setup

The XM3 power module was mounted to the KIT-CRD-CIL12N-XM CIL board. A CGD12HBXMP gate driver was attached to the module to control the gates. The CIL board was connected to the high voltage dc supply through the relay and control board to charge the DC link capacitors. Fig. 5 shows a simplified block diagram of the test setup.

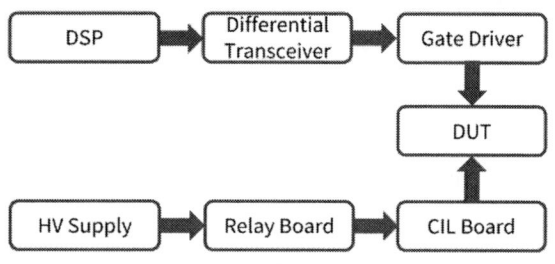

Fig. 5 Test Setup Signal Chain

Fig. 6 Shows the actual test setup with annotations describing each component.

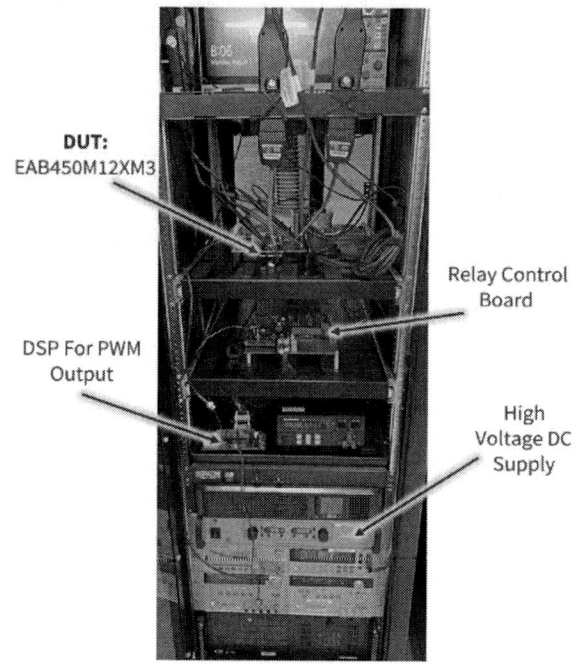

Fig. 6 Experimental Test setup

3.2 Test Sequence

The DC link capacitor was charged by a high voltage power supply through the relay control board. Once charged, the relays were opened, disconnecting the CIL board from any galvanic ground connections. The DSP initiates the test by sending the complementary PWM signals to the gate to discharge the bus. The entire discharge event is captured on the oscilloscope and the discharge time is measured Fig. 7.

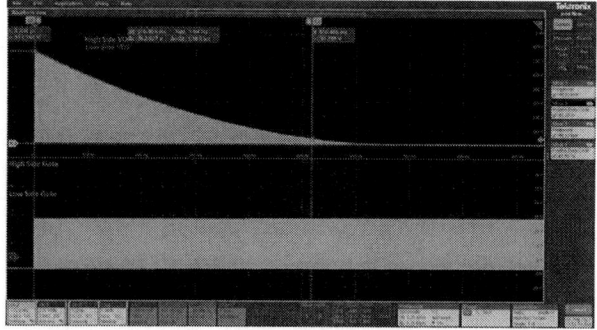

Fig. 7 Example of Discharge Event

The process was repeated for all conditions in the test matrix Table 1.

3.3 Experimental Results

The discharge times for each of the test conditions is plotted in Fig. 8. The discharge time reduces proportionally to the increased switching frequency as expected from the theoretical understanding of the discharge event.

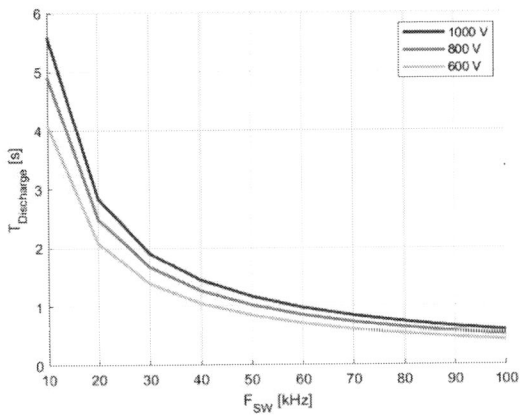

Fig. 8 DC Link Discharge Time vs. F_{SW}

4 Analytical Model

An analytical model was created to estimate the discharge time for different configurations of DC link capacitance and number of modules. The model was verified against the experimental test results and was proven to accurately predict the discharge time.

4.1 Model Operation

For each switching period, E_{OSS} of the module is calculated based on the C-V characteristics of the power module. The C-V characteristics of the EAB450M12XM3 power module is shown in Fig. 9. The initial energy stored in the DC link capacitor is calculated based on the current bus voltage (10). Four times E_{OSS} (2) is subtracted from the energy stored in the DC link capacitor.

$$E_{DC,Initial} = \frac{1}{2} \cdot C_{DC} \cdot V_{DC}^2 \qquad (10)$$

The new energy stored in the DC link capacitor is calculated with (11) and the new voltage of the DC link capacitor is calculated with (12).

$$E_{DC,New} = E_{DC,Initial} - 4 \cdot E_{OSS} \qquad (11)$$

The discharge time is increased by one period and the process repeats until the new DC link voltage is less than or equal to the target stop voltage (50 V).

$$V_{DC,New} = \sqrt{\frac{E_{DC,New}}{\frac{1}{2} \cdot C_{DC}}} \qquad (12)$$

The CIL board used to perform the experimental tests contained bleed resistors to discharge the DC link capacitor in the event of a test failure.

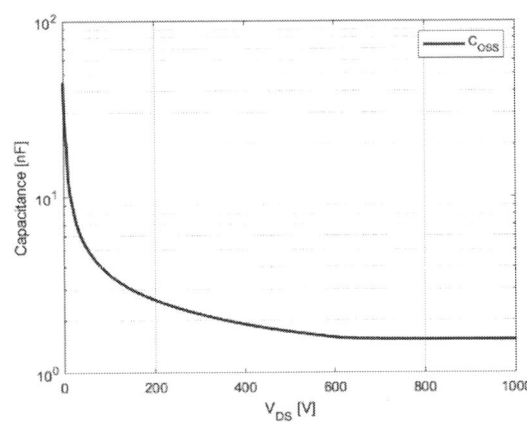

Fig. 9 C_{OSS} vs. V_{DS} for an EAB450M12XM3 power module

4.2 Model Validation

Average Percent Error	0.352
Standard Deviation of Percent Error	0.610
Variance of Percent Error	0.372

Table 2 Summary of Model Accuracy

To validate the model, the experimental test results for a starting DC link voltage of 800 V were compared to the simulated results produced by the analytical model.

The results for each of the ten test conditions are given in Table 1 and a summary of the results are given in Table 2 and Table 3.

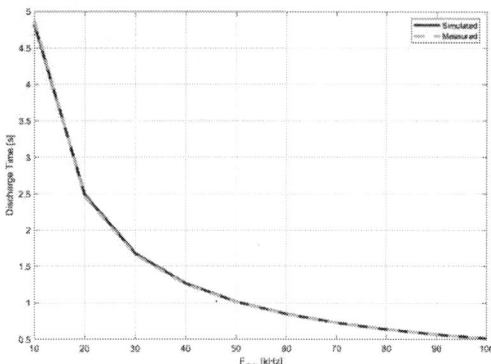

Fig. 10 Simulated Discharge vs. Measured Discharge time

The analytical model results in an average of less than one percent error across the test conditions. That error is likely to be due to extra capacitance in the CIL board not captured by the analytical expressions.

5 Conclusion

The active discharge method proposed in this paper operates the power modules in a mode that matches the intended use of the devices by switching completely from the negative V_{GS} to positive V_{GS}. The DC link capacitors can be fully discharged in under one second by increasing the switching frequency. The temperature rise of the power module is well within the operating limits of the power modules.

With the increasing demand to find a solution to discharge the DC link capacitance in electric vehicles, this method is viable.

Alternating between zero sequence states at elevated switching frequencies provides a low cost, predictable method for discharging the DC link capacitance. The analytical method described in this paper can be used to predict the discharge time for a given system.

References

[1] I. Aghabali, J. Bauman, P. J. Kollmeyer, Y. Wang, B. Bilgin and A. Emadi, "800-V Electric Vehicle Powertrains: Review and Analysis of Benefits, Challenges, and Future Trends," in IEEE Transactions on Transportation

F_{SW} (kHz)	$T_{Dis,sim}$ (ms)	$T_{Dis,exp}$ (ms)	Error (%)
10	4824	4872	0.99
20	2491	2468	-0.93
30	1679	1660	-1.17
40	1267	1257	-0.83
50	1017	1012	-0.46
60	849	844	-0.56
70	729	727	-0.21
80	639	637	-0.24
90	568	567	-0.23
100	512	512	0.12

Table 3 Results Comparison

Electrification, vol. 7, no. 3, pp. 927-948, Sept. 2021, doi: 10.1109/TTE.2020.3044938.

[2] K. Kumar, M. Bertoluzzo and G. Buja, "Impact of SiC MOSFET traction inverters on compact-class electric car range," 2014 IEEE International Conference on Power Electronics, Drives and Energy Systems (PEDES), Mumbai, India, 2014, pp. 1-6, doi: 10.1109/PEDES.2014.7042156.

[3] M. W. Kroll, D. Panescu, P. E. Perkins, M. Koch and C. J. Andrews, "Electrocution Risk of Capacitive Discharge Shocks: Application to Electric Vehicle Charging," 2022 44th Annual International Conference of the IEEE Engineering in Medicine & Biology Society (EMBC), Glasgow, Scotland, United Kingdom, 2022, pp. 1418-1422, doi: 10.1109/EMBC48229.2022.9871541.

[4] H. Yang, J. Yang and X. Zhang, "DC-Bus Capacitor Maximum Power Discharge Strategy for EV-PMSM Drive System With Small Safe Current," in IEEE Access, vol. 9, pp. 132158-132167, 2021, doi: 10.1109/ACCESS.2021.3112462.

[5] X. Wei, B. Yao, Y. Peng, Y. Sun, K. Wang and H. Wang, "An Improved Discharge Profile-Based DC-Link Capacitance Estimation for Traction Inverter in Electric Vehicle Applications," in IEEE Transactions on Power

Electronics, doi: 10.1109/TPEL.2024.3383153.

[6] C. Gong, Y. Hu, G. Chen, H. Wen, Z. Wang and K. Ni, "A DC-Bus Capacitor Discharge Strategy for PMSM Drive System With Large Inertia and Small System Safe Current in EVs," in IEEE Transactions on Industrial Informatics, vol. 15, no. 8, pp. 4709-4718, Aug. 2019, doi: 10.1109/TII.2019.2895317.

[7] Z. Ke, J. Zhang and M. W. Degner, "DC bus capacitor discharge of permanent magnet synchronous machine drive systems for hybrid electric vehicles," 2016 IEEE Applied Power Electronics Conference and Exposition (APEC), Long Beach, CA, USA, 2016, pp. 241-246, doi: 10.1109/APEC.2016.7467879.

PCIM Europe 2024, 11– 13 June 2024, Nuremberg
DOI: 10.30420/566262187

Study on Circulating Current in Parallel Operation of Transformerless UPS

Koji Kato[1] , Hisakatsu Igarashi[1] , Akira Sato[1]

[1] GS Yuasa International Ltd., Japan

Corresponding author: Koji Kato, koji.kato@jp.gs-yuasa.com
Speaker: Koji Kato, koji.kato@jp.gs-yuasa.com

Abstract

This paper reports on circulating currents in transformer less modular UPSs. There is concern that transformer less modular UPSs decline control performance and exceed component ratings due to circulating currents in the switching frequency component. Circulating currents in a transformer less UPS connected in parallel are even more complex. Therefore, this paper analyzes the circulating current focusing on the switching frequency component. The analysis method of the circulating current was confirmed by the experimental results.

1   Introduction

The number of power failures in Japan was 0.16 per household in FY2022[1], indicating that the power supply in Japan is very stable compared to other countries [2]. However, it is difficult to completely eliminate power failure due to power failure and instantaneous voltage drop caused by various factors such as lightning strikes, natural disasters, and human-caused accidents. Uninterruptible power supplies (UPS) [3] are widely used to continue power supply in the event of power failure or instantaneous voltage drop at data centers, telecommunications equipment, factory production facilities, and other customers that demand power quality.

However, if a UPS fails, it will not be able to supply power to critical loads, so redundant UPS system is used to increase the reliability of power supply to critical loads. The redundant UPS system consists of multiple UPS, and if UPS fails, it is disconnected from the redundant UPS system and the remaining UPS continue to supply power. There are two types of redundant UPS systems which is parallel redundant system and serial redundant system.

In the parallel redundant UPS system, the voltage difference or phase difference in the output voltage of each UPS causes an imbalance in load sharing among the UPSs. Several methods [4]-[7] have been proposed and put into practical use as countermeasures. These methods are realized by adding droop characteristics to the output voltage and

phase to balance the load current evenly. Conventional redundant UPSs have built-in trans-formers and individual UPSs have their own storage batteries, so there is no path for circulating current to flow. However, the parallel operation of conventional UPS is expensive due to the transformer and battery. Therefore, a modular UPS with transformerless and common batteries has been proposed[8].

In case of transformer less UPSs, there is no isolation by a transformer, so even a voltage difference caused by differences in switching timing generate circulating currents. The more units connected in parallel, the more complicated the circulating current becomes, which may degrade control performance and exceed component ratings.

The purpose of this paper is to analyze the circulating currents generated in parallel operation of transformer less UPSs and to clarify the conditions under which the circulating currents are generated. This paper focuses on the circulating current caused by the switching component, and especially formulates the circulating current that flows through the EMI filter and verifies it experimentally.

2   Configuration of UPS

2.1   Topology of UPS

Figure 1 shows a typical UPS configuration. The standby type UPS shown in Fig.1(a) supplies power to the load as is when the power system is normal. When the power system is abnormal, the

PCIM Europe 2024, 11– 13 June 2024, Nuremberg DOI: 10.30420/566262187

UPS supplies power to the load. The standby type UPS can be configured simply and inexpensively with a small number of components, but the reliability of the power supply is low because an instantaneous voltage drop occurs at the switching of the selector switch.

Figure 1(b) shows an on-line type UPS that always supplies power to the load from the inverter. The power quality is high because instantaneous voltage drop does not occur in the power failure. However, the UPS is expensive, including running costs, due to the large number of parts and losses caused by constant power conversion.

The type of UPS is determined by the power supply reliability and cost required for the load. The on-line type UPS is used for critical loads because it has the highest power supply reliability. A single-configuration UPS supplies commercial power from a bypass circuit when the UPS fails. However, commercial power supply is not acceptable for power supplies that require particularly high reliability, such as data centers. Therefore, a redundant configuration with multiple UPSs is used.

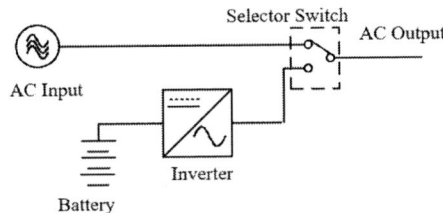

(a) Standby type UPS

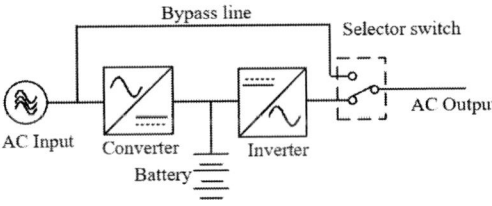

(b) On-line type UPS

Fig. 1 Topology of UPS

2.2 Redundant system of UPS

Figure 2 shows the redundant system configuration of the UPS. The serial redundant UPS system shown in Fig2(a) consists of two UPSs. The standby UPS and the permanent UPS are connected in series and switched to the standby UPS when the permanent UPS fails. However, this system is expensive because it requires two UPSs of the same capacity.

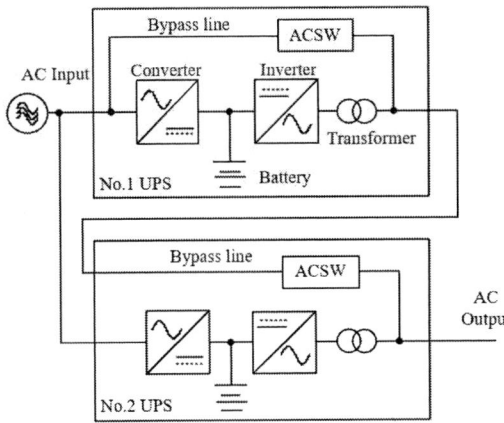

(a) Serial redundant UPS system

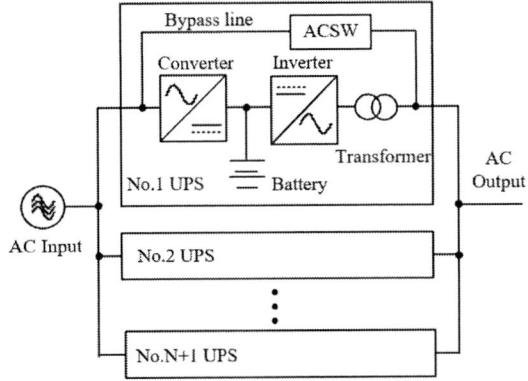

(b) Parallel redundant UPS system

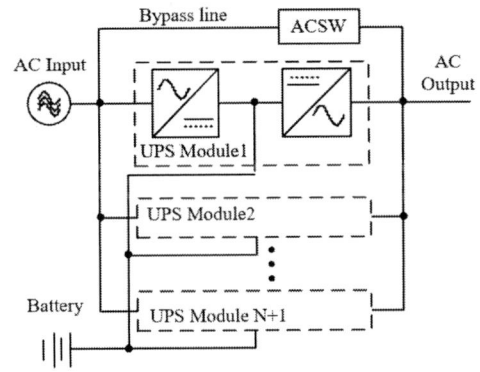

(b) Modular redundant UPS system

Fig. 2 Redundant UPS system

Figure 2(b) shows a parallel redundant UPS, in which UPSs are connected in parallel and each UPS shares the load current. If an N+1 redundant configuration is used, even if one UPS fails, the remaining UPS can continue operation. Parallel

1369

redundant UPSs can be redundant using UPSs with capacity smaller than the load capacity. Therefore, parallel redundant UPSs are more economical than standby redundant UPSs. However, a battery and isolation transformer are required for each UPS to avoid circulating current.

Figure 2(c) shows the UPS with a modular redundancy system. The bypass line and battery with a low failure rate are the common parts, and the UPS modules with a high failure rate are in parallel. As with parallel redundant UPSs, if one UPS module fails, the remaining UPS modules continue operation. Modular redundant UPSs are low-cost compared to parallel redundant configurations because the bypass line and battery are common. However, the modular redundant UPS generate circulating currents without isolation by transformers.

3 Modeling of circulating current

3.1 Unit redundant UPS without transformer

Figure 3(a) shows the circuit configuration of the parallel redundant UPS, with EMI filters at the input and output to reduce EMI and a Y capacitor for grounding. The transformer is built into the output section, and the input and output sections are insulated, thus blocking the path of circulating current. In parallel, the output transformer and battery are included in each UPS, so there is no circulating current path because there is no common part.

Figure 3(b) shows modular UPS without transformer. When there is a voltage difference between the input and output, circulating currents

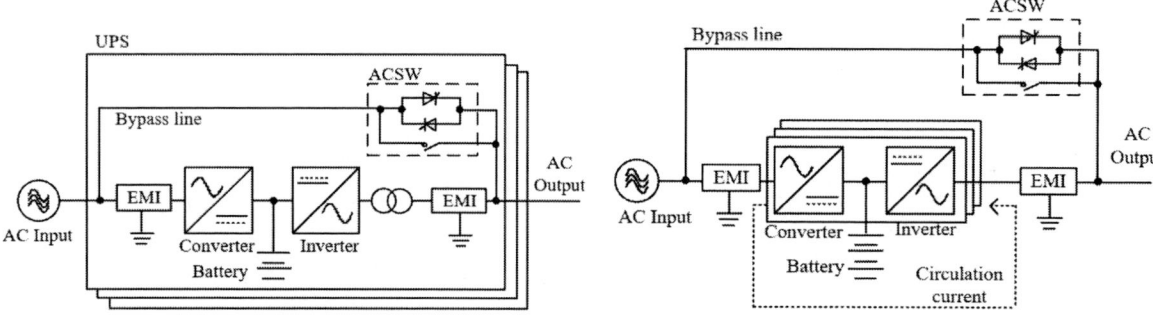

(a) Traditional parallel UPS   (b) Modular UPS without transformer

Fig. 3   Circuit diagram of parallel UPS

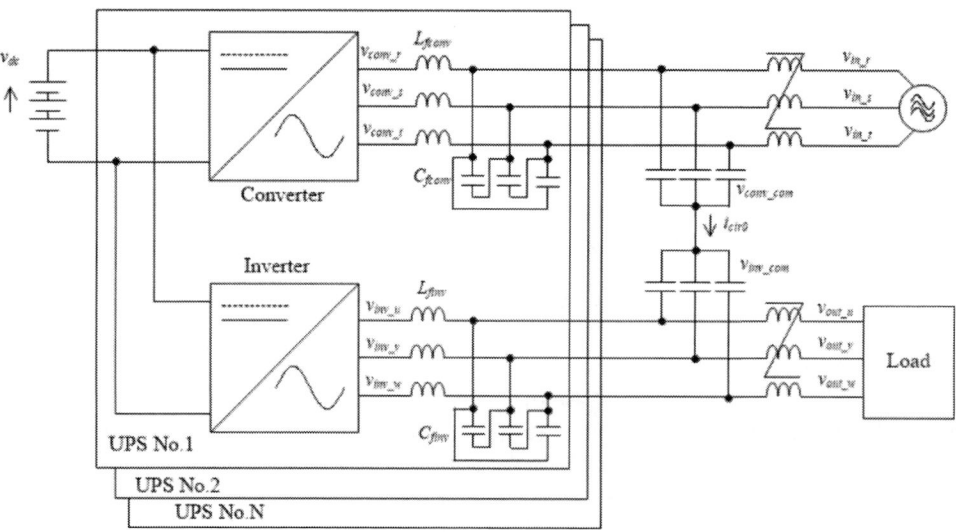

Fig. 4   Circulation Current flow of UPS without transformer

1370

also flow due to the common mode voltage difference between the inverter and converter via the Y capacitor of the EMI filter. Analysis of circulating currents is important because of concerns about deteriorating control performance and effects on component derating.

3.2 Circulating current analysis

Figure 3(a) shows the circulating current path through the Y capacitor of the transformer less modular UPS shown in Fig.4 as a common part connected to the converter and inverter viewed from the DC link voltage for clarity.

The current i_{cir0} flowing in the Y capacitor can be expressed by Eq. (1) based on the voltage difference v_{com} between the converter common voltage v_{conv_com} and the inverter common voltage v_{inv_com} and the impedance of the Y capacitor.

$$i_{cir_0} = \frac{v_{conv\_com} - v_{inv\_com}}{Z_y} = \frac{v_{com}}{Z_y} \tag{1}$$

The inverter common voltage and converter common voltage of one UPS module are shown in the following equations.

$$v_{conv\_com} = \frac{1}{3}\left(v_{conv\_r1} + v_{conv\_s1} + v_{conv\_t1}\right) \tag{2}$$

$$v_{inv\_com} = \frac{1}{3}\left(v_{inv\_u1} + v_{inv\_v1} + v_{inv\_w1}\right) \tag{3}$$

Using Fourier series expansion, the converter R-phase voltage is expressed as in Eq. (4).

where ω_{sconv} is the converter switching angular frequency.

$$v_{conv\_r} = v_{dc}\left\{ \frac{1}{2}v_{r1}^* + \sum_{n=1}^{\infty} \frac{2}{n\pi}\sin\left(\frac{n\pi}{2}(1 + v_{r1}^*)\right)\cos\omega_{sconv1}t \right\} \tag{4}$$

Using the converter voltage command v_{r1}^* shown in Eq. (5), the converter R-phase voltage v_{conv_r1} can be developed as in Eq. (6). where is the converter modulation rate a_{cnv1}, the input angular frequency ω_{in}, and the phase difference ϕ_1 with the inverter voltage.

$$v_{r1}^* = \alpha_{cnv1}\sin(\omega_{in}t + \phi_1) \tag{5}$$

$$v_{conv\_r1} = \frac{1}{2}\alpha_{conv1}\sin(\omega_{in}t + \phi_1)v_{dc}$$
$$+ \frac{1}{2}v_{dc}\sum_{n=1}^{\infty}\frac{2}{n\pi}\sin\left(\frac{n\pi}{2}(1 + \alpha_{cnv1}\sin(\omega_{in}t + \phi_1))\right)\cos n\omega_{sconv1}t \tag{6}$$

The S-phase and T-phase voltages also have a phase difference of $2/3\pi$ from the voltage command value. From Eq. (2) and Eq. (6), the converter common voltage can be expressed as in Eq. (7).

The inverter side common voltage can also be expressed by Eq. (9) using the inverter voltage command v_{u1}^* shown in Eq. (8). where is the converter modulation rate a_{inv1} and the output angular frequency ω_{out1}.

$$v_{u1}^* = \alpha_{inv1}\sin\omega_{out1}t \tag{8}$$

$$v_{conv\_com} = \frac{1}{3}\left(v_{conv\_r1} + v_{conv\_s1} + v_{conv\_t1}\right)$$
$$= \frac{1}{3}\left(\frac{1}{2}\alpha_{conv1}\sin(\omega_{in}t + \phi_1)v_{dc} + \frac{1}{2}v_{dc}\sum_{n=1}^{\infty}\frac{2}{n\pi}\sin\left(\frac{n\pi}{2}(1 + \alpha_{conv1}\sin(\omega_{in}t + \phi_1))\right)\cos n\omega_{sconv1}t \right.$$
$$+ \frac{1}{2}\alpha_{conv1}\sin\left(\omega_{in}t - \frac{2}{3}\pi + \phi_1\right)v_{dc} + \frac{1}{2}v_{dc}\sum_{n=1}^{\infty}\frac{2}{n\pi}\sin\left(\frac{n\pi}{2}\left(1 + \alpha_{conv1}\sin\left(\omega_{in}t - \frac{2}{3}\pi + \phi_1\right)\right)\right)\cos n\omega_{sconv1}t$$
$$\left. + \frac{1}{2}\alpha_{conv1}\sin\left(\omega_{in}t - \frac{4}{3}\pi + \phi_1\right)v_{dc} + \frac{1}{2}v_{dc}\sum_{n=1}^{\infty}\frac{2}{n\pi}\sin\left(\frac{n\pi}{2}\left(1 + \alpha_{conv1}\sin\left(\omega_{in}t - \frac{4}{3}\pi + \phi_1\right)\right)\right)\cos n\omega_{sconv1}t \right) \tag{7}$$

$$v_{inv\_com} = \frac{1}{3}\left(v_{inv\_u1} + v_{inv\_v1} + v_{inv\_w1}\right)$$
$$= \frac{1}{3}\left(\frac{1}{2}\alpha_{inv1}\sin\omega_{out1}t\, v_{dc} + \frac{1}{2}v_{dc}\sum_{n=1}^{\infty}\frac{2}{n\pi}\sin\left(\frac{n\pi}{2}(1 + \alpha_{inv1}\sin\omega_{out1}t)\right)\cos n\omega_{sinv1}t \right.$$
$$+ \frac{1}{2}\alpha_{inv1}\sin\left(\omega_{out1}t - \frac{2}{3}\pi\right)v_{dc} + \frac{1}{2}v_{dc}\sum_{n=1}^{\infty}\frac{2}{n\pi}\sin\left(\frac{n\pi}{2}\left(1 + \alpha_{inv1}\sin\left(\omega_{out1}t - \frac{2}{3}\pi\right)\right)\right)\cos n\omega_{sinv1}t$$
$$\left. + \frac{1}{2}\alpha_{inv1}\sin\left(\omega_{out1}t - \frac{4}{3}\pi\right)v_{dc} + \frac{1}{2}v_{dc}\sum_{n=1}^{\infty}\frac{2}{n\pi}\sin\left(\frac{n\pi}{2}\left(1 + \alpha_{inv1}\sin\left(\omega_{out1}t - \frac{4}{3}\pi\right)\right)\right)\cos n\omega_{sinv1}t \right) \tag{9}$$

From Eq. (7) and Eq. (9), the circulating current is generated by the difference between input frequency and output frequency $\omega_{in}-\omega_{out}$, the phase difference between the converter voltage command and inverter voltage command ϕ_1 and the difference in modulation rate $\alpha_{conv1} - \alpha_{inv1}$, the difference in switching frequency between inverter and converter $\omega_{conv1} - \omega_{inv1}$, and the phase difference.

The inverter common voltage and converter common voltage when N UPS module are in parallel can be expressed as Eq. (10) and Eq. (11) by extending Eq. (2) and Eq. (3).

$$v_{conv\_com} = \frac{1}{N}\sum_{N=1}^{N} \frac{1}{3}\left(v_{conv\_rN} + v_{conv\_sN} + v_{conv\_tN}\right) \quad (10)$$

$$v_{inv\_com} = \frac{1}{N}\sum_{N=1}^{N} \frac{1}{3}\left(v_{inv\_uN} + v_{inv\_vN} + v_{inv\_wN}\right) \quad (11)$$

Figure 5(a) shows the results of numerical calculations of the potential difference between the converter common voltage and the inverter common voltage. The results are obtained when the voltage command phase difference between the converter and inverter is varied from 0 to 2π with the input voltage of 200V/50Hz and the output voltage of 200V/50Hz and the DC voltage of 400V. The inverter and converter generate gate signals on the same carrier, therefore no difference in carrier frequency is assumed in the calculations. When the phase difference is 0°, 120°, and 240°, the inverter and converter voltages are perfectly matched, resulting in the lowest common voltage difference, and the phase reverses at 60°, 180°, and 300°, resulting in the highest common voltage difference. If there is a voltage difference between the converter and the inverter, the common voltage difference at 0°, 120°, and 240° increases.

Figure 5(b) shows the results when the phase difference between the inverter and converter switching angular frequency is varied from 0° to 180° for an input voltage of 200 V/50 Hz and an output voltage of 200 V/50 Hz with a DC voltage of 400 V. The common voltage difference is largest when the phase difference between the switching angular frequency of the inverter and converter is 180°.

4 Experimental results

Circulating currents were measured to confirm the validity of the analysis results. The UPS specifications are shown in Table 1. Three UPSs of 10 kVA each are operated in parallel, and the circulating current through the Y capacitor is measured when the phase difference between the converter

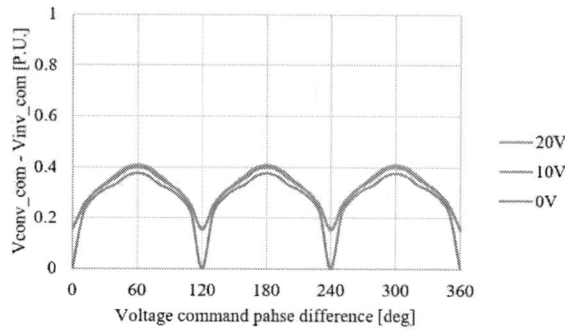

(a) Converter and inverter voltage command difference from 0V to 20V

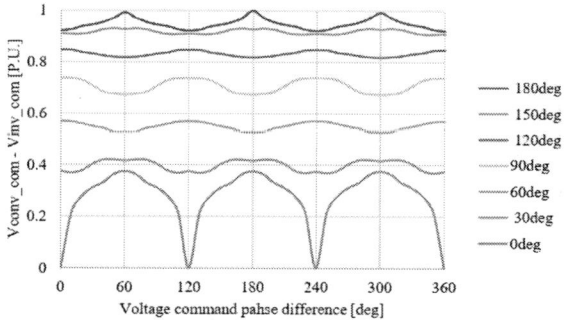

(b) Switching angular frequency phase difference from 0deg to 180deg

Fig. 5 Numerical calculation results of common voltage difference of parallel UPS

Table 1 Specific of prototype UPS

Rating power	10kVA
Input / output voltage	AC200V
Input /output frequency	50Hz
Switching frequency	20kHz
DC link voltage	400V
Number of Parallel UPS	3

and inverter is varied from 0 to 360° with the input voltage of 200 V/50 Hz and the output voltage of 200 V/50 Hz. The inverter and converter generate gate signals on the same carrier. Also, the parallel UPS carriers are synchronized, so there is almost no phase difference in switching angular frequency.

Figure 6 shows the circulating current waveforms. At a phase difference of 0° between the converter and the inverter, there is almost no high frequency circulating current, but at a phase difference of 60°, a high frequency current with a high peak is flowing.

Figure 7 shows the results of plotting the RMS values of the currents for phase differences between the converter and the inverter from 0° to 360°. The circulating current is maximum at phase differences of 60°, 180°, and 300°. Large circulating currents also flow around 150° phase difference, but this is considered to be due to leakage current to the ground side. Therefore, it can be said from these results that the analytical results and the test results are in general agreement.

5 Conclusion

The purpose of this study is to analyze the circulating current that occurs in parallel operation of the transformerless modular UPS and to clarify the conditions under which the circulating current occurs. By formulating the common voltage fluctuation, the circulating current in the Y capacitor was analyzed, and the dominant effect of the phase difference between the converter and the inverter was confirmed from the analysis and experimental results.

References

[1] Organization for Cross-regional Coordination of Transmission Operators Japan: "Annual Report FY2023", HP, https://www.occto.or.jp/en/index.html

[2] CEER: "7TH CEER-ECRB BENCHMARKING REPORT ON THE QUALITY OF ELECTRICITY AND GAS SUPPLY 2022", HP, https://www.ceer.eu/

[3] GS Yuasa: "UPS INDEX," HP, https://ps.gs-yuasa.com/products/, (2023)

[4] T. Kawabata and S. Higashino, "Parallel Operation of Voltage Source Inverters," IEEE Trans. on Industry Application, vol. 24, no. 2, March/April 1988, pp. 281-287.

[5] H. Hanaoka; M. Nagai; M. Yanagisawa, "Development of a novel parallel redundant UPS", Proc. of INTELEC 2003, pp.493-498

[6] J.M. Guerrero; N. Berbel; L.G. de Vicuna; J. Matas; J. Miret; M. Castilla, "Droop control method for the parallel operation of online uninterruptible power systems using resistive output impedance", Proc. of APEC '06., 19-23 March 2006, pp.1716-1722.

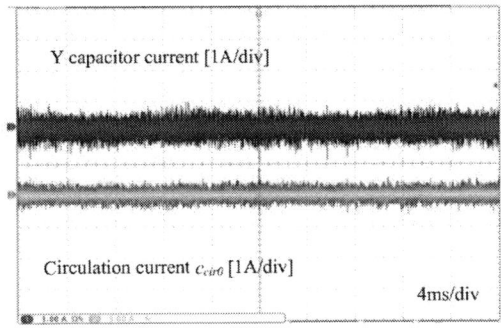

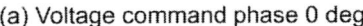

(a) Voltage command phase 0 deg

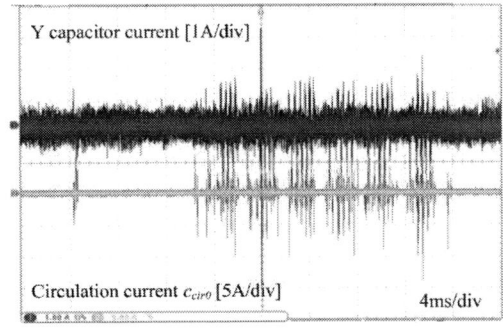

(b) Voltage command phase 60 deg

Fig. 6 Circulation current waveform

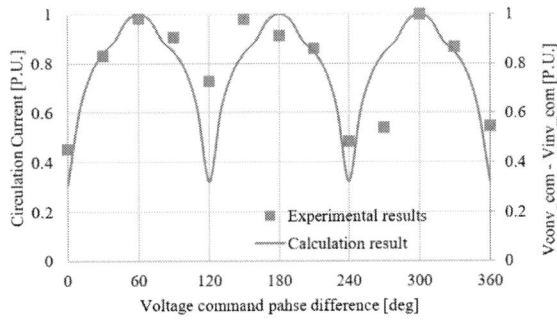

Fig. 7 Circulation Current

[7] Baoze Wei; Josep M. Guerrero; Juan C. Vásquez; Xiaoqiang Guo, "A Circulating-Current Suppression Method for Parallel-Connected Voltage-Source Inverters with Common DC and AC Buses", IEEE Trans. on Industry Applications, Vol.53, no. 4, July-Aug. 2017, pp.3758 – 3769.

[8] K.Kato, H.Igarashi, F.Nagano, T.kojima, A.Suzuki, A.sato, "Development of unit type UPS FULLBACK MLU Series", GS Yuasa Technical Report, Vol.20 No.2 December 25, 2023, pp.7-11

PCIM Europe 2024, 11– 13 June 2024, Nuremberg   DOI: 10.30420/566262188

Design Challenges and Considerations for Gate Drivers of SiC MOSFETS and their Testing

Niranjan Hegde[1], Shubha B[2], Srikrishna NH[3]

[1] Tektronix, India
[2] Tektronix, India
[3] Tektronix, India

Corresponding author: Niranjan Hegde, niranjan.hegde@tektronix.com
Speaker: Niranjan Hegde, niranjan.hegde@tektronix.com

Abstract

The testing of SiC gate drivers involves various parameters, including turn-on and turn-off delay times, switching losses, and short circuit protection. These tests help to ensure that the SiC-based power electronics devices are operating efficiently, reliably, and safely.

Silicon carbide typically has high dv/dt and di/dt and is accompanied with high-speed switching operations. This can cause EMI noise and stability issues. Gate drivers need to be immune to these noise problems. While designing high performance gate drive circuits, some of the unique characteristics of SiC, GaN MOSFETs like transconductance and negative gate voltage to fully discharge the gate are very important. Dynamic characteristics such as on–resistance, gate charge (Miller plateau) and over–current protection are also of high impact.

When designing gate drivers for silicon carbide (SiC) power MOSFETs, it is crucial to consider the negative impacts caused by parasitic inductance in the gate loop. These effects may include false turn-on, gate overstress, and decreased switching speed.

This paper talks about ways of testing driver parameters using oscilloscope and signal generator, which helps in signal visualization and accurate debugging.

1 Introduction

SiC (Silicon Carbide) gate drivers testing is a process of evaluating the performance and reliability of gate drivers used in SiC-based power electronics devices. This testing is crucial because SiC-based devices are known for their high switching frequency and high-temperature operation, which can make them prone to various issues such as gate oxide breakdown, gate threshold voltage shift, and voltage overshoot.

In recent years, SiC-based power electronics devices have gained significant attention due to their high efficiency, high power density, and high-temperature operation capabilities. Therefore, the testing of SiC gate drivers has become increasingly important to ensure the reliability and performance of SiC-based power electronics devices.

1.1 Significance of SiC (silicon carbide) power modules

Silicon carbide (SiC) power modules have revolutionized high-power electronics due to their exceptional properties. These modules offer higher efficiency by lowering switching losses and reducing conduction losses. SiC can operate at elevated temperatures, making it suitable for demanding environments. They can handle higher voltage levels, enabling compact designs. These modules allow downsizing of power converters while maintaining performance [8].

In summary, SiC power modules pave the way for more efficient, compact, and robust high-power systems.

1.2 Gate drivers for SiC MOSFETs

Gate drivers play a crucial role in the performance and reliability of Silicon Carbide (SiC) MOSFETs, especially due to their unique switching characteristics. SiC MOSFETs are known for their ability to switch at high frequencies and handle high voltages, which are beneficial for power electronics applications. However, these features also present challenges that gate drivers must address:

1374

Fast Switching Speeds: SiC MOSFETs can switch much faster than silicon devices, which requires gate drivers to manage very high dv/dt (rate of change of voltage) during switching. This can lead to issues like ringing, which can potentially damage the MOSFET.

Negative Voltage Handling: To prevent unwanted turn-on due to voltage spikes, gate drivers for SiC MOSFETs are designed to provide a negative voltage, which helps minimize these spikes and the chances of accidental turn-on.

Drive Strength: The low transconductance of SiC MOSFETs makes them challenging to drive. Gate drivers must be capable of delivering sufficient current to ensure efficient and reliable switching.

Protection Features: Gate drivers must include features like under-voltage lockout, over-current protection, and short-circuit protection to safeguard the SiC MOSFETs during abnormal operating conditions [10].

Thermal Management: Due to the high power densities involved, gate drivers must be designed to handle the thermal aspects of SiC MOSFET operation, ensuring that the devices do not overheat and fail. BTI is a serious reliability concern in power MOSFETs, including those made of SiC. It can cause significant changes (drifts) in the threshold voltage of the device. This drift can be influenced by various factors, including the gate stress bias and temperature [4].

Overshoot protection: Overshoot during start-up can trigger overvoltage protection, which can cause the system to fail to power up, cause it to restart constantly, or, even worse, result in damage to the next stage or load [3].

2 Design challenges

Designing gate drivers for Silicon Carbide (SiC) MOSFETs presents several challenges due to their unique properties. Some of the key challenges are highlighted below.

Switching Performance: SiC MOSFETs switch much faster than silicon devices, which requires gate drivers to handle higher dV/dt rates. This can lead to issues like ringing and potential damage to the MOSFET if not managed properly [9].

Thermal Management: The gate driver must be able to operate efficiently across a wide temperature range, sometimes up to 300°C, while maintaining the integrity of the switching cell.

Electromagnetic Interference (EMI): Fast switching speeds can induce EMI, which necessitates careful design to minimize common-mode currents and cross-talk.

Stray Inductances: Operating at higher switching speeds can result in overshoots of current and voltage due to stray inductances, which need to be managed to prevent losses and ensure reliable operation [1].

Integration: Achieving heterogeneous integration of the gate driver with the SiC power module is crucial. This involves matching the Coefficient of Thermal Expansion (CTE) and reducing parasitics inside the module [11].

Protection Features: Implementing features like voltage clamping, overcurrent, and short-circuit protection is essential to safeguard the SiC devices during abnormal conditions [7].

Power Supply Levels: Designing gate drivers that can work with different power supply voltage levels is necessary to accommodate various SiC module requirements [7].

3 Implementation Strategies

3.1 Modular Gate Driver Approach

Base Driver (BD) Board: This component is tailored to the specific electrical characteristics of the SiC module. It addresses the unique gate charge and threshold voltage requirements, ensuring that the gate driver is optimized for the particular SiC device it's paired with [5].

Core Driver (CD) Board: This board is designed to be universal for SiC power modules that share the same voltage operation level. It provides a standardized solution that can be used across various applications, promoting design uniformity and ease of integration [5].

Design Simplification: The design avoids the use of programmable elements to reduce complexity and cost. This approach leads to a more straightforward and reliable gate driver circuit, which is easier to produce and maintain.

3.2 Additional Considerations:

High-Temperature Operation: SiC devices often operate at high temperatures, so the gate driver must be designed to function reliably in such conditions. This includes selecting materials and components that can withstand the thermal demands without degradation.

Heterogeneous Integration: Integrating the gate driver within the SiC power module can significantly reduce parasitics and improve performance. This requires careful consideration of the thermal expansion coefficients and packaging techniques to ensure reliability and longevity.

EMI Management: Due to the high switching speeds of SiC devices, electromagnetic interference can be a concern. The gate driver design must include strategies to minimize EMI, such as shielding and proper layout practices [6].

Protection Features: Implementing robust protection features like voltage clamping and overcurrent protection is crucial to prevent damage to the SiC devices during fault conditions.

Drive Strength Control: The gate driver should have the capability to adjust the current drive strength. This is typically achieved through multiple pull-up and pull-down transistors in the output stage, which can be varied to match the load requirements.

Thermal Management: As the operating temperature increases, the peak drive current may also increase, necessitating a gate driver design that can handle these variations without compromising the output swing or the device's performance.

By addressing these considerations, the gate driver can be effectively implemented to harness the full potential of SiC power modules, leading to more efficient and reliable power electronic systems.

4 Testing gate drivers

Testing Silicon Carbide (SiC) gate drivers presents a unique set of challenges due to the advanced characteristics of SiC devices. These challenges are critical to address to ensure the reliability and performance of SiC-based power systems.

High-frequency operation: SiC and GaN devices are designed to operate at high frequencies, which can make it challenging.
The paper explains how some of these gate driver technical challenges can be measured usin g Oscilloscope and Function generators (AFG3 1000) in Fig. 4 test setup.
For SiC power MOSFETs and GaN devices are very important to test Miller clamping parameters and how drivers can be protected again st negative overvoltage stress.
We would like to highlight that switching turn O FF process can be accelerated, and by measu re gate loop oscillations, which will help to red uce switching loss.

Hence overall, testing gate drivers for SiC and GaN devices can be challenging due to the high-frequency operation, parasitic effects, power dissipation, voltage and current transients, and noise and EMI issues.

4.1 Double Pulse Test (DPT)

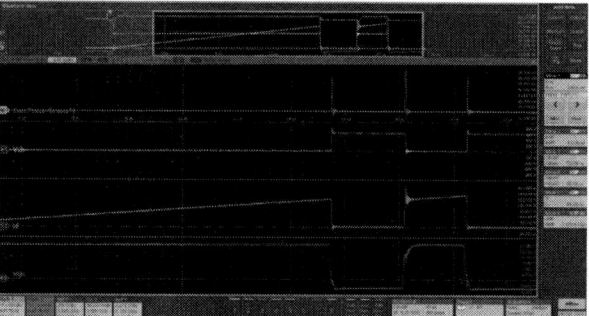

Fig. 1   Typical DPT measurement on oscilloscope

The double pulse test is a standard method for evaluating the switching performance of SiC MOSFETs and their gate drivers. It involves applying two short pulses to the gate of the device to turn it on and off twice in quick succession. This test is crucial because it allows engineers to observe the dynamic behaviors of the device, including turn-on and turn-off times, switching losses, and the effects of parasitic inductances and capacitances.

Typical DPT using switching measurements is shown in Fig. 1.

Challenges in DPT include the following.

High dv/dt and di/dt: SiC devices switch at very high speeds, leading to significant voltage and current rate changes. Capturing these rapid transitions requires specialized equipment with high bandwidth and fast sampling rates.

Thermal Management: The rapid switching can generate significant heat, which must be managed to prevent thermal runaway and ensure accurate measurements.

Electromagnetic Interference (EMI): The fast-switching speeds can induce EMI, which can affect measurement accuracy and require additional shielding or filtering.

4.2 Dead Time Measurement

Dead time is the brief period during switching transitions when both the high-side and low-side switches are off to prevent short circuits. Accurate measurement of dead time is vital for optimizing the efficiency of power converters and preventing shoot-through events [12].

A deadtime measurement using high and low side gate signals is shown in Fig. 2.

Some challenges in Measuring Dead Time are listed below.

Precision Timing: Measuring dead time requires precise timing instruments to capture the narrow time intervals between switch operations.

Variability: Dead time can vary with operating conditions such as temperature and load, making it necessary to test under a range of conditions.

Safety Margins: Ensuring that the measured dead time includes safety margins to account for device and driver variances is essential.

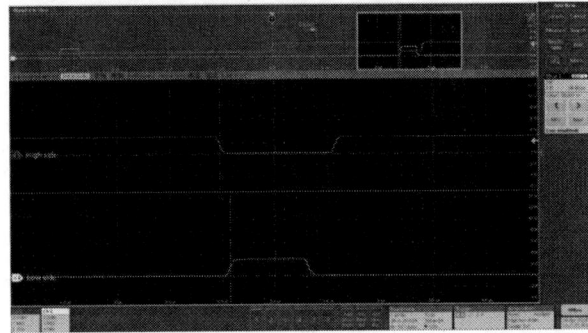

Fig. 2   Deadtime measurement

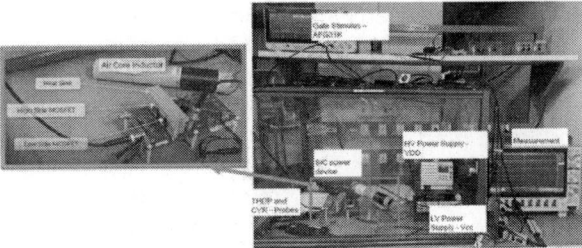

Fig. 3   DPT setup

4.3 Gate Charge Measurement

Gate charge (Qg) is a critical parameter that affects the driving requirements of SiC MOSFETs. It represents the total charge required to turn the device on and off.

Challenges in Measuring Gate Charge:

High Precision Instruments: Accurately measuring gate charge requires high-precision instruments capable of detecting small charge quantities.

Device Variability: Gate charge can vary from device to device, necessitating testing of multiple samples to obtain a representative measurement.

Voltage Levels: The gate charge measurement must be performed at different gate-source voltage levels to understand the device's behavior across its operating range.

4.4 Other Testing Considerations

Short-Circuit Testing: Evaluating the gate driver's ability to protect the SiC device during short-circuit conditions is crucial. This involves measuring the response time of the driver's protection features.

Common-Mode Transient Immunity (CMTI): High CMTI is essential for SiC gate drivers to handle the rapid voltage changes without malfunctioning. Testing for CMTI involves subjecting the driver to high common-mode voltage transients.

Long-Term Reliability: SiC devices are expected to operate reliably over a long period. Stress testing under accelerated aging conditions helps predict the long-term behavior of gate drivers.

4.5 Typical Test setup

In a typical double pulse test setup, the power semiconductor being tested is referred to as the Device Under Test (DUT). The DC Supply provides the necessary power for the test, and the DC Link Capacitor Bank, also known as a capacitive energy storage device, stores the DC energy for the test. The Load Inductor functions as a magnetic transducer. It should not saturate at the peak test currents and should limit the di/dt so that the switching times are not less than 10 μs [2]. The Gate Driver Circuit is used to control the gate of the DUT, and the Arbitrary Waveform Generator (AWG) is used to generate the required waveforms. Typical test setup is shown in Fig. 3.

4.6 Test setup considerations

Probe Selection: Probes that can handle high voltages and fast slew rates typical of SiC devices should be chosen. Differential probes are often preferred for their ability to measure small signals in the presence of high common mode voltages.

A schematic with appropriate probing is shown in Fig. 4.

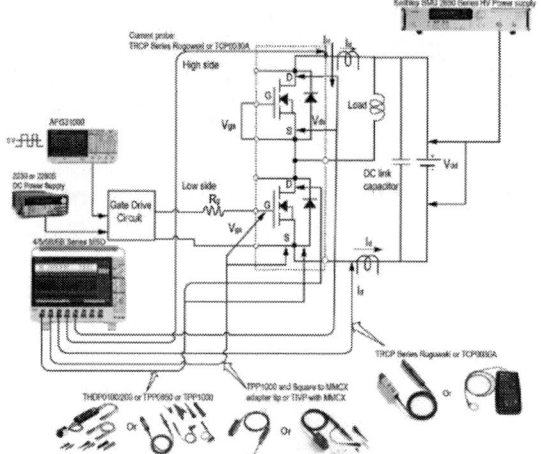

Fig. 4   Typical schematic for a DPT

Bandwidth and Attenuation: It should be checked that the probe has enough bandwidth to capture the high-frequency components of the signal and appropriate attenuation to protect the oscilloscope and maintain signal fidelity.

Parasitic Inductance: Parasitic inductance of the probes should be low to avoid distortions in the measured signal, especially important in high-speed switching applications.

Common Mode Rejection: High common mode rejection is vital to reduce the effects of noise and interference that can be present in high-power environments.

Isolation: Consider optically isolated probes, which offer superior isolation from the common mode voltage and noise, providing more accurate measurements.

Signal Fidelity: The probe should not affect the signal it is measuring. This means it should have low loading on the circuit to prevent altering the performance of the device under test.

Layout Considerations: Short, direct connections with minimal lead length help reduce additional inductance and noise coupling.

Calibration: Regular calibration of the probe and oscilloscope setup is necessary to ensure accurate measurements over time.

Safety: Always follow safety protocols when working with high voltages and currents to prevent damage to the equipment and ensure personal safety. Negative voltage handling is measured as shown in Fig. 4, and Fig. 5 shows overshoot measurement.

5   Results

Fig. 5 shows overshoot measurement, and negative voltage handling is measured as shown in Fig. 6.

Fig. 5   overshoot measurement

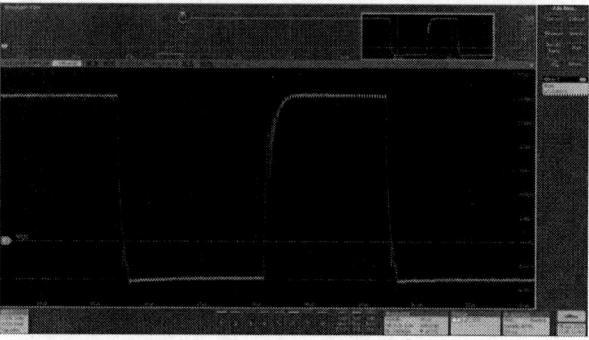

Fig. 4   Negative voltage handling

6   Conclusions

In conclusion, the design considerations and testing of SiC gate drivers are crucial for optimizing the performance and reliability of power electronic systems.

A modular gate driver design, divided into a base driver (BD) board and a core driver (CD) board, can be effective. The BD board depends on the electrical characteristics of the SiC module, while the CD board can be considered 'universal' for any SiC power module with the same voltage operation level.

Understanding and optimizing the gate drive circuitry has a profound effect on reliability and the overall switching performance that can be achieved. The proposed driving circuit ensures the reliability of the SiC MOSFETs and can solve the problem of the driving waveform oscillations efficiently, with a better anti-interference capability and lower loss.

In summary, careful design and thorough testing of SiC gate drivers can lead to significant improvements in the performance and reliability of power electronic systems.

References

[1] S. Hazra, K. Vechalapu, S. Madhusoodhanan, S. Bhattacharya and K. Hatua, "Gate driver design considerations for silicon carbide MOSFETs including series connected devices," 2017 IEEE Energy Conversion Congress and Exposition (ECCE), Cincinnati, OH,USA, 2017, pp. 1402-1409, doi: 10.1109/ECCE.2017.8095954.

[2] David Levett, Ziqing Zheng and Tim Frank, Infineon Technologies "Double Pulse Testing: The How, What and Why"

[3] Penn Zhang, Jason Song, Texas Instruments, "Avoid Start-up Overshoot of LDO".

[4] J. Berens, M. Weger, G. Pobegen, T. Aichinger, G. Rescher, C. Schleich and T. Grasser, "Similarities and Differences of BTI in SiC and Si Power MOSFETs"

[5] Alejandro Rujas, Víctor M. López, Luis Mir, Txomin Nieva, "Gate driver for high power SiC modules: design considerations, development and experimental validation"

[6] J. Shu, J. Sun, Z. Zheng and K. J. Chen, "Gate Driver Design for SiC Power MOSFETs With a Low-Voltage GaN HEMT for Switching Loss Reduction and Gate Protection," in IEEE Transactions on Power Electronics, vol. 39, no. 5, pp. 5558-5566, May 2024, doi: 10.1109/TPEL.2024.3353460.

[7] J. Ma et al., "Ultrafast Gate Driver With GaN HEMTs for ns-Pulse Generator Using SiC MOSFET," in IEEE Transactions on Plasma Science, vol. 51, no. 10, pp. 2771-2780, Oct. 2023, doi: 10.1109/TPS.2023.3289868.

[8] H. Obara, K. Wada, K. Miyazaki, M. Takamiya and T. Sakurai, "Active gate control in half-bridge inverters using programmable gate driver ICs to improve both surge voltage and switching loss," 2017 IEEE Applied Power Electronics Conference and Exposition (APEC), Tampa, FL, USA, 2017, pp. 1153-1159, doi: 10.1109/APEC.2017.7930841.

[9] D. Chatterjee and S. K. Mazumder, "Turn-on Switching Transition Control using a GaN-FET based Active Gate Drive," 2021 IEEE 12th International Symposium on Power Electronics for Distributed Generation Systems (PEDG), Chicago, IL, USA, 2021, pp. 1-7, doi: 10.1109/PEDG51384.2021.9494187.

[10] Y. Sukhatme, V. Krishna, G. P., K. Hatua and S. Bhattacharya, "A simple reduced order model for switching dynamics of Silicon Carbide MOSFET power module," 2018 IEEE International Conference on Power Electronics, Drives and Energy Systems (PEDES), Chennai, India, 2018, pp. 1-5, doi: 10.1109/PEDES.2018.8707647.

[11] M. Amer et al., "Fully Integrated Dual-Channel Gate Driver and Area Efficient PID Compensator for Surge Tolerant Power Sensor Interface," 2020 18th IEEE International New Circuits and Systems Conference (NEWCAS), Montreal, QC, Canada, 2020, pp. 166-169, doi: 10.1109/NEWCAS49341.2020.9159789.

[12] C. -K. Cheung and Z. Gao, "Driver-Integrated Silicon Carbide Based Power Module with Self-Optimized Current-Sensorless Temperature-Driven Deadtime Control," IECON 2023-49th Annual Conference of the IEEE Industrial Electronics Society, Singapore, Singapore, 2023, pp. 1-6, doi: 10.1109/IECON51785.2023.10312695.

PCIM Europe 2024, 11– 13 June 2024, Nuremberg   DOI: 10.30420/566262189

A Portable Efficiency Characterization Setup for Technology Demonstration of Power Modules

Sebastian Tengvall[1], Ariel Muszkat[1], Hoà Lê Thanh[1] ⓘ, and Ahmed Ammar[1] ⓘ

[1] Lotus Microsystems, Denmark

Corresponding author: Sebastian Tengvall, st@lotus-microsystems.com
Speaker: Sebastian Tengvall, st@lotus-microsystems.com

Abstract

This article presents an efficiency characterization setup used for technology demonstration and competitive analysis of low-mid power dc-dc converters. The setup is designed to provide for a high-level of portability, where all components can fit into a size of 30 cm x 10 cm x 10 cm. Two devices under test (DUTs) can be characterized at the same time for the same operational point. A sense circuit measures the input and output voltage and current for each device, and the analog signals are fed into a high-resolution analog-to-digital converter (ADC). A microcontroller (MCU) processes each device's efficiency and writes the data on an on-board LCD display. The presented setup has an accuracy of ± 1 % for efficiency measurement with currents higher than 100 mA, which provides a good trade-off between measurement accuracy and portability.

1   Introduction

1.1   Technology demonstration

Technology demonstration is a key pillar of introducing new technologies to the market. There is a need for accurate and robust demo setups that can also provide for a high-level of portability, allowing for illustrating the technology advantages for the different interest groups in a simple manner, with no requirements for large lab benchtop equipment and wiring. Such setups can be of high relevance when used in technology fairs and technical meetings.

Lotus Microsystems develops power modules based on a silicon substrate. Using a proprietary silicon interposer process, namely the Lotus Power Interposer™, highly integrated power modules that host the power management integrated circuit (PMIC) as well as passive devices can be realized. The Lotus Power Interposer™ is designed to carry higher currents and withstand higher voltages, compared to existing interposer technologies, enabling its use in power applications. Compared to the state-of-the-art packaging technologies for power modules, Lotus Microsystems solutions offer higher power density and superior thermal performance, with no loss in efficiency.

This paper presents the technology demo setup used to demonstrate the performance of power modules that use the Lotus Power Interposer™ vs. conventional substrates such as the laminate substrate and the copper lead frame.

2   Efficiency Characterization

2.1   Setup design

Figure 1 shows a block diagram of the demo setup which is used to characterize the efficiency of different power modules as DUTs. The setup can support two DUTs that are powered from a single source and loaded by similar loads. An LCD display shows the electrical characteristics of each device including the input and output voltage and current, where a thermal camera displays the difference in thermal performance.

The power source is a USB-C adapter that can support the fixed voltages of 5/9/15/20 V, as well as the programmable power supply (PPS) feature that enables a controlled input voltage within a range of 3.3 - 21 V. The power of the USB adapter can vary according to the characterization requirements, however commercial adapters can support up to 120 W in power delivery.

The load is comprised of a number of power resistors of different values mounted on a heat sink. The Arduino kit is responsible for setting the USB voltage of the source, compile the sense circuit output, calculating efficiency, and writing the information to the LCD display.

The interface board hosts the USB negotiation chip, the sensing circuit, and the LCD Display. In addition, it includes the termination for the source, load, DUT devices, and the Arduino kit. The connectors to the DUT boards provide power connections as well as 3 general-purpose I/O connections. Each GPIO is configured using a number of analog multiplexers on the interface board, providing connection to the main system bus voltages (3.3 V and 5 V) as well as analog/digital IOs on the Arduino. This ensures versatility with DUTs that has, among others, enable signal, power good signal, I²C and SPI interfaces. The sensing circuit is comprised of sense resistors and a sensing integrated circuit (IC), INA228 [1], for current sensing, as well as voltage sensing. This IC includes the voltage and current sensing amplifiers, voltage dividers and a 20-bit ADC. The values for the sense resistors and the gain of the current shunt amplifier are chosen according to the ADC full-scale voltage range (span) to maximize accuracy.

Figures 2 and 3 show a picture of the implemented setup and an illustration of the thermal performance difference between two DUTs, respectively. The figures demonstrate that for the same operational point and power loss for the two DUTs, the device employing the Lotus Power Interposer™ has superior thermal performance over the conventional counterpart based on a laminate substrate. That is thanks to the higher thermal conductivity of the silicon material over the different FR4 variants employed in laminate substrates.

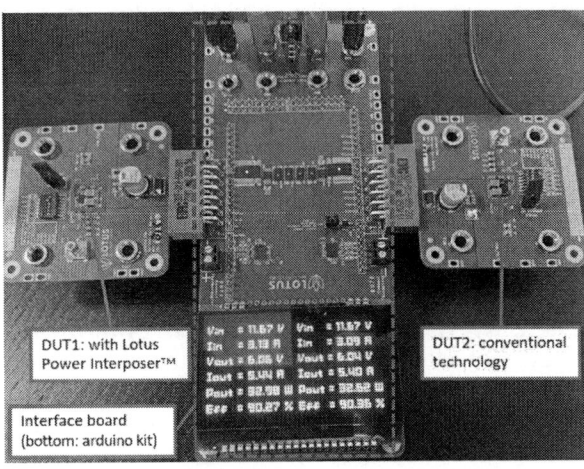

Fig. 2   Implemented characterization setup.

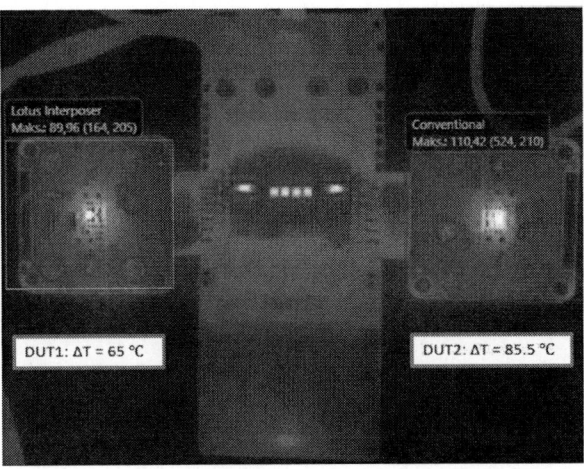

Fig. 3   Illustration of the thermal performance of two DUTs employing the Lotus Power Interposer™ and the conventional substrate technology using the implemented characterization setup.

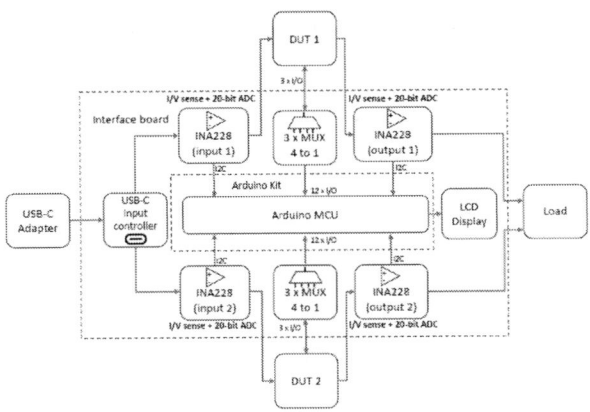

Fig. 1   Block diagram of the demo setup.

Component	Type/Part number
Arduino kit	Arduino MEGA 2560 Rev 3
Sensing circuit	INA228
Current sense resistors	Y14880R01000B9R 2 W 0.1 %
USB controller	AP33772
Analog MUX	TMUX1204
Load resistors	Wire-wound power resistor 50 W 5 %
Heat sink	Aluminum extrusion 300 x 100 x 34.5 mm

Table 1 Summary of the characterization setup bill of material.

Table 1 shows the simplified bill of material of the measurement setup, which is designed for efficiency characterization of low-mid power DUTs up to 100 W.

3 Efficiency Error Model

To address the question of the accuracy of the operational voltages, currents, and efficiency reported by the implemented setup, this section covers how the measurement error can be estimated, where the dominant sources of error are the current sense resistor tolerances and the sensing IC.

Equation (1) below refers to the measured efficiency, where V_{x_meas} and I_{x_meas} are the measured input and output voltage and current values respectively.

$$\eta_{meas} = \frac{V_{OUT\_meas} \cdot I_{OUT\_meas}}{V_{IN\_meas} \cdot I_{IN\_meas}} \quad (1)$$

Figure 4 shows a simplified schematic of the sensing circuit around the INA228 IC.

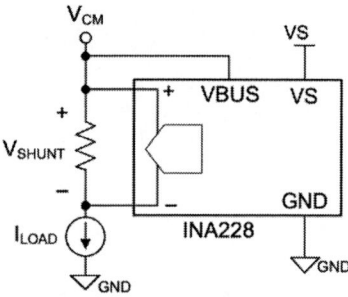

Fig. 4 Sensing circuit schematic [1].

3.1 Current sensing

Equation (2) defines the measured current where x is the direction of the current (input or output), I_{x_err} is the total current measurement error coefficient and I_x is the effective real current flowing in the circuit. Equation (3) defines the current measurement error coefficient in percentage where V_{OS_err} is the error generated by the offset voltage of the current sensing circuit, $CMRR_{err}$ and $PSRR_{err}$ are the errors related to the common mode rejection ratio (CMRR) and power supply rejection ratio (PSRR), respectively, G_{err} is the gain error from the amplifier in the current sensing circuit, and where R_{shunt_err} is the total error resulting from the shunt resistor [2]. I_{x_err} is calculated as a root sum square value (RSS) value instead of

a worst-case value where all the parameters contributing to the errors are at their worst at the same time. All the error sources are expressed in a percentage coefficient between 0 and 1.

$$I_{x\_meas} = I_x \left(1 + I_{x\_err}\right) \quad (2)$$

$$I_{x\_err} = \sqrt{\left(V_{OS\_err} + CMRR_{err} + PSRR_{err}\right)^2 + G_{err}^2 + R_{shunt\_err}^2} \quad (3)$$

Equation (4) shows V_{OS_err}, where V_{OSI} is the current sensing circuit offset in V, V_{OS_drift} is the offset drift in V/°C, ΔTa is the temperature difference between the temperature calibration value of the sensing circuit (datasheet room temperature) and its ambient operating temperature in °C, and where R_{shunt} is the nominal current sensing resistor value in Ω.

$$V_{OS\_err} = \frac{V_{OSI} + \left(V_{OS\_drift} \cdot | \Delta Ta |\right)}{I_x \cdot R_{shunt}} \quad (4)$$

Equation (5) shows the $CMRR_{err}$ definition where V_{CM} is the common mode voltage of the current sensing circuit in V and $CMRR_{min}$ is the worst-case CMRR expressed in dB at room temperature.

$$CMRR_{err} = \frac{|V_{CM} - 48| \cdot 10^{-\frac{CMRR_{min}}{20}}}{I_x \cdot R_{shunt}} \quad (5)$$

Equation (6) shows the $PSRR_{err}$ definition where V_S is the supply voltage of the current sensing circuit in V and where $PSRR_{min}$ is the worst case PSRR of the current sensing circuit expressed in V/V at room temperature.

$$PSRR_{err} = \frac{|V_S - 3.3| \cdot PSRR}{I_x \cdot R_{shunt}} \quad (6)$$

Equation (7) shows the G_{err} definition where AMP_{err} is the amplifier gain error of the current sensing circuit in V at room temperature and where AMP_{err_drift} is the worst-case amplifier gain drift expressed in ppm/°C.

$$G_{err} = AMP_{err} + \left(AMP_{err\_drift} \cdot |\Delta Ta|\right) \quad (7)$$

Finally, (8) shows the R_{shunt_err} definition where $SHUNT_{tol}$ is the current sensing resistor tolerance in percentage and $SHUNT_{temp\_coeff}$ is the temperature coefficient expressed in ppm/°C.

$$R_{shunt\_err} = SHUNT_{tol} + \left(SHUNT_{temp\_coeff} \cdot |\Delta Ta|\right) \quad (8)$$

Note that I_{x_err} depends on the real effective current I_x and the error grows exponentially as the current is reduced. Figure 5 shows an example of how the current measurement error varies with current.

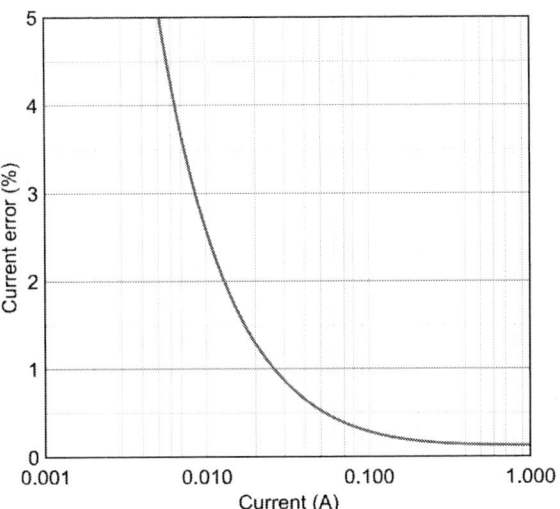

Fig. 5 Current error variation dependency on the measured current [2].

3.2 Voltage sensing

Equation (9) defines the measured voltage, where V_{x_err} is the total voltage measurement error coefficient, and V_x is the effective real voltage in the circuit. Equation (10) defines the voltage measurement error coefficient in percentage. V_{x_err} is also calculated as an RSS value instead of a worst-case value.

$$V_{x\_meas} = V_x \left(1 + V_{x\_err}\right) \quad (9)$$

$$V_{x\_err} = \sqrt{\left(V_{OS\_err} + CMRR_{err} + PSRR_{err}\right)^2 + G_{err}^2} \quad (10)$$

Equations (12-15) define the different error factors similar to (4-7), yet for voltage sensing instead of current.

$$V_{OS\_err} = \frac{V_{OSI} + \left(V_{OS\_drift} \cdot |\Delta Ta|\right)}{V_x} \quad (12)$$

$$CMRR_{err} = \frac{|V_{CM} - 48| \cdot 10^{-\frac{CMRR_{min}}{20}}}{V_x} \quad (13)$$

$$PSRR_{err} = \frac{|V_S - 3.3| \cdot PSRR}{V_x} \quad (14)$$

$$G_{err} = AMP_{err} + \left(AMP_{err\_drift} \cdot |\Delta Ta|\right) \quad (15)$$

The error parameters of the sensing IC and shunt resistor used in (2) to (15) are presented in table Table 2.

Parameter	Value	
	Current sense	**Voltage sense**
V_{OSI}	1 μV	2.5 mV
$V_{OS\ drift}$	10 nV/°C	20 μV/°C
$CMRR_{min}$	154 dB	154 dB
$PSRR$	0.5 μV/V	0.25 mV/V
AMP_{err}	0.05 %	0.05 %
$AMP_{err\ drift}$	20 ppm/°C	20 ppm/°C
$SHUNT_{tol}$	0.1 %	-
$SHUNT_{temp\_coeff}$	15 ppm/°C	-
V_S	5 V	5 V
T_A	30 °C	30 °C

Table 2 Summary of the characterization setup bill of material.

4 Measurement results

A set of 6 comparative efficiency measurements are conducted to compare the accuracy of the demo setup to a lab setup. The lab setup comprises 4 high-precision Keithley DMM6500 multimeters used to measure input and output currents and voltages, and efficiency is calculated. Table Table 3 shows the obtained results.

Table Table 4 shows the output power and efficiency calculations, as well as the difference between the efficiency measurements of the demo setup and the laboratory setup η_{err_meas}. Whereas η_{err_est} denotes the efficiency error calculated with (1) to (15).

Op.	Setup	V_{IN} (V)	I_{IN} (A)	V_{OUT} (V)	I_{OUT} (A)
1	Demo	14.2840	1.5700	6.0680	3.4162
	Lab	14.2920	1.5705	6.0722	3.4162
2	Demo	14.5600	0.5000	3.3700	1.9000
	Lab	14.5646	0.4989	3.3663	1.8958
3	Demo	14.6700	0.1500	1.8200	1.0200
	Lab	14.6714	0.1494	1.8152	1.0222
4	Demo	4.6740	0.1370	3.3060	0.1819
	Lab	4.6730	0.1373	3.3042	0.1821
5	Demo	4.6075	0.0806	3.509	0.1001
	Lab	4.6059	0.0809	3.508	0.1000
6	Demo	14.7590	0.0145	1.8190	0.0997
	Lab	14.7592	0.0152	1.8170	0.0990

Table 3 Voltage and current measurements obtained from the demo and lab setups.

Op.	Setup	P_{OUT} (W)	η (%)	η_{meas}^{err} (%)	η_{est}^{err} (%)
1	Demo	20.7295	92.4356	0.0172	0.3600
	Lab	20.7438	92.4184		
2	Demo	6.4030	87.9533	0.1252	0.3770
	Lab	6.3818	87.8281		
3	Demo	1.8564	84.3626	0.2895	0.4900
	Lab	1.8555	84.6522		
4	Demo	0.6014	93.9131	0.1346	0.5931
	Lab	0.6017	93.7785		
5	Demo	0.3513	94.5839	0.4793	0.8148
	Lab	0.3508	94.1046		
6	Demo	0.1814	84.7428	4.5596	1.8290
	Lab	0.1799	80.1833		

Table 4 Efficiency measurement results obtained with the demo setup relative to the lab setup and the worst-case error.

The table shows that the error in the efficiency measurement reported by the demo setup is less than 1% for measured currents of down to 100 mA. That is verified by the error in the obtained measurements relative to the lab setup. For lower measurement currents, the current error variation is substantially higher, as shown in fig. 5, and the obtained measurement from the demo setup is no longer accurate. Nevertheless, the presented setup offers a good trade-off between setup size and accuracy. At the same time, when comparing the two DUTs, the relative efficiency is of more significance than the absolute one, where the error

contributions are factored into the efficiency calculation of the two devices.

5 Conclusion

An efficiency characterization setup for technology demonstration of power modules is presented. Two devices under test (DUTs) can be characterized at the same time for the same operational point. A sense circuit characterizes the input and output voltage and current for each device where the analog signals are fed into a high-resolution ADC. A DSP processes each device efficiency and writes the info on an on-board LCD display. The presented setup with a size of size of 30 cm x 10 cm x 10 cm has an accuracy of ± 1% for efficiency measurement of currents above 100 mA, which provides for a good trade-off between setup size (portability) and accuracy.

References

[1] INA228 85-V, 20-Bit, Ultra-Precise Power/Energy/Charge Monitor with I2C Interface datasheet (Rev. A).

[2] M. Toa, P, Lliya, "Calculating and simulating error in current sensing systems" *TI Tech Days, Texas Instruments*, pp. 7–9, 2020.

PCIM Europe 2024, 11– 13 June 2024, Nuremberg   DOI:10.30420/566262190

Fast EME Characterization of Bare Die SiC-MOSFETs

Frederik Jülich[1,2], Robert Kragl[1], Karl Oberdieck[1], Konstantin Spanos[1], Rik W. De Doncker[2]

[1] Robert Bosch GmbH, Mobility Electronics, Germany
[2] Institute for Power Electronics and Electrical Drives (ISEA), RWTH Aachen, Germany

Corresponding author: Robert Kragl, robert.kragl@de.bosch.com
Speaker: Robert Kragl, robert.kragl@de.bosch.com

Abstract

This paper introduces a test setup to characterize the electromagnetic emissions (EMEs) of bare-die SiC MOSFETs. In the current design process of power electronics, the EME is solely measured in the product validation phase of the converter during electromagnetic compatibility (EMC) tests (CISPR-25). By measuring the EME of bare dies, the introduced setup enables an EME optimization of SiC MOSFETs during prototyping in the frequency range of 20 MHz to 150 MHz. Besides the introduction and the validation of the developed test setup, this paper shows measurement results of the EME noise spectra of two different SiC MOSFET prototypes. A clear difference can be detected in the two spectra, identifying the superior EME behavior of one prototype.

1   Introduction

The next generation of power semiconductors is greatly influenced by the measurement equipment used during development. In a classical development process, semiconductor prototypes, which differ in their design realization (doping profile, active area, ...), are measured and characterized in their static and dynamic behavior. Here, the measurement equipment for static device characteristics ($R_{DS,on}$, etc.) is a curve tracer, and for dynamic characteristics (E_{sw}, etc.) a double pulse test (DPT). The measured device characteristics are used as inputs for an optimization cost function, which evaluates every prototype. Afterwards, the best performing designs are further developed in the next optimization loop. As a result, modern power semiconductors are continuously improving their measured characteristics [1], [2].

However, there are important device characteristics, like the EME behavior, which are not measured during the semiconductor development. And, consequently, new device generations are not specifically improving in their noise characteristic. Instead, the standardized test for EME measurements is done at the very end of the product development, simultaneously with the measurements which should result in the EMC verification. As a result, the optimization for EME is mostly executed for system parameters and not for the semiconductor itself. This is disad-

vantageous as not all design degrees of freedom are utilized, especially since the semiconductor is the dominant noise source.

Therefore, this paper introduces a reference setup, to characterize the EME of bare-die SiC MOSFETs. Here, the characterization is used for a relative comparison of the devices, rather than defining absolute values for a later product release. This allows for considering the EME of a semiconductor during prototyping with the aim of an EME optimized chip design.

The paper is structured as follows. First, the developed test setup is introduced and important aspects for EMC conform measurements are highlighted. In the second section, the setup is validated and checked for accuracy, repeatability, and meaningfulness. Lastly, two different bare-die SiC MOSFET prototypes are compared in the setup and differences in their EME behavior are discussed.

2   Test Setup

The EMEs of the MOSFETs are measured in a setup similar to their later use in a power electronic system. A three-phase inverter as depicted in Fig. 1 is used. For the discussion of this paper only two phases are utilized in a H-bridge operation.

The EME noise spectra are measured at two Line Impedance Stabilization Networks (LISNs), which are connected at the two DC ports. The LISNs provide high-frequency decoupling of the external

PCIM Europe 2024, 11– 13 June 2024, Nuremberg   DOI:10.30420/566262190

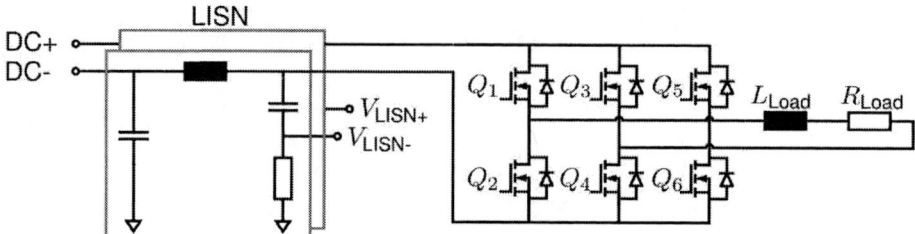

Fig. 1: Schematic overview of test setup. Two of three phases are used to realize a H-bridge configuration.

DC source and two measurement ports, which use a RC high-pass filter. The measurement ports of the LISNs (V_{LISN+} and V_{LISN-}) are connected to a 12-bit oscilloscope. The two voltages can be utilized to determine both the common mode and differential mode voltages of the EME. The seperation can be achieved either through an additional hardware setup, as described in [3], or by digital calculation in post-processing.

A fast Fourier transform (FFT) of the measured data is calculated in the post-processing to transfer the time-domain measurement of the oscilloscope into a frequency spectrum. However, to create a spectrum in accordance to an EMI receiver, a peak detection is used to consider the worst-case for each frequency. Furthermore, the FFT uses, in accordance to the CISPR-25 standard, different bandwidth resolutions between 150 kHz and 30 MHz (9 kHz) and between 30 MHz and 1 GHz (120 kHz) [4]. The post-processing steps, which create voltage spectra in accordance with an EMI receiver, are presented in detail in [5].

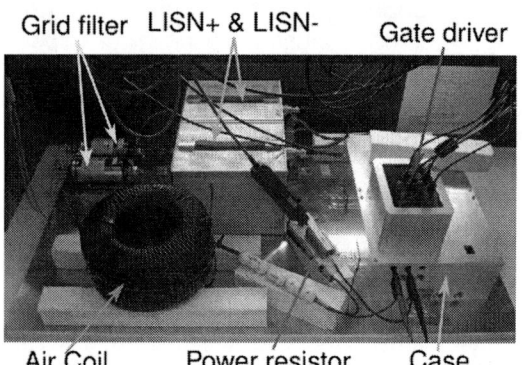

Fig. 2: Picture of the measurement setup.

The setup is developed for EMC conform measurements and a picture of it is shown in Fig. 2. To minimize non-linear effects aside the MOSFETs itself, an air coil is used as L_{Load}. Furthermore, the gate drivers are supplied by batteries to reduce cou-

pling by common ground. The gate drivers are fed by a PWM, generated by a FPGA, which operates in open-loop control. The power stage of the inverter is fully surrounded by a grounded aluminum case to avoid absorption or coupling of external noise.

In the first development stage of a power semiconductor, the devices are not mounted inside of a package or module. Instead, only bare dies are available. Therefore, an adapter PCB is used to mount the devices. The adapter PCB can be soldered on a D2PAK-7 footprint, thereby enabling a quick change of the DUTs. To be functional, the devices source and gate contacts are bonded and surrounded by gel for isolation reasons afterwards. A photograph of the used adapter PCB is shown in Fig. 4.

One challenge of this approach is the limited thermal conductivity of the adapter PCB, which is mounted on top of the main PCB in comparison to the base plate of a later used package or module. Hence, the devices should only be operated in a pulsed operation to avoid exceeding thermal limitations. Furthermore, a sinusoidal operation is beneficial for the EMC characterization of the device, because then the measurement takes all operating points into account. This behavior can be seen in the measurement results in Fig. 3 for some fixed operating points and a sinusoidal inverter operation. The noise spectrum of the operation as an inverter works as an envelope over all occurring operating points and thereby, providing the worst-case information for every frequency.

Here, it should be pointed out, that the step in the noise spectra at 30 MHz is a superposition of two effects. First, there is an actual peak in the emitted noise of the device. Second, as mentioned above, the noise spectra are generated in close accordance with the operation of an EMI receiver and hence, have a change in the used resolution bandwidth at 30 MHz from 9 kHz to 120 kHz.

1386

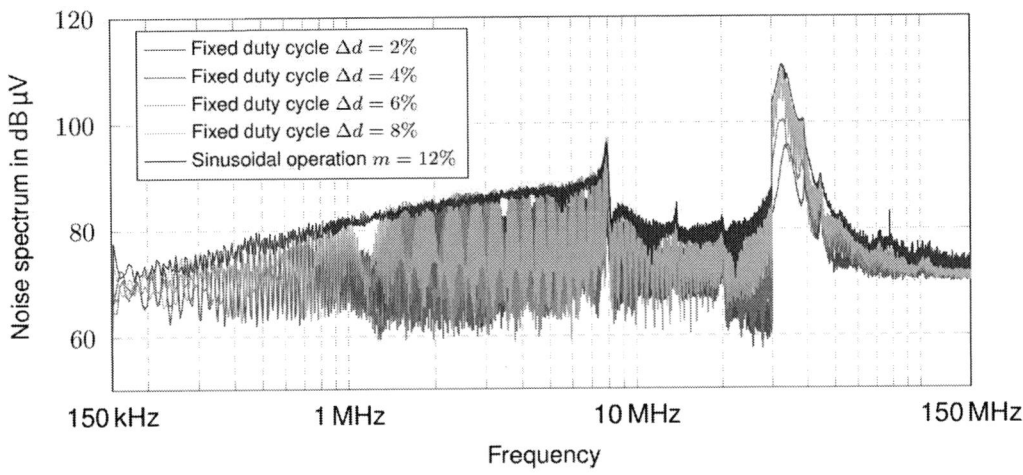

Fig. 3: Noise spectrum of multiple fixed operating points/duty cycles and sinusoidal inverter operation which works as envelope.

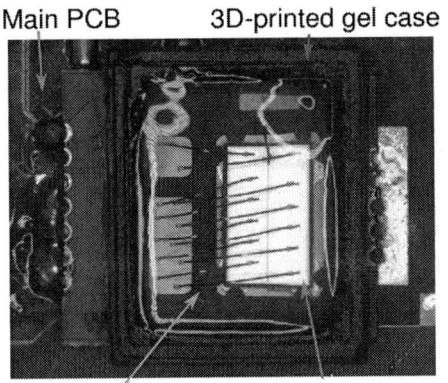

Fig. 4: Adapter PCB to utilize bare dies.

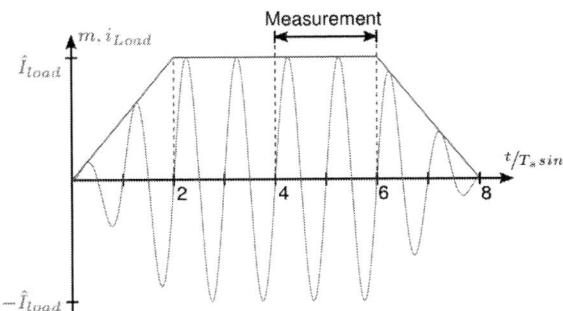

Fig. 5: Schematic siunsoidal, pulsed operation of measurement.

Hence, the inverter was operated as shown in Fig. 5. A sinusoidal load current is ramped up and after a settling time the measurement is triggered. Theoretically, one period is sufficient for measurement, but to increase accuracy multiple periods are taken into account. As the current investigations are executed for devices designed for the automotive market, the frequency is in the range of couple of hundred Hertz (measurement results for 200 Hz). Nevertheless, grid-tied applications with a fixed frequency of 50 Hz are also feasible.

2.1 Validation of Test Setup

This section aims to validate the introduced test setup through three control measurements. Firstly, a measurement assesses the reproducibility of measurements. Secondly, the noise spectrum of the adapter PCB is compared to a D2PAK-7 pack-

age. Lastly, a measurement is performed with a variation in the gate resistance to verify if the system behavior aligns with expected outcomes.

2.1.1 Reproducibility

The EME measurement results of the test-setup show a good reproducibility over multiple operating points. Fig. 6 shows the minimum, maximum, and mean values obtained from five repetitions of an identical test setup configuration (inverter operation with $U_{DC} = 800\,V$). It is evident that the minimum and maximum values closely align across the entire frequency range. This observation holds true for all the measurement results presented, thereby justifying the representation of the mean value for the remaining noise spectra of this paper.

2.1.2 Chip Packaging

The impact of the presented adapter PCB in comparison to a standard D2PAK-7 is investigated. Two devices of the same product were used for the mea-

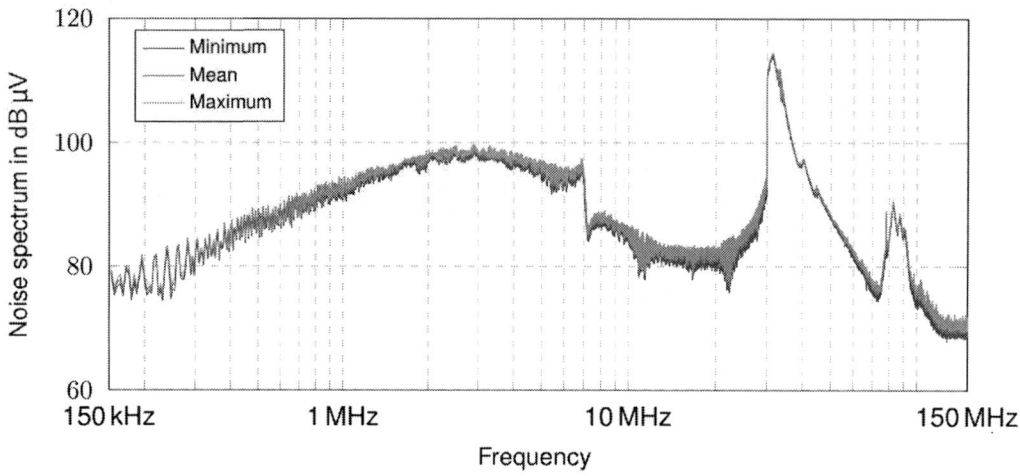

Fig. 6: Minimum, maximum and mean of five repetitive measurements to validate reproducibility of measurement. Operating point in inverter operation and U_{DC} = 800 V.

surement, with one device mounted on the adapter PCB and the other mounted in a D2PAK-7 package. The influence of the chip packaging on the noise spectra is depicted in Fig. 7. It can be observed that the different mounting options do not have any effect on the noise spectrum below 30 MHz. In this frequency range, the systems operation, including voltage and current values, as well as the switching frequency, dominate the spectrum.

Above 30 MHz the switching event becomes prominent and creates two peaks at 31 MHz and 100 MHz. The influence of the package on these two peaks differs. While there is only a slight change in the peak value at 31 MHz, the peak at 100 MHz is shifted to lower frequencies for the adapter PCB. This shift can be explained by an increased leakage inductance of the adapter PCB, which reduces the resonance frequency. The adapter PCB, in contrast to the package, does not have a continuous metal drain contact but instead has vias through the PCB, leading to the increased inductance. Additionally, the bonding wires of the source contact play a role. The D2PAK-7 package utilizes two 500 μm bond wires, whereas the adapter PCB employs four 125 μm bond wires.

2.1.3 Gate Resistance

In a last validation measurement, the same device is measured with a variety of gate resistors. This is done to cross-check the new test setup with reported EMC behavior from the literature. It is well known that the switching speed and hence the EME noise spectrum is increasing with decreasing gate resistance. This was among others reported in [6], [7].

The influence of the gate resistance on the noise spectra is presented in Fig. 8. As anticipated, the influence of the switching event and thus the effect of the gate resistor becomes noticeable at a frequencies above 30 MHz. Additionally, increasing noise voltage levels clearly correlate with decreasing gate resistances.

Based on these observations, the test setup can be deemed validated.

3 Comparison of Device Specimen

As a final step, this section discusses the EME noise spectra comparison of two bare-die SiC MOSFET prototypes. Both devices have the same voltage (1200 V) and current rating (175 A) but differ in their design realization. The used operating point is $I_{Load} = 30$ A, $U_{GS,on} = 15$ V. Measurement results for $U_{DC} = 450$ V and $U_{DC} = 800$ V are shown.

3.1 Double Pulse Test

In Fig. 9 the active turn-on and passive turn-off of a DPT are depicted. The test was executed at $U_{DC} = 800$ V and $I_{Load} = 40$ A. For a positive current, the low-side is utilized as active switch, whereas the high-side is turned off passively.

The DPT was executed in a specific DPT setup, which is the standardized setup for the characterization of datasheet values. Hence, the leakage inductance and thereby the switching event and the resulting spectra are different to measurements of the presented measurement setup. Nevertheless,

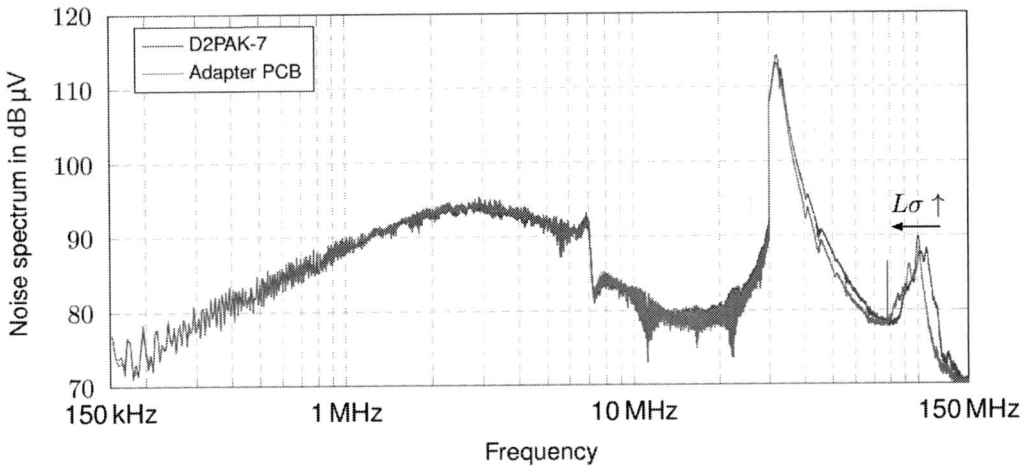

Fig. 7: Comparison of noise spectrum of adapter PCB with D2PAK-7.

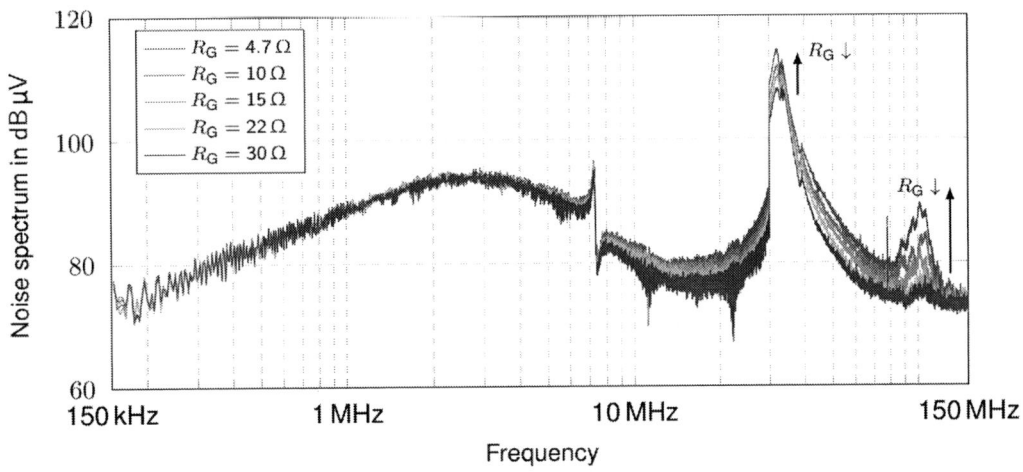

Fig. 8: Noise spectra for a variation of gate resistance.

the DPT setup provides a relative comparison of the prototypes in the time domain.

Prototype		A	B
	du/dt_{max}	−48.3 V/ns	−39.1 V/ns
Active On	di/dt_{max}	7.6 A/ns	6.2 A/ns
	$t_{u,10-90\%}$	33.6 ns	35.7 ns
	$t_{i,10-90\%}$	10.1 ns	13.8 ns
	du/dt_{max}	107.4 V/ns	82.5 V/ns
Passive Off	di/dt_{max}	−7.6 A/ns	−6.2 A/ns
	$t_{u,10-90\%}$	19.1 ns	32.2 ns
	$t_{i,10-90\%}$	9.0 ns	12.6 ns
Sw. Energy	E_{Sw}	1.10 mJ	1.29 mJ

Tab. 1: Comparison of parameters of switching behavior of Prototype A & B in DPT bench.

In the DPT measurements prototype A shows an increased switching amplitude, and shorter rise and fall times, which result in a smaller switching energy loss. The measurement characteristica are listed in Tab. 1.

For the discussion of this paper, the external circuit and the operation of the inverter remain unchanged for both chips. In future research, both prototypes will be operated for same du/dt_{max}, ... or E_{Sw}, and then the noise spectra will be compared. Since the change of one MOSFET design parameter affects multiple aspects of the switching characteristic simultaneously, no simple correlation is found yet. Regarding the EME behavior the DPT results alone are hard to interpret since EME is characterized in the frequency domain. There, it has to be distinguished between overall steepness of the noise in different frequency ranges and the position and

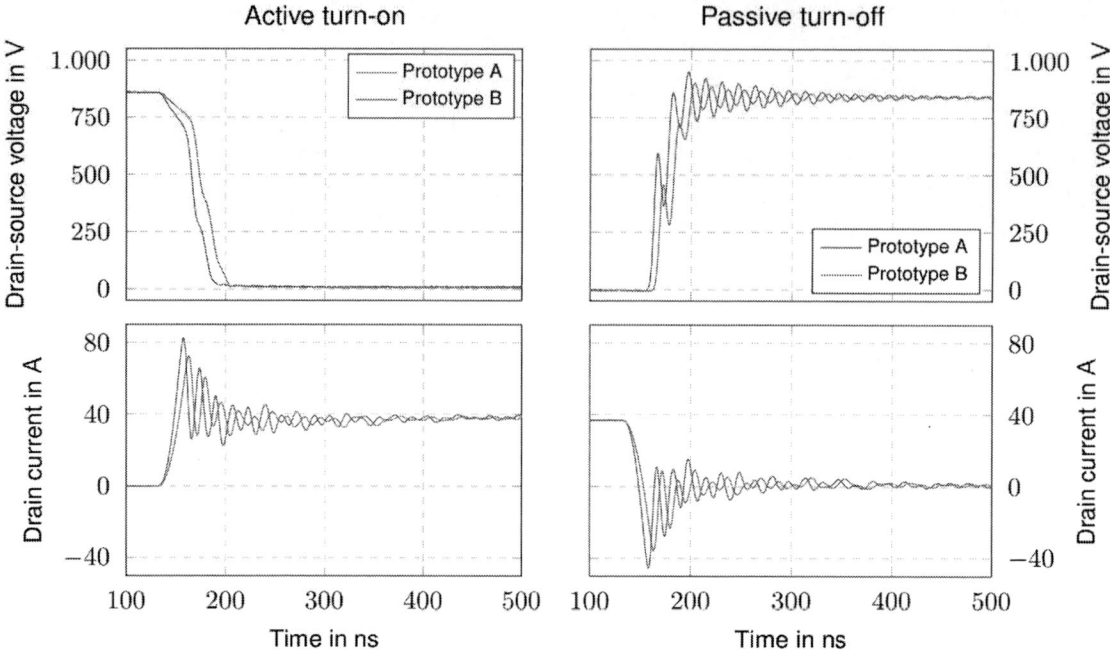

Fig. 9: DPT for prototypes A & B. The measurement is not executed in the presented test setup, but a dedicated DPT test bench.

amplitude of different resonances. Furthermore, as mentioned above, a DPT only shows one specific operating point, while in the proposed inverter operation an integration over all occurring operating points is shown in one graph.

3.2 Fast EMC Characterization

The measured noise spectra at the LISN+, plotted in Fig. 10 and 11, demonstrate superior EME behavior of prototype B across all frequencies, for both $U_{DC} = 450\,V$ and $U_{DC} = 800\,V$. However, the relative behavior of the two prototypes at specific frequencies differ for the two voltages, indicating a non-linear voltage behavior of the MOSFETs.

Comparing the noise spectra at $U_{DC} = 450\,V$ a couple of aspects should be noted. First, the general EME spectrum in between 150 kHz and 30 MHz is the same for both prototypes. The system behavior in this frequency range is solely defined by the operating point and the passives of the system.

Second, the increased voltage and current rise and fall times of prototype A in comparison to prototype B in the DPT translate to an overall increased broadband noise in the frequency spectrum above 20 MHz. In particular, at the frequencies of resonance at 31 MHz, 40 MHz and 45 MHz the amplitudes of prototype A are 4 dB higher than of prototype B. However, the frequency values remain the same for both. This shows that the inductive and capacitive parasitics are the same for both prototypes.

Lastly, the broadband noise in between 60 MHz and 90 MHz has the biggest difference for both prototypes. Here, the amplitude difference is up to 7 dB.

At $U_{DC} = 800\,V$, the EME levels are similar to the behavior at $U_{DC} = 450\,V$ except the EME at the first resonance at 31 MHz. Here, the two prototypes have the same voltages levels.

Overall, prototype B shows superior EME behavior over all frequencies. Thus, in the next optimization loop, prototype B contributes lower EME cost to the cost function. However, the further device developments do not solely depend on the EME cost but on all defined and measured costs.

4 Conclusion

This paper presents a measurement setup that is used to measure and analyze the electromagnetic emissions of bare-die SiC MOSFETs. The purpose of this setup is to enable a relative comparison between different SiC prototypes. This can be used during product development to select the best performing design realization for the next optimization loop.

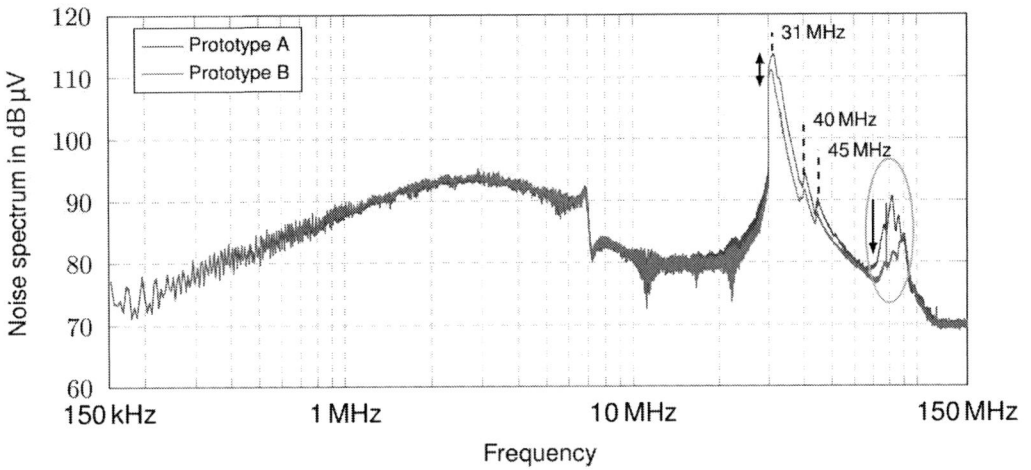

Fig. 10: Noise spectrum comparison of prototypes A & B for $U_{DC} = 450$ V.

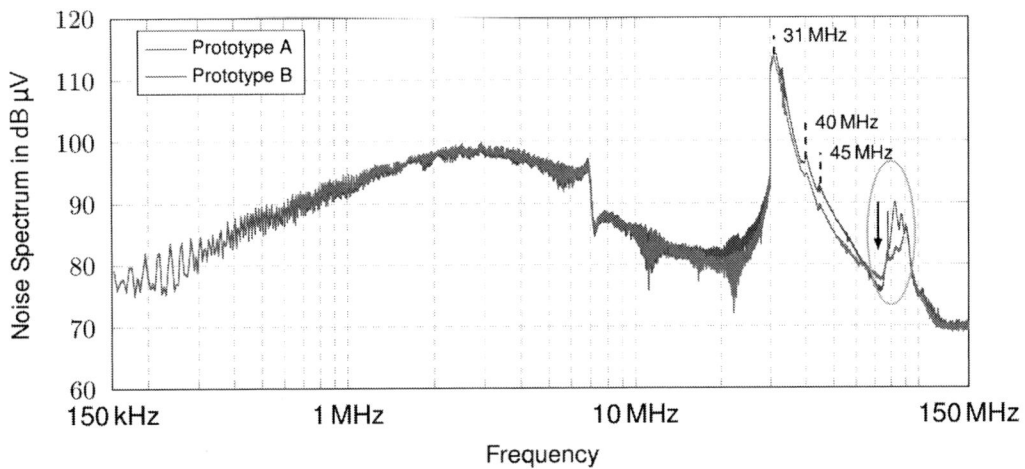

Fig. 11: Noise spectrum comparison of prototypes A & B for $U_{DC} = 800$ V.

The measurement, however, does not provide an estimation of the absolute noise levels emitted in the actual application system. It is important to test the behavior of the final chip design in both the application and the presented test-setup. It is expected that chips performing better in the test-setup will also perform better in the actual application.

The paper includes the measurement of noise spectra from two prototypes with the same voltage and current rating but different design implementations. In comparison, one prototype demonstrates superior EME behavior with up to 7 dB lower EME. This highlights the potential of this measurement test setup in developing a new generation of power semiconductors optimized for EME.

References

[1] S. Schwaiger, K. Heyers, A. Martinez-Limia, K. Oberdieck, and C. Foerster, "Advanced SiC Trench-MOS Technology for Automotive Application", in *International Exhibition and Conference for Power Electronics, Intelligent Motion*, Nuremberg, Oct. 2023, pp. 1–4. DOI: 10.30420/566091096.

[2] B. J. Baliga, "Silicon Carbide Power Devices: Progress and Future Outlook", *IEEE Journal of Emerging and Selected Topics in Power Electronics*, vol. 11, no. 3, pp. 2400–2411, Jun. 2023. DOI: 10.1109/JESTPE.2023.3258344.

[3] A. Nagel and R. De Doncker, "Separating Common Mode and Differential Mode Noise in EMI Measurements", *Taylor & Francis*, vol. 10, no. 2, pp. 27–30, 2000. DOI: 10.1080/09398368.2000.11463462.

[4] *CISPR 25*, en, Edition 4.0. Geneva, Switzerland: International Electrotechnical Commission, 2016, OCLC: 1039732887.

[5] C. Keller and K. Feser, "Fast Emission Measurement in Time Domain", *IEEE Transactions on Electromagnetic Compatibility*, vol. 49, no. 4, pp. 816–824, Nov. 2007. DOI: 10.1109/TEMC.2007.908282.

[6] X. Yuan, I. Laird, and S. Walder, "Opportunities, Challenges, and Potential Solutions in the Application of Fast-Switching SiC Power Devices and Converters", *IEEE Transactions on Power Electronics*, vol. 36, no. 4, pp. 3925–3945, Apr. 2021. DOI: 10.1109/TPEL.2020.3024862.

[7] D. Labrousse, B. Revol, and F. Costa, "Switching cell EMC behavioral modeling by transfer function", in *10th International Symposium on Electromagnetic Compatibility*, IEEE, 2011, pp. 603–606.

PCIM Europe 2024, 11– 13 June 2024, Nuremberg   DOI: 10.30420/566262191

Theoretical Comparison of Component-Related Measurement Methods of Photovoltaic Inverters for Long-Term Testing

Niclas Reitz[1], Sebastian Sprunck[1] , Ron Brandl[1] , Marco Jung[1,2]

[1] Fraunhofer Institute for Energy Economics and Energy System Technology, Germany
[2] Bonn-Rhein-Sieg University of Applied Sciences, Germany

Corresponding author: Niclas Reitz, niclas.reitz@iee.fraunhofer.de
Speaker: Niclas Reitz, niclas.reitz@iee.fraunhofer.de

Abstract

This paper presents a theoretical evaluation of different methods to measure relevant component parameters of photovoltaic (PV) inverters for the purpose of creating a digital twin. The main objective is to measure relevant data about component degradation during long-term testing without changing or influencing the inverter behavior. The paper provides a brief overview of different measurement opportunities and degradation indicators as well as a suggestion for the planned test bench. Chosen options and their implementation for long-term testing of the un-modified PV inverter out of several are described.

1   Introduction

PV inverters are playing an increasingly important role in today's and future energy supply. Therefore, it is becoming more important to ensure a stable and reliable power supply through these devices and to monitor and detect potential failures early enough to reduce and avoid unplanned downtime. Digital twins can play an important role in systems reliability through testing of different environmental and operational impacts and by developing and implementing control algorithms that extend the lifetime of sensitive components. To create a reliable digital twin, it is necessary to have meaningful data that represent the interaction between different parameters and the degradation processes of components over their lifetime. For this purpose, a test setup with different systems will be set up, which are starting at different times. The first test system is to be equipped with measurement technology that is easy to integrate in order to achieve the longest possible runtime during the project schedule, while subsequent systems are to be equipped with more complex measurement technology for more precise condition recording. The investigation is based on PV-inverter Blueplanet 92.0 TL3 from KACO new energy.

To collect necessary data, different measurement options are considered and the most suitable ones for a long-term test will be discussed in more detail. The main objective in this paper focuses on monitoring degradation of film capacitors, IGBTs and SiC-MOSFETs as well as the options for inductors. The above-mentioned components represent the power path for the investigated PV inverter. The goal behind this work is to create a long-term test bench that consists of multiple inverters which are based on the same serial product, but that are each equipped with different options on measurement equipment and components. These modifications should help to measure real-life component behavior in more detail, so that better physical components and more realistic digital twins can be developed for future devices.

2   Literature review

Monitoring the interaction of the various components of the inverter over a long period of normal operation involves greater challenges than is necessary for special lifetime tests for the individual components. Numerous approaches have been published which introduce options on how to measure the change of a certain parameter during defined operation under laboratory conditions. They often use specialized measurement circuits or specialized equipment which is unusable for inverters of serial production under field operation conditions. Fig. 1 shows the relevant components

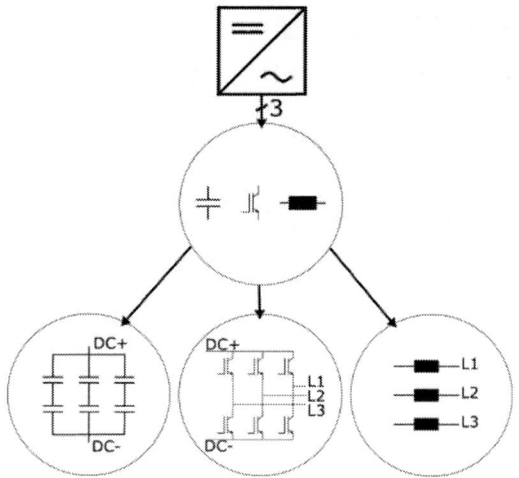

Fig. 1   Schematic overview of the relevant main components of a inverter power path

of the power path which are the main objects under investigation during the long-term test. According to [1] the capacitors possess the highest failure rate out of the three components, followed by the semiconductors. Together, they are responsible for around 50 % of all PV inverter failures [1]. The other 50 % are divided between different components and assemblies of the inverter. Therefore, these two components are the main objects of interest during the planned long-term tests. This investigation focuses mainly on online measurement techniques, since regular offline measurements would come with a high effort to disassemble the inverter and disconnect components, if that was possible without damaging or altering the system for further use. Corresponding offline measurements are planned for the end of the term.

2.1   Capacitors

Capacitors are categorized into three groups based on their mechanical structure: electrolytic capacitors (E-Caps), film capacitors (F-Caps) and ceramic capacitors (C-Caps). All three groups have different advantages what makes them more practical in specific applications. The main purpose in converter applications is to balance instantaneous power differences and minimize voltage variations. Therefore, electrolytic capacitors and film capacitors are commonly used due to their higher capacitance. Authors of [2] compare the different capacitor types in Fig. 2 which visualizes the electrical differences, which can be critical for specific applications.

The relevant capacitors in the investigated PV-inverter are film capacitors which are used in the dc-link as well as in the ac filter structure. Film capac-

itors typically consist of two sheets of polypropylene which are covered with a thin film of conducting material, often aluminum to serve as the electrode. The films are separated by an insulating material which ensures the physical distance and isolation [2]. An advantage of film capacitors in the construction is the so-called self-healing process which removes defects in the insulating material at the cost of a reduction in capacitance, but with the benefit of still retaining functionality.

During the degradation process of capacitors, the capacitance decreases and the equivalent resistance increases, which leads to a change in operation temperature and electrical characteristics of the system, e.g. worse equalization of voltage fluctuations.

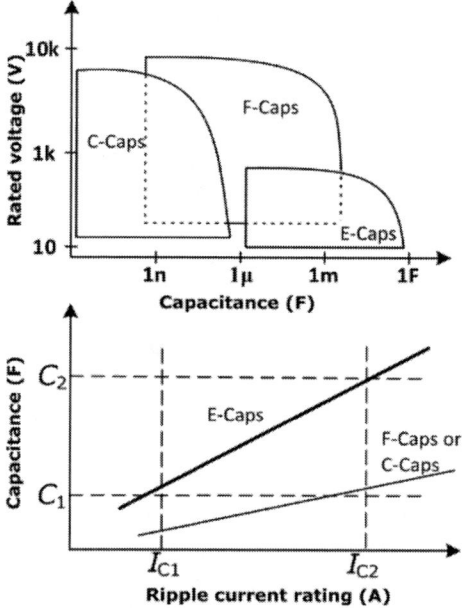

Fig. 2   Electrical capacitor comparison, based on [2]

LCR Measurement

For offline measurements, an LCR meter provides reliable measurements for capacitors and inductances if they can be disconnected from the circuitry. The LCR meter measures the impedance at different AC frequencies to calculate the corresponding parameters of the component. Based on the phase shift between voltage and current, it is possible to determine the capacitance or inductance of the component as well as the equivalent resistance and other parasitic properties. For components which are permanently mounted in the PV inverter or are under operation due to a multitude of currents paths and overlap of signals, the LCR meter is insufficient. Therefore, other measurement methods are required which can be used during operation and are easy to implement.

Ripple-Based Estimation

An often-considered method for monitoring the degradation of capacitors uses the ripple in current and voltage to calculate the capacitance (C) and the equivalent series resistance (ESR) of capacitors [3–5]. According to [5], the ESR of an electrolytic capacitor can be monitored by measuring the voltage and current of the capacitor by using a bandpass filter to reduce the measurement signal to the relevant elements in order to calculate the resistance. There are different approaches to the implementation for determining the resistance, which relate to an analogue or digital approach. In the analogue implementation, the cost of a sufficiently large bandpass filter is high, while for the digital implementation, the costs of a sufficiently fast A/D converter are higher. The proposed method of [5] is based on the power losses (1) and current of the component. The digital implementation requires a relatively quick calculation of power and current which are then summarized for the calculation of the ESR (2). This approach poses the drawback that the frequency dependency is lost. However, this error can be neglected since the frequency dependence of the ESR occurs only outside the normal working area.

$$P_L = \frac{1}{\tau} \int p_c \, dt = \frac{1}{\tau} \int u_{C,ac} i_{C,ac} \, dt$$
$$= R_{ESR} I_{C,rms}^{2} \tag{1}$$

$$R_{ESR} = \frac{P_L}{I_{C,rms}^{2}} = \frac{\frac{1}{\tau} \int u_{C,ac} i_{C,ac} \, dt}{\frac{1}{\tau} \int (i_{C,ac})^{2}} = \frac{\overline{p_c}}{(i_{C,ac})^{2}} \tag{2}$$

This method, as most others based on the ripple, requires the measurement of current of the capacitor, which can be difficult to realize on an already designed PCB without adding any additional components and without being able to monitor individual component parameters.

Dissipation Factor

A different approach is taken in [6] to monitor the degradation of electrolytic capacitors by observing the dissipation factor. There are two possibilities: The first is based on the capacitance and ESR, which makes the investigation using the dissipation factor needless because these values are describing the health status of the capacitor already. The second method is based on the angle between current and voltage at the capacitor and is mainly software-based. With this method, the real parameters of the capacitance and ESR are unknown and the change over time in the dissipation factor is instead used as a health indicator. The

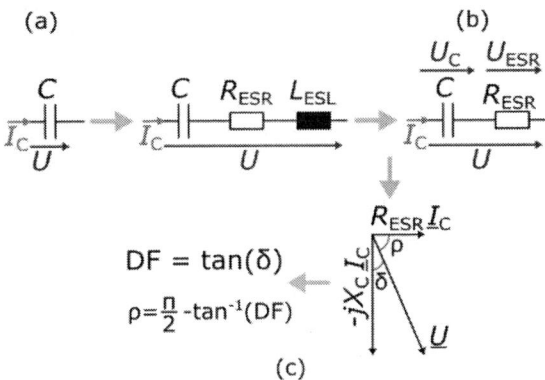

Fig. 3   (a) normal capacitor (b) simplified block diagram for capacitor behavior (c) dissipation factor, based on [6]

changing factor is calculated by dividing the current value by the initial value of the DF. It should be noted that for capacitors with a relatively small ESR compared to the reactance, the calculation error becomes significantly larger due to the characteristics of the tangent function (Fig. 3). To reduce the error, the impedance angle should be close to 45 degrees. Since it is impossible to change the capacitance or the ESR, the only changeable parameter for optimization is the frequency. Also, the research is based only on electrolytic capacitors, since the ESR of film capacitors is relatively small compared to the reactance, which certainly doesn't make it unsuitable as a health indicator, but requires further investigations. Furthermore, frequency and temperature are to be accounted for to reduce the overall error. As the overall error depends partly on the frequency, this method is only recommended if the frequency is adjustable or in a sector where the error is small enough and doesn't affect the reliability of the measurement. The authors suggest a change of 100 % in the DF as a good indication for the end of useful life (EUL) in correlation to the change of ESR and measurement results.

2.2   Semiconductor Switches

To monitor semiconductors, there are different degradation indicators to be considered between insulated gate bipolar transistors (IGBT) and metal oxide semiconductor field effect transistors (MOSFET), whereby some indicators are valid for both components as well as the causes of degradation.

IGBT

IGBT modules belong to the most stressed parts of an inverter due to their continuous exposition to temperature changes, mainly created through power losses [7]. Also, modules start to degrade during operation, which leads to further heat exposure for the devices. There are many mechanisms connected to aging, e.g. bond wire fatigue, die cracking, corrosion and voids. A main cause for degradation and the mentioned degradations option is the coefficient of thermal expansion, which is different between the different materials of a semiconductor and acts as an accelerator for the different processes.

For IGBT degradation monitoring, the on-state voltage ($U_{CE,on}$) is a widely used parameter [8–11]. A typical problem with online monitoring of this voltage is the voltage difference between blocking and conducting states, which as a result needs a high area of operation, while still retaining sufficient resolution to detect changes in the on-state voltage.

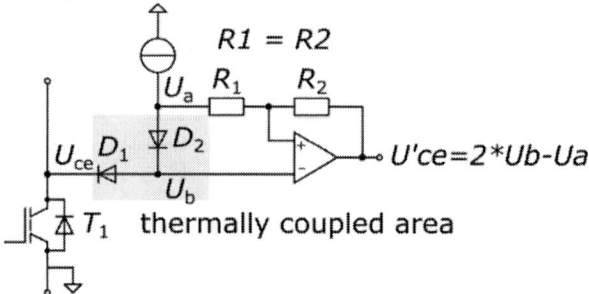

Fig. 4   Block diagram of measurement circuit for U_{CE}, based on [9]

Authors of [8] and [9] describe a circuit on how to measure $U_{CE,on}$ of high-power IGBTs and the anti-parallel diode during normal operation of an inverter for wear out prediction. $U_{CE,on}$ is an often used indicator for degradation of IGBTs, but is normally used in a quasi-offline measurement, since it is temperature dependent and increases with the corresponding junction temperature. Usually, a measurement current is chosen which doesn't generate heating in the device and is easily measurable due to lower requirements on the measurement equipment. The described circuit in Fig. 4 is based on a desaturation protection circuit, which is usually used to protect transistors against short circuit and overload. It allows the measurement of $U_{CE,on}$ and U_f. $U_{CE,on}$ is the voltage drop during conductive state and U_f. the voltage drop across the body diode. For a precise measurement, it is critically relevant that the current through the diodes is equal and that they are thermally coupled to each other, as well as having similar forward voltage drops. It is recommended to use Schottky diodes for the application. The circuitry got validated in a setup and compared to offline measurements by the authors of [8].

[12] gives an overview over recent results of detection methods and estimation processes for IGBTs with the purpose of finding a practical and cost-efficient solution to monitor the health status as well as proposing possible research topics. The recommended voltage measurement is already described in the section above and can be extended with a current measurement to complete an electrical black box of the semiconductor where just the parameters voltage and current are considered. Other influences are neglected to ensure the simplicity of the model. The current measurement can be divided into two main principles, the invasive and non-invasive measurement. Invasive measurements normally include a shunt resistor which can have a significant power loss but is also a relatively cheap option compared to a hall sensor. By tracking these two parameters, voltage and current, it is possible to create a black box model of the semiconductor in which the power loss is related to the junction temperature, while complex and hard to estimate coupling effects are neglect.

[10] introduces an FPGA-based online switching loss measurement, which uses the power loss as a degradation indicator in correlation to the junction temperature, which for example indicates bond wire lift-off and solder fatigue of semiconduc-

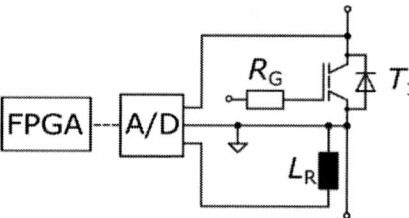

Fig. 5   Block diagram for measuring voltage and current via a Rogowski coil, based on [10]

tors. The evaluation of the measurement results is performed through an FPGA, which uses a voltage measurement and two current measurement options. For the collector current, two approaches are validated with the suggestion of using PCB integrated Rogowski coils over current measurement by parasitic inductors near the IGBT. The measurement signals are digitized by an A/D converter with a high-enough sampling rate before being evaluated by the FPGA. With the switching power loss, the junction temperature can be estimated, and the evaluating results correlate well with a comparison measurement done via an oscilloscope by the authors of [10].

In [7], a highly capable circuit for measuring the voltages $U_{CE,on}$ and U_{th} of an IGBT during the start and stop routine of an inverter is presented. The admissibility of this approach can be justified by the fact that the ageing process is not a short-term event but takes place over a longer period of time, which may end in a clear material failure. The parameter $U_{CE,on}$ indicates package related failures such as solder degradation or bond wire lift of which is measurable in the power path, whereas U_{th} indicates a degradation of the gate oxide retractable to the gate current. The junction temperature gets mentioned as well as a reliable aging indicator, yet is difficult to measure due to the enclosure and requires extensive calculations to achieve a trustworthy result due to the effects of the encapsulation material. The measurement circuit is visible in Fig. 6 and allows the separated measurement of $U_{CE,on}$ and U_{th}, while still being able to allow normal operation for the DUT. To measure the different parameters, a control algorithm is necessary for switching the different com-

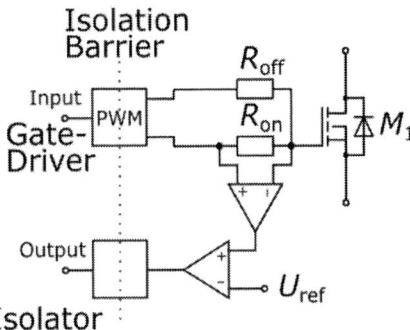

Fig. 7   Simplified block diagram for the measurement of the gate current, based on [13]

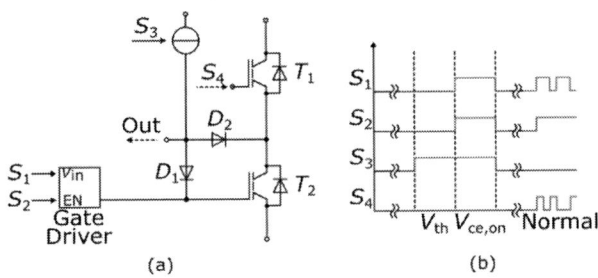

Fig. 6   (a) block diagram of the measurement circuit for $U_{CE,on}$ and U_{th} (b) timing diagram for measuring the different parameters and allowing normal operation, based on [7]

ponents. The proposed circuit is validated by the authors of [7] with a simulation as well as a prototype circuit by comparing the simulation results to a curve tracer.

MOSFET

MOSFETs show similar weak spots as IGBTs and even share similar mechanisms of degradation.

A widely-used parameter for monitoring MOSFET degradation is the on resistance ($R_{DS,on}$) [13–15]. Measuring $R_{DS,on}$ requires equipment which needs to be capable of handling high voltages during the blocking state while still being able to measure low voltages with high resolution during the conduction state as well. Another aspect to consider is the current measurement, which usually requires additional components in the power path that can lead to additional losses.

In [13], a degradation monitoring technique for SiC-MOSFETs based on the gate leakage current

is introduced, with the justification that it is easier to measure than the low $R_{DS,on}$, which tends to be considerably smaller in SiC-MOSFETs compared to Si-MOSFETs. Different parameters such as gate leakage current, drain source leakage current, threshold voltage and device capacitances can be monitored for degradation and were compared in a power cycling test to find the best usable one by the authors of [13]. Another advantage of this approach is the lack of high voltages or currents, easing requirements on probes and acquisition circuits as well as not having a significant temperature dependency of the measured value. The gate leakage current on a healthy device is very low, but significantly higher on an aged device and indicates a degradation of the gate oxide. To estimate the gate current, the voltage drop over the gate turn-on resistance is used to avoid additional components in the driver circuit. For implementation, a differential amplifier is suggested with an additional comparator (Fig. 7) to create a signal which clearly indicates degradation of a certain grade as a yes-or-no condition. For practical implementation, several remarks are given to ensure a reliable realization.

Also Authors of [16] recommend the gate leakage current as a reliable precursor for degradation of SiC-MOSFETs. The proposed circuit is suitable for higher voltages and differs from the one above, since their current measurement approach is

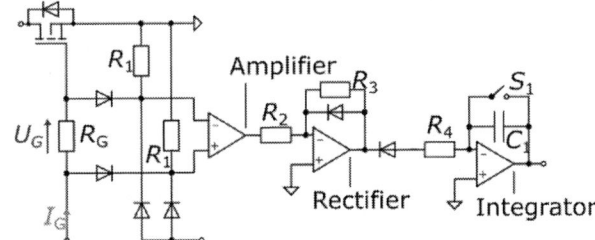

Fig. 8   Block diagram of the measurement circuit for the gate charge over the gate current, based on [16]

based on the positive gate charge. The gate current gets integrated over time to calculate the gate charge during on-state. Furthermore, the characteristics of the gate circuit are mentioned, which can make the current measurement more complicated due to higher currents caused by the gate capacitance during switching. The on-state gate leakage current is estimated over the differential voltage across the external gate resistance. The diode circuitry in necessary to avoid common mode voltage swings during the switching events and ensure a positive voltage while being integrated which is sensitive to the duty ratio. The integrator must be reset periodically to avoid saturation.

Notice that both mentioned options for degradation monitoring based on the gate current require additional circuits that need to be adapted and integrated into the existing circuit. For implementation in a existing product, this requires a modification of the main circuit board or the addition of a separate module, which in turn must be integrated without adding interfering signals to the main circuit.

Authors of [14] investigated the degradation of a MOSFET in a buck convert application to implement a model which takes degradation into account. The used degradation indicators are the $R_{DS,on}$ and the junction temperature, which are directly related and proportional to each other and describe the power loss in the device. The junction temperature is an often-used precursor of degradation which is difficult to measure due to the enclosure of the semiconductor. Therefore, the case temperature is suggested as a credible alternative with a clear correlation to the junction temperature. Furthermore, it was found that an increase in junction temperature is proportional to the increase of $R_{DS,on}$ which permits to use the case temperature to monitor degradation compared to the initial conditions.

Notice here that the investigated MOSFET is a discrete component and not embedded in a module. In a module, the case temperature is less strictly coupled to the junction temperature of one particular semiconductor, as the enclosure includes more components which work as a heat source and therefore make the correlation less precise. Furthermore, the correlation isn't always given between case temperature and junction temperature as it is often for discrete components. This requires a series of calibration measurements to determine the relationship between case temperature and heat source which can be each of the included semiconductors. The concept of heat flow and its dissipation is much more complex in the module, as there are more influencing factors to consider.

2.3 Inductors

The last considered main component in the power path is the choke, where limited studies on durability and online monitoring are available [17]. The cause for this lack of information likely lies in the high reliability of these components [18]. To estimate the parameters of an inductor, an LCR meter can be used during offline measurement as a reliable data source. For online measurements, the winding resistance or impedance can be used as a degradation indicator. Further investigations to identify ageing characteristics are planned.

3 Critical debate on implementation

The measurement methods considered were validated in special setups and in some cases followed the approach of being integrated into an inverter. However, the circuits should be considered at an early stage in the circuit design to enable problem-free integration. The introduction of measurement technology into an existing system should not disrupt the reliability and behavior of this system. The electrical parameters current and voltage are usually mandatory for the approaches considered for monitoring the health condition of capacitors. For the metrological realization, the challenge here is that the capacitors are usually installed in different circuit forms, as is typically the case with the DC link capacitor bank. The approach used here is that the different capacitors are considered as one large one. Alternatively, measurement taps must be created that allow the measurement of current and voltage without redesigning essential components of the inverter. This could be realized, for example, with PCB integrated Rogowski coils in a separated PCB and voltage taps on the capacitor connection pins. This approach involves a certain amount of effort, as various components have to be dismantled and PCBs redesigned.

The approaches considered for monitoring the condition of semiconductors usually also require the use of separate circuits. The circuits mentioned for monitoring IGBTs place high demands on dielectric strength, as these are connected to the power path. The protection and transmission of signals must be designed accordingly to prevent interference with the measuring equipment as well as ensuring the undisturbed transmission of the measurement signals in the long-term. Furthermore, the integration of measuring circuits must be

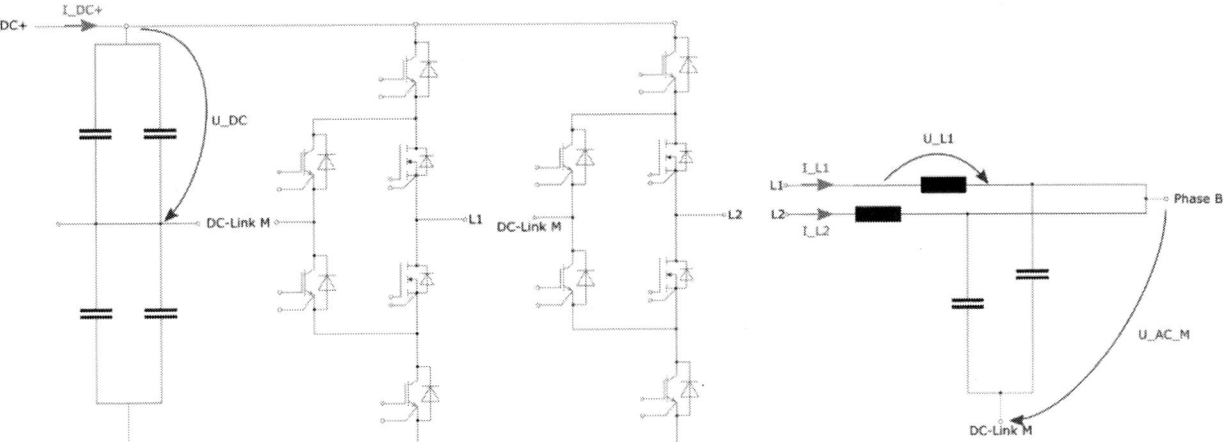

Fig. 9   Simplified single-phase block diagram of the PV inverter power path with implemented measurements for voltage and current

considered a challenge, as semiconductor modules are used in the inverter which make it difficult or impossible to tap at relevant points if they are not led out. The circuits for measuring the gate current are particularly critical here, as the switching behavior of the power semiconductor and thus the entire functionality can be affected. The realization of such measurement interventions must therefore be viewed critically. This can be improved if the measuring circuit can be connected directly to the existing PCB or can be integrated on the existing PCB. If a separate measuring circuit is necessary, special attention must be paid to this in order not to change the basic behavior. The implementation effort is correspondingly high as comparative measurements or simulations are recommended here. However, the costs for the further development and construction of new driver boards should not be underestimated. The measurement of the junction temperature can be used as a permissible alternative; this is possible to a good approximation using temperature sensors that can be integrated into the module and are positioned close to the individual semiconductor chips.

4   Realization of Measurements

The above-mentioned measurements all originate from dedicated component tests where the purpose was to describe and validate a certain method of measurement. The used setups were not necessarily included in inverters and not in serial products, which come with the added challenge of implementing these setups in a readily available product. Due to the reason that there are multiple PV inverters to be used throughout the project, the first one is equipped with minimal effort and mainly non-invasive measurement options.

Different components are measured periodically to monitor the change of state compared to the initial values. The considered components are the DC link capacitors, AC chokes and AC capacitors. The used product allows for separate offline measurements on the AC chokes and on the AC capacitors using an LCR meter by disconnecting corresponding screw terminals.

The DC link capacitors are measured as a group connected to the semiconductor module which can not be disconnected easily. It must be considered that the output voltage of the LCR meter doesn't exceed the threshold voltage of the diodes within the module, as otherwise the measurement would be incorrect due to a conducting path via these diodes. Furthermore, the parasitic effects of the PCB affect the measurement as well.

The measurements of chokes are performed at the beginning of the test as well as in regular time steps using an LCR meter to monitor changes of the electrical characteristics. A high resolution of measurements for the choke is not intended, as significant changes are not expected due to component-specific investigations.

The AC capacitors can be measured with an LCR-meter as well under the condition that both capacitors of one phase are considered as one due to the interleaved system and the reason that there is no opportunity to make a physical separation in the circuitry. Furthermore, it must be noted that through the connection to the PCB and conducting paths, higher parasitic effects are to be expected compared to a component-only measurement.

For online measurements, there are different non-invasive options chosen for the first test system. A wide range of temperature measurements is used to monitor the change of the temperature from different components compared to the initial value. A change in temperature indicates higher losses, which often are a precursor for degradation.

A large number of temperature sensors is distributed across the DC link capacitors, output capacitors, coils and output relays. Furthermore, the temperatures of the semiconductor modules are recorded via an already existing, integrated temperature sensor, as well as the temperature and humidity inside and outside of the inverter housing, serving as references. In order to avoid interfering with the circuits, differential voltage probes and Rogowski coils were used, at the positions as shown in Fig. 9 to record relevant variables and draw conclusions for the other systems.

5   Conclusion

The paper provided an overview of various measurement methods that can be used to monitor ageing. This reflects the current state of the art for integrating measurement sensors into an inverter. Various methods were explained, and their advantages and disadvantages discussed, as well as a possible implementation in a series product. It can be stated that the temperature of components is an important measurement parameter that can be measured relatively easy and can be used as an indicator of change. For capacitors, condition monitoring via the ripple current is recommended, although there are various implementation approaches here. For power semiconductors, a distinction must be made between the individual types, in this case between Si IGBTs and SiC-MOSFETs. The collector-emitter-voltage of IGBTs is cited as a good ageing indicator, but it requires the use of separate measuring circuits which must be integrated into the existing structure and can therefore be problematic. For MOSFETs on the other hand, the gate leakage current is mentioned as a reliable indicator. This also requires modifications or extensions to the current circuit. No clearly identifiable ageing characteristics are known for the inductors that would significantly influence their behavior. This is partly due to the generous design of the components. The first system under investigation was taken into operation using measurement equipment that took minimal effort to install. The research results of this paper are to be used for the following systems.

6   Acknowledgment

This work was supported by the German Federal Ministry for Economic Affairs and Climate Action (BMWK) under grant "PV4Life – Erhöhung der Stromrichter-Lebensdauer durch optimierte Komponenten-Optimierung mittels Digitaler KI Zwillinge" (FKZ: 03EE1165E). Only the authors are responsible for the contents of this publication.

References

[1] P. Sun, C. Gong, X. Du, Q. Luo, H. Wang, and L. Zhou, "Online Condition Monitoring for Both IGBT Module and DC-Link Capacitor of Power Converter Based on Short-Circuit Current Simultaneously," *IEEE Trans. Ind. Electron.*, vol. 64, no. 5, pp. 3662–3671, 2017, doi: 10.1109/TIE.2017.2652372.

[2] H.-L. Dang and S. Kwak, "Review of Health Monitoring Techniques for Capacitors Used in Power Electronics Converters," *Sensors (Basel, Switzerland)*, vol. 20, no. 13, 2020, doi: 10.3390/s20133740.

[3] M. W. Ahmad, A. Arya, and S. Anand, "An online technique for condition monitoring of capacitor in PV system," in *2015 IEEE International Conference on Industrial Technology (ICIT)*, Seville, 2015, pp. 920–925.

[4] Q. Luo, B. Luo, Y. Zhu, H. Wang, Q. Wang, and G. Zhu, "Condition Monitoring of DC-Link Capacitors by Estimating Capacitance and Real-time Core Temperature," in *2022 IEEE 13th International Symposium on Power Electronics for Distributed Generation Systems (PEDG)*, Kiel, Germany, 2022, pp. 1–5.

[5] M. A. Vogelsberger, T. Wiesinger, and H. Ertl, "Life-Cycle Monitoring and Voltage-Managing Unit for DC-Link Electrolytic Capacitors in PWM Converters," *IEEE Trans. Power Electron.*, vol. 26, no. 2, pp. 493–503, 2011, doi: 10.1109/TPEL.2010.2059713.

[6] M. Ghadrdan, S. Peyghami, H. Mokhtari, H. Wang, and F. Blaabjerg, "Dissipation Factor as a Degradation Indicator for Electrolytic Capacitors," *IEEE J. Emerg. Sel. Topics Power Electron.*, vol. 11, no. 1, pp. 1035–1044, 2023, doi: 10.1109/JESTPE.2022.3183837.

[7] S. H. Ali, X. Li, A. S. Kamath, and B. Akin, "A Simple Plug-In Circuit for IGBT Gate Drivers to Monitor Device Aging: Toward Smart Gate Drivers," *IEEE Power Electron. Mag.*, vol. 5, no. 3, pp. 45–55, 2018, doi: 10.1109/MPEL.2018.2849653.

[8] S. Beczkowski, P. Ghimre, A. R. de Vega, S. Munk-Nielsen, B. Rannestad, and P. Thogersen, "Online Vce measurement

method for wear-out monitoring of high power IGBT modules," in *2013 15th European Conference on Power Electronics and Applications (EPE)*, Lille, 2013, pp. 1–7.

[9] U.-M. Choi, S. Joergensen, and F. Blaabjerg, "Advanced Accelerated Power Cycling Test for Reliability Investigation of Power Device Modules," *IEEE Trans. Power Electron.*, p. 1, 2016, doi: 10.1109/TPEL.2016.2521899.

[10] T. Krone, L. Dang Hung, M. Jung, and A. Mertens, "On-line semiconductor switching loss measurement system for an advanced condition monitoring concept," in *2016 18th European Conference on Power Electronics and Applications (EPE'16 ECCE Europe)*, Karlsruhe, Germany, 2016, pp. 1–10.

[11] P. Ghimire, S. Beczkowski, S. Munk-Nielsen, B. Rannestad, and P. B. Thogersen, "A review on real time physical measurement techniques and their attempt to predict wear-out status of IGBT," in *2013 15th European Conference on Power Electronics and Applications (EPE)*, Lille, 2013, pp. 1–10.

[12] M. Huang, H. Wang, L. Bai, K. Li, J. Bai, and X. Zha, "Overview of recent progress in condition monitoring for insulated gate bipolar transistor modules: Detection, estimation, and prediction," *High Voltage*, vol. 6, no. 6, pp. 967–977, 2021, doi: 10.1049/hve2.12149.

[13] F. Erturk and B. Akin, "A method for online ageing detection in SiC MOSFETs," in *2017 IEEE Applied Power Electronics Conference and Exposition (APEC)*, Tampa, FL, USA, 2017, pp. 3576–3581.

[14] P. S. Kathribail and T. Vijayakumar, "Comprehensive Study of MOSFET Degradation in Power Converters and Prognostic Failure Detection Using Physical Model," *J. Inst. Eng. India Ser. B*, vol. 104, no. 1, pp. 305–317, 2023, doi: 10.1007/s40031-022-00814-7.

[15] F. Erturk, E. Ugur, J. Olson, and B. Akin, "Real-Time Aging Detection of SiC MOSFETs," *IEEE Trans. on Ind. Applicat.*, vol. 55, no. 1, pp. 600–609, 2019, doi: 10.1109/TIA.2018.2867820.

[16] P. Wang, J. Zatarski, A. Banerjee, and J. S. Donnal, "Condition Monitoring of SiC MOSFETs Based on Gate-Leakage Current Estimation," *IEEE Trans. Instrum. Meas.*, vol. 71, pp. 1–10, 2022, doi: 10.1109/TIM.2021.3137866.

[17] Z. Shen, M. Chen, H. Wang, X. Wang, and F. Blaabjerg, "EMI Filter Robustness in Three-Level Active Neutral-Point-Clamped Inverter," *IEEE Trans. Power Electron.*, vol. 37, no. 4, pp. 4641–4657, 2022, doi: 10.1109/TPEL.2021.3124282.

[18] X. Yu and A. M. Khambadkone, "Reliability Analysis and Cost Optimization of Parallel-Inverter System," *IEEE Trans. Ind. Electron.*, vol. 59, no. 10, pp. 3881–3889, 2012, doi: 10.1109/TIE.2011.2175670.

PCIM Europe 2024, 11– 13 June 2024, Nuremberg   DOI: 10.30420/566262192

Dynamic Transients and Reliability of High-Voltage Silicon & 4H-SiC Bipolar Junction Transistors Under Avalanche and Short-Circuits

Mana Hosseinzadehlish [1], Saeed Jahdi [1], Xibo Yuan [1], Jose Ortiz-Gonzalez [2], Olayiwola Alatise [2]

[1] University of Bristol, Bristol, UK
[2] University of Warwick, Coventry, UK

Corresponding author: Mana Hosseinzadehlish, m.hosseinzadeh@bristol.ac.uk
Speaker: Mana Hosseinzadehlish, m.hosseinzadeh@bristol.ac.uk

Abstract

This paper explores the dynamic characteristics and performance of two commercial bipolar junction transistors (BJTs) with the highest voltage ratings: an 800 V, 20 A Silicon BJT and a 1.7 kV, 15 A 4H-SiC BJT. The investigation focuses on their behavior under unclamped inductive switching and short circuit conditions. All the measurements have been done at room temperature. Clamped inductive switching tests have been conducted using various load inductors and switching rates, at 800 V and maximum collector current of 14 A. In the unclamped inductive switching test, the stress level is increased by either extending the pulse length or increasing the dc-link voltage until the device fails. As for the short circuit test, the dc-link voltage is gradually increased until the device reaches its failure point. The dynamic transient measurements reveal that 4H-SiC BJTs outperform their silicon counterparts. They exhibit higher current gain with a magnitude ten times higher. Notably, the Silicon BJT displays significant delays in both turn-ON and turn-OFF transitions when compared to the 4H-SiC BJT leading to increased losses.

1 Introduction

Within the domain of semiconductor devices, current-driven bipolar Junction Transistors (BJTs) have been introduced in two forms of lateral and vertical configurations. Among these, high voltage silicon vertical BJT has faced a challenge in the realm of power electronic applications. This challenge is primarily explained by its low DC current gain, leading to the requirement for complex base drivers [1]–[4]. Voltage-driven devices, such as metal-oxide-semiconductor field-effect transistors (MOSFETs) and insulated gate bipolar transistors (IGBTs), have gained popularity over silicon bipolar junction transistors (BJTs) due to their simpler gate drivers and the limitations of silicon BJTs including low current gain and second breakdown issues [5], [6]. However, SiC MOSFETs also face challenges related to gate oxide reliability due to the thinner gate oxide and higher electric field compared to their silicon counterparts [7]. This can lead to electron tunneling and significant degradation of the gate oxide. Consequently, 4H-SiC BJTs have been developed to overcome these limitations due to their several advantages

over silicon BJTs and SiC MOSFETs including absence of a gate oxide, lower ON-state resistance, simpler manufacturing process, and significantly higher dc current gain. The smaller base and drift region in 4H-SiC BJTs, achieved through the wider bandgap and higher critical electric field, enhance emitter injection efficiency and result in higher current gain [8]–[10]. Additionally, 4H-SiC BJTs exhibit higher switching speeds due to faster carrier mobility and lower minority carrier lifetime, making them promising for high-frequency applications [11]. Moreover, SiC BJTs possess exceptional conduction efficiency in comparison to unipolar devices, particularly in applications like boost converters where reverse conduction is not required [12].

The unclamped inductive switching (UIS) [13] and short-Circuit (SC) [14] are two important tests used to assess the reliability and ruggedness of the devices under electrothermal stress. The results of these tests are often used as important indicators in the selection of the devices for specific applications. Limited research has been conducted on UIS and SC reliability of the silicon and 4H-SiC NPN

BJTs in recent years [15]–[18]. This paper covers comprehensive investigation of the destructive reliability tests on the BJTs.

This paper is structured as follows. In section. 2 the experimental setup and test conditions are introduced. In section. 3, the dynamic performance of two recent generations of highest voltage devices: the 800 V Fairchild silicon and 1.7 KV GeneSiC 4H-SiC NPN power BJTs have been investigated through experimental measurements employing a wide range of base resistances (R_{Base} = 3.75 Ω to 11.75 Ω) and load inductors. In section. 4, destructive unclamped inductive switching test have been conducted through incremental increase in the dc-link voltage and pulse length in order to induce the electrothermal stress across the devices. Section. 5 focuses on the destructive short circuit of the devices and section. 6 presents the conclusion of this paper.

2 Experimental Set-up

Fig. 1 depicts the cross-sectional schematic of vertical power Silicon and 4H-SiC NPN BJTs.

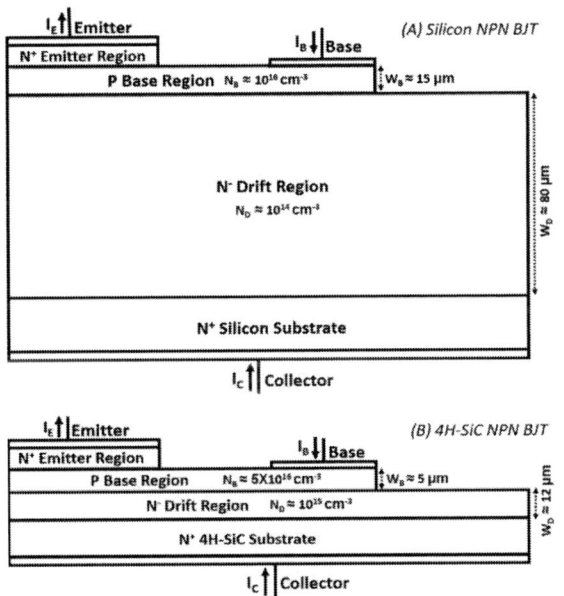

Fig. 1: Cross-sectional schematics of (top) Silicon & (bottom) 4H-SiC BJTs.

Fig. 2 illustrates the schematic of the double pulse circuit, while Fig. 3 shows the corresponding setup. The circuit design consists of a high-voltage ETPS power supply (HTP-HPp 40 757) with a maximum rating of 4 kV and 750 mA, connected

to a 5 mF DC-link capacitor to maintain a stable V_{DC}. On the board, the Fairchild silicon BJT (800 V) and GeneSiC 4H-SiC BJT (1700 V) are mounted, accompanied by a CREE SiC Schottky diode placed in antiparallel with the load. The devices are subjected to a voltage of 800 V, and a double pulse with durations of 35 μs and 8 μs is applied. The first pulse duration is set to ensure sufficient collector current increase for the given inductors. All tests are conducted at room temperature. The base resistance is varied from 3.75 Ω to 11.75 Ω by adding external base resistor up to 8 Ω to the base driver board, and load inductors of 2 mH and 4 mH are employed in the tests. Notably, a high-voltage decoupling capacitor of 100 nF is connected between the cathode of the Schottky diode and the emitter of the BJT in close proximity to the devices under test (DUTs). This capacitor serves to mitigate oscillations on the DC-link during switching, which is particularly crucial for the 4H-SiC power BJT due to its faster switching transient and susceptibility to oscillations.

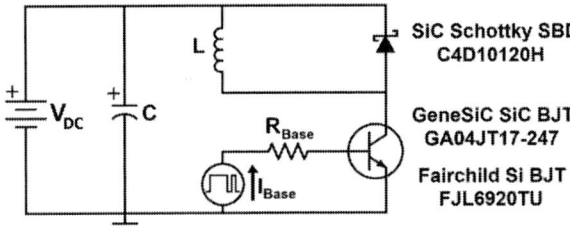

Fig. 2: Schematic of clamped inductive switching circuit and DUTs.

Fig. 3: Setup of double-pulse test: 1. DUTs, 2. Schottky Diode, 3. Inductor, 4. DC Capacitors, 5. Base driver, 6. Input Signal, 7. HV Supply, 8. Voltage Probes, 9. Rogowski Coil, 10. Current Probe.

In the unclamped inductive switching test, the clamping diode is removed from the circuit configuration, and an inductor of 2 mH is used in series with the DUT. The test has been conducted

1403

in two different scenarios. Firstly, the DC-link voltage is gradually increased while a fixed base pulse is applied to the device. Secondly, a fixed DC-link voltage is maintained while the duration of the applied base pulse is extended to increase the avalanche current. The same scenarios have been employed in the short circuit test where both the clamping diode and the load inductor are removed from the circuit configuration.

3 Dynamic transient in silicon & 4H-SiC BJT

Fig. 4 demonstrates the measured trends of turn-OFF delay in voltage of the Silicon NPN BJT switched at 800 V for various base resistances and $L_{Load} = 2\ mH$. Silicon BJT demonstrates a noticeable delay between base turn-OFF and collector-emitter voltage rise. This delay is attributed to the storage time of minority carriers, which significantly impacts the device's dynamic behavior. During turn-ON, minority carriers injected from the emitter region are stored in the base and drift regions. As a result, during turn-OFF, recombination and extraction of these minority charge carriers cause the observed delay.

The delay time is reduced in the second pulse due to its shorter duration, resulting in fewer charge carriers being stored in the base and drift regions. By increasing the base resistance, the delay time between the base current drop and the rise of collector-emitter voltage decreases. This reduction can be explained by a decrease in charge storage and the conductivity-modulated region within the drift region which occurs due to the decrease in base current.

Additionally, in the second pulse, it is observed that the collector-emitter voltage does not drop to zero during the turn-ON phase, which can be attributed to energy losses during the first and break pulse periods. The utilization of bank capacitors in the double-pulse test circuit is common to store energy and stabilize the DC-link voltage. However, the first pulse consumes energy from the bank capacitors, and losses occur due to external inductors and resistors in the circuit. Consequently, the bank capacitors are unable to sustain 800 V during the second pulse, leading to the collector-emitter voltage not reaching zero during the turn-on period. In our test, as the base-emitter junction is heavily forward-biased, the phenomenon of high-level injection significantly affects the switching

performance and dc current gain of the device. High-level injection occurs when a large number of minority carriers are injected into the base region, leading to reduced emitter injection efficiency and dc current gain. Additionally, high-level injection prolongs the turn-OFF time and increases the storage time in the silicon BJT. While increasing the base resistance may help reduce turn-OFF delay, its effectiveness in high-level injection conditions may be limited due to the continued injection of minority carriers into the base region.

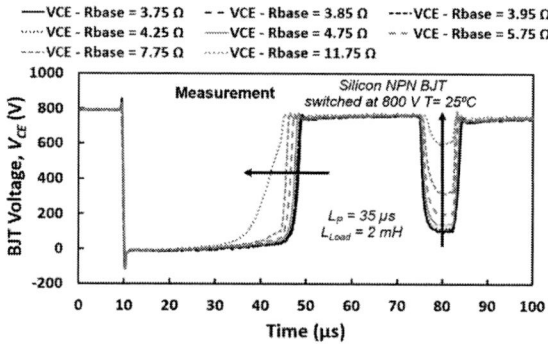

Fig. 4: The measured trends of turn-OFF delay in voltage of Silicon NPN BJT with various base resistances ranged from 3.75 Ω-11.75 Ω for $L_{Load} = 2\ mH$ at 25 °C.

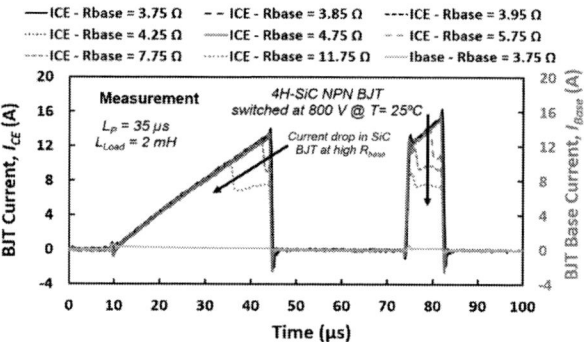

Fig. 5: The measured trends of the turn-OFF current drop in 4H-SiC NPN BJT with various base resistances ranged from 3.75 Ω-11.75 Ω for $L_{Load} = 2\ \mu H$ at 25 °C.

Fig. 5 shows the measurement results of the double-pulse test for the 4H-SiC NPN BJT. Unlike silicon BJT, the base and the collector current are turned off, concurrently. This behavior is due to the smaller dimensions of the 4H-SiC BJTs and lower minority carrier lifetime compared to their silicon counterparts.

An interesting finding is that the collector current of SiC BJT experiences a collapse when the base resistance is increased. Moreover, this behavior becomes more pronounced with higher base resistances. This phenomenon can be due to a decrease in current gain at high collector currents in SiC BJTs. When the base resistance is increased, it leads to significant power loss due to the resulting increase in collector voltage across the load. Consequently, increase of the base current beyond a certain point can lead to instability and degradation of SiC BJT.

4 Experimental Measurements of Avalanche UIS

UIS test is one of the most important reliability tests in order to evaluate the ruggedness of the devices under rough electrothermal stress conditions [13], [19]. During this test, as the base pulse is applied during the turn-ON, the inductor starts to store energy via the power supply through the DUT. During this time interval, the collector current reaches the maximum avalanche value which is specified by the size of the load inductor, base pulse length and the dc-link voltage. During the turn-OFF interval, as the base pulse drops, the device enters the avalanche mode. This results in a high voltage equal to the breakdown voltage being generated across the DUT. Additionally, a shoot-through current flows from the collector to the emitter of the DUT as the current of the load inductor cannot immediately reach zero. Avalanche collector-emitter voltage and load current of the silicon NPN BJT for various DC-link voltage are depicted in Fig. 6, a single pulse of $L_P = 80 \ \mu s$ is applied to the device. It can be seen by increasing the dc-link voltage, the collector-emitter voltage reaches its avalanche threshold and does not increase further. However, the avalanche load current continuous to increase. ultimately, silicon BJT withstands maximum avalanche current of 15 A before it fails at dc-link voltage of 320 V. It is worth mentioning that due to the significant minority carrier charge storage time in silicon BJT, there exists a distinct delay between turning off the base current and the device entering the avalanche mode.

Furthermore, we evaluated the impact of increasing the base pulse length once at a voltage before failure and once at the voltage where the DUT undergoes avalanche failure. The pulse length is

varied from 20 to 100 μs. Based on Fig. 7, as the pulse length increases the peak of avalanche current increases. By applying V_{DC} = 180 V and increasing the pulse duration to 100 μs, DUT does not enter the avalanche failure mode which indicates that the applied voltage and maximum avalanche current in the applied pulse length cannot induce enough critical avalanche energy across the device and enter it in failure mode. We could either keep the dc-link voltage fixed and increase the pulse length further or just increase the applied dc-link voltage to evaluate the ruggedness of the silicon BJT. So, we have increased the dc-link voltage to 320 V and based on Fig. 7, silicon BJT fails for pulse length of 80 μs. The failure in the silicon BJT occurs at the beginning of the avalanche process, when the peak electric field is at the maximum which suggests that high electric field density across the P-base N-drift junction is the failure reason.

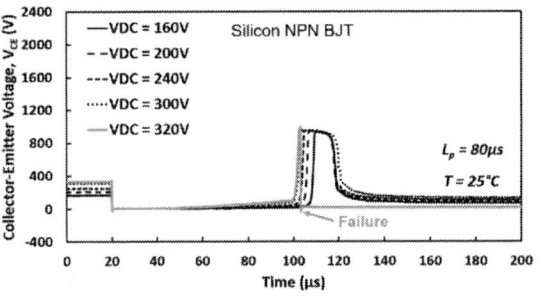

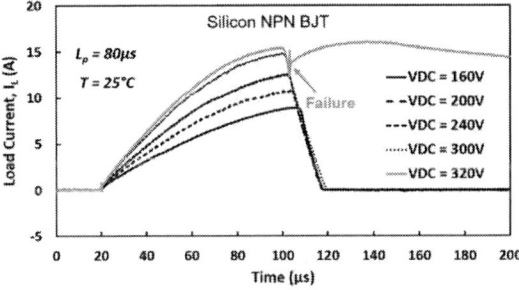

Fig. 6: Avalanche collector-emitter voltage and load current of the silicon NPN BJT for various DC-link voltage until the devices failed at 25 °C.

Furthermore, we have conducted an evaluation of the impact of increasing the base pulse length, both at a voltage before failure and at the voltage associated with DUT avalanche failure. The pulse length is varied from 20 to 100 μs. Based on the measurements depicted in Fig. 7, it is observed that as the pulse length increases, the peak avalanche current also increases. By applying a fixed

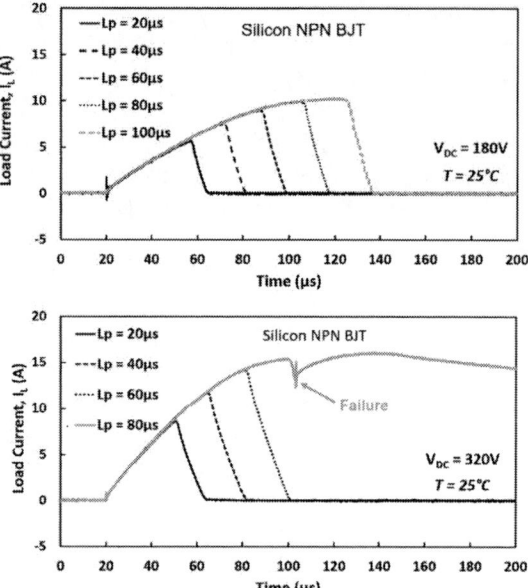

Fig. 7: Avalanche Load current of the silicon NPN BJT for various pulse lengths until failure at 25 °C.

V_{DC} = 180 V and extending the pulse duration event to 100 μs, the DUT does not undergo avalanche failure. This suggests that the applied voltage, along with the maximum avalanche current within the applied pulse length, cannot induce sufficient critical avalanche energy across the device to trigger failure mode. There are two possibilities to further evaluate the ruggedness of the silicon BJT: either increasing the pulse length while keeping the dc-link voltage constant, or increasing the applied dc-link voltage. In this study, we have increased the dc-link voltage to 320 V. As shown in Fig. 7, the silicon BJT failed at pulse length of 80 μs.

As illustrated in Fig. 9, the same test procedure as the silicon have been applied to the 4H-SiC BJT. Specifically, a fixed base pulse length of 100 μs is utilized, and the dc-link voltage is incrementally increased until the failure of the DUT occurs at room temperature. It can be seen that SiC BJT reaches its avalanche voltage threshold for a lower dc-link voltage compared to silicon one. SiC BJT are able to operate at higher voltages compared to the silicon ones due to their wider bandgap. Accordingly, they can withstand higher voltages before experiencing avalanche breakdown.

The impact of the base pulse length increase on avalanche reliability of the SiC BJT has been also investigated as shown in Fig. ref9, specifically at a voltage of 180 V. The pulse length is changed

from 20 to 100 μs. At the pulse length of 100 μs, DUT fails close to the termination of the avalanche process. The overlap of the high voltage across the PN-junctions and the high current density flowing through them results in formation of hot spots at those regions. In the case of SiC BJT that failure occurs towards the end of the avalanche process, the failure mechanism is thermal. Figs. 10 and 11 illustrate the avalanche voltage and current peaks of both silicon and SiC BJTs for different dc-link voltages and pulse lengths of 40 μs & 100 μs, respectively. In both devices and for both pulse lengths, it is observed that the peak avalanche current increases with an increase in the dc-link voltage. However, the avalanche voltage remains stable, as both devices reach their respective avalanche voltage thresholds.

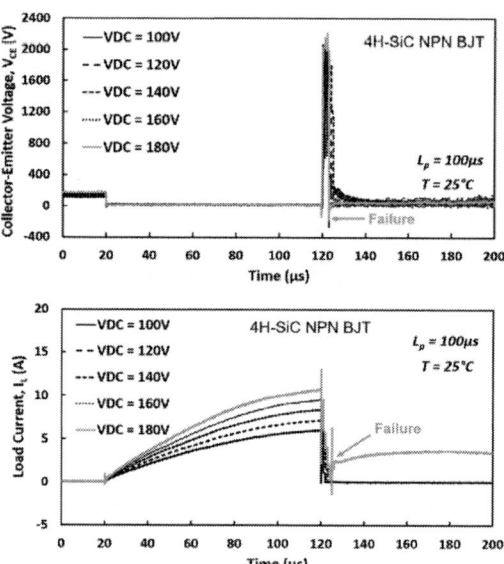

Fig. 8: Avalanche collector-emitter voltage and load current of the 4H-SiC NPN BJT for various DC-link voltage until the devices failed at 25 °C.

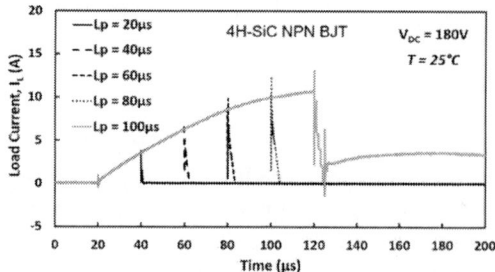

Fig. 9: Avalanche Load current of 4H-SiC NPN BJT for various pulse lengths until failure at 25 °C.

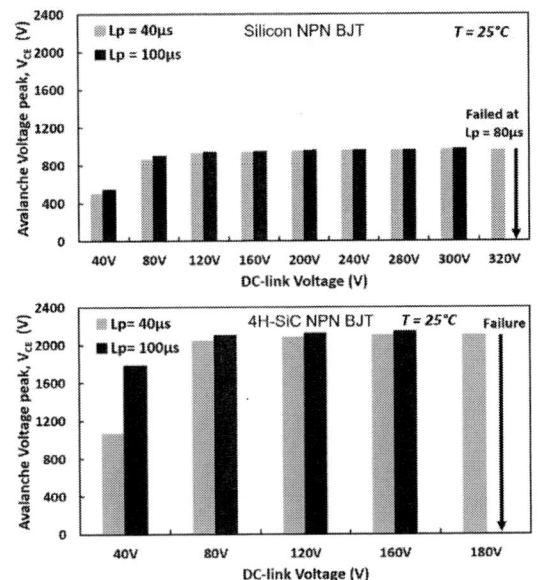

Fig. 10: Peak of avalanche Load current of the silicon & 4H-SiC NPN BJT for various DC-link voltages.

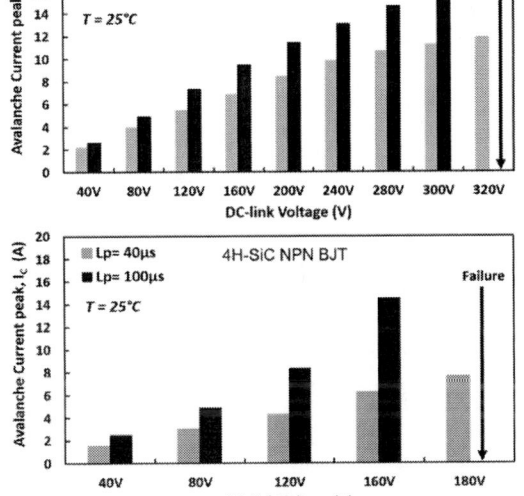

Fig. 11: Collector-emitter voltage peak of the silicon & 4H-SiC NPN BJT for various DC-link voltages.

5 Experimental Measurements of Short circuit

Short circuit is another important test to assess the reliability and ruggedness of power semiconductor devices [14], [20]. The impact of the dc-link voltage and short circuit duration on performance of the silicon and SiC BJTs have been investigated in this paper. The measured results of the short circuit tests conducted on the silicon BJT are presented in Figs. 12 and 13.

In Fig. 12, the short circuit duration is fixed at 200 μs, while the dc-link voltage is varied from 200 V to 800 V. The experimental results clearly demonstrate that as the short circuit occurs, the collector current of the device starts to rise. This increase in current is observed across all tested dc-link voltages. However, once the dc-link voltage is fully applied to the device, it enters the saturation mode of operation. In this mode, the device must withstand high voltage and current simultaneously. Accordingly, the temperature of the junction increases rapidly due to self-heating resulting in a reduction in the carrier mobility within the drift region and a negative slop of the short circuit current. Moreover, the short circuit current increases with the rise in the dc-link voltage. However, it is important to note that no failure events occurred under these specific test conditions since the junction temperature of the device is within the safe area.

In Fig. 13, the applied dc-link voltage is 800 V while the short circuit duration is increase from 100 μs to 400 μs. Under this test condition, silicon BJT experiences failure at 400 μs. As illustrated in Fig. 13, when the silicon BJT is turned off after 400 μs short circuit duration, the heat generation caused by high temperature leakage current exceeds a critical value and the internal thermal instability after device turn-off occurs after a delay time. This ultimately results in a thermal runaway and device failure indicated with a short between the base-emitter contacts of the device.

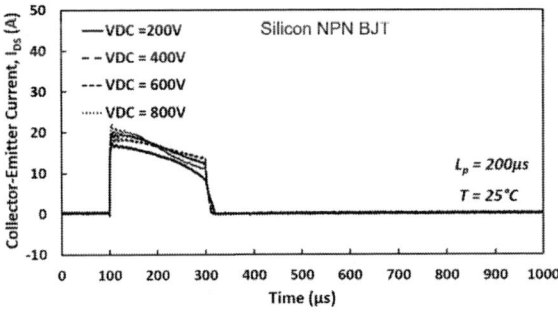

Fig. 12: Short circuit current of silicon NPN BJT for various dc voltages and $L_P = 200 \mu$s at 25°C.

The same short circuit strategy has been employed on SiC BJT. According to Fig. 15, SiC fail within 100 μs when the dc-link voltage reaches 480 V. It is worth mentioning that failure of the SiC BJT

PCIM Europe 2024, 11– 13 June 2024, Nuremberg   DOI: 10.30420/566262192

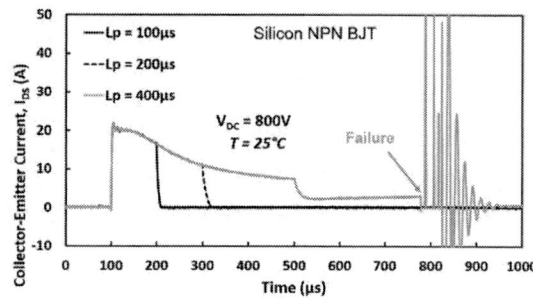

Fig. 13: Short circuit current of silicon NPN BJT for various pulse lengths and V_{DC} = 400 V until the device failed at 25 °C.

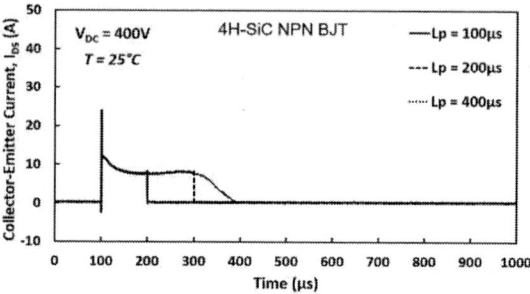

Fig. 15: Short circuit current of the 4H-SiC NPN BJT for various pulse lengths and V_{DC} = 400 V until the device failed at 25 °C.

happens at the start of the short circuit which can be due to the high current density. The SiC BJT has a much smaller (around 20 times) die size compared to the silicon one, that why the current density is SiC BJT is higher than in silicon one. Accordingly, SiC BJT can withstand shorter short circuit duration.

In order to evaluate the reliability of the SiC BJT under different short circuit durations, a series of tests have been conducted with a fixed dc-link voltage of 400 V while varying the pulse length from 100 μ to 400 μ. The measurement results, presented in Fig. 15 indicate that the DUT is able to withstand this specific short circuit condition.

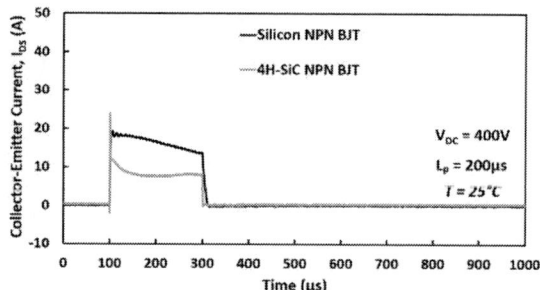

Fig. 16: Comparison of the Short circuit current of the silicon and 4H-SiC NPN BJT at a specific dc-link voltage and pulse length at 25 °C.

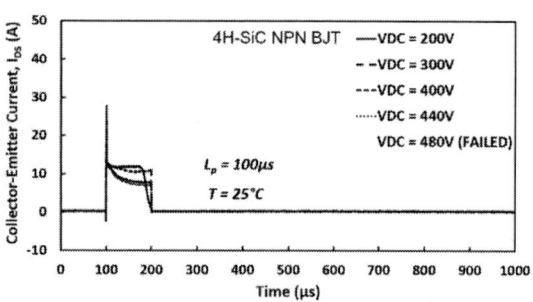

Fig. 14: Short circuit current of the 4H-SiC NPN BJT for various dc-link voltages and pulse length of 100μs until the device failed at 25 °C.

A comparison between the short circuit current of the silicon and 4H-SiC BJTs have been carried out at a specific dc-link voltage and short circuit duration as illustrated in Fig. 16 and 17. The short circuit current of the SiC BJT exhibits saturation at a lower value compared to that of silicon. This behavior can be related to the lower carrier mobility of SiC as 900 $\times$ cm^2/Vs for the 4H-SiC electrons as to 1350 $\times$ cm^2/Vs for Silicon [21].

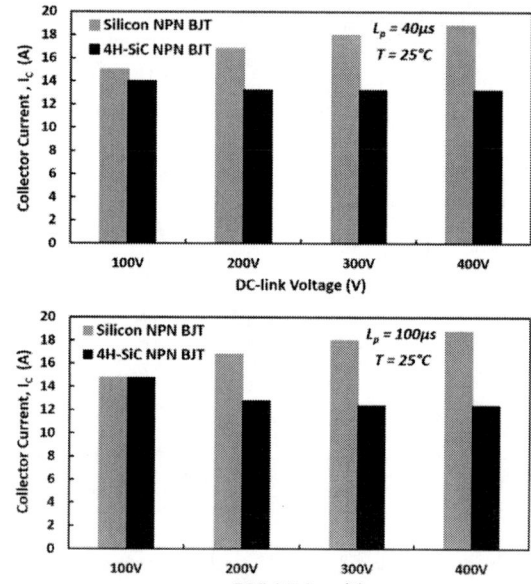

Fig. 17: Comparison of short circuit current of the silicon and 4H-SiC NPN BJT for various dc-link voltages and pulse length of (top) 40 μs & (bottom) 100 μs at 25 °C.

1408

6 Conclusion

In this paper, the dynamic transient and ruggedness of the high voltage silicon and 4H-SiC NPN BJTs under short circuit and UIS tests have been comprehensively investigated thorough experimental tests. It has been observed that the 4H-SiC BJT exhibits significantly higher current gain and faster switching capabilities compared to the silicon BJT which make the it well-suited for high-frequency applications. The impact of the incremental increase of the dc-link voltage and base pulse length have been examined in UIS and short circuit tests.

References

[1] B. J. Baliga, *Fundamentals of power semiconductor devices*. Springer, 2010.

[2] M. Hosseinzadehlish and et al., "Impact of temperature and base bias stress on the static characteristics of silicon and 4h-sic npn vertical power bjts," in *PCIM Europe 2023*, 2023, pp. 1–9.

[3] L. Liao and et al., "A new proportional base drive technique for sic bipolar junction transistor," *IEEE Transactions on Power Electronics*, vol. 32, no. 6, pp. 4600–4606, 2016.

[4] S. Sundaresan and et al., "Comparison of energy losses in high-current 1700 v switches," in *IEEE 3rd Workshop on Wide Bandgap Power Devices and Applications (WiPDA)*, 2015, pp. 180–183.

[5] C. Shen and et al., "Impact of carriers injection level on transients of discrete and paralleled silicon and 4h-sic npn bjts," *IEEE Open Journal of the Industrial Electronics Society*, vol. 3, pp. 65–80, 2022.

[6] S. M. Sze and et al., *Physics of Semiconductor Devices*, 3rd. Hoboken, NJ, USA: Wiley, 2007.

[7] S. Jahdi and et al., "Temperature and switching rate dependence of crosstalk in si-igbt and sic power modules," *IEEE Transactions on Industrial Electronics*, vol. 63, no. 2, pp. 849–863, 2016.

[8] R. Singh and et al., "Fulfilling the promise of high-temperature operation with silicon carbide devices: Eliminating bulky thermal-management systems with sjts," *IEEE Power Electronics Magazine*, vol. 2, no. 1, pp. 27–35, 2015.

[9] S. Jahdi and et al., "The impact of temperature and switching rate on dynamic transients of high-voltage silicon and 4h-sic npn bjts: A technology evaluation," *IEEE Transactions on Industrial Electronics*, vol. 67, no. 6, pp. 4556–4566, 2020.

[10] Y. Shi and et al., "Experimental comparison of sic mosfet and bjt," in *2018 IEEE International Power Electronics and Application Conference and Exposition (PEAC)*, 2018, pp. 1–6.

[11] X. Li and et al., "On the temperature coefficient of 4h-sic bjt current gain," *Solid-State Electronics*, vol. 47, no. 2, pp. 233–239, 2003.

[12] J. Rabkowski and et al., "Parallel-operation of discrete sic bjts in a 6-kw/250-khz dc/dc boost converter," *IEEE Transactions on Power Electronics*, vol. 29, no. 5, pp. 2482–2491, 2014.

[13] S. G. Sundaresan and et al., "Characterization of the stability of current gain and avalanche-mode operation of 4h-sic bjts," *IEEE Transactions on Electron Devices*, vol. 59, no. 10, pp. 2795–2802, 2012.

[14] R. Singh and et al., "Short circuit robustness of 1200 v sic switches," in *IEEE 3rd Workshop on Wide Bandgap Power Devices and Applications (WiPDA)*, 2015, pp. 1–4.

[15] B. Asllani and et al., "Demonstration of the short-circuit ruggedness of a 10 kv silicon carbide bipolar junction transistor," in *22nd European Conference on Power Electronics and Applications (EPE'20 ECCE Europe)*, 2020, pp. 1–10.

[16] C. Chen and et al., "Study of short-circuit robustness of sic mosfets, analysis of the failure modes and comparison with bjts," *Microelectronics Reliability*, vol. 55, no. 9, pp. 1708–1713, 2015.

[17] M. Domeij and et al., "Large area 1200 v sic bjts with beta 100 and rho(on) 3 m omega cm(2)," *Materials Science Forum*, vol. 717-720, pp. 1123–1126, May 2012.

[18] D.-P. Sadik and et al., "Comparison of thermal stress during short-circuit in different types of 1.2-kv sic transistors based on experiments and simulations," *IEEE Transactions on Industrial Electronics*, vol. 68, no. 3, pp. 2608–2616, 2021.

[19] S. Sundaresan and et al., "10 kv sic bjts — static, switching and reliability characteristics," in *25th International Symposium on Power Semiconductor Devices & IC's (ISPSD)*, 2013, pp. 303–306.

[20] R. Singh and et al., "Switching and robustness analysis of 10 kv sic bjts," in *IEEE 3rd Workshop on Wide Bandgap Power Devices and Applications (WiPDA)*, 2015, pp. 406–409.

[21] G. Kampitsis and et al., "Comparative analysis of the thermal stress of si and sic mosfets during short circuits," *Materials Science Forum*, vol. 856, pp. 362–367, May 2016.

PCIM Europe 2024, 11– 13 June 2024, Nuremberg   DOI: 10.30420/566262192

Power Cycling Test Optimization Toward Reliability Assessment of Sintered Power Modules.

Robert Graham

[1] Macdermaid Alpha Electronic Solutions, United States

Corresponding author: Robert Graham, Robert.Graham@macdermaidalpha.com
Speaker: Robert Graham, Robert.Graham@macdermaidalpha.com

Abstract

The optimization of power cycling is a multifaceted endeavor, contingent upon variables such as the specific power module, sintering and solder materials, and their respective specifications. Power cycling, serving as a crucial stress and reliability evaluation tool for power modules, plays a pivotal role in the end-of-life (EOL) assessment of these critical components. It encompasses a comprehensive examination of various facets within the module. This paper aims to elucidate the methodologies involved in optimizing these conditions to obtain precise power cycling data for accelerated EOL testing and the assessment of reliability in sintered power modules employing Alpha silver sinter paste and soldered power modules.

1   Experimental Setup

1.1   Introduction

The paper focuses on enhancing the reliability and performance evaluation of power modules exposed to challenging operational conditions, including temperature variations and voltage fluctuations. As accelerated testing gains prominence, meticulous test conditions are crucial, with specific emphasis on the materials used. Figure 1 shows the power tester procured from Siemens for testing. The study employs nano silver sintering, emphasizing its benefits over traditional solder materials. The meticulous steps in optimizing test conditions for nano silver sintered power modules are explored, encompassing calibration, cable selection, temperature settings, measurements, and adherence to temperature limits. It acknowledges the potential for variability when transitioning between materials and semiconductor devices.

Fig. 1   Siemens 2400A Power Tester being calibrated for Power Cycling.

1.2   EOL Testing and Defining Parameters

This section highlights the significance of evaluating the reliability of power modules, their components, and soldering materials. The test discussed here, including wire bonding, die attachment, thermal interfaces, and cable management. The focus is on the end-of-life (EOL) test, based on the equation used below

$$V_{TH} = \frac{2d}{\varepsilon_{ox}} \sqrt{\varepsilon_8 N_A kT ln\left(\frac{N_A}{n_i}\right) + \frac{2kT ln}{9}\left(\frac{N_A}{n_i}\right)}$$

Currently, two distinct test methods conform to AQG-324, namely IEC60749-34 and JESD22-A122A. This paper primarily focuses on the AQG-324 method for EOL testing, characterized by a 5% increase in forward voltage and a 20% increase in thermal resistance (ΔT) across the device [3]. The parameter of utmost interest to many is ΔT, which estimates the junction temperature of the power module, drawing upon various methodologies when assessing MOSFETs.

Along with defining the method at which we test the ΔT parameter, there are other important parameters to look at when power cycling.

- **V$_{GS}$** – Gate-Source voltage, the safe range at which the power module can operate without being stressed to failure fast.

- **I$_{DN}$** – Drain current, this defines the current going through the power module, most semiconductors have a maximum allowable current.

- **V$_{ON}$** – Voltage on Device, this defines the voltage flowing into the power module while it is on and running calculates the voltage drop as well.

- **ΔT** – Average Junction Temperature, this is the average temperature of the power module that is calculated from the Tj$_{(max)}$ and Tj$_{(min)}$

- **Tj$_{(min)}$** – Minimum junction temperature of the power module

- **Tj$_{(max)}$** – Maximum junction temperature of the power module

- **°C/W** – Thermal Resistance, a calculation of the resistance to heat flow for the whole power module

- **T$_{on}$** – The time that the device is turned on for

- **t$_{off}$** – The time that the device is turned off for

These are the basic parameters that will be mentioned throughout this paper as the reliability test relies on these units. One other important factor with power cycling these devices is the cycling method, for these power modules, it will be constant current that is applied during testing when the device is turned on.

1.3 Equipment and Materials Used

The power modules used in this study were supplied by a vendor and underwent modification to render them compatible with our silver sinter paste due to its limited adhesion to tin surfaces. Figure 2 shows a complete setup of the testing module. These adapted power modules serve as the Device Under Test (DUT) for the experiment. The materials and equipment employed include the Power Tester 2400A, as previously mentioned, in conjunction with a cold plate provided by Siemens for testing purposes. To facilitate efficient thermal management during cycling, Thermal Interface Materials (TIMs) were introduced. The TIMs being selected for this experiment are of different thermal conductivities and thickness, as well as having a high operating range of up to 200°C, three different TIMs will be used.

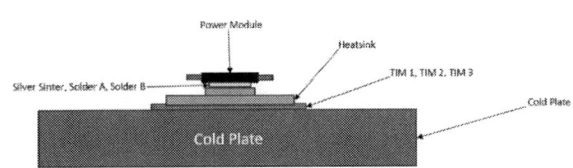

Fig. 2   Drawing of Power cycling setup for power module testing

It is imperative to underscore the importance of robust thermal conductivity during power cycling experiments, as this property greatly aids in the dissipation of heat from the heatsink into the cold plate. When considering the aspect of thermal management, it's crucial not to overlook the significance of the wires and cables used in power cycling. In this study, 70 mm^2 cables and 21.2mm^2 were employed, characterized by their ability to handle voltages of up to 600V and withstand temperatures of up to 150°C. It is worth noting that while smaller cables with identical specifications can be utilized, their physical size renders them incapable of accommodating sustained high current loads. Subsequently, we will present data illustrating the necessity of employing larger cables for high current testing.

1.4 High Temperature Calibration

The calibration methodology for the power module is primarily dictated by the manufacturer, predicated on a voltage and temperature curve observed during the calibration process. As current levels increase, so does the device's temperature, rendering it essential to accurately gauge the temperature emanating from the power module during calibration. To facilitate this, the Power Tester employs a K-factor calibration, a critical procedure aimed at characterizing the thermal cycling performance of the power module.

However, there exists a conundrum linked to the calibration current, specifically the milliampere (mA) value required for precise calibration of the power module. While many power modules permit calibration currents as high as 1A, such elevated currents can potentially introduce undue stress to the device, thereby jeopardizing its reliability. In certain instances, higher calibration currents may enhance calibration stability but can simultaneously intensify device stress during cycling. Figure 3 details what an calibration would look like and the linear structure of the devices.

The current calibration setting for the power module is -100mA, and observations indicate that within the range of -50mA to -200mA, no discernible alteration occurs during power module cycling. However, a perceptible change emerges in the range of -250mA to 400mA.

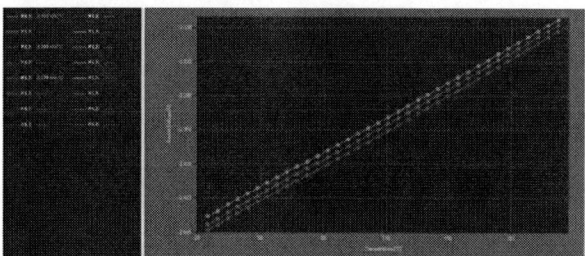

Fig. 3 Example of power modules being calibrataed

It is imparritive that the calibrated remains very linear with small deviation in the calibration temperature. The line is not completely linear when calibrating but the power tester accounts for that

1.5 VGS Range effect on Power Module

The gate-source voltage, as previously alluded to, demarcates the operational range of the device when initiating the power module. It is imperative to emphasize that the power module should exclusively function within this specified range to ensure accurate data acquisition and to uphold the reliability of the power module. For the power module employed in this study, the prescribed voltage range spans from 18V to -5V, a parameter explicitly derived from the specification sheet provided by the supplier in their technical document. Verification of this operating range's effectiveness with the Power Tester is accomplished through a short transient test conducted at room temperature.

The transient test serves a dual purpose: first, it assesses whether the manufacturer's specified range is compatible with the Power Tester, and second, it scrutinizes the cables' capacity to handle the current load without incurring adverse consequences. This examination leverages both 21.2mm² cables, provided by Siemens, and 70mm² welding cables, crafted in-house. Figures 4-7 show the difference between vgs ranges to verify the manufacturers settings.

The parameters of the transient test are defined as follows: a current of 150A is applied to both power modules, with a t_{on} time of 30 seconds and a t_{off} time of 40 seconds. The cold plate temperature is maintained at 30°C, and the TIM 1 is employed. During this test, the VGS + remains constant at 18V, while VGS (-) assumes values of -5V, -6V, -7V, and -8V. This variation in VGS - facilitates a comparative analysis of how each voltage range might impact the reliability and performance of the power module.

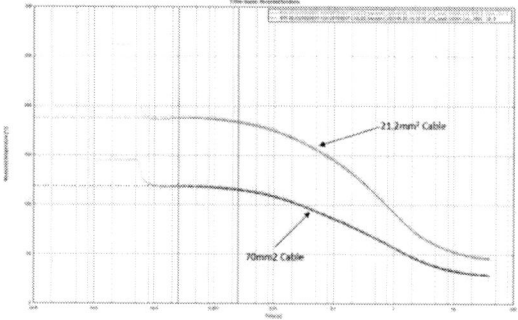

Fig. 4 VGS at 18V to -5V with both cables

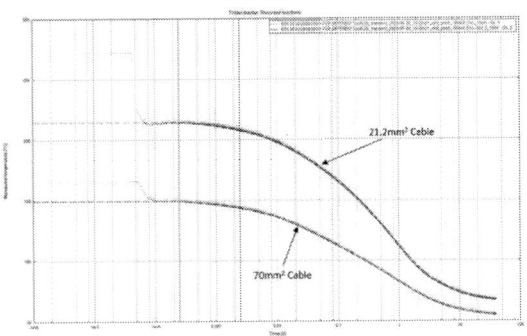

Fig. 5   VGS at 18V to -6V with both cables

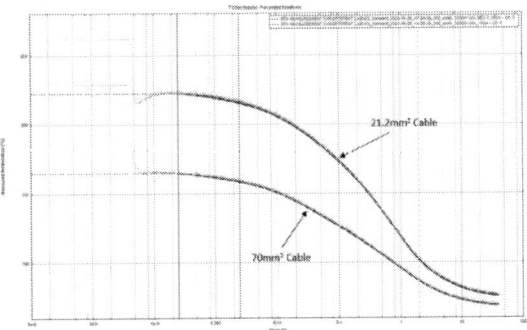

Fig. 6   VGS at 18V to -7V with both cables

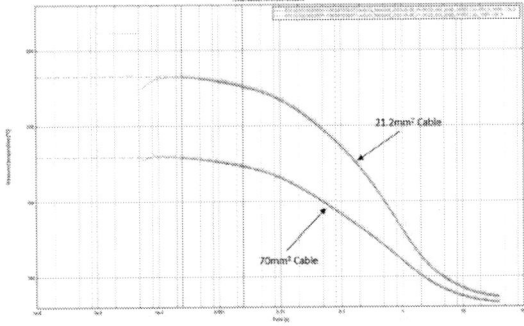

Fig. 7   VGA at 18V to -7V with both cables

The results of the transient test unequivocally affirm the merit of adhering to the voltage range recommended by the power module. Figures 4-7 clearly indicate that the device's performance exhibits an escalating thermal response as the gate-source voltage range expands. As the voltage increases, the power module displays a diminished capacity to return to the cold plate's ambient temperature, consequently generating a steeper temperature decrease curve, rather than a more gradual cooling cycle.

A noteworthy observation is the influence of cable size on the thermal response of the power module. Even within the -5V range, the 21.2mm² cables exhibit a protracted recovery period and do not return to room temperature. In contrast, the 70mm² cables prove superior in managing the thermal response, and this can be attributed to the larger cross-sectional area of the wires, which reduces electrical resistance, subsequently enhancing their thermal performance.

1.6 Stability of Various TIM Level 2 Materials Under Test Conditions

Once the appropriate VGS range is determined, a critical next step involves the assessment of Thermal Interface Materials (TIMs) designed to enhance heat dissipation from the heatsink. The choice of TIMs hinges on the power module's maximum operating junction temperature, a pivotal parameter that dictates the selection of a TIM with a matching maximum operating temperature. If the power module's operational junction temperature is set at 175°C, the selected TIM should possess a maximum operating temperature of 175°C or higher to effectively manage the heat emanating from the heatsink. Two formulas play an instrumental role in gauging the impact of the TIM:

Thermal conductivity $\quad k = \left(\dfrac{q}{A}\right)\left(\dfrac{\Delta z}{\Delta T}\right)\dfrac{w*m}{m^2*K} =$

$$\frac{W}{m*K}$$

Thermal Resistance* $\quad \dfrac{1}{k} = \left(\dfrac{A}{q}\right)\left(\dfrac{\Delta T}{\Delta z}\right)\dfrac{m^2*K}{w*m} =$

$$\frac{m*K}{W}$$

These formulas serve as indispensable tools in the evaluation of TIM performance, assisting in the informed selection of the most suitable material for efficient heat dissipation in accordance with the power module's specific operating conditions.

Thermal conductivity is the reciprocal of thermal resistance, and as per this formula, it can be deduced that an increase in thermal conductivity results in a decrease in thermal resistance, and vice versa. The objective behind the TIM selection process is to identify a TIM with high thermal conductivity to ensure optimal thermal management for heat dissipation from the heatsink, figures 8-10 help show the difference between TIMs and how they perform.

PCIM Europe 2024, 11– 13 June 2024, Nuremberg   DOI: 10.30420/566262192

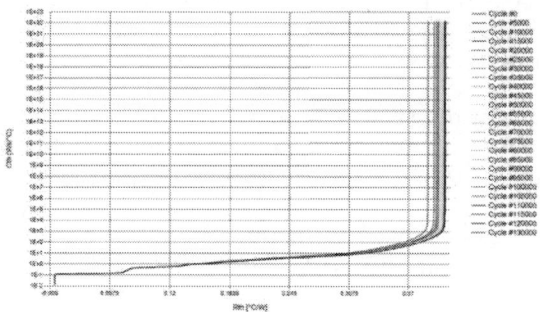

Fig. 8   TIM1 Thermal Resistance

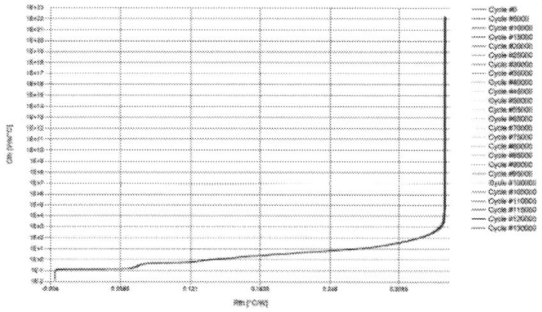

Fig. 9   TIM 2 Thermal Resistance

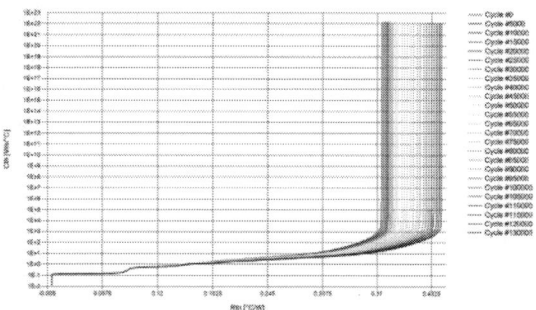

Fig. 10   TIM 3 Thermal Resistance

The figures presented highlight a crucial aspect of the experimentation, revealing that the choice of Thermal Interface Material (TIM) exerts a profound impact on the testing conditions for the power module. Particularly, when the power module is subjected to heavy loads and exposed to rigorous testing conditions, a TIM characterized by a favorable balance of thermal resistance and high thermal conductivity assumes paramount significance. Such a TIM not only contributes to the stability of the test but also plays a pivotal role in maintaining consistent and reliable testing conditions for both power modules and semiconductor dies.

1.7 Effect of ton and toff for modulating ΔT

The transient test results provide compelling evidence in favor of adhering to the VGS range recommended by the power module manufacturer, and this recommendation stands on solid grounds. The data, as depicted in Figures 8-10, unequivocally reveals that as the gate-source voltage range expands, the device experiences a notable escalation in temperature, diminishing its ability to return to the cold plate temperature. Furthermore, an intriguing observation surfaces concerning the time taken for cooling. Instead of a gradual, natural cooldown cycle, a more abrupt temperature decrease occurs, further underscoring the impact of the extended voltage range on device performance.

Notably, cable size emerges as a critical factor influencing thermal response. Even within the -5V range, the 21.2mm² cables exhibit a persistent temperature elevation, failing to return to room temperature. In stark contrast, the 70mm² cables exhibit superior thermal management, attributable to their larger wire cross-sectional area, which translates to lower electrical resistance. The ton and toff can effect the ΔT of the power module, the longer your ton generally the hotter your power module will become. So the T_{off} should be optimized to help bring the ΔT down back to the correct range.

2 Results

Incorporating these modifications prior to and during power cycling procedures enhances the precision of voltage and temperature measurements on the power module. Additional considerations encompass current and voltage levels, where higher current necessitates larger cables and adverse environmental conditions may demand alternative TIMs. A comprehensive understanding of power module specifications facilitates streamlined data collection and testing for both current and previous power module generations, as well as semiconductor dies. It is crucial to emphasize that an erroneous setup can yield inaccurate test data, potentially triggering unnecessary design alterations, impacting not only the power module but also the efficacy of the power testing apparatus. Hence, meticulous attention to the setup is imperative before embarking on substantial testing endeavors.

1414

2.1 Power Cycling with Solder Material

This investigation remains ongoing, with data collection efforts focused on soldered materials and the repercussions of the PCmin test on the integrity of the solder joint layer. The examination seeks to illuminate the intricate interplay between these factors and their potential consequences on the overall performance and reliability of the tested components.

The investigation scrutinizes the performance of OM340 and CVP390V solder pastes during power cycling, emphasizing their resilience to high temperature differentials and electrical currents. Aligning with previous optimizations in TIM selection, cable choice, VGS range, and calibration methods enhances the accuracy of power cycling data. Variations in solder paste composition constitute the sole discernible difference between power module configurations. External factors like tilting during reflow and voiding in solder pastes could influence power module performance. Even minor voiding can escalate ΔT during cycling, particularly if tilting occurs during reflow. Pre-cycling X-ray inspection aids in assessing potential performance deviations caused by solder joint issues. Comparative analysis of cycling data reveals solder A's superior stability over solder B, underscored by optimized calibration settings facilitating accurate performance assessment.

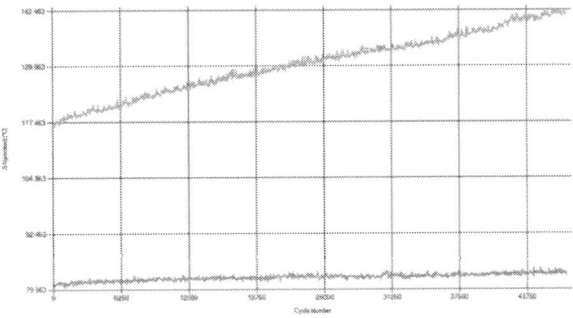

Fig. 13   Solder A is dark orange and Solder B is the orange line

The data presented in the table below highlights the substantial performance disparity between solder A and solder B, affirming solder A's superior suitability as a package attach material compared to solder B. This experiment was conducted twice to ensure the accuracy and consistency of the results, with meticulous verification of setup and testing parameters. The thorough examination of testing parameters instills confidence in the data's fidelity, providing a close approximation of how these solder materials would function in their intended application as package attach materials. This assertion is further corroborated by the comparative analysis depicted in the table below.

	Solder A			
Forward Voltage [V]	Thermal Resistance [°C/W]	Tj(max) Temperature [°C]	ΔT Average Junction Temperature [°C]	Forward Voltage [V]
1.33	0.401	135.726	105.65	1.33
Percent Change over 50k Cycles	Percent Change over 50k Cycles	Percent Change over 50k Cycles	Percent Change over 50k Cycles	Percent Change over 50k Cycles
1.17%	2.47%	3.02%	2.81%	1.17%

Table 1   Solder A performance over 50k cycles

	Solder B			
Forward Voltage [V]	Thermal Resistance [°C/W]	Tj(max) Temperature [°C]	ΔT Average Junction Temperature [°C]	Forward Voltage [V]
1.491	0.583	157.834	130.372	1.491

Percent Change over 50k Cycles	Percent Change over 50k Cycles	Percent Change over 50k Cycles	Percent Change over 50k Cycles	Percent Change over 50k Cycles
7.86%	11.80%	16.76%	18.73%	7.86%

Table 2 Solder B performance over 50k cycles

As evidenced in Table 1 and Table 2, the optimization of the system yields a notable performance advantage for solder A over solder B. Key metrics such as ΔT and forward voltage, aligned with EPCE standards, reveal that solder B exceeds the 5% threshold for forward voltage, indicating a failure in its testing regimen. By refining the testing methodology, a clearer comparison between different solder package attaches is facilitated, mitigating noise stemming from factors like incorrect calibration curves, wire thickness, and TIM selection. Replication of this test affirms the superior performance of solder A over solder B under identical testing conditions. This approach will be extended to the evaluation of silver sinter materials to discern their relative performance.

2.2 Power Cycling with Silver Sinter Material

This ongoing investigation is currently focusing on sintered materials and assessing the impact of the PCmin test on the integrity of the solder joint layer. The study aims to elucidate the intricate relationship between these variables and their potential ramifications on the overall performance and reliability of the components under examination.

The silver sinter material developed by Alpha Electronic Solutions exhibits notable distinctions from conventional solder materials. It not only withstands higher sintering conditions but also operates effectively at elevated temperatures. While typical solder materials have a limit of around 185°C, Alpha's silver sinter can endure temperatures up to 962°C, although such extremes are rarely encountered in power module applications. The experimental evaluation involves comparing two different sinter materials, namely sinter A and sinter B, to discern their performance characteristics.

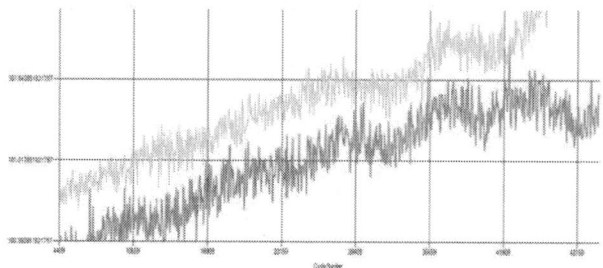

Fig. 14 Sinter A is the dark Orange and sinter B is the orange

As illustrated in Figure 14, the performance of the sinter package attach materials exhibits similarities, albeit with a slight advantage for sinter A over sinter B. While the initial ΔT mirrors that of solder material, the subsequent temperature change over time is notably reduced. This reduction is attributed to the advanced technology and process of silver sinter, which enhances thermal dissipation even under similar voltage and starting ΔT conditions. Comparatively, variations among different silver sinter formulations are generally minimal, irrespective of the application. The tables below provide further insights into these findings.

	Sinter A			
Forward Voltage [V]	Thermal Resistance [°C/W]	Tj(max) Temperature [°C]	ΔT Average Junction Temperature [°C]	Forward Voltage [V]
1.71	0.32	169.8	101.04	1.71
Percent Change over 50k Cycles	Percent Change over 50k Cycles	Percent Change over 50k Cycles	Percent Change over 50k Cycles	Percent Change over 50k Cycles
0.81%	0.60%	1.36%	1.33%	0.81%

Table 3   Sinter A performance over 50k cycles

	Sinter B			
Forward Voltage [V]	Thermal Resistance [°C/W]	T_j(max) Temperature [°C]	ΔT Average Junction Temperature [°C]	Forward Voltage [V]
1.71	0.33	166.27	101.59	1.71
Percent Change over 50k Cycles	Percent Change over 50k Cycles	Percent Change over 50k Cycles	Percent Change over 50k Cycles	Percent Change over 50k Cycles
1.19%	1.03%	1.83%	2.19%	1.19%

Table 4   Sinter B performance over 50k cycles

Utilizing the established methodology applied in the solder test, Tables 3 and 4 highlight the distinction observed when testing two distinct silver sinter materials. Employing the same optimization approach for both sinters, the data indicates that neither sinter approached the EPCE failure criteria, remaining well below the 5% threshold for forward voltage and the 20% threshold for junction temperature change. Contrasting solder with sinter, while it is acknowledged that silver sinter holds a thermal advantage over solder package attach, having empirical data to support this assertion marks a significant stride forward. Analyzing the average junction temperature growth over a consistent timeframe serves as a robust indicator of the superior thermal management afforded by silver sinter.

3   Conclusion

The advancements in solder and sinter technology have significantly enhanced the performance of power modules and their applications. While solder package attach materials remain efficient, they pale in comparison to the superior thermal management offered by silver sinter. Silver sintering confers a considerable advantage in extending power module lifetimes and maintaining optimal junction temperatures. By refining the power cycling processes for both solder and silver sinter package attach materials, obtaining accurate data becomes more feasible.

Further testing and experimentation are necessary to elucidate the thermal distribution dynamics during power cycling of power modules, particularly to discern the comparative effects of solder versus silver sinter applications. Streamlining the calibration process and equipment setup, as outlined in this paper, can reduce guesswork, and facilitate precise data collection. This optimization allows more focused analysis of the structural, functional, and package attach layer impacts on power module performance.

Continued experimentation and optimization efforts will explore additional silver sinter package attach materials and potentially novel solder alternatives. The ultimate objective is to establish a concise yet accurate process for selecting the most suitable application for power module use.

References

Please follow international scientific citation rules.

[1] C. Miraglia, Fundamentals of Heat Transfer in Thermal Interface Gap Filler Materials (2014).

[2] Siemens, Simcenter™ Micred™ Power Tester 2400A Measurement Control Software User Guide (2023).

[3] T. Harder, Qualification of Power Modules for Use in Power Electronics Converter Units in Motor Vehicles (2021).

PCIM Europe 2024, 11– 13 June 2024, Nuremberg   DOI: 10.30420/566262193

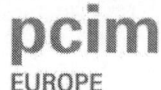

Real-Time Estimation and Sensitivity Analysis of Parasitic Capacitances in Electric Drive Systems

Mohammadreza Bagheribavaryani[1], Niklas Langmaack[1]

[1] Technische Universität Braunschweig, Germany

Corresponding author: Niklas Langmaack, N.Langmaack@tu-braunschweig.de
Speaker: Mohammadreza Bagheribavaryani, mohammadreza.bagheribavaryani@tu-braunschweig.de

Abstract

This study introduces a novel approach for real-time high-frequency modeling of Electric Drive Systems using an Unscented Kalman Filter. It accurately estimates the parasitic capacitance of AC cables, motors, and power switches under actual operational conditions while traditional offline models fail to capture the true high-frequency characteristics of the Electric Drive Systems, leading to errors in Electromagnetic Interference studies or unnecessary over-design of EMI filters. This over-design results in added weight and volume, compromising the efficiency of electric vehicles. Hence, real-time estimation of parasitic elements is crucial for precise EMI analyses and the optimization of system efficiency. Additionally, the paper conducts a sensitivity analysis on the impact of parasitic capacitances on EMI level of the system, guiding the development of targeted reduction strategies to comply with strict Electromagnetic Compatibility standards in automotive industry.

1   Introduction

The integration of Wide Band Gap (WBG) semiconductors into mobile applications, notably electric vehicles, has significantly enhanced Electric Drive Systems' (EDS) performance by enabling higher switching frequencies and reducing inverter size. Yet, this advancement introduces complex challenges, particularly in managing Electromagnetic Interference (EMI), which poses risks of malfunction or system failure if not adequately controlled [1]. The central role of AC cables, motors, and power switches in EMI propagation within EDS underscores the need for precise high-frequency modeling to address these EMI challenges effectively. Moreover, the inherent parasitic capacitances of the components in an EDS are sensitive to electrical stress, mechanical vibration, and environmental factors like humidity and temperature leading to dynamic changes in their high-frequency behavior [2], [3]. This variability complicates the accuracy of offline modeling efforts, which do not accurately reflect operational conditions, often resulting in EMI studies' errors or excessive EMI filter designs. Such over-designs unnecessarily increase the weight and volume of EDS components, directly impacting vehicle efficiency. Consequently,

a shift towards real-time modeling of system-wide parasitic capacitance is imperative for precise EMI analysis and optimizing EDS efficiency in mobile applications, especially when it comes to active filter design [4]. Recent research has spotlighted the critical impact of accurately estimating parasitic capacitance on the performance and reliability of EDS [5]–[7]. Traditional methods, such as the least-square approach combined with offline measurement setups, offer limited practicality due to their time-consuming nature and inability to adapt to real-time changes [8]. Recognizing these limitations, this paper pioneers a real-time estimation technique using an Unscented Kalman Filter (UKF) approach. This method overcomes the shortcomings of previous models by accounting for the complex interplay of multiple parasitic elements within the EDS, rather than simplifying them into a single capacitance value.

By leveraging the UKF's advanced capabilities in handling uncertainties, our method ensures accurate, real-time insight into parasitic capacitances, crucial for effective EMI mitigation and system optimization. Through MATLAB/Simulink simulations, we validate the model's effectiveness, showcasing its potential in condition monitoring and guiding the

PCIM Europe 2024, 11– 13 June 2024, Nuremberg   DOI: 10.30420/566262193

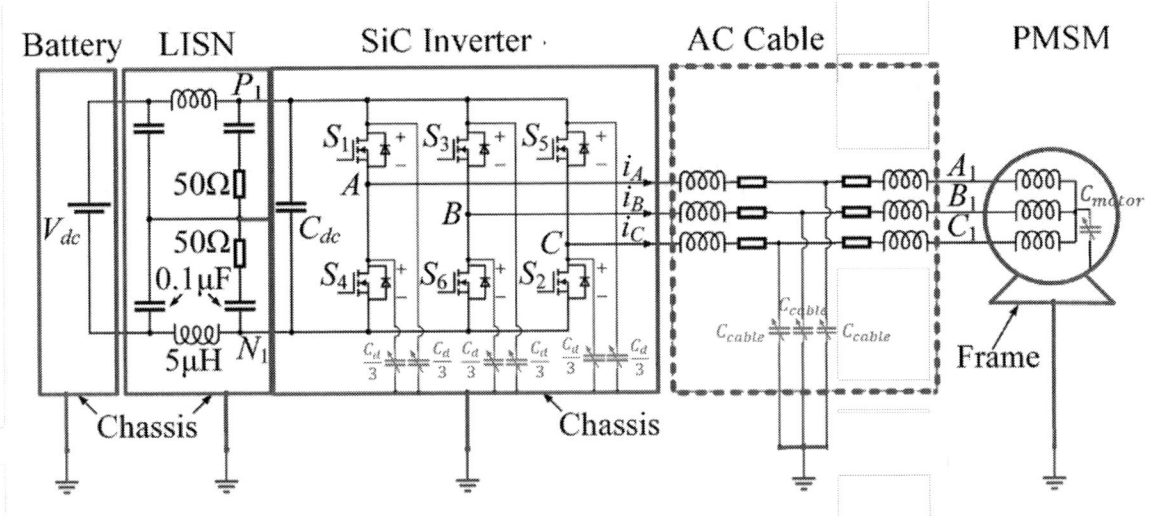

Fig. 1: An EDS with parasitic elements

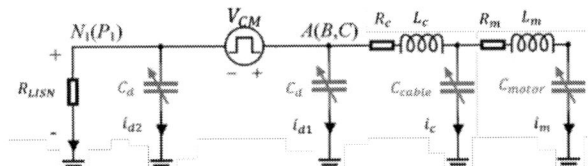

Fig. 2: Simplified model of the common mode current in an EDS

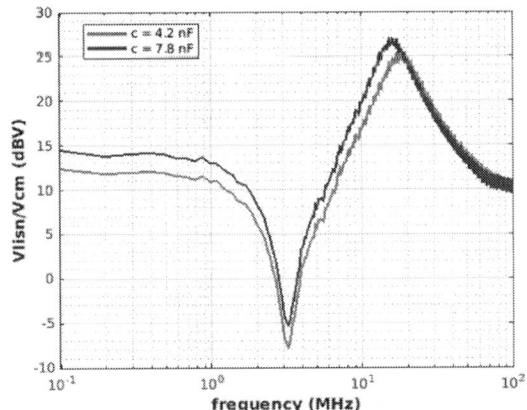

Fig. 3: Common mode voltage of the EDS influenced by variation of the value of the parasitic capacitance of the cable

design of more efficient EMI filters. This approach not only enhances the precision of EMI studies but also contributes to the development of lighter, more efficient electric vehicles by preventing over-design in EMI filtering solutions, specially in active filter approaches.

Moreover, a sensitivity analysis is also conducted on the system to elucidate the contribution of the parasitic capacitances of the EDS on the common-mode voltage of the system, giving the designer an insight into how different components influence the EMI of the EDS. This analysis is critical, as the common-mode voltage can have significant implications for the reliability and efficiency of the EDS, affecting everything from electromagnetic compatibility to the lifespan of the system components.

2 Common Mode Current Analysis in EV Powertrains

The study of electric vehicle (EV) powertrains at high frequencies necessitates an all-encompassing modeling approach to accurately capture the system's dynamics. A pivotal aspect of this study is the development of an equivalent high-frequency model that accurately represents the key components of an EV powertrain. This model, as illustrated in Figure 1, encompasses several critical elements, including the battery pack, the Line Impedance Stabilization Network (LISN), a Silicon Carbide (SiC) inverter, AC cables, and a Permanent Magnet Synchronous Machine (PMSM), along with their respective parasitic elements. These components are foundational in understanding the high-frequency behavior and the propagation of common mode (CM) noise within the system. In the realm of CM current analysis, a distinctive modeling approach is adopted. The lower switches of the SiC inverter are conceptualized as voltage sources, a simplification made possible by short-circuiting the positive and negative DC terminals. This is a direct consequence of the constant DC voltage provided by the bulk capacitor (C_{dc}), which effectively parallels each upper switch with its lower counterpart and their parasitic capacitances as a consequence, since bulk DC capacitor provides a very low impedance for high-frequency common mode current, ultimately making the capacitor of the DC link negligible in EMI studies. Furthermore, this setup permits the assumption that the positive and negative DC nodes are at an identical voltage potential, facilitating the short-circuiting of the LISN terminals. On the AC side, the analysis simplifies further by short-circuiting all three phase nodes, a step validated by the assumption of identical nature of CM components across all three phases. This in turn, results in the lumped model of each phase of the cable being identical. On the PMSM side, the parasitic capacitance, as shown, connects the neutral point of the motor to the grounded housing. By implementing the T-model for the AC cable in this study, the parasitic capacitance of the cable can be distinguished with that of the motor, thanks to the R-L impedance of the cable (and motor) on the motor side of the cable. Finally, the common mode model of the EDS can be simplified into the circuit in Figure 2. Here, the common mode voltage source, denoted as V_{CM} is mathematically defined as the average of the common mode voltages across the three phases, represented by the formula:

$$V_{\text{CM}} = \frac{V_{\text{AN}} + V_{\text{BN}} + V_{\text{CN}}}{3} \qquad (1)$$

Before delving into the UKF method, to emphasis on the effect of parasitic capacitance variations on system-level EMI, common mode voltage sensitivity to this variation is depicted in Figure 3, where the ratio of V_{LISN} to V_{CM} serves as a metric for assessing common mode interference. The comparison of this ratio across two different values of cable parasitic capacitance illuminates the sensitive nature of the system to a small variation in the value of the parasitic capacitance of the cable, which is inevitable in automotive applications, particularly

in terms of common mode noise intensity. Further, the analysis reveals the frequency-dependent response of the LISN to cable parasitic capacitance. Especially below 20 MHz, the LISN circuit's resonant frequency shifts to lower values with decreasing parasitic capacitance, manifesting a peak at 20 MHz and showcasing an anti-resonant effect at 3 MHz. This nuanced understanding underscores the critical influence of cable parasitic capacitance, as a show case, on the behavior of common mode current and the overall EMI profile of the EV powertrain system. Through this expanded discussion, the complex interplay of various components and their role in shaping the high-frequency behavior of EV powertrains, particularly concerning common mode current, is elucidated. This comprehensive model and the insights drawn from it are instrumental in advancing our understanding of EMI in electric vehicle powertrains, paving the way for more effective mitigation strategies and design optimizations.

3 Methodology

In this study, we propose to estimate the parasitic capacitances of the motor, AC cables, and switches in an EDS using the UKF method. To this end, we derive the non-linear state space equation of the system.

3.1 Non-Linear State space equations of the system

To implement the UKF method, we first need to define a state space model for the simplified CM EMI circuit of Figure 2. Taking into account the parasitic capacitance of motor, AC cables and power switches as the states number 1 to 3, respectively, and taking into account that the changes in the value of the parasitic capacitance of the EDS within the sampling time period of the UKF is negligible, one can write the first three state equations as follow:

$$\frac{dx_1}{dt} = 0 \qquad \frac{dx_2}{dt} = 0 \qquad \frac{dx_3}{dt} = 0 \qquad (2)$$

where x_1, x_2 and x_3 are the parasitic capacitance of motor C_m, parasitic capacitance of AC cable C_c and parasitic capacitance of the power switches C_d, respectively. Moreover, defining four more system states as $x_4 = V_m$, $x_5 = V_c$, $x_6 = i_m$, and $x_7 = i_c$ indicating the voltages and currents of the parasitic capacitances of the motor and cable respectively, the following state equations are driven:

$$C_m \frac{dV_m}{dt} = i_m \rightarrow \frac{dV_m}{dt} = \frac{i_m}{C_m} \rightarrow \frac{dx_4}{dt} = \frac{x_6}{x_1} \quad (3)$$

$$C_m \frac{dV_c}{dt} = i_c \rightarrow \frac{dV_c}{dt} = \frac{i_c}{C_c} \rightarrow \frac{dx_5}{dt} = \frac{x_7}{x_2} \quad (4)$$

Applying Kirchhoff's voltage law yields the sixth state equation as:

$$V_m + V_{R_m} + V_{L_m} = V_c$$

$$\rightarrow V_m + R_m i_m + L_m \frac{di_m}{dt} = V_c$$

$$\rightarrow \frac{di_m}{dt} = -\frac{V_m}{Lm} - \frac{R_m i_m}{Lm} + \frac{V_c}{Lm} \quad (5)$$

$$\rightarrow \frac{dx_6}{dt} = -\frac{x_4}{L_m} + \frac{x_5}{L_m} - \frac{R_m x_6}{L_m} \quad (6)$$

Also, in order to maintain the observability of the system, which will be explained more in detaile in section 3.2, we define the measured common mode voltage as the 8th state as $x_8 = V_{LISN}$, which is the voltage across the resistor of the LISN in the simplified equivalent circuit. By applying Kirchhoff's voltage law in another mesh including U_{CM} and V_{LISN} one may write:

$$V_{LISN} + U_{CM} = V_c + L_c \frac{d(i_c + i_m)}{dt}$$

$$+ R_c(i_c + i_m)$$

$$\xrightarrow{(5)} V_{LISN} + U_{CM} = V_c + L_c \frac{di_c}{dt} + \frac{L_c V_c}{L_m}$$

$$- \frac{L_c R_m i_m}{L_m} - \frac{L_c V_m}{L_m} + R_c i_c + R_c i_m$$

$$\rightarrow \frac{di_c}{dt} = \frac{V_m}{L_m} - (\frac{1}{L_c} + \frac{1}{L_m})V_c$$

$$+ (\frac{R_m}{L_m} - \frac{R_c}{L_c})i_m - \frac{R_c}{L_c}i_c$$

$$+ \frac{1}{L_c}V_{LISN} + \frac{1}{L_c}U_{CM} \quad (7)$$

By considering the V_{CM} as the first input to the system, the 7th state equation of the system can be written as:

$$\frac{dx_7}{dt} = \frac{x_4}{L_m} - (\frac{1}{L_c} + \frac{1}{L_m})x_5$$

$$+ (\frac{R_m}{L_m} - \frac{R_c}{L_c})x_6 - \frac{R_c}{L_c}x_7$$

$$+ \frac{1}{L_c}x_8 + \frac{1}{L_c}u_1 \quad (8)$$

In order to derive the state equation for the 8th state, by applying the Kirchhoff's current law on the ground node one may write:

$$i_m + i_c + i_{d1} + i_{d2} + i_{LISN} = 0$$

$$i_m + i_c + C_d \frac{d(U_{CM} + V_{LISN})}{dt}$$

$$+ C_d \frac{dV_{LISN}}{dt} + \frac{V_{LISN}}{R_{LISN}} = 0$$

$$\rightarrow \frac{dV_{LISN}}{dt} = -\frac{i_m}{2C_d} - \frac{i_c}{2C_d}$$

$$- \frac{V_{LISN}}{2R_{LISN}C_d} - \frac{1}{2}\frac{dU_{CM}}{dt} \quad (9)$$

By considering $\frac{dV_{CM}}{dt}$ as the second input to the system, which can be easily calculated in real time, the state equation for the 8th equation would be as follows:

$$\frac{dx_8}{dt} = -\frac{x_6}{2x_3} - \frac{x_7}{2x_3} - \frac{x_8}{2R_{LISN}x_3} - \frac{u_2}{2} \quad (10)$$

Moreover, as the final step for defining the system, the output of the system has to be derived, which in this case, is the measured voltage of the LISN:

$$y = V_{LISN}$$

$$y = x_8 \quad (11)$$

Having established the state-space model, we now understand the system's dynamics and the relationships between state variables. This sets the stage for employing state estimation techniques like the Unscented Kalman Filter (UKF). The effectiveness of such techniques is contingent upon the system's observability. Therefore, our next step involves assessing the system's observability to ensure all state variables can be estimated accurately. This assessment is crucial as it informs any necessary modifications to the state-space model to facilitate observability, guiding us towards achieving comprehensive state estimation and control.

3.2 Observability Analysis of the State-Space Model

Before delving into the estimation results obtained through the Unscented Kalman Filter (UKF) algorithm, it is paramount to ensure the system defined by the provided state-space equations is observable. Observability, a fundamental property of control systems, indicates the feasibility of inferring the complete state vector of a system from its output measurements. For the UKF algorithm to yield reliable and accurate state estimations in real-time, the system must be fully observable. This necessitates a thorough observability analysis, which not only validates the potential for real-time state estimation but also reinforces the integrity of the system's control strategy. To apply the observability evaluation method, we implement the method explained in [9] we construct an inference diagram that captures the dynamical correlation between the states as stipulated by these equations. This diagram will visually represent the system, indicating paths where information flow from one state to another is explicit. Steps in this methodology include:

1. **Drawing Straight Paths:** For each equation where a state variable's differential equation directly depends on another state variable (or an algebraic function involving other states), a straight path is drawn in the inference diagram. This path symbolizes the potential to infer the state at the path's destination from the state at its origin. For example, the equation $\frac{dx_4}{dt} = \frac{x_6}{x_1}$ suggests a path from x_4 to x_6 and x_1, indicating the existence of the information of x_6 and x_1 in the differential equation of x_4.

2. **Identifying Strongly Connected Components (SCCs):** The inference diagram is analyzed to identify SCCs, which are defined as maximal subgraphs where there is a path from each node to every other node within the subgraph. These SCCs highlight groups of state variables that are interdependent, suggesting that if one variable in the SCC can be observed, the others can also be inferred.

3. **Determining Root SCCs (RSCCs):** Among the SCCs, those with no incoming edges from outside the SCC are classified as RSCCs. The absence of incoming edges implies these state variables can be considered as starting points for observability analysis. They are pivotal in determining the system's overall observability.

4. **Evaluating Observability:** Based on the presence of RSCCs and the structure of the inference diagram, the system's observability can be ascertained. A system is deemed fully observable if there is at least one sensor node (measured state) in each RSCC of the system.

For the system in hand, the analysis of the system's dynamics through its inference diagram, illustrated in Figure 4, elucidates the structure of SCCs within

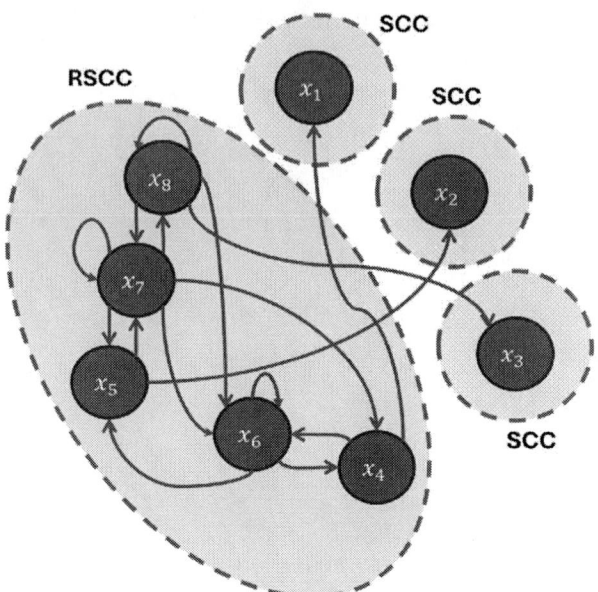

Fig. 4: Inference diagram of the system

the network, as shown in dashed circles. Notably, the interconnection between states x_4, x_5, x_6, x_7, and x_8 constitutes a significant SCC, indicating that the information from any one of these states could be utilized to deduce the others within this grouping. Additionally, the diagram highlights that the first three states, x_1, x_2, and x_3, each independently establish their own SCCs, as depicted by the dashed circles in Figure 4. A pivotal aspect of the observability analysis, as elaborated earlier, involves the identification of RSCCs, which are characterized by the absence of incoming edges from outside the component. The SCC that includes x_4, x_5, x_6, x_7, and x_8 qualifies as an RSCC due to its insular nature, as indicated by the shaded area in Figure 4. For the system to be deemed observable, it is imperative that at least one state within the RSCC is accessible for measurement. In this instance, state x_8 serves as such a sensor node, thereby confirming the system's observability and ensuring the unique estimation of its states. This observability criterion underscores the significance of judiciously selecting V_{LISN} as the eighth state, given its status as the only measurable state within this system.

4 UKF Implementation and Results

The UKF stands out as a powerful tool for estimating the states of nonlinear systems with a higher degree of accuracy than linear approximation methods like the Extended Kalman Filter (EKF). Its implementation involves a series of systematic steps, starting with the discretization of state-space equations, followed by the formulation specific to non-linear systems, initialization of the filter, and the iterative update process. Before implementing the UKF, the continuous state-space model needs to be discretized. This is because the UKF operates in discrete time steps, making it necessary to convert the continuous differential equations into discrete updates.

4.1 Discretization of State-Space Equations

The non-linear state-space equations, as presented in the system model, necessitate discretization for implementation within the UKF framework. One common method for achieving this transformation is Euler's method, which facilitates the transition from continuous to discrete representation. This process yields the general state-space equations of the system as follows:

$$\mathbf{x}_{k+1} = \mathbf{x}_k + \Delta t \cdot \mathbf{f}(\mathbf{x}_k, \mathbf{u}_k) + \mathbf{w}_k \quad (12)$$

$$\mathbf{y}_k = \mathbf{h}(\mathbf{x}_k) + \mathbf{v}_k \quad (13)$$

Here, $\mathbf{f}$ represents the transition function that models how the state of the system evolves over each time step from $\mathbf{x}_k$ to $\mathbf{x}_{k+1}$, considering the input $\mathbf{u}_k$. The function $\mathbf{h}$ denotes the measurement function, mapping the state vector $\mathbf{x}_k$ to the observed outputs $\mathbf{y}_k$. The terms $\mathbf{w}_k$ and $\mathbf{v}_k$ are the process and measurement noise at step k, assumed to be white and with covariance matrices Q_k and R_k, respectively. The voltage across the LISN, considered as output $\mathbf{y}$, is measurable through various techniques, effectively incorporating the effects of EMI as indicated by the input U_{CM}, the source of EMI. This discretized model forms the basis for the subsequent implementation of the UKF, enabling the algorithm to estimate the system's state from the measurable outputs and the known inputs, while accounting for the inherent uncertainties represented by the noise components $\mathbf{w}_k$ and $\mathbf{v}_k$.

4.2 UKF implementation

The UKF addresses the challenge of non-linear state estimation by employing a deterministic sampling technique known as the Unscented Transform (UT) to capture the mean and covariance of the state distribution. The general steps for implementing the UKF are as follows:

1. **Initialization:** Define the initial state estimate $\hat{\mathbf{x}}_0$ and covariance matrix P_0. Set the UKF scaling parameters α, β, and κ, which influence the spread of sigma points around the mean.

2. **Sigma Point Calculation:** Generate sigma points $\mathbf{X}_k^{(i)}$ around the current state estimate to capture the state distribution.

$$\mathbf{X}_k^{(0)} = \hat{\mathbf{x}}_k,$$
$$\mathbf{X}_k^{(i)} = \hat{\mathbf{x}}_k \pm (\sqrt{(n+\lambda)P_k})_i, \quad \text{for } i = 1, \dots, 2n$$

where $\lambda = \alpha^2(n+\kappa) - n$, and $(\sqrt{P})_i$ denotes the i-th column of the matrix square root of P_k. The term n is the dimension of the state vector.

3. **Prediction:** Estimate the state mean $\hat{\mathbf{x}}_{k|k-1}$ and covariance $P_{k|k-1}$ from the sigma points, incorporating process noise Q_k.

$$\hat{\mathbf{x}}_{k|k-1} = \sum_{i=0}^{2n} W_m^{(i)} f(\mathbf{X}_k^{(i)}, \mathbf{u}_k), \quad (14)$$

$$P_{k|k-1} = \sum_{i=0}^{2n} W_c^{(i)} \left(f(\mathbf{X}_k^{(i)}, \mathbf{u}_k) - \hat{\mathbf{x}}_{k|k-1} \right)$$
$$\times \left(f(\mathbf{X}_k^{(i)}, \mathbf{u}_k) - \hat{\mathbf{x}}_{k|k-1} \right)^T + Q_k \quad (15)$$

4. **Predict Measurement:** Predict the measurement $\hat{\mathbf{z}}_{k|k-1}$ and its covariance $P_{zz,k}$ from the sigma points and compute the cross-covariance between the state and the measurement.

$$\hat{\mathbf{z}}_{k|k-1} = \sum_{i=0}^{2n} W_m^{(i)} h(\mathbf{X}_k^{(i)}), \quad (16)$$

$$P_{zz,k} = \sum_{i=0}^{2n} W_c^{(i)} \left(h(\mathbf{X}_k^{(i)}) - \hat{\mathbf{z}}_{k|k-1} \right)$$
$$\times \left(h(\mathbf{X}_k^{(i)}) - \hat{\mathbf{z}}_{k|k-1} \right)^T + R_k, \quad (17)$$

$$P_{xz,k} = \sum_{i=0}^{2n} W_c^{(i)} \left(f(\mathbf{X}_k^{(i)}, \mathbf{u}_k) - \hat{\mathbf{x}}_{k|k-1} \right)$$
$$\times \left(h(\mathbf{X}_k^{(i)}) - \hat{\mathbf{z}}_{k|k-1} \right)^T \quad (18)$$

where R_k is the measurement noise covariance matrix and $h(\cdot)$ is the measurement function.

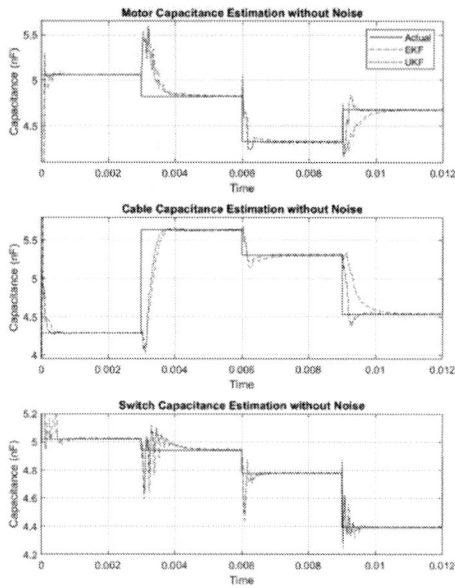 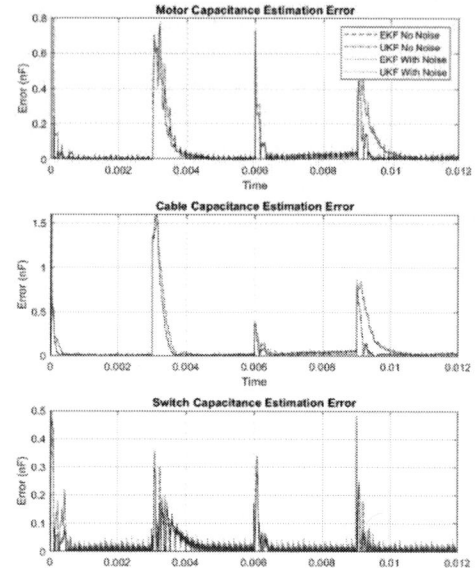

Fig. 5: Estimation results and the corresponding error

5. **Update:** With the actual measurement z_k, compute the Kalman gain K_k and update the state estimate $\hat{x}_k$ and the covariance matrix P_k for the next step.

$$K_k = P_{xz,k}P_{zz,k}^{-1}, \tag{19}$$

$$\hat{x}_k = \hat{x}_{k|k-1} + K_k(z_k - \hat{z}_{k|k-1}), \tag{20}$$

$$P_k = P_{k|k-1} - K_k P_{zz,k} K_k^T \tag{21}$$

The UKF iteratively repeats these steps, refining the state estimate and covariance with each new measurement, thus providing a robust framework for tracking the system dynamics under non-linear conditions and uncertainties. As we are particularly interested in the parasitic capacitance of the motor, AC cable, and the power switches, we can obtain their respective values in real-time by reading the first three states as the UKF outputs the estimations through $\hat{x}_k$.

4.3 Simulation Results

In this section, we embark on a detailed examination of the simulation results obtained from implementing the Unscented Kalman Filter (UKF) for the estimation of parasitic capacitances in motors, AC cables, and power switches within dynamic environments. This study primarily focuses on the UKF, chosen for its robustness and accuracy in

handling non-linear system dynamics typical of EV systems. However, to provide a comparative analysis and underscore the computational efficiency and performance trade-offs, results from the Extended Kalman Filter, a method known for its lower computational burden, are also presented.

The simulation incorporates scenarios with three random step changes in the parasitic capacitances values to simulate real-world fluctuations. These fluctuations aim to emulate the effects of mechanical vibrations or environmental changes on the parasitic capacitances which is inevitable throughout a driving cycle for an EV. This dynamic setup challenges the estimation methodologies to adapt and accurately track the system's state amidst sudden changes, thus providing a realistic test bed for their evaluation. The UKF's non-linear state estimation capability enables it to excel in this dynamic simulation, offering precise and timely tracking of the step changes in capacitance values.

Figure 5 shows the estimation results for UKF and EKF compared to the actual value of the parasitic capacitances of the EDS. As it can be observed from Figure 5, the UKF's performance illustrate its capacity to closely follow the actual capacitance values, even in the face of abrupt system variations in three different time instances throughout the time. As it can be observed, the parasitic capacitances of EDS undergo 3 random step changes within ±20% of the nominal value, and the estimation method follows the reference values in less than 180,000 sample time. This accuracy is particularly noteworthy under conditions of measurement noise, where the UKF demonstrates significant resilience, maintaining estimations within acceptable error margins. For a comprehensive performance evaluation, the EKF's estimations are juxtaposed with those of the UKF. Despite the EKF's lower computational demand, its linear approximation approach, as opposed to the non-linear transform (Unscented Transform) implemented by UKF method, exhibits a slight delay in adapting to the step changes, as seen in the comparative figures. This contrast highlights the trade-off between computational efficiency and the ability to handle nonlinear dynamics and measurement noise, with the UKF emerging as the more adept methodology under the conditions tested.

Under scenarios with measurement noise, both methodologies were scrutinized for their noise handling capabilities under same scenarios. The UKF's performance remained robust, underscoring its utility in real-world applications where sensor noise is unavoidable. Conversely, the EKF, while still producing viable estimations, encountered increased estimation errors in comparison to the UKF. Notably, the errors for both methodologies spiked at the instant of step changes but rapidly returned to acceptable levels, illustrating the filters' capacity to recover from sudden disturbances. The UKF consistently demonstrated lower estimation errors across all components, reaffirming its superiority in tracking capabilities, especially in noisy conditions. The transient increase in errors observed at the step changes for both filters underscores the challenge of abrupt system dynamics, yet the quick return to nominal error levels highlights the resilience and adaptability of the methodologies.

This simulation study accentuates the UKF's superior performance in dynamically changing environments and under the influence of measurement noise, making it exceptionally suitable for the estimation of parasitic capacitances in EVs. While the EKF offers a computationally less intensive alternative, with a computational time of about 30% less than that of the UKF, its performance in nonlinear conditions and its responsiveness to sudden changes are outmatched by the UKF.

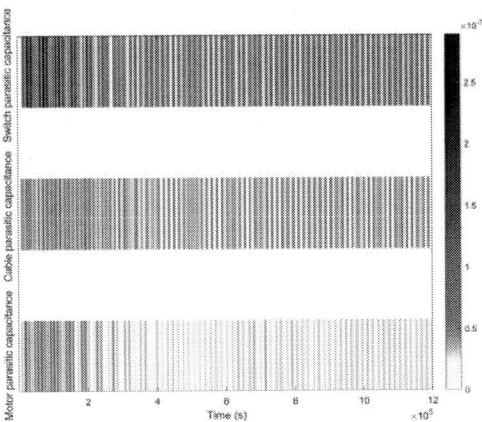

Fig. 6: Sensitivity of the Common mode voltage measured by LISN to the perturbance in each parasitic capacitances in the EDS

5 Sensitivity Analysis

Sensitivity analysis serves as a cornerstone in the assessment of electric drive systems, offering insight into how variations in system states impact the output. This study emphasizes the local sensitivity analysis (LSA) of the system's common mode voltage in response to perturbations in parasitic capacitances of the motor, cable, and switch.

5.1 Local Sensitivity Analysis Framework

The Local Sensitivity Analysis (LSA) framework, adapted to this electric drive system, is employed to evaluate the influence of small perturbations in the parasitic capacitances of the motor, cable, and switch on the common mode voltage (y). The sensitivity of the system's output to these perturbations is captured by numerically approximating the derivative of the output function with respect to each parameter of interest, employing a finite-difference method for estimation. The local sensitivity of the model output to the i th state of the system at the nominal value $\bar{x}$ can be defined as [10]:

$$\hat{S}_i(\bar{x}) = \frac{\bar{x}_i}{\bar{V}_i} \frac{[f(\bar{x}_1,..,\bar{x}_i + \Delta_i,..,\bar{x}_m) - f(\bar{x}_1,..,\bar{x}_i,..,\bar{x}_m)]}{\Delta_i} \tag{22}$$

where subscript i denotes the i th state, f is the nonlinear output function of the common mode voltage, $\bar{x}_i$ is used to adjust state on different scales to a common scale, and the average voltage $\bar{V}_i$ is used to achieve the normalization of sensitivity. Given the system's nonlinear dynamics and the complexity of its state space representation, the LSA provides a pragmatic approach to discern the relative impact of each parasitic capacitance on the common mode voltage. The perturbation size, Δ_i is chosen to be sufficiently small, typically less than 1% of the nominal parameter value, to ensure the analysis remains locally pertinent.

5.2 Analysis Results

The sensitivity analysis, performed on the system, uncovers that the common mode voltage is most significantly affected by perturbations in the parasitic capacitance of the switch, followed by the cable, and least by the motor. These insights are illustrated in Figure 6, depicting the relative sensitivity of the common mode voltage to changes in each of the parasitic capacitances.

This hierarchical sensitivity underscores the paramount importance of the switch's parasitic capacitance in influencing the electric drive system's performance. By identifying and quantifying the most influential factors affecting the common mode voltage, this analysis aids in the prioritization of design and optimization efforts, enhancing the system's efficiency and reliability.

6 Conclusion and Future Works

In this paper, we introduced a real-time approach for high-frequency modeling of Electric Drive Systems (EDS) through the accurate estimation of parasitic capacitances using an Unscented Kalman Filter (UKF). This method addresses the shortcomings of traditional offline models by offering precise EMI mitigation and system optimization, crucial for enhancing the performance and reliability of electric vehicles. Our methodology, validated through comprehensive MATLAB/Simulink simulations, demonstrates the UKF's superior ability to track dynamic changes in parasitic capacitances accurately, providing a robust solution for real-time EMI analysis. Looking ahead, several avenues for future research emerge from this study. One key area involves incorporating more detailed models for each component within the EDS. This enhancement would enable a finer-grained analysis of parasitic effects, potentially uncovering new insights into EMI mitigation strategies and equip the the implemented method to be valid for a wider range of frequency. Additionally, considering the skin effect's impact on high-frequency behavior of conductors could provide a more accurate representation of real-world conditions, improving the precision of parasitic capacitance estimates.

Further research could also explore the incorporating the influence of external factors into the model such as temperature and moisture on the accuracy of the proposed method. These environmental variables can significantly affect the electrical properties of parasitic elements, thereby influencing EMI characteristics. Understanding these effects could lead to more robust models capable of adapting to varying operational conditions.

Extending the proposed method to estimate all parasitic elements in the EMI circuit, including the stray inductances and resistors, in real-time represents another promising direction. This comprehensive approach would offer a holistic view of the EMI landscape within EDS, facilitating more effective mitigation strategies.

Moreover, experimental validation of the proposed method in real-world scenarios is essential. The authors are currently in the process of preparing a test bench setup aimed at conducting experimental verifications of the proposed method. By correlating simulation results with experimental data, we can further enhance the reliability and applicability of our model.

In conclusion, while our study marks an advance in real-time EMI analysis for electric drive systems, the path forward is rich with opportunities for further exploration. By addressing these future research directions, we can continue to refine our understanding and methodologies, pushing the boundaries of electric vehicle technology and contributing to the development of more efficient, reliable, and high-performing electric vehicles.

References

[1] V. Karakasli, A. Allioua, and G. Griepentrog, "Common-mode emi noise modeling of three-level t-type inverter for adjustable speed drive systems," in *2022 24th European Conference on Power Electronics and Applications (EPE'22 ECCE Europe)*, 2022, pp. 1–8. DOI: 10.23919/EPE22ECCEEurope.2022.1079042.

[2] M. S. Seyfried and L. E. Grant, "Temperature effects on soil dielectric properties measured at 50 mhz," *Vadose Zone Journal*, vol. 6, no. 4, pp. 759–765, 2007.

[3] P. Palta, P. Kaur, and K. S. Mann, "Dielectric behavior of soil as a function of frequency, temperature, moisture content and soil texture: A deep neural networks based regression model," *Journal of Microwave Power and Electromagnetic Energy*, vol. 56, no. 3, pp. 145–167, 2022. DOI: 10.1080/08327823.2022.2103630.

[4] S. Karimi, F. Naseri, E. Farjah, T. Ghanbari, and J.-L. Schanen, "Estimation of cm parasitic capacitances in front-end llc resonant dc-dc converters," in *2021 12th Power Electronics, Drive Systems, and Technologies Conference (PEDSTC)*, 2021, pp. 1–5. DOI: 10.1109/PEDSTC52094.2021.9405841.

[5] J. Luszcz, "Motor cable effect on the converter fed ac motor common mode current," in *2011 7th International Conference-Workshop Compatibility and Power Electronics (CPE)*, Tallinn, Estonia, 2011, pp. 445–450. DOI: 10.1109/CPE.2011.5942277.

[6] J. Yao, M. El-Sharkh, Y. Li, Z. Ma, S. Wang, and Z. Luo, "Investigation of radiated emi in non-isolated power converters with power cables in automotive applications," in *2019 IEEE Energy Conversion Congress and Exposition (ECCE)*, Baltimore, MD, USA, 2019, pp. 6957–6964. DOI: 10.1109/ECCE.2019.8912255.

[7] K. Choksi, Y. Wu, M. U. Hassan, F. Luo, B. Liu, and X. Wu, "Inspecting impact of cabling infrastructure on reflected wave and emi for more electric aircraft (mea) motor drives," in *2022 IEEE Transportation Electrification Conference & Expo (ITEC)*, Anaheim, CA, USA, 2022, pp. 529–533. DOI: 10.1109/ITEC53557.2022.9813941.

[8] N. Hadjigeorgiou and P. P. Sotiriadis, "Parasitic capacitances inductive coupling and high-frequency behavior of amr sensors," *IEEE Sensors Journal*, vol. 20, no. 5, pp. 2339–2347, 2020. DOI: 10.1109/JSEN.2019.2953113.

[9] Y.-Y. Liu, J.-J. Slotine, and A.-L. Barabasi, "Observability of complex systems," *Proceedings of the National Academy of Sciences*, vol. 110, no. 7, pp. 2460–2465, Jan. 2013. DOI: 10.1073/pnas.1215508110.

[10] V. C. Machado, G. Tapia, D. Gabriel, J. Lafuente, and J. A. Baeza, "Systematic identifiability study based on the fisher information matrix for reducing the number of parameters calibration of an activated sludge model," *Environmental Modelling and Software*, vol. 24, no. 11, pp. 1274–1284, 2009. DOI: https://doi.org/10.1016/j.envsoft.2009.05.001.

PCIM Europe 2024, 11– 13 June 2024, Nuremberg   DOI: 10.30420/566262194

Parasitic Component and Gate Resistance Effects of Internal and External Package Level on **Switching Performance of SiC Power Module**

Nguyen-Nghia Do[1], Sheng-Tsai Wu[1], Tai-Jyun Yu[1], Chen-Min Chen[1], Jing-Yao Chang[1], Wei-Zhong Huang[1], Zhen-Yi Liao[1], Yuan-Zhe Zhang[1]

[1] PowerX Semiconductor, Taiwan

Corresponding author: Nguyen-Nghia Do, nghia.do@powerx-semiconductor.com
Speaker: Nguyen-Nghia Do, nghia.do@powerx-semiconductor.com

Abstract

This paper investigates critical influences of parasitic components and gate resistances at both the internal and external package levels on the switching behavior and efficiency of Silicon Carbide (SiC) power modules. Internal package-level parasitics, emanating from the package design and interconnections, as well as external package-level parasitics, arising from the interaction with external circuits, are examined in-depth. The study proposes a simulation model and leverages advanced simulation techniques to analyze these parasitic elements and gate resistancess. The results reveal their profound impact on switching characteristics and subsequent efficiency. This research provides valuable insights into the optimization of SiC power module design in the V_{ds} = 1.2 kV class for enhanced performance and efficiency.

1   Introduction

In the realm of power electronics, SiC power modules have arose as a revolutionary technology, promising enhanced switching characteristics and efficiency in a variety of applications. Such modules, comprising SiC-based semiconductor devices, are gaining momentum in power conversion systems, renewable energy sources, electric vehicles, and more, owing to their superior material properties. By virtue of its smaller tail current compared to IGBT, the SiC device offers a significant reduction in switching losses. Furthermore, the SiC MOSFETs exhibit higher breakdown voltages, faster switching speeds, and lower conduction losses than traditional Silicon (Si) counterparts, which translates into increased efficiency, reduced heat generation, and a smaller form factor for power electronic systems [1-3]. However, the full realization of SiC power modules' potential is intricately linked to the understanding and mitigation of parasitic components at both the internal and external package levels. These parasitic components, often considered as unwanted byproducts, encompass a range of effects, including stray inductances, capacitances, and resistances [4-5]. As SiC devices can switch at considerably higher frequencies, the impact of these parasitic elements and gate resistances becomes even more pronounced, influencing the overall performance and efficiency of power modules [6-8].

This research delves into the critical aspects of parasitic components and gate resistances in the context of SiC power modules. Figure 1 describes the package of a half-bridge power module from the front and back views. Within the scope, the power module with multiple parallel SiC bare dies (chips) per switch on the same substrate was designed and fabricated. The SiC bare dies used in the power module are rated for the current of 130 A and the voltage of 1200 V. The current and power ratings can be scaled up or down by changing the quantity of the SiC chips.

Fig. 1   SiC half-bridge power module with front and back views

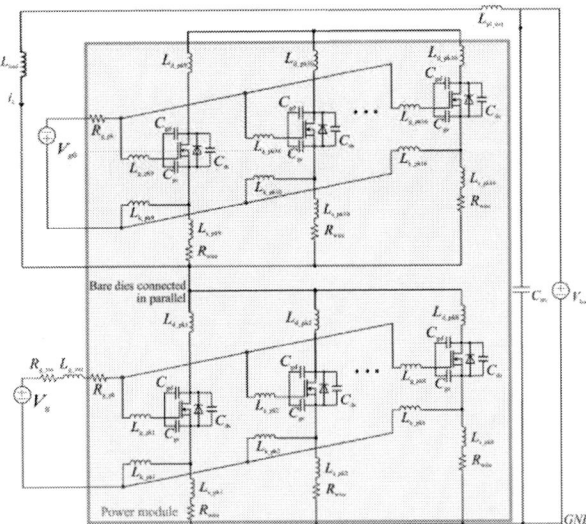

Fig. 2 Equivalent circuit for the DPT of SiC half-bridge power module

Furthermore, a model of the SiC power module is proposed illustrates the presence of both internal parasitic elements, attributed to the module packaging, and external parasitic elements associated with the surrounding circuits for double pulse test (DPT), as shown in Fig. 2. Experimental and electro-thermo-mechanical simulation results were compared and used to adjust and expand the simulation model with reduced parasitic inductances. In this experimental setup, the high-side MOSFET served as a freewheeling path, enabling current to flow through their channels. Meanwhile, the low-side chips were designated as the devices under test (DUT). The experiment involved applying two distinct pulses, including the first pulse lasting 20 μs and the second one lasting 20 μs, with applied V_{gs} of +15 V/ -4 V. Figure 3 shows the preliminary test results with waveforms of V_{gs}, V_{ds} and I_s. An inductor was connected in parallel to the high-side switch.

When multiple chips are connected in parallel, it is vital to optimize their placement to minimize discrepancies in current distribution. This ensures that each chip handles its fair share of the load, enhancing overall efficiency and reliability. Additionally, reducing differences in conduction paths between chips can help mitigate issues associated with uneven current distribution. By carefully managing the layout and design of parallel-connected chips, the performance and longevity of power modules can be enhanced, particularly in applications requiring high output currents. The analysis

in this paper not only minimizes switching losses but also maximizes the potential of SiC-MOSFETs in the power module. Section 2 investigates the ramifications stemming from external package-level components, focusing on power stray inductance and external gate resistance. Following this analysis, Section 3 of the paper presents an inclusive exploration of the influences exerted by internal package-level elements, including their associated parasitic inductances and internal gate resistance. The testing results and evaluations for some effect validations are provided in section 4 before concluding remarks are encapsulated in Section 5.

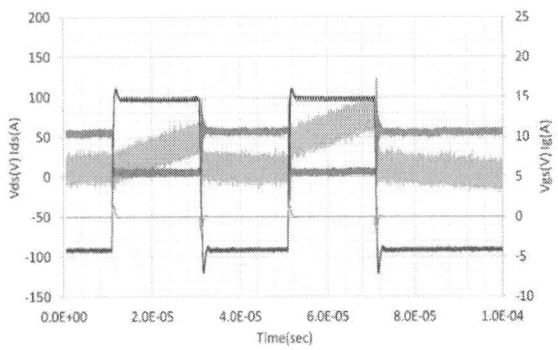

Fig. 3 Dynamic curves with typical double pulse test setup (blue: gate-source voltage V_{gs}, yellow: device current I_s, and orange: drain-source voltage V_{ds}).

2 External package-level elements

Understanding and managing the contained parasitic elements within conducting loops is imperative to harness the full capabilities of SiC technology and optimize system performance. The objectives are to characterize these parasitic elements, assess their impact on performance, and present effective mitigation strategies. This study focuses primarily on the power loop and gate loop. The gate loop encompasses the control path for the device, while the power loop comprises the critical current path that flows through the power semiconductor device. Power loop stray inductance L_{pl_ext} originating from wires and printed circuit board traces in the DPT circuit causes overvoltage and oscillation.

2.1 Power Stray Inductance

The circuit structure was subjected to simulation, and the parasitic inductances were accurately extracted using ANSYS-Q3D. These values were subsequently input into PSIM/LTSpice to model the electrical performance of the half-bridge. The creation of external package-level stray inductance arises from various factors. One key contributor is the wiring connections within the DPT circuit. These introduce several nanohenries to the current path, such that it establishes the stray inductance. Furthermore, the integration of current measurements within or around the commutation loop, while essential for sensing currents, also play a significant role in contributing to the external stray inductance. is In Fig. 2(a), when increasing power loop stray inductance L_{pl_ext} from 10 nH to 50 nH and 100 nH, the voltage stress and ringings on V_{ds} increase due to the resonant with the parasitic capacitance. The higher voltage overshoot with stray inductance results in higher turn-off loss. The abrupt surge of drain-source voltage ΔV_{ds} depends on the power loop stray inductance and the switch current slope dI_s/dt as in Eq. (1).

$$\Delta V_{ds} = L_{pl\_ext} \times \frac{dI_s}{dt} \qquad (1)$$

d

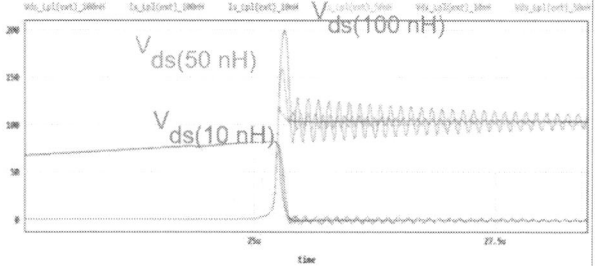

Fig. 2 Effects of outside the package from the power loop stray inductance

2.2 External gate resistor

As considering the gate loop, the inclusion of external gate resistance R_{g_ext} serves to lessen ringing and reduce overvoltage effects. However, it is important to acknowledge that while R_{g_ext} effectively addresses these issues, it introduces limitations on switching speed, consequently leading to an increase in unnecessary switching losses. As shown in Fig. 2(b), with 2 Ω, 5 Ω and 10 Ω respectively, when increasing gate resistances, noises

are suppressed but switching time becomes longer, thus switching loss is higher.

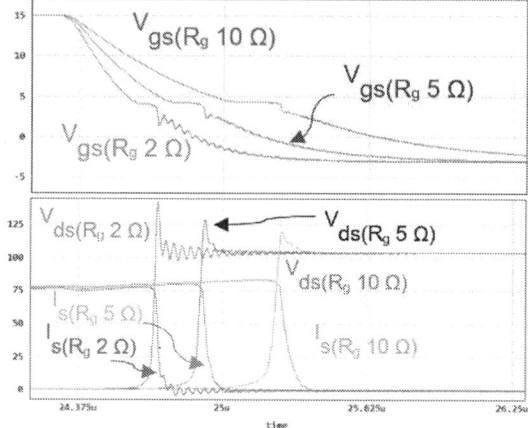

Fig. 3 Effects of outside the package from the external gate resistance

3 Internal package-level elements

3.1 Internal Parasitic Inductances

To achieve high output currents, connecting multiple chips in parallel is necessary. However, achieving uniform current distribution among parallel-connected chips can be challenging, particularly with high-speed switching devices like SiC-MOSFETs, which are sensitive to variations in conduction paths. Imbalanced current sharing in a power module occurs when the parasitic elements associated with the bare dies are unevenly distributed or exhibit differences in their electrical characteristics [3]. Proper placement of these parallel-connected chips is crucial to ensure equal current sharing under both static and dynamic conditions. The study presented two design schemes in the package layout and structure of the power module. In Alpha design scheme, as shown in the Fig. 4, the internal parasitic inductances emerge as relatively high and imbalanced. Specifically, L_{g_pk}, L_{k_pk}, L_{d_pk}, and L_{s_pk} denote the respective parasitic inductances from gate pad, Kelvin source pad, drain pad and source pad to the package terminals. There is a large variation among these internal inductances since each chip features a difference in the length of its conduction paths. The Beta design scheme is built with more balanced and smaller overall parasitic inductances. The overall internal inductances can be lowered 28.1% for L_{g_pk}, 4.5% for L_{k_pk}, 31.3% for L_{d_pk} and 3.1% for L_{s_pk} while the average gap of imbalance is markedly diminished from 22.85% to 6.1%.

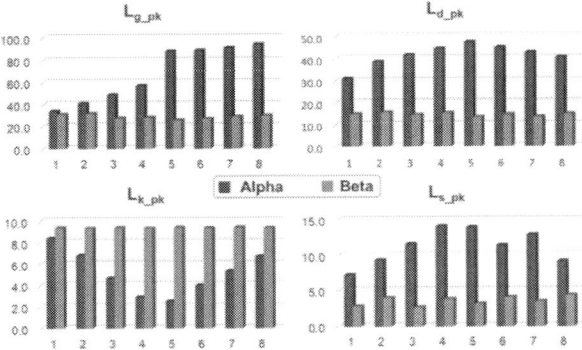

Fig. 4 Extracted parasitic inductances of each bare die in the SiC power module of two versions: Alpha design versus Beta design.

Figure 5 provides a comparative analysis of the analytical waveforms between two distinct design schemes operating under the identical condition. The first two frames show the gate-source voltage waveforms of each bare die for Alpha design version (above) and Beta design version (below). Concurrently, the last two frames illustrate the device current waveforms corresponding to the Alpha design version (above) and the Beta design version (below). By mitigating parasitic inductance values and dropping discrepancies among these parasitic elements across individual chips, the switching performance of the power module can be considerably improved, as shown in Fig. 3. Utilizing a power module with lower inductance also can effectively shorten the switching time, which is prominently observed in the I_s waveforms, thereby contributing to a further decrease in switching loss. The compatibility of the Beta design scheme with SiC technology lies in its ability to accommodate the need of balanced current distribution among multiple dies for the high switching speed. Moreover, ensuring balanced currents among parallel-connected chips is of paramount importance in high-power applications to prevent localized overheating and uneven wear. The enhancements of the Beta design version layout have fostered more equitable current sharing, thus contributing to enhanced reliability and longevity.

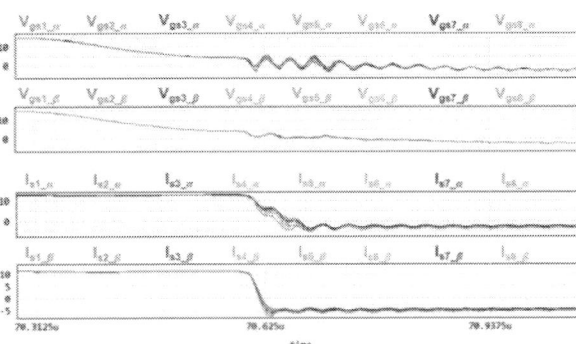

Fig. 5 Simulation results of two different design schemes for each parallel-connected SiC chips on the power module with V_{gs} and I_s respectively at the switching event

3.2 Internal gate resistance

In view of internal gate resistance of the package level, Figure 6(a) describes the modified power module model with individually connected gate resistance R_{g_pk} to each bare die. The significance of this modification becomes apparent when examining Fig. 6(b), where the consequence of placing a dedicated gate resistance per chip is visually demonstrated. The transient current sharing for each individual chip exhibits a notably flatter profile, indicating a more balanced distribution of current among the chips during dynamic operation. Consequently, the gaps observed among the current waveforms of the individual chips are significantly reduced compared to those observed when utilizing a common R_{g_pk} for the entire package, as depicted in the original model shown in Fig. 2.

This advancement in current sharing uniformity is influential in optimizing the performance of the power module. By ensuring that each chip receives a more consistent share of the current load, potential issues such as localized overheating or uneven stress distribution across the module are mitigated. Moreover, the implementation of individually connected gate resistances contributes to the achievement of faster switching characteristics. This improvement is attributed to the reduced turn-on and turn-off times facilitated by the optimized current sharing configuration.

PCIM Europe 2024, 11– 13 June 2024, Nuremberg   DOI: 10.30420/566262194

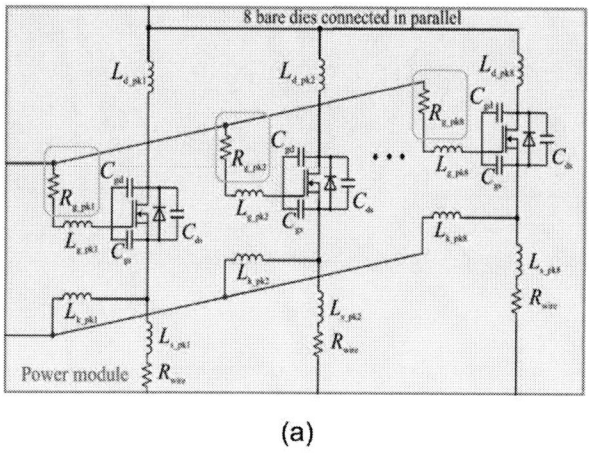

(a)

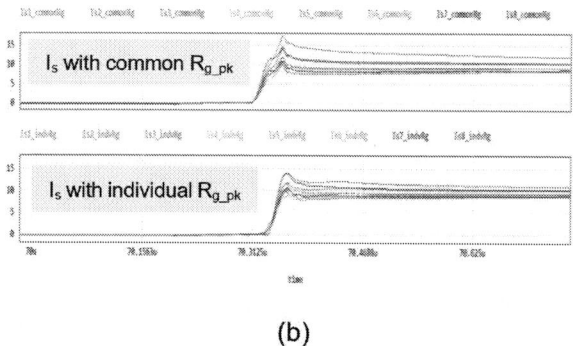

(b)

Fig. 6 (a) Power module model of individual gate resistance and (b) their simulation results at transient

4 Experimental Validation

The measurement procedure of DPT is implemented not only to verify the design and simulation model, but also to justify some effects analyzed above. As depicted in Fig. 7, the attenuation of ringings and overshoot is more evident in the current and voltage waveforms when utilizing a gate resistance of 10 Ω compared to using a gate resistance of 5 Ω. Nevertheless, this improvement comes at the expense of prolonged switching times, as can be seen in both switching on duration of Fig. 7(a) and switching off duration of Fig. 7(b). Consequently, the extended switching duration results in higher switching losses, diminishing the overall efficiency of the system. This trade-off between noise suppression and switching speed underscores the importance of carefully optimizing the selection of external gate resistance. This enables designers to strike a delicate balance between minimizing ringing and overvoltage effects while simultaneously mitigating the associated increase in switching losses.

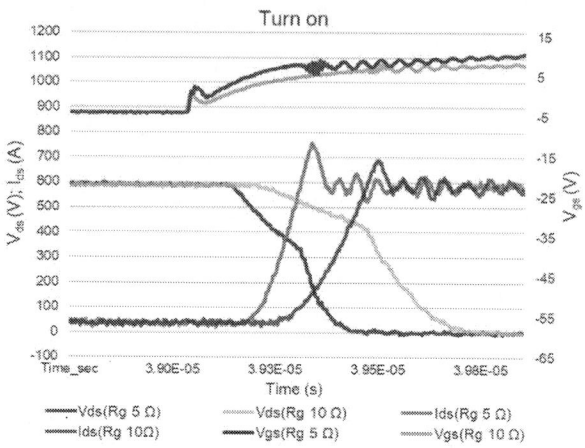

(a)

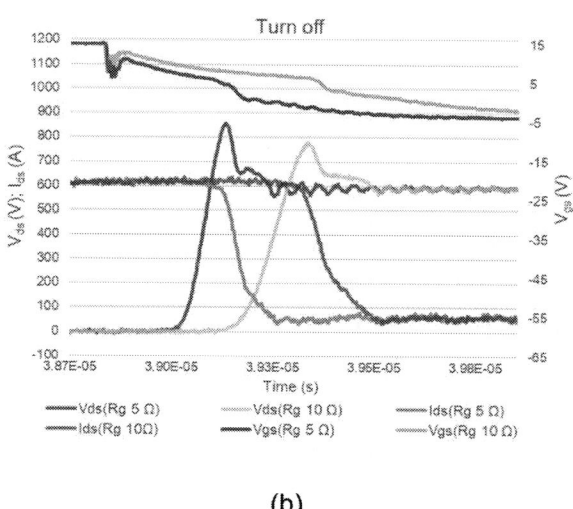

(b)

Fig. 7 Measured results in different R_{g_ext} of 5 Ω and 10 Ω with (a) turn on waveforms and (b) turn off waveforms

Furthermore, it's essential to consider the specific requirements and constraints of the application when determining the appropriate value of R_{g_ext}. While higher values may offer superior noise suppression, they also incur greater switching losses, which may not be acceptable in applications demanding rapid switching speeds and high efficiency. Therefore, a thorough evaluation of the trade-offs between noise mitigation, switching speed, and overall system performance is crucial in selecting the optimal value of external gate resistance for a given application.

1432

5 Conclusion

In conclusion, this research serves as a novel perspective into the parasitic element and gate resistance effects of interconnections inside and outside the package of the power module and the optimization of SiC power module design. By identifying and understanding the parasitic components and their effects, the study offers valuable insights that can be employed to enhance the efficiency and performance of SiC power modules in high-performance high-power applications. It is also worthwhile to note that the improved connections for low and equal stray inductances has significantly impacted transient current sharing among the bare dies within the package. Furthermore, the integration of individually connected gate resistances to each SiC chips inside the package represents a substantial enhancement in power module design, offering improved performance, reliability, and efficiency in high-speed switching applications.

Acknowlegment

The authors would like to acknowledge the measurement assistance from Nidec Advance Technology Taiwan Corporation and Nidec Advance Technology Inspection Instrument Division, Japan.

References

[1] S. Schwaiger, K. Heyers, A. Martinez-Limia, K. Oberdieck and C. Foerster, "Advanced SiC Trench-MOS Technology for Automotive Application," PCIM Europe 2023; International Exhibition and Conference for Power Electronics, Intelligent Motion, Renewable Energy and Energy Management, Nuremberg, Germany, 2023, pp. 1-4, doi: 10.30420/566091096.

[2] B. J. Baliga, "Wide Bandgap Semiconductor Power Devices Materials, Physics, Design and Appli-cations" Elsevier, ISBN: 9780081023068, 2018.

[3] J. Biela, M. Schweizer, S. Waffler and J. W. Kolar, "SiC versus Si—Evaluation of potentials for performance improvement of inverter and DC–DC converter systems by SiC power semiconductors," IEEE Transactions on Industrial Electronics, vol. 58, pp. 2872-2882, 2011.

[4] M. Meisser, M. Schmenger and T. Blank, "Parasitics in Power Electronic Modules: How parasitic inductance influences switching and how it can be minimized," Proceedings of PCIM Europe 2015, Nuremberg, Germany, 2015, pp. 1-8.

[5] M. Ostling, R. Ghandi and C. Zetterling, "SiC power devices—Present status, applications and future perspective," in International Symposium on Power Semiconductor Devices and ICs (ISPSD), 2011, pp. 10-15.

[6] D. -. Sadik, J. Colmenares, D. Peftitsis, Jang-Kwon Lim, J. Rabkowski and H. -. Nee, "Experimental investigations of static and transient current sharing of parallel-connected silicon carbide MOSFETs," in proceeding of European Conference on Power Electronics and Applications (EPE), 2013, pp. 1-10.

[7] Y. Cui, M. S. Chinthavali, F. Xu and L. M. Tolbert, "Characterization and modeling of silicon carbide power devices and paralleling operation," in International Symposium on Industrial Electronics (ISIE), 2012, pp. 228-233.

[8] H. Li et al., "Influence of Paralleling Dies and Paralleling Half-Bridges on Transient Current Distribution in Multichip Power Modules," in IEEE Trans. on Power Electron., vol. 33, no. 8, pp. 6483-6487, Aug. 2018, doi: 10.1109/TPEL.2018.2797326.

PCIM Europe 2024, 11– 13 June 2024, Nuremberg DOI: 10.30420/566262195

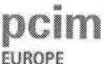

A Multi-Physics Iterative Approach for Temperature Estimation in SiC Power Module for Electric Vehicle

Ludovica Longo[1], Daniela Cavallaro[1], Marco Papaserio[1], Stefano Orlando[1]

[1] STMicroelectronics, Catania, Italy

Corresponding author: Stefano Orlando, stefano.orlando@st.com
Speaker: Stefano Orlando, stefano.orlando@st.com

Abstract

The objective of this paper is to describe the development of a single side cooled (SSC) silicon carbide based (SiC) power module for electric vehicle traction, with a specific focus on estimating the temperature of connection bars due to thermal stress. The proposed methodology involves a two-way coupling between an electrical model (EM) and a conjugate heat transfer analysis (CHT) to accurately evaluate the heat generated by internal parasitic resistances and by power dissipation. The iterative process involved evaluating the temperature of the device under operative conditions, based on the customer's mission profile, while considering both electrical and thermal properties as a function of temperature. This approach provides a comprehensive evaluation of the device's temperature and ensures that the connection bars are operating within safe temperature limits.

1 Introduction

The increasing current and switching frequency that a power module (PM) must handle leads to new package designs focusing on power terminal technology. The development of these terminals is concentrated on improving insulation, simplifying assembly systems and increasing reliability.

During the design phase of a PM considering the high currents involved, a proper size of busbars is necessary to ensure a reliable connection between the power module and the surrounding environment, such as the DC link and the motor. To evaluate the proper integration between the power module and the external system, it is crucial to estimate the temperature reached by the connection bars under operating conditions, due to the thermal stress induced by the current flow. Accurately estimating this temperature during the de-sign phase is essential because the plastic mate-rial of capacitors has a maximum allowable temperature that can be a significant constraint. This parameter is important to ensure optimal performance of the power module and its reliable operation under different conditions.

The management of electrothermal effects (ETM) in electronic devices involves monitoring the flow of current and the resulting heat generation, which causes electrical losses. An electrical simulation digital twin can effectively identify heat sources, calculate heat generation, and estimate electrical losses resulting from temperature variations in the model geometry.

The proposed Multiphysics methodology enables, in an iterative process, the concurrent evaluation of the losses caused by internal parasitic resistances and the associated heat generated by the Joule effect, as well as the power dissipated by the active dice working in parallel within a power module.

The case study involves three power modules mounted on a cooling system that uses a fluid to facilitate cooling, considering a specific current level based on the customer's mission profile. This integrated analysis in Ansys Electronic simulation environment enables a comprehensive evaluation of the thermal stress induced by the current flow on the power module, improving its integration with the external system, ensuring optimal performance and reliable operation under different conditions.

2 Electrothermal Model and Workflow

Electrothermal models are essential for predicting the operational behaviour of power modules. These models allow accurate and reliable analysis of power module performance by predicting the thermal and electrical behaviour of their components, the distribution of electric current and the heat flow within the PM. These factors are crucial in assessing the life and reliability of the power module and can influence its performance.

In addition, electrothermal models allow the optimisation of power module design by predicting its behaviour under different operating conditions and identifying any problems that might occur. This reduces development costs and improves the performance of the power module.

The methodology presented in this paper represents an effective solution because it involves electrothermal modelling of the entire system comprising the power module, including the connection bars, mounted on the cooler.

The ETM workflow methodology is a multi-step process used to analyse the behaviour of a power module. The simulation involves ANSYS simulation environment for thermal and electrical simulations and is conducted through an iterative feedback simulation until the simulated results converge. To determine the current flow and calculate the power losses associated to the passive parts of the PM, the first step of the analysis involves an electrical simulation analysed by Ansys Q3D. The volume losses generated by the electrical simulations are applied as a heating source for the thermal simulation, run in Ansys ICEPAK, which considers the heat transfer modes of conduction, convection, and radiation. In this model, a thermofluidic dynamic analysis is performed to estimate the temperature rise due to the power dissipated by the active dice operating simultaneously in parallel. The temperature information obtained from this analysis is then fed back into the electrical model, which incorporates temperature-dependent material properties that can induce further heat losses and, consequently, further electromagnetic losses. The advantage of this methodology, compared to the previously used one involving a single physical coupling [1], is that it also considers the variation of electrical properties of materials that change as a function of temperature.

In this iterative two-way process, the temperature determined by the thermal simulation is established as the reference point for the ensuing electrical simulation. The outputs of this electrical analysis are then fed back as inputs for the next round of thermal simulation. This cycle is repeated until the results achieve stability. In this way, the results will be more accurate than those of the single coupled simulation.

The workflow, resumed in figure 1, comprehends a fully coupled multi-physics simulation.

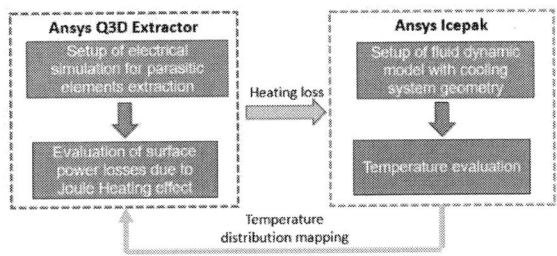

Fig. 1 Iterative simulation workflow

This methodology provides a comprehensive evaluation of the thermal and electrical behaviour of the power module, allowing for optimal performance and reliable operation under different conditions.

3 System Under Analysis

3.1 Overview of the System

The focus of this study is a moulded single-side indirectly cooled power module. Its design features a half-bridge topology with two switches, high-side and low-side, composed of parallel-operating silicon carbide dice. SiC technology has gained popularity among power semiconductor manufacturers in recent years and is expected to eventually replace silicon in power converter applications. Indeed, SiC MOSFETs offer lower on-state losses and higher switching speeds than silicon IGBT technology, making them attractive for applications where low weight and increased efficiency are essential, such as power converters for electric mobility. The use of silicon carbide MOSFETs also enables very high thermal and electrical performance, making SiC power modules suitable for electric vehicle drive inverters.

A fundamental aspect that cannot be overlooked when designing a power module for electric traction applications is the overall inductance. If the parasitic inductance is not kept very low, high switching frequencies lead to significant losses. The power terminal configuration of the module under analysis, shown in fig. 2, consists of the dc+ and dc- bars positioned on the same side one above the other, and the connection bar to the ac phase of the motor on the opposite side. This allows a current chaining that minimises the value of the final inductance.

Externally, the module has an exposed copper layer ready to be attached to the baseplate of the cooling system. The attachment techniques vary from soldering and sintering to the use of thermal interfaces (TIM). In the analysed system, one of these technical interfaces was used for the thermal and reliability performance required by a power module in an electric vehicle traction application.

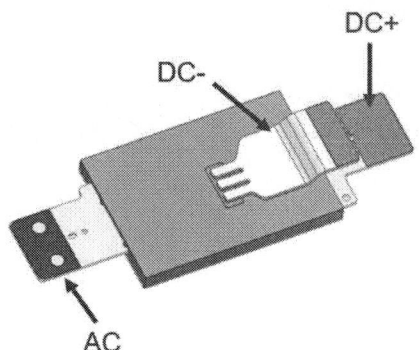

Fig. 2 Single side cooled power module and bars

3.2 Electrical Boundaries

To perform the simulations, the geometric configuration of the device was imported into the simulation environment, assigning the electrical material characteristics (such as relative permittivity, relative permeability, and conductivity) to the various model components. In particular, to enable the bidirectional interaction between the electrical and thermal simulation, a relative permeability of copper that varies as a function of temperature was used. Therefore, at the end of the first coupling cycle, the temperature obtained from the thermal simulation is adopted as the new reference point for the calculation of electrical losses (EM losses), and the material response will reflect the updated values based on this new thermal state.

To carry out the evaluation, it is necessary to define paths and networks, which consist of a set of conductive objects in contact separated by non-conductive materials or the background material. Each path must have a source and sink reference port assigned to emulate the current flow in the considered path. Since one of the main limitations in the design of connections between power modules and the power system is the maximum temperature that can be handled by the plastic capacitor, it is very important to provide a good prediction of the temperature of the dc busbar under operating conditions. For this reason, the analysis focuses on the electrical paths that include these busbars, considering in the study dc+ dc- and ac dc- paths.

The ports considered in the simulation are surfaces that represent the footprint of the connection between the DC link and the power module, in fact, to calculate the contributions of the parasitic elements of the device correctly, it is very important to reproduce the real connections.

In this case, the study was conducted in direct current considering the root mean square (RMS) current supplied, considering:

- 260 A in dc+ dc- path;
- 370 A in ac dc- path.

3.3 Thermal Boundaries

The second step of this methodology is a CHT simulation that uses the output of the electrical simulation, i.e. the surface power loss, for the thermal evaluation of the complete system composed by three power modules attached to the cooling system. This enables the simulation of thermal management aspects related to both conduction and radiation-convection effects.

To ensure the correct mapping of power losses in both simulation environments, the geometry used to extract EM losses is imported to perform the thermal simulation. For this analysis, it is necessary to define the thermal properties of the materials (density, thermal conductivity, and specific heat). To make the analysis as realistic as possible, some of these material properties are temperature dependent.

The investigated solution comprises three single-sided power modules, attached via a TIM onto a cooling system, as shown in figure 3. This system is made up of an aluminium heatsink embedded in a water jacket in which a water-glycol 50-50 flows. This cooling system, in which the module is not in direct contact with the coolant, is called indirect cooling.

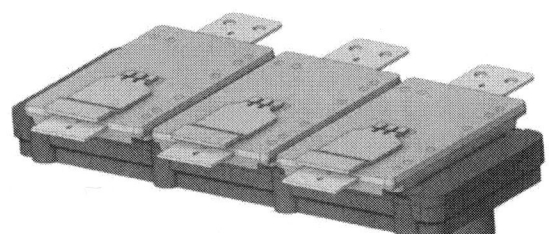

Fig. 3 Global system

The boundary conditions used in the present case are:

- Fluid flowrate 12 lpm.
- Ambient temperature $T_{amb}= 80°C$.
- Inlet temperature $T_{inlet}= 65°C$.
- Active power losses per switch $P_{switch}=500$ W.

4 Methodology in Practise

In the first stage, the electrical model of the module is developed. The geometry is imported, and the relevant materials are assigned to the components. Once the paths are defined, the simulation returns current density maps of the analysed paths.

The surface current density maps (A/cm²) resulting from the EM evaluation step are shown in Fig. 4 and in Fig. 5 for both analysed dc- ac and dc+ dc- paths. The total energy losses evaluated are 17.25 W and 12.95 W, respectively.

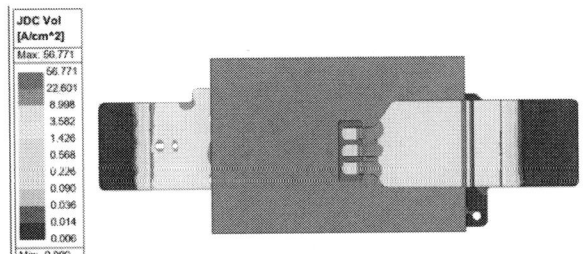

Fig. 4   Current density map at 1 A - ac dc- path

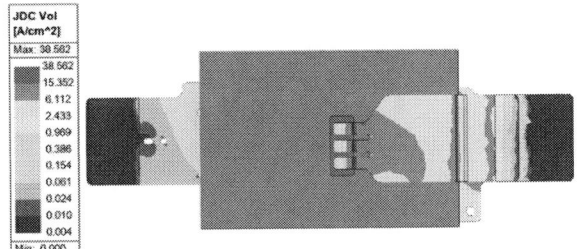

Fig. 5   Current density map at 1 A – dc+ dc- path

These losses due to the joule effect, since calculated by considering the forced current and parasitic resistance of the path, are then used as input for thermal simulation. In fact, they will be incorporated as thermal boundary conditions in the first iteration of the methodology.

Once the first fluid dynamics simulation is completed, the temperature information obtained from this analysis is fed back into the EM model.   Following the ETM methodology, the second iteration starts from the results of the first Multiphysics coupling, particularly considering the new reference temperature, at which the updated EM losses will be calculated. At each iteration, therefore, the electromagnetic losses associated with the working condition are reevaluated and used as a new input for the thermal simulation.

The resulting losses from this second iteration are 24.1 W for the ac dc- path and 18.36 W for the dc+ dc- path. The new losses, therefore, are increased by about 40% in both cases.

	One way coupling	Two-way coupling	ΔW %
dc- ac path	17.25 W	24.1 W	+ 39.7%
dc+ dc- path	12.95 W	18.36W	+ 41.8%

Table 1   Passive losses comparison

The updated losses are then considered as input into the fluid dynamic simulation to evaluate the average temperature of the power terminals. The results of the simulation are shown in Tab. 2 for the path ac dc- and in Tab. 3 for the path dc+ dc-.

To better highlight the difference between the single way and the two-way coupling, in the tables the average temperature of each power terminal in both cases are shown.

	One way coupling	Two-way coupling	ΔT %
dc+	90.35°C	90.9°C	+ 0.06%
dc-	101.04°C	104.73°C	+ 3.65%
ac	100.71°C	103.48°C	+ 2.75%

Table 2   Average temperature for dc- ac path

	One way coupling	Two-way coupling	ΔT %
dc+	93.18°C	93.65°C	+ 0.51%
dc-	96.81°C	98.61°C	+ 1.86%
ac	90.6°C	90.87°C	+ 0.03%

Table 3   Average temperature for dc- dc+ path

The findings indicate that the average temperature of the power terminals rises when subjected to the iterative process, compared with the temperatures obtained through a single-coupled analysis. Specifically, it is observed that the bar not involved in the current flow (dc+ in the first scenario and ac in the second) maintains a constant temperature. This suggests that the model accurately reflects the heat generation that is limited to the components traversed by the electric current.

The capacitor connected to the power module contains plastic components with an established maximum temperature tolerance. It is therefore critical to examine the temperatures that the terminals, particularly the dc+ and dc-, reach in contact with the capacitor under operating conditions. Literature sources indicate that these materials can handle temperatures up to 105°C.

The analysis conducted reveals that under maximum current load conditions, specifically 370A in the ac to dc path, current-induced thermal rise occurs. This results in an average temperature of about 104.8°C for the dc- bar.

The phase bar, which carries the full RMS current leading to an increase in temperature, is connected to the motor. This connection affords a less stringent limitation on the maximum permissible temperature compared to the section linked to the capacitor.

As previously outlined, this study used an RMS current during a steady state electrothermal simulation. The RMS value of a current denotes the magnitude of a direct current that would result in the same level of average power dissipation in a resistive load as compared to a time-varying alternating current. Consequently, the temperature derived from the simulation is representative of the average temperature during alternating current operation. In all current paths analysed and within the current levels examined, the maximum temperature virtually ascertained in the dc- bar remained consistently below 105°C, thus confirming that the simulated conditions met the prescribed safety thresholds.

5 Validation

As final step of the methodology the thermal model has been validated comparing simulation and measurement in a dedicated test vehicle, reproducing, by simulation, the same experimental operative condition.

Two different measurements and simulations trial have been performed on different device switched on.

The Tj has been evaluated in laboratory by using the forward diode voltage drop as thermo-sensitive electrical parameters (TSEP), during a cool down after reaching the steady state condition.

In the simulation to reproduce the experimental test, the following approach was taken by first turning on the HS and LS switches of two different legs of the inverter providing same power and flow rate used in the real measurement till the steady state condition is reached. In a second step, the switches have been turned off and the virtual Tj has been monitored till reaching the inlet temperature of the fluid.

In figure 6 is possible to observe one of the curves related to the Tj time history of experiment vs the simulation; the maximum delta temperature is observed at the starting point, then it drops with the reduction of temperature.

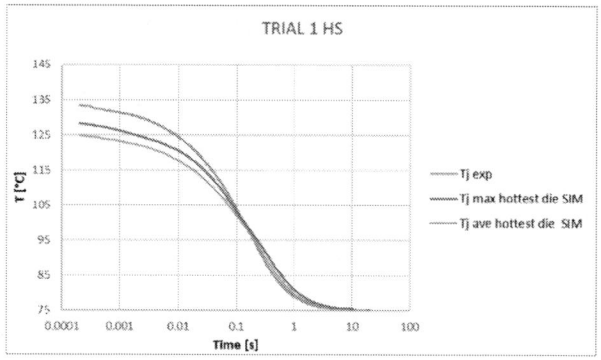

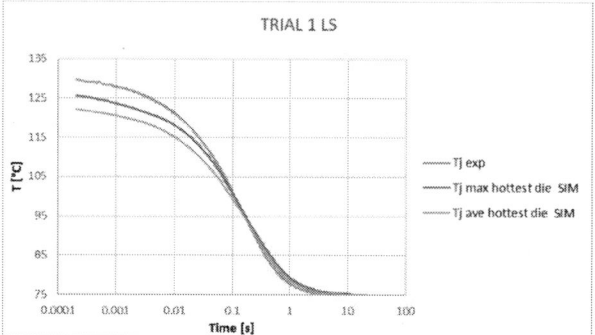

Fig. 6 Tj sim. vs Tj measured in trial 1

In table 4 and 5 it is possible to see the ΔT% between measurement and experiment observed in two different experimental trial.

	ΔT % at tstart	
	Trial 1 sim. vs Trial 1 exp.	
Simulated values	**HS switch**	**LS switch**
T_{Max}	2.75 %	1.88 %
T_{Ave} **hottest die**	5.6 %	4.89 %

Table 4 ΔT% between exp. and sim at t_{start} trial 1

| | ΔT % at tstart | |
| | Trial 2 sim. vs Trial 2 exp. | |
Simulated values	HS switch	LS switch
T_{Max}	0.46%	2.03%
T_{Ave} hottest die	3.38%	5.04%

Table 5 ΔT% between exp. and sim at t_{start} trial 2

Considering the maximum junction temperature (Tj max) derived from the simulations, the percentage temperature difference ΔT% is less than 3% in all tests conducted, indicating a high degree of correlation between the simulated model and the real system.

The validation presented here focuses on comparing the simulated results with experimental measurements of die temperature achieved under operating conditions, and thus the thermal model.

A full validation of the electrothermal model applied to the device in analysis is currently underway. This validation focuses on power terminal temperatures to corroborate the temperature predictions made by the electrothermal model outlined in this work. The details of this validation will be outlined in a forthcoming publication.

6   Conclusion

The Electrothermal Model workflow involves an electromagnetic simulation to determine the current flow and calculate the associated electromagnetic power losses. This is supplemented with thermal and fluid dynamic simulation to determine the temperature distribution in the model geometry. In this study, an iterative Multiphysics methodology of coupled electrical and thermal simulation was developed, providing a powerful tool for thermal analysis and optimization of high-power inverters. This methodology can improve the reliability and efficiency of these devices and reduce the risk of thermal failure.

The strength of the proposed methodology is that it allows to evaluate not only the temperature generated by the active power applied to the silicon carbide MOSFETs working in parallel inside the module switch but also to consider the temperature rise generated by the joule effect due to losses caused by the internal parasitic resistors of the power module. This allows a more accurate analysis of the thermal behaviour of the device, resulting in design optimization.

Compared with previous alternatives, the proposed FEM-based methodology offers the advantage of coupling the effects of electromagnetic and thermal phenomena in an iterative manner. This considers the increased power losses due to the variation of electrical parameters in the EM simulation, resulting in a more accurate analysis of the thermal performance of the device.

The electrical and thermal simulation coupling methodology is an integrated approach to analyse and optimize the behaviour of power modules under real operating conditions. In practice, this process is critical for predicting how heat generated by electrical currents will affect device performance and lifetime.

This methodology makes it possible to accurately predict potential problems and optimize the design of PMs prior to production, ensuring optimal reliability and performance during their life cycle.

The objective of the study was to monitor the temperatures of the power terminals during operating conditions, with particular attention to the capacitor interface terminal. This is because the capacitor is subject to temperature thresholds that must be respected to ensure the proper functionality of the entire system. This limit, according to the literature, is 105°C.

The operating conditions examined in this analysis revealed that the dc- terminal, even under the most severe conditions, reached an average temperature of around 104.8°C. This temperature is within the operating limits set by the capacitor's thermal constraints.

However, the analysis did not consider the influence of any additional external cooling mechanisms that could help manage the temperatures of the connections between the power module and the capacitor. The use of such cooling systems could allow for more demanding operating conditions while maintaining regulated temperatures at the connection points.

References

[1] D. Cavallaro, L. Longo, M. Papaserio, "SiC power modules for automotive traction inverters: a fully integrated electromagnetic-thermal modeling methodology for the estimation of busbar heating", *PCIM 2023.*

[2] J. Lim, J. Jeon, J. Seong, J. Cho, S. M. Cho, K. S. Kim, S. W. Yoon, "Iterative Electrical-Thermal Coupled Simulation Method of Automotive Power Module Used in Electric Power Steering System", IEEE Access, vol. 9, pp. 164712-164719, 2021.

[3] H. Gao et al., "Research on Fast Design Method for Power Module Terminal RMS Current Capacity via Thermal Equivalent Model,", *International Power Electronics and Application Conference and Exposition (PEAC)*, 2022.

[4] Adrian Plesca, "Thermal Analysis of Power Semiconductor Device in Steady-State Conditions", *MDPI Energies 2019*

[5] M. Szulborski, S. Łapczy and L. Kolimas "Thermal Analysis of Heat Distribution in Busbars during Rated Current Flow in Low-Voltage Industrial Switchgear", *MDPI Energies 2021.*

[6] G. Mauromicale, A. Cascio, M. Papaserio, D. Cavallaro, G. Bazzano, A. Messina, S. Patanè, M. Calabretta, A. Sitta, "Directly Cooled Silicon Carbide Power Modules: Thermal Model and Experimental Characterization", PCIM 2021

[7] G. Mauromicale, A. Raciti, S.A. Rizzo, G. Susinni, L. Abbatelli, S. Buonomo, D. Cavallaro, V. Giuffrida, "SiC Power Modules for Traction Inverters in Automotive Applications" *IECON 2019*

[8] G. Mauromicale, A. Sitta, M. Calabretta, S. M. Oliveri, G. Sequenzia, "Integrated Electromagnetic-Thermal Approach to Simulate a GaN-Based Monolithic Half-Bridge for Automotive DC-DC Converter", *MDPI Applied Science 2021.*

[9] C. Scognamillo et al., "Electrothermal Modeling, Simulation, and Electromagnetic Characterization of a 3.3 kV SiC MOSFET Power Module", *33rd International Symposium on Power Semiconductor Devices and ICs (ISPSD)*, 2021.

PCIM Europe 2024, 11– 13 June 2024, Nuremberg   DOI: 10.30420/566262196

Voltage Balancing Method for Series-Connection of 50 SiC MOSFETs

Antoine PHILIPPE[1,2], Nicolas GINOT[2], Guillaume PIQUET BOISSON[1], Anne-Sophie DESCAMPS[2], Van-Sang NGUYEN[1], Christophe BATARD[2]

[1] Univ. Grenoble Alpes, CEA, LITEN, Campus INES, Le Bourget du Lac

[2] University of Nantes, IETR, Nantes, France

Corresponding author: Antoine PHILIPPE, antoine.philippe@cea.fr
Speaker: Antoine PHILIPPE

Abstract

Silicon Carbide (SiC) MOSFETs above 3.3 kV are expensive and not yet commercially available. Partly because of this, medium voltage (MV) converters use multi-level architectures that are far more complex to design and maintain. In order to simplify the design of MV converters by using a two-level topology, this paper introduces a synthetic MV switch made up of 48 series-connected 1.2 kV SiC MOSFETs. As such series-connected switches suffer from dynamic voltage unbalancing issues; this paper discusses a new method for the voltage balancing of a series-connection of near 50 transistors, using a voltage clamping circuit and an associated command scheme for gate delay control.

1   Introduction

Nowadays, most of the largest photovoltaic power plants can generate hundreds of megawatts [1], and their connection to the grid is made possible by the use of inverters and step-up transformers. The latter are bulky and could be advantageously replaced by MV converters, made out of multi-level topologies. Compared to multi-level converters, two-level converters are well known and easier to design, but require transistors with high breakdown voltages to perform MV conversion. Regarding SiC MOSFET technology, the highest breakdown voltage a SiC MOSFET has reached is 15 kV [2] under laboratory conditions, and commercially available MOSFETs reach up to 3.3 kV.

To still benefit from the simple design of two-level converters for MV applications, a possible solution would be to synthesise MV switches by the serial-connection of SiC MOSFETs, as shown in Fig. 1.

By switching all of the serialised transistors simultaneously, the medium voltage is distributed among them in such a way that the series connection can act as a synthetic switch for a conventional two-level converter. This solution enables using widely available low voltage SiC MOSFETs. Connecting a voltage-source inverter to the 20 kV grid requires a 34 kV DC-bus (as the French grid is 20 kV$_{RMS}$±10%, and a 10% margin is kept for the controllability). Designing a synthetic switch for such a two-level inverter using 1.2 kV MOSFETs (with 40% additional margin) would require 48 SiC MOSFETs in series, as shown in Eq. (1)

$$\frac{34\ kV}{1.2\ kV \cdot (1 - 0.4)} = 48\ transistors \qquad (1)$$

Unfortunately, operating series-connected transistors suffers from static and dynamic voltage unbalancing issues. This voltage imbalance is due to latencies between the driving signals, parameter differences between the MOSFETs as well as parasitic common-mode currents [3]. To overcome this issue, mainly four voltage balancing techniques can be used [4]:

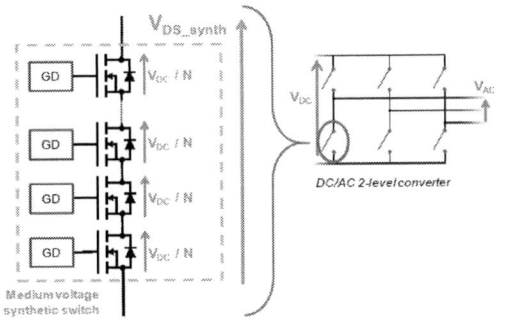

Fig. 1 Series-connected MOSFETs as a synthetic medium-voltage switch

- Passive voltage balancing [5] uses snubbers to slow down switching (thus increasing switching losses and hindering switching frequency).

- Active voltage clamping [6], where a clamping circuit is added between the drain and the gate of the MOSFET to protect each transistor from an overvoltage by means of a gate feedback above a defined voltage threshold.

- Gate timing control [7] compensates the timing mismatch by purposely adding a small time delay in chosen transistors commands.

- Active gate control [3] controls independently each MOSFET switching speed to slow down the fastest switching transistors during turn-off.

This paper proposes a new voltage balancing method applicable to a high count of SiC MOSFETs in series, as well as the control scheme applied to the transistors. This method will be described here through simulations and will lead to large-scale laboratory experiments in the future.

2 Voltage Balancing Method Based on Drain-Gate Voltage Clamping Detection

2.1 Introduction to Voltage Imbalance

In a commutation cell comprising two MOSFETs, both of them have the drain-source voltage (Vds) change between 0 V and V_{DC} (the bus voltage), at each switching event. With a switch made of series-connected MOSFETs this behaviour is applicable to the voltage of the series. In a perfectly balanced situation, the overall voltage is distributed equally along every transistor of the series, meaning that for N MOSFETs a voltage V_{DC}/N is applied to each one of them, as in Fig. 2 a).

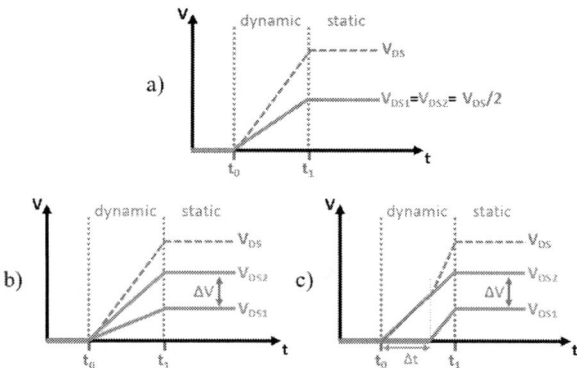

Fig. 2 Comparison for 2 ideal series-connected transistors between a) perfectly balanced voltage b) 2 transistors with a different switching speed and c) 2 transistors switched at a different time

Variability between components, command delays and parasitic capacitances lead to different switching speeds or delays in the command, resulting in a voltage imbalance as shown in Fig. 2 b) and Fig. 2 c).

These examples refer to turn-off switching, where the voltages rise to certain peak values (creating dynamic imbalance) and converge to certain voltages (resulting in a static imbalance). On the other hand, turn-on mainly has dynamic imbalance as the voltage rapidly cancels.

Voltage imbalance is an issue as the voltage across some of the MOSFETs could exceed their limits, causing additional losses through avalanche, reducing lifetime, or even causing permanent damages. As mentioned in the introduction, several voltage-balancing methods exist to overcome this issue and are represented in Fig. 3. Some reduce or adjust the switching speed (Passive voltage balancing, Active gate control), some add a delay to the fastest switching transistors (Gate timing control) and others limit the evolution of the drain-source voltage with a clamping circuit (Active voltage clamping).

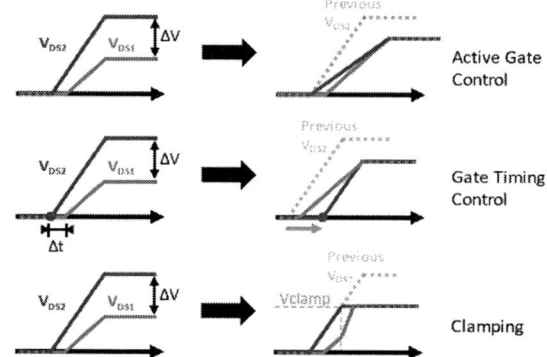

Fig. 3 Visual representation of three different voltage-balancing methods, from top to bottom: active gate control, gate timing control and voltage clamping

2.2 Proposed Voltage Balancing Method

As we plan to use up to 48 SiC MOSFETs in series, the gate driver circuits must stay as simple as possible to limit the number of components, for the sake of reliability and simplicity. The proposed voltage balancing method uses a hybrid approach of two of the previously presented balancing solutions: active voltage clamping combined with gate timing control, as shown in Fig. 4. Each MOSFET is equipped with a clamping circuit between drain and gate, mainly composed of a diode, a Zener diode (transient-voltage-suppression – TVS) and a resistor. A maximum

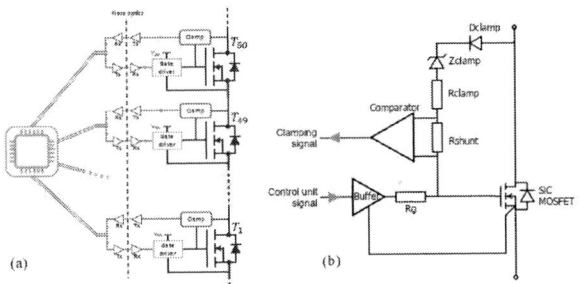

Fig. 4 a) Series connection of 50 SiC MOSFETs, b) simplified circuit for each unit.

threshold value is defined by the TVS diode Zclamp. Once the applied voltage exceeds the threshold, the TVS becomes conductive letting current flow through the clamp, changing the voltage between gate and source, thus the state of the MOSFET, limiting the evolution of the drain-source voltage. The clamp resistor is used to limit the current and define the feedback dynamics. This circuit is usually used as a protection circuit against drain-source overvoltage [8].

Applied in a series-connection circuit during a turn-off event, the voltage of the fastest MOSFETs rises higher than the others. This clamping circuit acts as a protection by reducing the MOSFETs resistivity, and in turn offloading more voltage to all other devices. With this approach the MOSFETs that are clamped are the fastest ones. This assessment can be used by relaying the clamping signals via optical fibres to the control unit, and adding delays to the fastest transistors. In our approach, the clamping time is measured and used to calculate a time delay for the next switching event. In a nutshell, the clamping circuit is used as a short-term protection, and the activation of this protection is used to fix the problem's origins on a longer term through gate timing corrections, to reduce as much as possible the dissipative clamping.

A similar balancing circuit was proposed in [9], with a series-connection of four IGBTs. As IGBTs work with lower dynamics than SiC MOSFETs, our paper studies the case of non-restrained SiC MOSFETs, with a series of 48 MOSFETs forming a synthetic switch, aiming at 20 kHz switching frequency in a megawatt-scale converter.

3 Control Scheme and Simulation of the Synthetic Switch

3.1 Delay Calculation Algorithm

During turn-off, where the drain-source voltage rises, the fastest or earliest MOSFET handles a larger share of the series voltage, thus activating its clamp circuit faster and longer than other MOSFETs. A longer delay must therefore be added to the turn-off command of this particular transistor. To quantify the required delays, the duration of the clamp is measured for each MOSFET in the series, all being subject to the described principles.

The clamp duration is defined as the row vector Tc for a number of MOSFETs N, where the ith MOSFETs clamp duration is noted Tc_i, as shown in Eq. (2).

$$Tc = [Tc_1, Tc_2, ..., Tc_N] \qquad (2)$$

The clamping durations are heavily dependent on the number of transistors, the bus voltage, the clamping thresholds, the temperature, the exact circuit and others. Therefore, it is challenging to directly define a timing delay to apply based on clamping durations. As the highest clamp duration requires a longer delay, a first variable is defined to represent the relative importance of adding a delay to each MOSFET turn-off. The row vector α, shown in Eq. (3), represents the percentages of each duration compared to the maximum clamp duration.

$$\alpha = [\alpha_1, \alpha_2, ..., \alpha_N] \text{ where } \alpha_i = \frac{Tc_i}{Tc_{max}} \qquad (3)$$

By multiplying each α_i coefficient by a time reference, a new row vector of incremental timing delays for the MOSFETs, D_{incr}, is defined in Eq. (4). The average clamp duration multiplied by an adjusting factor β, is chosen as a time reference. β is to be tweaked depending on system constants (N mainly) and possibly on operating conditions (bus voltage, load changes) to optimise voltage balancing.

$$D_{incr} = [D_{incr_1}, D_{incr_2}, ..., D_{incr_N}]$$

$$\text{where } D_{incr_i} = \alpha_i \cdot \beta \cdot \left[\frac{1}{N} \cdot \sum_{i=1}^{N} Tc_i \right] \qquad (4)$$

This incremental set of delays, calculated at each switching period is then added to the previously applied delay values T_{delay}, as shown in Eq. (5).

$$T_{delay} = [T_{delay_1}, T_{delay_2}, ..., T_{delay_N}]$$

$$\text{where } T_{delay_i} = T_{delay_{i,previous}} + D_{incr_i} \qquad (5)$$

The delay is kept from diverging by subtracting the minimum delay from all delays, Eq. (6). The purpose of this gate timing is conserved, as each transistor is still delayed according to the others.

$$T_{applied\_delay_i} = T_{delay_i} - \min(T_{delay}) \quad (6)$$

Regarding the turn-on event, the behaviour is reversed. During turn-on, the drain-source voltage across the fastest or earliest MOSFET changes from its previous static voltage (ideally V_{DC}/N) to zero. It is thus the slowest or latest transistors that handle the highest voltage. Therefore, new clamp durations are measured and every transistor must be delayed, except the slowest ones. To adapt the previously explained method to this situation, we simply use $(1-\alpha)$ instead of α, in otherwise identical equations, and a different β coefficient for on and off events.

3.2 Simulation Setup on PLECS

To simulate the behaviour of this voltage balancing method on a synthetic switch comprising up to 48 transistors in series, we use Plexim's PLECS simulator. The command scheme explained previously is applied in PLECS by implementing it in a "C-script block", representing the control unit. Based on an assessment of existing control units such as FPGAs, the incremental step for delays is chosen as 1 ns.

Most of PLECS components being considered ideal, with its library's MOSFET having a binary control, we chose to use the Spice netlist of an existing TO-247 1200 V 30 mΩ SiC MOSFET to create an equivalent circuit on PLECS, as shown in Fig. 5. This way we can use the electrical and thermal model of a real SiC MOSFET on PLECS simulator, with the main advantage of a realistic gate feedback loop simulation.

Next, we design the gate driver (GD) circuit, shown in Fig. 6, to control the MOSFET. The voltage applied between gate and source can vary between -4 V and 15 V. It contains a gate driving circuit with on and off gate resistors Rg_on and Rg_off, a gate protection circuit with two Zener diodes, and a pull-down resistor Rgs.

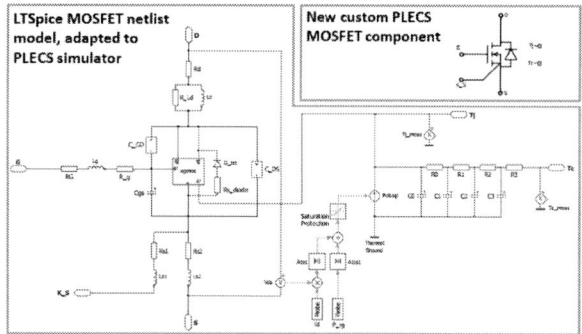

Fig. 5 Electrical and thermal Spice netlist model of a SiC MOSFET adapted for PLECS simulator

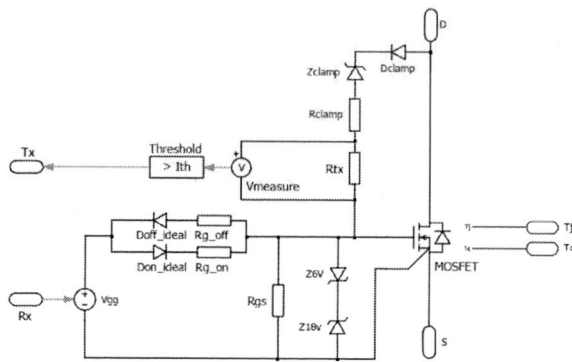

Fig. 6 Gate driver schematic connected to the SiC MOSFET model. It is composed of a clamp circuit, signal input/output (Rx/Tx) and gate resistors

The upper part of the gate driver contains the clamp circuit described in section 2. The series-connection is done by connecting the source of a MOSFET to the drain of another.

Voltage imbalance being caused by circuit imperfections, those have to be added manually in the simulations. Studies on internal equivalent capacitors of SiC MOSFETs show a variation of 10% to 30% compared to datasheet values [10]. For this simulation, random values are generated for each MOSFET, with ±10% for Cgs and Cds, and ±30% for Cdg; according to truncated Gaussian distributions centred on datasheet values, where either of these limits have a 2% chance of appearing.

The propagation time of the command from the control unit to the MOSFET (through optical transmitter/receiver, fibre and buffer) is also variable, electronically and through deviations in the implementation. Datasheet values for each component have been aggregated into a random delay between 0 and 150 ns for each transistor (according to a Gaussian distribution, centred between both limits, with a 2% chance of choosing either of the limits).

The resulting simulation has had to be lightened in some cases, the parasitic capacitors to ground being deleted in the 48-MOSFETs model. This trade-off between accuracy and simulation time does not hinder the demonstration, as the aforementioned variability creates enough imbalance to study our balancing method.

To begin, we simulate a block of 6 MOSFETs, allowing to demonstrate the feasibility of the balancing method with a fast simulation, followed by the 48 MOSFETs simulation (6*8 blocks). The synthetic switch is simulated in a Buck converter circuit.

3.3 Simulation Results

The aim is to apply the 34 kV DC bus voltage on 48 MOSFETs in series, with a current around 15 A. In this situation, the ideal balanced bus voltage should be around 700 V on each of the transistors. Following this approach, the DC bus voltage for 6 transistors must be 6×700 = 4200 V. As the chosen MOSFETs maximum voltage is 1200 V, the clamp value is chosen at 900 V, in-between the ideal voltage balance and maximum voltage.

Figure. 7 shows the simulation results of the first switching event with 6 transistors each having different initialisation parameters, and where no delay is applied yet. From top to bottom are represented the gate-source voltages (Vgs), the total drain-source across the series-connection (Total Vds), the drain-source voltage of each MOSFET (Each Vds) and the clamp states. Except for the total Vds, each colour matches a different MOSFET.

To visualise the effect of the applied delays after a few switching events, we chose to plot the clamping durations, the applied delay and the quasi-static drain-source voltage after switching oscillations, for different values of β and including a strong disturbance (sudden parameter change). The results are shown for turn-off in Fig. 8 and turn-on in Fig. 9. Finally, Fig. 10 represents a similar data measurement for 48 transistors in series.

3.4 Result Analysis

3.4.1 6 MOSFETs in series

In Fig. 7, we can see that the clamping circuit works as expected, by increasing the Vgs voltage, converging Vds voltage towards the threshold value and activating the clamping signal when it exceeds the threshold value. We can also note that the clamp isn't fully optimised as the Vds voltage exceeds the MOSFETs maximum voltage. This can be fixed by decreasing the voltage threshold value.

Regarding the turn-off correction in Fig. 8, as the intensity of the correction is controlled with the β_{OFF} coefficient, we can see that a too high β_{OFF} creates oscillations, as the applied delay reacts too sharply to the clamping durations. When β_{OFF} is too low, the clamping takes a longer time to disappear. With a value between 0.3 and 0.5 the command reacts quickly and the clamps disappear rapidly, even after a disturbance.

As the proposed method does not actually include a measure of the Vds voltage, the control unit can only assess if the voltage across a MOSFET is considered too high. As we can see in the simulations, the quasi-static voltage after turn-off is balanced differently depending on β_{OFF}: with a lower value, the delays are increased just enough to remove the clamps, without causing a sufficiently higher quasi-static voltage on the other switches, to correct the voltage imbalance. With a higher β_{OFF} value (but low enough not to cause

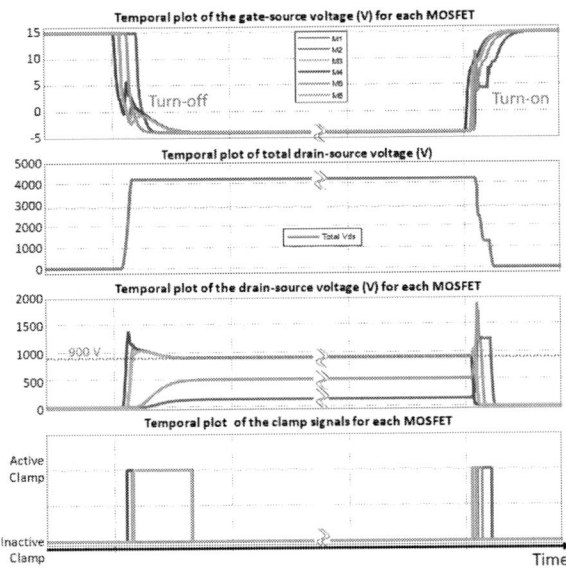

Fig. 7 Simulated temporal plot of a 6 MOSFETs series-connection, each comprising a clamping circuit

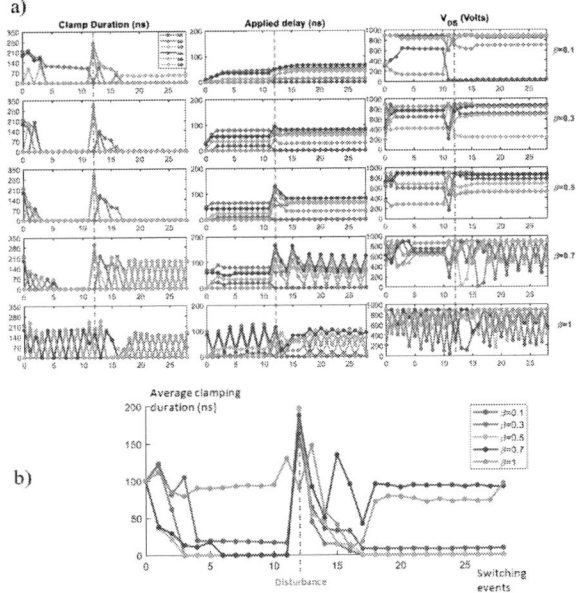

Fig. 8 Turn-off behaviour for 28 switching events, for different β$_{OFF}$ values, including a disturbance at the 12th iteration

an uncontrolled oscillating state), the delay steps are larger than strictly necessary, forcing some other MOSFETs to clamp, thus allowing a better voltage distribution by convergent oscillations. In this paper, we study the behaviour of the system for a constant β coefficient, but it would be possible to adjust it according to a closed loop control regulating the average clamp duration (shown in Fig. 8 b)).

Voltage imbalance also appears during turn-on before each Vds cancels. The reasons for this imbalance are similar to those of turn-off, as shown in Fig. 2, but are also due to quasi-static voltages resulting from the turn-off event. This makes it complicated to predict the behaviour of the clamping correction through iterative studies. For this, the turn-on clamp duration is measured for different β_{OFF} and β_{ON} values, as shown in Fig. 9. We can note that the β_{ON} coefficient has to be above 1 to work properly, meaning that a delay higher than the average clamp duration is necessary to remove the turn-on clamps. The clamps disappear with any properly working β_{OFF} regardless of β_{ON}. For instance, $\beta_{OFF} = 0.7$ causes turn-off instability, see Fig. 8, hence the turn-on clamp cannot stabilise either.

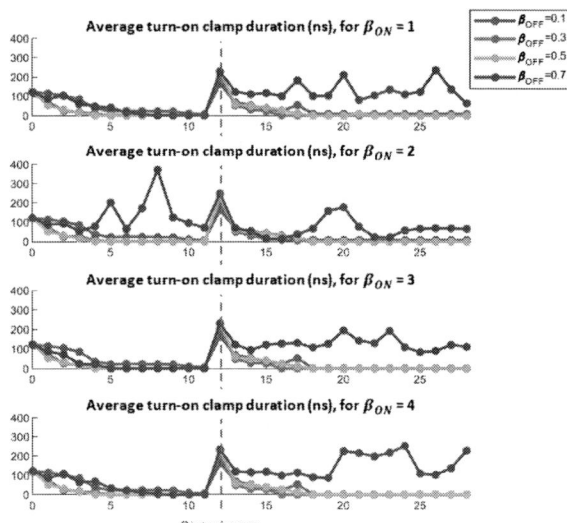

Fig. 9 Turn-on behaviour for 28 switching events, for different β$_{ON}$ and β$_{OFF}$ values, including a disturbance at the 12th iteration

3.4.2 48 transistors in series

Studying the behaviour of 48 MOSFETs is challenging as a simulation can take a very long time. We chose to use $\beta_{OFF} = 0.3$ and $\beta_{ON} = 3$. The results are shown in Fig. 10. With as much transistors in series, the clamp takes more time to disappear completely and the voltage doesn't

balance as much as for 6 MOSFETs in series within a limited duration. A longer simulation would need to be implemented. In addition, adapting the β values during the simulation can change the voltage balancing of these 48 transistors. As said previously this could be implemented in future works by using the average clamp duration or standard deviation for a closed loop control regulation.

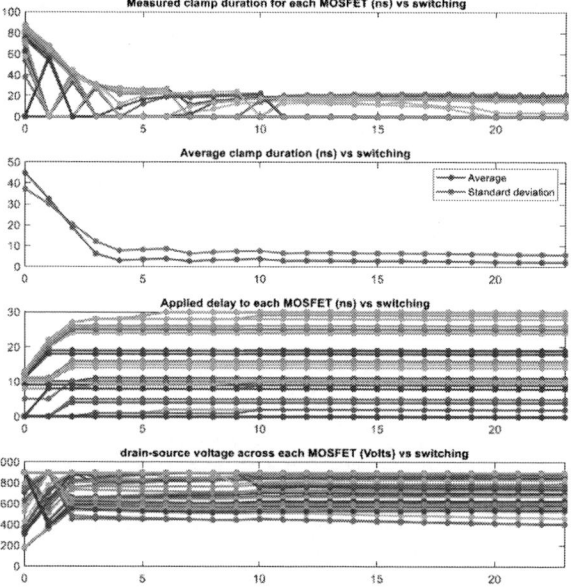

Fig. 10 Evolution of the clamping for 23 switching events for 48 MOSFETs in series each with a different color, with β$_{OFF}$ = 0.3

4 Implementation

The simulations show that the balancing method works for a small and large number of transistors in series. The implementation of such a synthetic switch is challenging, as many factors must be taken into account, starting with the DC bus, where the bus and decoupling capacitors need to sustain medium voltage. Power needs to be delivered to each gate driver, requiring high voltage insulation with the least possible capacitance to avoid disturbances due to high dv/dt. The overall inductance loop must be designed as small as possible to avoid oscillations caused by di/dt. The overall switch will be discussed in a future work.

To verify the proper operation of the clamp detection for one MOSFET and its GD, we set up a double pulse test (DPT) circuit, shown in Fig. 11, with a high (and adjustable) inductance loop, to generate pulses above the clamps threshold value. The aim is to observe how the clamping circuit reacts when an oscillation exceeds the

PCIM Europe 2024, 11– 13 June 2024, Nuremberg   DOI: 10.30420/566262196

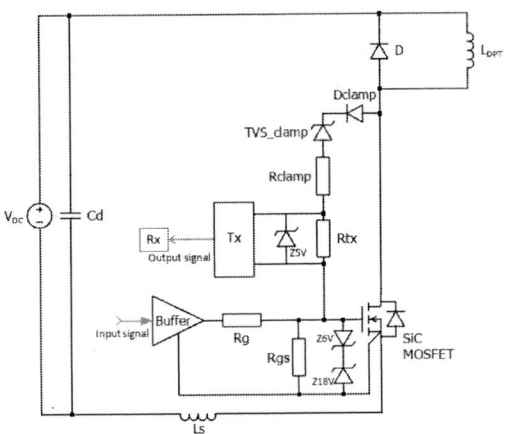

Fig. 11 Double pulse test schematic to confirm the active clamp operation and its detection

threshold clamping value, and if a clamping signal is sent through the optical fibre transmitter.

The clamp circuit is implemented by using four 180 V TVS diodes in series, resulting in a 720 V clamp threshold. The resistors are chosen to limit the current, and a 5 V Zener diode is placed across the Rtx resistor to limit the voltage (optical transmitter maximum input signal voltage). The Ls inductor connected to the source of the MOSFET has a high inductance value of 500 nH. It is important that the bus voltage does not exceed the threshold voltage, otherwise the clamping circuit could always be active, thus constantly closing the MOSFET. The bus voltage V_{DC} is set to 650 V, and the first pulse duration set to attain 13 A.

Figure. 12 shows the measured results, with and without the clamping circuit. On the first plot, we can see the gate-source voltage rise when the drain-source voltage is above the clamp threshold. The voltage across the Rtx resistor represents the current flowing through the clamp, and activates the optical transmitter when exceeding 2 V. The received signal has the same timing as the active clamp, with a small delay mainly due to fibre length. With the clamp activated, the maximum Vds voltage is reduced from 1016 V to 930 V, and the oscillations are more dampened.

This setup proves that the clamp detection works even for a very small clamp duration, but does not yet demonstrate how the clamp could react when adding transistors in series. In a next step, the aim will be to connect 6 MOSFETs in series at first, to make a small synthetic switch by taking into account the capacitive effects, the inductance loop, the cooling system... This will be followed by adding multiple of these blocks, in an appropriate arrangement minimizing inductance loop and parasitic capacitances, up to 8*6=48 MOSFETs in series.

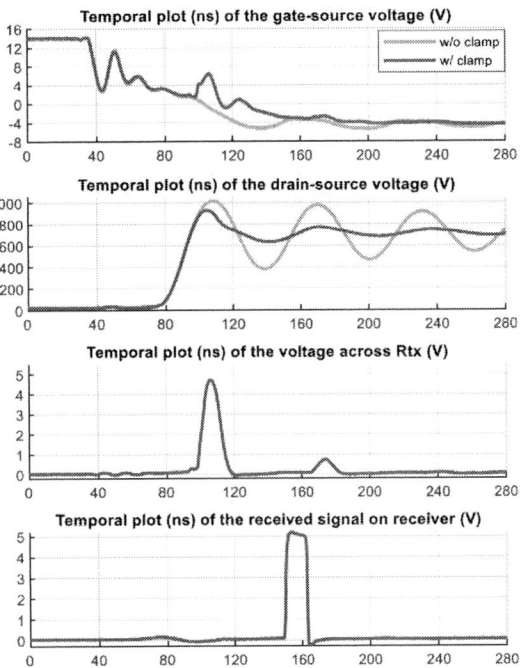

Fig. 12 Experimental results testing the clamp in a DPT circuit. Yellow curves is the behavior without clamp, and blue curves with clamp

5   Conclusion

In this paper, a new voltage-balancing method for a series-connection of up to 48 SiC MOSFETs was proposed. This was achieved by using clamping circuits, used both as protection and as inputs for a gate timing correction algorithm for each transistor. In addition, was proposed a control scheme using the clamping signal durations to generate timing delays, which are applied to the control of the transistors. The first steps of this work were carried out through simulations, using from 6 to 48 SiC MOSFETs in series. This method efficiently removes the clamps while balancing the voltages for a small amount of transistors. Further studies need to be done regarding the behaviour for a larger amount of transistors. Experimental setup was also made to verify the operation of the clamp signal transmission through optical fibre, validating the possibility of applying such a circuit around SiC MOSFETs.

The next experimental works will consist in creating such a synthetic switch, first as a single block of 6 SiC MOSFETs, up to 8 blocks, all in series.

1447

Acknowledgments

This work was supported by the French National Program "Programme d'Investissements d'Avenir – INES.2S" under Grant ANR-10-IEED-0014-01

References

[1] M. Rabiul Islam, A. M. Mahfuz-Ur-Rahman, K. M. Muttaqi and D. Sutanto, "State-of-the-Art of the Medium-Voltage Power Converter Technologies for Grid Integration of Solar Photovoltaic Power Plants," in IEEE Transactions on Energy Conversion, vol. 34, no. 1, pp. 372-384, March 2019

[2] V. Pala et al., "10 kV and 15 kV silicon carbide power MOSFETs for next-generation energy conversion and transmission systems," 2014 IEEE Energy Conversion Congress and Exposition (ECCE), Pittsburgh, PA, USA, 2014, pp. 449-454

[3] A. Marzoughi, R. Burgos and D. Boroyevich, "Active Gate-Driver With dv/dt Controller for Dynamic Voltage Balancing in Series-Connected SiC MOSFETs," in IEEE Transactions on Industrial Electronics, vol. 66, no. 4, pp. 2488-2498, April 2019

[4] V. U. Pawaskar, G. Gohil and P. T. Balsara, "Study of Voltage Balancing Techniques for Series-Connected Insulated Gate Power Devices," in IEEE Journal of Emerging and Selected Topics in Power Electronics, vol. 10, no. 2, pp. 2380-2394, April 2022

[5] K. Vechalapu, S. Bhattacharya and E. Aleoiza, "Performance evaluation of series connected 1700V SiC MOSFET devices," 2015 IEEE 3rd Workshop on Wide Bandgap Power Devices and Applica tions (WiPDA), Blacksburg, VA, USA, 2015, pp. 184-191

[6] R. W. Maier and M. -M. Bakran, "Active clamping method for SiC MOSFET high power modules - Benefits and Limits," 2020 22nd European Conference on Power Electronics and Applications (EPE'20 ECCE Europe), Lyon, France, 2020, pp. P.1-P.10

[7] M. Zhao, H. Lin and T. Wang, "An Adaptive Driving Signals Delay Control for Voltage Balancing of Multiple Series-Connected SiC MOSFETs," 2022 IEEE Applied Power Electronics Conference and Exposition (APEC), Houston, TX, USA, 2022, pp. 1619-1624

[8] R. W. Maier and M. -M. Bakran, "Active clamping method for SiC MOSFET high power modules - Benefits and Limits," 2020 22nd European Conference on Power Electronics and Applications (EPE'20 ECCE Europe), Lyon, France, 2020, pp. P.1-P.10

[9] S. Ji, F. Wang, L. M. Tolbert, T. Lu, Z. Zhao and H. Yu, "An FPGA-Based Voltage Balancing Control for Multi-HV-IGBTs in Series Connection," in IEEE Transactions on Industry Applications, vol. 54, no. 5, pp. 4640-4649, Sept.-Oct. 2018

[10] C. M. de Vienne, B. Asllani, B. Lefebvre, P. Lefranc and P. -O. Jeannin, "Model Parameter Extraction Tool for the Analysis of Series-Connected SiC-MOSFETs," PCIM Europe digital days 2021; International Exhibition and Conference for Power Electronics, Intelligent Motion, Renewable Energy and Energy Management, Online, 2021, pp. 1-8.

PCIM Europe 2024, 11– 13 June 2024, Nuremberg   DOI:10.30420/566262196

Voltage Balancing Method for Series-Connection of 50 SiC MOSFETs

Antoine Philippe[1,2], Nicolas Ginot[2], Guillaume Piquet Boisson[1], Anne-Sophie Descamps[2],
Van-Sang Nguyen[1], Christophe Batard[2]

[1] Univ. Grenoble Alpes, CEA, LITEN, Campus INES, Le Bourget du Lac
[2] University of Nantes, IETR, Nantes, France

Corresponding author: Antoine PHILIPPE, antoine.philippe@cea.fr
Speaker: Antoine PHILIPPE

Abstract

Silicon Carbide (SiC) MOSFETs above 3.3 kV are expensive and not yet commercially available. Partly because of this, medium voltage (MV) converters use multi-level architectures that are far more complex to design and maintain. In order to simplify the design of MV converters by using a two-level topology, this paper introduces a synthetic MV switch made up of 48 series-connected 1.2 kV SiC MOSFETs. As such series-connected switches suffer from dynamic voltage unbalancing issues; this paper discusses a new method for the voltage balancing of a series-connection of near 50 transistors, using a voltage clamping circuit and an associated command scheme for gate delay control.

1   Introduction

Nowadays, most of the largest photovoltaic power plants can generate hundreds of megawatts [1], and their connection to the grid is made possible by the use of inverters and step-up transformers. The latter are bulky and could be advantageously replaced by MV converters, made out of multi-level topologies. Compared to multi-level converters, two-level converters are well known and easier to design, but require transistors with high breakdown voltages to perform MV conversion. Regarding SiC MOSFET technology, the highest breakdown voltage a SiC MOSFET has reached is 15 kV [2] under laboratory conditions, and commercially available MOSFETs reach up to 3.3 kV.

To still benefit from the simple design of two-level converters for MV applications, a possible solution would be to synthesise MV switches by the serial-connection of SiC MOSFETs, as shown in Fig. 1.

By switching all of the serialised transistors simultaneously, the medium voltage is distributed among them in such a way that the series connection can act as a synthetic switch for a conventional two-level converter. This solution enables using widely available low voltage SiC MOSFETs. Connecting a voltage-source inverter to the 20 kV grid requires a 34 kV DC-bus (as the French grid is 20 kV$_{RMS}$±10%, and a 10% margin is kept for the controllability). Designing a synthetic switch for such a two-level inverter using 1.2 kV MOSFETs (with 40% additional margin) would require 48 SiC MOSFETs in series, as shown in Eq. (1)

$$\frac{34\ kV}{1.2\ kV \cdot (1 - 0.4)} = 48\ transistors \qquad (1)$$

Unfortunately, operating series-connected transistors suffers from static and dynamic voltage unbalancing issues. This voltage imbalance is due to latencies between the driving signals, parameter differences between the MOSFETs as well as parasitic common-mode currents [3]. To overcome this issue, mainly four voltage balancing techniques can be used [4]:

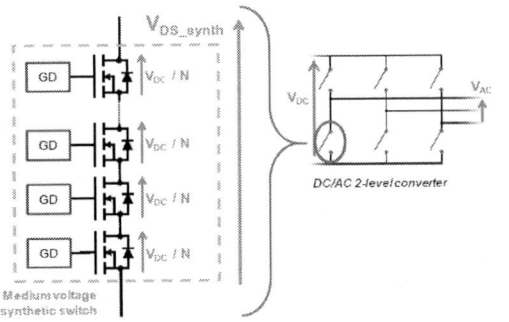

Fig. 1 Series-connected MOSFETs as a synthetic medium-voltage switch

- Passive voltage balancing [5] uses snubbers to slow down switching (thus increasing switching losses and hindering switching frequency).

- Active voltage clamping [6], where a clamping circuit is added between the drain and the gate of the MOSFET to protect each transistor from an overvoltage by means of a gate feedback above a defined voltage threshold.

- Gate timing control [7] compensates the timing mismatch by purposely adding a small time delay in chosen transistors commands.

- Active gate control [3] controls independently each MOSFET switching speed to slow down the fastest switching transistors during turn-off.

This paper proposes a new voltage balancing method applicable to a high count of SiC MOSFETs in series, as well as the control scheme applied to the transistors. This method will be described here through simulations and will lead to large-scale laboratory experiments in the future.

2 Voltage Balancing Method Based on Drain-Gate Voltage Clamping Detection

2.1 Introduction to Voltage Imbalance

In a commutation cell comprising two MOSFETs, both of them have the drain-source voltage (Vds) change between 0 V and V_{DC} (the bus voltage), at each switching event. With a switch made of series-connected MOSFETs this behaviour is applicable to the voltage of the series. In a perfectly balanced situation, the overall voltage is distributed equally along every transistor of the series, meaning that for N MOSFETs a voltage V_{DC}/N is applied to each one of them, as in Fig. 2 a).

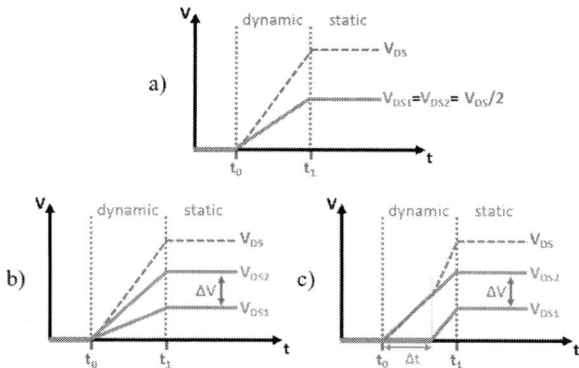

Fig. 2 Comparison for 2 ideal series-connected transistors between a) perfectly balanced voltage b) 2 transistors with a different switching speed and c) 2 transistors switched at a different time

Variability between components, command delays and parasitic capacitances lead to different switching speeds or delays in the command, resulting in a voltage imbalance as shown in Fig. 2 b) and Fig. 2 c).

These examples refer to turn-off switching, where the voltages rise to certain peak values (creating dynamic imbalance) and converge to certain voltages (resulting in a static imbalance). On the other hand, turn-on mainly has dynamic imbalance as the voltage rapidly cancels.

Voltage imbalance is an issue as the voltage across some of the MOSFETs could exceed their limits, causing additional losses through avalanche, reducing lifetime, or even causing permanent damages. As mentioned in the introduction, several voltage-balancing methods exist to overcome this issue and are represented in Fig. 3. Some reduce or adjust the switching speed (Passive voltage balancing, Active gate control), some add a delay to the fastest switching transistors (Gate timing control) and others limit the evolution of the drain-source voltage with a clamping circuit (Active voltage clamping).

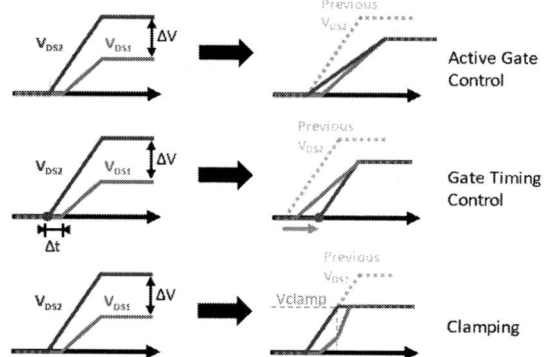

Fig. 3 Visual representation of three different voltage-balancing methods, from top to bottom: active gate control, gate timing control and voltage clamping

2.2 Proposed Voltage Balancing Method

As we plan to use up to 48 SiC MOSFETs in series, the gate driver circuits must stay as simple as possible to limit the number of components, for the sake of reliability and simplicity. The proposed voltage balancing method uses a hybrid approach of two of the previously presented balancing solutions: active voltage clamping combined with gate timing control, as shown in Fig. 4. Each MOSFET is equipped with a clamping circuit between drain and gate, mainly composed of a diode, a Zener diode (transient-voltage-suppression – TVS) and a resistor. A maximum

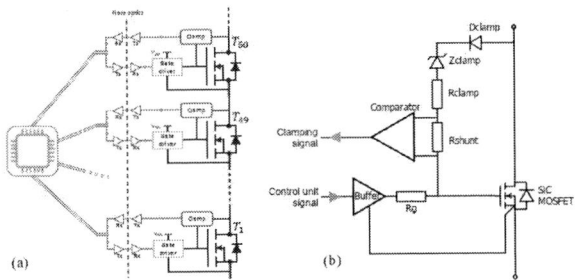

Fig. 4 a) Series connection of 50 SiC MOSFETs, b) simplified circuit for each unit.

threshold value is defined by the TVS diode Zclamp. Once the applied voltage exceeds the threshold, the TVS becomes conductive letting current flow through the clamp, changing the voltage between gate and source, thus the state of the MOSFET, limiting the evolution of the drain-source voltage. The clamp resistor is used to limit the current and define the feedback dynamics. This circuit is usually used as a protection circuit against drain-source overvoltage [8].

Applied in a series-connection circuit during a turn-off event, the voltage of the fastest MOSFETs rises higher than the others. This clamping circuit acts as a protection by reducing the MOSFETs resistivity, and in turn offloading more voltage to all other devices. With this approach the MOSFETs that are clamped are the fastest ones. This assessment can be used by relaying the clamping signals via optical fibres to the control unit, and adding delays to the fastest transistors. In our approach, the clamping time is measured and used to calculate a time delay for the next switching event. In a nutshell, the clamping circuit is used as a short-term protection, and the activation of this protection is used to fix the problem's origins on a longer term through gate timing corrections, to reduce as much as possible the dissipative clamping.

A similar balancing circuit was proposed in [9], with a series-connection of four IGBTs. As IGBTs work with lower dynamics than SiC MOSFETs, our paper studies the case of non-restrained SiC MOSFETs, with a series of 48 MOSFETs forming a synthetic switch, aiming at 20 kHz switching frequency in a megawatt-scale converter.

3 Control Scheme and Simulation of the Synthetic Switch

3.1 Delay Calculation Algorithm

During turn-off, where the drain-source voltage rises, the fastest or earliest MOSFET handles a larger share of the series voltage, thus activating its clamp circuit faster and longer than other MOSFETs. A longer delay must therefore be added to the turn-off command of this particular transistor. To quantify the required delays, the duration of the clamp is measured for each MOSFET in the series, all being subject to the described principles.

The clamp duration is defined as the row vector Tc for a number of MOSFETs N, where the ith MOSFETs clamp duration is noted Tc_i, as shown in Eq. (2).

$$Tc = [Tc_1, Tc_2, \dots, Tc_N] \tag{2}$$

The clamping durations are heavily dependent on the number of transistors, the bus voltage, the clamping thresholds, the temperature, the exact circuit and others. Therefore, it is challenging to directly define a timing delay to apply based on clamping durations. As the highest clamp duration requires a longer delay, a first variable is defined to represent the relative importance of adding a delay to each MOSFET turn-off. The row vector α, shown in Eq. (3), represents the percentages of each duration compared to the maximum clamp duration.

$$\alpha = [\alpha_1, \alpha_2, \dots, \alpha_N] \text{ where } \alpha_i = \frac{Tc_i}{Tc_{max}} \tag{3}$$

By multiplying each α_i coefficient by a time reference, a new row vector of incremental timing delays for the MOSFETs, D_{incr}, is defined in Eq. (4). The average clamp duration multiplied by an adjusting factor β, is chosen as a time reference. β is to be tweaked depending on system constants (N mainly) and possibly on operating conditions (bus voltage, load changes) to optimise voltage balancing.

$$D_{incr} = \left[D_{incr_1}, D_{incr_2}, \dots, D_{incr_N}\right]$$

$$\text{where } D_{incr_i} = \alpha_i \cdot \beta \cdot \left[\frac{1}{N} \cdot \sum_{i=1}^{N} Tc_i\right] \tag{4}$$

This incremental set of delays, calculated at each switching period is then added to the previously applied delay values T_{delay}, as shown in Eq. (5).

$$T_{delay} = \left[T_{delay_1}, T_{delay_2}, \dots, T_{delay_N}\right] \tag{5}$$

$$\text{where } T_{delay_i} = T_{delay_{i,previous}} + D_{incr_i}$$

The delay is kept from diverging by subtracting the minimum delay from all delays, Eq. (6). The purpose of this gate timing is conserved, as each transistor is still delayed according to the others.

$$T_{applied\_delay_i} = T_{delay_i} - \min(T_{delay}) \quad (6)$$

Regarding the turn-on event, the behaviour is reversed. During turn-on, the drain-source voltage across the fastest or earliest MOSFET changes from its previous static voltage (ideally V_{DC}/N) to zero. It is thus the slowest or latest transistors that handle the highest voltage. Therefore, new clamp durations are measured and every transistor must be delayed, except the slowest ones. To adapt the previously explained method to this situation, we simply use $(1 - \alpha)$ instead of α, in otherwise identical equations, and a different β coefficient for on and off events.

3.2 Simulation Setup on PLECS

To simulate the behaviour of this voltage balancing method on a synthetic switch comprising up to 48 transistors in series, we use Plexim's PLECS simulator. The command scheme explained previously is applied in PLECS by implementing it in a "C-script block", representing the control unit. Based on an assessment of existing control units such as FPGAs, the incremental step for delays is chosen as 1 ns.

Most of PLECS components being considered ideal, with its library's MOSFET having a binary control, we chose to use the Spice netlist of an existing TO-247 1200 V 30 mΩ SiC MOSFET to create an equivalent circuit on PLECS, as shown in Fig. 5. This way we can use the electrical and thermal model of a real SiC MOSFET on PLECS simulator, with the main advantage of a realistic gate feedback loop simulation.

Next, we design the gate driver (GD) circuit, shown in Fig. 6, to control the MOSFET. The voltage applied between gate and source can vary between -4 V and 15 V. It contains a gate driving circuit with on and off gate resistors Rg_on and Rg_off, a gate protection circuit with two Zener diodes, and a pull-down resistor Rgs.

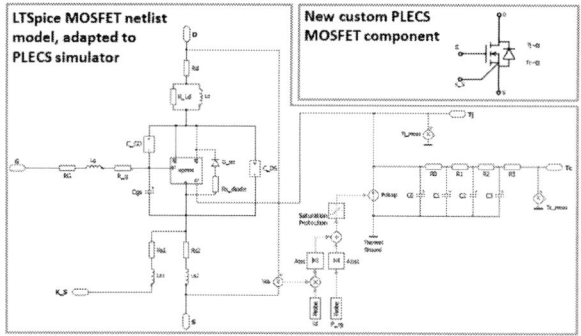

Fig. 5 Electrical and thermal Spice netlist model of a SiC MOSFET adapted for PLECS simulator

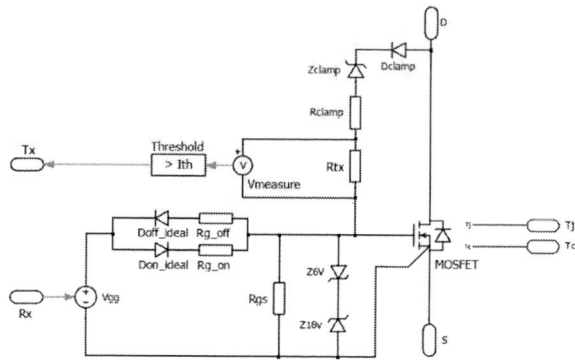

Fig. 6 Gate driver schematic connected to the SiC MOSFET model. It is composed of a clamp circuit, signal input/output (Rx/Tx) and gate resistors

The upper part of the gate driver contains the clamp circuit described in section 2. The series-connection is done by connecting the source of a MOSFET to the drain of another.

Voltage imbalance being caused by circuit imperfections, those have to be added manually in the simulations. Studies on internal equivalent capacitors of SiC MOSFETs show a variation of 10% to 30% compared to datasheet values [10]. For this simulation, random values are generated for each MOSFET, with ±10% for Cgs and Cds, and ±30% for Cdg; according to truncated Gaussian distributions centred on datasheet values, where either of these limits have a 2% chance of appearing.

The propagation time of the command from the control unit to the MOSFET (through optical transmitter/receiver, fibre and buffer) is also variable, electronically and through deviations in the implementation. Datasheet values for each component have been aggregated into a random delay between 0 and 150 ns for each transistor (according to a Gaussian distribution, centred between both limits, with a 2% chance of choosing either of the limits).

The resulting simulation has had to be lightened in some cases, the parasitic capacitors to ground being deleted in the 48-MOSFETs model. This trade-off between accuracy and simulation time does not hinder the demonstration, as the aforementioned variability creates enough imbalance to study our balancing method.

To begin, we simulate a block of 6 MOSFETs, allowing to demonstrate the feasibility of the balancing method with a fast simulation, followed by the 48 MOSFETs simulation (6*8 blocks). The synthetic switch is simulated in a Buck converter circuit.

3.3 Simulation Results

The aim is to apply the 34 kV DC bus voltage on 48 MOSFETs in series, with a current around 15 A. In this situation, the ideal balanced bus voltage should be around 700 V on each of the transistors. Following this approach, the DC bus voltage for 6 transistors must be 6×700 = 4200 V. As the chosen MOSFETs maximum voltage is 1200 V, the clamp value is chosen at 900 V, in-between the ideal voltage balance and maximum voltage.

Figure. 7 shows the simulation results of the first switching event with 6 transistors each having different initialisation parameters, and where no delay is applied yet. From top to bottom are represented the gate-source voltages (Vgs), the total drain-source across the series-connection (Total Vds), the drain-source voltage of each MOSFET (Each Vds) and the clamp states. Except for the total Vds, each colour matches a different MOSFET.

To visualise the effect of the applied delays after a few switching events, we chose to plot the clamping durations, the applied delay and the quasi-static drain-source voltage after switching oscillations, for different values of β and including a strong disturbance (sudden parameter change). The results are shown for turn-off in Fig. 8 and turn-on in Fig. 9. Finally, Fig. 10 represents a similar data measurement for 48 transistors in series.

3.4 Result Analysis

3.4.1 6 MOSFETs in series

In Fig. 7, we can see that the clamping circuit works as expected, by increasing the Vgs voltage, converging Vds voltage towards the threshold value and activating the clamping signal when it exceeds the threshold value. We can also note that the clamp isn't fully optimised as the Vds voltage exceeds the MOSFETs maximum voltage. This can be fixed by decreasing the voltage threshold value.

Regarding the turn-off correction in Fig. 8, as the intensity of the correction is controlled with the β_{OFF} coefficient, we can see that a too high β_{OFF} creates oscillations, as the applied delay reacts too sharply to the clamping durations. When β_{OFF} is too low, the clamping takes a longer time to disappear. With a value between 0.3 and 0.5 the command reacts quickly and the clamps disappear rapidly, even after a disturbance.

As the proposed method does not actually include a measure of the Vds voltage, the control unit can only assess if the voltage across a MOSFET is considered too high. As we can see in the simulations, the quasi-static voltage after turn-off is balanced differently depending on β_{OFF}: with a lower value, the delays are increased just enough to remove the clamps, without causing a sufficiently higher quasi-static voltage on the other switches, to correct the voltage imbalance. With a higher β_{OFF} value (but low enough not to cause

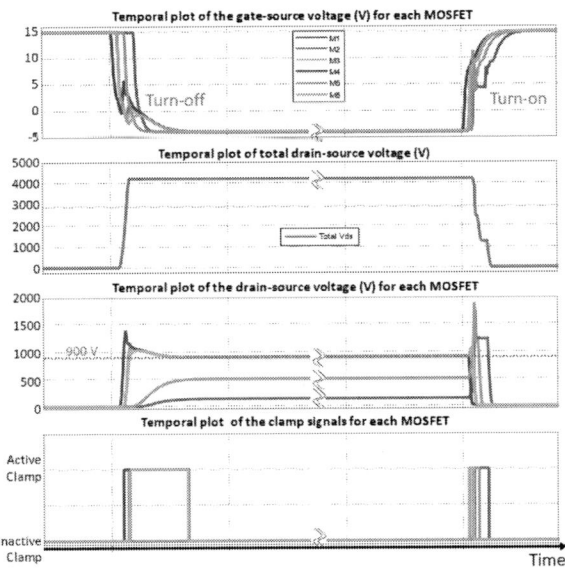

Fig. 7 Simulated temporal plot of a 6 MOSFETs series-connection, each comprising a clamping circuit

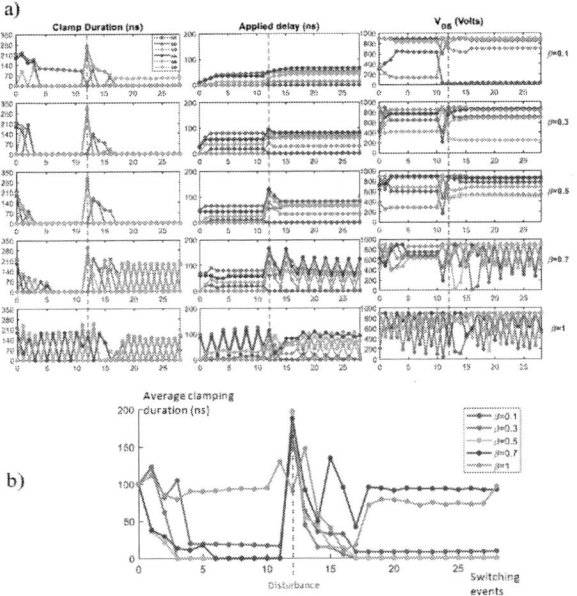

Fig. 8 Turn-off behaviour for 28 switching events, for different βOFF values, including a disturbance at the 12th iteration

an uncontrolled oscillating state), the delay steps are larger than strictly necessary, forcing some other MOSFETs to clamp, thus allowing a better voltage distribution by convergent oscillations. In this paper, we study the behaviour of the system for a constant β coefficient, but it would be possible to adjust it according to a closed loop control regulating the average clamp duration (shown in Fig. 8 b)).

Voltage imbalance also appears during turn-on before each Vds cancels. The reasons for this imbalance are similar to those of turn-off, as shown in Fig. 2, but are also due to quasi-static voltages resulting from the turn-off event. This makes it complicated to predict the behaviour of the clamping correction through iterative studies. For this, the turn-on clamp duration is measured for different β_{OFF} and β_{ON} values, as shown in Fig. 9. We can note that the β_{ON} coefficient has to be above 1 to work properly, meaning that a delay higher than the average clamp duration is necessary to remove the turn-on clamps. The clamps disappear with any properly working β_{OFF} regardless of β_{ON}. For instance, $\beta_{OFF} = 0.7$ causes turn-off instability, see Fig. 8, hence the turn-on clamp cannot stabilise either.

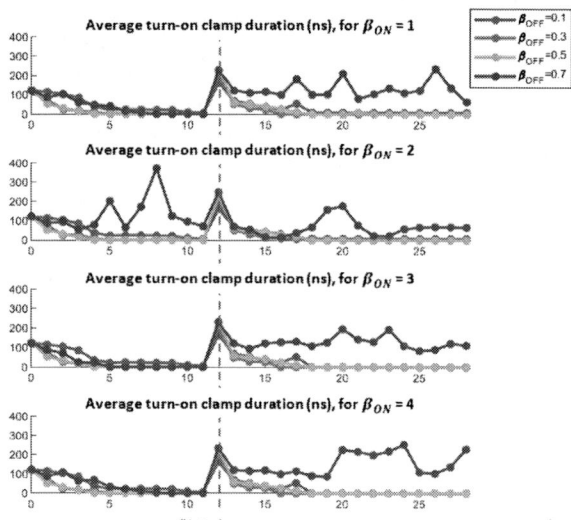

Fig. 9 Turn-on behaviour for 28 switching events, for different β_{ON} and β_{OFF} values, including a disturbance at the 12th iteration

3.4.2   48 transistors in series

Studying the behaviour of 48 MOSFETs is challenging as a simulation can take a very long time. We chose to use $\beta_{OFF} = 0.3$ and $\beta_{ON} = 3$. The results are shown in Fig. 10. With as much transistors in series, the clamp takes more time to disappear completely and the voltage doesn't

balance as much as for 6 MOSFETs in series within a limited duration. A longer simulation would need to be implemented. In addition, adapting the β values during the simulation can change the voltage balancing of these 48 transistors. As said previously this could be implemented in future works by using the average clamp duration or standard deviation for a closed loop control regulation.

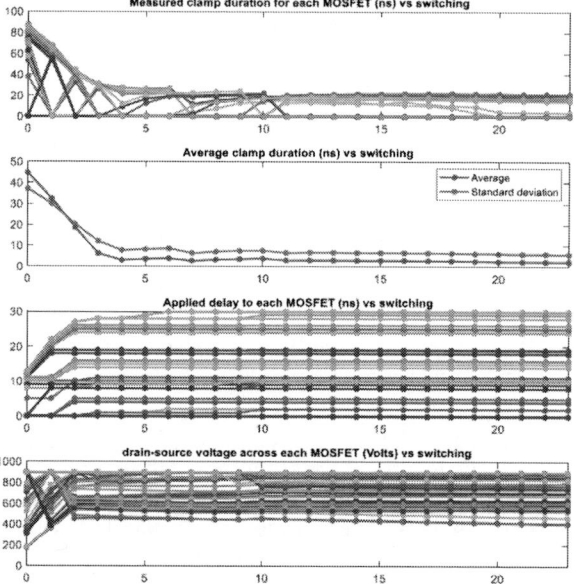

Fig. 10 Evolution of the clamping for 23 switching events for 48 MOSFETs in series each with a different color, with β_{OFF} = 0.3

4   Implementation

The simulations show that the balancing method works for a small and large number of transistors in series. The implementation of such a synthetic switch is challenging, as many factors must be taken into account, starting with the DC bus, where the bus and decoupling capacitors need to sustain medium voltage. Power needs to be delivered to each gate driver, requiring high voltage insulation with the least possible capacitance to avoid disturbances due to high dv/dt. The overall inductance loop must be designed as small as possible to avoid oscillations caused by di/dt. The overall switch will be discussed in a future work.

To verify the proper operation of the clamp detection for one MOSFET and its GD, we set up a double pulse test (DPT) circuit, shown in Fig. 11, with a high (and adjustable) inductance loop, to generate pulses above the clamps threshold value. The aim is to observe how the clamping circuit reacts when an oscillation exceeds the

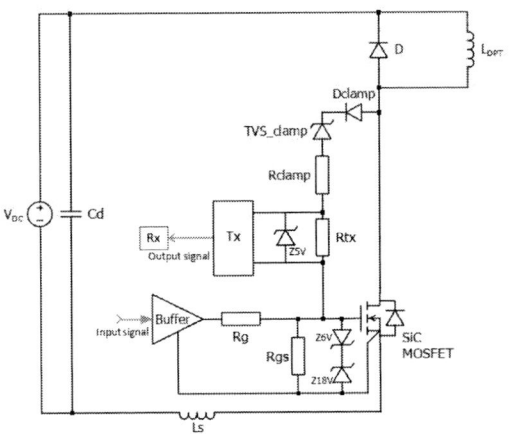

Fig. 11 Double pulse test schematic to confirm the active clamp operation and its detection

threshold clamping value, and if a clamping signal is sent through the optical fibre transmitter.

The clamp circuit is implemented by using four 180 V TVS diodes in series, resulting in a 720 V clamp threshold. The resistors are chosen to limit the current, and a 5 V Zener diode is placed across the Rtx resistor to limit the voltage (optical transmitter maximum input signal voltage). The Ls inductor connected to the source of the MOSFET has a high inductance value of 500 nH. It is important that the bus voltage does not exceed the threshold voltage, otherwise the clamping circuit could always be active, thus constantly closing the MOSFET. The bus voltage V_{DC} is set to 650 V, and the first pulse duration set to attain 13 A.

Figure. 12 shows the measured results, with and without the clamping circuit. On the first plot, we can see the gate-source voltage rise when the drain-source voltage is above the clamp threshold. The voltage across the Rtx resistor represents the current flowing through the clamp, and activates the optical transmitter when exceeding 2 V. The received signal has the same timing as the active clamp, with a small delay mainly due to fibre length. With the clamp activated, the maximum Vds voltage is reduced from 1016 V to 930 V, and the oscillations are more dampened.

This setup proves that the clamp detection works even for a very small clamp duration, but does not yet demonstrate how the clamp could react when adding transistors in series. In a next step, the aim will be to connect 6 MOSFETs in series at first, to make a small synthetic switch by taking into account the capacitive effects, the inductance loop, the cooling system... This will be followed by adding multiple of these blocks, in an appropriate arrangement minimizing inductance loop and parasitic capacitances, up to 8*6=48 MOSFETs in series.

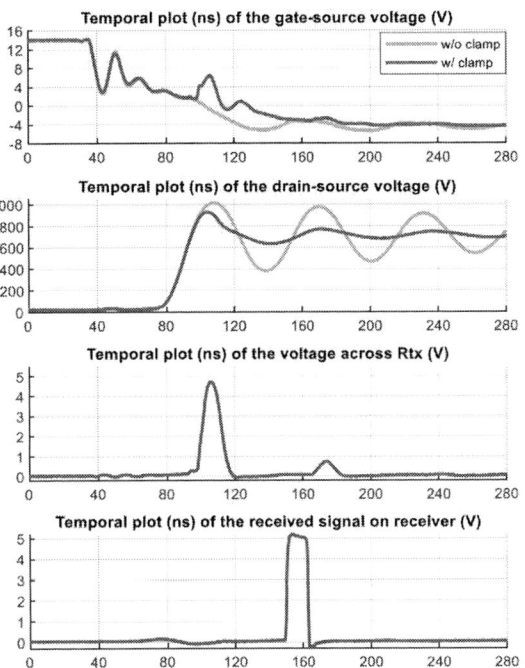

Fig. 12 Experimental results testing the clamp in a DPT circuit. Yellow curves is the behavior without clamp, and blue curves with clamp

5 Conclusion

In this paper, a new voltage-balancing method for a series-connection of up to 48 SiC MOSFETs was proposed. This was achieved by using clamping circuits, used both as protection and as inputs for a gate timing correction algorithm for each transistor. In addition, was proposed a control scheme using the clamping signal durations to generate timing delays, which are applied to the control of the transistors. The first steps of this work were carried out through simulations, using from 6 to 48 SiC MOSFETs in series. This method efficiently removes the clamps while balancing the voltages for a small amount of transistors. Further studies need to be done regarding the behaviour for a larger amount of transistors. Experimental setup was also made to verify the operation of the clamp signal transmission through optical fibre, validating the possibility of applying such a circuit around SiC MOSFETs.

The next experimental works will consist in creating such a synthetic switch, first as a single block of 6 SiC MOSFETs, up to 8 blocks, all in series.

Acknowledgments

This work was supported by the French National Program "Programme d'Investissements d'Avenir – INES.2S" under Grant ANR-10-IEED-0014-01

References

[1] M. Rabiul Islam, A. M. Mahfuz-Ur-Rahman, K. M. Muttaqi and D. Sutanto, "State-of-the-Art of the Medium-Voltage Power Converter Technologies for Grid Integration of Solar Photovoltaic Power Plants," in IEEE Transactions on Energy Conversion, vol. 34, no. 1, pp. 372-384, March 2019

[2] V. Pala et al., "10 kV and 15 kV silicon carbide power MOSFETs for next-generation energy conversion and transmission systems," 2014 IEEE Energy Conversion Congress and Exposition (ECCE), Pittsburgh, PA, USA, 2014, pp. 449-454

[3] A. Marzoughi, R. Burgos and D. Boroyevich, "Active Gate-Driver With dv/dt Controller for Dynamic Voltage Balancing in Series-Connected SiC MOSFETs," in IEEE Transactions on Industrial Electronics, vol. 66, no. 4, pp. 2488-2498, April 2019

[4] V. U. Pawaskar, G. Gohil and P. T. Balsara, "Study of Voltage Balancing Techniques for Series-Connected Insulated Gate Power Devices," in IEEE Journal of Emerging and Selected Topics in Power Electronics, vol. 10, no. 2, pp. 2380-2394, April 2022

[5] K. Vechalapu, S. Bhattacharya and E. Aleoiza, "Performance evaluation of series connected 1700V SiC MOSFET devices," 2015 IEEE 3rd Workshop on Wide Bandgap Power Devices and Applica tions (WiPDA), Blacksburg, VA, USA, 2015, pp. 184-191

[6] R. W. Maier and M. -M. Bakran, "Active clamping method for SiC MOSFET high power modules - Benefits and Limits," 2020 22nd European Conference on Power Electronics and Applications (EPE'20 ECCE Europe), Lyon, France, 2020, pp. P.1-P.10

[7] M. Zhao, H. Lin and T. Wang, "An Adaptive Driving Signals Delay Control for Voltage Balancing of Multiple Series-Connected SiC MOSFETs," 2022 IEEE Applied Power Electronics Conference and Exposition (APEC), Houston, TX, USA, 2022, pp. 1619-1624

[8] R. W. Maier and M. -M. Bakran, "Active clamping method for SiC MOSFET high power modules - Benefits and Limits," 2020 22nd European Conference on Power Electronics and Applications (EPE'20 ECCE Europe), Lyon, France, 2020, pp. P.1-P.10

[9] S. Ji, F. Wang, L. M. Tolbert, T. Lu, Z. Zhao and H. Yu, "An FPGA-Based Voltage Balancing Control for Multi-HV-IGBTs in Series Connection," in IEEE Transactions on Industry Applications, vol. 54, no. 5, pp. 4640-4649, Sept.-Oct. 2018

[10] C. M. de Vienne, B. Asllani, B. Lefebvre, P. Lefranc and P. -O. Jeannin, "Model Parameter Extraction Tool for the Analysis of Series-Connected SiC-MOSFETs," PCIM Europe digital days 2021; International Exhibition and Conference for Power Electronics, Intelligent Motion, Renewable Energy and Energy Management, Online, 2021, pp. 1-8.

PCIM Europe 2024, 11– 13 June 2024, Nuremberg   DOI: 10.30420/566262197

A Laboratory-Scale MMC-Based DC System with RCP and PHiL Simulation Capabilities

Marc René Lotz[1,2], Eugene Tinjinui Ndoh[3], Michael Kurrat[2], Martin Könemund[1]

[1] Ostfalia University of Applied Sciences, Germany
[2] Technische Universität Braunschweig, Germany
[3] German Aerospace Center (DLR), Germany

Corresponding author: Marc René Lotz, m.lotz@ostfalia.de
Speaker: Marc René Lotz, m.lotz@ostfalia.de

Abstract

A laboratory-scale DC system based on Modular Multilevel Converters (MMCs) with Rapid Control Prototyping (RCP) and Power-Hardware-in-the-Loop (PHiL) simulation capabilities is presented. After a system overview with the characterization of the components, and its application in the area of validating power electronics systems, two use cases are presented in detail. One is the validation of grid-tied converter controller designs, showing the RCP capabilities of the testbed. Another one is the PHiL simulation of an MMC, which shows that the number of converters and configurations, as well as the test and fault coverage, can be increased with feasible accuracy.

1   Introduction

Demonstration in the area of DC systems and their connected converters is an important topic of research in order to increase the Technology Readiness Level (TRL), defined by the European Commission and available, for example, in [1]. It closes the gap between plain simulation studies and full-scale prototype development. This increases credibility of research, and enables to give recommendations for the full-scale prototypes by validating the fundamental designs regarding their feasibility. The demand for such demonstrators is emphasized by analyzing latest calls for proposals, which "Components and interfacing for AC & DC side protection system - AC & DC grid: components and systems for grid optimisation" from the Horizon Europe Framework Programme clearly illustrates for prototypes of HVDC systems including converter operation.

Among utilizing HVDC systems for power transmission and renewables integration, described in [2], DC systems are also solutions for microgrids in general, as well as on-board power supply systems for aircrafts and ship vessels. Overviews are given in [3], [4], and [5], respectively. The applications share the main challenges in the areas of operation, control, and protection.

Especially for operation and control concepts, as well as controller designs, various demonstrators are implemented, like the one presented in [6]. In this work, a Modular Multilevel Converter (MMC)-based flexible laboratory environment is presented, which serves both as a Rapid Control Prototyping (RCP) and Power-Hardware-in-the-Loop (PHiL) simulation platform. It is used to demonstrate developments for power electronics and DC energy supply systems, as well as HVDC grids, in a scaled laboratory environment.

The benefit of utilizing real-time simulation is manifold, shown in [7] for power and energy system applications. Well-known principles of Hardware-in-the-Loop (HiL) and RCP allow the test of actual controller hardware, while the plant is simulated, and vice versa. With PHiL applications, not only power hardware can be tested, but it is also possible to connect simulated systems and systems in a laboratory environment together to analyze their interactions.

The outline of this work is as follows. In Section 2, an overview of the laboratory environment with its components and key characteristics will be given. After that, the advantages and challenges of incorporating PHiL simulations will be explained briefly in Section 3. Then, in Section 4, three use cases will be presented, that depict the testbed's capabilities for different applications, ranging from grid-tied converter control validation, over PHiL MMC simulations significantly enhancing the testbed's performance, to utilizing it for the validation of modern control concepts, here a Capacitor Voltage Estimation (CVE) algorithm for MMCs of ship on-board power supply systems. In Section 5, a conclusion is drawn reflecting on the use cases and the feasibility of the laboratory environment.

2 System Overview

The RCP and PHiL system consists of multiple components that can be connected to each other with variable topologies, depending on the desired application, thus creating a flexible, but still dedicated, laboratory environment. A schematic overview is shown in Fig. 1.

The system is based around an 11-level MMC (A), called OP1200 and manufactured by OPAL-RT, with 6 kVA rated power and a nominal DC voltage of 400 V. A RCP platform is included in order to implement control and protection algorithms via MATLAB/Simulink. It is CPU-based and allows complete freedom to define higher- and lower-level controls. The sample times are typically in the range of 20 to 50 microseconds. Although it is possible to generate the PWM swichting signals on the CPU as well, here, modulation references are sent to the dedicated FPGA unit in order to ensure high PWM resolution. In general, the PWM frequency can be varied up to $10\,\mathrm{kHz}$. Here, $f_{\mathrm{PWM}} = 4\,\mathrm{kHz}$ is used.

For a better understanding of the later sections, where a control architecture is presented, and PHiL MMC simulations demonstrated, the equivalent circuit is depicted in Fig. 2.

The real-time simulator (B) from OPAL-RT, called OP5707XG, is used for PHiL simulations of MMCs, their connected AC grids, and offshore wind farms. The aurora interface enables the exchange of set points and measurements with the two power amplifiers described later. Additional ethernet and analog interfaces are also utilizied, to control other components like the AC grid simulator, and the MMC.

One power amplifier (C) connected to the real-time simulator is a Triphase PM15 system. It is a four-quadrant switched inverter at $16\,\mathrm{kHz}$ with two operation modes: A four-wire grid-following AC current source, and a two-wire DC voltage source. Both modes can operate independently. It is possible

to implement user-defined control algorithms with MATLAB/Simulink similar to the MMC, as it also consists of a RCP platform. Among that, the aforementioned communication with the real-time simulator (B) makes PHiL simulations possible.

The second power amplifier (D) is a linear power amplifier from Spitzenberger and Spies, configured as a two-wire voltage source. It is capable of full four-quadrant operation in the power range of the MMC, while providing a high bandwidth necessary for PHiL simulations of DC systems and converters. One use case is the validation of control and protection principles of HVDC systems, for which a point-to-point connection with two MMCs is realized. For that, laboratory-scale cable models (E) have been developed, which represent $\pm 200\,\mathrm{kV}$ cables of 50 km length as pi-sections each.

The last component is a Chroma Regenerative Grid Simulator 61815 (F) used as a four-wire AC voltage source with the capability of simulating symmetric and asymmetric faults necessary for validation of Fault Ride Through (FRT) test cases.

3 The Role of PHiL Simulations

With PHiL simulations, the testbed's capabilities can be enhanced significantly. Among various application examples and considerations, as explained in [8], here the focus lies on presenting multiple possible configurations all depending on PHiL MMC simulations, while referring to the components presented in Fig. 1: A point-to-point connection with one actual MMC (A), and one MMC simulated in real-time (B, D), a three-terminal connection, which uses the remaining power amplifier (C), and a bipolar converter station.

In general, the number of relevant test cases is increased, as complex AC and DC systems, topologies, protection systems, and faults can be simulated in real-time, and interfaced with the actual hardware components. Drawbacks of the scaled laboratory environment, like inaccurate cable models or improper transformer characteristics, can be overcome as well due to the possibility of simulating these components.

Still, setting up a PHiL simulation is a complex task. Due to the connection of (discrete) real-time simulations with actual hardware via a power amplifier with its own dynamics, as well as communication and measurement delays, a stability analysis has to be performed beforehand, showing stability regions and if a proper interface design is necessary.

PCIM Europe 2024, 11– 13 June 2024, Nuremberg   DOI: 10.30420/566262197

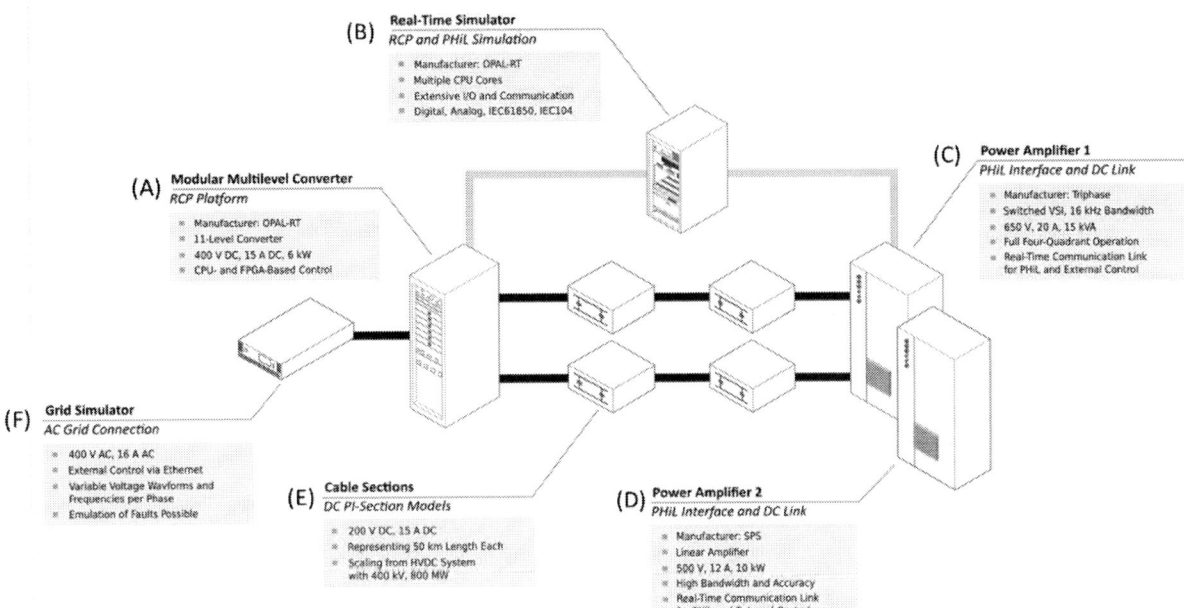

Fig. 1: System overview with highlighted power exchange (black) and communication (gray).

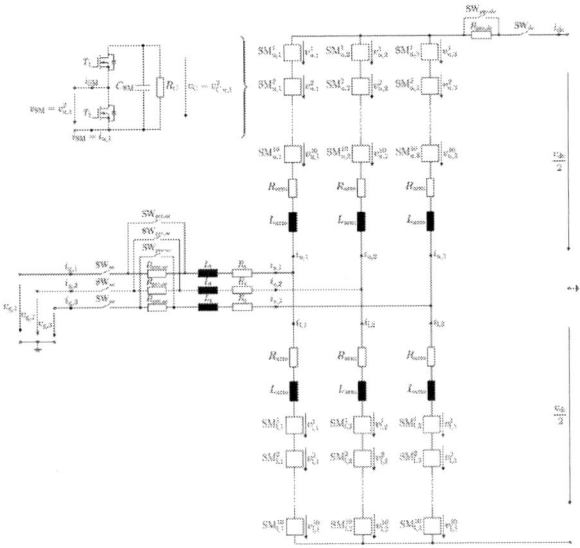

Fig. 2: MMC equivalent circuit.

Additionally, the simulation model complexity limits sample times, solver types, and thus the accuracy of the real-time simulation. Especially for DC systems and converters, with their high bandwidth, fast transients, and non-linearities, stability has to be analyzed in detail, even to the level of understanding the sequence of execution of real-time simulation and communication, see [9] for further information.

4 Use Cases

In this section, two use cases of the laboratory environment are shown in detail to illustrate the sufficiency and flexibility of the setup for control and protection system validation of DC systems and converters. After that, it will be shown that the setup is also beneficial for the validation of novel concepts, with a capacitor voltage estimation algorithm as an example.

4.1 Grid-Tied MMC Control Validation

As discussed in Section 2, the RCP platform of the MMC enables the implementation of user-definable control and protection algorithms on a CPU down to the voltage balancing and modulation reference generation. To illustrate the sufficiency of the platform, a simplified grid-tied control architecture of a HVDC system, which originates from [10] with controls based on [11] and [12], will be implemented. Fig. 3 shows a schematic overview of the control architecture.

Active and reactive power are controlled by controlling the Q- and D-frame currents, respectively, all with proportional-integral controllers. The PLL necessary for transforming the instantaneous voltages and currents is not depicted here. It should also be noted, that the original Park (DQ0) transformation is used, where the Q-component represents active currents or power, while the D-component

1459

PCIM Europe 2024, 11– 13 June 2024, Nuremberg   DOI: 10.30420/566262197

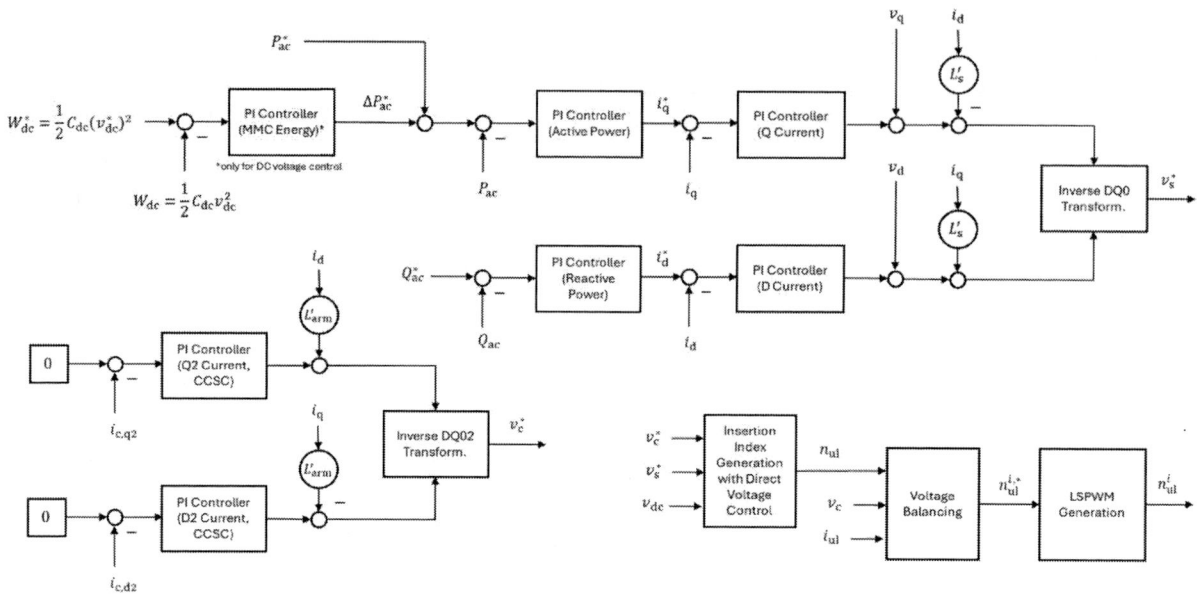

Fig. 3: Schematic control architecture implemented on the MMC RCP platform.

represents reactive currents or power.

Leaving the control architecture as described in the previous paragraph, circulating currents of double the fundamental AC grid frequency exist that lead to higher overall currents flowing in the MMC arms, and thus to higher capacitor voltage fluctuations. To suppress these currents, Circulating Current Suppression Control (CCSC) is implemented.

Here, the principles of CCSC will be discussed in more detail, while providing a controller design and illustrating actual measurements of the MMC and DC system. The circulating currents are expressed as $i_{c,k}$, with $k = 1, 2, 3$ and are assumed to exist as common-mode currents appearing in the upper and lower arm currents $i_{ul,k}$, among the grid currents $i_{g,k}$.

$$i_{u,k} = \frac{1}{2} \cdot i_{g,k} + i_{c,k}, \ i_{l,k} = -\frac{1}{2} \cdot i_{g,k} + i_{c,k} \quad (1)$$

Additionally, the arm submodule voltages are defined to be input voltages without their dynamics introduced by the capacitors. Then, the sum of arm submodule voltages is as follows.

$$v_{ul,k} = \sum_{i=1}^{N_{arm}} v_{ul,k}^i \quad (2)$$

The control signal needed for the CCSC is expressed as

$$v_{c,k} = \frac{v_{u,k} + v_{l,k}}{2}. \quad (3)$$

These assumptions simplify the MMC's dynamic equations so that it is possible to address the circulating currents isolated from the grid currents, emphasizing that there are two degrees of freedom for current control.

$$v_{c,k} = \frac{v_{dc}}{2} + R_{arm} \cdot i_{c,k} + L_{arm} \cdot \frac{di_{c,k}}{dt} \quad (4)$$

As discussed before, currents twice the fundamental frequency appear that do not contribute to active power transmission between MMC AC and DC side. In order to suppress them with proportional-integral controllers, the dynamic equation is transformed into a double fundamental frequency DQ0 coordinate system, denoted with the subscript "dq2". The 0-component will not be controlled and thus is omitted.

$$v_{c,d2} = R_{arm} \cdot i_{c,d2} + L_{arm} \cdot \frac{di_{c,d2}}{dt} - \omega \cdot L_{arm} \cdot i_{c,q2} \quad (5)$$

$$v_{c,q2} = R_{arm} \cdot i_{c,q2} + L_{arm} \cdot \frac{di_{c,q2}}{dt} + \omega \cdot L_{arm} \cdot i_{c,d2} \quad (6)$$

The cross-coupling between the D- and Q-axis will be tackled by adding feed-forward terms, as visible

1460

in Fig. 3. Then, the equations can be simplified, as they are principally the same for both axes.

$$v_{c,dq2} = R_{arm} \cdot i_{c,dq2} + L_{arm} \cdot \frac{di_{c,dq2}}{dt} \qquad (7)$$

A controller for this linear system will be developed in the Laplace domain, leading to

$$\frac{i_{c,dq2}(s)}{v_{c,dq2}(s)} = \frac{1}{s \cdot L_{arm} + R_{arm}} \qquad (8)$$

and the controller transfer function

$$G(s) = K_{P,c} + K_{I,c} \cdot \frac{1}{s} \qquad (9)$$

with $K_{P,c}$ and $K_{I,c}$ as proportional and integral gains. By comparing the closed-loop system with a first-order system

$$\frac{\alpha_c}{s + \alpha_c} = \frac{G(s) \cdot \dfrac{i_{c,dq2}(s)}{v_{c,dq2}(s)}}{1 + G(s) \cdot \dfrac{i_{c,dq2}(s)}{v_{c,dq2}(s)}} \qquad (10)$$

where α_c is the bandwidth, the controller gains can be calculated.

$$K_{P,c} = \alpha_c \cdot L_{arm}, \; K_{I,c} = \alpha_c \cdot R_{arm} \qquad (11)$$

The bandwidth is assumed to be smaller than the PWM frequency with $\alpha_c = 2\pi f_{PWM}/10$. After normalization, the controller is discretized using the exact Euler discretization. With the point-to-point topology shown in Fig. 1, the controller design is validated experimentally.

Fig. 4 depicts the normalized arm current $i_{u,1}$. Almost immediately after enabling the CCSC at $t = 0.1$ s, the double fundamental frequeny component is removed, reducing both the RMS and peak current.

This is also visible in the capacitor voltages, illustrated in Fig. 5 with normalized $v^1_{c,u,1}$. The fluctuations are decreased, reducing stress on the capacitors.

In general, the presented control architecture is working properly, but it lacks handling of asymmetric faults. The validation of FRT capabilities according to German grid codes, see [13], where reactive current injection in positive and negative sequence during faults is demanded, is examined in a future publication for laboratory-scale MV systems and the analysis of inverter-based influence on protection systems. There, proportional-resonant controls

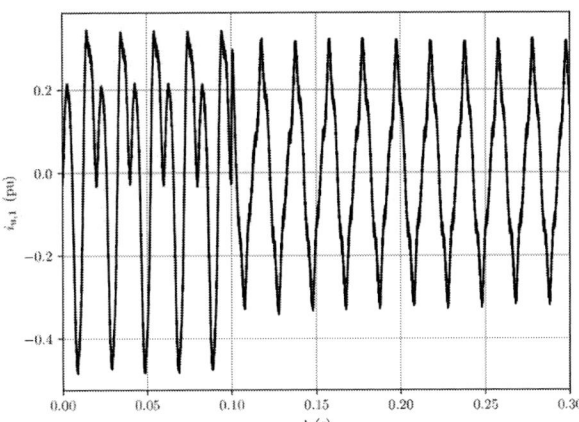

Fig. 4: CCSC enable at $t = 0.1$ s, arm current normalized with $I_{b,ac} = 19.63$ A.

are implemented, together with a robust PLL, and a positive and negative sequence current limiter.

4.2 PHiL MMC Simulation

To further enhance the setup's performance and capabilities regarding validation of laboratory-scale HVDC systems, as well as their control and protection, PHiL MMC simulations are incorporated. Referring to Fig. 1, one MMC is simulated on the real-time simulator (B), and power exchange with the actual hardware MMC (A) is realized with the linear power amplifier (D). This creates a more realistic point-to-point connection for validating the propagation of faults and the interaction of converter controls.

A limiting factor for the level of detail with which a MMC can be simulated is the computational performance of real-time simulation. The test cases that

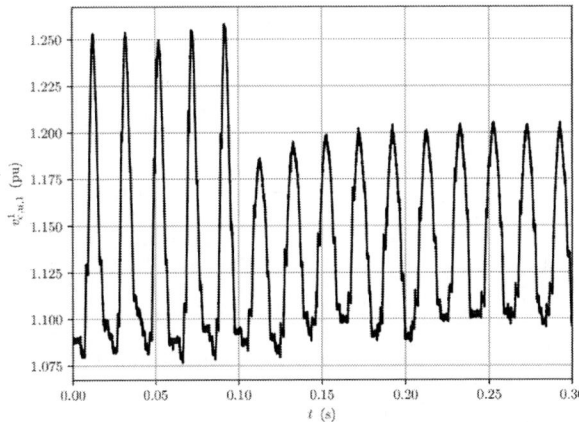

Fig. 5: CCSC enable at $t = 0.1$ s, capacitor voltage normalized with $V_{b,SM} = 33.97$ V.

can be performed depend on the type of MMC models used, while the model complexity and solver configuration limit the real-time simulation sample time. As explained in Section 3, the power amplifier and actual hardware dynamics, together with discretization and communication delays, affect stability and influence accuracy of the PHiL simulation, depending on the interface or filter designs needed. Different types of MMC models exist, where most models especially average the switching and submodule characteristics to reduce the computational burden. In [14] and [15], typical implementations are presented, ranging from averaging the whole MMC circuit, over to averaging each arm, or each submodule separately. The type of model used depends on the simulation studies. For higher-level controls and interactions with AC grids, simple averaged models are sufficient, but if lower-level controls, including voltage balancing, and the DC-side dynamics are of interest, the submodule characteristics also have to be considered.

Especially in [10], a typically used Type 4 simulation model is described, where the submodule's characteristics with simplified switching dynamics are considered, and separated from calculating the rest of the MMC and its connected AC and DC side by representing them as a Thévenin equivalent circuit for each arm. This reduces the computational burden, while still preserving access to the submodule capacitor voltages needed for analyzing lower-level controls including voltage balancing.

In this work, the Type 4 model is considered for performing PHiL MMC simulations. Fig. 6 shows the corresponding equivalent circuit of a submodule.

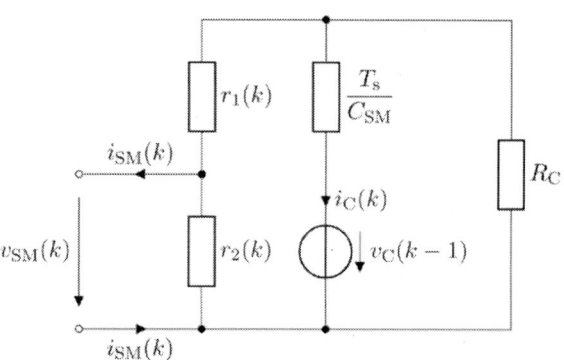

Fig. 6: Submodule equivalent circuit with Backward Euler.

The submodule transistors T_1 and T_2 with their anti-parallel diodes are simplified by variable resistors r_1

and r_2. During normal operation, where the MMC is deblocked, the transistors are switched on and off in an alternating manner. The on-state is modelled with an on-state resistance R_{on}, the off-state with R_{off}. The states are depicted in Tab. 1.

Tab. 1: Submodule resistances depending on DBLK and PWM states.

DBLK	PWM	r_1	r_2
0	X	$\to 0$	R_{off}
1	0	R_{off}	R_{on}
1	1	R_{on}	R_{off}

If arm-level blocking is considered, r_1 will be set close to zero, as the resistance is modelled by the diode's differential resistance instead.

The implementation of the Type 4 model also depends on the solver type. Instead of the typically used trapezoidal solver, Backward Euler is applied to increase computational efficiency and robustness. The continuous derivative

$$i_C(t) = C_{SM} \cdot \frac{dv_C(t)}{dt} \tag{12}$$

is discretized and approximated resulting in

$$i_C(k) = C_{SM} \cdot \frac{v_C(k) - v_C(k-1)}{T_s}. \tag{13}$$

Then, the capacitor can be modelled by a resistance and a voltage source that represents the capacitor voltage of the last time step $k-1$.

The submodule's characteristics are then summarized as a Thévenin equivalent circuit resistance r_{eq} and voltage v_{eq}.

$$r_{eq} = \frac{r_2 \cdot \left(r_1 + \frac{\frac{T_s}{C_{SM}} \cdot R_C}{\frac{T_s}{C_{SM}} + R_C} \right)}{r_2 + r_1 + \frac{\frac{T_s}{C_{SM}} \cdot R_C}{\frac{T_s}{C_{SM}} + R_C}} \tag{14}$$

$$v_{\text{eq}} = \frac{v_{\text{C}}(k-1) \cdot r_2 \cdot R_{\text{C}}}{\dfrac{T_{\text{s}}}{C_{\text{SM}}} \cdot (r_1 + r_2 + R_{\text{C}}) + (r_1 + r_2) \cdot R_{\text{C}}} \quad (15)$$

The equivalent circuits of the submodules are summarized into one equivalent circuit per arm, illustrated in Fig. 7. The submodules and the MMC are decoupled computation-wise, which significantly reduces the number of electrical nodes of the MMC circuit. Updating the capacitor voltage state variables can be performed after calculating the MMC circuit, as the arm current measurements then are available. The capacitor voltages can then also be used as measurements for control and voltage balancing.

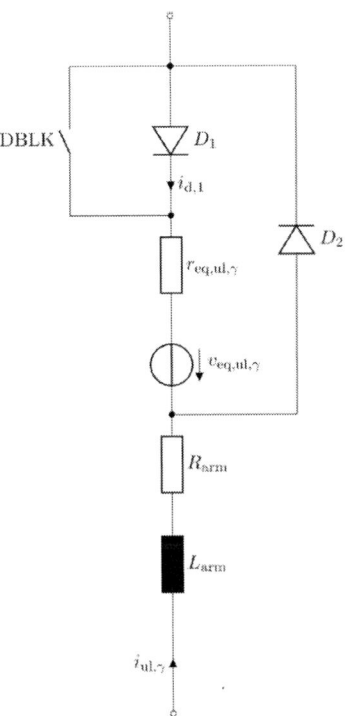

Fig. 7: One arm of the MMC Type 4 model with arm-level blocking.

In Fig. 7, blocking capabilities are also incorporated. Instead of being able to block each submodule individually, arm-level blocking is introduced as described in [15], leading to three non-linear elements per arm. The number of non-linear elements has a significant impact on the computation time due to the necessity of implementing a non-linear solver. The whole Type 4 MMC model including its control, an AC grid connection, and a transformer, has

been implemented on a real-time simulator. The computation times achieved are in the range of ten to twenty microseconds depending on the system state, meaning that a sample time of $T_{\text{s}} = 25\ \mu\text{s}$ is chosen. The simulation environment used is MATLAB/Simulink R2020b, with which the model is compiled and executed on the real-time simulator. It should be noted that the circuit is implemented with mathematical equations, without the use of typical toolboxes like Simscape Specialized Power Systems, due to higher values and fluctuations of the computation times which could not be reduced. The PHiL interface used is the Ideal Transformer Method (ITM). This means that the simulated MMC's DC side is connected to a current source. The simulated voltage drop across it is sent as a voltage set point to the power amplifier. The current flowing in the actual DC system is then measured by the amplifier and sent back into the real-time simulation, controlling the current source. The so created closed-loop enables operation of the whole system with its discretized simulation and actual hardware in the laboratory.

Fig. 8 now shows the complete PHiL setup, where the MMC Type 4 model is simulated in real-time on a real-time simulator (a), interfaced with a power amplifier (b), enabling power exchange with the actual hardware MMC (d) over DC cable sections (c).

By connecting the PHiL MMC simulation with the actual hardware MMC, the testbed's capabilities are significantly enhanced, as now, the interaction of converters can be studied. Here, one example will be given considering a simplified HVDC point-to-point connection. The offshore converter is typically operated in grid-forming mode, where the wind farms are injecting active power into the converter AC side. The onshore converter controls the DC voltage and reactive power. According to the German grid codes, see [19], this converter needs to inject reactive currents during AC faults both in positive and negative sequence via droop control.

A severe fault could lead to the converter losing its ability to control the DC voltage. If the offshore wind farms then still inject active power, the DC bus voltage will increase rapidly. One counter measure is to install a DC chopper at the onshore converter station, which dissipates the active power during a fault when a DC voltage threshold is reached. The principles of DC choppers for such applications, in-

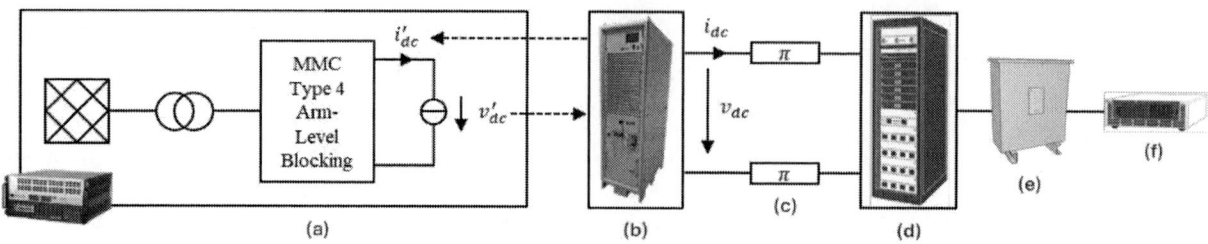

Fig. 8: Overview of the PHiL setup with adapted pictures from [16], [17], and [18].

cluding sizing of the resistor, is described in [20]. The chopper here is implemented and simulated on the real-time simulator, meaning that the simulated converter in Fig. 8 (a) represents the onshore converter with DC voltage control, while the actual MMC (d) controls active and reactive power to represent an offshore wind farm connected to it in a simplified manner.

Typically, the chopper is activated using hysteresis control or modulation. In this case, the chopper is modulated with the modulation reference depending on the DC voltage, if the DC voltage is above a certain threshold. An average model is used which consists of a controlled current source, as only CPU simulation is considered, opposed to including additional FPGA simulation as shown in [21].

During an AC fault at the converter that controls the DC voltage, reactive currents are injected to support the grid voltage. This is shown in Fig. 9 in the DQ0 reference frame for positive sequence currents, where the D-component corresponds to reactive, and the Q-component to active current.

DC voltage control in this case is not possible anymore, leading to a rising DC voltage, as the other converter with active power control is not affected by the fault. If the DC voltage is above a threshold, the chopper is modulated, leading to a stable operation point for the duration of the fault. This is visible in Fig. 10.

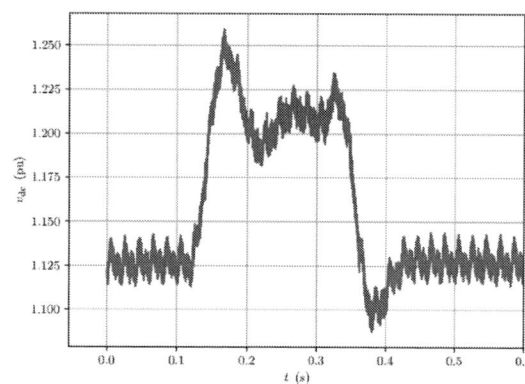

Fig. 10: DC voltage with chopper operation during an AC fault, normalized with $V_{\mathrm{b,dc}} = 339.7$ V.

In conclusion, the PHiL MMC simulation enhances the testbed's capabilities, as converter interactions can be analyzed, while providing the flexibility of implementing user-definable control and protection algorithms. Still, stability and accuracy constraints incorporated by connecting real-time simulations with actual hardware due to delays, discretization, and the power amplifier dynamics, need to be well-understood and handled accordingly, for example by a proper interface design.

4.3 Capacitor Voltage Estimation

The testbed will also be utilized to validate novel control concepts. As proposed in [22], the robustness of MMCs for ship energy supply systems can be increased by estimating the capacitor voltages with additional arm voltage measurements, if the

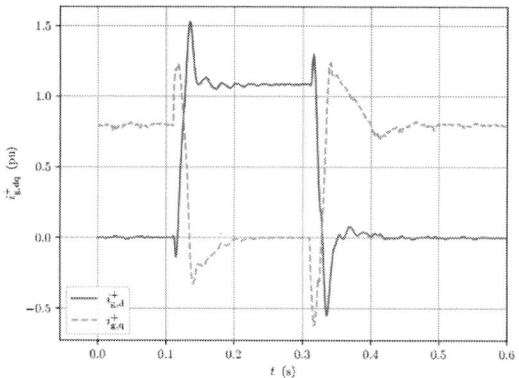

Fig. 9: Positive sequence active (Q-component) and reactive (D-component) currents injected into the grid during an AC fault, normalized with $I_{\mathrm{b,ac}} = 19.63$ A.

capacitor voltage sensors malfunction, adding an additional layer of safety. Fig. 11 shows simulative results of the CVE algorithm, where the sensor measuremets of the capacitor voltage are compared with the estimated values.

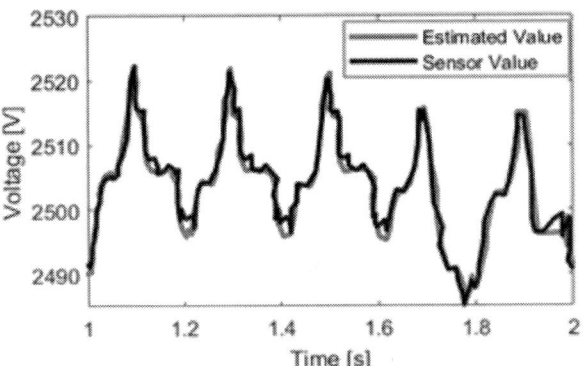

Fig. 11: Comparison of measured and estimated capacitor voltage from [22].

In principle, the algorithm works as follows. A change of the number of inserted submodules, together with the knowledge which submodule will be additionally inserted or bypassed, will be evaluated together with the arm voltage measurement. If a submodule is in its on-state, and has been additionally inserted, then, the difference between current arm voltage measurement and the previous measurement, should approximately correspond to the additionally inserted capacitor's voltage. Performance of this algorithm depends on multiple factors, like the voltage balancing algorithm, the modulation technique (NLM, PSPWM, LSPWM etc.), as well as availabiliy and access to hardware and signals to be measured.

As a next step, the CVE algorithm will be implemented on the MMC RCP platform presented in this work to validate its performance in a real-world application. Due to constraints imposed by the actual hardware and laboratory environment, the algorithm's robustness can also be analyzed regarding different MMC implementations and restricted access to the hardware. The adaptations necessary will also be evaluated.

5 Conclusion

In the introduction, the necessity of laboratory-scale demonstration has been emphasized as a means to increase the TRL from plain simulation studies to prototypes, bringing developments a step closer to full-scale implementations.

After that, the laboratory-scale MMC-based DC system has been presented. Due to the compatibility of the components voltage- and power-wise, and its RCP and PHiL simulation capabilities, it is well-suited for a variety of applications in the area of AC/DC converters, DC systems, control and protection.

The laboratory environment's capabilities have been illustrated by multiple use cases. A grid-tied control architecture has been developed and validated, while emphasizing the principals of CCSC. In a future publication, it will be shown how the controls need to be adapted in order to inject reactive currents in positive and negative sequence for grid conformance tests.

A second use case showed the advantages of utilizing PHiL MMC simulations to enhance the testbed's capabilities by analyzing converter interactions, here with a laboratory-scale HVDC system as an example. State-of-the-art mechanisms like activating DC choppers during AC faults to limit the increase of the DC voltage could be validated properly. Still, setting up a feasible PHiL simulation is complex, and especially accuracy and stability of the setup have to be analyzed.

The last use case depicted has shown that the laboratory environment will be utilized to validate modern control concepts, like the CVE algorithm to increase robustness and reliability of MMCs especially in ship energy systems. The implementation of such an algorithm with actual hardware and constrained access, for example to signals to be measured, can significantly contribute to improving robustness of the algorithm itself, among validating its performance in general.

In conclusion, the sufficiency of the MMC-based DC system with RCP and PHiL simulation capabilities for increasing TRL has been proven with different applications, and further progress will be made emphasizing on control and protection concepts for future DC systems.

References

[1] European Commission, "Technology readiness levels (trl)," *Horizon 2020 - Work Programme 2014 - 2015. General Annexes. Annex G*, 2015.

[2] M. Wang, T. An, H. Ergun, Y. Lan, B. Andersen, *et al.*, "Review and outlook of hvdc grids as backbone of transmission system," *CSEE Journal of Power and Energy Systems*, vol. 7, no. 4, pp. 797–810, 2021. DOI: 10.17775/CSEEJPES.2020.04890.

[3] F. S. Al-Ismail, "Dc microgrid planning, operation, and control: A comprehensive review," *IEEE Access*, vol. 9, pp. 36154–36172, 2021. DOI: 10.1109/ACCESS.2021.3062840.

[4] A. Barzkar and M. Ghassemi, "Electric power systems in more and all electric aircraft: A review," *IEEE Access*, vol. 8, pp. 169314–169332, 2020. DOI: 10.1109/ACCESS.2020.3024168.

[5] L. Xu, J. Guerrero, A. Lashab, B. Wei, N. Bazmohammadi, *et al.*, "A review of dc shipboard microgrids—part i: Power architectures, energy storage, and power converters," *IEEE Transactions on Power Electronics*, vol. 37, no. 5, pp. 5155–5172, 2022. DOI: 10.1109/TPEL.2021.3128417.

[6] E. Talon Louokdom, S. Gavin, D. Siemaszko, F. Biya-Motto, B. Essimbi Zobo, *et al.*, "Small-scale modular multilevel converter for multi-terminal dc networks applications: System control validation," *Energies*, vol. 11, no. 7, p. 1690, 2018. DOI: 10.3390/en11071690.

[7] X. Guillaud, M. O. Faruque, A. Teninge, A. H. Hariri, L. Vanfretti, *et al.*, "Applications of real-time simulation technologies in power and energy systems," *IEEE Power and Energy Technology Systems Journal*, vol. 2, no. 3, pp. 103–115, 2015. DOI: 10.1109/JPETS.2015.2445296.

[8] A. von Jouanne, E. Agamloh, and A. Yokochi, "Power hardware-in-the-loop (phil): A review to advance smart inverter-based grid-edge solutions," *Energies*, vol. 16, no. 2, 2023. DOI: 10.3390/en16020916.

[9] M. R. Lotz, "Insights on laboratory-scale dc system validation with phil and rcp," *Real-Time Simulation Workshop at Karlsruhe Institute of Technology*, 2023.

[10] Cigré Working Group B4.57, "Guide for the development of models for hvdc converters in a hvdc grid," 2014.

[11] F. Martinez-Rodrigo, D. Ramirez, A. Rey-Boue, S. de Pablo, and L. Herrero-de Lucas, "Modular multilevel converters: Control and applications," *Energies*, vol. 10, no. 11, p. 1709, 2017. DOI: 10.3390/en10111709.

[12] Z. Yan, H. Xue-hao, T. Guang-fu, and H. Zhi-yuan, "A study on mmc model and its current control strategies," in *The 2nd International Symposium on Power Electronics for Distributed Generation Systems*, IEEE, 2010, pp. 259–264. DOI: 10.1109/PEDG.2010.5545924.

[13] VDE-AR-N 4110:2023-11, *Technical requirements for the connection and operation of customer installations to the medium voltage network*, 2023.

[14] J. Peralta, H. Saad, S. Dennetiere, J. Mahseredjian, and S. Nguefeu, "Detailed and averaged models for a 401-level mmc–hvdc system," *IEEE Transactions on Power Delivery*, vol. 27, no. 3, pp. 1501–1508, 2012. DOI: 10.1109/TPWRD.2012.2188911.

[15] K. Sharifabadi, L. Harnefors, and H.-P. Nee, *Design, Control and Application of Modular Multilevel Converters for HVDC Transmission Systems* (Wiley - IEEE), 1. Aufl. s.l.: Wiley-IEEE Press, 2016.

[16] OPAL-RT, *Modular multilevel converter (mmc) test bench for rapid control prototyping and laboratory research*, 2024.

[17] Hammond Power Solutions, "Encapsulated distribution transformers: Section 8," 2024.

[18] ZES ZIMMER Electronic Systems GmbH, "Instrument family Img600: User manual," 2019.

[19] VDE-AR-N 4131:2019-03, *Technical requirements for grid connection of high voltage direct current systems and direct current-connected power park modules (tar hvdc)*, 2023.

[20] C. Nentwig, J. Haubrock, R. H. Renner, and D. van Hertem, "Application of dc choppers in hvdc grids," in *2016 IEEE International Energy Conference (ENERGYCON)*, IEEE, 2016, pp. 1–5. DOI: 10.1109/ENERGYCON.2016.7513951.

[21] L. Kou, Y. Gao, H. Gu, Y. Gao, J. Li, and F. Li, "A generic real-time simulation modeling method of modular dc chopper for vsc-based offshore wind farm connection system," in *2023 7th International Conference on Power and Energy Engineering (ICPEE)*, IEEE, 2023, pp. 323–328. DOI: 10.1109/ICPEE60001.2023.10453764.

[22] E. Ndoh, S. Byeon, M. Lotz, and S. Ehlers, "An improvement to analytical capacitor voltage estimation for modular multilevel converters," *4th International Conference on Modelling and Optimisation of Ship Energy Systems*, 2023.

PCIM Europe 2024, 11– 13 June 2024, Nuremberg   DOI: 10.30420/566262198

Film Capacitor Standard Series Digitalization: Electromagnetic & Thermal Modeling implementation in CLARA Web Tool.

Fernando Auñón[1], Fernando Rodríguez[1], Sergio Sepúlveda[1], David Olalla[2]
[1] TDK Electronic Components SAU, Spain
[2] TDK Electronics AG, Germany

Corresponding author: Fernando Auñón, fernando.aunon@tdk.com
Speaker: Fernando Auñón, fernando.aunon@tdk.com

Abstract

CLARA (Capacitor Life And Rating Application) is a web tool available on TDK Electronics' website. You can find it under Design Support > Design Tools > Film Capacitors. This tool digitalizes TDK's standard film capacitor series using electromagnetic and thermal modeling supported by simulation and confirmed by corresponding tests. It provides detailed information about the behavior of capacitors in a wide range of conditions, even those that are tough to evaluate. Electromagnetic modeling is the basis of this tool and involves the electromagnetic analysis of capacitors, the calculation of the 3D loss matrix, and virtual characterizations. The primary aim of CLARA is to provide users with comprehensive insights into the performance of capacitors under a wide array of operational conditions, including those that are traditionally challenging to assess. The foundation of this digitalization effort is a sophisticated electromagnetic analysis that allows for the detailed calculation of a three-dimensional loss matrix and facilitates the virtual characterization of capacitors. The outcome of this process is the creation of equivalent SPICE models, which are invaluable for conducting power converter simulations in various electronic design scenarios. Moreover, electromagnetic simulations play a crucial role in augmenting thermal modeling efforts, offering a realistic portrayal of loss distribution across different conditions. This synergy between electromagnetic and thermal simulations, underpinned by empirical testing, culminates in the development of CLARA. Its in-depth insights empower designers to accurately predict the thermal behavior of various capacitor types when deployed within specific application contexts.

1 Introduction to the digitalization of film capacitor standard series

1.1 CLARA: Capacitor Life And Rating Application

Capacitor Life And Rating Application (CLARA) is an online web tool for engineers to help select film capacitors based on the final application conditions. Currently, the tool includes PCB-mount and some screw-type film capacitors, and more capacitor types will be added soon.

CLARA is accessible through TDK Electronics' website in the Design Tools section. It is available for the standard product series of film capacitors within TDK's product line, whereby "standard series" refers to entirely standard, off-the-shelf products as specified in the respective datasheets.

CLARA is supporting engineers to search and identify film capacitors candidates for their power electronics design (PCB mounting and some screw options). Engineers can simulate up to four components under the desired application conditions. This reduces the time needed to select, test, and evaluate components that might otherwise be discarded due to unsuitability.

It includes a detailed parametric search: By dimensions, volume, reference standard, application, circuit position, etc. So, for example, a design engineer can search for a specific range of capacitance, voltage, current, and maximum dimensions that comply with his application requirements in only a few seconds. In the search results, clicking on the part number provides detailed information: electrical, mechanical, and reliability ratings; ESR vs. frequency curves; applicable SPICE models for the time and frequency domain; or 3D files for mechanical design considerations.

This tool is not only a detailed product database, but it also provides engineers with simulation tools to estimate the behavior of the capacitor under different operation conditions. The two main simulation modules in the tool are:

- Useful Life Simulator: This module calculates the lifetime parameters under different load conditions. The operating temperature is provided, along with safety indexes in voltages (DC and AC), current, and an estimation of the service life.
- Capacitor Banks [Parallel] Simulator: This module, detailed in section 3.2, estimates the average surface temperature of an array of PCB-mounted capacitors under certain electrical conditions and with forced convection with a specified air velocity.

1.2 Digitalization principles

With the increasing importance of Industry 4.0 and the creation of digital twins, there is a need to provide more information about electronic components.

Power electronic design projects require the joint operation of multiple components to provide a reliable and efficient product. Passive components play a significant role in electronic systems.

The usage of capacitors in power converters is varied: DC-link capacitors decouple the variable loads required by the power stages from the DC supply [1], input and output filtering capacitors remove noise from input/output power signals, EMC capacitors reduce electromagnetic interferences between components, etc.

1.3 Methodology

TDK adopts a methodology for digitalization that utilizes Finite Element Analysis (FEA) simulation capabilities, which facilitates faster processes and provides customers with product-specific information. To ensure the accuracy of models and simulations, a "Test - Simulation - Model" approach is applied based on the following principles:

- Test: Representative samples of the standard series are selected for testing. Electromagnetic or thermal (or both) tests and characterizations are performed under controlled conditions to ensure repeatability.
- Simulation: Comparative simulations are run under the same conditions as the tests. Results are cross-checked and the simulation model is adjusted to match testing results.
- Model: After the simulations are completed, a model is developed for the whole series, based on the application conditions and the required use.

1.4 Digitalization workflow

Figure 1 shows the workflow adopted to digitalize the capacitor standard series. This workflow is also applied to customized products and projects when the design team requires the support of simulations, or when a request is received from a customer.

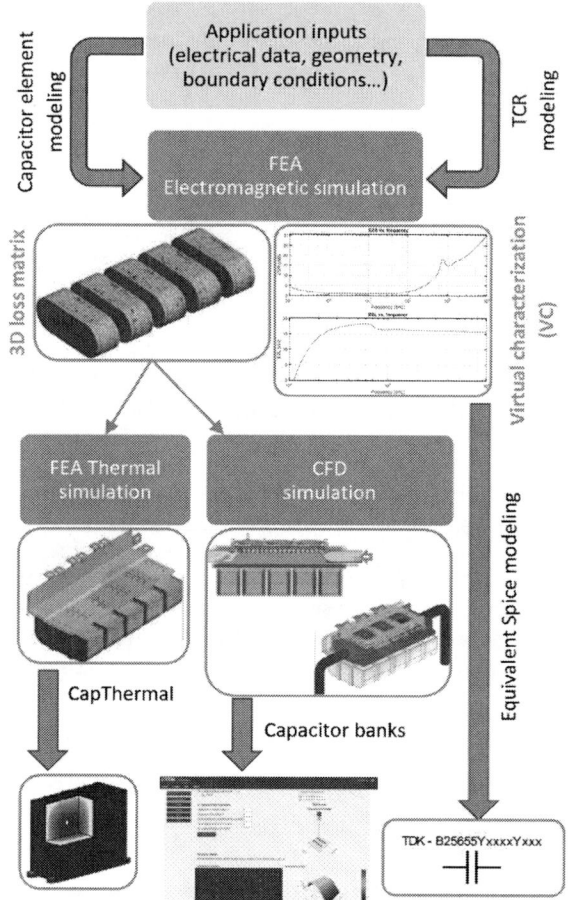

Fig. 1   Digitalization flowchart

The whole process starts with the input information: geometry, and conditions in the application (voltage, current, ambient temperature, etc.). With this information, **electromagnetic simulations** are run to obtain a virtual characterization *(VC)* of the capacitor (see section 2.1) and a loss distribution based on the electrical input (*3D loss matrix*).

The electromagnetic SPICE model is generated from these virtual characterizations, allowing design engineers to simulate the behavior of the capacitor within the overall converter simulation.

From the 3D loss matrix, **FEA thermal simulations** and **CFD (Computational Fluid Dynamics) simulations** are run, considering the complete loss distribution inside the capacitor and the boundary conditions, which play a crucial role in the system's behavior [2].

Finally, after results from thermal simulations are available, the different *thermal models* are developed. Thermal modeling is highly adapted to the final application of each product series and will provide the most relevant information in each case. *Capacitor Banks*, available within the CLARA tool, allows to estimate the self-heating of a matrix of m-by-n capacitors (PCB mounted) connected in parallel with a perpendicular air flow, for larger power capacitors, like ModCap and MKP standard series, the *CapThermal* tool was created to obtain a thermal map of a single capacitor under specific conditions faster than by simulation.

2 Electromagnetic modeling

Electromagnetic modeling is one of the fundamental pillars of the digitalization process. As shown in the workflow in Figure 1, the electromagnetic and thermal models are based on previous results from electromagnetic simulations.

Thermal simulation and modeling can be considered "straightforward" – thermal simulations are pretty standard under steady-state conditions and considering uniform convection conditions. However, electromagnetic simulations require a further understanding of the electromagnetic effects inside the capacitor.

This in-depth understanding of the behavior of the capacitive elements in the component allows complete electromagnetic simulations to be performed, including both the capacitive elements and the conductors.

2.1 Virtual characterization

The most direct output from electromagnetic simulations is virtual characterization, a complete and precise characterization of the capacitor's main electrical parameters.

With virtual characterizations, all the electrical parameters that could be measured by a traditional component analyzer (|Z|, C, ESR vs. frequency, ESL, etc.) can be obtained with the help of electromagnetic simulation software.

Figure 2 compares the ESR and ESL curves obtained by traditional measurement using a component analyzer and virtual characterization of a DC-link solution for an xEV inverter.

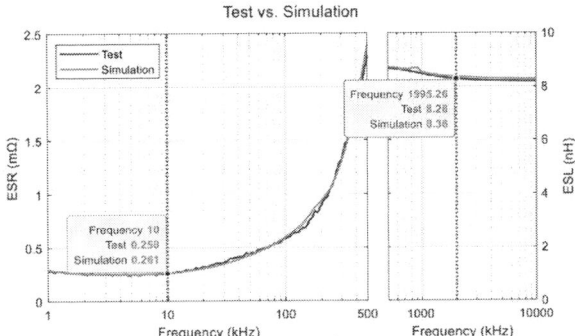

Fig. 2 Examples of ESR and ESL curves obtained by simulation (virtual characterization) and test

As shown in this figure, the electrical behavior of the system is accurately predicted by the virtual characterization. One of the main advantages of virtual characterization is the fast extraction of the electrical parameters in capacitors without the need for the time-consuming and resource-intensive process of manufacturing samples.

Additionally, virtual characterization can consider different manufacturing process tolerances and stages in the product's lifetime: film profile dimension tolerances, film metallization tolerances, EOL (End-of-life) characteristics, etc. This can help designers determine the system's behavior under nominal and worst-case conditions.

2.2 TCR modeling

Virtual characterization can also consider the TCR (*Thermal Coefficient of Resistance*) in addition to the calculation of the electrical behavior of a sample as if measured with an impedance analyzer.

The TCR represents the reduced electrical conductivity in metals when the temperature increases. While this effect is well known in the literature for the conductors inside the capacitor — they can be considered bulk materials —the main contributor to a film capacitor's ESR is the metallization film's square resistance. The value of the TCR for such thin metalized films is different from the value for the bulk resistivity of the material [3].

For this reason, TDK has characterized the TCR in the different elements of a capacitor to model the effect of temperature in virtual characterizations correctly. The impact of the TCR will differ depending on the various applications (DC-links, filters, PCB-mounted capacitors, etc.). This is due to the intrinsic characteristics of each application: operation temperatures, maximum currents, construction differences, etc.

Figure 3 shows representative values of the effect in total thermal losses generated for different film capacitor technologies.

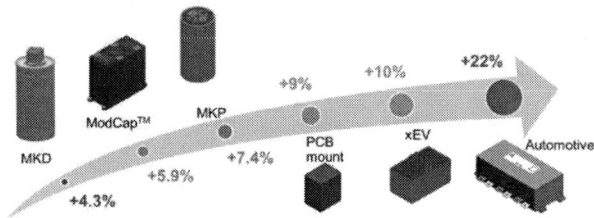

Fig. 3 Increase in losses caused by the TCR, considering representative application conditions (current spectrums) from ambient temperature to representative application temperatures

2.3 SPICE equivalent modeling

SPICE is an excellent tool for electronic designers to predict the behavior of their systems. The most accurate models of each system component are needed to make a reliable prediction. However, there is always a compromise between performance and computation time. The usual equivalent circuit of real capacitors is simplified to a single equivalent series resistance and inductance paired with the isolation resistance [3].

Equivalent SPICE models developed at TDK allow you to model the complete electrical behavior of the components. These models are composed of passive elements arranged to have the same electrical behavior as the real capacitor in both the time and frequency domains. The number of elements required to model the capacitor behavior has been selected to achieve a compromise between accuracy – more elements yield more accurate results – and computational requirements – more elements require more resources. The equivalent SPICE models are based on a mathematical model following capacitor electrical characteristics. This equivalent modeling is done for single capacitors. Still, it can also be obtained for complete DC-link systems, for assemblies of capacitors on a PCB (also considering the PCB behavior), and for different subsystem configurations.

Figure 4 shows a simplified schematic of the equivalent model implemented for TDK film capacitors.

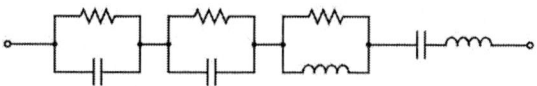

Fig. 4 Simplified capacitor model

The capacitor model behaves as the actual element in the time and frequency domain with this structure. Following the example shown in Figure 2, the SPICE equivalent model for the DC-link has been obtained and is shown in Fig. 5.

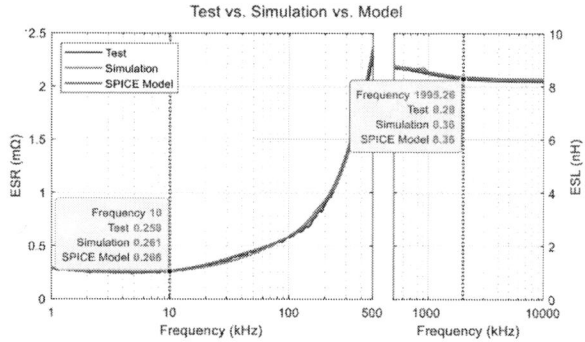

Fig. 5 Comparison of ESR and ESL curves between measurement, simulation, and equivalent spice modeling

TDK generates SPICE-equivalent models for all film capacitor standard series and uploads them to different product websites. They are also available on the CLARA tool. Models for customized products can be generated on demand. With these models, design engineers can directly incorporate the behavior of TDK capacitors in their electronic converter simulations.

2.4 Electromagnetic interactions

Another significant advantage of using electromagnetic simulation in film capacitors is the ability to detect and analyze electromagnetic interactions. These effects that can be seen using EM software are inhomogeneous impedances, skin effects, internal resonances, etc.

As an example, consider the effects of internal resonance. The subject of this analysis is a cylindrical capacitor with two capacitive elements, as shown in Fig. 6.

Fig. 6 Structure of the analyzed capacitor. The two capacitive elements are marked in blue (bottom) and brown (top)

The ESR was obtained using electromagnetic simulation and a frequency analysis. A total RMS current of 60 A was injected through the sample's terminals, and the current flowing through each capacitive element was measured.

Figure 7 shows the ESR curve and the current analysis results.

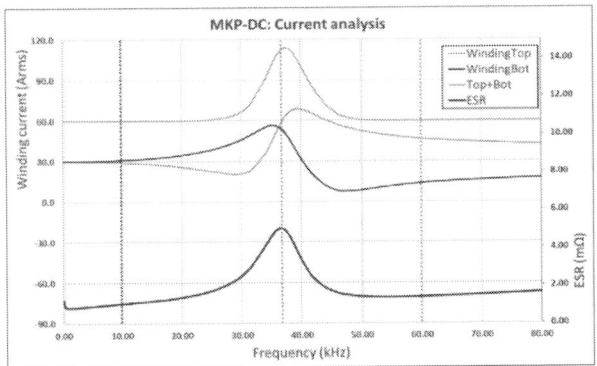

Fig. 7 ESR curve (bottom) and current analysis (top) of the analyzed capacitor

As shown in the figure, the peak in the ESR curve corresponds to a point where the capacitor is in resonance, and the total current flowing through the two capacitive elements (green curve) is higher than the injected current.

3 Thermal modeling

Thermal modeling follows the same philosophy as electromagnetic modeling described in 1.2: first, the test, then the simulation, and finally, the **model**. With these steps, the accuracy of simulations and models is validated through empirical tests.

In many cases, the conditions under which the capacitor operates can't be easily replicated in thermal testing. For this reason, thermal simulations are necessary to analyze and evaluate these elements. In these specific cases, thermal testing is conducted under conditions as close as possible to the actual application. Then, after the (simulation) model is adjusted, a thermal simulation is run under the specified boundary conditions.

Additionally, thermal simulations and models can be used to evaluate temperatures and the thermal behavior of specific points that are not easily measured.

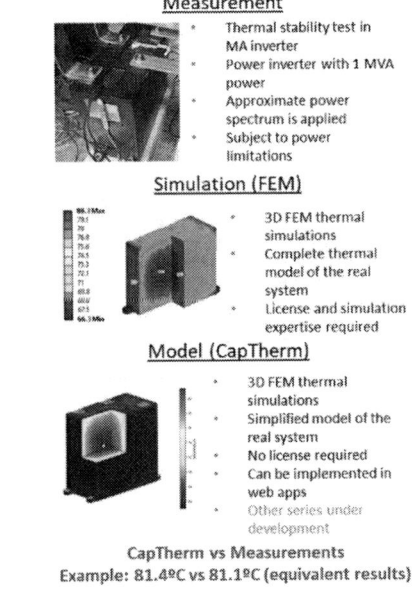

Fig. 8 Thermal test – simulation – model example

3.1 Thermal simulations supported by electromagnetic simulations

To correctly predict the capacitor behavior, thermal simulations must be supported by the corresponding electromagnetic simulation. This link between the electromagnetic and thermal simulation allows the loss distribution in each system element to be determined, not only the heat generation value (in watts) but also its distribution within the component.

Following the example introduced in section 2.4, where a cylindrical capacitor with two capacitive elements is analyzed, the thermal behavior of the sample can be studied based on the electromagnetic analysis. Considering the same boundary conditions (ambient temperature, terminals temperature, and current input), the capacitor has been simulated thermally at three different frequencies: 10 kHz (low frequency), 37 kHz (at the peak of the ESR curve), and 60 kHz.

Figure 9 shows the thermal results of the three simulations. This capacitor has a resonance at

37 kHz, the point at which the losses are dramatically increased (three times more than the others) with the same current input through the terminals.

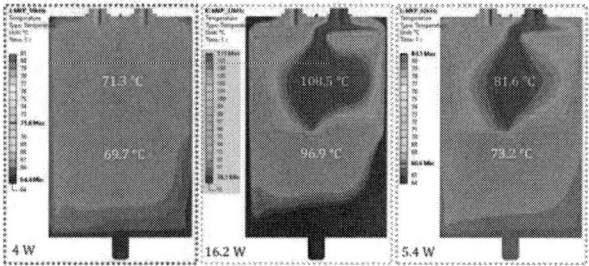

Fig. 9 Thermal simulation results for the analyzed capacitor at 10 kHz (left), 37 kHz (middle) and 60 kHz (right)

These temperature distribution results are only possible thanks to the electromagnetic simulation, from which the 3D loss matrix has been obtained, and it provides the thermal simulation software with a mapping of the loss distribution inside the analyzed sample.

3.2 Capacitor banks online thermal module

One of the developed thermal modeling tools is the so-called *"Capacitor Bank [Parallel] Simulator."* This application arises from the need to estimate the self-heating of film capacitors (for PCB mounting) when assembled in an array with forced cooling by a fan. This application is publicly available as a CLARA module on the TDK Electronics website under Design Tools—Film Capacitors.

This tool is based on a series of CFD simulations, supported by empirical tests, which consider different parameters:

- Capacitor characteristics (dimensions, electrical parameters)
- Matrix arrangement (number of capacitor rows and columns)
- Distanced between rows and columns (fixed for both directions)
- Mission profile (voltage and current spectrum)
- Airspeed
- Temperature difference between ambient and terminals

With this input information, the tool can estimate the average temperature rise of the surface of each capacitor inside the array. Figure 10 shows the results obtained for a 2-by-4 array of capacitors under certain conditions.

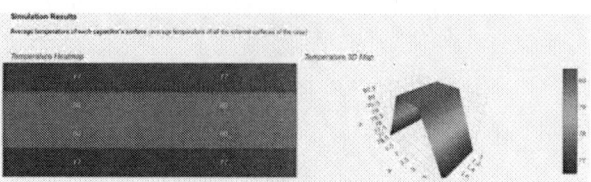

Fig. 10 Capacitor banks simulation result

3.3 CapThermal thermal software

CapThermal was born as a design tool based on modeling the thermal behavior of capacitors. Using boundary conditions (ambient temperature, convection condition, terminal temperature, etc.) and operating conditions (losses, spectrum, etc.), the thermal distribution inside the capacitors is calculated. This is an essential advantage because the user knows the hotspot temperature and the entire thermal distribution. With this information, the user can decide which capacitor is better suited to the specific application and determine the best conditions for it or how to maximize its use. The amount of information CapThermal provides is more significant and more realistic than theoretical calculations. Furthermore, it is easier than traditional simulation software, making simulations accessible to everyone.

In the end, CapThermal is a simulation emulator. Based on the inputs, it uses an algorithm (like any FEA software) to output the thermal distribution result.

Of course, as with any thermal model, it has been validated through accurate tests and simulations. The CapThermal simulation tool will soon be available on the TDK product website.

Fig. 11 CapThermal example simulation result

3.4 Thermal Integration: from components to subsystems

In cases where the application conditions for the capacitors are more restrictive, such as in automotive applications, more than the traditional approach of thermal simulation is required. In these applications, the constraints play a crucial role in the thermal behavior of the capacitor: cooling method and position, connection to the semiconductor modules, etc.

Under these circumstances, the new simulation trends focus on **thermal integration.** This is based on the need to consider a higher-level subsystem to accurately determine the operating conditions of the analyzed component, in this case, the capacitor.

Taking an automotive application as an example, the *minimum subsystem* to correctly evaluate the DC-link capacitor would be comprised of the DC-link capacitor itself, the semiconductor modules, and the cooling unit (including coolant flow).

Figure 12 shows the temperature of a DC-link system simulated using CFD, including the forced cooling unit and the semiconductor modules.

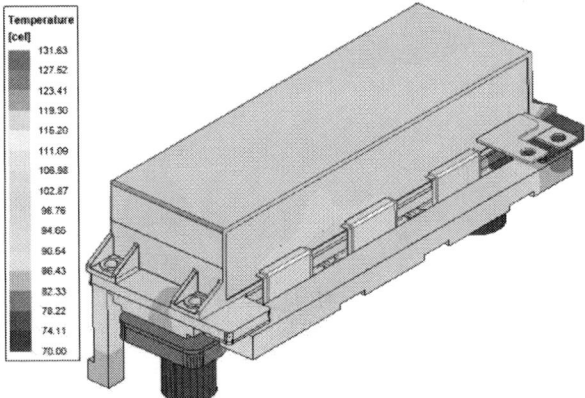

Fig. 12 Thermal integration simulation

The main advantage of including the components in direct contact with the DC-link capacitor in the simulation is that the heat transfer between each element is better represented.

Usually, the capacitor terminals temperature are fixed at the semiconductor module connection. However, the temperature is not set if these modules are included in the simulation. Instead, the temperature at the capacitor terminals reaches its steady state when the capacitor and the power modules are in thermal equilibrium.

Additionally, when considering the fluid flow in the cooler module, the temperature of the contact surface of the capacitor with the cooler is not fixed. Therefore, the heat transfer is better estimated.

4 Capacitor Standard Series Digitalization

At TDK, a comprehensive digitalization initiative is underway to transform all standard product series into digitally accessible information, enhancing the resources available to electronic designers during their design process. This initiative involves the creation and distribution of SPICE equivalent models for both standard and, upon request, customized products. These models are meticulously crafted and made readily available to facilitate a more informed design process.

These resources are designed to offer precise estimations of how TDK's diverse capacitor solutions will perform under the unique conditions specified by their intended applications. Among the resources currently accessible on the CLARA website are the Capacitor Banks and SPICE equivalent models, with plans to introduce further tools such as CapThermal soon.

The digitalization methodologies employed, encompassing electromagnetic simulations, SPICE equivalent modeling, thermal simulations, and thermal modeling, extend beyond individual products to encompass higher-level system integrations. This includes comprehensive solutions such as complete DC-link systems, PCB assemblies incorporating various capacitor components, and thermal management systems featuring cooling solutions. These techniques underscore TDK's commitment to offering advanced digital tools that can significantly enhance the design and implementation of electronic systems. For further information about simulation capabilities and support on specific projects, please contact CAPSimulations@tdk.com.

References

[1] S. Chowdhury, E. Gurpinar and B. Ozpineci, "Capacitor Technologies: Characterization, Selection, and Packaging for Next-Generation Power Electronics Applications," in IEEE Transactions on Transportation Electrification, vol. 8, no. 2, pp. 2710-2720, June 2022, doi: 10.1109/TTE.2021.3139806.

[2] H. Wang, R. Zhu, H. Wang, M. Liserre and F. Blaabjerg, "A Thermal Modeling Method Considering Ambient Temperature Dynamics,"

in *IEEE Transactions on Power Electronics*, vol. 35, no. 1, pp. 6-9, Jan. 2020, doi: 10.1109/TPEL.2019.2924723.

[3] V V R Narashimha Rao, S Mohan, and P J Reddy "Electrical resistivity, TCR and thermo-electric power of annealed thin copper films," in Journal of Physics D: Applied Physics, vol. 9, no. 1, 1976, doi: 10.1088/0022-3727/9/1/015

[4] M. F. Staniloiu, H. S. Popescu, B. I. Glod, and M. Iordache, "SPICE model of a real capacitor: Capacitive feature analysis with voltage variation," *2020 International Conference and Exposition on Electrical And Power Engineering (EPE)*, Iasi, Romania, 2020, pp. 333-338, doi: 10.1109/EPE50722.2020.9305554.

Accuracy Evaluation and Proposed Dynamic Tuning Procedure of a Compact SiC SPICE Model

Brian DeBoi[1], Blake Nelson[1], Austin Curbow[1]

[1] Wolfspeed, USA

Corresponding author: Brian DeBoi, Brian.DeBoi@Wolfspeed.com
Speaker: Austin Curbow, Austin.Curbow@Wolfspeed.com

Abstract

The commercialization of wide bandgap technology has increased demand for accurate circuit-level simulation models of devices such as Silicon-Carbide (SiC) MOSFET power modules. These models assist with design challenges such as minimizing overshoot and electromagnetic interference associated with wide bandgap switching speeds. However, it is challenging to create highly accurate models across all the conditions that the device may be operated in, such as varying gate resistance, temperature, operating voltage, and operating current. A major contributor to this problem is that the procedure for characterizing and modeling SiC MOSFETs is simplistic relative to their complexity in switching. In particular, the standard methods for characterizing and modeling the device capacitances are simplified, ignoring dependencies on voltage biases and frequency. However, increasing the model complexity to address these issues is generally not feasible because 1) the models must be efficient to converge and simulate quickly, and 2) the necessary characteristics are often impossible to measure. Rather than increasing the model complexity, this paper builds upon a previously presented compact behavioral model and applies a dynamic tuning procedure to improve alignment with empirically derived datasheet parameters. The procedure is applied to a half-bridge SiC MOSFET power module model and it is demonstrated that the overall dynamic accuracy of the model is increased by 50% across a wide range of double pulse test conditions. While this procedure further divorces the behavior model from physical reality, the tradeoff is acceptable given the purpose of this model: accurately predicting device behavior in application while minimizing computational complexity.

1 Introduction

Improvements to wide bandgap (WBG) semiconductors, such as Silicon-Carbide (SiC) MOSFETs, have increased the efficiency and power density of converter applications [1]. Multi-chip power modules (MCPMs) further these performance advantages by offering easy-to-implement packages that are optimized for their high edge rates and power levels. The performance advantages are enabling growth in new technologies, such as long-range electric vehicles [2]. Furthermore, despite higher up-front cost for the semiconductors, the reduction in the necessary passive components often yields an overall reduction in cost.

In order to take full advantage of the capabilities of WBG power devices, designers must optimize their system layout to mitigate the drawbacks of fast edge rates. Typically, this requires improving packaging, passive components, and system layout, which incurs additional development costs and presents new challenges for designers unfamiliar with WBG power electronics [2]. Simulation tools offer a method to reduce time and cost through virtual prototyping studies that inform hardware decisions or facilitate optimization. Compact behavioral models implemented in circuit-level solvers (such as SPICE) are particularly useful for optimization as they can predict the system behavior at specified conditions and can evaluate shifts in behavior caused by parasitic elements.

The SPICE models necessary for these simulations are usually provided by device manufacturers. These models are generated by measuring the "static" characteristics of the device and fitting parameters in equations to the measured data. Here, "static" refer to characteristics that are obtained during quasi-steady-state operation, such as the conduction characteristics, small-signal device capacitances, gate resistance, and parasitic elements. The models fit to these input characteristics are then used to predict how the device behaves when switched in a full system. However, this presents some fundamental issues with the approach. First, it assumes that these device characteristics are adequate for predicting switching, but the behavior of a MOSFET during switching is much more complex than what is observed during quasi-steady-state measurements. Second, the input dataset is substantially smaller than the number of conditions that the model may be evaluated in. It is unreasonable to expect that a model generated with static characteristics will accurately predict device behavior when switched at any arbitrary set of conditions.

These issues compound with the fact that the provided models must simultaneously have good convergence behavior, be computationally efficient, and accurately predict the device behavior in general circuits. However, this creates a tradeoff: increasing the accuracy requires a more complex model, but in turn a more complex model reduces the efficiency. More complex models also require larger input datasets, increasing the amount of testing required per part. Thus, the implementation of these models must be simple to ensure that they are able to be created and simulated.

Because it is unrealistic to develop a general model that can accurately predict switching for any arbitrary condition for all devices, another approach is to use the switching characteristics to tune the parameters of the device model. This paper presents such a dynamic tuning method that can improve its accuracy without increasing its complexity. The dynamic tuning procedure is applied to a CAB016M12FM3 half-bridge SiC MCPM (shown in Fig. 1) and demonstrates how the accuracy and utility of a model can be significantly improved by applying the approach without increasing the model complexity.

2   Target Model Overview

When selecting a SPICE model for dynamic tuning, it is important to consider the implementation of the SiC MOSFET. In general, MOSFET die-level models can be separated into three distinct categories: physics, semi-physics, and behavioral. Physics-based MOSFET models leverage the fundamental physical parameters of a device to predict its behavior and are generally the most complex [3], [4]. Semi-physics models also leverage device physics equations, but adjustments to the equations are made to improve performance at the cost of physical accuracy [5], [6]. Behavioral models are distinct from physics-based models in that they are decoupled from the physics of the device and can therefore be optimized for simplicity, computational efficiency, or accuracy [7], [8]. Behavioral models follow a "black-box" approach, where the implementation is arbitrary with intent only to provide the desired output. For dynamic tuning, this arbitrary design of behavioral models provides

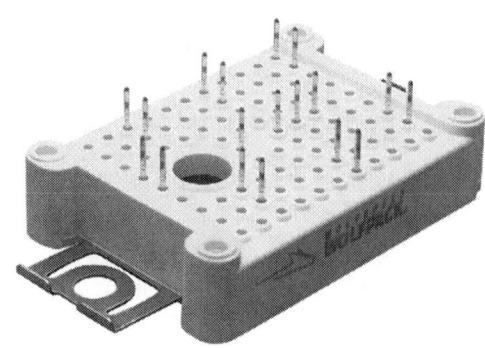

Fig. 1   Test subject for dynamic tuning procedure - CAB016M12FM3 power module

an advantage because the equations and parameters can be adjusted freely to improve the accuracy of the output. In addition, equations and parameters within behavioral models are usually specific to one aspect of behavior, and thus can be adjusted without affecting other model behaviors.

This paper applies the dynamic tuning approach to a previously presented robust and efficient behavioral compact SiC MCPM SPICE model [9]. The general architecture of the module model is shown in Fig. 2. The model is segmented into a package and die model. The package model connects the die to the module terminals and includes parasitic and thermal elements. The die model includes the electrical characteristics of the MOSFETs themselves. A brief description of the behavior and capabilities of the model is provided in Table 1 and is described in more detail below.

The forward conduction parameters are implemented in a lookup table with dependence on V_{GS}. An $R_{DS(ON)}$ and V_{GS} shift is used to model the temperature dependence. The equations and tables implemented provide excellent agreement with empirical data. The equations also include parameters for fitting the short-channel effects at high-voltage and high-current. These parameters are fit using empirically measured high-voltage high-current static characteristics [10]. This model can accurately predict conduction losses during normal operation and during short-circuit or surge events.

The conduction behavior between first and third quadrant is continuous through the origin, improving convergence and speed. The reverse conduction parameters are also implemented using a lookup table with V_{GS} dependence, and temperature dependence is achieved with a resistance and V_{GS} shift.

The device capacitances are implemented using behavioral current sources, with the capacitance

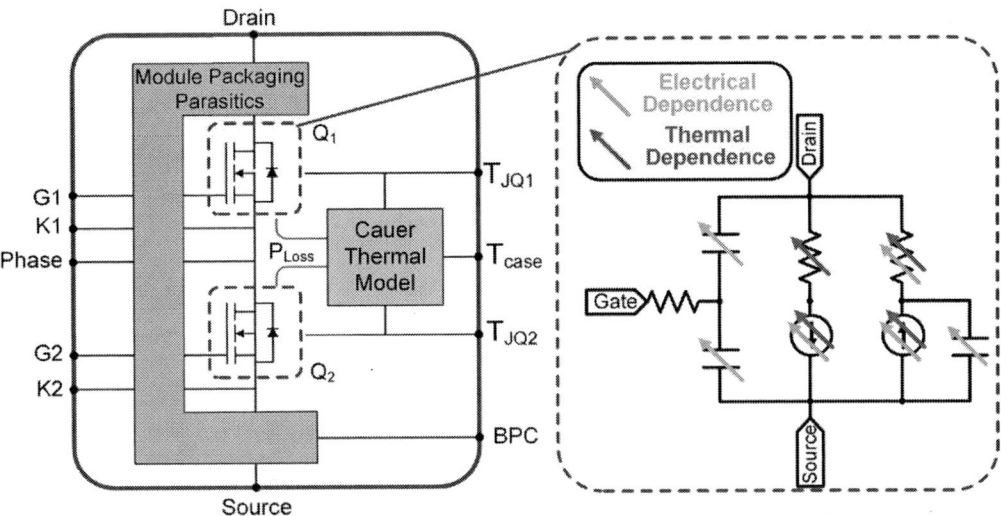

Fig. 2   Half-bridge power module model architecture [3]

at a given voltage bias provided by a lookup table. The capacitance across drain-to-source (C_{DS}) is dependent on the drain-source voltage (V_{DS}), the

Behavior	Model Implementation
Forward Conduction	Full dependence on V_{GS} (0 V – 18 V) and T_J (25°C – 175°C). V_{DS} biases include up to the bus voltage.
Reverse Conduction	Full dependence on V_{GS} (-5 V – 18V) and T_J (25°C – 175°C).
Device Capacitances	C_{GD} dependence on V_{GD}, C_{DS} dependence on V_{DS}, C_{GS} dependence on V_{GS}
Reverse Recovery	Modified lumped-charge reverse recovery model with dependence across temperature, derived from [12] and [13].
Parasitic Network	Per-terminal power terminal inductance, gate loop inductance, per-terminal baseplate capacitances.
Thermal Network	Closed-loop electrothermal behavior with a 3rd-order Cauer network.

Table 1   Brief description of leveraged model behaviour based on work in [3]

capacitance across gate-to-drain (C_{GD}) is dependent on across gate-drain voltage (V_{GD}), and the capacitance across gate-to-source is dependent on the gate-source voltage (V_{GS}). This method is simple to quantify, implement, and is efficient to simulate.

The reverse recovery model is based on a lumped-charge model first presented in [11] and modernized in [12]. The parameters are implemented in a lookup table with dependence on temperature. The reverse recovery parameters are decoupled from the reverse conduction parameters such that they can be tuned independently. This allows for the model to predict changes in reverse recovery charge, losses, and current overshoot across temperature.

The parasitic model includes the partial inductance from each terminal to the die (both power and gate), along with the baseplate capacitance for electromagnetic emission simulations. A 3rd-order Cauer network is included to model temperature changes in the die caused by power losses and heat exchange with the baseplate.

2.1   Determining Model Accuracy

The accuracy of a SPICE model depends on the conditions in which it is being evaluated in simulation. SPICE models can be used to predict various behaviors (i.e. conduction losses, short-circuit survivability, voltage overshoot, switching losses, and reverse recovery losses), and the accuracy of that prediction also depends on the conditions specified in the circuit. In general, the conduction losses are well predicted by the model of the forward and reverse static characteristics. However, the switching characteristics are more complex and more difficult to accurately predict. The transient behavior is also often an important consideration for designers implementing these devices into their systems. For these reasons, the focus in this work is on improving the accuracy of the transient characteristics, such as slew rates, switching losses, and reverse recovery.

There are two common methods for validating the transient characteristics of SPICE models. The first method, known as qualitative model analysis, uses overlain waveforms to observe agreement between the simulation and empirical data. Several models in the literature apply this approach to validate their models at a small number of operating conditions [13], [14]. However, this method has several disadvantages. First, it is a subjective metric that cannot be consistently applied across tests and models. Second, the model accuracy is sensitive to the operating conditions of the circuit; a model that is accurate at one operating condition will not necessarily be accurate at another. It is important to vary the operating conditions include the gate resistance (R_G), temperature (T_j), operating voltage (V_{bus}), and operating current (I_{load}). Because of the number of parameters that can be varied, the test matrix is often very large, and it is unreasonable to apply the method across many tests.

An alternative method, quantitative model analysis, uses a set of clearly defined metrics to quantify the transient characteristics of a DPT waveform. Each parameter is given a specific definition such that it can be calculated consistently through automated means across various operating conditions for both simulation and experimental data. This technique has been applied in the literature, but the number of operating conditions considered are often limited [15], [16], [17].

The quantitative metrics chosen for this analysis are described in Fig. 3. The selected parameters encompass only a small subset of the overall device behavior but describe important features of switching dynamics. In Fig. 3 (a), the V_{DS} and I_{DS} waveforms at turn-off and turn-on are parsed to obtain the slew rates for each transition. In this work, the slew rates are calculated at the average slope between 10% and 90% of the nominal operating point. For example, dv/dt(off) would for a DPT at a bus voltage of 600 V would be calculated between 60 V - 540 V. In Fig. 3 (b), the instantaneous power waveforms are integrated to obtain the switching energies, E_{off} and E_{on}. The switching energies are integrated between 10% V_{DS} and 10% I_{DS}.

Any waveform metric that can be programmatically calculated on a waveform can be considered. Some examples include the reverse recovery energy, peak reverse recovery current, and the turn-

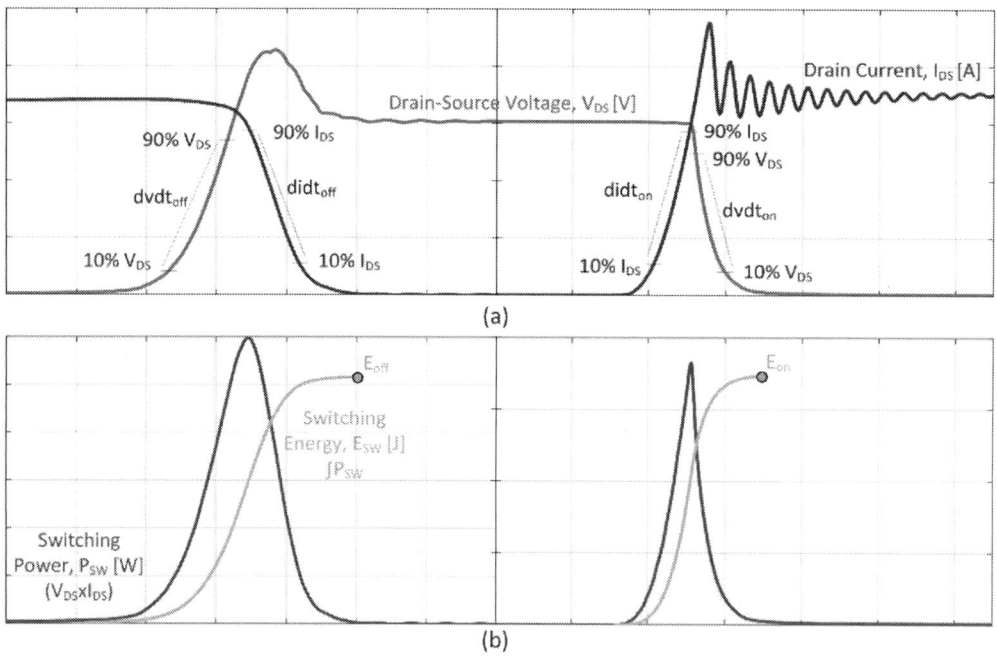

Fig. 3 Description of quantitative model metrics, (a) V_{DS} and I_{DS} waveforms for slew rate calculation, and (b) instantaneous power and integrated energy waveforms for SW energy calculations

off and turn-on delay. These metrics should be selected based on which characteristics are important for the model application.

With the quantitative metrics selected, the method for determining model accuracy is as follows. First, a model of the CIL test circuit is implemented in a SPICE solver software. Second, the simulation is run at a set of conditions that match the empirical data. Third, both the empirical data and the simulation waveforms are analyzed to obtain the quantified values for each metric. Finally, this information is used to generate an error for each metric that is a function of the operating conditions. These error functions can be used to generate heat maps or histograms to visually understand the error or can be combined to generate a total sum of error that can be input into fitting functions.

3 Dynamic Tuning Procedure

The procedure for the dynamic tuning approach is outlined in Fig. 4. For the process described here, MATLAB is used to control, quantify, and tune while and LTspice was selected for model simulation, but any high-level language and SPICE solver could be used.

First, a set of DPT conditions appropriate for the device are selected. The conditions should vary across R_G, T_J, V_{bus}, and I_{load} within the recommended bounds of operation for the part. However, while many tests can be used for accuracy validation, the number of tests used for fitting should be limited to 10-20 conditions. If too many conditions are selected, the required time for tuning will become excessive because the solver must simulate the circuit at all conditions for each iteration.

Second, MATLAB calls LTspice to run the simulation at each specified condition. The LTspice waveforms are processed to obtain the quantitative metrics from Fig. 3. The error for a single DPT operating condition is calculated by equation (1), where *emp* is the quantitative metric value for empirical data and *sim* is the quantitative metric value for simulation data. The error for each metric is summed to obtain the total error for a specific DPT condition. Finally, the total error for each DPT condition is summed to obtain a total error between the simulation and the model. This process can be changed as necessary, such as by applying weights to specific conditions or tests or by using other methods to sum the errors. It is only necessary that the result is a single error value that can be input to a fitting function.

$$Error = abs\left(\frac{emp - sim}{emp}\right) \quad (1)$$

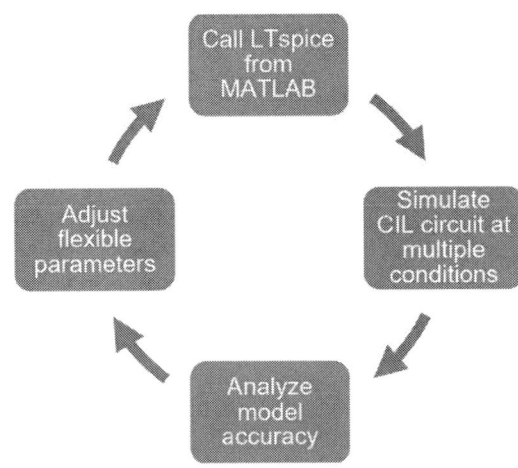

Fig. 4   Dynamic tuning process

Finally, the error term is input into a fitting function that adjusts the model parameters to decrease the error, and the cycle repeats until a certain number of iterations or tolerance is reached. The fit function will balance the adjustment of parameters to minimize the error across all of the provided conditions, resulting in a robust and balanced model that is useful and accurate across the various conditions the model may be used in.

3.1 Selecting Flexible Tuning Parameters

The model parameters that are adjusted by the fitting function should meet a set of criteria. First, the transient behavior should be sensitive to changes in the parameter value; otherwise, the tuning algorithm may change the parameter arbitrarily and cause unintended effects. Second, changing the parameter to improve transient behavior should minimally affect other behaviors of the model. For example, if a parameter affects the on-state resistance, changes to it may increase conduction loss error. Finally, it should be reasonable to change the parameter within certain bounds. If a parameter has a well-defined and strict definition, then it is not suitable for tuning. For example, the commutation inductance of a module package can be well-defined by impedance analysis measurements. Significantly adjusting this parameter in the model can make it non-physical.

Considering the above criteria, the most flexible parameters for tuning are those related the device capacitances of the SiC MOSFET and the internal gate resistance: C_{DS}, C_{GS}, C_{GD}, and R_{Gi}. These parameters do not affect conduction losses and primary purpose in a model is to control the transient behavior. Most importantly, the standard measurement and modeling procedures for these characteristics are oversimplified. They are generally represented as having a single voltage dependence in datasheets and device models. However, it has been shown repeatedly that the device capacitances are highly dependent on both V_{DS} and V_{GS} biases [18], [19], [20], [21], [22]. There is also evidence that they are sensitive to the stimulus frequency at which capacitances are measured [10]. Quantifying a simultaneous V_{DS} and V_{GS} bias is challenging because applying DC current will flow through the device while V_{GS} is above the threshold voltage, leading to device failure. Attempts to solve this issue have been made, such as using a custom pulsed setup that applies a pulsed V_{GS} and V_{DS} bias and performs a small-signal measurement of the capacitances within this window [22] or leveraging physics-based simulation software [20]. Implementing these processes to characterize and model the physical device capacitance behavior is unrealistic for creating SPICE models. Instead, the dynamic tuning procedure described herein can adjust the simplified device capacitance model to match the transient behavior without increasing model complexity.

The internal gate resistance is similar to the device capacitances and has additional complexity that is ignored in modeling. While this parameter is modeled as a lumped resistance, it is actually a distributed resistance that is both frequency and temperature dependent. The transient behavior is also highly sensitive to the internal gate resistance making it suitable for tuning.

The selected tuning parameters and their descriptions are provided in Table 2. The gate-drain capacitance is multiplied by A_{GD} and has an added offset O_{GD}, as shown in equation (2). The C_{GD} term is V_{GD}-dependent, and the scaling terms are applied equally at all biases. O_{GD} can be negative, but is bounded such that $C_{GD,effective}$ is not negative. The gate-source capacitance is scaled by A_{GS} but does not include an added offset, as shown in (3). This is because, while C_{GS} is a largely constant term, the C_{GD} capacitance decreases by several orders of magnitude from low V_{GD} to high V_{GD}. As such, O_{GD} has a larger effect at high V_{GD}, while A_{GD} has a larger effect at low V_{GD}, yielding more tuning control. Finally, the internal gate resistance is scaled by A_{RGI}, as shown in equation (4). Scaling of C_{DS} is omitted because it can cause errors with overpredicting reverse recovery and Q_{OSS} losses.

$$C_{GD,effective} = A_{GD} * (C_{GD}(V_{GD}) + O_{GD}) \quad (2)$$

$$C_{GS,effective} = A_{GS} * C_{GS}(V_{GS}) \quad (3)$$

$$R_{GI,effective} = A_{RGI} * R_{GI} \quad (4)$$

Parameter	Description
A_{GS}	Scalar multiplier to gate-source capacitance equation
A_{GD}	Scalar multiplier to gate-drain capacitance equation
O_{GD}	Added offset to gate-drain capacitance equation
A_{RGI}	Scalar multiplier to internal gate resistance

Table 2   Description of parameters selected for dynamic tuning

3.2 Dynamic Tuning Results

To demonstrate its effectiveness, dynamic tuning procedure was applied to a CAB016M12FM3 power module SPICE model using the metrics described in Fig. 3 and tuning the parameters described in Table 2. For the initial conditions, the scalar parameters were set to 1 and the offset parameter was set to 0 (meaning that the parameters have no effect on the model). The empirical CIL data was collected on a KIT-CRD-CIL12N-FMA Wolfspeed® evaluation kit shown in Fig. 5. Over 1,000 CIL test conditions were evaluated on this setup. However, for tuning, the smaller subset of conditions shown in Table 3 was considered. The

Fig. 5   KIT-CRD-CIL12N-FMA CIL evaluation kit for evaluating CAB016M12FM3 power module

T_J (°C)	R_G (Ω)	V_{bus} (V)	I_{load} (A)
25	0	800	30
25	0	800	90
25	0	800	150
25	3	800	30
25	3	800	90
25	3	800	150
25	10	800	30
25	10	800	90
25	10	800	150
150	4	800	30
150	4	800	90
150	4	800	150

Table 3 DPT conditions selected for dynamic tuning

selected conditions consider a wide range of gate resistance, load current, and temperature conditions. A single bus voltage is used because most module operation will occur at 800 V. More conditions can be included but doing so increases the tuning time.

As mentioned previously, the error between the simulated DPT and the empirical measurements for each of the quantified metrics in Fig. 3 is calculated for each condition in Table 3 to yield the total model error. The mean error across all conditions for each metric before and after tuning is shown in Fig. 6. For this model, each metric shows a significant reduction in error. For example, the error in dvdt$_{Off}$ decreased from 48.9% to 18.5%, and the error for E$_{Off}$ decreased from 26.4 % to 12.9 %. To achieve this reduction in error, A$_{GS}$ decreased to 0.68, A$_{GD}$ increased to 1.47, O$_{GD}$ increased to 5.8 pF, and A$_{RGI}$ decreased to 0.84.

It should be noted that Fig. 6 represents only a small subset of the overall dataset and potential operating conditions. Thus, an additional heat map that compares the empirical DPT data and SPICE simulation model across a wider set of conditions is provided in Fig. 7. Each subplot shows the error between the empirical data and the simulation model for one of the six parameters defined in Fig. 3. The color of the surface at each X and Y coordinate indicates the error at an R$_G$ and I$_{load}$ condition for V$_{bus}$ = 800 V and T$_j$ = 25°C. Warm colors indicate more error, and cool colors indicate less error. Before tuning in Fig. 7 (a), the model error is quite high, especially for the turn-on event. The turn-on loss error exceeds 100% in many cases and is no lower than 40%; dvdt$_{On}$ exceeds 50% at

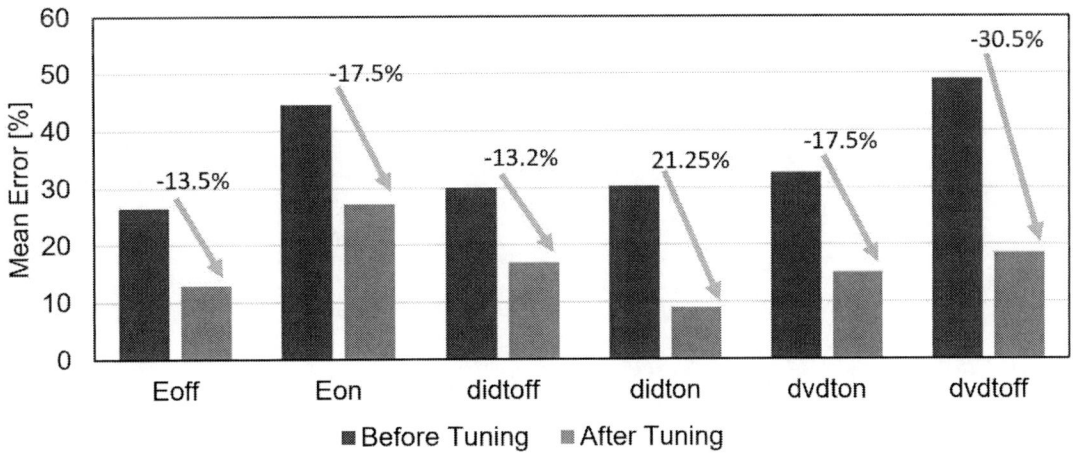

Fig. 6 Reduction of individual error metrics before and after tuning

most conditions, and the $didt_{On}$ is no lower than 40% at any condition.

After applying the dynamic tuning procedure in Fig. 7 (b), the error for each metric improves considerably. The E_{On} error decreases to less than 40% in all conditions, and under 10% error below a 2 Ω gate resistance. The $dvdt_{on}$ error decreased to below 30% at all conditions, and the $didt_{On}$ error decreased from above 40% across all conditions to less than 10%. Overall, the model demonstrates

extremely good agreement with the empirical measurements across a wide range of operating conditions after tuning. This improves the overall utility of the model as it and can more accurately predict the slew rates and switching losses.

The improvement in model accuracy can be observed in waveform overlays as well. Fig. 8 shows an overlay between the empirical measurements and simulation results before and after tuning at 800 V, 25°C, 70 A, and 2 Ω. While the turn-off

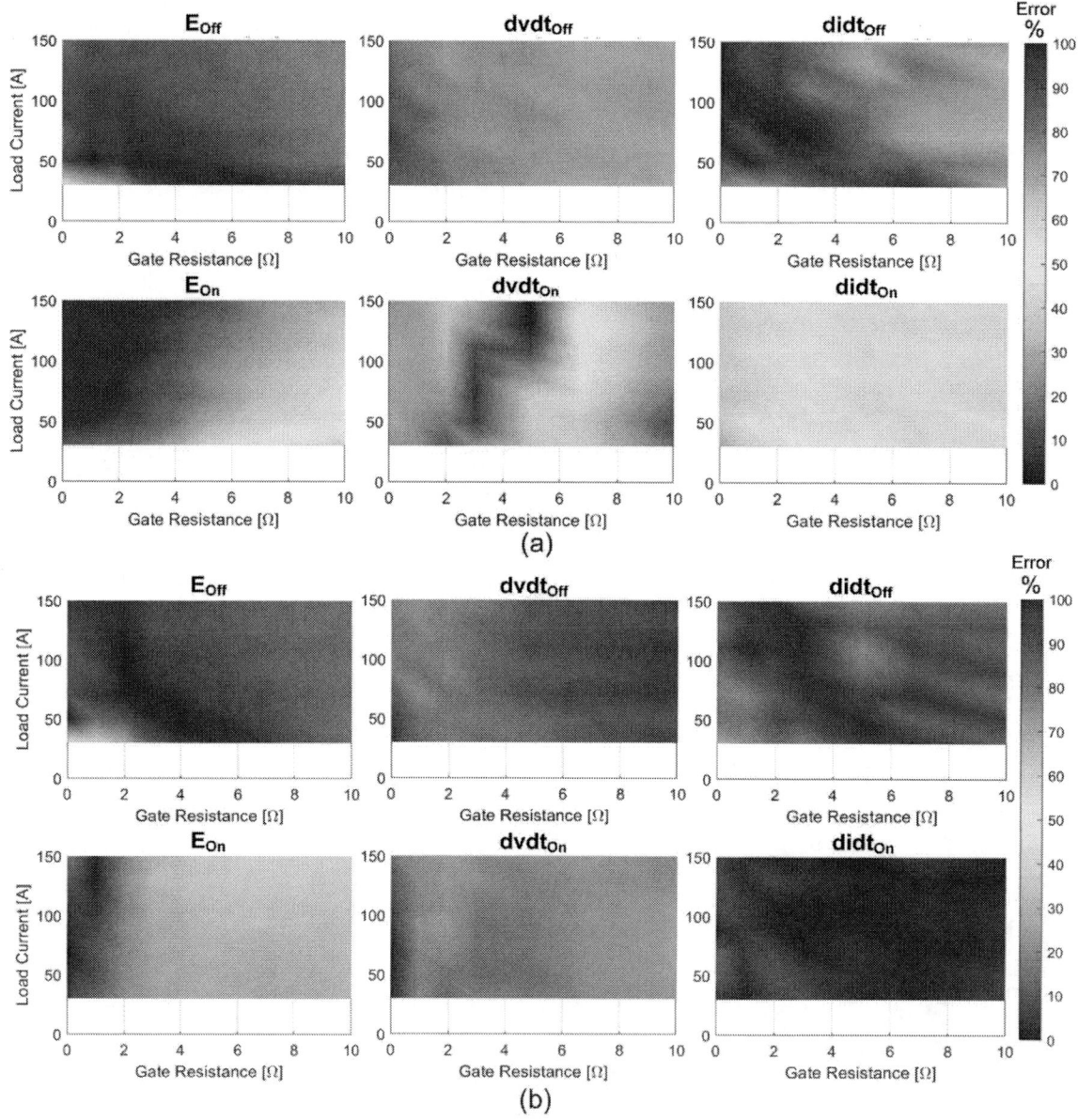

Fig. 7　CAB016M12FM3 model error across gate resistance and load current at 800 V and 25 °C (a) before dynamic tuning and (b) after dynamic tuning. Subplot titles indicate error parameter. Warmer colors indicate higher error. Color at Y-axis and X-axis coordinate indicate the conditions that the error is observed.

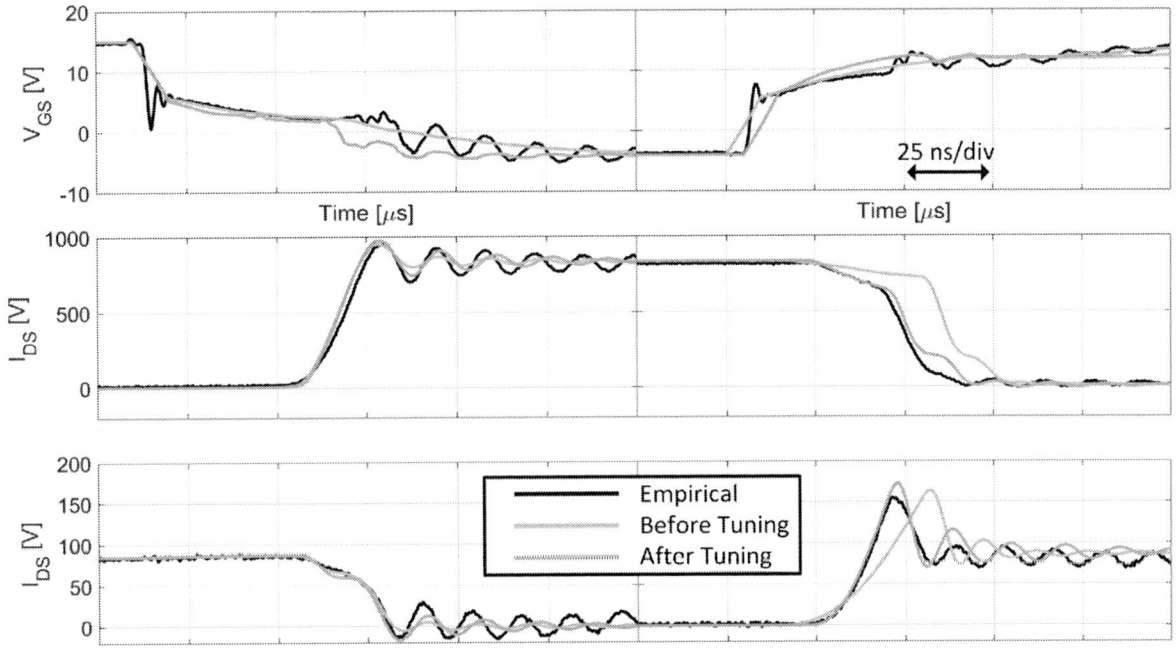

Fig. 8 Turn-off and turn-on DPT waveform overlay comparisons for CAB016M12FM3 SPICE model before and after applying dynamic tuning procedure. DPT conditions are 800 V, 25°C, 70 A, and 2 Ω.

waveforms are similar between all three, the post-tuning results agree much better with the empirical measurements than before. This demonstrates that not only are the quantified metrics easier to use for a tuning algorithm, but the end results provide good agreement between overlain waveforms as well.

4 Conclusion

Circuit-level simulation models face a distinct challenge in that they are expected to operate efficiently, converge in general circuits of varying complexity, and provide accurate predictions of fast switching edges across a wide range of operating conditions. Developing modelling and characterization techniques that simultaneously satisfy these conditions is extremely challenging and impractical.

Rather than focusing on implementing new model behaviours and developing new measurement techniques, this paper proposes an output-based dynamic tuning approach that uses the measured CIL waveforms of a device to tune its parameters. The final result is a model that more accurately predicts the transient behaviour while retaining its simplicity such that it can operate efficiently in general circuits. This method is easily applicable to device models and provides a general approach to

improve transient accuracy without redeveloping the model itself.

References

[1] X. She, H. Q. Alex, L. Oscar and O. Burak, "Review of Silicon Carbide Power Devices and Their Applications," IEEE Transactions on Industrial Electronics, vol. 64, no. 10, pp. 8193-8205, 2017.

[2] K. Kumar, M. Bertoluzzo and G. Buja, "Impact of SiC MOSFET traction inverters on compact-class electric car range," 2014 IEEE International Conference on Power Electronics, Drives and Energy Systems (PEDES), Mumbai, India, 2014, pp. 1-6, doi: 10.1109/PEDES.2014.7042156.

[3] B. W. Nelson, "Computationally Efficient Design and Implementation of SiC MSOFET Models in SPICE," Dissertation, University of Alabama, Tuscaloosa, AL, USA, 2020.

[4] M. Mudholkar, S. Ahmed, N. M. Ericson, S. S. Frank, C. L. Britton and H. A. Mantooth, "Datasheet Driven Silicon Carbide Power MOSFET Model," IEEE Trans. Power Electron., vol. 29, pp. 2220-2228, 2014.

[5] A. P. Arribas, F. Shang, M. Krishnamurthy and K. Shenai, "Simple and Accurate Circuit Simulation Model for SiC Power MOSFETs,"

IEEE Trans. Electron Devices, vol. 62, pp. 449-457, 2015.

[6] A. AlHoussein, H. Alawieh, Z. Riah and Y. Azzouz, "A New Modeling Approach for Predicting the Static and Dynamic Behavior of SiC Power MOSFETs," in 2018 International Symposium on Electromagnetic Compatibility, 2018.

[7] B. W. Nelson, "Computational Efficiency Analysis of SiC MOSFET Models in SPICE: Static Behavior," IEEE Open Journal of Power Electronics, Vols. 499-512, p. 1, 2020.

[8] Y. Mukunoki, "Characterization and Modeling of a 1.2-kV 30-A Silicon-Carbide MOSFET," IEEE Trans. Electron Devices, vol. 63, pp. 4339-4345, 2016.

[9] B. T. DeBoi, B. W. Nelson, A. Curbow, T. McNutt and A. N. Lemmon, "Computational Efficiency Analysis of a Compact Behavioral SiC SPICE Model," in PCIM Europe 2023; International Exhibition and Conference for Power Electronics, Intelligent Motion, Renewable Energy and Energy Management, Nuremberg, Germany, 2023. 10.1109/WiPDA.2018.8569185

[10] B. T. DeBoi, "CHARACTERIZATION AND MODELING OF SIC MULTI-CHIP POWER MODULES," The University of Alabama, Tuscaloosa, 2022.

[11] P. O. Lauritzen and C. L. Ma, "A simple diode model with reverse recovery," in IEEE Transactions on Power Electronics, vol. 6, no. 2, pp. 188-191, April 1991, doi: 10.1109/63.76804.

[12] Zaikin, Denys. (2021). Practical implementation of diode SPICE model with reverse recovery. 10.36227/techrxiv.14915139.v1.

[13] M. Mudholkar, S. Ahmed, M. N. Ericson, S. S. Frank, C. L. Britton, and H. A. Mantooth, "Datasheet Driven Silicon Carbide Power MOSFET Model," IEEE Trans. Power Electron., vol. 29, no. 5, pp. 2220–2228, 2014, doi: 10.1109/TPEL.2013.2295774.

[14] A. Endruschat, C. Novak, H. Gerstner, T. Heckel, C. Joffe, and M. Marz, "A Universal SPICE Field-Effect Transistor Model Applied on SiC and GaN Transistors," IEEE Trans. Power Electron., vol. 34, no. 9, pp. 9131–9145, 2019, doi: 10.1109/TPEL.2018.2889513.

[15] H. Li, X. Zhao, K. Sun, Z. Zhao, G. Cao, and T. Q. Zheng, "A Non-Segmented PSpice Model of SiC mosfet with Temperature-Dependent Parameters," IEEE Trans. Power

Electron., vol. 34, no. 5, pp. 4603–4612, 2019, doi: 10.1109/TPEL.2018.2865611.

[16] H. Sakairi, T. Yanagi, H. Otake, N. Kuroda, and H. Tanigawa, "Measurement Methodology for Accurate Modeling of SiC MOSFET Switching Behavior over Wide Voltage and Current Ranges," IEEE Trans. Power Electron., vol. 33, no. 9, pp. 7314–7325, 2018, doi: 10.1109/TPEL.2017.2764632.

[17] P. Sochor, A. Huerner, R. Elpelt and I. T. Ag, "A Fast and Accurate SiC MOSFET Compact Model for Virtual Prototyping of Power Electronics Circuits," in PCIM Europe 2019; International Exhibition and Conference for Power Electronics, Intelligent Motion, Renewable Energy and Energy Management, Nuremberg, Germany, 2019.

[18] Y. Mukunoki et al., "An Improved Compact Model for a Silicon-Carbide MOSFET and Its Application to Accurate Circuit Simulation," in IEEE Transactions on Power Electronics, vol. 33, no. 11, pp. 9834-9842, Nov. 2018, doi: 10.1109/TPEL.2018.2796583.

[19] T. Funaki, N. Phankong, T. Kimoto and T. Hikihara, "Measuring Terminal Capacitance and Its Voltage Dependency for High-Voltage Power Devices," in IEEE Transactions on Power Electronics, vol. 24, no. 6, pp. 1486-1493, June 2009, doi: 10.1109/TPEL.2009.2016566.

[20] R. Stark, A. Tsibizov, N. Nain, U. Grossner and I. Kovacevic-Badstuebner, "Accuracy of Three Interterminal Capacitance Models for SiC Power MOSFETs Under Fast Switching," in IEEE Transactions on Power Electronics, vol. 36, no. 8, pp. 9398-9410, Aug. 2021, doi: 0.1109/TPEL.2021.3053330.

[21] S. Jimenez, A. Lemmon, B. Nelson and B. Deboi, "Comprehensive Characterization of MOSFET Intrinsic Capacitances," 2021 IEEE Applied Power Electronics Conference and Exposition (APEC), 2021, pp. 1524-1530, doi: 10.1109/APEC42165.2021.9487289.

[22] C. Salcines, B. Holzinger and I. Kallfass, "Characterization of Intrinsic Capacitances of Power Transistors Under High Current Conduction Based on Pulsed S-Parameter Measurements," 2018 IEEE 6th Workshop on Wide Bandgap Power Devices and Applications (WiPDA), 2018, pp. 180-184, doi:

PCIM Europe 2024, 11– 13 June 2024, Nuremberg   DOI: 10.30420/566262200

Investigation of Use-Case-Dependent Modeling Approach for Switched-Mode Power Converters for LVDC Grid Evaluation

Melanie Lavery ©[1], Raffael Schwanninger ©[1], Martin März ©[1]

[1] Institute for Power Electronics, Friedrich-Alexander-University Erlangen-Nuremberg, Germany

Corresponding author: Melanie Lavery, melanie.lavery@fau.de
Speaker: Melanie Lavery, melanie.lavery@fau.de

Abstract

In this paper, the methodology for creating models of power electronic converters based on a recently presented simulation approach for LVDC grids is discussed. This analysis-use-case dependent modeling approach defines three complexity levels for the component models to be used for specific evaluation cases. The theoretical basics and conventions for the three modeling levels of switched-mode converters are presented in this paper. The converter components are collated into functionally distinct subsystems to explore the modeling conventions and analytical descriptions. The parameters required to create models of buck, boost and buck-boost converters are derived and explained. Finally the model behavior for the three model levels is briefly studied and discussed. Based on this in-depth description, a detailed investigation of model accuracy and its implications on the results of the grid simulation can be carried out.

1 Introduction

Switched-mode power converters are key for a sustainable power supply employing renewable energy sources [1]. One of the many reasons is their high efficiency particularly when being used to form DC and AC/DC hybrid grids [2]. DC-grids offer many advantages for power distribution on medium voltage (MVDC) [3] and high voltage (HVDC) [4] levels. These advantages are also evident - and maybe even more prominent - in microgrids with low voltage DC (LVDC). LVDC and AC/DC microgrids hold promise for industrial production plants due to their potential to boost efficiency, availability, and reliability [5]. But to achieve this potential, the correct layout of the grid and design of the individual components need careful consideration. Simulations of individual components and entire systems are often used and sometimes even required to ensure a suitable layout [6], [7]. Even though the interest in DC-grids has increased dramatically, such tools are not yet readily available for DC and hybrid AC/DC systems. This is partly due to the lack of suitable modeling approaches and subsequently a general DC-grid component library is yet to be developed. One of the many challenges in DC grid simulation is the non-linear, time-variant behavior of switched-mode components. Modeling the converters switching behavior can require a lot of computing resources and may take a long time. This can be mitigated by using averaged linearized modeling techniques, that return a time-invariant system description. These analytical approaches, however, are only valid for one specific operating point of the converter [8], [9] but can be expanded and included in a generalized model as presented in [10].

Such a method is used in the novel LVDC system modeling approach presented in [11]. It provides a versatile modeling technique with a standardized interface for various component types, promising a reduction of computing resources. However this work does not describe the modeling procedure of the components in full. Given the complexity of converter modeling, it is necessary to explain the methodology in more detail to explore the proposed benefits of the approach.

2 General Conventions

The general definitions and assumptions of the modeling approach are described in [11]. To summarize, the presented approach offers a standardized modeling methodology to reduce computation time by adjusting model complexity. The models are categorized into the following three complexity level groups:

- Level 1: No dynamic behavior

- Level 2: Linear dynamic behavior

- Level 3: Full dynamic behavior

Depending on the scope of the grid analysis, the complexity of the individual component models is varied accordingly. The approach uses time-dependent analytical equations to describe the physical behavior of the individual components. These equations need to be derived for specific input signals $u(t)$ and output signals $y(t)$ of the particular component. As can be easily deducted, the input-to-output transfer functions depend on the type of input and output signal. For example the transfer function $G = \frac{y(t)}{u(t)}$ for a current input signal $u(t) = i(t)$ of a resistor is its resistance $G_R = \frac{v(t)}{i(t)} = R$. If the input signal is a voltage $u(t) = v(t)$ the transfer function is its conductance $G_R = \frac{i(t)}{v(t)} = \frac{1}{R}$. Hence, to make interconnection of individual blocks feasible, all models are created employing a standardized interface. This interface requires all component models to be partitioned into one of the following three input type groups:

- *Current-Input*: All input signals are currents, all output signals are voltages

- *Voltage-Input*: All input signals are voltage, all output signals are currents

- *Mixed-Input*: One *port* requires input voltages and outputs current, the other port outputs voltages and requires current inputs

In the context of this paper *input* refers to the signal $u(t)$ fed from a connected component and *output* refers to the required signal $y(t)$. The term port is used to designate the power flow direction. To designate the respective ports of a converter in regard to the direction of the power flow, the terms *in-port* and *out-port* are used. From a grid point-of-view, the in-port of the converter acts as a load $P_{in} > 0$, and the out-port acts as a source $P_{out} < 0$. A port encompasses all the signals of the terminals required for connecting one component to another. Ports must always have at least one input $u(t)$ and one output signal $y(t)$ but may have more. Components may also have more than one port.

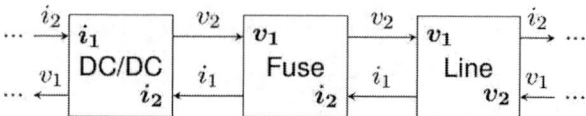

Fig. 1: Example of possible interconnection of component models using the standardized model interface. The current output of *Voltage-Input* blocks is used as input for *Current-Input* blocks and vice-versa.

An example of a possible interconnection of these three different input block types and their ports is displayed in Fig. 1. The *Current-Input* DC/DC block is connected to the *Mixed-Input* fuse block which in turn is connected to the *Voltage-Input* line block. The primary port of the *Mixed-Input* block requires a voltage input and current output, the secondary port requires a current input and voltage output.

The standardized modeling interface defines the reference sign system of the currents. For *Current-Input* blocks, all currents are counted positive if they are directed towards the terminal. Naturally, for converters the out-port current i_{out} will be negative due to the power flow being designated out of the converter. To mark all the signals relating to a specific port, the ports are either named (as will be done in this paper) or numbered. This designator is then used as an index for the port signal i.e. $v_{\text{in}}(t)$ to indicate an in-port voltage or $i_2(t)$ for a secondary port current.

The resulting output voltages $v_n(t)$ are calculated in reference to the respective component's ground potential $\phi_{\text{gnd},n}(t)$. For a converter with three terminals L+, L− and GND the voltages at L+ and L− are calculated via (1) and (2):

$$v_{+,n}(t) = \phi_{+,n}(t) - \phi_{\text{gnd},n}(t) \qquad (1)$$
$$v_{-,n}(t) = \phi_{\text{gnd},n}(t) - \phi_{-,n}(t) \qquad (2)$$

In this paper, the modeling method will generally be explained assuming a monopolar TN-grid with a low impedance grounding scheme. This leads to $\phi_{\text{gnd},n}(t) = 0$ and hence the terminal L− is directly connected to GND.

Time-dependent voltages v and currents i will be represented by lower case letters. To improve legibility, the (t) indicating a function is time-dependent will be omitted, i.e. $v(t) = v$. Upper case letters signify constant parameters such as inductance L and capacitance C. Bold font is used to mark input signals, e.g. the converter input in-port current $\mathbf{i_{\text{in}}}$.

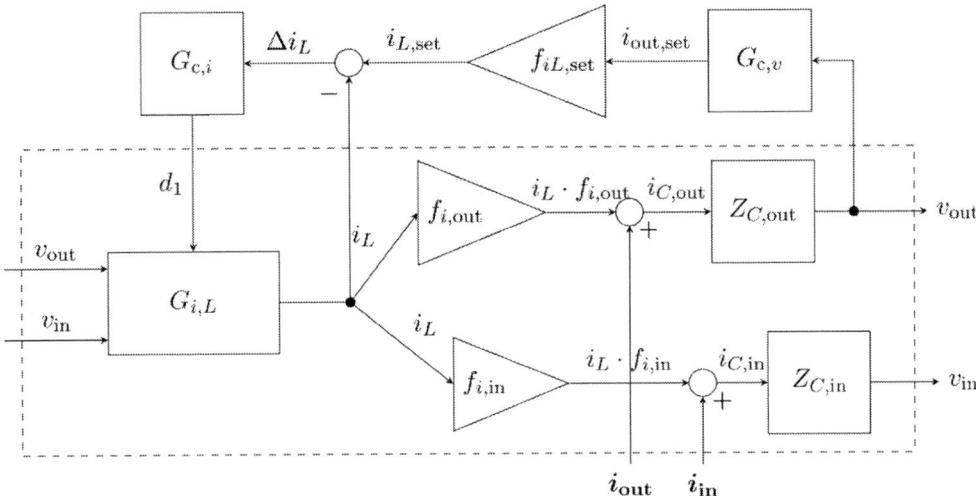

Fig. 2: Generalized block diagram of Level 3 converter model using blocks to represent functionally related subsystems. The blocks inside the dashed rectangle form the power stage of the converter.

As mentioned, DC-DC converters are modeled as *Current-Input* blocks. Therefore, the input signals u of the converter models will be the currents at the ports of the converter. Using all of these conventions and assumptions, the system of equations for a two-port converter given in (3) is derived:

$$\begin{pmatrix} v_{\text{in}} \\ v_{\text{out}} \end{pmatrix} = \begin{pmatrix} Z_{\text{in}} & Z_{\text{in-to-out}} \\ Z_{\text{out-to-in}} & Z_{\text{out}} \end{pmatrix} \cdot \begin{pmatrix} i_{\text{in}} \\ i_{\text{out}} \end{pmatrix} \quad (3)$$

To keep this first introduction to component modeling as simple as possible, the method will be explained for lower-order non-isolating converters such as buck, boost and buck-boost converters. However this methodology can of course also be used for isolating and higher-order systems. An example of a state-space description of such a converter can be found in [12].

The most complex Level 3 models are explained thoroughly first. Based on this, the simplifications and abstractions of Level 2 and Level 1 models are then derived.

3 Design of Level 3 Models

Organizing the converter's individual systems into functionally related groups simplifies the explanation of the modeling methodology. This results in the configuration given in Fig. 2 and consists of:

- the switched-mode components that determine the inductor current i_L, i.e. transistors and diodes but also the inductances, are represented by the block $G_{i,L}$,

- the inductor current scaling functions $f_{i,\text{in}}$ and $f_{i,\text{out}}$, as well as the set-point scaling function $f_{iL,\text{set}}$,

- the dynamic behavior of the in-port and out-port capacitors is represented by $Z_{C,\text{in}}$ and $Z_{C,\text{out}}$ respectively,

- the voltage control transfer function $G_{c,v}$ as well as inductor current control transfer function $G_{c,i}$.

The most complex subsystems of the modeling approach are the converter power stage which will be explained first. The modeling of the converter control and the output signal dynamics of the in- and out-port voltages will be explained subsequently.

3.1 Converter Power Stage Modeling

The generalized equivalent circuit of the power stage of a lower-order converter is displayed in Fig. 3. The switched-mode components of the converter are represented by $G_{i,L}$ in Fig. 2 and 3. The applied modeling method omits the switching dynamics and transforms the time-variant system into a time-invariant system by using averaging methods. It must be noted, that these resulting models are not linear. However, simulation tools such as *MATLAB Simulink* are able to solve non-linear systems of equations using numerical solvers [13]. If need be the systems can also be linearized for further analysis [14].

The theoretical basics of this method are described in various works such as [8], [9], [15]–[17]. These

PCIM Europe 2024, 11– 13 June 2024, Nuremberg   DOI: 10.30420/566262200

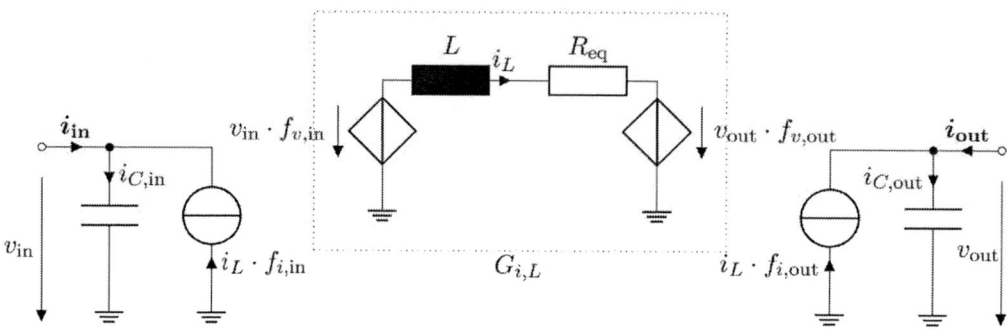

Fig. 3: Generalized equivalent circuit for the power stage of Level 3 converter models.

approaches generally focus on finding and solving a system of differential equations of the converter's state-space parameters to find the transfer functions. The equations for the inductor current are derived using $\frac{di_L}{dt} = \frac{v_L}{L}$. As it is shown in (4), in this approach the inductor voltage v_L is used to directly determine the inductor current i_L via integration:

$$\int \frac{di_L}{dt} dt = \int \frac{v_L}{L} dt \qquad (4)$$

The averaged voltage $\bar{v}_L$ can be found using Kirchhoff's voltage law and applying scaling functions to represent the effects of switching. The resulting equation to determine the inductor voltage v_L is given in (5):

$$\bar{v}_L = v_{in} \cdot f_{v,in} - v_{out} \cdot f_{v,out} - i_L \cdot R_{eq} - V_{eq} \qquad (5)$$

The voltage scaling functions $f_{v,in}$ and $f_{v,out}$ emulate the changing voltage levels of the inductor during a switching cycle. These scaling functions, as given in (6) and (7), depend on the converter topology and the operating mode and are determined using the on- and off-time duty cycles d_1 and d_2:

$$f_{v,in} = \begin{cases} d_1, & \text{if buck} \\ d_1 + d_2, & \text{if boost} \\ d_1, & \text{if buck-boost} \end{cases} \qquad (6)$$

$$f_{v,out} = \begin{cases} d_1 + d_2, & \text{if buck} \\ d_2, & \text{if boost} \\ -d_2, & \text{if buck-boost} \end{cases} \qquad (7)$$

The equivalent loss voltage V_{eq} represents the sum of all effective forward voltages V_f of transistors and diodes. Assuming a transistor T_1 with a forward voltage of $V_{f,1}$, and a diode D_2 with a forward voltage

$V_{f,2}$, the equivalent voltage V_{eq} can be calculated using (8):

$$V_{eq} = V_{f,1} \cdot d_1 + V_{f,2} \cdot d_2 \qquad (8)$$

Similarly, the equivalent resistance R_{eq} defined in (9) is the sum of all resistances and ohmic losses effective during switching:

$$R_{eq} = R_L + f_{i,in} \cdot R_{on,1} + f_{i,out} \cdot R_{on,2} \qquad (9)$$

Due to the averaging of the switching behavior, "switched" resistances need to be scaled using the functions $f_{i,in}$ and $f_{i,out}$. They represent effect of the relative switched inductor current on the outport and in-port output voltages. These functions depend on the converter topology and its current operating mode. They are dependent on the on-time duty cycle d_1 and off-time duty cycle d_2. The current scaling functions for buck, boost and buck-boost converters are given in (10) and (11):

$$f_{i,in} = \begin{cases} -d_1, & \text{if buck} \\ -1, & \text{if boost} \\ -d_1, & \text{if buck-boost} \end{cases} \qquad (10)$$

$$f_{i,out} = \begin{cases} 1, & \text{if buck} \\ d_2, & \text{if boost} \\ -d_2 & \text{if buck-boost} \end{cases} \qquad (11)$$

As shown in Fig. 2, the on-time duty cycle d_1 is generated directly by the current control block $G_{c,i}$. Focusing on the power stage alone, it is viewed as an input parameter. The determination the off-time duty cycle d_2 however depends on the operating mode of the converter. In continuous conduction mode (CCM) d_2 depends solely on the on-time duty cycle, see (12). For discontinuous conduc-

1488

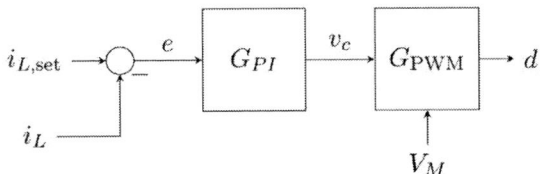

Fig. 4: Block diagram of converter current control of the subsystem $G_{c,i}$.

tion mode (DCM) however, d_2 can either be calculated employing a reduced-order approach using the voltage-second-balance of the inductor or a full-order approach based on the demagnetization time of the inductor [18]. This modeling method proposes the use of the full-order modeling for d_2 to retain the inductor dynamics during DCM. This full-order off-time duty cycle d_2 is determined by the inductor on-time voltage $v_L^{|1}$, the averaged inductor current $\bar{i}_L$, the on-time duty cycle d_1, as well as the converter inductance L and switching frequency $f_S = \frac{1}{T_S}$. The two operating mode dependent equations used to determine d_2 are given in (12):

$$d_2 = \begin{cases} 1 - d_1, & \text{if CCM} \\ \frac{\bar{i}_L \cdot L}{v_L^{|1} \cdot d_1 \cdot T_S} - d_1, & \text{if DCM} \end{cases} \quad (12)$$

3.2 Current Control Modeling

As can be seen in Fig. 2, the converter has a cascaded current and voltage control loop. Since there is a myriad of different converter and grid control mechanisms, it is not possible to demonstrate the modeling approach for all of them. A generalized block diagram of a current control loop based on a PI-controller is displayed in Fig. 4.
The output-to-input transfer function $G_{c,i}$ of the current control can be calculated via (13):

$$G_{c,i} = \frac{d}{i_L} = G_{PI} \cdot G_{PWM} \quad (13)$$

The general transfer function of a PI-controller in the Laplace-domain is given in (14):

$$G_{PI}(s) = K_P(1 + \frac{1}{s \cdot T_I}) \quad (14)$$

In *MATLAB-Simulink* this transfer function can be used directly without transforming it to the time domain. If the time-dependent description is however required, the analytical description can be found for the input error e and the output value v_c resulting in (15). This equation represents an idealized transfer function of a PI-controller. The output signal

v_c represents the carrier voltage produced by the PI-controller.

$$v_c = e \cdot K_P + K_P \int e \mathrm{d}t \quad (15)$$

Next, the transfer function of the Pulse Width Modulator (PWM) and delays caused by sampling and computation are also included. According to [19], the transfer function of the PWM is a constant factor determined by the peak value of the modulator saw tooth voltage V_M. The transfer function, as given in (16), when multiplied with the carrier voltage v_c leads to the time dependent duty cycle d_1:

$$G_{PWM} = \frac{d_1}{v_c} = \frac{1}{V_M} \quad (16)$$

The PWM inherently has a sampling characteristic, which introduces phase shifts due to delays. In DCM additional latencies are introduced which can lead to instabilities. The theoretical basics of this behavior is described in detail in [20]. However, the effects and implications of this on the converter modeling approach are yet to be evaluated. As of now, the PWM-transfer function given in (16) is used.

3.3 Voltage Control Modeling

The modeling of the voltage or grid control of a converter will be explained assuming a droop-control for the converter. The functionality and usage especially for DC grids, as well as the layout of droop-control is discussed in [21]. Generally, droop-control uses either the out-port voltage or current as the control input signal. The corresponding output signal $i_{\text{out,set}}$ of a voltage droop-control, as given in (17), is the set-value of the out-port current:

$$i_{\text{out,set}} = \begin{cases} i_{\text{out,Droop}}, & \text{if } V_{\min} < v_{\text{out}} < V_{\max} \\ I_{\max}, & \text{if } v_{\text{out}} < V_{\min} \\ I_{\min}, & \text{if } V_{\max} < v_{\text{out}} \end{cases} \quad (17)$$

As can be seen in (17), the behavior of the droop-control is not linear but a piecewise-function. The voltages $V_{\min}$ and $V_{\max}$ define the levels at which the converter switches from a proportional control method to a current or voltage control mode. The maximum and minimum out-port currents $I_{\max}$ and $I_{\min}$ are defined during system layout. The output current $i_{\text{out,Droop}}$ generally is determined by (18):

$$i_{\text{out,Droop}} = I_{\max} - v_{\text{out}} \cdot G_{\text{Droop}} \quad (18)$$

Using this method, the converter acts as a voltage source with an internal resistance. The transfer function G_{Droop} can simply be a constant virtual resistance, as is given in (19), but can also be much more complex [22].

$$G_{\text{Droop}} = \frac{1}{R_{\text{Droop}}} \tag{19}$$

The droop-control inherently designates the set-value for the out-port current i_{out}. The current control however regulates the inductor current i_L. The averaged out-port current and inductor current are only the same for converters where the inductor is connected directly to the out-port without any switching-elements in between. Therefore an out-port to inductor current scaling function $f_{iL\text{set}} = \frac{i_{L,\text{set}}}{i_{\text{out,set}}}$ is required. Similarly to the current scaling functions (10) and (11), the function defined in (20) is adjusted depending on converter topology:

$$f_{iL,\text{set}} = \begin{cases} 1, & \text{if buck} \\ \frac{1}{d_2}, & \text{if boost} \\ \frac{1}{d_2}, & \text{if buck-boost} \end{cases} \tag{20}$$

3.4 Out- and In-Port Dynamics

The in-port and out-port voltages of the converters are found by applying Kirchhoff's voltage and current laws at the in- and out-port capacitor nodes. The voltages are determined via the sum of the capacitor voltages and the voltage drop across parasitic serial resistances of C_{in} and C_{out}, resulting in (21) and (22):

$$v_{\text{in}} = \int \frac{i_{C,\text{in}}}{C_{\text{in}}} \mathrm{d}t + i_{C,\text{in}} \cdot R_{C,\text{in}} \tag{21}$$

$$v_{\text{out}} = \int \frac{i_{C,\text{out}}}{C_{\text{out}}} \mathrm{d}t + i_{C,\text{out}} \cdot R_{C,\text{out}} \tag{22}$$

The capacitor currents $i_{C,\text{in}}$ and $i_{C,\text{out}}$ are calculated via the sum of all currents at the connection node of the respective in- and out-ports, leading to (23) and (24):

$$i_{C,\text{in}} = i_L \cdot f_{i,\text{in}} + i_{\text{in}} \tag{23}$$

$$i_{C,\text{out}} = i_L \cdot f_{i,\text{in}} + i_{\text{out}} \tag{24}$$

The in-port and out-port currents are both considered positive for a current-flow towards the capacitor nodes and thus the resulting equations (23) and (24) have a similar form.

3.5 Required Model Parameters

For the modeling certain parameters required. The minimum set of required parameters for a buck, boost or buck-boost converter for Level 3 models is given in Tab. 1. The parameters are either component values of the passive components or have been defined during the layout of the converter.

Passive Components	L, C_{in}, C_{out}
Parasitics	R_L, $R_{C,\text{in}}$, $R_{C,\text{out}}$, $V_{\text{f},1}$, $V_{\text{f},2}$, $R_{\text{on},1}$, $R_{\text{on},2}$
Switching Parameters	f_S, V_M
Current Control	T_I, K_P
Voltage Control	R_{Droop}, $I_{\max}$, $I_{\min}$, $V_{\max}$, $V_{\min}$

Tab. 1: Minimum set of required parameters for Level 3.

4 Design of Level 2 Models

After having established the basics of the modeling approach for Level 3 models, the description for Level 2 will focus on the differences in modeling. The biggest change from Level 2 and Level 3 models is the complexity of the switched-stage $G_{i,L}$. A block diagram of the subsystems of the Level 2 model is given in Fig. 5.

4.1 Converter Power Stage and Current Control Modeling

A generalized equivalent circuit representing the function of the simplified power stage of Level 2 models is given in Fig. 6. As can be seen, the inductor dynamics are not included in Level 2 models. The closed current control loop is simplified in such a way that the resulting transfer function G_{conv} is based on the out-port current i_{out}. The switched converter currents $i_{\text{conv,in}}$ and $i_{\text{conv,out}}$ are thus not based on the inductor current i_L. The equations to find these currents are given in (25) and (26):

$$i_{\text{conv,in}} = i_{\text{conv}} \cdot f_{\text{conv,in}} \tag{25}$$

$$i_{\text{conv,out}} = i_{\text{conv}} \cdot f_{\text{conv,out}} \tag{26}$$

This closed loop transfer function for the averaged current G_{conv} is approximated by a first-order low pass. Level 2 models hence do not distinguish between DCM and CCM and the duty cycles d_1 and d_2 need not be determined. The out-port to inductor current scaling function $f_{iL,\text{set}}$ used in Level 3

PCIM Europe 2024, 11– 13 June 2024, Nuremberg   DOI: 10.30420/566262200

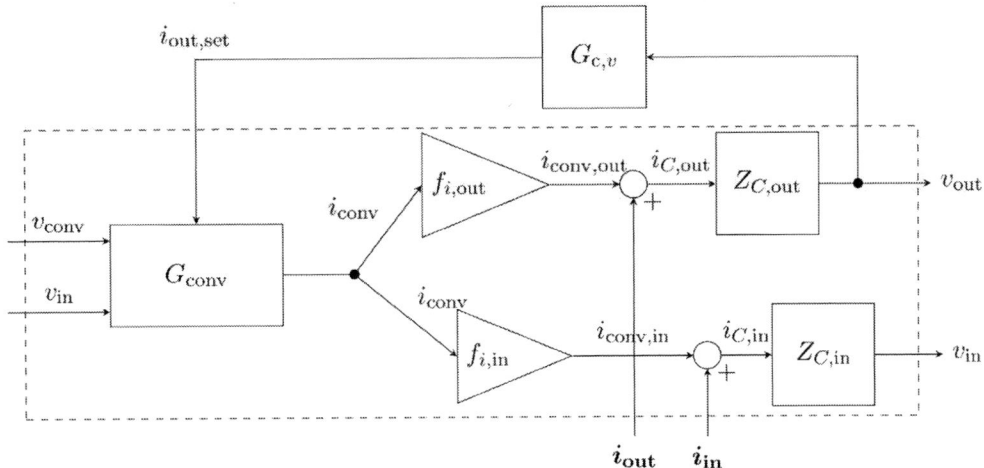

Fig. 5: Generalized block diagram of a Level 2 converter model using blocks to represent functionally related subsystems. The blocks inside the dashed rectangle form the power stage of the converter.

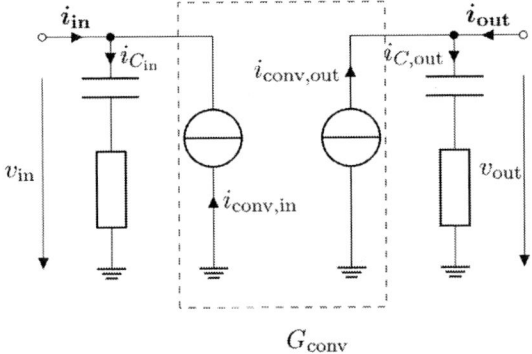

Fig. 6: Generalized equivalent circuit for the power stage of Level 2 converter models.

models is not required. Therefore the converter current control loop transfer function $G_{\mathrm{conv}}(s)$ results in (27):

$$G_{\mathrm{conv}}(s) = \frac{i_{\mathrm{out}}}{i_{\mathrm{out,set}}} = \frac{1}{1 + s \cdot T_{\mathrm{LP}}} \qquad (27)$$

The time constant T_{LP} of the transfer function $G_{\mathrm{conv}}(s)$ is equal to the inverse of the crossover frequency of the current control loop [23].
The current scaling functions $f_{i,\mathrm{in}}$ and $f_{i,\mathrm{out}}$ in (10) and (11) no longer depend on the converter's topology. The out-port current scaling function for buck, boost or buck-boost is simply 1, as given in (28):

$$f_{i,\mathrm{out}} = 1 \qquad (28)$$

The in-port current is calculated using the converter transmission ratio M, resulting in (29):

$$f_{i,\mathrm{in}} = -\frac{1}{M} \qquad (29)$$

The transmission ratio M is not a linear function, since the converter's inherent step-down or step-up as well as the voltage limiting behavior need to be included. M is determined using the in-port and out-port current i_{in} and i_{out} and is a unitless factor ranging from 0 to 1, as given in (30):

$$M = \begin{cases} \frac{i_{\mathrm{in}}}{i_{\mathrm{out}}}, & \text{if } 0 \leq \frac{i_{\mathrm{in}}}{i_{\mathrm{out}}} \leq 1 \\ 1, & \text{if } 1 \leq \frac{i_{\mathrm{in}}}{i_{\mathrm{out}}} \\ 0, & \text{if } \frac{i_{\mathrm{in}}}{i_{\mathrm{out}}} \leq 0 \end{cases} \qquad (30)$$

It must be noted, that this method for modeling the converter current control does not include any parasitic resistances from the switching stage. The currently defined use-cases for the grid modeling approach do not require this level of accuracy for Level 2 converter models.

4.2 Voltage Control and Out- and In-Port Dynamics Modeling

The voltage control modeling approach for Level 2 does not differ from Level 3. As can be seen in Fig. 5 and 6 the calculation of the capacitor currents $i_{C,\mathrm{in}}$ and $i_{C,\mathrm{out}}$, as well as the dynamic behavior of the in- and out-port capacitors $Z_{C,\mathrm{in}}$ and $Z_{C,\mathrm{out}}$ is equal to that of Level 3. The grid or voltage control $G_{c,v}$ also does not differ and can be copied from Level 3 models.

4.3 Required Model Parameters

Concluding the modeling approach for Level 2, the minimum set of parameters required for modeling a

1491

buck, boost or buck-boost converter is given in Tab. 2. All these parameter are either component values of the passive components or have been defined during the layout of the converter.

Passive Components	C_{in}, C_{out}
Parasitics	$R_{C,\mathrm{in}}$, $R_{C,\mathrm{out}}$
Current Loop	T_{LP}
Voltage Control	R_{Droop}, I_{max}, I_{min}, V_{max}, V_{min}

Tab. 2: Minimum set of required parameters for Level 2.

5 Design of Level 1 Models

Level 1 models are the most basic converter models in terms of dynamic complexity. The converter control and power stage are combined and cannot be separated into functionally independent blocks. It therefore is not necessary to derive a block diagram as for the previous levels. Instead, an equivalent circuit describing the model function for Level 1 models as given in Fig. 7 will suffice.

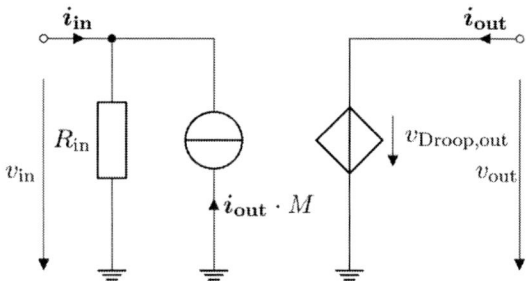

Fig. 7: Generalized equivalent circuit for the power stage of Level 1 converter models.

5.1 Out-Port Voltage

The modeling approach for Level 1 models differs in many aspects. The biggest difference is the physical disconnection of the in-port and out-port signals. The out-port voltage v_{out} is dependent only on the voltage control and the out-port current i_{out}, as described in (31):

$$v_{\mathrm{droop,out}} = \begin{cases} v_{\mathrm{Droop}}, & \text{if } I_{\mathrm{min}} < i_{\mathrm{out}} < I_{\mathrm{max}} \\ V_{\mathrm{max}}, & \text{if } I_{\mathrm{out}} < i_{\mathrm{out}} \\ V_{\mathrm{min}}, & \text{if } I_{\mathrm{max}} < i_{\mathrm{out}} \end{cases} \quad (31)$$

The droop voltage v_{Droop} is determined in a similar manner to Level 3. As can be seen in (32) however,

for Level 1 the input signal of the voltage control is not the out-port voltage v_{out} but its current i_{out}:

$$v_{\mathrm{Droop}} = V_{\mathrm{max}} - i_{\mathrm{out}} \cdot G_{\mathrm{Droop}} \quad (32)$$

As previously explained for Level 3 models, the droop-control transfer function G_{Droop} can be either a constant value or a more complex function.

5.2 In-Port Voltage

The in-port voltage v_{in} is determined by the in- and out-port currents i_{in} and i_{out}, the transmission ratio M, and an internal equivalent resistance R_{in} as given in (33). This virtual resistance R_{in} has a very high value and does not represent any actual converter component but is merely a modeling utility required to mitigate convergence problems.

$$v_{\mathrm{in}} = (i_{\mathrm{in}} - i_{\mathrm{out}} \cdot M) \cdot R_{\mathrm{in}} \cdot \frac{1}{\eta} \quad (33)$$

Losses are represented via the converter efficiency η which is a unitless factor ranging from $0\,\%$ to $100\,\%$. The efficiency η used for simulation can be a constant factor or operating point dependent.

As explored in (30), the converter transmission ratio M is not a linear, but a piece-wise function. Since M is used to determine v_{in} in (33) the transmission ratio is determined using the in- and out-port voltages resulting in (34):

$$M = \begin{cases} \frac{v_{\mathrm{out}}}{v_{\mathrm{in}}}, & \text{if } 0 \leq \frac{v_{\mathrm{out}}}{v_{\mathrm{in}}} \leq 1 \\ 1, & \text{if } 1 \leq \frac{v_{\mathrm{out}}}{v_{\mathrm{in}}} \\ 0, & \text{if } \frac{v_{\mathrm{out}}}{v_{\mathrm{in}}} \leq 0 \end{cases} \quad (34)$$

5.3 Required Model Parameters

Concluding the modeling approach for Level 1, the minimum set of required parameters for modeling buck, boost and buck-boost converters is summarized in Tab. 3.

Passive Components	R_{in}
Parasitics	η
Voltage Control	R_{Droop}, I_{max}, I_{min}, V_{max}, V_{min}

Tab. 3: Minimum set of required parameters for Level 1.

6 Model Behavior

Finally, after having established the theoretical basics for the converter modeling, a brief example of the converter model behavior is explored. In Fig. 8 and Fig. 9, the out-port voltages and currents for the different levels of a buck converter model are shown. To demonstrate the differing dynamic behavior, a low-impedance short-circuit with $R_{sc} = 1 \text{ m}\Omega$ is introduced at the out-port of the converter. Before the fault onset, the out-port voltages and currents are similar for all three model levels, as would be expected. However after fault onset, the differences in modeling become apparent:

- Level 1: The out-port voltage is limited by the voltage control to $V_{out} = 360$ V after fault onset. The out-port current rises greatly due to the change of the out-port voltage and the low impedance of the fault.

- Level 2: The out-port voltage drops to nearly zero almost immediately. 1.5 ms after fault onset the out-port current is limited to a steady-state value of $-I_{out} = -80$ A.

- Level 3: The out-port voltage and current have a similar albeit slower dynamic than Level 2. After fault onset, there are some oscillations at the out-port before the current rises to the steady state value.

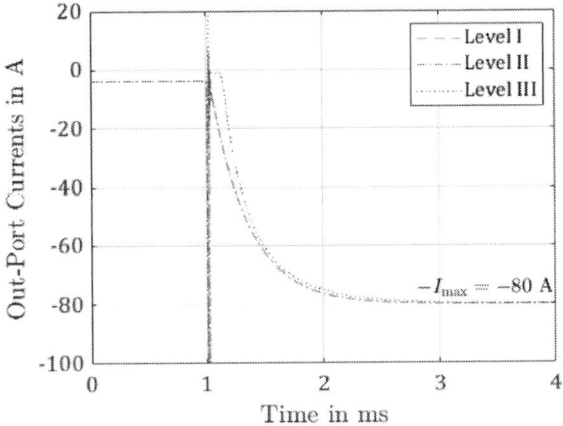

Fig. 8: Out-port currents for a buck converter with a low-resistance fault at $T_{sc} = 1$ ms.

The low-resistance short-circuit, causes the out-port voltage to drop quickly for all three converter models. As can be seen in Fig. 8 there is little difference in out-port current dynamics for Level 2 and 3. The out-port current for Level 3 oscillates for a short time at the onset of the fault before dropping

to zero for 0.1 ms before finally rising to the steady state value I_{max} determined by the droop-control.

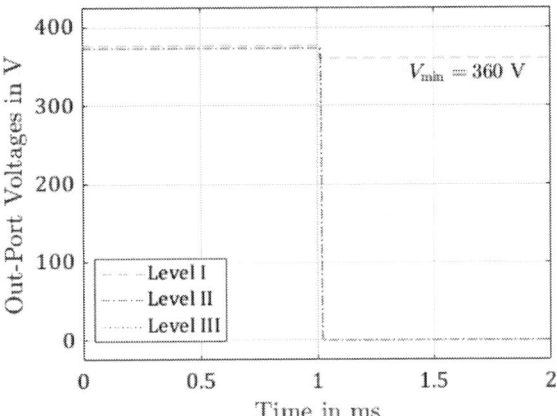

Fig. 9: Out-port voltages for a buck converter with a low-resistance fault at $T_{sc} = 1$ ms.

7 Conclusion and Further Work

In this paper, the theoretical basics and steps required for converter modeling for on a newly presented grid modeling approach are explored and explained in detail. The methodology is explained starting with the most detailed Level 3 converter models. The converter components are grouped into functionally linked subsystems. Based on Level 3, the simplifications and abstractions for the less complex Level 2 and Level 1 models are derived. For Level 2 models the dynamic behavior of the closed current control loop is reduced to a first-order low pass. The modeling of the voltage control and the capacitances is equal to Level 3. The modeling of the Level 1 models is fundamentally different - the out-port voltage is determined only by the out-port current. This physically disconnects the in-port and out-port behavior of the converter. Using these detailed descriptions, a detailed investigation of component model accuracy and its implications on the results of the grid simulation can be discussed in the future. The proposed reduction of computing-resources during grid simulation as well as the model accuracy will need to be investigated in the future.

Acknowledgement

The research for this work was greatly supported by the German Federal Ministry for Economic Affairs and Climate Action within the research project "DC | hyPAMod".

References

[1] K. Shenai and K. Shah, "Smart dc micro-grid for efficient utilization of distributed renewable energy," pp. 1–6, DOI: 10.1109/EnergyTech.2011.5948505.

[2] M. März, B. Wunder, and L. Ott, "Lvdc networks - challenges and perspectives," *Bauelemente der Leistungselektronik und ihre Anwendungen 2017 - 7. ETG-Fachtagung 2017*, 2017.

[3] A. Clerici, R. Chiumeo, and C. Gandolfi, "Mvdc multi-terminal grids: A valid support for distribution grids improvement," pp. 1–6, DOI: 10.23919/AEIT50178.2020.9241183.

[4] O. Saadeh, B. A. Sba, Z. Dalala, and A. Bashaireh, "Comparative performance analysis of hvdc and hvac transmission systems in the presence of pv generation: A case study using the ieee-5-bus network," pp. 1–5, DOI: 10.1109/AEITHVDC58550.2023.10178994.

[5] M. Barth, B. Gutwald, E. Russwurm, J. Franke, M. Lavery, *et al.*, "Simulation-based planning and design of hybrid ac/dc energy grids for production systems: A holistic approach: 20. asim fachtagung simulation in produktion und logistik 2023, p. 31," 2023. DOI: 10.22032/dbt.57898.

[6] L. Ott, J. Kaiser, K. Gosses, Y. Han, B. Wunder, *et al.*, "Model-based fault current estimation for low fault-energy 380vdc distribution systems," *2016 IEEE International Telecommunications Energy Conference (INTELEC)*, pp. 1–8, DOI: 10.1109/INTLEC.2016.7749126.

[7] D. Gutina, A. Bärmann, G. Roeder, M. Schellenberger, and F. Liers, "Optimization over decision trees: A case study for the design of stable direct-current electricity networks," *Optimization and Engineering*, 2023. DOI: 10.1007/s11081-023-09788-x.

[8] S. Cuk and R. D. Middlebrook, "A general unified approach to modelling switching dc-to-dc converters in discontinuous conduction mode," vol. 1977, pp. 36–57, DOI: 10.1109/PESC.1977.7070802.

[9] R. D. Middlebrook and S. Cuk, "A general unified approach to modelling switching-converter power stages," vol. 1976, pp. 18–34, DOI: 10.1109/PESC.1976.7072895.

[10] T. Suntio, "Unified average and small-signal modeling of direct-on-time control," *IEEE Transactions on Industrial Electronics*, vol. 53, no. 1, pp. 287–295, 2006. DOI: 10.1109/TIE.2005.862221.

[11] M. Lavery, R. Schwanninger, and M. März, "Use-case-dependent modeling approach for computer simulation of hybrid ac-dc grids," pp. 1–6, DOI: 10.1109/ICDCM54452.2023.10433620.

[12] X. Yang, S. Pohlenz, R. Schwanninger, N. Schleipmann, B. Wunder, and M. März, "Modelling and design of a droop-based cascaded controller for llc resonant converter," *2023 IEEE International Conference on DC Microgrids (ICDCM)*,

[13] V. F. Pires and J. Silva, "Teaching nonlinear modeling, simulation, and control of electronic power converters using matlab/simulink," *IEEE Transactions on Education*, vol. 45, no. 3, pp. 253–261, 2002. DOI: 10.1109/TE.2002.1024618.

[14] MathWorks, *Linearize simulink model at model operating point*, 2024.

[15] W. M. Polivka, P. R. Chetty, and R. D. Middlebrook, "State-space average modelling of converters with parasitics and storage-time modulation," pp. 119–143, 1980. DOI: 10.1109/PESC.1980.7089440.

[16] P. T. Krein, J. Bentsman, R. M. Bass, and B. L. Lesieutre, "On the use of averaging for the analysis of power electronic systems," *IEEE Transactions on Power Electronics*, vol. 5, no. 2, pp. 182–190, 1990. DOI: 10.1109/63.53155.

[17] D. Maksimovic, A. M. Stankovic, V. J. Thottuvelil, and G. C. Verghese, "Modeling and simulation of power electronic converters," *Proceedings of the IEEE*, vol. 89, no. 6, pp. 898–912, 2001. DOI: 10.1109/5.931486.

[18] J. Sun, D. M. Mitchell, M. F. Greuel, P. T. Krein, and R. M. Bass, "Averaged modeling of pwm converters operating in discontinuous conduction mode," *IEEE Transactions on Power Electronics*, vol. 16, no. 4, pp. 482–492, 2001. DOI: 10.1109/63.931052.

[19] R. W. Erickson and D. Maksimović, *Fundamentals of Power Electronics*. New York: Springer Science + Business Media and Kluwer Academic Publishers, 2004. DOI: 10.1007/b100747.

[20] S. Matlok, "Digitale regelung bidirektionaler gleichspannungswandler," Ph.D. dissertation, Erlangen.

[21] B. Wunder, L. Ott, J. Kaiser, K. Gosses, M. Schulz, *et al.*, "Droop controlled cognitive power electronics for dc microgrids," *2017 IEEE International Telecommunications Energy Conference (INTELEC)*, pp. 335–342, 2017. DOI: 10.1109/INTLEC.2017.8214158.

[22] R. Wang, Q. Sun, Y. Gui, and D. Ma, "Exponential-function-based droop control for islanded microgrids," *Journal of Modern Power Systems and Clean Energy*, vol. 7, no. 4, pp. 899–912, 2019. DOI: 10.1007/s40565-019-0544-3.

[23] L. Dixon, "Switching power supply control loop design," *Unitrode Power Supply Seminar Handbook*, no. Seminar 800, 1991.

PCIM Europe 2024, 11– 13 June 2024, Nuremberg   DOI: 10.30420/566262201

Averaged Model with Blocking Capability for Solid-State Transformers

Ahmed Meligy [1,2], Rafael Coelho-Medeiros [1], Ilknur Colak[1], Seddik Bacha[2]

[1] Schneider Electric, 38000 Grenoble, France
[2] Univ. Grenoble Alpes, CNRS, Grenoble INP, G2ELab, 38000 Grenoble, France

Corresponding author: Ahmed Meligy, Ahmed.meligy@se.com
Speaker: Ahmed Meligy, Ahmed.meligy@se.com

Abstract

This study develops a hybrid time-domain averaged model capable of reproducing the blocked state of power modules to address the computational challenges associated with modular Solid-State Transformer (SST) modeling. The proposed model provides a valuable tool for examining external dynamics within SST-based systems, encompassing aspects such as power flow and stability analysis, operation during external fault scenarios as well as startup and shutdown procedures. This study validates the effectiveness of the proposed model by conducting a comprehensive analysis of various test cases compared to an equivalent switched time-domain model. The results showcase its capability to speed up simulation processing time up to 9 fold, maintain marginally less than 1% error during steady-state operation and accurately depict the external dynamics of modular SSTs.

1 Introduction

The Solid-State Transformer (SST) is a versatile and efficient power converter, finding diverse applications encompassing railways, transmission and distribution grids, microgrids, and the integration of renewable energy sources[1]–[7]. Modern SST have demonstrated advanced technological readiness levels, showcasing efficiencies exceeding 97% for AC/AC conversion, over 98% for AC/DC conversion, and surpassing 99% for DC/DC conversion, with notable power densities reaching up to 500 kW/m^3 for a complete system [8]. In Medium-to-Low-Voltage (MV/LV) conversion applications, a modular approach is often adopted with SSTs using multiple fundamental converters, here referred to as power cells, each rated for an arbitrary power. These power cells are interconnected in series on the Medium-Voltage (MV) side and in parallel on the Low-Voltage (LV) side, thereby facilitating the scalability of voltage and power requirements for MV systems. This configuration gives rise to the Input-Series Output-Parallel (ISOP) topology, which has gained widespread adoption in the field [8]. Modeling modular topologies with numerous cells can impose significant computational challenges

due to the numerous nonlinear switches that are commonly present. This complexity is further exacerbated in multi-SST systems, particularly simulations that capture electromagnetic transient behavior [9]. An established solution to address these challenges is using averaged models, a practice widely applied and validated for Modular Multilevel Converters (MMCs) in high-voltage direct current [10], [11] and static compensation applications [12]. Averaged models effectively capture the low-frequency response of switching devices corresponding to the frequencies where power is transferred [13], significantly reducing the model's complexity. These models can be implemented as time-dependent, often integrated into circuit-based simulations, or as time-independent models when applied within a stationary framework, such as the Harmonic State Space method. Averaged models are well-suited for analyzing external dynamic behaviors, including interactions between the converter and the grid or between multiple converters under normal and fault conditions [10].

Multiple studies have explored time-dependent averaged models for SSTs [14]–[20], employing a common approach of approximating the switching functions of AC/DC converters by their fundamental values. While these models have been validated for

1495

various operating conditions, none have addressed the blocked state of the power modules composing the SST, a mode important for fault ride-through and pre-charging. Although the prior models could be mathematically modified to capture diode bridge behavior, this introduces a complex non-linear Jacobian matrix in ordinary differential equation (ODE) solvers. To overcome this challenge, this article presents a hybrid time-domain averaged model for SSTs, specifically addressing the blocking state of the power cells. This model can be efficiently integrated into circuit-based simulation software, facilitating the study of external dynamics in SST-based systems. This work extends the model from [9] for the ISOP SST configuration.

The remainder of this article is structured as follows. In section 2, we introduce the baseline system and provide an overview of the system and control strategy used in this study. In Section 3, we present the development of the proposed averaged model, designed to handle blocking scenarios for the ISOP SST. In section 4, we compare and validate the proposed model against a switched model under three test cases: normal operation, system startup, and fault conditions. In Section 5, we conclude this article.

2 Baseline System

2.1 System Configuration

The schematic of the considered SST is illustrated in Fig. 1. It comprises a cascaded configuration of N power cells per phase connected to an 11 kV AC grid. The MVAC grid is modelled by an equivalent three phase source $v_{g_{a,b,c}}$ in series with its equivalent impedance L_g, R_g. In the remainder of this article, the phase indices (a, b, and c) are omitted from the variables for simplicity. The ISOP SST is connected to the grid through a line inductor with inductance L_f and resistance R_f. The grid current and modulated voltage are denoted i_g and v_m^Σ, respectively. Whereas the current and voltage at the LVDC side are denoted i_L and v_L, respectively.

Each cell of the SST deploys a Full-Bridge Voltage-Source Converter (FB-VSC) at each conversion stage, as presented in Fig. 2. The Active Front End Converter (AFEC), with modulated voltage $v_m^{(k)}$ and $k \in [1..N]$, interfaces the AC grid and the primary-side DC bus of the Isolated DC/DC converter. The isolated DC/DC converter can be operated as a Dual-Active Bridge (DAB) [21] or Single-Active Bridge (SAB) [22]. Both of which

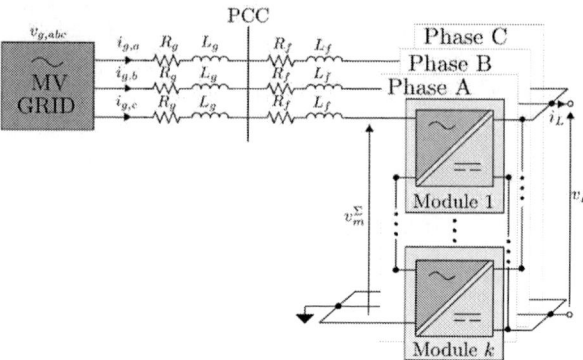

Fig. 1: MVAC-LVDC ISOP SST.

dynamics could be emulated by the proposed hybrid averaged model. The medium frequency link depicted in Fig. 2 contains a medium frequency transformer and may also incorporate passive components to replicate the characteristics of resonant converters. At the input and output of the DC/DC converter are DC-link capacitors represented by C, p and C, s and their corresponding current and voltage measurements are denoted as $i_{C,p}^{(k)}$, $v_{C,p}^{(k)}$ and $i_{C,s}^{(k)}$, $v_{C,s}^{(k)}$, respectively. Finally, the remaining variables $i_{dc,1-4}^{(k)}$, $u_{1-6}^{(k)}$, $i_p^{(k)}$ and $i_s^{(k)}$, $v_{m,p}^{(k)}$ and $v_{m,s}^{(k)}$ depicted in Fig. 2, refer to the DC side currents per module, the gating signals of the switches, the medium frequency AC-link currents on the primary and secondary side of the MFT, and the modulated voltage on either side of the MFT, respectively.

2.2 Adopted Control Strategy

The control strategies implemented in ISOP SST topologies are primarily deployed to ensure DC-link voltage balancing and module power dispatch balancing. Voltage balancing strategies fall into two primary categories: (1) Employing the AFEC [23] or (2) via the Isolated DC/DC stage [24]–[29]. The former approach entails conventional multi-level inverter control for the AFEC, which regulates the primary DC-link voltage by employing a cascaded outer voltage inner current loop alongside a voltage balancing algorithm. Simultaneously, the DC/DC stage governs the secondary DC-link voltage while integrating a power-balancing algorithm. Conversely, the latter approach integrates the voltage balancing controller within the DC/DC control, compensating for deviations in the primary DC-link voltage. A variation of the latter proposed by [30], simplifies the control structure and reduces the communication requirements. This control con-

PCIM Europe 2024, 11– 13 June 2024, Nuremberg   DOI: 10.30420/566262201

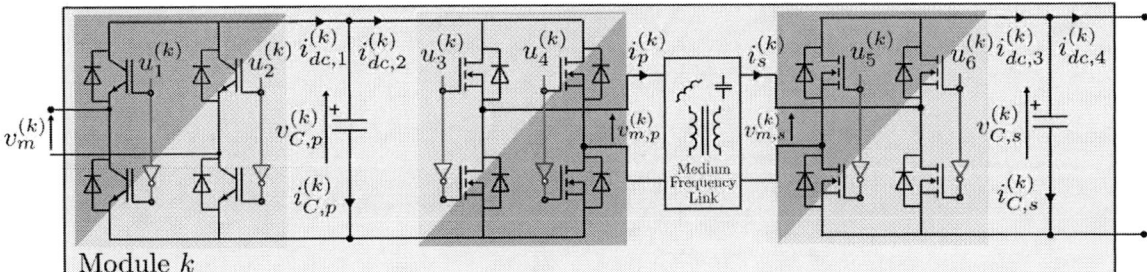

Fig. 2: Single AC-DC/DC module.

figuration, illustrated in Fig. 3 for DAB operation of the DC/DC stage, entails the AFEC controlling the secondary DC voltage, while the DC/DC converters regulate the primary-side DC voltage. Consequently, in scenarios involving equivalent modules, power flow automatically balances out among the modules. This control arrangement offers additional advantages for SAB operation, whereby maintaining a constant duty cycle causes the primary-side voltage to adjust automatically to accommodate the power to be transferred by each module. While the averaged model proposed in this paper can accommodate any of the described control structures, the simplicity of the strategy proposed by [30] makes it the preferred choice for this study.

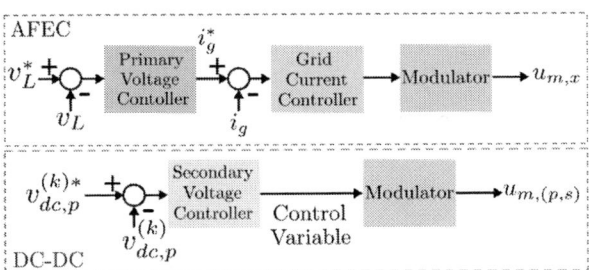

Fig. 3: Control structure proposed by [30].

3 Averaged Model with Blocking Capability

To deploy an equivalent averaged model of the SST, we condense all modules of each phase into a single equivalent circuit in which we deploy controlled voltage and current sources to emulate the converter's behavior without modeling each switching component.

The proposed model allows for simplification in the context of voltage and power balancing, which is achieved through a two-fold approach. Firstly, we assume a perfect voltage balance by considering all primary and secondary side capacitors' voltages

of the cells of a given phase are the same. This assumption is mathematically translated as follows.

$$v_{C,p}^{(k)} \triangleq v_{C,p} \qquad v_{C,s}^{(k)} \triangleq v_{C,s} \qquad (1)$$

Secondly, we assume uniform utilization among all modules considering the same primary and secondary-side modulation functions, this assumption yeilds

$$v_m^\Sigma = u_m^\Sigma v_{C,p} \quad u_{m,p}^{(k)} \triangleq \frac{u_{m,p}^\Sigma}{N} \quad u_{m,s}^{(k)} \triangleq \frac{u_{m,s}^\Sigma}{N} \qquad (2)$$

$$i_p^{(k)} \triangleq i_p \qquad i_s^{(k)} \triangleq i_s \qquad i_{dc,4}^{(k)} \triangleq \frac{i_L}{N} \qquad (3)$$

where u_m^Σ, $u_{m,p}^\Sigma$, and $u_{m,s}^\Sigma$ are the sum of the modulating functions of all AFEC, primary-side DC/DC converter, and secondary-side DC/DC converter within a phase. Taking into account the aforementioned assumptions presented in Eqs. (1) to (3), the capacitors' dynamics can be deduced as

$$C_p \frac{d}{dt} v_{C,p} = \frac{u_m^\Sigma}{N} i_g - \frac{u_{m,p}^\Sigma}{N} i_p \qquad (4)$$

$$N C_s \frac{d}{dt} v_{C,s} = u_{m,s}^\Sigma i_s - i_L \qquad (5)$$

The proposed averaged model simplifies the whole AFEC using the Fundamental Harmonic Approximation (FHA), where only the fundamental component of the switching function is considered while modulation harmonics are neglected. This allows for accurate modeling of AC/DC converters where the first-order harmonic prevails [13]. Thus, the sum of modulating functions of AFECs in a phase is approximated as

$$u_m^\Sigma = \sum_{k=1}^{N} u_m^{(k)} \approx N M \cos\left(2\pi f_0 t + \theta_m\right) \qquad (6)$$

where M and θ_m are the AFEC modulation index amplitude and phase, respectively, and f_0 is the grid fundamental frequency.

1497

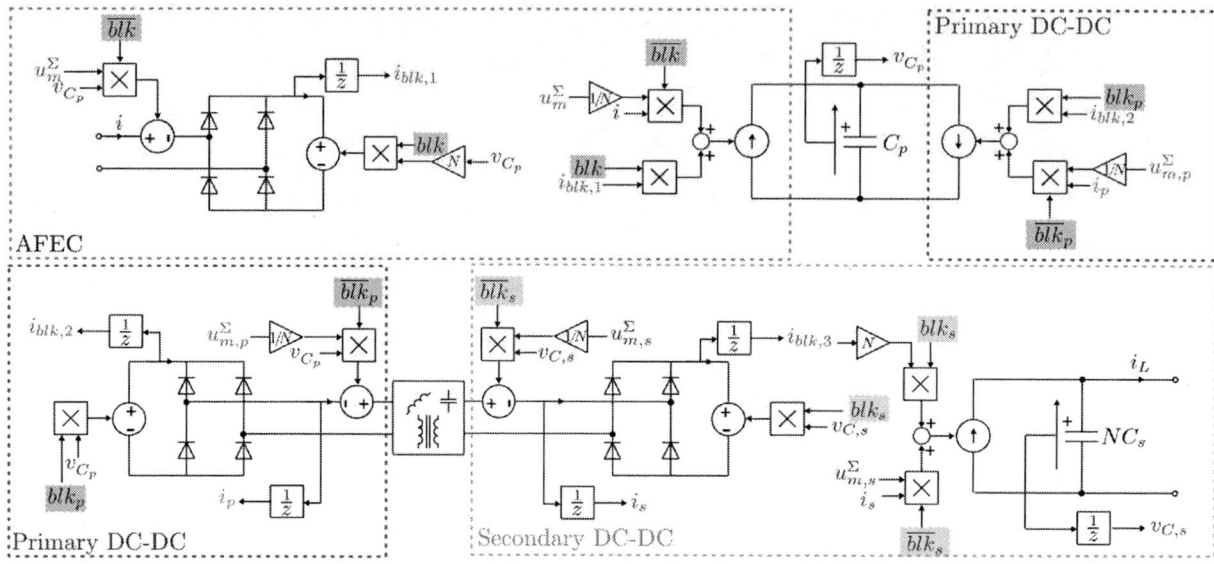

Fig. 4: SST hybrid averaged model.

Whereas for the Isolated DC/DC converter, we consider the first N_h harmonics of the modulating function. Including higher-order harmonics introduces harmonic power transfers, which influence the DC voltage output. As more harmonics are included, the steady-state error decreases. In this work, we consider up to the 9th harmonic, going above this threshold has negligible effect on the converters output [31]. The modulating function as presented in Fig. 5 considers three degrees of freedom, which allows for different modulations: (1) phase-shift $\varphi_{(p,s)}$, (2) duty cycle $\alpha_{(p,s)}$, and (3) switching frequency f_{sw}.

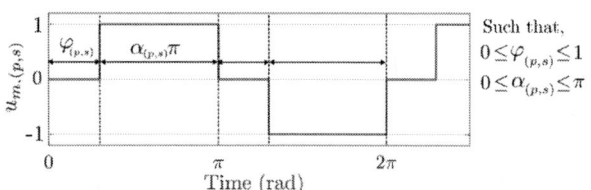

Such that,
$$0 \leq \varphi_{(p,s)} \leq 1$$
$$0 \leq \alpha_{(p,s)} \leq \pi$$

Fig. 5: DC/DC converter modulated voltage.

The modulation function of the DC/DC converter can be written in terms of Fourier series as

$$u_{m,(p,s)} = \sum_{n=-\infty}^{\infty} U_{m,(p,s)}^{(k)} e^{j2\pi n f_{sw} t} \qquad (7)$$

The Fourier coefficients corresponding to the modulated voltage depicted in Fig. 5 are

$$U_m^{(k)} = \begin{cases} 0 & \text{, if } n = 0 \\ \frac{j}{2\pi n}\left(e^{-jn\left(\varphi_{(p,s)}+\alpha_{(p,s)}\pi\right)} \right. \\ \quad - e^{-jn\left(\pi+\varphi_{(p,s)}+\alpha_{(p,s)}\pi\right)} \\ \quad + e^{-jn\left(\varphi_{(p,s)}+\pi\right)} \\ \left. \quad - e^{-jn\left(\varphi_{(p,s)}\right)} \right) & \text{, if } n \neq 0. \end{cases}$$

$$(8)$$

Thus, the switching function of an arbitrary FB-VSC in the Isolated DC/DC converter is given by

$$u_{m,(p,s)} = \sum_{n=1}^{N_h} \frac{4}{n\pi} \sin\left(\frac{n\pi}{2}\alpha_{(p,s)}\right)$$
$$\cos\left(n\left(2\pi f_{sw}t + \frac{\varphi_{(p,s)}}{} + \frac{\pi}{2}\alpha_{(p,s)}\right)\right)$$
$$(9)$$

Finally, to accurately capture the blocking capability of the modules, the equivalent circuit should reproduce the behavior of the anti-parallel diodes when power cell are blocked. This feature is achieved by integrating the diodes bridge with the averaged model of each FB-VSC as illustrated in Fig. 4. Thus, the converters within each module must be capable of transitioning between two operating conditions: the simplified harmonic behaviour of the actively switched bridge and its blocked state. This transition between states is accomplished via a trigger signal denoted blk, which sets the respective voltage source representing the behaviour of an actively switched bridge to zero when the module is blocked ($blk = 1$). As a result, the equivalent model of each FB-VSC is realized by two voltage sources and four diodes on the AC side and a current source on the DC side. Applying the same analogy to each FB-VSC within a power cell alongside the aforementioned simplifications, yields the hybrid averaged model presented in Fig. 4.

4 Model Validation

To verify the proposed averaged model, time-domain simulations of both the switched model (denoted $Full$) and the hybrid averaged model (denoted Avg) were conducted in MATLAB/Simulink. These simulations were carried out using the parameters outlined in Tab. 1. The DC/DC converter of the tested SST integrates a symmetrical resonant tank at its medium frequency link and operates in SAB operation. In addition, a fixed time-step solver with a time step of $T_{sim} = 0.1~\mu s$ is deployed in both models, while the sampling rate is set to $T_{sys} = 10~T_{sim}$. Three distinct test cases are explored to verify the accuracy of the averaged model in representing the dynamics of the ISOP SST: 1) Active Power Step, 2) SST start-up procedure, and 3) Fault Operation. These simulation models and the test cases were conducted on a Windows 10 Pro system with an 11th Gen Intel(R) Core(TM) i7-11850H @ 2.50 GHz processor. Additionally, it is essential to note that the tuning of the control loops remains consistent when transitioning between models and/or test cases.

Tab. 1: System parameters

Network Parameter	Value	Units
Grid ph-ph RMS Voltage (V_g)	11	kV
Nominal Apparent Power	6.32e6	VA
Nominal Active Power (P_{nom})	6e6	W
Fundamental Frequency (f_0)	50	Hz
Grid Resistance (R_g)	58.1e-3	Ω
Grid Inductance (L_g)	7.48e-4	H
Line Filter Resistance (R_f)	1e-4	Ω
Line Filter Inductance (L_f)	15e-3	H
Number of Modules per phase (N)	12	-
Primary Capacitance (C_p)	10e-3	F
Secondary Capacitance (C_s)	0.8e-3	F
AFEC switching frequency ($f_{sw,AFE}$)	500	Hz
Isolated DC/DC switching frequency ($f_{sw,DC}$)	10e3	Hz
Isolated DC/DC Resonant frequency (f_{res})	11.11e3	Hz
Transformer leakage inductance (L_σ)	1.1e-4	H
Transformer magnetizing inductance (L_m)	14.4e-3	H
Transformer turns-ratio	1	-
Nominal DC-link voltages ($V_{dc,p}, V_{dc,s}$)	0.95	kV

4.1 Power Step Response

At t = 0.5 s, a step change of $P = 0.8~P_{nom}$ is introduced into the system. The results of this step change are gathered in Fig. 6, illustrating the modulated voltage and grid current of phase a, along with the voltages of the primary and secondary DC-link capacitors, respectively. The results demonstrate a comparable transient response between the two models, with a 1% marginal difference during the transient state of the DC-link voltages and approximately 0.5% error in the steady-state values. This observed error is attributed to the reduced number of harmonics of the averaged model.

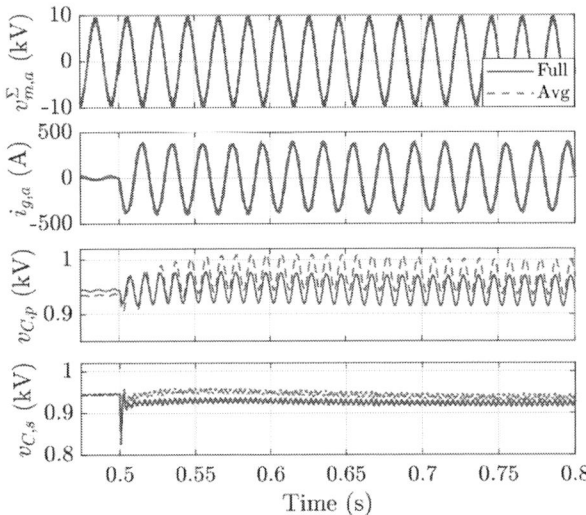

Fig. 6: System dynamics with respect to $0.8~P_{nom}$ step change.

The simulation runtime for the power step response is summarized in Tab. 2. In this test case, the averaged model runtime was approximately 9 times faster than the switched model.

Tab. 2: Elapsed real-time per simulation fundamental period: Power Step Response

	Switched	Averaged
Time (s)	70.9094	7.8773

4.2 SST Start-up Routine

The start-up routine for the SST in this study follows a similar routine used for MMCs as described in [32]. Initially, the control system is down while AC current is provided from the grid, charging the primary DC-link capacitors via pre-insertion resistors (PIRs) and diode rectifiers. Subsequently, the primary bridge of the DC/DC converter is activated to enable current flow to the secondary-side capacitors. In this study, the duty cycle undergoes a gradual ramp-up to its maximum value within 0.45 s to mitigate rapid transients. Once the voltage stabilizes, the PIRs are bypassed further increasing the capacitors' voltage. Following this, the control system is activated, incrementally adjusting the capacitors' voltages to align with their specified references through the voltage control loops. Once transient behaviors reach a stable state, the SST is put into operation with the initiation of a load step. These procedural stages were implemented for both the switched and averaged models, with outcomes illustrated in Fig. 7. Although the transient behavior in both models is nearly identical, a small steady-state error around 0.5% is evident between them, as highlighted more distinctly in the outcomes of the previous test case illustrated in Fig. 6.

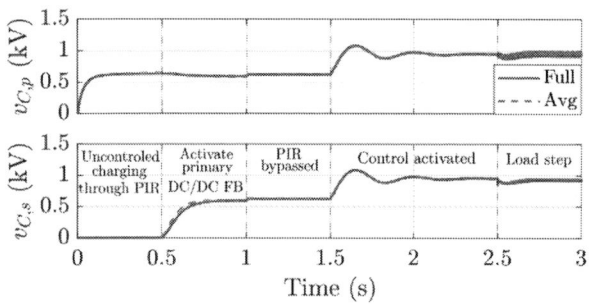

Fig. 7: Primary and secondary DC capacitors voltage during start-up routine.

The simulation runtime for the SST start-up routine is summarized in Table 3. In this test case, the averaged model runtime was approximately 8 times faster than the switched model.

Tab. 3: Elapsed real-time per simulation fundamental period: Start-up routine

	Switched	Averaged
Time (s)	82.2407	10.3325

4.3 Fault-operation

Two tests were performed on the SST to examine the behavior of the averaged model under blocked operation conditions in the event of a fault. The location and nature of the faults are indicated in Fig. 8 and are discussed in the following subsec-

tions. Both faults are introduced into the system during an operation at $0.8 P_{nom}$.

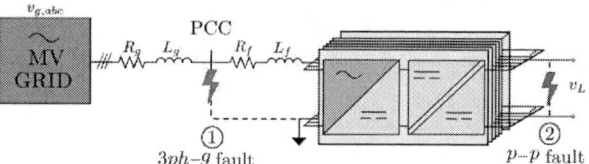

Fig. 8: Investigated fault scenarios.

4.3.1 AC-Grid Fault

At t=0.3 s, a 3-phase AC fault to ground is introduced into the system, with the assumption that the fault is detected within 1 ms, prompting the converter to transition into blocked operation. As shown in Fig. 9, when the fault occurs, the AC current increases rapidly and equally in both models reaching a surge value of $i_{fault} = 54.5$ kA. Once the SST switches into blocked operation, the path of power flow between the primary and secondary bridge of the DC/DC converter is blocked, Thus, the d-component of the grid current $i_{g,d}$ drops to zero. This leads to the load being entirely supplied by the secondary capacitor, causing the voltage $v_{C,s}$ to discharge to zero.

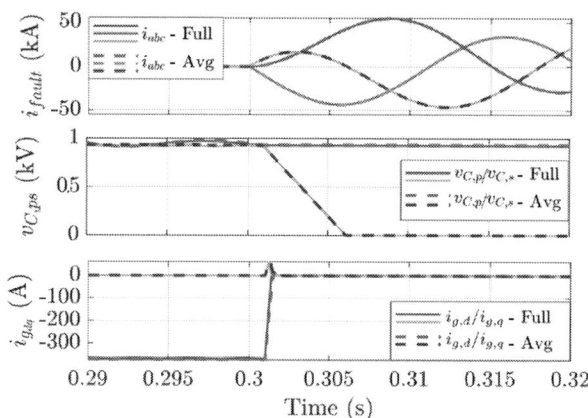

Fig. 9: Three-phase to ground fault.

The simulation runtime for the AC-grid fault is summarized in Table 4. In this test case, the averaged model runtime was approximately 6.5 times faster than the switched model.

Tab. 4: Elapsed real-time per simulation fundamental period: AC-Grid Fault

	Switched	Averaged
Time (s)	56.0141	8.6153

4.3.2 DC-Grid Fault

At t=0.3 s, a pole-to-pole DC fault is introduced in the system, with blocked operation activated 1 ms following the occurrence of the fault for self-protection. As depicted in Fig. 10, the fault current attains the approximately the same surge value ($\approx$ 86 kA) in both models, and their system dynamics show consistency. Unlike the AC-phase fault scenario, where the discharge of the secondary DC-link voltage occurs only after entering the blocked state, in this case, the secondary-side capacitors promptly discharge into the fault. Furthermore, the grid current experiences a rapid increase in its peak value as presented by $i_{g,d}$ in response to the rapid rise in load seen by the AFEC. These findings demonstrate how the averaged model adeptly reproduces the resultant transient behavior of the DC fault.

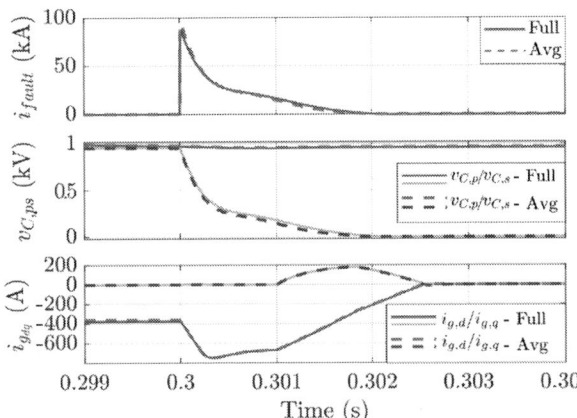

Fig. 10: Pole-to-pole DC fault.

The simulation runtime for the DC-grid fault is summarized in Table 5. In this test case, the averaged model runtime was approximately 6.9 times faster than the switched model.

Tab. 5: Elapsed real-time per simulation fundamental period: DC-Grid Fault

	Switched	Averaged
Time (s)	55.0075	8.0153

5 Conclusion

This paper presents the fundamental concept and methodologies for developing a hybrid time-domain averaged model of modular SSTs with blocking capability. The primary objective of the model is to reduce the computational burden associated with switched models while retaining a portion of the power electronic device-level information. The obtained results demonstrated the high accuracy of the averaged model in capturing the external dynamics of the SST. Furthermore, the results indicate that the internal dynamics of the SST remain unaffected by the balancing controller in the implemented test cases. However, it is noteworthy that in fault scenarios, this may not hold true if the SST fails to enter the blocked state, thereby leading to differences in internal dynamics. Whereas in terms of steady-state analysis, the averaged model waveforms exhibited an error margin of approximately 0.5% in compared to the switched model. Furthermore, the proposed model enhances the overall runtime of the modular SST system, up to a factor of 9, and paves way for system-level simulation aimed at power flow, control, and stability analysis, thereby facilitating simulations involving multiple SSTs.

References

[1] M. Brenna, F. Foiadelli, and H. J. Kaleybar, "The evolution of railway power supply systems toward smart microgrids: The concept of the energy hub and integration of distributed energy resources," *IEEE Electrification Magazine*, vol. 8, no. 1, pp. 12–23, Mar. 2020, Conference Name: IEEE Electrification Magazine. DOI: 10.1109/MELE.2019.2962886.

[2] S. Cui, R. Marquardt, and R. W. de Doncker, "Modular multilevel DC-DC converters interconnecting high-voltage and medium-voltage DC grids," ISBN: 9783942789677 Number: RWTH-2019-05892, Ph.D. dissertation, E.ON Energy Research Center, RWTH Aachen University, 2019. DOI: 10.18154/RWTH-2019-05892.

[3] J. E. Huber and J. W. Kolar, "Applicability of solid-state transformers in today's and future distribution grids," *IEEE Transactions on Smart Grid*, vol. 10, no. 1, pp. 317–326, Jan. 2019, Conference Name: IEEE Transactions on Smart Grid. DOI: 10.1109/TSG.2017.2738610.

[4] R. Pillon Barcelos and D. Dujić, "Direct current transformer impact on the DC power distribution networks," *IEEE Transactions on Smart Grid*, vol. 13, no. 4, pp. 2547–2556, Jul. 2022, Conference Name: IEEE Transactions on Smart Grid. DOI: 10.1109/TSG.2022.3162310.

[5] X. Yu, X. She, X. Zhou, and A. Q. Huang, "Power management for DC microgrid enabled by solid-state transformer," *IEEE Transactions on Smart*

Grid, vol. 5, no. 2, pp. 954–965, Mar. 2014, Conference Name: IEEE Transactions on Smart Grid. DOI: 10.1109/TSG.2013.2277977.

[6] H. A. B. Siddique and R. W. De Doncker, "Evaluation of DC collector-grid configurations for large photovoltaic parks," *IEEE Transactions on Power Delivery*, vol. 33, no. 1, pp. 311–320, Feb. 2018, Conference Name: IEEE Transactions on Power Delivery. DOI: 10.1109/TPWRD.2017.2702018.

[7] G. Abeynayake, G. Li, J. Liang, and N. A. Cutululis, "A review on MVdc collection systems for high-power offshore wind farms," in *2019 14th Conference on Industrial and Information Systems (ICIIS)*, ISSN: 2164-7011, Dec. 2019, pp. 407–412. DOI: 10.1109/ICIIS47346.2019.9063352.

[8] R. Medeiros and I. Colak, "Solid-state transformers: How far have we come?" In *PCIM Europe 2023; International Exhibition and Conference for Power Electronics, Intelligent Motion, Renewable Energy and Energy Management*, May 2023, pp. 1–12. DOI: 10.30420/566091329.

[9] H. Saad, S. Dennetière, and J. Mahseredjian, "On modelling of MMC in EMT-type program," in *2016 IEEE 17th Workshop on Control and Modeling for Power Electronics (COMPEL)*, Jun. 2016, pp. 1–7. DOI: 10.1109/COMPEL.2016.7556717.

[10] R. Coelho-Medeiros, B. Džonlaga, J.-C. Vannier, J. Dai, L. Queval, and P. Egrot, "A comparison between different models of the modular multilevel converter," in *2020 22nd European Conference on Power Electronics and Applications (EPE'20 ECCE Europe)*, Sep. 2020, P.1–P.10. DOI: 10.23919/EPE20ECCEEurope43536.2020.9215698.

[11] A. Zama, B. Seddik, A. Benchaib, D. Frey, and S. Silvant, "A novel modular multilevel converter modelling technique based on semi-analytical models for hvdc application," *Journal of Electrical Systems*, vol. 12, Nov. 2016.

[12] A. Meligy, T. Qoria, and I. Colak, "Assessment of sequence extraction methods applied to mmc-sdbc statcom under distorted grid conditions," *IEEE Transactions on Power Delivery*, vol. 37, no. 6, pp. 4923–4932, 2022. DOI: 10.1109/TPWRD.2022.3162959.

[13] S. Bacha, I. Munteanu, and A. I. Bratcu, "Generalized averaged model," in *Power Electronic Converters Modeling and Control: with Case Studies*, ser. Advanced Textbooks in Control and Signal Processing, London: Springer, 2014, pp. 97–147. DOI: 10.1007/978-1-4471-5478-5_5.

[14] Y. Jiang, "Modeling of solid state transformer for the FREEDM system demonstration," 2014.

[15] J. A. Martinez-Velasco, S. Alepuz, F. González-Molina, and J. Martin-Arnedo, "Dynamic average modeling of a bidirectional solid state transformer for feasibility studies and real-time implementation," *Electric Power Systems Research*, vol. 117, pp. 143–153, Dec. 1, 2014. DOI: 10.1016/j.epsr.2014.08.005.

[16] J. A. Mueller and J. W. Kimball, "Generalized average modeling of DC subsystem in solid state transformers," in *2017 IEEE Energy Conversion Congress and Exposition (ECCE)*, Oct. 2017, pp. 1659–1666. DOI: 10.1109/ECCE.2017.8095992.

[17] G. Nayak and A. Dasgupta, "Full order averaged modelling for modular solid state transformer," in *2020 IEEE International Conference on Power Electronics, Drives and Energy Systems (PEDES)*, Dec. 2020, pp. 1–6. DOI: 10.1109/PEDES49360.2020.9379573.

[18] Z. Yu, R. Ayyanar, and I. Husain, "A detailed analytical model of a solid state transformer," in *2015 IEEE Energy Conversion Congress and Exposition (ECCE)*, ISSN: 2329-3748, Sep. 2015, pp. 723–729. DOI: 10.1109/ECCE.2015.7309761.

[19] T. Zhao, J. Zeng, S. Bhattacharya, M. E. Baran, and A. Q. Huang, "An average model of solid state transformer for dynamic system simulation," in *2009 IEEE Power & Energy Society General Meeting*, ISSN: 1932-5517, Jul. 2009, pp. 1–8. DOI: 10.1109/PES.2009.5275542.

[20] H. Cheng, Y. Gong, and Q. Gao, "The research of coordination control strategy in cascaded multilevel solid state transformer," in *2015 IEEE International Conference on Mechatronics and Automation (ICMA)*, ISSN: 2152-744X, Aug. 2015, pp. 222–226. DOI: 10.1109/ICMA.2015.7237486.

[21] A. Meligy, R. Coelho-Medeiros, I. Colak, and S. Bacha, "Efficiency-driven parameter selection for dual active bridge converters," in *2023 25th European Conference on Power Electronics and Applications (EPE'23 ECCE Europe)*, Sep. 2023, pp. 1–8. DOI: 10.23919/EPE23ECCEEurope58414.2023.10264507.

[22] S. Deshmukh (Gore), A. Iqbal, S. Islam, I. Khan, M. Marzband, *et al.*, "Review on classification of resonant converters for electric vehicle application," *Energy Reports*, vol. 8, pp. 1091–1113, Nov. 1, 2022. DOI: 10.1016/j.egyr.2021.12.013.

[23] T. Zhao, G. Wang, S. Bhattacharya, and A. Q. Huang, "Voltage and power balance control for a cascaded h-bridge converter-based solid-state transformer," *IEEE Transactions on Power Electronics*, vol. 28, no. 4, pp. 1523–1532, Apr. 2013, Conference Name: IEEE Transactions on Power Electronics. DOI: 10.1109/TPEL.2012.2216549.

[24] H.-J. Yun, H.-S. Kim, M.-H. Ryu, J.-W. Baek, and H.-J. Kim, "A simple and practical voltage balance method for a solid-state transformer using cascaded h-bridge converters," in *2015 9th International Conference on Power Electronics and ECCE Asia (ICPE-ECCE Asia)*, 2015, pp. 2415–2420. DOI: 10.1109/ICPE.2015.7168109.

[25] S. Pugliese, M. Andresen, R. Mastromauro, G. Buticchi, S. Stasi, and M. Liserre, "Voltage balancing of modular smart transformers based on dual active bridges," in *2017 IEEE Energy Conversion Congress and Exposition (ECCE)*, 2017, pp. 1270–1275. DOI: 10.1109/ECCE.2017.8095935.

[26] G. Wang, X. She, F. Wang, A. Kadavelugu, T. Zhao, *et al.*, "Comparisons of different control strategies for 20kva solid state transformer," in *2011 IEEE Energy Conversion Congress and Exposition*, ISSN: 2329-3748, Sep. 2011, pp. 3173–3178. DOI: 10.1109/ECCE.2011.6064196.

[27] J. Liu, J. Yang, J. Zhang, Z. Nan, and Q. Zheng, "Voltage balance control based on dual active bridge DC/DC converters in a power electronic traction transformer," *IEEE Transactions on Power Electronics*, vol. 33, no. 2, pp. 1696–1714, Feb. 2018, Conference Name: IEEE Transactions on Power Electronics. DOI: 10.1109/TPEL.2017.2679489.

[28] S. Pugliese, M. Andresen, R. A. Mastromauro, G. Buticchi, S. Stasi, and M. Liserre, "A new voltage balancing technique for a three-stage modu-lar smart transformer interfacing a DC multibus," *IEEE Transactions on Power Electronics*, vol. 34, no. 3, pp. 2829–2840, Mar. 2019, Conference Name: IEEE Transactions on Power Electronics. DOI: 10.1109/TPEL.2018.2840961.

[29] S. Pugliese, G. Buticchi, R. A. Mastromauro, M. Andresen, M. Liserre, and S. Stasi, "Soft-start procedure for a three-stage smart transformer based on dual-active bridge and cascaded h-bridge converters," *IEEE Transactions on Power Electronics*, vol. 35, no. 10, pp. 11 039–11 052, 2020. DOI: 10.1109/TPEL.2020.2977226.

[30] P. Jang, H.-p. Park, J. Baek, D.-U. Kim, and S. Kim, "A simple control method without voltage balance algorithm for modular solid-state transformer," in *2022 IEEE Energy Conversion Congress and Exposition (ECCE)*, ISSN: 2329-3748, Oct. 2022, pp. 1–6. DOI: 10.1109/ECCE50734.2022.9948182.

[31] S. Ghosh, D. Das, B. Singh, S. Janardhanan, and S. Mishra, "Frequency-domain modeling of dual-active-bridge converter based on harmonic balance approach," *IEEE Journal of Emerging and Selected Topics in Industrial Electronics*, vol. 3, no. 1, pp. 166–176, Jan. 2022. DOI: 10.1109/JESTIE.2021.3051591.

[32] S. Dennetière, P. Rault, T. Priebe, S. Beckler, H. Saad, *et al.*, "Guide for electromagnetic transient studies involving VSC converters," *e-cigre*, Apr. 16, 2021.

PCIM Europe 2024, 11– 13 June 2024, Nuremberg   DOI: 10.30420/566262203

Surfactant-Modified Nanocomposite Thin-Film Capacitors

Bartosz Gackowski[ID], Shova Neupane[ID], Luciana Tavares[ID], Thomas Ebel[ID], William Greenbank[ID]

Centre for Industrial Electronics, University of Southern Denmark, Alsion 2, 6400 Sønderborg, Denmark

Corresponding author: William Greenbank, greenbank@sdu.dk
Speaker: Bartosz Gackowski, gackowski@sdu.dk

Abstract

This paper discusses the potential and importance of interface engineering of barium titanate nanoparticles for dielectric capacitors. The $BaTiO_3$ nanoparticles were dispersed in a polypropylene gel with the aid of two types of surfactants: an ionic sodium dodecyl sulfate (SDS) and a nonionic Triton X-100. Optical microscopy and light scattering analyses revealed a notable decrease in nanoparticle agglomeration within the polypropylene gel when surfactants were employed, which could reach the diameter of a single nanoparticle. Dielectric capacitors were manufactured through multi-layer spin coating. The capacitor comprised top and bottom layers of neat polypropylene with three intermediary nanocomposite layers. Comparative analysis showcased a 37% increase in the dielectric constant of capacitors utilizing the ionic surfactant compared to the nonionic alternative, and a 23% increase over the devices without any surfactant. Conversely, the incorporation of the nonionic surfactant resulted in a 140% surge in dielectric strength compared to devices with the ionic surfactant. Overall, devices with the nonionic surfactant exhibited an energy density exceeding four times that of the ionic surfactant, and double that of biaxially oriented or spin-coated polypropylene devices. Thus, interface engineering emerges as a promising and efficient strategy for enhancing nanoparticle dispersibility in polypropylene, consequently elevating the energy density of film capacitors.

1 Introduction

Capacitors can temporarily store and release electrical energy; thus, they can smooth out power supply fluctuations, filter noise in electronic circuits, and block direct current, among others. Therefore, they are commonly used in power electronics for power transmission and distribution, motors and drives, power supplies, adapters, and converters, responsible for 31%, 23%, and 13% of the overall capacitor market, respectively. Film capacitors are the most widely used type of capacitors and are responsible for 50% of the power capacitor market, followed by ceramic (30%), aluminium electrolytic capacitors (15%), tantalum (1%), and carbon supercapacitors (4%) [1]. Film capacitors are composed of two conductive electrodes separated by an insulating dielectric material. In most cases, the dielectric is made of polypropylene (PP), which offers high dielectric strength, self-healing, low leakage current, and reliability. However, they suffer from poor temperature tolerance and low volumetric capacitance. That means improvements in the energy density of polypropylene could lead to smaller and lighter capacitors, which is particularly important for e-mobility.

Nanocomposite dielectrics are a promising way forward to solve the problem. Ceramic nanoparticles (NPs), such as silicon dioxide or barium titanate, have high thermal stability and dielectric constants. However, once mixed with the thermoplastic resin, the nanoparticles tend to cluster which may result in a short circuit or greatly decrease the breakdown strength. As the capacitive energy density of a dielectric depends on the square of breakdown strength (E_b), any decline in that property can cancel out any gains from the higher dielectric constant (ε_r) of polypropylene nanocomposites, as shown in Equation 1, where ε_0 is the constant vacuum permittivity.

$$U = \frac{1}{2}\varepsilon_r \varepsilon_0 E_b^2 \qquad (1)$$

The agglomeration of nanoparticles has many origins, including chemical incompatibility between the nanoparticles and thermoplastics, high surface area and surface energy exhibited by nanoparticles, and related non-covalent interactions, such as van der Waals forces [2-4]. Thus, obtaining a uniform blend of nanoparticles within a polymer matrix is challenging [5], and the dispersion level will dictate the thermal, electrical, and mechanical properties of the nanocomposite [6]. Moreover, the most common polymer used for film capacitors is biaxially oriented polypropylene [7], which is hydrophobic [8]. On the other hand, many nanoparticles, such as barium titanate [9] and silicon dioxide [10] exhibit hydrophilic properties. The difference in polarity and surface energies restrict dispersion and impede achieving a homogenous composite [11].

On the other hand, it has been recently proposed that dielectric capacitors can be manufactured from a polypropylene gel in a mixture of solvents [12-14]. The advantage of this method is the lower viscosity than in melt mixing, which aids in separating nanomaterials. However, this approach alone does not tune the complex interfacial phenomena between the nanoparticles and polymer matrix. That can be overcome by chemical functionalization of nanoparticles, which may include using surfactants with polar and non-polar groups [15]. Thus, surfactants can tune the interfacial interaction [16], and lead to a reduced agglomeration of nanoparticles in a liquid medium [17]. The hydrophilic part of surfactants may either be electrically neutral (nonionic) or carry charge groups (for example negative in anionic surfactants) [18]. This work aims to determine if surfactants are an efficient surface treatment for barium titanate nanoparticles in polypropylene matrix and explore the impact of either anionic or nonionic surfactants on the dispersibility of nanoparticles in the gel and the ensuing dielectric properties.

2 Methodology

2.1 Materials

Polypropylene pellets (amorphous), o-xylene, and sodium dodecyl sulfate (SDS) were supplied by Merck, Denmark. Toluene and ethanol were obtained from VWR, Denmark. Barium titanate nanoparticles with a diameter of 50 nm were purchased from US Research Nanomaterials, USA. Triton X-100 was obtained from Fisher Scientific, Denmark.

2.2 Fabrication of dielectric capacitors

First, a polypropylene gel was prepared by dispersing approx. 4.6 wt.% of polypropylene pellet in a mixture of toluene and xylene (50:50, by weight). It was subsequently heated to 140°C and then rapidly cooled down by submerging it in water, which resulted in a semi-transparent liquid. NPs (5 wt.%) and a desired wt.% of either SDS or Triton X-100 (0.5 wt.%, 1 wt.%, 3 wt., or 5 wt.%) were dispersed in ethanol through probe sonication (Fisherbrand™ Model 505) for one hour. Subsequently, the dispersion of NPs in ethanol was

diluted in the PP gel so that it created a 5 wt.% (0.8 vol.%) composite layer after evaporation of the solvents.

The composite dielectric layers were manufactured by spin-coating on a glass substrate with thin layers of titanium (10 nm) and silver (200 nm). A total of five layers were spin-coated on the substrate: the first and last one was the neat PP gel and the three middle layers contained the PP, NPs, and surfactants. After drying, a top electrode made of aluminium (200 nm) was deposited on top of the dielectric layers using Cryofox Explorer 600 thin film deposition system in a cleanroom environment (see Figure 1). More details about the PP gel and fabrication of dielectric capacitors by spin-coating are available elsewhere [12, 13].

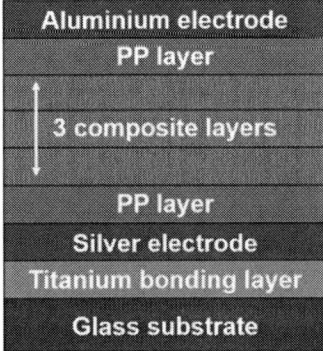

Fig. 1 A schematic of the structure of a dielectric capacitor.

2.3 Characterization

Capacitance was measured using E4980AL Precision LCR Meter from Keysight, USA. Breakdown testing was conducted by applying a DC voltage ramp from a Vitrek 951i Electrical Safety Compliance Analyzer. The thickness of the dielectric layers was determined with a Veeco Dektak 150 profilometer. Scanning Electron Miscroscopy (SEM) and Energy-Dispersive X-ray spectroscopy (EDX) were carried out using Hitachi S-4800. 3D cross-correlation dynamic light scattering (3D-DLS) was performed using LS Instruments LS Spectrometer, equipped with a DP SS 660 nm, 100 mW laser, and a Julabo CF31 external temperature control unit. For these measurements, the 5 wt.% suspensions of NPs in ethanol were diluted in the PP gel to 0.01 wt.%. The diluted suspension was subsequently homogenised using an IKA Ultra Turrax digital homogeniser for 10 minutes and poured into glass test tubes for testing.

3 Results and discussion

3.1 Optical microscope

Optical microscope images of PP layers containing NPs are shown in Figure 2. The non-functionalized NPs form agglomerates scattered throughout the dielectric layer, which can exceed 10 μm in length. As their size may be larger the thickness of the dielectric layer, they would cause a short circuit by creating a conductive junction between the top and bottom electrodes in a capacitor. The addition of either anionic or nonionic surfactants to the suspension of NPs can reduce the extent of agglomeration. However, the images suggest that the ionic surfactant still possessed agglomerates with a diameter of a few microns, although less than the reference sample without surfactants. Thus, the nonionic surfactant was more efficient in dispersing the nanoparticles in PP.

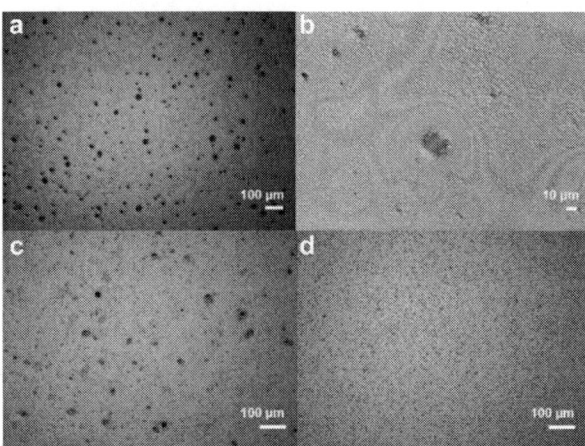

Fig. 2 Optical microscopy images showing the dispersion levels of NPs in PP without surfactants (a-b), and after the addition of SDS (c) and Triton X-100 (d).

3.2 Dynamic light scattering

The quality of dispersion was characterized by light scattering on NPs suspended in the PP gel (see Figure 3). Commercial, neat NPs showed a peak at 1079 nm and a tail that started from 2.7 μm to over 100 μm. The NPs were first dispersed in ethanol and subsequently in the polypropylene gel using both high shear mixing and ultrasonication. The scattering data indicate that these physical methods were not sufficient to overcome the agglomeration forces.

The addition of 0.5 wt.% of nonionic surfactant reduced the size of detected agglomerates to 178-292 nm and 624-1762 nm, while 1 wt.% diminished it further to peaks that ranged between 121-188 nm and 362-1762 nm. On the other hand, both 3 wt.% and 5 wt.% concentrations of the surfactant led to peaks at 63 nm, which is close to the diameter of the barium titanate nanoparticles - 50 nm, and therefore likely represents isolated nanoparticles.

A similar trend can be observed in the polypropylene suspensions with nanoparticles and anionic surfactant. The primary peaks can be observed at 91 nm, 63 nm, and 60 nm after the addition of 0.5 wt.%, 1 wt.% and both 3 and 5 wt.% of the surfactant to the gel, respectively. The secondary peaks reached the highest intensity at 574 nm (0.5 wt.%), 275 nm (1 wt.%), 398 nm (3 wt.%) and 442 nm (5 wt.%), which is significantly lower than the non-functionalized nanoparticles.

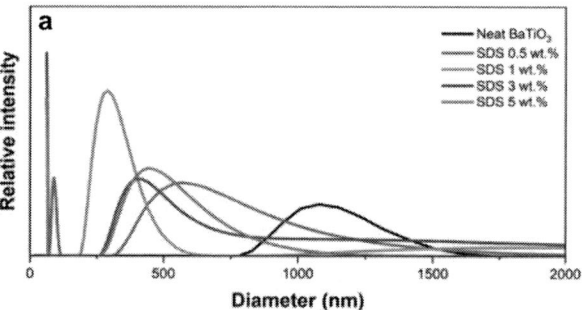

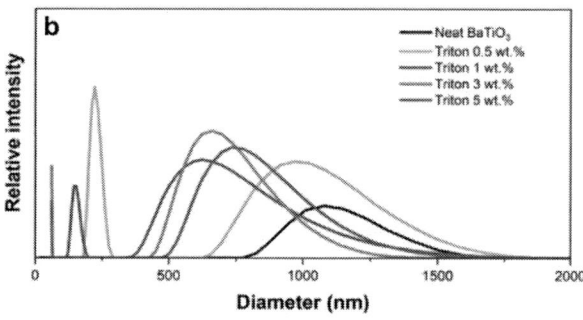

Fig. 3 Light scattering spectra of PP gels with nanoparticles indicating the change in dispersion levels with increasing concentrations of anionic SDS (a) or nonionic Triton X-100 (b) surfactants.

3.3 SEM-EDX

The electron microscopy image of a cross-sectional area of a nanocomposite dielectric capacitor is shown in Figure 4. The thickness of the dielectric layer is below 1 micron, which is in good agreement with the value measured with the profilometer (see section 3.4). Numerous cracks are visible throughout the dielectric material, which might have originated from the sample preparation process that involved freezing the sample with liquid nitrogen and subsequent breaking. The spherical-like masses that are scattered in the PP matrix are most likely the NPs with the surfactant. The light scattering experiments showed that the surfactant leads to peaks at around 50 nm and between 500-1000 nm. The NP agglomerates visible on the SEM image are mostly between 100-200 nm, which highlights the effectiveness of the surfactant in improving dispersion of NPs in PP. An EDX spectrum indicates that most of the material is composed of silica (21 wt.%) and oxygen (37 wt.%), which may come from the glass substrate. Carbon atoms made 25 wt.% of the elemental composition, which might originate from PP and the surfactant. The measurement also revealed the presence of silver (16 wt.%) as it was used as the bottom electrode of the capacitor. The last two elements were titanium (0.45 wt.%), which thin layer (10 nm) was deposited on glass to help bind silver to the substrate. Titanium together with barium (0.25 wt.%) are the elements that form the NPs.

The thickness of five dielectric layers of PP nanocomposite with unmodified NPs spin-coated on the silver electrode was 876 ± 47 nm. After the addition of the anionic or nonionic surfactants to the suspension, the thickness deviated by 100 nm and amounted to 761 ± 27 nm and 930 ± 40 nm, respectively. The capacitance of the devices with neat NPs was measured to be 163 ± 2 pF, which increased to 201 ± 14 pF when the NPs were dispersed with the anionic surfactant. On the other hand, the capacitance decreased to 147 ± 14 pF when the nonionic surfactant was used (see Figure 5a). The frequency sweep data indicated that devices with either anionic or nonionic surfactants show a minimal reduction in capacitance up to 10^6 Hz, but they undergo a rapid decrease to around 100 pF at higher frequencies. The resistance of all devices decreased from approx. 10^5 Ω to 10^1 Ω. After simulating a circuit using the impedance-frequency data, the equivalent series resistance (ESR) of capacitors with neat NPs amounted to 4.64 Ω. The use of either anionic or nonionic surfactants led to an increase in ESR to 8.8 Ω or 16 Ω, respectively.

Assuming the vacuum permittivity is equal to $8.85 \cdot 10^{-12}$ F/m, the dielectric constant of PP composites with NPs dispersed without surfactant was 2.98. The use of SDS increased the dielectric constant by 7% to 3.2, while Triton X-100 reduced it to 2.85 (see Figure 5b). For reference, the dielectric constant of unmodified BOPP is typically cited as 2.2 [19].

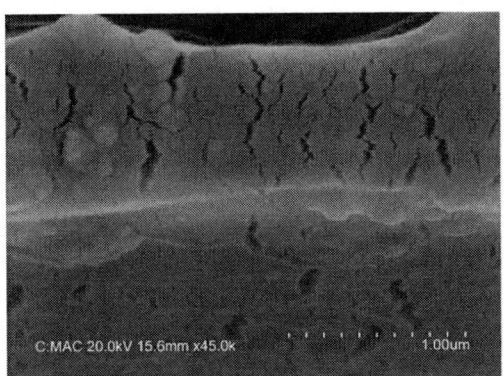

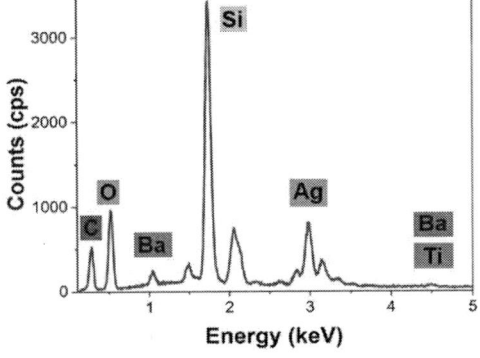

Fig. 4 SEM image of a PP-BaTiO₃-Triton X-100 nanocomposite capacitor (a) and an EDX spectrum (b).

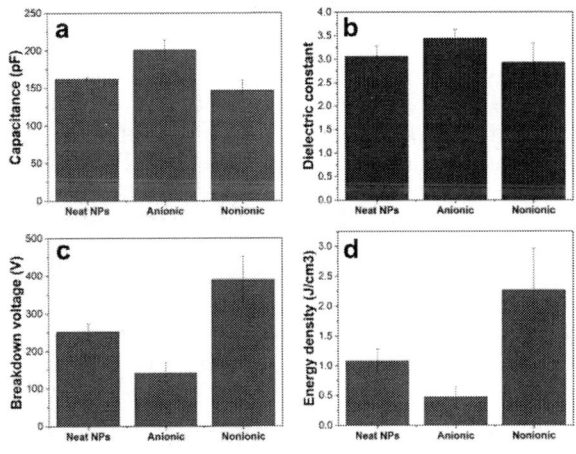

Fig. 5 The comparison of capacitance (a), the dielectric constant (b), breakdown voltage (c), and energy density (d) of nanocomposite capacitors.

3.4 Dielectric properties

The average breakdown voltage of devices with unmodified NPs was 251 ± 22 V. The use of the anionic surfactant in the NP dispersion decreased the breakdown voltage to 141 ± 27 V, while the devices with the nonionic surfactant increased it by 55% to 390 ± 61 V. The representative curves recorded during the experiments are shown in Figure 6.

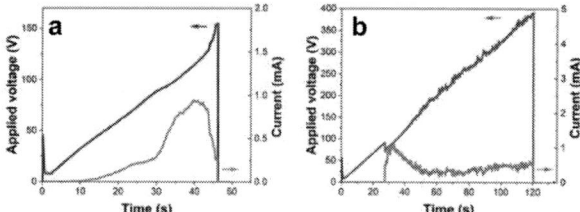

Fig. 6 Breakdown voltage measurements of PP composites with NPs dispersed with anionic SDS (a) or nonionic Triton X-100 (b).

It was reported in previous studies that fabricating PP devices through a similar process led to capacitance of 144 pF and the dielectric constant of 2.03 at 676 nm of thickness [12, 13]. Substituting three layers of PP with the nanocomposite led to an increase in capacitance by 14% and the dielectric constant by 51%. The capacitance of devices containing Triton X-100 was at a similar level as PP, but those with the anionic surfactant exhibited an increase in capacitance by 40%. On the other hand, the dielectric strength of PP devices was within 200-350 MV m⁻¹, and decreased to 100-300 MV m⁻¹ after introducing 0.77 vol.% (5.2 wt.%) of NPs into the capacitor [12]. This work applied a similar volume fraction of NPs (0.8 vol.% or 5 wt.%), which resulted in the dielectric strength of 276 ± 10 MV m⁻¹. The addition of the anionic surfactant to the devices decreased it to 171 ± 38 MV m⁻¹, while the nonionic surfactant improved it to 412 ± 44 MV m⁻¹. Therefore, the results suggest a certain paradigm: the anionic surfactant can improve the capacitance, but it entails a reduction in breakdown strength, while the nonionic surfactant increases the dielectric strength that is far more important for energy density. Thus, the composite capacitors with Triton X-100 can reach 2.2 J cm⁻³ in energy density, which is over twice as much as their counterparts without surfactant (1 J cm⁻³), and over four times more (0.5 J cm⁻³) than the devices with SDS. It is possible that the inclusion of charged species in the dielectric (as in the case of anionic SDS) increases the polarizability of the dielectric (and therefore its dielectric constant), but this also results in the formation of sub band-gap states and/or charge injection into the conductionband of the dielectric, resulting in a decreased dielectric strength [20, 21]. However, identifying the exact physical mechanism goes beyond the scope of this study. The results highlight, though, the importance of tuning the interfacial properties in nanocomposite capacitors, and the critical role that the choice of surfactant chemistry can play in determining device performance.

4 Conclusions

The results presented in this paper led to the following conclusions:

- The addition of nanoparticles to the polypropylene gel led to agglomeration that exceed 10 microns in a dry film or 1-100 µm in a composite dispersion.
- Anionic SDS and non-anionic Triton X-100 surfactants improve the dispersion of nanoparticles in polypropylene as observed through optical microscopy and light scattering.
- The capacitance of devices with SDS was higher by 37% and 23% than devices with Triton X-100 or without surfactant, respectively.
- The average dielectric strength of devices with Triton X-100 was 140% times higher than the counterparts with the SDS surfactant.
- The overall energy density of devices with neat nanoparticles, and those with SDS or Triton X-100 amounted to 1.09 J cm⁻³, 0.5 J cm⁻³ and 2.2 J cm⁻³, respectively.

Acknowledgements

This work was generously supported by the ECPE European Center for Power Electronics through the Joint Research Programme project 2022/PC07 entitled: *Nanoparticle Surface Engineering for Next Generation, High Energy Density Film Capacitors*. Parts of this work were done in the cleanroom facility belonging to the SDU NanoSYD and the Mads Clausen Institute. The authors wish to thank Jacek Fiutowski for his help with light scattering measurements.

References

[1] "Passive Electronic Components: World Market Outlook: 2020-2025," Paumanok Publications, Inc., 2020.

[2] X. Liu *et al.*, "Interface modified BTO@PS-co-mah/PS composite dielectrics with enhanced breakdown strength and ultralow dielectric loss," *RSC Advances*, 10.1039/D2RA06524J vol. 13, no. 2, pp. 1278-1287, 2023, doi: 10.1039/D2RA06524J.

[3] Y. Zare, K. Y. Rhee, and D. Hui, "Influences of nanoparticles aggregation /agglomeration on the interfacial /interphase and tensile properties of nanocomposites," *Composites Part B: Engineering*, vol. 122, pp. 41-46, 2017/08/01/ 2017, doi: https://doi.org/10.1016/j.compositesb.2017.04.008.

[4] S. Shrestha, B. Wang, and P. Dutta, "Nanoparticle processing: Understanding and controlling aggregation," *Advances in Colloid and Interface Science*, vol. 279, p. 102162, 2020/05/01/ 2020, doi: https://doi.org/10.1016/j.cis.2020.102162.

[5] Y. Haldorai and J.-J. Shim, "Manufacturing Polymer Nanocomposites," in *Rheology and Processing of Polymer Nanocomposites*, 2016, pp. 29-67.

[6] T. Hassan *et al.*, "Functional nanocomposites and their potential applications: A review," *Journal of Polymer Research*, vol. 28, no. 2, p. 36, 2021/01/08 2021, doi: 10.1007/s10965-021-02408-1.

[7] Z. Ran *et al.*, "Significantly improved high-temperature capacitive performance in polypropylene based on molecular semiconductor grafting," *Materials Today Energy*, vol. 38, p. 101429, 2023/12/01/ 2023, doi: https://doi.org/10.1016/j.mtener.2023.101429.

[8] G. Tao, A. Gong, J. Lu, H.-J. Sue, and D. E. Bergbreiter, "Surface functionalized polypropylene: synthesis, characterization, and adhesion Properties," *Macromolecules*, vol. 34, no. 22, pp. 7672-7679, 2001/10/01 2001, doi: 10.1021/ma010941b.

[9] S.-J. Chang, W.-S. Liao, C.-J. Ciou, J.-T. Lee, and C.-C. Li, "An efficient approach to derive hydroxyl groups on the surface of barium titanate nanoparticles to improve its chemical modification ability," *Journal of Colloid and Interface Science*, vol. 329, no. 2, pp. 300-305, 2009/01/15/ 2009, doi: https://doi.org/10.1016/j.jcis.2008.10.011.

[10] B. Xu and Q. Zhang, "Preparation and properties of hydrophobically modified Nano-SiO_2 with hexadecyltrimethoxysilane," *ACS Omega*, vol. 6, no. 14, pp. 9764-9770, 2021/04/13 2021, doi: 10.1021/acsomega.1c00381.

[11] M. Gilbert, "Surface treatments for particulate fillers in plastics," in *Plastics Additives: An A-Z reference*, G. Pritchard Ed. Dordrecht: Springer Netherlands, 1998, pp. 590-603.

[12] W. Greenbank and T. Ebel, "Layer-by-layer printable nano-scale polypropylene for precise control of nanocomposite capacitor dielectric morphologies in metallised film capacitors," *Power Electronic Devices and Components*, vol. 4, p. 100025, 2023/03/01/ 2023, doi: https://doi.org/10.1016/j.pedc.2022.100025.

[13] W. Greenbank, P. Gupta, J. Fiutowski, and T. Ebel, "Layer-by-layer printed dielectrics: scalable nanocomposite capacitor fabrication for the green transition," in *PCIM Europe 2022; International Exhibition and Conference for Power Electronics, Intelligent Motion, Renewable Energy and Energy Management*, 10-12 May 2022 2022, pp. 1-7, doi: 10.30420/565822052.

[14] Greenbank, W., & Ebel, T., "Composite Dielectric Material, Capacitor and Methods for producing said Composite Dielectric Material and said Capacitor" (Patent No. WO 2023/131491 A1). 2023. European Patent Office.

[15] Staszak, K., Wieszczycka, K. and Tylkowski, B, Chemical Technologies and Processes. Berlin, Boston: De Gruyter, 2020.

[16] S. M. Shaban, J. Kang, and D.-H. Kim, "Surfactants: recent advances and their applications," *Composites Communications*, vol. 22, p. 100537, 2020/12/01/ 2020, doi: https://doi.org/10.1016/j.coco.2020.100537.

[17] D. O. Zelentsov *et al.*, "Influence of anionic surfactant on stability of nanoparticles in aqueous solutions," *Chimica Techno Acta*, vol. 10, no. 3, p. 202310302, 2023.

[18] T. Tadros, "Surfactants," in *Encyclopedia of Colloid and Interface Science*, T. Tadros Ed.

Berlin, Heidelberg: Springer Berlin Heidelberg, 2013, pp. 1242-1290.

[19] J. Dong *et al.*, "Scalable high-permittivity polyimide copolymer with ultrahigh high-temperature capacitive performance enabled by molecular engineering," *Advanced Energy Materials,* vol. 14, no. 9, p. 2303732, 2024, doi: https://doi.org/10.1002/aenm.202303732.

[20] V. Bordo and T. Ebel, "Theory of electrical breakdown in a nanocomposite capacitor," *Applied Sciences,* vol. 12, no. 11, p. 5669, 2022, doi: https://www.mdpi.com/2076-3417/12/11/5669.

[21] X. He *et al.*, "Combining good dispersion with tailored charge trapping in nanodielectrics by hybrid functionalization of silica," *e-Polymers,* vol. 21, no. 1, pp. 897-909, 2021, doi:10.1515/epoly-2021-0054.

PCIM Europe 2024, 11– 13 June 2024, Nuremberg   DOI: 10.30420/566262204

Increasing Energy Storage Capabilities of Powder Cores by Adapting the Winding and the Use of Fringing Flux

Paul Winkler, Bhartindu Kumar Bunny, AnhTuan Luong, Wulf Günther

Acal BFi Germany GmbH, Germany

Corresponding author: Paul Winkler, paul.winkler@acalbfi.de
Speaker: Paul Winkler, paul.winkler@acalbfi.de

Abstract

Soft magnetic powder cores with a permeability of 14 to 160 are widely used in the design of storage chokes and coupled inductors. Their main advantage is the absence of a discrete air gap, which keeps the fringing minimized and equally distributed around the whole core. Placing the winding in a specific manner can help to design chokes with higher inductance and so higher energy storage capabilities. The theory of how to place the winding to get the desired effect is explained and measurements on practical examples are shown as well as the influence on parasitic effects.

1   Soft-magnetic Powder-Cores

Powder-cores (like MPP, Sendust or HighFlux cores) consist of ground and pressed soft magnetic particles (Figure 1). The particles themselves are high permeability metal alloys. A binder in which the particles are embedded and/or their non-magnetic oxidized surfaces form a distributed air gap, which reduces the overall relative permeability of this cores (14 to approximately 160).

The effective permeability of these cores depends on the particle material, their size and density.

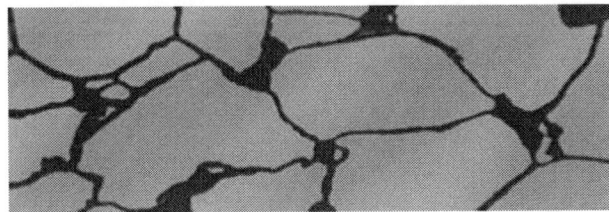

Figure 1: Microscopic Structure of a Powder Core [1]

The airgap is already distributed within the material itself. So there is no need for a bulk air-gap to get down the inductance and make the core able to store energy. The distributed air-gap leads to equally distributed fringing flux, when the core is excited with a magnetic force.

2   Idea

When prototyping for new developments a design-engineer will experience the influence of the position of the winding/turns on the coils inductance. In some cases this effect is strong enough to lead to inductance values, which exceed the tolerance of the cores inductance given by the core-manufacturer. Indeed this variation of inductance is not a "problem" of the core, but of the position of the winding.

The effect can be used to create chokes with higher inductance and so higher energy storage capabilities, using the same core and the same number of turns. This means a higher energy density and so less required space, but no additional labor- or material-cost.

3   Using Fringing to Increase Inductance

3.1   Core Cross Section and AL-value

The A_L-value of the core and so the inductivity L of the choke are depending on the core material permeability μ_r as well as on its size and geometry.

$$L = A_L \cdot N^2 \tag{1}$$

$$A_L = \mu_0 \cdot \mu_r \cdot \frac{A_{core}}{l_{mag}} \qquad (2)$$

where A_{core} is the cross section of the core and l_{mag} is its mean magnetic path length.

The A_L-value of the core expresses its magnetic conductivity. Similar to the ohmic conductivity of a (bigger) wire, a higher cross-section of the soft-magnetic core will increase its capability to handle more (magnetic) flux and lead to a higher inductance.

There is also a portion of the magnetic flux forced by the electrical current in the winding, which is leaking out of the core. Placing the turns of a winding in a manner to encompass this leakage flux, will lead to an enlarged cross section and so to a higher inductance, see Equation (2).

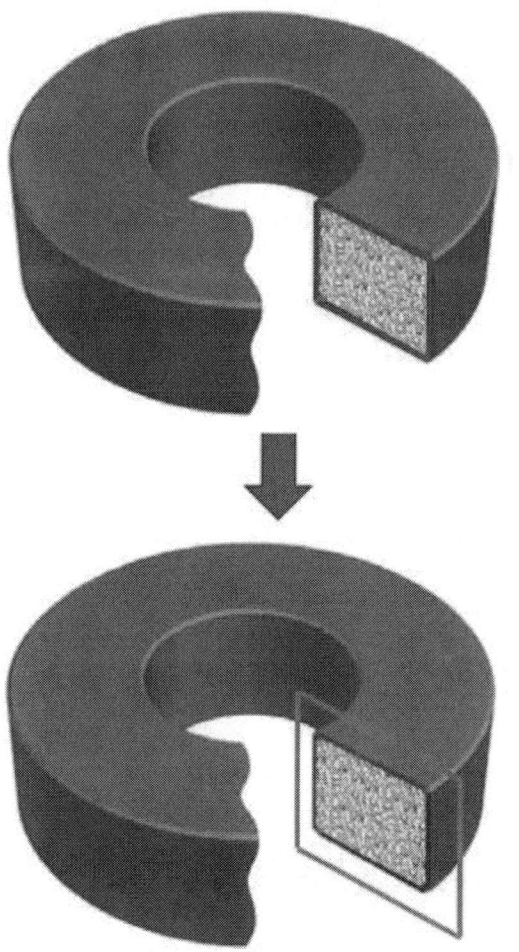

Figure 2: Using the flux leaking outside of the core increases its cross section.

3.2 Sectional Winding to Force and Encompass Fringing Flux

If a toroidal core is regularly and fully covered with a single layer of turns, the winding is covering the flux inside the core (Figure 3). If only a section of the core is used for winding, the winding is also covering some of the fringing flux, which is outside of the core in the section of the core, which is not covered with turns (Figure 4).

Both, Figure 3 and Figure 4, are based on a FEMM-simulations using the same following setup:

- Core material: KoolMµ powder with relative permeability of 60
- Core size: od=46mm, id=28mm, h=15mm
- N = 50 turns.
- I=2A

Both pictures use 40 flux lines to represent the flux and the same color-legend for the flux density, which is shown in Figure 5.

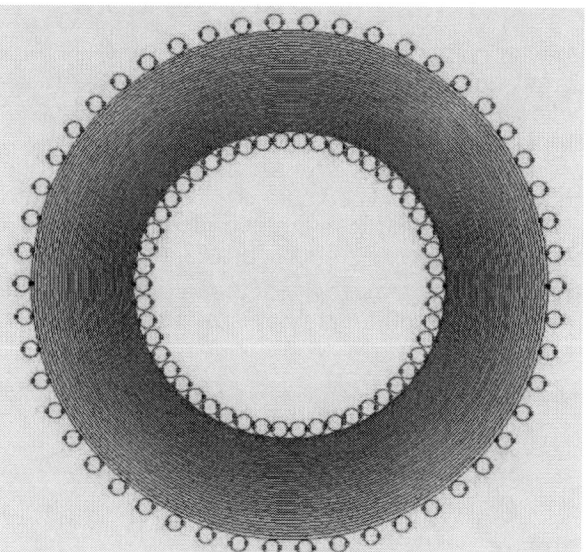

Figure 3: Flux-lines in a powder toroid with evenly spread winding

On the picture in Figure 3 all turns are placed in one layer and are equally distributed around the whole core. The flux distribution is similar everywhere, with a bit higher concentration in the inner diameter of the core. All 40 flux lines are inside the toroid.

On the picture in Figure 4 the winding is placed on just one core half. In this section the flux concentration inside the core is significantly higher than in the upper core-halve. A part of the flux (three flux lines in Figure 4) leaks out of the toroid and uses the space around the core in the section without turns (upper core-half), before entering the core and the winding again.

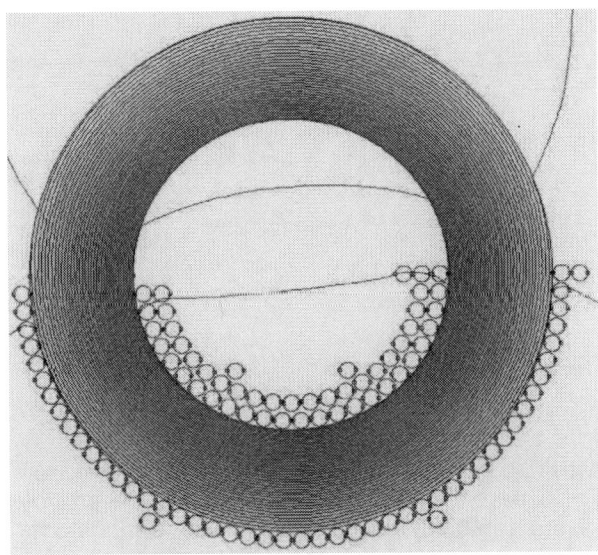

Figure 4: Flux-lines in a powder toroid with unevenly spread windings (all turns just on one core-half)

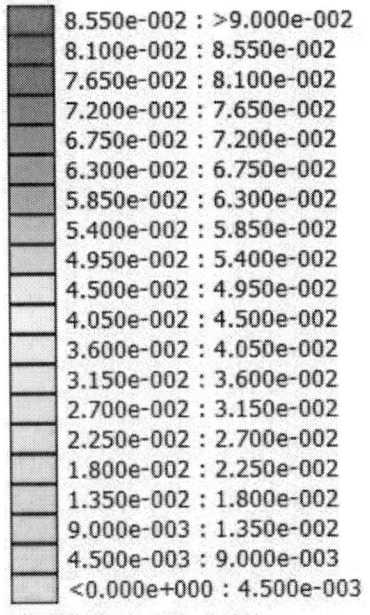

Density Plot: |B|, Tesla

Figure 5: Flux-Density-Legend for **Figure 4** and Figure 5

3.3 Measurement

To prove the theoretical findings, several chokes were realized, to check the effect at different permeabilities. This prototypes are done accordingly to the setting in the FEMM simulation, presented in section 3.2, beside that different permeabilities were used. To get reliable results, the equalness of the used cores was ensured by the following method:

- Both cores come from the same production batch.
- Both cores were at first wound with all turns in one layer and equally spread around the core as shown in Figure 6.
- If both inductors have a $\Delta L < 1\%$, one of the windings will be removed from the core again and an new sectorial winding will be applied to the core as shown in Figure 7

This procedure was applied to cores in the same size as mentioned in section 3.2 and with permeabilities between 1 and 147. Permeability of 1 was realized with dummy core made by 3D-printing, while all other cores use KoolMμ-powder material.

Figure 6: Prototype: N=50, equally distributed on the core

Figure 7: Prototype: N=50, wind on just one core-half

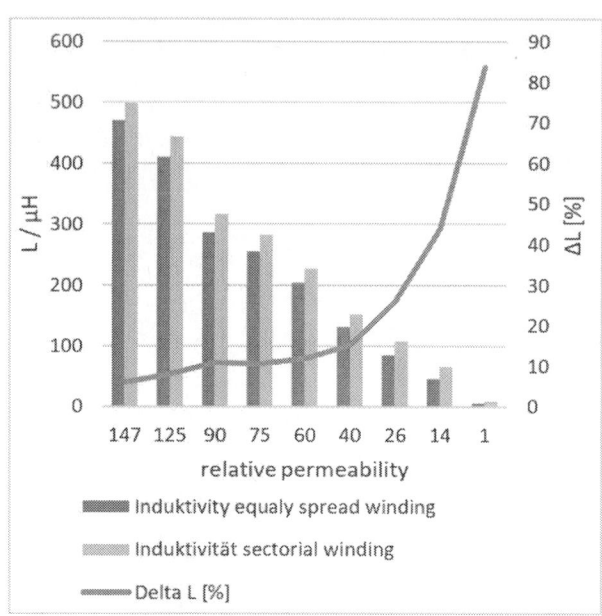

Figure 8: Comparing inductivity of different winding placements

In Figure 8 the result of the measurements of the prototypes is shown. It can be seen that on all permeabilities the inductance of the sectorial wound chokes is higher than on the one with equally distributed turns, also the number of turns is the same. The total difference in inductivity drops with the permeability of the core, while the relative difference increases at lower permeabilities, as shown in the diagram. Especially at permeability 26 and 14 the increase of inductance just due to different winding placement is high: 27% respectively 45%. In case of a dummy core with no soft-magnetic material, the inductance is almost double.

4 Increase of Energy-Storing-Capability

Storage Chokes, like PFC-, Buck-, Boost, Input-Output- or differential-mode-filter chokes, are storing energy in their magnetic field. The amount of stored energy is proportional to the inductivity of the choke and to the square to its operational current.

$$L = \frac{1}{2} L \cdot I^2 \qquad (3)$$

In general the choke has to become bigger, if it should store more energy, meaning it should operate at higher current or have a higher inductivity. This normally results in higher cost for material and heavier parts.

With the winding technique presented, it is possible to reach a significant increase in the energy saving capability of a choke, without changing its size or cost. Also the required number of turns stays the same.

On the cores described in section 3.2 (KoolMµ, $\mu_r = 60$) the saturation curve was measured. Based on this saturation curves the energy stored at different currents was calculated for the choke with the equally distributed winding and the choke with the half-core-sectional winding using equation (3). The result is shown in Figure 9.

At zero current there is no magnetic energy in the choke. At low currents the difference in energy increases with increasing currents, as more electrical current means more energy in the magnetic field. This increase is stopped, when the inductance of the powder core starts dropping due to saturation effects. With dropping inductance the amount of additional saved energy in the choke is also dropping. At higher currents the inductivity of the sectional wound choke gets more equal to the one with the equally distributed winding. This is most probably to a higher flux concentration in the core section where the winding is placed and so a (relatively) faster saturation (see also Figure 3 and Figure 4). For the currents typical for the wire- and choke-size used, the increase in energy is in it highest range of around 10% to 15%.

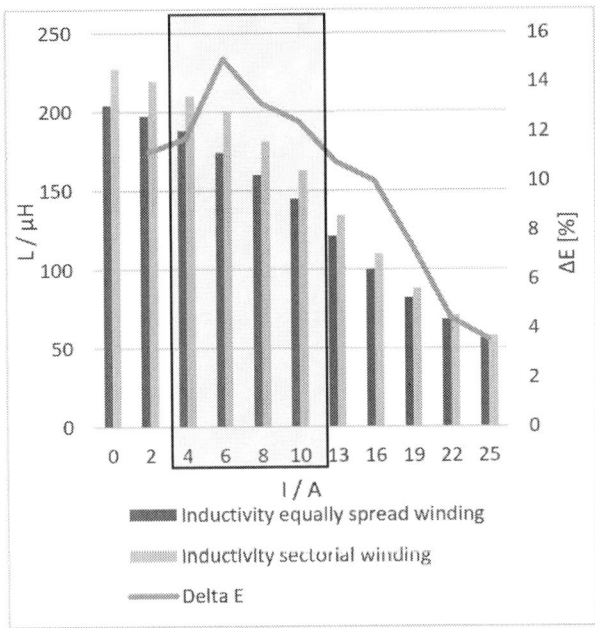

Figure 9: Inductance and stored energy over current for core KoolMµ, µr=60 with N=50 and different winding placements

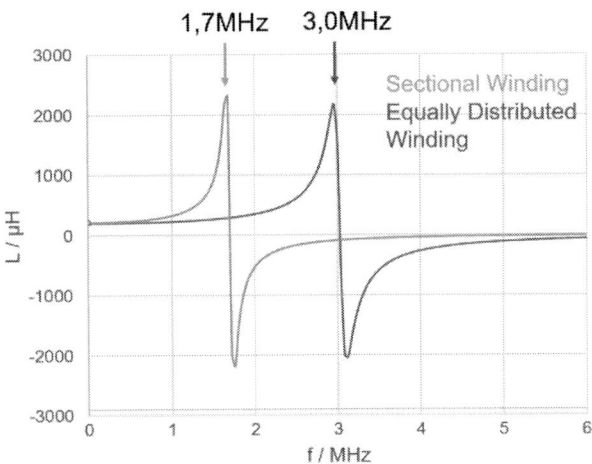

Figure 10: Inductance versus frequency for different winding placement (choke as described in section 3.2)

5 Effect on Parasitic Capacitance and Resonant Frequency

The shown increase in inductivity and the related increase in energy-storage-capability of the sectional wound choke come along with an increase of the inner-winding capacitance C of this chokes. This capacitance is mainly driven by turns touching each other and creating micro capacitors within the winding. This effect and also the increased inductance lead to a lower resonant frequency f_r of the chokes with the sectional winding, according to Equation (4).

$$f_r = \frac{1}{\sqrt{L \cdot C}} \tag{4}$$

Using again the prototypes with the setting described in section 3.2 the reason frequency of the two chokes is shifted from around 3 MHz to 1,7MHz.

Using this resonant frequencies and rewriting Equation (4) to

$$C = \frac{1}{L \cdot f_r^2} \tag{5}$$

results into the parasitic capacitances shown in the Table 1.

Table 1: Calculating parasitic capacitances from measurement

	L / µH	f_r / MHz	C / pF
Spread winding	200	3,0	0,6
Section winding	220	1,7	1,7

As shown in the table the parasitic capacitance is highly increased due to the sectional winding.

6 Conclusion

In this paper we presented a method to place a winding around a soft magnetic powder core, which can be used to increase the inductivity and energy-saving-capability of storage chokes, without changing the number of turns or the core size and without increasing the required space or manufacturing cost.

Depending on the permeability an increase of 40 percent in inductance can be achieved, resulting in the same percentage of increase of the capability to store magnetic energy.

When using the proposed method to use the space more efficient, some side effects like increased parasitic capacitance and lower resonant frequency have to be considered.

References

[1] Chang Sung Corporation, Magnetic Powder Cores Catalogue. V. 13.

PCIM Europe 2024, 11– 13 June 2024, Nuremberg   DOI: 10.30420/566262205

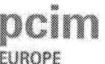

PEEC-Based Thermal Modeling of Passive Components

Sascha Langfermann[1] , Michael Owzareck[1] , Lukas Fräger[1] , Regine Mallwitz[2]

[1]BLOCK Transformatoren-Elektronik GmbH, Germany,   [2]TU Braunschweig, Germany

Corresponding author: Sascha Langfermann, Sascha.Langfermann@block.eu
Speaker: Sascha Langfermann, Sascha.Langfermann@block.eu

Acknowledgements

The research leading to this publication has received funding by the German Federal Ministry of Education and Research under grant number 16ME0355 – CODAPE

Abstract

This Paper presents a Partial Element Equivalent Circuit (PEEC) thermal model to calculate the temperature of inductors and transformers. With a modular concept different cores can be modeled by using a unique meshing concept with rectangular and cylindrical cells. The model consists of conduction, radiation to the ambient as well as between elements and heat dissipation like free and forced convection. The separate calculation methods are compared to FEM-Models. The final model is compared to a measurement.

1   Introduction

When designing magnetic components, you need to consider not only the magnetics, but among other things the electrical, mechanical, and thermal design. The thermal design is just as important as the others since the size of the component depends on it. The complexity of the model can vary greatly.

Often a first estimation is based on how much losses per area the component can dissipate. For small standard ferrite cores, it is not unusual that a thermal resistance to the ambient is given in the datasheets, which can be used to estimate the temperatures. Also, simple formulas based on experiments can be found [9]. Often a thermal network with the analogy to the electrical circuit is used. This network can also vary in its complexity by using quite small networks [8] or larger networks for a more detailed modelling [7]. The thermal networks are often static and tailored to a certain geometry. For a more detailed modeling, finite element methods (FEM) and computational fluid dynamics (CFD) can be used [13]. While those models are highly flexible, since usually geometries can be built up easily or imported from computer aided design (CAD) files, the computational

efforts are quite high, and the meshing is time consuming and must be done with care.

The Partial Element Equivalent Circuit (PEEC) method based on a thermal network offers a balanced alternative. The geometry can be modeled in greater detail, while using a simplified calculation approach. With a clever designed model, a fast parameter study can be made for an optimization. In this Paper a dynamic PEEC thermal model is proposed for different kinds of magnetic components.

2   PEEC-Based Thermal Model

The thermal model consists of rectangular legs and yokes, that automatically connect, creating a modular approach.

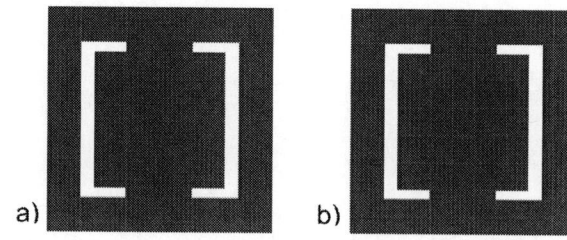

Fig. 1:   Methodology of Subdividing an EE Core with windings for PEEC Modeling; Front view. a) Coarse Subdivision b) Finer subdivision

The model subdivides the yokes and legs into rectangles. On each leg a winding can be placed. The windings are not separately modeled, but as a homogenized block. The yokes are segmented following the limbs to ensure an alignment of the rectangles. This way the interconnected mesh is the same size resulting in a correctly calculated thermal resistance even when a coarser mesh is used. A cylindrical element is used for the round corners of the winding.

In [2], [14] a PEEC-thermal model has been developed for a specific core and winding geometry for a planar PCB Magnetic component. It uses a similar principle, but lacks the flexibility of the developed approach, since different types of cores and windings for inductors and transformers can be used with the same model.

2.1 Conduction

From the center of the rectangular and cylindrical mesh-cell the thermal resistances are calculated in each direction as shown in Fig. 2. This way anisotropic materials like electrical steel or the windings, which are radially wound around the core can be considered.

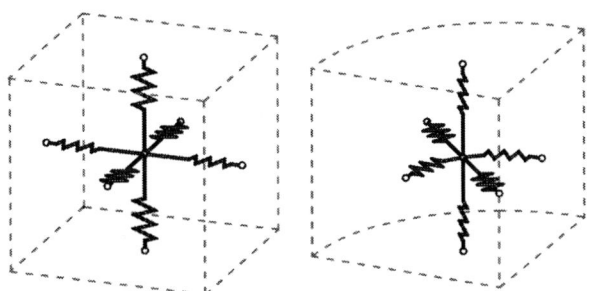

Fig. 2: Rectangular and cylindrical mesh-cell

The windings are not separately modeled, but as a homogenized block. While this approach reduces the accuracy, it offers the advantage of increased computational speed. This has already been proven in previous studies [4], [5], [6], [12].

With the combination of the cylindrical and rectangular element, the windings can be modeled with only a few mesh-elements as seen in Fig. 3 in comparison to only using rectangular or even triangular mesh elements.

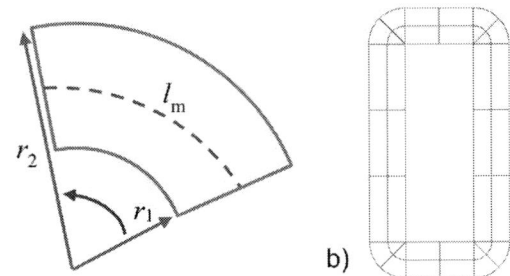

Fig. 3: a) definition of the cylindrical element
b) subdivision of the windings with cylindrical and rectangular mesh-elements

With the mesh-elements defined, the heat flux q, can be expressed with the material's thermal conductivity k, the cross-sectional area A, the temperature T, and the pathing along x.

$$\frac{dq}{dt} = -k \cdot A \cdot \frac{dT}{dx} \qquad (1)$$

The thermal resistance in one direction can be calculated with the following equation:

$$R_{con} = \frac{\Delta x}{k \cdot A} \qquad (2)$$

For the cylindrical element, half of the mean path length l_m, as shown in Fig. 3 a), can be used to calculate the resistance along it. For the heat transfer from the centrum to the exterior, the following equation can be used, where H is the height of the cylindrical element [10]:

$$R_{cond} = \frac{ln\frac{r_2}{r_1}}{2\pi Hk} \pi r^2 \cdot \frac{360}{\theta} \qquad (3)$$

One important factor for the thermal conduction for the core geometry are the air gaps in the legs, since the thermal conductivity usually is low to very low, depending on the material of the core and the material used for the air gaps in comparison to the core. The air gaps can be defined in quantity, size, and location at each leg. They are not modeled separately but are included in the direction dependent calculated conduction resistance. The mesh refinement is not dependent on the air gaps resulting in an optional faster calculation time. For a more refined calculation, the mesh size can be increased in the legs.

2.2 Convection

External heat dissipation is modeled using a resistance to the ambient. The resistances, where h is the heat transfer coefficient, is defined as:

$$R_{conv} = \frac{1}{h_{conv} \cdot A} \qquad (4)$$

The coefficient can be either calculated for a vertical plane, horizontal plane, for a cooling channel in free convection, or for forced convection [1]. Alternatively, the heat transfer coefficient can be looked up for different use cases. Each outer element can be set differently with an equation or a fixed heat transfer coefficient in the model.

2.3 Radiation

Radiative heat transfer from a body can be directed towards other objects or the surrounding environment and is described by the Stefan-Boltzmann law. This law quantifies the power, denoted as q_{rad}, radiated per unit area based on the temperature difference between the body, with temperature T_1 and its target, which could be an adjacent object or the ambient environment with temperature T_2. The equation considers the emissivity ε of the material and the Stefan-Boltzmann constant σ and the view factor F, capturing the geometrical relationship between the radiating and receiving surfaces. It is given by [1]:

$$q_{rad} = F \varepsilon \sigma (T_1^4 - T_2^4) \qquad (5)$$

The view factor is calculated between the different surfaces of the legs, yokes, and windings numerically in MATLAB. The equation is given by [1]:

$$F_{A \to B} = \frac{1}{A} \int_A \int_B \frac{\cos \theta_A \cos \theta_B}{\pi r^2} \, dA \, dB \qquad (6)$$

The view factor is calculated for all visible surfaces from a given surface from a mesh element. For each surface-to-surface view factor, the radiation exchange through a resistance can be calculated.

$$R_{rad,A \to B} = \frac{1}{A} \frac{T_1 - T_2}{F_{A \to B} \varepsilon \sigma (T_1^4 - T_2^4)} \qquad (7)$$

To determine the radiation to the ambient for a particular surface, the sum of all view factors is subtracted from one, yielding the view factor to the ambience.

$$R_{rad,A \to Amb} = \frac{1}{A} \frac{T_1 - T_2}{\left(1 - \sum_i^n F_{A \to i}\right) \varepsilon \sigma (T_1^4 - T_2^4)} \qquad (8)$$

In the modular PEEC model, an algorithm has been created which checks all the surfaces in between the legs and windings of the component, including the yokes.

2.4 Model Construction and Calculation

To solve the PEEC method, the nodal voltage potential approach is used. The equation (9) can be constructed with the temperature difference ΔT, the losses P in each node, and the admittance matrix.

$$
\begin{bmatrix} P_1 \\ P_2 \\ \vdots \\ P_n \end{bmatrix} = \begin{bmatrix} Y_{11} & Y_{12} & \cdots & Y_{1n} \\ Y_{21} & Y_{22} & \cdots & Y_{2n} \\ \vdots & \vdots & \ddots & \vdots \\ Y_{n1} & Y_{n2} & \cdots & Y_{nn} \end{bmatrix} \begin{bmatrix} \Delta T_1 \\ \Delta T_2 \\ \vdots \\ \Delta T_n \end{bmatrix} \qquad (9)
$$

For the admittance matrix the thermal network must be constructed with the previous calculated resistances The resistances must be correctly placed in the admittance matrix. The radiation to ambient resistance is parallel to the convection resistance. Between the mesh elements there is either conduction or radiation. This principle is shown in Fig. 4.

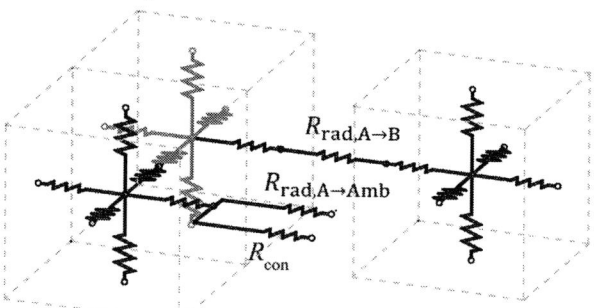

Fig. 4: Resistance network between the mesh elements

In the diagonal matrix, the summation of all admittances adjacent to the nodes is done. The connection between the mesh elements is done by placing the negative admittance on the indices of the matrix where the two nodes connect. The temperature difference can be calculated with the inverse admittance matrix.

$$\vec{\Delta T} = Y^{-1} \vec{P} \qquad (10)$$

For the whole algorithm an iteration must be done due to temperature dependencies. A base admittance matrix with the thermal conduction is created since the thermal dependency can be neglected.

The view factor can also be calculated in beforehand. The resistances due to convection and radiation, are however highly temperature dependent, as well as the core losses and the conduction losses in the windings.

3   Algorithm and Industrial Use

The separate calculations have been shown in the previous section. The calculation order and how the algorithm works is shown in the following Fig. 5. First, the geometry must be defined. The core configuration options include single-phase cores (EE/EI and UU/UI), three-phase 3UI cores, and four-legged chokes equipped with differential and an additional common-mode inductance. For each configuration different core materials like electrical steel, nanocrystalline, powder and ferrite cores can be chosen.

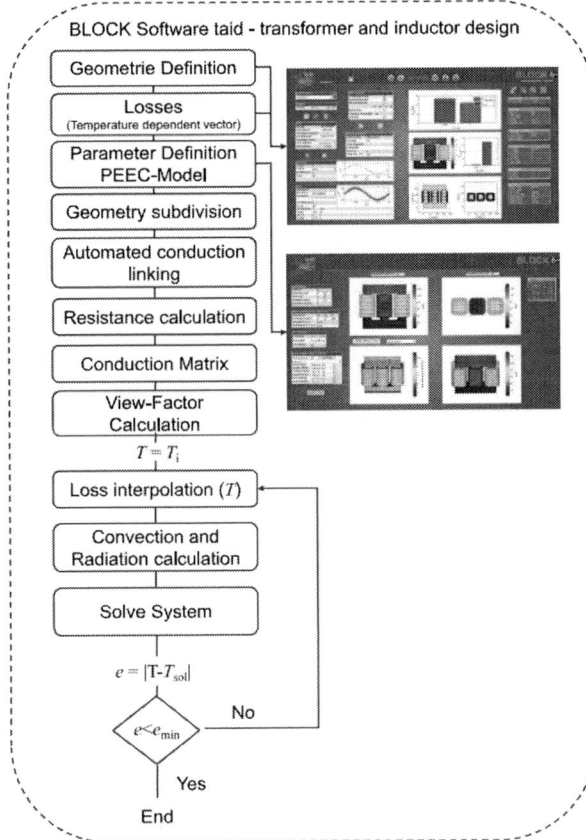

Fig. 5: Algorithm of the PEEC-Model

For each core, air gaps can be set. The winding configuration can be defined for single or multi winding configurations, depending on the application. The losses in the windings and in the core are calculated temperature dependent and can later be interpolated for the temperature calculation. This can be done separately and can be fed into the PEEC thermal model. In this case all of this is done in the BLOCK developed Software taid – transformer and inductor design. Afterwards, the Parameters for the PEEC-Model can be defined. The mesh size, thermal conductivity, and external heat dissipation can be chosen for each outside area separately. Once all the boundary conditions are set, the geometry is divided according to the previously described methodology. Based on the geometry, an automated conduction linking between the legs and windings is done for the conduction. The conduction is calculated for each element in all directions and set in the matrix according to the automated conduction linking. This Matrix is set as the base matrix for conduction. The view factor is calculated between the elements. Now a temperature iteration starts with an initial temperature T_i. The losses are interpolated for the given temperature. The resistances are calculated for convection and radiation and added to the base conduction matrix. After solving the system of equations, the new temperature is compared to the previously calculated temperature. The iteration is complete once the temperature difference is smaller than the chosen error.

4   Model Verification

It is possible to directly compare the result of the model with a real measurement, however, to rule out any errors, it is possible to do a comparison for the different calculated parts. The conduction and radiation will be compared with FEM simulations.

4.1   Conduction and matrix generation

To verify the accurate generation of the matrix for heat conduction, an identical geometry was constructed in COMSOL for comparison with an FEM simulation. To confirm the functionality of all anisotropic resistances, multiple test cases were executed. One such case is depicted in Fig. 6, where losses are introduced in the upper yoke, and a heat transfer coefficient is applied to the front area of the bottom yoke. Different subdivision sizes were tested and are visibly represented to illustrate the varying resolution of the model. The thermal conductivity works as intended. With a finer mesh size for the PEEC-Model, the resolution of the model and the temperature distribution approaches the FEM simulation. The temperature is

PCIM Europe 2024, 11– 13 June 2024, Nuremberg   DOI: 10.30420/566262205

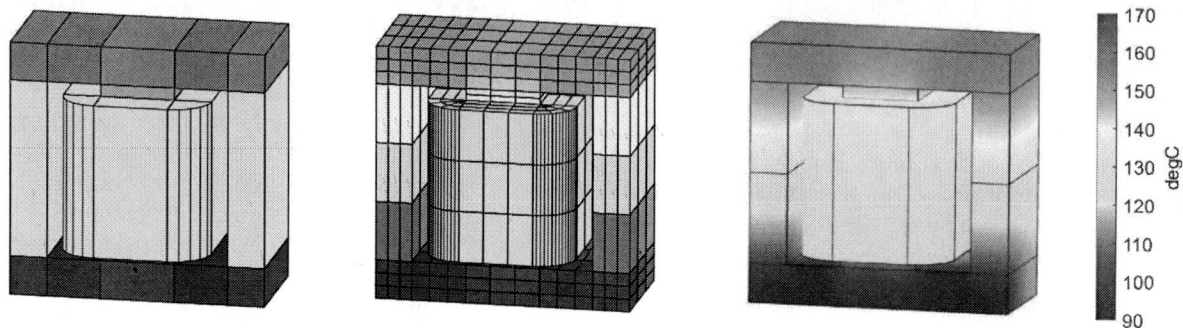

Fig. 6: PEEC Model: a) 23 elements, b) 603 elements, c) FEM (Comsol, Meshsize: 186830), EE 42/20, 4 W in upper yoke, 200 W/(m²K) front section of lower yoke. 5 W/(mK) for the core and windings

higher for a lower mesh since the losses are fed into the centrum of the mesh element. Because of the larger element, the thermal resistance to the next Block increases, resulting in a higher temperature. mesh size.

To check if the anisotropic behavior works properly, different properties have been tested as seen in Fig. 7. In this test scenario, losses were put in the lower yoke and a heat transfer to the ambient is placed on the front winding. With this configuration the special anisotropic behavior of the windings can be tested properly.

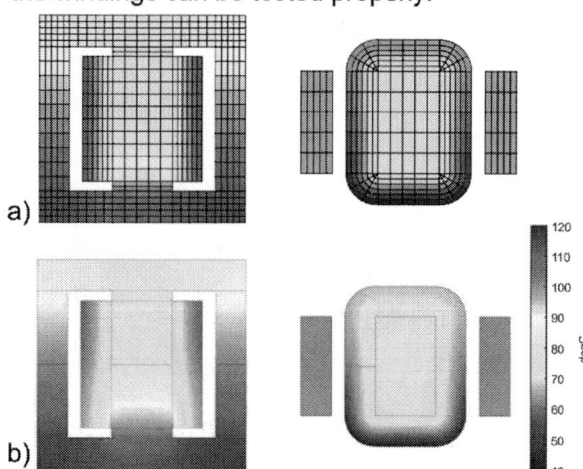

Fig. 7: Comparison of anisotropic thermal conductivity. a) PEEC Model 4020 elements b) FEM Comsol. Core xyz (10;4;2 W/(mK)). Winding ryphi (0.5;2;20 W/(mK)) 4 W in upper yoke, 200 W/(m²K) front section of winding

In Comsol curvilinear coordinates coordinates with the adaptive method has been used to model the anisotropic thermal behavior of the windings correctly. The temperature minimum and maximum temperatures as well as the distribution only has a small difference of under 2% because of the larger

4.2 Radiation

The solution for the parallel plates gets complex once the surface normal vectors are not aligned, since there is no direct solution. It is possible by creating new plates, subdividing them, and using the equation to get the view factor for each area. By appropriate addition or subtraction of the view factors a solution can be found.

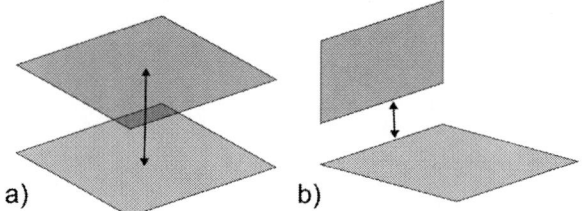

Fig. 8: View Factor calculation. a) parallel Plates b) orthogonal plates

The view factor is calculated numerically with equation (6). Two different test cases as shown in Fig. 8, parallel plates, and orthogonal plates with varying distances, are shown in this paper which are compared to FEM simulations.

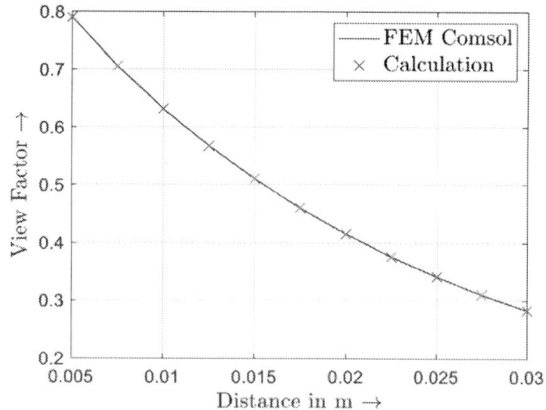

Fig. 9: View Factor calculation for parallel Plates. Length 40 mm, Width 40 mm

1520

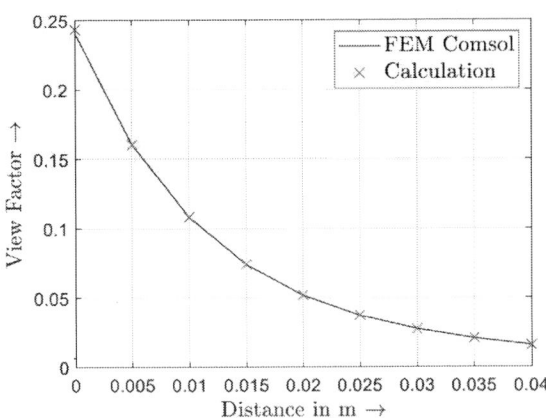

Fig. 10: View Factor calculation for orthogonal Plates. Length 40 mm, Width 20 mm

Both test cases shown in Fig. 9 and Fig. 10 show good results with only a small error compared to the FEM simulation. Different test cases with additional offsets of the plates have also been tested with similar results.

The view factor calculation has been implemented into the PEEC model. The calculation is only done for surface areas which are visible to each other. Therefore, vector calculations are done in beforehand to check if surfaces are blocking the view.

5 Comparison to Measurements

For a measurement comparison a 50 Hz Transformer has been used. The losses are easily measured and can be directly used to compare the thermal model directly.

For the conduction of the electrical steel, 25 W/(mK) has been used in the rolling direction and 4 W/(mK) along the stacks [11]. The windings are round wires, resulting in a good thermal conduction of 150 W/(mK) in the winding direction. For the other directions 1 W/(mK) has been used according to [12]. For the bobbin 0,25 W/(mK) has been used, as well as a small contact resistance between the bobbin and the core. The core losses are 16 W and the winding losses 23 W. The result can be seen in Fig. 11. The temperature distribution is very similar between the measurement and the PEEC-model. In the PEEC-model the bobbin is only between the winding and the core which results in a higher visual temperature. When looking at the Table 1 the temperature inside the winding is very similar to the model.

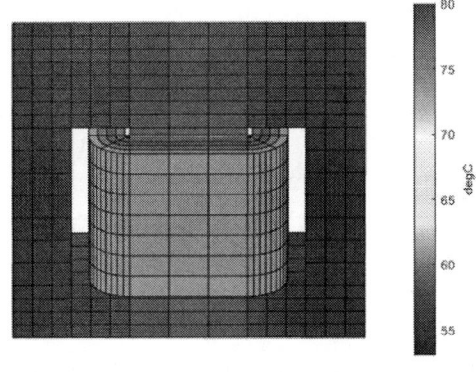

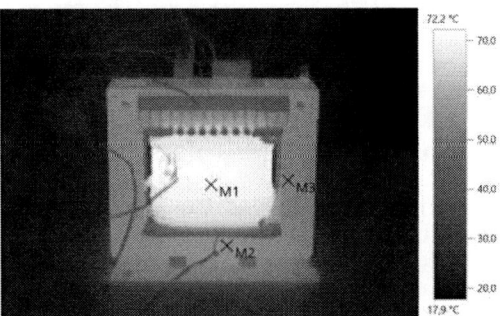

Fig. 11: Calculation vs Measurement of a 50 Hz Transformer.

In the model accessories such as electrical terminals and mounting brackets for the core are not considered resulting in a slightly different temperature.

Table 1: Temperature comparison of the Measurement and PEEC-Model

	Measurement in °C	PEEC-Model in °C
Ambient Temp.	21,8	21,8
M1	71,5	73,2
M2	56,0	57,7
M3	53,1	54,6
Sensor Winding	79,8	79,3

The heat transfer coefficient to the outside could be tuned for the winding and the core to get a closer result.

6 Conclusion

In this paper a PEEC thermal model has been developed with a modular concept. Different core shapes can be calculated with the same model. The mesh consists of rectangular elements and cylindrical elements for the corners of the windings to reduce further elements. The conduction of the model and the view factor for the radiation is verified through FEM models. For the outside heat dissipation, equations from the Literature can be used, as well as set heat transfer coefficients by the user. The model is compared to a measured 50 Hz Transformer showing good results with a maximum temperature difference of 1,7 K.

7 References

[1] VDI-Gesellschaft Verfahrenstechnik und Chemieingenieurwesen (Ed.), VDI-Wärmeatlas, 12th Edition, Springer, 2019

[2] L. C. Ordonez, A. D. Exposito, P. A. Cervera, M. Bakic and T. Wijekoon, "Fast and Accurate Analytical Thermal Modeling for Planar PCB Magnetic Components," in *IEEE Transactions on Power Electronics*, vol. 38, no. 6, pp. 7480-7491, June 2023, doi: 10.1109/TPEL.2023.3259064

[3] K. Nakamura, H. Yoshida and O. Ichinokura, "Electromagnetic and thermal coupled analysis of ferrite orthogonal-core based on three-dimensional reluctance and thermal-resistance network model," in *IEEE Transactions on Magnetics*, vol. 40, no. 4, pp. 2050-2052, July 2004, 10.1109/TMAG.2004.832497

[4] M. Jaritz, A. Hillers and J. Biela, "General Analytical Model for the Thermal Resistance of Windings Made of Solid or Litz Wire," in *IEEE Transactions on Power Electronics*, vol. 34, no. 1, pp. 668-684, Jan. 2019, doi: 10.1109/TPEL.2018.2817126.

[5] P. A. Kyaw, M. Delhommais, J. Qiu, C. R. Sullivan, J. -L. Schanen and C. Rigaud, "Thermal Modeling of Inductor and Transformer Windings Including Litz Wire," in *IEEE Transactions on Power Electronics*, vol. 35, no. 1, pp. 867-881, Jan. 2020, doi: 10.1109/TPEL.2019.2914661

[6] G. Salinas López, A. D. Expósito, J. Muñoz-Antón, J. Á. O. Ramírez and R. P. López, "Fast and Accurate Thermal Modeling of Magnetic Components by FEA-Based Homogenization," in *IEEE Transactions on Power Electronics*, vol. 35, no. 2, pp. 1830-1844, Feb. 2020, doi: 10.1109/TPEL.2019.2921160.

[7] M. Mogorovic and D. Dujic, "Thermal modeling and experimental verification of an air cooled medium frequency transformer," *2017 19th European Conference on Power Electronics and Applications (EPE'17 ECCE Europe)*, Warsaw, Poland, 2017, pp. P.1-P.9, doi: 10.23919/EPE17ECCEEurope.2017.8099176.

[8] S. Langfermann, M. Owzareck and L. Fräger, "Design Space Optimization of a SiC Drive Inverter with an Integrated All-Pole Sine Filter," *2021 23rd European Conference on Power Electronics and Applications (EPE'21 ECCE Europe)*, Ghent, Belgium, 2021, pp. P.1-P.10, doi: 10.23919/EPE21ECCEEurope50061.2021.9570448.

[9] McLyman, Wm. T.: Transformer and Inducto Design Handbook. Boca Ra- ton, London, New York: CRC Press: Taylor & Francis Group. 2011

[10] Baehr, Hans Dieter and Karl Stephan, "Convective heat and mass transfer. Flows with phase change" in Heat and Mass Transfer, Berlin, Heidelberg:Springer, pp. 443-543, 2011.

[11] S. Yamazaki, Y. Kurosaki, and T. Wakisaka, "Anisotropic Thermal Conductivity of Lamination Stacks of Non-oriented Electrical Steel," in Tetsu-to-Hagané, Journal of the Iron and Steel Institute of Japan, vol. 107, no. 2, pp. 121-127, 2021, 10.2355/tetsutohagane.TETSU-2020-066

[12] R. Wrobel, S. Ayat and J. L. Baker, "Analytical methods for estimating equivalent thermal conductivity in impregnated electrical windings formed using Litz wire,*" 2017 IEEE International Electric Machines and Drives Conference (IEMDC)*, Miami, FL, USA, 2017, pp. 1-8, doi: 10.1109/IEMDC.2017.8002003.

[13] S. Langfermann and M. Owzareck, "Design of a 40 kHz Inverse Coupled Inductor for an Inverter System," *PCIM Europe digital days 2020; International Exhibition and Conference for Power Electronics, Intelligent Motion, Renewable Energy and Energy Management*, Germany, 2020, pp. 1-7.

[14] L. C. Ordonez, A. D. Exposito, P. A. Cervera, M. Bakic and T. Wijekoon, "Optimized Thermal Modelling of High Power Planar PCB Magnetics," *2022 IEEE Energy Conversion Congress and Exposition (ECCE)*, Detroit, MI, USA, 2022, pp. 1-8, doi: 10.1109/ECCE50734.2022.9948122.

PCIM Europe 2024, 11– 13 June 2024, Nuremberg   DOI: 10.30420/566262206

Galvanically Isolated Power Supply for Gate Drivers in High Voltage Applications

Priyanka Ghosh[1], Michael Meissner[1], Klaus F. Hoffmann[1]

[1] Helmut Schmidt University, Germany

Corresponding author: Priyanka Ghosh, priyanka.ghosh@hsu-hh.de
Speaker: Priyanka Ghosh, priyanka.ghosh@hsu-hh.de

Abstract

This work presents a feasibility study of power supply solutions for isolated gate drivers meant for use in high voltage applications. At high potential differences, maintaining the standard specified creepage and clearances increases the size and volume of the integrated system. Therefore, this becomes a bottleneck in space-constrained applications. In this work, the galvanically isolated power supply for gate drivers designed to drive a modular multilevel converter [1] is realized by using a common primary voltage source to supply multiple gate driver units over individual isolation transformers. Two configurations of isolation have been investigated, and two further power supply arrangements have been constructed and tested in nominal operation. All arrangements have their own advantages and disadvantages, and therefore, a preferred selection is dependent on application, and the available volume to integrate the solution. A comparison of the configurations in terms of their performances, robustness against partial discharges and the associated construction effort concludes this paper.

1 Introduction and motivation

Isolated gate driver ICs have been readily available in the market. While the transient isolation voltage rating of these devices goes as high as 5 kV or even more, the continuous rating seldom exceeds 1 kV. This feature renders them unsuitable in certain applications with higher requirement of continuous isolation voltage. An established method to achieve this is to transfer power inductively over an isolation barrier, with the windings sufficiently isolated from each other and from the core, where needed. In addition, it is further necessary to isolate the associated digital circuitry.

In previous literature, different methods of achieving isolated power supply to gate drivers for medium voltage applications have been published. They include using optical power transducer technology [2], magnetic coupling with discrete wire-wound magnetic cores [3], and wireless power transfer between printed coils [4], [5]. A current transformer based approach is presented in [6], which is designed for half bridges. In this work voltage transformers are configured for up to four isolated driver units, aiming at reduction of total space consumption. The two main configurations studied in this paper for realizing galvanic isolation are presented in Fig. 1.

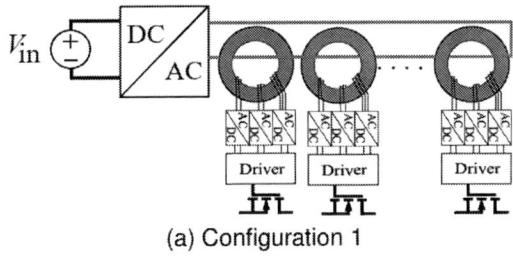

(a) Configuration 1

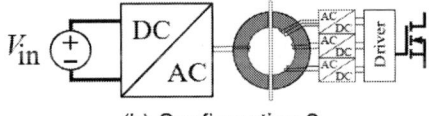

(b) Configuration 2

Fig. 1: Transformer configurations

2 Principle of operation

The voltage isolation transformer for power transfer is driven with a switching frequency of about 300 kHz derived from an oscillator, followed by a corresponding driver IC. Any DC bias from the transformer input is removed by a series capacitor placed in the path. The transformer in configuration 1 uses a toroid core with a high voltage insulated cable for the primary side, and thin, lacquer-coated

1523

PCIM Europe 2024, 11– 13 June 2024, Nuremberg   DOI: 10.30420/566262206

wires for the secondary high voltage side. For configuration 2, two toroid cores are split in two parts, stacked and held together across a teflon sheet barrier. All windings use thin, lacquer-coated wires, thus reducing the occupied window area. The transformer has two windings on the secondary side to provide the positive and negative power supply for driving the gate, while a third winding provides a stable 5 V supply for the digital circuit. On the secondary side, the voltages at the transformer outputs are rectified, and regulated to desired levels with the support of adjustable linear regulators.

3   Test setup and results

The hardware realization of the configurations depicted in Fig.1 is presented in Fig.2. In the first case, a single turn primary side winding is used, that causes the primary inductance to be low, and the magnetizing current to be high. The core material

is supplied by 15 V and −5 V, derived from the isolation transformer voltages on the secondary side. Each driver unit is meant to drive one power switch, and is loaded with a corresponding capacitance of 200 nF for the tests conducted. The switching frequency chosen is 150 Hz, defined by the application requirement.

3.1   Configuraion 1: Single turn primary winding

In configuration 1, two toroid cores [7] of T35 material [8] are stacked for each unit. The tests were performed with up to four units connected in series. With an increasing number of units the secondary turns were also proportionally increased, according to the voltage transformation relation $\frac{V_p}{N_p} = \frac{V_s \cdot x}{N_s}$, where x is the number of units connected in series and p and s refer to the primary and secondary sides, respectively. The gate driver outputs for four connected units are shown in Fig. 3a. The primary

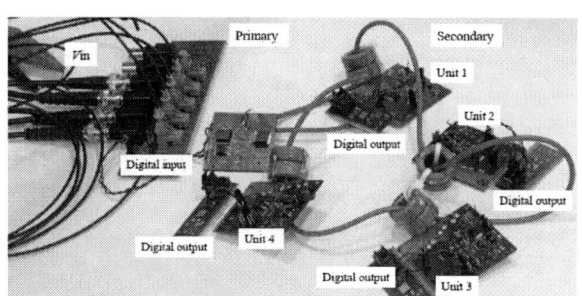

(a) Configuration 1: Single turn winding

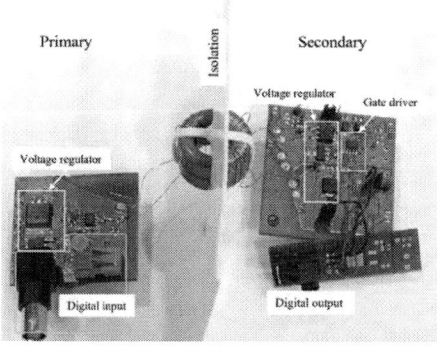

(b) Configuration 2: Split core

Fig. 2: Test setup of the two isolating configurations

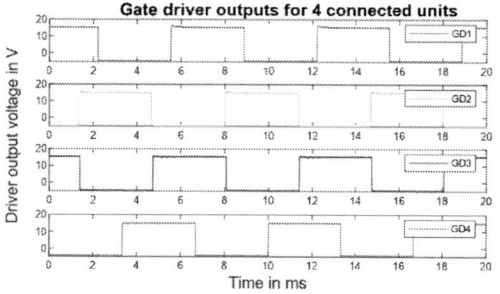

(a) Gate driver output signals

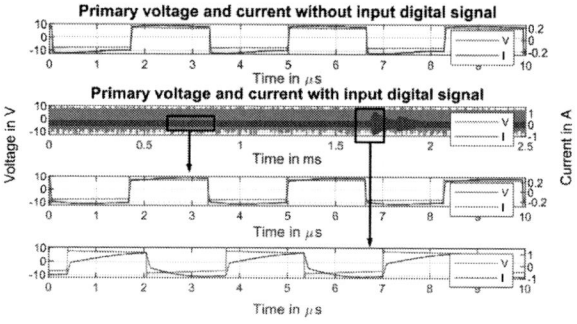

(b) Primary voltage and current

Fig. 3: Driver characteristics with four units connected in configuration 1

is chosen with appropriate A_L value such that the driver IC used on the primary side is able to provide the required current continuously within its thermal performance limit. The digital signals are provided by waveform generators via discrete fiber optic components. On the secondary side, the gate driver

current and voltage are illustrated in Fig. 3b, where the first and second subplots are obtained before and after providing digital input signals respectively. Peaks appear in the transformer primary current at the switching instants of the gate driver. This is marked in the envelope with black rectangles and

1524

shown magnified in the third and fourth subplots.

3.2 Configuration 2: Split core, single unit

In the first step, configuration 2 was tested with one secondary driver unit as depicted in Fig. 2b. Two T35 toroid cores split in the middle, and separated by a 1.5 mm thick teflon sheet were used. Differing from configuration 1, in this case, six primary turns were used to reduce the primary current peak, which can be further decreased with a higher number of turns. This is compared in Fig.4 between transformers using 6 and 10 turns on their primary windings. To maintain the turns ratio, the secondary turns must be proportionally increased. This is restricted in space by the available window area in the secondary side of the core that has three windings. The current shapes obtained for configuration 1 in

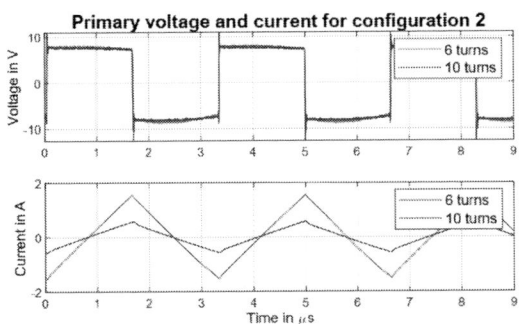

Fig. 4: Performance with split core arrangement

Fig. 3b and for configuration 2 in Fig.4 are different. This is governed by the dominance of either the load current or the magnetizing current. The primary inductances were measured to be 48 μH for configurations 1, and 12 μH for configuration 2 with ten (10) primary turns, so that the magnetizing current is more significant in the second case, leading to a prominent triangular current shape for an imposed rectangular voltage on the transformer primary winding. Following the transformer secondary windings, there are rectifiers and linear voltage regulators on the gate driver units. Therefore, the load seen by each secondary winding can be approximated to be resistive. This causes the current to assume a somewhat rectangular shape, according to the shape of the induced winding voltage. Consequently, the primary current shown in Fig.3b reflects a similar rectangular shape. The triangular magnetizing current is small in this case, and has negligible influence on the current shape.

This split-core arrangement was extended to supply

multiple driver units in parallel, and further, in series connection, and have been categorized under configurations 2a and 2b respectively, and elaborated in the next sections.

3.3 Supplying multiple units using transformer with split core

As shown in Fig. 4, higher number of turns on the primary winding can be used to decrease the primary current demanded from the driver. This advantage can be utilized to supply multiple units in parallel, taking into consideration the continuous current rating of the gate driver IC on the primary side. It is trivial, that a parallel connection of units will draw a proportional amount of current from the primary driver. A series connection of the same units is expected to reduce this requirement due to the higher number of turns on the transformer primary winding. Consequently, the number of secondary turns needs to be proportionally higher in this case, to achieve the same induced voltage level on the secondary side. This demands a larger window area, and subsequently, the toroid cores used for this purpose were cut in a ratio of 1:3 by height for easier accommodation of the secondary turns. This is seen in Fig. 5 on the left, while the core on the right is split into equal parts. A mate-

Fig. 5: Toroid cores split 1:3 (left) and 1:1 (right)

rial thickness of 1 mm was removed by the used water jet cutting machine, for both geometries. For parallel operation of the driver units, either core geometry can be used, because their magnetic properties do not change between the two alternatives. A schematic is drawn in Fig.6 to represent both serial and parallel connections of several gate driver units. The two combinations do not operate simultaneously but are depicted together for conciseness. Operations with these two arrangements are discussed in the next subsections.

3.3.1 Configuration 2a: Driver units supplied in parallel

In this configuration, the primary winding of the transformer on each unit has multiple turns, and are supplied in parallel from a common source, as depicted in Fig.6 on the right. This was realized by using 1:3 split toroid cores, separated by a teflon

PCIM Europe 2024, 11– 13 June 2024, Nuremberg   DOI: 10.30420/566262206

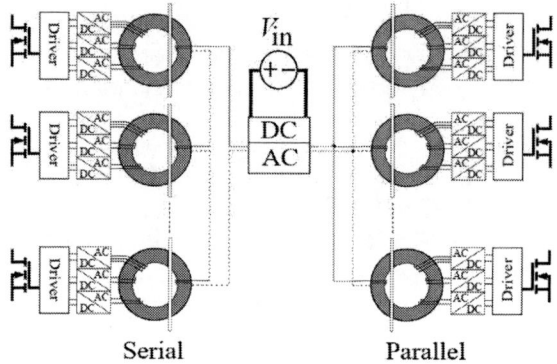

Fig. 6: Series and parallel connected driver units

sheet isolation barrier 1.5 mm thick. Four such units were prepared and operated in parallel, as can be seen in Fig. 7, together with the primary driver circuit.

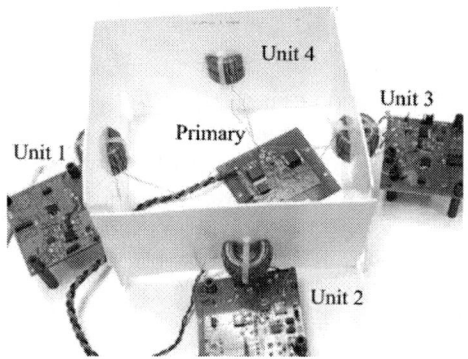

Fig. 7: Setup of four parallel driver units

The primary current drawn by each of the four units is plotted in Fig.8 along with the total demand, which peaks at around 5 A. This magnitude of current is undesirable and can be lowered by either increasing the number of primary turns, or reducing the number of connected units. Following a verifica-

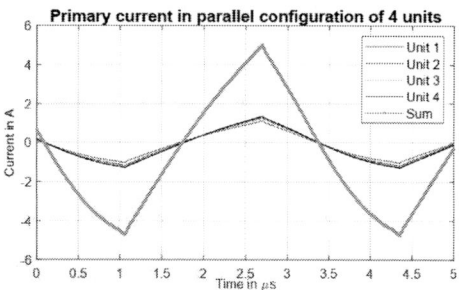

Fig. 8: Primary current demand of four parallel units

tion of normal switching behavior of all units, a thermal measurement was additionally performed while

operating a single unit, with the primary current peak of 1.2 A. The driver IC, the 5 V and 15 V voltage regulators on the primary side are labeled as 'Gate Driver', 'VR5' and 'VR15' respectively in the image in Fig. 9. The highest temperature reached was 30.8 °C, appearing at the 15 V voltage regulator owing to the voltage drop appearing from its input to output. The measurement was not repeated for the parallel connection of 4 units, which due to higher primary current was operated for a shorter time duration that was insufficient for the temperature to reach a steady-state value.

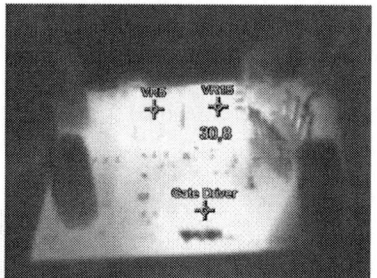

Fig. 9: Temperature profile of primary side driver with one unit

3.3.2 Configuration 2b: Driver units supplied in series

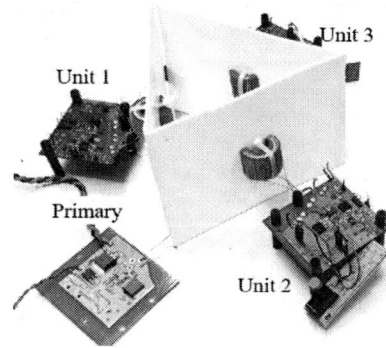

Fig. 10: Setup of three series connected driver units

In this section, three driver units were supplied in series, set up as in Fig. 10. As explained previously, when supplying several units in series the number of turns in the secondary windings has to be scaled by the number of units, because the induced voltage on the secondary side is distributed among the connected units. Alternatively, the same turns ratio can be maintained by keeping the secondary turns fixed and reducing the primary winding turns proportionally, but at the cost of higher primary current. The gate driver IC [9] in the primary side provides

1526

an alternating square wave output varying between 15 V and 0 V. The DC blocking capacitor placed in its path translates it to a signal varying between 7.5 V and –7.5 V, which is proportionally distributed among the three series connected gate driver units, as seen in Fig.11. With an increment in the number of primary turns from two (2) to four (4), the primary winding current peak decreased from 3.8 A to 1.2 A, as compared in the same figure. A thermal mea-

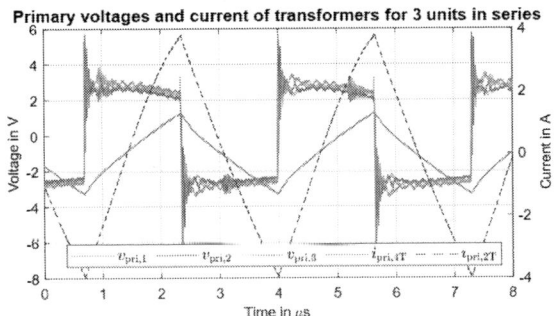

Fig. 11: Primary voltage distribution of three series connected units with four primary winding turns

surement was additionally performed while operating in this configuration, in which the primary side gate driver IC and voltage regulators are shown marked in Fig. 12, corresponding to a peak current of 1.2 A. The highest temperature reached was 44.6 ℃, appearing at the 5 V voltage regulator marked as 'VR5'. The label 'VR15' corresponds to the 15 V linear voltage regulator. The temperature rise is expected to improve with a higher copper area under the device for better heat dissipation, but it is not dealt with in this work.

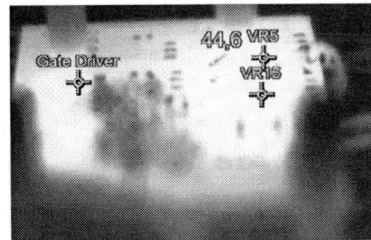

Fig. 12: Temperature profile of primary side driver with three series connected units

3.3.3 Inference

It was possible to drive multiple units with their output switching between 15 V and –5 V at 150 Hz, for all the aforementioned configurations. The primary current has a higher amplitude in the parallel connection compared to that in the series connection,

when the same number of primary turns and secondary units are used. On the other hand, the number of transformer secondary turns in the series arrangement has to be scaled proportionally with the number of units. Keeping the primary turns constant, the augmentation achieved in the induced voltage by increasing the secondary turns gradually becomes smaller when the number of turns is large, as is the case in series configuration. Furthermore, with a weak coupling across the isolation barrier, and the geometrical variation among similarly split cores, non-uniformity across the units is more prominent in the series arrangement compared to the parallel one. With this in mind, a new combination utilizing the identified advantages of both the configurations will be investigated in the future, but is outside the scope of this paper. A comparison of the results is presented in Tab. 2. In the next section, the galvanic isolation provided by the designed transformers is studied.

4 Validation for high voltage isolation applications

The power supply solution for the gate driver units should provide the desired voltage isolation between the primary and secondary side of the transformer, specified up to 2.5 kV for the target application. Additionally, the insulation resistance (IR) measurement of the transformer windings across the isolation barrier should provide a high ohmic value, when subjected to the high potential difference, specified by application. In the following subsections, the corresponding results are presented.

4.1 Insulation resistance measurement

Measurement of insulation resistance is performed to test the insulation quality of an electrical conductor, to check for its suitability in application, and protection provided against electrical shocks and subsequent damage arising from leakage currents. IR tests were carried out on both the configurations with measurements done as shown in Fig.13. Operation at an imposed voltage of 10 kV with a test time of 20 s resulted in 889 GΩ for configuraton 1, and 18 GΩ for configuraton 2. These values can be accepted to be sufficient for the developed prototypes.

4.2 Partial Discharge (PD) measurement on transformer configurations

Localized electrical discharges occuring internally in an insulating material or even on the material

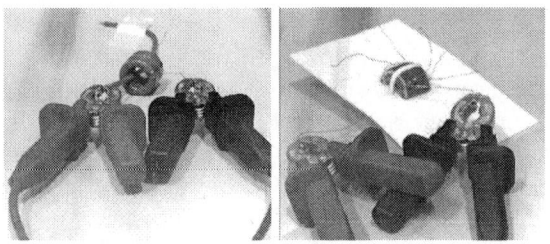

Fig. 13: Insulation resistance measurement

surface caused by high electrical field strength are known as partial discharges. This effect can cause deterioration of the insulation, and might create a bridge between the insulator and the conducting material over time. Therefore, it is necessary to observe the partial discharge behavior of the test object under parameters specified by application. For the transformer arrangements discussed in this paper, the insulation medium is a combination of airgap and cable insulation for configuration 1, and teflon sheet for configuration 2. As per standard IEC 60270 [10], the following test procedure was adopted.

4.2.1 Measurement procedure

After calibration of the measurement device, a 50 Hz sinusoidal voltage was imposed across the test terminals comprising of one end each of the primary and secondary windings of the transformer. The voltage at which the charge present at the test terminals exceeds 10 pC is defined as the inception voltage, where partial discharge begins to occur. The observed value of the inception voltage (RMS) multiplied by a factor of 1.5, was then applied across the terminals and sustained for approximately 40 s. Following this, the applied voltage was successively reduced till partial discharge peaks could no longer be observed, and this voltage was held for 25 s. The peak voltage at which the existing electric charge goes lower than 10 pC provides the value of extinction voltage. The peak value of this 50 Hz sinusoid during PD measurement is used to set the maximum equivalent DC operation voltage of the power switches, and that in turn determines the galvanic isolation level necessary on the gate driver units. A safety factor of 1.2 is further considered to identify the maximum permissible operating voltage. Therefore, for an isolation requirement of 2.5 kV, corresponding to the same level of operating voltage, an extinction voltage of 3 kV is necessary to achieve.

The obtained results have been summarized in

Tab.1, with a brief description of the configurations, corresponding to each item as marked in Fig.14. For items U1-U3, uncut cores as used in trans-

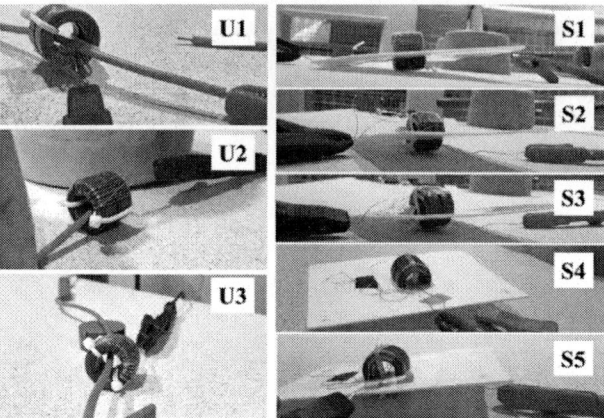

Fig. 14: Configurations tested for Partial Discharge

former configuration 1 were tested, with the worst case scenario with U1 and the best with U2. The transformer of configuration 2b was used in item S1, and that of configuration 2a was used for items S2 through S5. The test winding in the secondary side of item S1 has twice the number of turns compared to items S2 to S5. Therefore, the winding area occupied is larger for S1, where the turns come in close proximity to the insulation sheet. This condition should be avoided, because the electric field present between the windings on either side of the isolation barrier is stronger with smaller distance, and might lead to an earlier onset of partial discharges. This condition was not present in items S2 to S5, due to a more compact arrangement of their turns.

4.2.2 Measurement results and inference

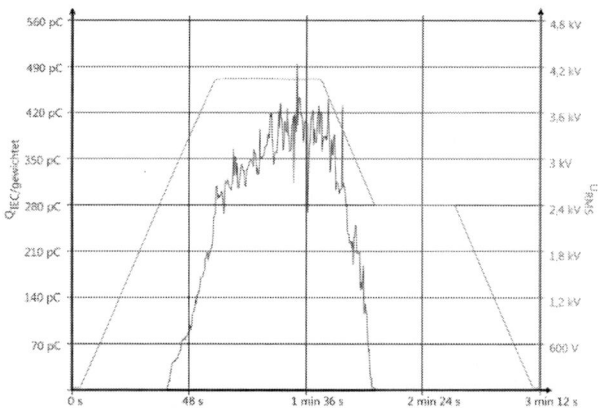

Fig. 15: Partial Discharges observed for item S1

Acceptance criteria: Peak Extinction voltage $\geq$ 3 kV					
Item	Description	Inception voltage (RMS, kV)	Imposed voltage (RMS, kV)	Extinction voltage (Peak, kV)	Remark
U1	HV insulated primary winding in contact with insulated secondary winding	2.35	3.59	2.86	Not Acceptable
U2	HV insulated primary winding farthest from insulated secondary winding, and in contact with transformer core	3.7	3.7	4.52	Safe operation
U3	HV insulated primary winding passing through the middle of transformer core	2.9	4.1	3.82	Safe operation
S1	Secondary windings occupy entire core area and come in contact with teflon sheet isolation barrier	2.72	4	3.4	Safe operation
S2	Secondary windings spread over core area (lesser turns than in S1)	2.74	4.06	3.64	Safe operation
S3	Secondary windings arranged more compactly, and brought closer to center, and farther from isolation barrier	2.79	4	3.7	Safe operation
S4	Test terminals brought closer and separated by 2.5 cm	1.88	2	2.66	Not Acceptable
S5	Test terminals separated by 8 cm	2.5	3.8	3.51	Safe operation

Tab. 1: Summary and evaluation of the Partial Discharge test results

The voltage profile applied across the test terminals, and the localized charge are shown in Fig.15 for item S1. The inception voltage was reached at 2.7 kV, following which the imposed RMS voltage was increased up to around 4 kV, and then gradually reduced as seen. Partial discharge was no longer observed at approximately 3.4 kV peak extinction voltage, which exceeds the 3 kV threshold, and therefore this result can be accepted in operation.

Item S3 with its turns wound closer together has a slightly higher extinction voltage, and higher inception voltage than S2, and is therefore the preferred configuration. With a more compact arrangement of the windings, improvements in inception and extinction voltage can be expected. For all the items mentioned above, the measurement terminals were kept farthest from each other, separated by approximately 10 cm. For item S4, the terminals were brought closer, and separated by 2.5 cm. The peak extinction voltage was obtained at 2.66 kV, indicating the presence of partial discharges below the target of 3 kV. Therefore the minimum separation between the two winding terminals should be increased further, while an upper threshold for separation can be obtained from item S5, which has a peak extinction voltage at 3.51 kV, indicating safe operation.

For items U1 through U3 using uncut toroid cores, the measurement was done between one end of the high voltage (HV) isolation cable and one end of the thin multi-turn windings. Firstly, they were placed in contact, and separated by their insulation, as seen for U1, resulting in a peak extinction voltage of 2.86 kV, which being lower than 3 kV, is not an acceptable arrangement. In the next step, the HV cable was placed farthest from the multi-turn winding, and in contact with the inner wall of the toroid core, as seen in item U2. The peak extinction voltage obtained at 4.52 kV was a significant improvement from the previous case. Finally, an intermediate condition was simulated by adjusting the HV cable to pass approximately through the center of the toroid cavity. The peak extinction voltage in this case was 3.82 kV, a value in between those observed for U1 and U2, as expected. As seen, items U2 and U3 indicated safe operation, unlike item U1.

Configuration	Description	Turns ratio	Advantages	Disadvantages
1	Uncut cores in series	1:4:4:12	a) Strong coupling between primary to secondary side b) Low construction effort	a) Number of primary turn limited by window area and cable thickness b) Sufficient clearance between primary and secondary windings must be maintained
2	Split core, one unit	6:10:11:21	a) More window area available to increase turns b) Medium construction effort	a) Weak coupling, so more turns needed for each secondary winding b) Higher non-uniformity between units arising from splitting of cores
2a	Split core in parallel	7:12:13:30	a) Lower number of turns compared to series connection b) Medium construction effort	a) Higher primary current b) Might cause worse thermal performance
2b	Split core in series	4:28:30:60	Lower primary current	a) Number of secondary turns multiply with number of units b) High construction effort

Tab. 2: Summary of investigated transformer configurations

4.2.3 Inference from PD tests

The following can be concluded from the partial discharge measurements conducted. For the uncut cores, the primary and secondary windings should be kept apart by at least the length of radius of the used cores. Higher separation distance, where feasible, is better. For the split core, closer arrangement of winding turns is better than spreading them across the core width. Contact of the winding with the isolation sheet should be avoided, because this increases the electrical field present between the primary and secondary side of the transformer, and results in a lower extinction voltage. In summary, sufficient clearance between the two sides of isolation, and from the isolating medium itself should be provided in the gate driver units to ensure an operation free of partial discharges.

5 Conclusion and future prospective

Within the scope of this work, two main solutions of isolated power supply for gate drivers in high voltage application are studied as proof-of-concept. The solution with an physical isolation barrier is further extended to two combinations, and their performances are compared. Furthermore, partial discharge tests are carried out on the two main configurations, and recommendations are made regarding their arrangement based on the results. Finally, a comparison is tabulated for the transformer arrangements highlighting their benefits and drawbacks. In conclusion, the prototypes show promising results, and can be expected to be successfully implemented in the concerned high voltage application. Other different combinations are possible to realize, and might lead to a more efficient and compact solution, for example, a combination of configurations 2a and 2b. These need to be further investigated, and will be reported in a future publication.

6 Acknowledgement

This research as part of the project DEFINE is funded by dtec.bw – Digitalization and Technology Research Center of the Bundeswehr. dtec.bw is funded by the European Union – NextGenerationEU.

The authors are grateful to Prof. Dr.-Ing. Rainer Marquardt and Mr. Martin Rasch for their valuable technical suggestions. The authors are also thankful for the support received in conducting the Partial Discharge measurements at the Laboratory for High Power Electronic Systems at University of the Bundeswehr, Munich, Germany.

References

[1] R. Marquardt, "Modular Multilevel Converter: An universal concept for HVDC-Networks and extended DC-Bus-applications," Sapporo, Japan: IEEE, 2010, pp. 502–507. DOI: 10.1109/IPEC.2010.5544594.

[2] M. Ishigaki, S. Fafard, D. P. Masson, M. M. Wilkins, C. E. Valdivia, and K. Hinzer, "A new optically-isolated power converter for 12 V gate drive power supplies applied to high voltage and high speed switching devices," Tampa, FL, USA: IEEE, 2017, pp. 2312–2316. DOI: 10.1109/APEC.2017.7931022.

[3] F. Teng, H. Feng, and S. Lukic, "Gate Driver Power Supply with Air-gapped Transformer for Medium Voltage Converters," Houston, TX, USA: IEEE, 2022, pp. 451–456. DOI: 10.1109/APEC43599.2022.9773678.

[4] V. T. Nguyen, G. V. Bharath, and G. Gohil, "Design of Isolated Gate Driver Power Supply in Medium Voltage Converters using High Frequency and Compact Wireless Power Transfer," Baltimore, MD, USA: IEEE, 2019, pp. 135–142. DOI: 10.1109/ECCE.2019.8912184.

[5] V.-T. Nguyen, V. U. Pawaskar, and G. Gohil, "Isolated Gate Driver for Medium-Voltage SiC Power Devices Using High-Frequency Wireless Power Transfer for a Small Coupling Capacitance," *IEEE Transactions on Industrial Electronics*, vol. 68, pp. 10 992–11 001, 11 2021. DOI: 10.1109/TIE.2020.3038095.

[6] J. Gottschlich, M. Schäfer, M. Neubert, and R. W. D. Doncker, "A galvanically isolated gate driver with low coupling capacitance for medium voltage SiC MOSFETs," Karlsruhe, Germany: IEEE, 2016, pp. 1–8. DOI: 10.1109/EPE.2016.7695608.

[7] TDK, "Ferrites and accessories, May 2023, Toroids (ring cores) R 25.3 x 14.8 x 10.0, Type B64290L0618."

[8] TDK, "Ferrites and accessories, February 2023, SIFERRIT material T35."

[9] IXYS, "Gate Treiber 12.5 V 5-pin TO-263 MOSFET driver, 30 A, 2017, IXD 630."

[10] International Electrotechnical Commission, "IEC 60270: High-voltage test techniques - Partial discharge measurements, Third edition 2000-12."

PCIM Europe 2024, 11– 13 June 2024, Nuremberg
DOI: 10.30420/566262207

Fabrication Technique for Novel Nanocrystalline Cores with High Saturation Polarization and Low Losses

Merlin Thamm[1], Inge Lindemann-Geipel[1], Christoph Höhnel[2], Thomas Weißgärber[1,2]

[1]Fraunhofer Institute for Manufacturing Technology and Advanced Materials IFAM, Germany
[2]Technische Universität Dresden, Faculty Mechanical Engineering, Institute of Materials Science, Chair Powder Metallurgy, Germany

Corresponding author: Merlin Thamm, merlin.thamm@ifam-dd.fraunhofer.de
Speaker: Merlin Thamm, merlin.thamm@ifam-dd.fraunhofer.de

Abstract

Nanocrystalline toroidal cores of the Nanomet alloy are produced by FAST/SPS. Flakes are prepared by milling starting from an amorphous ribbon. An oxidation layer is formed by heat treatment for electrical insulation between the flakes in the compacted toroid. A layer of sol gel is applied to further improve the insulation. The compacted toroidal cores have a density of about 80 %. Low losses of only 50 W/kg at 1 T and 1 kHz can be achieved by using a pressing aid.

1   Introduction

Nanocrystalline Fe-based alloys are excellent soft magnetic materials. These alloys are fabricated by melt spinning as amorphous ribbons with a thickness of around 20 μm [1]. By a heat treatment over the crystallization temperature, crystal grains below 30 nm are precipitated in an amorphous matrix. Due to the amorphous phase and the small crystallite size, the crystal anisotropy is neglectable, resulting in a low coercivity usually below 10 A/m [2]. Nanocrystalline cores are particularly suitable for high frequency applications due to the low ribbon thickness and higher electrical resistivity compared to conventional electrical steel. However, there is only one nanocrystalline alloy on the market, namely Fe-Cu-Nb-Si-B with the trade name Finemet or Vitroperm. This alloy was invented by Yoshizawa, Oguma and Yamauchi [3]. It has one of the lowest coercivity of all soft magnetic materials with $H_c = 0.5$ A/m, but the saturation polarization of 1.24 T is only in the middle range [4]. Therefore, nanocrystalline alloys with higher saturation polarization are currently being developed. Makino, Men, Kubota, Yubuta and Inoue [5] invented 2009 the alloy $Fe_{83.3-84.3}Si_4B_8P_{3-4}Cu_{0.7}$ called Nanomet, which is the first type of new subgroup within the nanocrystalline soft magnetic materials. These materials, referred to as high-B alloys, exhibit saturation polarization between 1.5 and 2.0 T. The high saturation of the new nanocrystalline alloy allows a wide range of applications in modern power electronics. Petzold [6] mentions in particular power transformers, pulse power cores and DC/DC-converters, which require high saturation polarization and low losses. The main disadvantage of the new high-B alloys is the required high heating rate of larger than 100 K/min to avoid grain growth and deterioration of the soft magnetic properties. The required heating rate can be realized with FAST/SPS as special hot-pressing technique. This fabrication method was already used successfully for uncoated Nanomet-powder, leading to high losses for high induction levels and frequencies [7–9]. The aim of this work is to analyze the fabrication route of soft magnetic Nanomet cores with high saturation polarization and low losses by insulation the powder before compaction. Two approaches are compared with each other on the one hand the powders are compacted with a resin and on the other hand a thin insulating layer is applied on the flakes before compaction in order to achieve the highest possible stacking factor.

2   Experimental

Amorphous ribbons of Nanomet alloy (NMAQ-W60, $Fe_{85}Si_2B_8P_4Cu_1$ in at.-%, TOHOKU MAGNET INSTITUTE Co., Ltd.) were crushed into flakes with a size of several millimeters. Two different approaches were chosen in order to reduce the eddy currents in the compacted component. Firstly, the flakes were insulated by a resin similar to powder cores. To do that the flakes were mixed with various amounts of silicone resin for 10 minutes. To crosslink the resin on the surface of the flakes, the wet flakes were heat treated in a drying oven at 180°C for 30 minutes. The second approach aims for thin layers by thermal oxidation. Ribbons were thermally oxidized at different temperatures up to 300°C and different dwell times up to 1 h to form a homogeneous oxide layer. The fracture behavior was observed as a function of the annealing temperature. The flakes in this approach were annealed at 250°C for up to 1 h to form an oxide layer without becoming brittle. An additional coating layer based on silicon oxide was applied to the flakes using the sol-gel process. The coated flakes were mixed with and without distearylethylene diamide (Crodamide® EBS, Croda Chemicals Europe Ltd, UK) and compacted by FAST/SPS to toroidal cores at 125 MPa and 100 K/min up to 380°C. The outer diameter and inner diameter were 30 mm and 20 mm, respectively. The height was around 2 mm. The magnetic properties of the toroidal cores were analyzed using a B-H analyzer (Remagraph-Remacomp C 705, MAGNET-PHYSIK Dr. Steingroever GmbH). Coercivity and maximum permeability of the toroidal cores were determined under quasi-isostatic conditions.

3   Results and discussion

3.1   Insulation by resin

The relative density ρ_{rel} and electrical resistivity ρ_{el} are shown in Fig. 1 for the toroidal cores based on the resin coated Nanomet flakes. By increasing the amount of resin in the specimen, the relative density unexpectedly decreases and the porosity increases despite a low viscosity resin being used. Possibly, the friction between the uncoated flakes is lower than that of the resin-coated flakes, so the resin hampers the rearrangement of the particles during pressing. The electrical resistivity increased as expected due to the higher insulation content. The higher porosity supports this trend. As a result of the porosity, the saturation polarization of the samples decreases. The saturation polarization of the Nanomet alloy used is 1.76 T. With a pore content of 20 %, it is reduced to 1.41 T, with 40 % pores to 1.06 T. This trend can also be seen in the induction level at 10 kA/m.

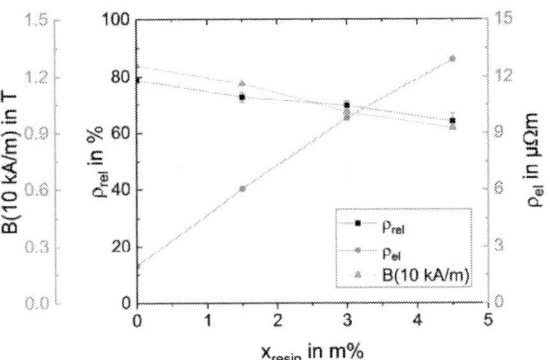

Fig. 1   Relative density ρ_{rel} and specific electrical resistivity ρ_{el} of compacted Nanomet flakes depending on the resin content x_{resin}.

Figure 2 shows the losses at 1 kHz for the Nanomet alloy with different additions of silicon resin up to 1 T. The losses of the coated samples are significantly lower than the losses of the uncoated sample. However, the difference between 3.0 and 4.5 wt% resin is small. Therefore, it can be assumed that an increasing amount of resin will not reduce the losses any further. As a result of the increasing porosity with increasing resin content, the possible polarization decreases, so the loss at 1 kHz and 1 T could no longer be measured for the specimens with 3.0 and 4.5 wt% resin content.

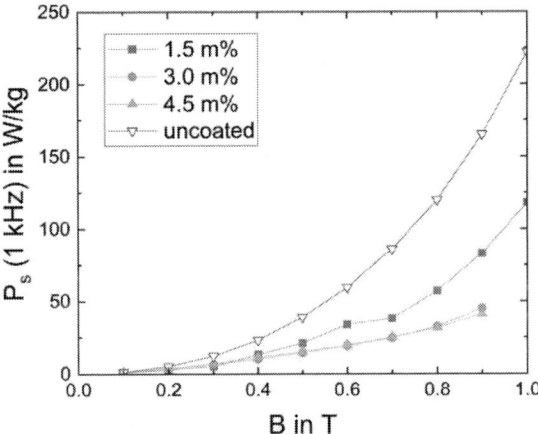

Fig. 2   Power losses depending on the induction level for compacted flake cores with different resin contents.

The cross-sections are shown in Fig. 3. The addition of the resin hampers a dense packing of the flakes. Large pores can be recognized. Furthermore, vortex structures form with a high resin content. It is assumed that the resin cannot be applied homogeneously to the flakes. Therefore, the aim of high polarization with low losses cannot be achieved with this approach, which is why an isolation approach with thin layers will be considered in the next chapter.

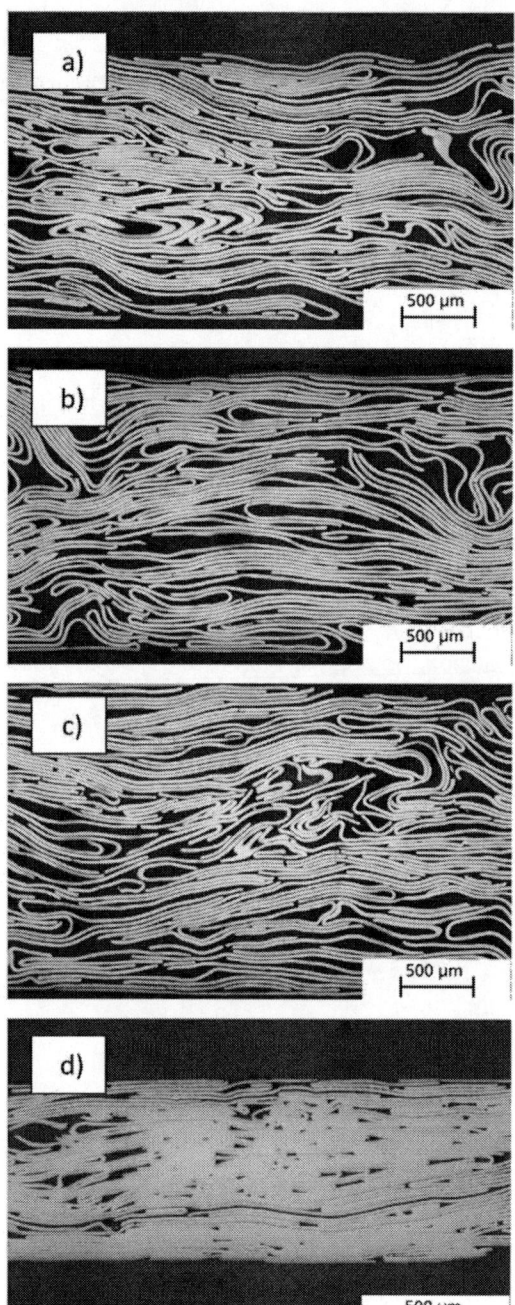

Fig. 3 Cross-section of compacted Nanomet flake cores with different mass contents of resin a) 1.5 wt%, b) 3.0 wt% and c) 4.5 wt%. d) is the uncoated reference.

3.2 Insulation by thermal oxidation and sol-gel coating

Higher relative density up to 80 % is achieved by the compaction of oxidized and sol gel coated Nanomet flakes. It is comparable to the density achieved with uncoated flakes.

The hysteresis loops of the compacted uncoated flakes, resin coated flakes and the compacted oxidized and sol gel coated flakes are displayed in Fig. 4. Due to the thin insulation layer the shearing of the hysteresis loop is less pronounced and the permeability is higher in comparison to the resin flake core.

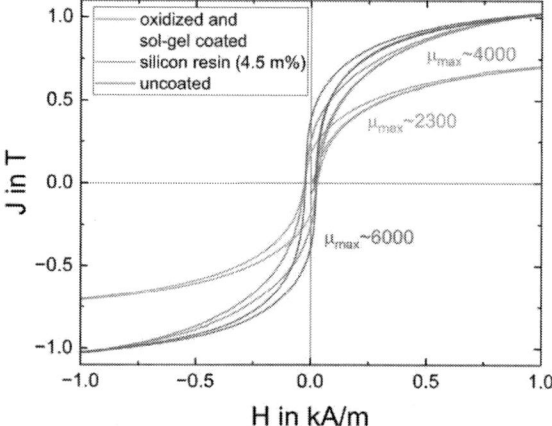

Fig. 4 Hysteresis loops of compacted Nanomet flakes under DC conditions.

To illustrate the potential of this approach, the minimum possible loss should be estimated. The power loss can be estimated by calculating the hysteresis loss and eddy current loss using the Eq. 1 to 3. Equation (1) of the calculation of the power loss according to the Bertotti model is depending on the hysteresis loss P_h, the eddy current loss P_e and the anomalous loss P_a.

$$P = P_h + P_e + P_a \tag{1}$$

The hysteresis loss can further be estimated by Eq. (2) from the coercivity H_c, the maximum induction B, the density ρ and the frequency f.

$$P_h = \frac{4 H_c \widehat{B}}{\rho} f \tag{2}$$

The eddy current loss is depending on the sheets thickness d, the electrical resistivity, and similar to the hysteresis loss, on the induction, frequency and density. The formula is given by Eq. (3).

$$P_e = \frac{(\pi d B)^2}{\rho_{el}\rho} f^2 \tag{3}$$

The anomalous loss cannot be calculated easily, so it is neglected at this point.

The hysteresis loss is estimated by the coercivity of 20 A/m. The lowest eddy current loss appears if the flakes are perfectly electrical insulated. Therefore, the flakes have a thickness of 26 μm and the specific electrical resistivity of 0.74 μΩm [16]. Assumed minimal power loss at 1 T and 1 kHz is shown in Fig. 5. The fabricated cores have five times higher power losses than the lowest possible power loss if the anomalous loss is neglected. It can be assumed that the losses of the Nanomet flake cores can be further reduced by optimizing the electrical insulation layer.

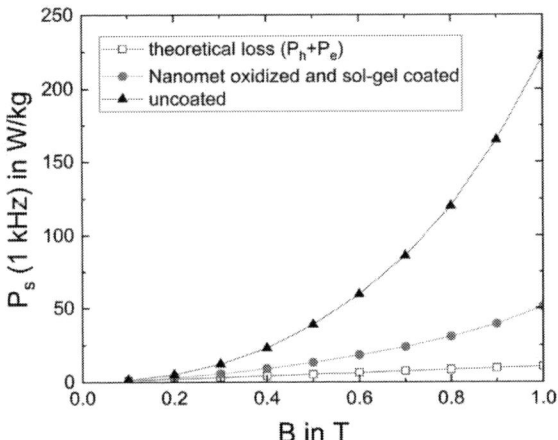

Fig. 5 Calculation of the minimum loss of the belt using the Eq. 1 to 3, H_c =20 A/m, ρ_{el}=0,74 μΩm and d=26 μm.

3.3 Comparison with state-of-the-art materials

The specific power losses P_s at 1 T and 1 kHz and maximum permeability μ_{max} are shown in Fig. 5 for different insulation approaches of Nanomet flake cores. In addition, commercially available products are also categorized. The resin flake cores are missing in Fig. 6 because the induction of 1 T was not reached due to the low permeability and the low staking factor. The permeability decreases with the number of insulation layers applied. At the same time, the losses are also reduced due to the better electrical insulation of the flakes from each other, which prevents eddy currents. Somaloy 130i 5P is a soft magnetic composite (SMC) from Höganäs. Pure iron powder is pressed with an insulating layer of iron phosphate, among other layers and pressing aid. The good insulation results in very low eddy currents between the powder particles. At the same time, the hysteresis curve is strongly sheared due interruptions acting as air gaps for the magnetic flux, so that the coercivity is larger than 100 A/m resulting in a high hysteresis loss. Therefore, the permeability of the Somaloy

130i 5P is lower than that of the fabricated Nanomet flakes cores. Due to the higher electrical resistivity of the nanocrystalline alloy the eddy current losses in the flakes are lower than in the iron particles.

The conventional electrical steel M270-35A has a higher permeability than the Nanomet core. The lower losses of the Nanomet flake core are lower than the losses of the electrical steel due to the lower thickness and insulation of the Nanomet flakes.

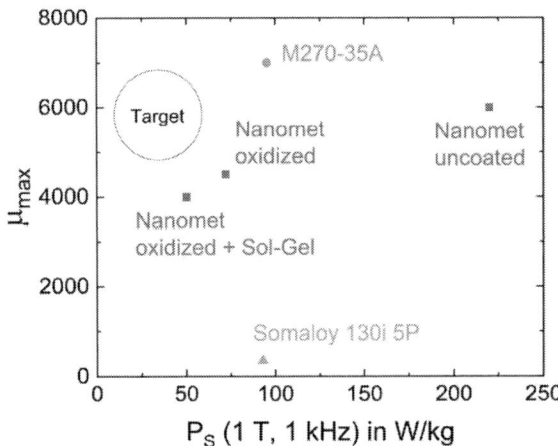

Fig. 6 Classification of the manufactured specimens into commercially available materials as electrical steel M270-35A and soft magnetic composite Somaloy 130i 5P. High permeability and low losses are desirable for many applications.

Electrical steel sheets are used in radial flux machines and transformers. The low losses of the nanocrystalline alloys with high saturation polarization would make them very interesting for motor construction. Some prototype motors have already been built from laminated ribbons [10–12]. Due to the brittleness of nanocrystalline alloys, the ribbons cannot be laminated easily without breaking [13]. Furthermore, the conventional manufacturing process is especially expensive for those thin individual ribbons. Nonaka and Makino [14] cold compacted nanocrystalline Nanomet flakes with a binder in order to fabricate stator cores. However, the density was not homogeneous and the nanocrystalline flakes broke during the compaction leading to small particle with high coercivity (>100 A/m).

The Nanomet flakes cores compacted by FAST/SPS overcome the brittle issues as the flakes are initially amorphous at the start of compaction. Due to the additional low losses, the cores produced in this way can be used in stators of radial flux motors or in transformers.

SMCs are suitable for axial flux motors due to their isotropic properties and for DC/DC converters due to their low eddy current losses. Due to their lamellar structure, the Nanomet flake cores can be used in radial flux motors in a similar way to electrical sheets. Due to the pressing process, the favorable orientation of the flakes cannot be combined with the required core shape of the axial flux machines. However, the Nanomet flake cores can be used in DC/DC converters. Petzold [6] recommends for DC/DC converters high saturation polarization, low losses and stability against DC-premagnetization. In addition, the work of Nuetzel, Rieger, Wecker, Petzold and Mueller [15] is mentioned in which flakes of the conventional nanocrystalline Vitroperm alloys are compacted in to powder cores. However, the Vitroperm alloy (Fe-Si-B-Cu-Nb) has just a saturation polarization of 1.24 T. Therefore, the here introduced Nanomet flake cores would be an improvement in power density.

4 Conclusions

Toroidal cores with a density of around 80 % relative density were fabricated from oxidized Nanomet flakes by FAST/SPS. The high heating rate of 100 K/min allows a compaction with small grains leading to low coercivity and low static losses. Using the pressing aid EBS, the insulation layers remain intact and decrease the dynamic losses. Because of the high saturation polarization of Nanomet with 1.76 T and the low losses, this fabrication technique could be used for processing nanocrystalline alloys for power transformers and DC/DC-converters. To keep up with the conventional materials the density as well as the stacking factor has to be increased. However, by increasing the pressure or optimizing the pressing die for better particle stacking a higher density can be perhaps accomplished.

5 Acknowledgement

This work was carried out as a part of the project "WeiMag" (600420) within the research grant Fraunhofer Attract and the DFG-FhG-project "NanoKompakt".

6 References

[1] S. Tumanski, "Modern magnetic materials-the review," *Organ*, vol. 4, p. 10, 2010.

[2] G. Herzer, "Modern soft magnets: Amorphous and nanocrystalline materials," *Acta Materialia*, vol. 61, no. 3, pp. 718–734, 2013, doi: 10.1016/j.actamat.2012.10.040.

[3] Y. Yoshizawa, S. Oguma, and K. Yamauchi, "New Fe-based soft magnetic alloys composed of ultrafine grain structure," *Journal of Applied Physics*, vol. 64, no. 10, pp. 6044–6046, 1988, doi: 10.1063/1.342149.

[4] VACUUMSCHMELZE GMBH & CO. KG, *Datenblatt Vitroperm 800*. [Online]. Available: www.vacuumschmelze.com

[5] A. Makino, H. Men, T. Kubota, K. Yubuta, and A. Inoue, "New Fe-metalloids based nanocrystalline alloys with high Bs of 1.9T and excellent magnetic softness," *Journal of Applied Physics*, vol. 105, no. 7, 07A308, 2009, doi: 10.1063/1.3058624.

[6] J. Petzold, "Applications of nanocrystalline softmagnetic materials for modern electronic devices," *Scripta Materialia*, vol. 48, no. 7, pp. 895–901, 2003, doi: 10.1016/S1359-6462(02)00624-3.

[7] Y. Zhang, P. Sharma, and A. Makino, "Sintered magnetic cores of high B s Fe 84.3 Si 4 B 8 P 3 Cu 0.7 nano-crystalline alloy with a lamellar microstructure," *Journal of Applied Physics*, vol. 115, no. 17, 17A322, 2014, doi: 10.1063/1.4865324.

[8] Y. Zhang, P. Sharma, and A. Makino, "Sintered powder cores of high Bs and low core-loss Fe 84.3 Si 4 B 8 P 3 Cu 0.7 nano-crystalline alloy," *AIP Advances*, vol. 3, no. 6, p. 62118, 2013, doi: 10.1063/1.4811465.

[9] Y. Zhang, P. Sharma, and A. Makino, "Fe-Rich Fe–Si–B–P–Cu Powder Cores for High-Frequency Power Electronic Applications," *IEEE Trans. Magn.*, vol. 50, no. 11, pp. 1–4, 2014, doi: 10.1109/TMAG.2014.2316543.

[10] N. Nishiyama, K. Tanimoto, and A. Makino, "Outstanding efficiency in energy conversion for electric motors constructed by nanocrystalline soft magnetic alloy "NANOMET ® " cores," *AIP Advances*, vol. 6, no. 5, p. 55925, 2016, doi: 10.1063/1.4944341.

[11] T. Nonaka, S. Zeze, S. Makino, and M. Ohto, "Research on motor with nanocrystalline soft magnetic alloy stator cores," *Electr Eng Jpn*, vol. 211, 1-4, pp. 55–62, 2020, doi: 10.1002/eej.23260.

[12] R. Parsons and K. Suzuki, "Nanocrystalline soft magnetic materials produced by continuous ultra-rapid annealing (CURA)," *AIP Advances*, vol. 12, no. 3, p. 35316, 2022, doi: 10.1063/9.0000274.

[13] Z. Wang, Y. Enomoto, R. Masaki, K. Souma, H. Itabashi, and S. Tanigawa, "Development

of a high speed motor using amorphous metal cores," in *8th International Conference on Power Electronics - ECCE Asia*, Jeju, Korea (South), 2011, pp. 1940–1945.

[14] T. Nonaka and S. Makino, "Research on Stator Core with Crushed pieces of Nanocrystalline Soft Magnetic Alloy," *IEEJ Journal IA*, vol. 10, no. 6, pp. 755–760, 2021, doi: 10.1541/ieejjia.21001263.

[15] D. Nuetzel, G. Rieger, J. Wecker, J. Petzold, and M. Mueller, "Nanocrystalline soft magnetic composite-cores with ideal orientation of the powder-flakes," *Journal of Magnetism and Magnetic Materials*, 196-197, pp. 327–329, 1999, doi: 10.1016/S0304-8853(98)00736-7.

[16] A. Makino, "Nanocrystalline Soft Magnetic Fe-Si-B-P-Cu Alloys With High B of 1.8–1.9T Contributable to Energy Saving," *IEEE Trans. Magn.*, vol. 48, no. 4, pp. 1331–1335, 2012, doi: 10.1109/TMAG.2011.2175210.

PCIM Europe 2024, 11– 13 June 2024, Nuremberg   DOI: 10.30420/566262208

Excitation-Dependent Temperature Behavior of the Quasi-static Hysteresis Loss Energy Density of N87 Ferrite Material

Jeremias Kaiser [1], Erika Stenglein [2], Thomas Dürbaum [1]

[1] Optoelectronics, Friedrich-Alexander-Universität Erlangen-Nürnberg
[2] Siemens Energy Global GmbH & Co. KG

Corresponding author: Jeremias Kaiser, jeremias.kaiser@fau.de
Speaker: Jeremias Kaiser, jeremias.kaiser@fau.de

Abstract

This paper presents a study of the temperature behavior of the quasi-static hysteresis loss energy density of the ferrite material N87. First, the material data is measured by use of a small toroidal core with the volt-amperometric method. To adjust the temperature of the device under test, the core is dived into an oil bath of a temperature calibrator. The excitation frequency of 1 kHz ensures the quasi-static behavior. The measurements are performed for various excitation levels of the magnetic flux density. Second, the measured data is approximated by various fit formulas. A model that uses a polynomial fit of the temperature dependence for different excitation levels of the magnetic flux density shows the most promising results. Additionally, the model uses a further fit of the polynomial coefficients. This allows the prediction of the loss energy density for a wide temperature range by using only a few parameters.

1 Introduction

Many power conversion applications contain a magnetic component as an integral part to store energy or to transform voltage. While these components play a key role, they are often bulky and suffer from significant losses that contribute a substantial amount to the losses of the entire power electronic system. This makes it all the more important to predict losses as accurately as possible. In literature exist various approaches to calculate the losses in ferrite material [1]–[4]. Even though these models are quite accurate, a magnetic component represents a multi-physical problem because the losses feed back to the temperature, which in turn affects the losses. Besides an accurate thermal simulation, an adequate description requires the temperature dependence of the losses. Additionally, the excitation dependence of the losses changes with temperature. The description of this behavior of ferrite material is missing in current models.

The recently published core loss model in [2] is based on the loss separation approach. This means, that the total power losses can be described by a quasi-static and a dynamic component. The quasi-static part is the loss energy density of one hysteresis loop and characterized, here by a frequency of 1 kHz. Due to the low frequency, the losses are independent from the time course between two extreme values of the flux density as long as it is monotonic [5]. The dynamic part of the power losses describes the rate dependent behavior at higher frequencies. It is considered by an effective frequency that is influenced by an integral over the absolute value of the second derivative of the flux density [2]. These two loss mechanisms are characterized by different measurement setups according to [6] and [7].

This paper deals with the quasi-static part of the losses, the quasi-static energy loss density of one hysteresis loop. Measurements are performed for different temperatures and excitation levels of the flux density. The influence of DC premagnetization of the material is not subject of this study.

The paper is structured as follows: Section 2 describes the measurement setup based on the volt-amperometric method and presents its results. The measured data is fitted in section 3 in different ways: the polynomial approach of [5], [6] is fitted for every measured temperature value and the coefficients of this model were fitted over temperature again (Section 3.1). For N87, this can alternatively be done by using an exponential approach according to the steinmetz equation [5] (Section 3.2). Additionally, by using the value of the existing model at 25 °C in [5], the temperature progression is fitted for every measured excitation (Section 3.3).

2 Measurement of the quasi-static hysteresis

The measurement setup of this paper is based on the volt-amperometric method to determine the flux density and the magnetic field within a material probe. This method is a common approach to obtain hysteresis loops and to deduce the magnetic losses. Due to its simplicity, it enables the operation with almost any signal shape.

This paper deals with the quasi-static properties, that define the low frequency behavior of the material. They depend only on the materials history (e.g. the initial value in the B-H trajectory) and the extreme values of the flux density. The shape between the extreme values is not relevant, as long as the course is monotone. The investigated ferrite material N87 has a quasi-static behavior up to approximately 1 kHz. [5]

2.1 Measurement setup

The measurement setup is based on the standards DIN EN 60404-6 and DIN EN 62044-3. A detailed description of the setup is given in [5], [6] and [8]. Figure 1 shows the circuit diagram of the volt-amperometric measurement principle.

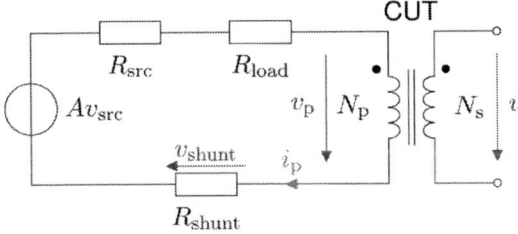

Fig. 1: Circuit diagram of the measurement principle.

The material probe is a small R16 × 9.60 × 6.30 toroidal core made of N87 ferrite material [9]. To set a defined temperature, this core-under-test (CUT) is dived into the oil bath of a WIKA CTM9100-150 temperature calibrator. The CUT has two windings with N_p turns on the primary and N_s turns on the secondary side, respectively. The magnetic field strength in der material probe is proportional to a linear combination of the currents in these two windings. When using a measuring instrument with a high ohmic termination for the secondary voltage v_s,

the secondary current can be assumed zero. So the magnetic field strength in core material is adjusted by feeding a current i_p in the primary winding. It calculates by

$$H(t) = \frac{N_p i_p(t)}{l_e} = \frac{N_p v_{shunt}(t)}{l_e R_{shunt}} \quad (1)$$

with the effective core length l_e. The exact current i_p and field strength H is obtained by measuring the voltage drop v_{shunt} across the shunt resistor R_{shunt} with an high resolution digitizer.

The setup is fed by a voltage amplifier, represented by a voltage source with the small source impedance R_{src}. An arbitrary function generator creates the voltage v_{src} that is amplified with the gain A. To obtain the required current source characteristic with respect to the CUT, a high ohmic load resistor R_{load} is placed in series. Because this linear resistance dominates the total impedance that is fed by the voltage source $A v_{src}$, the current i_p is nearly proportional to the generator voltage v_{src}

$$i_p(t) = \frac{A v_{src}(t) - v_p(t)}{R_{load} + R_{shunt} + R_{src}} \approx \frac{A v_{src}(t)}{R_{load}}. \quad (2)$$

The secondary induced voltage is measured by a second channel of the digitizer. Due to its high input impedance, the voltage v_s is proportional to the derivation of the magnetic flux density

$$v_s(t) = N_s A_e \frac{dB(t)}{dt} \quad (3)$$

with the effective core area A_e. The flux density is calculated by an integration

$$B(t) = \frac{1}{N_s A_e} \int_0^t v_s(\tau) d\tau + B_0. \quad (4)$$

The initial flux density value B_0 depends on the material's history and is set to zero by a demagnetization procedure.

Plotting the measured flux density B over the field strength H yields the hysteresis loop. Subsequently, the hysteresis loss density is the enclosed area of the B-H loop multiplied by the frequency.

2.2 Measurement results

The hysteresis loss energy of one hysteresis loop per volume is measured for temperatures from 25 °C to 105 °C in 5 °C steps by preheating the oil bath to the corresponding value. The amplitude of the flux density excitation ΔB is adjusted from

PCIM Europe 2024, 11–13 June 2024, Nuremberg   DOI: 10.30420/566262208

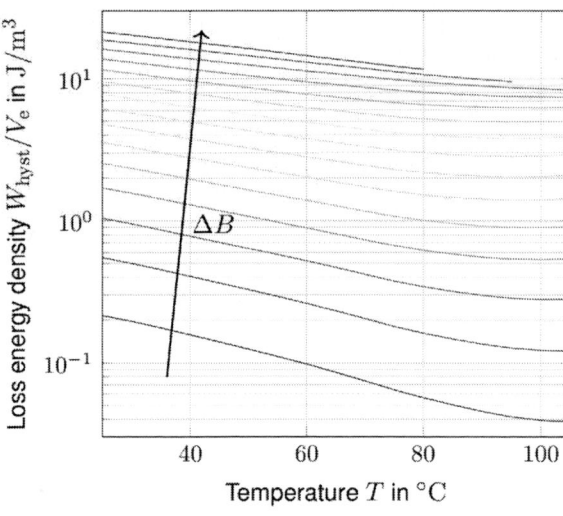
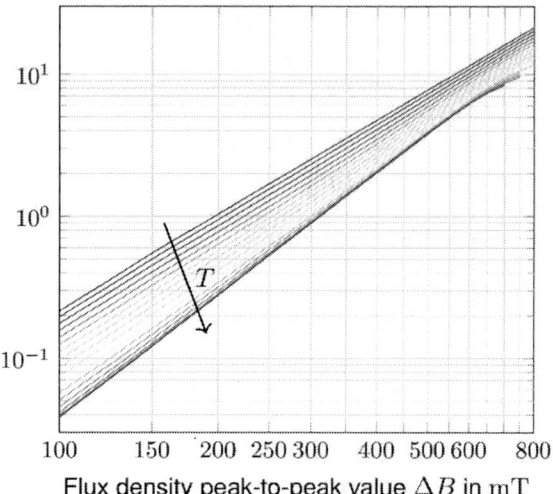

Fig. 2: Measurement of the loss energy density at temperatures T from 25 °C to 105 °C in 5 °C steps and and for a peak-to-peak flux density ΔB from 100 mT to 800 mT in 50 mT steps. Left against the temperature T, right against the flux density amplitude ΔB.

100 mT to 800 mT in 50 mT steps. To avoid a DC premagnetization of the material, the measuring object is demagnetized and carefully balanced during measurement. The measuring frequency is 1 kHz. Figure 2 shows the measured loss energy of one hysteresis loop in two different ways.

The left diagram in Figure 2 plots the measured loss energy W_{hyst} per volume for different peak-to-peak flux densities ΔB against the temperature T. The results at high amplitudes and simultaneously high temperatures are lacking because of the onset of saturation effects. The right diagram in Figure 2 depicts the same results of the measured loss energy W_{hyst} but against the flux density amplitude ΔB for different temperatures T.

3 Approximation of the measurements using various fit formulas

Developers need simple models to accurately describe losses in magnetic material. These models are usually simple formulas with parameters determined empirically by measurements. The most prominent example is the power equation of steinmetz [10], which uses an exponential approach for the excitation and frequency dependence. There are newer extensions to cover nonsinusoidal waveforms and to extend the accuracy [1], [3], [4]. Besides that, a recent model [5] uses loss separation in a quasi-static and a rate dependent part with a polynomial approach for the quasi-static loss energy. The following sections show three approaches to

describe the measurement data using simple analytical relationships from known models. The resulting coefficients are further fitted by polynomials.

3.1 Fit of the excitation dependence for different temperatures by a polynomial approach

The model in [5], [6] describes the quasi-static hysteresis energy per volume at 25 °C with a polynomial approach. This model also includes a DC bias of the magnetization, which is not the subject of this paper. Without a DC part in the flux density, a third-degree polynomial for the dependence on the amplitude of the flux density remains

$$\frac{W_{\mathrm{hyst}}}{V_{\mathrm{e}}} = a_1\left(\frac{\Delta B}{B_{\mathrm{sat}}}\right) + a_2\left(\frac{\Delta B}{B_{\mathrm{sat}}}\right)^2 + a_3\left(\frac{\Delta B}{B_{\mathrm{sat}}}\right)^3.$$
(5)

According to [5], the three coefficients for the tested sample of the material N87 at 25 °C are $a_1 = -5.46 \cdot 10^{-1}$, $a_2 = 7.49$ and $a_3 = 5.68 \cdot 10^{-1}$ with the saturation flux density $B_{\mathrm{sat}} = 490\,\mathrm{mT}$ as a normalization variable. This section attempts to approximate the loss energy for each temperature using this equation. The normalization variable is left unchanged at the 25 °C value of $B_{\mathrm{sat}} = 490\,\mathrm{mT}$, even if the saturation flux density depends on the temperature. The blue curves in Figure 3 show the change in the three parameters a_1, a_2 and a_3 against the temperature.

The parabolic trends indicate that an analytical fit using quadratic equations may be possible. The

1540

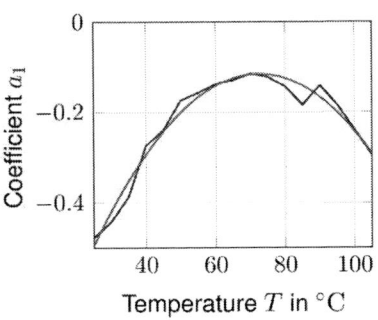 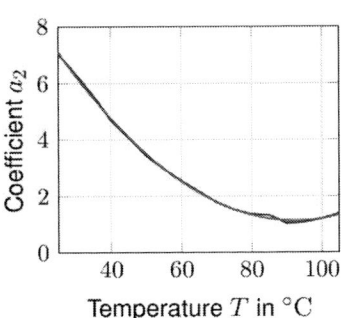 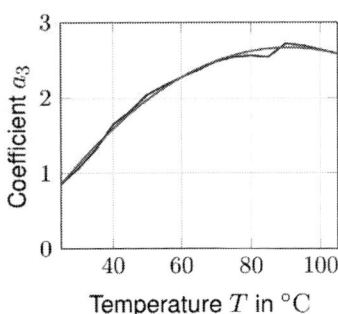

Fig. 3: Coefficients a_1, a_2 and a_3 of the polynomial approach in equation (5) over the temperature T. The blue curve is the fit of equation (5) to the measurements. The red curve is the fit of the coefficients according to equation (6), (7) and (8).

red curves in Figure 3 are approximations using the quadratic approaches

$$a_1 = a_{12}T^2 + a_{11}T + a_{10} \qquad (6)$$

$$a_2 = a_{22}T^2 + a_{21}T + a_{20} \qquad (7)$$

$$a_3 = a_{32}T^2 + a_{31}T + a_{30} \qquad (8)$$

with the coefficients according to Table 1.

a_{12}	$-1.6811 \cdot 10^{-4}$
a_{11}	$2.4389 \cdot 10^{-2}$
a_{10}	$-9.9953 \cdot 10^{-1}$
a_{22}	$1.3139 \cdot 10^{-3}$
a_{21}	$-2.4287 \cdot 10^{-1}$
a_{20}	12.3511
a_{32}	$-4.1968 \cdot 10^{-4}$
a_{31}	$7.6123 \cdot 10^{-2}$
a_{30}	$-7.8555 \cdot 10^{-1}$

Tab. 1: Coefficients for the fit of a_1, a_2 and a_3 according to equation (6), (7) and (8).

This approximation can be used to reproduce the measurement results from Figure 2. Figure 4 depicts the loss energy W_{hyst} per volume against the temperature T. The continuous lines represent the proposed model, whereas the measured data points are marked with $\times$.

For a medium excitation, the model shows a good agreement to the measurement results over the whole temperature range. With high levels above 600 mT (close to saturation) and high temperatures, the model predicts a slightly too high loss energy. Likewise, the prediction is too low for very small excitation levels at high temperatures. This may result

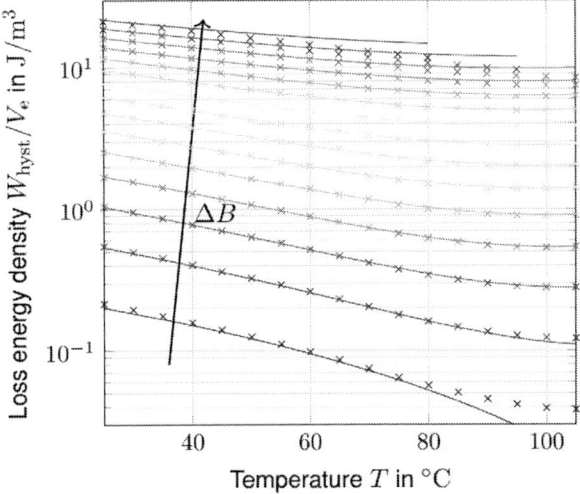

Fig. 4: Measurement (marked with $\times$) and proposed fit (continuous) of loss energy density over the temperature for ΔB from 100 mT to 800 mT in 50 mT steps.

from a problem of the cubic polynomial approach, which is not necessarily monotonically increasing in contrast to an exponential approach. Nevertheless, adequate agreement can be achieved by expressing a curve by polynomials of sufficient order.

3.2 Fit of the excitation dependence for different temperatures by an exponential approach

The next fit is based on the exponential approach for calculating core losses according to steinmetz's power equation [2]

$$\frac{P_{\mathrm{spec}}}{V_{\mathrm{e}}} = C_{\mathrm{m}} \cdot \left(\frac{f}{1\,\mathrm{Hz}}\right)^x \left(\frac{\Delta B}{1\,\mathrm{T}}\right)^y. \qquad (9)$$

Referring to [5], this exponential approach provides for the materials N87, N49 and 3F3 similar quality

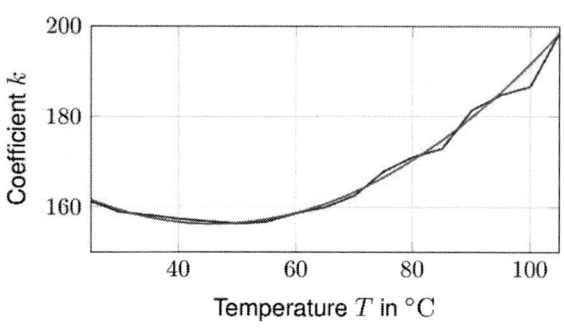
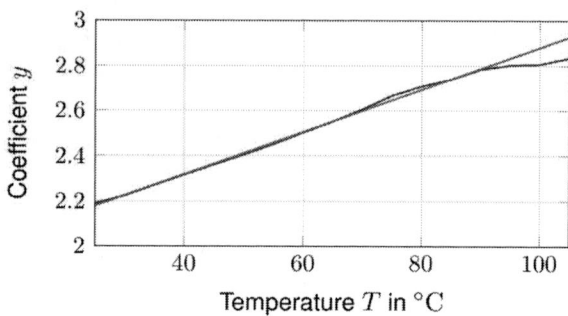

Fig. 5: Coefficients k and y of the exponential approach in equation (10) over the temperature T. The blue curve is the fit of equation (10) to the measurements. The red curve is the fit of the coefficients according to equation (11) and (12).

of results as the polynomial approach. For the materials 3C90 and 3C97, the polynomial approach achieves a higher accuracy. For the study of N87, the power equation (9) can therefore also be used. Analogous to the previous section 3.1, the following attempts to approximate the measurement results over the excitation for each measured temperature value using the exponential approach

$$\frac{W_{\text{hyst}}}{V_{\text{e}}} = k \cdot \left(\frac{\Delta B}{1\,\text{T}}\right)^{y}. \tag{10}$$

The blue curves in Figure 5 show the resulting coefficients k and y against the temperature T for a fit of equation (10) to the measurements. The coefficient k has a parabolic trend again, whereas y increases almost linearly.

Simple polynomial equations for the parameters k and y against the temperature are

$$k = k_2 T^2 + k_1 T + k_0 \tag{11}$$

$$y = y_1 T + y_0. \tag{12}$$

With the coefficients of 2, they define the red curves in Figure 5.

k_2	$1.2071 \cdot 10^{-2}$
k_1	-1.1071
k_0	181.6467
y_1	$9.3574 \cdot 10^{-3}$
y_0	1.9446

Tab. 2: Coefficients for the fit of k and y according to equation (11) and (12).

This model is compared again to the measurements from Figure 2. The dependence on the temperature

is obtained by inserting equations (11) and (12) into equation (10). Figure 6 shows the loss energy W_{hyst} per volume for all amplitudes ΔB against the temperature T.

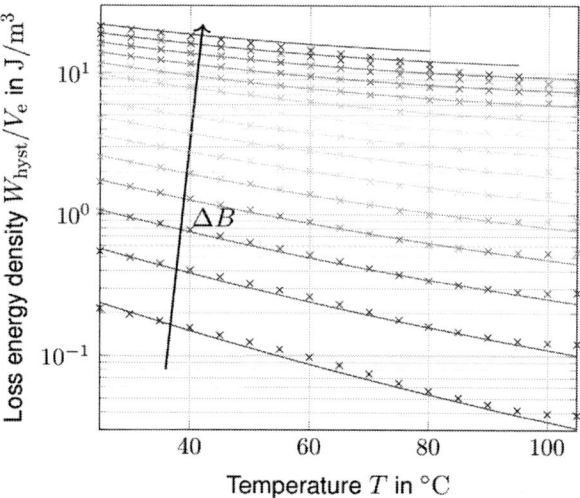

Fig. 6: Measurement (marked with ×) and proposed fit (continuous) of loss energy density over the temperature for ΔB from 100 mT to 800 mT in 50 mT steps.

The deviations in the case of average levels appear slightly larger in this model than in the model in section 3.1. The polynomial approach is better able to reproduce the characteristic when the temperature changes, but the significant deviation at low levels no longer occurs with the exponential approach.

3.3 Fit of the temperature dependence for different excitation levels by a polynomial approach

The quasi-static hysteresis loss energy density at 25 °C can be calculated by the polynomial approach

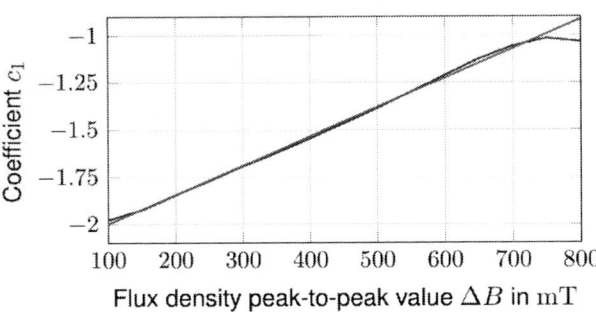
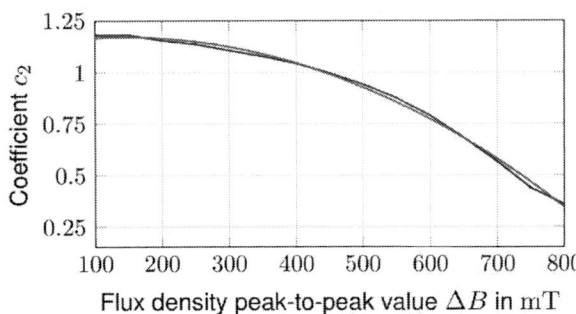

Fig. 7: Coefficients c_1 and c_2 for the temperature factor in equation (14) against the flux density peak-to-peak value ΔB. The blue curve is the fit of equation (13) to the measurement. The red curve is the fit of the coefficients according to equation (15) and (16).

in equation (5) according to [5], [6]. To describe the temperature dependence of power loss in ferrites, Mulder proposed in [11] a quadratic polynomial as a factor. Using the losses at 25 °C as reference, the loss energy calculates by

$$\frac{W_{\text{hyst}}}{V_e} = \frac{W_{\text{hyst,25}}}{V_e} \cdot C(\tau) \tag{13}$$

with the temperature factor

$$C(\tau) = c_2 \tau^2 + c_1 \tau + 1 \quad \text{with} \quad \tau = \frac{T - 25\,°C}{100\,°C}. \tag{14}$$

The variable τ is the temperature change to 25 °C, normalized to 100 °C. Due to the already known loss density $W_{\text{hyst,25}}/V_e$ at 25 °C, the two variables c_1 and c_2 are left to match this quadratic equation to the measurement results. They are fitted separately to the measurements for the flux density peak-to-peak values from 100 mT to 800 mT. The blue curves in Figure 7 show the results.

The coefficients are approximated again by polynomials. This paper simplifies by using a linear behavior for c_1 and a quadratic for c_2. The red curves in Figure 7 represent a fit with the formulas

$$c_1 = c_{11} \cdot \Delta B + c_{10} \tag{15}$$
$$c_2 = c_{22} \cdot (\Delta B)^2 + c_{21} \cdot \Delta B + c_{20}. \tag{16}$$

Their coefficients are given in Table 3.

Inserting these calculated coefficients back to equation (14) and (13) yields with the loss energy density of (5) the continuous lines in Figure 8.

This model fits best over a wide amplitude range. In comparison to the other models, the deviation is smaller at very high flux density levels. In addition, the progression over the temperature is well reproduced.

c_{11}	1.5495
c_{10}	−2.1546
c_{22}	−1.919
c_{21}	$5.5165 \cdot 10^{-1}$
c_{20}	1.1307

Tab. 3: Coefficients for the fit of c_1 and c_2 according to equation (15) and (16).

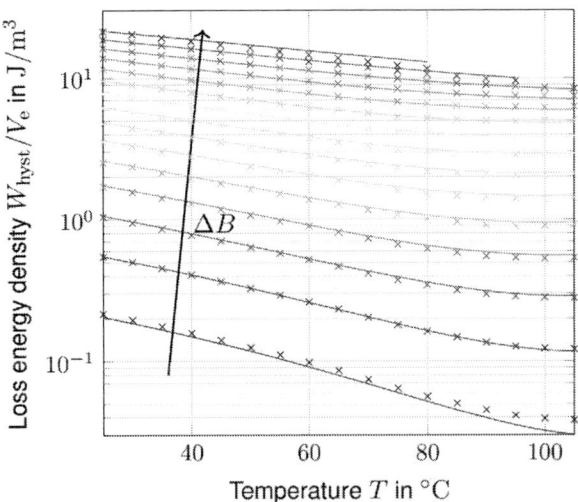

Fig. 8: Measurement (marked with ×) and proposed fit (continuous) of loss energy density over the temperature for ΔB from 100 mT to 800 mT in 50 mT steps.

4 Conclusion

Measurements of the low frequency hysteresis loss of the ferrite material N87 are presented in dependence of the temperature. To calculate the losses for different excitation levels, three models are proposed. The first and second are based on common core loss models whose coefficients are fitted by polynomials via the temperature. The third model uses a quadratic polynomial as a factor to match the temperature dependence of hysteresis loss for a given peak-to-peak flux density. Due to a clear trend of the coefficients, they can be described again with simple formulas. Further investigations should concern the rate dependent losses at higher frequencies, since these also depend strongly on the temperature, but show a behavior that differs from the quasi-static one.

References

[1] M. Albach and T. Dürbaum, "Core losses in transformers with an arbitrary shape of the magnetizing current," in *6th European Conference on Power Electronics and Applications (EPE)*, 1995.

[2] E. Stenglein and T. Dürbaum, "Core loss model for arbitrary excitations with DC bias covering a wide frequency range," *IEEE Transactions on Magnetics*, vol. 57, no. 6, pp. 1–10, Jun. 2021. DOI: 10.1109/TMAG.2021.3068188.

[3] J. Li, T. Abdallah, and C. Sullivan, "Improved calculation of core loss with nonsinusoidal waveforms," in *2001 IEEE Industry Applications Conference*, vol. 4, Chicago, IL, USA, 2001, pp. 2203–2210. DOI: 10.1109/IAS.2001.955931.

[4] J. Muhlethaler, J. Biela, J. W. Kolar, and A. Ecklebe, "Improved core-loss calculation for magnetic components employed in power electronic systems," *IEEE Transactions on Power Electronics*, vol. 27, no. 2, pp. 964–973, Feb. 2012. DOI: 10.1109/TPEL.2011.2162252.

[5] E. Stenglein, "Messtechnische Charakterisierung und Vorhersage der Kernverluste bei weichmagnetischen Ferriten," Doktorarbeit, Friedrich-Alexander-Universität Erlangen-Nürnberg (FAU), 2021, 276 pp.

[6] E. Stenglein, D. Kübrich, M. Albach, and T. Dürbaum, "Novel fit formula for the calculation of hysteresis losses including DC-premagnetization," in *Power Conversion and Intelligent Motion Conference (PCIM)*, May 7, 2019, pp. 1328–1335.

[7] E. Stenglein, B. Kohlhepp, D. Kubrich, M. Albach, and T. Durbaum, "GaN-half-bridge for core loss measurements under rectangular AC voltage and DC bias of the magnetic flux density," *IEEE Transactions on Instrumentation and Measurement*, vol. 69, no. 9, pp. 6312–6321, Sep. 2020. DOI: 10.1109/TIM.2020.2972140.

[8] E. Stenglein, D. Kübrich, M. Albach, and T. Dürbaum, "Guideline for hysteresis curve measurements with arbitrary excitation: Pitfalls to avoid and practices to follow," in *Power Conversion and Intelligent Motion Conference (PCIM)*, Jun. 5, 2018, pp. 1328–1335.

[9] TDK Electronics AG, *EPCOS Data Book 2017: Ferrites and Accessories*. 2017.

[10] C. P. Steinmetz, "On the law of hysteresis," *Transactions of the American Institute of Electrical Engineers*, vol. IX, no. 1, pp. 1–64, Jan. 1892. DOI: 10.1109/T-AIEE.1892.5570437.

[11] S. A. Mulder, "Fit formulae for power loss in ferrites and their use in transformer design," in *PCIM'93 Europe: official proceedings of the twenty-ninth international Intelligent Motion Conference*, Nürnberg, Germany, Jun. 1993, pp. 345–359.

PCIM Europe 2024, 11– 13 June 2024, Nuremberg   DOI: 10.30420/566262210

Passive Methods Limiting Leakage Current in Metal-Oxide Varistor Used as Voltage Clamping Circuit in Low Voltage DC Semiconductor Circuit-Breakers

Kenan Askan[1], Radek Klof[2]

[1] Eaton Industries GmbH, Austria
[2] Eaton European Research Center, Czech Republic

Corresponding author: Kenan Askan, kenanaskan@eaton.com
Speaker: Kenan Askan, kenanaskan@eaton.com

Abstract

In this work, two passive methods are proposed to limit the leakage current of metal-oxide varistor (MOV) used as the most common voltage clamping circuit in semiconductor circuit-breakers (SCCBs) known as hybrid and solid-state circuit breakers during standby operation defined in the draft standard for SCCBs. The first method is placing MOV in series with silicon bidirectional SYDAC (Silicon Thyristor for Alternating Current). The second connects MOV in series with GDT (Gas Discharge Tube). Both methods are characterized by experiments and compared with each other, and other passive methods. Results enable utilization of low cost-high energy dissipation capable of MOV with limited leakage current and low breakdown voltage semiconductor switches in SCCBs used for low voltage DC applications enabling standby operation of breakers.

1 Motivation

Voltage clamping circuit (VCC), used in semiconductor circuit-breakers (Fig. 1), has a function of clearing remaining energy stored in line and stray inductance (L_{Line}) after switching off power semiconductor switch (PSS) under normal and fault current conditions. The VCC is connected parallel to the PSS and its clamping voltage is selected to be lower than breakdown voltage of semiconductor switches used in the PSS. Moreover, the clamping voltage shall be significantly higher than the line voltage (V_{Line}) so that the remaining energy can be cleared fast. According to the definition in the draft standard for low voltage semiconductor circuit breakers [1], these breakers (Semiconductor Circuit Breaker – SCCB and Semiconductor Hybrid Circuit Breaker – SCHCB) need to provide mechanical switching contacts (MSCs) connected in series for safe isolation purposes like for thermo-magnetic circuit-breakers [2]. Therefore, the VCC cannot stay under continuous line voltage once the breaker is in the steady-state open position (open-state). After the remaining energy has been cleared, there is still a leakage current flowing through the VCC for a few tens of milliseconds until the MSC opens its

contacts. This leakage current is then switched off by the MSC. There is also some leakage current flowing through the PSS; however, this leakage current is negligible compared to the leakage current of the VCC.

The draft standard for the low-voltage semiconductor circuit-breakers [1] defines a new position (state) called standby. At the standby position, the MSC is closed, the PSS is in off-state, and for the SCHCB, parallel switching contacts (PSCs) are open. In the standby position, the leakage current measured through each pole, shall not be larger than 6 mA when $1.1 \times U_{\text{e}}$ (rated operational voltage) is applied as a test voltage. One of the reasons behind using the semiconductor circuit-breakers in standby position is to use the breaker as a solid-state relay without opening the MSC to achieve a larger switching operation under non-fault conditions (e.g. during frequent load connection and disconnection).

One of the main application areas of the semiconductor circuit-breakers is low voltage direct current (LVDC) microgrid application [3]. DC microgrids usually consist of bidirectional power electronics converters having filter capacitors at their input and output (e.g. 25 µF/kW). Due to this fact, these

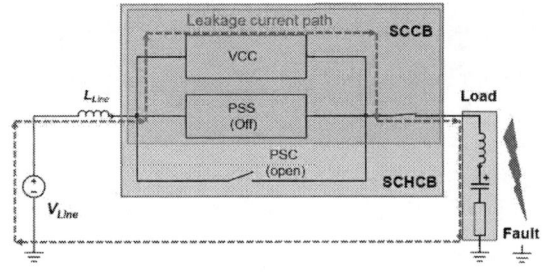

Fig. 1: Single pole circuit topology of DC SCCB and SCHCB.

Parameters	Values	Definition
U_e	900 V	Rated operational voltage
I_n	100 A	Rated current
I_i	400 A	Instantenous trip current
SCHCB delay time (t_{delay})	650 µs	Delay in interruption of current after fault detection time (t_{det})
SCCB delay time (t_{delay})	5 µs	Delay in interruption of fault current after detection time (t_{det})

Tab. 1: Example rated values of semiconductor circuit breaker [4,5].

$$\tau = \frac{L}{R}, \tag{1}$$

$$R = \frac{U_e}{I_{SC}}, \tag{2}$$

$$I_{peak} = I_{SC} \times 0.632, \tag{3}$$

$$t_{det} = \frac{\tau}{I_{peak}} \times I_i, \tag{4}$$

$$t_{int} = t_{det} + t_{delay}, \tag{5}$$

$$I_{Line} = i(t_{int}) = \frac{U_e}{R} \times \left(1 - e^{-\frac{t_{int}}{\tau}}\right), \tag{6}$$

$$E = \frac{1}{2} \times L \times I_{Line}^2. \tag{7}$$

grids have the characteristics of a capacitive grid. The circuit-breakers are usually placed between sources and loads to protect the cables and connected equipment. In Fig. 1, in the case of the standby position, the leakage current through the VCC may charge up the capacitor of the load when the standby operation lasts for a significantly long time. After a certain time, the leakage current is limited because the line voltage is divided by the capacitor of the load and the VCC. The capacitor can be charged up above 25% of the line voltage. The energy stored in the capacitor can be significantly large which might create potential safety issues and inadequate operation of the converters.

Nominal voltage-current rating, instantaneous trip current, and fault current interruption time for the semiconductor circuit-breakers are given in Tab. 1 as an example [4,5]. Values of fault test currents (I_{SC}) and corresponding time constants (τ) are given in Tab. 2 [1]. The corresponding peak of line current (I_{Line}) and fault energy are calculated based on Tab.1 with Eq. 1 − 7 and given parameters in Tab. 2. For instance, in the case of a test circuit having I_{SC} = 20 kA and τ = 10 ms, the energy which is stored in the L_{Line} for SCCB is 90 J and for SCHCB is 764 J. This energy (E) needs to be dissipated within the VCC (Eq. 7). It can be seen that the energy dissipation capability of the VCC in the case of the SCHCB is significantly larger than that of the SCCB.

A Metal-Oxide Varistor (MOV), which is a voltage-dependent resistor (also known as non-linear resistor), is the most commonly used voltage clamping device within the VCC of SCCB and SCHCB because of its high energy dissipation capability and low cost. Due to its low continuous operational voltage under DC voltage (high leakage current) and high clamping voltage characteristics, it requires higher breakdown voltage semiconductor switches. To increase peak current and energy handling capability, multiple MOVs can be connected in parallel; however, this increases the total leakage current. An alternative solution is using a transient voltage suppressor (TVS) diode. The TVS diode has a higher continuous operational voltage, lower leakage current, and lower clamping voltage compared to the MOV. Nevertheless, TVS diodes are usually extremely expensive and have significantly lower energy dissipation capabilities compared to MOVs. In addition, TVS diodes are only available in limited voltage and energy dissipation levels. To obtain the required continuous voltage and energy dissipation capability, multiple TVS diodes need to be connected in series and parallel. This results in a bulky and expensive solution. Apart from the solution with TVS diodes, there are other voltage clamping devices such as snubber networks (C, RC, RCD) and these snubber networks with MOV. One of the methods to limit the leakage current of the MOV under continuous source voltage is to use an active switch in series with MOV [6,7]. The MOV is connected in series with semiconductor switches such as MOSFET, IGBT, and JFET. The leakage current under continuous voltage is limited by driving the semiconductor switches to the off state after the current switching event. This solution increases cost and complexity. Also, the semiconductor devices need to be able to conduct high-pulse current during the energy clearing.

In this paper, in addition to the state-of-the-art voltage clamping circuits used in the semiconductor circuit-breakers, two additional novel passive methods are proposed and discussed to limit the MOV's leakage current in the standby position of the breaker. The first method is placing MOV in series with silicon bidirectional SYDAC (Silicon Thyristor for Alternating Current) shown in Fig. 2 - a. SYDAC stays nonconducting until the voltage rises above its breakdown voltage. Once the voltage is above the breakdown voltage, SYDAC goes into conduction mode. Therefore, MOV starts to conduct the current and dissipates the remaining energy as heat. Once the current is below the hold-on current of SYDAC, it naturally turns off. Similar to this method, in [8] electronic MOV (EMOV) is proposed. EMOV depends on breakover-diode, thyristor, and driver circuit. This requires multiple components to be interconnected which results in a bulky solution. The second method is placing MOV in series with GDT (Gas Discharge Tube) shown in Fig. 2 - b. The GDT stays nonconducting until the voltage rises above its DC sparkvoltage. Once the voltage is above this level, the gas ionizes, and the GDT starts conducting current. Hence, MOV begins conducting current. Once the remaining energy is cleared, GDT stops conducting. This method is already proposed in [9] for medium voltage circuit-breaker applications. Here in this paper, it is evaluated for low-voltage DC. These two methods can increase the maximum holding voltage of the MOV up to 50% by using the same breakdown voltage semiconductors (leakage current $\leq$ 1 mA).

The proposed methods are characterized to figure out their leakage current characteristics under standby mode at different DC voltage and ambient temperatures, and under short-circuit fault events. Results are compared with the VCC consisting of only MOV and only TVS diodes.

2 Semiconductor Circuit-Breaker Topologies

According to [1], the low voltage semiconductor circuit-breakers are classified under two groups:

Fault Circuit		SCHCB		SCCB	
I_{SC} (kA)	τ (ms)	I_{Line} (A)	E (J)	I_{Line} (A)	E (J)
6	5	1259	594	606	138
10	5	1758	695	623	87
20	10	1842	764	633	90
50	15	2723	1001	645	56
65	15	3360	1172	651	44
100	15	4845	1584	664	30
200	15	9087	2787	698	16

Tab. 2: Values of fault test current and corresponding time constant [1], and calculated peak of line current and fault energy according to Tab. 1.

2.1 Semiconductor Circuit-Breaker (SCCB)

A Semiconductor Circuit-Breaker (SCCB), also known as a solid-state circuit breaker (SSCB) in literature, has a power semiconductor switch (PSS) and voltage clamping circuit (VCC) and, to provide a safe isolation function, it has a mechanical switching contact (MSC) connected in series (Fig. 1). It can interrupt a fault current in a few tens of μs once the fault has been detected by an electronic current measurement unit [4]. Even though SCCBs can interrupt faults much faster than traditional thermal-magnetic circuit breakers, SCCBs have much lower overcurrent and thermal limits.

2.2 Semiconductor Hybrid Circuit-Breaker (SCHCB)

SCHCB, also known as a hybrid circuit breaker (HCB) in literature, has additional low-ohmic parallel switching contact (PSC) to the PSS (Fig. 1). In the closed position of the breaker, the current flows through the low-ohmic mechanical switch. The PSC opens its mechanical contacts to generate a micro-electrical arc so the current commutates to the PSS [5]. The overcurrent and thermal limits of the SCHCBs are comparable with the traditional thermal-magnetic circuit breakers; however, the interruption speed of SCHCBs is much slower than SCCBs.

3 Proposed Passive Methods

3.1 SYDAC-MOV Arrangement

The first proposed method is using MOV in series with silicon bidirectional SYDAC (Silicon Thyristor for Alternating Current) shown in Fig. 2 - a. The SYDAC, also known as SIDACtor [10], is a semiconductor device that has four layers of alternating P-type and N-type materials.

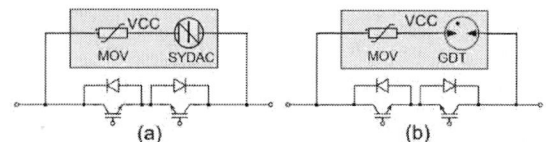

Fig. 2: Proposed topologies: (a) SYDAC-MOV arrangement, (b) GDT-MOV arrangement.

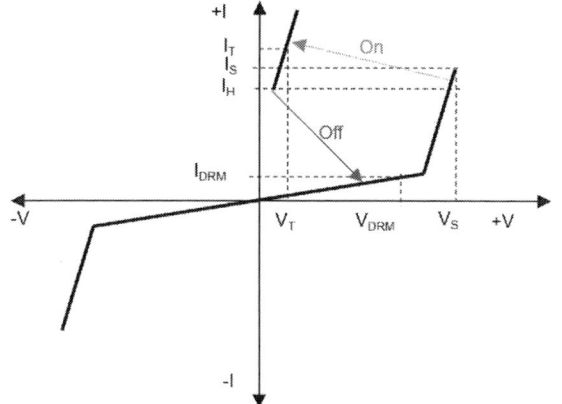

Fig. 3: Basic operational principle of SYDAC [10].

Key parameters for SYDAC are V_{DRM}, I_{DRM}, V_S, I_H, and V_T (Fig. 3). V_{DRM} is the repetitive peak off-state voltage rating of the device (also known as stand-off voltage) and is the continuous peak DC voltage that may be applied to the SYDAC in its off-state condition. I_{DRM} is the maximum value of leakage current that results from the application of V_{DRM}. Switching voltage (V_S) is the maximum voltage that subsequent components may be subjected to during a fast-rising voltage transient. Holding current (I_H) is the minimum current required to maintain the device in the on-state. On-state voltage (V_T) is the maximum voltage across the device during on-state.

As the voltage across the device increases and exceeds the V_{DRM}, the electric field across the center junction reaches a value sufficient to cause avalanche multiplication. As avalanche multiplication occurs, the impedance begins to decrease, and current flow starts to increase until the device's current gain exceeds unity (Crowbaring). Once unity is exceeded, the device switches from a high impedance (measured at V_S) to a low impedance (measured at V_T) until the current flowing through the device is reduced below the I_H. If the peak pulse current (I_{PP}) rating is exceeded, the device typically turns into a permanent short-circuit.

The SYDAC connected in series with MOV operates like a switch. In the off-state, the line voltage is divided by the junction capacitance of the MOV and SYDAC. When the voltage drop on the SYDAC is smaller than V_{DRM}, the leakage current is only a few µA. Therefore, the leakage current of the MOV is limited to almost zero. During the on-state of the power semiconductor switch (PSS), the voltage drop is only due to the on-state resistance of the semiconductor switches in the case of SCCB and on-state contact resistance in the case of SCHCB which is significantly small. During the PSS turns off, due to the energy stored in line inductance (L_{Line}), a transient voltage appears on the VCC. As the junction capacitance of SYDAC is much smaller than MOV's junction capacitance ($C_{SYDAC} \ll C_{MOV}$), most of the transient voltage appears on the SYDAC. As the transient voltage exceeds the device's V_{DRM}, the device begins to enter its protective mode with characteristics similar to an avalanche diode. Finally, the transient voltage appears on the MOV. Until the transient voltage reaches the clamping voltage, only leakage current flows through the VCC. This leakage current is larger than I_S which results in the SYDAC turning from off-state to on-state. While SYDAC operates in on-state, the MOV clamps the transient voltage to protect the PSS and clear the remaining energy in the L_{Line}. Once the leakage current of the VCC (MOV) is below I_H, the SYDAC resets, returning to its off-state. If the leakage current stays above the I_H, the SYDAC will not reset. In this case, the mechanical switch contact (MSC) shall open the circuit to avoid overheating the MOV.

3.2 GDT-MOV Arrangement

The first proposed method is placing MOV (Metal-Oxide Varistor) in series with silicon bidirectional GDT (Gas Discharge Tube) shown in Fig. 2 - b. The GDT remains at high impedance off-state condition (e.g. 1000 MΩ at 200 V) until the voltage appears on the GDT rises above its rated sparkover voltage (V_{SO}). During this phase, the leakage current (I_{leak}) is almost 0 µA. The V_{SO} is measured at a change of rate of voltage (dV/dt) between 100 V/s and 2000 V/s. With rising voltage on its terminals, the GDT will go into its glow voltage (V_{GLR}) region. This region is where gas inside the tube begins ionizing because of the charge developed across. During the glow region the rise of the current flow will result in an avalanche effect in the gas ionization that will result in the transition of the GDT into a virtually short-circuit mode (Crowbaring) and current

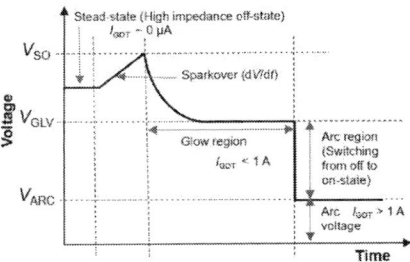

Fig. 4: Basic operational principle of GDT [11].

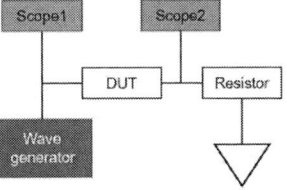

Fig. 5: Block diagram of setup for measurement of junction capacitance.

depending on the impedance of the line will flow between two conductors. The voltage across the GDT during the short-circuit is called arc-voltage (V_{ARC}). The V_{ARC} is usually around $10-15$ V. The GDT will reset to off-state once there is insufficient energy to maintain the arc condition.

The GDT connected in series with MOV operates like a switch. In the off-state of the PSS, the line voltage is divided by the junction capacitance of the MOV and GDT. When the voltage drop on the GDT is smaller than V_{SO}, I_{leak} is almost equal $0\,\mu$A (Fig. 4). Therefore, the leakage current of the MOV is limited to $0\,\mu$A. During the on-state of the power semiconductor switch (PSS), the voltage drop is only due to the on-state resistance of the semiconductor switches in the case of SCCB and on-state contact resistance in the case of SCHCB which is significantly small. During the PSS turning off, due to the energy stored in line inductance (L_{Line}), a transient voltage appears on the voltage clamping circuit (VCC). As the junction capacitance of GDT is much smaller than MOV's junction capacitance ($C_{\mathrm{GDT}} \ll C_{\mathrm{MOV}}$), most of the transient voltage appears on the GDT. As a transient voltage ($\mathrm{d}V/\mathrm{d}t$) exceeds the device's V_{SO}, the GDT goes from high impedance off-state to the glow region. As the V_{GLR} is smaller than the V_{SO}, more voltage appears MOV which results in a significant amount of the leakage current flowing to start the arc-condition in the GDT. Finally, the GDT goes from the glow region to the arc region. Consequently, the complete transient voltage appears on the MOV. Until the transient voltage reaches the MOV clamping voltage, there is a large leakage current flowing through the MOV which maintains the arc-condition in the GDT. While the GDT operates in on-state with the V_{SO} appearing on its terminals, the MOV clamps the transient voltage to protect the PSS and clears the remaining energy in the L_{Line}. Once the MOV's current is below the minimum at which energy is insufficient to maintain the arc in the GDT, the GDT resets and returns to its high-impedance off-state. If the leakage current through the MOV still drives the arc, then in this case, the mechanical switch contact (MSC) shall open the circuit to avoid overheating the MOV.

4 Experimental Results

Experiments are conducted on different types of devices (MOV, TVS, SYDAC, and GDT). The list of protection devices used during experiments and their comparison is given in Tab. 3 according to their datasheets.

4.1 Junction Capacitance Measurement

The junction capacitance of the selected devices is given in their datasheets at different frequencies. To determine the capacitance of the DUT (Device under test) at the same frequencies, the test setup shown in Fig. 5 is used. Each device is connected in series with a highly precise resistor. A wave generator is used to feed the circuit. The impedance of the DUT and the resistor and the phase shift between them are measured by voltage using voltage probes and an oscilloscope (Scope1, Scope2). The measurement results are given in Fig. 6 from 0.5 MHz to 5 MHz. The obtained results are almost identical to the values given in their datasheets. The junction capacitance of the MOV is nearly 4 times the selected SYDAC and 1000 times the selected GDT.

4.2 Leakage Current Measurement

The leakage current (I_{leak}) measurement is conducted on the MOV, two TVS diodes connected in series, the MOV connected in series with the SYDAC, and the MOV connected in series with GDT forming the VCC. The VCC is placed in a thermal chamber and measurements are done at different positive ambient temperatures and at applied different continuous DC voltages.

As it can be seen in Fig. 7 that the SYDAC-MOV and GDT-MOV arrangements show the lowest I_{leak} at continuous voltage up to 1000 V DC and ambient

Parameters		MOV	TVS	SYDAC	GDT
Designation		TDK S25K460E4R12	Bourns PTVS3-430C-TH	Littlefuse SYDACtor P3500 MEL	Bourns 2047-350
Type of Protection		Clamping	Clamping	Crowbar	Crowbar
Continues DC voltage / Stand-off voltage / DC Sparkover voltage		750 V ($I_{leak} < 1$ mA) at 25 ℃ 615 V ($I_{leak} > 1$ mA) at 85 ℃	430 V ($I_{leak} = 10\,\mu$A)	320 V ($I_{leak} < 5\,\mu$A) at 25 ℃	350 V ($I_{leak} = 0\,\mu$A)
Working Principle		Zinc-oxide grain structure	Avalanche	Controllable silicon	Gas ionization
Clamping Voltage / Switch Voltage / Impulse Sparkover		1250 V at 150 A	580 V	400 (100 V/µs)	475 V / 550 V (100 V/µs) 725 V / 800 (1000 V/µs)
Residual Voltage (During on-state)		>1250 V at >150 A (non-linear)	465 V	4 V at 2.2 A	12 V at 2 A
Maximum Peak Current		20 kA	3 kA	5 kA	40 kA
Maximum Energy Dissipation		720 J	22 J [4]	NA	NA
Number of Operations		Low	High	High	Low

Tab. 3: Selected devices for experiments (According to datasheets available online).

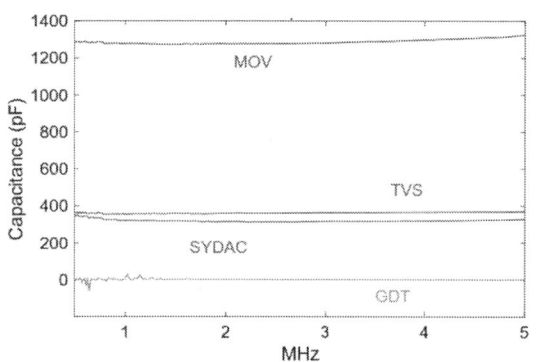

Fig. 6: Measured junction capacitance vs. frequency: MOV, TVS, SYDAC and GDT.

temperature up to 125 ℃, compared to only MOV and only TVS diodes arrangement. For the GDT-MOV arrangement, the I_{leak} is almost the $0\,\mu$A and the I_{leak} is below $500\,\mu$A in the case of the SYDAC-MOV arrangement. Due to the positive temperature of the silicon, the maximum holding voltage of the TVS diode increases with the temperature.

4.3 Dynamic Operation

Test setup given in Fig. 8 is used to characterize the VCC consisting of only MOV, only TVS diodes, the SYDAC-MOV and GDT-MOV arrangements under a short-circuit event (Fig. 9). A capacitor bank (C_{Line}) is first of all charged with a voltage source through a current limiting resistor (R). The line inductance (L_{Line}) during the short-circuit event is set to 150 µH. The power semiconductor switch (PSS), consisting of a bidirectional IGBT module, is switched on into the short-circuit and switched off after a certain time delay.

4.3.1 VCC consisting of only MOV

Fig. 10 shows current and voltage waveforms of the VCC consisting of only one single MOV. Before the PSS switches on into the shorted output,

the line voltage (V_{Line}) of 750 V appears on the VCC parallel connected to the PSS, which represents the standby mode of the breaker. At t_0, the PSS turns on. The peak of the short-circuit current rises to 950 A in 200 µs (approx. 4.75 A/µs). At t_1, the PSS turns off and the line current ($I_{Line} = I_{PSS} + I_{MOV}$) commutates to the MOV. The MOV clamps the voltage at around 1245 V and the peak voltage appearing on the PSS is approx. 1290 V. The PSS current (I_{PSS}) decreases from 90% to 10% in approx. 1.1 µs. In approximately 375 µs the remaining energy in the L_{Line} is cleared (dissipated by the MOV as heat) and at t_2, the breaker switches back to standby.

4.3.2 VCC consisting of TVS Diodes

Fig. 11 shows current and voltage waveforms of the VCC consisting of only two TVS diodes connected in series (Fig. 9 - a). Before the PSS switches into the shorted output, the V_{Line} of 840 V appears on the VCC parallel connected to the PSS, representing the breaker's standby mode. At t_0, the PSS turns on. The peak of the short-circuit current rises to 1050 A in 200 µs (approx. 5.25 A/µs). At t_1, the PSS turns off and I_{Line} commutates to the VCC consisting of TVS diodes. The VCC clamps the transient voltage at around 1095 V. The I_{PSS} decreases from 90% to 10% in 0.9 µs. Even though the TVS diodes are successfully clamping the voltage and the PSS is turned off, due to insufficient energy dissipation capability of the TVSs the VCC faces thermal run-aways and switches back to conductive mode. The total energy stored in the L_{Line} is approx. 83 J (Eq.7). On the other hand, a single TVS diode can dissipate only approx. 22 J [4]. At least four TVSs are required to successfully clear the remaining energy in the fault circuit, resulting in bulky and expensive VCC.

PCIM Europe 2024, 11– 13 June 2024, Nuremberg   DOI: 10.30420/566262210

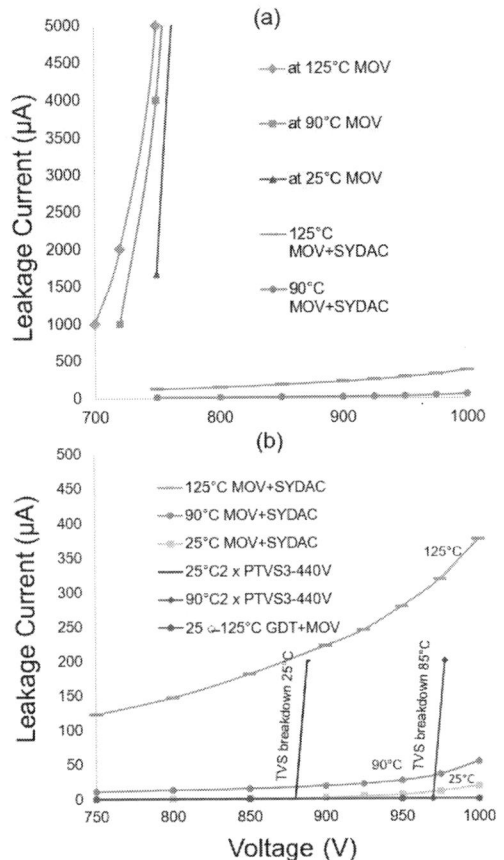

Fig. 7: Comparison of Leakage Current of different VCCs.

4.3.3 SYDAC-MOV Arrangement

Fig. 12 - a shows current and voltage waveforms of the VCC consisting of the MOV connected in series with the SYDAC (Fig. 9 - b). Before the PSS switches on into the shorted output, the V_{Line} of 840 V appears on the VCC parallel connected to the PSS, representing the standby mode of the breaker. During the standby mode (Fig. 12 - b), the 840 V is divided by the MOV (485 V) and SYDAC (355 V). The PSS is turning on at t_0. Because the PSS is now shorting the junction capacitance

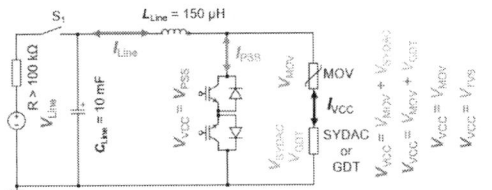

Fig. 8: Experimental setup for evaluation of VCCs under fault.

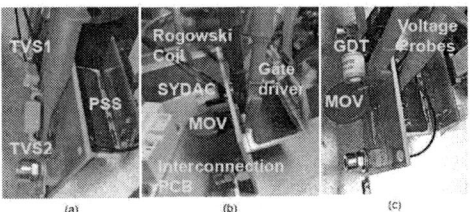

Fig. 9: Image of VCC: (a) TVS diodes, (b) SYDAC-MOV, (c) GDT-MOV.

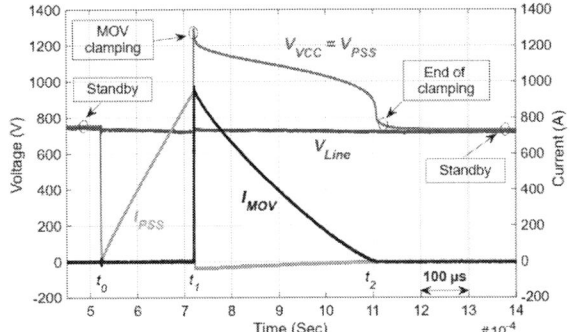

Fig. 10: Only MOV: Short-circuit experimental waveforms at 750 V.

of the MOV and the SYDAC, the voltage probes measure a very short time a polarized voltage on the MOV and the SYDAC. The short-circuit current (Fig. 12 - c) rises to approx. 1040 A in 200 μs (5.2 A/μs). At t_1, the PSS starts with turning off, which results in a transient voltage on the VCC due to the energy stored in the L_{Line} (Fig. 12). As the junction capacitance of the SYDAC is smaller than the MOV's capacitance, the transient voltage appears on the SYDAC. Once the transient voltage rises to the switch voltage of the SYDAC (V_S), the SYDAC switches on and the transient voltage appears now only on the MOV. However, the transient voltage is not yet sufficient to be clamped by the MOV, there is no sufficient large switch current (I_S) to keep the SYDAC on. Therefore, the SYDAC goes from on-state to off-state, resulting in the transient voltage re-appearing on the SYDAC, and the SYDAC re-turns on. Finally, the transient voltage reaches a sufficient level so that MOV can clamp the voltage which results in the rise of the current through the MOV, which is sufficient to keep the SYDAC in on-state. The MOV clamps the voltage at around 1220 V. The peak voltage of the VCC reaches up to 1395 V. The PSS current decreases from 90% to 10% in approx. 1.2 μs.

Finally, at t_2 the fault energy is cleaned and there

1551

PCIM Europe 2024, 11– 13 June 2024, Nuremberg   DOI: 10.30420/566262210

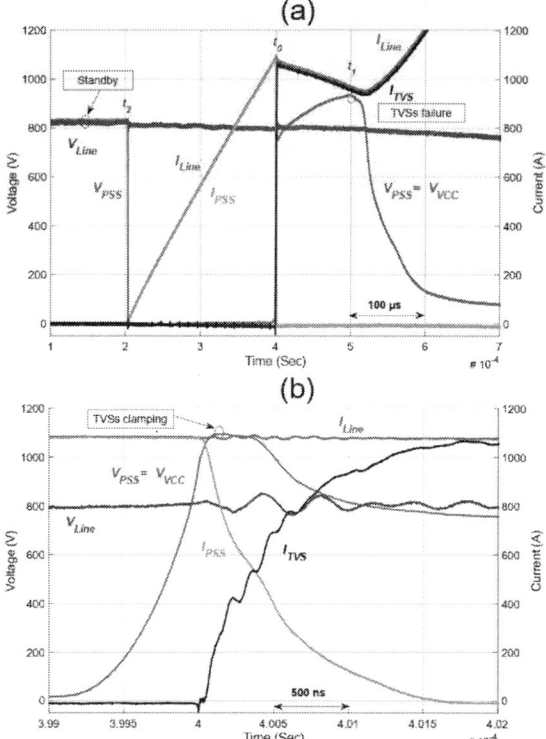

Fig. 11: TVS diodes: Fault current experimental waveforms at 840 V, (a) Complete event, (b) Zoomed
view showing fault current turn-off event t_1.

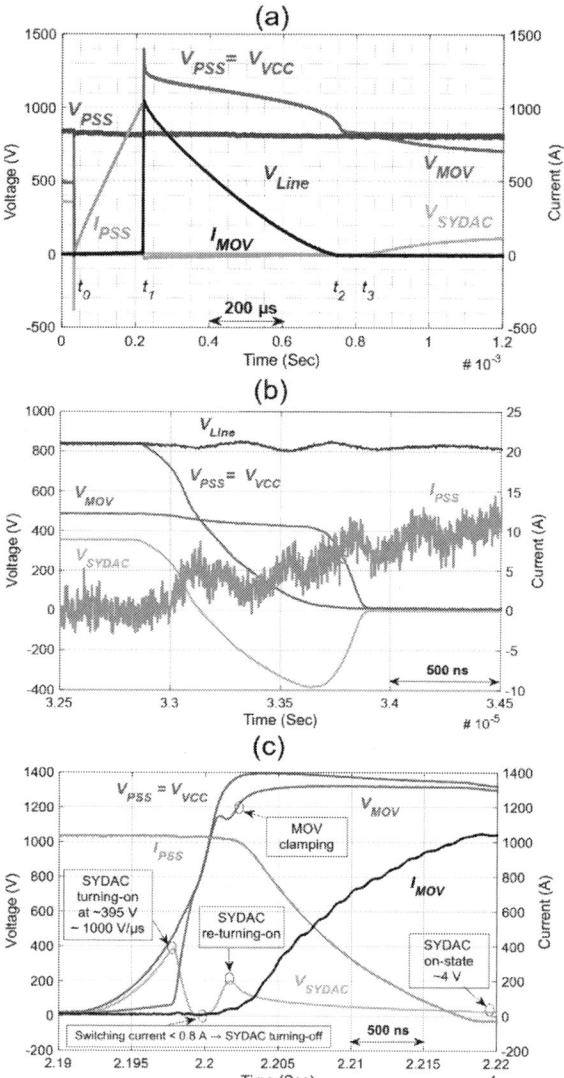

Fig. 12: SYDAC-MOV: Fault current experimental waveforms at 840 V, (a) Complete event $t_0 - t_3$, (b)
Zoomed view showing turn-on into fault event
t_0, (b) Zoomed view showing fault turn-off event
t_1.

is only the I_{leak} flowing through the VCC due to
the MOV (Fig. 12 - a). At t_3, once I_{leak} is below
the holding current (I_{H}) of the SYDAC, the SYDAC
switches to off-state and the V_{Line} is again divided
by the MOV and the SYDAC, and the breaker goes
back the standby position.

4.3.4 GDT-MOV Arrangement

Fig. 13 shows current and voltage waveforms of
the VCC consisting of the MOV connected in series
with the GDT (Fig. 9 - c). Before the PSS switches
on into the shorted output, the V_{Line} of 840 V appears on the VCC parallel connected to the PSS,
representing the standby mode of the breaker. During the standby mode (Fig. 13 - b), the 840 V is
divided by the MOV (520 V) and the GDT (325 V).
The PSS is turning on at t_0. Because the PSS now
is shorting the junction capacitance of the MOV
and the GDT, the voltage probes measure for a
very short time a polarized voltage on the MOV and
the GDT. The short-circuit current rises to approx.
1050 A in 200 µs (approx. 5.25 A/µs) (Fig. 13 -
c). At t_1, the PSS starts to turn off which results in
a transient voltage on the VCC due to the energy

1552

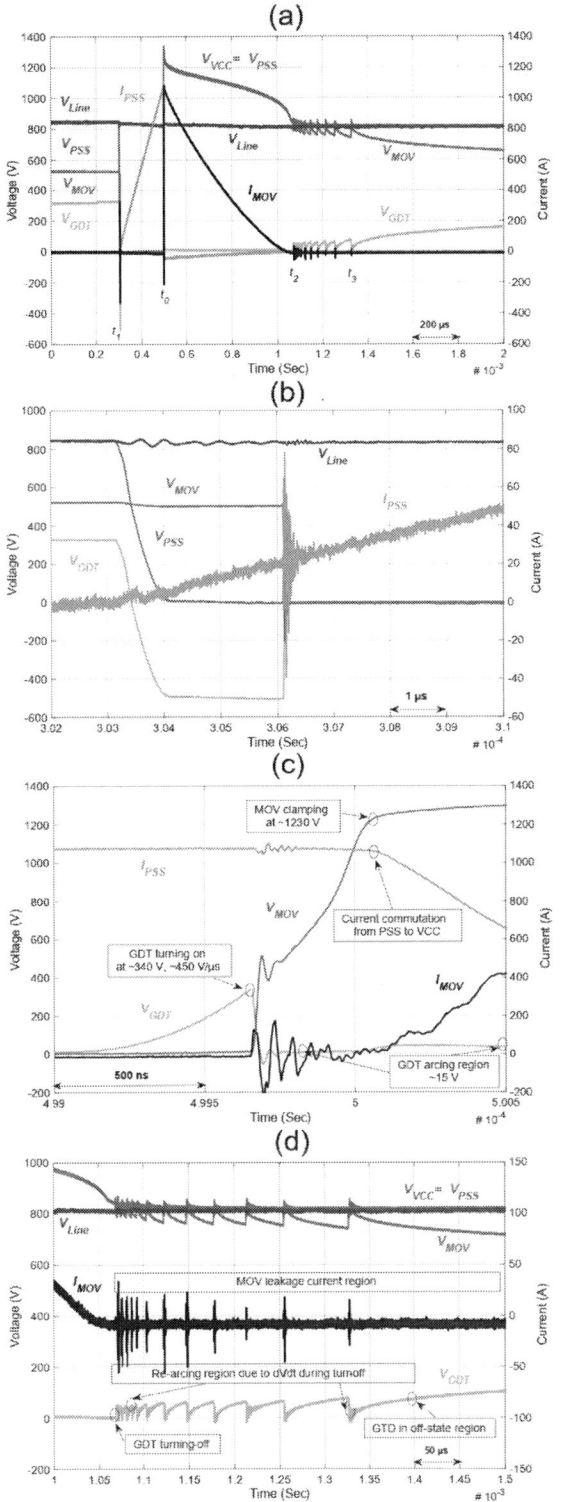

Fig. 13: GDT-MOV: Fault current experimental waveforms at 840 V, (a) Complete event $t_0 - t_3$, (b) Zoomed view showing turn-on into fault event t_0, (c) Zoomed view showing fault turn-off event t_1, (d) Zoomed view transition to standby $t_2 - t_3$.

stored in the L_{Line}. As the junction capacitance of the GDT is smaller than the MOV's capacitance, the transient voltage appears on the GDT. Once the transient voltage rises to the sparkover voltage of the GDT (V_{SO}), the GDT goes from the high-impedance off-state region to the on-state arcing region. The transient voltage appears now only on the MOV. The leakage current flowing through the MOV is sufficient to keep the arcing condition for the GDT. Finally, the transient voltage on the MOV rises to the clamping voltage and the MOV clamps the voltage at approx. 1230 V (Fig. 13 - c). The peak voltage of the VCC reaches up to 1340 V. The PSS current decreases from 90% to 10% in approx. 1.2 µs. Once the complete line current flows through the VCC, there is around 15 V arc-voltage (V_{ARC}) measured on the terminals of the GDT at 1070 A. In average 10 V of the V_{ARC} is measured while the current decreases from 1070 A (t_1) to 0 A (t_2) in approx. 400 µs. At t_2, the leakage current of the MOV is not sufficient to keep the arcing conditions in the GDT. The GDT tries to turn off; however, the transient voltage that appears on the GDT forces the GDT to go from high-impedance off-state to re-arcing. As the current is insufficient to realize the arc condition, the GDT tries to turn it off again. After around 600 µs duration of the oscillation, at t_3, the temperature of the MOV and the GDT go down and the leakage current decreases further, and it is not sufficient anymore to realize arcing conditions. Finally, the GDT goes completely off-state and the line voltage is again divided by the MOV and GDT (Fig. 13 - d).

4.4 Comparison According to Experimental Results

The VCC consisting of the two TVS diodes connected in series has a peak voltage lower than 1200 V and turn-off time smaller than 1 µs which results in utilizing 1200 V breakdown voltage (V_{BR}) components (e.g. SiC MOSFET and JFET) and reduced turn-off losses (Tab.4). However, due to the low energy dissipation capability of TVS and high cost, the utilization of TVS diodes results in a bulky and expensive solution. Arrangements of MOV with SYDAC and GDT result in only approx. 8% increase in the peak of the voltage and turn-off time. This requires semiconductor switches with a V_{BR} larger than 1200 V (e.g. 1700 V IGBT and diode). By choosing MOV with a lower clamping voltage, even 1200 V components can be used. In the case of SYDAC-MOV, after the energy clearing phase of the

VCC	Peak I_{Line}(A)	Peak voltage (V)	PSS turn-off time (µs)
MOV	950	1290	1.1
TVS	1050	1095	0.9
SYDAC-MOV	1040	1395	1.2
GDT-MOV	1070	1340	1.2

Tab. 4: Comparison of VCCs consisting of different devices under fault event.

switching (such as under normal and fault events), I_{Leak} of the chosen MOV shall be smaller than I_H of chosen SYDAC so that the SYDAC switches back to off-state. For the case of using GDT with MOV, the leakage current of the chosen MOV shall be sufficiently small to extinguish the arc in the GDT.

5 Conclusion

In this paper two novel voltage clamping circuit (VCC) topologies are presented to be able to utilize low voltage DC semiconductor circuit-breakers (SCCB) also in standby operation. Existing topologies based on only MOV and TVS diodes are compared with the SYDAC-MOV and GDT-MOV arrangements according to the leakage current under continuous DC voltage and dynamic operation under fault events. These arrangements allow using MOV with lower clamping voltage and therefore semiconductor switches with lower breakdown voltage. This enables placing, for instance, 1200 V SiC MOSFET / JFET without using expensive TVS diodes for 1000 V DC applications. MOV-GDT arrangement enables the lowest leakage current at 1000 V DC and 125 °C ambient temperature compared to MOV-SYDAC arrangement. Both proposed arrangements enable the highest energy dissipation and highest continuous voltage holding capable of the SCCBs with lower cost and allow standby operation of hybrid and solid-state DC circuit breakers with the lowest leakage current without opening the mechanical switching contacts (MSCs) connected in series for safe isolation purposes for non-fault operation conditions.

References

[1] IEC 60947-10 ED1, Low-voltage switchgear and controlgear – Part 10: Semiconductor Circuit-Breakers (121A/565e/CD).

[2] IEC 60947-2, Low-voltage switchgear and controlgear – Part 2: Circuit-Breakers (Edition 5.1, 2019).

[3] DC-INDUSTRIE System Concept, https://odca.zvei.org .

[4] K. Askan, M. Bartonek and F. Stueckler, "Bidirectional Switch Based on Silicon High Voltage Superjunction MOSFETs and TVS Diode Used in Low Voltage DC SSCB," PCIM Europe 2019; International Exhibition and Conference for Power Electronics, Intelligent Motion, Renewable Energy and Energy Management, Nuremberg, Germany, 2019, pp. 1-8.

[5] K. Askan and P. Schasfoort, "Variable Voltage IGBT Gate Driver for Low Voltage Hybrid Circuit Breaker," 2021 IEEE Fourth International Conference on DC Microgrids (ICDCM), Arlington, VA, USA, 2021, pp. 1-7, doi: 10.1109/ICDCM50975.2021.9504654.

[6] X. Song, Y. Du and P. Cairoli, "Survey and Experimental Evaluation of Voltage Clamping Components for Solid State Circuit Breakers," 2021 IEEE Applied Power Electronics Conference and Exposition (APEC), Phoenix, AZ, USA, 2021, pp. 401-406, doi: 10.1109/APEC42165.2021.9487424.

[7] L. Qiao, T. Liu, J. Ying and D. Zhang, "Research and Application of Solid State DC Circuit Breaker Based on SiC Series and Parallel," PCIM Asia 2020; International Exhibition and Conference for Power Electronics, Intelligent Motion, Renewable Energy and Energy Management, Shanghai, China, 2020, pp. 1-8.

[8] L. Ravi, D. Zhang, D. Qin, Z. Zhang, Y. Xu and D. Dong, "Electronic MOV-Based Voltage Clamping Circuit for DC Solid-State Circuit Breaker Applications," in IEEE Transactions on Power Electronics, vol. 37, no. 7, pp. 7561-7565, July 2022, doi: 10.1109/TPEL.2022.3149757.

[9] K. Liu, X. Zhang, L. Qi, X. Qu and G. Tang, "A Novel Solid-State Switch Scheme With High Voltage Utilization Efficiency by Using Modular Gapped MOV for DC Breakers," in IEEE Transactions on Power Electronics, vol. 37, no. 3, pp. 2502-2507, March 2022, doi: 10.1109/TPEL.2021.3115254.

[10] Littelfuse, Application Note, High Power Semiconductor Crowbar Protector for AC Power Line Applications.

[11] Bourns, Whitepaper, First Principle of Gas Discharge Tube (GDT) Primary Protector.

ESD Solutions for Discrete 650V Normally-off AlGaN/GaN HEMTs

Plinio Bau, Thanh Hai Phung, Bernard Bancal, Dominique Bergogne

Wise-Integration, France

Corresponding author: Plinio Bau, plinio.bau@wise-integration.com

Abstract

The 650V AlGaN/GaN Schottky gate technology presents a gate structure that can be damaged by ESD discharges. Therefore, discrete power GaN devices can have ESD protection circuits to be used in diverse applications. The use of diodes in series for clamping negative voltages in the gate of HEMT power transistors is discussed. The disadvantage of those solutions is the large area required by ESD diodes and the low breakdown value achieved. To increase the performance of ESD circuits while keeping the circuit consuming low area for a cost-effective solution, two circuits are presented in this paper.

1 Introduction

ESD (Electrostatic Discharge) protection circuits monolithically integrated on GaN-on-Si (Gallium Nitride on Silicon) technology represent a crucial advancement in ensuring the reliability and robustness of GaN-based devices. These integrated protection circuits serve as a shield against potential damage caused by electrostatic discharge events, a significant concern in electronic systems. By incorporating ESD protection directly onto GaN-on-Si chips, these circuits effectively safeguard the sensitive components from high-voltage transients and electrostatic discharges, preserving the integrity and longevity of the devices.

Fig. 1 is a schematic of the HMB ESD (Human Body Model for Electrostatic Discharge) test setup for a GaN power transistor (M_{PWR}).

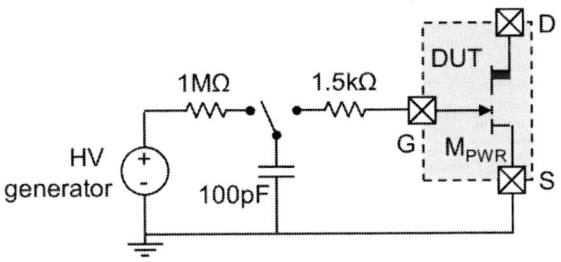

Fig. 1. HBM ESD test bench. The Device Under Test (DUT) is a 650V GaN HEMT. High voltage (HV) pulses both positive and negative polarity should be used to test.

The monolithic integration of ESD protection circuits on GaN-on-Si technology signifies an improvement towards more robust and reliable electronic systems, helping the widespread adoption of GaN devices across industries such as consumer electronics, power supplies for data centers and automotive.

Fig. 2 presents two examples of recommended ESD protection circuits for 650V GaN HEMTs. The option illustrated in **Fig. 2** (a) has the disadvantage of preventing the utilization of negative voltage in normal operation for increased dv/dv robustness in a switching leg topology. Another disadvantage is the large area required by the diode-mounted transistors and therefore the price associated. To reduce the area required by the ESD protection circuit, power clamp circuits are also reported in the literature for GaN transistors [1-3] similarly to CMOS [4-6] and in the patents databases of GaN manufacturers.

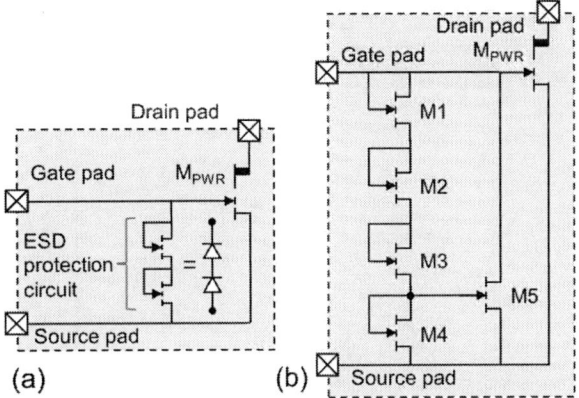

Fig. 2. ESD protection circuits for the gate of GaN Schottky p-type technology. Commonly recommended solutions are either (a) diodes-mounted transistors or with (b) power clamp topology.

The disadvantage of this second solution is a small current consumption in normal operation and the impossibility of driving with negative PWM waves for the dv/dt robustness.

In this paper, two other solutions are presented to obtain good performance in a small chip area for a cost-effective solution.

2 Diode-counted Transistors for ESD Circuits

For pGaN AlGaN/GaN Schottky technology, ESD robustness was reported previously in the literature that depends on many factors, for example, UV light and epitaxial quality. [7,8]. It was also reported that the gate of the transistors has a leakage current proportional to the voltage applied in the gate and creates a relatively robust behavior in direct polarity regarding positive ESD discharges [9]. However, the gate structure is fragile in the inverse polarity. By only placing one diode (diode-mounted transistor), it is possible to clamp the gate voltage by deviating the current into the diode. A simple solution to increase the ESD withstand voltage is obtained by placing one diode-mounted transistor connected the cathode to the gate and anode to the source of a GaN transistor.

2.1 TLP I-V Characteristics of Diode-connected Transistors

For sizing one ESD diode, the I-V characteristics of a diode-connected transistor should be observed. The ESD current is set by the HBM resistor of 1.5kΩ. The increase in the gate node voltage of a power device is then translated by the I-V characteristics of the diode. Examples of I-V curves depending on the size of the ESD diode are presented in **Fig. 3**.

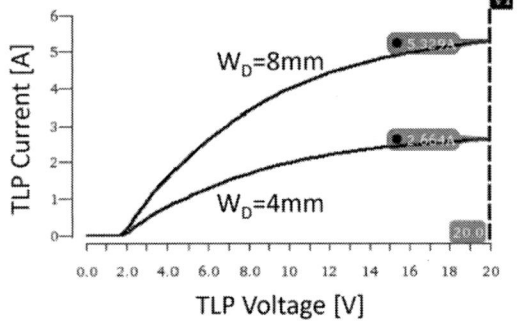

Fig. 3. I-V direct characteristics of diode made of 12V GaN transistors of size with width W=4 and 8 mm.

TLP stands for transmission line pulses and is a method to measure the I-V curve by making short enough voltage and current pulses to not heat the device under test (DUT). A DC simulation (static behavior) should match TLP-measured data if the pulses are short enough to avoid self-heating.

2.2 Design Trade-off for ESD Diodes

If one ESD protection circuit is composed of many diodes in series, it becomes possible to drive the transistor with negative waveform to increase immunity to high dv/dt in a switching leg commonly used topology. However, by stacking up 4 diodes or more, the voltage across each diode can overpass the maximum 12 V maximum recommended value during ESD discharge and the required surface is also too high (11% added area for a 100 mΩ transistor), making a solution implemented by using them economically disadvantaged. The option of two diodes in series presented in **Table I** below also corresponds to the schematic of **Fig. 2(a)**.

Table I. Number of ESD diodes in series between gate and source of a GaN power device.

How many diodes	TLP current	TLP Voltage	Vg @2kV ESD discharge	ESD area
1	1A	1.25 V	3.5 V	2.5 %
2	1A	2.7 V	7 V	5.3 %
3	1A	4.0 V	10 V	7.8 %
4	1A	5.6 V	14 V	11 %
5	1A	7 V	17.4 V	13 %

Therefore, simple stacking up diodes is not the best recommended solution for mass production of GaN power transistors.

3 New Solution for ESD Circuit for Negative ESD Pulses

A typical solution to avoid using area consuming large diodes is using only one large size transistor for clamping the voltage above a certain level and using a trigger circuit connected to its gate. Similar topologies to **Fig. 2** (b) can be made with different circuit branches to trigger the large transistor M5.

The circuit presented in **Fig. 4** below is one way to achieve an area effective ESD protection circuit. All the devices in this circuit work as a trigger for the large transistor M11. M11 is a large dimension clamp transistor sized to provide above 1A of ESD current when triggered by negative ESD spikes of 2kV. M10 and M9 are also non-minimum size transistors to drive the gate of the clamp transistor M11. Transistors M6 to M8 act as clamps to prevent overvoltage damage to the gate of M10. M5 is placed for its capacitive effect only (C_{OSS}). Resistor R1 acts as a current limiter and R2 to R4 acts as pull up logic. Transistors M2 to M4 form a branch for activation (threshold) for static negative signals below 4 times the gate-to-source threshold voltage (Vth).

In normal operation, a small capacitive current will also trigger the clamp, however, the current i_{M11} will achieve a peak below 2 mA in the moment the driver is providing above 1 A to switch on and off the power device (M_{PWR}), therefore this circuit presents a negligible current consumption during normal operation.

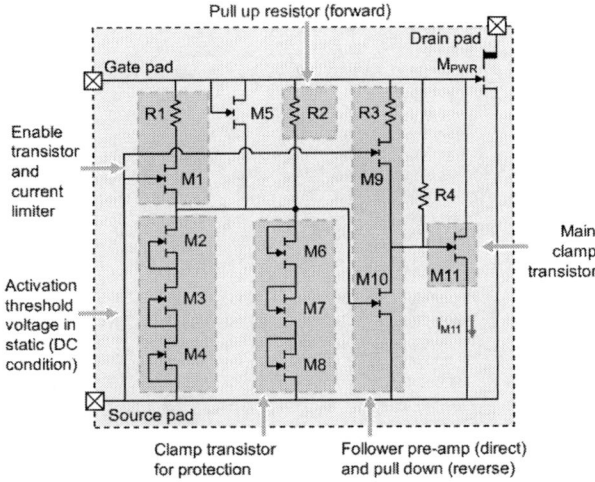

Fig. 4. ESD circuit 1 for monolithic GaN technology. The clamp transistor M11 will provide a peak current during negative ESD discharges to clamp the voltage in the gate of the power transistor (M_{PWR}).

Transient simulation during an ESD pulse discharge of -2kV is presented in **Fig. 5.** The gate-to-source voltage of M11 should trigger this transistor that will conduct a current above 1A and will prevent the gate-to-source voltage value of the power device from staying below -15V. Notice the gate of the power device will stay between -15 and -10V for less than 1µs. After this period, the gate of the power device stays at a safe (static) value above -10V. The disadvantage

of this circuit is it is only triggered by negative ESD spikes.

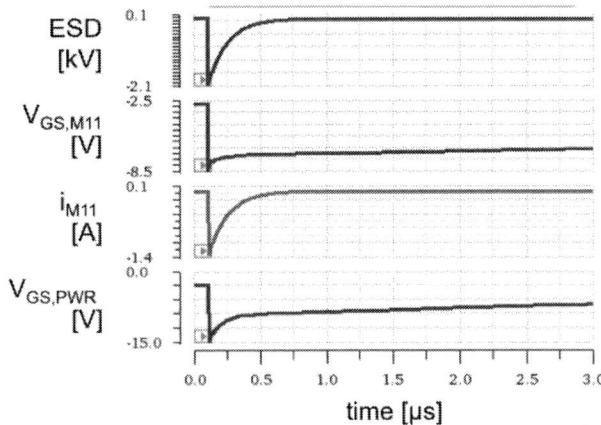

Fig. 5. Simulated waveforms during negative ESD discharge pulses. For positive ESD pulses, the leakage current of the gate of the power transistor (M_{PWR}) should self-protect the device.

4 New Solution for ESD Circuit for Positive and Negative ESD Pulses

To be able to trigger the ESD clamp transistor with both positive and negative ESD spikes in static condition, a solution is presented in **Fig. 6.** In this topology when the gate signal is above 5 times the threshold value of a 12V enhanced mode GaN transistor, the clamp M10 and M11 are switched on. In case the gate node voltage is below 5 times the threshold value of the GaN transistors it will also switch on the clamp transistors.

In this topology, the source of M11 is above its drain and is connected to the source of M10 for better symmetry in the layout.

Another alternative topology for this circuit is connecting R1 only between gate and source of M10 and similarly to R2, connecting only between gate and source of M11.

The threshold value to trigger the clamp transistors M10 and M11 is 8V for 1 mA. The simulated DC curve is presented in **Fig. 7.**

PCIM Europe 2024, 11– 13 June 2024, Nuremberg   DOI: 10.30420/566262211

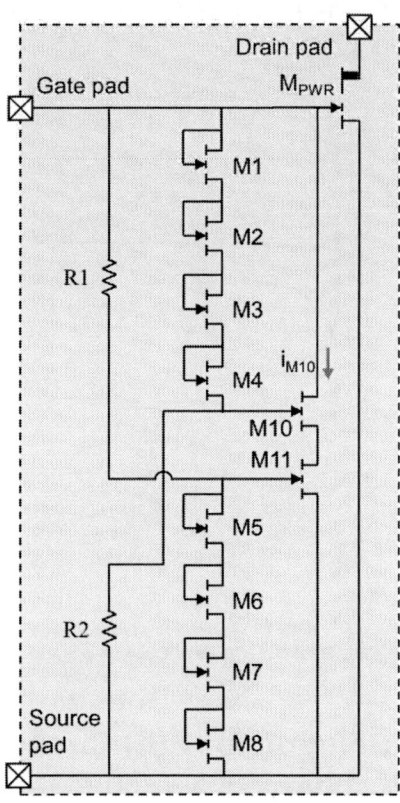

Fig. 6. Schematic of proposed monolithic GaN ESD solution. Transistors M10 and M11 act as clamp transistors for both positive and negative ESD pulses.

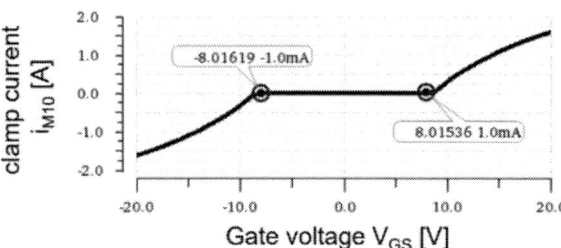

Fig. 7. Static DC simulated waveforms to clamp the ESD at a desired value level for the proposed solution.

5   Test Bench Setup

The HBM ESD experimental test consists of two main parts. In the first part, the GaN transistor (DUT) previously soldered in a mini-PCB is connected in the ESD generator to apply the high voltage between gate and source. The second part consists of removing the mini-PCB from the ESD generator and connecting it to another

board to test if the DUT operates normally. The pass criteria to evaluate the device in this work is to see if the drain-to-source waveshape presents a square shape in the oscilloscope while using a resistive load of 20Ω. See **Fig. 8** for the test bench equipment and printed circuit boards (PCBs).

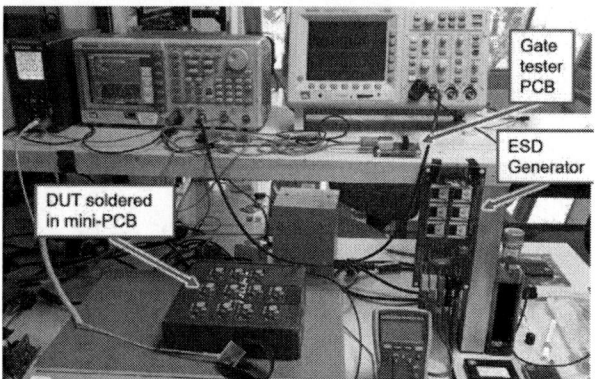

Fig. 8. Test bench with ESD generator capable of providing up to 4 kV.

Some oil is applied around the DUT to avoid an electrostatic discharge in the air between the package pins where the ESD pulse is applied.

The ESD generator is made using the DC/DC high voltage converter with reference AF40P-5. To send ESD pulses a mechanical switch made of rhodium triggered by a magnet is used (reference 2202-1513-060 Asemtech). A voltage divider bridge is made with high voltage resistors to be able to measure the applied voltage using a simple multimeter. To verify the magnetic switch dielectric strength, some measures are performed with output in open circuit and short circuit.

A second motherboard PCB is used as a gate tester board. This board consists of simply having connectors to access the 3 terminals of the GaN (DUT) in the smaller daughter board. When a square wave of 20Hz and 6V is applied to the gate, the oscilloscope shows the drain-to-source V_{DS} waveshape of the DUT connected to a 20Ω resistor to a 20V DC bus. When the V_{DS} waveshape is square, it indicates the device is capable of switching 1A and it is assumed no permanent damage to the device. The breakdown voltage is obtained when after the ESD pulse, the waveshape becomes rounded. See **Fig. 9**.

The test procedure is described in **Fig. 10** and consists basically of applying 400V more than the previous voltage value until it reaches the breakdown value.

1558

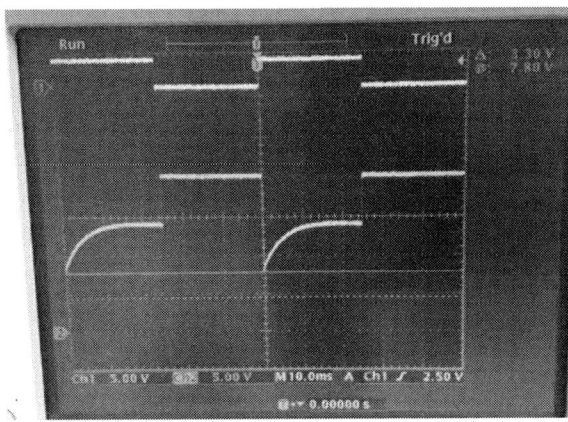

Fig. 9. Example of waveshape of a damaged device after ESD test. Instead of a square wave, a round wave shape is obtained after the breakdown voltage is applied.

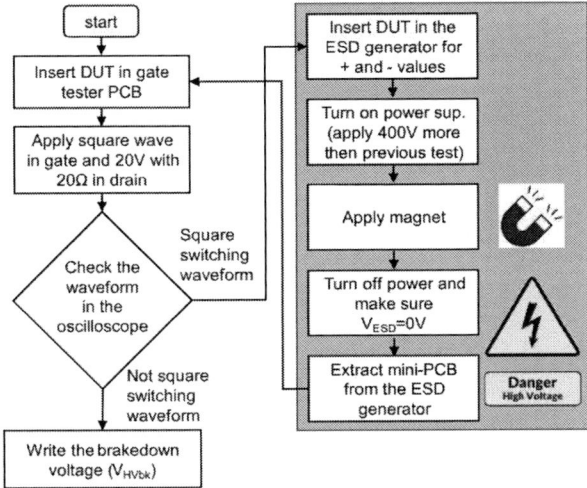

Fig. 10. Flowchart of the test procedure to obtain the breakdown voltage value the DUT can withstand before destructive irreversible damage.

6 Experimental Results and Discussion

The breakdown voltage values obtained for 3 parts for different designs are presented in **Table II** below. The first line of the table is about a power transistor without any ESD circuit. The second line contains one diode-mounted transistor with cathode in the gate and anode in the source of the power transistor.

Table II. Breakdown voltage measured for the HBM ESD test for different chip designs.

GaN design	Obtained ESD breakdown values for 3 parts			Unit
without ESD	-1.2	2	-1.6	kV
with 1 diode	2.4	2	1.6	kV

Without any ESD protection, the lowest measured breakdown voltage applied between gate and source of the GaN power transistors is -1.2kV. Notice this value is negative. This value differs from the values reported in scientific literature and the reason is attributed to the pass criteria of this work. The pass criteria in this work is by waveform analysis and therefore does not detect partial and irreversible ESD damage. With a simple diode added between gate and source, the measured breakdown voltage increases to positive 1.6kV. The two new proposed solutions were not characterized yet due to the time needed to fabricate the prototypes.

7 Conclusion

The presented ESD circuits tailored for GaN technology showcase a robust and efficient solution to safeguard sensitive electronic components from electrostatic discharges. Two different ESD circuits are presented to overcome the disadvantage of only using one diode for ESD protection of GaN devices. Using a simple method to measure ESD breakdown voltage it is possible to have acceptable repeatability to evaluate the performance of ESD circuits to evaluate the viability of these circuits in industrial applications, highlighting their potential to enhance the reliability and longevity of GaN-based electronic systems.

8   Reference

[1] S. -Y. Lin et al., "Monolithic GaN-Based Gate Driver with On-Chip Adaptive On-Time Controller and Negative Current Slope Detector to Prevent Shoot-Through," in *IEEE Solid-State Circuits Letters*, vol. 6, pp. 217-220, 2023.

[2] J. -H. Lee *et al.*, "Incorporation of a Simple ESD Circuit in a 650V E-Mode GaN HEMT for All-Terminal ESD Protection," *2022 IEEE International Reliability Physics Symposium (IRPS)*, Dallas, TX, USA, pp. 2B.3-1-2B.3-6, 2022.

[3] C. Zhou *et al.*, "On-Chip Gate ESD Protection for AlGaN/GaN E-Mode Power HEMT Delivering >2kV HBM ESD Capability," *2019 IEEE 7th Workshop on Wide Bandgap Power Devices and Applications (WiPDA)*, Raleigh, NC, USA, pp. 175-176, 2019.

[4] Z. Shen, Y. Wang, X. Zhang and Y. Wang, "Area-efficient Power-rail ESD Clamp Circuit with False-trigger Immunity in 28nm CMOS Process," *2022 6th IEEE Electron Devices Technology & Manufacturing Conference (EDTM)*, Oita, Japan, pp. 271-273, 2022.

[5] J. Zhang *et al.*, "Modeling Injection of Electrical Fast Transients Into Power and IO Pins of ICs," in *IEEE Transactions on Electromagnetic Compatibility*, vol. 56, no. 6, pp. 1576-1584, Dec. 2014.

[6] Ming-Dou Ker, Jia-Huei Chen and Kuo-Chun Hsu, "Self-substrate-triggered technique to enhance turn-on uniformity of multi-finger ESD protection devices," *IEEE VLSI-TSA International Symposium on VLSI Technology, 2005. (VLSI-TSA-Tech).*, Hsinchu, Taiwan, pp. 17-18, 2005.

[7] E. Canato et al., "ESD-failure of E-mode GaN HEMTs: Role of device geometry and charge trapping" Microelectronics Reliability, Volumes 100–101, 2019.

[8] B. Shankar, et al. "ESD Behavior of AlGaN/GaN Schottky Diodes," in *IEEE Transactions on Device and Materials Reliability*, vol. 19, no. 2, pp. 437-444, June 2019.

[9] S. -H. Chen et al., "HBM ESD Robustness of GaN-on-Si Schottky Diodes," in *IEEE Transactions on Device and Materials Reliability*, vol. 12, no. 4, pp. 589-598, Dec. 2012.

[10] P. -Y. Kuei et al., "Improvement of ESD robustness in gallium nitride-based flip-chip HEMT by introducing metal-insulator-metal capacitor," *2014 International Symposium on Electromagnetic Compatibility*, Tokyo, Japan, pp. 721-724, 2014.

[11] B. Shankar and M. Shrivastava, "Unique ESD behavior and failure modes of AlGaN/GaN HEMTs," *2016 IEEE International Reliability Physics Symposium (IRPS)*, Pasadena, CA, USA, pp. EL-7-1-EL-7-5, 2016.

[12] B. Shankar, S. Raghavan and M. Shrivastava, "Distinct Failure Modes of AlGaN/GaN HEMTs Under ESD Conditions," in *IEEE Transactions on Electron Devices*, vol. 67, no. 4, pp. 1567-1574, April 2020.

[13] Y. Xin *et al.*, "Electrostatic Discharge (ESD) Behavior of p-GaN HEMTs," *2020 32nd International Symposium on Power Semiconductor Devices and ICs (ISPSD)*, Vienna, Austria, 2020.

[14] Y. Shi *et al.*, "A Comparative Study on G-to-S ESD Robustness of the Ohmic-Gate and Schottky-Gate p-GaN HEMTs," in *IEEE Transactions on Electron Devices*, vol. 70, no. 5, pp. 2229-2234, May 2023.

[15] P. Bau, S. Gavira-Duque, F. Rothan, C. Reymond and D. Bergogne, "Voltage Reference and Zero Current Detector Monolithically Integrated on p-GaN Technology Designed for Process Corners Compensation," 2023 IEEE Applied Power Electronics Conference and Exposition (APEC), Orlando, FL, USA, , pp. 673-677, 2023.

[16] C. Reymond, L. Mistre, P. Bau, T. H. Phung and D. Bergogne, "Digital GaN 300W AC/DC Power Supply," *PCIM Europe 2023; International Exhibition and Conference for Power Electronics, Intelligent Motion, Renewable Energy and Energy Management*, Nuremberg, Germany, pp. 1-6, 2023.

[17] P. Bau, S. Hariharan, S. Gaviria-Duque and D. Bergogne, "Digital Control for Efficient Switching Over a Wide Range of Supply Voltages," *PCIM Europe 2023; International Exhibition and Conference for Power Electronics, Intelligent Motion, Renewable Energy and Energy Management*, Nuremberg, Germany, pp. 1-6, 2023.

PCIM Europe 2024, 11– 13 June 2024, Nuremberg   DOI: 10.30420/566262212

A Simulative Study of Measurement Errors During Double Pulse Testing of GaN Devices

Severin Klever[1], Rik W. De Doncker[1]

[1] RWTH Aachen University, Institute for Power Electroncis and Electrical Drives (ISEA), Germany

Corresponding author and speaker: Severin Klever, post@isea.rwth-aachen.de

Abstract

The accuracy of double pulse tests of wide-bandgap semiconductor devices is a common issue. This is particularly true for Gallium Nitride devices, where non-ideal characteristics such as bandwidth limitations and parasitic elements of the probes can falsify the measurement of the switching loss. Although the individual sources of error are well known, there is little literature on their overall impact on the measurement accuracy. This study aims to improve the understanding of the interaction of the individual parameters and to derive recommendations. Statistical statements were derived using a simulative approach based on a design of experiments with over 15 000 setups. The results can assist future users in estimating the magnitude of unavoidable measurement errors.

1 Introduction

The accuracy of double pulse tests (DPT) performed on wide-bandgap (WBG) devices is a common issue [1]. The very short rise times place high demands on the measurement techniques that are used, which is particularly critical for Gallium Nitride (GaN) devices. Therefore, methods are continuously researched to improve the measurement accuracy for the switching energies E_{off} and E_{on} [2], [3]. The accuracy is influenced by several effects such as limited measurement bandwidths or parasitic elements of the probes [4], [5]. Although the individual sources of error are well known, there is little research on how their interaction affects the overall quality of the characterization result. Therefore, it is difficult to evaluate an existing setup and to predict the magnitude of the measurement error. The aim of this work is to provide a better understanding of the individual parameters and their impact on the overall measurement error. Section 2 reviews the primary sources of error, followed by section 3, which presents the simulation model that allows the evaluation of a large number of DPTs. Based on a design of experiments with over 15 000 setups, the error can be statistically evaluated. Finally, section 4 analyzes the results and establishes requirements and recommendations for the measurement that are summarized in section 5.

2 Sources of Error in Double Pulse Testing

The following section describes the sources of error and their effects on switching behavior. To accurately measure the fast switching behavior of GaN with rise times in the nanosecond range, the appropriate measurement equipment must be carefully selected. The major criteria for selection are limited bandwidth, parasitic elements, and deskewing, which can significantly impact the measurement results.

1. Limited bandwidth: In order to sense transient signals correctly, it is necessary that both current and voltage probes have a sufficiently high bandwidth. Since most probes act as lowpass filters, the minimum measurable rise time t_{rise} is directly related to the 3 dB cutoff frequency of the probe, which is an approximation valid for 1st-order low-pass filters (1). In addition, the measured rise time is also influenced by the oscilloscope and might be less than specified in the data sheet of the probe (2) [6]. Figure 1a shows an example of how a reduced bandwidth affects the measurement of the switching transient v_{ds} of a GaN device.

$$f_{\text{c}} = \frac{0.35}{t_{\text{rise}}} \tag{1}$$

$$t_{\text{rise,meas}} = \sqrt{t_{\text{r,signal}}^2 + t_{\text{r,probe}}^2 + t_{\text{r,scope}}^2} \tag{2}$$

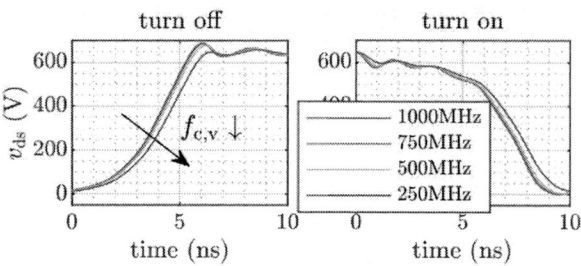

(a) Limited Bandwidth

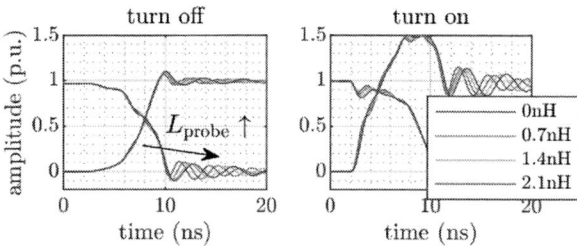

(b) Parasitic inductance of current probe (L_{probe})

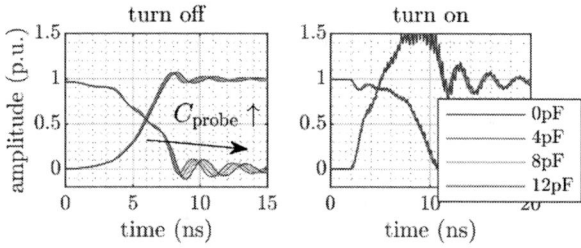

(c) Parasitic capacitance of voltage probe (C_{probe})

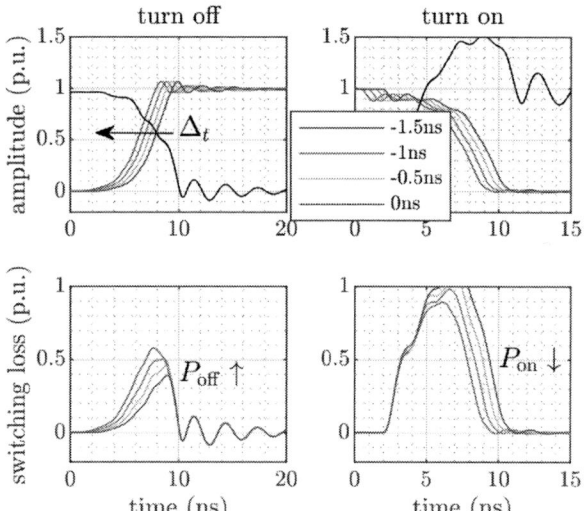

(d) Deskewing error (negative shift of v_{ds})

Fig. 1: Influence of the most significant error sources on the switching transients (simulation at $V_{\mathrm{DC}} = 650\,\mathrm{V}$, $I_{\mathrm{sw}} = 40\,\mathrm{A}$, $T_{\mathrm{j}} = -40\,^{\circ}\mathrm{C}$)

2. Parasitic elements: An ideal and non-invasive measurement of electrical quantities is usually not possible. When characterizing WBG devices, the parasitic capacitance of the voltage probe and the parasitic inductance of the current sensor will show an impact on the switching behavior of the device under test (DUT). Figure 1b and Fig. 1c give examples on this effect. It can be seen that the switching transients are slowed down and resonance frequencies are shifted.

3. Deskewing error: To calculate switching energies, voltage and current are measured simultaneously. Therefore, it is necessary to compensate for different group delays between the recorded waveforms. The total group delay is determined by the phase response of the probe and the length of the attached cable. Figure 1d gives an example and shows, how shifting the v_{ds} waveform will affect the calculation of the instantaneous power. In this case, the turn-off energy would increase while the turn-on energy would decrease. With an deskew error in the other direction, this effect would be reversed [7].

4. Oscilloscope: Another source of error is the analog-to-digital conversion (ADC). The data is usually acquired by an oscilloscope, which has a limited sampling rate and a limited vertical resolution expressed by the effective number of bits. In this work, it is assumed that the horizontal and vertical resolution is sufficient, since modern oscilloscopes offer sampling rates above 10 GS/s and 12 bit ADC.

In addition, there are other sources of error, such as common-mode and differential-mode currents that are caused by inductive or capacitive coupling. As these errors are very specific and depend on the test setup, they are not analyzed in this study. It is necessary to take suitable countermeasures, such as galvanically isolated gate drivers, floating DC link capacitors and ferrite beads [8]. In the best case, optically isolated voltage probes are used which have a very high common mode transient immunity in contrast to regular differential probes [9]. Finally, temperature changes can also affect the accuracy and repeatability of the measurement results.

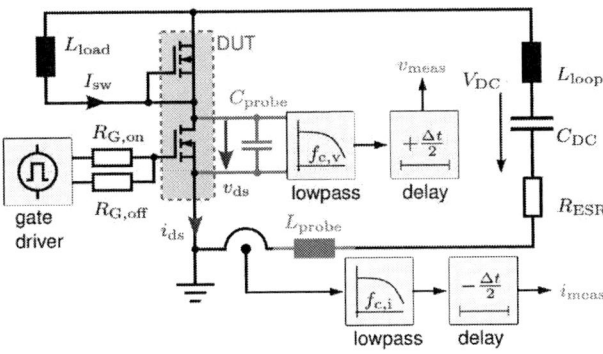

Fig. 2: Circuit diagram of DPT and measurement setup

parameter	min.	max.	steps
V_{DC}	260 V	650 V	4
I_{sw}	10 A	40 A	4
T_j	−40 °C	125 °C	4
L_{probe}	0 nH	2.1 nH	4
C_{probe}	0 pF	12 pF	4
$f_{c,i}$	0.1 GHz	1.7 GHz	5
$f_{c,v}$	0.1 GHz	1.7 GHz	5
t_{skew}	−100 ps	100 ps	5

Tab. 1: Simulation parameters being varied

3 Setup of the Simulation Model and Design of Experiments

The selected device under test (DUT) is the commercial GaN HEMT *GS66508T* from with a voltage rating of $V_{DS} = 650\,V$ and a nominal current of $I_D = 30\,A$. The manufacturer provides a *LTSpice* behavioral model which is suitable for simulating the switching behavior over the entire operating range. The DPT is modeled according to the circuit shown in Fig. 2. The DPT is performed with a constant ramp time $t_{ramp} = 1.7\,\mu s$. To achieve the desired current I_{sw}, the load inductance L_{load} is dimensioned according to (3). The DC-link capacitor is dimensioned according to (4) as a function of the operating point, whereby the energy balance from [10] is applied (4). It is assumed that the energy ΔE_C taken from the capacitor C_{DC} is completely used to magnetize L_{load}. At the time when I_{sw} is reached, the DC-link voltage should maintain at 99 % of its initial value, from which C_{DC} can be calculated (5).

$$L_{load} = t_{ramp} \cdot \frac{V_{DC}}{I_{sw}} \qquad (3)$$

$$\Delta E_L = \Delta E_c \Rightarrow \frac{L_{load}I_{sw}^2}{2} = \frac{C_{DC}V_{DC}^2}{2}\left(1 - 0.99^2\right)^2 \qquad (4)$$

$$\Leftrightarrow C_{DC} \approx 50 \cdot \frac{t_{ramp}I_{sw}}{V_{DC}} \qquad (5)$$

The parameters in the first section of Tab. 1 determine the investigated operating point. These are the switching voltage V_{DC}, the switching current I_{sw} and the junction temperature T_j. The entire temperature range of the device is considered, as very fast switching is to be expected, especially at low temperatures [11]. The second section shows the parameters of the measurement setup. L_{probe} describes the parasitic inductance of the current sensor and C_{probe} the parasitic capacitance of the voltage probe. Further parameters are the 3 dB cutoff frequencies of voltage and current probes ($f_{c,v}$ and $f_{c,i}$) and the time offset t_{skew} to simulate the error from improper deskewing. t_{skew} is defined as the relative shift of v_{ds} in relation to i_{ds}. For all parameters, it was ensured that practice-oriented values were selected. The maximum value of C_{probe} corresponds to the input capacitance of a passive probe and the maximum value of L_{probe} corresponds to the parasitic inductance of a coaxial shunt [9], [12], [13]. Each parameter is swept linearly between its minimum and maximum values for the given number of steps, resulting in a full-factorial design of experiments with $N_{sim} = 5^4 \cdot 3^5 = 151\,875$ combinations.

The remaining parameters are listed in Tab. 2 and are kept at a constant level. These are the power loop inductance L_{loop}, the equivalent series resistance of the switching cell R_{ESR}, the gate resistors ($R_{G,on}$, $R_{G,off}$) and the gate driver output levels ($V_{G,on}$, $V_{G,off}$).

For each combination, the switching energies E_{off} and E_{on} are calculated from the waveforms v_{ds} and i_{ds}. In addition, the "measured" energies $E_{off,meas}$ and $E_{on,meas}$ are calculated using v_{meas} and i_{meas} (see Fig. 2). All integrals are calculated over the

param.	value	param.	value
L_{loop}	2 nH	$R_{G,on}$	10 Ω
R_{ESR}	1 Ω	$R_{G,off}$	2 Ω
$V_{G,on}$	6 V	$V_{G,off}$	0 V

Tab. 2: Simulation parameters at a constant level

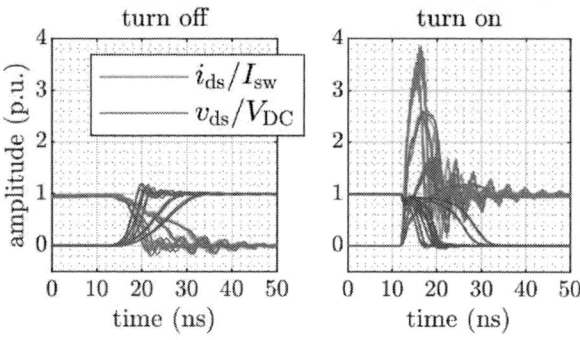

Fig. 3: Superimposed view of the simulated switching transients at the parameter limits ($N = 32$)

full switching event, as the definition from [14] does not take oscillations into account and therefore only represents a portion of the losses. To evaluate a single simulation, the relative error $\varepsilon_{\mathrm{rel}}$ (6) is used and will be calculated for both the turn-on and turn-off energy. For a set of N simulations, the mean percentage error (mpe) (7) is used as a benchmark.

$$\varepsilon_{\mathrm{rel}} = 100\,\% \cdot \frac{E_{\mathrm{meas}} - E_{\mathrm{sim}}}{E_{\mathrm{sim}}} \qquad (6)$$

$$\varepsilon_{\mathrm{mpe}} = \frac{1}{N} \cdot \sum_{i=1}^{N} \varepsilon_{\mathrm{rel,i}} \qquad (7)$$

Figure 3 shows a superimposed view of the simulated switching transients. This view shows only those setups in which the parameters are set to their respective minimum or maximum. This allows verification of proper simulation of all switching events. All switching events take place in a time window of approx. 20 ns and converge accordingly. In some cases there is a high relative overshoot of the current at turn-on. The cause of this is the E_{qoss} loss mechanism, which is a discharge of the parasitic capacitance of the upper HEMT [15]. Since this current peak is independent of I_{sw}, this leads to a high relative overshoot for small values of I_{sw}. Figure 4a shows the distribution of non-falsified switching losses across all N_{sim} combinations. The accurate measurement of E_{on} is especially important, since E_{on} accounts for a larger share of the total losses. Figure 4b shows the distribution of the errors according to (6). The error of E_{off} has a distribution shifted in the positive direction and can exceed 20 %. The error of E_{on} is normally distributed and ranges between –10 % and 10 %. The mean value μ is close to 0 % in both cases, which is due to the fact that the individual sources

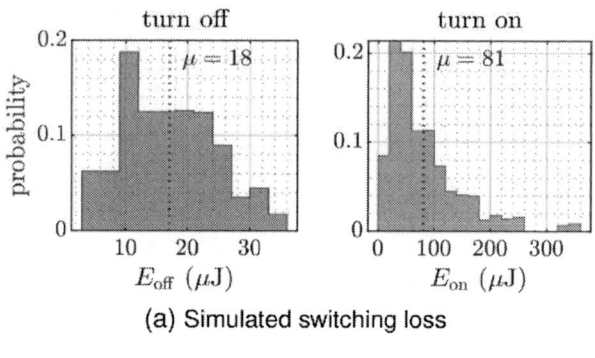

(a) Simulated switching loss

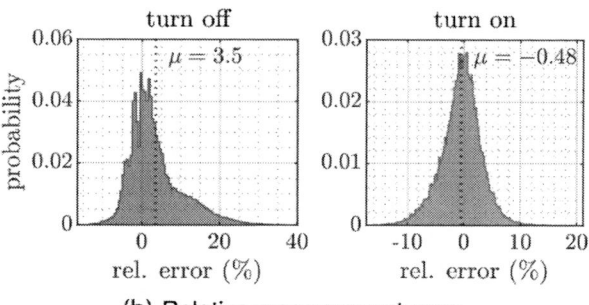

(b) Relative measurement error

Fig. 4: Distribution of switching losses E_{off} and E_{on} and relative measurement errors across all simulation setups

of error partially compensate each other, as will be determined in the following section.

4 Interpretation of the Simulation Results

The methodology used to analyze the simulation results is the evaluation of main effect diagrams, which is a standardized method to evaluate the influence of individual parameters [16]. The effects considered are the influence of the operating parameters (V_{DC}, I_{sw}, T_{j}) on the absolute values and the influence of the measurement parameters (L_{probe}, C_{probe}, $f_{\mathrm{c,v}}$, $f_{\mathrm{c,u}}$, t_{skew}) on the relative errors. In each subplot of Fig. 5 and Fig. 6, all N_{sim} simulations are grouped according to the value of the corresponding parameter on the x-axis. Depending on the considered parameter, either the mean value of the absolute switching energy or the mean percentage error $\varepsilon_{\mathrm{mpe}}$ (7) is plotted on the y-axis. In addition the overall mean value μ and the difference between the first and last mean value is indicated. This gives the magnitude and direction of the effect so that a quantitative comparison between the parameters is possible. The individual plots are interpreted below.

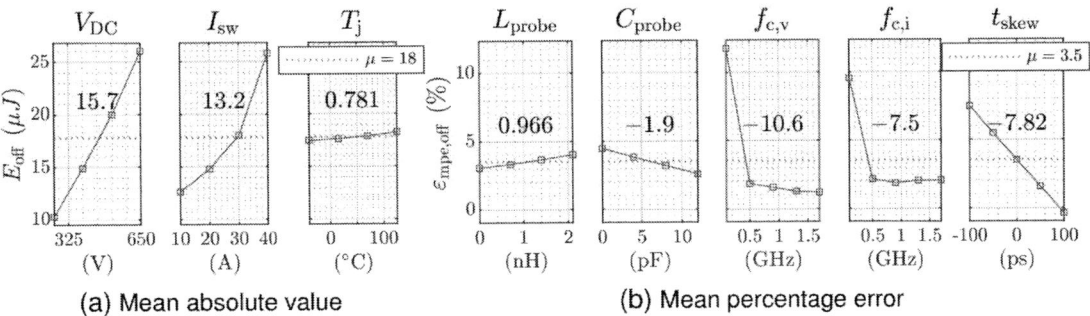

(a) Mean absolute value   (b) Mean percentage error

Fig. 5: Main effect diagrams for E_{off} measurement of all N_{sim} experiments

4.1 Turn-off Energy (E_{off})

Figure 5a shows the main effect plot for the absolute switching loss E_{off}, whose distribution was shown in Fig. 4a. It can be seen that V_{DC} and I_{sw} both show a very strong influence on E_{off} in the positive direction. The temperature T_{j}, on the other hand, shows almost no influence, which is confirmed by measurements in [11].

Figure 5b shows the main effect plot for the relative measurement error of E_{off}. With regard to the parasitic elements L_{probe} and C_{probe}, slight trends can be seen. Increasing L_{probe} increases the measurement error in the positive direction, which means that the inductance of the current sensor increases the turn-off energy. For C_{probe}, this effect is reversed and E_{off} is reduced. Since there is an average positive measurement error of 3.5 %, the measurement error is therefore getting closer to zero C_{probe}. Both parameters show a linear behavior.

There is a very clear correlation between the two bandwidths $f_{\text{c,v}}$ and $f_{\text{c,i}}$. A low bandwidth leads to a large positive measurement error for both parameters. However, this curve is flattening. Increasing the bandwidths above 500 MHz only has a very small effect on the measurement error. For this reason, it can be assumed that this is the minimum required bandwidth.

The deskewing error t_{skew} shows a linear behavior and is another very significant parameter. Negative values increase the measured E_{off}, resulting in a positive measurement error (see Fig. 1d) With a positive value, this effect is reversed and E_{off} is reduced. By averaging all experiments, a measurement error of 0 % is achieved by coincidence for $t_{\text{skew}} = 100$ ps. This is due to the fact that the positive measurement error from the bandwidths is compensated in this case by false deskewing.

4.2 Turn-on Energy (E_{on})

Analogous observations can be made for the measurement of E_{on}. Figure 6a shows the main effect plot for the absolute switching loss E_{on}, whose distribution was also shown in Fig. 4a. Again, V_{DC} and I_{sw} both show a very strong influence in the positive direction. However, the junction temperature T_{j} also has an effect on the losses and shows a non-linear influence.

Figure 6b shows the main effect plot for the relative error of E_{on}, which is on average –0.48 %. The most significant influence is L_{probe}. Increasing this parameter increases the error in the negative direction, which means that the parasitic inductance of the current sensor reduces the turn-on energy. As described in [17], this is caused by an additional voltage drop of v_{ds}. As also observed in Fig. 5b, C_{probe} counteracts the effect of L_{probe}. The reason for this is the discharge of this parasitic capacitance at turn-on, which results in a higher current peak, which in turn increases E_{on}. Because the two effects overlap, a resulting error of 0 % is possible by coincidence.

With regard to the bandwidth of the voltage probe, it is noticeable that $f_{\text{c,v}}$ shows almost no influence on the measurement error. For $f_{\text{c,i}}$, the curve is similar to Fig. 5b. The effect is initially very strong and creates a positive error, but flattens out at 500 MHz and converges to a stationary value.

The effect of t_{skew} acts in the positive direction for E_{on}, which corresponds to Fig. 1d. This effect is less significant than with E_{off} and is of the same order of magnitude as the influence of L_{probe}.

The available data set also makes it possible to identify interactions between the individual parameters [16]. The analysis of first-order interactions showed no irregularities for either E_{off} or E_{on}.

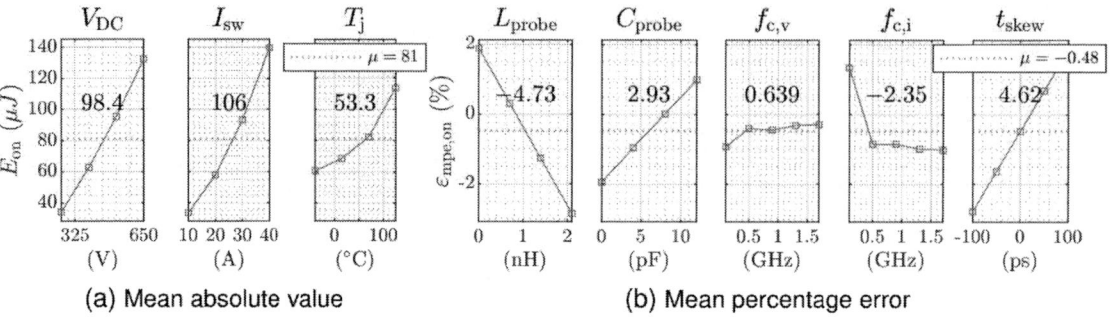

(a) Mean absolute value   (b) Mean percentage error

Fig. 6: Main effect diagrams for E_{off} measurement of all N_{sim} experiments

5 Recommendations for Measurements

Based on the evaluated simulation results, some recommendations can be given for the measurement of E_{off} and E_{on} of the selected 650 V GaN device. The recommendations may also apply to devices with similar rise time and make it possible to evaluate existing setups.

1. Bandwidth: it has been shown that too low bandwidths of the current and voltage measurement have a strong influence on the measurement error. the selected DUT has a rise-time of 3.1 ns specified in the data sheet, which leads to a minimum bandwidth of 94.6 MHz according to eq. (1). However, this is not a sufficient value and it was observed that at least 500 MHz is required. An increase above this value improves the accuracy only slightly.

2. Parasitics: The parasitics of the probes have a significant influence on the measurement result, as they affect the switching behavior. It is noticeable that the influences of L_{probe} and C_{probe} behave linearly and in opposite directions. In particular, the measurement of E_{on}, which has the higher proportion of the total losses, is strongly influenced. For current measurement, this means that devices such as coaxial shunts should not be used. With regard to the voltage measurement, passive probes with a high input capacitance should be avoided.

3. Deskewing: Even a small offset of 100 ps causes a large measurement error. For an oscilloscope with a sampling rate of 10 GS, this corresponds to a single sampling step.

Therefore, special measures must be developed. With the right methodology, it is possible to fully compensate for this source of error in post processing [2], [18].

6 Conclusion

In this work, the problem of hardly predictable measurement errors in double pulse tests with WBG semiconductors is addressed. Using a simulation framework based on design of experiments, it was possible to analyze the individual effects of the most significant measurement parameters on the switching energies E_{off} and E_{on}. The sources of error investigated are the limited bandwidth of the probes, the parasitic capacitance of the voltage probe, the parasitic inductance of the current sensor and the time offset between the waveforms. Using a commercial 650 V GaN HEMT as an example, it was shown that the measurement error of the switching energies can exceed 10 % if the measurement setup is inadequate. By evaluating main effect plots, it was possible to derive recommendations for accurate measurements and quantify the impact of the individual parameters. It was shown that at least 500 MHz of bandwidth is required and that the rule of thumb $0.35/t_{\text{rise}}$ is invalid for this application. In order to achieve a low measurement error, the focus should be on reducing the parasitic elements of the probes and, above all, on accurate deskewing. Even a small time offset of 100 ps can invalidate the measurement. The method presented can be applied to other sources of error and devices and is a versatile tool for evaluating the measurement error of double pulse tests. Future studies must investigate the effects of even faster switching semiconductors such as GaN HEMTs in the 100 V class.

References

[1] Z. Zeng, J. Wang, L. Wang, Y. Yu, and K. Ou, "Inaccurate switching loss measurement of sic mosfet caused by probes: Modelization, characterization, and validation," *IEEE Transactions on Instrumentation and Measurement*, vol. 70, pp. 1–14, 2021. DOI: 10.1109/TIM.2020.3024356.

[2] M. Dong, H. Li, S. Yin, Y. Wu, and K. Y. See, "A postprocessing-technique-based switching loss estimation method for gan devices," *IEEE Transactions on Power Electronics*, vol. 36, no. 7, pp. 8253–8266, 2021. DOI: 10.1109/TPEL.2020.3043801.

[3] H. Dymond, Y. Wang, S. Jahdi, and B. Stark, "Probing techniques for gan power electronics: How to obtain 400+ mhz voltage and current measurement bandwidths without compromising pcb layout," in *PCIM Europe 2022*, Aug. 2022, pp. 1–10. DOI: 10.30420/565822010.

[4] S. Sprunck, C. Lottis, F. Schnabel, and M. Jung, "Suitability of current sensors for the measurement of switching currents in power semiconductors," *IEEE Open Journal of Power Electronics*, vol. 2, pp. 570–581, 2021. DOI: 10.1109/OJPEL.2021.3127225.

[5] S. Sprunck, M. Koch, C. Lottis, and M. Jung, "Suitability of voltage sensors for the measurement of switching voltage waveforms in power semiconductors," *IEEE Open Journal of Power Electronics*, pp. 1–12, 2022. DOI: 10.1109/OJPEL.2022.3201952.

[6] C. Mittermayer and A. Steininger, "On the determination of dynamic errors for rise time measurement with an oscilloscope," *IEEE Transactions on Instrumentation and Measurement*, vol. 48, no. 6, pp. 1103–1107, 1999. DOI: 10.1109/19.816121.

[7] Z. Zhang, B. Guo, F. F. Wang, E. A. Jones, L. M. Tolbert, and B. J. Blalock, "Methodology for wide band-gap device dynamic characterization," *IEEE Transactions on Power Electronics*, vol. 32, no. 12, pp. 9307–9318, Dec. 2017. DOI: 10.1109/TPEL.2017.2655491.

[8] G. Sergentanis, Y. R. De Novaes, L. De Lillo, L. Empringham, and M. C. Johnson, "Dynamic characterization of 650v gan hemt transistors," in *2023 IEEE 8th Southern Power Electronics Conference and 17th Brazilian Power Electronics Conference (SPEC/COBEP)*, 2023, pp. 1–7. DOI: 10.1109/SPEC56436.2023.10407647.

[9] Teledyne LeCroy, Inc., *Oscilloscope probes and accessories (catalog)*, Dec. 5, 2023.

[10] J. Schweickhardt, K. Hermanns, and M. Herdin, *Tips & tricks on double pulse testing*, Rohde & Schwarz Application Note, Mar. 2021.

[11] R. Ren, H. Gui, Z. Zhang, R. Chen, J. Niu, *et al.*, "Characterization of 650 v enhancement-mode gan hemt at cryogenic temperatures," in *2018 IEEE Energy Conversion Congress and Exposition (ECCE)*, 2018, pp. 891–897. DOI: 10.1109/ECCE.2018.8557868.

[12] Z. Xin, H. Li, Q. Liu, and P. C. Loh, "A review of megahertz current sensors for megahertz power converters," *IEEE Transactions on Power Electronics*, vol. 37, no. 6, pp. 6720–6738, 2022. DOI: 10.1109/TPEL.2021.3136871.

[13] W. Zhang, Z. Zhang, and F. Wang, "Review and bandwidth measurement of coaxial shunt resistors for wide-bandgap devices dynamic characterization," in *IEEE Energy Conversion Congress and Exposition (ECCE)*, Sep. 2019, pp. 3259–3264. DOI: 10.1109/ECCE.2019.8912750.

[14] *Semiconductor devices - Discrete devices - Part 8: Field-effect transistors (IEC 60747-8:2010)*, International Standard, International Electrotechnical Commission, Dec. 2010.

[15] R. Hou, J. Lu, and D. Chen, "Parasitic capacitance eqoss loss mechanism, calculation, and measurement in hard-switching for gan hemts," in *2018 IEEE Applied Power Electronics Conference and Exposition (APEC)*, Mar. 2018, pp. 919–924. DOI: 10.1109/APEC.2018.8341124.

[16] J. Antony, *Design of Experiments for Engineers and Scientists*. Elsevier Science, 2023. DOI: 10.1016/C2022-0-01075-8.

[17] S. Gao, J. Tian, X. Fu, Y. Li, B. Wang, and L. Zhao, "Complete loss distribution model of gan hemts considering the influence of parasitic parameters," *Journal of Power Electronics*, Oct. 2023. DOI: 10.1007/s43236-023-00710-3.

[18] S. Yin, Y. Liu, Y. Gu, X. Xin, S. Deng, *et al.*, "Automatic v - i alignment for switching characterization of wide band gap power devices," in *2018 1st Workshop on Wide Bandgap Power Devices and Applications in Asia (WiPDA Asia)*, May 2018, pp. 75–78. DOI: 10.1109/WiPDAAsia.2018.8734681.

PCIM Europe 2024, 11– 13 June 2024, Nuremberg DOI: 10.30420/566262213

Parallel Connection of GaN FETs: an Experimental Investigation Approach

Marco Palma[1] , Vincenzo Barba[2] , Salvatore Musumeci[2]
[1] Efficient Power Conversion, Volpiano (TO), Italy
[2] DENERG PEIC - Politecnico di Torino, Italy

Corresponding author: Marco Palma, marco.palma@epc-co.com
Speaker: Marco Palma, marco.palma@epc-co.com

Abstract

The paper investigates the parallel connection of the GaN FETs in inverters for low-voltage motor control applications. Several simulation results and experimental tests show the main device parameters and layout constraints involved in the effectiveness of parallel connection arrangement. A dedicated measurement board setup has been developed for the experimental evaluation. The experimental tests enable the empirical prediction of the application's maximum phase load current amplitude, considering the effects of the spread of the parameters among the parallel-connected GaN FETs. Furthermore, an inverter board with four devices in parallel is designed and assembled to validate the advanced GaN FETs parallel operation in an actual application, increasing the output phase current value for high-current motor drive applications.

1   Introduction

Nowadays, power converter applications demand increasing amounts of power in small volumes. In high-current, low-voltage applications such as motor drives for mild hybrid automotive applications, high-performance robotic systems, or Unmanned Aerial Vehicle (UAV) [1-3], the Gallium Nitride devices (GaN FET) are more attractive compared to Si MOSFET devices. The GaN FETs are chosen because of their high-performance characteristics, such as high-speed switching and low conduction losses. This makes them suitable for motor control applications [4], [5]. For thermal management purposes, using power devices in parallel is worthwhile, especially with insulated gate devices with a positive temperature coefficient that allows a current balance in steady state [6].

In low-voltage applications (less than 200V), connecting multiple devices in parallel configurations becomes essential to expand the utilization of GaN FETs in high-power applications. However, the parallel connection of devices is challenging for the PCB design phase and the devices' parametric spread management [7]. Parameter mismatch among the parallel circuit legs significantly affects the dynamic characteristics and thermal management of GaN FETs. To mitigate the influence of these parasitic components, alternative packaging techniques for power devices have been proposed [8].

The main objective of this paper is to assess the feasibility and effectiveness of using GaN FETs in parallel within an inverter setup to control low-voltage AC motors. The paper examines device parameters and layout constraints that impact the effectiveness of parallel connections. This could include the number of devices to connect in parallel, gate driving control, thermal management, and circuit layout considerations.

The paper involves both simulation and experimental testing. Simulation results are used to understand how GaN FETs behave in parallel configurations, predicting and making them better understand the outcomes before experimental tests in actual operative conditions. The experimental evaluation setup is designed to provide accurate and controlled conditions for testing the GaN FETs in parallel configuration and to validate the simulation results. Furthermore, the paper discusses an inverter circuit with four paralleled GaN FETs to demonstrate how GaN FETs can be used in parallel to increase the output phase current, which is crucial for high-current motor drive applications.

2 Design considerations of parallel GaN FETs and simulation results

Parallel connections of GaN FET devices demand careful consideration of various static and dynamic factors to achieve uniform current sharing and reliable operation. Several factors impact device temperature differences. The junction temperature variation may result in dynamic or static current-sharing unbalances depending on device characteristics. The behavior of the conduction resistance $R_{DS(on)}$ and the threshold voltage V_{GSth} of the GaN FET as a function of the junction temperature is shown in Fig.1a and Fig.1b, respectively.

The parameters' spread plays a crucial role in the parallel connection of devices, both during conduction and switching transients [9].

In GaN FETs' parallel connections, RDS(on) significantly affects static current sharing. However, the positive temperature coefficient aids current self-balancing. On the other hand, dynamic current sharing during turn-on and turn-off is mainly influenced by the spread of V_{GSth}. The lower threshold voltage V_{GSth} leads to an earlier turn-on, higher switching current/loss, and positive feedback [10]. Also, the transconductance (g_m) plays a role in dynamic current sharing.

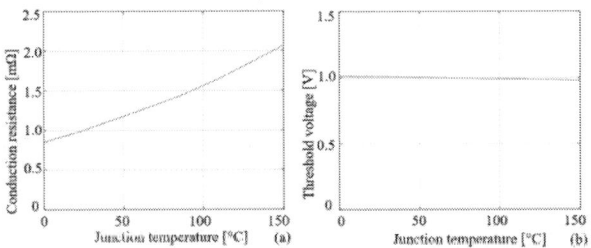

Fig. 1 Dependency of GaN FET parameters on junction temperature: a) conduction resistance, b) threshold voltage.

A balanced circuit layout is crucial for dynamic current sharing and parallel operation stability. This is especially critical for high-speed power switches like GaN/SiC. In addition to synchronized device control, layout symmetry is pivotal.

Simulations are carried out to understand the impact of the V_{GSth} spread and the parasitic components of the circuit layout when several GaN FETs are connected in parallel. Fig.2 shows the electrical schematic of the system composed of N GaN FETs in parallel connection and connected to an inductive load with a freewheeling diode in parallel.

In Fig. 2, the GaN FETs gates are driven by a pulse voltage source V_q=5 V, with a series gate resistance of 1 Ω. Eight (N=8) GaN FETs (EPC2302) are connected in parallel. In the simulation test circuit shown in Fig.2, the devices are arranged in rows considering the equivalent parasitic inductances for each GaN FET. The output capacitance $C_{OSS}= C_{DS}+C_{GD}$ is represented in parallel to each GaN FET. The GaN FET model is described in [11]. L_D=0.3 nH represents the drain inductance due to the parallel connection. Shunt resistors R_S=100 mΩ are used for sensing the GaN FET currents. The equivalent capacitance of the PCB is C_{PCB}=11 nF. The bus voltage is V_{DC}= 24V, and the load is L_{Load}=90 µH. The freewheeling diode has a parasitic inductance of L_{diode}=1 nH. Moreover, the load is connected to the GaN FETs through the drain-board inductance L_{DB}=1 nH.

These stray inductances and C_{PCB} values come from actual measurements of a PCB mounting twelve GaN FETs in parallel.

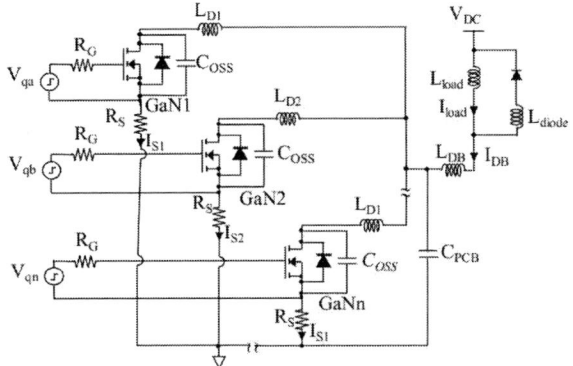

Fig. 2 N GaN FETs in parallel connected to an inductive load with a freewheeling diode in parallel.

Different simulations are carried out by turning on only 2, 4, 6, or 8 GaN FETs at a time while the others remain in the off-state. The load current is I_{load}=150 A. Only the single GaN FET that turns on has a low-value threshold voltage, $V_{th\ low}$=0.95 V. The other GaN FETs have a higher threshold voltage of $V_{th\ high}$=1.65 V. The device's current with $V_{th\ low}$ is shown in Fig. 3a for the turn-on and in Fig. 3b for the turn-off. The current waveforms in the cases of turning on 2, 4, 6, or 8 GaN FETs are indicated with the respective number. The current I_{DB} entering the common drain of the GaN FETs in parallel is depicted in black.

After the transient, during the conduction time, the I_{load} is equally shared between the GaN FETs in the on-state. A higher number of devices in the on-state results in a lower current conducted by each GaN FET because the simulation always used the same load current. The different threshold voltage causes the device with $V_{th\,low}$ to switch on earlier and switch off later than the other GaN FETs with typical V_{th}.

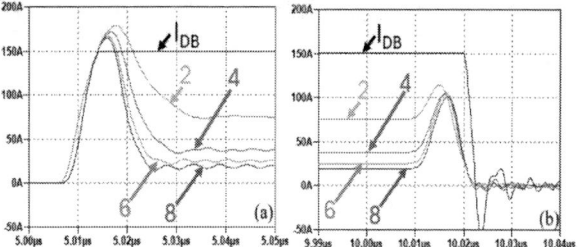

Fig. 3 Current of the GaN FET with $V_{th\,low}$. Number of GaN FETs turning on: 2 (yellow), 4 (red), 6 (green), 8 (blue). The current I_{DB} entering the common drain of the GaN FETs in parallel is shown in black. a) turn-on. b) turn-off. I=50 A/div. Time scale = 10 ns/div.

At turn-on and turn-off, the current peak is higher if the GaN FET has $V_{th\,low}$. In this simulation, I_{DB} rises to the I_{load} value at turn-on and falls to 0 A quickly because of the low L_{diode} and L_{DB} values.

At turn-on, the GaN FET with $V_{th\,low}$ conducts I_{load}. Moreover, the current shows a peak higher than I_{load} due to the discharging parasitic capacitances in the circuit, C_{OSS}, and C_{PCB}.

Similarly, during the turn-off, the GaN FET with V_{thlow} has a peak lower than that at turn-on. For both turn-on and turn-off, the peak is higher in the case of only two GaN FETs turning on because of the higher current conducted at steady state.

To separate the current peak due to the parasitic capacitance from the load current (I_{DB}) rise, it is necessary to slow down the I_{DB} dynamic by L_{diode} and L_{DB} with higher values than the interconnecting inductances.

3 PCB with eight GaN FETs in parallel

The experimental evaluation is carried out with the board prototype equipped with up to twelve GaN FETs in parallel. The board is shown in Fig.4.

Each GaN FET has an independent gate driver in the prototype to decide how many devices supply the load. Moreover, a turn-on and turn-off mismatch between the GaN FETs can be set by adding a proper delay on the different control signals V_{PWM} applied to the GaN FETs. The control signal

delay allows us to test and emulate the different operative conditions and V_{GSth} spread. Shunt resistors R_S=0.1 Ω are used to sense each GaN FET current.

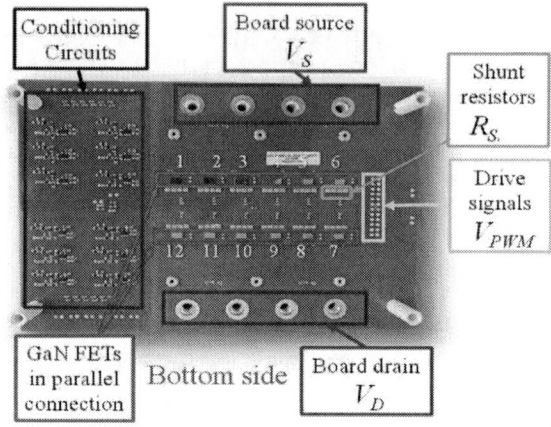

Fig. 4 Picture of the experimental parallel-GaN FETs test board.

The experimental test is carried out by connecting the board as depicted in Fig.5. The load is composed by L_{Load}=90 μH and R_{Load}=0.1 Ω connected in parallel with a freewheeling diode. The DC voltage is V_{DC}=24 V.

Fig. 5 Electric schematic of the connection of the

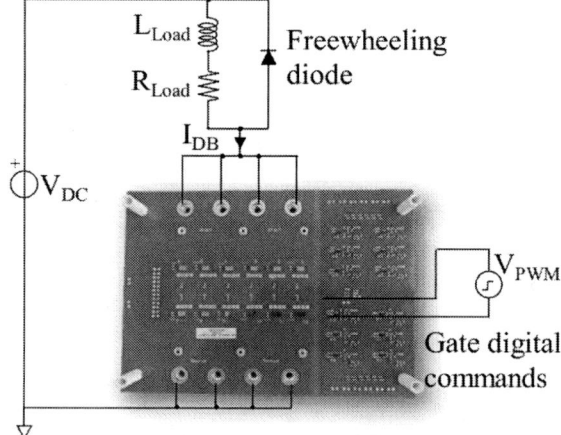

parallel-GaN FETs board for the experimental tests.

In the experimental tests, two GaN FETs are driven by two V_{PWM} with an established delay between them. The gate driving of GaN$_2$ is delayed with respect to GaN$_1$. All the other GaN FETs (GaN$_n$) remain in off-state.

Fig. 6 depicts the waveforms of the GaN FETs' current I_S, the current I_{DB} entering the board, and the drain-source voltage V_{DS} of the GaN FETs in parallel, which are measured during the turn-on and turn-off transients.

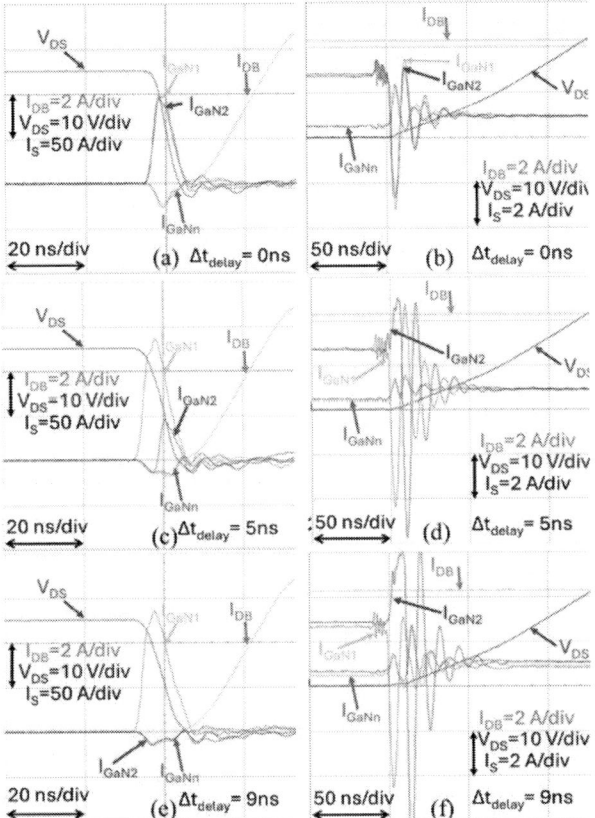

Fig. 6 Switching waveforms with GaN$_1$ driven earlier than GaN$_2$ that is delayed. a) turn-on delay=0 ns. b) turn-off delay=0 ns. c) turn-on delay=5 ns. d) turn-off delay=5 ns. e) turn-on delay=9 ns. f) turn-off delay=9 ns.

The switching waveforms obtained with delays of 0 ns, 5 ns, and 9 ns are shown respectively in Fig. 6a, Fig. 6c, and Fig. 8e for the turn-on and in Fig. 6b, Fig. 6d and Fig. 8f for the turn-off.

At turn-on, the I_S peak amplitude depends on the equivalent C_{OSS} of the GaN FETs in parallel and the C_{PCB}. I_{load} does not influence the peak because the high value of $L_{diode} + L_{DB}$ (50 nH circa) lowers the I_{DB} slow rate.

Looking at the turn-on transient, with null delay, the I_S peak is minimized by splitting the total charge of the parasitic capacitances between GaN$_1$ and GaN$_2$ (Fig.6a). When GaN$_2$ has a delay, GaN$_1$ has a higher peak that lasts longer (Fig.6c). The area underneath the I_S curve is maximized for delay

equal to 9ns (Fig.6e) or longer. It represents the total charge of the parasitic capacitances. The current peak in Fig.6e is I_S=135 A, which is only due to parasitic capacitances. It is the maximum achievable considering the number of GaN FETs mounted in the board and the designed PCB.

At turn-off, in the case of null delay, there is no current peak, and the I_S waveforms of GaN$_1$ and GaN$_2$ are congruent (Fig.6b). When GaN$_2$ turns off later than GaN$_1$, GaN$_2$ shows an I_S peak (Fig.6d). In this example, the I_S turn-off peak of GaN$_2$ is maximum for delay=9 ns because the I_S reaches the amplitude of I_{load}. There is no contribution of the parasitic capacitances on the turn-off peak.

Overall, in a board with GaN FETs in parallel, the I_S peak due to the spread of parameters is maximum at turn-on in the condition of the low value of $L_{diode} + L_{DB}$ (e.g., less than 2 nH circa in Fig.3a).

These results obtained with the board prototype featuring a high value of $L_{diode} + L_{DB}$ are significant when it is intended to design a board with the same number of GaN FETs. Typically, the GaN FET-based inverter has a low value of $L_{diode} + L_{DB}$ because another high-side device substitutes the freewheeling diode. In this scenario, the I_{DB} slew rate is steep enough to reach the I_{load} amplitude before the peak due to the parasitic capacitances.

The chosen EPC2302 GaN FET allows a pulse current of up to 400 A, while the parasitic capacitances of the parallel connection layout have been demonstrated to cause an IS of up to 135 A. Hence, theoretically, a hypothetical inverter board designed in similarity to the tested prototype can work seamlessly even with extreme parameters spread, as in the case of this paper, with currents I_{load} up to 265 A.

4 Experimental evaluation of an inverter with four GaN FETs in parallel

An inverter board with four parallel GaN FETs per each switching device has been designed using the experience from the 12 devices' parallel board characterization.

The top view of the experimental board picture is shown in Fig. 7a. In Fig. 7b, a half-bridge layout arrangement is highlighted.

Each inverter leg is driven by a single gate driver that supplies the eight GaN FETs: four on the high side (H) and on the low side (L).

The inverter was tested by powering a 5 kW three-phase BLDC motor with a DC-Bus voltage of V_{DC}=60 V. A DC-Bus capacitance of 720 µF was used. The switching frequency was f_{sw}=20 kHz.

The phase current I_0 is measured and depicted in Fig. 8.

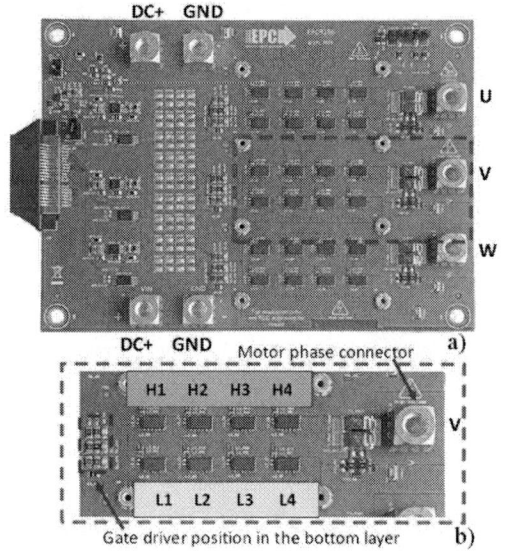

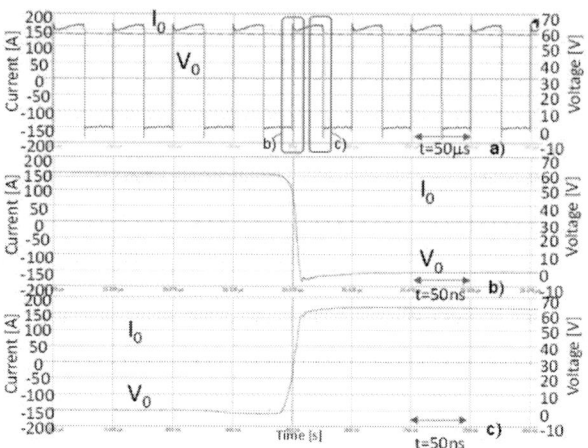

Fig. 7   a) Top view of EPC9186 inverter board. b) Zoomed view of the half-bridge layout arrangement of phase V.

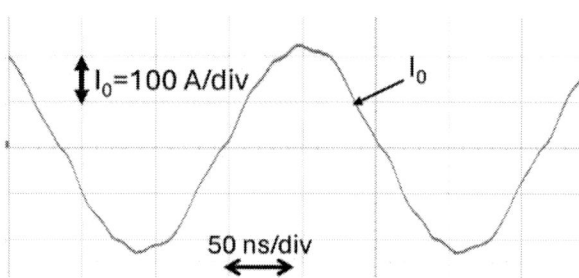

Fig. 8 I_0=150 A_{RMS} output phase (U) current obtained using the inverter board with four GaN FETs in parallel per switch. f_{sw}=20 kHz. V_{DC}=60 V.

The experimental test shows that the designed inverter can supply the motor with a phase current of I_0=150 A_{RMS}.

Fig. 9a depicts the switching waveforms related to a motor phase at a current level of I_0=150 A_{RMS}. Fig. 9 b, and Fig. 9c depict zoomed views of the node voltage at turn-on and turn-off, respectively.

Looking at Fig. 9, the switching voltage does not ring during commutations, and no overvoltage spikes are observed. This demonstrates that the layout's stray inductance is minimized and that the DC-link filter is well-designed.

Fig. 9   a) Output phase (U) switching voltage V_0 and current I_0 waveforms. f_{sw}=20 kHz. V_{DC}=60 V, I_0=150 A, b) zoomed view of the turn-on switching waveforms, c) zoomed view of the turn-off switching waveforms.

5   Conclusions

The paper investigates the advantages and issues of parallel-connected GaN FETs within inverter applications. The study discusses the effect of the spread of device parameters and layout considerations through a combination of simulation and practical experimentation through an explicitly developed board. This breakthrough promises to cater to the escalating demands of high-current motor drive applications, offering a more efficient and high-performance solution for the industry.

In simulations, the parameters spread, particularly the threshold voltage, has been revealed to cause a timing mismatch between the GaN FETs connected in parallel and, consequently, current peaks in the device that turns on earlier and turns off later. The strategy of separating the contribution of the parasitic capacitances to the rise in load current, as presented in this work, allows for empirical prediction of the maximum applicable load current level. Moreover, the design of the GaN FET-based inverter with parallel devices has followed the proposed analysis procedure and has demonstrated its capability to supply an electrical machine operating with a phase current of more than 150 A_{RMS}.

Future studies should focus on thermal analysis and heat spreading issues when connecting in parallel GaN FETs with extreme parameters spread.

References

[1] B. Li, W. Qin, Y. Yang, Q. Li, F. C. Lee and D. Liu, "A High Frequency High Efficiency GaN Based Bi-Directional 48V/12V Converter with PCB Coupled Inductor for Mild Hybrid Vehicle," 2018 IEEE 6th Workshop on Wide Bandgap Power Devices and Applications (WiPDA), Atlanta, GA, USA, 2018, pp. 204-211, doi: 10.1109/WiPDA.2018.8569067.

[2] S. Musumeci, V. Barba, "Gallium Nitride Power Devices in Power Electronics Applications: State of Art and Perspectives," Energies, 16(9), 3894 2022. https://doi.org/10.3390/en16093894.

[3] G. Faraci, A. Raciti, S. A. Rizzo, G. Schembra,. "Green wireless power transfer system for a drone fleet managed by reinforcement learning in smart industry," Applied Energy, Elsevier, 2020, 259, 114204. https://doi.org/10.1016/j.apenergy.2019.114204vol.

[4] M. Palma, S. Musumeci, F. Mandrile and V. Barba, "GN Devices for Motor Drive Applications," 2021 IEEE 8th Workshop on Wide Bandgap Power Devices and Applications (WiPDA), Redondo Beach, CA, USA, 2021, pp. 146-151, doi: 10.1109/WiPDA49284.2021.9645113.A.

[5] Lidow, Ed. GaN Power Devices and Applications. El Segundo, CA, USA: Power Conversion Publication, 2022. pp. 1-4.

[6] F. Chimento, A. Raciti, A. Cannone, S. Musumeci and A. Gaito, "Parallel connection of super-junction MOSFETs in a PFC application," 2009 IEEE Energy Conversion Congress and Exposition, San Jose, CA, USA, 2009, pp. 3776-3783, doi: 10.1109/ECCE.2009.5316419.

[7] M. Zhang and W. Zhang, "Current Sharing Analysis of Paralleled GaN HEMT Via Feedback Theory," 2018 1st Workshop on Wide Bandgap Power Devices and Applications in Asia (WiPDA Asia), Xi'an, China, 2018, pp. 94-99, doi: 10.1109/WiPDAAsia.2018.8734638.

[8] Y. Zhang, J. Li and J. Wang, "Investigations on Driver and Layout for Paralleled GaN HEMTs in Low Voltage Application," in IEEE Access, vol. 7, pp. 179134-179142, 2019, doi: 10.1109/ACCESS.2019.2957190.

[9] V. Barba, S. Musumeci, M. Palma, R. Bojoi, "Maximum Peak Current and Junction-to-ambient Delta-temperature Investigation in GaN FETs Parallel Connection," Power Electronic Devices and Components, 5, 100035, 2023, https://doi.org/10.1016/j.pedc.2023.100035.

[10] F. Luo, Z. Chen, L. Xue, P. Mattavelli, D. Boroyevich and B. Hughes, "Design considerations for GaN HEMT multichip halfbridge module for high-frequency power converters," 2014 IEEE Applied Power Electronics Conference and Exposition - APEC 2014, Fort Worth, TX, USA, 2014, pp. 537-544, doi: 10.1109/APEC.2014.6803361.

[11] V. Barba, L. Solimene, M. Palma, S. Musumeci, C. S. Ragusa and R. Bojoi, "Modelling and Experimental Validation of GaN Based Power Converter for LED Driver," 2022 IEEE International Conference on Environment and Electrical Engineering and 2022 IEEE Industrial and Commercial Power Systems Europe (EEEIC / I&CPS Europe), Prague, Czech Republic, 2022, pp. 1-6, doi: 10.1109/EEEIC/ICPSEurope54979.2022.9854660

PCIM Europe 2024, 11– 13 June 2024, Nuremberg   DOI: 10.30420/566262214

Repetitive Short-circuits on 650 V GaN

Adrien Lambert[1,2], Hervé Morel[1], Dominique Planson[1], Luong Viêt Phung[1],
Dominique Tournier[1], Pascal Bevilacqua[1], Laurent Guillot[2]

[1] INSA Lyon, Université Claude Bernard Lyon 1, Ecole Centrale de Lyon, CNRS, Ampère, UMR5005, France
[2] STMicroelectronics, France

Corresponding author: Adrien Lambert, Adrien.lambert@st.com
Speaker: Adrien Lambert, Adrien.lambert@st.com

Abstract

The study consists in analyzing the robustness of 650 V GaN HEMT components in case of short-circuits. The objective is to develop a test bench for repetitive short-circuits to a component. Electrical characterizations will be carried out before and after the short-circuit cycles to identify the possible parameters drift of the tested devices, to identify the failure mechanisms. Parametric variations have been observed.

1   Introduction

Wide-band gap (WBG) components, and specifically gallium nitride (GaN) High Electron Mobilty Transistor (HEMT) , are expected to be highly promising in the field of power electronics. Indeed, these transistors have enviable performances with reduced conduction and switching losses and thus a high cutting frequency, up to a few MHz. Nevertheless, a proven robustness and reliability are essential for their commercialization on a large scale, in particular in the automotive field with traction control system of electric vehicles or in the aeronautics field. With simultaneous high-voltage and high-current, short-circuits represent one of the most undesirable events due to the significant amount of energy to be dissipated [1]. Two types of short-circuit are mainly distinguished, type I (a hard switching fault, the component is gated on under full tension) and type II (a fault under load, the fault appears while the component is gated on). Type I, considered as the most frequent and de facto preferred for testing, is studied here.

The study of short-circuits began a few years ago, with initial work observed in 2013 [2]. However, it remains limited in several respects. The normally-off technology of p-GaN and cascode components has been studied almost exclusively. Current studies of enhanced mode (e-mode) components based on p-GaN components have revealed that the highest voltages could only be achieved at the cost of concessions (overheated environment, increased gate resistance) ([1], [2], [3,], [4], [5], [6], [7]). To the best of our knowledge, very few, if any, papers focus on normally-on technology depletion mode (d-mode). What is more, current studies are mainly concerned with the robustness of components to a single short-circuit and the time to failure.

The study presented here differs in several ways: normally-on components, MIS-HEMT technology (insulated gate HEMT[1]) and normally-off (p- GaN[1]) components are chosen. P-GaN technology is chosen as this structure seems to be preferred in the current industrial development.

Component	Vth	Id	Rds_ON
D-mode	-12.5 V	10 A	250 mΩ
p-GaN Batch 1	1.8 V	15 A	120 mΩ
p-GaN Batch 2	1.7 V	11 A	150 mΩ

Table 1   Different types of studied components

In addition, it has been chosen to study the impact of repetitive, non-destructive short-circuits, as the severe damage of the component complicates the study. What is more, in a real-life application, the aim is to have only non-destructive short-circuits, repeated a few times in the life of the converter.

1574

These short-circuits are interspersed with electrical characterizations to identify parametric variations and gain a better understanding of degradation and failure mechanisms. The study of induced damage is therefore a key point and it will contribute to the design of more short-circuits robust chips.

2 Test Set-Up

2.1 Technical Choices

Previous short-circuit studies has often reported high overvoltage (>100 V) limiting the drain-source voltage range during short-circuit testing [1]. Because of the very fast current fronts typical of GaN technology, any parasitic inductance in the PCB power loop leads to a significant overvoltage at the component level and turns out to be deleterious.

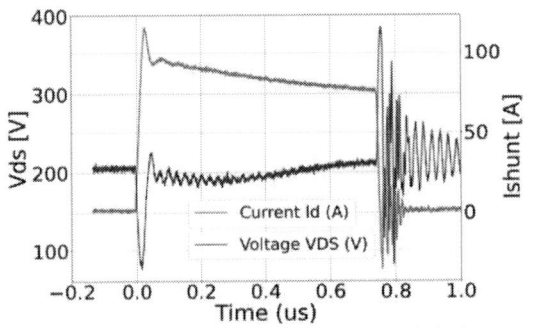

Fig. 1 Typical drain-source voltage and drain current measured at short-circuit

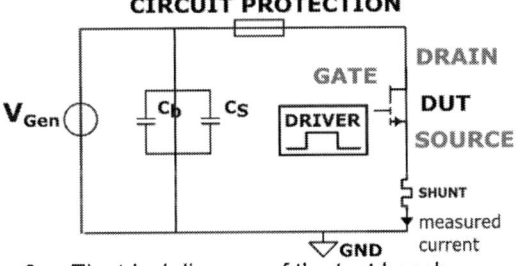

Fig. 2 Electrical diagram of the test bench

For the studied DUTs, with a short-circuit current of around 50 A (low estimate) and a typical switching time of 20 ns, a parasitic inductance of 80 nH causes 200 V of overvoltage! The circuit has therefore been designed to reduce the wiring inductance as much as possible. The protection element, often active switch like IGBT, considered highly inductive, is replaced here by an SMD fuse, which likely preserves the circuit before its destruction. To reproduce short-circuit conditions, and in particular the rapid rise of a high-current, while minimizing the voltage drop, two capacitors

of 2 µF (ceramic with ultra-low ESR, C_s) and 133 µF (polypropylene, C_b) respectively are connected in parallel. The test bench consists of three key elements: the *power* mother board with capacitors, fuse, and current measurement shunt; the "device" daughter board with the transistor encapsulated in its sarcophagus, and the driver section for -24 V / 0 V or -3 V / + 6 V control.

Fig. 3 *Device* Daughter Board and *Power* mother Board

Fig. 4 3-D printed sarcophagus

To reduce component handling, which could affect these electrical properties, the daughter board has been designed to be easily connected to the Keysight B1505A characteristic tracer. In addition, to avoid soldering/unsoldering of components on this daughter board, and thus cause significant thermal stress, a 3D-printed sarcophagus was designed to maintain the component and ensure pressure conduction (**Fig. 4**).

Fig. 5 Daughter Board with Attached Driver

Similar electrical characterizations with a brazed component ensured that the sarcophagus did not interfere with the component electrical properties.

2.2 First Optimisation of the Test-Bench Set-up with D-mode components

2.2.1 First Short-circuits

Particular care has been taken to reduce parasitic inductances in advance. This involves careful routing and optimised compactness, as well as surface-mounted components with reduced ESL (Equivalent Series Inductances).

Despite this, the first short-circuits caused significant disturbance and the bench had to be adjusted. The first short-circuit was conducted for drain-source voltage V_{DS} of 10 V. Non-negligible overvoltages are observed at turn-off on V_{GS} and V_{DS} signals. For V_{DS} = 50 V, the signals oscillate at the risk of damaging the component or even causing it to break. Solutions therefore had to be found.

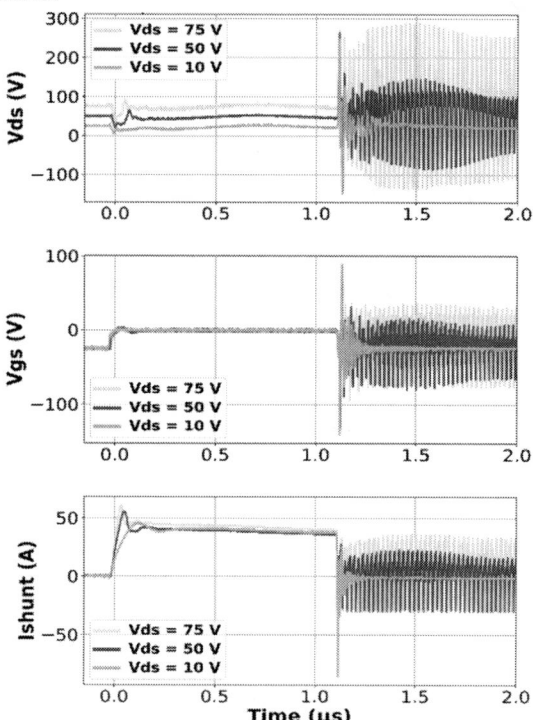

Fig. 6 First short-circuit with unstable behaviour

2.2.2 Impact of Voltage Probes (V_{DS}, V_{GS})

The observed overvoltages and oscillations necessarily raise the impact of the probes to the extent it can be questioned if the observed perturbations are an artifact of the probes or are truly undergone by the component. For this purpose, a spice simulation of the experimental bench has been developed.

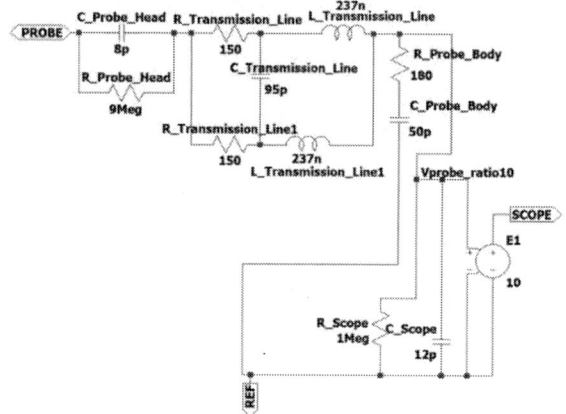

The spice simulation presents the significant interest of being able to access the voltages within the component to compare them to the voltages observed through the probes. A passive probe model is therefore developed

Fig. 7 Probe Spice model inspired by [8], [9], [10]

According to the results of simulation, the probes do not necessarily represent the reality of the signals observed on the oscilloscope. Different types of probes are tested. The V_{GS} signal is considered to be the most critical, so the study of the impact of the probes is focused on this signal. Three types of probes were studied: passive, differential and optically isolated differential. The different types of probes are summarized in the table below.

Type	Manufacturer	Reference
Passive Probe	Tektronix	TPP0500B
Differential Probe	Tektronix	THDP0100
Optically isolated differential probe	Tektronix	TIVM1

Table 2 Different types of used probes

Several tests are carried out at a reduced voltage V_{DS} = 10 V. In the case of differential probes, V_{GS} signals are directly obtained while it is calculated by subtracting the V_G and V_S signals when using passive probes.

Fig. 8 shows the full sequence of short-circuit while Fig. 9 focuses on the turn-off, the most-critical judged moment.

The differential probe, instead of improving the measurement, exacerbates the noise. The differences in the V_{GS} signal depending on the type of probe used are clearly visible. The differential probe is surprisingly the most disturbed, perhaps because its use requires an electronic box, potentially disturbed by the magnetic field radiated by the experimental bench. Passive probes therefore turn out not to be the worst. The least noisy signal comes from the optically isolated probe.

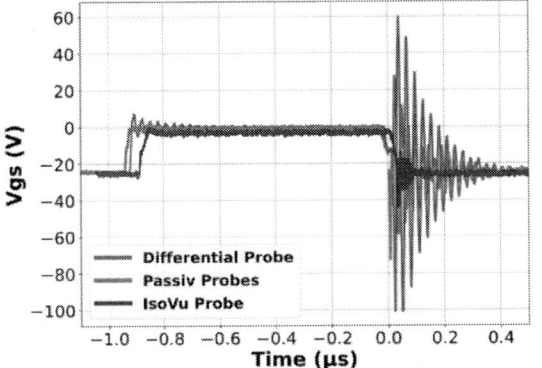

Fig. 8 Comparison of probes during short-circuit (full sequence)

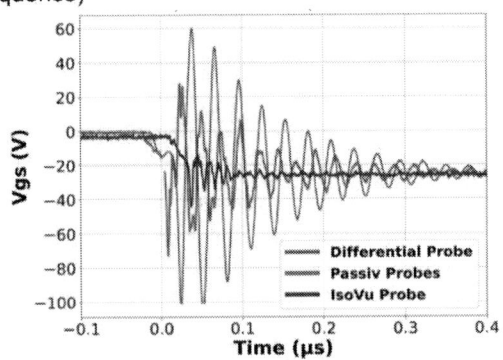

Fig. 9 Comparison of probes during short-circuit (zoom on Turn-OFF)

2.2.3 Impact of Current Sensing (I$_S$)

The shunt has its own inductance, and adding it to the board causes parasitic inductances. It was therefore investigated to what extent the shunt could have an impact.

It was studied whether the direction of the shunt ground position had an influence. Different test configurations were carried out. For config. 1, the ground leg of the shunt was soldered to the ground side of the high-voltage supply (in red on the diagram below), while for config. 2, the shunt ground was soldered to the source side of the device (in blue on the diagram below). The reduction of the loop formed by the shunt legs was also studied, to reduce the parasitic inductance caused. This corresponds to the optimized case and to configurations 4 and 5. Config. 3 corresponds to the absence of a shunt.

Type of config.	Technical choices
Config. 1	shunt mass side of PS ground
Config. 2	shunt mass side of DUT
Config. 3	no shunt
Config. 4	optimized shunt, mass side PS
Config. 5	optimized shunt, mass side DUT

Table 2 Different configurations to analyse the impact of shunt

Fig. 10 shows the entire duration of the short-circuit, while Fig. 11 shows a zoom of V$_{DS}$ during the turn-off. It can be seen that, although the direction of the shunt is of little importance in terms of overvoltages on V$_{DS}$ with very similar overvoltage between config 1 and 2, reducing the parasitic inductance generated by the shunt loop significantly reduces the overvoltage on V$_{DS}$ from 75 V.

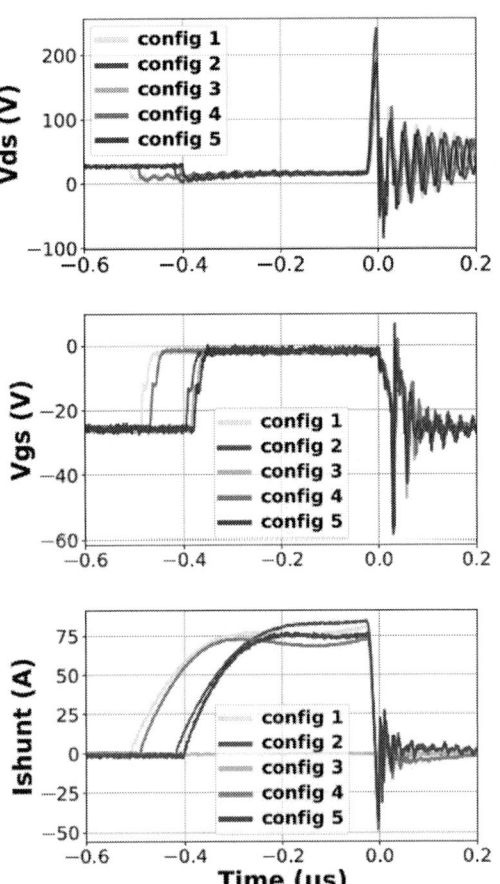

Fig. 10 Short-circuits measures considering several configurations

PCIM Europe 2024, 11– 13 June 2024, Nuremberg   DOI: 10.30420/566262214

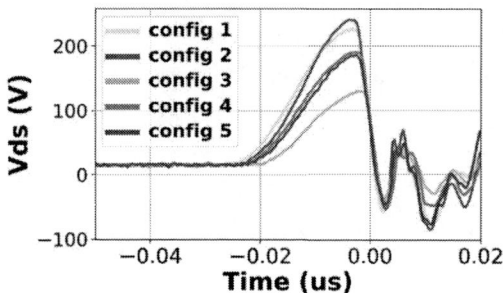

Fig. 11   Short-circuits (zoom on Turn-OFF)

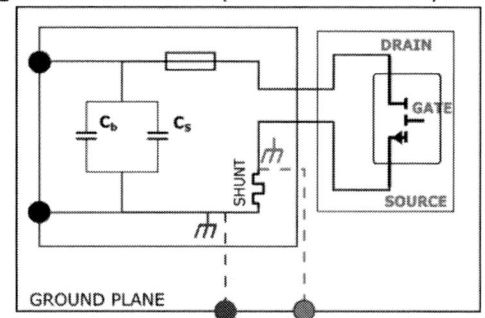

Fig. 12   Configuration 1 (blue) or 2 (red)

Reducing the parasitic inductances of the shunt significantly reduced the power loop and V_{DS} oscillations during Turn-OFF. A gradual rise in voltage could then be achieved for normally-ON components. Config.1 has been maintained. Tests for longer conduction times were also carried out. Tests for longer conduction times were also carried out.It can be noted in Fig.13 that the overvoltage is constant, what underlines it only depends current and inductance.

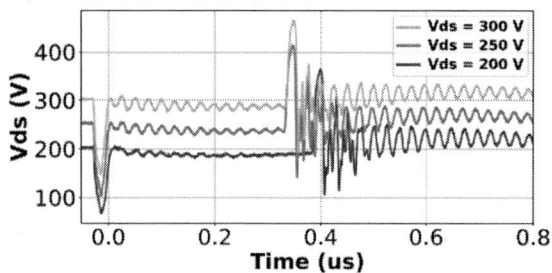

Fig. 13   Short-circuit for a gradually increased V_{DS}

2.3 Second Optimisation for N-OFF Components

The study is extended to normally-OFF devices, with p-GaN devices from two manufacturers (called batch 1 and batch 2) with similar characteristic. While the tests on the batch 1 did not pose any problems for drain-source biases of 100 V and 200 V (**Fig. 14**), it turns out to be different for batch 2.   Various ways of improvement have been explored.

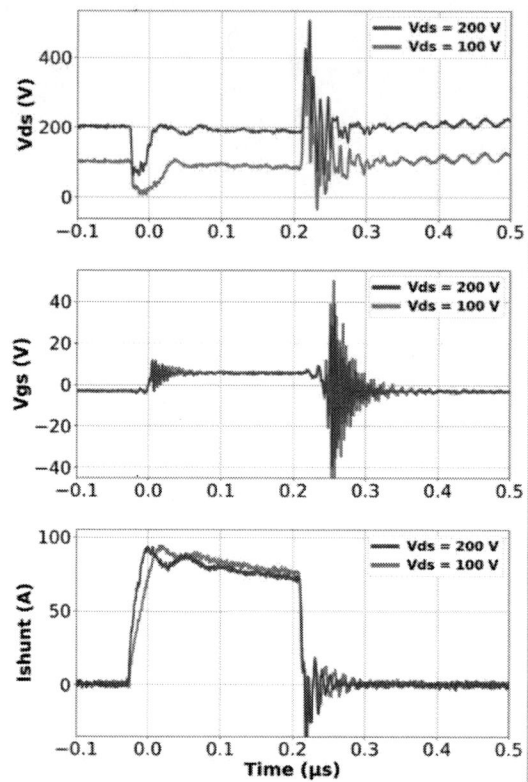

Fig. 14   Short-circuit for batch 1 component

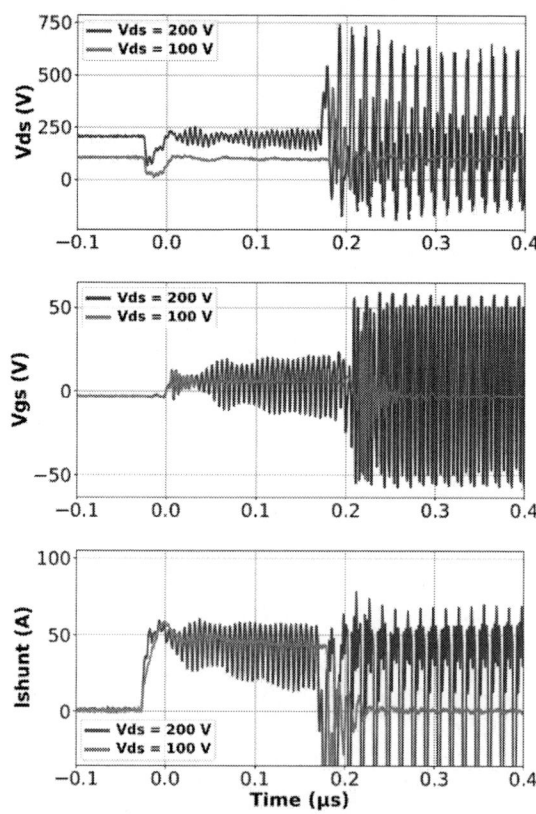

Fig. 15   Short-circuit for batch 2 component

1578

2.3.1 Impact of the Gate-resistance

Increasing the gate-resistor was the first idea.

Indeed, by increasing gate-resistance, the authors [11] have observed an improvement in signals that had previously been disturbed by short-circuiting. It was tested to increase the R_{gOFF} resistance to 33 Ω compared with 22 Ω previously used. It did not improve the shape of signals at both V_{DS} = 100 V and 200 V. Given the lack of improvement and the fact that R_{gOFF} cannot be increased indefinitely, as the industry favours low value of gate-drive resistance to minimise losses, the impact of R_{gOFF} is not studied further.

2.3.2 Impact of Common Mode Current

Considering the absence of any improvement to the experimental bench and in view of the very high disturbance of the V_{GS} signal, it was decided to study the common-mode currents at the driver. The study focuses on the wire connection between the driver power supply board and the driver board.

Fig. 16 Driver Power Supply (left) and driver Boards (right)

The initial configuration includes four connection wires (V+Drive, V-Drive, +5V opto, GND) between the driver power supply card and the driver. A first attempt at improvement was made by removing the wires by pinning the cards via the white connectors shown in the photo above. This first solution is unsatisfactory, and even degrading.
Ferrite torus are then inserted into the wire connection

Fig. 17 Insertion of Ferrite Torus in the driver wires

The wires are first looped twice around the ferrite cores.

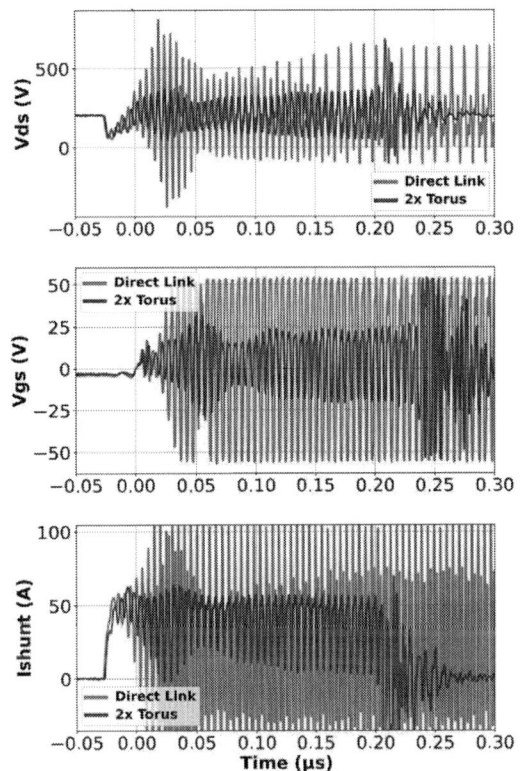

Fig. 18 Comparison between a direct connection without or with ferrite torus (looped twice)

The signals shown in **Fig. 17** demonstrate a appreciable reduction of disturbance. Considering that it was decided to increase the looping through the torus from two to five times. The first improvement is noticeable at V_{DS} = 100 V. At V_{DS} = 200 V, shown in **Fig. 18** , the results were indisputable with a significant stabilization.

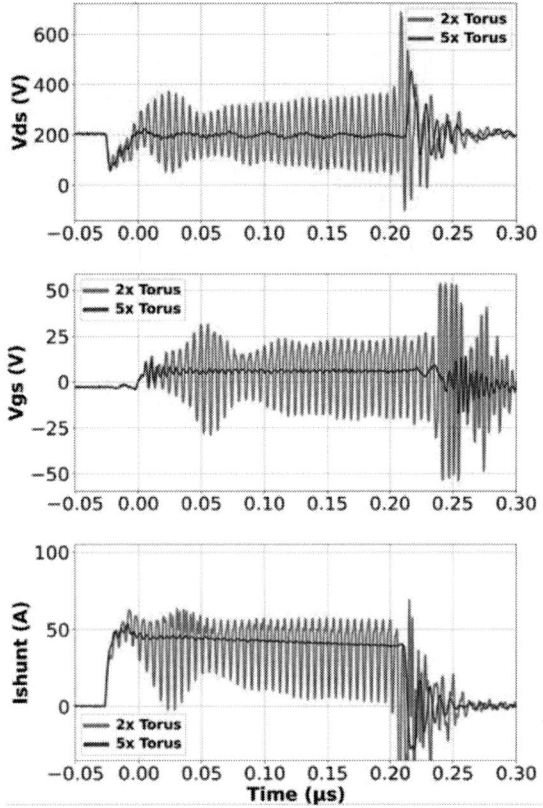

Fig. 19 Comparison between a connection with ferrite torus looped twice or five times

3 Results

The bench has been optimized to allow repetitive short-circuits of variable duration and V_{DS} polarization. Progressive voltage increases are achieved.

A distinction is made in this study between normally-ON, or D-mode, components from manufacturer 1 and normally-OFF or P-GaN components from manufacturers 1 and 2.

The components are subjected to a series of short-circuits of variable duration and variable V_{DS} polarization. Before and after each series of short-circuits, the components are characterized using a B1505 electrical tracer to record parameters such as Threshold Voltage (V_{th}) and on-state resistance (RdsON). Each initial series of measurements includes the "Triple-Sens" [12] V_{th} reset protocol, allowing a reliable initial V_{th} value to be taken into account.

3.1 D-mode

Optimization of the test bench enables the voltage to be ramped up to the breaking point with a

maximum polarization V_{DS} = 325 V.

The B01505 characterizations reveal a variation in the threshold voltage. Figures ... and ... show respectively the variation of V_{th} in value or in relative terms to its initial value before short-circuit. This V_{th} variation is shown as a function of the energy accumulated during successive short-circuits. The study was continued for components in batch B, for polarization V_{DS} = 200 V and conduction times varying from 400 ns to 1 µs .

Fig. 20 shows the last three consecutive short-circuits experienced by component 17B. The failure before the end of the last short-circuit demonstrates the fatigue of the component due to the accumulation of successive short-circuits.

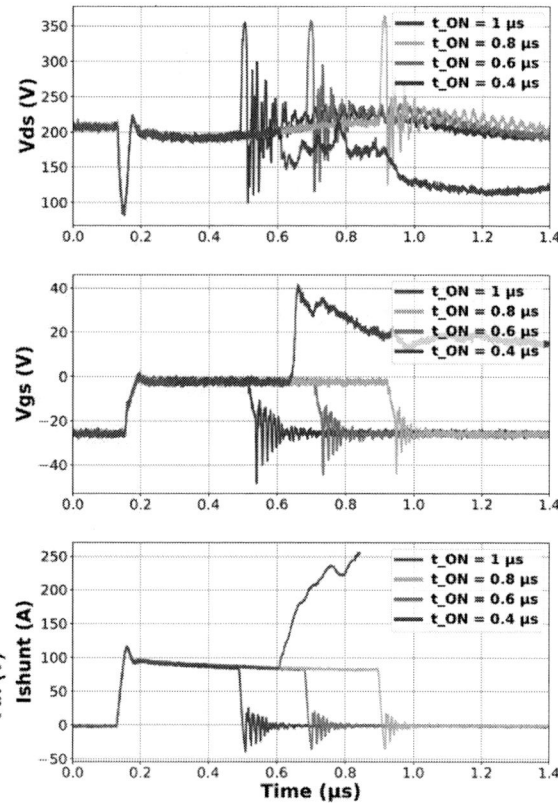

Fig. 20 Breakage of batch B component before the end of short-circuit

3.1.1.1 Static Variation

Fig. 21 and Fig. 22 illustrate the influence of repetitive short-circuits on the threshold voltage, in some cases with a variation of more than 10%.

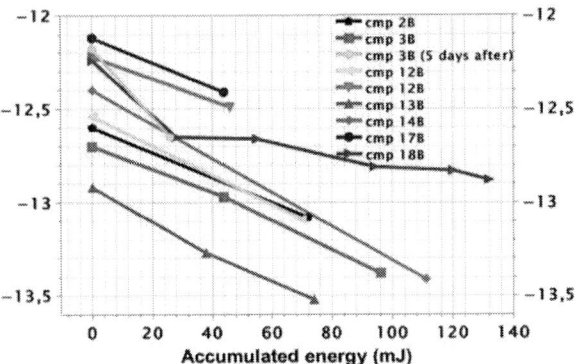

Fig. 21 V_th variation (in values) for D-mode Batch B

It is interesting to note the overall trend emerging from the multiple tests on the components in batch B, with a general reduction in threshold voltage to values even more negative than the initial value.

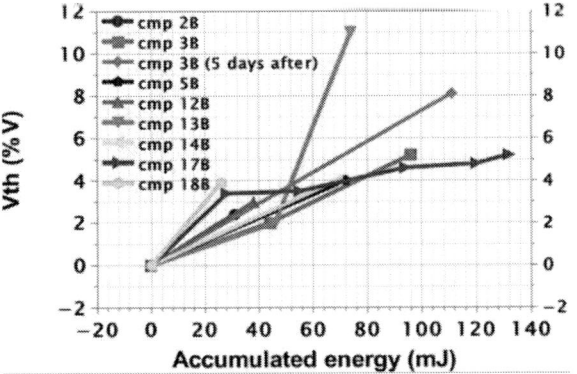

Fig. 22 V_th variation (in %) for D-mode Batch B

3.1.1.2 Dynamic Variation

As it is represented in the **Fig. 23**, an increase of the drain current was noted during the repetitve short-circuits on a part.

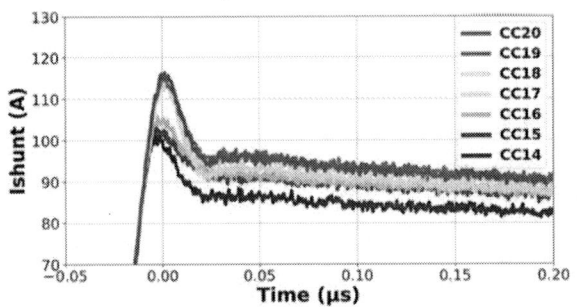

Fig. 23 Zoom on the drain current of component 17B during a series of consecutive short-circuits.

3.2 Normally-Off

3.2.1 P-GaN Batch 1

The p-GaN components in batch 1 are highly robust, which means that conduction times can be extended to as little as 10 µs. It was also possible to carry out a voltage rise, with a recorded break for V_{DS} = 400 V and 2 µs of conduction as shown in **Fig. 24**. The progressive degradation of the V_{GS} and V_{DS} signals at turn-ON can be seen as the voltage increases in **Fig. 24**.

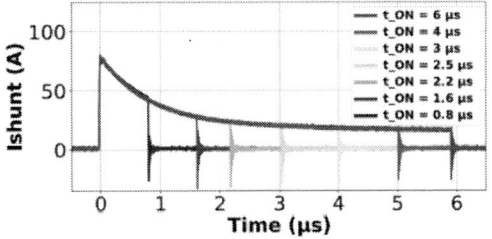

Fig. 24 Drain current signal for a progressive Increase of the conduction time for p-GaN component

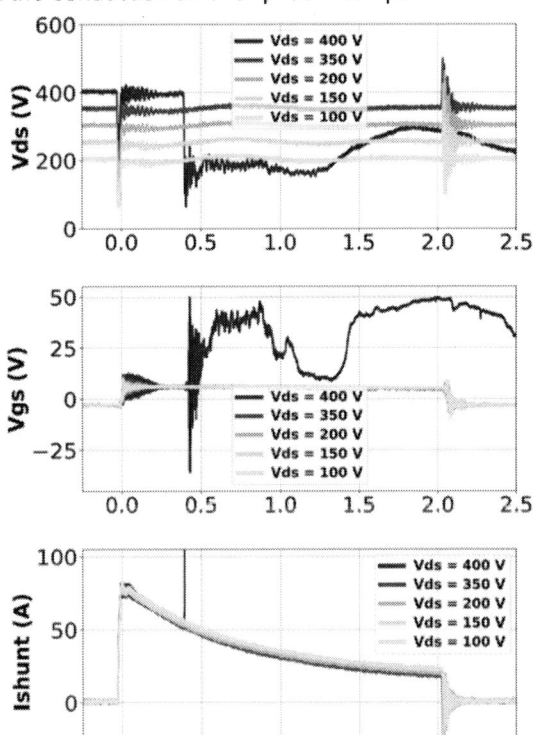

Fig. 25 V_{DS}, V_{GS} and drain current signals during a Rise in V_{DS} for p-GaN component with a breakage at V_{DS} = 400 V

3.2.1.1 Vth Variation

Significant V_{th} variations have been observed as shown in **Fig. 26** and **Fig. 27**.

To make these measurements reliable, witness component number 34, known as the "Golden Device" is used. This golden device is not short-circuited, but is characterized at the same time as other component. Small variation in Vth for the Golden device compared to short-circuited components demonstrates the impact of short-circuits on the variation in threshold voltage. It would be interesting to develop a measurement protocol which includes a temperature calibration.

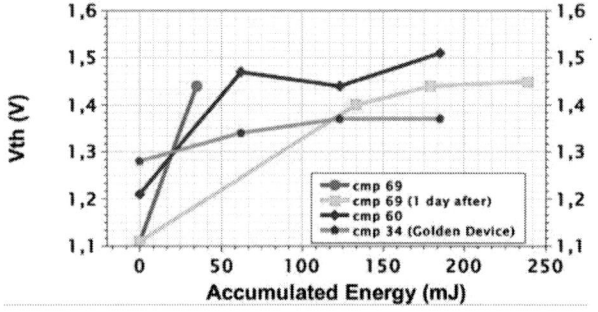

Fig. 26 V_{th} variation (in values) for p-GaN Batch 1

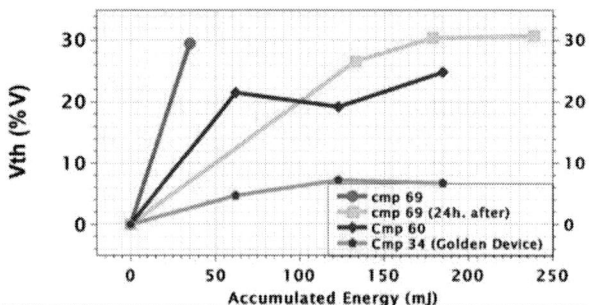

Fig. 27 V_{th} variation (in %) for p-GaN Batch 1

3.2.1.2 Rds-ON Variation

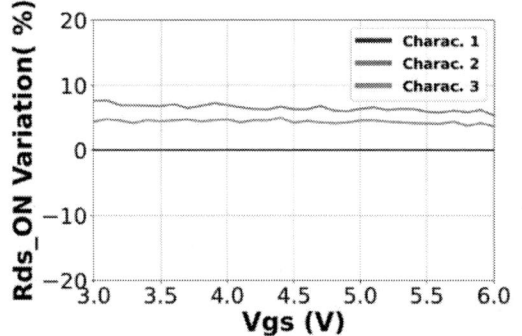

Fig. 28 Rds-ON variation for p-GaN batch 1

In view of the large variation in threshold voltage, an analysis of the possible variation in Rds-ON is carried out. In the same way as for the V_{th} study, a component is taken as golden device. With a variation of Rds-ON below 1% between each characterization, it demonstrates that variation observed in short-circuited component is imputable to this specific stress. **Fig. 28** shows a variation of Rds-ON between 5 and 10% for a short-circuited component.

3.2.2 P-GaN Batch 2

Although the p-GaN components in Batch 2 are also more robust than the D-mode components studied, they broke more easily than p-GaN batch 1 with a maximum V_{DS} of 300 V. Component failure does not always occur during conduction or at turn-off, as shown in the **Fig. 29** below, which illustrates component failure AFTER turn-off.

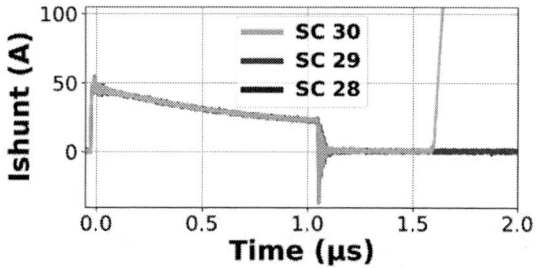

Fig. 29 Breakage after turn-off for p-GaN batch 2

3.2.2.1 Vth Variation

The potential variation of V_{th} is also studied and gives results considered similar to p-GaN batch 1, although slightly lower in variation and leads to Rds_on study.

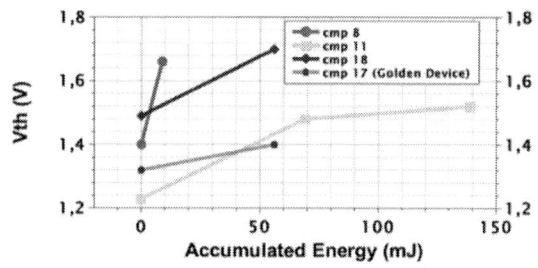

Fig. 30 V_{th} variation (in values) for p-GaN batch 2

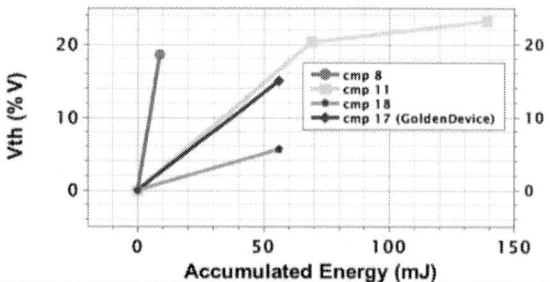

Fig. 31 V_{th} variation (in %) for p-GaN batch 2

3.2.2.2 Rds-ON Variation

As the batch 1, the variation of Rds-ON is studied. After a series of short-circuits, a variation of more than 10% is noted in **Fig. 32** in comparison to the initial situation, before the short-circuits. In the same way as batch 1, a golden device was taken, with a variation of less than 1% in the absence of short circuit, what demonstrates the impact of short-circuit.

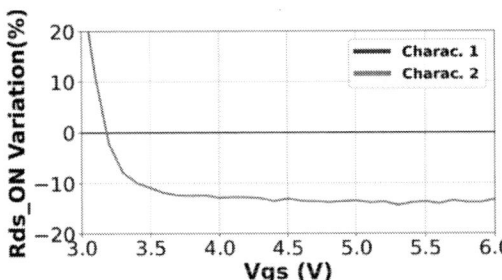

Fig. 32 Rds-ON variation for p-GaN batch 2

4 Conclusion

By optimizing the test bench, repetitive short-circuits were applied to normally-ON and normally-OFF components. Degradations were observed with non-negligible drift of static parameters such as the threshold voltage V_{th} and the on-state resistance. In the case of normally-ON components, a few dynamic variations of the short-circuit drain current were observed.

References

[1] M. Landel, « Étude de la robustesse de transistors GaN en régime de court-circuit », p. 263.

[2] C. Abbate, « Thermal instability during short-circuit of normally-off AlGaN/GaN HFETs », *Microelectron. Reliab.*, 2013.

[3] H. Li *et al.*, « Robustness of 650 V Enhancement-Mode GaN HEMTs under Various Short-circuit Conditions », p. 9.

[4] T. Oeder, « Electrical and thermal failure modes of 600V p-gate GaN HEMTs », *Microelectron. Reliab.*, p. 6, 2017.

[5] M. Riccio *et al.*, « Experimental analysis of electro-thermal interaction in normally-off pGaN HEMT devices », p. 6.

[6] M. Fernández *et al.*, « Short-Circuit Study in Medium Voltage GaN Cascodes, p-GaN HEMTs and GaN MISHEMTs », *IEEE Trans. Ind. Electron.*, p. 10, 2017.

[7] P. J. Martínez, « Unstable behaviour of normally-off GaN E-HEMT under short-circuit », *Semicond Sci Technol*, 2018.

[8] J. S. de Oliveira, « A Methodology for designing SiC and GaN device based converters for automotive applications ».

[9] K. Ammous et H. Morel, « Analysis of Power Switching Losses Accounting Probe Modeling », *IEEE Trans. Instrum. Meas.*, vol. 59, n° 12, 2010.

[10] K. Mbaitelbe, « Inductance dans son environnement : Caractérisation des inductances planaires intégrées dans les conditions d'utilisation de l'électronique de puissance ».

[11] P. Xue, L. Maresca, M. Riccio, G. Breglio, et A. Irace, « Experimental Study on the Short-Circuit Instability of Cascode GaN HEMTs », *IEEE Trans. ELECTRON DEVICES*, vol. 67, n° 4, p. 7, 2020.

[12] T. G. Bade, H. Hamad, A. Lambert, H. Morel, et D. Planson, « Threshold Voltage Measurement Protocol ``Triple Sense'' Applied to GaN HEMTs », 2023.

PCIM Europe 2024, 11– 13 June 2024, Nuremberg   DOI: 10.30420/566262215

Comparison of Switching Losses and Dynamic On Resistance of 600 V-Class GaN HEMTs

André Thönnessen [1], Joshua Baumgärtner[1], Carsten Fronczek[1], Rik W. De Doncker [1]

[1] Institute for Power Electronics and Electrical Drives (ISEA), RWTH Aachen University, Germany

Corresponding author: André Thönnessen, post@isea.rwth-aachen.de
Speaker: André Thönnessen, post@isea.rwth-aachen.de

Abstract

Semiconductor devices based on GaN promise lower switching losses and thus enable higher switching frequencies and higher power densities. However, measuring the switching losses is challenging due to the fast switching transients and places high demands on the sensor technology. In this paper, a high-bandwidth and low-inductance current shunt is used to evaluate and compare switching losses of five GaN devices and one SiC MOSFET as a benchmark. Furthermore, the dynamic on resistance of the semiconductors is measured and compared. By measuring all semiconductor devices with an identical measurement setup, a neutral comparison of the different power switches is achieved.

1 Introduction

Due to the increasing requirements regarding power density and efficiency, wide-bandgap devices (WBG) play a crucial role in today's power electronics. Gallium nitride (GaN) high-electron-mobility transistors (HEMT) are particularly attractive with very low gate and output charges as well as almost zero reverse recovery charge, which enable high switching frequencies and power densities. These advantages are achieved due to material properties, such as higher electron mobility, bandgap voltage and higher breakdown field strength compared to silicon carbide (SiC) or silicon (Si) [1].

However, the semiconductor devices suffer from trapping and detrapping effects, which lead to an increase in on-state resistance R_{on}, which decays after turning on. This effect varies depending on the structure of the GaN HEMTs and is often not considered in data sheets [2]. In this work, the dynamic on-state resistance is measured in the first 10 μs after turn on of the devices, since this period is of major interest in high-frequency power electronic converters.

Since GaN HEMTs are inherently depletion-mode (d-mode) devices, additional actions must be taken to achieve enhancement-mode (e-mode) behavior, which is desired for voltage source converters for safety reasons [3]. Manufacturers follow different approaches to accomplish this.

1) Nexperia and Transphorm use a cascode configuration consisting of a d-mode GaN HEMT and a series-connected low-voltage Si MOSFET for the devices considered in this work [4], [5]. With this configuration, an ordinary gate driver for gate voltages of Si MOSFETs can be used. This advantage comes at the price of a reverse-recovery charge caused by the series-connected Si MOSFET, which degrades the switching behavior especially at turn on [6].

2) VisIC also uses a cascode configuration, however with a continuously turned on Si MOSFET during operation, which only turns off, if the supply voltage of the gate driver is falling below a threshold (under-voltage lockout). As a result, the reverse recovery charge of the Si MOSFET does not impact the switching behavior. The d-mode GaN HEMT technology is called D³GaN and can be driven by a 0 V/+15 V gate driver, but it must be connected in a way such that the gate-source voltage is 0 V when turned on and –15 V when turned off. Thus the normally-off behavior is obtained [3].

3) An e-mode HEMT is achieved by Infineon formerly GaN Systems through a p-type GaN layer underneath the gate called GaN Systems E-HEMT [7]. However, the low threshold voltage of typically 1.7 V and the low maximum gate

1584

PCIM Europe 2024, 11– 13 June 2024, Nuremberg   DOI: 10.30420/566262215

Tab. 1: Overview of the measured devices. The SiC device is used for the sake of comparability of SiC and GaN.

Device	Manufacturer	Technology	Cont. Current Rating	Package
GS66516T	GaN Systems Inc.	GaN Systems E-HEMT	60 A	SMD
IGOT60R070D1	Infineon Technologies AG	GIT	31 A	SMD
V22TC65S1A1	VisIC Technologies	D³GaN	100 A	SMD
TP65H035G4WS	Transphorm Inc.	cascode	46.5 A	TO247-3
GAN041-650WSB	Nexperia	cascode	47.2 A	TO247-3
C3M0025065K	Wolfspeed Inc.	SiC	120 A	TO247-4

voltage of 7 V are disadvantages in terms of robustness [8]. The manufacturer of this device is referred to hereafter as GaN Systems in order to distinguish it from the next device investigated.

4) The e-mode gate injection transistor (GIT) from Infineon uses an ohmic gate contact and a pGaN gate, which leads to a non-isolated gate structure with a diode-like input characteristic. This creates a robust gate against over-voltage, but it must be driven with a continuous gate current [9].

These different approaches not only result in different on-state but also different switching behaviors. GaN HEMTs switch extremely fast, which is challenging for measurement equipment. To overcome this, low-impedance voltage probes and high-bandwidth coaxial current shunts are used to evaluate the switching losses in this work. The used coaxial current shunts have less than 100 pH parasitic inductance and thus distort the switching behavior significantly less than state-of-the-art current shunts [10], [11].
Table 1 presents the devices investigated, manufacturer and used technology, the continuous current rating at 25 °C as well as the package. All devices are rated for 600 to 650 V blocking voltage and offer top-side cooling, which allows better heat dissipation than bottom-side cooling. So far, literature shows only comparisons of two GaN HEMTs in terms of dynamic conduction losses [12] or switching losses at low power levels (<12 A) [13]. Other publications show comparisons of total losses based on individual converter topologies [14] but in this work, switching and dynamic conduction losses of more than two GaN HEMTs are measured at high currents (>30 A) with an identical setup, which establishes a neutral comparison between the devices.

2 Measurement Setup

To ensure a fair comparison, all circuit designs were individually optimized for the respective GaN HEMTs. This includes the selection of the gate driver, its circuitry and the auxiliary voltage supply of the secondary side of the gate driver.
The gate resistors of the driver circuits influence the switching losses, but also the voltage and current gradients as well as the over- and undershoot during switching operations [15]. In order to

Tab. 2: Voltage slopes and gate resistances for the turn-on and turn-off events

Device	Turn-on	Turn-off
GaN Systems	$-14.7\,\text{V/ns}$ @ 8.2 Ω	33.1 V/ns @ 15 Ω
Infineon	$-16.0\,\text{V/ns}$ @ 37 Ω	46.6 V/ns @ 15 Ω
VisIC[1]	$-1.47\,\text{V/ns}$ @ 33 Ω	41.1 V/ns @ 3.3 Ω
Transphorm	$-14.9\,\text{V/ns}$ @ 56 Ω	40.2 V/ns @ 15 Ω
Nexperia	$-15.8\,\text{V/ns}$ @ 47 Ω	36.1 V/ns @ 1 Ω
Wolfspeed	$-6.17\,\text{V/ns}$ @ 47 Ω	15.0 V/ns @ 1 Ω

[1] Limited turn-on slope of the device from VisIC to prevent parasitic turn off

achieve comparable results, the gate resistors of the individual circuits were trimmed to similar voltage slopes between the GaN HEMTs. When sizing the gate resistors, it is usually necessary to find a trade-off between voltage slope and switching losses, as voltage slope has a decisive influence on the size, e.g., of the filters of a dc-dc converter and can lead to insulation breakdowns in the windings of motors [16]. The selected gate resistors and the resulting voltage slopes are listed in Tab. 2. The aim was to trim the voltage slopes to the same values. All measurements were taken at 400 V and 50 A. To avoid a parasitic turn-off event, the turn-on voltage slope of the VisIC device had to be limited.
The voltage slope was evaluated at turn-off event as the average from 10 % to 90 % of the rated

1585

voltage. At turn on, the start point was chosen when the current reached 90 % of its rated value and the endpoint when the voltage decreased below 10 % of its rated value. This prevents averaging from starting well before the end of current commutation for the turn-on event.

When designing the measuring boards, special attention was paid to a low-inductive power and gate loop. A four-layer PCB was used for the circuit in accordance to [11] to minimize the parasitic inductance. CeraLink capacitors with a total capacitance of 3 µF are used as dc-link capacitance on all boards. All measurements have been conducted with a double-pulse test bench according to [17].

In addition to the switching losses, the dynamic on resistance was analyzed. The resistance is evaluated by measuring the drain-source voltage and the current. A clamping circuit is required for precise measurement of the on-state voltage, as the voltage scale changes from several hundred volts to a few volts by turning the device on.

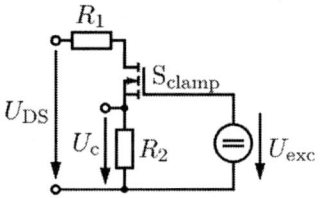

Fig. 1: Clamping circuit for measuring the drain-source voltage of the turned-on HEMTs

The clamping circuit was designed and constructed based on [18] and is shown in Fig. 1. A 180 Ω and 10 kΩ resistor is used for R_1 respectively R_2. Component *STN1NK60ZL* from STMicroelectronics is used for MOSFET S_{clamp}. The gate of the clamping MOSFET S_{clamp} is tied to a constant excitation voltage source U_{exc}. The MOSFET follows the behavior of the device under test (DUT) and is in conduction mode whenever the DUT turns on. In this case, the drain-source voltage U_{DS} can be calculated based on the clamped voltage U_c with Eq. 1. When the DUT turns off, the clamping MOSFET starts blocking so that the clamped voltage U_c is limited. With means of the clamping circuit, the oscilloscope input can be configured to the on-voltage scale of the DUT, thus achieving a high resolution.

$$U_c = U_{DS} \cdot \frac{R_2}{R_1 + R_2 + R_{S,clamp}} \qquad (1)$$

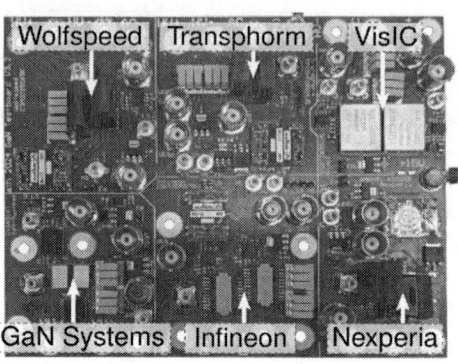

Fig. 2: Measurement PCB for all six semiconductors

Figure 2 shows the PCBs used for the measurements. The PCB is divided into six sections, one for each semiconductor. BNC connectors are used as interface to the oscilloscope. The gate signals are transmitted via SMB connectors. The double-pulse test bench is contacted via screw connections.

Fig. 3: Coaxial current shunt consisting out of 16 resistors with 0402 package and 1 Ω each and an SMA connector

The current shunts are designed according to [11] and shown in Fig. 3. Its small size and the coaxial current flow results in low parasitic inductance while ensuring a high bandwidth.

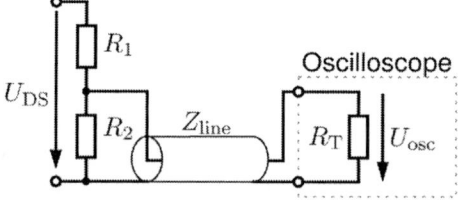

Fig. 4: Schematic of the low-impedance voltage probes.

Figure 4 shows the schematic of the low-impedance voltage probes. These probes are used to measure the drain-source voltage with high bandwidth during the switching events. The resistor

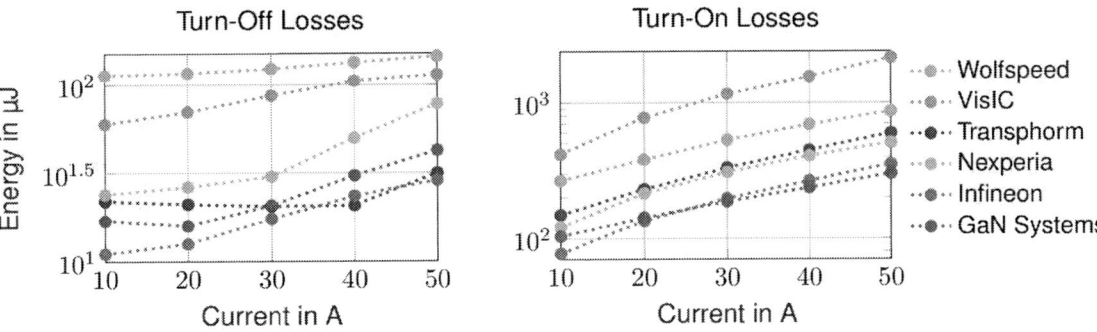

Fig. 5: Measured switching losses at 400 V dc-link voltage.

$R_2 = 50\,\Omega$ is selected to match the line impedance Z_{line} and the terminating resistor of the oscilloscope R_T in order to avoid reflections. To limit the voltage of the oscilloscope to the maximum input voltage, the resistor R_1 is set to 3.15 kΩ.

3 Measurement Results

Figure 5 shows the double-pulse switching losses at 400 V dc-link voltage and 25 °C ambient temperature. Integration limits of the turn-on losses start at 10 % of rated current and end at a voltage of less than 10 % of rated voltage — and vice versa for the turn-off losses. The limits are shown in Fig. 6 as an example for the turn-on event of the device from GaN Systems at 400 V and 50 A.

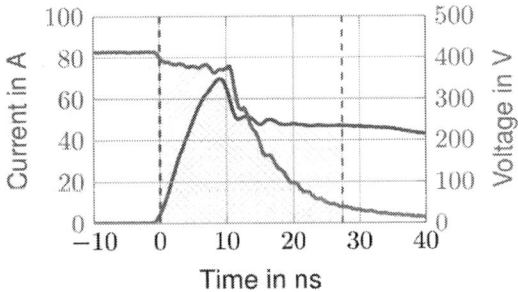

Fig. 6: Voltage and current waveforms of the GaN HEMT from GaN Systems for the turn-on event at 400 V and 50 A. The integration limits of the switching losses are shown with dashed lines.

All measurements of the GaN HEMTs show lower turn-off losses in comparison to the SiC MOSFET. In contrast to SiC MOSFETs, GaN HEMTs generally have no reverse recovery losses and the switching processes are faster [19]. However there are also major differences in the measurements between the GaN HEMTs. The semiconductor device from VisIC in particular has higher turn-off losses

than the other GaN HEMTs. The differences are even greater at turn on. Depending on the operating point, the semiconductor from VisIC has 4 to 7 times higher losses compared to the semiconductor from GaN Systems. However, the HEMT from VisIC is also designed for higher currents than the others (Tab. 1). A higher current carrying capacity requires more chip area, which means a larger output capacitance and output charge. Both add up to the total switching losses [20]. Also the voltage slope of this component had to be limited by an order of magnitude more than the others to avoid parasitic turn off during switching.

The HEMT from Infineon has the lowest total losses for currents of 10 to 20 A and the HEMT from GaN Systems for currents of 30 to 50 A as can be seen in Fig. 7. Both devices come in an SMD package with low parasitic inductance, which enables fast commutation of the current.

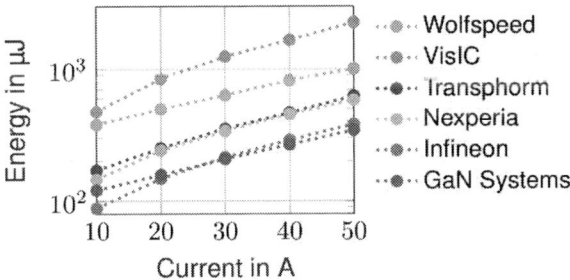

Fig. 7: Total switching losses at 400 V dc-link voltage

The measurements of the dynamic on resistance at 50 A are shown in Fig. 8 normalized to the static on resistances 500 µs after turn on of the respective devices. The devices were previously turned off for 2.5 µs at 400 V dc-link voltage. All measurements were taken with deskewed probes.

It is evident that the GaN HEMTs from GaN Sys-

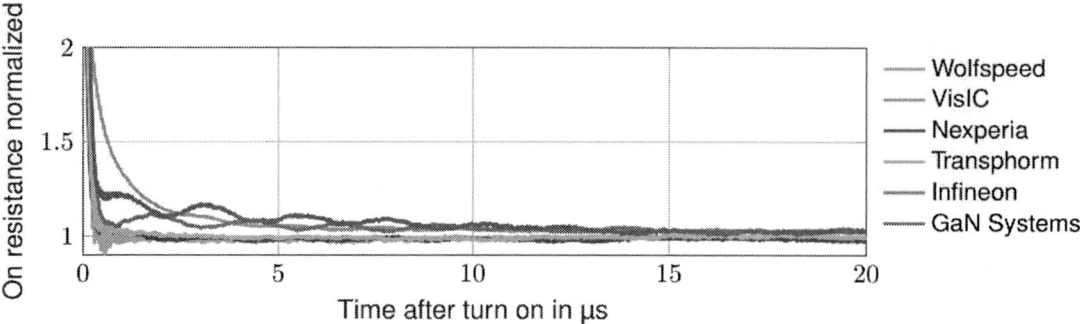

Fig. 8: Measurement of the dynamic on resistance after turn on at 50 A normalized to the measured on resistance 500 µs after turn on of the respective device at 25 °C ambient temperature

tems, Infineon and VisIC do not reach the static resistance in the first 10 µs. Only the GaN HEMTs from Nexperia and Transphorm (both cascode configuration) as well as the SiC device from Wolfspeed reach the static value almost immediately after turn on. Oscillations are caused by fluctuations in the dc-link voltage that couple into the gate voltage of the devices.

4 Conclusion

GaN HEMTs can push the limits of efficiency and power density in power electronic converters. However, detailed knowledge of the devices and their operating principles is required to achieve the best possible result. The measured GaN HEMTs are optimized for various parameters such as turn-on losses, turn-off losses, total losses, current capacity and short-circuit withstand time. When selecting, it is important to find the best possible compromise for the corresponding application.

The measurements show that all but one GaN HEMT exhibit lower total switching losses than the SiC semiconductor. In particular, the GaN HEMTs in SMD package have lower total switching losses compared to the others. However, the voltage slope of the GaN HEMT with the highest losses had to be limited more than the others in order to prevent parasitic switching off. The losses of these devices are therefore not directly comparable at the operating point under consideration.

For the on resistance of the HEMTs, it has been shown that the static data sheet values differ from their dynamic counterparts and are not suitable for calculating conduction losses for all devices, since the resistance is significantly increased shortly after turn on and takes up to several microseconds to decay to its final value.

Acknowledgment

Funded by the Federal Ministry of Education and Research (BMBF, FKZ03SFO596A), Flexible Electrical Networks (FEN) Research Campus.

References

[1] Udrea, F., "ICeGaN™ – a novel technology for integrated power GaN," Cambridge GaN Devices Limited, White Paper, 2022.

[2] Meneghini, M.; Vanmeerbeek, P.; Silvestri, R.; Dalcanale, S.; Banerjee, A., et al., "Temperature-dependent dynamic R_{ON} in GaN-based MIS-HEMTs: Role of surface traps and buffer leakage," *IEEE Transactions on Electron Devices*, vol. 62, no. 3, pp. 782–787, 2015. DOI: 10.1109/TED. 2014.2386391.

[3] Smith, K. V., "VisIC D3GaN reliability whitepaper," VisIC Technologies, White Paper, 2020.

[4] Chowdhury, D., "Power gallium nitride technology: The need for efficient power conversion," *IEEE Electrification Magazine*, vol. 8, no. 2, pp. 6–10, 2020. DOI: 10.1109/MELE.2020.2985480.

[5] Parikh, P.; Wu, Y.; Shen, L.; Barr, R.; Chowdhury, S., et al., "GaN power commercialization with highest quality-highest reliability 650V HEMTs-requirements, successes and challenges," in *2018 IEEE International Electron Devices Meeting (IEDM)*, 2018, pp. 19.7.1–19.7.4. DOI: 10.1109/ IEDM.2018.8614579.

[6] Chen, J.; Du, X.; Luo, Q.; Zhang, X.; Sun, P., and Zhou, L., "A review of switching oscillations of wide bandgap semiconductor devices," *IEEE Transactions on Power Electronics*, vol. 35, no. 12, pp. 13 182–13 199, 2020. DOI: 10.1109/TPEL. 2020.2995778.

[7] Di Maso, P. and Chen, D., "System level considerations with GaN power switching," presented at the APEC Industry Session (San Antonio, Texas, Mar. 17, 2018), 2018.

[8] *GS66516T top-side cooled 650 V e-mode GaN transistor datasheet*, Rev. 210727, GaN Systems Inc., 2021.

[9] Varajao, D. and Zojer, B., "Gate drive solutions for CoolGaN™ GIT HEMTs: Exploiting the full potential of GaN," White Paper, Nov. 2021. DOI: 10.13140/RG.2.2.14900.27526/1.

[10] Klever, S.; Thönnessen, A., and De Doncker, R. W., "Characterization of conventional and advanced current measurement techniques suitable for WBG semiconductor devices," in *2022 24th European Conference on Power Electronics and Applications (EPE'22 ECCE Europe)*, 2022, pp. 1–10.

[11] Klever, S. and De Doncker, R. W., "A high-bandwidth and low-inductive sensor for measuring the commutation current of WBG devices," in *2023 25th European Conference on Power Electronics and Applications (EPE'23 ECCE Europe)*, 2023, pp. 1–8. DOI: 10.23919/EPE23ECCEEurope58414.2023.10264674.

[12] Li, K.; Evans, P. L., and Johnson, C. M., "Characterisation and modeling of gallium nitride power semiconductor devices dynamic on-state resistance," *IEEE Transactions on Power Electronics*, vol. 33, no. 6, pp. 5262–5273, 2018. DOI: 10.1109/TPEL.2017.2730260.

[13] Abdullah, Y.; Li, H., and Wang, J., "Evaluation of 600 V direct-drive GaN HEMT and a comparison to GaN GIT," in *2017 IEEE 5th Workshop on Wide Bandgap Power Devices and Applications (WiPDA)*, 2017, pp. 273–276. DOI: 10.1109/WiPDA.2017.8170559.

[14] Saglam, B.; Aksit, M. H., and Tamyurek, B., "A comparison of GaN-based cascode and e-mode HEMTs using bridgeless totem pole PFC," in *2022 IEEE Energy Conversion Congress and Exposition (ECCE)*, 2022, pp. 1–6. DOI: 10.1109/ECCE50734.2022.9947785.

[15] Henn, J.; Lüdecke, C.; Laumen, M.; Beushausen, S.; Kalker, S., *et al.*, "Intelligent gate drivers for future power converters," *IEEE Transactions on Power Electronics*, vol. 37, no. 3, pp. 3484–3503, Mar. 2022. DOI: 10.1109/tpel.2021.3112337.

[16] Grau, V.; Wienhausen, A. H.; Kossek, M., and Doncker, R. W. D., "GaN-based full-bridge converter with digitally adjustable voltage slopes for characterization of interwinding insulation properties of magnetic high-frequency power components," *IEEE Transactions on Industry Applications*, vol. 57, no. 6, pp. 6288–6294, Nov. 2021. DOI: 10.1109/tia.2021.3112508.

[17] Gottschlich, J.; Kaymak, M.; Christoph, M., and De Doncker, R. W., "A flexible test bench for power semiconductor switching loss measurements," in *International Conference on Power Electronics and Drive Systems (PEDS)*, IEEE, Jun. 2015, pp. 442–448. DOI: 10.1109/PEDS.2015.7203495.

[18] Fritz, N.; Kamp, T.; Polom, T. A.; Friedel, M., and De Doncker, R. W., "Evaluating on-state voltage and junction temperature monitoring concepts for wide-bandgap semiconductor devices," *IEEE Transactions on Industry Applications*, vol. 58, no. 6, pp. 7550–7561, 2022. DOI: 10.1109/tia.2022.3191632.

[19] Buffolo, M.; Favero, D.; Marcuzzi, A.; De Santi, C.; Meneghesso, G., *et al.*, "Review and outlook on GaN and SiC power devices: Industrial state-of-the-art, applications, and perspectives," *IEEE Transactions on Electron Devices*, vol. 71, no. 3, pp. 1344–1355, 2024. DOI: 10.1109/TED.2023.3346369.

[20] Hou, R.; Lu, J., and Chen, D., "Parasitic capacitance E_{qoss} loss mechanism, calculation, and measurement in hard-switching for GaN HEMTs," in *2018 IEEE Applied Power Electronics Conference and Exposition (APEC)*, 2018, pp. 919–924. DOI: 10.1109/APEC.2018.8341124.

PCIM Europe 2024, 11– 13 June 2024, Nuremberg   DOI: 10.30420/566262216

Performance Evaluation of Deadtime and Gate Resistance for Parallel Connected GaN HEMTs

Junhyeok Jegal[1] , Minho Kwon[1] , Dongsul Shin[1] , Jong-Pil Lee[1]

[1] KERI(Korea Electrotechnology Research Institute), Republic of Korea

Corresponding author: Jong-Pil Lee, jplee@keri.re.kr
Speaker: Junhyeok Jegal, jgjh0421@keri.re.kr

Abstract

Gallium nitride (GaN) is gaining attention in the modern power electronics industry due to its superior performance. However, the lower current ratings of GaN devices serve as a bottleneck for scaling up power conversion system (PCS). While parallel connection of devices presents a straightforward approach to address this issue, comprehensive research in this area remains insufficient. This study aims to address several issues related to the parallel connection of GaN. Firstly, compared to Si devices, GaN has large losses due to high forward voltage during reverse conduction, so reasonable deadtime can reduce unnecessary losses. Secondly, it provides criteria for selecting resistance values from the perspective of driver ratings when using a single gate driver in the parallel structure of GaN. Finally, experiments were conducted to explore the efficiency limits achievable in the operation of three parallel units using the selected deadtime and gate resistance value. These research findings are expected to provide insights for researchers employing GaN-based PCS, particularly in setting limits for gate resistance selection and determining deadtime values when using GaN in parallel.

1   Introduction

The current trend in power electronics is to increase efficiency and reduce volume. However, as conventional commercial power converters based on silicon (Si) are gradually reaching their theoretical performance limits, it is difficult to improve efficiency and density [1, 2]. To overcome these challenges, wide-bandgap (WBG)-based devices are being developed, and among them, gallium nitride enhancement-mode high electron mobility transistors (GaN HEMTs) are attracting attention as a possible replacement for conventional Si MOSFET devices due to their superior switching performance, low ON-state resistance $R_{ds,on}$, and small package, as shown in Fig. 1 [3]. However, despite these advantages, commercial GaN devices have been mainly used in small-scale applications because their current rating is up to 60 A (@25°C). Therefore, it is essential to connect the devices in parallel in order to realize large-scale power conversion system.

Parallelization has been studied for a long time in order to increase the capacity of power conversion devices. For SiC MOSFETs, parallelization has

been studied for the past decade, while parallelization research for GaN devices has been slower. Nevertheless, there have been studies on high di/dt, dv/dt, and parasitic effects [4-6], on reducing losses under light load conditions [7], on parallelization based on positive temperature coefficient (PTC) characteristics [5, 8], on protection circuits in GaN with short SCWT for reliability [9], and on four-parallel studies using 60 A, the highest rating among currently commercialized devices [6, 10, 11].

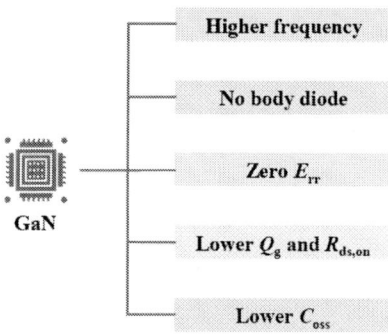

Fig. 1   Advantages of GaN HEMTs devices

As mentioned earlier, the superior characteristics of GaN devices have led to the expectation of high-efficiency, high-density power converters, but there are still several challenges in moving toward parallel connection-based, high-volume applications. First, optimization of fixed deadtime during complementary operation between switches is required. In the past, when using IGBTs, it was sufficient to set the deadtime longer than required. However, unlike MOSFETs, GaN devices do not have a body diode, as shown in Fig. 2, so the high forward voltage of GaN devices during the deadtime window causes unnecessary losses. As a result, modeling studies have been conducted to find the optimal deadtime value for GaN, but they are limited to single-switch, light-load conditions [12, 13]

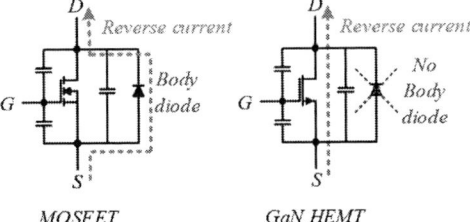

Fig. 2   Current flow in reverse conduction for

MOSFETs and GaN HEMTs

In addition, a key characteristic of GaN devices is PTC, which is an important factor in parallel structures. This is a characteristic that allows the temperature of GaN devices to be balanced between devices during parallel operation. However, there is a problem that the PTC characteristic does not appear during the reverse operation, which is the dead time period, as shown in Fig. 3, with a temperature difference of up to 41°C. Therefore, it is necessary to optimize the deadtime period when current flows in the reverse direction. The PTC characteristic causes a dominant effect on temperature sharing even if the parallel devices are temperature balanced [10].

Furthermore, there is a lack of research on parallel structures utilizing a single driver for GaN parallel structures and the optimal number of parallelism. In [14], a four-parallel drive that minimizes the loop is studied. This enables simultaneous turn-on, but it is a disadvantage in terms of driver price and density. Also, since there are limitations of GaN parallelism when utilizing a single driver, it is necessary to select the number of parallelisms considering the limitations. Therefore, research on deadtime and external gate resistance based on a single gate driver is needed to maximize the superior performance of GaN devices.

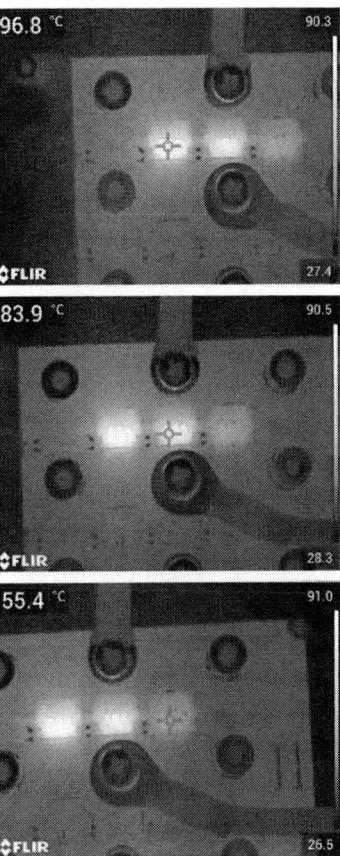

Fig. 3   Temperature imbalance due to reverse conduction in GaN parallel structures

In this paper, the reasonable driving conditions are selected by performing performance evaluation based on deadtime and gate resistance to utilize the advantages of GaN devices and to apply large-scale parallelization. For this purpose, an experimental evaluation was performed, and the case temperature Tc of the GaN device was operated below 100°C for the experiment. Unlike SiC devices, GaN devices have a relatively low T_j (T_j<150°C) and losses increase rapidly at higher temperatures, so temperature management is critical for parallel operation. The variables considered in this study have a direct impact on loss and temperature. In the case of deadtime, the experiments were conducted considering the load-dependent efficiency over the entire deadtime range, and a fixed deadtime was selected to account for the long transient time at the light side. In addition, for gate resistance, a single driver was utilized for high density, and the optimal gate resistance and number of parallel can be selected by considering the output current of the driver, and three parallel was finally selected.

2 Deadtime Performance

This section describes the losses that can occur in the deadtime region when utilizing GaN devices. This is a significant loss compared to conventional MOSFETs and should be minimized, especially at light load conditions, and the deadtime should be set based on the long transients.

2.1 Deadtime Principles

Figure 4 shows the schematic of the half-bridge converter used in this research. The high-side switch, Q_H, works as the control switch for operation in buck mode, while the low-side switch, Q_L, works as the rectifier switch. The GaN device used for the performance evaluation is GaN Systems' GS66516B, a 650 V device rated at 60 A at 25°C and 47 A at 100°C.

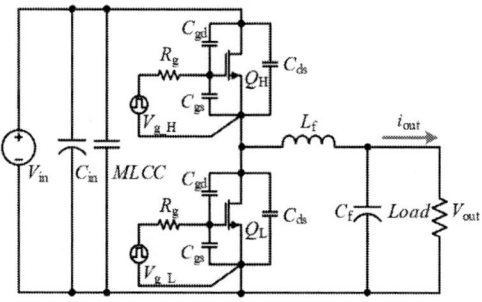

Fig. 4   Half-bridge converter prototype for parallel operation of GaN devices

A synchronous buck converter in half-bridge form, as shown in Fig. 4, is more efficient than an asynchronous buck converter when operating at high load and low duty conditions because it conducts through the switches during the diode conduction section. However, to avoid shoot-through or too large conduction losses in a synchronous buck converter, deadtime is essential for the gate signals of the complementary switches. Fig. 5 shows the deadtime as a function of the complementary gate voltages in a switching cycle. Here, a deadtime t_{don} is applied before the turn-on of the top switch Q_H and a deadtime t_{doff} is applied before the turn-on of the bottom switch Q_L. The losses incurred during the deadtime period can be simply expressed as Eq. (1), and it is necessary to minimize the conduction losses during this period by reducing the time of deadtime. However, if the deadtime is too short, both switches will be in the turn-on state due to overlap during some transient periods, and shoot-through current will flow during this period, resulting in a sharp decrease in efficiency.

$$P_{dead} = V_{ds} \cdot i_L \cdot f_{sw} \cdot (t_{don} + t_{doff}) \qquad (1)$$

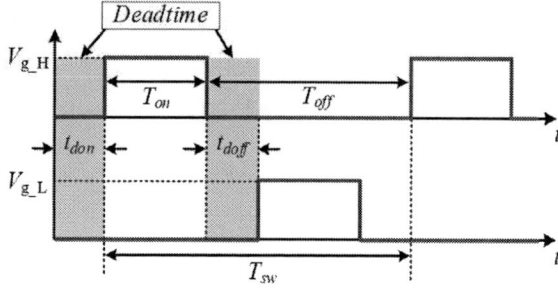

Fig. 5   Deadtime sequence of gate driver voltages

2.2 Consideration for Evaluation

Especially for deadtime evaluation, it is important to consider that the reasonable deadtime value varies depending on the output power. The experimental results in Fig. 6 show the transient voltage waveforms of the top and bottom switches under light and heavy loads. It can be seen that the transient section is longer than the heavy load in the light load condition. This is because the current flowing into the inductor becomes smaller when the current flowing into the load becomes smaller in the light load condition. As a result, the current charged into the output capacitance is smaller, so the dv/dt in the light load condition becomes smaller by Eq. (2). In this way, the transient slope changes depending on the load, so it is necessary to consider this and select the deadtime based on light load when hard-switching.

$$\frac{dv_c}{dt} = \frac{V_{in} \cdot i_{out}}{C_{oss}} \qquad (2)$$

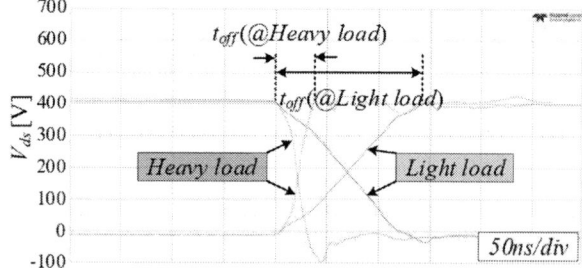

Fig. 6   Transient duration based on load conditions

3 External gate resistance performance

This section describes the parallelization of GaN devices using a single driver for external gate resistance selection. The gate resistance can be selected by considering the output current of the driver, which can improve the density of the driver part. However, when GaN devices are connected in parallel, the gate slope decreases and losses increase rapidly when utilizing a single driver due to the increase in input capacitance, so it is necessary to select the number of parallel devices considering this.

3.1 Single driver with Parallel GaN

When driving GaN devices in parallel, the external gate resistance must also be considered as it has a significant impact on the performance of the converter. Basically, the switching speed depends on the value of the gate resistance, increasing the gate resistance value will slow down the switching speed and increase the switching loss. Conversely, reducing the gate resistance value increases the switching speed but may cause the device to burn out due to surge voltage between the drain and source terminals due to wire stray inductance and other factors. In addition, the external gate resistance limits noise and ringing in the gate driver path and determines losses in the switching transient region, so the selection of optimal gate resistance is key to high-performance design. Therefore, it is necessary to select the optimal gate resistance in parallel GaN structures through performance evaluation based on gate resistance.

This can be illustrated by the configuration of the gate driver and GaN device in Fig. 7. The gate driver has a positive voltage V_{DD} for turn-on and a negative voltage V_{SS} for turn-off, and internal resistors R_{OH} and R_{OL} inside the driver. Then, the GaN devices in parallel can be simply represented as a single GaN device, and the device internals can consist of a parasitic capacitor and internal resistors. Finally, the parallel GaN device can be driven by an external gate resistor connecting the driver and the device. In this study, a gate driver (Si8271GB-IS) from Silicon Labs and a GaN device (GS66516B) from GaN Systems were used, with specifications as shown in Table 1 and Table 2. In conventional parallel operation, individual drivers are used due to current rating limitations and noise issues of the driver, but in this study, a single driver is used for performance evaluation according to GaN parallel structure and limitations of GaN device parallelization.

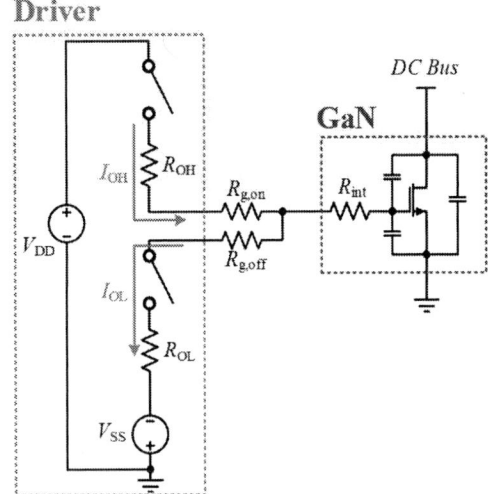

Fig. 7 Gate driver and GaN configuration

Description	Value
High output transistor R_{OH}	2.7Ω
Low output transistor R_{OL}	1.0Ω
High level peak output current I_{OH}	1.8A
High level peak output current I_{OL}	4.0A
On / off state gate voltage V_g	6 / -3V

Table 1 Key parameters of the gate driver (Si8271GB-IS)

Description	Value
Internal gate resistance R_{int}	0.3Ω
Drain-to-source blocking voltage V_{ds}	650V
Continuous drain current (T_c = 25°C)	60A
Continuous drain current (T_c = 100°C)	47A
Drain-to-source on resistance $R_{ds,on}$	25mΩ

Table 2 Key parameters of the GaN (GS66516B)

3.2 Consideration for Evaluation

In this study, a single-driver based parallel GaN was applied to evaluate the performance as a function of external gate resistance. Since a single driver is utilized, the gate resistance can be selected by considering the output current of the driver. Since the specifications of the selected driver are maximum source current I_{OH} = 1.8 A and maximum sink current I_{OL} = 4 A, the external gate resistor can be calculated from Eq. (3) and (4) by considering the margin as a value smaller than the driver output current, and the remaining variables can be obtained from the datasheet.

$$i_{OH} > \frac{V_{DD} - V_{SS}}{R_{OH} + R_{g,on} + R_{int}} \quad (3)$$

$$i_{OL} > \frac{V_{DD} - V_{SS}}{R_{OL} + R_{g,off} + R_{int}} \quad (4)$$

3.3 Simulation results

Based on the previous equation, the final external gate resistor values of $R_{g,on}$ = 3.3 Ω and $R_{g,off}$ = 1.5 Ω were selected, taking into account the driver peak current margin, depending on the current value allowed by the gate driver. The selected resistor-based three-parallel circuit can be represented by PSIM as shown in Fig. 8. The input capacitance of the GaN device was utilized from the datasheet. From this, Fig. 9 shows the simulation results for the gate voltage and current when the GaN devices are configured in parallel (1, 2, 3). For the same gate resistance value, as the number of parallel increases, the gate slope decreases as the input capacitance of the GaN device increases, which leads to switching losses. Therefore, based on the trade-off between switching losses and large capacity as the number of parallel increases, three parallelisms were finally configured.

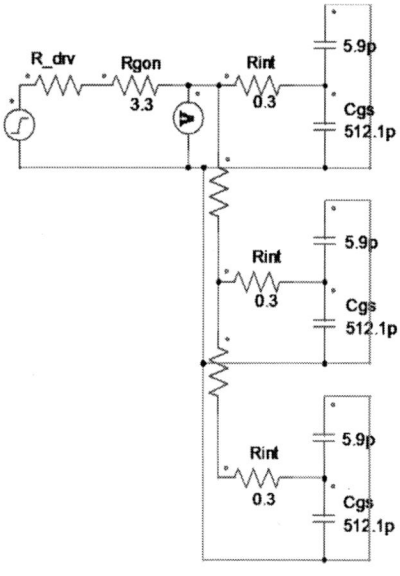

Fig. 8 PSIM schematic for comparison based on GaN device parallel configuration

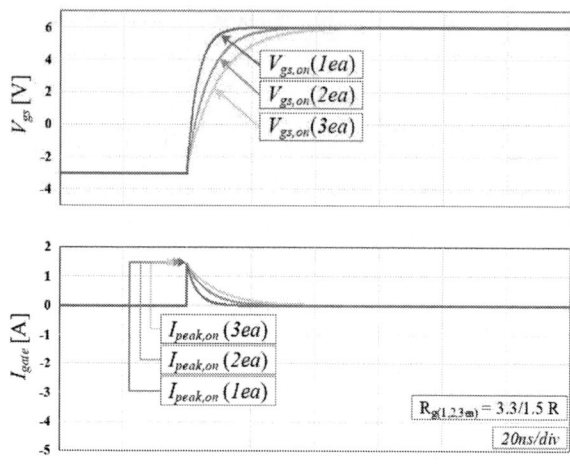

Fig. 9 Turn-on gate voltage and gate current waveforms as the number of parallels increases

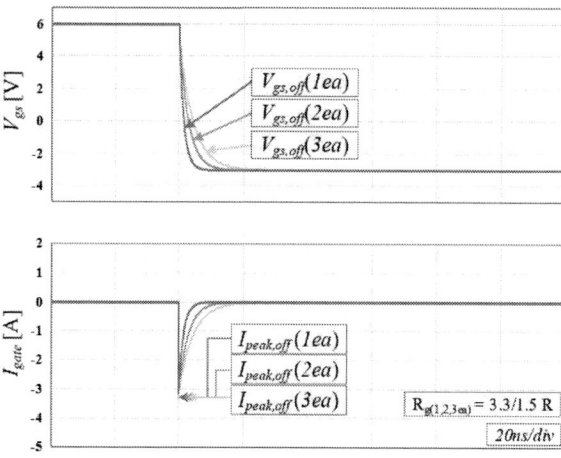

Fig. 10 Turn-off gate voltage and gate current waveforms as the number of parallels increases

4 Experimental Results

In this section, hardware capable of up to five parallel operations was fabricated for experiments on the two variables described above, and finally, the performance of the half-bridge-based converter was experimentally evaluated for the main variables using three parallel operations.

The test bench of the synchronous buck converter to evaluate the performance under parallel operation based on the previously considered deadtime and external gate resistor is shown in Fig. 10. YOKOKAWA (WT5000) power analyzer was used for efficiency measurements. The main specifications for the experiments are summarized in Table

3. The input voltage V_{in} was set to 400 V to account for the GaN voltage rating, and the output voltage V_{out} was limited to 140 V due to the voltage specification of the DC electric load (V_{out} < 150 V). Better efficiency can be expected by improving the specifications of the electronic load in the future. In addition, for excellent heat transfer performance, we minimized thermal resistance by using a metal PCB and a high-performance (18W/m-K) thermal interface material (TIM).

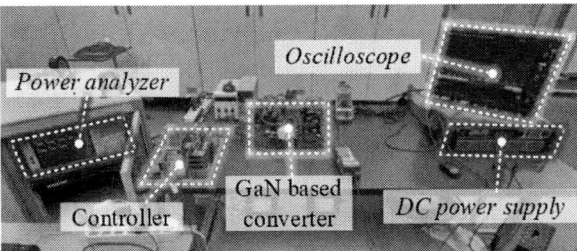

Fig. 11 Test bench for GaN-based parallel experimentation

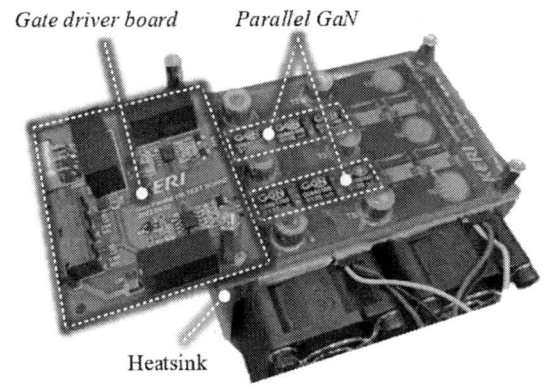

Fig. 12 Half-bridge converter prototype for parallel operation of GaN devices

Description	Value
Input voltage V_{in}	400V
Output voltage V_{out}	140V
Inductor L_f	340µH
Number of GaN	1~3ea

Table 3 Key Parameters of the gate driver (Si8271GB-IS)

4.1 Deadtime evaluation

For deadtime, the value of deadtime mentioned in the paper refers to the digital gating signal, not the output of the physical gate driver. Therefore, the actual gate driver output will have transients and will be shorter than the applied deadtime value of the digital gating signal. From this, the measured efficiency of the converter as a function of deadtime (0 ns to 8000 ns) for four different load conditions of different sizes (0.5 to 2.0 kW) can be plotted as shown in Fig. 13. Here, the white marks represent the actual experimental data and the solid lines represent the results by interpolation. The experimental results show that in the dead time region, which represents the antiparallel conduction time of GaN devices, the efficiency decreases as the conduction losses increase with increasing time. At high loads, such as 2 kW (blue circle), there is a limit to increasing the dead time due to the temperature rise of GaN. The bottom zoom-in in Fig. 13 shows that as the dead time approaches 0 ns, the efficiency decreases rapidly due to shoot-through during the transient between the top and bottom switches of the converter. Especially under light load conditions with small output current, it takes more time to charge the output capacitance C_{ds} during the transient, so sufficient deadtime is required to avoid the short-circuit region. Therefore, the evaluation was performed with a fixed deadtime (about 100 ns) based on a light load of about 0.5 kW. Also, negative deadtime was not tested, but it is expected that the efficiency will drop more sharply from the results of this experiment.

4.2 Gate resistance evaluation

For parallel driving of GaN devices, the external gate resistor was selected based on a single gate driver. Figure 14 shows that, similar to the previous simulation result (Figure 9), the slope of the gate voltage decreases as the number of parallel GaN devices increases with the same gate resistor value, and the input capacitance is experimentally confirmed to increase. In the case of three-parallel, the slope of the GaN devices is small, so the switching losses of all parallel devices increase rapidly. Therefore, three-parallel was selected based on the trade-off between the density improvement by utilizing only one driver and the efficiency of GaN. The efficiency measurement experiment utilized gate resistance values selected according to the number of parallel GaN devices (1-3ea) as shown in Fig. 15, and the efficiency was measured while increasing the load based on the device's case temperature T_c < 100°C.

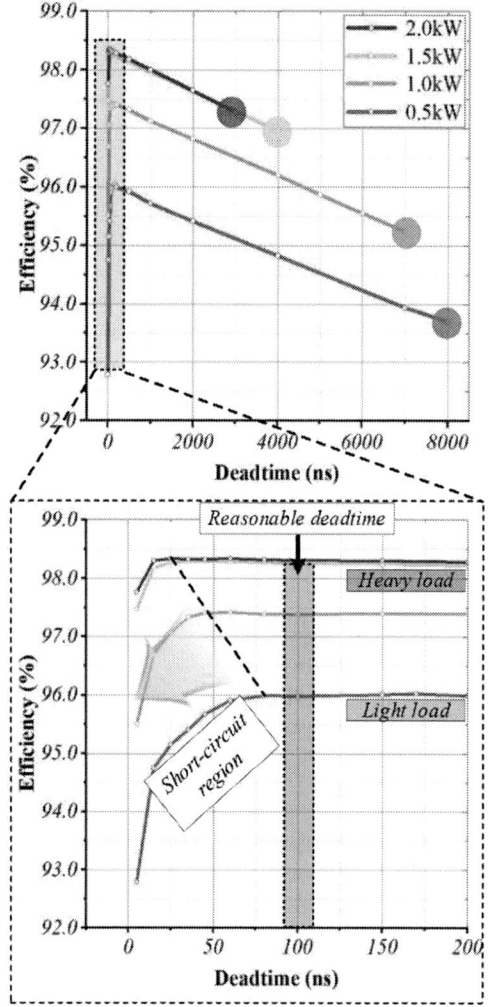

Fig. 13 Efficiency graph as a function of deadtime by load

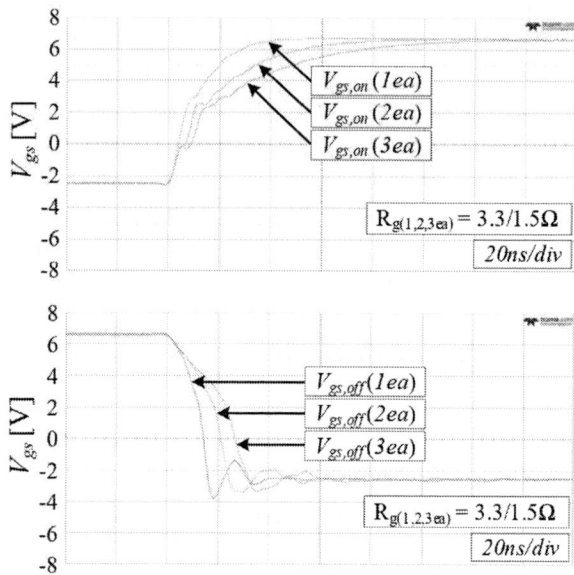

Fig. 14 Experimental results of gate voltage slope change with increasing the number of parallel

Efficiency measurements show that even at low duty (D=0.35) peak 98.7 % is achieved, which is better than existing papers with similar specifications [3]. Furthermore, the device can output up to 3.9 kW (28 A output current) when a single GaN device is configured. However, it can be seen that the output power does not increase proportionally with the number of parallel GaNs. This is because the parallel configuration utilizes a single gate driver, which causes large switching losses as the gate voltage decreases rapidly, as shown in Fig. 14. These losses cause the temperature of the GaN devices to rise, which also limits the allowable power as the number of parallels increases. In addition, parallel GaN configurations are more prone to thermal interference between switches than single GaN configurations, which can affect the allowable power.

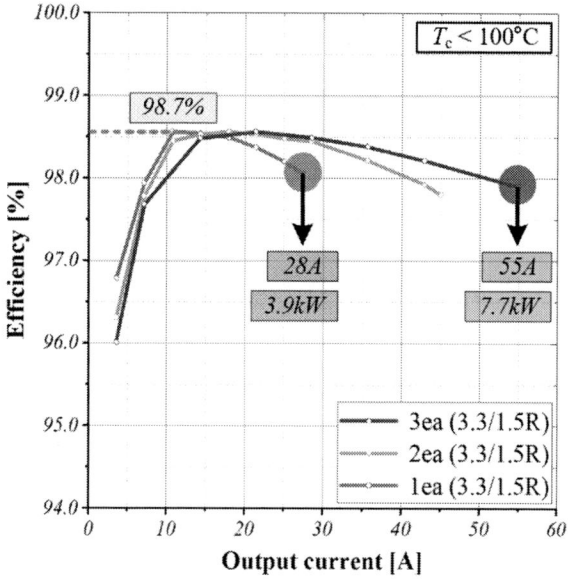

Fig. 15 Efficiency graph as a function of load by number of GaN devices

5 Conclusion

In this paper, GaN-based high-capacity converters were studied. For this purpose, deadtime and single driver-based external gate resistance were selected using GaN in a three-parallel configuration, and their performance was experimentally evaluated. For deadtime, the efficiency increases with shorter deadtime, but the efficiency decreases sharply when overlapping in the deadtime transient. Therefore, deadtime around 100 ns was selected. However, the optimal value may vary depending on the structure and device of the hardware, so related modeling studies are needed. For the external gate resistor, an external gate resistor was selected that considers the output peak current by utilizing a single gate driver. Experimental results show that an efficiency of up to 98.7% is possible, but switching losses increase rapidly due to the increase in input capacitance as the number of parallelism increases, so more than three parallelisms are irrelevant. From this study, it can be expected that other GaN-based converters with high capacity and high density in parallel operation can be achieved.

6 Acknowledgement

This work was supported by Korea Institute of Energy Technology Evaluation and Planning(KETEP) grant funded by the Korea government(MOTIE)(24A02026, Advanced Control & Protection Platform for Multi-terminal MVDC System and Engineering Design Protocol)

References

[1] C.-T. Ma and Z.-H. Gu, "Review of GaN HEMT Applications in Power Converters over 500 W," Electronics, vol. 8, no. 12, 2019.

[2] J.-H. Jegal, M. Kwon, C.-Y. Oh, K. Kim, and J.-P. Lee, "Implementation of three-phase four-leg inverter using SiC MOSFET for UPS applications," Journal of Power Electronics, vol. 21, no. 1, pp. 103-112, 2020.

[3] P. Prajapati and S. Balamurugan, "Leveraging GaN for DC-DC Power Modules for Efficient EVs: A Review," IEEE Access, vol. 11, pp. 95874-95888, 2023.

[4] Z. Liu, X. Huang, F. C. Lee, and Q. Li, "Package Parasitic Inductance Extraction and Simulation Model Development for the High-Voltage Cascode GaN HEMT," IEEE Trans. Power Electron., vol. 29, no. 4, pp. 1977-1985, 2014.

[5] J. L. L. D. Chen, "Paralleling GaN E-HEMTs in 10kW–100kW systems," 2017 IEEE Applied Power Electronics Conference and Exposition (APEC), 2017.

[6] P. P. Das, S. Satpathy, S. S. Shah, S. Bhattacharya, and V. Veliadis, "Paralleling of Four 650V/60A GaN HEMTs for High Power Traction Drive Applications," 2021 IEEE Energy Conversion Congress and Exposition (ECCE), pp. 5269-5276, 2021.

[7] Y. Shen, L. Shillaber, H. Zhao, Y. Jiang, and T. Long, "Desynchronizing Paralleled GaN HEMTs to Reduce Light-Load Switching Loss," IEEE Trans. Power Electron., vol. 35, no. 9, pp. 9151-9170, 2020.

[8] Y. Zhang, J. Li, and J. Wang, "Investigations on Driver and Layout for Paralleled GaN HEMTs in Low Voltage Application," IEEE Access, vol. 7, pp. 179134-179142, 2019.

[9] F. Karakaya, O. S. Alemdar, and O. Keysan, "Layout-Based Ultrafast Short-Circuit Protection Technique for Parallel-Connected GaN HEMTs," IEEE Journal of Emerging and Selected Topics in Power Electronics, vol. 9, no. 5, pp. 6385-6395, 2021.

[10] J. Lu, R. Hou, and D. Chen, "Loss Distribution among Paralleled GaN HEMTs," 2018 IEEE Energy Conversion Congress and Exposition (ECCE), 2018, 2018.

[11] J. Lu, Q. Tian, K. Bai, A. Brown, and M. McAmmond, "An Indirect Matrix Converter based 97%-efficiency On-board Level 2 Battery Charger Using E-mode GaN HEMTs " 2015 IEEE 3rd Workshop on Wide Bandgap Power Devices and Applications (WiPDA), 2015.

[12] D. Han and B. Sarlioglu, "Deadtime Effect on GaN-Based Synchronous Boost Converter and Analytical Model for Optimal Deadtime Selection," IEEE Trans. Power Electron., vol. 31, no. 1, pp. 601-612, 2016.

[13] Z. Qi et al., "An Accurate Datasheet-Based Full-Characteristics Analytical Model of GaN HEMTs for Deadtime Optimization," IEEE Trans. Power Electron., vol. 36, no. 7, pp. 7942-7955, 2021.

[14] F. Stalleicken, S. Dieckerhoff, K. Handt, and S. Nielebock, "Analysis of current sharing in the parallel connection of GaN transistors," 2022 24th European Conference on Power Electronics and Applications (EPE'22 ECCE Europe), 2022.

PCIM Europe 2024, 11– 13 June 2024, Nuremberg

DOI: 10.30420/566262217

Reaching Beyond 1200 V: Lateral GaN HEMTs for High-Reliability EV and Industrial Applications

Kamal Varadarajan[1], Alexei Ankoudinov[1], Robert Yang[1], Alexey Kudymov[1], Bhawani Shankar[1], Karthick Murukesan[1], Sorin Georgescu[1], John Rongavilla[1], Doug Kang[1]

[1] Power Integrations, USA

Corresponding author: Kamal Varadarajan, kamal.varadarajan@power.com
Speaker: Kamal Varadarajan, kamal.varadarajan@power.com

Abstract

The industry's first, true 1250 V-rated lateral GaN HEMT allows significant transient overvoltage capability making it ideal for high-voltage, high-reliability EV and industrial applications. Based on the PowiGaN™ technology platform of Power Integrations, the 1250 V GaN cascode device shows stable off-state leakage beyond 2000 V with a typical breakdown voltage of 2300 V. Reliability qualification data is presented confirming the advantage of the 1250 V device for such applications. In this work, we have also demonstrated high-efficiency power supply operation at high-input voltages using a 1250 V flyback switcher IC, confirming its readiness for real-world use.

1 Introduction of InnoSwitch with PowiGaN

GaN-based power semiconductor devices are ideally suited for high-efficiency power converters due to their superior material properties [1]. Lateral GaN HEMTs have been adopted into many commercial power supply designs covering a broad range of applications [2]. Devices available from various manufacturers typically offer voltage ratings <200 V and 650 V. Beyond 650 V, only a few manufacturers have released lateral GaN HEMTs with a rated voltage of 900 V [3-5]. Because of this, applications needing wide-bandgap power devices with 1200 V rating and beyond have been constrained to using SiC switches. However, cost is still a significant barrier to SiC adoption [6]. Besides improved efficiency and higher switching frequency, GaN offers the advantage of a significantly lower cost when compared to SiC [7], making it a compelling alternative even at 1200 V and beyond.

While 1200 V lateral GaN HEMTs have recently been reported as a cost-effective, high-performance platform for EV and industrial applications [8], the off-state drain to source leakage characteristics are shown to only be stable up to 1200 V, which leaves no margin for transient overvoltage conditions. Similarly, another lateral GaN HEMT device rated at 1200V has an actual breakdown of only 1500V leaving no margin for overvoltage capability [9]. Rapidly increasing off-state drain-to-source leakage current beyond 1000 V is also observed in a reported 1200 V GaN Polarisation Superjunction (PSJ) HFET [10]. Significant dynamic $R_{DS(ON)}$ increases ranging between 40% and 50% were observed even at a 600 V switching condition on the reported 1200 V GaN FET [11] making it unsuitable for applications that require a true 1200 V rating. In addition, none of these works have reported reliability qualification test results, which are essential to demonstrate readiness for field deployment.

Surge-robust, high-performance flyback power supplies with 750 V GaN-based InnoSwitch™ ICs operate reliably through surge events without any hard failure or change in converter efficiency over 90 VAC to 264 VAC line conditions [12]. Subsequently, flyback switcher ICs with 900 V GaN targeting industrial applications and 400 V-system automotive power supplies up to 100 W were introduced [4]. The AEC-Q100-qualified product family by Power Integrations is ideal for EVs based on 400 V bus systems where the 900 V GaN switch provides more power and increased design margin with enhanced efficiency over silicon-based converters. The efficiency improvement is valued both for range extension and thermal management.

PCIM Europe 2024, 11– 13 June 2024, Nuremberg

DOI: 10.30420/566262217

In this work, we present device level characterization, reliability qualification and power supply operation results of a 1250 V lateral GaN HEMT cascode device, also built on the PowiGaN technology platform. Development of the 1250 V-rated device now allows us to take advantage of the proven benefits of lateral GaN HEMTs in power conversion applications at a higher voltage. With this device, power supply designers can reliably specify an operating peak V_{DS} of 1000 V, while allowing for industry-standard 80% de-rating from the 1250 V device rating.

For industrial applications, this provides significant headroom and is particularly beneficial in challenging power grid environments where robustness is an essential defense against grid instability, surge, and other power perturbations. For EV, a reliable 1250 V lateral GaN technology can provide significant cost benefits over SiC in high-power 800 V system bus applications such as in an OBC.

2 Device structure and electrical characteristics

The 1250 V GaN is a normally-on, depletion mode device built using PowiGaN technology with a proprietary field plate architecture optimized to control peak electric fields across the device. It is connected in series with a low-voltage silicon MOSFET in a cascode configuration to achieve effective normally-off operation, which is essential for safe operation of power electronic systems. The circuit schematic for the cascode switch with the 1250 V GaN and low-voltage silicon MOSFET is shown in Fig. 1.

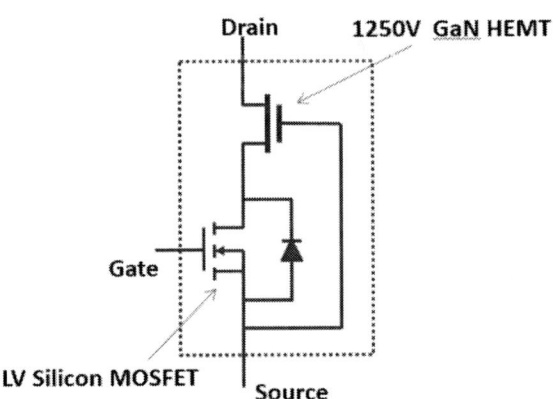

Fig. 1 Circuit schematic for the 1250V GaN cascode switch.

Pulsed output characteristics of a typical 330 mOhm, 1250 V GaN cascode device are shown in Fig. 2, illustrating high current capability.

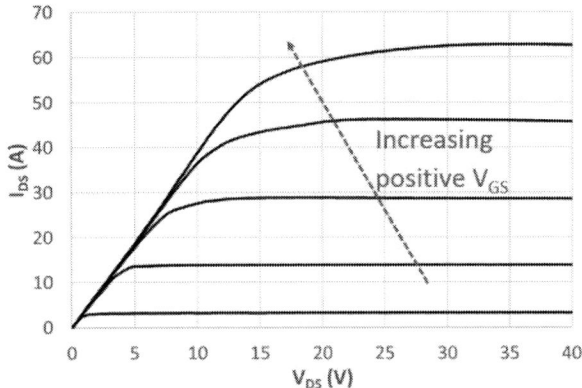

Fig. 2 Output characteristics of a 330 mOhm, 1250 V GaN cascode device.

Off-state characteristics of a typical 1250 V cascode device are shown in Fig. 3 illustrating stable leakage behavior to beyond 2000 V. This ensures that the device has excellent transient overvoltage capability, and demonstrates significant margin compared to silicon or SiC devices with similar voltage ratings.

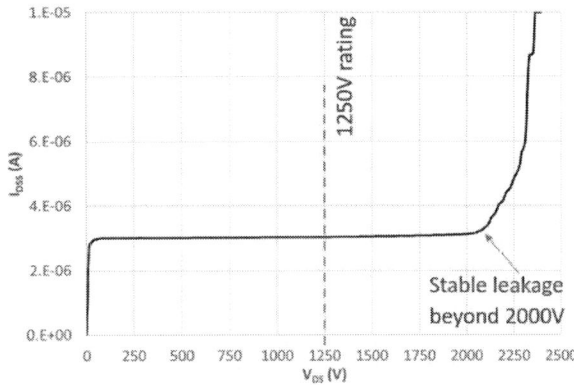

Fig. 3 Typical off-state characteristics demonstrating stable leakage beyond 2000 V.

At the product level, the 1250 V GaN cascode has been incorporated into an integrated flyback switcher IC belonging to the InnoSwitch™3-EP product family [13]. This InnoSwitch IC includes the 1250 V GaN cascode as the high-voltage primary-side switch, a primary-side controller, and a

1599

secondary-side controller for synchronous rectification as shown in Fig. 4. Inside the IC, secondary-side generated primary-switch timing requests are transferred across the safety isolation barrier using proprietary high-speed digital FluxLink™ technology [14].

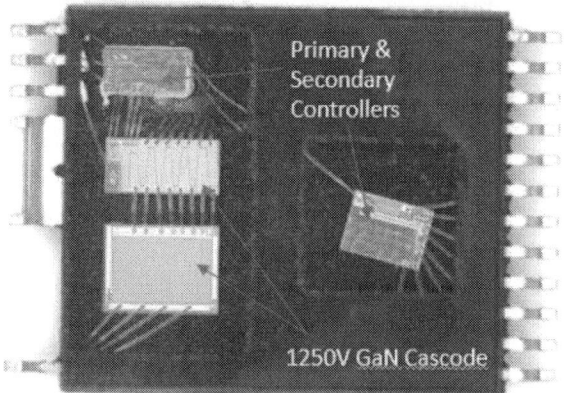

Fig. 4   InnoSwitch flyback switcher IC with the 1250V GaN cascode, primary, and secondary controllers.

3   Device reliability evaluation

In this section, results from key reliability qualification tests performed on the 1250 V GaN device in both the off-state and switching modes of operation are described.

High Temperature Reverse Bias (HTRB) is an off-state reliability stress test that evaluates the long-term stability of the power device under high drain-to-source bias and is intended to accelerate thermally activated failure mechanisms under high E-fields over an extended period. An HTRB stress test of the 1250 V GaN cascode device was performed in a 150°C ambient environment and stressed with an off-state V_{DS} of 1000 V (80% of 1250 V rating) for 1000 hrs with passing results in accordance with the specifications laid out in JEP198 [15]. JEP198 is the JEDEC guideline for reverse bias reliability evaluation procedures for gallium nitride power conversion devices. One of the most important parameters monitored during the course of the HTRB stress test is the off-state drain to source leakage current. The excellent device leakage stability is shown in Fig. 5, validating its suitability for high-reliability applications at high voltage.

In addition to the HTRB qualification test described above, intrinsic off-state failure rates were exposed by running tests on a large number of units under accelerated V_{DS} conditions (2100 V to 2200 V) across multiple temperatures (80°C to 120°C) as shown in Table 1.

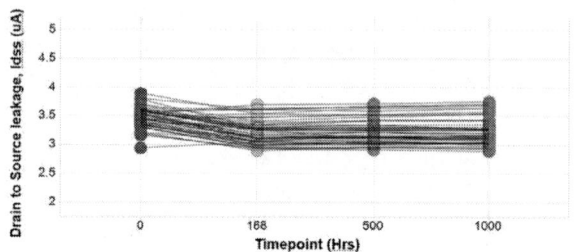

Fig. 5   Stable off-state drain leakage through 1000 hrs of HTRB stress at 1000 V / 150°C.

Voltage / Temperature	80°C	100°C	120°C
2100V		✓	
2150V	✓	✓	✓
2200V		✓	

Table 1 Accelerated evaluation conditions for exposing intrinsic failure rate in the off-state.

Based on the time to failure Weibull distribution obtained from this set of experiments, voltage and temperature acceleration factors were extracted and the projected failure rate under typical use conditions was calculated. As shown in Fig. 6, the model predicts a cumulative failure rate of 1 ppm in >15 thousand years of operation at 1000 V / 100°C indicating a significant built-in reliability margin for the 1250 V GaN cascode power device.

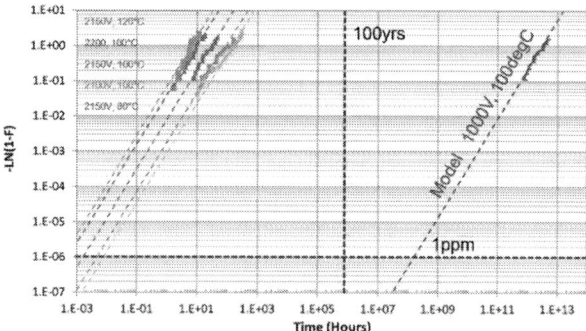

Fig. 6   Weibull distributions indicating time to failure under multiple accelerated stress conditions and the projected distribution at 1000 V / 100°C.

Switching reliability at the product level was evaluated using the 1250 V InnoSwitch flyback switcher IC with a Dynamic High Temperature Operating Life (DHTOL) test, conditions for which are specified in JEP 180 [16], which is the JEDEC guideline for switching reliability evaluation procedures for gallium nitride power conversion devices. A custom test-bed was developed to evaluate multiple units in parallel at 125°C (Fig. 7). The most stringent operating conditions possible were chosen — hard-switched turn-on and turn-off at 1000 V (80% of 1250 V). The hard-switching waveform seen in the evaluation is illustrated in Fig. 8.

Fig. 7 Custom test-bed used to perform DHTOL on multiple 1250V InnoSwitch ICs in parallel at 125°C.

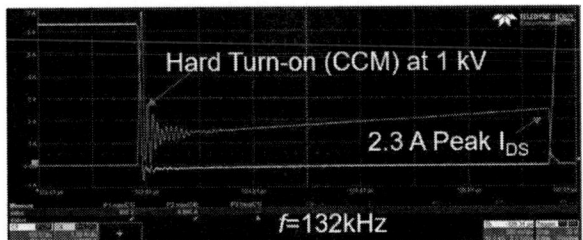

Fig. 8 Hard-switching waveform used during DHTOL evaluation of the 1250 V InnoSwitch IC.

For lateral GaN HEMT power devices, dynamic $R_{DS(ON)}$ during high-voltage switching is of particular interest. An improperly designed device will likely exhibit a significant increase in $R_{DS(ON)}$ during high-voltage switching transitions due to additional electron trapping, which can have a detrimental effect on the converter efficiency.

Fig. 9 Stable $R_{DS(ON)}$ observed during 1000 hrs of

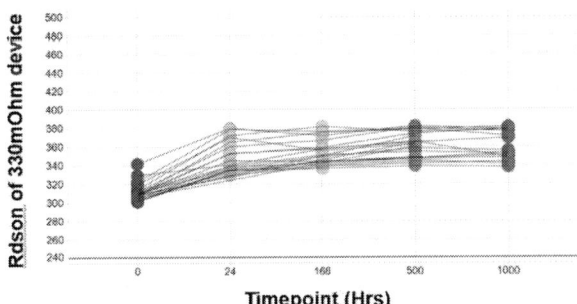

DHTOL with minimal shift (<20%).

$R_{DS(ON)}$ of the 1250 V GaN cascode monitored over the course of 1000 hrs of DHTOL stress at 1000 V / 2.3 A / 125°C is shown in Fig. 9, and shows stable performance from the earliest timepoint with a minimal shift of <20%. This result confirms the hard-switching capability of the Power Integrations

1250 V lateral GaN HEMT device and demonstrates its robustness to meet all potential high-reliability industrial and EV application use cases.

4 Power supply evaluation

The 1250 V InnoSwitch IC, introduced in Sec. 2, was evaluated in an automotive flyback power supply design using a 60 W, 24 V / 2.5 A test set-up described in RDR-919Q [17]. The schematic for the power supply design, detailed in RDR-919Q, is shown in Fig. 10, along with the top-side photograph of the circuit board used in the evaluation (Fig. 11).

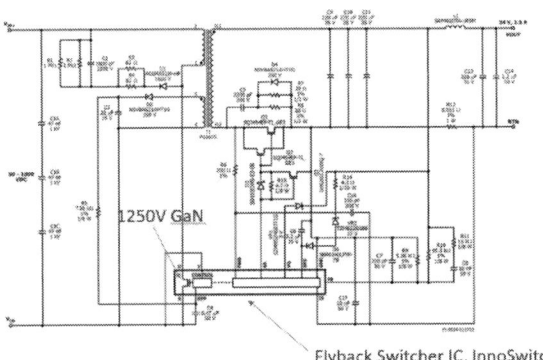

Fig. 10 Circuit schematic of the power supply design used to evaluate the 1250 V GaN flyback switcher IC.

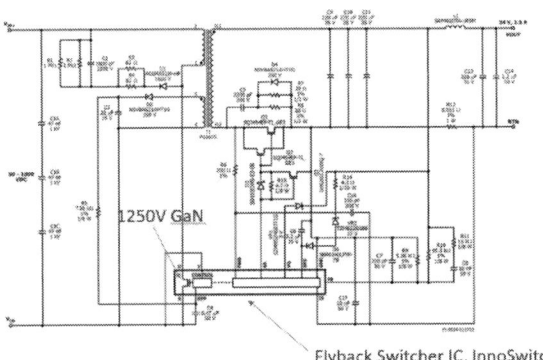

Fig. 11 Top-side photograph of the power supply design used in the evaluation.

Steady-state operation of the flyback power supply with a peak V_{DS} of 1000 V (80% of 1250 V) was evaluated using a DC input of 585 V. The corresponding switching waveforms are shown in Fig. 12. Temperatures measured after 2 hours of stable operation at this condition showed a maximum temperature of 54.6°C for the GaN HEMT, indicating a temperature rise of below 30°C compared to the 25°C ambient (Fig. 13).

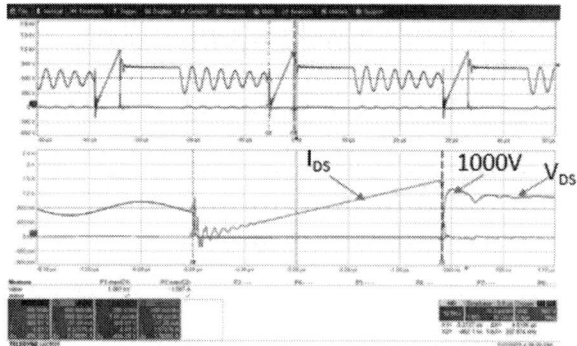

Fig. 12 Steady-state switching waveform illustrating peak V_{DS} of 1000 V.

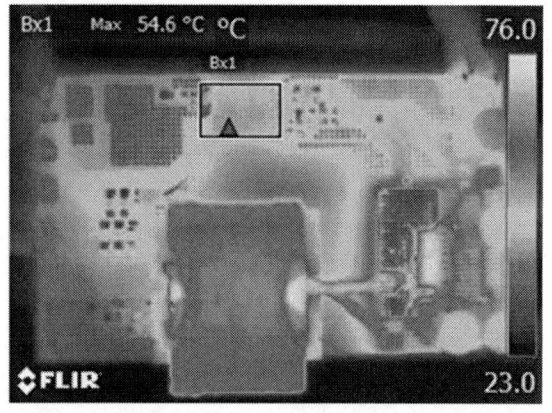

Fig. 13 Steady-state thermals indicating temperature rise below 30°C during continuous switching and a $V_{DS(PK)}$ of 1000 V.

While the GaN cascode is designed to operate with peak V_{DS} below 1000 V in a continuous mode, it can accommodate momentary increases in supply voltage up to 1250 V. This transient voltage capability of the GaN cascode allows it to continue to switch through the surge event for short periods without suffering a hard failure. This capability of the device was demonstrated by increasing the input voltage until the peak V_{DS} during switching reached the maximum rated value of 1250 V without any issues. An illustrative waveform showing the 1250 V V_{DS} stress is shown in Fig. 14.

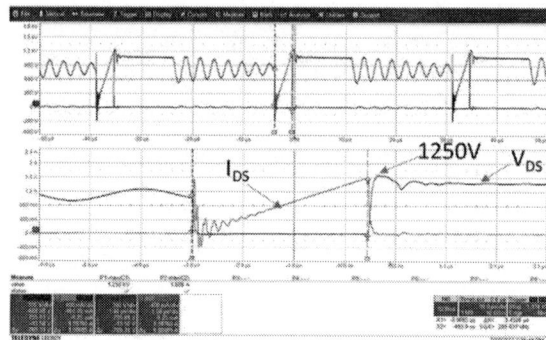

Fig. 14 Switching waveform illustrating transient overvoltage capability up to 1250 V.

Besides being extremely robust, the flyback power supply designed with the 1250 V InnoSwitch IC achieves best-in-class full-load efficiency exceeding 92% at 700 V input, as illustrated in Fig. 15.

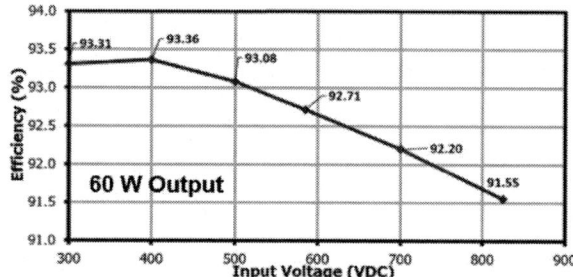

Fig. 15 Efficiency vs. input voltage

This power supply evaluation demonstrated the capability of 1250 V lateral GaN HEMTs as a credible alternative wide-bandgap technology to SiC in the 1200 V space and beyond. When higher switching frequency, improved efficiency and lower costs compared to SiC are factored in, lateral GaN HEMTs will become compelling in high-reliability EV and industrial applications.

5 Conclusion

This work introduces an industry-first, high-reliability 1250 V-rated lateral GaN HEMT and its readiness for real-world use at very high input voltage is illustrated by means of a high voltage flyback power supply design using a 1250 V InnoSwitch flyback switcher IC.

References

[1] M. Menegheni, G. Meneghesso, E. Zanoni, "Power GaN Devices: Materials, Applications and Reliability", Springer International Publishing, 2017.

[2] P. Waurzyniak, "GaN Application Base Widens, Adoption Grows", Semiconductor Engineering, 2021. https://semiengineering.com/gan-application-base-widens-adoption-grows/.

[3] Transphorm: 900 V GaN FETs; c2024. https://www.transphormusa.com/en/products/900v/.

[4] M. Di Paolo Emilio, "Power Integrations launches 900 V GaN flyback switcher ICs" Power Electronics News, 2023. https://www.powerelectronicsnews.com/power-integrations-launches-900v-gan-flyback-switcher-ics/.

[5] GaNPower International Discrete GanFET 900V Series. https://iganpower.com/gan-hemts

[6] M. Ahmad, "Silicon carbide's wafer cost conundrum and the way forward", EDN Asia, 2023. https://www.ednasia.com/silicon-carbides-wafer-cost-conundrum-and-the-way-forward/

[7] M. Beheshti, "Wide-bandgap semiconductors: Performance and benefits of GaN versus SiC", Analog Design Journal. Texas Instruments. 2022. https://www.electronicdesign.com/21256962

[8] G. Gupta, M. Kanamura, B. Swenson, C. Neufeld, T. Hosoda, P. Parikh, R. Lal, U. Mishra, "1200 V GaN Switches on Sapphire: A low cost, high-performance platform for EV & industrial applications", Proc. IEEE IEDM, 2022.

[9] 1200 V Enhancement-Mode GaN Power Devices Challenge SiC. https://www.how2power.com/newsletters/1912/products/H2PToday1912_products_GaNPower.pdf?NOREDIR=1

[10] Y. Du, H. Yan, P. Luo, X. Tan, E.M. Shankara Narayanan, H. Kawai, S. Yagi, H. Narui, "Investigation on Shift in Threshold Voltages of 1.2kV GaN Polarisation Superjunction (PSJ) HFETs", IEEE Trans. on Electron Devices. Vol. 70; No. 1, pp 178-184, January 2023.

[11] S. Li, Y. Ma, W. Lu, M. Li, L. Wang, Z. Zhang, T. Zhu, Y. Li, J. Wei, L. Zhang, S. Liu, W. Sun, "1200V E-mode GaN Monolithic Integration Platform on Sapphire with Ultra-thin Buffer Technology", Proc. IEEE IEDM, 2023.

[12] K. Varadarajan, S. Singamaneni, S. Kappala, "Surge-Robust Flyback Power Supplies with GaN", Proc. IEEE APEC, 2020.

[13] M. Di Paolo Emilio, "Power Integrations Releases Innovative 1,250 V GaN Switcher IC", Power Electronics News, 2023. https://www.powerelectronicsnews.com/power-integrations-releases-innovative-1250v-gan-switcher-ic/.

[14] L O'Toole, "The Ultimate Flyback: Power Integrations' FluxLink Conversion Technology", EEWeb. 2014. https://www.eeweb.com/the-ultimate-flyback-power-integrations-fluxlink-conversion-technology/

[15] JEP 198 JEDEC Standard, Guideline for Reverse Bias Reliability Evaluation Procedures for Gallium Nitride Power Conversion Devices. 2023.

[16] JEP180 JEDEC Standard, Guideline for Switching Reliability Evaluation Procedures for Gallium Nitride Power Conversion Devices. Ver 1.0. 2023.

[17] Power Integrations Applications Engineering Department. Reference Design Report for a 60 W Isolated Flyback Power Supply Using InnoSwitch3-AQ INN3949CQ. High Input Voltage for Automotive Application. 2023. https://www.power.com/sites/default/files/documents/rdr-919q_60w_high_input_voltage_psu_automotive_innoswitch3-aq-1700v_sic.pdf

PCIM Europe 2024, 11– 13 June 2024, Nuremberg   DOI: 10.30420/566262218

SmartSiC™ 150 & 200mm engineered substrate: increasing SiC power device current density up to 30%

Eric Guiot[1] , Frédéric Allibert[1], Jürgen Leib[2], Tom Becker[2], Oleg Rusch[2], Alexis Drouin[1], Walter Schwarzenbach[1]

[1] Soitec, France,
[2] Fraunhofer IISB, Germany

0009-0008-3971-3805

Corresponding author:	Eric Guiot, eric.guiot@soitec.com
Speaker:	Eric Guiot, eric.guiot@soitec.com

Abstract

The Smart Cut™ technology enables the integration of high quality SiC layer transfer for device yield optimization, combined with a low resistivity handle wafer (below 5mOhm.cm) to lower device conduction and/or switching losses both for 150mm and 200mm wafers diameter. Based on material characterisation, we anticipate a benefit of up to 15% or 30% in terms of R_{DSon} for state of the art 1200V SiC MOSFET and JFET. 1200V SiC diodes and MOSFETs have been fabricated by Fraunhofer IISB. 1200V diodes (JBS and MPS) with voltage drop improvement by 12% at rated current have been demonstrated.

1 Introduction

Silicon Carbide (SiC) technology has emerged as a crucial component in the realm of power electronics, playing a pivotal role in propelling electric mobility forward and enhancing the harnessing of renewable energy sources. The surging demand for SiC in the market has placed significant pressure on power semiconductor companies to rapidly scale up their production capabilities. Despite notable advancements in the quality and availability of 4H-SiC material, the quest for low defect density wafers to ensure high yields persists as a main challenge.

In response to this pressing need, a groundbreaking SiC engineered substrate has been introduced to address the industry's requirements [1-4].

2 Substrate description

The Smart Cut™ technology (fig.1) is a significant advancement in the field of semiconductor manufacturing, particularly for Silicon Carbide (SiC) devices. Here's a detailed breakdown of its features and benefits:

1.High-Quality SiC Layer Transfer: The technology enables the integration of high-quality SiC layer transfer, which is crucial for optimizing device yield.

2.Low Resistivity Handle Wafer: By incorporating a low resistivity handle wafer with a resistivity of less than 5mOhm.cm, the technology enhances device conduction and/or reduces switching losses. This feature is essential for improving the overall efficiency and performance of SiC-based devices.

3.Compatibility with Different Wafer Diameters: The Smart Cut™ technology is scalable to both 150mm and 200mm wafer diameters, providing flexibility in manufacturing processes and meeting various application requirements.

4.SmartSiC™ Engineered Substrate Composition (fig.2): The engineered substrate consists of a thin layer (400 to 800nm) of high-quality 4H-SiC bonded on top of a 350µm thick polycrystalline SiC (pSiC) handle wafer. This composition ensures structural integrity and performance while contributing to the reliability and efficiency of SiC devices.

5. Efficient Utilization of SiC Boule Materials: The reusability of initial single crystal donor wafers significantly enhances the efficiency of SiC boule material usage. Compared to conventional wafering technology, which typically yields a maximum of 50 wafers per boule, the Smart Cut™ technology enables the preparation of up to 500 wafers from the same boule. This substantial increase in yield represents a significant cost-saving and resource-efficient solution for SiC device fabrication.

Overall, the Smart Cut™ technology offers a comprehensive solution for optimizing SiC device manufacturing processes, enhancing device performance, and maximizing the utilization of SiC boule materials. These benefits make it a valuable advancement in semiconductor technology, particularly for applications requiring high-performance SiC devices.

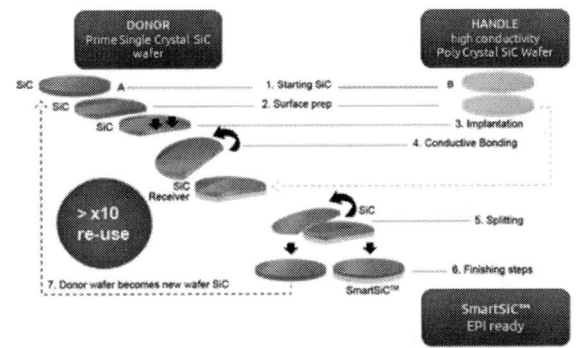

FIG. 1. SMART CUT™ TECHNOLOGY ADAPTED TO SILICON CARBIDE

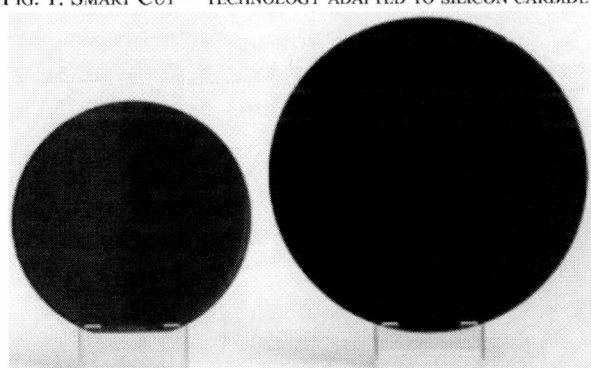

FIG. 2. 150mm & 200mm SmartSiC™ substrate ready for SiC drift epitaxy

Through the utilization of high material doping, the electrical resistivity of the pSiC substrate is significantly reduced by a factor of at least 4, up to 10, in comparison to conventional

single crystal SiC (mSiC) wafers. It guarantees an electrical resistivity of less than 5 mOhm.cm, with typical values falling within the range of 1 to 2 mOhm.cm. Employing CVD technology ensures consistent performance across wafers for substrate diameters of both 150mm and 200mm. This uniformity can be replicated using various material sources, which is a critical aspect for industrial applications, as depicted in figure 3.

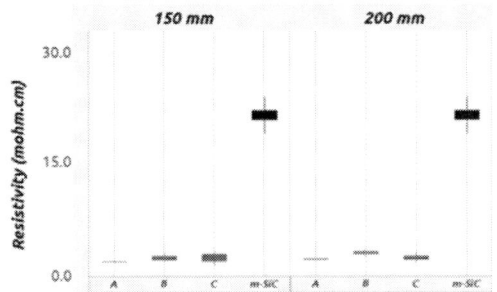

FIG. 3. ELECTRICAL RESISTIVITY DISTRIBUTION OF pSiC (A, B, C SOURCES) COMPARED TO mSiC SUBSTRATE.

The electrical conductivity achieved through high material doping remains consistent from room temperature (25°C) to the maximum operating temperature of devices (175°C), as indicated in Table 1. This table compares the electrical resistivity of single crystal SiC, pSiC, and the SmartSiC™ wafer bonding interface across the entire temperature range [5].

Temperature	25°C	175°C
4H-SiC bulk [mOhm.cm]	$15 - 25$	$15 - 25$
Bonding Interface [mOhm.cm^2]	$0.003 - 0.006$	$0.002 - 0.006$
Handle pSiC bulk [mOhm.cm]	$1.5 - 2.2$	$1.7 - 2.5$

TABLE 1: ELECTRICAL RESISTIVITY FOR EACH SmartSiC LAYER, FROM 25°C TO 175°C

Polycrystalline SiC handle wafers are prepared through the chemical vapor deposition (CVD) technique. This process is much more energy-efficient than the conventional physical vapor transport (PVT) used to manufacture mSiC wafers, cutting CO_2 emissions of the final SmartSiC™ product by at least 70% compared to conventional single crystal SiC.

3 Device experimental conditions

The fabrication of 0.09mm² 1200V JBS diodes and MPS diodes (with areas of 2.5 and 6.12mm²: see figure 4) has been performed in collaboration with Fraunhofer IISB. In parallel

PCIM Europe 2024, 11– 13 June 2024, Nuremberg   DOI: 10.30420/566262218

1200V SiC planar MOSFETs are also being fabricated.

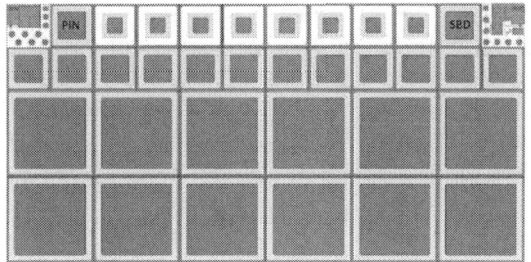

FIG. 4: TEST MASK FOR SiC DIODES FABRICATION.

Diode devices were fabricated with both commercial 150 mm 4H-SiC wafers and 150mm SmartSiC™ engineered substrate processed in parallel for both epitaxy and device fabrications. The drift layer is an n-type epitaxial layer of 11μm thickness and 1.10^{16}cm^{-3} doping concentration. This epitaxial layer is designed for a 1200 V blocking capability. The die thickness was reduced down to 340μm. The rated current ranges from 1A for 0.09mm² JBS diodes, 4A and 9A for respectively 1.6x1.6mm and 2.5x2.5mm MPS diodes. P+ doped layers are 500nm deep with a width of 2.5μm and 2.0μm respectively for JBS and MPS diodes.

For the front side metallization (fig.5), we deposited (through evaporation) 80 nm Ti and 300 nm Al on the p+-regions and annealed via RTP (980°C during 2min under Argon). On top (and on the n- region) we sputtered the power metal (50 nm Ti, 3500 nm Al and 20 nm Ti), which also builds the Schottky contact on the not-implanted epi layer.

For the back side metallization, we deposited (through sputtering) 60 nm NiAl$_{2.6\%}$ and use laser annealing on 4H-SiC to create the ohmic contact, on top of that we sputter 2000 nm Al and then evaporate the solder stack – 100 nm Cr, 1000 nm Ni and 1000 nm Ag. For the SmartSiC substrate, no laser annealing was performed.

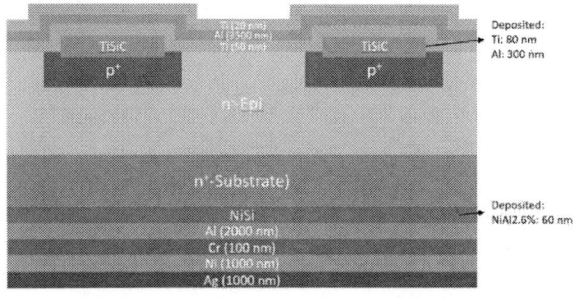

FIG. 5. METAL STACKS FOR FRONT AND BACK SIDE.

4   Device results

4.1   JBS diodes

For 1200V JBS diodes, we have performed wafer level 300x300μm dies forward (see fig.6) and reverse characteristics (see fig.7). No critical change in the reverse characteristics is observed. Forward characterization is showing forward voltage drop lowering at the rated voltage.

Reverse I-V characteristics of SmartSiC™ based SiC diodes are shown in Fig.8 for JBS diodes. Despite some device failures below the targeted voltage, the tested diodes reached the targeted voltage with a current leakage below the specifications.

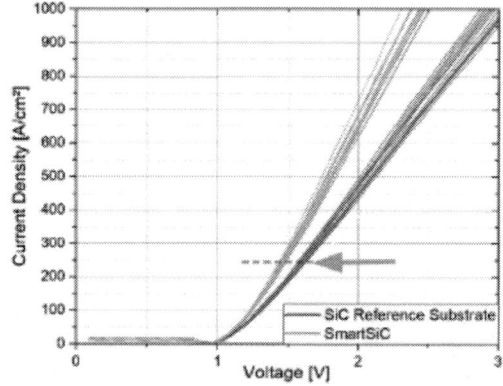

FIG. 6. FORWARD RESISTANCE OF RESPECTIVELY STANDARD SiC AND SMARTSiC™ FOR JBS DIODES.

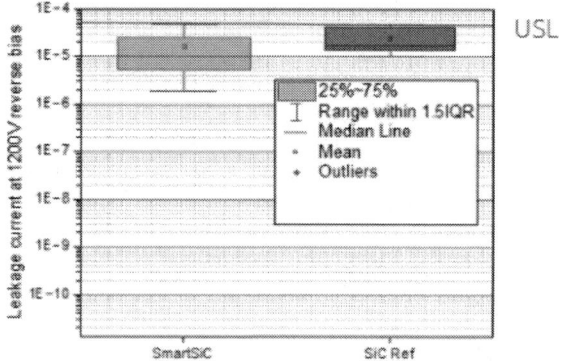

FIG. 7. REVERSE BIAS LEAKAGE AT 1200V OF RESPECTIVELY STANDARD SiC AND SMARTSiC™ FOR JBS DIODES.

1606

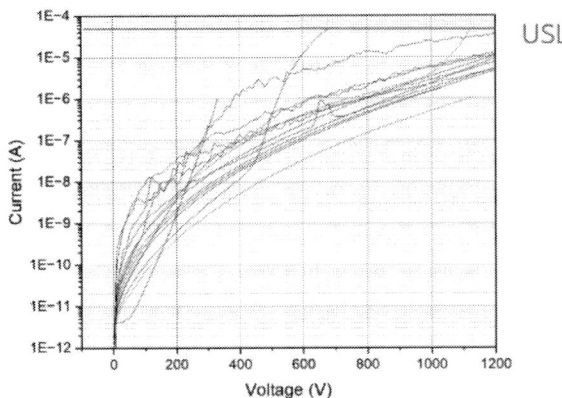

FIG. 8. REVERSE BIAS LEAKAGE OF SMARTSIC™ BASED JBS DIODES.

4.2 MPS diodes

For 1200V MPS diodes, we first report wafer level dies forward characteristics. Voltage drop lowering at the respectively rated current of 4 and 9A for 1.6x1.6mm (see fig.9) and 2.5x2.5mm (see fig.10) MPS diodes, is around 12%. A first extraction of the dynamic on resistance of the forward regime (2 to 10A) of the MPS diodes leads to a benefit (linked to SmartSiC™ transition) around 0.9mOhm.cm². This is even beyond the expected gain linked to the improved material resistivity: around 0.77 mOhm.cm². This further gain is supposed to be linked to the lower contact resistance of the back side contact coming from the high level of doping in the pSiC [3].

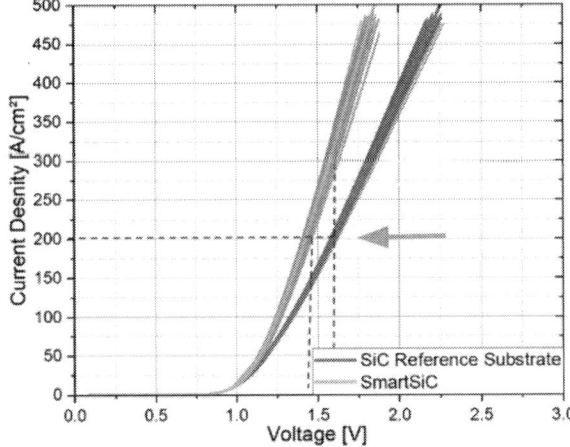

FIG. 9. FORWARD RESISTANCE OF STANDARD SiC AND SMARTSIC™ FOR 1.6x1.6mm 6A rated MPS DIODES.

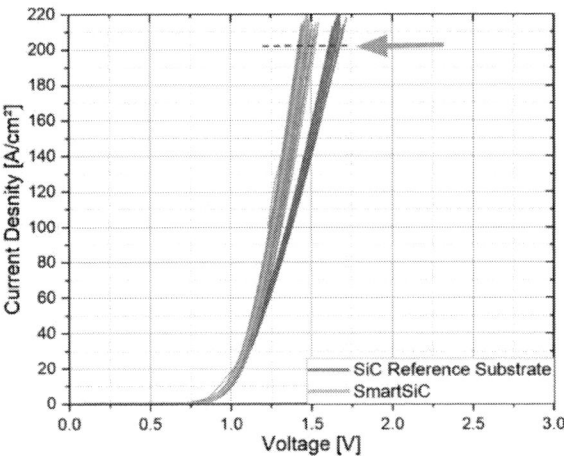

FIG. 10. FORWARD RESISTANCE OF STANDARD SiC AND SMARTSIC™ FOR 2.5x2.5mm 10A rated MPS DIODES.

Reverse bias leakage at 1200V (see fig.11) is slightly higher for the SmartSiC™ samples . This is attributed to the 20% higher doping level which lower the breakdown voltage.

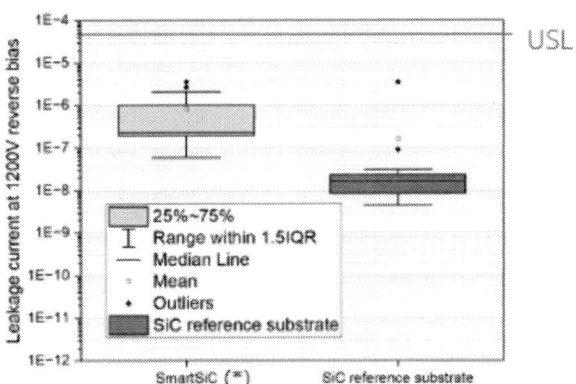

(*) 20% higher n- doping concentration

FIG.11. REVERSE BIAS LEAKAGE AT 1200V OF RESPECTIVELY STANDARD SiC AND SMARTSIC™ FOR MPS DIODES.

Reverse I-V characteristics of SmartSiC™ based SiC diodes are shown in Fig. 12 for PIN diodes. Again, despite some device failures below the targeted voltage, the tested diodes reached the targeted voltage with a current leakage below the specifications. This is showing that SmartSiC™ engineered substrate is fully compatible with SiC power device fabrication and reaches the targeted device leakage levels.

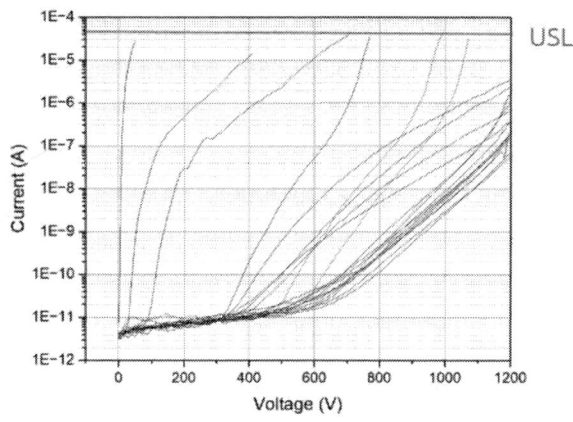

FIG. 12. REVERSE BIAS LEAKAGE OF SMARTSIC™ BASED PIN DIODES.

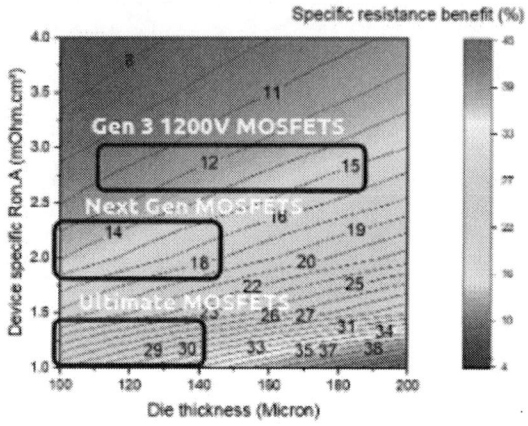

FIG. 13. RON.A BENEFIT ENABLED BY SMARTSIC™ USE FOR 1200V SiC MOSFET.

BIPOLAR DEGRADATION BENEFITS OF SMARTSIC™ MATERIAL

5 Device performance perspectives

SmartSiC™ enables an increase of the current density at the device level thanks to both the polycrystalline SiC material's low resistivity and the lowered back side contact resistance. Compared to standard single crystal SiC substrate with a material electrical resistivity around 20mOhm.cm, the polycrystalline SiC material can reach material resistivity as low as 2mOhm.cm. In parallel, thanks to the high doping level of polycrystalline SiC, the contact resistance is lowered around 10-50 µOhm.cm². Associated with substrate selection and/or specific surface preparation prior to Smart Cut™ that will guarantee the best yield, the technology will enable a lowering of the total cost of ownership of power devices.

The benefit of the electrical parameters can be calculated as a function of device specific resistance (Ron.A) at room temperature and die thickness. We clearly see that for Ron.A improvement up to 15% for state of the art 1200V SiC MOSFETs and up to 18% for next generation 1200V SiC MOSFETs, can be envisioned (see fig.13). Considering ultimate SiC MOSFET or state of the art JFET, improvement up to 30% is anticipated.

Bipolar degradation of post epitaxy SiC substrates without diode processing was carried out on reference and engineered SiC substrates using the E-V-C technique developed by ITES, Co. (Japan) [8]. The results suggest that the SmartSiC™ design possesses an inherent advantage over bulk material in terms of robustness against bipolar degradation. This characteristic was previously evaluated through a forward-current stress test conducted on a 4H-SiC epitaxial layer subjected to proton irradiation [7]. It appears that both the number of SSFs and their typical size are lower in the case of SmartSiC™ compared to bulk. We are currently assessing the above characterized MPS diodes under high current density with short pulses to directly check the bipolar degradation [9].

6 Conclusion

In conclusion, we are demonstrating that the SmartSiC™ engineered substrate solution available in 150 and 200mm diameters will bring a higher current density (up to 30%) both for state of the art SiC MPS diodes and MOSFETs (either planar or vertical). Besides modeling, SiC devices have been fabricated and diodes (JBS and MPS) with voltage drop improvement by 12% at rated current have been demonstrated. Finally we have demonstrated post epitaxy robustness with regards to bipolar degradation. This will be further assessed at device level.

Acknowledgements

This work is supported by the H2020 - ECSEL JU programme of the European Union under the grant of the TRANSFORM project 'Trusted European SiC Value Chain for a greener Economy' (ECSEL JU Grant No. 101007237).

References

[1] https://www.eetimes.eu/soitecs-smartsic-to-hit-the-road-in-2024/

[2] S. Rouchier et al., in Materials Science Forum 1662-9752, Vol. 1062, pp 131-135

[3] T. Shimono et al., Intl. Conf. Silicon Carbide and Rel. Mat. 2021

[4] E. Guiot et al., in Materials Science Forum Vol. 1092, pp 201-207

[5] H. Biard et al., in Materials Science Forum Vol. 344, pp 47-52

[6] Y. Igarashi et al., Defect and Diffusion Forum. 425. 75-82.

[7] Harada et al., Scientific Reports. 12. 13542.

[8] A. Drouin et al, ICSCRM 2023, to be published in Material Science Forum

[9] S.Laha et al., CIPS 2024, Milliseconds Power Cycling (PC_{msec}) driving bipolar degradation in Silicon Carbide Power Devices.

PCIM Europe 2024, 11– 13 June 2024, Nuremberg DOI: 10.30420/566262219

Dynamic Transients in High-Voltage Silicon and 4H-SiC NPN Bipolar Junction Transistors

Mana Hosseinzadehlish[1], Saeed Jahdi[1], Xibo Yuan[1], Jose Ortiz-Gonzalez[2], Olayiwola Alatise[2]

[1] University of Bristol, Bristol, UK
[2] University of Warwick, Coventry, UK

Corresponding author: Mana Hosseinzadehlish, m.hosseinzadeh@bristol.ac.uk

Topic: A Power Semiconductors, A05 SiC Devices
Preferred presentation form: Oral presentation

Abstract

In this paper, dynamic characteristics of two bipolar junction transistors as 800 V, 20 A Silicon BJT and a 1.7 KV, 15 A 4H-SiC BJT have been explored. Our investigation covers dynamic transients at high temperatures and different base resistances. The experimental measurements conducted at 800 V with a collector current of 7 A and temperatures reaching up to 175 °C reveal that 4H-SiC BJTs outperform their silicon counterparts. They exhibit higher current gain and maintain temperature-invariant dynamic transients. We also employ Silvaco TCAD simulations to analyze the dynamic response of NPN SiC BJTs under double-pulse transients. Notably, the Silicon BJT displays significant delays in both turn-ON and turn-OFF transitions when compared to the 4H-SiC BJT leading to increased switching energy and losses.

1 Introduction

Within the domain of semiconductor devices, current-driven bipolar Junction Transistors (BJTs) have been introduced in two forms of lateral and vertical configurations. Among these, high voltage silicon vertical BJT has faced a challenge in the realm of power electronic applications. This challenge is primarily explained by its low DC current gain, leading to the requirement for complex base drivers [1]. Consequently, 4H-SiC BJTs have been developed by GeneSiC to overcome this limitation due to their considerably higher current gain. This can be attributed to the reduced dimensions of the base and drift regions, resulting from the wider bandgap and higher critical electric field when compared to silicon counterparts, which, in turn, leads to a higher level of emitter-collector injection efficiency [2], [3]. Fig. 1 depicts the cross-sectional schematic of vertical power NPN BJTs in both Silicon and 4H-SiC and their equivalent simulated TCAD models. It can be seen that emitter region is highly doped (in range of $\times 10^{19}$ cm^{-3}) in 4H-SiC BJT in order to increase the efficiency of carrier injection in PN junction of emitter-collector to achieve a higher current gain This paper investigates the performance of 4H-SiC BJTs compared to vertical silicon BJTs, focusing on dynamic switching transients across a wide temperature range (25 °C to 175 °C) and various base resistances through both experimental tests and Silvaco TCAD modelling analysis.

2 Test setup and experimental results

Fig. 2 illustrates the double pulse test circuit and setup, respectively. A double pulse with charging pulse length of 35 μs and switching pulse length of 8 μs is applied to the silicon and 4H-SiC NPN BJTs in order to investigate the impact of temperature rise and switching rate on their transient behavior. Duration of the charging pulse is adequate to increase the current to 7 A for the load inductor size of 4 mH. The applied voltage to the devices is equal to 800 V. Collector and base currents of both devices in the double-pulse test are shown in Fig. 3 at different temperatures. It is obvious that there is a significant turn-OFF delay between the base current and drop of collector current in silicon BJT when compared to the 4H-SiC BJT as a result of higher storage time, a phenomenon that further intensifies with increasing temperature.

1610

PCIM Europe 2024, 11– 13 June 2024, Nuremberg   DOI: 10.30420/566262219

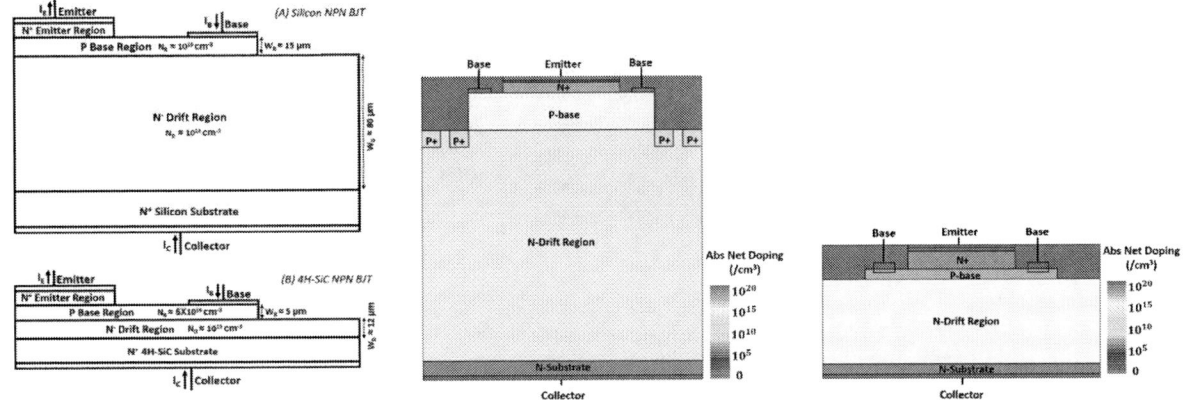

Fig. 1: Cross-sectional schematics, and TCAD models of (center) Silicon & (right) 4H-SiC BJTs.

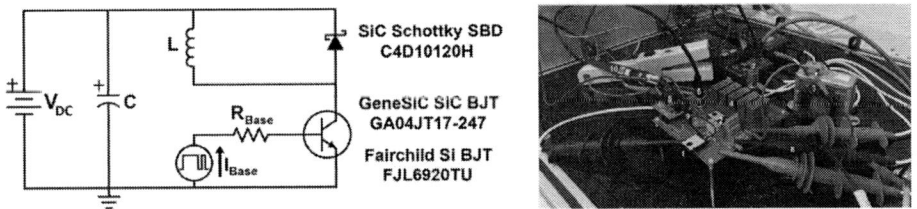

Fig. 2: Setup of double-pulse test: 1. DUTs, 2. Schottky Diode, 3. Load Inductor, 4. DC-link Capacitors, 5. Base driver, 6. Input Signal, 7. Power Supply, 8. Voltage Probes, 9. Rogowski Coil, 10. Current Probe.

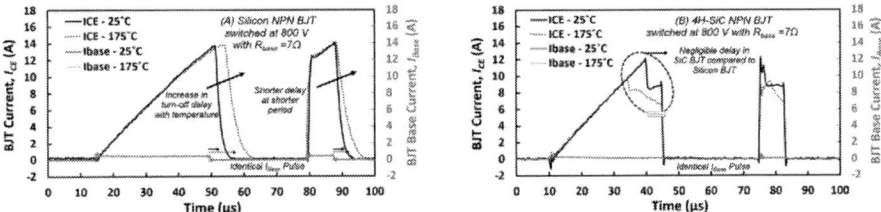

Fig. 3: Collector and base current of (A) Silicon BJT and (B) 4H-SiC NPN BJT with the R_{Base}= 3.75 Ω indicating the delay influenced by temperature under the same base current.

3 TCAD Validation of Experimental Measurements

To validate the TCAD simulation models of NPN BJTs, a comparisons have been done between the static electrical characteristics of the TCAD models and the real devices provided in their datasheets. Proper definition of doping profiles and device dimensions is crucial to achieve a good match between simulation and real devices. Fig. 4 illustrates a comparison between the TCAD Modelling and measurement of IV characteristics of Silicon and 4H-SiC NPN BJT as per datasheets which are almost close supporting further analysis of dynamic performance.

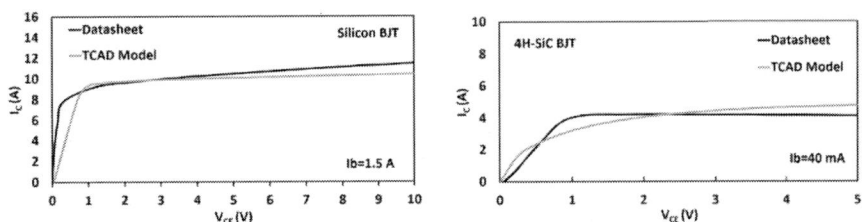

Fig. 4: TCAD model of IV characteristics of (left) Silicon NPN BJT for I_b= 1.5 A & (right) 4H-SiC NPN BJT for I_b= 40 mA, both compared with datasheet.

1611

4 TCAD Simulation Analysis

Fig. 5 demonstrates a comparison of the current and voltage of the modelled 4H-SiC NPN BJT under double-pulse test and the experimental measurements at 25 °C.

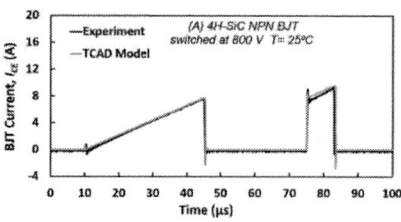 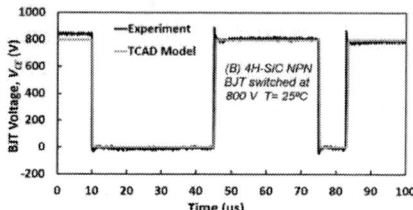

Fig. 5: Comparison of Modeled and Measured collector (A) Current and (B) Voltage of 4H-SiC NPN BJT under clamped inductive switching test at 800 V & 25 °C.

TCAD model and experimental measurements of the turn-OFF and turn-ON transients of 4H-SiC BJT are also compared in Fig. 6 and Fig. 7, respectively which demonstrate a clear alignment between the results.

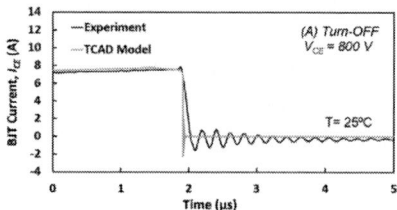 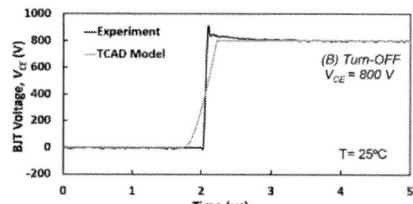

Fig. 6: Comparison of Modeled and Measured Turn-OFF transient of 4H-SiC BJT with pulse length of 35 μs with for the BJT collector (A) current and (B) Voltage at 25 °C.

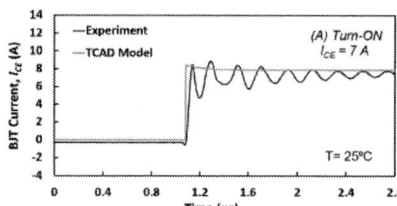 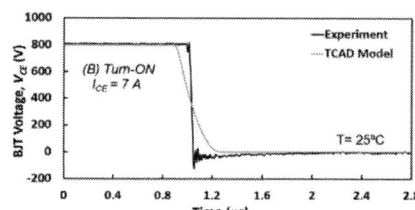

Fig. 7: Comparison of Modeled and Measured Turn-ON transient of 4H-SiC BJT with pulse length of 35 μs with for the BJT collector (A) current and (B) Voltage at 25 °C.

5 Scope of full paper

The full paper will include the results of complete range of experimental double-pulse measurements on high voltage Silicon and 4H-SiC NPN BJTs, and discuss the dynamic transient of the tested Silicon BJT and 4H-SiC BJT for different temperatures and switching rates. The transient times of current and voltage in 4H-SiC BJT will be shown to be significantly smaller in SiC compared to the silicon case, with turn-off delay almost non-existent. To evaluate these measurements, the full paper will also include comprehensive results of the TCAD modellings of both the silicon and 4H-SiC NPN BJTs performed on dynamic transients, enabling analysis of the trends seen.

References

[1] B. J. Baliga, *Fundamentals of power semiconductor devices*. Springer, 2010.

[2] C. Shen and et al., "Impact of carriers injection level on transients of discrete and paralleled silicon and 4h-sic npn bjts," *IEEE Open Journal of the Industrial Electronics Society*, vol. 3, pp. 65–80, 2022.

[3] S. M. Sze and K. K. Ng, *Physics of Semiconductor Devices*, 3rd. Hoboken, NJ, USA: Wiley, 2007.

PCIM Europe 2024, 11– 13 June 2024, Nuremberg   DOI: 10.30420/566262220

An Advanced Multi-Aspect Performance Analysis of Planar-Gate 1.2 kV SiC Power MOSFETs

Anja Katerina Brandl [1], Salvatore Race [1], Ivana Kovacevic-Badstuebner [1], Bhagyalakshmi Kakarla [1], Michel Nagel[1], Ulrike Grossner [1]

[1] Advanced Power Semiconductor Laboratory, ETH Zurich, Switzerland

Corresponding author: Anja Katerina Brandl, brandl@aps.ee.ethz.ch
Speaker: Anja Katerina Brandl, brandl@aps.ee.ethz.ch

Abstract

This paper introduces a parameterized SiC power MOSFET model enabling to find optimal design solutions by solving 2-D numerical device simulations. Namely, it allows exploring the design space of hundreds of MOSFET designs while accounting for electro-thermal and dynamic effects to evaluate the application-relevant device figure-of-merits (FOM). The analysis of a large MOSFET design space shows that short-circuit (SC) ruggedness of planar-gate SiC power MOSFETs featuring the same SC peak current is defined by the saturation current at high temperatures beyond 175°C, which is then directly correlated to the temperature-dependent on-state resistances. Furthermore, a comprehensive analysis of different MOSFET designs points to the impact of circuit layout parasitics on the dynamic device performance.

1 Introduction

The design of Silicon Carbide (SiC) power Metal-Oxide-Semiconductor Field-Effect Transistor (MOSFETs) to comply with specified application requirements in terms of performance, reliability, and ruggedness is necessary in order to facilitate the development of high-performance power electronic converters in various applications. This, however, requires the exploration of an extensive MOSFET design parameter space while addressing various coupled performance aspects and trade-offs. The considerable performance variability observed among commercially available SiC power MOSFETs highlights the complexity of SiC power MOSFET design optimization.

The examination of several commercially available 1.2 kV 4H-SiC power MOSFETs under short-circuit (SC) operation demonstrated a clear trade-off between the short-circuit withstand time ($SCWT$) and the specific on-state resistance $R_{ds(on,sp)}$ as illustrated in Fig. 1. In previous work, it has been shown that device parameters such as the design of the channel region [1], [2], the p-base [3], [4], and the JFET design [5], [6] impact the SC, switching, and conduction behavior of SiC power MOSFETs. Fur-

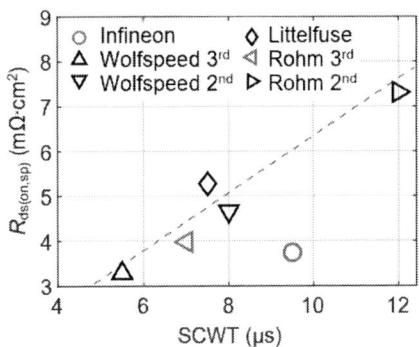

Fig. 1: Trade-off relationship between measured specific on-state resistance $R_{ds(on,sp)}$ and experimentally characterized $SCWT$ at $V_{DC} = 600\,\text{V}$ for commercial $1.2\,\text{kV}$ SiC power MOSFETs.

thermore, the impact of striped [5], hexagonal, and squared [6], [7] cell layouts of $1.2\,\text{kV} - 3.3\,\text{kV}$ SiC power MOSFETs were analyzed with respect to the aforementioned performance metrics.

Practical validation of design concepts has traditionally relied on the fabrication of a few MOSFET prototypes which is both time consuming and costly. Technology Computer Aided Design (TCAD) based numerical simulations of device designs can therefore be used for exploring different optimization strategies [8]. This paper introduces a fully parametrized, 2-dimensional (2-D) TCAD electrothermal model, enabling the variation of MOSFET cell design parameters and the evaluation of their impact on the device performance and state-of-the-art figure-of-merits (FOMs). The proposed model incorporates the complete manufacturing procedure of a planar-gate 1.2 kV power MOSFET. Furthermore, the developed simulation framework facilitates automated electrical characterization of static and dynamic properties, as well as the evaluation of application-relevant performance metrics such as short-circuit withstand time $SCWT$ and switching characteristics. The potential of this simulation framework is demonstrated on the example of a set of hundreds of SiC power MOSFET designs, generated with variations of the JFET width and doping, and the channel length and peak doping, with the aim to investigate the trade-offs between static and dynamic performance metrics as well as the SC ruggedness.

2 Figure-of-Merits (FOM)

Depending on the power electronic application, various requirements are imposed towards power semiconductor devices. A device comparison is typically performed based on performance evaluation metrics, commonly referred as Figure-of-Merits (FOMs), proposed to select the most suitable device for a given application.

The well-established Baliga's FOM (B-FOM) merges the breakdown voltage BV or the critical electric field E_{crit}, respectively, with the on-state resistance $R_{\mathrm{ds(on)}}$ to a single metric (Eq. (1)), enabling effective comparison across semiconductor materials, doping concentrations or layer thicknesses [9].

$$\text{B-FOM} = \varepsilon_{\mathrm{SiC}} \cdot \mu_{\mathrm{n}} \cdot E_{\mathrm{crit}}^3 = \frac{4 \cdot BV^2}{R_{\mathrm{on(sp)}}} \qquad (1)$$

The High-Frequency FOM (HF-FOM), cf. Eq. (2), serves as a suitable metric to compare the dynamic performance of different device designs [5], [10].

$$\text{HF-FOM} = R_{\mathrm{ds(on)}} \cdot C_{\mathrm{gd}} \qquad (2)$$

The SC-FOM is not clearly defined in literature. The SC ruggedness of a SiC power MOSFET has been so far associated to its saturation current $I_{\mathrm{ds(SAT)}}$, i.e. $SCWT$ is inversely proportional to $I_{\mathrm{ds(SAT)}}$ [8], [11]. Generally, a higher $R_{\mathrm{ds(on)}}$ corresponds to a lower on-state current, and thereby results in a prolonged endurance of the device under SC condition.

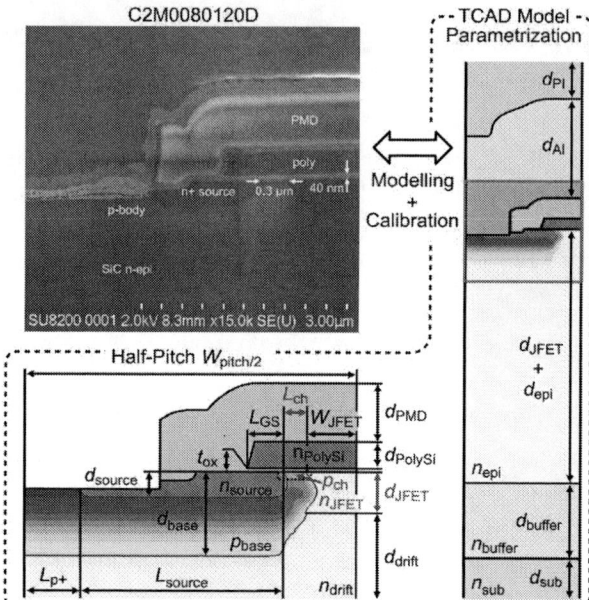

Fig. 2: SEM cross-section of the commercially available device C2M0080120D, and of the corresponding parameterized TCAD model.

This paper addresses the HF-FOM and SC-FOM with respect to their applicability for selecting SiC power MOSFETs to achieve a better SC and/or switching performance.

3 Device Modeling and Simulation Methodology

A 2-D parameterized TCAD model of a n-channel, planar-gate 4H-SiC power MOSFET was developed based on the reference 2-D TCAD model of an 80 mΩ, 1.2 kV 4H-SiC power MOSFET presented in [12], as shown in Fig. 2. In the next step, a simulation framework was implemented based on this parameterized TCAD to provide a complete, fully automated design flow as illustrated in Fig. 3. Based on process-related input parameters, for example implantation energies and fluences, etch rates or thermal oxidation parameters, the device fabrication procedure is simulated using Sentaurus^{TM} Process (S-Process). By varying the input parameters, a set of SiC power MOSFET designs is created. Subsequently, relevant geometrical properties and doping profiles are extracted for every device structure during runtime. The generated device structures are characterized using quasi-stationary and transient (mixed-mode) simulations in Sentaurus^{TM} Device (S-Device), as presented in detail in the following sections. Mixed-mode TCAD simulations allow time-domain simulations of electrical circuits consisting of Spice-based models of electrical components and 2-D TCAD, i.e. numerical, device models. The boundary conditions for the generated SiC power MOSFET design space are a minimum breakdown voltage BV of 1.2 kV, maximum on-state resistance $R_{ds(on)}$ at room temperature of 100 mΩ, a threshold voltage V_{th} in a range of (2 V, 4 V), a fixed active area of $A = 6.25\,\text{mm}^2$, and a fixed pitch size of 9.1 μm.

3.1 I-V/C-V Simulations

Quasi-stationary TCAD simulations are performed to extract the datasheet device parameters, such as BV, V_{th}, $R_{ds(ON)}$, and transconductance g_m. Namely, the current-voltage (I-V) characteristics are extracted by using a quasi-stationary dc-bias sweep. The evaluated BV corresponds to the drain-source voltage V_{ds} at a drain current I_{ds} of 100 mA and zero gate bias $V_{gs} = 0$ V. V_{th} is extracted in accordance with the *JEDEC standard* (JEDEC JEP183, Version 1.03). The on-state resistance $R_{ds(ON)}$ is evaluated at $V_{ds} = 50$ mV with $V_{gs} = 20$ V. Additionally, the $R_{ds(ON)}$ components, namely channel R_{ch}, JFET R_{JFET}, epi R_{epi} and substrate R_{sub} resistances are extracted, where R_{sub} includes the resistances within the drain contact, buffer and substrate, cf. Fig. 2. The sum $R_{JFET} + R_{epi} + R_{sub}$ is then denoted as a drift resistance R_{drift} [13]. The g_m is extracted from the quasi-stationary $I_{ds} - V_{ds}$ curve at $V_{gs} = 20$ V, $V_{gs} = V_{ds}$, and $I_{ds} = 20$ A.

The two-voltage dependent capacitance-voltage (C-V) characteristics, the gate-drain, $C_{gd}(V_{gs}, V_{ds})$, the gate-source $C_{gs}(V_{gs}, V_{ds})$, and the drain-source $C_{ds}(V_{gs}, V_{ds})$, are extracted from TCAD simulations by applying quasi-stationary dc-bias sweeps and a small ac signal frequency of 100 kHz [14].

3.2 Short-Circuit Simulations

Short-circuit (SC) ruggedness is one of the key evaluation parameters of SiC power MOSFETs (i.e. the publicly available experimental data base is rather large), and it represents a weak point in comparison to their counterpart Si-IGBTs. Here, SC behavior is simulated for the SiC power MOSFET design space for evaluation of the suggested methodology. An exemplary short-circuit curve is given in Fig. 4. The SC capabilities are evaluated by mixed-mode electrothermal SC simulations based on an extracted value t_{SC} for the *SCWT*. The DC-link voltage $V_{dc(SC)}$ as well as the SC pulse duration τ_p are defined as input parameters. The t_{SC} of the

generated devices is estimated according to two major failure mechanisms of SiC power MOSFETs under SC operations: the fail-to-open (FTO) and fail-to-short (FTS). FTS is caused by thermal runaway typically at $V_{dc(SC)} \geq 0.5 \cdot BV$, and it is accompanied by a steep increase of the device current as shown in Fig 4. Namely, three regions of the I-V SC waveforms can be distinguished: (I) the device turn-on causes a rapid increase of the current accompanied by a device temperature increase, followed by (II) an increase of on-state resistance due to self-heating effects, and a consequent decrease of the device current I_d, while temperature and carrier generation increase, which can result finally in a (III) parasitic latch-up/thermal runaway for longer τ_p. Therefore, t_{SC} for higher $V_{dc(SC)}$ is approximated by a time instance when the device current during a SC event I_{SC} reaches its minimal value, as shown in Fig. 4.

On the other hand, as the first degradation leading to FTO is aluminum melting, t_{SC} for lower $V_{dc(SC)}$ is approximated by a time instance when the maximum simulated temperature of the source contact reaches the Aluminium melting point $T_{melt(Al)} = 660\,°C$.

3.3 Switching Simulations

The dynamic performance of the simulated SiC power MOSFETs was assessed by the time-domain transient simulations of Double Pulse Test (DPT) measurements. The switching simulations are performed for the user-defined test voltage $V_{ds(DPT)} = 800$ V and test current $I_{ds(DPT)} = 20$ A, the turn-on and -off gate voltages $V_{gs(DPT,ON/OFF)} = -5/+20$ V, and an external gate resistance $R_{g(ext)} = 2.5\,\Omega$. The simulations were performed in the Simetrix (Spice-based) circuit simulation environment, using a lumped MOSFET model based on the look-up-tables (LUTs) [15], representing the $I(V_{gs}, V_{ds})$, $C_{gs}(V_{gs}, V_{ds})$, $C_{gd}(V_{gs}, V_{ds})$ and $C_{ds}(V_{gs}, V_{ds})$ MOSFET characteristics. The LUT-models are implemented in Verilog-A with the I-V and C-V data extracted from the 2-D TCAD device simulations described in Section 3.1. The simulated DPT circuit includes the broadband frequency-dependent model of the PCB layout parasitics extracted by ANSYS Q3D for a 4-layer PCB described in [16], and TO-247-3 package parasitics. The lumped MOSFET model allows us to distinguish the channel and displacement currents and hence, more accurately evaluate the turn-on E_{on} and turn-off E_{off} switching losses. The E_{on}, E_{off} as well as

PCIM Europe 2024, 11– 13 June 2024, Nuremberg   DOI: 10.30420/566262220

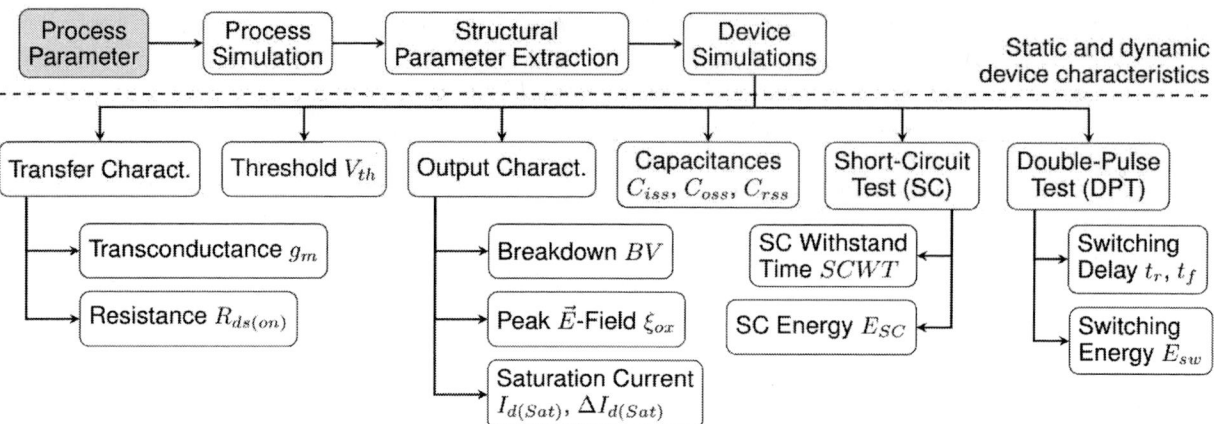

Fig. 3: Visualisation of the established fully-automated tool flow with a selection of extracted device parameters. Gray rectangles indicate the user input instance.

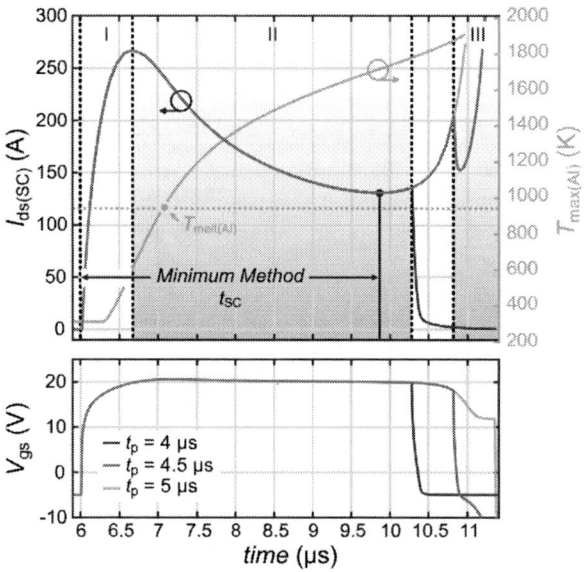

Fig. 4: Exemplary SC current, voltage and temperature waveforms at $800\,\text{V}$ DC-link bias for three SC pulse duration times τ_p with three distinguished regions.

the total switching losses E_{sw} are evaluated for the device comparison.

4   Simulation Study: SC Performance Trade-offs

A simulation study was performed to evaluate the design trade-offs of SiC power MOSFETs focusing on the SC ruggedness. In total, 106 different MOSFET designs were created by varying channel length L_{ch}, channel peak doping p_{ch}, JFET region width L_{JFET}, and JFET doping n_{JFET}, as summarized in table 1. These properties are ex-

tracted $2\,\text{nm}$ below the SiC/SiO$_2$-interface. The peak concentrations within the horizontal cutline is taken for doping concentrations. The base depth $d_{\text{base}} = 1.17\,\mu\text{m}$ and the peak doping $p_{\text{base}} = 7.9 \times 10^{18}\,\text{cm}^{-3}$ are kept equal for all simulated devices, cf. Fig. 2. It should be also noted that the channel length variation was achieved by changing mask properties in the Sentaurus S-Process, affecting also the poly-Si gate width w_{g}. The w_{g} on the other hand affects the internal gate resistance $R_{\text{g(int)}}$, which was then approximated by scaling the $R_{\text{g(int,ref)}}$ of the reference $80\,\text{m}\Omega$, $1.2\,\text{kV}$ SiC power MOSFET shown in [12], with $w_{\text{g(ref)}} = 3.6\,\mu\text{m}$, i.e. $R_{\text{g(int)}} = R_{\text{g(int,ref)}} \cdot w_{\text{g(ref)}}/w_{\text{g}}$. Additionally, the same SiC-SiO$_2$ interface quality is assumed for all devices, which was modeled by a reference density of interface traps [12] $D_{\text{it1}} = 2 \times 10^{13}\,\text{eV}^{-1}\text{cm}^{-2}$.

Device Parameter	Value Range
W_{JFET}	$834.65 \pm 352.06\,\text{nm}$
n_{JFET}	$\{1 \cdot 10^{16}, 2.8 \cdot 10^{16}\}\,\text{cm}^{-3}$
L_{ch}	$327.18 \pm 83.72\,\text{nm}$
p_{ch}	$(1.42 \pm 0.44) \cdot 10^{17}\,\text{cm}^{-3}$

Tab. 1: Range of parameters varied to create a set of 106 SiC power MOSFET designs.

All generated devices fulfill the boundary conditions specified in Section 3.1, i.e. a blocking capability between $1.4\,\text{kV}$ and $1.8\,\text{kV}$, and a $R_{\text{ds(ON)}}$ below $100\,\text{m}\Omega$. Furthermore, all designs showed a threshold voltage above 2.99V. The specific on-state resistance $R_{\text{ds(ON,sp)}}$ is plotted against BV in Fig. 5 for all simulated MOSFET designs as well as for the selected commercial SiC power

1616

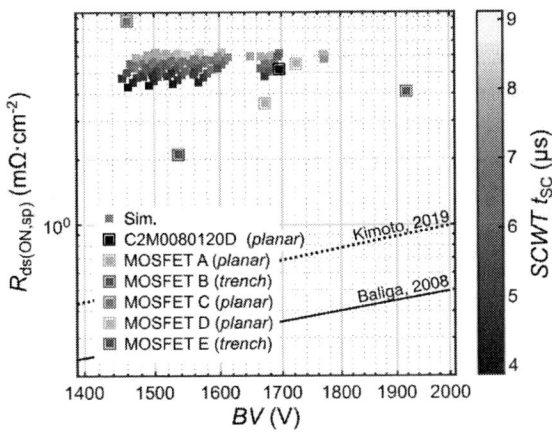

Fig. 5: Trade-off relationship between specific on-state resistance $R_{ds(ON,sp)}$ and breakdown voltage BV of commercial and simulated devices including two unipolar limits [9], [17].

MOSFETs. The $SCWT$ extracted as described in Section 3.2 is also shown for the simulated SiC power MOSFETs. The simulated MOSFET designs exhibit comparable performance as commercially available planar-gate SiC power MOSFETs. The selected commercially available trench-gate SiC power MOSFETs show a better $R_{ds(ON,sp)}$-BV trade-off relationship that can be explained by smaller cell pitch and JFET resistance.

Dev.	W_{JFET} (μm)	n_{JFET} (cm^{-3})	L_{ch} (nm)	p_{ch} (cm^{-3})
D_4	1.184	$2.8 \cdot 10^{16}$	296.18	$1.22 \cdot 10^{17}$
D_5	0.484	$2.8 \cdot 10^{16}$	244.96	$1.15 \cdot 10^{17}$
D_6	0.866	$1.0 \cdot 10^{16}$	268.08	$1.50 \cdot 10^{17}$
D_7	1.184	$2.8 \cdot 10^{16}$	245.45	$1.15 \cdot 10^{17}$

Tab. 2: Properties of the selected planar-gate SiC power MOSFETs extracted from TCAD simulations.

For all generated devices, the SC simulations were performed as described in Section 3.2, while the switching energies E_{sw}, E_{on} and E_{off} are computed as described in Section 3.3.

4.1 Short-Circuit Performance Space

Four MOSFETs marked D_4 to D_7 are selected for an in-depth analysis of their SC capabilities as well as static and dynamic performance with the aim to evaluate device design trade-offs and the state-of-the-art FOMs. Namely, the MOSFETs D_4, D_5 and

D_6 were selected due to their comparable SC peak current $I_{ds(SC,peak)}$, whereas the devices D_5 and D_7 are compared in order to explore the impact of the design parameters on switching behavior. The main parameter differences between D_4-D_7 MOSFETs and an overview of the corresponding performance metrics are summarized in table 2 and in table 3, respectively.

The SC capabilities were evaluated by the t_{SC} at $V_{dc(SC)} = 800$ V, with the turn-on $V_{gs} = 20$ V and a long pulse duration $\tau_p = 20\,\mu s$. A commonly used SC-FOM of SiC power MOSFETs is the saturation current $I_{ds(SAT)}$, i.e. smaller $I_{ds(SAT)}$ leads to prolonged $SCWT$ [11]. Accordingly, the relationship between $SCWT$ and the saturation current at room temperature $I_{ds(SAT,300K)}$ is demonstrated for all simulated MOSFET designs in Fig. 6(a), for both the reference high density of interface traps D_{it1} [12], and a low $D_{it2} = 2 \times 10^{11} eV^{-1}cm^{-2}$, which corresponds to a Si-SiO$_2$ oxide quality. The same trend was observed independently of D_{it}. However, three selected MOSFET designs D_4, D_5 and D_6 in Fig. 6(a) with similar $I_{ds(SAT,300K)}$ demonstrate that a significant improvement of $SCWT$ can be achieved without a reduction of $I_{ds(SAT,300K)}$. The corresponding SC currents $I_{ds(SC)}$ and the maximum device temperatures $T_{SiC(max)}$ during the SC operation of D_4, D_5 and D_6 are shown in Fig. 6(b). The devices D_4, D_5 and D_6 lead to comparable peak $I_{ds(SC)}$ at around 170 A, while D_5 exhibits a $\approx 2\,\mu s$ longer $SCWT$, but has a higher $R_{ds(ON)}$ than D_4 and D_6. The prolonged $SCWT$ of D_5 is explained by a sharper decrease of $I_{ds(SC)}$ with time once the peak current is reached. This confirms that $I_{ds(SAT)}$ of SiC power MOSFETs features a strong temperature-dependence, which has a high impact on the SC performance.

This is further confirmed by the I-V simulations conducted over a wide temperature range T from 300 K to 1000 K, as shown in Fig. 6 (c). Among four selected devices, D_5 demonstrates a stronger temperature-dependence of $I_{ds(SAT)}$ compared to D_4 and D_6, which can be attributed to the relative contribution of the drift resistance R_{drift} component to overall on-state resistance $R_{ds(ON)}$, cf. Fig. 2, where $R_{drift} = R_{JFET} + R_{epi} + R_{sub}$. Fig. 6(d) shows the temperature-dependent R_{drift} and channel R_{ch} resistances extracted from the I-V TCAD simulations for D_4, D_5 and D_6 MOSFETs. The R_{drift} exhibits a strong positive temperature coefficient and becomes the dominant part of

	BV (kV)	V_{th} (V)	g_{m} (S)	$R_{\text{ds(ON)}}$ (mΩ)	$I_{\text{ds(Sat)}}$ (A)	$I_{\text{ds(SC)}}$ (A)	$SCWT$ (μs)	E_{sw} (μJ)	E_{on} (μJ)	E_{off} (μJ)
D_4	1.46	3.51	6.87	80.19	150.98	170.78	5.55	435.3	428.3	7.0
D_5	1.77	3.00	5.70	92.07	162.37	170.26	7.38	421.8	414.4	7.4
D_6	1.60	3.62	6.86	87.67	153.08	170.70	5.86	432.8	426.6	6.2
D_7	1.46	3.00	7.89	69.15	237.00	245.66	3.87	402.5	394.4	8.1

Tab. 3: Summary of performance metrics of the selected MOSFETs D_4 to D_7.

$R_{\text{ds(ON)}}$ at high temperatures typically reached during SC operation. This behavior contributes to a more substantial reduction of $I_{\text{ds(SC)}}$ with time during SC operation for higher R_{drift} resistance. It should be noted that D_5 corresponds to a design with a narrow JFET region, which leads to a larger contribution of R_{drift} in $R_{\text{ds(ON)}}$, and hence, an improved $SCWT$.

4.2 HF-FOM Comparison

As the HF-FOM is typically defined by the gate-drain capacitance at high voltage $C_{\text{gd(HV)}}$ and $R_{\text{ds(ON)}}$, a plot $C_{\text{gd(HV)}}$ vs. $R_{\text{ds(ON)}}$ for simulated MOSFET designs is shown in Fig. 8, together with the corresponding simulated $SCWT$. It can be seen that D_7 and D_5 are close to the Pareto-optimal device designs. Namely, D_7 features the highest $C_{\text{gd(HV)}}$, while D_5 features the lowest $C_{\text{gd(HV)}}$, for the investigated parameter space.

In the next step, the correlation between the HF-FOM and the switching waveforms is evaluated.

4.3 Switching Performance

It was observed that the 4-layer PCB setup and $R_{\text{g(ext)}} = 2.5\,\Omega$ leads to a fast decrease of channel current at turn off for all MOSFETs so that E_{off} is negligible in comparison to E_{on} as shown on the example of D_5 and D_7 in Fig. 7. Furthermore, larger $g_{\text{m,D7}}$ and smaller $R_{\text{g(int,D7)}}$ lead to a steeper and earlier current rise at turn on, however due to smaller $C_{\text{gd(HV,D5)}}$ at turn ON, the oscillations due to layout parasitics are less damped, leading to a 5 % higher $E_{\text{on(D5)}}$ than $E_{\text{on(D7)}}$. A higher $C_{\text{gd(LV,D7)}}$ on the other hand leads to a delayed V_{ds} rise in comparison to D_5, and hence higher $E_{\text{off,D7}}$ However, D_5 has a higher HF-FOM than D_7, which should suggest a better dynamic performance. By using a larger $R_{\text{g(ext)}} > 10\,\Omega$, the E_{off} becomes non-negligible and the PCB layout parasitic effects less pronounced, so that $E_{\text{sw(D7)}} > E_{\text{sw(D5)}}$, which is then in agreement to HF-FOM. Accordingly, the

V_{ds}	C_{gd} (pF) LV	HV	C_{gs} (nF) LV	HV	C_{ds} (pF) LV	HV
D_4	1136.87	7.39	1.06	1.15	854.72	50.84
D_5	377.29	1.39	0.99	1.04	973.00	57.05
D_6	692.17	4.42	1.05	1.10	853.75	53.99
D_7	1139.09	7.41	1.03	1.11	854.52	50.82

Tab. 4: Low (LV) and high (HV) voltage inter-terminal capacitances of the selected SiC power MOSFETs, with $V_{\text{ds(LV)}} = 10\,\text{mV}$ and $V_{\text{ds(HV)}} = 800\,\text{V}$ and $V_{\text{gs}} = 0\,\text{V}$.

4-layer PCB layout parasitics strongly affect the dynamic device performance at fast transients, which is not considered by HF-FOM.

5 Conclusion

This paper presents a simulation framework based on a fully-parameterized TCAD model of a planar-gate 4H-SiC Power MOSFET, which enables a comprehensive device analysis considering multiple performance aspects all together. The presented results are highly valuable for optimizing the SiC power MOSFET with respect to the application-required SC performance and the switching performance. The developed parameterized 2-D TCAD model will enable an evaluation of Pareto-optimized SiC power MOSFET designs and can easily be extended to other designs of power devices.

References

[1] G. Romano, M. Riccio, L. Maresca, *et al.*, "Influence of design parameters on the short-circuit ruggedness of SiC power MOSFETs," in *Proc. of 28th International Symposium on Power Semiconductor Devices and ICs (ISPSD)*, 2016, pp. 47–50.

[2] L. Knoll, A. Mihaila, S. Wirths, *et al.*, "Planar 1.2kV SiC MOSFETs with retrograde channel profile for enhanced ruggedness," in *Proc . of 31st International Symposium on Power Semiconductor Devices and ICs (ISPSD)*, 2019, pp. 211–214.

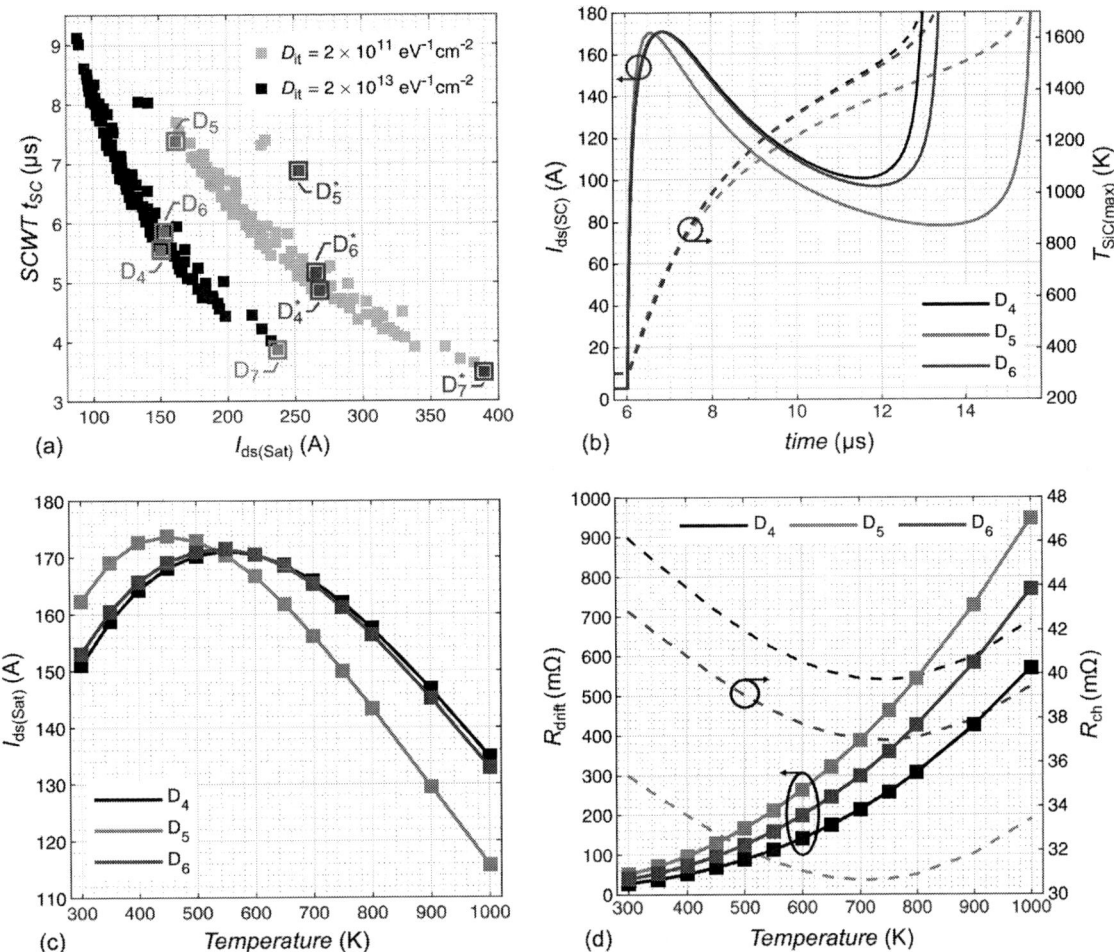

Fig. 6: Electrothermal analysis of the selected MOSFETs D_4, D_5 and D_6: (a) the relation between the $SCWT$ extracted as shown in Fig. 4 for $800\,\text{V}$ DC-link bias, and the saturation current $I_{ds(SAT,300K)}$, extracted at room temperature, $V_{ds} = 800\,\text{V}$, and $V_{gs} = 20\,\text{V}$. D_i^* correspond to the design D_i with a low D_{it}, (b) SC current $I_{ds(SC)}$ and maximum SiC lattice temperature $T_{SiC(max)}$ curves under the SC operation, (c) temperature-dependence of saturation current $I_{ds(Sat)}$ at $V_{ds} = 800\,\text{V}$, and (d) the temperature-dependence of MOSFET resistance components,i.e. the JFET and drift layer resistance R_{JFET} and R_{drift}, and the channel resistance R_{ch}.

[3] B. Kakarla, S. Nida, J. Mueting, *et al.*, "Trade-off analysis of the p-base doping on ruggedness of SiC MOSFETs," *Microelectronics Reliability*, vol. 76-77, pp. 267–271, 2017.

[4] D. Kim and W. Sung, "Improved Short-Circuit Ruggedness for 1.2kV 4H-SiC MOSFET Using a Deep P-Well Implemented by Channeling Implantation," *IEEE Electron Device Letters*, vol. 42, no. 12, pp. 1822–1825, 2021.

[5] C. Lin, N. Ren, H. Xu, *et al.*, "Performance and Short-Circuit Reliability of SiC MOSFETs With Enhanced JFET Doping Design," *IEEE Transactions on Electron Devices*, vol. 70, no. 5, pp. 2395–2402, 2023.

[6] X. Chen, X. Li, Y. Wang, *et al.*, "Different JFET Designs on Conduction and Short-Circuit Capability for 3.3 kV Planar-Gate Silicon Carbide MOSFETs," *IEEE Journal of the Electron Devices Society*, vol. 8, pp. 841–845, 2020.

[7] C. Lin, N. Ren, H. Xu, *et al.*, "1.2-kV Planar SiC MOSFETs With Improved Short-Circuit Capability by Adding Plasma Spreading Layer," *IEEE Transactions on Electron Devices*, vol. 70, no. 9, pp. 4730–4736, 2023.

[8] C. Leendertz, M. Hell, G. Zeng, *et al.*, "CoolSiC trench mosfet chip design for the 3.3 kV class," in *Proc. of International Exhibition and Conference for Power Electronics, Intelligent Motion, Re-*

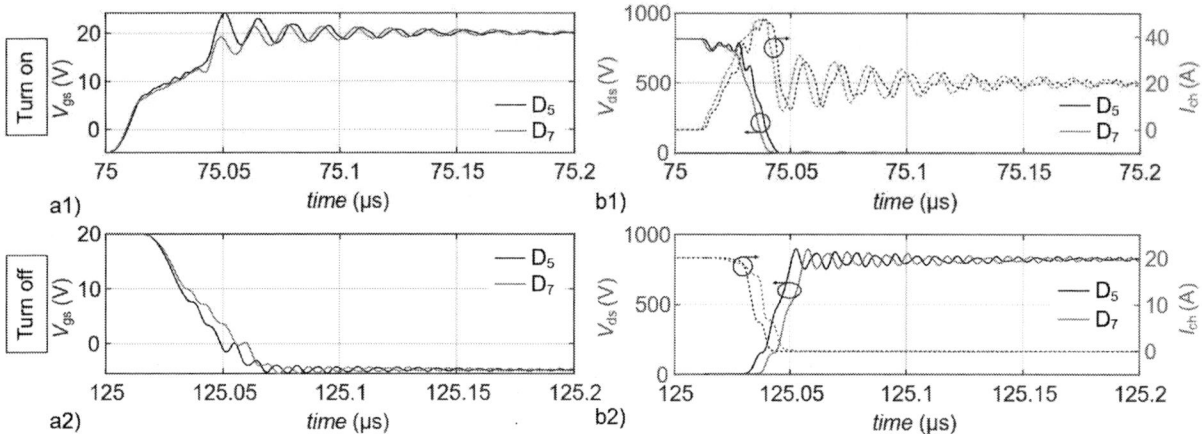

Fig. 7: Simulated turn on- and off-switching waveforms of the SiC power MOSFETs D_5 and D_7 with a 4-layer PCB and TO-247-3 packages for $V_{ds(DPT)} = 800\,\text{V}$, $I_{ds(DPT)} = 20\,\text{A}$, $V_{gs(DPT,ON/OFF)} = -5/+20\,\text{V}$, and $R_{g(ext)} = 2.5\,\Omega$: a) the internal V_{gs}, b) the channel current I_{ch} and the internal V_{ds}.

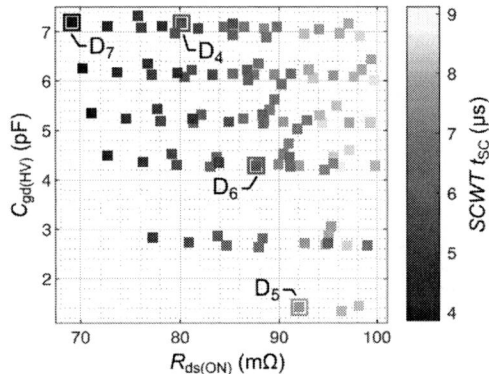

Fig. 8: $C_{gd(HV)}$ vs. $R_{ds(ON)}$ for simulated MOSFET designs together with the corresponding simulated $SCWT$.

newable Energy and Energy Management (PCIM Europe), 2023, pp. 1–6.

[9] B. J. Baliga, "Fundamentals of Power Semiconductor Devices," in Boston, MA: Springer US, 2008.

[10] A. Agarwal and B. J. Baliga, "Implant Straggle Impact on 1.2 kV SiC Power MOSFET Static and Dynamic Parameters," *IEEE Journal of the Electron Devices Society*, vol. 10, pp. 245–255, 2022.

[11] K. Han, A. Kanale, B. J. Baliga, *et al.*, "New Short Circuit Failure Mechanism for 1.2kV 4H-SiC MOSFETs and JBSFETs," in *Proc. of IEEE 6th Workshop on Wide Bandgap Power Devices and Applications (WiPDA)*, 2018, pp. 108–113.

[12] A. Tsibizov, I. Kovacevic-Badstuebner, B. Kakarla, *et al.*, "Accurate temperature estimation of SiC power MOSFETs under extreme operating conditions," *IEEE Transactions on Power Electronics*, vol. 35, no. 2, pp. 1855–1865, 2020.

[13] R. Stark, A. Tsibizov, S. Race, *et al.*, "Temperature dependence of the channel and drift resistance of SiC power MOSFETs extracted from I-V and C-V measurements," *Materials Science Forum*, vol. 1092, pp. 165–170, Jun. 2023.

[14] R. Stark, A. Tsibizov, I. Kovacevic-Badstuebner, *et al.*, "Gate capacitance characterization of silicon carbide and silicon power MOSFETs revisited," *IEEE Transactions on Power Electronics*, vol. 37, no. 9, pp. 10 572–10 584, 2022.

[15] R. Stark, A. Tsibizov, N. Nain, *et al.*, "Accuracy of Three Interterminal Capacitance Models for SiC Power MOSFETs Under Fast Switching," *IEEE Transactions on Power Electronics*, vol. 36, no. 8, pp. 9398–9410, 2021.

[16] S. Race, M. Nagel, I. Kovacevic-Badstuebner, *et al.*, "Towards Digital Twins for the Optimization of Power Electronic Switching Cells with Discrete SiC Power MOSFETs," in *Proc. of International Exhibition and Conference for Power Electronics, Intelligent Motion, Renewable Energy and Energy Management (PCIM Europe)*, 2022.

[17] T. Kimoto, "Updated trade-off relationship between specific on-resistance and breakdown voltage in 4H-SiC{0001} unipolar devices," *Japanese Journal of Applied Physics*, vol. 58, no. 1, p. 018 002, 2018.

PCIM Europe 2024, 11– 13 June 2024, Nuremberg   DOI: 10.30420/566262221

SiC MOSFET Die Sorting and Parallel for Optimal Module Design

Zhong Ye, Yi Yang, Qing Wang, Yandong Zhou, Zhaoqiang Zhao

InventChip, China

Corresponding author: Zhong Ye, zhong.ye@inventchip.com.cn
Speaker: Zhong Ye, zhong.ye@inventchip.com.cn

Abstract

SiC MOSFET power modules often have multiple dies connected in parallel inside the packages. Variation or mismatch of the dies' Vth and Ron can degrade the power module performance, including output current capability and even reliability. This article proposes a new die sorting methods to arrange dies in Vth value sequence and screen out outlier dies in any designated consecutive die number. This sorting method and its die output arrangement facilitate module manufacturing. Die screen criteria are also discussed for optimal module design. HPD modules are built with this new die selection scheme and tests are designed to investigate the performance. The tests show that proper die sorting and combination can achieve a good current sharing and thermal balance.

1   Introduction

Power modules dominate high voltage and high power applications, such as electric vehicles, power grids and railway transportation, etc. Compared with Si IGBTs, SiC MOSFETs have larger parameter variation, including Vth (MOSFET gate threshold voltage) and Ron (MOSFET turn-on resistance) due to semiconductor process limitation and fabrication discrepancies. How to select SiC MOSFET dies for parallel connection is an eternal topic for module manufacturers. Current imbalance among the MOSFET dies is the major concern as to a module performance and reliability[1,2]. Static current sharing is determined by the MOSFETs' Ron while dynamic current is affected by the MOSFET Vth and gate drive circuit. Since SiC MOSFET's Ron typically presents a positive temperature coefficient at high temperature, static current sharing can be achieved naturally to a certain degree[3]. However, Vth has negative temperature coefficient. A MOSFET with a lower Vth can hog the current at both turn-on and turn-off switching edges, which can result in more switch-

ing loss. The more the switching loss is, the higher the device temperature is, which can in turn causes the Vth to decrease further and even device damage at the end in the parallel application. The worst case scenario is that the Vth mismatch and DBC leakage inductance-induced voltage cause the MOSFET current oscillation. The oscillation is one of the lethal factors causing module failure[4]. To mitigate this issue, manufacturers often sort and group dies into bins by Vth values, which limits the Vth variation in each bin. However, to make it feasible for manufacturing, this process can become quite tricky. CP (Circuit Probe) tests can provide fairly accurate MOSFET Vth, while to obtain more accurate dies' Ron ,a KGD (Known Good Device) machine is still needed. After CP test, a new map needs to be generated with bin information. The KGD machine picks the good dies by bin sequence and load to a new frame after the test. A second map is generated by excluding the dies whose Ron exceeds resistance limits of the bins. This new map is then used for assembly machines to pick dies for final assembly. The issue is that the die numbers in each bin may not be the integer multiple of the paralleled dies' number, which can result in die waste. A new sorting method is proposed in this article to facilitate the process and achieve better module performance. In Section 2, the detail of the new sorting method is presented; and Section 3 and 4 discusses the criteria of paralleled die selection, and Section 5 provides validation test data. Conclusions are given in Section 6 at the end.

2   A New Die Sorting Method

The terminology "Bin" is often used in die sorting. We can still use the bin concept, but each bin contains only one die in this new sorting method afterward. In this way, the dies can actually be sorted with continuous Vth values. A new map can be generated by this concept after the CP test, as shown in Fig. 1. The new map excludes bad dies and assigns bin numbers to the good dies based on Vth value sequence from the minimum value to the maximum value. A KGD can then test the dies in the designated sequence, and load them to a new frame one by one in CP's Vth value sequence, as shown in Fig. 2. The KGD can test die parameters more accurately and isolate out any defect dies. On the new frame, the dies are arranged in the same sequence of a module assembly machine's pick-&-place and a second new map is generated, as shown in Fig.3.. Based on the module assembly machine's requirement, KGD can load the good dies to the new frame in S or Z shape. S shape means that the value sequence of odd lows is opposite to that of the even lows, while Z shape means that all lows' value sequences are the same. Therefore, for the Z-shape frame input, the module assembly machines have to move the die-picking nozzle to the beginning of new lows after finishing one low's die placement; for the S shape frame input, the machines pick dies by alternating head-to-toe and toe-to-head sequences low by low. The S-shape die arrangement is accepted by most module assembly machines currently.

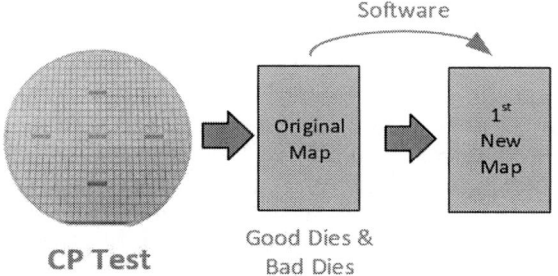

Fig. 1 CP test and the 1ˢᵗ new map generation.

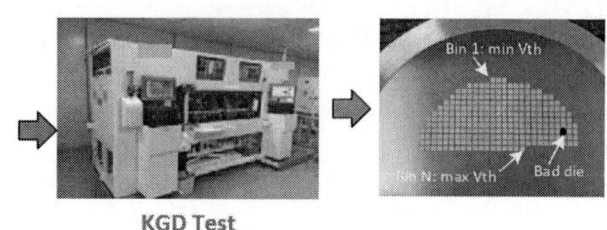

Fig. 2 KGD test and die loading on a new frame.

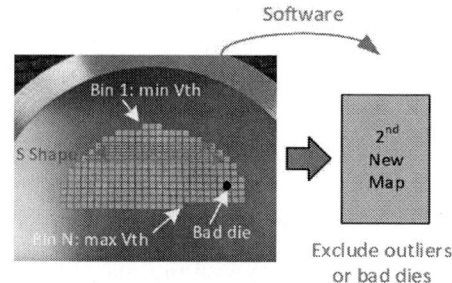

Fig. 3 2ⁿᵈ new map generation.

The second new map can be generated by excluding the dies with Ron apart too far from the adjacent dies' average value. In this way, the frame can be used for any module build, disregarding paralleled die numbers, and the paralleled dies can have minimum Vth difference for the modules to achieve optimal performance.

3   Die Sorting Limit Setting

In order to perform die sorting, die parameter limit setting is important. There are many parameters can be used for die sorting, but the most important two parameters are Vth and Ron. This section will focus only on these two parameters and discuss the sorting limit setting.

To sort Vth, the Vth data must be screened for outliers. The criteria for outliers are firstly consistent with the PAT rules adopted by automotive-grade wafer products. The upper and lower limit algorithms are: Mean ± n × Robust σ (the default value of n is 6). These limits isolate out the dies with Vth too high and too low, which are not suitable for module production. For module build and to achieve good current balancing among the paralleled dies, the Vth variation of adjacent dies is critical. A method of rolling mean is used for the outlier screen. Since most automotive power modules use 6 or 8 dies in parallel, 8 dies with consecutive Vth values or bin numbers are used for the parameter limit calculation here. The upper and lower limit algorithms are: Mean(i; i+N) ± ΔVth; where "i" is No.i die (bin) and N is the rolling bin number; N is 8 here for the following data plot. Mean(i; i+N) is the Vth mean value of the dies from bin-i to bin-i+N; ΔVth is the maximum allowed Vth difference from the Mean among the N dies. The same algorithms is used for Ron limit setting. ΔVth = 0.05V and ΔRon=0.425mΩ are used in the following limit calculation.

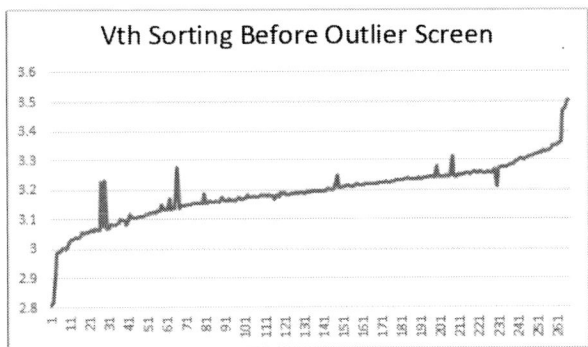

Fig. 4   Vth sorting before outlier screen.

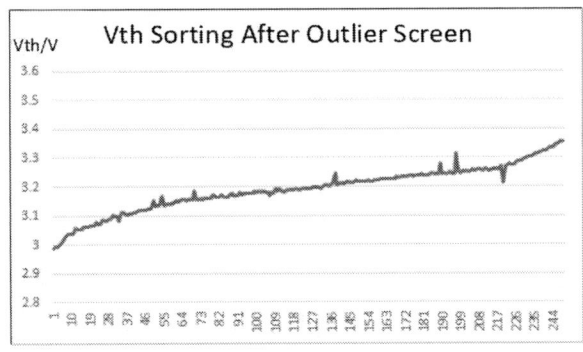

Fig. 5   Vth sorting after outlier screen.

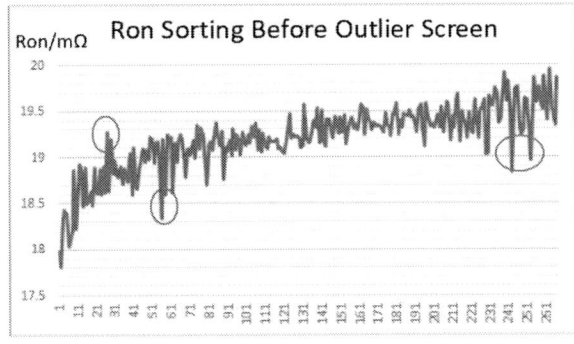

Fig. 6   Ron sorting before outlier screen.

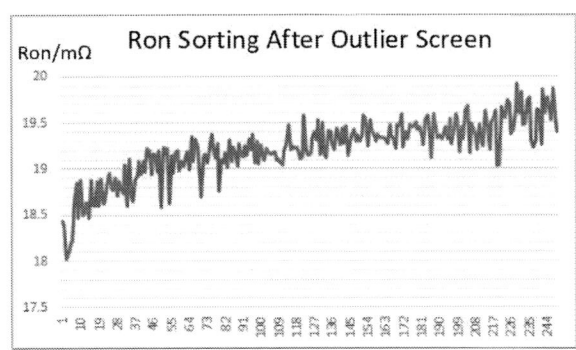

Fig. 7   Ron sorting after outlier screen.

Fig. 4,5,6 and 7 show Vth sorting and Ron sorting before and after outlier screen. It can be seen that after outlier screen any 6 or 8 consecutive bins (dies) can meet both Vth and Ron selection criteria. If a module needs more dies in parallel, for example 10 dies in parallel, N should be 10 in the Mean calculation.

4   ΔRon and ΔVth Criteria Setting

There are no standard criteria of ΔRon and ΔVth selection. For different dies and module structures, different ΔRon and ΔVth should be used to achieve targeted current balancing and thermal balancing goals. The most common goals of dynamic and static current balancing variation and die temperature variation are 10% and 5°C respectively. To achieve the goals, ΔRon and ΔVth are pre-selected to screen and sort dies for module build. Since it is difficult or not feasible to measure dies' current in the module level, discrete MOSFETs with the same dies can be used instead and tested on a specially designed double-pulse board for current balancing measurement, as shown in Fig. 8 and 9.

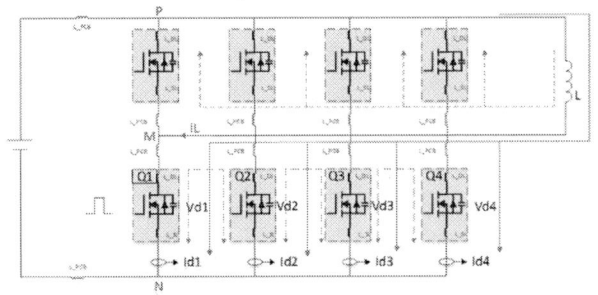

Fig. 8   A double pulse test with 4 MOSFETs in parallel.

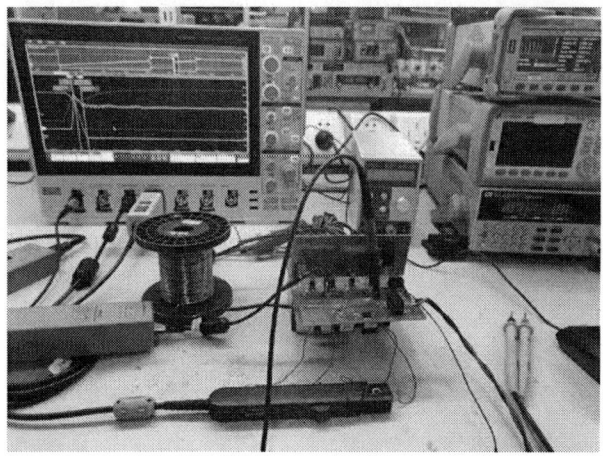

Fig. 9 Double pulse test setup.

	A (ΔVth=0.35V)	B (ΔVth≤0.05V)	C (ΔVth≤0.05V)	D (ΔVth≤0.05V)
Eoff	0.505mJ	0.646mJ	0.670mJ	0.629mJ
Idsoff	60.8A	63A	62A	61A
Eon	1.218mJ	1.440mJ	1.453mJ	1.412mJ
Idson	51A/82.4A	63A/94A	65.4A/92.6A	63.7A/93A

Fig. 10 Switching loss test results.

Eight 17mΩ 1200V MOSFETs IV2Q12017T4Z were used for the current balancing test. The double pulse was tested at 600V and 240A. It can be seen that current balancing or switching loss can meet the 10% design goal with a good margin if ΔVth is limited to 0.05V.

For the ultimate goal of maximum 5°C die-to-die temperature variation, it is estimated that ΔRon should be less than 0.6 mΩ for a HPD module, and 0.425 mΩ was used for Ron screen for the following validation tests. Note this criteria setting is a try-and-adjust process. A real module must be built and tested. If the goal can not be achieved, ΔRon and ΔVth limits must be shrunk, vice versa. Tests show that dynamic current balancing reaches near perfect condition when ΔVth is reduced to 0.1V. Further reduction of ΔVth has little effect on the current balancing improvement. Due to this reason, in the following validation tests, we focus more on ΔRon.

5 Module Validation Test

The proposed sorting and screen method is validated by building and testing a HPD module. The process elaborated above can be depicted as Fig.11.

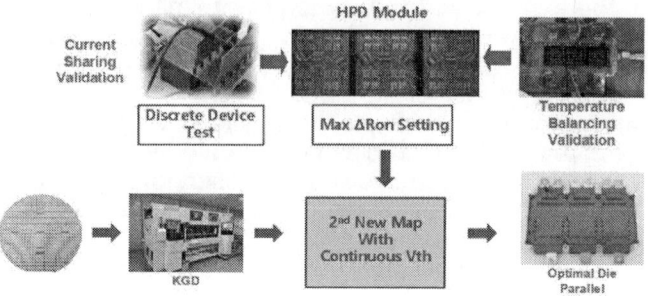

Fig. 11 Die sorting with maximum ΔRon setting.

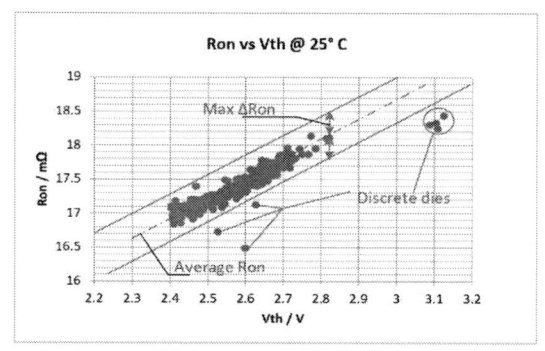

Fig. 12 Vth and Ron distribution of one wafer.

A new wafer is used for the module build. The wafer's Vth and Ron distribution is plot in Fig. 12. It is clear that outlier dies often exist. There is no doubt the outlier dies can cause current or thermal balance issues if they are used for module build. After excluding the outliers, modules can be built with the die optimal combination process aforedescribed. The modules' average die Ron values are plot in Fig. 13. Each "dot" represents a die's average Ron of a switch. A module has 6 switches, namely 6 "dots". Three complete modules were built with the dies.

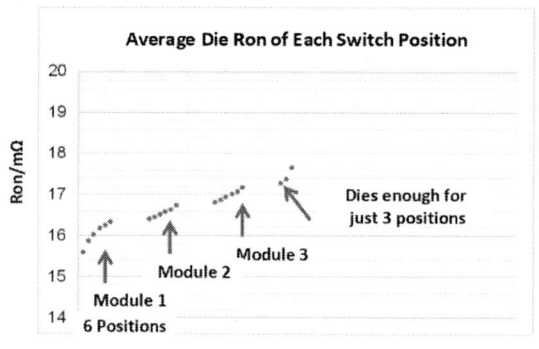

Fig. 13 Average die Ron of HPD modules.

One of the modules was used for validation tests. A special test setup was designed to test static and dynamic thermal balancing separately. Three bridges of the module were connected in series. All switches were turned on and a DC current passed the switches for static thermal balancing test. The schematic diagram and thermal image are shown in Fig. 14 and 15.

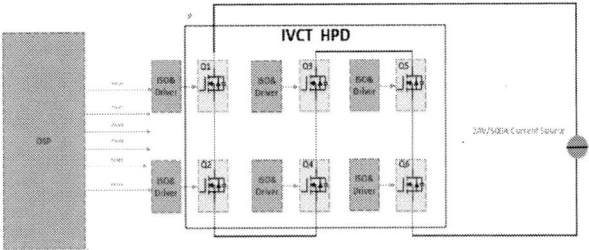

Fig. 14 Static current/thermal balancing test.

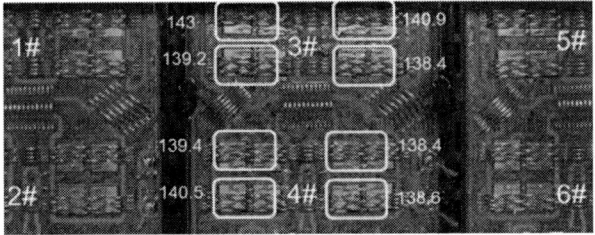

Fig. 15 Static thermal Balancing data.

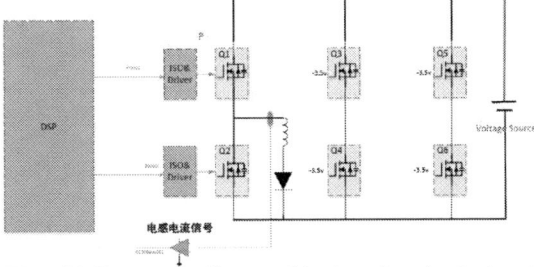

Fig. 16 Dynamic thermal balancing test circuit.

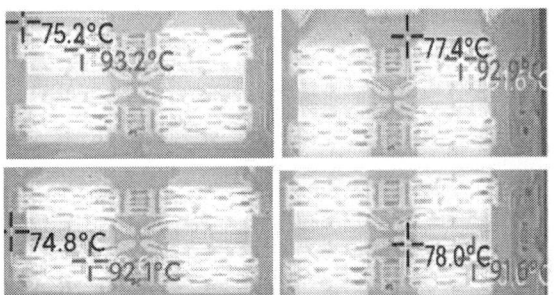

Fig. 17 Dynamic current/thermal balancing test.

Fig. 16 shows the circuit diagram for the dynamic thermal balance test. The test was conducted on Phase A which has switches Q1 and Q2. Since the phase output is shorted to the negative input by a series connected inductor and diode directly, to regulated a constant inductor current, Q1 has a very narrow duty cycle. In this way, Q1 has essentially only switching loss. Its conduction loss is negligible. Fig. 17 shows the thermal image of dynamic current balance. The test results in Fig. 15 and 17 show that the temperature difference caused by static and dynamic current imbalance can be within 4°C and 1.6°C respectively. The static current balance test was conducted at 100% duty cycle. The thermal imbalance will reduce with the decrease of the duty cycle. It agreed with the system level test result and maintained the temperature difference within 5°C.

6 Conclusions

A new die sorting method was proposed to facilitate module manufacture and optimize the paralleled dies' current sharing and thermal balance. The new die sorting method screening out outlier dies is an important step to ensure modules' high performance and excellent reliability. The new sorting method enables flexible module manufacturing disregard the die-paralleling number. The concept was validated with a HPD module. The thermal balance related to both static and dynamic current sharing was tested and the concept was validated. It shows that a targeted thermal balance can be achieved by properly setting die screen criteria.

References

[1] Helong Li et al, "Parallel Connection of Silicon Carbide MOSFETs—Challenges, Mechanism, and Solutions," IEEE TRANSACTIONS ON POWER ELECTRONICS, VOL. 38, NO. 8, AUGUST 2023.

[2] T. Bertelshofer, et al, "Derating of Parallel SiC MOSFETs Considering Switching Imbalances," PCIM Europe 2018; International Exhibition and Conference for Power Electronics, Intelligent Motion, Renewable Energy and Energy Management.

[3] Yizhuo Dong et al, "Effects of On-State Resistance Temperature Effect on The Static Current Balancing Capability of SiC MOSFET Power Module," 2022 23rd International Conference on Electronic Packaging Technology (ICEPT).

[4] Zhou Dong et al, "Instability Issue of Paralleled Dies in an SiC Power Module in Solid-State Circuit Breaker Applications," IEEE Transactions on Power Electronics (Volume: 36, Issue: 10, October 2021).

PCIM Europe 2024, 11– 13 June 2024, Nuremberg   DOI: 10.30420/566262222

Simulation Approach for Radiated Electro-Magnetic Field Estimation on Acepack™ Drive SiC Power Module

Andrea Cusumano[1], Debora Crimi[1], Salvatore Oliveri[1], Alessandra Manzitto[1]

[1] STMicroelectronics, Italy

Corresponding author: Andrea Cusumano, andrea.cusumano@st.com
Speaker: Andrea Cusumano, andrea.cusumano@st.com

Abstract

Power electronics solutions generate electromagnetic fields in the environment leading to possible interferences in both near and far field conditions. A prediction of radiated and conducted electromagnetic pollution allows the evaluation of the related issues that can cause device failures and malfunctioning. For this purpose, the radiated emissions in different working points are evaluated through electromagnetic simulations based on Ansys HFSS tool. The goal of this work is the presentation of a simulation approach to perform the electromagnetic analysis of the Acepack™ Drive SiC power module. This method estimates the radiated emission levels with the aim of seeking to reduce potential electromagnetic compatibility problems.

1   Introduction

The trend of using wide band-gap devices in modern high-power electronics applications have enabled the possibility to increase the switching frequencies of the power switches; this allows the reduction of the size of the power converters and the increase of the power density and power efficiency [1]. The drawback is that all the passive components behave differently at high frequency and especially at MHz range the parasites might be the dominant factor [2]. The high dv/dt and di/dt give rise to radiated emission issues; in this way the switching power devices get coupled to other auxiliary circuits such as the gate drivers and other control circuits. [3]. The high-power density converters are more widely used in applications such as electric vehicles. Many International Standards such as CISPR 25 are adopted to regulate the radiated emissions [1] and the circuit designers must make their devices electromagnetically compatible, satisfying that EMC Regulatory Standards. Conventionally the EMC compliance test is carried out through testing of the first prototype in semi-anechoic chamber; this method is rather iterative and time-consuming. Therefore, it is important to predict the radiated emissions of the electronic equipments during the design stage before the first prototype is fabricated. Electromagnetic field simulations have become one of the common tools for designers to validate analytical designs and investigate details that are complex to consider analytically. For this reason, an accurate modelling is necessary to simulate and analyze the performance of the power converter during the design process [4]. The market offers many simulation' software to solve the Maxwell Equations; some of them are based on analytical solvers, others on FEM based quasi-static analysis or full-wave analysis applied to 3D objects. Finite element method is a well-known approach to calculate field distribution, losses and to extract parasitic components which can directly deteriorate the accuracy of the model if not considered [2]

In this paper the radiated electromagnetic field evaluation of the Acepack™ Drive traction power module is performed by means of the FEM based simulation software Ansys HFSS (High Frequency Structure Simulator). It solves the Maxwell equations using a full-wave solver without any simplifying assumptions and it allows to simulate 3D objects, accounting for materials composition and shaped/geometries of each object. This tool is used for antenna design and to simulate complex electronic circuit elements including filters, transmission lines and packaging. [5]

The paper is organized as follows. In the first section, an overview of the power module is reported, with a description of its electromechanical characteristics. The second and third sections are dedicated to the modelling of the power module and to

1627

the experimental model validation and losses evaluation. The last section is focused on the radiated emissions analysis in different working points.

2 Power Module Overview

The Acepack™Drive is a three-phase inverter power module optimized for hybrid and electric vehicles traction. This power module features switches based on Silicon Carbide Power MOSFET 3rd generation, characterized by very low R_{DSon} and very limited switching losses. A copper base plate with pin-fin base structure makes direct fluid cooling available for this power module minimizing thermal resistance. A dedicated pin-out has been developed to get the best switching performance and press-fit pins will ensure optimal connection with driving board [6].

2.1 Specifications

The power module under investigation is the Acepack™ Drive ADP480120W3 (Fig.1); it is a three-phase topology with three parallel legs. Each switch is constituted by eight $25.55\ mm^2$ SiC Power MOSFETs and $5\ x\ 15\ mils$ bond wires for each die. The internal layout is shown in Fig.2

Fig. 1 Acepack™Drive package with pin-fins detail.

The main features of the PM are [7]:

- 1200 V blocking voltage;
- 1.9 $m\Omega$ typical $R_{DS\,on}$;
- Maximum operating junction temperature $Tj = 175°C$;
- Very low switching energy;
- Low inductive compact design to guarantee high power density;
- Si_3N_4 AMB substrate to improve thermal performances;
- SiC power MOSFET chip sintered to substrate for improved lifetime;
- 1.2 kV_{DC} 1s insulation;

- Directly liquid cooled base plate with pin-fins;
- Three integrated NTC temperature sensors.

Fig. 2 Internal layout of the Acepack™ Drive.

One of the most important figures of merit for the power modules is the equivalent inductance of the path between $DC+$ and $DC-$, called stray inductance. It's important to evaluate the L_{stray} during the design process because high values may cause voltage overshoot, reduced efficiency, and high frequency oscillation during switching transients. The typical stray inductance of the ADP480120W3 is around 10 nH; small shifts from this value are due to the manufacturing process. The stack of materials of the structure are listed in table I.

Material	Object	Relative Permittivity	Relative Permeability	Bulk Conductivity
Copper	Lead, Top layer, Bottom layer	1	1	58e6 Si/m
Solder	Solder	1	1	7e6 Si/m
Al2O3	Dielectric Layer	9.8	1	0
Cu-DHP	Pin holder	1	0.999991	45e6 Si/m
CuSn6	Pin	1	1	9e6 Si/m
Aluminum	Bondwire	1	1.000021	38e6 Si/m

Table 1 Stack of materials.

From the thermal point of view, the power module has a heat exchange surface equipped with pin fins to maximize the thermal exchange with the fluid. This detail is shown in Fig.1.

2.2 Electrical Modeling

The virtual prototyping of the system concerns the modeling of all the components which constitutes the prototype to validate the design. The main involved components are:

- The behavioral SPICE models of the dies;
- The SPICE parasitic matrix of the layout;
- The models of DC link, load and driving circuits.

The behavioral SPICE model of the MOSFET is developed from its characterization with curve tracer. This model, thanks to a set of mathematical expressions, can accurately reproduce the behavior of the device, in all allowed working points and with respect to changing temperatures. It is also a self-heating model, its temperature changes during the functioning and its parameters change with the temperature. The SPICE parasitic matrix of the layout is derived from Ansys HFSS starting from the 3D CAD of the power module. By assigning all materials and nets, the broadband parasitic matrix (up to $1\,GHz$) is extracted. Such a matrix is represented as a block scheme with its ports interconnected between them and with the external components (dies, DC link; load and gate driving) to reproduce the real interconnections of the system. The DC link is modeled as a DC voltage source with its real equivalent series resistance ($5\,m\Omega$) and inductance ($25\,nH$) coming from the laboratory experiments. At the same manner the gate driving circuit parasites are modeled. In particular, $10\,nH$ lumped inductance is used to model the loop inductance of the driving of the active switch and $5\,nH + 1\Omega$ for the Miller loop of the passive switch (to emulate the active Miller clamp). The square wave voltage generator gives in output the two voltage levels requested for the SiC driving, $-5\,V$ and $18\,V$. The $R_{G\,on}$ and $R_{G\,off}$ are chosen for each working point to set the correct switching speeds. Also, the RL load is set in each case to establish the load current level.

3 Ansys-Based Model Validation

Once the model of the system is defined, the Double Pulse Test has been performed to analyze the switching behavior of the power module. In particular, the switching losses have been estimated and compared to the energy levels get in measurements. One simulation-experimental waveforms comparison is showed to evidence the alignment.

3.1 Switching Behaviour of the PM

The defined model has been simulated through Ansys Electronics Circuit tool. The time step of the simulation ($T_s = 0.1\,ns$) is chosen to spot several characteristics, as potential high frequency oscillations (several MHz) during commutations. The matching between the waveforms coming from measurements and simulations has been conducted at $V_{bus} = 800\,V$ and different load currents and switching speeds. In Fig. 3 and 4, a turn off and turn on waveforms are shown to evidence the alignment between measurements and simulations. The result is the good behaviour of the model in both commutations. The turn off is ana-

lysed at $25°C$, because it represents the worst operating condition due to the lower overall resistance of the PM.

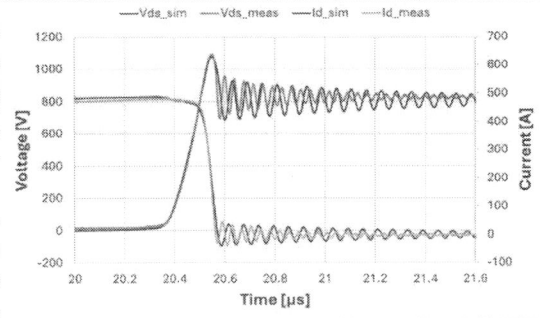

Fig. 3 Current and voltage comparison between measurements and simulations during turn-off commutation. $V_{bus} = 800V$, $I_{LOAD} = 840A$.

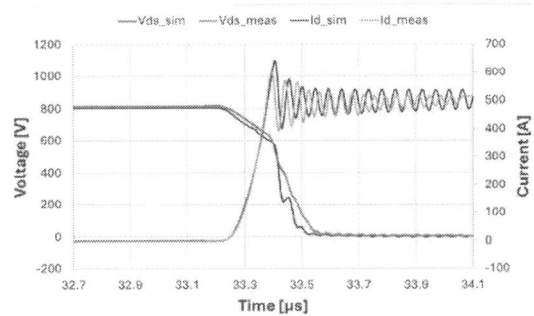

Fig. 4 Current and voltage comparison between measurements and simulations during turn-on commutation. $V_{bus} = 800V$, $I_{LOAD} = 840A$.

The voltage peak on the active switch is monitored during this commutation and it must be lower than $1200\,V$, that is the breakdown voltage of the MOSFETs. This voltage peak value is related to the stray inductance of the power module and to the switching speed through the equation (1).

$$V_{peak} = V_{DC} + L_{stray}\,{di}/{dt} \tag{1}$$

The turning on is evaluated at $175°C$, because higher is the working temperature of the die, higher is the slope of the SiC transfer characteristic that leads to higher switching speed. The ground unbalance phenomenon is usually investigated during the turn-on because it concerns the effect of the switching speed on the passive switch. If the V_{gs} of the passive switch is higher than V_{th}, for a certain interval related to the die characteristics, it could lead to unwanted turn-on. This phenomenon could cause a short circuit of the leg with a consequent permanent damaging of the power module. This investigation is not treated in this paper.

3.2 Losses Evaluation

To achieve good switching performance of the power module a compromise between the switching speeds and the switching losses is necessary: The higher the speeds, the lower the losses. Unfortunately, the speeds are limited by the system, especially by parasitic inductances and by the specific application. The power losses evaluation is a fundamental step, and it must be carried out at different temperatures and working conditions to ensure the goodness of the model. To verify its predictive capabilities, the simulated energies are compared to the measured ones (Fig. 5 and 6).

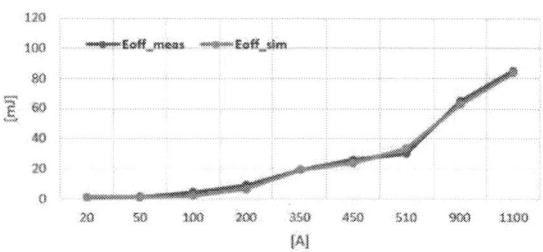

Fig. 5 Energy comparison for different load current and constant R_G during turn-off commutation at 25°C.

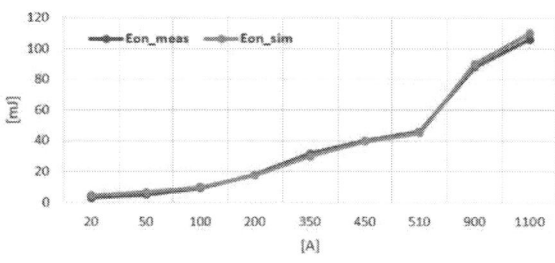

Fig. 6 Energy comparison for different load current and constant R_G during turn-on commutation at 175°C.

The different working points have been chosen maintaining the same R_G setting and varying the load currents from $20\,A$ to $1100\,A$; this implies the variation in the switching speeds and in the overvoltage value. The optimum alignment and the goodness of the simulated model is proven, and it can be used to esteem the PM behavior in the further and even more complex analysis.

4 Ansys HFSS Radiated Field Evaluation

The proposed EMC/EMI simulation workflow is a multi-step process that begins with the reproduction of the system's 3D geometry to be simulated. This involves a 3D structure that encompasses dies, bonding wires, pins, connectors, vias, solder balls, signal traces, power, and ground planes. The structure is elaborated in HFSS, that is a full-wave 3D simulator. All the system's input and output ports are defined for the extraction of S, Y, Z parameters. The time-domain waveforms coming from the simulations described in the section 3 are then converted into electrical spectra by applying the Fourier transformation and are pushed for the final radiated electromagnetic field simulation. The simulation workflow is depicted in Fig. 7.

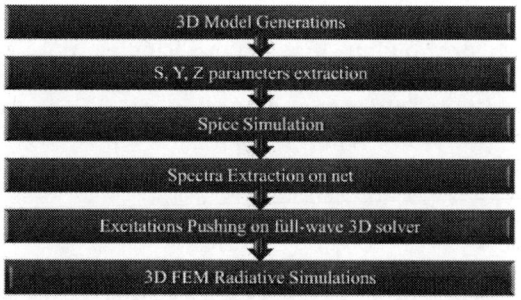

Fig. 7 Proposed simulation flow for radiated emissions estimations.

4.1 3D FEM Radiative Simulation Setup Definition

Before enforcing the excitations for the final radiative simulation, is necessary to import the complete structural model into HFSS via a dynamic link. It allows a dynamic relation between the transient circuital simulation and the FEM analysis. To establish the tetrahedral curvilinear mesh, the software employs the finite-element method (FEM); the automatic mesh method was selected for its efficiency and effectiveness. The resulting mesh is in Fig.8.

Fig. 8 Final Mesh of the entire structure.

Boundary conditions are then defined to specify the field behavior into a proper radiation boundary, in which the software solves Maxwell's equations to calculate the electromagnetic fields. The software sets it automatically depending on the 3D model of the DUT. The solution type chosen for this study is the Terminal Modal one. It allows to investigate the structure in terms of voltages and currents; another possibility is to set the driven modal solution type which is related to the incident and reflected waves. The solution frequency is set at $100\ MHz$ to guarantee the accuracy of the results into a wide range of frequencies; the discrete frequency sweep is configured in the following way:

- Linear Count: $0.001\ Hz \div 1 Hz$ with 2 points;
- Log Scale: $1\ Hz \div 50\ kHz$ with 25 samples;
- Linear Step: $50\ kHz \div 1MHz$ with a step size of $50\ kHz$;
- Linear Step: $1\ MHz \div 1GHz$ with a step size of $1\ MHz$.

4.2 Excitation pushing of full-wave 3D solver

Once the setup is defined, as described in the previous section, each port is connected to its corresponding input within the HFSS subcircuit. These inputs are the spectra coming from the experimental validated waveforms and are the V_{bus}, v_{ds} and v_{gs} of the active and passive switches. They are implemented through frequency dependent source data linked to the ports defined for the model. A frequency sweep ($0\ Hz$ to $1GHz$ with a step of $100\ kHz$) was introduced to facilitate communication between the HFSS block and the frequency-dependent sources. Once all the parameters and excitations are implemented, a push excitation approach is executed, by means of a dynamic link between Ansys Circuit and Ansys HFSS (Fig 9).

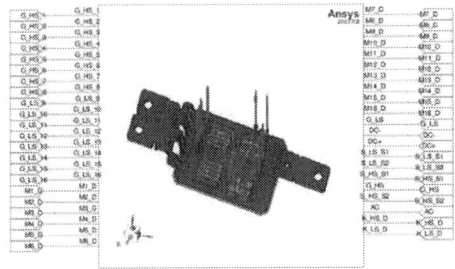

Fig. 9   HFSS dynamic link.

4.3 Simulation Results

Once the simulation is concluded, it's possible to plot the results. These can be displayed in the form of electromagnetic radiation as a function of the frequency harmonics, current density maps, electric and magnetic field behaviors. In this study, the electromagnetic energy plot is evaluated in two different working points, and the electric and magnetic field behaviors are reported for the case of the maximum radiated energy, in the frequency range higher than the commutation frequency ($50\ kHz$). The electromagnetic field emissions, coming from the power module under test, have been evaluated in two different conditions in which the active switch is the low side one:

- $V_{Bus} = 800\ V,\ I_{Load} = 840\ A,\ f = 50\ kHz$;
- $V_{Bus} = 400\ V,\ I_{Load} = 840\ A,\ f = 50\ kHz$

The commutation speeds are maintained in the two different cases and are equal to *15.5 V/ns* during the turning off and *6.8 A/ns* during the turning on. The electromagnetic energy plot resulting from this analysis is in Fig. 10; the vertical axis is expressed in *dBμV/m* as requested by the EMI standards (such as CISPR). Analyzing the EM energy plot, it can be noted that in the high frequency range, the high bus voltage contribution is dominant. On the contrary, in low frequency range, the low bus voltage contribution is higher than the other. This result is due to higher peak to peak oscillations after the commutations, as in Fig. 11 and 12. These oscillations, both during the turn off and turn on are due to internal resonances of the circuit and to the voltage-dependent capacitance of the dies.

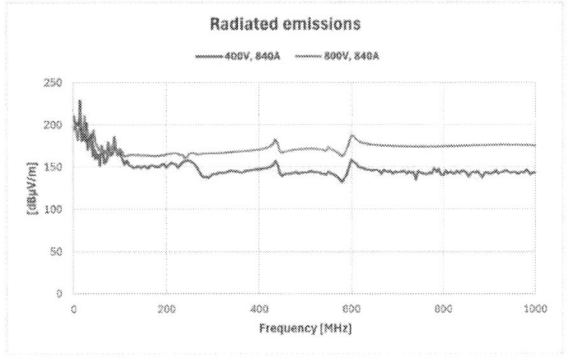

Fig. 10 Electromagnetic field plot evaluated in 1 m sphere in red $V_{Bus} = 800V$, in blue $V_{Bus} = 400V$.

The focus is on these higher amplitude peaks at low frequency:

- 14 MHz with 228 $dB\mu V/m$ for $V_{bus} = 400V$;

- 16 MHz with 211 $dB\mu V/m$ for $V_{bus} = 800V$.

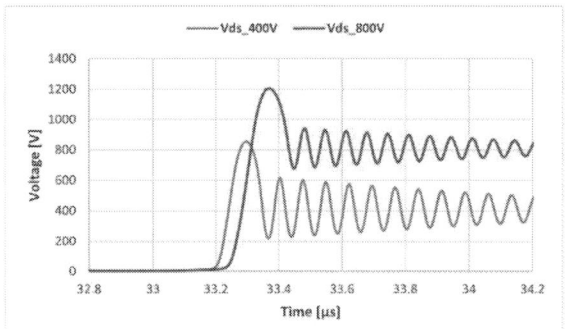

Fig. 11 Comparison between V_{ds} waveforms of the active switch for $V_{bus} = 400\,V$ (in blue) and $V_{bus} = 800\,V$ (in red).

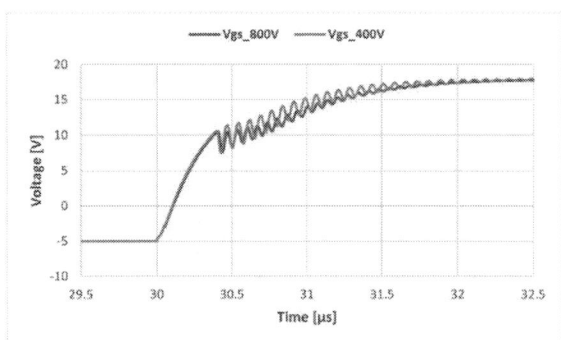

Fig. 12 Comparison between V_{gs} waveforms of the active switch for $V_{bus} = 400\,V$ (in blue) and $V_{bus} = 800\,V$ (in red).

The surface electric (Fig. 13 and 14) and magnetic (Fig.15 and 16) fields are traced respectively, for the two cases, at 14 and 16 MHz. As expected, the field levels reflect the results of the electromagnetic energy emission plot. The effect of the oscillations is more evident in the magnetic field map, in which is possible to distinguish the higher field concentration in the gate connections of the active switch due to the gate signal oscillations during the commutation. The same holds for the power bars which are affected by the iso-frequency oscillations in the power loop.

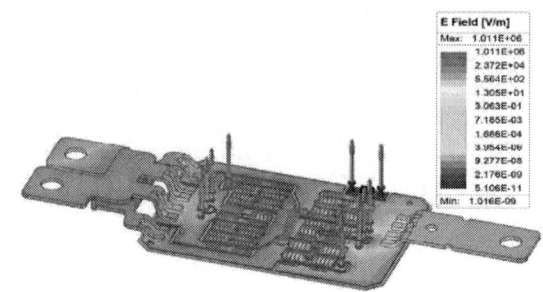

Fig. 13 Electric field behaviour on the PM surface at 14 MHz, in the case $V_{bus} = 400\,V$.

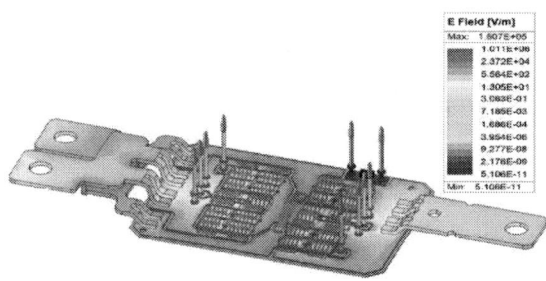

Fig. 14 Electric field behaviour on the PM surface at 16 MHz, in the case $V_{bus} = 800\,V$.

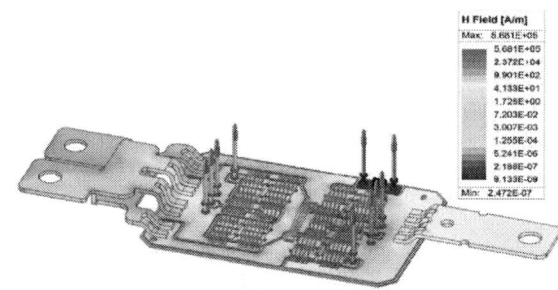

Fig. 15 Magnetic field behaviour on the PM surface at 14 MHz, in the case $V_{bus} = 400\,V$.

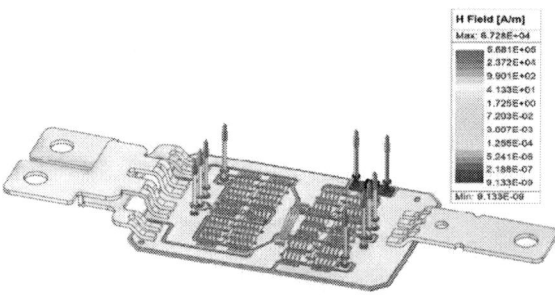

Fig. 16 Magnetic field behaviour on the PM surface at 16 MHz, in the case $V_{bus} = 800\,V$.

5 Conclusions

SPICE-based electrical model of the Acepack™ Drive ADP480120W3 housing 16 SiC MOSFETs has been developed and validated by experimental tests. Looking at the discussed results, the developed model can be suitably exploited to predict the module behavior for other operating conditions, keeping same simulation settings of the measurements-validated one. This complex behavioral system has been exploited to perform the estimation of the irradiated electric and magnetic fields in different operating conditions. It allows us to discover and predict, during the design phase, any crosstalk or system compatibility issues in a cheaper and less-time consuming way, compared to the physical tests. Some layout modifications could be performed during this phase, analyzing their effects on the emission spectra to satisfy the regulatory standards. In this work a method for the radiated emission estimation, for traction power modules, has been presented and it could be used also for the EM analysis of the whole system, integrating the real model of DC link, gate driver, motors, etc. and other electrical equipments that could interfere with the system under test.

References

[1] Y. Lai, Y. Yang, S. Wang and Z. Luo, "A Novel Low-Frequency Radiated Emissions Prediction Technique for the Inductor of a Non-Isolated Power Converter," *2022 IEEE Energy Conversion Congress and Exposition (ECCE)*, Detroit, MI, USA, 2022, pp. 1-8.

[2] A. KhakparvarYazdi, M. Mostafavi, A. Safaee and S. A. Khajehoddin, "An Alternative Method to Accurately Model Magnetic Components Using Ansys HFSS 3D," 2023 IEEE Applied Power Electronics Conference and Exposition (APEC), Orlando, FL, USA, 2023, pp. 1557-1564.

[3] A. M. Sayegh and M. Z. M. Jenu, "Estimation of Radiated Emissions from microstrip PCB using neural network model," *2016 IEEE Asia-Pacific Conference on Applied Electromagnetics (APACE)*, Langkawi, Malaysia, 2016, pp. 211-216.

[4] F. Lopez, A. Ramanujan, Y. V. Gilabert, C. Arcambal, A. Louis and B. Mazari, "A Radiated Emission Model Compatible to a Commercial Electromagnetic Simulation Tool," *2009 20th International Zurich Symposium on Electromagnetic Compatibility*, Zurich, Switzerland, 2009, pp. 369-372.

[5] https://www.ansys.com.

[6] A. Cusumano, D. Crimi, A. Raffa, G. Bazzano, A. Manzitto and L. Longo, "SPICE-based model validation for 1200V AcepackTM Drive traction power module," *2023 International Symposium on Electromagnetic Compatibility – EMC Europe*, Krakow, Poland, 2023, pp. 1-5.

[7] StMicroelectronics, Automotive-grade ACE-PACK DRIVE power module, sixpack topology 1200 V, 1.9 mΩ typ. SiC MOSFET gen.3 based, ADP480120W3 datasheet, Rev 3 - October 2022.

PCIM Europe 2024, 11– 13 June 2024, Nuremberg   DOI: 10.30420/566262223

Exact Analysis of a Control-to-Output Transfer Function of PWM-Converters - A Comparison of Two Methods

Sophia Roesel[©], Daniel Breidenstein[©], Thomas Duerbaum[©]

Friedrich-Alexander-Universität Erlangen-Nürnberg, Germany

Corresponding author: Daniel Breidenstein, daniel.breidenstein@fau.de
Speaker: Daniel Breidenstein, daniel.breidenstein@fau.de

Abstract

Generally, the design of control loops for power converters requires a transfer function. Commonly, averaging models are used to approximately describe the small signal behavior. This paper compares two alternative methods based on time discrete modeling, resulting in an exact prediction. Two perturbation methods will be explained and discussed with their advantages and disadvantages. The new methods are exemplarily demonstrated for a boost-converter. The paper reveals that with a typical design averaging methods correspond to the proposed exact methods. Furthermore, for critical designs given in literature, the exact methods predict the correct result, in contrast to the averaging method.

1   Introduction

There are three challenges that have to be faced in switched mode power supplies (SMPS) design. One of those is that the design of a converter has to guarantee the performance at the highest possible efficiency. This can be achieved with the proper component selection. Furthermore, the converter has to perform within the electromagnetic compatibility standards. Lastly, controlling the converter is necessary to keep the output voltage constant under varying line and load conditions.

Cost pressure commonly favors analog control-ICs for DC/DC-applications. In case of pulse width modulation (PWM) - converters, the pulse width of the gate signal typically results from natural sampling, which compares the control signal to a sawtooth with a repetition rate equal to the intended switching frequency. This paper focuses on the analysis of the control-to-output transfer function, which represents the important starting point for the controller design. Several papers describe different approaches to determine the small signal transfer function of PWM-converters. One way of getting the transfer function is state space averaging [1]. Other possibilities are average switch modeling (ASM) [2, 3], generalized switch inductor model (GSIM) [4], sample data modeling [5], or time discrete modeling. In case of an existing hardware, the transfer function can also be measured with a gain-phase analyzer [6].

All the theoretical approaches – except the exact time domain modeling – employ certain approximations and, therefore, introduce some errors. Especially the averaging methods have been shown to deliver incorrect transfer functions for certain designs, as depicted in [7]. Thus, validating the simplified method requires an exact transfer function received by time discrete modeling using, in contrast to sampled data modeling, waveforms with fine resolution in time. This paper will introduce and compare two different methods to obtain transfer functions. The first method emulates the measurement procedure by exciting the DC/DC converter with a small sine wave superimposed on the steady-state duty cycle. The corresponding simulation delivers amplitude and phase of the desired transfer function at one single frequency point. This procedure must be executed several times to obtain the complete transfer function. The second method circumvents this drawback by using a rectangle to disturb the steady-state. The multi-frequency content of the rectangle allows calculating the transfer function by one single simulation run.

The paper is structured as the following. Section 2 shows the standard method of getting a transfer function with ASM, with a boost-converter as an example. Section 3 introduces the time discrete model based on the augmented state-vector method [8]. Section 4 discusses the two proposed methods of perturbation. Section 5 depicts the obtained results. The paper ends with a conclusion and an outlook.

2 Boost-converter with ASM

Because of simplicity, for both ASM and perturbation in time domain, a boost-converter with synchronous rectifier is used. Replacing the diode with another switch (T1) enables the operation in FCCM (Forced CCM). Using a synchronous rectifier achieves a higher efficiency due to the reduction of conduction losses. Fig. 1 shows the boost-converter.

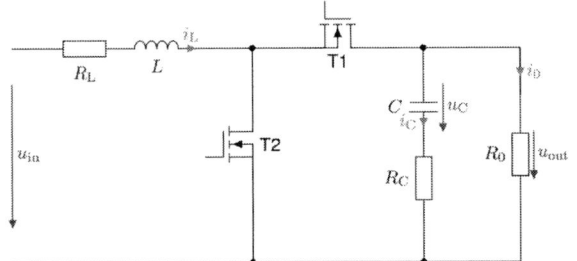

Fig. 1: Boost-converter with synchronous rectifier capable of operation in FCCM

Using FCCM allows a negative inductor current, while avoiding the discontinuous conduction mode (DCM). Furthermore, in FCCM, only two intervals occur. This resembles continuous conduction mode (CCM). However, in CCM, only a positive inductor current occurs. In the first interval with a duration of the duty cycle d, T2 is turned on. In the second interval with the duration corresponding to the complementary duty cycle, T1 is turned on. Fig. 2 shows the resulting inductor current waveform for different load conditions. With high load, the converter operates in normal CCM, and with low load, the operation is in FCCM. T_s is the switching period.

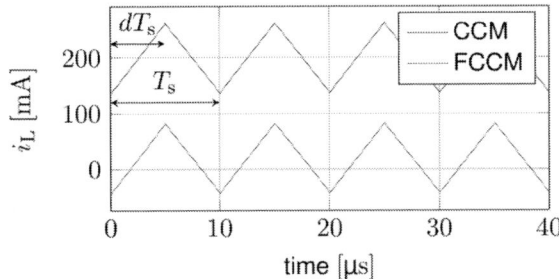

Fig. 2: Inductor current under FCCM and normal CCM operation

Since SMPSs are non-linear and time-variant, obtaining a transfer characteristic of such a device is rather difficult. To get rid of the time variance, the standard procedure applies averaging methods, such as ASM, GSIM, or state space averaging. ASM replaces the switch with a controlled voltage source and the diode with a controlled current source. The averaging over a switching period removes the high-frequency contents of the time waveforms, and therefore, the SMPS becomes time-invariant. The ASM of a boost-converter is shown in Fig. 3. This circuit diagram can be implemented in a tool like LTspice. With an AC-sweep, the control-to-output transfer function can be generated quickly. The parameters for the boost-converter used in this paper are summarized in Tab. 1.

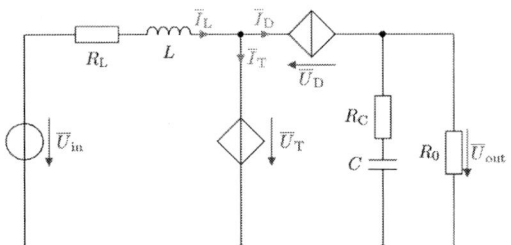

Fig. 3: Average switch model of the boost-converter

Fig. 4 depicts the magnitude and phase of the control-to-output transfer function received with the averaging model. It can be seen in [2] that the transfer function of the boost-converter with ASM agrees with the measurements. Therefore, a reasonably dimensioned boost-converter modeled with ASM can be taken as a reference point to verify the methods used in this paper for receiving the transfer function.

Tab. 1: Parameters of the used boost-converter

Parameters	Value
R_L	$100\,\text{m}\Omega$
R_C	$10\,\text{m}\Omega$
R_0	$10\,\Omega$
L	$200\,\mu\text{H}$
C	$220\,\mu\text{F}$
U_{in}	$5\,\text{V}$
d	0.5
f_s	$100\,\text{kHz}$

PCIM Europe 2024, 11– 13 June 2024, Nuremberg   DOI: 10.30420/566262223

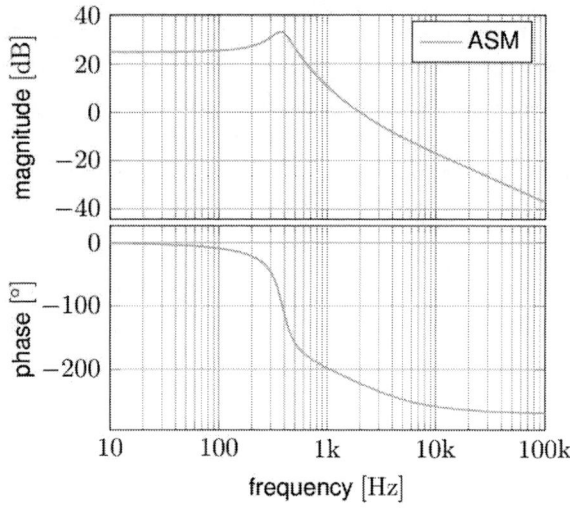

Fig. 4: Magnitude and Phase of control-to-output transfer function generated with ASM

However, as shown in [7], ASM does not always result in the correct transfer function. This is the case when the sizing of the components is done in a way that results in a very high amplitude of the output capacitor ripple. While ASM does not always correctly predict the transfer function, time discrete modeling achieves an accurate result, also for critical designs, like the one presented in [7].

3 Time discrete modeling

Time discrete modeling is used in power electronics to describe the stationary and dynamic behavior of a system. It allows fast simulation in time domain and is, therefore, beneficial during power circuit design.

This paper uses the time discrete modeling to acquire the small-signal response. According to the possible combinations of switching stages, a number of sub-circuits can be generated, which are de-

scribed by first-order differential equation systems as seen in (1). The index i clearly characterizes the sub-circuits. While the boost-converter generally exhibits three sub-circuits, the description of the FCCM only requires two sub-circuits. The vector $\vec{u}$ contains the values of the voltage sources and respectively of the current sources. For the boost-converter $\vec{u}$ equals u_{in}. Furthermore, the state-vector $\vec{x}(t)$ contains the inductor currents and the capacitor voltages, as shown in (2).

$$\frac{\mathrm{d}\vec{x}(t)}{\mathrm{d}t} = \mathbf{A}_{\text{i}}\vec{x}(t) + \mathbf{B}_{\text{i}}\vec{u} \qquad (1)$$

$$\vec{x}(t) = \begin{pmatrix} i_{\text{L}} \\ u_{\text{C}} \end{pmatrix} \qquad (2)$$

Every interval is described by a matrix $\mathbf{A}$ and $\mathbf{B}$. Fig. 5 lists the matrices used for the boost-converter in this paper. For simulation of the switching operation, the used matrices have to be interchanged when switching at dT_{s} and T_{s}.

With the matrices, the steady-state of the system can be calculated preferably with the augmented method, which is explained in detail in [8]. The known time duration of each interval in case of FCCM allows a faster calculation of the steady-state.

The augmented method extends the state-vector to (3) and forms a new matrix $\hat{\mathbf{A}}_{\text{i}}$ by combining the matrices $\mathbf{A}_{\text{i}}$ and $\mathbf{B}_{\text{i}}$ from Fig. 5 into a single one [8]. This is shown in (4). It has to be noted that $\vec{u}$ has to be constant.

$$\hat{\vec{x}}(t) = \begin{pmatrix} i_{\text{L}} \\ u_{\text{C}} \\ \hline 1 \end{pmatrix} \qquad (3)$$

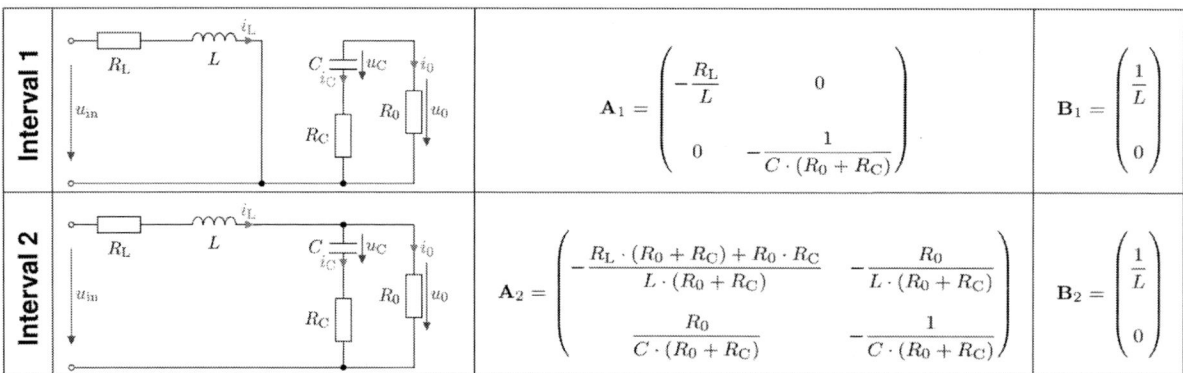

Fig. 5: State space representation of a synchronous boost-converter

1636

$$\hat{\mathbf{A}}_i = \left(\begin{array}{c|c} \mathbf{A}_i & \mathbf{B}_i \cdot \vec{u} \\ \hline 0 & 0 \end{array} \right) \qquad (4)$$

The waveforms of the state-vector in the time-domain can now be calculated with (5) [8]. One way to obtain the steady-state is to sequentially calculate (5), until reaching the steady-state condition given by (6). α_i represents the normalized duration of the corresponding interval.

$$\hat{\vec{x}}(t_i) = e^{\hat{\mathbf{A}}_i \alpha_i T_s} \cdot \hat{\vec{x}}(t_{i-1}) \qquad (5)$$

$$\hat{\vec{x}}(0) = \hat{\vec{x}}(T_s) \qquad (6)$$

Since (5) can be applied for any time instant within the corresponding interval by just changing the entry within the matrix exponent, it is also possible to oversample the high-frequency period by replacing $\alpha_i T_s$ with an arbitrary sampling time. This results in waveforms of the switching period over time. Fig. 6 shows an example of the inductor current and the capacitor voltage with an oversampling rate of 100 points per switching period over five switching periods. The other parameters are the same as in Tab. 1.

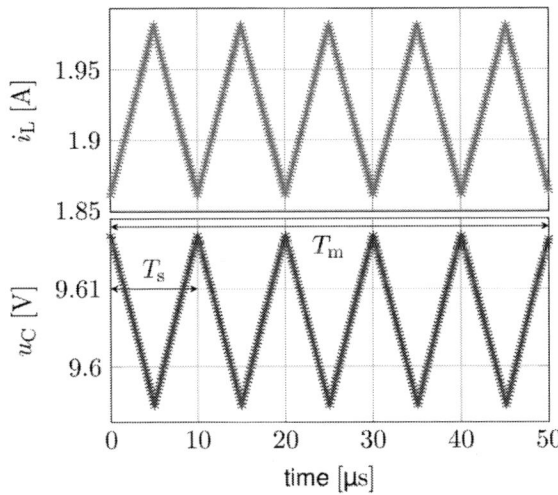

Fig. 6: Inductor current and capacitor voltage in time domain

With the time discrete modeling, the steady-state could be calculated. Apart from using the time discrete modeling for component dimensioning, it can also be applied to small-signal-analysis, shown in the next section.

4 Methods of Perturbation

To determine the control-to-output transfer function, the duty cycle needs to be perturbed in time domain, resulting in a corresponding disturbance of the output voltage. The perturbation amplitude needs to be within the small-signal range to avoid unwanted influences of non-linearities. The system can be perturbed with two different methods - with a sinusoidal signal (single frequency) or with a multi-frequency signal, e.g. in form of a rectangle. The implementation of both methods and the advantages and disadvantages will be described in this section.

4.1 Perturbation with sine wave

In section 3, the steady-state of the converter was calculated. After this step, a small-signal-perturbation is superpositioned on the steady-state, and a new steady-state, including the perturbation, must be calculated that differs from the original one. To receive amplitude and phase of the transfer function for a single frequency point, the duty cycle can be perturbed with a sine at the corresponding frequency. In order to obtain a steady-state with perturbation, the possible frequencies are limited by the condition that m perturbation periods T_m have to contain n switching periods T_s. With m and n being integer numbers, as shown in (7).

$$T_m = \frac{n}{m} \cdot T_s \qquad n, m \in \mathbb{N} \qquad (7)$$

The system reaches a new steady-state, including the perturbation that can be calculated by sequentially applying (5) until the steady-state condition (8) is met. However, instead of (6) the state vector must be periodic after $n \cdot T_s$.

$$\vec{x}(0) = \vec{x}(n \cdot T_s) \qquad (8)$$

For fast calculations, m can be set to one. If the control-to-output transfer function has too few values at higher frequencies, higher values of m allow more frequencies at intermediate points.

Section 1 mentions that analog control-ICs are typically used. In order to resemble the operation of standard control ICs, the duty cycle is calculated with natural sampling. Fig. 7 shows the principle, where $m = 1$ and $n = 4$.

It has to be noted that the amplitude of the perturbed control signal (in blue) is exaggerated for demonstration purposes and not in the range of a small signal perturbation. The sine equals the signal at one perturbation frequency. The straight lines represent a sawtooth signal that is compared to the sine, resulting in the PWM-signal. The black dots represent the turn-off instant for the corresponding switching period.

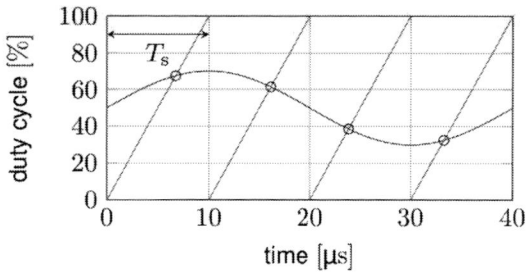

Fig. 7: Natural sampling

Commonly, the DC-value of the duty cycle, obtained by the unperturbed steady-state, is superpositioned with a sine. The amplitude of the perturbation sine is set in this example to $0.1\,\%$. The sine excites the system, due to the small amplitude, only at the perturbation frequency. Therefore, the result of the fast Fourier transformation (FFT) of the capacitor voltage only needs to be evaluated at the modulation frequency f_m, which is shown for $f_\mathrm{m} = 1\,\mathrm{kHz}$ in Fig. 8. It can be seen, that the non-linearities generate disturbances at a frequency of $2f_\mathrm{m}$ and higher frequencies, but due to the small-signal excitation their level is sufficiently small.

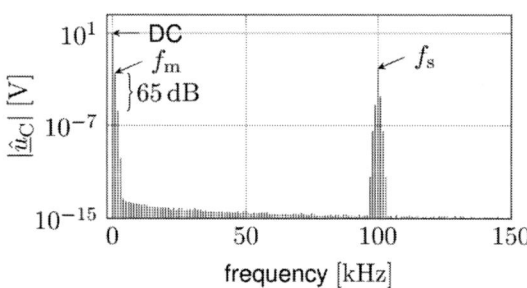

Fig. 8: Magnitude of the frequency spectrum of the sinusoidal perturbed capacitor voltage for $n = 100$

Magnitude and phase at this frequency produce one point of the control-to-output transfer function. This means that many perturbation frequencies have to be simulated to get a reasonable amount of points in the frequency domain for the control-to-output transfer function.

4.2 Perturbation with rectangle

Alternatively, the system can be perturbed with a rectangle. This yields the control-to-output transfer

function much faster than the sinusoidal perturbation because a rectangular signal superimposed on the steady-state duty cycle has multiple harmonics, as shown in (9). This has the effect that the system needs to be perturbated with only a few rectangles with different modulation frequencies, or even only with one.

$$\mathrm{rect} = \frac{A}{\pi} \sum_{k=1,3,\dots}^{\infty} \frac{1}{k} \cdot \sin(k \cdot \omega t) \qquad (9)$$

Of course, the perturbation with a rectangle is only allowed, if its amplitude (A) is in the small signal range, in order to suppress the non-linearities. Again, the perturbed steady-state given by (8) is calculated, the capacitor voltage is oversampled and an FFT is performed, which is shown in Fig. 9 for a fundamental perturbation frequency of $1\,\mathrm{kHz}$, which now can be evaluated at all uneven multiples of f_m up to below half the switching frequency.

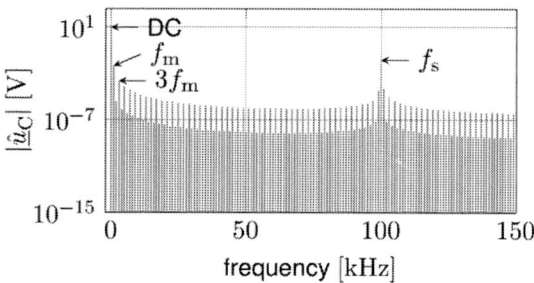

Fig. 9: Magnitude of the frequency spectrum of the rectangular perturbed capacitor voltage for $n = 100$

As seen in (9), the magnitude of the rectangle declines with higher frequencies. Therefore, the system needs to be perturbed with a few rectangles at different perturbation frequencies to avoid small amplitudes. If the magnitude of the FFT at the highest frequency, that is evaluated, is still large enough compared to noise, it is even possible to perturb with a rectangle at only one frequency. Needing less perturbation frequencies shortens the computing time drastically. Since the signal is a rectangle, only two values occur for the duty cycle within a modulation period. Contrarily, with a sinusoidal perturbation, each duty cycle value has to be calculated with natural sampling separately.

It has to be taken into account that perturbing the system with a rectangle leads to different aliasing-effects. There are different options to cancel the effects out to get a congruent control-to-output transfer function comparable to the perturbation with a sine. One possibility is to generate the oversampled control signal of the switch, perform an FFT, and extract the amplitudes and phases at the needed frequencies that correlate with the frequencies from the FFT of the capacitor voltage. The amplitude of the control-to-output transfer function results from the division of the magnitude of the capacitor voltage by the amplitudes of the modulation signal at the corresponding frequency. Additionally, the phase of the modulation signal is subtracted from the phase of the capacitor voltage. An alternative to the switch control signal is the discrete Fourier transformation (DFT) of the discrete duty cycle values for each switching period. Note that when using this option, the phase needs to be adjusted by the phase φ_{cor} after the subtraction, which is given by (10).

$$\varphi_{\mathrm{cor}} = 2\pi \cdot d \cdot f_{\mathrm{m}} \cdot T_{\mathrm{s}} \qquad (10)$$

4.3 Comparison of both perturbation methods

An example frequency of $100\,\mathrm{Hz}$ is taken to compare the computing time of the two perturbation methods. The results show roughly the same computational effort ($1.45\,\mathrm{s}$ for sine and $1.5\,\mathrm{s}$ for rectangle). However, as mentioned in 4.1, the computing time is needed for only one point of the transfer function in case of a sinusoidal perturbation. On the other hand, a single simulation with a rectangular excitation delivers the complete transfer function as explained in section 4.2. This shows that the computing time is much faster, when taking the rectangle for perturbation, because of the advantage of the multi-frequency excitation.

5 Results

Firstly, this section shows that the new methods can be used to verify the results obtained by ASM. After a short comparison of the two excitation methods the last part demonstrates that for special designs where ASM delivers misleading results, the new methods still yield, as expected, accurate transfer functions.

5.1 Comparison with ASM

The comparison of the sine wave excitation with ASM is done for the specification of the boost converter given in Tab. 1. Fig. 10 shows that using the time discrete model results in the same control-to-output transfer function that the ASM method

implemented in LTspice yields. This demonstrates that both ASM and the exact method presented in this paper result in the correct control-to-output transfer function for standard designs (see also [2] for validation of ASM against measurement).

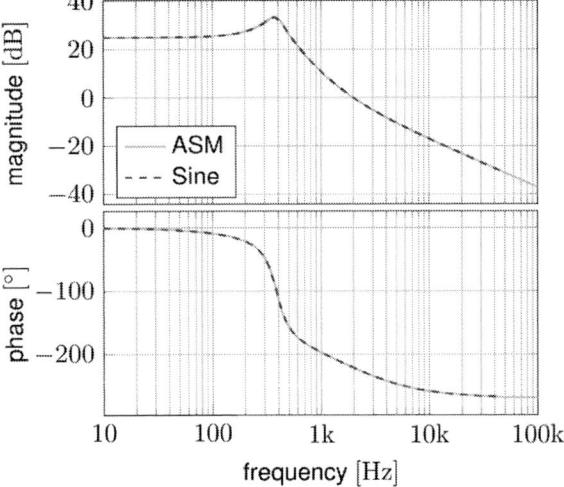

Fig. 10: Magnitude and Phase of control-to-output transfer function generated with ASM and time discrete modeling with sinusoidal perturbation

5.2 Comparison of the excitation methods

Since it is now verified that using time discrete modeling with a sinusoidal perturbation provides the correct result, it is further shown in Fig. 11 that perturbing with a rectangle also delivers the correct control-to-output transfer function.

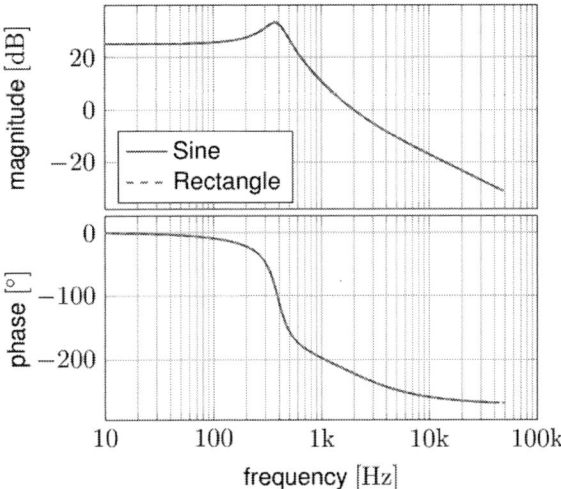

Fig. 11: Comparison of Magnitude and Phase of control-to-output transfer function between sinusoidal and rectangular perturbation

Note that with the time discrete modeling, the control-to-output transfer function can only be evaluated up to half the switching frequency. In section 4.3, the computing time of the two perturbation methods were presented. The comparison showed that using a rectangular perturbation results in a faster computing time. This is why this perturbation method is recommended when simulating the transfer function.

To further prove the accuracy of the presented method, a different duty cycle than the previous $d = 0.5$ is used. In Fig. 12, a duty cycle of $d = 0.6$ is used and again compared with ASM. It can be seen that both control-to-output transfer functions are congruent.

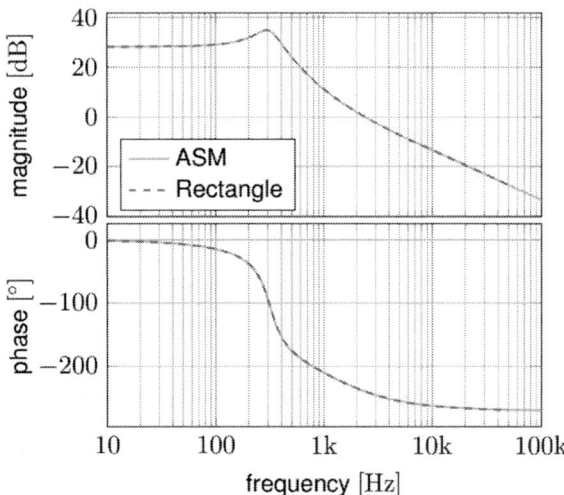

Fig. 12: Magnitude and phase of the control-to-output transfer function for a duty cycle of $d = 0.6$

5.3 Results for critical designs

In [7], it was shown that especially for small capacitance values with large ripple on the output voltage, ASM does not correctly predict the control-to-output transfer function anymore. However, it is expected that the presented exact methods determine the correct small signal transfer function. When using the dimensions of the component values, which were given in [7] ($L = 100\,\mu\text{H}$, $C = 4.4\,\mu\text{F}$, $R_\text{L} = 8\,\Omega$, $f_\text{s} = 10\,\text{kHz}$) a difference in the control-to-output transfer function with ASM and time discrete modeling can be seen in Fig. 13.

This is in agreement with Fig. 9 in [7]. The values calculated with the time discrete modeling result in a larger difference than the one presented in [7]. However, since the capacitor voltage with the high ripple in time domain calculated with the time dis-

crete modeling matches with Fig. 8 in [7], it can be concluded that the time discrete modeling results in the exact result for the control-to-output transfer function.

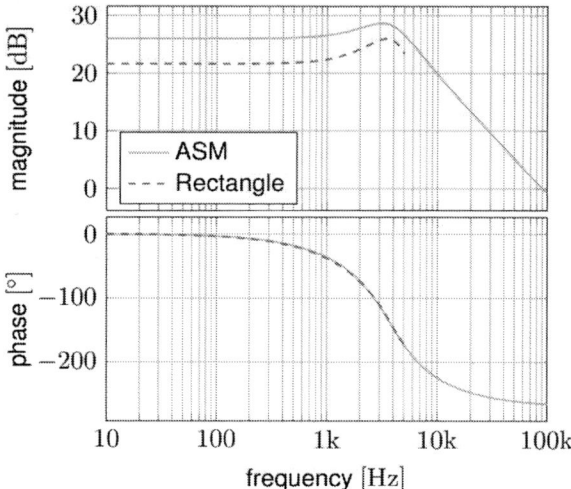

Fig. 13: Deviation between ASM and time discrete modeling

6 Conclusion

In this paper, the standard method for receiving a transfer function, which is ASM, was compared to a perturbation in time domain using time discrete modeling. Two ways of perturbation were explained. The results showed that the exact method met with ASM when the dimensioning of the converter values was reasonable. Additionally, there are some advantages of the exact method compared to ASM. It calculates an accurate result even in case of large capacitor voltage ripple.

The simulation of the transfer function when perturbed with a sine takes longer because the steady-state needs to be calculated for each perturbation frequency. Therefore, a multi-frequency stimulation was presented. A rectangle was used for perturbing because keeping the perturbation in the small signal scale can easily be guaranteed since its amplitude represents the maximum deviation from the steady-state duty cycle.

As explained, both methods were shown for a boost-converter operation in FCCM. This can be further extended to the discontinuous conduction mode. Additionally, the method can be applied to other converter topologies.

References

[1] R. D. Middlebrook and S. Cuk, "A general unified approach to modelling switching-converter power stages," in *1976 IEEE Power Electronics Specialists Conference*, Cleveland, OH: IEEE, Jun. 1976, pp. 18–34. DOI: 10.1109/PESC.1976.7072895.

[2] G. Wester and R. Middlebrook, "Low-Frequency Characterization of Switched dc-dc Converters," *IEEE Transactions on Aerospace and Electronic Systems*, vol. AES-9, no. 3, pp. 376–385, May 1973. DOI: 10.1109/TAES.1973.309723.

[3] A. Ayachit and M. K. Kazimierczuk, "Averaged Small-Signal Model of PWM DC-DC Converters in CCM Including Switching Power Loss," *IEEE Transactions on Circuits and Systems II: Express Briefs*, vol. 66, no. 2, pp. 262–266, Feb. 2019. DOI: 10.1109/TCSII.2018.2848623.

[4] "Generalized switched inductor model (GSIM): Accounting for conduction losses," *IEEE Transactions on Aerospace and Electronic Systems*, vol. 38,

no. 2, pp. 681–687, Apr. 2002. DOI: 10.1109/TAES.2002.1008997.

[5] A. R. Brown and R. Middlebrook, "Sampled-data modeling of switching regulators," in *1981 IEEE Power Electronics Specialists Conference*, Boulder, Colorado, USA: IEEE, Jun. 1981, pp. 349–369. DOI: 10.1109/PESC.1981.7083659.

[6] R. D. Middlebrook, "Measurement of loop gain in feedback systems," *International Journal of Electronics*, vol. 38, no. 4, pp. 485–512, Apr. 1975. DOI: 10.1080/00207217508920421.

[7] V. Caliskan, O. Verghese, and A. Stankovic, "Multi-frequency averaging of DC/DC converters," *IEEE Transactions on Power Electronics*, vol. 14, no. 1, pp. 124–133, Jan. 1999. DOI: 10.1109/63.737600.

[8] H. Visser and P. van den Bosch, "Modelling of periodically switching networks," in *PESC '91 Record 22nd Annual IEEE Power Electronics Specialists Conference*, Cambridge, MA, USA: IEEE, 1991, pp. 67–73. DOI: 10.1109/PESC.1991.162655.

PCIM Europe 2024, 11– 13 June 2024, Nuremberg   DOI: 10.30420/566262224

Three-Level Flying Capacitor Multilevel Topology with Delta-Sigma Modulation

Jannik Maier[1], Philipp Czerwenka[1], Eckhard Hennig[1], Ertugrul Sönmez[1], Gernot Schullerus[1]

[1] Electronics & Drives, Reutlingen University, Germany

Corresponding author: Jannik Maier, jannik.maier@reutlingen-university.de
Speaker: Jannik Maier, jannik.maier@reutlingen-university.de

Abstract

This paper proposes a novel modulation concept for a 3-level flying capacitor multilevel system using an asynchronous delta-sigma modulator (ADSM). The benefits of delta-sigma modulation, including noise shaping, spread-spectrum qualities, and a lower average switching frequency while maintaining the same signal quality as traditional pulse-width modulation (PWM) methods, can be exploited in multilevel power converters. A control circuit is introduced which determines the flying capacitor voltage from the output switching terminal and keeps the flying capacitor at a constant voltage. In this paper, we present the concept of the modulator and the control circuit to control the flying capacitor voltage. We confirm the modulation method by simulations and measurements.

1 Introduction

Multilevel inverter topologies increase the possible operating supply voltage for power devices with a given maximum voltage rating, enabling higher-power systems [1]. Increasing the voltage results in higher output power for a given maximum current [2]. Multilevel topologies for high-voltage systems up to 800 V are an emerging application in modern drive inverters due to weight savings and faster charging time, when it comes to battery-electric vehicles [3]. For such systems, a standard half-bridge leg requires semiconductor switches with breakdown voltages higher than the supply voltage V_{DC}. Gallium nitride (GaN) semiconductors offer advantages such as high-frequency switching and higher efficiency compared to other wide bandgap materials [4] but are limited to breakdown voltages of 650 V and can therefore not be used in an 800 V half-bridge topology. New circuit topologies are therefore necessary. Multilevel topologies allow the use of semiconductors with breakdown voltages lower than the supply voltage. The traditional multilevel topologies are the neutral point clamped (diode clamped), the cascaded H-Bridge and the flying capacitor topology [5].
The focus of this paper is on a 3-level flying capacitor topology shown in Fig. 1(b), which consists of

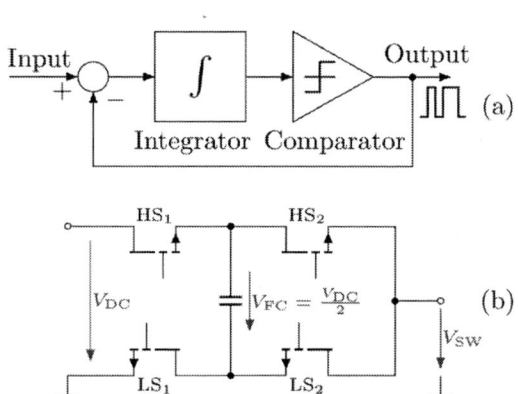

Fig. 1: Basic topologies (a) analog delta sigma modulator, (b) flying capacitor bridge-leg

the four switches HS_1, HS_2, LS_1, LS_2 and the flying capacitor (FC) charged to half the supply voltage. This ensures that the drain-source voltage of each switch is limited to $V_{DC}/2$. Such systems produce three output levels [V_{DC}, $V_{DC}/2$, 0 V] [6] where the middle level $V_{DC}/2$ is used for charging and discharging the capacitor [7]. Precautions must be taken to prevent excessive charging or discharging of the capacitor. In [8] various modulation schemes are presented which keep the flying capacitor at its target value. These modulation schemes often use pulse-width modulation (PWM), which modulates the output signal by varying the duty cycle of a fixed frequency signal. In the new approach presented

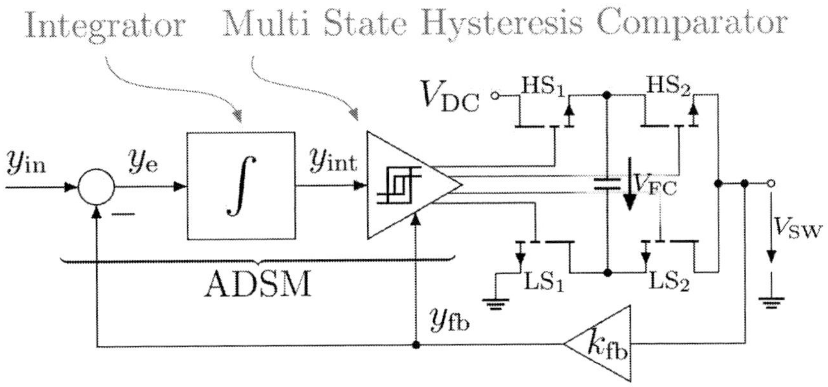

Fig. 2: Structure of 3-level analog delta-sigma modulator

in this paper, we propose an asynchronous delta-sigma modulator with a charge controller. The standard asynchronous delta-sigma modulator is shown in Fig. 1(a) and consists of an integrator and a comparator. In contrast to a PWM modulator, a delta-sigma modulator has a spread-spectrum feature that extends the switching frequency into a switching band. While maintaining the same signal quality, delta-sigma modulation reduces the average switching frequency as compared to pulse-width modulation methods [9], resulting in lower switching losses. During the modulation, the flying capacitor is charged and discharged, so the flying capacitor voltage V_{FC} changes slightly, causing the output voltage V_{SW} to deviate from $V_{DC}/2$. To reduce the error caused by the voltage ripple on the flying capacitor, it is important to introduce a feedback of the switching-node voltage V_{SW}. A delta-sigma modulator can inherently handle this feedback and is therefore well suited for this application. In the new approach, we combine the delta-sigma modulator and the flying capacitor multilevel topology.

The structure of the paper is as follows: First we describe the concept for a 3-level flying capacitor multilevel system using an asynchronous delta-sigma modulator in Section 2. This will also include design criteria for the flying capacitor. In Section 3 we present simulation results which are substantiated by measurements on a prototype in Section 4. Section 5 concludes this paper.

2 Concept of 3-level flying capacitor system using ADSM

The proposed concept for a 3-level flying capacitor multilevel system using an asynchronous delta-sigma modulator (ADSM) is shown in Fig. 2. The system consists of an asynchronous delta-sigma modulator, which generates the gate signals for the switches (HS_1, HS_2, LS_1, LS_2), and the flying capacitor multilevel topology. The ADSM subtracts the feedback signal y_{fb} from the input signal y_{in}, where y_{fb} depends on the switching-node voltage via the downscaling factor k_{fb}. The resulting error signal y_e is then integrated by the integrator and fed forward (y_{int}) to the multi-state hysteresis comparator (MSHC), while a standard delta-sigma modulator uses a conventional comparator at this point. In addition to controlling the switching-node voltage, the voltage of the flying capacitor is also controlled. To do so, the signal y_{fb} is used as an input signal for the MSHC. The four output signals of the MSHC are connected to the gates of the switches (HS_1, HS_2, LS_1, LS_2). A detailed description of the MSHC with the working principle is given in Section 2.1. Figure 3 illustrates the four possible output levels of the flying capacitor multilevel topology. In order to describe the different levels, we assume that the flying capacitor is charged to $V_{FC} = V_{DC}/2$.

In the high level ③, both high-side switches are ON, resulting in a switching-node voltage $V_{SW} = V_{DC}$. The node between the two switches, LS_1 and LS_2, is pulled to $V_{DC}/2$ by the flying capacitor. As a result, the drain-source voltage of both low-side switches equals $V_{DC}/2$.

In the first middle level ②c, the switches HS_1 and LS_2 are ON, resulting in a switching-node voltage $V_{SW} = V_{DC}/2$. According to the current direction, the flying capacitor is charged by the positive current i_{out}. The drain-source voltage of the switches HS_2 and LS_1 equals $V_{DC}/2$.

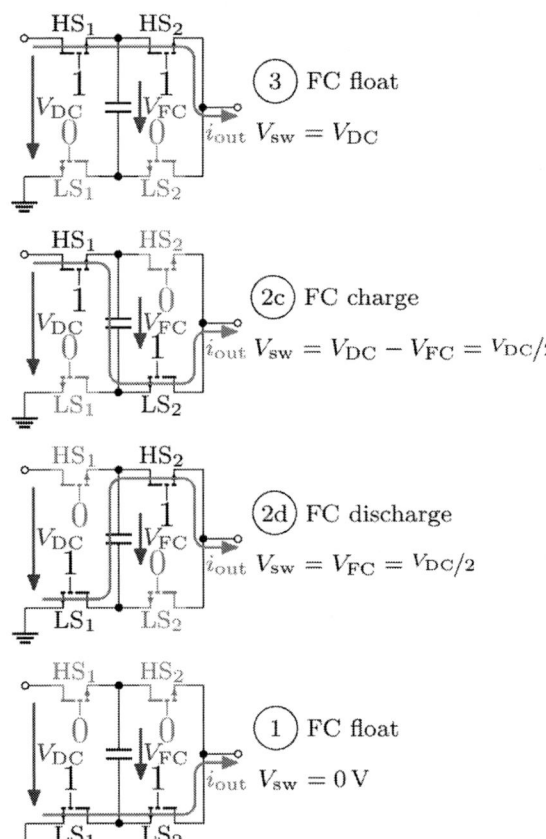

Fig. 3: Output level of the flying capacitor multilevel topology

In the second middle level ②d, the switches HS_2 and LS_1 are ON, resulting in a switching-node voltage $V_{SW} = V_{DC}/2$. According to the current direction, the flying capacitor is discharged by the positive current i_{out}. The drain-source voltage of the switches HS_1 and LS_2 equals $V_{DC}/2$.

In the low level ①, both low-side switches are ON, resulting in a switching-node voltage $V_{SW} = 0$. The node between the two switches, HS_1 and HS_2, is pulled to $V_{DC}/2$ by the flying capacitor. As a result, the drain-source voltage of both high-side switches equals $V_{DC}/2$.

The discussion given above demonstrates that three different switching-node voltages

$[V_{DC}, V_{DC}/2, 0\,V]$ are generated by the four output levels, while the drain-source voltage of each switch must not exceed $V_{DC}/2$.

2.1 Multi-State Hysteresis Comparator (MSHC)

The multi-state hysteresis comparator is the key component in the new concept presented in this paper. It can be divided into the signal processing part (Fig. 4) and the state machine (Fig. 5).

The signal processing part in Fig. 4(a) generates the clock signal (Clk) and the two transition statements c_1 and c_{th}. It consists of two hysteresis

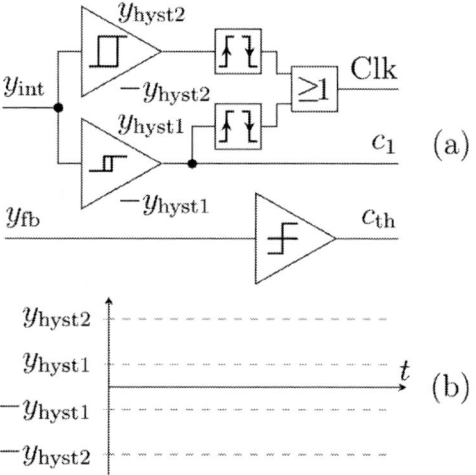

Fig. 4: MSHC (a) signal processing, (a) thresholds arrangement

comparators which compare the signal y_{int} with the inner thresholds $\pm y_{hyst1}$ (comparator 1) and the outer thresholds $\pm y_{hyst2}$ (comparator 2). The arrangement of the thresholds is shown in figure Fig. 4(b). As soon as the signal y_{int} reaches one of the four thresholds, a clock signal is generated and the state machine changes to a new state. The signal c_1 is the output of comparator 1 which indicates whether one of the positive thresholds (y_{hyst1}, y_{hyst2}) or one of the negative thresholds ($-y_{hyst1}$, $-y_{hyst2}$) is reached when a clock signal is generated. To detect whether a positive or a negative threshold is reached, the signal c_1 is sufficient due to the arrangement of the thresholds.

In state ②d and ②c, the switching-node voltage and the flying capacitor voltage are equal ($V_{SW} = V_{FC}$), so the feedback can be used to determine the voltage of the flying capacitor. The signal c_{th} evaluates if the flying capacitor voltage is above or below the middle voltage $V_{DC}/2$ and therefore needs to be charged or discharged.

If the signal c_{th} equals one ($c_{th} = 1$) in state ②d, the flying capacitor voltage is above $V_{DC}/2$ and must be discharged. If $c_{th} = 0$ the flying capacitor voltage is below $V_{DC}/2$ and needs to be charged.

If the signal c_{th} equals one ($c_{th} = 1$) in state ②c, the flying capacitor voltage is below $V_{DC}/2$ and must be charged. If $c_{th} = 0$ the flying capacitor voltage is above $V_{DC}/2$ and needs to be discharged.

The internal signals Clk, c_1 and c_{th} are the input signals for the multi-state hysteresis comparator state machine shown in Fig. 5. For each state the

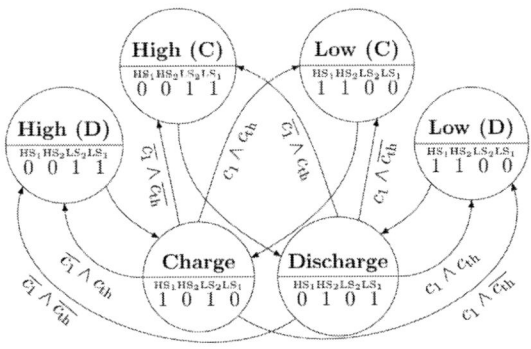

Fig. 5: MSHC state machine

name and the gate signals are specified. Note that level ① and ③ are represented by two states each, since the following state is also stored by only one possible exit path. The states are:

State	Output Level	Possible Following States
Low (C)	① $\cong 0\,V$	**Charge**
Low (D)	① $\cong 0\,V$	**Discharge**
Charge	②c $\cong V_{DC}/2$	**Low (C), Low (D), High (C), High (D)**
Discharge	②d $\cong V_{DC}/2$	**Low (C), Low (D), High (C), High (D)**
High (C)	③ $\cong V_{DC}$	**Charge**
High (D)	③ $\cong V_{DC}$	**Discharge**

Tab. 1: Switching states

The state evolution of the MSHC in Fig. 5 is governed by the clock signal events and the level of the signals c_1 and c_{th}. When a clock signal event occurs and one of the positive thresholds is reached ($c_1 = 1$), the MSHC switches to a state where the output voltage is increased by $V_{DC}/2$. If $c_1 = 0$ when a clock signal event is received, one of the negative thresholds is reached and the MSHC switches to a state where the output voltage is decrease by $V_{DC}/2$.

The two **High** states result in the highest possible output Voltage ($V_{out} = V_{DC}$) while the two **Low** states result in the lowest possible output voltage ($V_{out} = 0$). These states are always followed by one of the middle states (**Charge**, **Discharge**). In contrast, the middle states are always followed by a **High** or **Low** state.

The charge control is active in the two middle states. At the end of each middle state, a decision is made whether the voltage across the flying capacitor is above or below its target value. This determines whether it should be charged or discharged in the next middle state. If the flying capacitor should be charged, the following state can only be **High (C)** or **Low (C)** because both states are followed by the **Charge** state. If the flying capacitor should be discharged, the following state can only be **High (D)** or **Low (D)** because both states are followed by the **Discharge** state. This behavior ensures that, depending on the voltage across the flying capacitor, the middle states are chosen to keep V_{FC} at the desired voltage $V_{DC}/2$. In Fig. 6 two different switching cycles are shown, where it is assumed that the voltage on the flying capacitor does not chance significantly. Figure 6(a) shows a sequence in which the switching-node switches between the low level and the middle level, with the signal y_{int} staying between $\pm\, y_{hyst1}$. During the interval of duration t_1, the error signal is negative and therefore y_{int} decreases to $-y_{hyst1}$. During t_2, the error signal is positive and therefore y_{int} increases until y_{hyst1} is reached. The switching frequency for this cycle can be calculated with Eq. (1) where k_i is the proportional gain factor of the integrator.

$$f_{1,2} = \frac{1}{T_{1,2}} = \frac{1}{\frac{2\,y_{hyst1}}{k_i\,(y_{fb,t1}-y_{in})} + \frac{2\,y_{hyst1}}{k_i\,y_{in}}} \tag{1}$$

The maximum frequency is reached when y_{in} is halfway between ②c and ① which is a typical

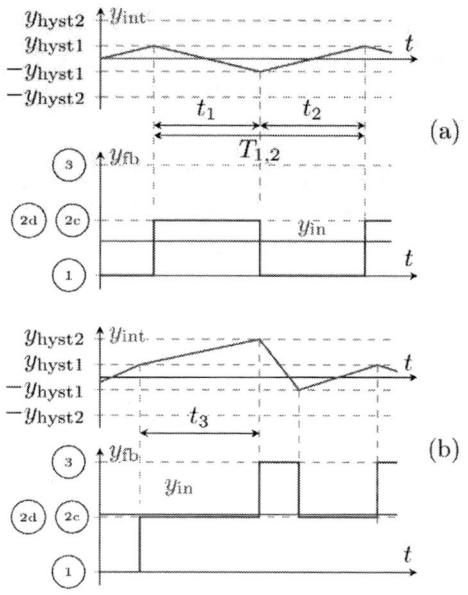

Fig. 6: Switching cycles (a) between low level and middle level, (b) input signal rises above the middle state

characteristic of delta-sigma modulators. Assume that y_{in} is in the range from zero to $\hat{y}_{in}$ and k_{fb} is designed with Eq. (2).

$$k_{fb} = \frac{\hat{y}_{in}}{V_{DC}} \qquad (2)$$

As a consequence, the signals y_{fb} and y_{in} are in the same range and the calculation of the maximum switching frequency can be done with Eq. (3).

$$f_{max} = \frac{k_i \hat{y}_{in}}{16 \, y_{hyst1}} \qquad (3)$$

The minimum ON/OFF time is reached, when y_{in} is close to its maximum or its average. The minimum time depends on y_{in}, the thresholds $\pm y_{hyst1}$ and the maximum feedback voltage $\hat{y}_{in}$ and can be calculated according to Eq. (4).

$$t_{min} = \frac{4 \, y_{hyst1}}{k_i \hat{y}_{in}} \qquad (4)$$

Figure 6 (b) illustrates the case when the input signal rises above the middle state for the first time, coming from the low state. After reaching y_{hyst1} at the beginning of t_3, the MSHC changes to a middle level. The signal y_{int} continues to rise because the error signal is still positive. The next clock signal is generated when y_{hyst2} is reached. At this

point, $c_1 = 1$ and the MSHC changes to one of the **High** states while y_{int} returns between $\pm y_{hyst1}$. This state only occurs when y_{in} exceeds the middle state.

2.2 Flying Capacitor Design Criteria

An important design parameter is the capacitance of the flying capacitor. The main design criterion is the maximum flying capacitor voltage ripple ΔV_{FC}, which is influenced by the maximum current, the maximum switching frequency and the longest ON time. The design criteria are discussed below.

Maximum Current: With a resistive load at the switching-node, the highest current is reached at the same time the input signal reaches the maximum. At this point the ON time is the shortest possible and C_{FC} can be calculated with Eq. (5), where $\hat{i}_{out}$ is the maximum output current and ΔV_{FC} the maximum allowed voltage change of the flying capacitor.

$$C_{FC} \geq \frac{\hat{i}_{out} \, t_{min}}{\Delta V_{FC}} \qquad (5)$$

Maximum switching frequency: The ON time during the maximum switching frequency can be calculated with Eq. (6).

$$t_{f,max} = \frac{1}{2 f_{max}} \qquad (6)$$

The capacitance for the maximum switching frequency can be calculated with Eq. (8). For a resistive load, the output current at this point is calculated in Eq. (7).

$$i_{f,max} = \frac{3 \, \hat{i}_{out}}{4} \qquad (7)$$

$$C_{FC} \geq \frac{i_{f,max} \, t_{f,max}}{\Delta V_{FC}} \qquad (8)$$

Longest ON time: Figure 7 shows a worst-case scenario where the middle state (t_{middle}) is ON for the longest possible time. To calculate the capacitance C_{FC}, the change in the flying capacitor voltage must be taken into account. During t_{middle}, the voltage of the flying capacitor changes due to the current i_{out} flowing through it. In t_4 the feedback voltage is higher than the input voltage resulting in a positive error signal. In the worst case, the positive error signal is integrated during t_4 as long as the upper threshold y_{hyst2} is just not reached when y_{fb}

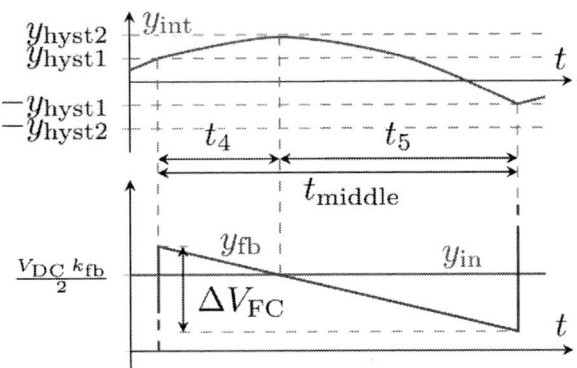

Fig. 7: Switching cycle of longest ON time

falls below y_{in} and the error signal becomes negative. As a result, the integrator output decreases for the longest possible time until $-y_{hyst1}$ is reached. With Eq. (9) and Eq. (10), the time t_{middle} and ΔV_{FC} can be calculated for a given C_{FC}.

$$t_{middle} = \frac{\sqrt{\dfrac{4\,k_i\,i_{out}\,y_{hyst2}}{C_{FC}\,k_{fb}}}}{\dfrac{k_i\,i_{out}}{C_{FC}\,k_{fb}}} \tag{9}$$

$$\Delta V_{FC} = \frac{i_{out}\,t_{middle}}{C_{FC}} \tag{10}$$

A design criterion for the flying capacitor given in Eq. (11) can be derived from Eq. (9) and Eq. (10).

$$C_{FC} \geq \frac{4\,i_{out}\,k_{fb}\,y_{hyst2}}{\Delta V_{FC}^2\,k_i} \tag{11}$$

If the assumption that the switching frequency is significantly greater than the frequency of the input signal is not sufficiently satisfied, a constant input signal is no longer sufficient and the input signal y_{in} must be approximated with a first order function. The value of C_{FC} can be influenced by the choice of the thresholds.

3 Simulation

The modulation strategy developed in Section 2 is compared by simulation with a PWM method, i.e. the alternative phase opposition disposition (APOD) method from [10]. Due to the charge and discharge state, the flying capacitor voltage oscillates around 400 V in Fig. 8(a). The input signal Fig. 8 (c) with a frequency of 100 Hz and peak-to-peak value of 90 % and the switching-node voltage, switching between all three output levels Fig. 8(b) are also presented. Figure 8(d) shows the frequency spectrum

of the switching-node signal. The harmonics of the fundamental frequency as well as the switching frequency of 97.25 kHz are clearly visible.

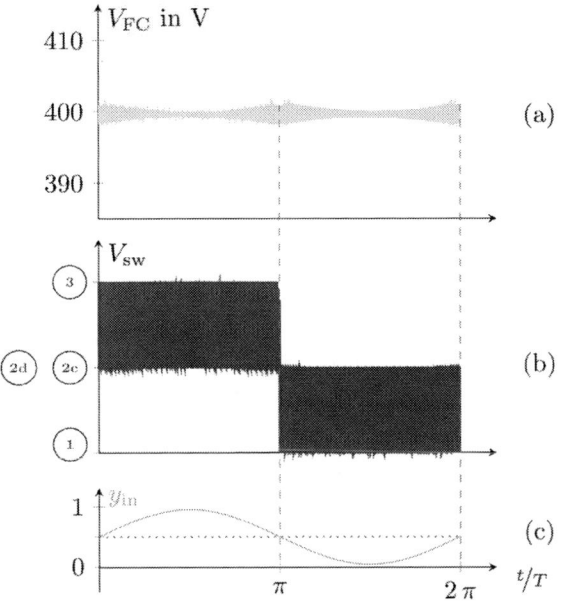

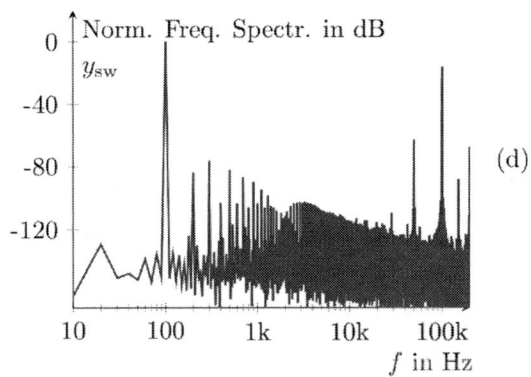

Fig. 8: Simulation of 3-level PWM (APOD) (a) flying capacitor voltage, (b) switching-node voltage, (c) input signal, (d) spectrum of the switching-node voltage

Figure 9 shows a simulation using the same input signal and an asynchronous delta-sigma modulator with the system parameters given in Tab. 2. The simulation indicates that the flying capacitor is stabilized at its target value with a deviation of ± 20 V Fig. 9(a). The voltage drop is due to the longest ON time and the worst case value can be calculated with Eq. (10). The output voltage switches between all three levels Fig. 9(b), and the frequency spectrum of V_{SW} exhibits the typical noise shaping

Tab. 2: Simulation parameters

f_{in}	1 kHz
V_{DC}	800 V
y_{hyst1}	$\pm\,2\,\text{mV}$
y_{hyst2}	$\pm\,3\,\text{mV}$
k_{i}	2×10^3

characteristic of an first-order ADSM Fig. 9(c) is also shown. Table 3 shows the difference in signal quality between a PWM and the ADSM. It is evident that the harmonics of the PWM are higher than the harmonics of the ADSM. To evaluate the signal quality, the total harmonic distortion (THD) was calculated for the first five harmonics. To compare the results, the average switching frequency was determined for the ADSM. For the same signal quality as a PWM modulator, an ADSM with a lower average switching frequency can be used. This results in lower switching losses.

Average switching frequency	PWM (THD)	ADSM (THD)
97.25 kHz	74.44 dB	77.69 dB
143.29 kHz	82.96 dB	85.2 dB
189.24 kHz	77.31 dB	86.68 dB

Tab. 3: Signal quality for pulse-width modulator and delta-sigma modulator

4 Measurements

To verify the simulation, measurements of the ADSM were performed with the results shown in Fig. 10. For safety reasons, the measurement was carried out at 100 V. The signal processing part of the multi-state hysteresis comparator was built on a printed circuit board (PCB), and the state machine was implemented on a microcontroller. The measurements were performed at lower switching frequencies due to the limitations of the microcontroller. The prototype on which the measurements

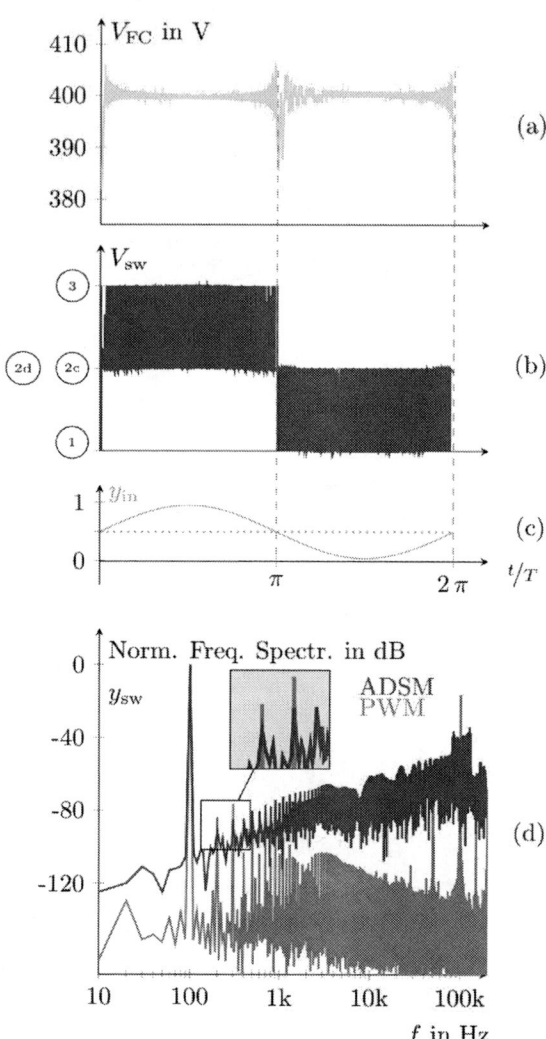

Fig. 9: Simulation of 3-level ADSM, (a) flying capacitor voltage, (b) switching-node voltage, (c) input signal, (d) spectrum of the switching-node voltage

were made is shown in Fig. 10(d). It consists of a DC-Link capacitor, the flying capacitor bridge-leg, the signal processing board, and the state machine. The input signal in Fig. 10(c) was chosen to 10 Hz with a peak-to-peak value of 90 % and a resistive load connected to the switching-node. The measurement confirms that the system exhibits the desired behavior alternating between all output levels in Fig. 10(b) and it is also evident that V_{FC} is maintained at its target value in Fig. 10(a). The voltage rise at $t = \pi$ is due to the longest ON time. In Fig. 10(d), the spectrum of the switching node is shown exhibiting the typical noise shaping characteristic of a first-order analog delta-sigma modu-

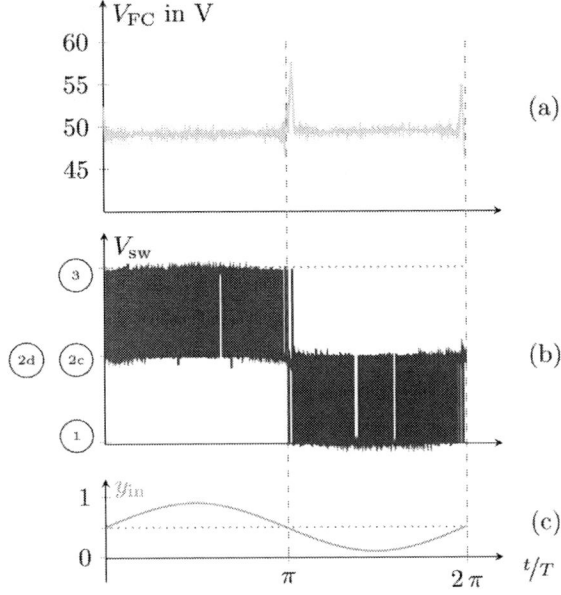

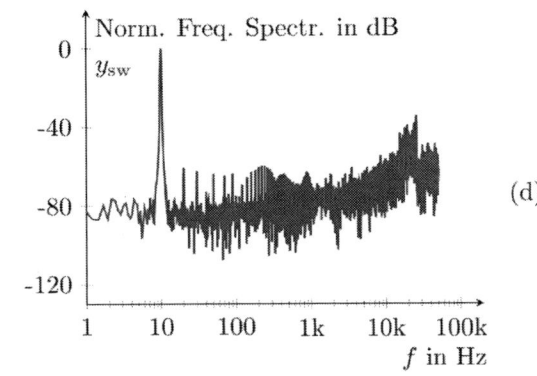

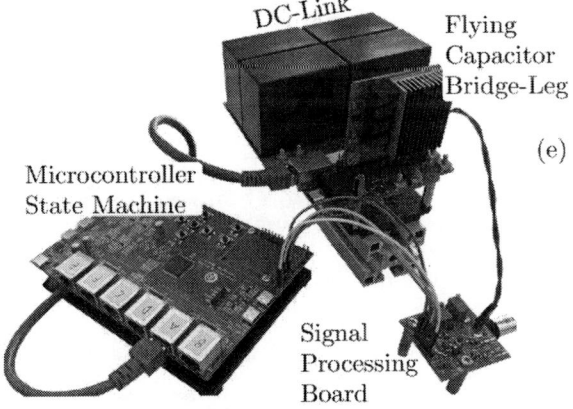

Fig. 10: Measurement of 3-level ADSM, (a) flying capacitor voltage, (b) switching-node voltage, (c) input signal, (d) spectrum of the switching-node voltage (e) measurement setup

lator with the switching band between 15 kHz and 25 kHz.

5 Conclusions

The paper presents a stable 3-level flying capacitor multilevel system controlled by an asynchronous delta-sigma modulator with a charge controller. The proposed topology can be used for motor applications since one system can be used for each motor winding. For the same switching frequency, delta-sigma modulation increases signal quality by up to 9.3 dB compared to pulse-width modulation methods. The proposed concept was verified by simulations and measurements.

Acknowledgment

This work was supported by the German Federal Ministry of Education and Research, FH-Kooperativ Project 13FH063KX1 "Clean Motor Supply".

References

[1] N. Choi, J. Cho, and G. Cho, "A general circuit topology of multilevel inverter," in *PESC '91 Record 22nd Annual IEEE Power Electronics Specialists Conference*, Cambridge, MA, USA, 1991, pp. 96–103. DOI: 10.1109/PESC.1991.162660.

[2] A. K. Vijayan, S. Chakkalakkal, and A. Emadi, "Efficiency evaluation of 800v electric vehicle powertrain using two-level voltage source inverter with different modulation techniques," in *2023 IEEE Transportation Electrification Conference & Expo (ITEC)*, Detroit, MI, USA, 2023, pp. 1–6. DOI: 10.1109/ITEC55900.2023.10186895.

[3] C. Jung, "Power up with 800-v systems: The benefits of upgrading voltage power for battery-electric passenger vehicles," *IEEE Electrification Magazine*, vol. 5, no. 1, pp. 53–58, 2017. DOI: 10.1109/MELE.2016.2644560.

[4] L. Spaziani and L. Lu, "Silicon, gan and sic: There's room for all: An application space overview of device considerations," in *2018 IEEE 30th International Symposium on Power Semiconductor Devices and ICs (ISPSD)*, Chicago, IL, USA, 2018, pp. 8–11. DOI: 10.1109/ISPSD.2018.8393590.

[5] A. Singh, S. Bhandari, and J. Kumar, "A comparative study on multilevel inverters with reduced number of components – a review," in *2022 1st International Conference on Sustainable Technology for Power and Energy Systems (STPES)*, SRINAGAR, India, 2022, pp. 1–6. DOI: 10.1109/STPES54845.2022.10006427.

[6] T. Meynard and H. Foch, "Multi-level conversion: High voltage choppers and voltage-source inverters," in *PESC '92 Record. 23rd Annual IEEE Power Electronics Specialists Conference*, Toledo, Spain, 1992, 397–403 vol.1. DOI: 10.1109/PESC.1992.254717.

[7] S.-G. Lee, D.-W. Kang, Y.-H. Lee, and D.-S. Hyun, "The carrier-based pwm method for voltage balance of flying capacitor multilevel inverter," in *2001 IEEE 32nd Annual Power Electronics Specialists Conference (IEEE Cat. No.01CH37230)*, vol. 1, Vancouver, BC, Canada, 2001, 126–131 vol. 1. DOI: 10.1109/PESC.2001.954006.

[8] B. McGrath and D. Holmes, "Multicarrier pwm strategies for multilevel inverters," *IEEE Transactions on Industrial Electronics*, vol. 49, no. 4, pp. 858–867, 2002. DOI: 10.1109/TIE.2002.801073.

[9] A. Mertens, "Performance analysis of three-phase inverters controlled by synchronous delta-modulation systems," *IEEE Transactions on Industry Applications*, vol. 30, no. 4, pp. 1016–1027, 1994. DOI: 10.1109/28.297919.

[10] D. Krug, "Vergleichende Untersuchungen von Mehrpunkt-Schaltungstopologien mit zentralem Gleichspannungszwischenkreis für Mittelspannungsanwendungen," *Dissertation, Technische Universität Dresden*, 2017.

PCIM Europe 2024, 11– 13 June 2024, Nuremberg DOI: 10.30420/566262225

Model Based Controlled Power Converter Test Platform

Dawid Koczy[1], Alexander Ernst[1], Wilfried Holzke[1], Bernd Orlik[2]

[1] IALB, University of Bremen, Germany
[2] IALB, University of Bremen, Germany until 01.04.2023
Corresponding author: Dawid Koczy, dkoczy@ialb.uni-bremen.de

Abstract

In order to investigate the cause of failure of power converters for e. g. wind energy plants, a 10 MW test rig and a climate chamber were set up at IALB. This configuration makes it possible to test entire switch cabinets under realistic climatic conditions and with varying electrical loads. To emulate the behaviour of a wind turbine generator a model-based approach was chosen. The shaft angle, for the field oriented based current control of the Device Under Test (DUT), is extracted from the calculation model. The machine angle is transferred with a low power inverter, a servomotor and a suitable encoder to the controller of the DUT.

1 Introduction

1.1 Topology

For the development of models and test scenarios, a 300 kVA pre-test platform was set up. The platform is a smaller replica of the 10 MW test rig. The system shown in Fig. 1 consist of two frequency converters, three transformers and a servo assembly. The power in this topology can be driven in a loop, only the loses are resupplied from the grid via the transformer T1. The taps on the secondary sides of T2 and T3 provide the flexibility to match the voltages required by the DUT.

A model-based control has been implemented in the frequency converter of the "load unit". The model consists of an electrically exited synchronous generator and a rotor which is driven by a generated wind field. The DUT is controlled like a typical wind energy plant converter with a current controller.

The servo assembly provides the mechanical angle of the generator shaft which is extracted from the generator model (see Fig. 2).

1.2 Generator model

The machine has been set up in MATLAB Simulink as an electrically excited synchronous generator [1]. The model is a normalized 12-pole generator with current inputs (see Fig. 3). Changing of parameter sets allows the adaptation to the DUT. This solution enables the investigation of wind turbine converters from different manufacturers. There is no need to purchase additional hardware.

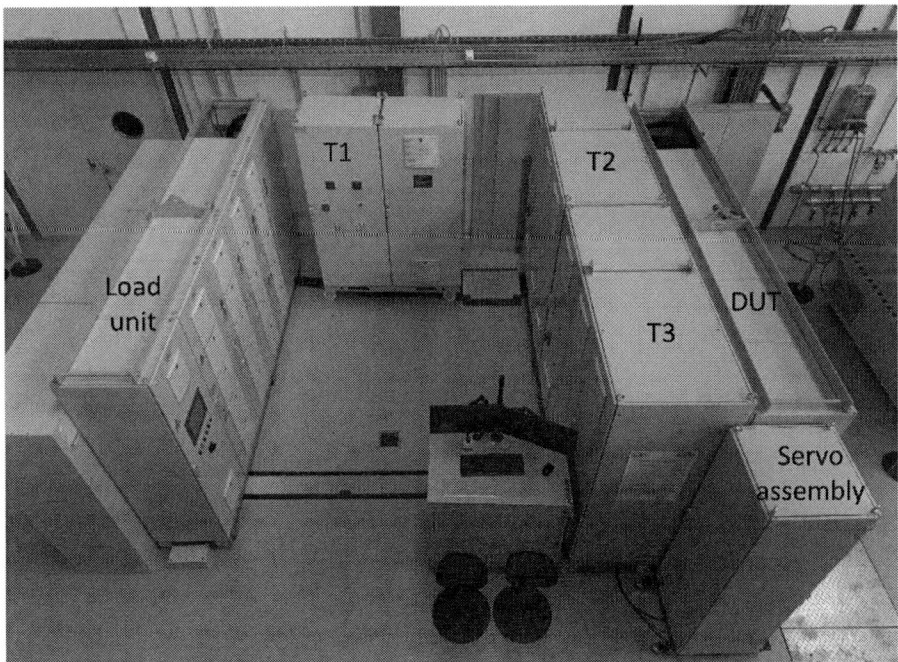

Fig. 1 Pre-test platform

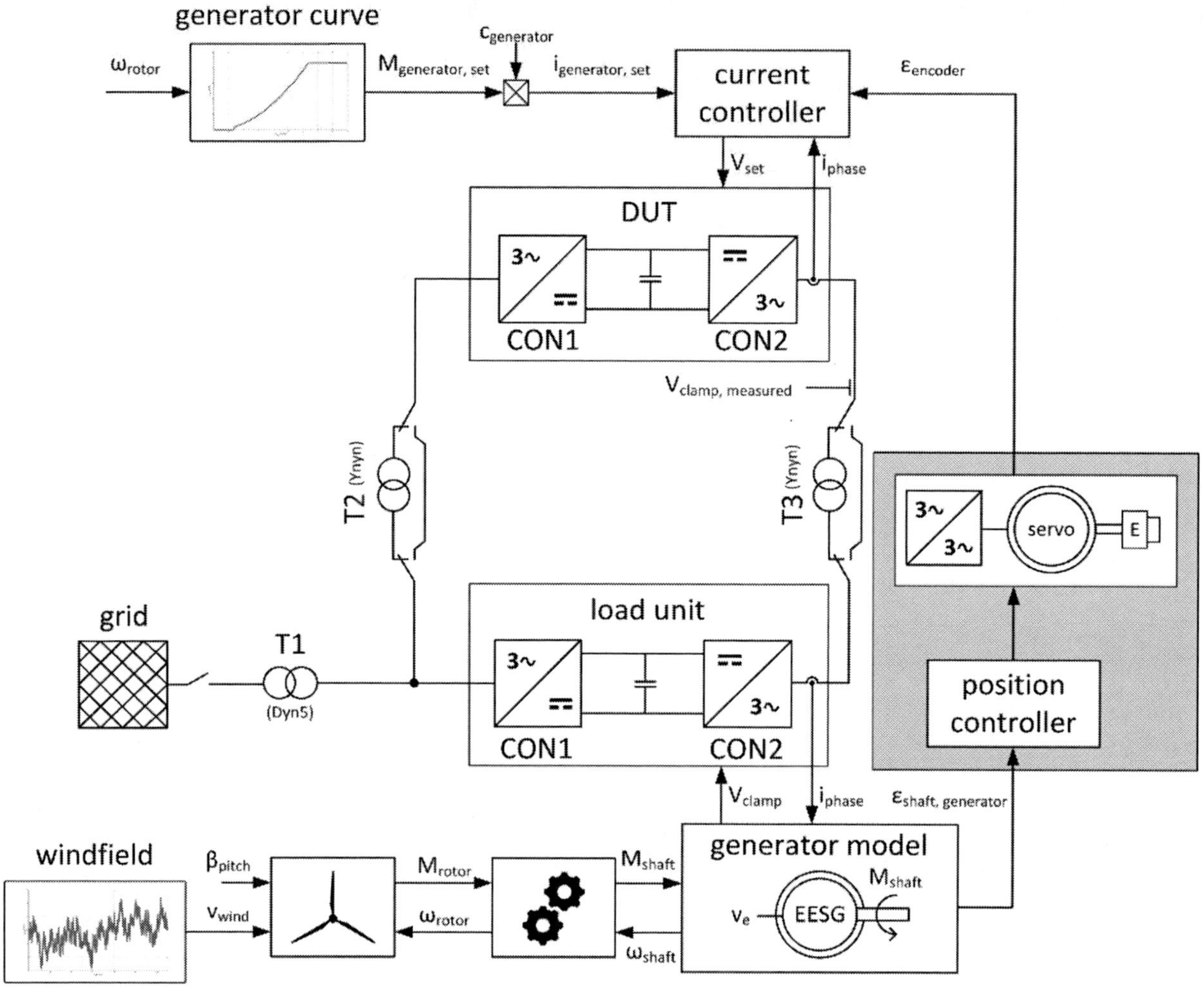

PCIM Europe 2024, 11– 13 June 2024, Nuremberg   DOI: 10.30420/566262225

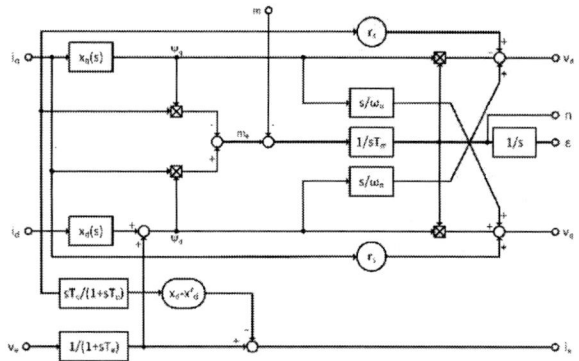

Fig. 2 System topology

2   Problems of the test bench

The investigation of the measurements shown in [2] revealed several issues which need to be addressed in order to test the entire system in dynamic conditions.

Fig. 3 Generator model

2.1   Encoder resolution

Figure 4 shows a Phase Lock Loop (PLL) [3] and the shaft angle of the generator measured on the DUT side. The shaft angle signal has a step-like form. The width of the steps is 1 ms, which corresponds to 4 cycles of the Programmable Logic Controller (PLC). In this configuration, the voltage setpoint is calculated four times, using the same angle in the inverse Park-Transformation [3]. This can cause an additional voltage difference and, consequently, an additional current flow. The shape of the encoder [4] signal and the time difference between the values indicate that the input refresh rate is too low.

2.2   Angle calibration

The angle calibration on the DUT side is coupled with homing of the servo [5] on the load unit side.

1652

PCIM Europe 2024, 11– 13 June 2024, Nuremberg   DOI: 10.30420/566262225

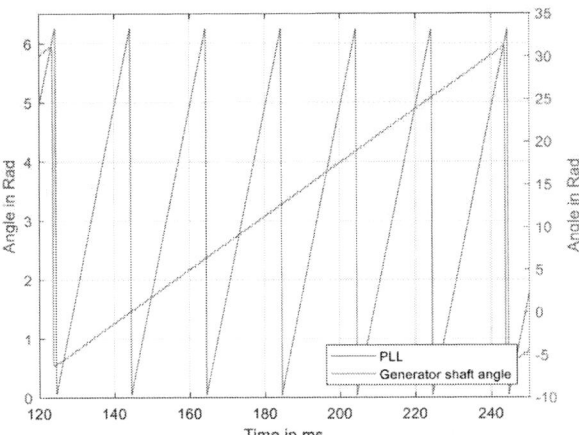

Fig. 4 PLL and the generator shaft angle with six-fold gradient

The homing command takes the actual position and set it as the position with the angle 0°. After the homing command on the load unit side has been executed, the angle can be adjusted on the DUT side. Just like on the load unit side the homing command reads the actual position of the encoder and subtracts it from the actual position. This allows a perfect hardware alignment but provides difficulties during the comparison of the PLL angle and the generator shaft angle provided by the encoder. Due to the latency in the measuring system the time difference between the PLL and the shaft angle need to be determined and subtracted from the shaft angle. The

comparison of the PLL and the mechanical angle read from the encoder was difficult due to the period and the range of the signal. The PLL changes its values between 0 and 2π, the period is equal to the period of the voltage. The mechanical angle changes the values between 0 and 12π with an additional offset. The range is six times bigger due to the machine model (12-pol), the additional offset is referred to the calibration. One rotation of the generator contains six PLL periods what can be referred as previous to the machine model.

2.3 Current peaks

Current peaks have been measured under load. The problems mentioned in 2.1 and 2.2 could force an additional current flow but not those current peaks. A time difference of approximately 48 s can be measured between two current peaks.

Figure 5 displays an analysis of the current peak causes. The first plot shows the measured voltage and current. It is evident, that the current begins to rise gradually from t1=29,84 s, and then increases more steeply after t2=29,86 s. The voltage drops after the current increases, which is a typical response to a change in load. The PLL and the measured voltage are shown in the second plot. You cannot see any deviation of the PLL from the output voltage. In the next plot (see Fig. 5, row 3) you can see the angle of the drive and the PLL. The servo angle leads, which can cause a

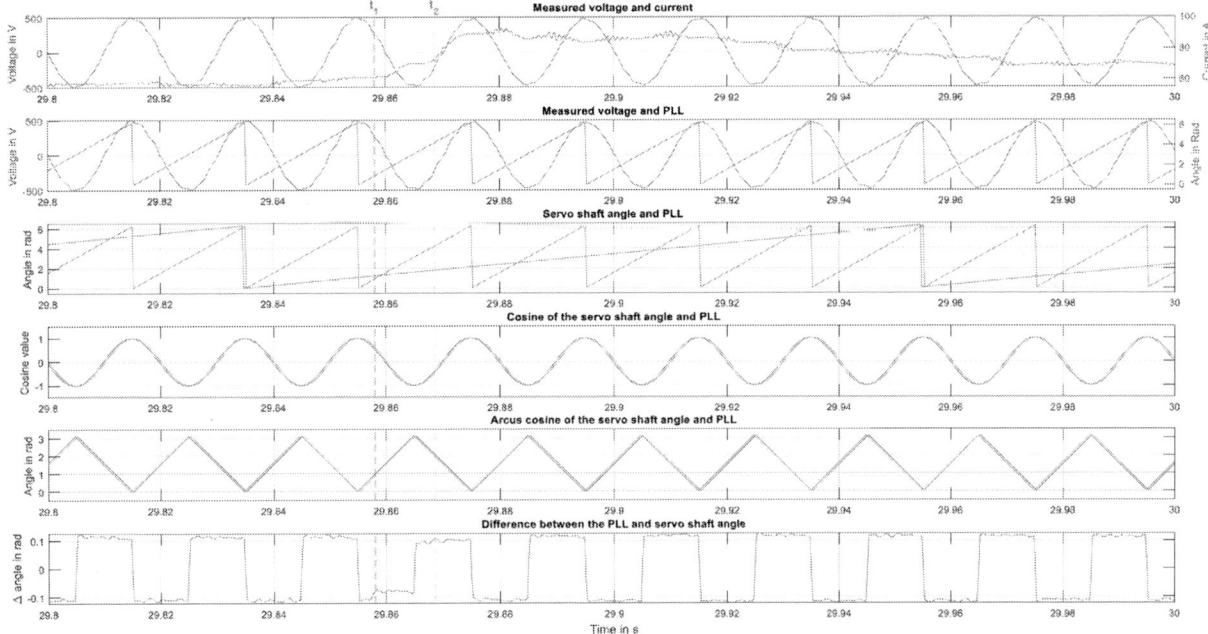

Fig. 5 Investigation of current peaks

1653

PCIM Europe 2024, 11– 13 June 2024, Nuremberg   DOI: 10.30420/566262225

constant current, but does not explain the current peaks. In order to be able to better examine the curve between the reset and the servo angle, the two variables have been represented as cosine values (see Fig. 5, row 4). The blue curve represents the cosine value of the servo angle, the orange one the PLL. Since there was not much difference the fifth and sixth plot have been added. Row five displays the inverse cosine values of the cosine waves from row four (blue is the servo position angle and orange is the PLL). The last curve in Fig. 5 displays the difference between the angle of the servo shaft (blue) and of the PLL (orange). The curve is similar to a square wave. The error fluctuates around two values, - 0,11 rad when the cosine of the PLL is rising and 0,11 rad when its falling. Between t=29,858 s and t=29,879 s the typical form of the signal is deformed. The cause can be a delay by the output of the voltages on the load unit or DUT side.

3   Problem solving

3.1   Improvement of the encoder resolution

Due to hardware limitations the position of the encoder cannot be read faster than 1 ms.

3.2   Angle adjustment

The angle calibration described in 2.2 was difficult due to the different ranges and periods of the PLL and mechanical angle signal. In order to force the signals to the same level, the signals were multiplied with a sine function. The result is a sine wave with an amplitude from -1 to 1 and a period equal to the output voltage period. In the second step the sine waves have been multiplied with an

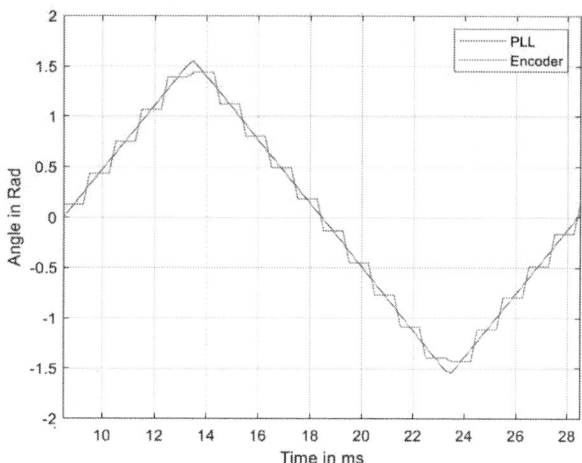

Fig. 7 Comparison of the PLL and encoder angle with calibration

inverse sinus function. The resulting signal has a triangle form in range from -0,5π to 0,5π. In order to automate the procedure, the angle from the PLL and from the encoder are compared. The difference is then added to the encoder angle to align the two curves (see Fig. 6 and Fig. 7).

3.3   Current peaks

The cause of the current peaks was identified in a few steps. To ensure that the load unit is working correctly an idling operation has been tested for several minutes. The observation of the deviation between the PLL and the actual position of the servo did not show any significant errors. The value of the error is fluctuating between 0,0049 and 0,011 rad (see Fig. 8).

During the second step, the DUT was incorporated into the test. The DUT's current setpoint was

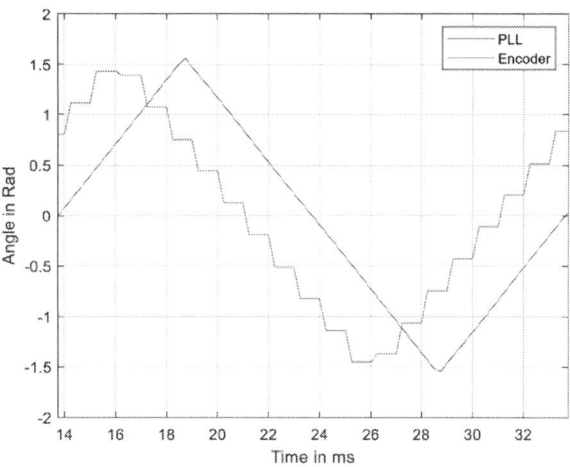

Fig. 6 Comparison of the PLL and encoder angle without calibration

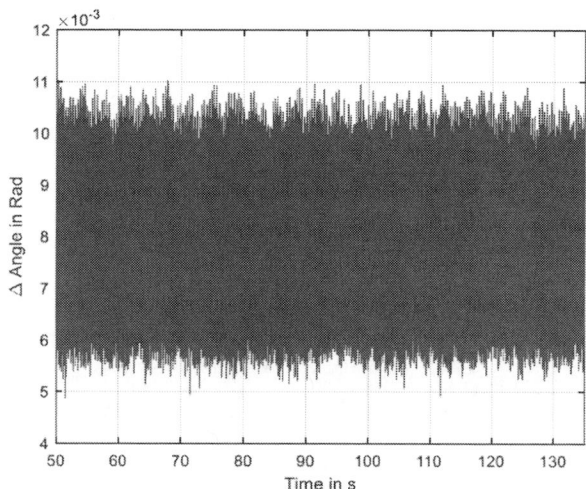

Fig. 8 Difference between the PLL and servo angle at idle

1654

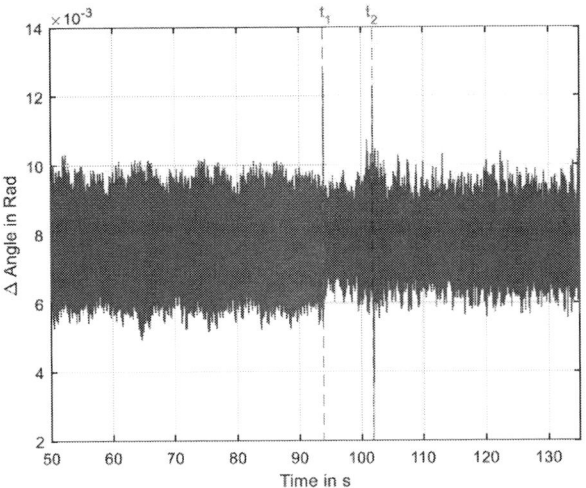

Fig. 9 Difference between the PLL and servo angle at current setpoint at 0 A

adjusted to 0 A. The measurement procedure described in the first step was then repeated (refer to Fig. 9). The left side of the plot shows similarities to the measurement taken at idle. However, at t_1=94 s and t_2=102 s, peaks in the difference between the PLL and the servo angle were observed. The deviation of the angle error decreased slightly and gained an offset after the peaks.

Figure 10 displays the current measured at the load unit side when the current setpoint is set to 0 at the DUT. The current curve also shows the impact of the previously described peaks. The average load unit current up to t_1=94 s was approximately 69 A, after t_2=102 s the value dropped to 58 A. The typical idle current of the load unit (due to the output filter) is approximately

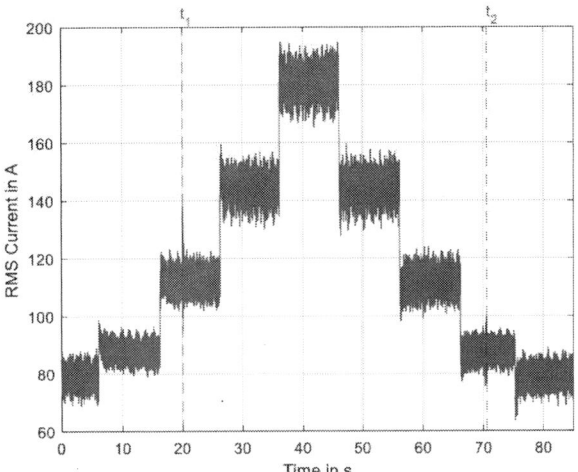

Fig. 11 RMS Current between the load unit and DUT

70 A. At this point the cause of the error current peaks can be identified as being on the DUT side. The investigation has been started with the current controller. A closer look at the PI-controller [6] values has revealed that the I-Part of the controller is not constant at constant current setpoint but is rising and resetting approx. every 48 s causing a current peak. The precise causes are still under investigation.

4 Measurements under load

4.1 Steady state

To confirm, that improvements described in 3 have reduced the current peaks from [2] a similar steady state measurement with a few current setpoints have been taken (see Fig. 11).

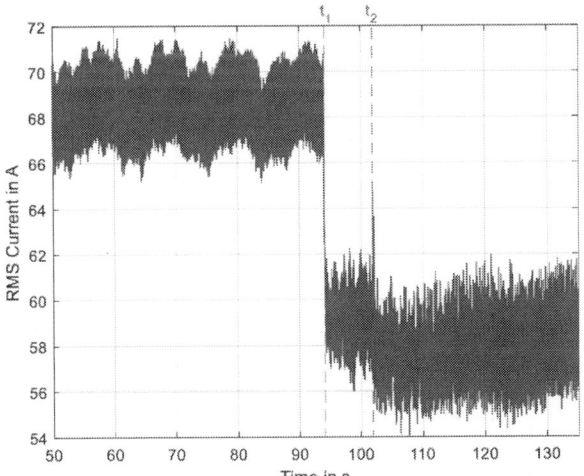

Fig. 10 Current peaks investigation. DUT current setpoint at 0 A

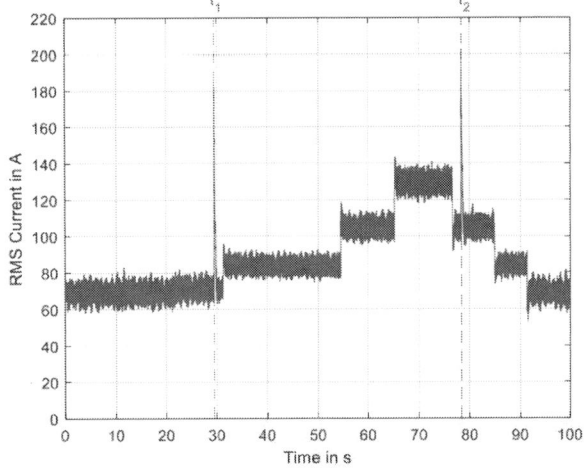

Fig. 12 RMS current between the load unit and DUT before optimisation

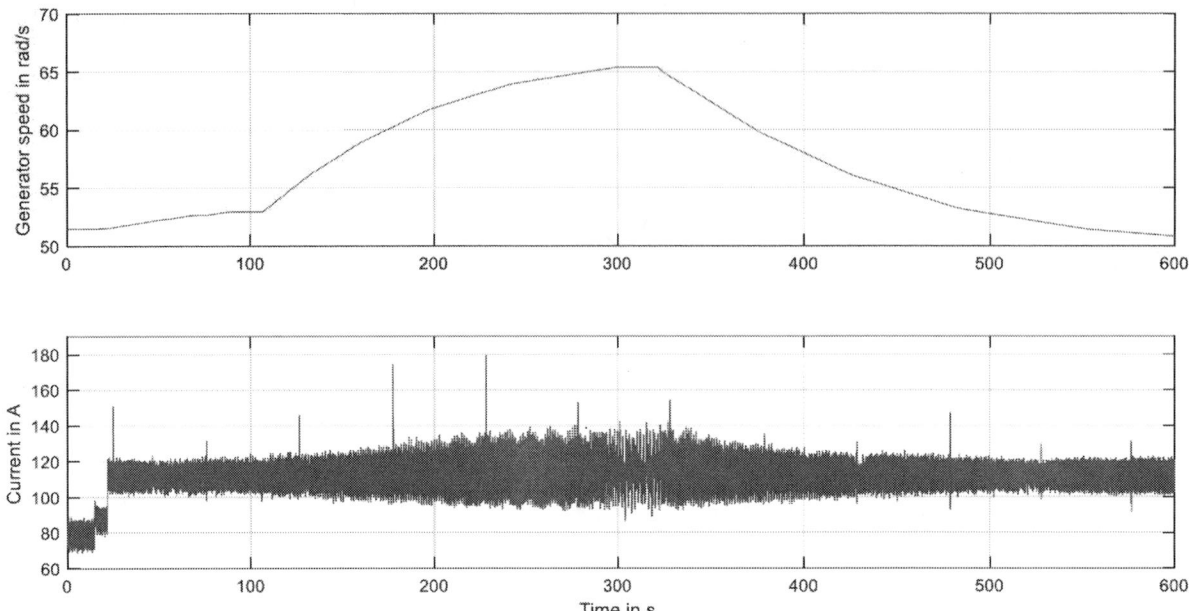

Fig. 13 Dynamic state

The current peaks have been significantly reduced as compared to the measurement presented in older publications (refer to Fig. 12).

4.2 Dynamic states

A final test was conducted to measure the dynamic state. Figure 13 shows the generator speed and the current drawn by the DUT from the load unit. The generator model was accelerated to the nominal speed of 52.3 rad/s and then increased to 62.8 rad/s. As shown in Fig. 13, the speed of the generator exceeded the setpoint of 62.8 rad/s due to the parameters of the model's speed controller. The voltage period at t=300 s reached Δt=16 ms

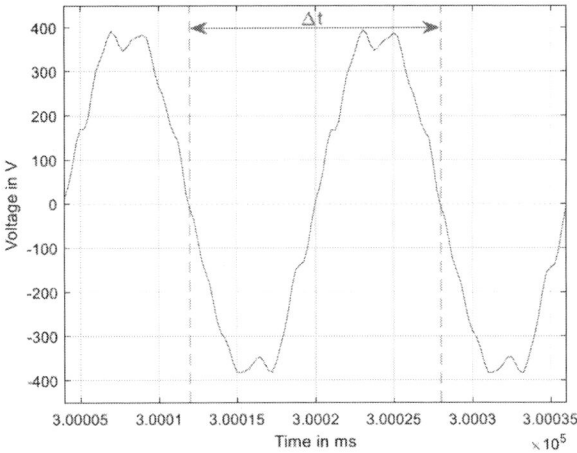

Fig. 14 Voltage at 62,5 Hz

which is equal to 62,5 Hz (see Fig. 14). There was some audible resonance during this phase, so the speed setpoint was reduced to the nominal value. At 21 s, the current setpoint of the DUT was set to 110 A. The current ripple increases with the speed of the generator and at 300 s, the previously mentioned resonance is also visible in the current. The error current peaks are significantly higher during acceleration or deceleration of the generator. This behaviour may indicate that the error current peaks are not only dependent on the PI-controller but also on determining the position of the generator shaft.

The current curve up to the setpoint is opposite and has a different offset to the current curve in Fig. 10. The difference comes from the point at which the current is measured. The current from Fig. 10 was measured on the load unit side and Fig. 13 displays the current measured on the DUT side. The level of the current at current setpoint equal to 0 is higher due to the transformer T3 (see Fig. 2) connecting the CON2 of the load unit and CON2 of the DUT.

5 Conclusion

This paper and its predecessor demonstrate the evolution of the test stand. A model-controlled load unit was developed and extended to include the generator angle transfer via a servo. The DUT is controlled by a field-oriented control that uses the angle from the encoder coupled with the servo. Transformers T2 and T3 are used to adjust the voltage level so that the DUT can be changed if necessary.

The aim was to optimise and test the pre-test platform in dynamic states. After investigating the current peaks, it was concluded that the load unit is functioning correctly and the DUT is responsible for the current peaks. This may depend on the workload of the PLC controlling the DUT. The code and angle adjustment of the DUT have been optimised, resulting in an improvement in the behaviour of the pre-test platform under load (current peaks have been reduced). The dynamic test has been completed successfully. The generator's speed, and therefore the frequency of the output voltage, has been increased to 65,4 rad/s, which is equivalent to 62,5 Hz for this generator model.

The control of the DUT will be enhanced to prevent any occurrence of error current peaks. Additionally, the generator model will be transferred to the 10 MW test rig, and a second servo assembly will be attached to the rig to transfer the machine angle.

References

[1] H. Bühler, *Einführung in die Theorie Geregelter Drehstromantriebe*. Basel: Birkhäuser Basel, 1977. doi: 10.1007/978-3-0348-5940-0.

[2] D. Koczy, A. Ernst, W. Holzke, und B. Orlik, „Optimisation of a Model Based Controlled Power Converter Test Platform", in *PCIM Europe 2023; International Exhibition and Conference for Power Electronics, Intelligent Motion, Renewable Energy and Energy Management*, Mai 2023, S. 1–7. doi: 10.30420/566091323.

[3] D. Schröder, *Elektrische Antriebe - Regelung von Antriebssystemen*. Berlin, Heidelberg: Springer Berlin Heidelberg, 2015. doi: 10.1007/978-3-642-30096-7.

[4] S. Basler, *Encoder und Motor-Feedback-Systeme*. Wiesbaden: Springer Fachmedien Wiesbaden, 2016. doi: 10.1007/978-3-658-12844-9.

[5] U. Probst, *Servoantriebe in der Automatisierungstechnik: Komponenten, Aufbau und Regelverfahren*. Wiesbaden: Springer Fachmedien Wiesbaden, 2022. doi: 10.1007/978-3-658-37423-5.

[6] H. Lutz und W. Wendt, *Taschenbuch der Regelungstechnik: mit MATLAB und Simulink*, 12., Ergänzte Auflage. in Edition Harri Deutsch. Haan-Gruiten: Verlag Europa-Lehrmittel, Nourney, Vollmer GmbH & Co. KG, 2021.

PCIM Europe 2024, 11– 13 June 2024, Nuremberg   DOI: 10.30420/566262226

Educational Hardware Trainer for Teaching the Dual Active Bridge in a DC Grid

S.J.C. Koning[1], D.C. Zuidervliet[1], P.J. van Duijsen[1]

[1] THUAS, DC-Lab, Delft, The Netherlands

Corresponding author: Peter van Duijsen, p.j.vanduijsen@hhs.nl
Speaker: Peter van Duijsen, p.j.vanduijsen@hhs.nl

Abstract

Two hardware trainers, along with design and simulation tools, are suggested as a method to educate individuals on the operational concept of the Dual Active Bridge. Explaining the operational principle of DAB and its modulation techniques can be made easier by utilizing design tools that provide an overview and assistance in setting the main parameters, along with simulation and visualization tools that aid in comprehending the impact of the modulation methods. In order to witness this in a real-life scenario, a hardware trainer is necessary to display these wave patterns on an oscilloscope, together with added control systems. What matters most is observing firsthand how altering parameters impacts the waveforms and power transfer. While simulation may demonstrate this concept, a hardware trainer has the capability to directly display the waveforms as parameters are adjusted either manually or through a control system. This paper proposes the use of a hardware trainer that can be utilized in power electronics courses at both the undergraduate and graduate levels.

1   Introduction

The Dual Active Bridge (DAB) [1], [2] is undeniably emerging as a key element within the DC Grid infrastructure. The primary characteristics of this converter design include the capability for bidirectional power flow control, incorporating galvanic isolation between input and output, as well as voltage scaling.

Typically, control of the converter is accomplished by adjusting the phase difference between the primary and secondary high-frequency AC signals. Several modulation techniques are available in addition to various operational modes of the converter, such as voltage control, startup of one output, and power flow management [3]–[5]. Educating undergraduate and graduate students on these principles necessitates educational resources like simulation and animation software, as well as hardware trainers.

This paper introduces two hardware trainers that allow for testing of different control methods with different parameters. By utilizing a digital replica in a simulation, the control technique is initially tested, allowing for the estimation of parameters. One can observe voltage and current waveforms and analyze power flow according to the modulation principles. In the hardware trainer, the real waveforms can be observed and compared with the simulated waveforms.

Section 2 briefly introduces the topology and an example of a control method of the DAB. The DAB together with the control method is put in a simulation in section 3, to calculate the exact waveforms together with Zero Voltage Switching. The proposed hardware trainers are presented in section 4, where the outcomes from the hardware instructors are showcased.

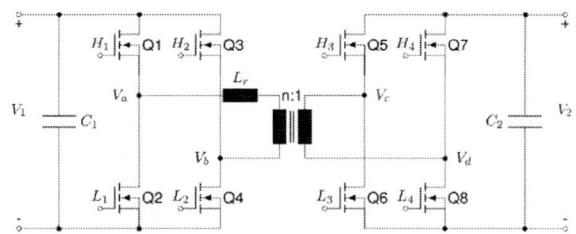

Fig. 1: Topology of a Dual Active Bridge with leakage inductance shown next to the transformer.

2 Topology and teaching framework

The DAB, as seen in Fig. 1, contains two full bridges that are coupled via a high-frequency transformer. Since the leakage inductance L_r plays an important role in the operation of the DAB, it is shown as external inductance in the schematics. In reality it will be the leakage inductance of the coupled inductors.

Each bridge has a constant DC voltage and depending on the modulation method, a pulsed waveform U_{Lr} is created across the coupled inductors. Since we consider the leakage inductance externally of the coupled inductors, the voltage is directly across the leakage inductor and the resulting current through the leakage inductance.

Controlling the phase shift and duty cycle of each bridge, defines the voltage across the leakage inductor and thereby the current through the inductor as demonstrated in Fig. 2. Power transfer is defined by the shape and amplitude of the current and if it is leading or legging the bridge voltages.

3 Simulation and Animation

Before measuring waveforms in the hardware trainer, it is wise to first study these waveforms in a simulation. In that case, the students learn what they can expect when they do the measurements and they can compare the measurement results with the expected simulation results. There are three important aspects that have to teached via simulation, the modulation method, the principle of bidirectional power flow, and the requirements for Zero Voltage Switching(ZVS) [6]. The first two, modulation and power flow, can be simulated and visualized using a simplified simulation model, see section 3.1. For ZVS, a detailed power electronics model is required and animation can visualize the modes of operation, required to obtain ZVS, see section 3.2.

3.1 Simplified simulation model

The goal of the simplified simulation model is to show the bidirectional operation, without going into detail in the power electronics. Therefore, in the simulation, the two bridges are replaced by two controlled voltage sources, and the magnetic coupling is replaced by the leakage inductance only. Using this simplified model, the voltage and current waveforms are simulated as well as the bidirectional power flow, see Fig. 3.

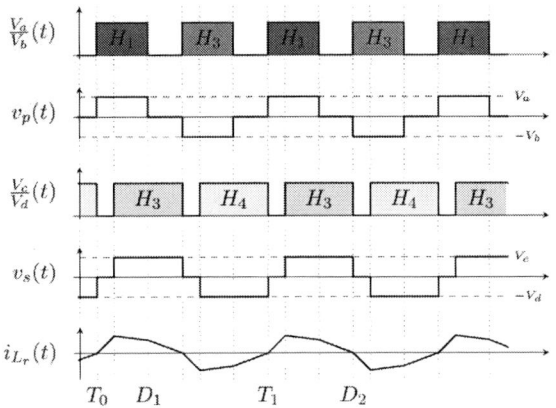

Fig. 2: Control of the inductor current $i_L(T)$ through phase shift and duty cycles. Shows the highside Mosfet states, full bridge voltages and the current through inductor L_r.

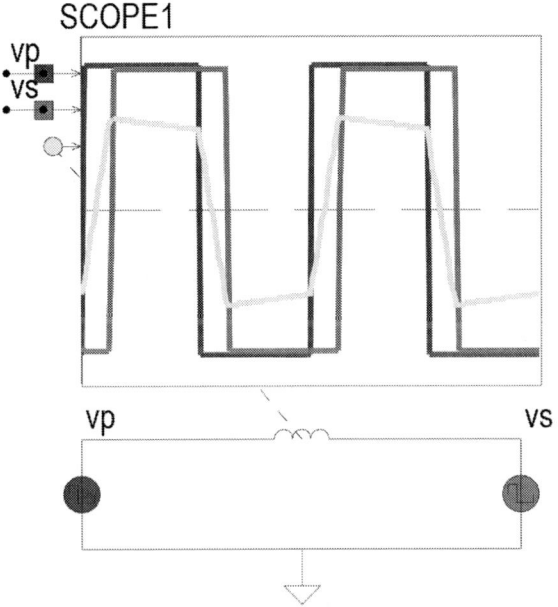

Fig. 3: Bridges modeled by square waveforms, inductive coupling replaced by the leakage inductance L_r. Scope shows v_p(Blue), v_s(Red) and leakage inductor current i_{L_r}(Lightblue).

PCIM Europe 2024, 11– 13 June 2024, Nuremberg   DOI: 10.30420/566262226

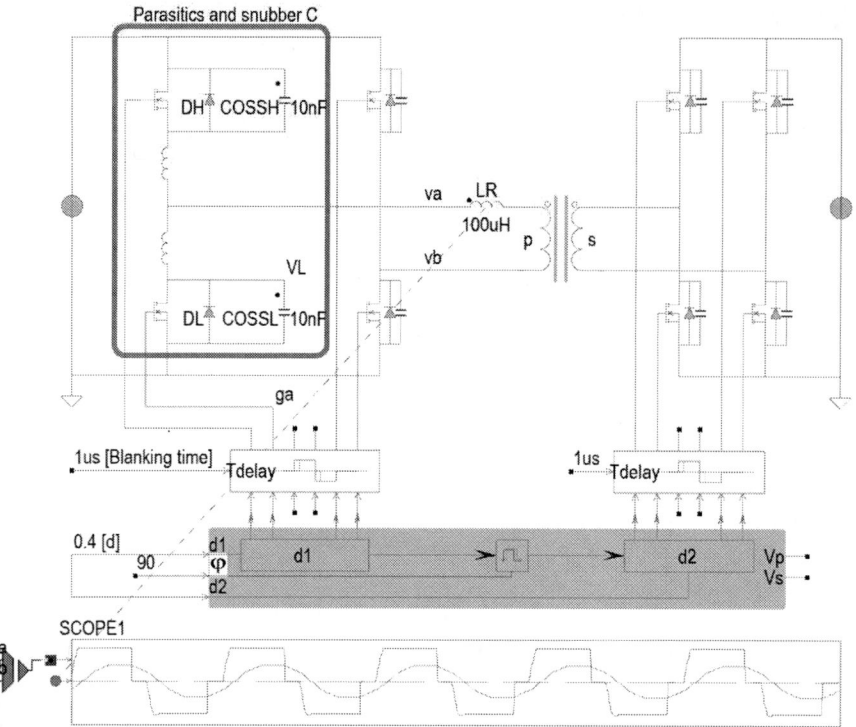

Fig. 4: Simulation of a Dual Active Bridge in Caspoc [7]. First leg in the left bridge is modeled with parasitic components and snubbber capacitors. Phase-Shift modulation $\varphi = 90^o$, with $d = 0.4$ and $1\mu s$ blanking time. Scope shows primary voltage V_{ab} and leakage inductor current i_{Lr}.

3.2 Power electronics simulation model

To simulate Zero Voltage Switching (ZVS), a detailed power electronics model is simulated in Caspoc [7], see Fig. 4. Here the interaction between the output capacitance of the Mosfets C_{oss}, and the leakage inductance is included. In the simulation, the principle of ZVS can be explained and demonstrated and the detailed operation can be visualized using animation of the current flow and Mosfet status, during the simulation.

The simulation includes 4 Bridges, where only the first leg of the left bridge is modeled with parasitic components. The switching frequency equals $Fs = 50kHz$ and the blanking time equals $1\mu s$. Because the low-side Mosfet turns on during the freewheeling of the anti-parallel diode D_L, the voltage V_{DS} across the low-side Mosfet is close to zero during turn-on. Therefore, there are nearly no losses at turn-on, as they mainly come from the discharging of the output capacitor C_{oss}. However the voltage across this capacitor is also zero, there is no discharge of the output capacitor C_{oss} through the low-side Mosfet, hence there are negligible turn-on losses.

In Fig. 5 the waveforms during turn-on of the low-side Mosfet are shown. At $t = 10\mu s$, the high-side Mosfet is turned off. At $t = 10\mu s$, the low-side Mosfet voltage V_{DS} is still equal to the DC link voltage of around 400 volt. The complete voltage drop of V_{DS} is outside of the scaling of the figure.

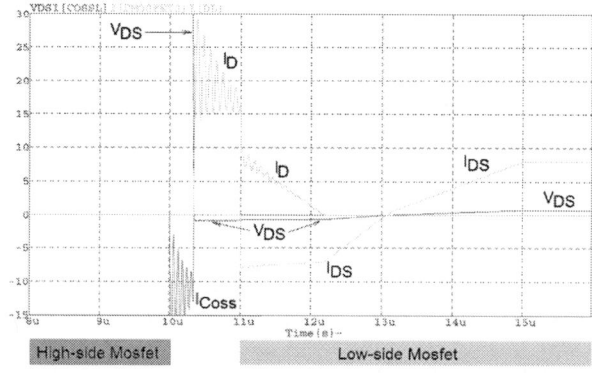

Fig. 5: Waveforms during turn-on of the low-side Mosfet. The high-side Mosfet turns off at time $t = 10\mu s$, the low-side Mosfet turns on at $t = 11\mu s$. Mosfet voltage v_{DS}[Blue], Capacitor current i_{Coss}, low-side diode current i_D and low-side Mosfet current i_{DS}

1660

The voltage V_{DS} over the low-side Mosfet, drops to zero, but this takes time, due to the snubber capacitors $C_{oss} = 10nF$. At $t = 10.4\mu s$ the freewheeling low-side diode D_L takes over the output current (flowing through the leakage inductor L_r), from the output capacitor C_{oss}. This happened, because V_{DS} became negative, and the diode becomes forward biased. The diode D_L carries the output current, until turn-on of the low-side Mosfet at $t = 11\mu s$. The Mosfet can then sustain the output current through the parasitic inductor L_r, and is displayed as I_{DS}. Between $t = 11\mu s$ and $t = 12.2\mu s$, both the low-side Mosfet and low-side diode D_L share the current through the leakge inductor L_r. At $t = 12.2\mu s$, the current I_D through the low-side diode D_L is reduced to zero, and the output current is from $t > 12.2\mu s$ only supported by the low-side Mosfet.

4 Hardware trainers

The principle schematic of the hardware trainer for the DAB is shown in Fig. 6. The idea is that the student can change the switching frequency F_s and dutycycles of each DAB, as well as the phase shift between the two bridges. In this way nearly all modulation methods, such as SPS and DPS can be demonstrated. The influence of these parameters on the trapezoidal current waveform is then directly visible on the oscilloscope.

4.1 Low voltage trainer

Fig. 6: Low voltage hardware trainer, where switching frequency, duty cycles and phase shift are controlled by the C2000 microcontroller.

The hardware trainer [8] for safe low voltage levels is shown in Fig. 6 using the U4L [9], [10]. Here the C2000 [11] is applied. Through software can the four parameters be adjusted to control the four parameters.

Different transformer models can be connected with the option to add additional inductors to increase the leakage inductance. Modulation techniques can then manually be applied to teach the differences between the modulation principles. Students can then directly apply the modulation techniques they simulated.

A typical result from a measurement is shown in Fig. 7. Here the maximum power transfer is measured occurring at a phase shift of $\varphi = 90^o$ degrees. In this case is the SPS modulation technique demonstrated on the trainer.

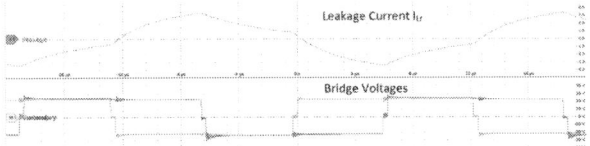

Fig. 7: Open loop example of a SPS modulation technique in a DAB. Upper trace, leakage inductor current (Ipp=6.6Ampere), and lower trace, bridge output voltages (Vpp=48v), of the DAB with phase difference of $\varphi = 90^o$ degrees, operating at 40kHz, Vdc=24v.

4.2 High voltage trainer

Fig. 8: High voltage hardware trainer, where switching frequency, duty cycles and phase shift are controlled by the C2000 microcontroller.

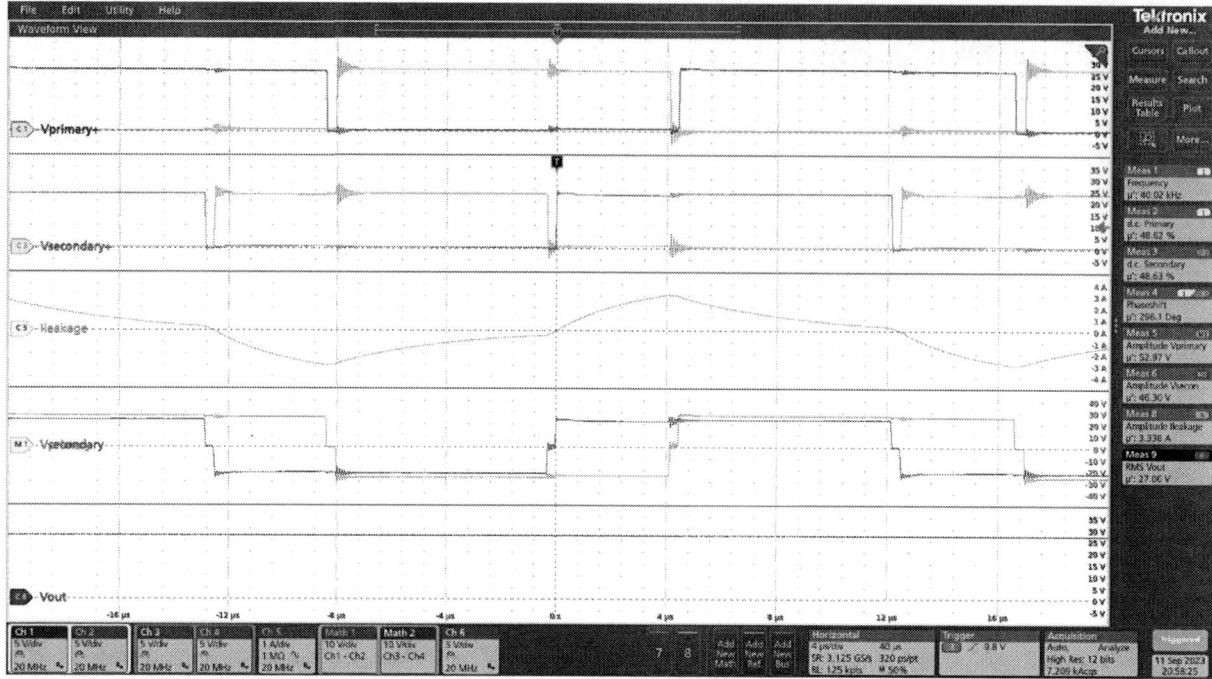

Fig. 9: Voltage controlled start up example on V_1 set on 27v. Upper group, half bridge output voltages(Vpp=24v) Va and Vb, mid up group, half bridge output voltages(Vpp=27v) Vc and Vd, middle group, leakage inductor current (Ipp=6.6Ampere), mid down group, bridge output voltages (Vpp=48v primary, Vpp=54v secondary), and lowest group, output voltage, of the DAB with phase difference of $\varphi = 64^o$ degrees, operating at 40kHz, Vdc=24v input.

The prototype of the hardware trainer for high voltage [12] is shown in Fig. 8. Again, here the C2000 is applied. A control system is added which measures the voltages on both inputs and depending on the direction can this voltage be used as to create a voltage on one end. Another way the measured voltages can be used is as part of a constant current constant voltage control system for the charging of batteries.

Having a higher voltage rating, this model is also suitable to emulate situations having a higher voltage or power rating normally seen in a lab environment. As example can it demonstrate the workings of a charger for an electric vehicle.

Another typical result is seen in Fig.9 where 27V is made on the output from a 24V source. A phase shift of 64 degrees is applied through a voltage controlled controller. The switching frequency is set to $40kHz$.

Conclusion

To educate undergraduate and graduate students in power electronics courses about the Dual Active Bridge, two hardware trainers have been de-

signed. The primary goal of these trainers is to demonstrate how the modulation method impacts the power transfer, current waveform, and efficiency of the Dual Active Bridge. To support the design process and gain insights into parameters within the DAB, design tools are utilized to compute waveforms and power transfer. Simulation demonstrates the similarity between a detailed and simplified model, illustrating that operational aspects and power transfer can be determined using the simplified model. Moreover, intricate principles like Zero Voltage Switching can also be simulated using the detailed simulation model. The waveforms from the design tool, simulation and hardware trainer are in accordance to each other, which clearly demonstrates the modulation methods to students. Understanding the DAB becomes easier for students since they can focus on the key elements, compared to understanding all the details of the converter.

References

[1] R. De Doncker, D. Divan, and M. Kheraluwala, "A three-phase soft-switched high power density dc/dc converter for high power applications," in

Conference Record of the 1988 IEEE Industry Applications Society Annual Meeting, vol. 1, 1988, pp. 796–805. DOI: 10.1109/IAS.1988.25153.

[2] R. De Doncker, D. Divan, and M. Kheraluwala, "A three-phase soft-switched high-power-density dc/dc converter for high-power applications," *IEEE Transactions on Industry Applications*, vol. 27, no. 1, pp. 63–73, 1991. DOI: 10.1109/28.67533.

[3] A. K. Jain and R. Ayyanar, "Pwm control of dual active bridge: Comprehensive analysis and experimental verification," in *2008 34th Annual Conference of IEEE Industrial Electronics*, 2008, pp. 909–915. DOI: 10.1109/IECON.2008.4758074.

[4] F. Krismer and J. W. Kolar, "Closed form solution for minimum conduction loss modulation of dab converters," *IEEE Transactions on Power Electronics*, vol. 27, no. 1, pp. 174–188, 2012. DOI: 10.1109/TPEL.2011.2157976.

[5] J. Liu and N. Zhao, "Improved fault-tolerant method and control strategy based on reverse charging for the power electronic traction transformer," *IEEE Transactions on Industrial Electronics*, vol. 65, no. 3, pp. 2672–2682, 2018. DOI: 10.1109/TIE.2017.2748032.

[6] R. Steigerwald, R. De Doncker, and H. Kheraluwala, "A comparison of high-power dc-dc soft-switched converter topologies," *IEEE Transactions on Industry Applications*, vol. 32, no. 5, pp. 1139–1145, 1996. DOI: 10.1109/28.536876.

[7] Simulation-Research, *Caspoc Simulation and Animation for Power Electronics and Electric Drives — caspoc.com*, https://www.caspoc.com/, [Accessed 08-04-2024].

[8] S. Koning, D. Zuidervliet, and P. van Duijsen, "Educational laboratory demonstrator for teaching dual active bridge control principles," in *2022 22nd International Symposium on Electrical Apparatus and Technologies (SIELA)*, 2022, pp. 1–4. DOI: 10.1109/SIELA54794.2022.9845754.

[9] P. J. van Duijsen and D. C. Zuidervliet, "Universal power electronics hardware trainer for teaching the dc grid," in *2022 International Conference on Electrical, Computer, Communications and Mechatronics Engineering (ICECCME)*, 2022, pp. 1–6. DOI: 10.1109/ICECCME55909.2022.9988666.

[10] P. Van Duijsen and D. Zuidervliet, "Laboratory setup for teaching dc grid droop control and protection," in *2021 44th International Convention on Information, Communication and Electronic Technology (MIPRO)*, 2021, pp. 1587–1592. DOI: 10.23919/MIPRO52101.2021.9596738.

[11] *C2000 real-time microcontrollers — TI.com — ti.com*, https://www.ti.com/c2000, [Accessed 08-04-2024].

[12] A. Drop, D. Zuidervliet, and P. van Duijsen, "The development of an universal six leg inverter for electrical drives laboratory experiments," in *2023 18th Conference on Electrical Machines, Drives and Power Systems (ELMA)*, 2023, pp. 1–4. DOI: 10.1109/ELMA58392.2023.10202552.

PCIM Europe 2024, 11– 13 June 2024, Nuremberg   DOI: 10.30420/566262227

Study on the Operating Performance of a FCS-MPC-Controlled Matrix-Converter for PMSM at Different Frequency Ratios

Robert Zipprich[1], Bünyamin Tekir[1], Jan Winter[1], Marcus Ziegler[1]

[1] Electrical Machines and Drives, University of Kassel, Germany

Corresponding author: Robert Zipprich, r.zipprich@uni-kassel.de
Speaker: Robert Zipprich, r.zipprich@uni-kassel.de

Abstract

In this study, the system behavior of a model-predictive controlled matrix converter as a motor converter of a permanent magnet synchronous machine (PMSM) is analyzed at different input and output frequencies. In addition to the variation of the frequencies, a differentiation of the chosen basis vectors takes place in order to present the differences by using rotating vectors. For the investigation, the Finite Control Set Model Predictive Control (FCS-MPC) algorithm was derived for PMSM in grid applications with and without rotating voltage vectors and these different operating parameters were subjected to a parameter study.

1   Introduction

Compared to classic indirect topologies, direct converters such as the matrix converter offer the advantage of a single-stage conversion of the supply network into a desired output network. In principle, this conversion is more efficient than indirect topologies, as there is no need for a lossy storage element. This has a positive effect on construction volume, weight and service life [1],[2], so that the matrix converter is in the interest of the industry and is already being used in several application variants in classic grid applications [2]. As an object of research, the direct matrix converter is the target for testing and validating novel control and regulation approaches. A promising control approach for the matrix converter is a finite control set model predictive control [3],[4],[5], in which the system behavior is predicted for each switching state using current variables such as grid voltage and machine currents. The subsequent switching state is selected on the basis of a defined cost function so that there is no fixed switching frequency. When using classic indirect converter topologies, the frequencies of the incoming grid are completely decoupled from the outgoing grid. However, due to the lack of storage elements, a matrix converter must interact with the directly connected grid in order to form the desired output behavior. Due to the possible long-term direct connection of the input-side grid to the output-side grid, the current shape of the feeding grid can also

be used to model the output-side characteristics [6]. Based on this fact, there is a direct dependence of the supply frequency of the matrix converter when modelling the system behavior.

The aim of the present work is to show the influence shown when using classical FCS-MPC approaches for the matrix converter without explicit adaptations of the algorithm and to what extent these can already be used to improve the system behavior. The basis of the investigation is a modelling of the overall structure in Matlab/Simulink with subsequent parameter study to determine the weighting factor for the cost function and simulation of various operating states.

2   Drivetrain with Direct Matrix Converter

The basic structure of a drive train with a matrix converter is shown in Figure 1. Direct matrix converters consist of an $n \times m$ switching matrix, where n describes the number of input phases and m the number of output phases. For classic industrial applications on existing networks and existing machines, this results in a three-phase application and the following applies: $n = m = 3$. The machine to be controlled is connected directly to the matrix converter on the output side. On the input side, the matrix converter requires an LC-filter for operation and to comply with industrial EMC standards. The supply network is idealized by three voltage sources.

1664

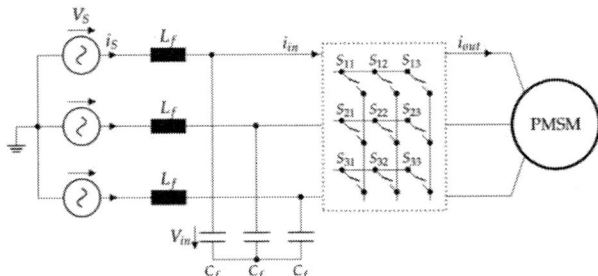

Fig. 1 Block diagram of the 3-phase drivetrain

2.1 PMSM

A permanent magnet synchronous motor (PMSM) is assumed as the load machine to be controlled in this work. This electrical machine, which has been widely discussed in the literature, has been dealt with in depth and is sufficiently described mathematically using the classic fundamental wave model. Other machine types are directly applicable by adapting the prediction model.

2.2 Switching Matrix

The switching matrix of the direct converter for the present case of a 3x3 phase system is to be constructed from a total of nine bidirectional switches. These semiconductor switches allow a bidirectional controlled current flow and open the possibility of elaborate commutation processes. Due to the abstinence of monolithically integrated semiconductor switches for industrial purposes, the bidirectional switches must be described in a discrete design [7].

The mathematical description of the semiconductor matrix is carried out by means of a matrix S [5]. The matrix elements Sm,n of the matrix S encode the complete bidirectional conduction state of each of the nine bidirectional switches, Eq. (1).

$$S_{m,n} = \begin{cases} 1, & Switch\ S_{m,n}\ is\ closed \\ 0, & Switch\ S_{m,n}\ is\ open \end{cases} \quad (1)$$

With the aid of the switching matrix S, the output voltage V_{out} of the matrix converter is determined by right-hand multiplication of the switching matrix S by the vector of the input voltages V_{in}, Eq. (2).

$$\begin{bmatrix} V_{out1} \\ V_{out2} \\ V_{out3} \end{bmatrix} = S \cdot \begin{bmatrix} V_{in1} \\ V_{in2} \\ V_{in3} \end{bmatrix} = \begin{bmatrix} S_{11} & S_{21} & S_{31} \\ S_{12} & S_{22} & S_{32} \\ S_{13} & S_{23} & S_{33} \end{bmatrix} \begin{bmatrix} V_{in1} \\ V_{in2} \\ V_{in3} \end{bmatrix} \quad (2)$$

Analogue to the determination of the output voltage, the input currents of the matrix converter are calculated by a right-sided multiplication of the transposed switching matrix S with the output currents i_{out}, Eq. (3).

$$\begin{bmatrix} i_{in1} \\ i_{in2} \\ i_{in3} \end{bmatrix} = S^T \cdot \begin{bmatrix} i_u \\ i_v \\ i_w \end{bmatrix} = \begin{bmatrix} S_{11} & S_{12} & S_{13} \\ S_{21} & S_{22} & S_{23} \\ S_{31} & S_{32} & S_{33} \end{bmatrix} \begin{bmatrix} i_{out1} \\ i_{out2} \\ i_{out3} \end{bmatrix} \quad (3)$$

Taking all permutations into account, the description of the switching state of the semiconductor matrix results in $2^9 = 512$ possible switching states. However, only the switching states that prevent an input-side short-circuit of the input voltages and an output-side current break of the output currents are technically relevant, which reduces the number of relevant switching states to 27.

The 27 switching states are grouped accordingly when the alpha/beta transformation Eq. (4),[6] of the output voltage of each switching state is considered. The name is derived from the behavior of the pointer in the $\alpha\beta$-plane.

$$g = \frac{2}{3} \cdot (g_U + g_V \cdot a + g_W \cdot a^2)\ , a = e^{j\frac{2\pi}{3}} \quad (4)$$

The defined switching states 1-18 are each characterized by a fixed angle of the output voltage pointer with variable amplitude. This group is defined below as voltage-forming group V. In addition, only 2 of the three input phases are used to form the output voltage, shown by the grey areas in Table 1. The defined switching states 19-24, on the other hand, have a time-variable angle with a constant amplitude. This group of switching states utilizes all three input phases to form the output voltage, whereby only a modification of the input voltages takes place regarding the phase sequence. Consequently, this special group of switching states is referred to as rotating vectors (R-V) R and represents one core topic of the work. Finally, the last switching state group 25-27 is analyzed. Due to the use of only one input phase, these switching states do not generate a voltage difference at the output of the switching matrix, which is why this group is referred to as zero pointer Z.

Nr.	S_{11}	S_{12}	S_{13}	S_{21}	S_{22}	S_{23}	S_{31}	S_{32}	S_{33}	G
1	1	0	0	0	1	1	0	0	0	V
2	1	1	0	0	0	1	0	0	0	V
3	0	1	0	1	0	1	0	0	0	V
4	0	1	1	1	0	0	0	0	0	V
5	0	0	1	1	1	0	0	0	0	V
6	1	0	1	0	1	0	0	0	0	V
7	0	0	0	1	0	0	0	1	1	V
8	0	0	0	1	1	0	0	0	1	V
9	0	0	0	0	1	0	1	0	1	V
10	0	0	0	0	1	1	1	0	0	V
11	0	0	0	0	0	1	1	1	0	V
12	0	0	0	1	0	1	0	1	0	V
13	0	1	1	0	0	0	1	0	0	V
14	0	0	1	0	0	0	1	1	0	V
15	1	0	1	0	0	0	0	1	0	V
16	1	0	0	0	0	0	0	1	1	V
17	1	1	0	0	0	0	0	0	1	V
18	0	1	0	0	0	0	1	0	1	V
19	1	0	0	0	1	0	0	0	1	R
20	0	1	0	0	0	1	1	0	0	R
21	0	0	1	1	0	0	0	1	0	R
22	1	0	0	0	0	1	0	1	0	R
23	0	1	0	1	0	0	0	0	1	R
24	0	0	1	0	1	0	1	0	0	R
25	1	1	1	0	0	0	0	0	0	Z
26	0	0	0	1	1	1	0	0	0	Z
27	0	0	0	0	0	0	1	1	1	Z

Table 1   Switching States of the Matrix Converter

2.3   LC-Input-Filter

A LC-filter required for the matrix converter must be located on the input side before the switching matrix [8],[9]. The design as a second-order LC filter represents an optimum in terms of filter properties, installation space and costs. In addition to the pure filtering property for suppressing the switching frequencies and compliance with the EMC guidelines, the input filter provides the required switching energy per switching pulse as well as the additional creation of required freewheeling branches when using the voltage-forming vectors and the zero vectors.

When dimensioning the components, it must be ensured that the respective system frequencies do not violate the equation Eq. (5).

$$f_{Grid} < f_{Filter} \ll f_{switching} \qquad (5)$$

Another key point of the filter design is the voltage drop that occurs during a switching operation. The size of the capacitance and inductance must be dimensioned in such a way that the voltage distortion falls within a tolerable range.

Cascaded or damped filter structures can be used to further improve the operating behavior [10].

3   FCS-MPC-Control

As a control approach for electrical machines, the model predictive control approaches of the FCS-MPC family represent promising options [1],[3]. By predicting the system behavior at time $[k + 1]$ for each available switching state and evaluating it based on a defined cost function, it is possible to determine the switching state that most precisely fulfils the defined control objectives while taking operating limits into account.

1. modelling of qualitative d-q currents for the PMSM
2. modelling of qualitative input phase currents without harmonics
3. minimization of the input-side reactive power

These goals are achieved by high-frequency prediction of the system behavior based on an input and output model with final minimization of the cost function. The target currents of the PMSM are specified by a superimposed speed control cascade.

3.1   Prediction Models

The basis of an FCS-MPC algorithm is a sufficiently detailed description of an input or output model for predicting the system behavior. Based on Figure 1, a model for the input-side LC filter and the output-side PMSM must therefore be created.

3.1.1   PMSM

A classic representation of the operating behavior of a PMSM is given by the fundamental wave model. By assuming the torque development by only the fundamental wave and the transformation of the three-phase equivalent circuit diagram into the dq -plane, a sufficient modelling of the PMSM [5] is described for the technical application. The resulting differential equations Eq. (6) and Eq. (7) thus form the basis of the prediction model of the PMSM, where R_s represents the winding resistance and L_s winding inductance, ψ describes the occurring permanent flux linkage and ω_{el} the angular frequency of the rotor. The currents i_d and i_q represent the variables to be regulated, while the voltages u_d and u_q represent the inputs for the regulation.

$$u_d(t) = R_s i_d(t) + L_d \frac{di_d}{dt} - \omega_{el} L_q i_q(t) \quad (6)$$

$$u_q(t) = R_s i_q(t) + L_q \frac{di_q}{dt} + \omega_{el} L_d i_d(t) \\ + \omega_{el} \psi_{pm} \quad (7)$$

To enable application within the FCS-MPC, the corresponding differential equations must be reformulated using the explicit Euler method into a discretized prediction according to $[k+1]$ based on current values at time $[k]$. Using the defined prediction period T_s, the prediction equations of the PMSM Eq. (8) and Eq. (9) result.

$$i_{d_p,1\ldots27}[k+1] = \left(1 - \frac{T_S R_s}{L_d}\right) i_d[k]$$
$$+ \frac{T_S}{L_d} u_{d,1\ldots27}[k] \quad (8)$$
$$+ \frac{T_S L_q}{L_d} \omega_{el}[k] i_q[k]$$

$$i_{q_p,1\ldots27}[k+1] = \left(1 - \frac{T_S R_s}{L_q}\right) i_q[k]$$
$$+ \frac{T_S}{L_q} u_{q,1\ldots27}[k]$$
$$- \frac{T_S L_d}{L_q} \omega_{el}[k] i_d[k] \quad (9)$$
$$- \frac{\psi_{pm} T_S}{L_q} \omega_{el}[k]$$

The required input voltages u_d and u_q result from the voltages across the filter capacitors that are currently present at time $[k]$, which are calculated into the corresponding output voltages according to Eq. (2) and subsequent dq-transformation, depending on the switching state to be predicted.

Once the respective predicted currents i_d and i_q have been determined, these must be converted into the corresponding three-phase input currents using Eq. (3).

3.1.2 Input-LC-Filter

For the prediction model of the input-side LC filter, the circulation and account equation must first be determined [8],[9].

$$V_s = L_f \cdot \frac{di_s}{dt} + i_s \cdot R_f + V_{in} \quad (10)$$

$$i_s = i_{in} + C_f \cdot \frac{dV_{in}}{dt} \quad (11)$$

Based on the equations Eq. (10) and Eq. (11), a state space representation is given by Eq. (12) and Eq. (13).

$$\dot{x}(t) = \underbrace{\begin{bmatrix} 0 & \frac{1}{C_f} \\ -\frac{1}{L_f} & -\frac{R_f}{L_f} \end{bmatrix}}_{A_c} x(t) + \underbrace{\begin{bmatrix} 0 & -\frac{1}{C_f} \\ \frac{1}{L_f} & 0 \end{bmatrix}}_{B_c} u(t) \quad (12)$$

$$\dot{x}(t) = \begin{bmatrix} V_{in}(t) \\ i_s(t) \end{bmatrix} und \ u(t) = \begin{bmatrix} V_s(t) \\ i_{in}(t) \end{bmatrix} \quad (13)$$

The continuous state space representation is to be discretized for use on corresponding control systems, whereby a zero-order hold discretization is to be applied. Considering the prediction period T_s, this results in the discretized model according to Eq. (14).

$$x[k+1] = A_d x[k] + B_d u[k] \quad (14)$$

with:

$$A_d = e^{A_c \cdot T_S} \cong \begin{bmatrix} a_{d_1} & a_{d_3} \\ a_{d_2} & a_{d_4} \end{bmatrix} \quad (15)$$

$$B_d = \int_0^{T_S} e^{B_c \cdot (T_S - \tau)} \cong \begin{bmatrix} b_{d_1} & b_{d_3} \\ b_{d_2} & b_{d_4} \end{bmatrix} \quad (16)$$

Based on the discretized prediction model and resolution according to the searched term of the source current, the prediction of the input currents results according to Eq. (17)

$$i_{s,1\ldots27}[k+1] = a_{d_2} V_{in}[k] + a_{d_4} i_s[k] \\ + b_{d_2} V_s[k] + b_{d_4} i_{in,1\ldots27}[k] \quad (17)$$

The values of the source voltage and the source currents as well as the input voltage of the matrix converter are measured. The term $i_{in}[k]$ is determined by the reverse transformation of the predicted dq currents of the PMSM and the equation Eq. (3). If the predicted value of the source current is available, this must be used to calculate the input-side reactive power. Assuming that the change in input voltage is insignificant, the reactive power Q in step $[k+1]$ is determined by Eq. (18).

$$Q_{1\ldots27}[k+1] \approx Imag\{V_s[k+1] \\ \cdot \overline{i_{s,1\ldots27}}[k+1]\} \\ \approx Imag\{V_s[k] \\ \cdot \overline{i_{s,1\ldots27}}[k+1]\}$$ (18)

3.2 Cost function

If the prediction results of the dq currents and the input reactive power of each switching state are available, the calculation and minimization of the cost function enables the selection of the next switching state to be assumed.

For this purpose, the corresponding error terms of the cost function must first be defined as e.g. squared deviation of the prediction from the set-point, Eq. (19), Eq. (20) and Eq. (21).

$$e_{i_d,1\ldots27}[k+1] = \left(i_d^* - i_{d_p,1\ldots27}[k+1]\right)^2$$ (19)

$$e_{i_q,1\ldots27}[k+1] = \left(i_q^* - i_{d_q,1\ldots27}[k+1]\right)^2$$ (20)

$$Q_{1\ldots27}[k+1] = (Q^* - Q_{1\ldots27}[k+1])^2$$ (21)

The required cost function J is then obtained by summing the respective error terms and explicitly weighting the error term of the input reactive power, Eq. (22).

$$J = e_{i_d}[k+1] + e_{i_q}[k+1] + \lambda \cdot Q[k+1]$$ (22)

Based on the weighting factor λ, the control target must be shifted between the input reactive power and the currents of the PMSM and optimized according to the desired system behavior. A simulative parameter study is a suitable means of determining the weighting factor λ, but there are other ways of determining it.

The next switching state SwS_{next} is now selected by determining a global minimum across all switching states, Eq. (23).

$$SwS_{next} = \min(J_{1\ldots27})$$ (23)

3.3 Algorithm Structure

Based on the prediction models of the input and output behavior of the topology and the formulated cost function, the description of the holistic control algorithm is given in Figure (2). Starting with the receipt of the trigger signal from an external clock encoder, the required process variables of the system must be measured. The actual prediction algorithm processes the mathematical operations

sequentially within a counter-based loop structure and stores the analyzed cost function of each switching state in an array. This memory is then used to perform a minimization by searching for the smallest element, whereby the next switching state is determined. The switching state found is then converted into the respective switch states for the simulation using the table Tab. (1) and the signals are fed to the simulation model of the switch matrix.

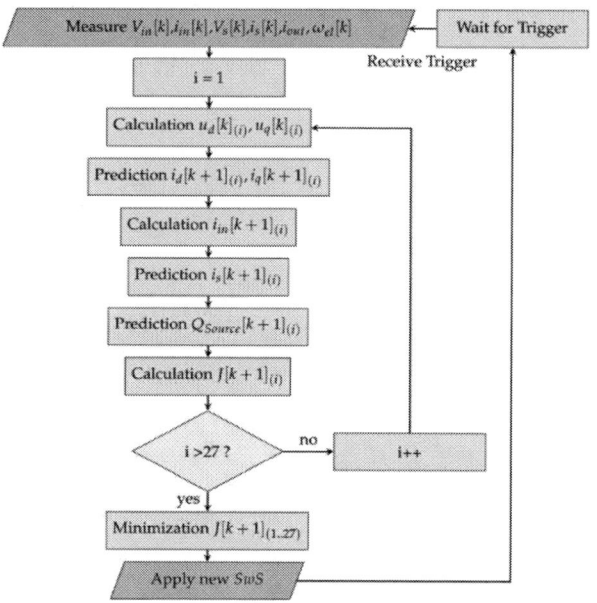

Fig. 2   Flowchart FCS-MPC-Algorithm

4   Experimental examination

The system behavior at different frequency ratios is investigated by simulating the topology. Modelled using Matlab/Simulink and the "Simscape-Electrical" library, the topology from Fig. (1) is to be mapped equivalently. The algorithm from Fig. (2) is to be implemented in a triggered subsystem according to the representation. The translation between the calculated next switching state and the ideal switches of the switching matrix is carried out using a look-up table.

Influences caused by non-linearities of the power electronics or sensors as well as influences of the commutation method are not taken into account in the analysis.

For the simulations, Table Tab (2) should be used to parameterize the model.

Parameter	Value
Voltage Source	3 Phase 264.5 VAC
R_f	0.5 Ω
L_f	150e-6 H
C_f	33e-6 F
FCS-MPC-Frequency	50 kHz
R_s	79e-3 Ω
L_d	1.13e-3 H
L_q	1.24e-3 H
ψ_{PM}	143.76e-3 Wb
Z_p	10
Rated Current	24.7 A RMS
J	68056e-6 kg*m²
Friction-Torque	1 Nm
Rated Torque	71.5 Nm
Rated Speed	2000 rpm

Table 2 Simulation parameters

The machine torque is controlled purely by regulating the q current, while the d current is to be regulated to zero, no MTPA- procedure are used for the investigation. An ideal stiff supply network consisting of three star-connected voltage sources is assumed as the source for the topology.

4.1 Determination of the weighting factor

A parameter study must be used to determine the required weighting factor λ of the cost function Eq. (22). For this purpose, the factor λ is increased incrementally while simultaneously recording the generated THD [11] of the source currents during the entire operation and the generated THD in the machine currents when loaded with nominal torque. The factor λ is assumed to be ideal if the sum of the generated THD is minimal. For better comparability of the approaches, the parameter study is to be carried out both for classic FCS-MPC approaches without the rotating vectors and for the FCS-MPC approach with rotating vectors. The mains frequency is defined as 50 Hz.

4.2 Basis for evaluation

The following criteria are to be used as a basis for evaluation:

1. achieved PMSM speed
2. generated THD of the source currents
3. generated THD of the load currents
4. average switching frequency
5. generated torque ripple

Based on the first point, the successful operation of the algorithm must be assessed under the given circumstances. Points 2-3 provide information about the subsequent operating behavior and allow an estimation of the EMC of the system as well as a quantitative evaluation of the currents. Point 4 is to be considered from the point of view of system efficiency. The last point 6 characterizes the system behavior with regard to the connected PMSM.

A simple load trajectory is provided to assess the topology based on the points mentioned. This can be divided into four sectors:

S1. Standstill of the PMSM
S2. Acceleration to target speed
S3. PMSM at target speed, load-free
S4. PMSM to target speed, target load

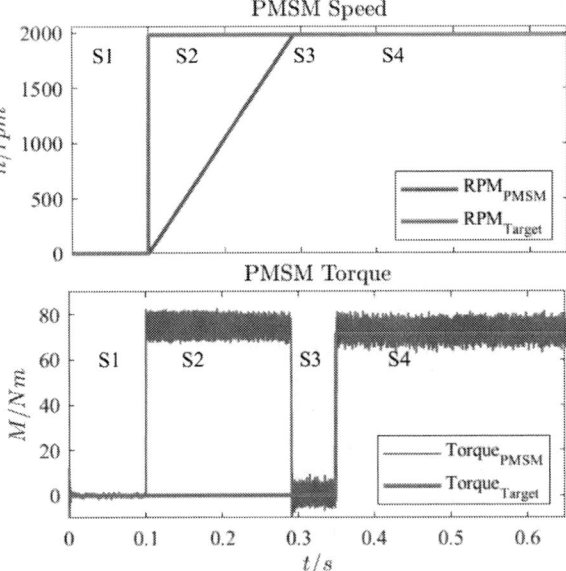

Fig. 3 Example trajectories used for evaluation (1980 rpm Target and rated Torque)

A parameter study must be carried out to determine the frequency influence. For this purpose, the frequency of the supply network is set from $10\,Hz$ to $330\,Hz$ in $10\,Hz$ steps with applied load steps from 0 to the rated torque of the machine in $10\,\%$ steps at the corresponding speed of the PMSM. This is to be carried out for the target speed $1980\,rpm$ (frequency ratio input/output 0.03 ... 1) and $990\,rpm$ (frequency ratio input/output 0.03 ... 2) of the PMSM. The reactive power target for all simulations is $0\,Var$.

4.3 Simulation Results and Discussion

4.3.1 Reached Machine Speed

To evaluate successful control of the PMSM, the machine speed achieved is decisive (point 1). If the operating speed of $1980\,rpm$ is the target of the control, a failure of the topology can be recognized from $120\,Hz$ supply frequency at nominal load. If the rotating vectors are used, this point can be delayed up to $160\,Hz$ supply frequency.

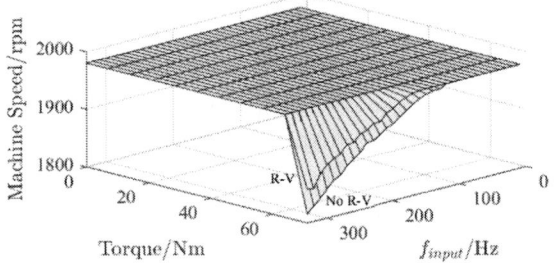

Fig. 4 Reached PMSM-Speed for Target 1980 rpm and Frequency ratio 0,03…1

Due to the constant component sizes of the input filter, the increasing input reactive power of the LC combination causes a shift in focus of the FCS-MPC algorithm when the supply frequency is increased. This effect is intensified by the reduced voltage reserve of the control. The results when the control target is missed are not to be considered for further evaluation but are shown for the sake of completeness, see Fig.4 .

If an operating speed of $990\,rpm$ is the control target, this is achieved over the entire operating range due to the greater voltage reserve and no difference between the classic vector basis and the extension by the rotating vectors can be seen.

4.3.2 THD Currents

The evaluation of the generated THD of the source currents (points 2 and 3) does not reveal any significant influence of the supply frequency on the power quality in the topology presented. While the load scenarios of the PMSM show no difference in this consideration, there is a marginal reduction in the input-side THD due to the use of rotating vectors [6] in all load cases.

4.3.3 Average Switching Frequency

When looking at the average switching frequency achieved by the FCS-MPC algorithm (point 3), there are clear dependencies on the supply frequency and the load torque for both speed variants. If the PMSM is at a speed of 1980 rpm (frequency ratio 0.03 ... 1), the range of 10Hz-100Hz represents a qualitatively constant area over the entire load range, see Fig.5. The unloaded PMSM exhibits a slight increase in the switching frequency when the supply frequency is increased. At nominal load, however, there is a significant reduction in the switching frequency when the supply and machine frequencies are equalized, and a convex surface is created. At a frequency of 330 Hz, for example, this results in a reduction of 16.85 % compared to no-load operation at a load of 35.75 Nm and 11.42 % at a load of 64.35 Nm (maximum power achieved).

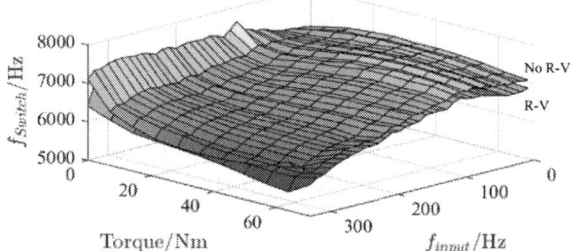

Fig. 5 Average Switching Frequency for Target 1980 rpm and Frequency ratio 0,03…1

The situation is similar at a machine speed of 990 rpm (frequency ratio 0.03 ... 2) , see Fig.6. If the frequency is increased, there is again an increase in the switching frequency in idling operation, whereas there is a reduction in nominal operation. The maximum possible reduction in the switching frequency increases further in the partial load range and produces a minimum at 330 Hz supply frequency and half the nominal load with a switching frequency reduction of 20.02 %. At nominal load, a reduction of 7.9 % is recorded.

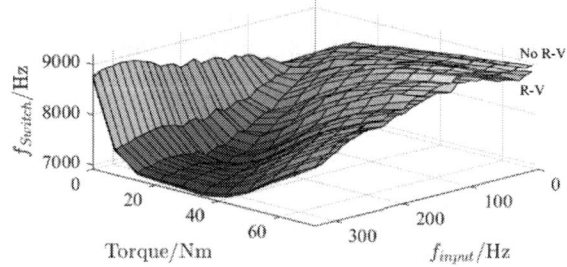

Fig. 6 Average Switching Frequency for Target 990 rpm and Frequency ratio 0,03…2

A comparison of both target speeds at a frequency ratio of 1 shows similar maximum reductions in the switching frequency at the minimum occurring. For a target speed of 990 rpm and a supply frequency of 160 Hz, the minimum switching frequency is located at 14.3 Nm with a percentage reduction of 12.2 %.

The simulations show a significant influence of the frequency ratio of input to output frequency. Across all simulations, a reduction in the switching frequency can be observed when using the rotating vectors. A significant influence of the rotating vectors on the switching frequency at different frequency ratios could not be determined.

The simulations also show that there is a general reduction in the switching frequency at similar voltage levels of the supplying grid and the PMSM.

4.3.4 Torque Quality

The basis for evaluating the torque behavior is the RMS value of the deviation of the generated torque from the load torque during load phase S4. The evaluation of the simulations with the control target of 1980 rpm shows in Fig.7 an increase in the key figure at higher load and higher supply frequency with a slight decrease at full load.

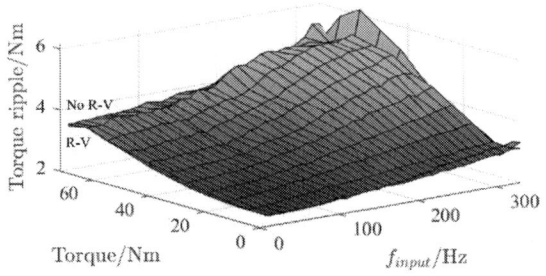

Fig. 7 Torque-Ripple for Target 1980 rpm and Frequency ratio 0,03...1

A similar picture applies to the control target of 990 rpm. The evaluation is similar for the lower supply frequency range up to approx. 100 Hz, but the maximum of the characteristic number shifts from the full load to the partial load range with a further increase in the supply frequency, Fig 8.

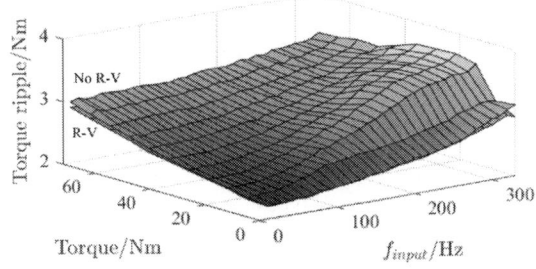

Fig. 8 Torque-Ripple for Target 990 rpm and Frequency ratio 0,03...2

Over the entire operating range, a significantly lower coefficient can still be determined. The use of the rotating vectors reduces the defined coefficient across most valid operating points and improves the torque behavior of the topology. Only with a frequency ratio of 1 and the target speed of 1980 rpm is no algorithm variant advantageous over the entire load range.

5 Conclusion

The simulations carried out show that the feeding frequency is a significant factor for the operation of the topology from Fig. (1). By increasing the input frequency, the average switching frequency can be significantly reduced during operation over a wide load range. At the same time, reducing the switching frequency leads to a deterioration in torque quality due to increased ripple.

The findings can contribute to an improvement when considering the entire system of network, converter, load machine and mechanics. If an increase in torque ripple can be dealt with due to the mechanical structure, a reduction in converter losses caused by the switching frequency can be achieved by selecting a PMSM with an operating frequency close to or below the mains frequency.

6 References

[1] Siti Hajar Yusoff, Nur Shahida Midi, Sheroz Khan, Majdee Tohtayong, "Predictive Control of AC/AC Matrix Converter" in International Journal of Power Electronics and Drive System (IJPEDS) Vol. 8, No. 4, December 2017, pp. 1932~1942 ISSN: 2088-8694, DOI: 10.11591/ijpeds.v8i4.pp1932-1942

[2] T. Friedli, J. W. Kolar, J. Rodriguez and P. W. Wheeler, "Comparative Evaluation of Three-Phase AC–AC Matrix Converter and Voltage DC-Link Back-to-Back Converter Systems" in IEEE Transactions on Industrial Electronics, vol. 59, no. 12, pp. 4487-4510, Dec. 2012, doi: 10.1109/TIE.2011.2179278.

[3] J. Rodriguez et al., "State of the Art of Finite Control Set Model Predictive Control in Power Electronics" in IEEE Transactions on Industrial Informatics, vol. 9, no. 2, pp. 1003-1016, May 2013, doi: 10.1109/TII.2012.2221469.

[4] J. Zhang, L. Li and D. G. Dorrell, "Control and applications of direct matrix converters: A re-view" in Chinese Journal of Electrical Engineering, vol. 4, no. 2, pp. 18-27, June 2018, doi: 10.23919/CJEE.2018.8409346.

[5] P. Gamboa, S. F. Pinto, J. F. Silva and E. Margato, "A flywheel energy storage system with Matrix Converter controlled Permanent Magnet Synchronous Motor", 18th International Conference on Electrical Machines, 2008, pp. 1-5, doi: 10.1109/ICELMACH.2008.4799861.

[6] R. Zipprich, J. Winter, B. Tekir and M. Ziegler, "Analyzing the Influence of Rotating Voltage Vectors in Case of FCS-MPC-Controlled Matrix-Converter for PMSM," PCIM Europe 2023; International Exhibition and Conference for Power Electronics, Intelligent Motion, Renewable Energy and Energy Management, Nuremberg, Germany, 2023, pp. 1-8, doi: 10.30420/566091326.

[7] Resutík, P.; Kaščák, S. Compact 3 × 1 Matrix Converter Module Based on the SiC Devices with Easy Expandability. Appl. Sci. 2021, 11, 9366. https://doi.org/10.3390/app11209366.

[8] A. Popovici and R. Mutiu, "Analysis of the Input Filter Parameters for a Power Matrix Converter," 2021 IEEE 27th International Symposium for Design and Technology in Electronic Packaging (SIITME), Timisoara, Romania, 2021, pp. 366-368, doi: 10.1109/SIITME53254.2021.9663721.

[9] Muñoz-Castillo, J.; Muñoz-Hernández, G.A.; Portilla-Flores, E.A.; Vega-Alvarado, E.; Calva-Yáñez, M.B.; Mino-Aguilar, G.; Niño-Suarez, P.A. Design of the Input and Output Filter for a Matrix Converter Using Evolutionary Techniques. Appl. Sci. 2020, 10, 3524. https://doi.org/10.3390/app10103524.

[10] H. She, H. Lin, X. Wang and L. Yue, "Damped input filter design of matrix converter," 2009 International Conference on Power Electronics and Drive Systems (PEDS), Taipei, Taiwan, 2009, pp. 672-677, doi: 10.1109/PEDS.2009.5385684.

[11] "IEEE Standard Definitions for the Measurement of Electric Power Quantities Under Sinusoidal, Nonsinusoidal, Balanced, or Unbalanced Conditions," in IEEE Std 1459-2010 (Revision of IEEE Std 1459-2000), vol., no., pp.1-50, 19 March 2010, doi: 10.1109/IEEESTD.2010.5439063.

PCIM Europe 2024, 11– 13 June 2024, Nuremberg   DOI: 10.30420/566262228

Enhancing Reactive Power Capacity in Battery-fed Power Conditioning Systems

Joseph K. Banda [1], Lucas S. Araujo [1], Elisabetta Tedeschi [1,2]

[1] Department of Electric Power Engineering, Norwegian University of Science and Technology (NTNU), Norway

[2] Department of Industrial Engineering, University of Trento, Italy

Corresponding author: Lucas S. Araujo, lucas.s.de.araujo@ntnu.no
Speaker: Lucas S. Araujo, lucas.s.de.araujo@ntnu.no

Abstract

In isolated electrical grids, battery energy storage systems (BESS) are crucial for increasing the integration of renewable energy sources and reducing the reliance on gas turbine generators. However, isolated systems often face power quality issues, especially when heavy motors are direct online started, requiring a large amount of reactive power. The BESS power conditioning system (PCS) is often oversized when high reactive power demands are to be met. To avoid oversizing and related additional costs, this paper proposes a novel strategy for enhancing the reactive power capacity of a single-stage battery-fed PCS. The simulation results show that employing the proposed strategy increases the reactive power capacity by approximately threefold when the battery bank's state-of-charge is at 20%, as compared to the scenario without the strategy.

1   Introduction

To increase the integration of renewable energy sources and reduce the reliance on gas turbine generators (GTGs) in isolated electrical grids, implementing energy storage systems such as battery energy storage system (BESS) is crucial. BESS offer essential services, including peak shaving, frequency regulation, and support for spinning reserves [1], [2]. Additionally, there is a growing trend for services such as energizing power transformers and aiding direct online start of heavy motors [3]. Notably, these services demand a large amount of reactive power. Hence, the BESS power conditioning system (PCS) can support with some, or potentially all, of the reactive power demand, thereby improving the power factor, helping with voltage regulation, and relieving the GTGs.

Reactive power support can be achieved through proper control of connected converters, e.g., statcoms or active-power filters, with or without the inclusion of BESS. However, considering BESS-fed PCS, the supply of reactive power is impaired by the variation in the batteries state-of-charge (SoC). This happens because a converter's reactive power supply capacity is mainly related to the dc bus and grid voltage levels.

Although there are advancements in modular architecture with multi-stage converters for BESS [4], BESS-fed PCS are usually single-stage converters [5], due to lower initial costs [6]. Single-stage PCS is also more compact, which is especially important in space-constrained locations, such as offshore oil and gas platforms (OOGPs). Hence, when the SoC of a BESS with a single-stage PCS is low, the dc bus voltage will also reflect into a lower voltage level, as they are directly connected in parallel. Once the available dc bus voltage for the PCS is lower, its reactive power supply capacity is impaired.

Common approaches to deal with this issue are: *i)* oversizing the BESS-PCS converter; *ii)* implementing double-stage arrangements for BESS-PCS; or *iii)* changing the tap changer position of the BESS connection transformer. While oversizing a converter and implementing a double-stage arrangement can provide additional reactive power support, they come with cost, efficiency, space, and complexity trade-offs. Additionally, transformers typically have a limited number of taps, so changing the tap might offer a discrete set of options for voltage adjustment.

In recent years, substantial efforts have been dedicated to refining the control of BESS-PCS systems to provide both active and reactive power, aiming to address power quality issues. In [7], [8], a single-stage BESS-PCS is used to provide active and reactive power and regulate the grid frequency and voltage. In [9], a BESS is used as a statcom, curtailing its active power momentarily and providing a large amount of reactive power when there is a voltage disturbance. However, in these works the BESS SoC and its related voltage value is not considered into the variability of reactive power capacity.

In [2], fluctuations in voltage and frequency within an isolated marine grid are smoothed through the utilization of grid-supporting BESS embedded of data-driven control when providing active and reactive power. In grid-connected applications, as outlined in [10], low-voltage distributed energy resources can be coordinated to provide controlled reactive power to the upstream grid, thereby supporting medium-voltage voltage regulation. Regarding OOGPs applications, in [3], active front-end variable frequency drives, used for motors applications, are explored to provide reactive power to regulate the power factor at the GTG terminals, and to enhance voltage regulation under a direct online start of a heavy motor. In [11], a similar approach is used but to mitigate voltage fluctuation caused by wind turbines connection. These services could also be handled by BESS-fed PCS, however the reactive power capacity could be limited by the BESS SoC. Therefore, recognizing the current gap in the literature regarding effective solutions for addressing the reactive power capacity constraint in single-stage BESS-fed PCS, this paper presents a novel strategy that enhances their reactive power capacity.

2 System Architecture - Isolated OOGP with BESS

To exemplify the application of the proposed strategy, it is considered an OOGP in the North Sea. Fig. 1 depicts its single-line diagram. Multiple gas turbine driven synchronous generators are lumped together as one generation unit, which is connected to the platform loads through a step-up transformer T_1 (11 kV/ 13.8 kV). An offshore wind farm with multiple Type 4 wind turbines is connected at the 13.8 kV bus through an ac collection system. Also, some heavy induction motors are connected to the 13.8 kV bus. These motors are started directly

online by closing the contactors (C_1, C_2, and C_3). For simplicity, other platform loads are represented lumped as fixed loads, and are connected to the same 13.8 kV bus through transformer T_3 (13.8 kV/6 kV).

A grid-scale BESS is employed at the 13.8 kV bus through transformer T_2 (600 V/ 13.8 kV) for providing backup power and/or peak shaving services. A simplified single-line diagram of the BESS is shown in Fig. 2. The BESS mainly comprises of a battery bank, a single stage dc-ac PCS, and a control system. A 2-level voltage source converter (VSC) with main reactor, L_f, and its resistance, R_f, filter capacitor, C_f, and transformer equivalent winding inductance referred to low voltage (LV) side, L_{LFT}, form the power circuit of the PCS, depicting a LCL filter. A dc contactor, $B1_{dc}$, connects the VSC to the battery bank, and an ac contactor, $K1_{ac}$, connects T_2 to the grid. The battery bank is realized with a series and parallel combination of several battery cells to obtain a 1100 V dc bus, and energy support of 5 MW for 15 min. The PCS parameters are tabulated in Table 1.

Tab. 1: Parameters of single-stage BESS PCS.

Parameter	Symbol	Value	Per unit
Base apparent power	S_{base}	5 MVA	1.0 pu
Base voltage ac	V_{base}	600 V	1.0 pu
Base current ac	I_{base}	4811 A	1.0 pu
Base frequency ac	f_{base}	50 Hz	1.0 pu
Voltage dc	V_{dc}	1100 V	
Current dc	I_{dc}	4545 A	
Transformer ratio		13.8 kV / 600 V	
Transf. winding inductance	L_{LFT}	18.3 µH	0.08 pu
Transf. winding resistance	R_{LFT}	0.360 mΩ	0.005 pu
Main reactor inductance	L_f	67.4 µH	0.293 pu
Main reactor resistance	R_f	0.720 mΩ	0.01 pu
Shunt capacitance	C_f	2.4 mF	0.055 pu
LCL resonance	f_{res}	850 Hz	
Dc capacitance	C_{dc}	15 mF	
Switching frequency	f_{sw}	3000 Hz	

3 Enhancing Reactive Power Capacity in BESS-fed PCS

3.1 Background of PCS Reactive Power Capability

Single-stage PCS are preferred over double-stage PCS for grid-scale BESS whose range exceeds 1 MVA. However, the battery's variable SoC results in a varying dc bus voltage. Thus, it is essential to quantify the dynamic variation in the dc bus voltage,

PCIM Europe 2024, 11– 13 June 2024, Nuremberg   DOI: 10.30420/566262228

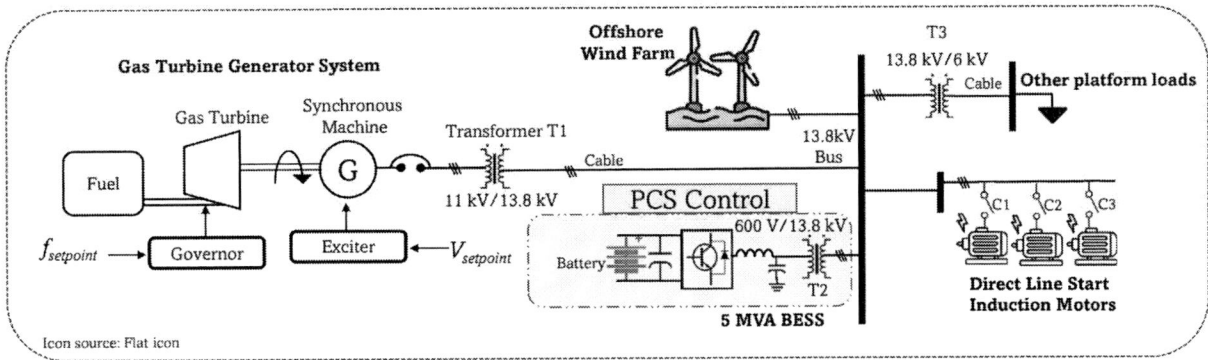

Fig. 1: Considered isolated OOGP power system partly electrified with an offshore wind farm and BESS.

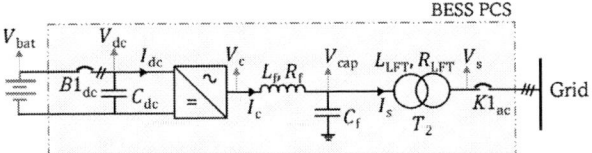

Fig. 2: Single-line diagram of single-stage BESS PCS.

and its resultant variation in the active and reactive power capacity of the PCS. The magnitude of the ac and dc voltages at the terminals of the PCS and the electrical ratings (voltage and current) of the semiconductor devices would decide its power capability. To better understand the limitations on the active and reactive power capabilities for the BESS PCS depicted in Fig. 2, mathematical expressions are derived to evaluate the capabilities of the VSC at its connection point to the grid.

$$Z_{LFT} = R_{LFT} + j\omega_n L_{LFT}; \tag{1}$$

$$Z_f = R_f + j\omega_n L_f; X_{Cf} = \frac{1}{\omega_n C_f} \tag{2}$$

$$v_{cdq} = R_f i_{cdq} + L_f \frac{d}{dt} i_{cdq} + j\omega_n L_f i_{cdq} + v_{capdq} \tag{3}$$

$$i_{sdq} = \frac{v_{cdq} - v_{sdq}\left(1 + j\frac{Z_f}{X_{Cf}}\right)}{Z_{LFT} + Z_f + j\frac{Z_{LFT} Z_f}{X_{Cf}}} \tag{4}$$

where Z_{LFT} represents the impedance comprising the leakage resistance R_{LFT} and the leakage inductance L_{LFT}; ω_n denotes the grid frequency in radians per second; Z_f signifies the impedance of the inductor of LC filter, consisting of the inductance L_f and the filter resistance R_f. X_{Cf} represents the capacitive reactance of LC filter, consisting of a shunt capacitance C_f. The variables

v_{cdq}, v_{sdq}, v_{capdq}, i_{cdq}, and i_{sdq} correspond to the instantaneous fundamental rotating vectors of the converter voltage, grid source voltage (referred to the LV side), capacitor voltage, converter current, and grid source current, respectively. These quantities are transformed from a stationary reference frame to a rotating reference frame dq. The angle for this transformation is acquired from a Phase-locked loop (PLL) which ensures the alignment of direct axis with the measured capacitor voltage.

Using the measured values of v_{sdq}, and the computed value of i_{sdq}, as in Eq. (4), the three-phase instantaneous active, P_{pcs}, and reactive, Q_{pcs}, power of BESS PCS at the point of connection to the grid can be calculated as:

$$P_{pcs} = 1.5 \times \text{Real}\left(v_{sdq} \cdot i_{sdq}^*\right) \tag{5}$$

$$Q_{pcs} = 1.5 \times \text{Imag}\left(v_{sdq} \cdot i_{sdq}^*\right) \tag{6}$$

Where i_{sdq}^* is the complex conjugate of i_{sdq}. Notably, both P_{pcs} and Q_{pcs} are constrained by two limits established during the design phase of the PCS:

1. PCS Voltage Limit: For a modulation index of 1.0 per unit, the peak value of the converter voltage ($|v_{cdq}|$) in Eq. (3) is limited to $1.15 \times 0.5 \times V_{dc}$ for a Sine Pulse Width Modulation (SPWM) with third harmonic injection [12].

2. PCS Current Limit: The peak value of the converter current ($|i_{cdq}|$) in Eq. (3) is constrained by the peak value of the PCS current ($I_{base} \times sqrt(2)$).

Thus, the calculated active and reactive power capabilities in Eqs. (5) and (6) are constrained by these two limits, shown in Eq. (7).

1675

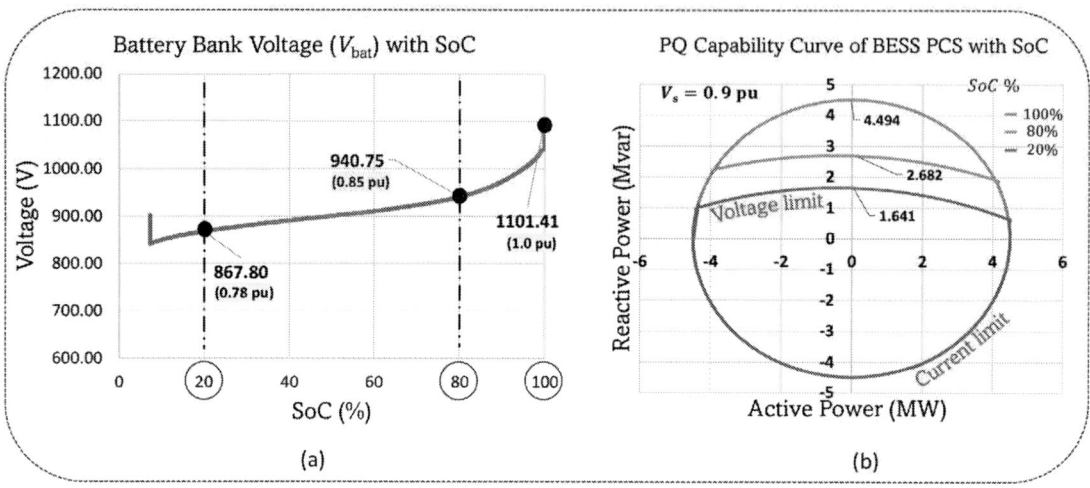

Fig. 3: (a) Battery bank voltage variation with SoC and (b) Corresponding capability curve.

$$P_{pcs} = 1.5 \times \text{Real} \left(v_{sdq} \cdot i_{sdq}^* \right)$$

$$Q_{pcs} = 1.5 \times \text{Imag} \left(v_{sdq} \cdot i_{sdq}^* \right)$$

$$\text{subject to:} \quad |v_{cdq}| \leq (1.15 \times 0.5 \times V_{dc})$$

$$|i_{cdq}| \leq (I_{base} \times \sqrt{2}) \tag{7}$$

Detailed theoretical derivations, sensitivity, and validation of active and reactive power capacities for a generic dc-ac converter are discussed in [13].

The dynamic variations in active and reactive capacities are particularly intriguing for the considered BESS PCS due to two main factors. First, the parallel connection of the battery with the dc bus, in which the voltage varies with the batteries SoC. Second, the isolated weak grid on the ac side is susceptible to voltage fluctuations.

Fig. 3a shows the battery bank voltage variation in relation to SoC values, and its corresponding capability curve for the considered BESS PCS in Fig. 3b. The dc bus voltage varies from 0.85 pu for SoC = 80% to 0.78 pu for SoC = 20%, i.e., for a base voltage of 1100 V, the dc bus voltage varies between 867 V and 940 V. These values are highlighted in Fig. 3a as most battery management systems regulate the SoC between 20% and 80%, by limiting charging and discharging operations, to increase the battery lifespan. In Fig. 3b, the grid voltage is maintained at a constant level of 0.9 pu, indicating a slight voltage dip. Notably, the reactive power capability is drastically affected by dc voltage variation. Reactive power values range from 1.6 Mvar to 4.5 Mvar for SoCs of 20% to 100%, respectively. This means that the theoretical capability curve

obtained with the assumption of dc bus voltage being constant is not valid for single-stage BESS-fed PCS. This issue must be addressed in applications that expect BESS-PCS to provide reactive power support as a priority service.

3.2 Proposed strategy

Fig. 4 shows the proposed BESS-PCS control block diagram for reactive power capacity enhancement. Essentially, it is proposed to add a dc contactor, $B1_{dc}$, between the battery and dc bus capacitor, C_{dc}, which is digitally controlled by the PCS control. Also, it is necessary to have measurements of battery voltage and SoC within PCS control. The proposed strategy is highlighted inside a dashed area with green-colored blocks added on top of a conventional cascaded control system of PCS in grid following mode. This strategy should be enabled when there is a need for reactive power contribution from BESS-PCS, e.g., for voltage support during heavy motor start, and the BESS SoC is low, with the compromised reactive power capacity. Rather than relying on the time-consuming process of recharging the BESS, the proposed strategy offers a significantly faster alternative.

When the proposed strategy is enabled, the state flow of the control algorithm is as follows:

1. Active and reactive power capacity curves are computed dynamically with inputs V_{bat}, SoC_{bat} and V_s.

2. Outer loops in PCS control, which are responsible for active and reactive power regulation, are first handled. Their references are set to

PCIM Europe 2024, 11– 13 June 2024, Nuremberg   DOI: 10.30420/566262228

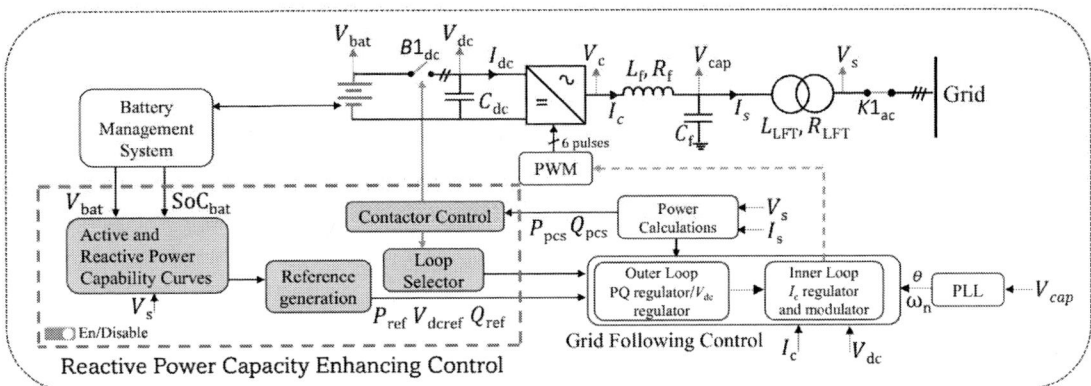

Fig. 4: Block diagram of BESS-PCS control with the proposed reactive power capacity enhancement.

zero regardless of the polarity and magnitude prior to enabling the proposed strategy.

3. Checkpoint to evaluate if the measured powers (both active and reactive) are tracked to zero. If this check passes, no active power flows from the battery to the PCS or viceversa. Thus, the Contactor control block opens the contactor $B1_{dc}$ at zero dc current.

4. Once the contactor is opened, the dc bus voltage is no longer attached to the battery voltage. Thus, the Loop Selector block switches the PCS control mode from active power regulation to V_{dc} regulation; the reactive power loop is unchanged.

5. Finally, the dc bus voltage is regulated to its nominal value. Thus, the BESS PCS maximum reactive power capacity shifts from the blue curve to the green curve, achieving ≈ 2.7 times of enhancement in reactive power, as seen in Fig. 3b.

4   Simulation Results

To verify the proposed strategy's performance, an isolated electric power system composed of a GTG, a 5 MVA BESS-PCS connected to the grid through a 600 V/13.8 kV transformer, and a load is simulated using Matlab/Simulink, and the results are shown in Fig. 5. Consider the situation when contactor $B1_{dc}$ is closed and the battery is discharged, i.e., SoC = 20%, while providing active power, P_{pcs}, of 1 MW.

The black solid line in Fig. 5-d captures the grid voltage referred to the LV side. At 4.0 s, a grid voltage, V_s, of 0.9 pu is sensed and identified as a voltage sag. Thus, BESS-PCS, which is originally providing active power, ramps up the reactive power in response to the voltage sag event. The reactive power reference is set to 1.5 Mvar, close to the maximum capability when battery SoC = 20%, as seen in the red solid line in Fig. 5-b. The measured reactive power, Q_{pcs}, tracks the set reference as desired, as seen in the purple solid line in Fig. 5-b. One can also note that the PCS is at its maximum operating limit, as the modulation index (MI) reaches its boundary, 1.0 pu, when the PCS provides its maximum reactive power, i.e., from 4.5 s to 9.0 s, as shown in Fig. 5-d. Thus, the grid voltage has improved only to 0.95 pu.

At 9.0 s, the proposed reactive power capacity enhancement strategy is enabled, as seen in the solid blue line in Fig. 5-a. When the proposed strategy is enabled, both outer loop regulators (active and reactive power) set their references to zero, and the measured powers track them accurately, as shown in Fig. 5-b. The battery bank current, I_{bat}, is function of PCS active power, and also drops to zero, as seen in Fig. 5-d, as a solid green line. Once the battery bank current is ensured at zero, the contactor $B1_{dc}$ opens at 12.0 s, and the PCS control mode switches from active power regulation to dc bus voltage regulation. As the dc bus is disconnected from the battery, it is then regulated to its nominal value of 1100 V, as shown in Fig. 5-c. When the dc bus voltage is regulated, the modulation index is observed to drop to 0.68, as shown in Fig. 5-d from 12 s to 14 s. Thereby, the PCS is relieved, making it possible to provide maximum reactive power of 4.4 Mvar. As a result, the grid voltage is improved to 1.0 pu, and the modulation index reaches the voltage limit of 1.0 pu, fully utilizing the enhanced

1677

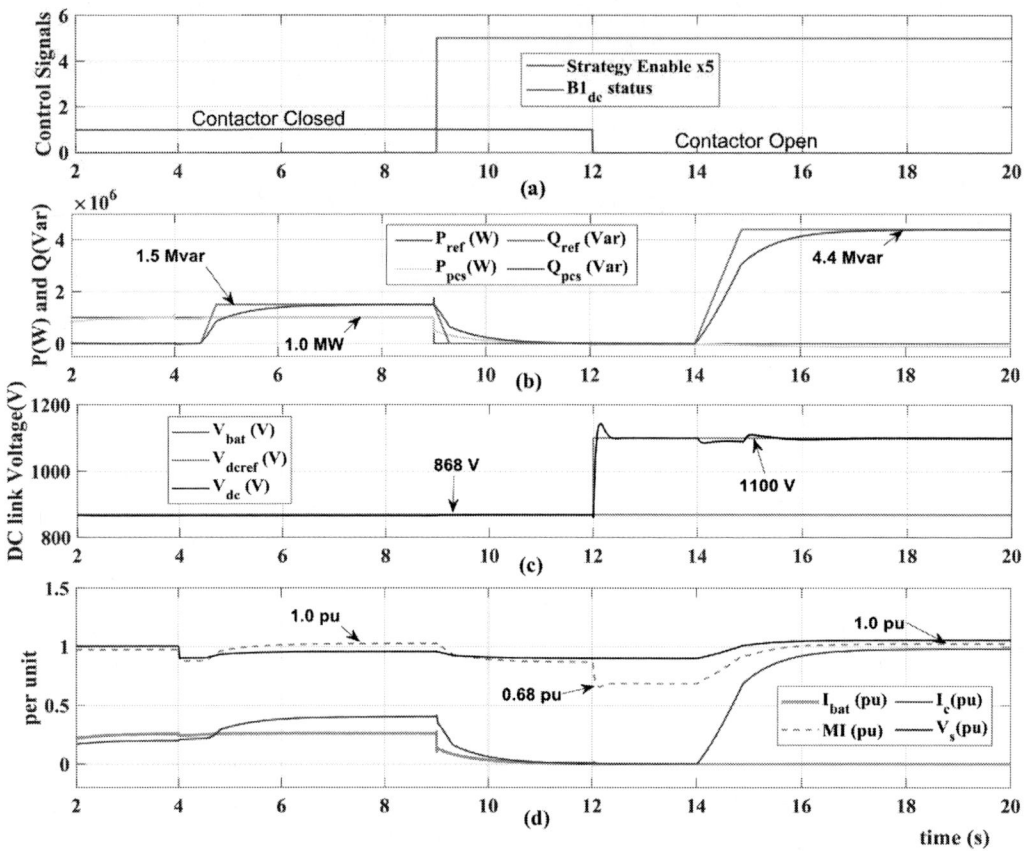

Fig. 5: Simulation Results: a) Control strategy enable signal, $B1_{dc}$ contactor status; b) active and reactive power reference and measured power at the grid; c) Battery and dc bus capacitor voltages; d) Battery and converter currents, converter modulation index, and grid voltage, all represented in per unit.

reactive power capacity. Therefore, it is shown that the proposed strategy increases the reactive power capacity by approximately three times for a BESS at 20% of SoC when compared to the scenario without the strategy implemented.

5 Conclusions

This paper develops a novel strategy to enhance the reactive power capacity of a single-stage battery-fed power conditioning system connected to an electrical grid. It is proposed to disconnect the dc bus capacitor from the battery when the battery SoC is low, thus allowing the dc bus voltage to be regulated to its nominal value. The results show that the proposed control strategy effectively increases the reactive power capacity approximately threefold for the scenario when the battery SoC is 20%, compared to the case without the strategy. By increasing the reactive power capacity, the need

for an oversized single-stage PCS can be avoided, saving significantly on costs and space. This is especially valuable in space-constrained applications where BESS-PCS is installed to provide reactive power support, such as when motor loads are directly online-started.

6 Acknowledgements

This research was supported by the program PETROMAKS2 of the Research Council of Norway, within the project "Smart Platform" (project number 308735).

References

[1] M. S. Pilehvar and B. Mirafzal, "Frequency and voltage supports by battery-fed smart inverters in mixed-inertia microgrids," *Electronics*, vol. 9, no. 11, p. 1755, 2020.

[2] D. J. Ryan, R. Razzaghi, H. D. Torresan, A. Karimi, and B. Bahrani, "Grid-supporting battery energy storage systems in islanded microgrids: A data-driven control approach," *IEEE Transactions on Sustainable Energy*, vol. 12, no. 2, pp. 834–846, 2021. DOI: 10.1109/TSTE.2020.3022362.

[3] J. M. S. Callegari, L. A. Vitoi, and D. I. Brandao, "Vfd-based coordinated multi-stage centralized/decentralized control to support offshore electrical power systems," *IEEE Transactions on Smart Grid*, vol. 14, no. 4, pp. 2863–2873, 2023.

[4] L. S. Araujo, N. T. D. Fernandes, D. I. Brandao, and B. J. Cardoso Filho, "Smartbattery: An active-battery solution for energy storage system," in *2019 IEEE 15th Brazilian Power Electronics Conference and 5th IEEE Southern Power Electronics Conference (COBEP/SPEC)*, 2019, pp. 1–6.

[5] S. Ponnaluri, G. Linhofer, J. Steinke, and P. Steimer, "Comparison of single and two stage topologies for interface of bess or fuel cell system using the abb standard power electronics building blocks," in *2005 European Conference on Power Electronics and Applications*, 2005, 9 pp.–P.9. DOI: 10.1109/EPE.2005.219502.

[6] L. S. Xavier, C. V. De Sousa, H. A. Pereira, and V. F. Mendes, "Design and performance comparisons of power converters for battery energy storage systems," *International Journal of Circuit Theory and Applications*, vol. 51, no. 7, pp. 3146–3166, 2023.

[7] I. Serban and C. Marinescu, "Control strategy of three-phase battery energy storage systems for frequency support in microgrids and with uninterrupted supply of local loads," *IEEE Transactions on power electronics*, vol. 29, no. 9, pp. 5010–5020, 2013.

[8] J.-T. Gao, C.-H. Shih, C.-W. Lee, and K.-Y. Lo, "An active and reactive power controller for battery energy storage system in microgrids," *IEEE*

Access, vol. 10, pp. 10490–10499, 2022. DOI: 10.1109/ACCESS.2022.3145009.

[9] R. K. Varma, M. Ahmadi, and C. Arpino, "Novel control of battery energy storage system (bess) as statcom (bess-statcom) for stabilization of a critical motor," in *2023 IEEE Canadian Conference on Electrical and Computer Engineering (CCECE)*, 2023, pp. 261–267. DOI: 10.1109/CCECE58730.2023.10289096.

[10] L. S. Araujo, D. I. Brandao, S. M. Silva, and B. J. Cardoso Filho, "Reactive power support in medium voltage networks by coordinated control of distributed generators in dispatchable low-voltage microgrid," in *2019 IEEE 28th International Symposium on Industrial Electronics (ISIE)*, 2019, pp. 2603–2608. DOI: 10.1109/ISIE.2019.8781376.

[11] L. F. da Rocha, D. I. Brandao, K. d. S. Medeiros, M. S. Dall'asta, and T. B. Lazzarin, "Coordinated decentralized control of dynamic volt-var function in oil and gas platform with wind power generation," *IEEE Open Journal of Industry Applications*, vol. 4, pp. 269–278, 2023. DOI: 10.1109/OJIA.2023.3307299.

[12] S. N. Vukosavic, *Grid-Side Converters Control and Design: Interfacing Between the AC Grid and Renewable Power Sources* (Power Electronics and Power Systems), 1st ed. Springer Cham, 2018, p. 266. DOI: 10.1007/978-3-319-73278-7.

[13] J. K. Banda, D. D. S. Mota, and E. Tedeschi, "A physics-informed scaling method for power electronic converters in power hardware-in-the-loop test beds," *IEEE Open Journal of Industry Applications*, vol. 5, pp. 1–14, 2024. DOI: 10.1109/OJIA.2024.3349480.

PCIM Europe 2024, 11– 13 June 2024, Nuremberg   DOI: 10.30420/566262229

Pulse Sharing: Achieving High Efficiency and Excellent Regulation in Multi-Output Flyback Power Supplies

Xingda Yan[1], Toine Werner[1]

[1] Power Integrations, United Kingdom

Corresponding author: Xingda Yan, xingda.yan@power.com
Speaker: Xingda Yan, xingda.yan@power.com

Abstract

This paper introduces a pulse sharing control strategy which is a key innovation employed in the InnoMux™-2 IC family of multi-output flyback converters from Power Integrations. InnoMux-2 ICs enable a unique flyback architecture, enabling a multi-output design that can adjust to changes in line and load without affecting the other outputs and eliminate the effects of cross regulation. Pulse sharing eliminates audible noise and reduces fluctuations in power output by distributing energy between outputs in each switching cycle. This article will explore the design challenges associated with this strategy, it will cover how the approach effectively accommodates sudden changes in power demand (load transients) and reduces EMI by adding variations in operating frequency (jitter).

1   Introduction of InnoMux-2 IC family

Traditional multiple output converter designs commonly rely on multiple DC-DC converter stages, such as buck/boost converters following the flyback converter to achieve accurate regulation. The InnoMux-2 IC-based flyback topology, shown in Fig. 1, offers a highly integrated solution featuring up to three individually regulated outputs. This innovative approach increases the efficiency of multi-output converter design and eliminates cross-regulation issues [1].

By independently regulating and protecting each output the InnoMux-2 IC family eliminates downstream conversion stages reducing the bill of materials (BOM).

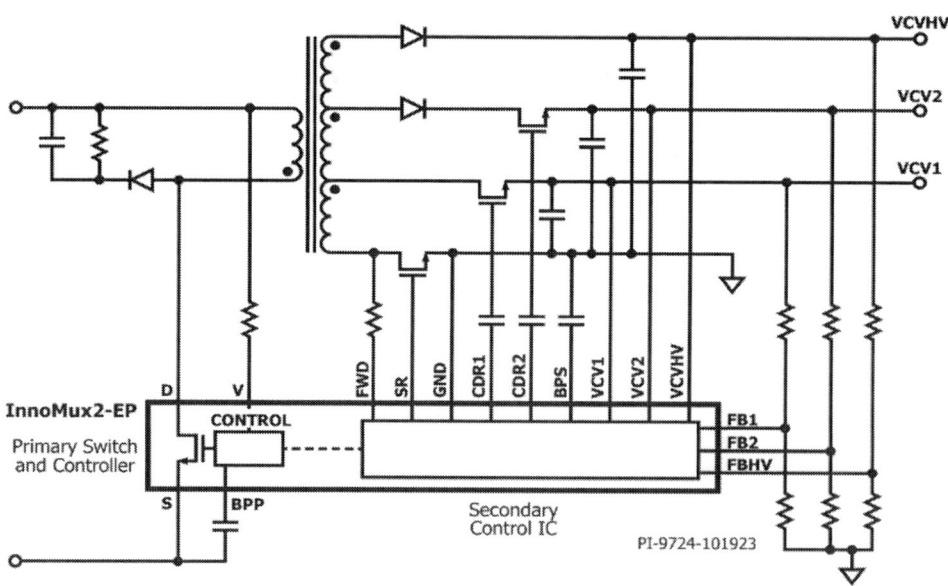

Fig. 1   InnoMux-2 architecture

1680

The InnoMux-2 combines a high-voltage power switch, along with primary-side and secondary-side controllers in one device. The InnoMux-2 architecture incorporates a proprietary inductive coupling feedback scheme using the package lead frame and bond wires to provide a safe and reliable means to accurately communicate switching requests from the secondary controller to the primary side. This eliminates the need for optocouplers.

The control mechanism on InnoMux-2 is a quasi-resonant (QR) flyback controller that has the ability to operate in continuous conduction mode (CCM) and discontinuous mode (DCM). The controller uses a variable current control scheme and consists of a receiver circuit magnetically coupled to the secondary controller, a current limit controller, an audible noise reduction engine, a lossless input line sensing circuit, current limit selection circuitry, overvoltage protection, secondary output diode / SR MOSFET short protection circuit, and a 650 V / 725 V silicon or 750 V PowiGaN™ power switch.

The secondary controller consists of a transmitter circuit that is magnetically coupled to the primary receiver, a multi-output controller for regulating up to three outputs independently, synchronous rectifier (SR) MOSFET driver, high-side MOSFET drivers, shunts to prevent individual output current from rising in abnormal loading conditions, single string LED driver, timing functions, and a host of integrated protection features.

1.1 An example: PSU with two constant voltage outputs

One typical application of InnoMux-2 is a two constant voltage outputs power supply as shown in

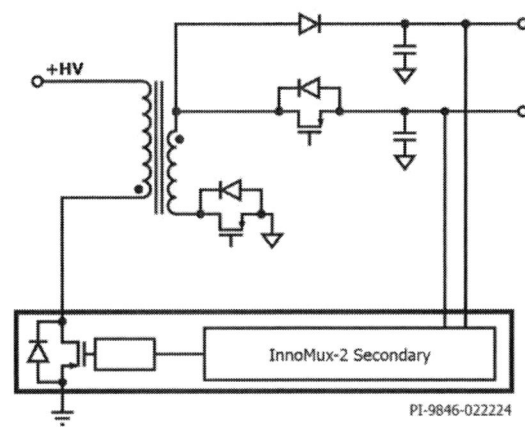

Fig. 2   Power supply with 2 CV outputs.

simplified form in Fig. 2. The main circuit consists of an InnoMux-2 controller, an isolation transformer with one primary winding, and a single secondary winding, an SR MOSFET on the return side of the output, and a selection MOSFET for the 5 V output plus a diode for 12 V output.

When a switching cycle is required, the secondary side requests the primary controller to energize the primary winding. This occurs whenever an output voltage dips below its set reference level. Once the inductor is charged to a pre-determined level (controlled by the rate that switching requests are received), the primary switch is turned off, and the flow of energy to the secondary side begins. The direction of this energy to the appropriate output stage is achieved by the state of the Selection FET. When the Selection FET is on, energy is di-

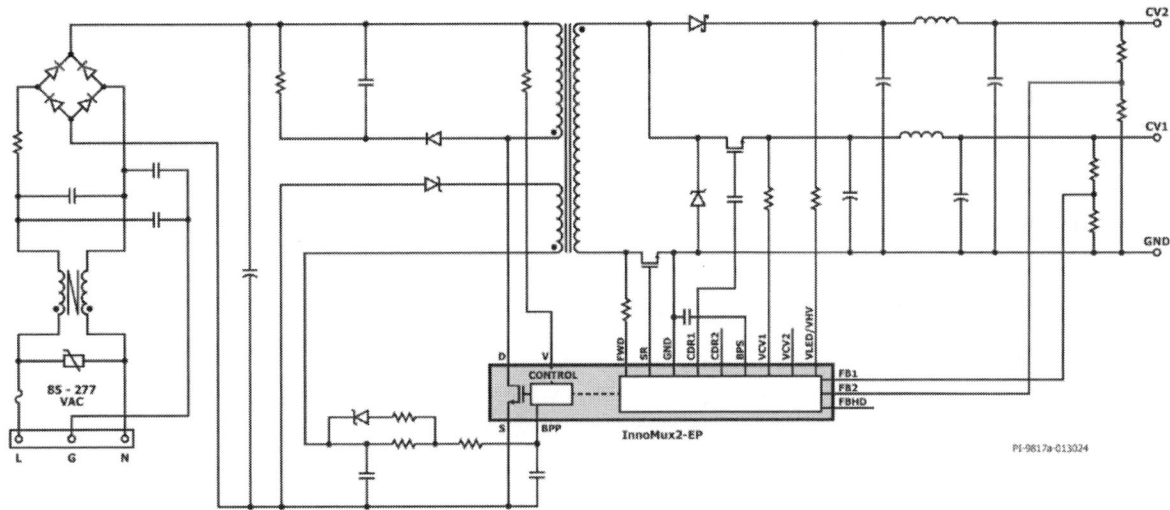

Fig. 3   Reference design showing dual output design (+5 V and +12 V outputs).

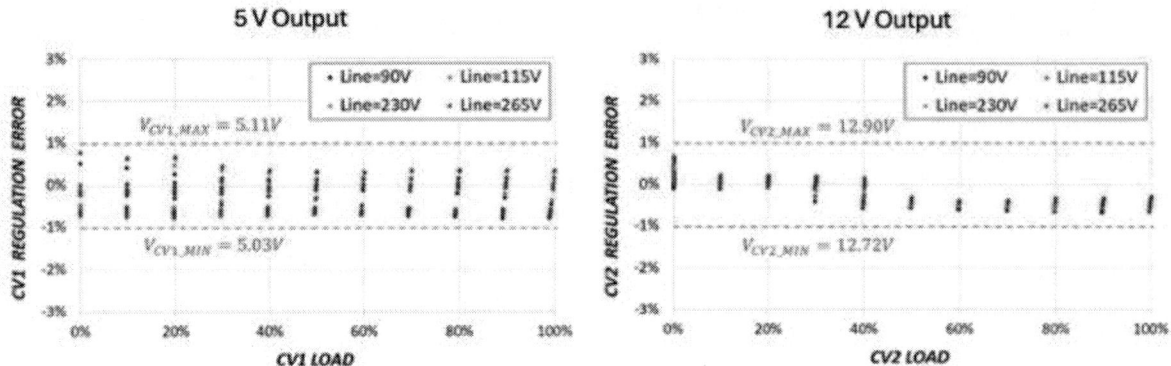

Fig. 4 Effect of line voltage, output load and loading on other outputs on output accuracy for each output of a dual output power supply with +5 V and +12 V outputs.

rected solely to the 5 V output capacitor. If the Selection FET is off, the energy is directed to the capacitor across the 12 V output. The SR (Synchronous Rectification) FET and a diode in the circuit for the 12 V output work together to block any backward flow of current. A specialized control strategy for the SR FET within the secondary control system is implemented to regulate the SR FET's operation, optimizing efficiency during the flyback discharge phase. The control of the Selection FET ensures that each output receives the correct amount of energy, maintaining stable voltage levels and high efficiency.

A primary switching cycle request is only triggered when one of the outputs drops below an output voltage threshold. By steering the energy provided between outputs, it is possible to ensure extremely accurate regulation for each output. Figure 4 illustrates output accuracy across line-, load-, and cross-regulation in a typical reference design (Fig. 3). Because output voltage is measured directly, transformer tolerances and other variability in power train performance between units is eliminated, ensuring that excellent load regulation will be retained across production.

Figure 4 shows the effect of line voltage, output load and loading on other outputs on accuracy for each output of a dual output power supply with +5

V and +12 V outputs. Note that regulation performance is achieved down to zero load without the need for dummy-load resistors on the output.

1.2 Audible noise and sub-harmonics

A typical switching pattern is shown in Fig. 5. The 5 V and 12 V outputs receive different numbers of energy packets. The main control scheme effectively eliminates cross-regulation effects, where loading on one output would influence the other output(s). However, a notable downside of this approach is the creation of audible noise. With each cycle, a pulse of energy is sent to one of the outputs, and since each output has a different reflected voltage, the speed at which magnetic energy changes in the transformer's core also changes based on which output is receiving the energy. This change in magnetic energy will induce a subharmonic transformer excitation frequency, which is lower than the main switching frequency. The nature of this subharmonic frequency depends on the load distribution between the two outputs. If this subharmonic frequency falls within the audible range, between approximately 1 kHz and 25 kHz, it's likely to produce a sound that can

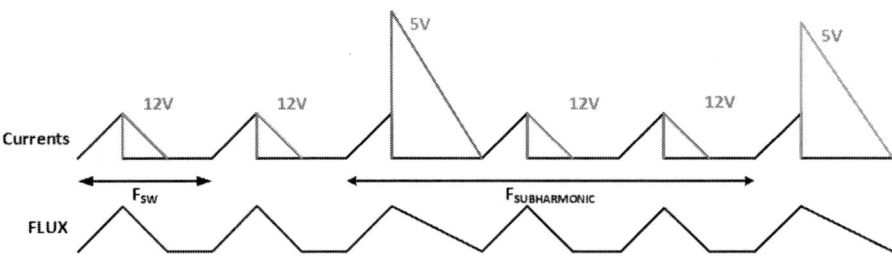

Fig. 5 Multi-output control switching pattern (2CV example).

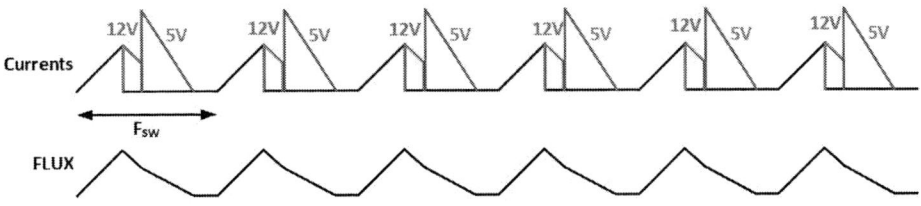

Fig. 6 InnoMux-2 switching pattern (2CV example) showing pulse sharing.

be heard. The magnetostriction effect will be amplified by the resonant frequency of the transformer mass which will typically also be in this region. This audible noise is a byproduct of the way the switching pattern operates under certain conditions.

2 Pulse sharing

A new control strategy — pulse sharing — solves this problem. The purpose of pulse sharing is to mitigate audible noise by distributing energy to multiple outputs during each cycle. Instead of delivering a complete pulse to a single output, each pulse is divided and shared between the two outputs (PWM divided for each to match output requirements). This approach ensures that every cycle is more uniform, eliminating the presence of subharmonics on the primary side. This also supports the system's capability to quickly adapt to changes in load without causing a disruption to the output voltage levels, ensuring reliable operation during dynamic load conditions.

2.1 Steady-state operation

Figure 6 provides a visual representation of the pulse sharing technique for the same two CV applications (5 V and 12 V). The process of pulse sharing involves sending the discharge pulse to

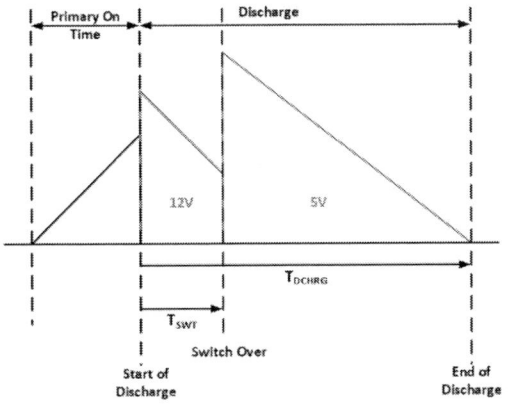

Fig. 7 Pulse sharing during each cycle.

the top-side 12 V output initially and then later in the discharge phase activating the selection MOSFET to enable the second part of the discharge to be directed to the 5 V output. By adjusting the duty cycle, regulation is maintained.

As shown in Fig. 7, the discharge period starts with the 12 V output. The amount of time discharged to the top-side output is defined as T_{SWT}. The secondary controller will regulate this value to match the operating conditions.

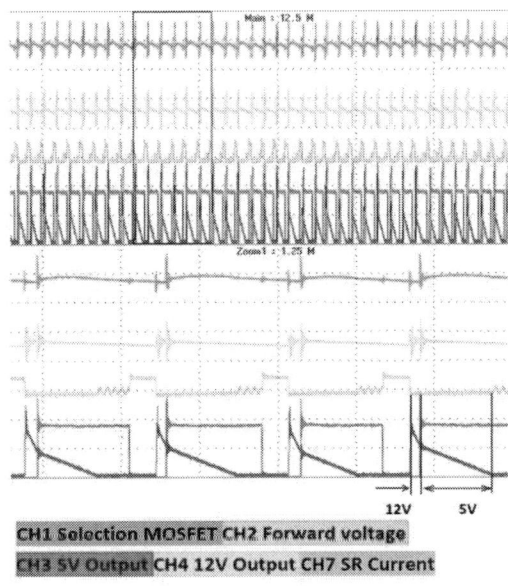

CH1 Selection MOSFET CH2 Forward voltage
CH3 5V Output CH4 12V Output CH7 SR Current

Fig. 8 Steady-state operation.

Figure 8 displays a screenshot from an oscilloscope when the two constant voltage (CV) application is functioning under nominal load conditions. Here, the switching frequency appears to be consistent, as shown by the very similar waveforms of inductor current from one cycle to the next. This indicates well-balanced operation.

2.2 Load transients

Operation must be maintained during disturbances caused by changes in the load on any of the outputs.

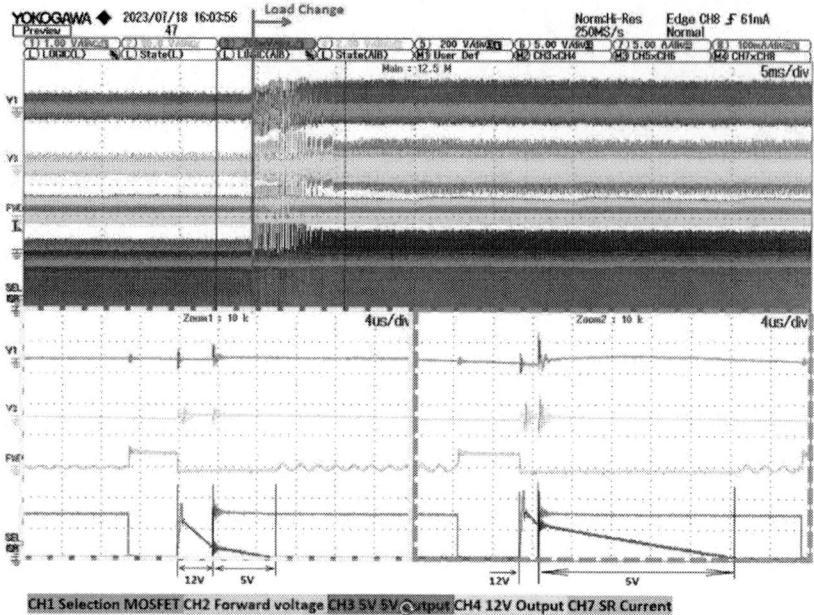

Fig. 9 Load change of bottom output from half-load to full-load with zoom.

A change in the load on either output leads to a change in the distribution of required energy between the outputs. This necessitates an adjustment in T_{SWT}, the period designated for top-side discharge.

In Fig. 9, the effect of a change in load is shown. A step change in the 5 V load is introduced. This causes a transient response from the system, during which the outputs experience a minor fluctuation (within the regulation limits). The fast response of the controller prevents significant overshoot or undershoot on any output. Within ~ 5 milliseconds, the system adjusts to a new balance point. At this new state, the period allocated for top-side discharge, T_{SWT}, is shorter, but there is an increase in the maximum current from the primary side. This illustrates the system's capability to quickly adapt to changes in load without causing disruption to the output voltage levels, ensuring reliable operation during dynamic load conditions.

2.3 Benefits of pulse sharing

The immediate benefit of pulse sharing is the reduction in audible noise. The acoustic performance of this two CV application (5 V, 12 V) is shown in Fig. 10. The graph above is the measurement without pulse sharing, whereas the lower plot shows the audible-noise measurement with the pulse sharing feature enabled. This clearly demonstrates the dramatic reduction of the audible noise – by 13 dBA in this example.

Pulse sharing allows the topology to work efficiently in CCM, which was previously limited by the

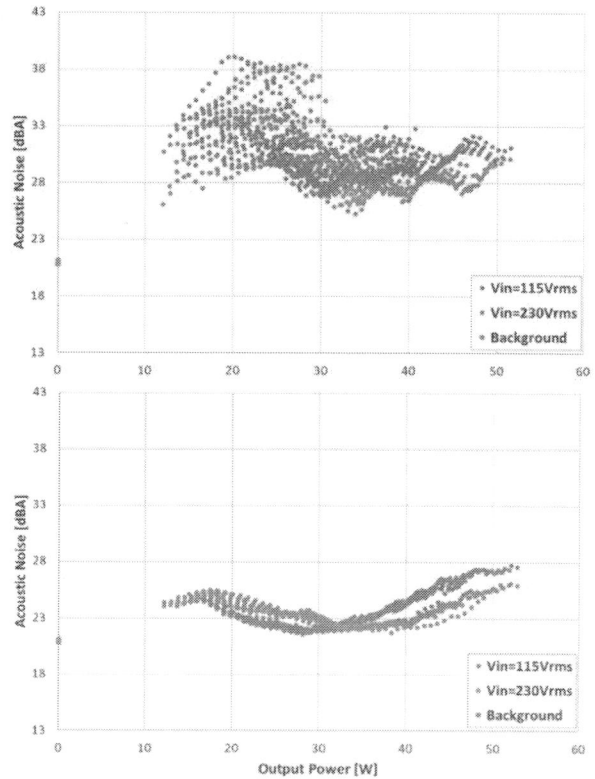

Fig. 10 Audible noise measurement (top: without pulse sharing; bottom: with pulse sharing)

high diode recovery losses for the top output if operated in DCM only.

Pulse sharing provides additional advantages, one of which is the reduction in the RMS current in the secondary winding, leading to a decrease in conduction losses. Moreover, this technique contributes to an increase in the operating frequency for each output, thereby reducing output ripple when using the same output filter capacitor.

3 Conclusion

This paper introduces an innovative control mechanism — pulse sharing — to improve the operation of multi-output power delivery and addresses the challenge of audible noise associated with varying discharge rates of the transductor in multi-output flyback converters. By distributing energy across multiple outputs in each cycle, the pulse sharing technique not only mitigates the issue of audible noise but also enhances the overall robustness of the system while increasing efficiency and improving transient response.

References

[1] Motte-Michellon D, Cogitore B, Lembeye Y and Ramdane B. A Study on the Influence of the Transformer on Cross-Regulation in DCM Multi-Output Flybacks. PCIM Europe 2022: Intelligent Motion, Renewable Energy and Energy Management: 1-8. VDE.

PCIM Europe 2024, 11– 13 June 2024, Nuremberg   DOI: 10.30420/566262230

Reliability-Optimized Space Vector Modulation (RO-SVM) for Semiconductors Lifetime Enhancement

Amin Rezaeizadeh [1], Silvia Mastellone [1]

[1] Institute of Electric Power Systems, University of Applied Science Northwest Switzerland, Windish, Switzerland

Corresponding author: Amin Rezaeizadeh, amin.rezaeizadeh@fhnw.ch
Speaker: Amin Rezaeizadeh, amin.rezaeizadeh@fhnw.ch

Abstract

This paper presents a reliability-optimized space vector modulation (SVM) method for controlling a three phase power converter while reducing the damage on semiconductor switches. Simulation results on IGBTs and SiC MOSFETs switches demonstrate that the thermal stress is significantly alleviated, leading to improved power device lifetime and overall system reliability.

1   Introduction

The increasing adoption of electronic power converters across different industries and critical applications has led to stronger requirements for reliable operation of those power converters. The inherent vulnerability of the high-power semiconductor switches raises reliability concerns in power converters. Specifically, power semiconductor devices, such as IGBTs and MOSFETs, are responsible for approximately 21% of inverter failures [1], [2]. Any fault in a power electronic component can lead to an unexpected system shut down, posing safety risks and increasing financial losses. Therefore, reliability analysis and operation of power semiconductor devices is a critical step in system lifetime management.

Thermal stress is the major cause of reduced lifetime for power switching devices, and a significant amount of research has been devoted to the electrical-thermal analysis and reliability assessment of these devices[3]–[5]. Several studies have proposed fault-tolerant control strategies to improve the safety and reliability of the system in case of failure[6], [7]. Additionally, techniques have been developed to extend the converter's lifetime by actively controlling the power module's temperature[8], [9]. Other solutions to reduced thermal stress are based on a topology with redundant switches [10]. Finally, other approaches design a switching strategy to control the temperature of the stressed power module and avoid overheating condition [11]–[13]. These methods use redundancy in the space vector modulation (SVM) to relocate power losses from the stressed switches to other switches, potentially causing more damage to other switches.

In this paper, we exploit the redundancy of SVM and formulate an optimization problem to minimize the damage experienced while providing the required voltage levels. The optimization algorithm computes the duty-cycles online for each power device in a two-level three-phase inverter topology (see Fig. 1). The objective of the proposed optimization is to reduce the total damage of the inverter by equalizing the thermal stress across all power semiconductor devices. The proposed switching strategy, which we call RO-SVM (Reliability-optimized SVM), is compared with a conventional SVM method developed in MATLAB/SIMULINK's SVM (2-level topology) block that generates a pattern with minimal switching losses.

The paper is structured as follows: Section 2 describes the principle of the conventional SVM and presents the lifetime model of the power semiconductor switches. The proposed SVM method is introduced in Section 3. Section 4 describes the simulation results, and finally, the conclusion is given in Section 5.

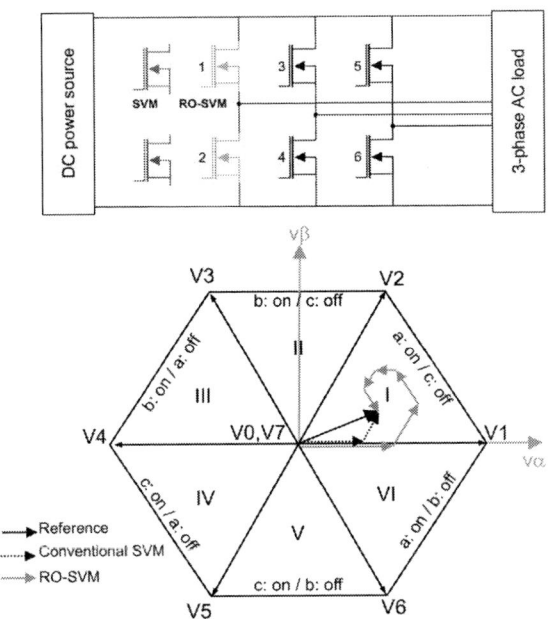

Fig. 1: Schematic of the reliability-optimized SVM (RO-SVM) method applied to a two-level three-phase inverter. With this method the overall damage is reduced by equalizing the thermal stress across all semiconductor switches.

2 System Description

2.1 Principle of SVM

The space vector modulation (SVM) is a widely employed modulation technique for power converters used to generate the desired voltage at the converter output[14], [15]. In the conventional SVM method, the desired voltage vector is generated by fast switching of the two closest space vectors (for example, V_1 and V_2 in Fig. 1), and a null vector (V_7 or V_0). In general, the desired output voltage vector V^* can be expressed as a weighted average of the space vectors V_i. The SVM algorithm determines the dwell times of each vector such that the average of the output voltage over the switching period, $\bar{V}$, approximates the desired voltage. That is,

$$V^* \approx \bar{V} = \frac{1}{T_{sw}} \sum_{i=0}^{7} T_i V_i \quad \text{with} \quad \sum_{i=0}^{7} T_i = T_{sw}, \quad (1)$$

where T_{sw} denotes the switching period, and T_i is the dwelling time of the space vector V_i.

According to (1), there are generally multiple combinations of space vectors that lead to the same V^*. This degree of freedom can be exploited to reach different objectives. The conventional SVM

schemes compute a switching pattern by following some predefined rules. For example, by appropriately selecting the null vectors, it is possible to generate a pattern that minimizes switching losses, similar to the implementation in SIMULINK 's SVM (2-level topology) block, which will be used as a benchmark in this study. However, none of these methods take into account the reliability of semiconductor switches.

The averaged output vector can be also expressed in terms of the switches duty-cycles:

$$\bar{V} = \frac{2}{3} V_{DC}(d_1 + d_3 e^{j\frac{2\pi}{3}} + d_5 e^{j\frac{4\pi}{3}}), \quad (2)$$

where V_{DC} denotes the input DC voltage, and d_1, d_3 and d_5 are, respectively, the duty-cycles of the upper switches in legs a, b, and c (see Fig. 1 for numbering of switches). The duty-cycles of the upper and the lower switch of each leg complement each other. In other words,

$$d_i + d_{i+1} = 1, \quad \text{for} \quad i = 1, 3, 5. \quad (3)$$

2.2 Lifetime model

Temperature fluctuations in the power semiconductor devices increases stress at the bonding interface between the wire and the silicon, which can eventually lead to disconnection of bond wires and open-circuit failures[16]–[18]. The lifetime expectancy of power devices is usually expressed in terms of the number of cycles that can perform effectively. In [19], the authors derived an empirical extension to the Coffin-Manson lifetime model for power semiconductor devices. According to this model, the number of cycles to failure, N_f, of a power semiconductor device can be estimated as

$$N_f = A_0 . A_1^\beta . \Delta T^{\alpha-\beta} . \exp\frac{E_a}{\kappa_B T_j} . \frac{C + t_{on}^\gamma}{C + 2^\gamma} . k_{thick},$$
$$(4)$$
$$\text{with } \beta = \exp\frac{-(\Delta T_j - T_0)}{\lambda},$$

where ΔT_j is the junction temperature fluctuation, T_j is the median junction temperature, t_{on} is the duration of the cycle, E_a and κ_B are, respectively, the activation energy and the Boltzmann constant, k_{thick} is the chip thickness factor, which is 1 for IGBTs and 0.33 for SiC MOSFETs. The other parameters A_0, A_1, T_0, λ, α, C, γ are constants and are given in [19].

It can be seen from 4 that the main factors that contribute to the degradation of power semiconductor devices are the junction temperature T_j, the temperature fluctuation ΔT_j, and the cycling time t_{on}. Using the parameter values given in [19], we observe an inverse relationship between these stress factors and the number of cycles to failure. In other words, reducing thermal stress, including T_j and ΔT_j, can significantly enhance the lifetime of power devices. On the other hand, longer thermal cycles allow the temperature fluctuation to affect critical layers of the assembly, strongly decreasing N_f.

The linear damage accumulation, proposed by Palmgren and Miner [20], [21], is considered es as a well-established technique for calculating the total damage accumulated in a device. This assumption states that with k different stress magnitudes within a spectrum S_i (where $1 \leq i \leq k$), and a total of $n_i(S_i)$ cycles within each of these stress levels, while the number of cycles to failure at a constant stress is denoted as $N_i(S_i)$, the overall damage can be expressed as follows:

$$D = \sum_{i=1}^{k} \frac{n_i(S_i)}{N_i(S_i)}. \tag{5}$$

For the semiconductor lifetime model, the total damage is computed as follows

$$D = \sum_{\Delta T_{j,min}}^{\Delta T_{j,max}} \sum_{T_{j,min}}^{T_{j,max}} \sum_{t_{on,min}}^{t_{on,max}} \frac{n(\Delta T_j, T_j, t_{on})}{N_f(\Delta T_j, T_j, t_{on})}, \tag{6}$$

where the number of cycles, $n(\Delta T_j, T_j, t_{on})$, is obtained from the Rainflow-counting algorithm.

Furthermore, we assume a linear expression to calculate the total damage across all devices. That is,

$$D_{\text{total}} = \sum_{n=1}^{6} D_n, \tag{7}$$

where D_n is the accumulated damage at each device as calculated per 6. The overall damage at the point of failure is often expressed as $D = 1$ [22].

3 Reliability-optimized SVM

The proposed modulation approach to extend the lifetime of the semiconductors and hence of the converter leverages the inherent redundancy of the space vectors for generating the desired voltage vector. This flexibility is exploited to reduce the total damage of all the devices. Since the thermal stress of semiconductor devices are directly dependent on their power loss, the proposed RO-SVM relies on optimizing the power loss distribution across all devices. Note that minimizing the distribution is not equivalent to minimizing the total power losses, and hence in some cases optimizing for thermal stress might lead to increased total power losses. With this optimized heat distribution, the temperature profile across all devices is equalized, alleviating the overall system damage.

The power dissipated at each device, P_i, is a function of the duty-cycle d_i. We formulate the problem as a convex optimization that compute the duty-cycles for each semiconductor device:

$$\min \quad \sum_{1}^{6} P_i^2 \tag{8a}$$

$$\text{s.t.} \quad P_i = d_i f(I_i) + P_{i,sw}, \text{ for } i = 1, \ldots, 6 \tag{8b}$$

$$\eta_{min} \leq d_i \leq 1 - \eta_{min}, \text{ for } i = 1, \ldots, 6 \tag{8c}$$

$$d_i + d_{i+1} = 1, \text{ for } \quad i = 1, 3, 5 \tag{8d}$$

$$|V^* - \frac{2}{3}V_{DC}(d_1 + d_3 e^{j\frac{2\pi}{3}} + d_5 e^{j\frac{4\pi}{3}})| \leq \epsilon \tag{8e}$$

where d_i is the duty-cycle of device i, $P_{i,sw}$ is the switching loss, $f(I_i)$ denotes a current-dependent term used to calculate the conduction loss, η_{min} is the minimum value for the duty-cycle, and ϵ is an error bound on the generated output voltage.

The objective of the above optimization problem is to minimize the temperature fluctuations in the semiconductor switches. Constraint 8b captures the impact of phase currents on power loss, constraint 8c keeps the duty-cycles within a given range, constraint (8d) ensures that the duty-cycles of devices on each leg sum up to one, and finally constraint 8e guarantees the desired output vector.

According to the power loss constraint in (8b), all devices are assumed to have switching losses. Therefore, duty-cycles cannot take values of 0 or 1, as limited by constraint (8c). This will eliminate the trivial case with zero switching losses. Because the switching loss is dependent on the duty-cycle, the optimization problem (8) is not convex in the general case. Therefore in the following we consider separately the limit cases of duty-cycles $d = 0$ and $d = 1$ where the problem can be solved algebraically.

3.1 Duty-cycle limit cases

According to Fig. 1, for any given voltage sector, there are two ways to generate the desired voltage without any change in the state of any of the legs, i.e. $d = 0$ or $d = 1$. For example, if the reference voltage is located inside of the sector 1 (as depicted in Fig. 1), either leg A can be on, i.e., switch 1 is conducting during the entire switching period (equivalently, switch 2 is open), or leg C can be off, i.e., switch 5 is fully open and switch 6 is conducting. These two cases correspond to using the vectors $\{V_1, V_2, V_7\}$ and $\{V_1, V_2, V_0\}$, respectively, to generate the reference vector. The analysis for other voltage sectors can be carried out similarly. For the two cases where a device is either on or off, the duty-cycles can be uniquely determined from the complex equation (2), that result in two algebraic equations. Therefore, in this limit cases, the problem can be solve algebraically without utilizing the optimization (8).

At the beginning of each switching period T_{sw}, the phase currents and the reference vector are captured and the cost function in (8a) is continuously minimized to reduce the total damage of the switches. The minimum cost value is achieved either by solving (8) or by the cost value obtained from the other two switch-free cases. The duty-cycles that achieve the minimum cost are then applied to the gate signals. Algorithm 1 summarizes the modulation scheme.

Algorithm 1: The RO-SVM Algorithm

Data: I_i, V^*

Result: d_i's for $i = 1, ..., 6$

Step 1: Solve (8), and store the cost value
$J_1 = \sum_{i=1}^{6} P_i^2$

Step 2: For the two non-switch cases:
Compute the duty-cycles according to (2), and the corresponding cost values, J_2 and J_3

Step 3: Find the minimum cost value in J_1, J_2, J_3 and output the corresponding duty-cycles

---

4 Simulation Results

In this section, the simulation results obtained by applying the RO-SVM to a two-level three-phase inverter are presented. The simulation is carried out in MATLAB and the RO-SVM method is compared with the pulse patterns generated by the SIMULINK 2-level SVPWM block. With the patterns generated by this block, the state of one of the legs in the inverter stays constant for the entire switching period, i.e. one semiconductor switch remains completely

either on or off. The specific SiC MOSFET [23] and Si IGBT [24] modules used in the simulation are manufactured by Semikron. The power loss associated to each module is calculated based on the parameters given in the datasheet. The corresponding Foster thermal models are retrieved using PLECS software [25]. The CVXGEN [26] fast QP solver is used in this work with the computation time of approximately 100us.

Without loss of generality of the proposed method, we use 3-phase balanced current signals as given by

$$I_a(t) = I_m \cos\left(2\pi ft + \phi\right), \tag{9a}$$
$$I_b(t) = I_m \cos\left(2\pi ft - 120° + \phi\right), \tag{9b}$$
$$I_c(t) = I_m \cos\left(2\pi ft + 120° + \phi\right) \tag{9c}$$

where I_m is the output current amplitude and ϕ is the output current phase angle.

The simulation is performed for one period of the fundamental frequency, f, and the current magnitude, phase angle and the modulation index are all are kept constant. The simulation parameters are gathered in Table 1.

I_m	200 A	Current magnitude
f_{sw}	10 kHz	Switching frequency
f	50 Hz	Signal frequency
T_a	70 °C	Ambient (heatsink) temperature
η_{min}	0.01	The minimum duty-cycle (1%)
ϵ	0.01	Voltage error threshold

Tab. 1: Simulation parameters.

Figure 2a compares the resulting power loss and the junction temperature for the Sic MOSFET number 1 (top device in leg A). We consider various voltage angles spanning from 0 to 360 degrees, with the voltage frequency set at 50 Hz. As illustrated in the figure, the peak of the power loss has been notably reduced when applying the RO-SVM method. However, the total power loss of all 6 devices with this methods is slightly increased by 3%. The resulting junction temperature profile is plotted in Fig. 2b Comparing the two methods reveals that the temperature cycle range with the RO-SVM approach is approximately 8 degrees, while the conventional SVM method (with minimal switching loss pattern) yields a range of 12 degrees. To compare

PCIM Europe 2024, 11– 13 June 2024, Nuremberg   DOI: 10.30420/566262230

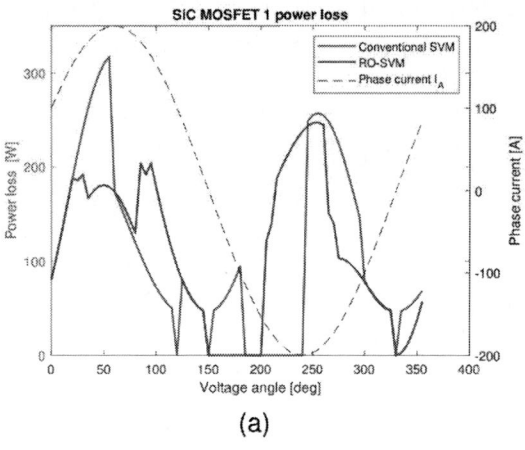
(a)

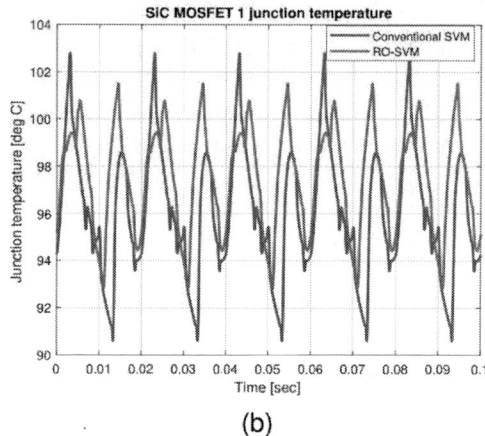
(b)

Fig. 2: Simulation results for the SiC MOSFET number 1 with the modulation index, $m = 0.4$ and the phase offset, $\phi = 30°$: (a) the power loss, and (b) the junction temperature versus different voltage phase.

the lifetime resulting by applying the two methods, we employ the damage model given in 4. The significant contributing factors that differs between the two methods are the temperature fluctuation, ΔT_j, and the mean temperature, T_j. Using the provided damage model, the RO-SVM method yields an increase of 400 times improvement in damage compared to the conventional SVM method. Figure 3 illustrates a similar comparison for an IGBT-type device. The RO-SVM method, in this case, improves the lifetime by a factor of 9, with overall 2% less power loss.

4.1 Varying m and ϕ

In the first simulation, we vary the modulation index between 0.1 and 0.8, and the phase offset, ϕ, is kept constant at $30°$. Figure 4 depicts the power loss of MOSFET 1 for different voltage angles and modulation indices. As we can see, with the RO-SVM method, especially at lower modulation indices, the power loss tends to be more distributed over the entire circle, resulting in lower temperature swing. At lower modulation indices, it is worth noting that the RO-SVM method results in a relatively higher power loss, approximately 16%. However, the significant improvement in lifetime, by a factor of 2×10^6, outweighs this drawback. For higher modulation indices the performance of the two methods converges and becomes similar. Similar result is achieved with the IGBT, as illustrated in Fig. 5, where the maximum lifetime improvement is 18 at the modulation index of 0.1 with almost no impact on the total power loss.

In the next simulation, we let the phase offset, ϕ,

vary from $-45°$ to $45°$, while the modulation index sweeps from 0.1 to 0.8, as before. Figure 6a provides the (total) lifetime enhancement for different phase offsets and modulation indices using SiC MOSFETs. The result of the same simulation but with the IGBTs is shown in Fig. 6b. Depending on the phase offset between the inverter voltage and the current, the lifetime improvement can vary significantly for the IGBT case, however, this is not the case with the MOSFETs as they can pass reverse currents.

5 Conclusion

This paper proposes a space vector modulation scheme that generates the reference output voltage, and by exploiting the redundancy of the space vectors, reduces the total damage of the semiconductor devices. The simulation results for Si IGBTs and SiC MOSFETs demonstrate the efficacy of the method to extend the lifetime of the devices by several orders of magnitude.

References

[1] E. Wolfgang, "Examples for failures in power electronics systems," *ECPE tutorial on reliability of power electronic systems, Nuremberg, Germany*, pp. 19–20, 2007.

[2] S. Yang, A. Bryant, P. Mawby, D. Xiang, L. Ran, and P. Tavner, "An industry-based survey of reliability in power electronic converters," in *2009 IEEE Energy Conversion Congress and Exposition*, 2009, pp. 3151–3157. DOI: 10.1109/ECCE.2009.5316356.

1690

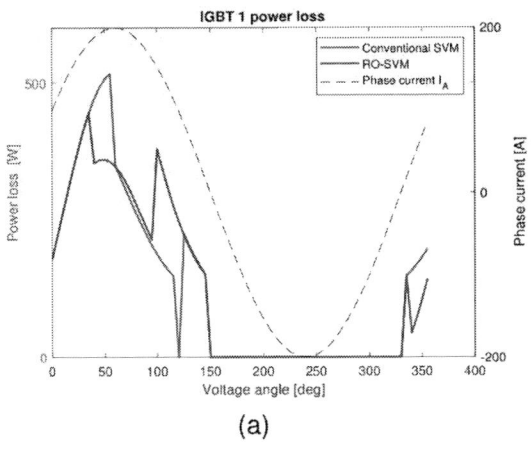
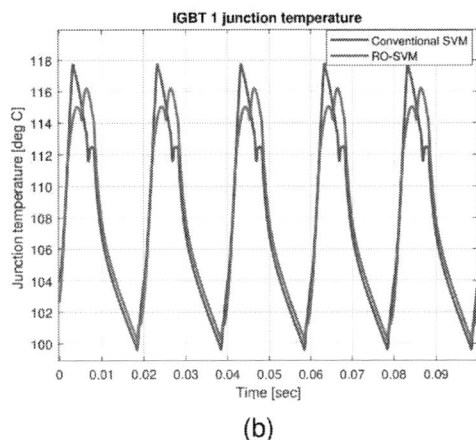

(a)   (b)

Fig. 3: Simulation results for the Si IGBT number 1 with the modulation index, $m = 0.4$ and the phase offset, $\phi = 30°$: (a) the power loss, and (b) the junction temperature versus different voltage phase.

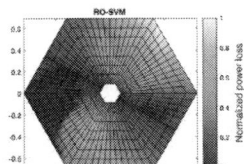
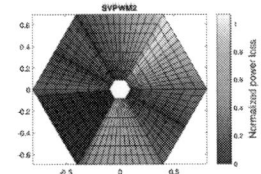

Fig. 4: The MOSFET 1 power loss for different modulation index and voltage phases: (left) with the RO-SVM method, (right) with the conventional SVPWM method.

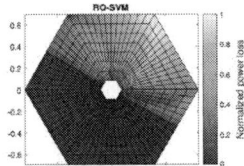
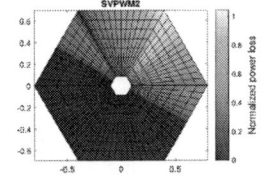

Fig. 5: The IGBT 1 power loss for different modulation index and voltage phases: (left) with the RO-SVM method, (right) with the conventional SVPWM method.

[3] H. Wang, K. Ma, and F. Blaabjerg, "Design for reliability of power electronic systems," in *IECON 2012 - 38th Annual Conference on IEEE Industrial Electronics Society*, 2012, pp. 33–44. DOI: 10. 1109/IECON.2012.6388833.

[4] A. Sangwongwanich and F. Blaabjerg, "Monte carlo simulation with incremental damage for reliability assessment of power electronics," *IEEE Transactions on Power Electronics*, vol. 36, no. 7, pp. 7366–7371, 2021. DOI: 10.1109/TPEL.2020. 3044438.

[5] M. Hernes, S. D'Arco, A. Antonopoulos, and D. Peftitsis, "Failure analysis and lifetime assessment

of igbt power modules at low temperature stress cycles," *IET Power Electronics*, vol. 14, no. 7, pp. 1271–1283, 2021.

[6] D. Kumar, R. K. Nema, and S. Gupta, "Investigation of fault-tolerant capabilities of some recent multilevel inverter topologies," *International Journal of Electronics*, vol. 108, no. 11, pp. 1957–1976, Nov. 2021. DOI: 10.1080/00207217.2020. 1870752.

[7] M. di Benedetto, A. Lidozzi, L. Solero, F. Crescimbini, and P. J. Grbović, "Reliability and real-time failure protection of the three-phase five-level e-type converter," *IEEE Transactions on Industry Applications*, vol. 56, no. 6, pp. 6630–6641, 2020. DOI: 10.1109/TIA.2020.3019358.

[8] C. Li, D. Jiao, J. Jia, F. Guo, and J. Wang, "Thermoelectric cooling for power electronics circuits: Modeling and active temperature control," *IEEE Transactions on Industry Applications*, vol. 50, no. 6, pp. 3995–4005, 2014. DOI: 10.1109/TIA. 2014.2319576.

[9] D. Murdock, J. Torres, J. Connors, and R. Lorenz, "Active thermal control of power electronic modules," *IEEE Transactions on Industry Applications*, vol. 42, no. 2, pp. 552–558, 2006. DOI: 10.1109/ TIA.2005.863905.

[10] A. Ghazanfari and Y. A.-R. I. Mohamed, "A resilient framework for fault-tolerant operation of modular multilevel converters," *IEEE Transactions on Industrial Electronics*, vol. 63, no. 5, pp. 2669–2678, 2016. DOI: 10.1109/TIE.2016.2516968.

[11] M. Aly, E. M. Ahmed, and M. Shoyama, "Developing new lifetime prolongation svm algorithm for multilevel inverters with thermally aged power devices," *IET Power Electronics*, vol. 10, no. 15, pp. 2248–2256, 2017.

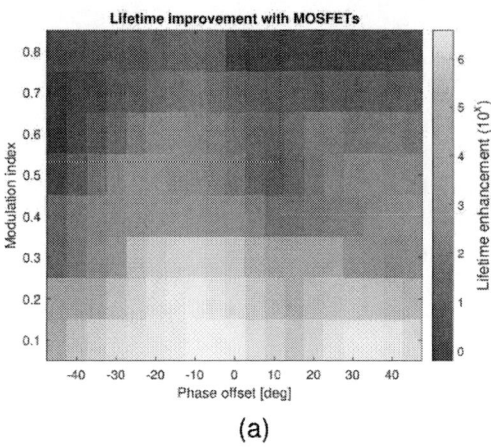 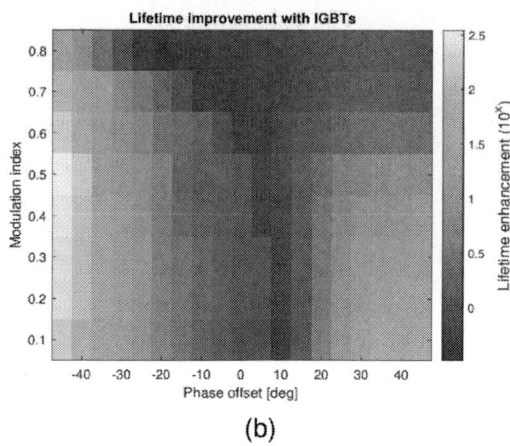

(a)   (b)

Fig. 6: Lifetime enhancement of the RO-SVM method compared to the conventional SVPWM2 method versus different modulation indices and phase offset values: (a) using SiC MOSFETs and (b) using Si IGBTs.

[12] T.-M. Phan, N. Oikonomou, G. J. Riedel, and M. Pacas, "Pwm for active thermal protection in three level neutral point clamped inverters," in *2014 IEEE Energy Conversion Congress and Exposition (ECCE)*, 2014, pp. 3710–3716. DOI: 10.1109/ECCE.2014.6953905.

[13] T.-M. Phan, G. J. Riedel, N. Oikonomou, and M. Pacas, "Active thermal protection and lifetime extension in 3l-npc-inverter in the low modulation range," in *2015 IEEE Applied Power Electronics Conference and Exposition (APEC)*, 2015, pp. 2269–2276. DOI: 10.1109/APEC.2015.7104665.

[14] Z. Yu, "Space-vector pwm with tms320c24x/f24x using hardware and software determined switching patterns," *Texas Instruments Application Report SPRA524*, pp. 24–28, 1999.

[15] P. Handley and J. Boys, "Space vector modulation: An engineering review," in *1990 Fourth International Conference on Power Electronics and Variable-Speed Drives (Conf. Publ. No. 324)*, IET, 1990, pp. 87–91.

[16] H. Wang, M. Liserre, F. Blaabjerg, P. de Place Rimmen, J. B. Jacobsen, *et al.*, "Transitioning to physics-of-failure as a reliability driver in power electronics," *IEEE Journal of Emerging and Selected Topics in Power Electronics*, vol. 2, no. 1, pp. 97–114, 2014. DOI: 10.1109/JESTPE.2013.2290282.

[17] M. Ciappa and W. Fichtner, "Lifetime prediction of igbt modules for traction applications," in *2000 IEEE International Reliability Physics Symposium Proceedings. 38th Annual (Cat. No.00CH37059)*, 2000, pp. 210–216. DOI: 10.1109/RELPHY.2000.843917.

[18] W. Wu, G. Gao, L. Dong, Z. Wang, M. Held, *et al.*, "Thermal reliability of power insulated gate bipolar transistor (igbt) modules," in *Twelfth Annual IEEE Semiconductor Thermal Measurement and Management Symposium. Proceedings*, 1996, pp. 136–141. DOI: 10.1109/STHERM.1996.545103.

[19] A. Wintrich and U. Scheuermann, "Power cycle model for IGBT product lines," Semikron Danfoss, Nuremberg, Germany, Application Note AN21-001, 2021.

[20] A. Czechowski and A. Lenk, "Miner's rule in mechanical tests of electronic parts," *IEEE Transactions on Reliability*, vol. R-27, no. 3, pp. 183–190, 1978. DOI: 10.1109/TR.1978.5220318.

[21] M. A. Miner, "Cumulative damage in fatigue," *Journal of Applied Mechanics*, vol. 12, no. 3, A159–A164, 1945.

[22] J. Slavič, M. Boltezar, M. Mrsnik, M. Cesnik, and J. Javh, *Vibration Fatigue by Spectral Methods: From Structural Dynamics to Fatigue Damage–Theory and Experiments*. Elsevier, 2020.

[23] *Half-bridge (full SiC)*, 24922580, Rev. 0.6, SEMIKRON, Jan. 2021.

[24] *Trench IGBT modules*, 22890045, SEMIKRON, Oct. 2009.

[25] *User Manual of PLECS Blockset Version 4.8.* February 2024, https://www.plexim.com/sites/default/files/plecsmanual.pdf.

[26] J. Mattingley and S. Boyd, "CVXGEN: A code generator for embedded convex optimization," *Optimization and Engineering*, vol. 12, no. 1, pp. 1–27, 2012.

PCIM Europe 2024, 11– 13 June 2024, Nuremberg   DOI: 10.30420/566262231

Analysis and Optimization of Internal Coupling Interference in Integrated SiC Power Module Based on DBC

Chenhang Zeng[1] , Yiyang Yan[1] , Heng Zhang[1] , Yue Wu[2] , Zhipeng He[2] , Yong Kang[1] , Cai Chen[1]

[1] Huazhong University of Science and technology, China
[2] State Key Laboratory of HVDC, Electric Power Research Institute, CSG, China

Corresponding author: Yong Kang, ykang@hust.edu.cn
Speaker: Chenhang Zeng, ch_z@hust.edu.cn

Abstract

Silicon carbide devices have higher switching speeds than traditional silicon-based devices and can operate at higher switching frequencies. The packaging structure of the integrated gate driver can significantly reduce the parasitic inductance in the drive circuit, reduce the drive resistance value, and increase the switching speed of SiC devices. Integrating the driver on the DBC can further improve the integration and heat dissipation of the driver chip. However, due to the shared DBC, the dv/dt during the switching process will be coupled to the signal input side of the driver chip through the parasitic capacitance between DBC layers, causing interference. The existing analysis of this phenomenon has problems such as incomplete consideration of parameters and the introduction of redundant parameters, and the accuracy is not high. This paper establishes a coupling interference model that considers the complete parasitic parameters in the DBC integrated power module, and proposes an optimized structure of the copper layer at the bottom of the DBC based on this model. The experimental results show that the proposed improved coupling interference model has high accuracy, and the calculation The error is within 8%. At the same time, the optimized structure of the copper layer at the bottom of the DBC proposed can increase the switching speed of the integrated SiC power module by more than 50%.

1   Introduction

Silicon carbide (SiC) devices have higher switching speeds than traditional silicon-based devices and can operate at higher switching frequencies, further reducing the volume of passive devices and achieving high power density [1]. However, high switching speed makes SiC devices very sensitive to parasitic inductance. In the driving circuit, high parasitic inductance will cause gate-source oscillation of the device. The driving resistance needs to be increased to provide damping, which limits the increase in switching speed. In order to reduce the parasitic inductance of the driving circuit, the packaging structure of integrated gate driver has been proposed in recent years, this type of structure is often called intelligent power module or integrated power module [2]-[3].

Depending on the connection method between the power circuit and the drive circuit and the substrate material, integrated power modules usually include PCB embedded structure, PCB/DBC hybrid packaging structure and DBC integrated structure

[4]. As shown in Fig. 1, the PCB embedded packaging structure has the smallest parasitic inductance, but is limited by the high thermal resistance of the FR4 material. The heat dissipation and reliability issues of the module need to be further verified, and the processing is difficult; The PCB/DBC hybrid package has relatively better heat dissipation than the PCB embedded structure, and can still maintain a low parasitic inductance. However, this structure needs to be realized by combining the PCB production process on the DBC substrate, and the processing is difficult.

The DBC integrated structure has the simplest process, the lowest cost and excellent thermal performance. However, since the driver chip and the power chip share the same DBC, during the switching process, dv/dt is easily coupled to the input side of the driver chip through the interlayer parasitic capacitance of DBC, causing coupling interference and affecting the normal drive waveform. Therefore, it is necessary to analyze the internal coupling interference phenomenon of the integrated power module based on DBC.

1693

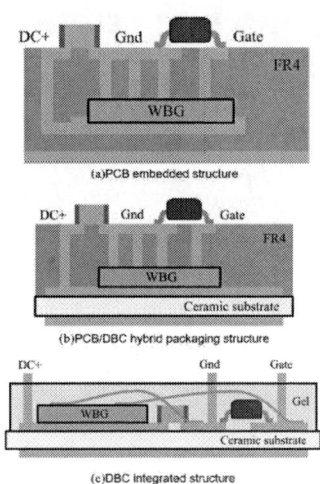

Fig. 1 Several packaging structures of integrated power modules

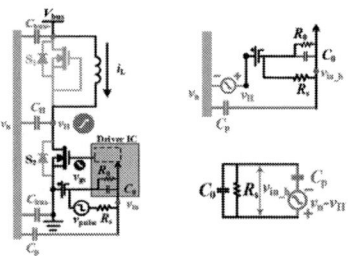

Fig. 2 Existing coupling interference model of DBC integrated power module

However, the existing analysis of internal coupling interference in DBC-based integrated power modules only considers a small part of the parasitic parameters in the path [5], as shown in Fig. 2, and does not consider the loop parasitic inductance. Moreover, additional redundant resistors are introduced in the interference loop, such as the series resistor in the signal input loop of the driver chip, which has a great impact on the analysis. However, in actual applications, this resistor is rarely connected in series. As a result, the accuracy of existing models in guiding actual module design is insufficient, and the parameters that have a greater impact on coupling interference phenomena in module design are still unclear. Therefore, it is necessary to establish a more accurate coupled interference model that considers the complete parasitic parameters in the path to provide theoretical guidance for the design of integrated power modules and the suppression of coupled interference in practical applications.

This paper proposes an improved interference model for the coupling interference phenomenon

of DBC integrated power modules. The model considers the complete parasitic parameters in the path, eliminates redundant parameters in the existing model, and proposes two optimizations based on this model structure. The full paper is organized as follows. Section 2 models the coupling interference path inside the DBC integrated power module. Section 3 analyzes the influence trend of parasitic parameters on the coupling interference voltage amplitude through parameter scanning and discrete numerical solution, and extracts Key parasitic parameters that have a significant impact,

Section 4 proposes two optimized structures for key parasitic parameters and conducts simulation verification. Section 5 processes and tests typical integrated power module structures and optimized structures.

2 DBC Integrated Power Module Coupling Interference Modeling

As shown in Fig. 3(a), a typical DBC integrated power module is built based on the S4661 (1200V/130A@25℃) SiC MOSFET die produced by Rohm and the UCC27531 (5A peak current) driver chip produced by Texas Instruments. Fig. 3(b) is its equivalent circuit. The module layout refers to the commercial dual-in-line IPM. The drive terminals and power terminals are distributed on both sides. Considering that the drive circuit area should match the power circuit, the chip integrated inside the module adopts a non-isolated form, and the volume of UCC27531 is only 3*3mm², which meets the requirements of integration. The isolation and power supply of the drive are implemented by external circuits.

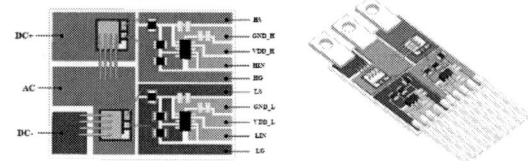

(a) typical DBC integrated power module

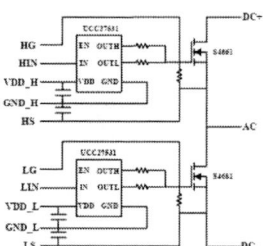

(b) equivalent circuit of integrated power module

Fig. 3 Three-dimensional structure and equivalent circuit of integrated power module

Consider the parasitic capacitance inside the module and the parasitic inductance in the loop, eliminate the series resistance in the signal input loop of the driver chip, and consider the output impedance of the isolation chip. First, take the low-side MOSFET as an example, the improved coupling interference model of the DBC integrated power module is established as shown in Fig. 4.

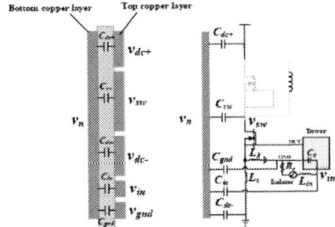

Fig. 4 Improved coupling interference model of the DBC integrated power module

C_{sw} is the parasitic capacitance of all half-bridge midpoint potential copper layers to the bottom copper layer of DBC, including the AC power copper layer, the high-side Kelvin source copper layer and the high-side gate copper layer; C_{dc+} is the parasitic capacitance of the DC+ power copper layer to the bottom copper layer of DBC; C_{dc-} is the parasitic capacitance of the DC-potential copper layer to the copper layer under DBC, including the DC-power copper layer, the low-side Kelvin source copper layer and the low-side gate copper layer; C_{in} is the parasitic capacitance between the copper layer on the signal input side of the low-side driver chip and the copper layer under DBC; C_{gnd} is the parasitic capacitance between the GND copper layer of the low-side driver chip and the copper layer under DBC; L_k and L_s are the driving Kelvin source inductance and power source inductance respectively; L_{in} is the parasitic inductance of the driving chip signal input circuit from the IN terminal to the GND terminal; C_0 is the capacitance to ground at the internal input pin of the driver chip, R_o is the output impedance of the isolation chip, and is an inherent parameter of the device used. V_{dc+} and V_{dc-} are the positive and negative bus potentials, v_{sw} is the half-bridge midpoint potential, v_{in} and v_{gnd} are the driver chip input and GND potentials, and v_n is the bottom copper layer potential of DBC.

The following analysis is divided into two situations, the DBC bottom copper layer is floating and the DBC bottom copper layer is grounded.

When the DBC is connected to the heat sink through insulating materials such as thermal grease or through solder and the heat sink itself is not grounded, the potential of the bottom copper layer is floating. Due to the existence of parasitic capacitance between DBC copper layers, when the half-bridge midpoint potential v_{sw} changes, it will cause the bottom copper layer potential v_n to change, which is equivalent to the series and parallel voltage division of the parasitic capacitance on the circuit. Since the area of the copper layer corresponding to C_{in} and C_{gnd} is relatively small, its parasitic capacitance value can basically be ignored. According to the AC equivalent, the positive and negative bus voltages can be equivalent to a short circuit at the transient moment when the midpoint potential changes. The relationship between v_n and v_{sw} can be established as shown in Fig. 5

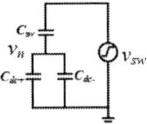

Fig. 5 Relationship between v_n and v_{sw}

$$v_n = \frac{C_{sw}}{C_{sw} + C_{dc+} + C_{dc-}} \cdot v_{sw} = \lambda \cdot v_{sw} \quad (1)$$

Equation (1) describes the relation of v_n and v_{sw}, where λ is $C_{sw}/(C_{sw}+ C_{dc+} +C_{dc-})$, which is the proportion of the parasitic capacitance of the copper layer corresponding to the midpoint potential of the half-bridge in the parasitic capacitance of the copper layer of the power circuit. After obtaining the expression of the copper layer potential v_n at the bottom of the DBC, using v_n as the interference source, a coupling interference loop is obtained, as shown in Fig. 6(a).

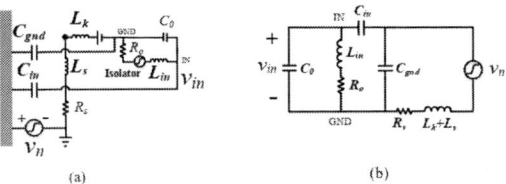

Fig. 6 Improved low-side coupling interference equivalent circuit

During the switching transient process, since the transmission delay of the driver has passed, the signal output by the isolation chip maintains a constant level at this time, so it can be equivalent to a separate output impedance R_o, If negative voltage driving is used, there will be a constant negative voltage between the GND pin of the driver chip and the source of the MOSFET, which can be regarded as a short circuit. The simplified equivalent interference circuit of the driver chip input interference voltage v_{in} is further obtained, as shown in Fig. 6(b).

The obtained interference simplified equivalent circuit is a fifth-order dynamic circuit. It is necessary to deduce the relationship between the interference voltage v_{in} and the bottom copper layer potential v_n. Due to the high order of the circuit, it is not easy to obtain the analytical expression of the interference voltage v_{in}. Therefore, we consider turning to find its numerical solution, and use charts and other more intuitive methods to express the waveform and amplitude of v_{in}.

Next, a mathematical model of the fifth-order circuit is established, and the state equation and output equation of the circuit are derived. First, the characteristic physical quantities of the independent dynamic components in the circuit are selected as state variables, that is, the capacitor voltage and the inductor current. As shown in Fig. 7, the five state variables are selected as the voltages on C_0, C_{in}, and C_{gnd} and the currents on L_{in} and L_k+L_s, which are recorded as x_1, x_2, x_3, x_4, and x_5 respectively. Their first-order descriptors are $\dot{x}_1$、 $\dot{x}_2$、 $\dot{x}_3$、 $\dot{x}_4$、 $\dot{x}_5$. x_1 is the voltage across C_0, which is the input interference voltage v_{in} of the driver chip.

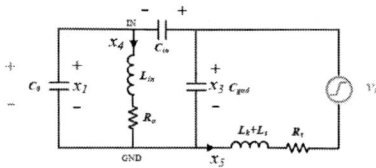

Fig. 7 Interference model equivalent circuit state variable diagram

$$C_{in}\dot{x}_2 = C_0\dot{x}_1 + x_4$$

$$C_{gnd}\dot{x}_1 + (C_{in} + C_{gnd})\dot{x}_2 = x_5$$

$$x_1 = L_1\dot{x}_4 + R_{s1}x_4 \tag{2}$$

$$x_1 + x_2 + (L_k + L_s)\dot{x}_5 + R_{s2}x_5 = v_n$$

Equation (2) describes the node current equation and loop voltage equation. After further sorting, the state equation can be obtained as shown in Equation (3). The A and B matrices are shown in Equation (4) and Equation (5).

$$\dot{x} = Ax + Bv_n = Ax + \lambda \cdot B \frac{dv_{sw}}{dt} t$$

$$x = [x_1 \ x_2 \ x_4 \ x_5]^T \tag{3}$$

$$\dot{x} = \frac{dx}{dt}$$

$$A = \begin{bmatrix} 0 & 0 & -\dfrac{C_{in}+C_{gnd}}{C_{in}C_{gnd}+C_{in}C_0+C_0C_{gnd}} & \dfrac{C_{in}}{C_{in}C_{gnd}+C_{in}C_0+C_0C_{gnd}} \\[2mm] 0 & 0 & \dfrac{C_{gnd}}{C_{in}C_{gnd}+C_{in}C_0+C_0C_{gnd}} & \dfrac{C_0}{C_{in}C_{gnd}+C_{in}C_0+C_0C_{gnd}} \\[2mm] \dfrac{1}{L_1} & 0 & \dfrac{-R_{s1}}{L_1} & 0 \\[2mm] -\dfrac{1}{L_k+L_s} & -\dfrac{1}{L_k+L_s} & 0 & \dfrac{-R_{s2}}{L_k+L_s} \end{bmatrix} \tag{4}$$

$$B = \begin{bmatrix} 0 & 0 & 0 & \dfrac{1}{L_k+L_s} \end{bmatrix}^T \tag{5}$$

The above analysis focuses on the coupling interference to the low-side driver input when the midpoint potential changes. In addition, variation in the midpoint potential v_{sw} will also cause corresponding interference problems at the input of the high-side drive chip. Fig. 8 shows the coupling interference equivalent circuit at the input end of the high-side driver chip. The interference source that causes the high-side input signal to change is the potential difference between the DBC bottom copper layer potential and the AC copper layer. The coupling interference of the high-side MOSFET can also be analyzed using the state equation, and will not be described again here.

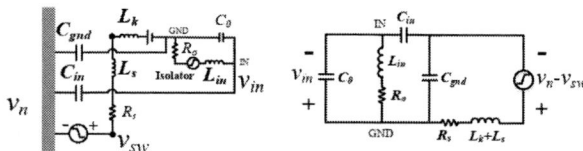

Fig. 8 Improved high-side coupling interference equivalent circuit

When the copper layer at the bottom of the DBC is grounded, its potential remains constant. At this time, changes in midpoint potential no longer cause potential changes on the bottom copper layer. Therefore, according to the improved model, the interference source of the low-side drive input becomes zero. As shown in Fig. 9(a), the low-side drive input no longer has coupling interference. On the other hand, at this time, the interference source of the high-side drive input becomes $-v_{sw}$, which is larger than when the bottom copper layer floats. Therefore, when the copper layer under the DBC is grounded, although it can suppress the coupling interference of the low-side drive input, it increases the coupling interference of the high-side drive input.

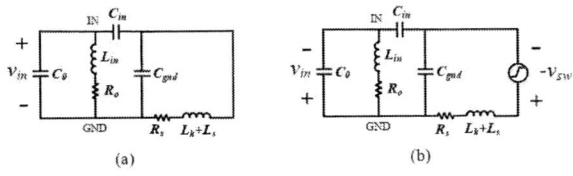

Fig. 9 Coupling interference equivalent circuit when the bottom copper layer is grounded

3 Influence of Parasitic Parameters on Coupling Interference Voltage Amplitude

The parasitic parameters in the interference loop are divided into three categories. The first category is the parasitic capacitance on the DBC power side, including C_{sw}, C_{dc+}, and C_{dc-}; the second category is the parasitic capacitance on the DBC drive side, including C_{in} and C_{gnd}; the third category is the parasitic inductance in the interference loop, including L_k, L_s, and L_{in}.

The parasitic capacitance on the power side mainly affects the potential change of the bottom copper layer through capacitive voltage division, which further affects the interference source in the equivalent circuit of driving coupling interference. When the potential of the bottom copper layer floats, in order to ensure consistent influence on the high-side and low-side switches, the proportion λ of the parasitic capacitance of the copper layer associated with the midpoint potential of the half-bridge in the parasitic capacitance of the power side should be close to 0.5; when the copper layer at the bottom of the DBC is grounded, the parasitic capacitance on the power side has no effect on coupling interference.

Next, the proposed improved coupled interference mathematical model is numerically solved. The typical values and approximate ranges of each parasitic parameter are given based on the actual module structure, and the corresponding DBC interlayer parasitic capacitance and loop parasitic inductance of the structure shown in Fig. 3 are extracted through ANSYS Q3D Extractor.

As shown in Table 1, After extracting the typical values of the parasitic parameters, they can be directly substituted into the obtained mathematical model equation (3)-(5). As for the intrinsic parameter values of the device in the interference path, the values are determined directly according to the datasheet. The input capacitance C_0 of the driver chip is 20pF according to the UCC27531 datasheet. The output impedance R_o of the isolation chip is 30Ω according to the Si8610 datasheet. The line resistance is 10mΩ. Under different half-bridge midpoint potential variation rates dv_{sw}/dt, the numerical solution of the coupling interference voltage amplitude(v_{in_max} & v_{in_min}) of low-side driver input end in typical module structure is shown in Fig. 10.

Parasitic capacitance	Value	Parasitic inductance	Value
C_{dc+}	14.9pF	L_{in1}(Inside the module)	30.2nH
C_{dc-}	20.6pF	L_{in2}(Outside the module)	5nH
C_{sw}	36.5pF	L_k	6.4nH
C_{in}	2.5pF	L_s	0.8nH
C_{gnd}	3.5pF		

Table 1 Parasitic parameter values of typical integrated power module structures

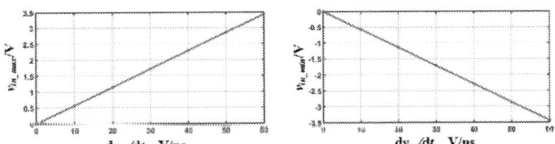

Fig. 10 v_{in_max} when the midpoint potential rises and v_{in_min} when the midpoint potential decreases

The amplitude of the interference voltage is proportional to the midpoint potential change rate, and under the same voltage change rate, there is little difference in the amplitude of the positive and negative interference voltages. According to Fig. 10, when the midpoint potential change rate dv_{sw}/dt of this typical three-dimensional structure is greater than 36V/ns, the coupling interference voltage amplitude on the driving side exceeds the threshold 2V, which may cause misleading conduction of the low-side MOSFET.

The parasitic inductance and parasitic capacitance are analyzed separately below. When analyzing the influence of the driving side parasitic capacitance, the parasitic inductance is set to a fixed value within the range determined based on the typical value. At this time, the parameter variable is only the parasitic capacitance value, and the analysis object becomes two-dimensional. Matlab can be used to perform a discrete numerical solution of two-dimensional parametric scanning of the driving side parasitic capacitance values C_{in} and C_{gnd}. After the dimensionality reduction process, the relationship between the target value and the parameters can be expressed more clearly using a three-dimensional diagram.

When the parasitic inductance takes a typical value, the variation trend of the interference voltage amplitude with the parasitic capacitance value under different dv_{sw}/dt is shown in Fig. 11.

C_{in} has a more significant impact on the coupling interference voltage v_{in} amplitude. The v_{in} amplitude increases with C_{in}. maintains an upward trend, while C_{gnd} has a small impact on the amplitude of the coupling interference voltage v_{in}.

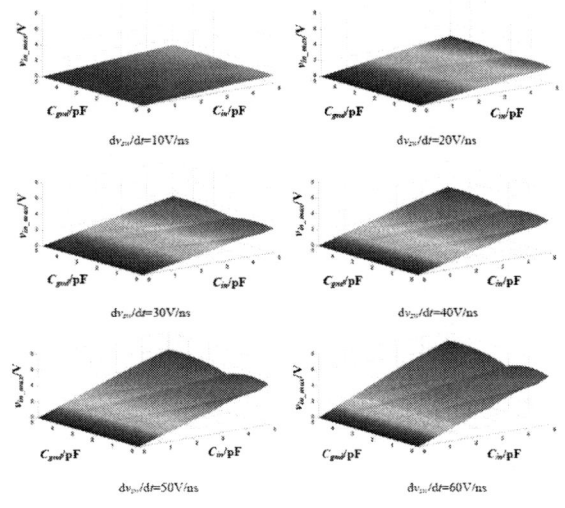

Fig. 11 Variation of interference voltage amplitude with parasitic capacitance value under different dv_{sw}/dt

As shown in Fig. 12, it is the intersection projection of the trend surface of the interference voltage amplitude under different voltage change rates and the plane $v_{in\_max}=2$. It can be clearly obtained that the value range of the parasitic capacitance where the interference voltage amplitude does not exceed 2V When dv_{sw}/dt is 10V/ns, the coupling interference voltage amplitude caused by the combination of parasitic capacitances within 5pF will

not exceed 2V, and there is no need to consider the problem of misdirection; When dv_{sw}/dt increases to 60V/ns, the parasitic capacitance C_{in} of the driver chip input end to the copper layer at the bottom of the DBC needs to be controlled below 1.2pF to ensure that the coupling interference voltage amplitude is less than 2V.

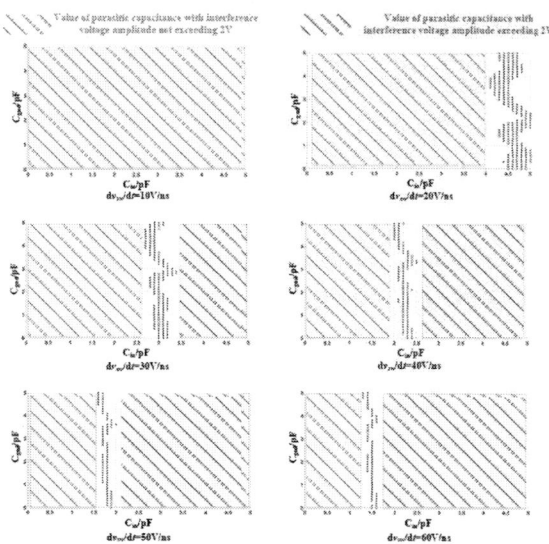

Fig. 12 Parasitic capacitance value range under different dv_{sw}/dt

Similarly, when analyzing the influence of parasitic inductance, set the parasitic capacitance to a fixed value, perform two-dimensional parameter scanning and discrete numerical solution for L_{in} and L_k+L_s. When the parasitic capacitance takes a typical value, the variation trend of the interference voltage amplitude with the parasitic inductance value under different dv_{sw}/dt is shown in Fig. 13, the coupling interference voltage amplitude increases with the increase of L_{in} and L_k+L_s, among which L_{in} has a more significant influence on the coupling interference voltage amplitude, and when L_{in} is small, the interference voltage will have an oscillation peak.

The value range of parasitic inductance is shown in Fig. 14. When the midpoint potential change rate dv_{sw}/dt is 10V/ns, the coupling interference voltage amplitude caused by the combination of parasitic inductances within 50nH will not exceed 2V, and there is no need to consider the problem of misleading conduction; When dvsw/dt increases to 60V/ns, it is necessary to control the parasitic inductance L_{in} of the driver chip input loop to 10nH, and the sum of Kelvin source inductance and power source inductance L_k+L_s to be controlled within 6nH to ensure that the generated coupling interference voltage does not affect the driving side.

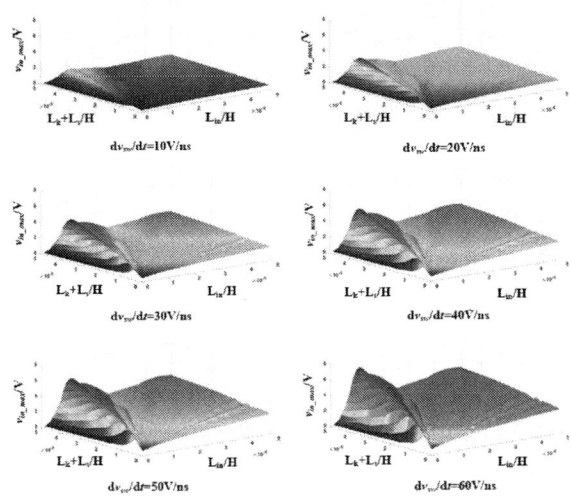

Fig. 13 Variation of interference voltage amplitude with parasitic inductance value under different dv_{sw}/dt

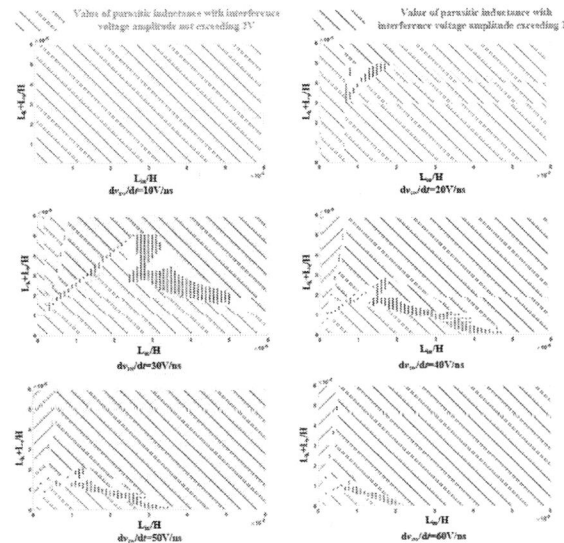

Fig. 14 Parasitic inductance value range under different dv_{sw}/dt

Through numerical calculation results, the parasitic parameter values that have a significant impact on the coupling interference voltage were discovered, which provided theoretical guidance for subsequent module structure optimization and verification.

4 Module Structure Optimization

According to the above analysis, among the parasitic parameters included in the interference loop, the parasitic capacitance C_{in} of the copper layer at the signal input end of the driver chip to the copper layer at the bottom of the DBC and the parasitic inductance L_{in} of the signal input circuit of the driver chip have a greater impact on the amplitude of the coupling interference voltage. Obviously, the goal of module structure optimization is to adjust the above key parasitic parameter values to achieve the effect of reducing the interference voltage amplitude and achieving coupled interference suppression.

The sources of parasitic inductance in the input signal loop of the driver chip mainly include three types of interconnect structures: upper surface copper layer wiring on the DBC substrate, driver terminals and external PCB lines. Although the structure and shape are different, in essence, the interconnection is a combination of multiple sections of rectangular cross-section flat conductors, and the flow cross-section of each section of conductor is rectangular. For a single-segment flat conductor, its parasitic inductance can be calculated using equation (6). The conductor length has a more significant impact on the parasitic inductance relative to the flow area, so it is shortened in practical applications. Conductor length reduces parasitic inductance more effectively.

$$L = \frac{\mu_0 l}{2\pi}(\ln\frac{2l}{t+a}+\frac{1}{2}) \qquad (6)$$

Analyze the parasitic inductance composition of the drive input loop in the structure of Fig.3, and extract several parts of the DBC substrate copper layer wiring, drive terminals and external PCB lines through Q3D. The parasitic inductance of the DBC copper layer is limited to the driving side copper layer wiring and cannot be effectively reduced. The same is true for the parasitic inductance of the external PCB line. However, the parasitic inductance of the driving terminal is not limited by the module layout. Therefore, consider reducing the parasitic inductance of the drive signal input loop by shortening the terminal length.

The three-dimensional dimensions of the drive terminal in a typical structure are shown in Fig. 15(a). The length is 15mm, the cross-sectional area is $0.8*0.8mm^2$, and its parasitic inductance is 9.2nH. According to Equation(6), the relationship between terminal parasitic inductance and length is shown on the right side of Fig. 15(b).

Considering that a part of the length needs to be reserved for connection with external circuits, the length of the drive side terminals can be adjusted to 5mm, and the parasitic inductance of the drive terminals is reduced to 1.9nH.

After adjustment, the parasitic inductance Lin value of the drive signal input loop is 21nH. Comparing the coupling interference voltage amplitude of the original structure and the structure after adjusting the drive terminal at different voltage change rates is shown in Fig. 16.

It can be seen that although adjusting the parasitic inductance of the signal input loop of the driver chip can alleviate the coupling interference phenomenon to a certain extent, the effect is relatively limited.

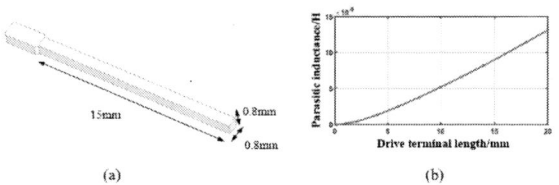

(a) (b)

Fig. 15 Driving terminal size and parasitic inductance changes with length chart

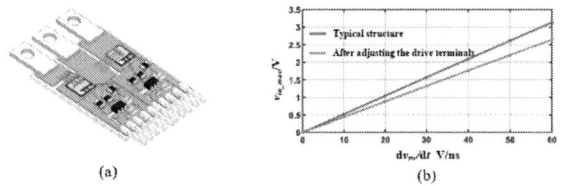

(a) (b)

Fig. 16 Optimized structure of drive terminal

On the other hand, the copper layer of the signal input end of the driver not only needs to be soldered to the driver chip pins and driver terminals for external connection, but the actual design may also be limited by the wiring and layout of the copper layer. The area of the copper layer itself is difficult to control very small. The corresponding reduction in parasitic capacitance value is also limited. Therefore, this paper proposes an optimized structure for the copper layer at the bottom of DBC.

Only the DBC bottom copper layer with the same pattern directly below the driver input copper layer is etched, which not only reduces the corresponding parasitic capacitance, but also minimizes the impact on the thermomechanical characteristics of the module.

The corresponding parasitic capacitance C_{in} value of the copper layer at the input terminal of the driver chip with a typical structure is 2.5pF, and the parasitic capacitance value extracted through Q3D after pattern etching of the corresponding position of the bottom copper layer is 0.9pF.

Fig. 17(a) shows the optimized structure of the bottom copper layer. Fig. 17(b) shows the relationship between the coupling interference voltage amplitude and the midpoint potential change rate before and after etching the bottom copper layer. After etching the bottom copper layer with the corresponding pattern, the ability to suppress coupling interference is greatly enhanced. When the midpoint potential change rate dv_{sw}/dt=80V/ns, the coupling interference voltage amplitude is 1.8V. It can suppress the internal coupling interference phenomenon of integrated SiC power modules at higher switching speeds.

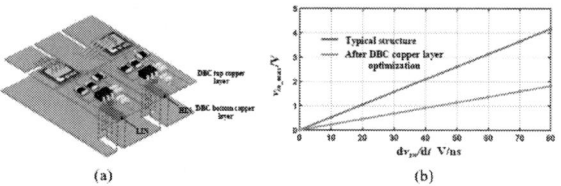

(a) (b)

Fig. 17 Optimized structure of copper layer at the bottom of DBC

5 Integrated Power Module Processing and Testing

The processing process of the integrated power module is shown in Fig. 18. The power chip and driver chip are consistent with the typical three-dimensional structure described above.

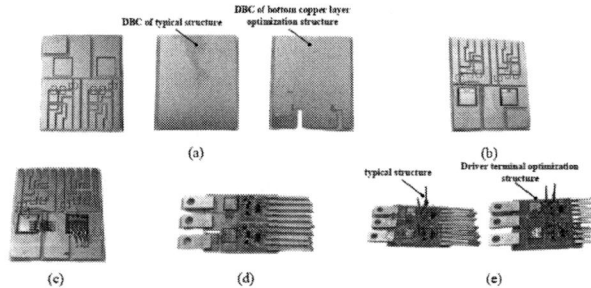

Fig. 18 Processing process of integrated power module: (a) Customized typical structure and bottom copper layer optimization DBC (b) First reflow soldering chip (c) Wire bonding (d) Terminal and driving component soldering (e) Silicone gel potting

Experiments are designed to verify the accuracy of the proposed coupling interference model and the effectiveness of the module structure optimization measures. The interference voltage at the input end of the driver chip and the gate-source voltage of the SiC MOSFET of the typical structure and the two optimized structures were tested under different voltage change rates. The basic hardware of the experimental platform is divided into modules to be tested, external power circuits, drive power supply and isolation circuits.

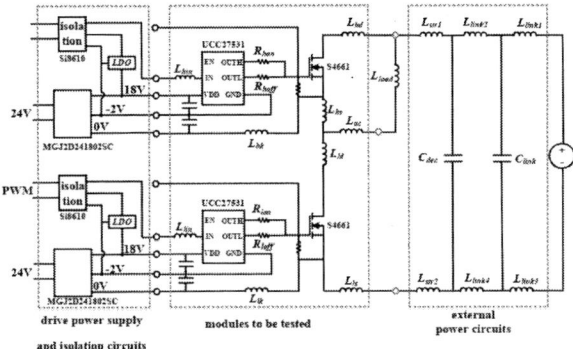

Fig. 19 Dynamic test schematic diagram

During testing, different modules use the same drive power supply and isolation circuit to ensure that the parasitic inductance introduced by the external PCB is the same, and only the parasitic parameters inside the module change. The dynamic test schematic diagram is shown in Fig. 19. The interfaces between the power side, the drive side and the external circuit of the power module to be tested are all marked in the figure.

The actual picture of the dynamic test platform is shown in Fig. 20.

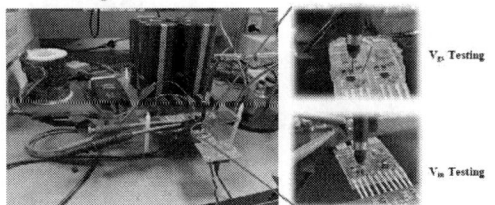

Fig. 20 Dynamic test platform

In the double-pulse test experiment, it is necessary to measure the drain-source voltage V_{ds}, drain current i_d, gate-source voltage V_{gs} and the input signal V_{in} of the driver chip of the low-side MOSFET, and observe the interference voltage at the input terminal of the driver chip during the switching process of the integrated power module. LeCroy's original differential probe HVD3106 is used to measure the drain-source voltage V_{ds}, and a CP9000S model Rogowski coil is used to measure the low-side MOSFET drain current i_d. Considering that the

gate-source voltage and the driver chip input signal amplitude are both small, it is more appropriate to use a passive probe for measurement, and because it uses a negative voltage drive, there is a 2V voltage difference between the source of the power chip and the GND of the driver chip. It is impossible to use two passive probes on the same oscilloscope at the same time for measurement, Therefore, the gate-source voltage V_{gs} and the driver chip input signal V_{in} can only be measured twice under the same working conditions, using Lecroy's original passive probe PP026 for measurement. The above signals are then captured by the Lecroy WaveSurfer 4104 oscilloscope with a 1GHz bandwidth.

First test a typical structure without optimization, the drain-source voltage V_{ds}, drain current i_d, gate-source voltage V_{gs} and the input signal V_{in} of the driver chip under different voltage change rates at the turn-off moment are shown in Fig.21.

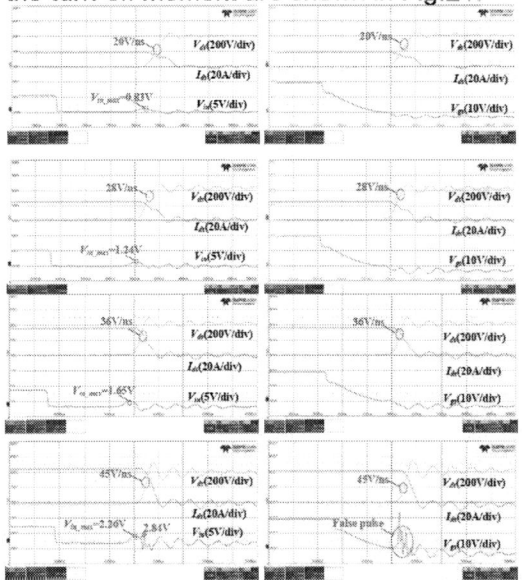

Fig. 21 Low-side MOSFET turn-off transient waveform of typical structure

When the turn-off speed reaches 45V/ns, the low-side gate-source voltage V_{gs} waveform has obvious false pulses. Compare the test results with the calculation results of the improved model. As shown in Fig.22(a), the experimental test results are about 0.2V lower than the calculation results of the improved model. Note that the capacitance from the internal input terminal of the driver chip to GND is 20pF, while the capacitance of the passive probe PP026 used for testing is 10pF. It can be regarded as a 10pF capacitor connected in parallel to the driver input terminal, which has a certain impact on the test results.

1701

Consider compensating the input capacitance C_0 of the driver chip in the improved interference model. The compensation value is the built-in capacitance value of the passive probe 10pF. The comparison of the compensated results is shown in Fig. 22(b). The difference between the two is reduced to within 0.08V, which can basically verify the accuracy of the proposed improved coupling interference model.

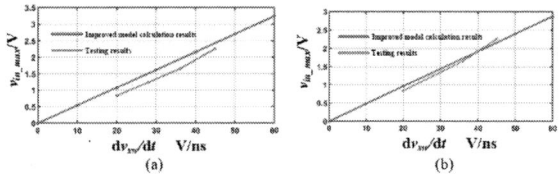

Fig. 22 Comparison of test results and model calculation results (a) before compensation (b) after compensation

Next, two optimized structures are tested. The test results of the optimized structure of the drive terminal are shown in Fig. 23. The amplitude of the drive input coupling interference voltage at a switching speed of 45V/ns is 1.8V, indicating that reducing the inductance of the drive chip signal input loop can alleviate coupling interference.

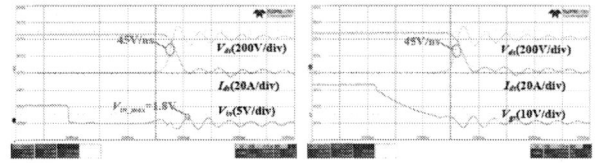

Fig. 23 Test results of the optimized structure of the drive terminal

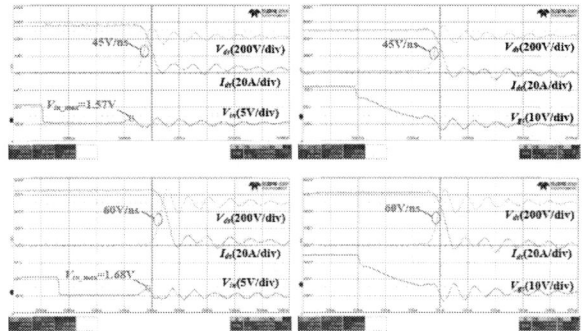

Fig. 24 Test results of the optimized structure of the DBC bottom copper layer

Then test the optimized structure of the copper layer at the bottom of the DBC. As shown in Fig. 24, the driver input coupling interference voltage amplitude is also at a switching speed of 45V/ns. At this time, the interference voltage amplitude is

1.57V, which further decreases. When dVds/dt increases to 60V/ns, the interference voltage amplitude at this time is only 1.68V, indicating that the optimized structure of the copper layer at the bottom of the DBC can effectively suppress the coupling interference phenomenon, the switching speed can be increased by more than 50%

6   Conclusion

In view of the internal coupling interference phenomenon of integrated power modules based on DBC, this article proposes an improved analysis model, and obtains the key parasitic parameters that have a greater impact on this phenomenon through numerical calculation. Based on this, two module optimization structures are proposed and verified through experiments. The model calculation difference is less than 8%, and the switching speed can be increased by more than 50%.

References

[1] J. Lv, C. Chen, B. Liu, Y. Yan and Y. Kang, "A Dynamic Current Balancing Method for Paralleled SiC MOSFETs Using Monolithic Si-RC Snubber Based on a Dynamic Current Sharing Model," in IEEE Transactions on Power Electronics, vol. 37, no. 11, pp. 13368-13384, Nov. 2022.

[2] L. Zhang, P. Liu, A. Q. Huang, S. Guo and R. Yu, "An improved SiC MOSFET-gate driver integrated power module with ultra low stray inductances," 2017 IEEE 5th Workshop on Wide Bandgap Power Devices and Applications (WiPDA), Albuquerque, NM, USA, 2017, pp. 342-345.

[3] S. Guo, L. Zhang, Y. Lei, X. Li, W. Yu and A. Q. Huang, "Design and application of a 1200V ultra-fast integrated Silicon Carbide MOSFET module," 2016 IEEE Applied Power Electronics Conference and Exposition (APEC), Long Beach, CA, USA, 2016, pp. 2063-2070.

[4] A. B. Jørgensen, S. Bęczkowski, C. Uhrenfeldt, N. H. Petersen, S. Jørgensen and S. Munk-Nielsen, "A Fast-Switching Integrated Full-Bridge Power Module Based on GaN eHEMT Devices," in IEEE Transactions on Power Electronics, vol. 34, no. 3, pp. 2494-2504, March 2019.

[5] Z. Dong, X. Wu and K. Sheng, "Suppressing Methods of Parasitic Capacitance Caused Interference in a SiC MOSFET Integrated Power Module," in IEEE Journal of Emerging and Selected Topics in Power Electronics, vol. 7, no. 2, pp. 745-752, June 2019.

PCIM Europe 2024, 11– 13 June 2024, Nuremberg   DOI: 10.30420/566262232

Multispectral Electroluminescence Sensing of SiC MOSFETs for Junction Temperature and Current Extraction

Lukas A. Ruppert ©, Rik W. De Doncker ©

Institute for Power Electronics and Electrical Drives, RWTH Aachen University, Germany

Corresponding author: Lukas A. Ruppert, lukas.ruppert@isea.rwth-aachen.de
Speaker: Lukas A. Ruppert, lukas.ruppert@isea.rwth-aachen.de

Abstract

The growing demand for reliability in power electronic systems leads to an increasing need for real-time condition monitoring. In this context, previous research demonstrated how the electroluminescence (EL) of SiC MOSFETs can be used for galvanically-isolated and high-bandwidth temperature and current sensing. However, existing EL-based sensing concepts exhibit poor sensitivity and accuracy when measuring at low currents and low temperatures. Thus, this work identifies key characteristics of the EL spectrum and proposes an improved EL-based sensing technique that measures multiple spectral sensitivities of the EL with optical sensors. The sensing approach is validated by measurements on a SiC MOSFET.

1 Introduction

Junction temperature monitoring is a crucial technique for tracking the state-of-health of semiconductor devices and identifying critical degradation [1]–[8]. For this, recent research has identified temperature sensitive optical parameters (TSOPs) of SiC MOSFETs, i.e. the EL, as promising research path [9], [10]. The EL, that is emitted when the intrinsic body diode is energized as shown in Fig. 1, exhibits both device temperature and current information [11]. Exploiting these spectral characteristics, various EL-based monitoring techniques have been proposed in literature that capture the EL with optical sensors to extract the device temperature and/or current information[11]–[18]. However, existing real-time EL-sensing solutions for simultaneous current and temperature sensing suffer from poor sensitivity and accuracy when measuring at low currents and low temperatures. Thus, this work presents an advanced EL-based sensing method that provides a higher sensing accuracy at such operating conditions. It utilizes multiple optical sensors and optical filters, which allows to separate three distinct characteristics of the EL spectrum. In this way, the sensing method is able to extract real-time current and temperature information over the entire operating range of SiC MOSFETs.

This paper is organized as follows: The noise ro-

Fig. 1: Electroluminescence of a SiC power module [18]

bustness of these sensing solutions is analyzed to identify which spectral characteristics should be utilized for a good estimation accuracy, also at low currents. Based on these results, a dynamic sensing circuitry is presented that is able to measure all key characteristics of the EL spectrum. Finally, the sensing approach is experimentally validated by double-pulse measurements on a commercially available power module.

2 Electroluminescence Sensing of SiC MOSFETs

MOSFETs emit light when the body diode is forward biased [10], which typically occurs when the MOSFET is operated in the third quadrant ($i_{SD} > 0$) such that minority carriers are injected in the internal p-n junction. The EL spectrum of 4H-SiC MOSFETs exhibits two major peaks in the visible

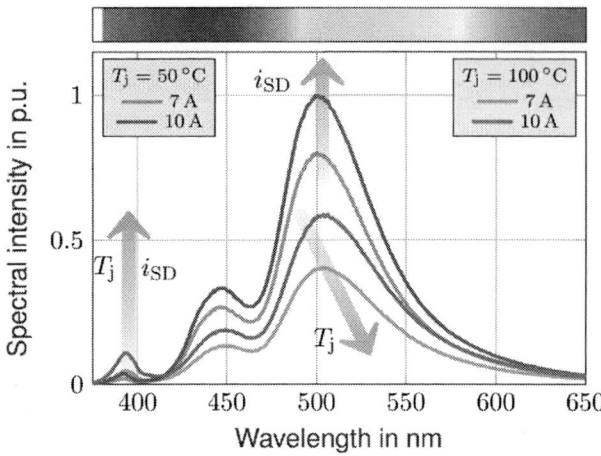

Fig. 2: EL spectra of a SiC MOSFET body diode for different device currents and temperatures [16]

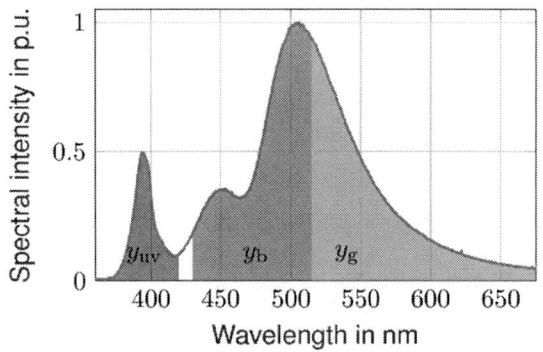

Fig. 3: Integration of the measured EL spectra to calculate the spectral areas

range [10], [19] as shown in Fig. 2. The first peak is at around 390 nm in the ultraviolet region, whereas the second peak is located in the blue-green region around 500 nm [19]–[22].

This sections first considers the basic characteristics of the EL spectrum and discusses how these are utilized for the extraction of current and temperature in existing monitoring concepts. Then, based on static EL measurements with a spectrometer, this section analyzes the noise robustness for different sensor configurations to show that all available characteristics of the EL spectrum should be utilized to achieve the best sensing accuracy.

2.1 Spectral Sensitivities

The spectrum shown in Fig. 2 exhibits three distinct spectral characteristics [20], [21], [23], [24]. First, the two peaks exhibit an opposing dependency on the junction temperature: The blue-green

peak decreases and the ultraviolet peak increases with rising temperature. Second, the device current has a uniform influence on both peaks, with the intensity increasing with rising current. Third, the blue-green peak exhibits a temperature-dependent redshift as this peak is shifted from lower to higher wavelength with rising temperature, as indicated by the down-right pointing arrow in Fig. 2 [16], [25].

In order to investigate these spectral characteristics more quantitatively, the EL of a SiC power module was characterized by means of static spectroradiometric measurements using an fiber-coupled spectrometer in [16]. Subsequently, the spectral intensities $a(\lambda)$, such as the one in Fig. 2, were integrated over different wavelength regions to separate the respective spectral areas y:

$$y = \int_{\lambda_1}^{\lambda_2} a(\lambda)\, d\lambda. \tag{1}$$

To analyze not only the temperature and current dependencies of the peak but also the temperature-dependent redshift of the blue-green peak, three spectral areas are calculated according to Fig. 3: y_{uv}, y_{bg} and y_{g}, whereby y_{bg} is the sum of y_{b} and y_{g}. The resulting integrated intensities shown in Fig. 4 prove the three aforementioned spectral characteristics, as y_{uv} and y_{bg} show an opposing temperature and a uniform current dependency. The temperature-dependent redshift can be recognized by comparing the different slopes of y_{bg} and y_{g}. Mathematically, the current and temperature dependency of any spectral area can be expressed by the function f_{i}

$$y_{\mathrm{i}} = f_{\mathrm{i}}(\boldsymbol{x}), \boldsymbol{x} = (i_{\mathrm{SD}}, T_{\mathrm{j}})^{\mathsf{T}}, \tag{2}$$

whereby the input $\boldsymbol{x}$ consists of the device current and temperature whereas the output y_{i} is the measured spectral area.

Also, it is evident from these plots that the intensity and sensitivity of the ultraviolet peak is poor at low temperatures and currents. This poses a challenge for EL-based current and temperature extraction and, thus, must be taken into account when designing the sensor system, as will be discussed in the following.

2.2 Unified Current and Temperature Estimation

Utilizing these spectral characteristics, different monitoring concepts for simultaneous current and

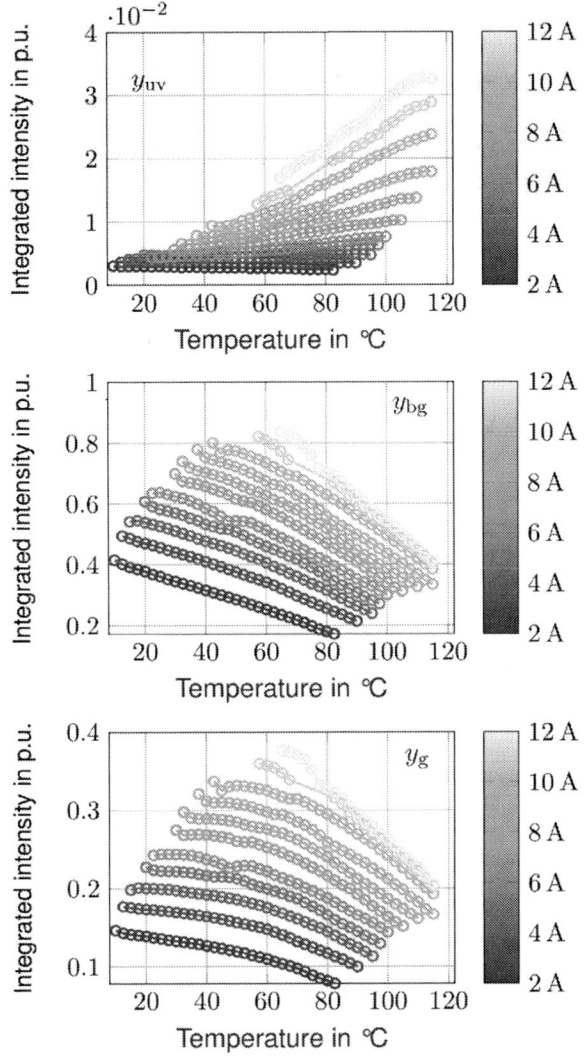

Fig. 4: Integrated intensities as a function of junction temperature and device current: y_{uv} (top), the y_{bg} (middle) and y_g (bottom) [16]

temperature extraction [13], [15] have been proposed by measuring different spectral areas of the EL. The general principle of these monitoring concepts is shown in Fig. 5. The emitted light of SiC MOSFETs is captured with multiple optical sensors, i.e. photodiodes, with distinct spectral characteristics to measure the spectral intensity of different wavelength regions. Subsequently, the measured signals are post-processed via e.g. look-up tables, analytical expressions or artificial neural networks to calculate the device current and junction temperature. From a mathematical point of view, the output in Eq. (2) is extended to a vector y when measuring multiple spectral regions simultaneously. Thus, the

output vector y now contains the measurement of each spectral region, such that a non-linear system of equations f is formed:

$$y = (y_1, \ldots, y_n)^\mathsf{T}, x = (i_{SD}, T_j)^\mathsf{T} \tag{3}$$

$$y = \begin{pmatrix} y_1 \\ \vdots \\ y_n \end{pmatrix} = f(x) = \begin{pmatrix} f_1(x) \\ \vdots \\ f_n(x) \end{pmatrix}. \tag{4}$$

To extract current and temperature information from the spectral measurements y, the post-processing calculates the inverse of f to find x:

$$x = f^{-1}(y). \tag{5}$$

For the simultaneous extraction of both the junction temperature and the device current, at least two linearly independent spectral characteristics and, thus, at least two optical sensors must be utilized ($n \geq 2$). For instance, [13] separated the ultraviolet peak y_{uv} and blue-green peak y_{bg} of the spectrum with optical filters and measured their respective intensity with two optical sensors ($n = 2$), thus utilizing the opposing temperature characteristic and the uniform current characteristic of the spectrum. However, the spectral intensity of the ultraviolet peak is poor at low currents and temperatures, as discussed in section 2.1, such that it can barely be used for solving Eq. (5) for x. At these operating points, effectively only one linearly independent measurement remains, which is not sufficient to extract two unknown quantities, i.e. i_{SD} and T_j, and may lead to inaccurate current and temperature estimates.

To achieve better sensor performance also at low currents and temperatures, this work proposes to utilize all three characteristics of the spectrum. For this, a third optical sensor is used to measure only the green part y_g of the spectrum, as depicted in Fig. 5. The simultaneous measurement of y_g and y_{bg} allows to evaluate the temperature-dependent redshift, thus providing an additional spectral characteristic. This significantly increases the estimation accuracy of x as shown in the following.

2.3 Noise Robustness Analysis

To evaluate the estimation accuracy of the EL-based sensing approach, as formulated in Eq. (5), this section analyses the robustness of the current and temperature extraction against noise. For this, it is assumed that the spectral measurement y is disturbed by a small noise component y_{noise}, which

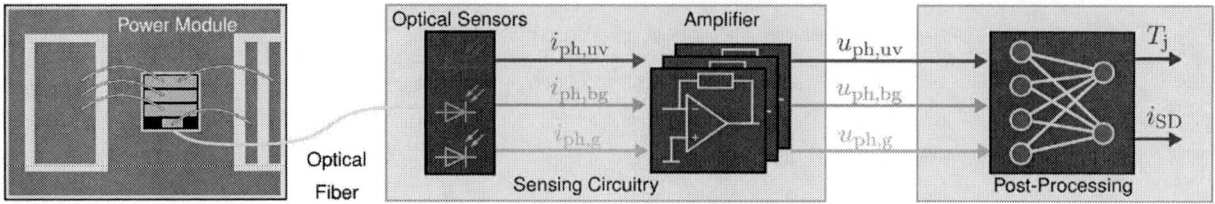

Fig. 5: Schematic representation of the proposed temperature sensing approach

leads to an estimation error of x_{err}:

$$y + y_{\mathrm{noise}} = f(x + x_{\mathrm{err}}). \quad (6)$$

The upper bound of the estimation error x_{err} is given by the product of the absolute condition number κ_{abs} of f^{-1} and the noise component y_{noise} [26]:

$$x_{\mathrm{err}} <= \kappa_{\mathrm{abs}}(x_0) \cdot y_{\mathrm{noise}} \quad (7)$$

$$\kappa_{\mathrm{abs}}(x_0) = \left|\left| \mathbf{J}^{-1}(x_0) \right|\right| = \left|\left| \left(\frac{\partial f(x, x_0)}{\partial x} \right)^{-1} \right|\right|, \quad (8)$$

where $\mathbf{J}(x_0)$ is the Jacobi matrix of f at a given operating point x_0. Consequently, the absolute condition number is an adequate performance indicator to evaluate the robustness of EL-based sensing approach against noise: The higher κ_{abs}, the higher the susceptibility to noise. It should be noted that if the number of inputs and outputs of f is not equal ($n \neq 2$), compare Eq. (3), the inverse of $\mathbf{J}$ does not exist and, thus, must be replaced by the Moore-Penrose pseudoinverse $\mathbf{J}^+$.

In the following, the noise robustness of three sensing concepts are compared:

$$\begin{aligned} y_\alpha &= (y_{\mathrm{uv}}, y_{\mathrm{bg}})^\mathsf{T} \\ y_\beta &= (y_{\mathrm{uv}}, y_{\mathrm{bg}}, y_{\mathrm{g}})^\mathsf{T} \quad (9) \\ y_\gamma &= (y_{\mathrm{uvb}}, y_{\mathrm{bg}})^\mathsf{T} . \end{aligned}$$

The first concept, expressed by y_α, measures the two spectral areas y_{uv} and y_{bg}, thus utilizing two spectral characteristics: the opposing temperature dependency of the ultraviolet peak and the blue-green peak and the uniform current dependency of both peaks. The second concept, expressed by y_β, additionally measures y_{g} such that it uses also the temperature-dependent redshift of the blue-green peak, which is given by the different temperature sensitivity of y_{bg} and y_{g}. Consequently, this sensing concept exploits all three spectral characteristics. The third concept, expressed by y_γ, also exploits

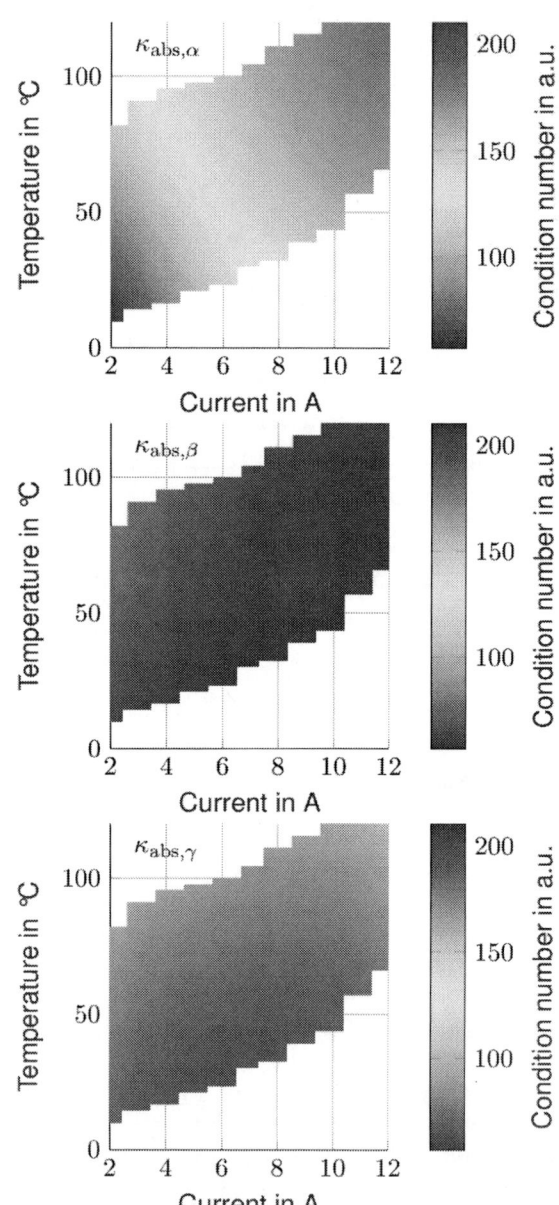

Fig. 6: Absolute condition number $\kappa_{\mathrm{abs},\alpha}$ (top) $\kappa_{\mathrm{abs},\beta}$ (middle) and $\kappa_{\mathrm{abs},\gamma}$ (bottom) of the three sensing scenarios y_α, y_β and y_γ

all three spectral characteristics, but by measuring only two spectral areas: y_{bg} and y_{uvb}, defined as the sum of y_{uv} and y_b. At first glance, it might appear unreasonable to measure three independent spectral characteristics with only two sensor values. However, as the two spectral areas overlap in the blue region, the temperature-dependent redshift of the blue-green peak is again captured by this sensing configuration, especially at low currents when the ultraviolet peak hardly contributes to y_{uvb}.

To derive the absolute condition numbers for the three scenarios, the static measurement results in Fig. 4 are fitted to bivariate polynomial regressions. This yields differentiable expression for the functions f_α, f_β and f_γ that are used to calculate the Jacobi matrix for a given operating point in Eq. (8). The resulting condition numbers for the three scenarios are depicted in Fig. 6 as a function of current and temperature. The absolute condition number $\kappa_{\text{abs},\alpha}$ of the first sensing method y_α, with one sensor each for the ultraviolet and the blue-green peak, significantly increases at low currents. Thus, this sensing approach suffers from high susceptibility to noise. In contrast, the sensing scenario y_β with the additional measurement of y_g shows a significantly lower condition number $\kappa_{\text{abs},\beta}$, especially at lower currents. The third sensing scenario y_γ also considerably reduces the condition number at low current compared to the scenario y_α due to the overlapping spectral areas that exploit an additional spectral characteristic, i.e. the temperature-dependent redshift. However, this sensing scenario cannot reach the low condition numbers of scenario y_β. At higher currents when additional information can be obtained from the ultraviolet peak, all the three condition values tend to converge.

Thus, utilizing all three spectral characteristics improves the estimation accuracy particularly at low current levels. Even though the three characteristics can theoretically be measured with only two sensors, the best overall noise robustness is achieved with three sensors.

3 Multispectral Sensing Circuitry

In the previous section, it was shown that all three characteristics of the EL spectrum should be utilized to improve the estimation accuracy of the temperature and current estimation at low current conditions. Previous publications mainly focused on two spectral characteristics and presented dynamic sensing circuits that allow measuring these charac-

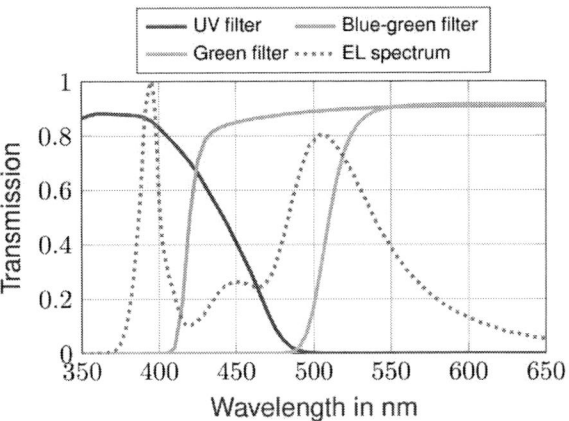

Fig. 7: Transmission profile of the optical filters

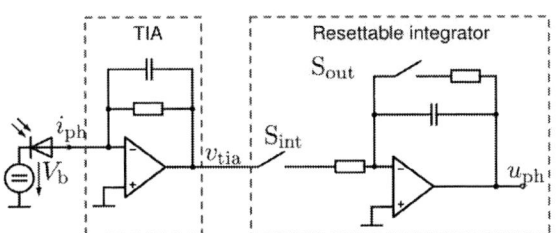

Fig. 8: Equivalent circuit of the amplifier circuitry for the SiPM

teristics during operation. However, the third spectral characteristic, i.e. the temperature-dependent redshift was only analyzed so far by means of static measurements using an optical spectrometer [16] but not during dynamic operation. Thus, this section presents a dynamic circuitry that is able to measure all three characteristics during operation, which is subsequently validated by experimental measurements.

3.1 Dynamic Sensing Circuitry

The general structure of the developed sensing circuitry is shown in Fig. 5. For the optical sensors, three silicon photomultiplier (SiPM) are used due to their high sensitivity, in particular in the ultraviolet and blue-green region of the visible spectrum. Each SiPM is equipped with an additional optical filter whose transmission profiles are depicted in Fig. 7. Due to these distinct transmission profiles, the optical sensors separate the ultraviolet, blue-green and green part of the EL spectrum, as suggested in section 2. It should be noted that the optical filters used in this work do not provide a perfect separation of the spectral areas, especially the ultraviolet filter. However, this does not limit the proposed sensing approach as all key characteristics of the spectrum are still measured.

As the EL in SiC MOSFETs typically only occurs during a short dead time in a hard-switched application, the output of the SiPM, i.e. the photocurrent i_{ph}, must be amplified with high bandwidth. This amplification is done by a transimpedance amplifier (TIA) for each SiPM that translates the photocurrent into a voltage signal, as depicted in Fig. 8. Subsequently, the output of the TIA is fed into a resettable integrator in order to improve the signal-to-noise ratio. Finally, this signal is captured by an 12-bit ADC of a Texas Instruments development board LAUNCHXL-F28379D that is also used for the post-processing.

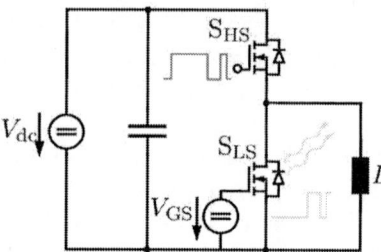

Fig. 10: Equivalent circuit of the double-pulse setup used to generate EL pulses in the LS switch S_{LS}

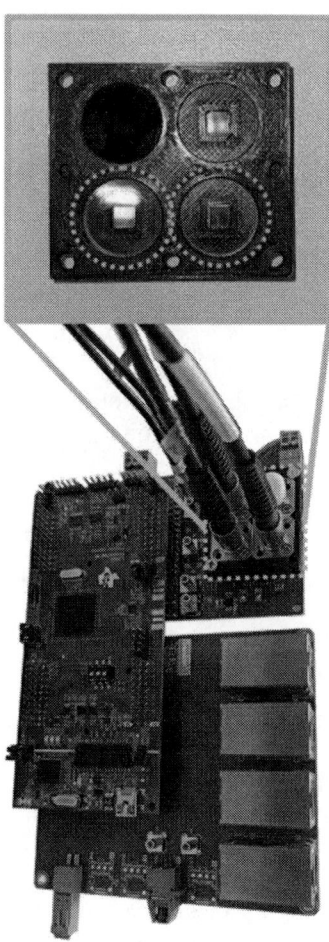

Fig. 9: Sensing circuitry with optical sensors

The overall setup of the sensing circuitry is shown in Fig. 9. In contrast to the schematic representation in Fig. 5, each sensor is equipped with a separate optical fiber that all measure the light emission of the same MOSFET die. Using multiple optical fiber for a single die simplifies the prototyping of different sensor configuration. Alternatively, any other method can be used to distribute the light from a single optical fiber to several sensors.

3.2 Experimental Validation

To validate that all of the key spectral characteristics, in particular the temperature-dependent redshift, can be measured by the dynamic sensing circuitry, the circuit was characterized experimentally on a commercially available SiC half-bridge power module. For this purpose, the double-pulse setup in Fig. 10 generated short light pulses of $3\,\mu s$ emitted by the low-side switch of the power module. The characterization was performed over a current range of $2\,A$ to $30\,A$ and a temperature range of $0\,°C$ to $120\,°C$, using a controlled heating plate.

The recorded sensor values u_{ph}, measured by the ADC, are given in Fig. 11. Considering the measurements of the ultraviolet sensor $u_{ph,uv}$ and blue-green sensor $u_{ph,bg}$, it is evident that the developed sensing circuitry captures the opposing temperature and uniform characteristic of the EL spectrum. Moreover, the third characteristic, i.e. the temperature-dependent redshift, is visible from the different slopes of $u_{ph,bg}$ und $u_{ph,g}$. To highlight the different temperature dependencies of the sensor signals, the ratio of the signals is calculated according to:

$$r_{ph} = \frac{u_{ph,bg}}{u_{ph,g}} \qquad (10)$$

The corresponding results in Fig. 12 show that the ratio r_{ph} exhibits a high temperature and only a low current dependency, thus proving that temperature-dependent redshift of the spectrum can be captured by means of dynamic measurements. Consequently, all three spectral characteristics are measured by the developed dynamic sensing circuitry.

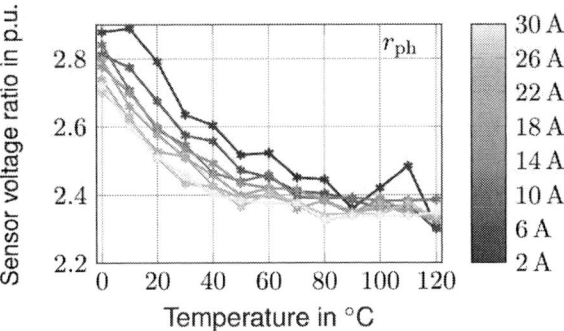

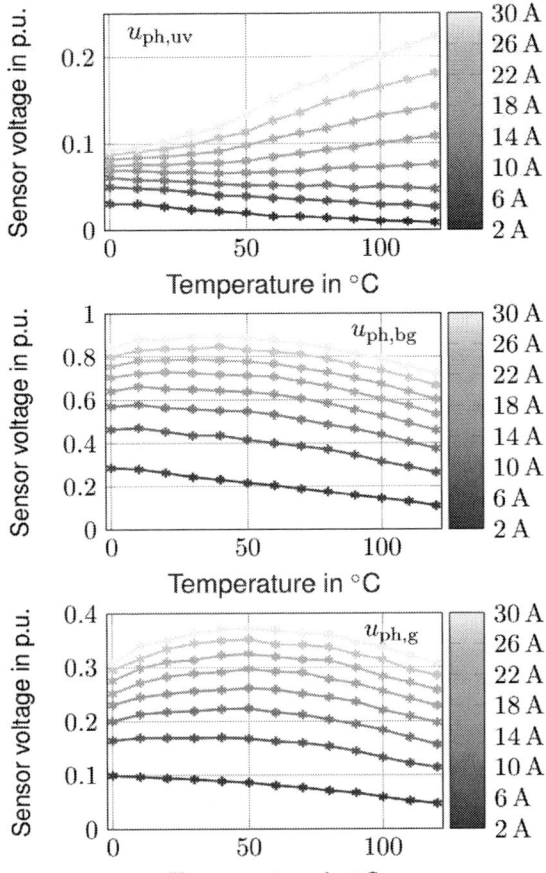

Fig. 12: Ratio of the measured sensor voltages r_{ph}

Fig. 11: Measurement results of double-pulse experiment of the ultraviolet sensor (top), the blue-green sensor (middle) and the green sensor (bottom)

4 Conclusion

The contributions and key associated conclusions of this paper are summarized as follows:

– Analyzing the noise robustness of EL-based sensing methods indicated that all three spectral characteristics should be exploited to achieve good sensor performance, especially at low current conditions when the ultraviolet peak suffers from low intensity. These characteristics are: the opposing temperature dependency of the ultraviolet peak and the blue-green peak, the uniform current dependency of both peaks as well as the temperature-dependent redshift of the spectrum.

– Even though only two optical sensors can already exploit all three spectral characteristics,

using three sensors yields the best overall noise robustness and sensing accuracy.

– A dynamic sensing circuitry was presented that measures three distinct areas of the EL spectrum. Double pulse measurements on an industry-standard SiC power module proved that this sensing circuitry is able to capture all three spectral characteristics. Thereby, it was shown for the first time that the temperature-dependent redshift can be measured dynamically during operation.

Acknowledgment

This research was funded by the by the German Research Foundation (DFG) within the project EluSense (BR 6266/2-1).

References

[1] M. Andresen, M. Liserre, and G. Buticchi, "Review of active thermal and lifetime control techniques for power electronic modules," in *2014 16th European Conference on Power Electronics and Applications*, Aug. 2014. DOI: 10.1109/EPE.2014. 6910822.

[2] Z. Ni, X. Lyu, O. P. Yadav, B. N. Singh, S. Zheng, and D. Cao, "Overview of real-time lifetime prediction and extension for sic power converters," *IEEE Transactions on Power Electronics*, vol. 35, no. 8, Aug. 2020. DOI: 10.1109/tpel.2019.2962503.

[3] N. Fritz, T. Kamp, T. A. Polom, M. Friedel, and R. W. De Doncker, "Evaluating on-state voltage and junction temperature monitoring concepts for wide-bandgap semiconductor devices," *IEEE Transactions on Industry Applications*, vol. 58, no. 6, 2022. DOI: 10.1109/tia.2022.3191632.

[4] S. Kalker, C. H. van der Broeck, and R. W. De Doncker, "Online junction-temperature extraction method for SiC MOSFETs utilizing turn-on delay," in *2021 IEEE 8th Workshop on Wide Bandgap Power Devices and Applications (WiPDA)*, IEEE, Nov. 2021. DOI: 10.1109/wipda49284.2021. 9645104.

[5] I. Austrup, T. B. Albert, C. H. van der Broeck, and R. W. de Doncker, "Identifying superimposed degradation effects in power electronic modules," in *PCIM Europe 2023; International Exhibition and Conference for Power Electronics, Intelligent Motion, Renewable Energy and Energy Management*, VDE VERLAG GMBH, May 2023. DOI: 10. 30420/566091218.

[6] J. O. Gonzalez and O. Alatise, "Challenges of Junction Temperature Sensing in SiC Power MOS-FETs," in *2019 10th International Conference on Power Electronics and ECCE Asia (ICPE 2019 - ECCE Asia)*, ISSN: 2150-6086, May 2019.

[7] S. Kalker, L. A. Ruppert, C. H. van der Broeck, J. Kuprat, M. Andresen, *et al.*, "Reviewing thermal-monitoring techniques for smart power modules," *IEEE Journal of Emerging and Selected Topics in Power Electronics*, Mar. 2021. DOI: 10.1109/ jestpe.2021.3063305.

[8] S. Kalker, J. Holz, I. Austrup, C. H. van der Broeck, and R. W. De Doncker, "Degradation diagnosis during active power cycling via frequency-domain thermal impedance spectroscopy," in *2023 IEEE International Conference on Power Electronics - ECCE Asia (ICPE)*, May 2023. DOI: 10.23919/ ICPE2023-ECCEAsia54778.2023.10213859.

[9] J. Winkler, J. Homoth, and I. Kallfass, "Electroluminescence in power electronic applications: Utilization of p-n junctions in power semiconductors as unintentional light emitting diodes for current and temperature sensing," Oct. 2018.

[10] J. Winkler, J. Homoth, and I. Kallfass, "Utilization of Parasitic Luminescence from Power Semiconductor Devices for Current Sensing," in *PCIM Europe 2018; International Exhibition and Conference for Power Electronics, Intelligent Motion, Renewable Energy and Energy Management*, Jun. 2018.

[11] L. A. Ruppert, S. Kalker, C. H. van der Broeck, and R. W. De Doncker, "Analyzing Spectral Electroluminescence Sensitivities of SiC MOSFETs and their Impact on Power Device Monitoring," in *PCIM Europe digital days 2021; International Exhibition and Conference for Power Electronics, Intelligent Motion, Renewable Energy and Energy Management*, May 2021.

[12] J. Winkler, J. Homoth, and I. Kallfass, "Electroluminescence-Based Junction Temperature Measurement Approach for SiC Power MOSFETs," *IEEE Transactions on Power Electronics*, vol. 35, no. 3, Mar. 2020. DOI: 10.1109/TPEL.2019.2929426.

[13] H. Luo, J. Mao, C. Li, F. Iannuzzo, W. Li, and X. He, "Online Junction Temperature and Current Simultaneous Extraction for SiC MOSFETs With Electroluminescence Effect," *IEEE Transactions on Power Electronics*, vol. 37, no. 1, Jan. 2022. DOI: 10.1109/TPEL.2021.3094924.

[14] S. Kalker, C. H. van der Broeck, and R. W. De Doncker, "Utilizing electroluminescence of SiC MOSFETs for unified junction-temperature and current sensing," in *2020 IEEE Applied Power Electronics Conference and Exposition (APEC)*, IEEE, Mar. 2020. DOI: 10.1109/apec39645.2020. 9124517.

[15] S. Kalker, C. H. van der Broeck, L. A. Ruppert, and R. W. De Doncker, "Next Generation Monitoring of SiC MOSFETs via Spectral Electroluminescence Sensing," *IEEE Transactions on Industry Applications*, 2021. DOI: 10.1109/TIA.2021.3062773.

[16] L. A. Ruppert, S. Kalker, and R. W. De Doncker, "Junction-Temperature Sensing of Paralleled SiC MOSFETs Utilizing Temperature Sensitive Optical Parameters," in *2021 IEEE Energy Conversion Congress and Exposition (ECCE)*, ISSN: 2329-3748, Oct. 2021. DOI: 10.1109/ECCE47101.2021. 9595734.

[17] L. A. Ruppert, B. Wirsen, S. Kalker, and R. W. De Doncker, "Utilizing electroluminescence of silicon IGBTs for junction temperature sensing," in *2023 11th International Conference on Power Electronics and ECCE Asia (ICPE 2023 - ECCE Asia)*, IEEE, May 2023. DOI: 10.23919/icpe2023-ecceasia54778.2023.10213620.

[18] L. A. Ruppert, M. Laumen, and R. W. De Doncker, "Utilizing the electroluminescence of sic mosfets as degradation sensitive optical parameter," in *2022 24th European Conference on Power Electronics and Applications (EPE'22 ECCE Europe)*, Sep. 2022.

[19] M. Anikin, A. Lebedev, N. Poletaev, A. chuk, A. Syrkin, and V. Chelnokov, "Deep centers and blue-green electroluminescence in 4H-SiC," Oct. 1993.

[20] A. M. Strel'chuk, E. V. Kalinina, and A. A. Lebedev, "Temperature dependence of the band-edge injection electroluminescence of 4h-sic pn structure," in *Silicon Carbide and Related Materials 2012*, ser. Materials Science Forum, vol. 740, Trans Tech Publications Ltd, Mar. 2013. DOI: 10.4028/ www.scientific.net/MSF.740-742.569.

[21] S. G. Sridhara, L. L. Clemen, R. P. Devaty, W. J. Choyke, D. J. Larkin, *et al.*, "Photoluminescence and transport studies of boron in 4h sic," *Journal of Applied Physics*, vol. 83, no. 12, 1998. DOI: 10.1063/1.367970.

[22] S. Bishop, C. Reynolds, J. Molstad, F. Stevie, D. Barnhardt, and R. Davis, "On the origin of aluminum-related cathodoluminescence emissions from sublimation grown 4h-sic(1 1 2⁻ 0)," *Applied Surface Science*, vol. 255, no. 13, 2009. DOI: https://doi.org/10.1016/j.apsusc.2009.02.036.

[23] A. M. Strel'chuk, Y. S. Kuz'michev, and K. F. Shtel'makh, *Features of the Band-Edge Injection Electroluminescence in 4H-SiC pn Structures*, 2015. DOI: 10.4028/www.scientific.net/MSF.821-823.289.

[24] A. A. Lebedev, "Deep level centers in silicon carbide: A review," *Semiconductors*, vol. 33, no. 2, Feb. 1999. DOI: 10.1134/1.1187657.

[25] A. Yang, K. Murata, T. Miyazawa, T. Tawara, and H. Tsuchida, "Time-resolved photoluminescence spectral analysis of phonon-assisted DAP and e-a recombination in n+b-doped n-type 4h-SiC epilayers," *Journal of Physics D: Applied Physics*, vol. 52, no. 10, Jan. 2019. DOI: 10.1088/1361-6463/aaf8e9.

[26] L. N. Trefethen and D. Bau, *Numerical Linear Algebra, Twenty-fifth Anniversary Edition*. Society for Industrial and Applied Mathematics, Jan. 2022. DOI: 10.1137/1.9781611977165.

PCIM Europe 2024, 11– 13 June 2024, Nuremberg   DOI: 10.30420/566262233

SiC-IPM for Compact and Energy Efficient Low-Power Motor Drives

Jong-Mu Lee[1], Jintae Kim[1], Jun-Ho Lee[1], Bum-Seok Suh[1]

[1]Alpha and Omega Semiconductor, Republic of Korea

Corresponding author: Jong-Mu Lee, jm.lee@kr.aosmd.com
Speaker: Jong-Mu Lee, jm.lee@kr.aosmd.com

Abstract

The demand on higher energy efficiency is on the rise even in low-power motor drive applications and it is expected that SiC will play an important role in saving the energy as well as reducing the inverter board size. This paper presents a new SiC-FET IPM (intelligent power module) which enables higher efficient and more compact inverter. All the performance including EMI and noise that designers must face will be directly compared to RC-IGBT IPM inverter which has been generally and wisely used.

1   Introduction

Nowadays the cost-efficient inverter design in low-power motor drives due to the higher adoption of energy efficient BLDC motors or PMSM is required in ceiling and various fan, refrigerator, pump and dish washer applications. In such applications RC-IGBT inverter are mainly chosen with heatsink because of thermal management issue. Therefore, designed PCB size and its shape can be limited and additional costs, which are heatsink and its assembly process, are required. This paper describes the concept of using surface-mount device integrated power module (SMD-IPM) employing thermal-enhanced direct-bonded copper (DBC) substrate, which can be crucial to effective power management and adopting Silicon Carbide (SiC) MOSFETs as inverter power devices, which can significantly reduce the power losses and improve the inverter efficiency [1].

It can remove the heatsink attachment by thermal spread along with PCB trace and reduce PCB footprint for compact and higher power density.

Fig. 1 shows the external view of propose SMD IPM (dimension: 7.5mm x 18mm) in which bottom side DBC substrate is placed. The DBC is designed with a structure to be electrically isolated from any internal circuits so that the substrate of IPM can be directly soldered to thermal pad on printed circuit board (PCB).

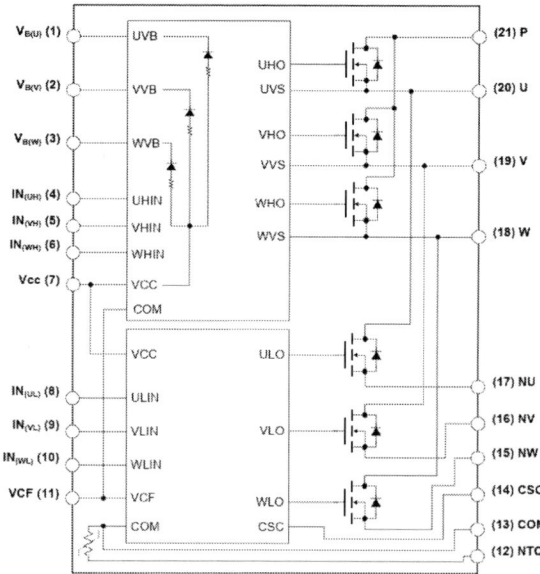

Fig. 1   Proposed SiC-IPM.

Fig. 2   Internal Block Diagram of new SiC-FET IPM.

Fig. 2 shows internal block diagram of the proposed SMD-IPM, which is composed of 500 mΩ SIC-MOSFETs, bootstrap-integrated HVIC, functional LVIC and NTC thermal sensor. The estimated power losses comparison indicates the inverter efficiency can be improved up to 69% comparing with conventional 3A IGBT IPM.

2 Electrical Characteristics

The Silicon Carbide MOSFET has several advantages over the RC-IGBT device. It has a very low 'ON'-state voltage drop particularly in lower current area and shows very low Qrr during turn-on behavior due to majority carrier device characteristics. The conduction, forward blocking and reverse blocking capabilities of the SiC-FET are superior to the RC-IGBT because of wide-band gap device.

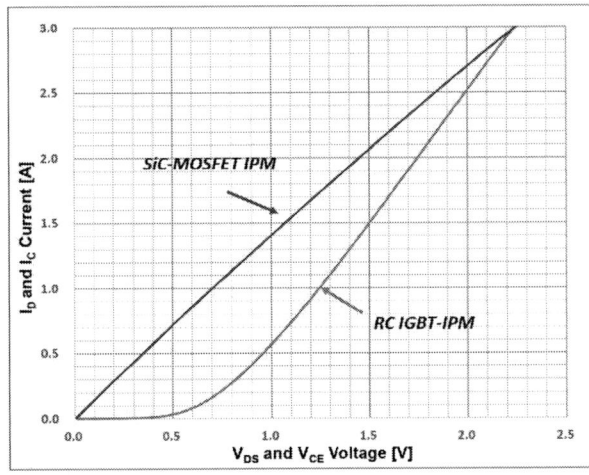

Fig. 3 Forward DC Characteristics at T$_J$=125 °C.

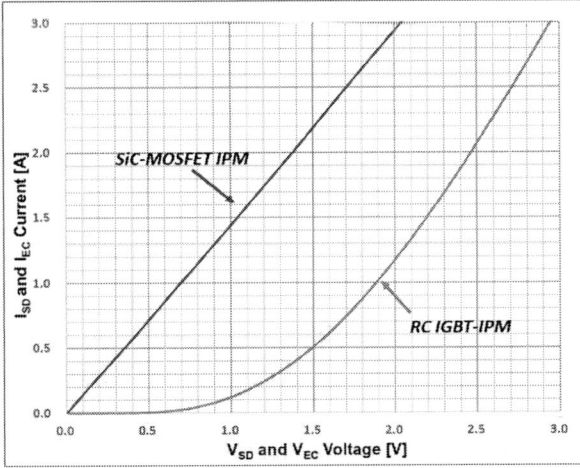

Fig. 4 Reverse DC Characteristics at T$_J$=125 °C.

The SiC-FET integrated in SMD-IPM is 1200V device, nevertheless, die size of 500 mΩ SiC-FET is smaller than 600 V RC-IGBT die. Fig. 3 shows the forward DC characteristics and Fig. 4 shows the reverse DC characteristics. In case of PWM inverter system, PWM is complementarily operated so that SiC-FET is driven by synchronous rectifying mode while flowing reverse current. Accordingly, much smaller conduction loss is generated at low current region.

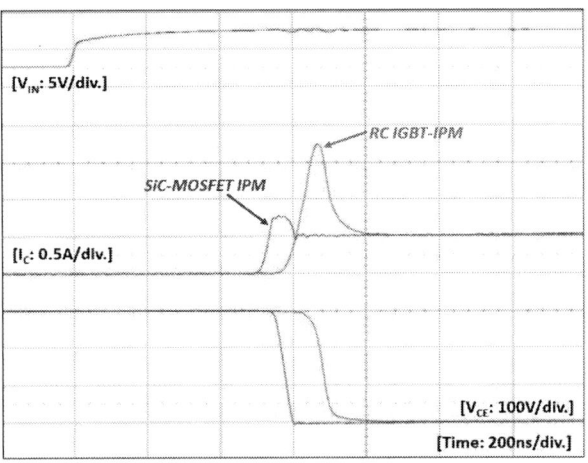

Fig. 5 Comparison of Turn-on Behavior.
I$_C$=0.5 A and T$_J$=125 °C

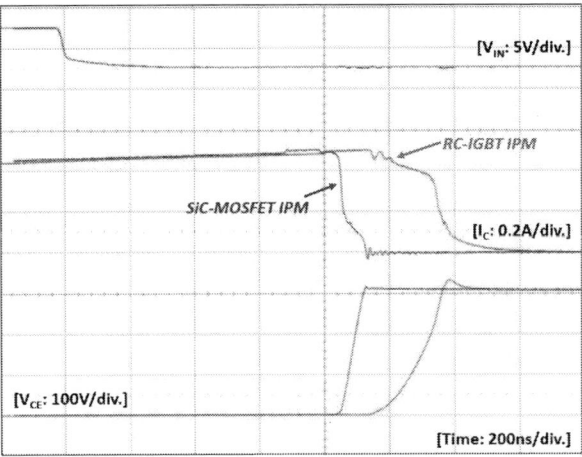

Fig. 6 Comparison of Turn-off Behavior.
I$_C$=0.5 A and T$_J$=125 °C

When a RC-IGBT is switched on, the large reverse recovery current is caused by Recovery Diode (FRD) of RC-IGBT body. However, SiC-FET has a schottky barrier diode (SBD) as a body diode of FET and the displacement current is much smaller than the reverse recovery of the conventional FRD. Accordingly, turn-on switching losses of both Eon and Err of SiC-FET at 0.5 A shows just about 15.4% in comparison with RC-IGBT ones which is measured in the waveforms of Fig. 5. From low reverse recovery current, low EMI characteristic would be expected. Upon comparison of waveforms during switch-off, a SiC-FET has no tail current theoretically, and it can have faster dv/dt behavior, which can provide extremely smaller turn-off switching loss (Eoff). Fig. 6 illustrates the Eoff of the SiC-IPM is just about 10% comparing to IGBT-IPM at 0.5 A switching operating condition.

3 Power Efficiency and Short-Circuit Ruggedness

From AC and DC electrical characteristics, the power losses comparisons can be estimated considering motor drive operation.

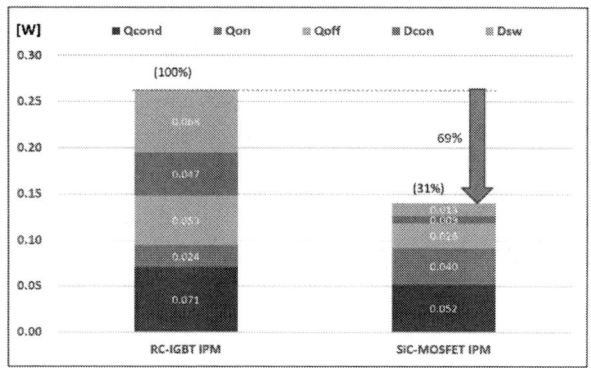

Fig. 7 Comparison of Power Losses at V_{DC}=300 V, F_{SW}=15 kHz, MI=0.8, PF=0.8, I_D=0.5 A, T_J=125 °C.

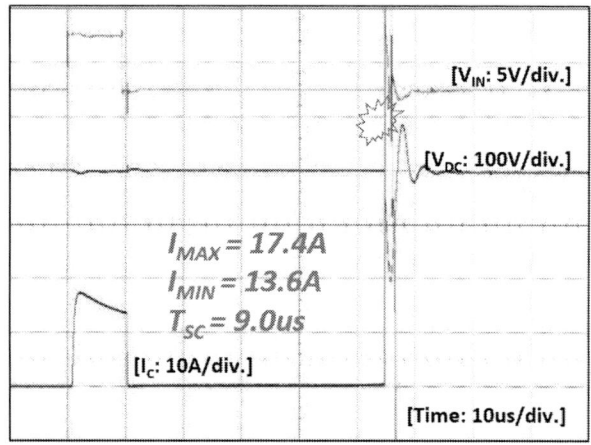

(a) Short-Circuit Waveform of RC-IGBT IPM

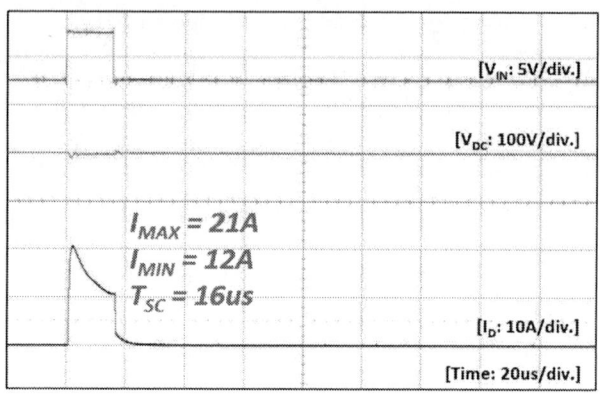

(b) Short-Circuit Waveform of SiC-FET IPM

Fig. 8 Short-Circuit Waveforms at T_J=125 °C.

In general, fan motor application drives under 15 kHz frequency. Assuming 60 W motor drive, following power losses are calculated as shown in Fig. 7. The operating temperature of SiC-FET is basically higher than that of RC-IGBT. It proves that the junction temperature of Si IGBT is limited by 200 °C, while SiC-FET can still operate in a high temperature of 600 °C [2]. Accordingly, SiC-FET shows short-circuit ruggedness is also much better performance than RC-IGBT as shown in Fig. 8, although die size and thermal impedance is smaller.

4 Thermal Performance

As is known to all, the efficiency of an IGBT device will never be 100% because it will unavoidably lose energy mainly by heat generation when operating. In addition, the heat energy will make temperature rise up gradually and the device will work in a higher temperature, thus some properties of the device will be influenced, especially the electrical characteristics. In addition, the on-state voltage drop of semiconductor increases with the increasing temperature. It shows that the on-state voltage drop of IGBT increases about 20% at 125 °C in general field-stop IGBT [3]. Thus, a positive feedback between the device voltage drop and its junction temperature will occur when such device is conducting significant current in reality, which causes further increases in both. As a result, increasing on-state voltage drop in high temperature environment will bring out a huge on-state power loss. On the other hand, SiC-FET has little change in characteristics with temperature and shows good characteristics even at high temperatures. To compare both devices, Fig. 9.

shows evaluation PCB and thermal pattern for SMD-IPM. To improve cooling down the SMD-IPM by PCB pattern, it is applied to optimized thermal vias and copper plate design [4]. And thermal comparison result of both devices is shown in Fig. 10. Even RC-IGBT IPM is triggered by temperature shut down (TSD) at 0.5 A.

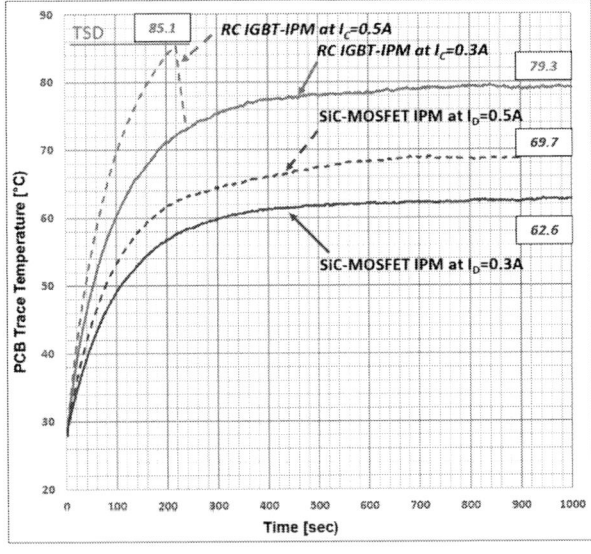

(a) Top side   (b) Bottom Side

Fig. 9 PCB photograph for the RC-IGBT IPM and SiC-IPM.

Fig. 10 Thermal Performance Comparison.
V_{DC}=300 V, V_{CC}=15 V, F_{SW}=15 kHz, MI=0.8, PF=0.8, I_D/I_C=0.3 A and 0.5 A, Ta=27 °C.

5 EMI Performance

Low Qrr and slow switching behavior makes a low noise from EMI point of view. From AC analysis, SiC-FET IPM shows low Qrr and slow dV_{DS}/dt, comparing to RC-IGBT switching behavior and it will result in lower EMI noises. To evaluate the EMI performance, 60 W motor demonstration board is used to drive inverter with sinusoidal waveforms

as show in Fig. 11. At 1000 rpm, Fig. 12(a) shows Common-Mode EMI noise level of RC-IGBT IPM and Fig. 12(b) shows that of SiC-FET IPM. Overall, the average level of SiC-FET IPM is lower about 1.0 dB and at more than 10 MHz region, it shows much lower than 3.0 dB.

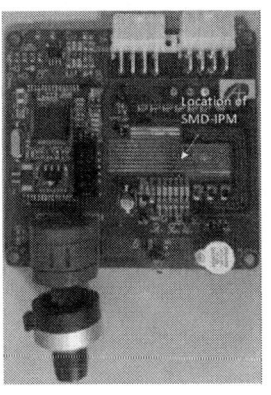 

(a) Top side   (b) Bottom Side

Fig. 11 60W Fan Motor Demonstration PCB Board using SMD-IPM.

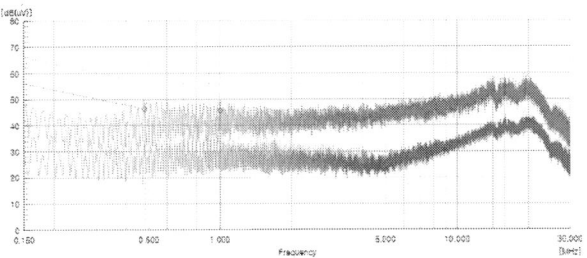

(a) Common-Mode EMI of 3A RC-IGBT IPM

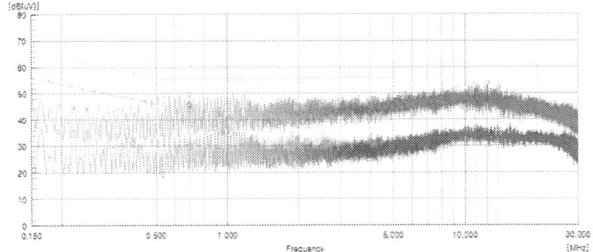

(a) Common-Mode EMI of SiC-FET IPM

Fig. 12 Comparison of CM EMI Test Results.

6 Conclusion

This paper deals with the DBC-based SMD type of full-functional SiC-IPM comparing with widely used RC-IGBT one for motor drives. In most applications with a supply voltage over 300 V, SiC-

FET with their inherent lower voltage drop will increase the efficiency, in particular when operating at high temperatures. The thermal impedance is optimized by DBC structure, PCB traces and thermal vias design. It becomes to reach into a higher current density. The test result without heatsink shows approximately junction temperature 86 °C under switching frequency at 15 kHz and 0.5 A. And 69% of power losses were reduced. Due to low Qrr and smooth switching behaviors characteristics, common-mode EMI noise level was also improved more than 1.3 dB margin. It is expected that the SMD-SiC-IPM line-up will be expanded to provide much higher efficiency and bigger power density in low-power motor drive applications.

7 Reference

[1] Jong-Mu Lee, Jintae Kim, Jun-ho Lee, Alex Niu, Bum-Seok Suh, "New SMD-IPM: Explore How to Make Low-Inverter Design More Compact and Cost-efficient," PCIM 2023.

[2] Jiawei Wang, "A Comparison between Si and SiC MOSFETs," IOP Conf. 2020.

[3] "IGBT vs. MOSFET – Determining the Most Efficient Power Switching Solution," White paper by BOURNS 2022.

[4] "A Guide to Board Layout for Best Thermal Resistance for Exposed Packages," AN-1520 by Texas Instrument.

PCIM Europe 2024, 11– 13 June 2024, Nuremberg   DOI: 10.30420/566262234

Concept for a GaN-Based Intelligent Motor Controller with Integrated Failure Prediction for the Inverter and the Drive

Martin Schellenberger[1] , Maximilian Hofmann[1] , Christoph Blechinger[1] , Michael Mensing[2] , Marc Gensch[2] , David Westerhoff[2], Patrick Barylla[3] , Holger Kapels[2] , Alexander Stanitzki[3], Lukas Krupp[3] , André Lüdecke[3]

[1] Fraunhofer Institute for Integrated Systems and Device Technology IISB, Germany
[2] Fraunhofer Institute for Silicon Technology ISIT, Germany
[3] Fraunhofer Institute for Microelectronic Circuits and Systems IMS, Germany

Corresponding author: Martin Schellenberger, martin.schellenberger@iisb.fraunhofer.de
Speaker: Christoph Blechinger, christoph.blechinger@iisb.fraunhofer.de

Abstract

Power electronics based on gallium nitride (GaN) or silicon carbide enable more efficient and compact drives for industry and mobility - with particularly demanding requirements for dynamics and reliability. In the joint research project "PowerCare", three Fraunhofer institutes combine their expertise to research a novel, miniaturized motor controller with integrated real-time failure prediction, yielding increased time-to-market for novel power electronics in mission critical applications. The custom controller combines integrated fault prediction for both the inverter electronics and the motor, utilizing a domain-specific RISC-V processor.

This paper gives an insight into the system concepts for modelling the degradation of power electronics and the motor in electrical drive systems. Additionally, opportunities and application scenarios of vertical GaN MOSFETs and intelligent motor controllers are highlighted, and initial development results are provided.

1   Motivation

From permanently operated production lines to electrified transportation – electric drives are the system's centerpiece and therefore critical to failure as well as primarily responsible for efficiency. Accounting for 30% of global electricity consumption, industrial electric motors have a significant share in greenhouse gas emissions (~4 gigatons CO_2e per year) [1] [2].

Gallium nitride (GaN) as a novel wide bandgap (WBG) semiconductor material offers enormous potential for more compact and efficient power modules for electric machines due to its higher power density compared to Silicon (Si) or Silicon-Carbide (SiC). GaN transistors in field-oriented controlled (FOC) motors can achieve

- a higher switching speed from 20 kHz to 60 kHz with ~5% increase in system efficiency (mechanical output vs. electrical input) [3],
- lower power losses (saving >20% over SiC and >50% over Si) leading to relaxed thermal design [4],

- less harmonics, lower torque ripple, and less acoustic noise [3].

If used on a widespread basis, savings of gigatons of greenhouse gases per year are possible [5]. Recent research demonstrates that with GaN, inverter efficiencies can be increased from 90-93% (silicon-based) up to 96% [6]. By using vertical GaN power semiconductors in industrial motor controls, PowerCare has the potential to save approx. 124 megatons of CO_2e per year (equivalent to Kenya's CO_2e emissions in 2022) [7].

However, for widespread industrial application of such novel power electronics, there are two major challenges - reliability and maintenance costs [8]. The most failure-critical components in converter systems are power semiconductors and capacitors, accounting for more than 50% of unplanned maintenance costs [9] [10] [11].

To get closer to the requirement for absolute reliability of electrical drives and the power semiconductors used to drive them, intelligent controllers require the capability to monitor and interpret the state of the system components in real time. In the

1717

strategic research project "PowerCare" [12], Fraunhofer will combine new, highly efficient GaN devices and a multi-channel pulse width modulator (PWM) controller with embedded AI to develop and demonstrate an inverter and motor controller with integrated failure prediction. The failure prediction will cover both the inverter electronics and the motor itself - where feasible, without additional sensors.

2 Concept Development and Initial Results

The overall solution concept (as depicted in Figure 1) is based on the Institutes' respective background and pre-developments. In the following sections, an insight will be given into the concept development and initial results in the three main target areas of the project:

1. Novel, vertical GaN trench MOSFETs and their behavioral models,
2. Embedded AI models for failure prediction of electric motors and GaN power semiconductors,
3. System demonstration of GaN MOSFETs and intelligent motor control.

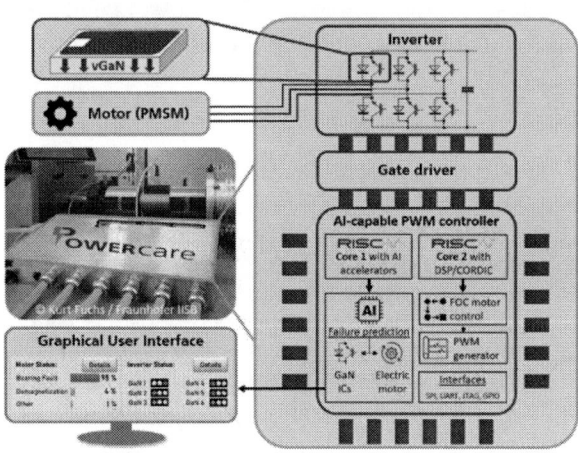

Fig. 1 Solution concept of the "PowerCare" project.

2.1 Vertical GaN Trench MOSFETs

Besides commercial GaN HEMTs, vertical GaN trench MOSFETs based on ceramic 8" GaN-on-QST substrates will be developed and utilized. While lateral GaN HEMTs are rapidly being adopted in many application areas, vertical GaN is still being scaled to commercial viability, currently focusing on high voltage applications, and its envisaged benefits are still to be verified. Here, a

voltage-scalable power device platform is being established, that offers significant advantages in area current density, cost and, in mid-term, inverter efficiency, compared to currently employed lateral GaN devices. Discrete components are subjected to reliability and stress tests, assembled into power modules, which are in turn modelled and characterized.

Before production of the components began, numerous process simulations were carried out to verify the process chain. Device simulations were also carried out to estimate the properties of the components. Among other things, blocking, transfer and output characteristics were investigated for this purpose. These were carried out for single and multiple core cell arrangements with edge terminations. An exemplary output characteristic is shown in Figure 2a. Figure 2b shows the current density distribution within a core cell while the transistor is switched on. The layout of the entire MOSFET is shown in Figure 2c as an example for a 4 mm^2 transistor. Based on the simulations, the exact dimensions for the edge termination, the gate dielectric thicknesses and the trench spacing could be determined for the process flow.

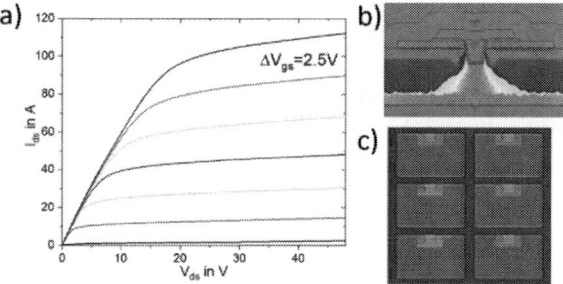

Fig. 2 a) TCAD simulations of an output characteristic of the transistor core cell. Current values are normalized to the active gate length to estimate the absolute current carrying capacity of the component. Alternative conservative estimates based on $R_{DS(on)}$ and heat distributions result in a current carrying capacity of about 50 A. b) Core cell for the MOSFET in the on-state. c) Exemplary layout for a size of 4 mm^2.

2.2 Embedded AI Models for Failure Prediction

Two failure models – for inverters and for connected electric motors – are developed and ported to a RISC-V based power module for in situ execution.

The basis for the failure models for electric motors is a hybrid motor model consisting of an application-specific pre-processing of the sensor data (e.g., FFT, filtering), a compact model component and a ML-based model, based on the "Cognitive

Power Electronics" (CPE) approach developed by Fraunhofer IISB [13]. The model will consider changes in load current and, optionally, other sensor data (vibration, acoustics, instantaneous rotation rate) that can be observed due to impending failures. Starting with the detection of bearing damage, the development will cover further faults, such as demagnetization or winding faults.

The failure models of the transistors and inverters are developed based on data from life tests and parameter measurements and enable the training of the failure prediction. The failure prediction of the GaN-based semiconductors and inverters will be trained using SPICE-based analogue circuit simulations to keep the generation of larger training sets practical and to include system-level effects. The calibrated models are then used to predict usage and failure modes in the inverter, which are in turn empirically validated and readjusted.

At this stage, various model approaches for the semiconductor and motor model were tested. The evaluation of the models include

- the capability for anomaly detection and failure classification, as well as
- the porting to a memory-constrained MCU (RISC-V and ARM-based).

For development and porting of the artificial neural network (ANN) models, Fraunhofer IMS' opensource edge AI framework "AIfES" [14] is used that provides inference and on-device training for resource constrained devices. Herewith, even training from scratch is possible, avoiding transfer of training data to a more powerful and energy consuming device. Hence, energy is saved without sharing the raw training data, and privacy increases. To ensure that the system cannot run out of memory during inference or training of an ANN, AIfES provides scheduler functions to calculate the required memory size beforehand based on the ANN structure. This is particularly important in safety-critical applications such as monitoring electric drives in public transportation. As a benchmark to commercially available solutions, AIfES outperforms, e.g., TensorFlow Lite for Microcontrollers in both execution time and memory consumption for fully connected neural networks (FCNNs). When using CNNs, AIfES reduces memory consumption by up to 54% [14].

2.2.1 Failure Models for Electric Motors

One focus of the research project is on the reliable detection of failures in electric motors. The approach chosen here is to forego additional sensors and to utilize the current sensors already integrated into the inverter for recording the current data. According to the CPE-concept, the current signal is then used to perform condition monitoring [13]. Furthermore, the AI algorithms must allow for implementation on corresponding hardware with respect to their memory requirements and the cache requirements of the data to be evaluated.

The basis for motor failure detection is shown in Figure 3: After the raw current signal has been measured and preprocessed, a transformation from the time domain to the frequency domain is performed by applying spectral analysis. To uncover nonlinear correlations and gain a better understanding of the data, the dimensionality is reduced in the subsequent step. Finally, the motor's condition is predicted using an appropriate algorithm for anomaly detection. [15] [16] [17] Should the described procedure not work on the developed hardware, alternative approaches for classification via deep learning may be considered.

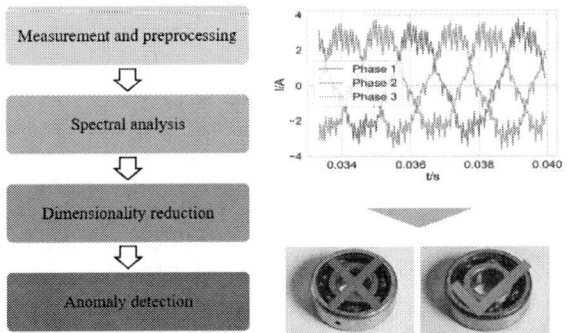

Fig. 3 Pipeline approach for the preparation and transformation of the data [15] [16] [17].

Given that the previously described approach pertains to anomaly detection and thus represents a semi-supervised method, the model is trained exclusively with data measured from healthy motors. Consequently, the model detects deviations from the normal, healthy state. The evaluation of the trained models is then based on both healthy and damaged data. To assess performance, a confusion matrix and corresponding metrics will be employed to ensure the lowest possible false-positive rate.

Ultimately, the trained algorithms are ported to C for integration on the RISC-V-based motor controller. At the current stage of development, the existing model for current-based anomaly detection for bearing faults could be optimized to run on edge devices. Initially, the ML model used

- 7.9 MB of RAM (variable data at runtime), and
- 79.7 MB of flash memory (unchangeable data).

After optimization towards CNNs and conversion to C using AIfES, the first executable model was ported to an MCU (STM32) using

- 320 KB of RAM, and
- 388 KB of flash memory.

From a future-oriented perspective, the maintenance of the algorithms can be realized through regular retraining of the models, e.g., via a cloud connection.

2.2.2 Failure Models for Transistors and Inverter

To develop the modelling methodology and generate initial training data, commercially available HEMTs are used, which could be potentially exchanged by the later developed vertical GaN devices. In parallel, a HEMT-based motor drive is built up evaluate real-world sensor data such that synthetic training data sets can be calibrated regarding their signal-to-noise, sampling rate and quantization properties.

To determine how the transistor degradation can be monitored, a motor control model was implemented with the simulation software QSPICE. The model includes the inverter, an equivalent circuit model of a permanent-magnet synchronous motor (PMSM) as well as the field orientated control unit (FOC). Figure 4 shows the schematic of the simulation model.

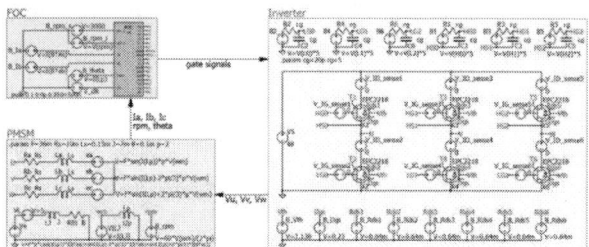

Fig. 4 QSPICE model schematic.

The sub-circuit model for the GaN-transistors of the inverter includes a thermal network for modelling the junction temperature, so that self-heating dependent behavior is included in the simulation. Additionally, the transistor model allows for the emulation of degraded states via external parameters. These currently include $R_{DS(on)}$ and leakage current degradations.

The motor model, based on the equivalent circuit proposed in [18], includes an electrical and a mechanical model. The electrical model consists of a star circuit of the motor phases, including the phase resistance, phase impedance and back EMF. The mechanical model makes use of math-

ematical equivalences to replace mechanical variables such as inertia or friction with electrical variables such as inductance or resistance.

The FOC is realized with C-code. It uses the Park and Clark transformation to transform the phase currents into the direct and quadrature current. These currents, as well as the rotational speed of the rotor are then regulated with PI controllers, under consideration of current and voltage limits. The output signals of the FOC are the PWM gate signals controlling the inverter.

Figure 5 shows the rotational speed of the rotor, the phase currents, and the junction temperatures of the transistors of the inverter, as simulated with the SPICE model. Modelling the whole system in an electrical context allows for simulating the transistors under realistic conditions typical for motor control systems. Thereby, the effects of transistor degradation on known variables, such as phase currents in relation to voltages set by the FOC, can be evaluated.

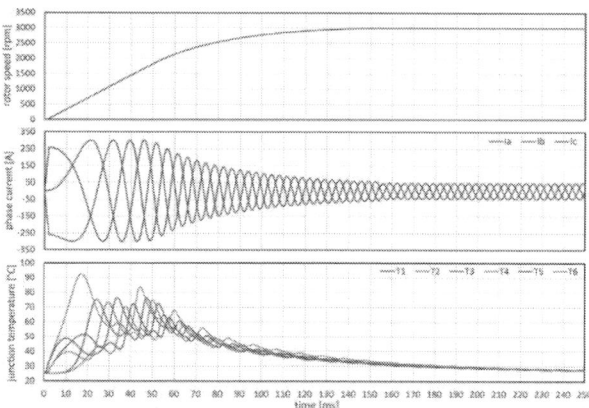

Fig. 5 Rotor speed, phase currents and junction temperature during motor start.

As reliability reports have shown, the $R_{DS(on)}$ increase is the most significant electrical characteristic, indicating a progressed degradation state of GaN power transistors. To observe the $R_{DS(on)}$ values during normal operation, the total phase impedance R_p for any given time can be obtained via equation 1, using the following known parameters: The quadrature current I_q calculated from the phase currents via Park and Clark transformation, the quadrature voltage V_q set by the FOC, the number of pole pairs p, the magnetic flux ϕ, and the rotational speed ω_m of the PMSM.

$$R_p = \frac{V_q - p \cdot \phi \cdot w_m}{I_q} \tag{1}$$

With equation 1 the back EMF is subtracted from the quadrature voltage, so that the effective voltage responsible for the quadrature current is divided by the quadrature current to obtain the quadrature resistance, which equates to the phase resistance.

As increase of the degradation dependent $R_{DS(on)}$ can be expected to be exponential and not simultaneous for the transistors of the inverter, the phase resistance shows a higher variability when the $R_{DS(on)}$ of a transistor has increased significantly due to critical degradation. The phase resistance can be impacted by other effects, such as temperature. However, for the steady state these effects can be viewed as locally static, so that the local deviation ΔR_P from the average phase resistance provides feedback about the degradation state of the transistors. Figure 6 shows the simulated phase resistance deviation ΔR_P over time with pristine transistors in blue. For red, one resistor is critically degraded, leading to a higher variability of ΔR_P.

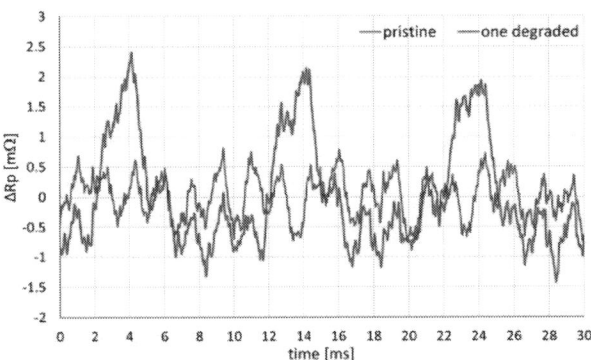

Fig. 6 Relative phase impedance for pristine and degraded HEMTs within the inverter.

The simulation results suggest that monitoring ΔR_P with AI pattern recognition can provide information on degradation states of the transistors. Benefit of the AI-based approach are an increased resilience towards noisy sensor data and shifting impedances due to e.g. thermal effects or load changes.

2.3 System Demonstration of GaN MOSFETs and Intelligent Motor Control

2.3.1 AI Capable PWM Controller

A GaN-based intelligent power module consisting of a power section and an AI section will be built, into which the trained failure models are integrated and executed locally, with the current and sensor data from the inverter and motor serving as input parameters.

For this purpose, an existing RISC-V-based PWM controller [19] is extended by additional hardware accelerators so that the resulting controller is optimized for real-time requirements. To effectively execute both the motor control and the AI models, a dual-core system implementation is favored: One RISC-V core (AI core) exclusively calculates the failure models for the electric motor and the inverter electronics and is extended by dedicated AI hardware accelerators so that an effective and high-performance calculation of even complex neural networks can be performed locally on the embedded system. Among other things, a part of the RISC-V SIMD extension ("P-Extension"), a reconfigurable logic part and additional activation functions [20] will be integrated. In addition, the AI framework "AIfES" [14] is ported to the PWM controller for inference and training on the device for the execution of hybrid models (e.g., physics guided neural networks).

The second RISC-V core (FOC core) is reserved exclusively for motor control. To increase performance and effectiveness, this core is also extended by dedicated hardware accelerators for the effective calculation of filtering and trigonometric functions, for example. Depending on the system requirements, the FOC core (or both cores) can also be implemented as a safety core with corresponding extensions for functional safety (the AI-RISC core used [21] is also available as an "ASIL-D ready" version).

In the first version, the PWM controller is implemented as a soft core on an FPGA, which can then be used as the basis for a later ASIC realization. This approach has the advantage that modified or new requirements due to changes in the AI models can still be flexibly implemented on the hardware side.

2.3.2 Demonstrator of vGaN Based Inverter

Based on the vertical GaN trench MOSFETs developed within the project, an inverter system for running 3-phase electric machines will be developed and realized as hardware prototype in the project. Based on the specific chip data, a nominal inverter voltage of 48 V and a maximum power of 20 kW for the highest output power have been defined. A 2-level B6 topology was chosen as internal inverter topology with three half bridge power modules.

A DCB (Direct Copper Based) based semiconductor packaging approach with ceramic insulated metal substrates is used to achieve the best performance also in terms of overall commutation cell inductance. The vertical GaN semiconductors will be soldered to the substrate and bond wires are used for the topside die attach. With this design, highest switching speeds (dv/dt) are realizable and therefore highest inverter efficiencies. Also, the thermal resistance from the chip to the coolant water/glycol can be reduced to achieve highest current densities in the semiconductors. The inverter will be characterized at a motor testbench together with a synchronous electric motor.

One of the primary goals of the testing procedure is to compare the operation behavior, for example the system efficiency, to state-of-the-art GaN HEMT or Si-MOSFET based inverter system realizations. This motor and inverter test setup will also be available for the experimental testing of condition monitoring methods and algorithms.

3 Summary and Outlook

It is expected that the research on the three levels described above (GaN MOSFETs, embedded AI, system demonstration), results in a technological leap in the intelligence of power modules, which equally includes the connected electric drive.

The first obvious area of application is the monitoring of the system's operating status, which is used here for anomaly detection in the event of impending component failure. A lean AI takes on the task of interweaving the complex phenomena that have so far been realized mainly by observing the thresholds of sensors or operating parameters. In a further development of the approach, it could also be possible to increase the efficiency of the system by dynamically adapting the control of the power module based on the load data recorded in real time.

Regarding potential application scenarios, permanently operated conveyor systems with high demands on system availability and reliability are considered as a first step. Other worthwhile use cases for intelligent drive systems with monitored safety are drones and electric aircrafts, cobots in production and logistics environments, as well as medical robots. In these applications, the new compound semiconductors enable drives with great precision and at the same time compact and low-maintenance designs due to their high maximum switching frequency.

References

[1] Maslin, M. (2021). Bill Gates, How to avoid a climate disaster: Solutions we have and the breakthroughs we need. Society (New Brunswick, N.J.), 58(2), 156–158. https://doi.org/10.1007/s12115-021-00581-z.

[2] Industrial drives: overview and main trends. (2023, November 16). https://community.infineon.com/t5/Training/Industrial-drives-overview-and-main-trends/ba-p/485216.

[3] M. Wattenberg, E. A. Jones and J. Sanchez, "A Low-Profile GaN-Based Integrated Motor Drive for 48V FOC Applications," PCIM Europe digital days 2021; International Exhibition and Conference for Power Electronics, Intelligent Motion, Renewable Energy and Energy Management, Online, 2021, pp. 1-8.

[4] Hesener, A. (2024). Reducing System Cost with GaN HEMTs in Motor Drive Applications. Applied Power Electronics Conference 2024, Industry Session 5.2.

[5] Sustainability Report 2021 | Navitas. (n.d.). https://navitassemi.com/sustainability-report-2021/.

[6] EU research: Energy-saving chips made of gallium nitride boost efficiency and CO2 savings. (2022, November 23). Fraunhofer Institute for Applied Solid State Physics IAF. https://www.iaf.fraunhofer.de/en/media-library/press-releases/ultimategan-end-of-project.html.

[7] EDGAR (Emissions Database for Global Atmospheric Research) Community GHG Database, a collaboration between the European Commission, Joint Research Centre (JRC), the International Energy Agency (IEA), and comprising IEA-EDGAR CO2, EDGAR CH4, EDGAR N2O, EDGAR F-GASES version 8.0, (2023) European Commission, JRC (Datasets).

[8] J. Endrenyi and G. J. Anders, "Aging maintenance and reliability: Approaches to preserving equipment health and extending equipment life", *IEEE Power Energy Mag.*, vol. 4, no. 3, pp. 59-67, May 2006.

[9] S. Yang, A. Bryant, P. Mawby, D. Xiang, L. Ran and P. Tavner, "An industry-based survey of reliability in power electronic converters", IEEE Trans. Ind. Appl., vol. 47, no. 3, pp. 1441-1451, May 2011.

[10] H. Wang, F. Blaabjerg, K. Ma and R. Wu, "Design for reliability in power electronics in renewable energy systems: Status and future", *Proc. 4th Int. Conf. Power Engineering Energy*

and Electrical Drives, pp. 1846-1851, May 2013.

[11] J. Falck, C. Felgemacher, A. Rojko, M. Liserre and P. Zacharias, "Reliability of power electronic systems: An industry perspective", *IEEE Ind. Electron. Mag.*, vol. 12, no. 2, pp. 24-35, June 2018.

[12] This work is supported by the Fraunhofer Internal Programs under Grant No. PREPARE 40-06175.

[13] M. Schellenberger, V. R. H. Lorentz, B. Eckardt, " Cognitive Power Electronics – An Enabler for Smart Systems," in PCIM Europe 2022: International Exhibition and Conference for Power Electronics, Intelligent Motion, Renewable Energy and Energy Management, Nuremberg, Germany, 2022: VDE Verlag GmbH, pp. 24-28, doi: 10.30420/565822006.

[14] L. Wulfert, J. Kühnel, L. Krupp, J. Viga, C. Wiede, P. Gembaczka, A. Grabmaier, "AIfES: A Next-Generation Edge AI Framework," in IEEE Transactions on Pattern Analysis and Machine Intelligence, doi: 10.1109/TPAMI.2024.3355495.

[15] G. Roeder, X. Liu, M. Hofmann, M. Schellenberger, F. Hilpert and M. März, "Cognitive Power Electronics for Intelligent Drive Technology," 2020 10th International Electric Drives Production Conference (EDPC), Ludwigsburg, Germany, 2020, pp. 1-8, doi: 10.1109/EDPC51184.2020.9388182.

[16] T. Huf, G. Roeder, M. Schellenberger, V. R. H. Lorentz and H. -F. Steinmetz, "Cognitive Power Electronics for Smart Drives in Unmanned Aerial Vehicles," PCIM Europe 2022; International Exhibition and Conference for Power Electronics, Intelligent Motion, Renewable Energy and Energy Management, Nuremberg, Germany, 2022, pp. 1-8, doi: 10.30420/565822007.

[17] C. Blechinger, D. Walch, M. Schellenberger, M. Hofmann and H. -F. Steinmetz, "Cognitive Power Electronics for Detection of Demagnetization in Electric Drives," 2023 13th International Electric Drives Production Conference (EDPC), Regensburg, Germany, 2023, pp. 1-7, doi: 10.1109/EDPC60603.2023.10372150.

[18] E. K. Beser, "Electrical equivalent circuit for modelling permanent magnet synchronous motors" Journal of electrical engineering, 72(2021), 3, 176–183,

[19] M. Richter, A. Lüdecke, Y.-C. Lee, A. Stanitzki, A. Utz, G. Grau, H. Kappert, R. Kokozinski, "A RISC-V-based System on Chip for High-Speed Control in Safety-Critical 650 V GaN-Applications", SMACD / PRIME 2021, International Conference on SMACD and 16th Conference on PRIME, 2021, pp. 1-4.

[20] I. Hoyer, A. Utz, A. Lüdecke, M. Rohr, C. H. Antink, and K. Seidl, "Inference runtime of a neural network to detect atrial fibrillation on customized RISC-V-based hardware" in *Current Directions in Biomedical Engineering*, vol. 8, no. 2, pp. 703–706, 2022. [Online]. Available: https://doi.org/10.1515/cdbme-2022-1179.

[21] AIRISC System-On-Chip Design, https://www.airisc.de, [Online; accessed 09-April-2024]

PCIM Europe 2024, 11– 13 June 2024, Nuremberg   DOI: 10.30420/566262235

Introducing the New 1200 V CIPOS Maxi IM817 Intelligent Power Module for Motor Drive Applications

Lee Kihyun[1], Kim Jinhyeok[1], Han Soohyuk[1], Song Bokkeun[1], Kang Kyoungpil[1] and Lee Minsub[1]
[1] Infineon Technologies Korea, South Korea

Corresponding author: Lee Kihyun, kihyun.lee@infineon.com
Speaker: Lee Kihyun, kihyun.lee@infineon.com

Abstract

This paper presents Infineon's new generation CIPOS™ 1200 V IM12BxxxC1 (old nomenclature: IM817) intelligent power modules (IPM) ranging up to 20 A in a dual in-line package with transfer molding. These IPMs can be for 3-phase AC motors and permanent magnet motors in variable speed motor drives applications. They combine the features of TRENCHSTOP™ IGBT7 S7 and emitter-controlled EC7 diodes with optimized silicon-on-insulator gate drivers and a direct copper bonding substrate. They offers the smallest package size with high power density. This paper provides an overview of the module as well as the technology behind the semiconductors, the electrical characteristics, the package, and the thermal performance.

1   Introduction

In recent years, the need for efficient power conversion has increased dramatically in response to global demands for decarbonization and improved natural resource sustainability.

Most inverter systems on the global industrial motor market are dominated by low-voltage motor drives and servo drives. Additionally, modern rectifier systems adopt a 3-phase pulse width modulation (PWM) rectifier method, instead of the tradiitional diode bridge rectifier method, to overcome the harmonic problem of variable speed drives. In these applications, the need for higher power density, lower cost, and higher efficiency along with improved reliability creates a greater need for power switches with lower power dissipation, improved robustness, and optimized packaging.

Keeping these requirements in mind, Infineon has launched the new CIPOS™ 1200 V IM12BxxxC1 (old nomenclature: IM817) development suite featuring high-speed variants of the innovative 7th generation IGBT and diode technology. It mainly combines the key advantages of 1200 V-rated TRENCHSTOP™ IGBT7 S7 and emitter-controlled EC7 Rapid diode technology with lower conduction and switching power losses, and smooth and stable rectification. The 1200 V TRENCHSTOP™ IGBT7 S7 technology has also been optimized for applications that require the short circuit safe operating area (SCSOA) functionality. Thus, improving the efficiency of inverter systems significantly.

In this regard, the trade-off between the SCSOA performance and power consumption of IGBTs is generally well known. SCSOA-rated IGBTs have higher collector-emitter saturation voltages and total switching energy losses than equivalent IGBTs without SCSOA ratings. Therefore, SCSOA functionality is very important if the target application requires it, but counterproductive if it is not required [1]. This product family also offers exceptional robustness under harsh conditions that require SCSOA, and a unique portfolio of current ratings ranging from 10 A to 20 A.

Fig. 1   External outline overview (size: 36 x 22.7 mm²).

This paper is organized as follows: Section 2 describes key features of the internal components and the package structure. Section 3 introduces the static characteristics, dynamic characteristics, and short-circuit characteristics. Thermal performance for low and high switching frequencies are presented in Section 4.

2 Feature of the new CIPOS™ Maxi IM12BxxxC1 IPM series

2.1 Electrical internal circuit and components

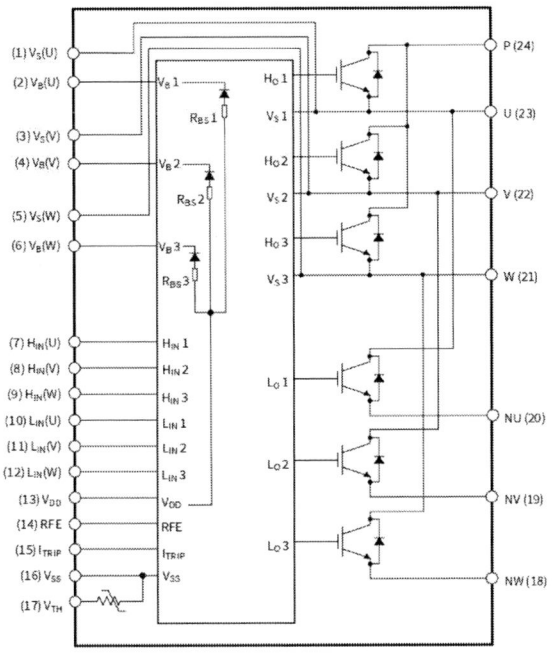

Fig. 2 Internal block diagram of the new Maxi

IM12BxxxC1 IPM product series

The new 1200 V Maxi IM12BxxxC1 IPM has a built-in gate driver IC with 3-phase inverter circuitry and control, as shown in Fig. 2.

The built-in gate drive IC is designed using the rugged silicon-on-insulator (SOI) technology with stability against transient and negative voltage. This gate drive IC has a built-in integrated bootstrap functionality. Temperature monitoring is done through a built-in, UL-certified negative temperature coefficient (NTC) thermistor and an open-emitter pin configuration is adopted for the low side emitter. Table 1 shows all the products in this new IPM product family and their current rating with target applications.

Product	Current	Target applications
IM12B10CC1	10 A	Motor drives for industrial applications
IM12B15CC1	15 A	
IM12B20EC1	20 A	

Table 1 Product line-up and target applications

2.2 Semiconductor

2.2.1 The 1200 V TRENCHSTOP™ IGBT7 S7 and EC7 diode technology

The TRENCHSTOP™ IGBT7 1200 V S7 and EC7 technology are, respectively, the 7th generation insulated gate bipolar transistor and emitter-controlled diodes, the cross-sections of which are shown in Figure 3.

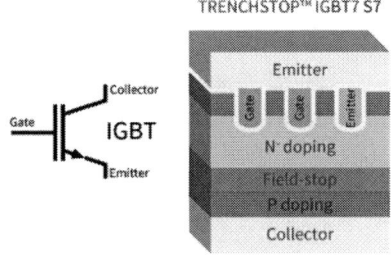

Fig. 3 Cross-section of the TRENCHSTOP™

IGBT 1200 V S7 and EC7 diode

The IGBT technology, based on micro-pattern trenches (MPT) with sub-micron mesas, has been specially designed to provide strongly reduced static losses ($V_{CE(sat)}$), a high level of controllability (dV/dt), and a short-circuit withstand time capability (t_{SC}). The EC7 diode field-stop region has been designed to provide enhanced softness without compromising the blocking voltage even at high V_{DC}, high temperature and very low current to reduce the oscillatory behavior to a minimum even under harsh conditions. Thanks to the combination of the aforementioned features, TRENCHSTOP™ IGBT7 S7 and EC7 diode portfolio is the best fit for electric power converters requiring e.g., moderate switching frequencies (≤ 20 kHz), relatively slow switching events (≤ 10 kV/μs), and short-circuit robustness [2][3].

2.2.2 1200 V SOI single-gate drive IC

The 1200 V SOI single-gate driver with integrated bootstrap circuit is used to achieve higher levels of integration, reliability, and performance in the CIPOS™ Maxi IPM. This SOI gate driver IC disables leakage or latch-up current between structurally adjacent devices. It prevents the latch-up effect even in case of high dV/dt switching and surge under elevated temperature [3]. This gate driver provides several protection functions such as cross-conduction prevention, undervoltage lockout, overcurrent detection, and enable input.

CIPOS™ Maxi IPM series provides an integrated fault output with sleep function and adjustable fault clear time through the RFE pin. There are two situations that can cause the driver IC to report a fault via the RFE pin. The first being an undervoltage condition of V_{DD} and the second when the I_{TRIP} function recognizes a fault. After the fault condition occurs, the RFE pin is internally pulled to V_{SS} and the fault-clear timer is activated. When the built-in fault-clear timer expires, the voltage on the RFE pin returns to its external pull-up voltage. The output remains disabled, and the fault condition is maintained until the voltage on the RFE pin charges up to enable threshold voltage. The charging characteristics are dictated by the RC time constant attached to the RFE pin.

2.3 Package

The CIPOS™ Maxi IPM package is designed with the smallest package size (36 mm x 22.7 mm x 3.1 mm) without dummy pins by optimizing internal PCB and Direct Bonding Copper (DBC) structures, as shown in Figure 4. The gate driver IC and thermistor are placed on the internal PCB.

The IGBTs and diodes are placed on a DBC for better thermal performance. It adopts Al wire bonding technology for electrical connections between the PCB and the DBC as well as the DBC and the lead frame.

The CIPOS™ Maxi IPM has an independent V_{TH} pin that is connected to the thermistor inside the package, which monitors the temperature. The case temperature is tracked by the NTC thermistor. As the temperature relationship between the NTC and the case can vary depending on the heat dissipation conditions defined by the users, the relationship between the two must be measured in a relevant experimental environment.

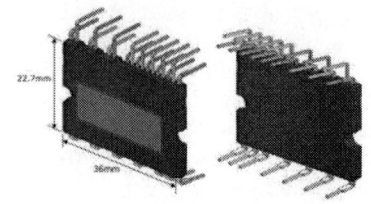

(a) External view of the Maxi IPM package

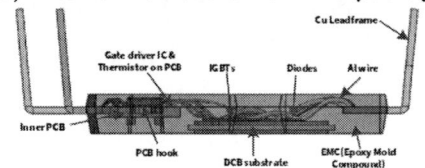

(b) Internal structure of the Maxi IPM package

Fig. 4 External view and Internal structure of CIPOS™ Maxi IPM

In addition, the IPM's internal body is encapsulated using transfer molding technology and meets all international industry standards, such as clearance for insulation distance and creepage distance. The corresponding insulation distance between two pins and between a pin and the DBC is shown in Figure 5 [4].

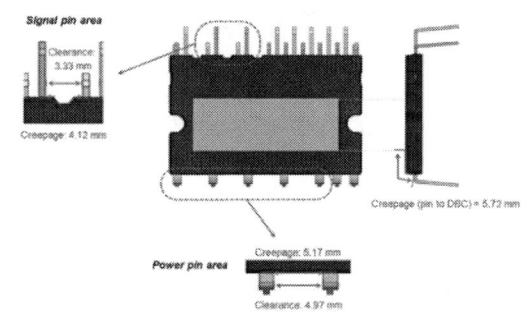

Fig. 5 Clearance and creepage distance of the CIPOS™ Maxi IPM package

3 Electrical characteristics

3.1 Static characteristics

Figures 6 and 7 show a comparison between the static DC characteristics of the 10 A and 20 A devices in the new CIPOS™ Maxi 1200 V IM12BxxxC1 IPM series with that of competitors, at room temperature and high temperature conditions. Specifically, the IM12B10CC1 at 10 A nominal current and 150°C junction temperature shows at least 300 mV lower $V_{CE(Sat)}$ (see Fig. 6 (a)) and at least 200 mV lower V_F (see Fig. 6 (b)).

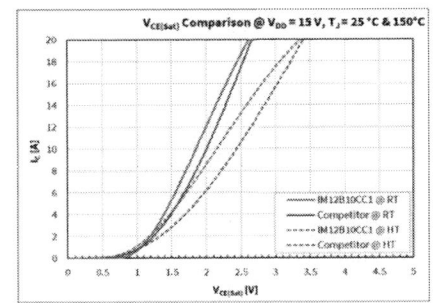

(a) IGBT collector-emitter voltage, $V_{CE(Sat)}$ [V]

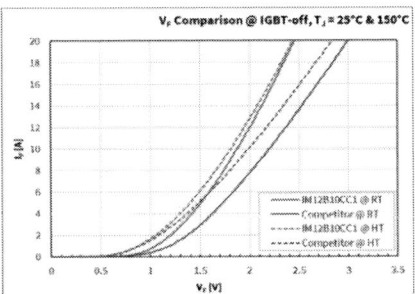

(b) Diode forward voltage, V_F [V]

Fig. 6 DC characteristics of the Maxi 1200 V 10 A IPM (IM12B10CC1)

As shown in Fig. 7 (a), the IM12B20EC1 at 20 A nominal current and 150°C junction temperature shows at least 140 mV lower $V_{CE(Sat)}$.

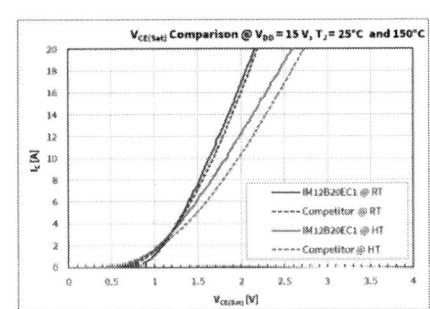

(a) IGBT collector-emitter voltage, $V_{CE(Sat)}$ [V]

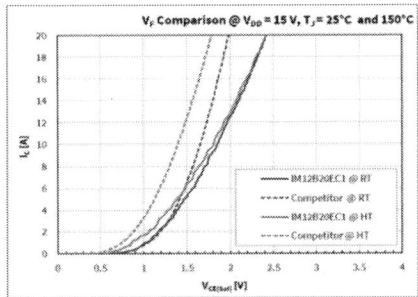

(b) Diode forward voltage, V_F [V]

Fig. 7 DC characteristics of the Maxi 1200 V 20 A IPM (IM12B20EC1)

3.2 Dynamic characteristics

Figures 8 and 9 show the switching-on and switching-off waveforms of 10 A and 20 A-rated devices, respectively, at $V_{DC\text{-link}}$ = 600 V, V_{DD} = 15 V, and T_J ≤ 150°C.

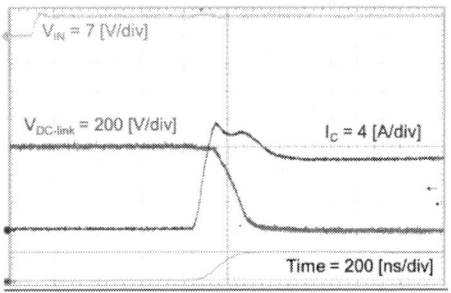

(a) Switching-on waveforms (T_J = 150°C)

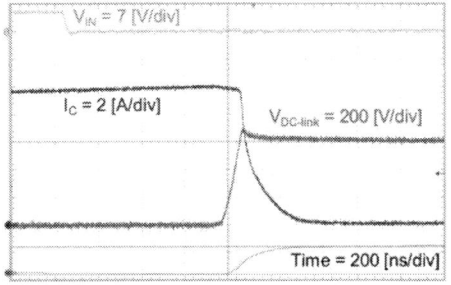

(b) Switching-off waveform (T_J = 150°C)

Fig. 8 DC characteristics of the Maxi 1200 V 10 A IPM (IM12B10EC1)

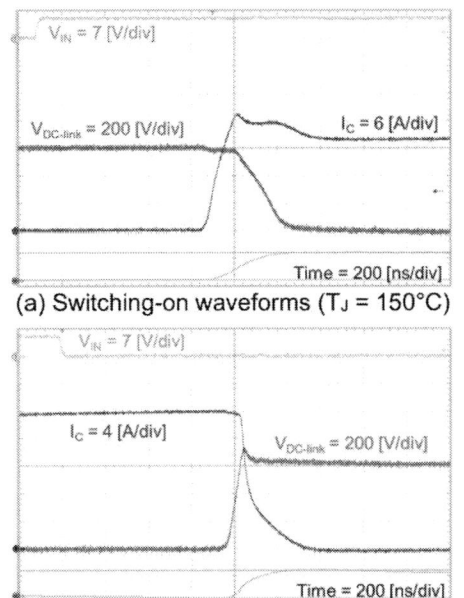

(a) Switching-on waveforms (T_J = 150°C)

(b) Switching-off waveform (T_J = 150°C)

Fig. 9 DC characteristics of the Maxi 1200 V 20 A IPM (IM12B20EC1)

3.3 Performance

Figure 10 shows the dynamic loss characteristics of the new Maxi 1200 V 10 A and 20 A products for $V_{DC\text{-}link}$ = 600 V, V_{DD} = 15 V, and inductive load conditions at T_J = 25°C and T_J = 150°C. These characteristics lead to highly improved switching characteristics.

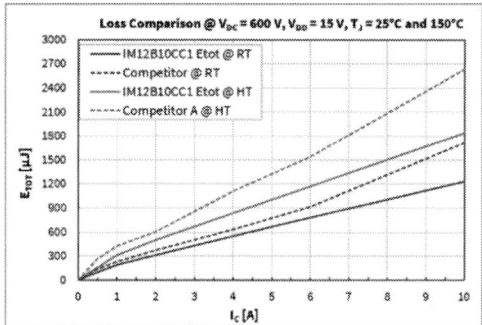

(a) Loss characteristics of the Maxi 1200 V 10 A IPM

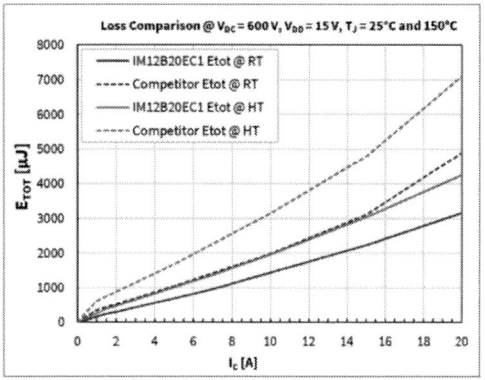

(b) Loss characteristics of Maxi 1200 V 20 A IPM

Fig. 10 Dynamic loss characteristics of CIPOS™ Maxi 10 A and 20 A products (IM12B10CC1 and IM12B20EC1)

The 1200 V 10 A product (IM120B10CC1) with nominal current of 10 A offers approximately 30% lower total loss than its competitors, especially at high temperature. The 1200 V 20 A product (IM12B20EC1) with a nominal current of 20 A offers 40% lower total loss at high temperature than its competitors. This makes the devices very suitable for the target applications.

3.4 SCSOA characteristics

The short-circuit withstand time of the CIPOS™ Maxi 1200 V IM12BxxxC1 product series is specified as 5 µs under the following conditions: $V_{DC\text{-}link}$ = 800 V, V_{DD} = 15 V, and $T_J \leq$ 150°C. Figure 11 shows the measured test waveform of IM12B10CC1 for $V_{DC\text{-}link}$ = 800 V, V_{DD} = 15 V, and T_J = 150°C.

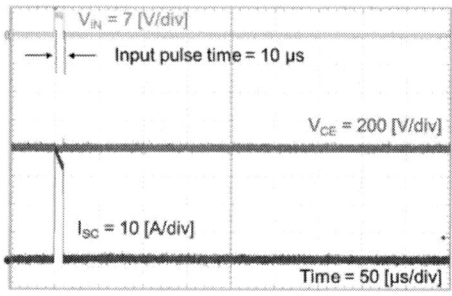

(a) Normal waveform

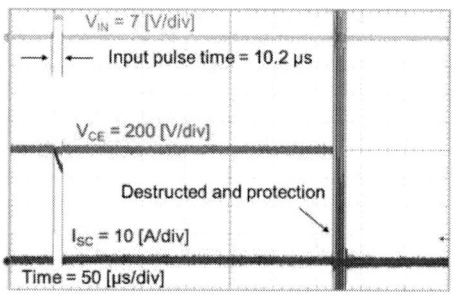

(b) Shutdown waveform

Fig. 11 Short-circuit test waveform of IM12B10CC1

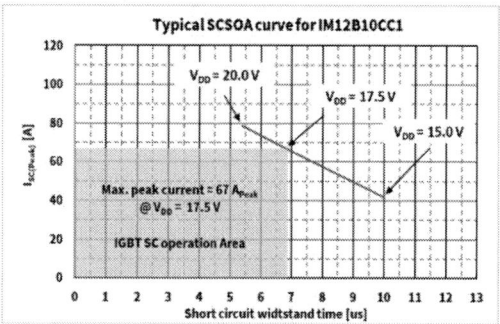

(a) SCSOA of the Maxi 1200 V 10 A IPM

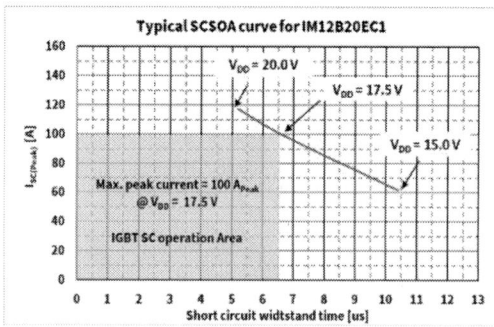

(b) SCSOA of the Maxi 1200 V 20 A IPM

Fig. 12 SCSOA characteristics of IM12B10CC1 and IM12B20EC1

Figure 12 show the typical (not guaranteed) SCSOA performance graphs of the 10 A and 20 A devices for $V_{DC\text{-link}}$ = 800 V, $V_{PN(surge)}$ < 900 V, and T_J = 150°C at different supply voltages.

4 Thermal performance

4.1 Test conditions

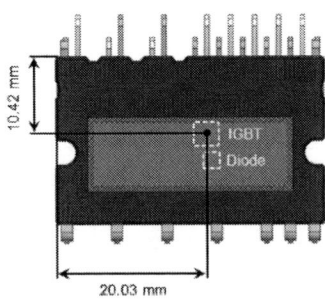

Fig. 13 Point for measuring the case temperature

to test thermal performance

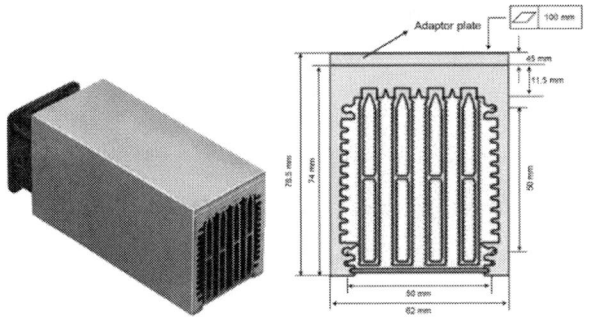

Fig. 14 Heatsink information

Measuring the case temperature can help confirm thermal performance. The points on the IGBT chip shown in Fig. 13 represent the hottest points on the module. The case temperature (T_C) is the low-side, U-phase IGBT temperature in the CIPOS™ Maxi 1200 V products. Figure 14 shows the heatsink used for testing the thermal performance at case-to-ambient temperature.

4.2 Thermal characteristics of the Maxi 1200 V 10 A IPM (IM12B10CC1)

Figure 15 shows the thermal performance at case-to-ambient temperature (ΔT_{CA}) for the new CIPOS™ Maxi 1200 V 10 A products at low switching and high switching frequencies. Their thermal performance was evaluated using a 3-phase inverter system with R-L load under the following

test conditions: $V_{DC\text{-link}}$ = 600 V, V_{DD} = 15 V, and switching frequency = 6 kHz and 15 kHz, output frequency = 60 Hz. The maximum ΔT_{CA} of the CIPOS™ Maxi 1200 V 10 A IPM (IM12B10CC1) was about 23°C at f_{SW} = 6 kHz and I_O = 10 A_{peak}. This was 2.5°C lower than that of the competitors. Its maximum ΔT_{CA} was approximately 40°C at f_{SW} = 15 kHz and I_O = 10 A_{peak}, which was 6.6°C lower than that of the competitor.

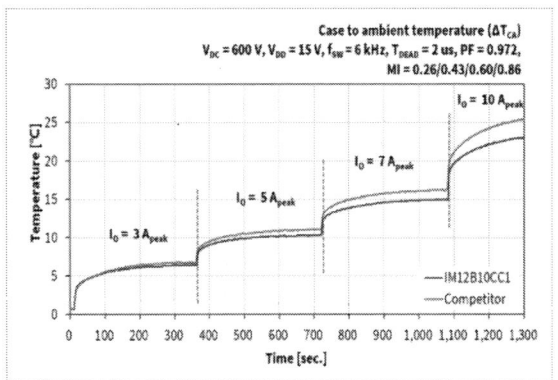

(a) f_{SW} = 6 kHz

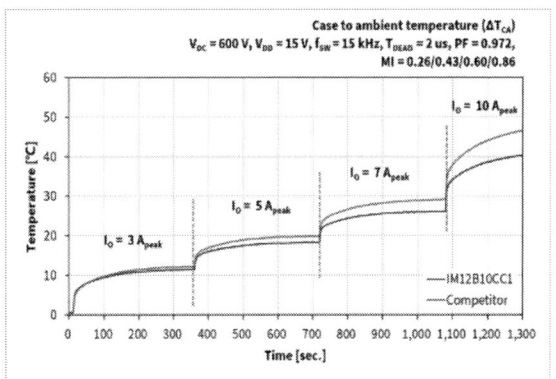

(b) f_{SW} = 15 kHz

Fig. 15 Results of case-to-ambient temperature measurement of IM12B10CC1 (1200 V, 10 A)

4.3 Thermal characteristics of the Maxi 1200 V 15 A IPM (IM12B15CC1)

Figure 15 also shows the thermal performance at case-to-ambient temperature (ΔT_{CA}) for the new CIPOS™ Maxi 1200 V 15 A products at low switching and high switching frequencies. The thermal performance was evaluated using a 3-phase inverter system with R-L load under the following test conditions: $V_{DC\text{-link}}$ = 600 V, V_{DD} = 15 V, switching frequency = 6 kHz and 15 kHz, and output frequency = 60 Hz. The maximum ΔT_{CA} of the

CIPOS™ Maxi 1200 V 15 A IPM (IM12B15CC1) was about 37°C at f_{SW} = 6 kHz, and I_O = 15 A_{peak}. Its maximum ΔT_{CA} was about 55°C at f_{SW} = 15 kHz and I_O = 15 A_{peak}.

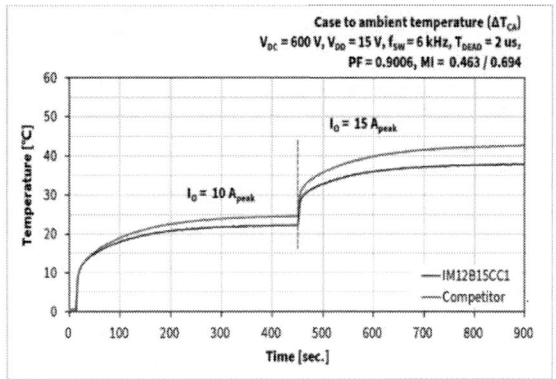

(a) f_{SW} = 6 kHz

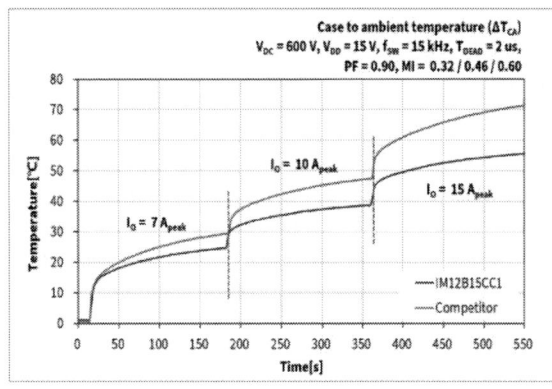

(b) f_{SW} = 15 kHz

Fig. 15 Results of case-to-ambient temperature measurement of IM12B15CC1 (1200 V, 15 A)

4.4 Thermal characteristics of the Maxi 1200 V 20 A IPM (IM12B20EC1)

Figure 16 show the thermal performance with case-to-ambient temperature (ΔT_{CA}) for the new CIPOS™ Maxi 1200 V 20 A IPM products at low switching and high switching frequencies. The thermal performance was evaluated using a 3-phase inverter system with R-L load under the following test conditions: $V_{DC\text{-link}}$ = 600 V, V_{DD} = 15 V, switching frequency = 6 kHz and 15 kHz, and output frequency = 60 Hz. The maximum ΔT_{CA} of the CIPOS™ Maxi 1200 V 20 A IPM (IM12B20EC1) was about 36°C at f_{SW} = 6 kHz, and I_O = 10 A_{peak}, which was 1°C lower than that of the competitor. Its maximum ΔT_{CA} was approximately 40°C at f_{SW} = 15 kHz and I_O = 15 A_{peak}, which was 18°C lower than that of the competitor.

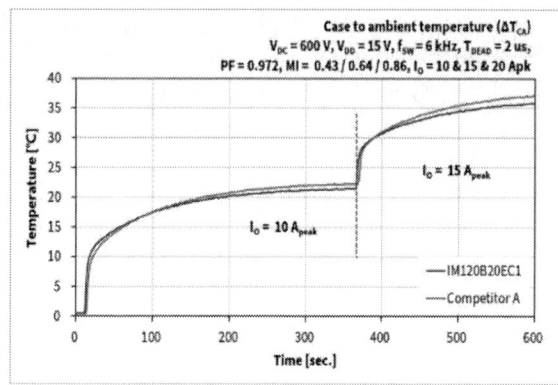

(a) f_{SW} = 6 kHz

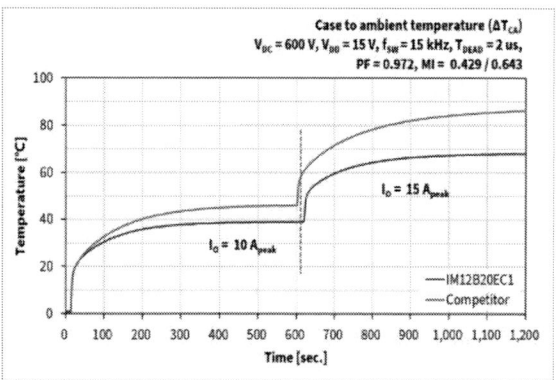

(a) f_{SW} = 15 kHz

Fig. 16 Results of case-to-ambient temperature measurement of IM12B20EC1 (1200 V, 20 A)

5 Conclusion

The new CIPOS™ Maxi 1200 V IPM product series (IM12BxxxC1) that uses Infineon's progressive 1200 V rating TRENCHSTOP™ IGBT7 S7 IGBTs and EC7 technology with C5SOI gate drive ICs was introduced in this paper. This product enables the development of highly energy-efficient appliances by offering higher efficiency in a small package that also adequately addresses the trend of system miniaturization.

References

[1] A. Piccioni, "Comprehensive performance evaluation of discrete 1200V IGBT7 S7 for drives application," PCIM Europe 2022, pp. 1951 – 1954.

[2] J. Cerezo, and A. K. Sekar, "1200 V TRENCHSTOPTM IGBT7 H7 and Emitter-Controlled EC7 Rapid Diode Technologies

Define an Enhanced Benchmark for Improved Energy-Efficient, Fast-Switching Inverter Applications," PCIM Europe 2023, pp. 2189 - 2195.

[3] Application Note, AN-2023-01, "The 1200 V TRENCHSTOPTM IGBT 7 S7 and EC7 technology, portfolio and applications," Infineon Technologies AG.

[4] R. Keggenhoff, Z. Liang, A. Arens, P. Kanschat, and R. Rudolf. "Novel SOI Driver for Low Power Drive Applications", Power Systems Design Europe, Nov. 2005.

[5] M. Lee, M. Baek, J. Lee, and D. Chung, "A New Smallest 1200V Intelligent Power Module for Three Phase Motor Drives", IPEC 2018, pp. 1141 – 1144.

PCIM Europe 2024, 11– 13 June 2024, Nuremberg   DOI: 10.30420/566262236

Thermal Performance of Infineon's New 600 V CIPOS Micro IM241 IPM for Low-Power Motor Drive Systems Without Heatsink

Jo David[1] ⓘ, Kim Jinhyeok[1] ⓘ, Song Hyunsoo[1] ⓘ, Song Bokkeun[1] ⓘ, Choo Byoungho[1] ⓘ, and Beaurenaut Laurent[2] ⓘ

[1] Infineon Technologies Korea, South Korea

[2] Infineon Technologies AG, Germany

Corresponding author: Jo David, David.Jo@infineon.com
Speaker: Jo David, David.Jo@infineon.com

Abstract

This paper presents the thermal performance of Infineon's new 600 V CIPOS™ Micro IM241 intelligent power modules (IPM) for low-power motor drive systems without heatsink. The CIPOS™ Micro IM241 IPM series combines the features of six Infineon TRENCHSTOP™ 2nd generation reverse conducting drive IGBTs (RC-D2 IGBT) and three half-bridge gate drive ICs. These days the demand for high-efficiency inverter motor driving systems for small home appliances and industrial motor applications has increased due to environmental concerns and to save energy. Additionally, heatsinks are not being used to reduce the size of the inverter motor drive system and manufacturing costs. Therefore, to implement a low-power motor driving system without a heatsink, devices with lower power loss are required. An optimal thermal analysis of power devices is required to avoid damaging the IPMs under low-power motor driving conditions without any heatsink. This paper provides an overview of the CIPOS™ Micro IM241 IPM series, defines the thermal measurement method for systems without a heatsink, and compares its thermal performance with competitor IPMs.

1   Introduction

In recent years, the demand for highly efficient inverter motor driving systems for low-power motor drive applications has increased in the global market to meet new efficiency regulations. These applications include fans, pumps, and low-power home appliances such as residential air conditioner (RAC) fans, hair dryers, refrigerators, dish washer and small washing machines [1]. In addition, there is also an increasing demand to reduce the size of low-power inverter motor driving systems to reduce the overall system cost. Removing the heatsink is considered a good way of making the low-power motor drive system highly efficient and compact, and for reducing manufacturing costs [2].

To meet the requirements of the low-power motor application market worldwide, Infineon has introduced a new 600 V CIPOS™ Micro IM241 IPM series. It is an optimized, highly efficient power semiconductor with high thermal performance and a compact package size. It features the latest TRENCHSTOP™ RC-D2 600 V IGBT technology with monolithically integrated diodes [3]. The CIPOS™ Micro IM241 IPM series comes not only in a through hole device (THD) package but also a surface mounted devices (SMD) package, so an

auto assembly process is possible. Due to the compact package size, it is easier to design the PCB. The overall module size is smaller than that of a discrete IGBT and the competitor IPM.

Figure 1 shows the new 600 V CIPOS™ Micro IM241 IPM series in DIP and SMD packages.

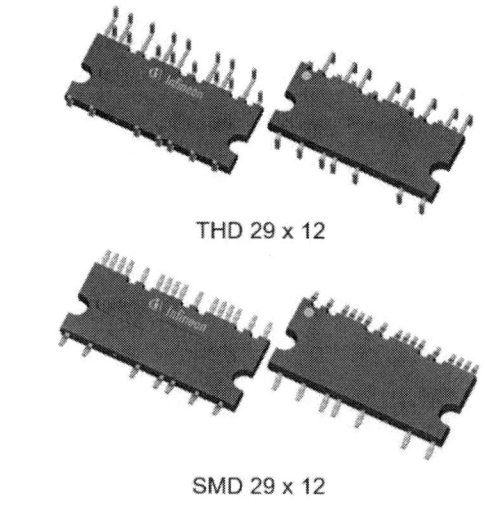

THD 29 x 12

SMD 29 x 12

Fig. 1   External outline overview (size: 36 x 22.7 mm²)

1732

To reduce the system cost and size, customers the world over tend to remove the heatsink from low-power motor driving applications. Thus, to prevent the IPM from being damaged by heat in low-power motor applications without a heatsink, especially 30 W to 160 W motor drives, optimal power loss and thermal analysis are required.

This paper provides an overview of the internal components and package structure of the CIPOS™ Micro IM241 IPM series. It also discusses the results of thermal simulation of the package without heatsink, and compares its package size, power loss, and thermal performance with that of competitor IPMs.

2 Feature of the new CIPOS™ Micro IM241 IPM series

2.1 Electrical internal circuit and components

The new CIPOS™ Micro IM241 IPM series provides a minimized package with low power consumption. These products are composed of six TRENCHSTOP™ 2nd generation reverse conducting drive IGBTs (RC-D2 IGBT), and three high-voltage, junction-isolated, half-bridge gate drive ICs with an integrated bootstrap field effect transistor (FET) and negative temperature coefficient (NTC) thermistor for temperature monitoring, as shown in Fig. 2 and Fig. 3.

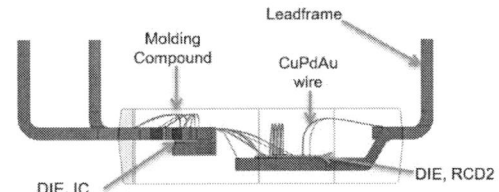

Fig. 3   Internal structure of the Micro IPM package

Table 1 lists the products in the CIPOS™ Micro IM241 IPM series and their current ratings.

Product	Package type	Target applications
IM241-x6T2y	THD 29x12	Low-power motor
IM241-x6S1y	SMD 29x12	drives (fan and
		pump)

x = S (2A) or M (4A) or L (6A)
y = B (fast speed for low losses; for x = S, M, L) or J (slow speed for low EMI; for x = S, M)

Table 1   Product line-up and target applications

2.2 Package

The CIPOS™ Micro IPM is designed with the smallest package size (29 mm x 12 mm x 3.15 mm). The Micro IPM package is roughly 54% smaller than the existing CIPOS™ Mini IPM used in low-power motor applications, as shown in Fig. 4. Compared to the size of the competitor IPMs, the CIPOS™ Micro IPM is the smallest, as listed in Table 2.

Generally, the PCB of low-power motor applications is small. In the limited space, it is easier to design using the CIPOS™ Micro IPM than with the competitor IPMs.

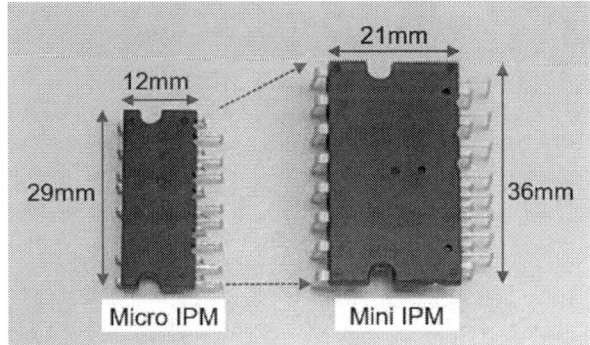

Fig. 4   Package size comparison with the existing CIPOS™ Mini IPM

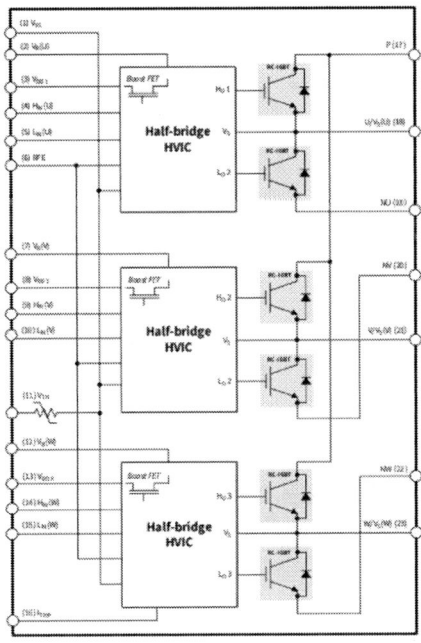

Fig. 2   Internal block diagram of the Micro IPM

Product	Package size
Micro IPM	29 mm x 12 mm x 3.15 mm
Mini IPM	36 mm x 21 mm x 3.1 mm
Competitor A	32.8 mm x 18.8 mm x 3.6 mm
Competitor B	36 mm x 14.8 mm x 4.0 mm

Table 2 Package size comparison with competitor IPMs in the market

3 Package thermal simulation

To define the thermal measurement method for systems without a heatsink, the correlation between the RC-D2 IGBT junction temperature (T_J) and the CIPOS™ Micro IPM module's case temperature (T_C) was simulated using the Ansys Icepak simulation tool, as shown in Fig. 5. The thermal simulation conditions for the package were: power loss of single RC-D2 IGBT = 0.7 W, T_A = 25°C, PCB copper thickness = 1 oz., without heatsink. The simulation results showed that the T_C of the CIPOS™ Micro IPM was similar to the T_J of the hottest RC-D2 IGBT, which was about 93°C. Figure 6 and Table 3 show the results of the correlation simulation between the T_J of the RC-D2 IGBT and the T_C of the CIPOS™ Micro IPM.

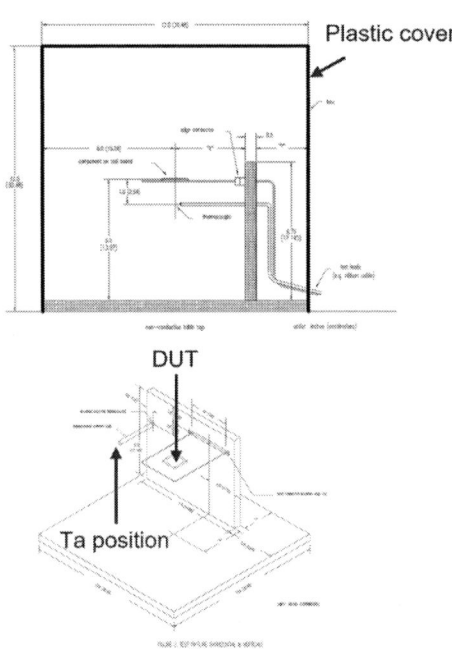

(a) Dimensions of the still air chamber

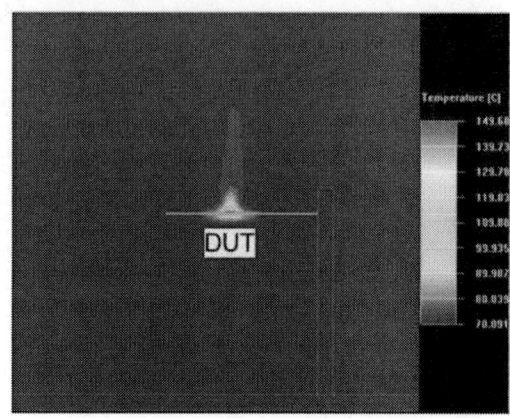

- Temperature distribution

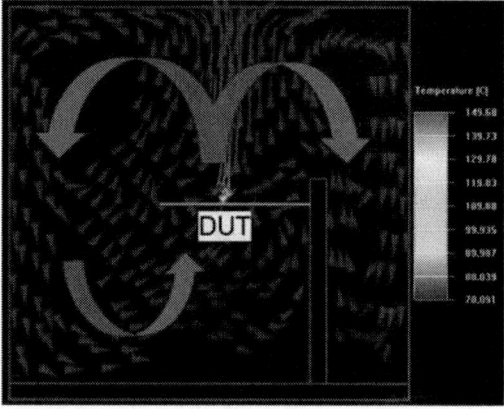

- Air velocity distribution

(b) Simulation model

Fig. 5 Simulation setup for package thermal analysis using JEDEC51-02 guidelines

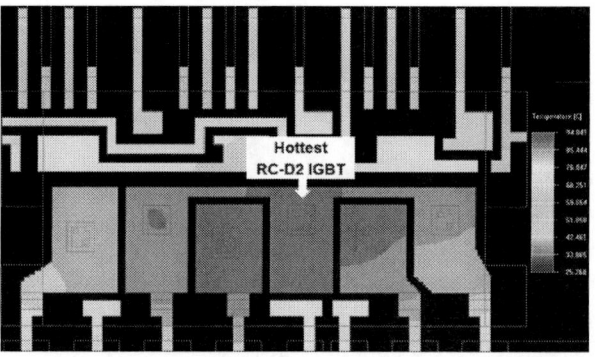

(a) Layout view of the Micro IPM

PCIM Europe 2024, 11– 13 June 2024, Nuremberg   DOI: 10.30420/566262236

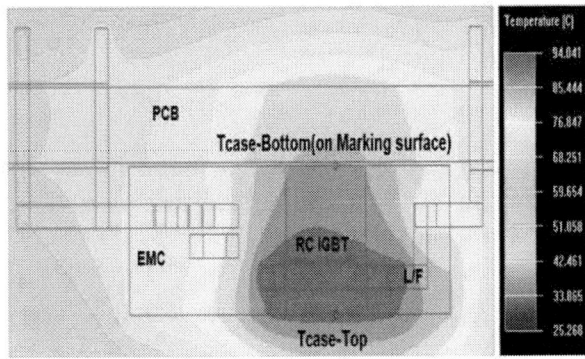

(b) Cross-section view of the Micro IPM

Fig. 6   Package thermal simulation results

Single chip power loss [W]	T_J [°C]	T_{case_top} [°C]	$T_{ambient}$ [°C]	R_{thja} [°C/W]
0.67	93.3	92.2	26.5	117.6

Table 3   Results of package thermal simulation

4   Power loss simulation

Figure 7 shows the comparison between the RC-D2 IGBT losses of 600 V, 2 A and 4 A Micro IPM and a competing 600 V, 5 A IPM used in 40 W and 120 W low-power motors without a heatsink. The simulation conditions were: V_{DC} = 300 V, F_{SW} = 16 kHz, SVPWM, T_A = 24°C, MI = 0.4 and 0.7, PF = 0.9, I_O = 0.31~0.52 Arms, and P_O = 40 W and 120 W, without heatsink. Under low-power motor drive conditions, the phase current was as small as 0.31 – 0.52 Arms, so the RC-D2 IGBT conduction loss of the Micro IPM 600 V, 2 A and 4 A products was worse than that of the competitor's 600 V, 5 A IPM. However, this does not account for a significant portion of the single chip's total loss. In addition, since the switching frequency condition was high at 16 kHz under the low-power motor drive conditions, the proportion of switching loss in the single chip's total loss was large, and the RC-D2 IGBT switching loss of the Micro IPM 600 V, 2 A and 4 A products was much lower than that of the competitor's 600 V, 5 A IPM. Thus, the overall RC-D2 IGBT loss of the Micro IPM 600 V, 2 A and 4 A products was superior to that of competitor's IPM.

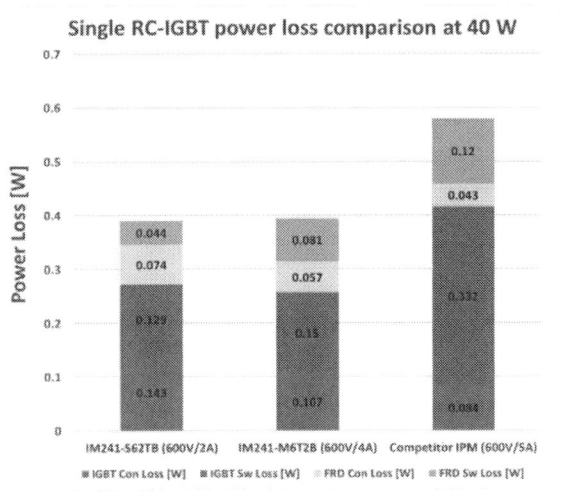

(a) Power loss simulation at 40 W

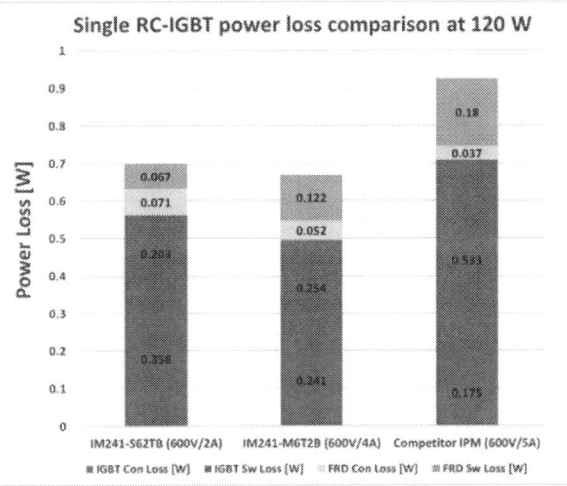

(b) Power loss simulation at 120 W

Fig. 7   Results of power loss simulation

5   Thermal performance

5.1   Test setup

The thermal analysis simulation of the CIPOS™ Micro IPM package without a heatsink showed that T_J of the power chip and T_C of the IPM were similar. The T_C of CIPOS™ Micro IM241 product and a competitor's IPM was measured using an infrared (IR) camera as shown in Fig. 8 and Fig. 9.

1735

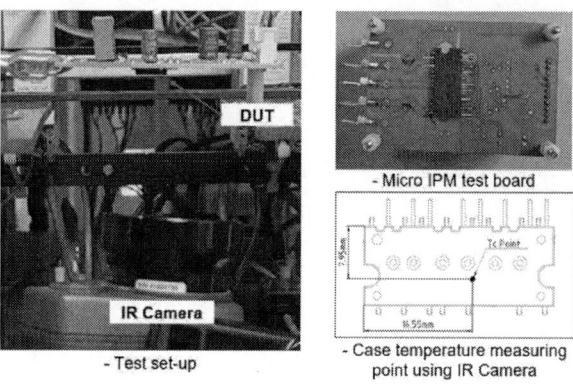

- Test set-up

- Micro IPM test board

- Case temperature measuring point using IR Camera

Fig. 8 Thermal test setup using an IR camera

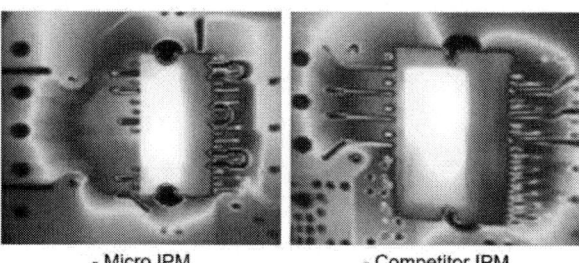

- Micro IPM

- Competitor IPM

Fig. 9 Thermal photo from an IR camera

5.2 Thermal performance comparison

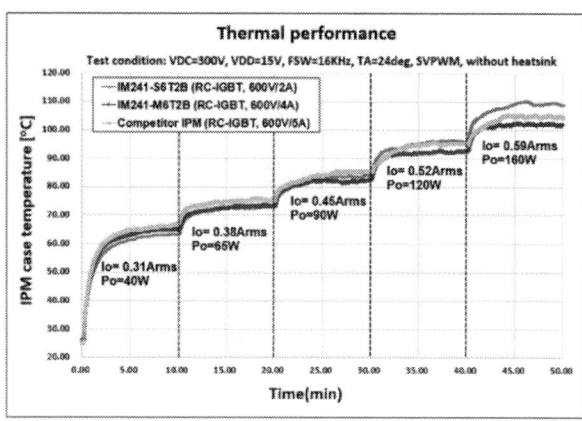

Fig. 10 Comparison between the module case temperatures of different IPMs

Figure 10 shows the comparison between the thermal performance of Infineon's CIPOS™ Micro IM241 product and a competitor's IPM for low-power motor drive applications. The thermal performance was evaluated using a 3-phase inverter system with R-L load under the following test con-

ditions: V_{DC} = 300 V, V_{DD} = 15 V, switching frequency = 16 kHz, T_A = 24°C, SVPWM, P_O = 40~160 W, without heatsink.

The test results showed that at P_O = 40 – 65 W, the T_C of CIPOS™ Micro IPM 600 V, 2 A and 4 A product was better than that of the competitor's IPM. At P_O = 90 – 160 W, the T_C of CIPOS™ Micro IPM 600 V, 4 A (IM241-M6T2B) was better than that of the competitor's IPM.

In addition, considering the RC-D2 IGBT guaranteed temperature of T_J = 150°C, it was verified that the temperature margin of the CIPOS™ Micro IPM 600 V, 4 A (IM241-M6T2B) was about 48°C compared to the T_C at P_O = 160 W.

6 Conclusion

This paper compared and analyzed the thermal measurement method, power loss, and thermal performance of the new CIPOS™ Micro IPM with RC-D2 IGBT under 40 – 160 W low-power motor application conditions without a heatsink. In particular, it was confirmed through the thermal analysis simulation of the IPM package without a heatsink that the RC-D2 IGBT junction temperature and the module case temperature were the same. It was also verified that the total loss of the Micro IPM RC-D2 IGBT is lower than that of its competitor IPMs, i.e., its thermal performance is excellent. Therefore, because of their high efficiency and excellent thermal performance without a heatsink, the Micro IPM can help realize the compact, highly efficient systems required by the low-power motor driving application market.

References

[1] K. Lee, T. Lee, D. Jo, and B. Song, "Introducing the New 600 V CIPOS™ Tiny IM323 Intelligent Power Module for Motor Drive Applications", PCIM Europe 2023, pp. 1445 – 1450.

[2] S. S. Arefin, C. Villani, and S. Ruzza, "The new CIPOSTM Micro intelligent power module with reverse conducting IGBT technology for home appliances", PCIM Asia 2022, pp. 299 – 304.

[3] K. Lee, B. Song, J. Lee, and T. Kwon, "A New Generation of 600V CIPOS™ Tiny IM323 Intelligent Power Module for Home Appliance Motor Drive Application", PCIM Asia 2022, pp. 99 – 103.

PCIM Europe 2024, 11– 13 June 2024, Nuremberg   DOI: 10.30420/566262236

Introducing the New 1200 V CIPOS™ Maxi IM12BxxxC1 Intelligent Power Module for Motor Drive Applications

Lee Kihyun[1] ⓘ, Kim Jinhyeok[1] ⓘ, Han Soohyuk[1] ⓘ, Song Bokkeun[1] ⓘ, Kang Kyoungpil[1] ⓘ and Lee Minsub[1] ⓘ
[1] Infineon Technologies Korea, South Korea

Corresponding author: Lee Kihyun, kihyun.lee@infineon.com
Speaker: Lee Kihyun, kihyun.lee@infineon.com

Abstract

This paper presents Infineon's new generation CIPOS™ 1200 V IM12BxxxC1 (old nomenclature: IM817) intelligent power modules (IPM) ranging up to 20 A in a dual in-line package with transfer molding. These IPMs can be for 3-phase AC motors and permanent magnet motors in variable speed motor drives applications. They combine the features of TRENCHSTOP™ IGBT7 S7 and emitter-controlled EC7 diodes with optimized silicon-on-insulator gate drivers and a direct copper bonding substrate. They offers the smallest package size with high power density. This paper provides an overview of the module as well as the technology behind the semiconductors, the electrical characteristics, the package, and the thermal performance.

1   Introduction

In recent years, the need for efficient power conversion has increased dramatically in response to global demands for decarbonization and improved natural resource sustainability.

Most inverter systems on the global industrial motor market are dominated by low-voltage motor drives and servo drives. Additionally, modern rectifier systems adopt a 3-phase pulse width modulation (PWM) rectifier method, instead of the traditional diode bridge rectifier method, to overcome the harmonic problem of variable speed drives. In these applications, the need for higher power density, lower cost, and higher efficiency along with improved reliability creates a greater need for power switches with lower power dissipation, improved robustness, and optimized packaging.

Keeping these requirements in mind, Infineon is developing a new CIPOS™ 1200 V IM12BxxxC1 (old nomenclature: IM817) development suite featuring high-speed variants of the innovative 7th generation IGBT and diode technology. It mainly combines the key advantages of 1200 V-rated TRENCHSTOP™ IGBT7 S7 and emitter-controlled EC7 Rapid diode technology with low conduction and switching power losses, and smooth and stable rectification. The 1200 V TRENCHSTOP™ IGBT7 S7 technology has also been optimized for applications that require the short circuit safe operating area (SCSOA) functionality. Thus, improving the efficiency of inverter systems significantly.

In this regard, the trade-off between the SCSOA performance and power consumption of IGBTs is generally well known. SCSOA-rated IGBTs have higher collector-emitter saturation voltages and total switching energy losses than equivalent IGBTs without SCSOA ratings. Therefore, SCSOA characteristic is very important if the target application requires it, but counterproductive if it is not required [1]. This product family also offers exceptional robustness under harsh conditions that require SCSOA, and a unique portfolio of current ratings ranging from 10 A to 20 A.

Fig. 1   External outline overview (size: 36 x 22.7 mm²).

This paper is organized as follows: Section 2 describes key features of the internal components and the package structure. Section 3 introduces the static characteristics, dynamic characteristics, and short-circuit characteristics. Thermal performance for low and high switching frequencies are presented in Section 4.

Product	Current	Target applications
IM12B10CC1	10 A	Motor drives for indus-
IM12B15CC1	15 A	trial applications
IM12B20EC1	20 A	

Table 1   Product line-up and target applications

2   Feature of the new CIPOS™ Maxi IM12BxxxC1 IPM series

2.1   Electrical internal circuit and components

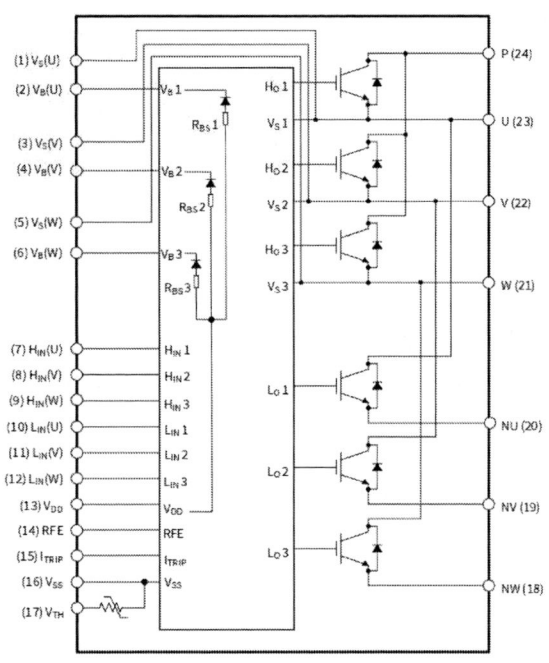

Fig. 2   Internal block diagram of the new Maxi

IM12BxxxC1 IPM product series

The new 1200 V Maxi IM12BxxxC1 IPM has a built-in gate driver IC with 3-phase inverter circuitry and control, as shown in Fig. 2.

The built-in gate drive IC is designed using the rugged silicon-on-insulator (SOI) technology with stability against transient and negative voltage. This gate drive IC has a built-in integrated bootstrap functionality. Temperature monitoring is done through a built-in, UL-certified negative temperature coefficient (NTC) thermistor and an open-emitter pin configuration is adopted for the low side emitter. Table 1 shows all the products in this new IPM product family and their current rating with target applications.

2.2   Semiconductor

2.2.1   The 1200 V TRENCHSTOP™ IGBT7 S7 and EC7 diode technology

The TRENCHSTOP™ IGBT7 1200 V S7 and EC7 technology are, respectively, the 7th generation insulated gate bipolar transistor and emitter-controlled diodes, the cross-sections of which are shown in Figure 3.

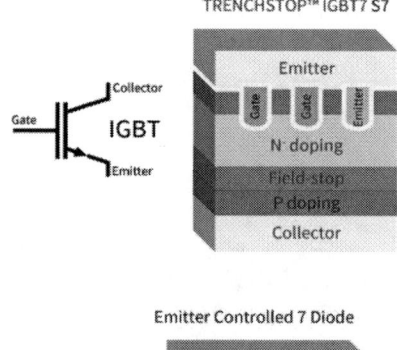

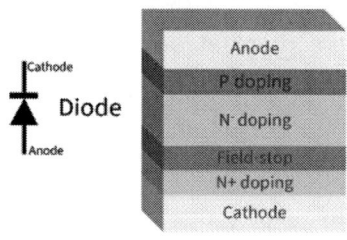

Fig. 3   Cross-section of the TRENCHSTOP™

IGBT 1200 V S7 and EC7 diode

The IGBT technology, based on micro-pattern trenches (MPT) with sub-micron mesas, has been specially designed to provide strongly reduced static losses ($V_{CE(sat)}$), a high level of controllability (dV/dt), and a short-circuit withstand time capability (t_{SC}). The EC7 diode field-stop region has been designed to provide enhanced softness without compromising the blocking voltage even at high V_{DC}, high temperature and very low current to reduce the oscillatory behavior to a minimum even under harsh conditions. Thanks to the combination of the aforementioned features, TRENCHSTOP™ IGBT7 S7 and EC7 diode portfolio is the best fit for electric power converters requiring e.g., moderate switching frequencies (≤ 20 kHz), relatively slow switching events (≤ 10 kV/μs), and short-circuit robustness [2][3].

2.2.2 1200 V SOI single-gate drive IC

The 1200 V SOI single-gate driver with integrated bootstrap circuit is used to achieve higher levels of integration, reliability, and performance in the CIPOS™ Maxi IPM. This SOI gate driver IC disables leakage or latch-up current between structurally adjacent devices. It prevents the latch-up effect even in case of high dV/dt switching and surge under elevated temperature [4]. This gate driver provides several protection functions such as cross-conduction prevention, undervoltage lockout, overcurrent detection, and enable input.

CIPOS™ Maxi IPM series provides an integrated fault output with sleep function and adjustable fault clear time through the RFE pin. There are two situations that can cause the driver IC to report a fault via the RFE pin. The first being an undervoltage condition of V_{DD} and the second when the I_{TRIP} function for over-current protection recognizes a fault. After the fault condition occurs, the RFE pin is internally pulled to V_{SS} and the fault-clear timer is activated. When the built-in fault-clear timer expires, the voltage on the RFE pin returns to its external pull-up voltage. The output remains disabled, and the fault condition is maintained until the voltage on the RFE pin charges up to enable threshold voltage. The charging characteristics are dictated by the RC time constant attached to the RFE pin.

2.3 Package

The CIPOS™ Maxi IPM package is designed with the smallest package size (36 mm x 22.7 mm x 3.1 mm) by optimizing internal PCB and Direct Bonding Copper (DBC) structures, as shown in Figure 4. The gate driver IC and thermistor are placed on the internal PCB.

The IGBTs and diodes are placed on a DBC for better thermal performance. It adopts Al wire bonding technology for electrical connections between the PCB and the DBC as well as the DBC and the lead frame.

The CIPOS™ Maxi IPM has an independent V_{TH} pin that is connected to the thermistor inside the package, which monitors the temperature. The case temperature is tracked by the NTC thermistor. As the temperature relationship between the NTC and the case can vary depending on the heat dissipation conditions defined by the users, the relationship between the two must be measured in a relevant experimental environment.

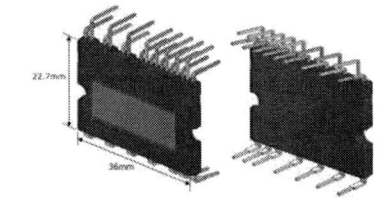

(a) External view of the Maxi IPM package

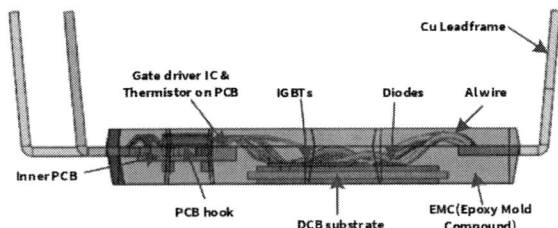

(b) Internal structure of the Maxi IPM package

Fig. 4 External view and Internal structure of CIPOS™ Maxi IPM

In addition, the IPM's internal body is encapsulated using transfer molding technology and meets all international industry standards, such as clearance for insulation distance and creepage distance. The corresponding insulation distance between two pins and between a pin and the DBC is shown in Figure 5 [5].

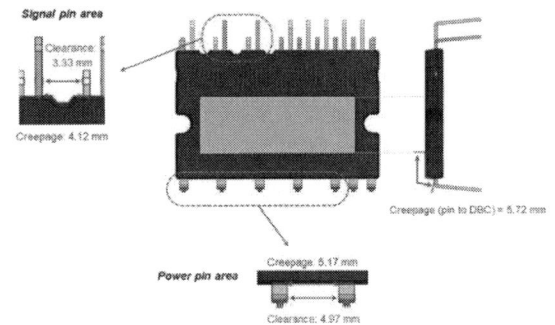

Fig. 5 Clearance and creepage distance of the CIPOS™ Maxi IPM package

3 Electrical characteristics

3.1 Static characteristics

Figures 6 and 7 show a comparison between the static DC characteristics of the 10 A and 20 A devices in the new CIPOS™ Maxi 1200 V IM12BxxxC1 IPM series with that of competitors, at room temperature and high temperature conditions. Specifically, the IM12B10CC1 at 10 A nominal current and 150°C junction temperature shows at least 300 mV lower $V_{CE(Sat)}$ (see Fig. 6 (a)) and at least 200 mV lower V_F (see Fig. 6 (b)).

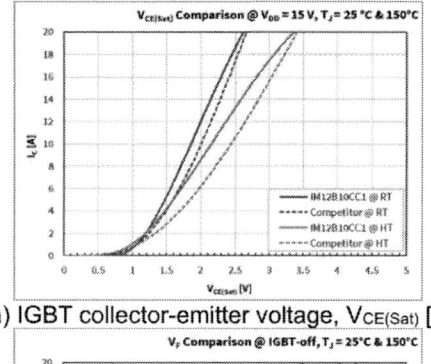

(a) IGBT collector-emitter voltage, $V_{CE(Sat)}$ [V]

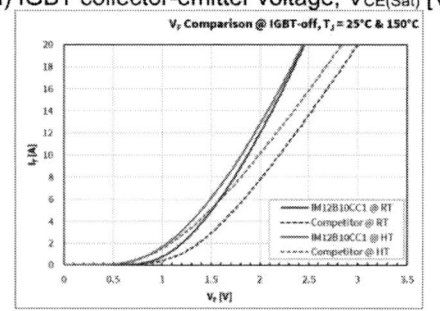

(b) Diode forward voltage, V_F [V]

Fig. 6 DC characteristics of the Maxi 1200 V 10 A IPM (IM12B10CC1)

As shown in Fig. 7 (a), the IM12B20EC1 at 20 A nominal current and 150°C junction temperature shows at least 140 mV lower $V_{CE(Sat)}$.

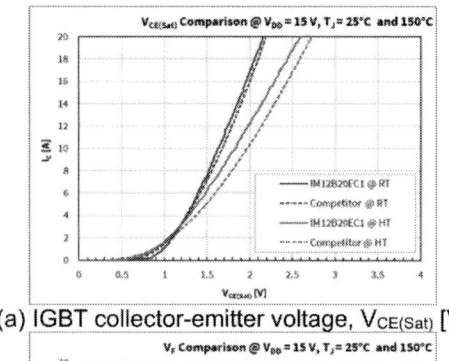

(a) IGBT collector-emitter voltage, $V_{CE(Sat)}$ [V]

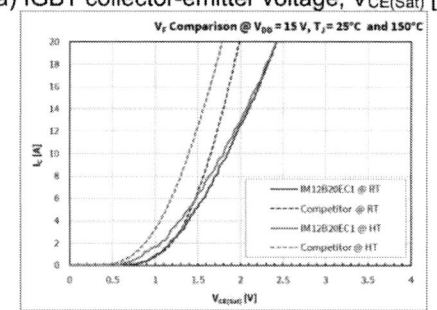

(b) Diode forward voltage, V_F [V]

Fig. 7 DC characteristics of the Maxi 1200 V 20 A IPM (IM12B20EC1)

3.2 Dynamic characteristics

Figures 8 and 9 show the switching-on and switching-off waveforms of 10 A and 20 A-rated devices, respectively, at $V_{DC\text{-link}} = 600$ V, $V_{DD} = 15$ V, and $T_J \leq 150$°C.

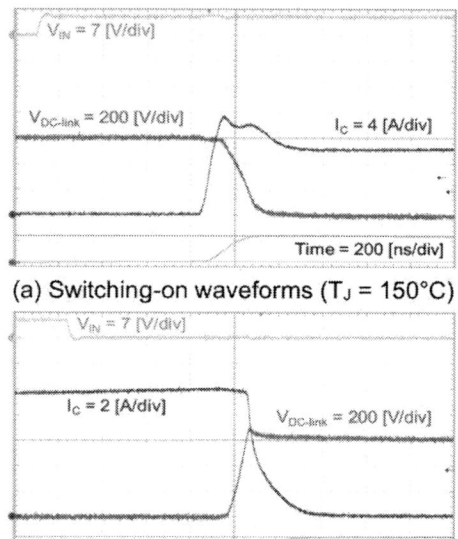

(a) Switching-on waveforms ($T_J = 150$°C)

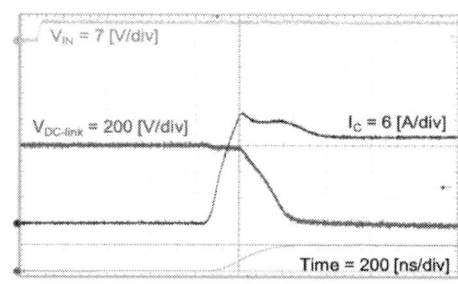

(b) Switching-off waveform ($T_J = 150$°C)

Fig. 8 DC characteristics of the Maxi 1200 V 10 A IPM (IM12B10EC1)

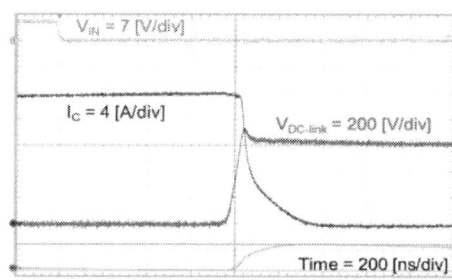

(a) Switching-on waveforms ($T_J = 150$°C)

(b) Switching-off waveform ($T_J = 150$°C)

Fig. 9 DC characteristics of the Maxi 1200 V 20 A IPM (IM12B20EC1)

3.3 Performance

Figure 10 shows the dynamic loss characteristics of the new Maxi 1200 V 10 A and 20 A products for $V_{DC-link}$ = 600 V, V_{DD} = 15 V, and inductive load conditions at T_J = 25°C and T_J = 150°C. These characteristics lead to highly improved switching characteristics.

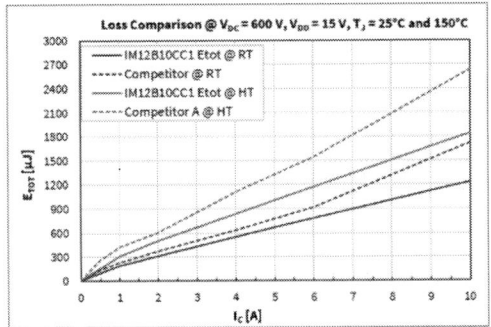

(a) Loss characteristics of the Maxi 1200 V 10 A IPM

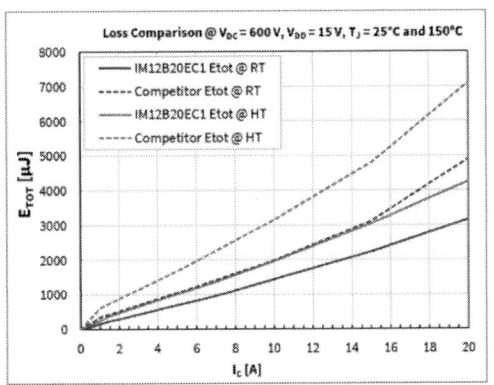

(b) Loss characteristics of Maxi 1200 V 20 A IPM

Fig. 10 Dynamic loss characteristics of CIPOS™ Maxi 10 A and 20 A products (IM12B10CC1 and IM12B20EC1)

The 1200 V 10 A product (IM120B10CC1) with nominal current of 10 A offers approximately 30% lower total loss than its competitors, especially at high temperature. The 1200 V 20 A product (IM12B20EC1) with a nominal current of 20 A offers 40% lower total loss at high temperature than its competitors. This makes the devices very suitable for the target applications.

3.4 SCSOA characteristics

The short-circuit withstand time of the CIPOS™ Maxi 1200 V IM12BxxxC1 product series is specified as 5 µs under the following conditions: $V_{DC-link}$ = 800 V, V_{DD} = 15 V, and $T_J \leq$ 150°C. Figure 11 shows the measured test waveform of

IM12B10CC1 for $V_{DC-link}$ = 800 V, V_{DD} = 15 V, and T_J = 150°C.

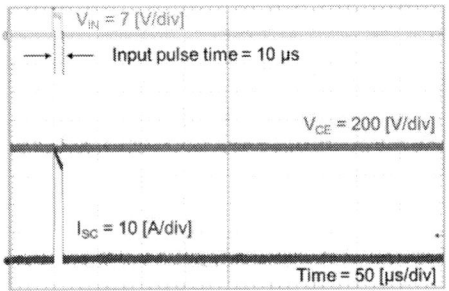

(a) Normal waveform

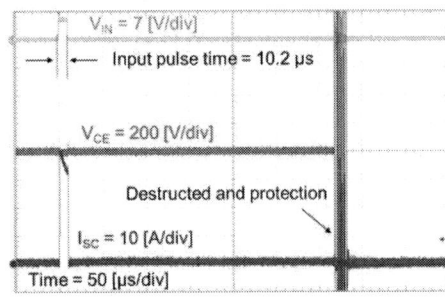

(b) Shutdown waveform

Fig. 11 Short-circuit test waveform of IM12B10CC1

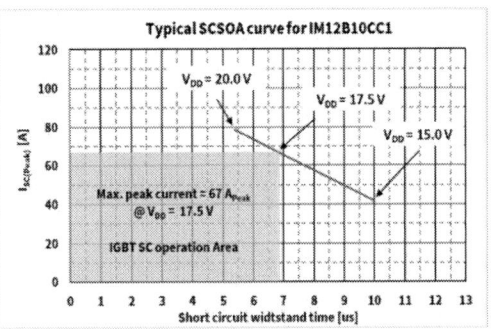

(a) SCSOA of the Maxi 1200 V 10 A IPM

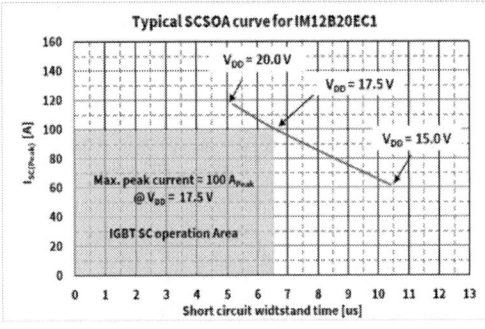

(b) SCSOA of the Maxi 1200 V 20 A IPM

Fig. 12 SCSOA characteristics of IM12B10CC1 and IM12B20EC1

1741

Figure 12 show the typical (not guaranteed) SCSOA performance graphs of the 10 A and 20 A devices for $V_{DC\text{-link}}$ = 800 V, $V_{PN(surge)}$ < 900 V, and T_J = 150°C at different supply voltages.

4 Thermal performance

4.1 Test conditions

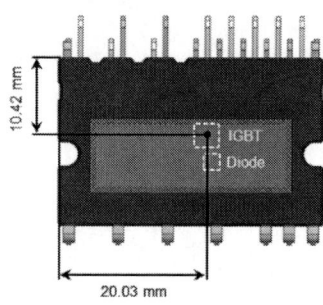

Fig. 13 Point for measuring the case temperature to test thermal performance

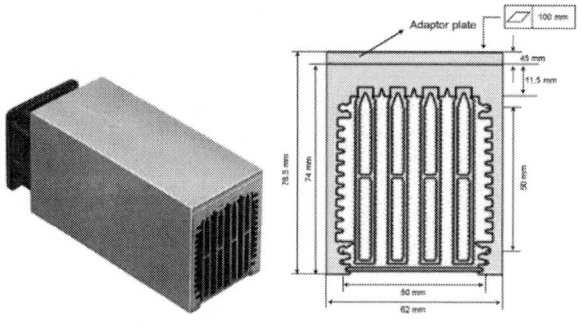

Fig. 14 Heatsink information

Measuring the case temperature can help confirm thermal performance. The points on the IGBT chip shown in Fig. 13 represent the hottest points on the module. The case temperature (T_C) is the low-side, U-phase IGBT temperature in the CIPOS™ Maxi 1200 V products. Figure 14 shows the heatsink used for testing the thermal performance at case-to-ambient temperature.

4.2 Thermal characteristics of the Maxi 1200 V 10 A IPM (IM12B10CC1)

Figure 15 shows the thermal performance at case-to-ambient temperature (ΔT_{CA}) for the new CIPOS™ Maxi 1200 V 10 A products at low switching and high switching frequencies. Their thermal performance was evaluated using a 3-phase inverter system with R-L load under the following

test conditions: $V_{DC\text{-link}}$ = 600 V, V_{DD} = 15 V, and switching frequency = 6 kHz and 15 kHz, output frequency = 60 Hz. The maximum ΔT_{CA} of the CIPOS™ Maxi 1200 V 10 A IPM (IM12B10CC1) was about 23°C at f_{SW} = 6 kHz and I_O = 10 A_{peak}. This was 2.5°C lower than that of the competitors. Its maximum ΔT_{CA} was approximately 40°C at f_{SW} = 15 kHz and I_O = 10 A_{peak}, which was 6.6°C lower than that of the competitor.

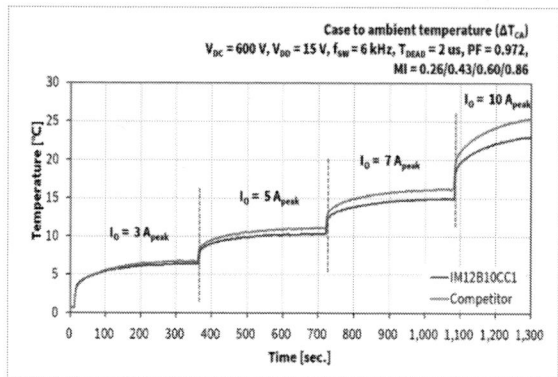

(a) f_{SW} = 6 kHz

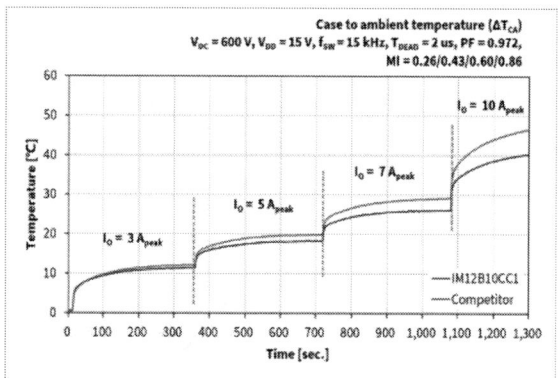

(b) f_{SW} = 15 kHz

Fig. 15 Results of case-to-ambient temperature measurement of IM12B10CC1 (1200 V, 10 A)

4.3 Thermal characteristics of the Maxi 1200 V 15 A IPM (IM12B15CC1)

Figure 15 also shows the thermal performance at case-to-ambient temperature (ΔT_{CA}) for the new CIPOS™ Maxi 1200 V 15 A products at low switching and high switching frequencies. The thermal performance was evaluated using a 3-phase inverter system with R-L load under the following test conditions: $V_{DC\text{-link}}$ = 600 V, V_{DD} = 15 V, switching frequency = 6 kHz and 15 kHz, and output frequency = 60 Hz. The maximum ΔT_{CA} of the

CIPOS™ Maxi 1200 V 15 A IPM (IM12B15CC1) was about 37°C at f_{SW} = 6 kHz, and I_O = 15 A_{peak}. Its maximum ΔT_{CA} was about 55°C at f_{SW} = 15 kHz and I_O = 15 A_{peak}.

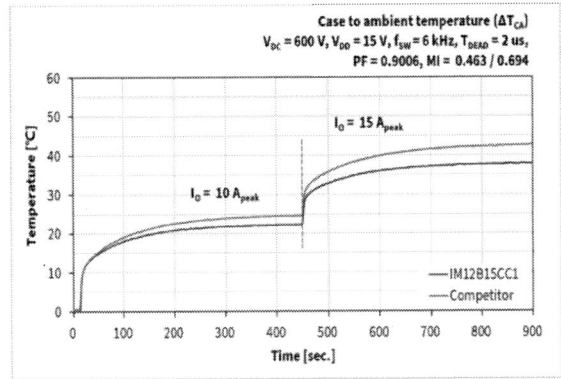

(a) f_{SW} = 6 kHz

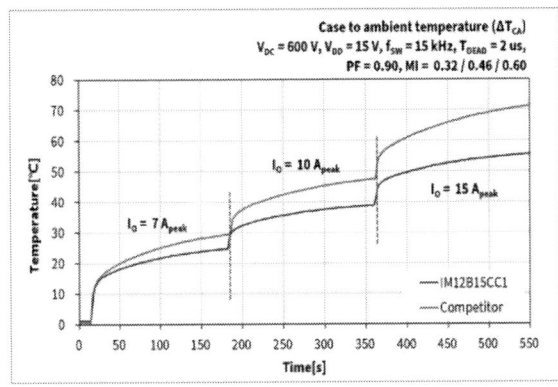

(b) f_{SW} = 15 kHz

Fig. 15   Results of case-to-ambient temperature measurement of IM12B15CC1 (1200 V, 15 A)

4.4   Thermal characteristics of the Maxi 1200 V 20 A IPM (IM12B20EC1)

Figure 16 show the thermal performance with case-to-ambient temperature (ΔT_{CA}) for the new CIPOS™ Maxi 1200 V 20 A IPM products at low switching and high switching frequencies. The thermal performance was evaluated using a 3-phase inverter system with R-L load under the following test conditions: $V_{DC\text{-}link}$ = 600 V, V_{DD} = 15 V, switching frequency = 6 kHz and 15 kHz, and output frequency = 60 Hz. The maximum ΔT_{CA} of the CIPOS™ Maxi 1200 V 20 A IPM (IM12B20EC1) was about 36°C at f_{SW} = 6 kHz, and I_O = 10 A_{peak}, which was 1°C lower than that of the competitor. Its maximum ΔT_{CA} was approximately 40°C at f_{SW} = 15 kHz and I_O = 15 A_{peak}, which was 18°C lower than that of the competitor.

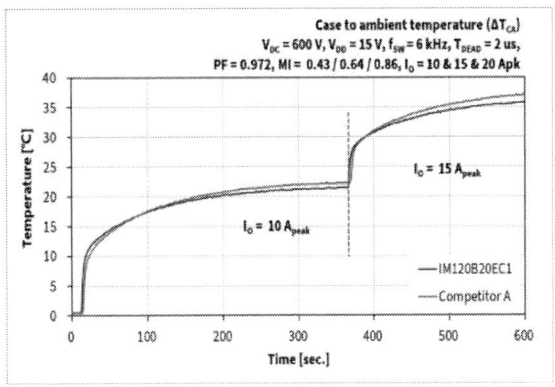

(a) f_{SW} = 6 kHz

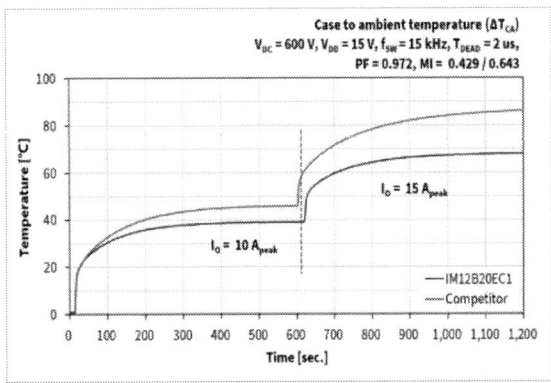

(a) f_{SW} = 15 kHz

Fig. 16   Results of case-to-ambient temperature measurement of IM12B20EC1 (1200 V, 20 A)

5   Conclusion

The new CIPOS™ Maxi 1200 V IPM product series (IM12BxxxC1) that uses Infineon's progressive 1200 V rating TRENCHSTOP™ IGBT7 S7 IGBTs and EC7 technology with C5SOI gate drive ICs was introduced in this paper. This product enables the development of highly energy-efficient appliances by offering higher efficiency in a small package that also adequately addresses the trend of system miniaturization.

References

[1] A. Piccioni, "Comprehensive performance evaluation of discrete 1200V IGBT7 S7 for drives application," PCIM Europe 2022, pp. 1951 – 1954.

[2] J. Cerezo, and A. K. Sekar, "1200 V TRENCHSTOPTM IGBT7 H7 and Emitter-Controlled EC7 Rapid Diode Technologies

Define an Enhanced Benchmark for Improved Energy-Efficient, Fast-Switching Inverter Applications," PCIM Europe 2023, pp. 2189 - 2195.

[3] Application Note, AN-2023-01, "The 1200 V TRENCHSTOPTM IGBT 7 S7 and EC7 technology, portfolio and applications," Infineon Technologies AG.

[4] R. Keggenhoff, Z. Liang, A. Arens, P. Kanschat, and R. Rudolf. "Novel SOI Driver for Low Power Drive Applications", Power Systems Design Europe, Nov. 2005.

[5] M. Lee, M. Baek, J. Lee, and D. Chung, "A New Smallest 1200V Intelligent Power Module for Three Phase Motor Drives", IPEC 2018, pp. 1141 – 1144.

PCIM Europe 2024, 11– 13 June 2024, Nuremberg   DOI: 10.30420/566262237

An Adaptive Dead Time Control based on the Switch-Node-Voltage Derivative

Lukas Knappstein[1], Niklas Falkenberg[1], Martin Pfost[1]

[1] Chair of Energy Conversion, TU Dortmund University, Germany

Corresponding author: Lukas Knappstein, Lukas.Knappstein@tu-dortmund.de
Speaker: Lukas Knappstein, Lukas.Knappstein@tu-dortmund.de

Abstract

With an increasing demand for higher switching frequencies in power electronic applications, the switching losses of semiconductor devices become more crucial. This can be avoided by operating the device in soft switching, which usually imposes restrictions on the dead time. This study proposes an active dead time controller that ensures soft switching by adjusting the dead time as a function of in situ measurements. For this purpose, the switch-node-voltage derivative is measured with a high-pass filter to determine the switching-transient length for the controller to set the ideal dead time for soft switching. The proposed controller can quickly adapt to changing operating conditions without the need for extensive offline training procedures.

1 Introduction

Modern power electronic applications demand higher energy densities, which can be achieved by higher switching frequencies. Modern semiconductors such as GaN HEMTs allow high-frequency operation but suffer from increased reverse-conduction losses compared to conventional Si MOSFETs [1]. However, a high-frequency operation can result in extensive switching losses that can be circumvented by soft switching. This usually asks for increased dead times, at least to a level where hard switching is prevented. On the other hand, too long dead times can result in excessive reverse-conduction losses, so the dead time should be kept as close as possible to the soft switching limit.

For that purpose, a dead time control is required that determines the optimal dead time for low switching and reverse conduction losses. Typically, this is done with an adaptive control utilising look-up tables [2][3] or analytical transistor models [4][5], which require tedious training procedures. This study proposes an active dead time controller that utilises the derivative of the switch-node-voltage to regulate the dead time for an optimal soft switching.

2 Method Overview

The proposed method is designed for being used in a half bridge application. Therefore, an LTSpice simulation utilising GaN Systems GS66516B that operate at V_{in} = 250 V is set, cf. Fig. 1. In order to investigate the impact of the dead time on the switching performance, a simulation is done for a positive inductor current i_q = 2 A during the negative switching transient. The simulation results for different dead times are shown in Fig. 2. Here, the dead time interval t_D is defined as the duration between the off-state of the high-side transistor HS and the on-state of the low-side transistor LS. At the beginning of the switching transition, the voltage v_{sw} is high. If a positive current i_q is flowing through L_{out} in the meantime, it will discharge the output capacitance C_{OSS} of the GaN HEMTs and v_{sw} will drop. Now, applying a too short dead time t_D results in the occurrence of hard switching and undesired switching losses, see Fig. 2(a). On the other hand, for too long t_D, v_{sw} falls below 0 V, leading to reverse operation and thus increased conduction losses. Consequently, tracking v_{sw} is crucial to determine the optimal dead time, which is done by its derivative. Note that v_{sw} is constant before and after a switching event, corresponding to a derivative of 0. Consequently, t_{meas} is defined

1745

by the negative interval of $\frac{dv_{sw}}{dt}$ during the switching event.

With those measurements, the dead time controller compares t_{meas} with t_D each switching cycle to either decrease or increase t_D for the next pulse. For example, if t_D is shorter than t_{meas}, hard switching occurs as shown in Fig. 2(a), thus t_D needs to be increased. A digital determination of the derivative by the direct measurement of v_{sw} has the disadvantage that a large measuring voltage range and a high resolution would be required. Consequently, this study investigates an analogue method for measuring the derivative of v_{sw} utilising a first-order high-pass filter: The frequency response of a derivative is given by the transfer function

$$G_{deriv}(j\omega) = j\omega. \tag{1}$$

A comparison of the first-order high-pass filter transfer function, which is given by

$$G_{HP}(j\omega) = \frac{j\omega}{j\omega + 2\pi f_0}, \tag{2}$$

to that of a first-order low-pass filter, which is

$$G_{LP}(j\omega) = \frac{1}{j\omega + 2\pi f_0}, \tag{3}$$

it is important to note that the high-pass filter acts like a derivative with a low-pass component that limits the bandwidth to f_0:

$$G_{HP}(j\omega) = j\omega \cdot \frac{1}{j\omega + 2\pi f_0} \tag{4}$$

$$= G_{deriv}(j\omega) \cdot G_{LP}(j\omega) \tag{5}$$

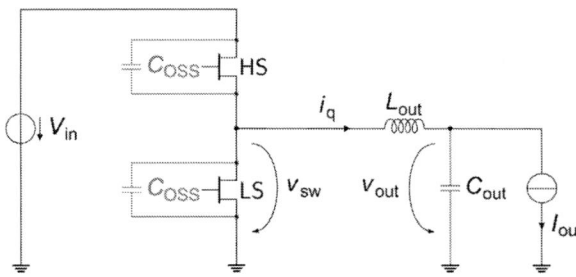

Fig. 1: Overview of the simulation setup utilizing a half bridge in a synchronous converter.

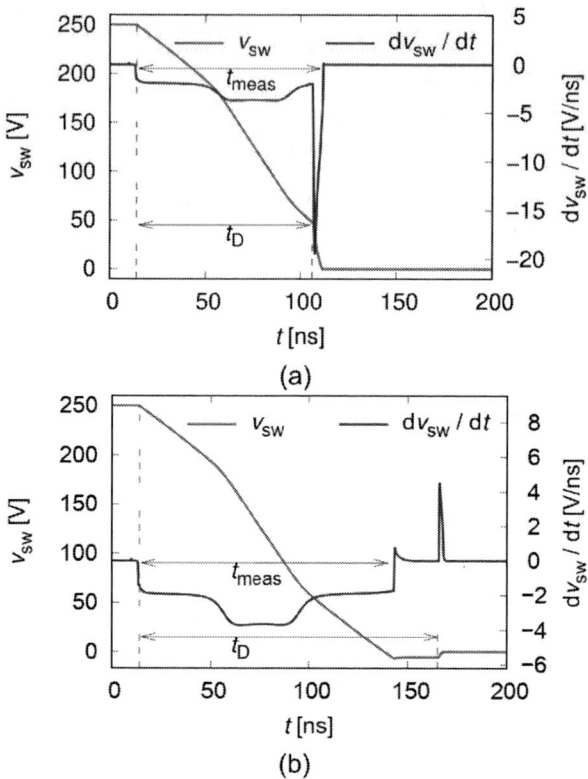

Fig. 2: Simulated switch node voltage v_{sw} and its derivative dv_{sw}/dt with (a) a too short dead time and (b) a too long dead time, both resulting in increased switching losses. Note that the simulation is done for a positive current $i_q = 2\,A$ during the negative switching event, leading to a negative transition.

3 System Overview

To experimentally investigate the proposed method, it is equpieed in a test bench with a synchronous converter that utilises GaN Systems GS66516B in a half bridge application. Here the switching frequency is set to $f_s = 500\,kHz$ and the duty cycle to 50%. An electrical power supply is set with $V_{in} = 250\,V$ and an electrical load (EL) is used to operate the synchronous converter at several operating points. An overview of the test bench circuit is shown in Fig. 3, the physical test bench is shown in Fig. 5.

The centrepiece of the proposed method is the derivative measuring circuit, which consists of the derivative high-pass filter whose signal is transmitted to a 250MSa/s. It is integrated directly next to the half bridge and constructed with $R_{HP} = 50\,\Omega$ and $C_{HP} = 1\,pF$ as shown in Fig. 7, resulting in

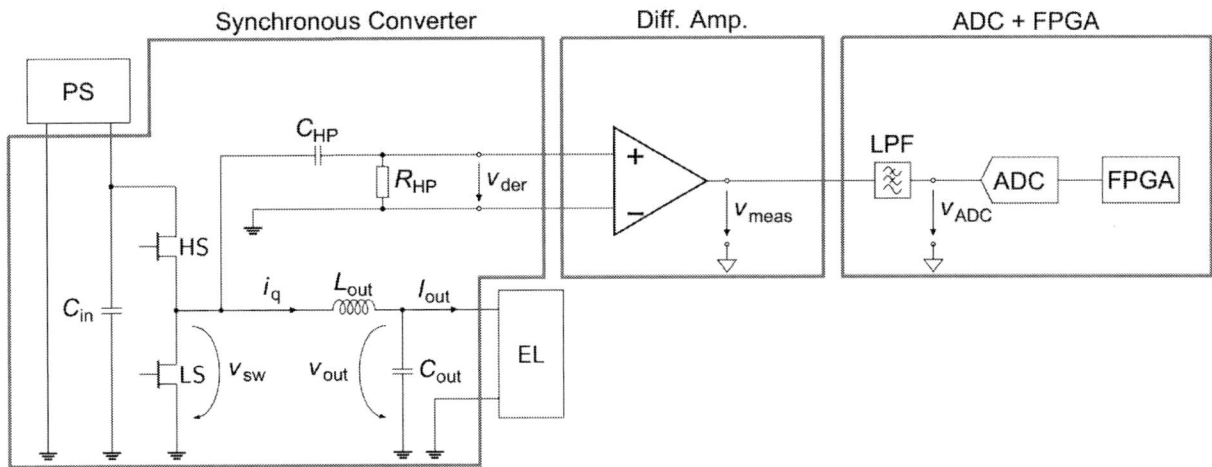

Fig. 3: Circuit of the test bench consisting of a synchronous converter and the measurement circuit. The derivative high-pass filter is implemented with C_{HP} and R_{HP}.

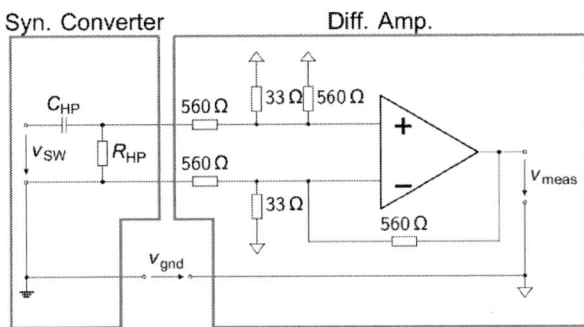

Fig. 4: Detailed differential amplifier circuit.

a $\tau_{HP} = R_{HP} \cdot C_{HP} = 50$ ps. Ensuring that this high-pass time constant is lower than the sample rate of the ADC is crucial for providing an instantaneous derivative signal for every ADC sample. Oscillations of v_{gnd} occur across the power ground and the signal ground due to rapid transients in v_{sw}. To mitigate these oscillations, a high-speed operational amplifier (op-amp) is used in a fully differential circuit to pass v_{der} to the utilised ADC. This op-amp is set as a subtraction amplifier by combining an inverting op-amp circuit and a non-inverting op-amp circuit. The full circuit of this differential op-amp is shown in Fig. 4 and the physical design in Fig. 6. As oscillations of v_{gnd} might surpass the op-amp supply voltage, the incoming measurement voltages from the half bridge are initially divided by the voltage dividers of 1 : 18. The resulting divided voltages act on the op-amp as if they had an impedance of 31.16 Ω, which is used to calculate the amplification of the inverting ampli-

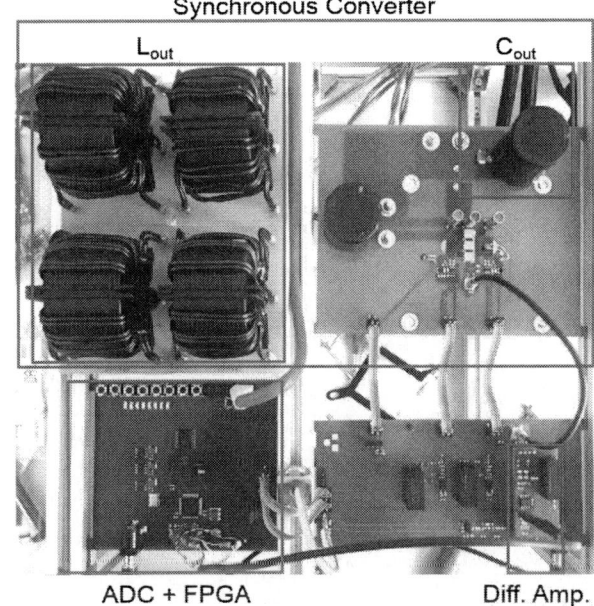

Fig. 5: Physical synchronous converter with the full measurement circuit.

Fig. 6: Detailed physical differential amplifier.

fier by 18. It is worth noting that the amplification of the non-inverting amplifier would be 19, which

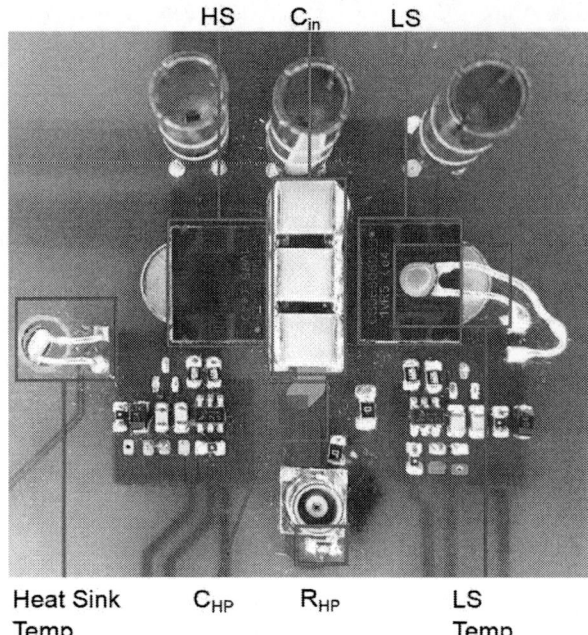

Fig. 7: Detailed physical half bridge and the derivative high-pass filter, C_{HP} is realized in PCB as plate capacitor.

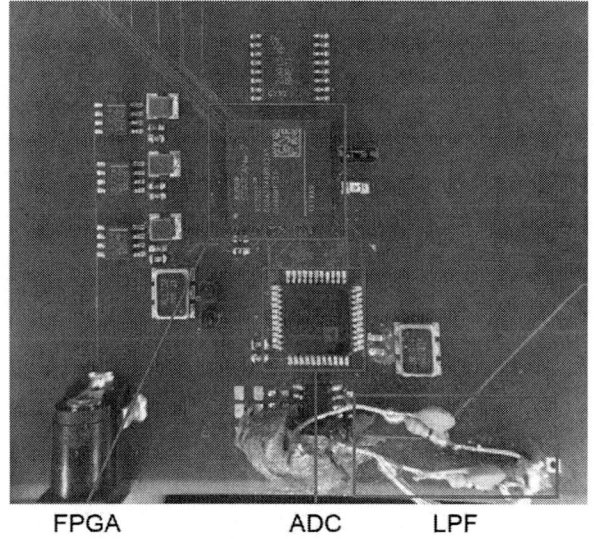

Fig. 8: Detailed physical ADC with its affiliated LPF and the FPGA.

is slightly above the inverting amplifier component. Therefore a fully subtraction behaviour will only be achieved if this path is grounded by $560\,\Omega$. The overall gain of the resulting differential circuit is 1, with the $33\,\Omega$ resistors primarily serving to enhance the suppression against v_{gnd}.

Before the resulting amplified signal $v_{meas.}$ is being sent to the ADC, it is filtered with a low-pass filter (LPF) of $125\,\mathrm{MHz}$ to comply with the Nyquist theorem. Finally, the ADC signal is transited to the FPGA, which implements the proposed dead time controller. The pcb carrying the FPGA and the ADC with its affiliated LPF is shown in Fig. 8.

4 Experimental Results

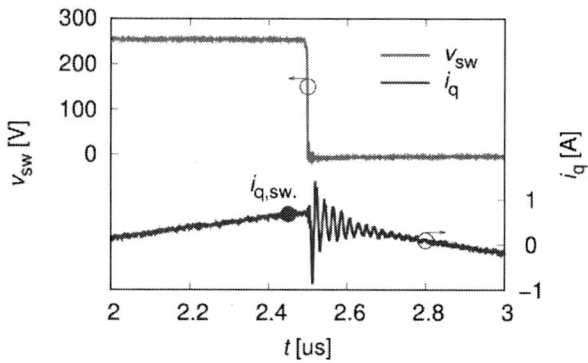

Fig. 9: Capturing i_q just before a switching event to be able to validate the proposed method precisely for specific values of i_q.

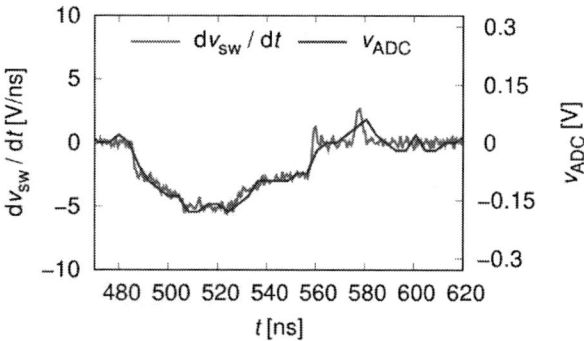

Fig. 10: Comparison of the captured ADC signal v_{ADC} with the belonging switch node derivative dv_{sw}/dt

The validation of the proposed method occurs by precisely defining the inductor current i_q of the synchronous converter during a switching event. This is important as it is responsible for draining C_{OSS} of the GaN-HEMTs and thus directly correlates with the optimal dead time. Consequently i_q is measured just before a switching event as shown in Fig. 9 by a current probe, which is directly placed on L_{out}. Any differences between the desired and

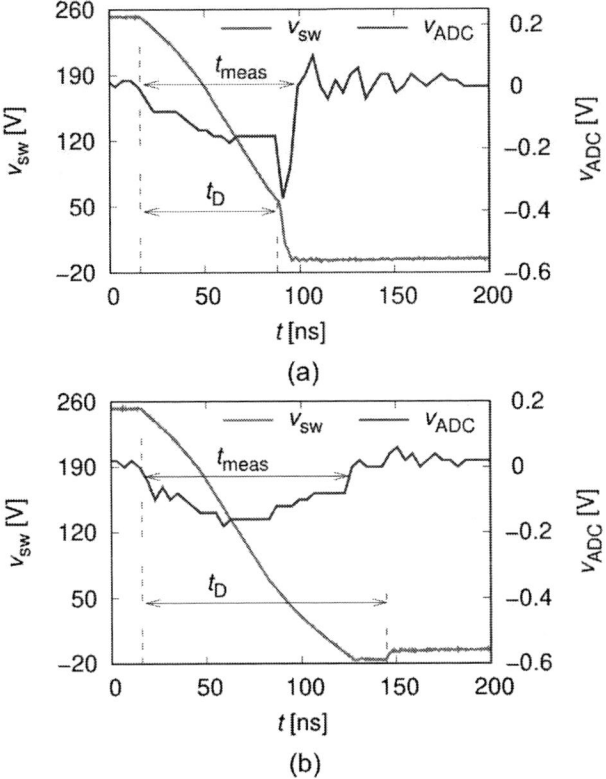

(a)

(b)

Fig. 11: Results for the measurement as v_{meas} and the switch node v_{sw} with (a) too short dead time and (b) too long dead time, which both result in increased switching losses.

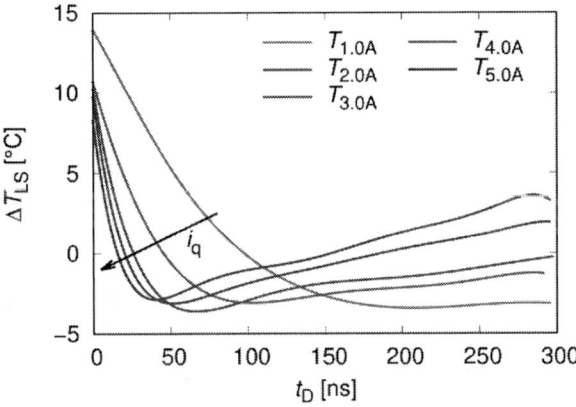

Fig. 12: Finding the optimal dead time visually by minimizing the operating temperature of LS

actual will be added to the EL by the test bench controlling software.

The output of the proposed derivative measurement circuit is shown in Fig. 10. Here, v_{sw} was measured for a switching event with $i_q = 3\,\text{A}$ and

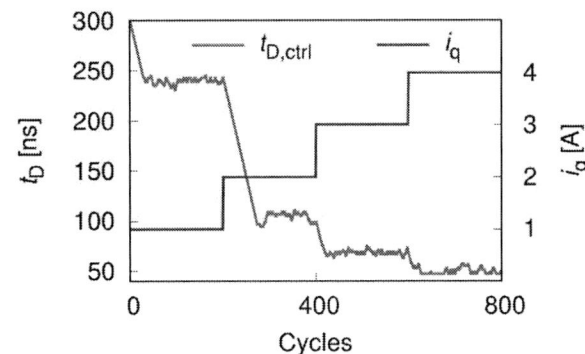

Fig. 13: Demonstration of the proposed controller varying the dead time for varying i_q.

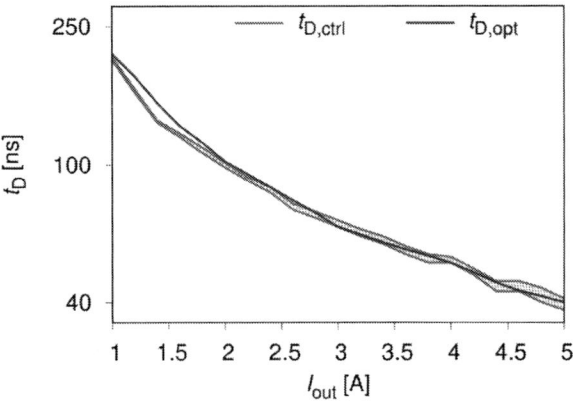

Fig. 14: Resulting predicted dead times $t_{D,ctrl}$ compared to the optimal dead times $t_{D,opt}$.

$t_D = 80\,\text{ns}$ and derived mathematically for comparison, showing an excellent agreement.

Testing the synchronous converter with the same operating points for i_q and t_D as in Fig. 2 shows an excellent agreement between the measurement circuit and the simulation, cf. Fig. 11. In this case the controller defines the beginning of t_{meas} when v_{meas} reaches values below $-0.05\,\text{V}$ and ends the interval, as v_{meas} raises above 0. This allows for hardware testing the proposed method with the synchronous converter.

With i_q set to a desired level, the dead time controller starts to regulate the dead time by altering t_D in steps of 2 ns each negative switching transient. The operation of the controller was tested for $i_q = 1\,\text{A}$ to $i_q = 4\,\text{A}$ with steps of 1 A and a starting value of 300 ns. The regulator operates in a stable manner as seen in Fig. 13.

To find the optimal dead time, the switching losses of LS must be measured separately, which is not electrically possible due to the compact half bridge design. As an alternative, the temperature difference between LS and the heat sink is measured, cf Fig. 7. Fig. 12 shows dead time sweeps from $t_D = 0$ ns to $t_D = 300$ ns each for currents from $i_q = 1.0$ A to $i_q = 5.0$ A. The optimal dead time for each i_q is given by the minimum of its temperature curve. Subsequently to evaluate the performance of the controller, the synchronous converter has been run for out flowing currents from $i_q = 1$ A to $i_q = 5$ A with steps of 0.2 A. Each current was run for 100 cycles, with minimum and maximum dead times recorded for the last 50 cycles. In Fig. 14 the resulting controller dead times $t_{D,ctrl}$ are shown in comparison to the optimal dead times, as well as the optimal dead times $t_{D,opt}$. Here, the predicted dead time $t_{D,ctrl}$ is compared to an empirically determined optimal dead time $t_{D,opt}$ that refers to minimised switching losses.

5 Conclusion

For power electronic application operation with various loads, soft switching of the semiconductor devices can not be guaranteed if static dead times are used. In this study, an active dead time control is proposed, which uses in situ measurements to readjust the dead time within each switching pulse. This controller can significantly reduce switching losses by adjusting the dead times towards an optimum level. It determines the optimal dead time as total switching-transient length, measured by a switch-node-voltage derivative utilising a simple high-pass filter with an ADC and an FPGA. Consequently, the active dead time controller is shown to accurately predict the optimal dead time for a GaN-HEMT-based half bridge with a load variation of 1 A up to 5 A. Furthermore the controller can be implemented as an extension for gate drivers, with no need for extensive training procedures.

The proposed derivative high-pass filter could also be used in other approaches: For example, as a cost-effective alternative the high-frequency ADC could be replaced by a simple DC measurement. The measurement signal could then be captured by a peak detector. Subsequently, the DC measurement could be used as an indication of the switching transient steepness and thus indirectly the switching transient length.

References

[1] R. Reiner, P. Waltereit, B. Weiss, R. Quay, and O. Ambacher, "Investigation of GaN-HEMTs in Reverse Conduction," in *PCIM Europe 2017*, May 2017, pp. 1–8.

[2] T. Krigar and M. Pfost, "Adaptive Dead-Time Control in a Resonant Wireless Power Transfer System," in *2022 24th European Conference on Power Electronics and Applications (EPE'22 ECCE Europe)*, Sep. 2022, pp. 1–8.

[3] B. Kohlhepp, D. Kübrich, M. Tannhäuser, and T. Duerbaum, "Adaptive dead time in high frequency GaN-Inverters with LC output filter," in *The 10th International Conference on Power Electronics, Machines and Drives (PEMD 2020)*, vol. 2020, Dec 2020, pp. 372–377.

[4] Y. Zhang, H. Peng, C. Chen, Y. Xie, T. Liu, and Y. Kang, "A High-Efficiency Dynamic Inverter Dead-Time Adjustment Method Based on an Improved GaN HEMTs Switching Model," *IEEE Transactions on Power Electronics*, vol. 37, no. 3, pp. 2667–2683, March 2022.

[5] Q. Huang, A. Q. Huang, W. Yu, and R. Yu, "Adaptive zero-voltage-switching control and hybrid current control for high efficiency GaN-based MHz Totem-pole PFC rectifier," in *2017 IEEE Applied Power Electronics Conference and Exposition (APEC)*, March 2017, pp. 1763–1770.

PCIM Europe 2024, 11– 13 June 2024, Nuremberg   DOI: 10.30420/566262238

Coupling Coil Design and Positioning Optimization on New High Power Semiconductor Module for Fast Short Circuit Detection

Yannick Dumollard[1], Vincent Escrouzailles[1], Emmanuel Batista[1], Damien Tisné-Grimaud[1],

[1] Alstom SA, France

Corresponding author: Yannick Dumollard, yannick.dumollard@alstomgroup.com
Speaker: Yannick Dumollard, yannick.dumollard@alstomgroup.com

Abstract

This study explores the integration of coupling coils within Power Semiconductor Modules for sensing applications in railway traction converters. It focuses on the new high-power module devices generation called Low Voltage Module (LVM) in Roll2Rail specification [R2R16]. The research delves into coupling coil in scenarios involving short circuits and switching phases, emphasizing di/dt measurement. Integrated sensors on driver circuit demonstrate efficient short-circuit detection, and this study introduces sensor optimization through numerical simulation. Di/dt measurements through High Voltage (HV) main terminals to enable rapid detection during operational phases. A comparison of coupling coil technologies, by opposition of well-known Rogowski coils place around a conductor, highlights the possibility of using it in the region of DC minus terminals and make integration easier. This solution lowers the concerns about electrical isolation and this work highlights the potential of coupling coils in power electronics, suggesting different ways to explore for further research and development.

1   Introduction

Power semiconductor module manufacturers have been proposing a new standard packaging that is now preferred to integrate the newest semiconductor chips including silicon IGBT and SiC MOSFET technologies. They are called respectively LV100, XHP2, LinPak, nHPD2, etc. All these packages share similar interfaces defined in the Roll2Rail specification [R2R16] and are called Low-Voltage Modules (LVM) in the document. In railway applications, protection systems for semiconductors during short-circuits are challenging and rapid detection is key (µs range). Therefore, we propose to explore the addition of a coupling coil on gate driver unit to improve short-circuit detection speed as it has been evaluated for some discrete package already [RP23]. It should be also possible in a second time to integrate the signal and get current information, this part will not be addressed in the paper. Finite element electromagnetic simulations will be used to identify the optimal coupling coil placement area and to explore the immunity over various power connections and busbar geometries. This article focuses on optimizing the detection principle using coil's signals taking into account system integration and patristics effect of the environment connection configurations. The optimized positioning of the coupling coil improves current detection and can be applied for different purposes such as short-circuit protection for power components and current measurement during switching. The study will also cover the tests carried out with two coupling coils placed in the previously determined zones. Our study highlights the importance of power connection geometry on current detection performance and positioning of coupling coil.

2   Theoretical Concepts

2.1   Finite element method (FEM)

Finite Element Method (FEM) has become an essential tool for the analysis and modelling of complex electromagnetic environments based on Maxwell's equations and Ohm's law. FEM can provide a detailed understanding of the electromagnetic behaviour issues encountered in power modules and power busbars. FEM simulations are effective in predicting the electromagnetic performance of systems, as well as the phenomena induced by changes in environmental conditions such as voltage, load, and switching. The accuracy of simulations also helps in identifying and resolving potential problems before their occurrence, thereby reducing costs and associated risks.

Rogowski current sensors are a particularly interesting choice for measuring power current, especially in complex environments with power busbars used in power converters. To maximize the coupling of Rogowski sensors, it is crucial to optimize the placement in the studied environment, as well as their electromagnetic behaviour. FEM simulations coupled with criteria optimization based algorithm support the design in iterating designs parameters . It will results in the identification of best compromise between different parameters like rotation angle for the coil, coupling surface and positioning around a dedicated area.

2.2 Power module MOSFET LVM

The recent power modules, namely LV100, nHPD², LinPak, XHP2, and HPnC, have similar external interfaces. Although the internal layout of copper tracks, high voltage (HV) and low voltage (LV) terminals may differ between suppliers, the magnetic field signatures are similar, thus, one can define some general rules on field lines mapping between normal operation of the module and internal short circuit.

During low side switching, the current flows through the phase and distributes within the component, before exiting through the negative terminals. Significant current concentration occurs in the lower arm chips. The same phenomenon occurs during high side switching but with incoming current from the positive terminals and exiting through the phase terminal, as illustrated in Figure 1.

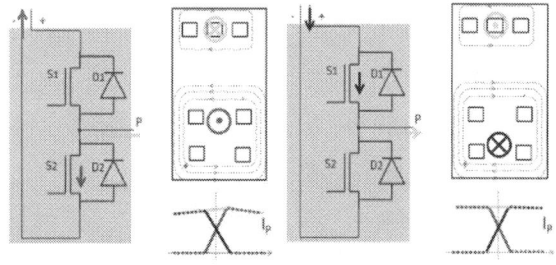

a) Bottom switching   b) Top switching

Figure 1: Schematic diagram of commutation of power module LVM

The magnetic field generated by the main connectors exhibits opposing polarities and intensities that depend on the inductance of the conductors. Due to the higher inductance of the phase terminal and low Di/dt, the magnetic field may pose challenges in detecting short circuits since the coupling coil can capture both short-circuit current and current generated during switching phases.

During an internal short-circuit event internally of the module, where one or more chips from a normally open stage are switched on, current flows directly from the positive terminals to the negative terminals. It results in a magnetic field centered

around these terminals forming two loops with opposing directions intersecting between these terminals. The highest concentration of the magnetic field occurs along the crossing points of these loops, as illustrated in Figure 2.

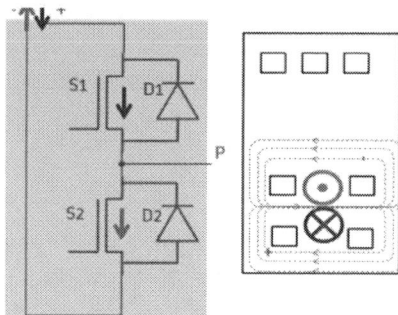

Figure 2: Schematic diagram of short-circuit of power module LVM

To maximize the capture of the magnetic field by the coupling coil during short-circuit phases, it is advisable to position it in the zone where the generated magnetic flux is most significant [YD21].

3 Feasibility Study of Di/dt Measurement using a Coupling Coil in HV main Terminals for Power Semiconductor Protection:

This section aims to investigate the feasibility of measuring di/dt through a central coupling coil positioned between High Voltage (HV) main terminals. During switching phases, the magnetic field concentration in this area is high, which allows to measure di/dt without requiring dedicated auxiliary terminals to sense a parasitic inductance inside the module [YD24].

Compared to a coil placed on the wings of the module, positioning the coupling coil in the center of the connections- [RXMS20] offers more compact integration in the converter, particularly when multiple modules are connected in parallel. To simulate its electromagnetic behaviour under testing conditions, we utilized numerical simulation tools to recreate and investigate the coupling coil, as shown in Figure 3.

Simulation followed two objectives. First, we aim to estimate measurable voltage at the coil output according values of di/dt during normal switching and short-circuit phases by placing the coil in the center of the connections. Second, we aim to use the numerical simulation methodology to optimize the sensor positioning and   validate the optimal placement that leads maximize at turn Turn-on the voltage level gap considering normal switching and module's internal short circuit event. Before to start optimization process by simulation, it have been performed a validation of our numerical results versus real measurements. Results are described just after.

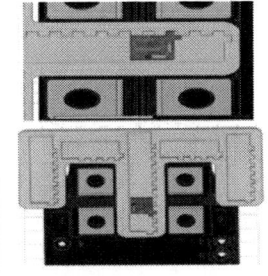

| a) Picture of the coupling coil on the power module | b) 3D model of the coupling coil on power module |

Figure 3: Positioning of the coupling coil

In this study, we conducted switching tests at both 600V and 1800V with three different current levels: 50A, 600A, and 1200A. We used the measured current from these tests as input in our numerical simulation to account for potential parasitic phenomena arising from the setup. Our primary focus was on comparing the results between the tests and simulation conducted at 1200A switched on the top of the power module supplied with Vds 1200V. Figures 4 and 5 show the switched current for both the turn-on and turn-off phases, along with the voltage measured at the coupling coil's terminals, and the voltage obtained through numerical simulation.

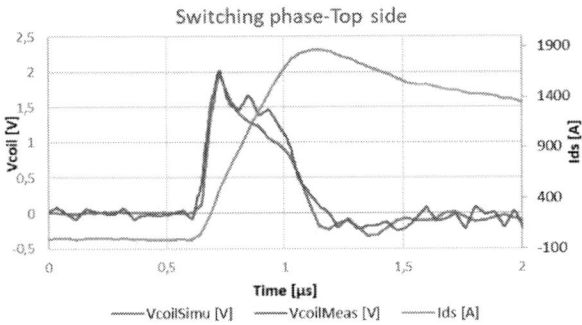

Figure 4: Measurement during a Turn-on

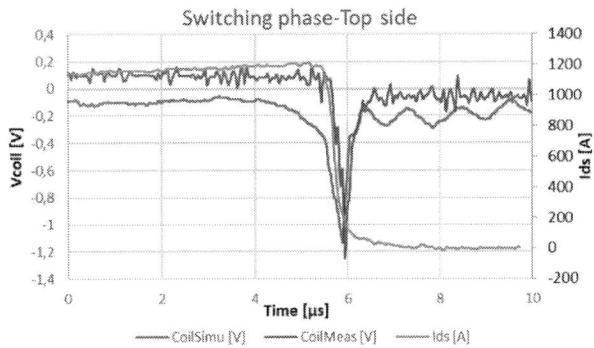

Figure 5: Measurement during a Turn-off Top switch

As shown, the numerical simulation results are correlated with the measured voltage during the switching phase, with a maximum value of 2V measured and 1.96V obtained through simulation during the Turn-on phase, and -1.25V measured and -1.13V simulated during the Turn-off phase.

It is noteworthy that the simulation signal waveform closely matches the measured signal waveform, despite a more complex external environment present during experiments than in our simulation.

Next, we replicate the process for the short-circuit phase by turning on the bottom switch in applying a constant gate voltage of 20V. We then switch the top component by sending a 3 µs pulse of 15V, as given by the supplier's datasheet, to the top gate.

In Figure 6, we can observe the short-circuit current reaching 2770 A, in response to this di/dt (~5kA/µs). The coupling coil measured a voltage of 1.45V at its terminals.

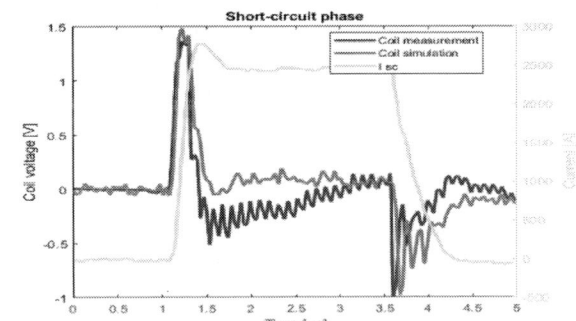

Figure 6: Measurement during a short-circuit.

We can observe that the coupling coil captures the negative di/dt. These results were accurately reproduced by the simulation and validated the simulation methodology applied to this sensor, with a maximum value of 1.47V simulated and 1.35V measured.

However, it was observed that the measured voltage during the normal switching phase was higher than during the short circuit. This can be explained by a higher di/dt during switching, 6kA/µs compared to 5kA/µs during the short circuit. Under these conditions, a positioning study has been performed based on numerical simulation (FEM) to determine the areas where the magnetic field is more concentrated during the short-circuit phase than during switching phases. The study will be presented in the next section.

4 Analysis and Optimization of Magnetic Field Distribution in Power Modules during Normal Switching and Short Circuit event Depending on Power Connection Design

In this section, we have examined the distribution of the magnetic field in the plan of driver's PCB during the high and low switching phases, as well as during the short-circuit phase. We will compare these different distributions of the magnetic field according to different external busbar layouts used for positive and negative terminals connection in a power converter. The objective is to identify an area where the field remains as constant as possible regardless of the connections' type.

Optimal positions were determined by coupling simulation results with an NSGA2 (Non-dominated Sorting Genetic Algorithm 2) algorithm, which maximizes the field present in a surface under short-circuit conditions compared to the field calculated under high switching conditions. NSGA2 [DPAM02], [MAJ19], [YD21] is a genetic algorithm used for multi-objective optimization problems. The basic algorithm is a research technique inspired by Darwinian natural selection [CD1859]. The objective is to find a compromise among several conflicting objectives.

NSGA2 is a multi-objective optimization method based on the non-dominated sorting algorithm. This algorithm sorts of candidate solutions based on their dominance over others. Non-dominant solutions are placed in the Pareto front, which represents a set of optimal non-dominant solutions, and the process is repeated until solutions in the Pareto front can no longer be improved.

NSGA2 also utilizes selection, crossover, mutation, and reinsertion techniques in the population to improve and maintain diversity among candidate solutions. The advantage of NSGA2 is that it provides a set of efficient solutions, closely located to the Pareto frontier.

The simulation results have identified three main zones, as shown in Figure 7.

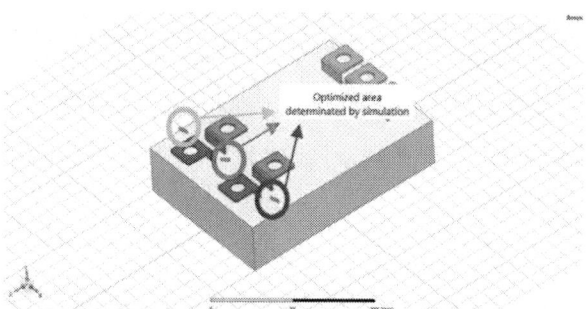

Figure 7: Best sensing area determinate by simulation (**Left**, middle and right)

Two areas on the sides of the power module and one in the center of the power connectors have been identified. It is important to note that these were calculated in an environment without busbars in order to simplify optimization [Y18]. The surfaces utilized correspond to the surface of a turn of the coupling coil used in these tests (16mm²). The surface in the center is positioned at an angle of 45 degrees and those on the sides at an angle of 15 degrees. It appears interesting to study the impact of different busbars in these areas during a short-circuit phase and a high switching phase. In Figure 8, the field level present in the three areas and its distribution in the plan define previously during the short-circuit phase and high switching phase have been observed.

negative terminals means that the field remains focused around these connectors. The simulation results using a standard busbar shows a shift of the field to the right. In terms of the calculated values in the three zones, it appears only the middle zone experiences higher magnetic field during the short-circuit phase compared to the switching phase. With the busbar, this field level even doubles (from 4.1mT to 8.5mT). We will now observe these variations in the calculated field levels in the three surfaces for busbars with different shapes in order to determine their impact on measuring di/dt during switching and short-circuit phases. The results are shown in Figure 9.

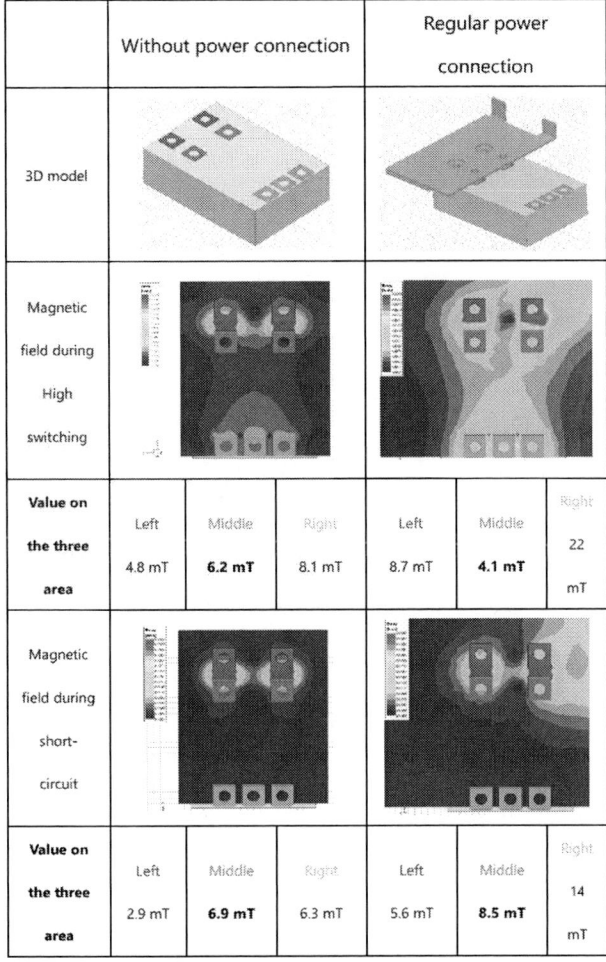

Figure 8: Compilation of simulation results without power connection and with a regular power connection.

During the switching phase, the power module without a power busbar generates a magnetic field as described in Figure 1, a). A symmetry of the magnetic field is observed between the left and right sides of the power module. The field levels are mainly concentrated around the positive electrodes' connectors and the AC output terminals. In the simulation considering a standard busbar, a shift of the magnetic field is observed towards the input of the HT power supply of the busbar. This imbalance is reflected in the calculated field level on each surface. Indeed, the central zone experiences less magnetic field (it goes from 6.2mT to 4.1mT) when the side zones are increased, specifically on the right side of the module, where its value more than doubles (from 8.1mT to 22mT). Regarding the short-circuit phase, the power module without a busbar gets a magnetic field distribution as described in Figure 2. The current entering via the positive terminals and exiting through the

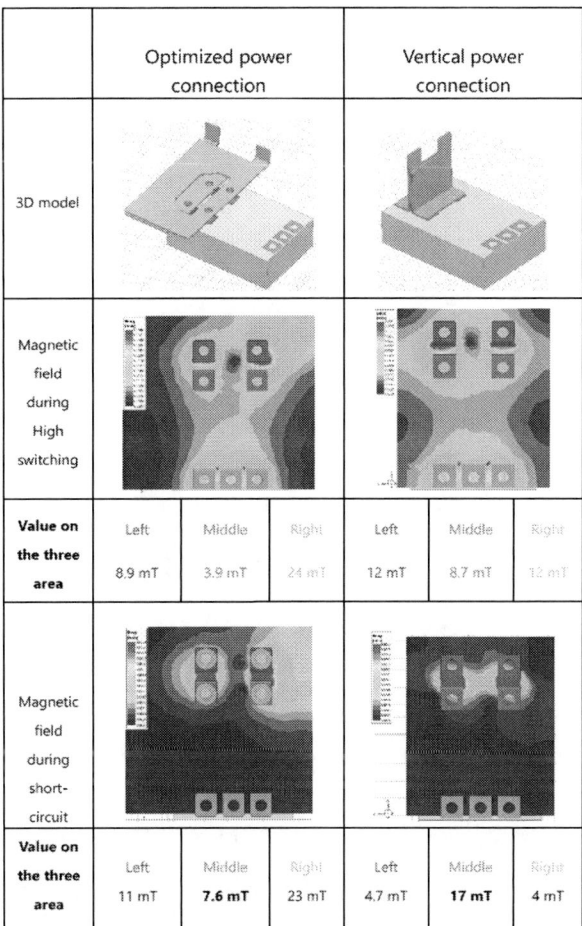

Figure 9: Compilation of simulation results with an optimized power connection and with a vertical power connection

Concerning the optimized power connection busbar, the field map remains similar to the standard connector. However, it is notable that the values for the central surface still vary twofold (3.9mT to 7.6mT). With the vertical power connector, the field distribution is slightly more symmetrical for switching and short-circuit phases. The calculated field values for the left and right zones are higher during the switching phase than during the short circuit. On the central surface, the field value varies significantly between the field calculated during switching and during short-circuit (8.7 mT to 17 mT). In conclusion, the central surface shows promising results in terms of the reproducibility of the results over different events and different busbar layouts. In other words, the field level in this zone doubles during the short-circuit phase compared to the calculated field during the switching phase. This observation is not only valid for the power module alone, but in a power converter environment, and for electrical design reasons, there will be no case without a busbar-type power connector. Based on these results and the various constraints related to power modules and their design, we are proposing to dedicate the areas presented in Figure 10 to the implementation of coupling coils to protect against short-circuits encountered during the life of the power module.

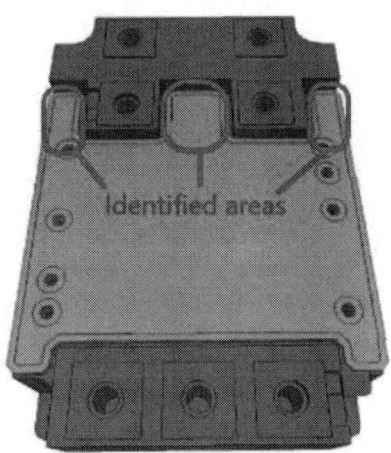

Figure 10: Identified areas for placement of coupling coils on gate driver PCB.

Among the planned zones, the study will focus on the center and right zones. The study will involve testing by placing two coupling coils in these zones and analyzing the different signatures depending on the high switching phase and short-circuit phase. The objective here is to provide the quicker protection for power modules against short-circuits using simple and very fast comparator threshold logic.

5 Measurement and Simulation on the selected area

In this study, the coupling coils are positioned lower than the surface determined by simulation for the purpose of integration into the zones identified in Figure 10. The middle coupling coil has an angle of 45 degrees and the one at the side has an angle of 50 degrees in order to follow the curvature of the magnetic field lines on the new positioning as presented in Figure 11.

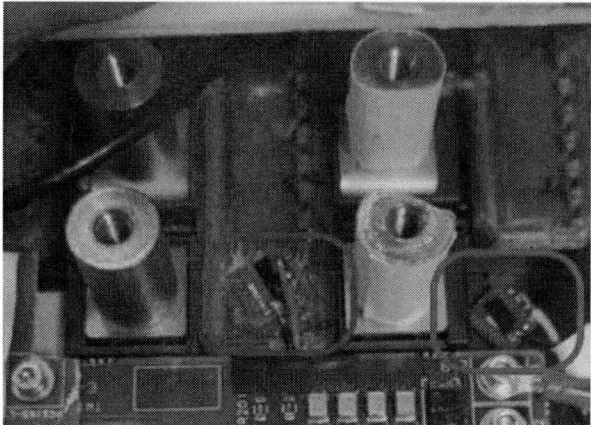

Figure 11: Coupling coil positioned on the optimizing position (Middle coupling coil, lateral coupling coil).

Initially, the tests were carried out by switching with a supply voltage Vds of 1800V and a switched current of 1200A. During the switching of the upper switch (Turn-on), the current increases in this switch and decreases in the diode of the lower switch. Figure 12 shows the measurement of the two coupling coils during the Turn-on event on the top switch.

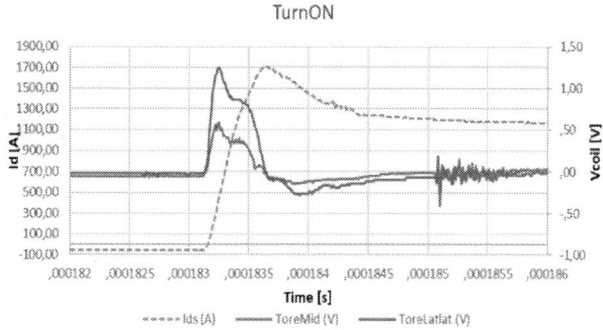

Figure 12: Coupling coil measurement during Turn-on event.

It can be observed that the switched current during Turn-on rises to 1700A before stabilizing at 1200A. The response to this di/dt (4.8kA/µs) is described by the blue and red curves. On the middle coupling coil, a measurement of 0.56V was obtained, and on the lateral coupling coil, one of 1.18V was obtained. The Turn-off results are described by Figure 13.

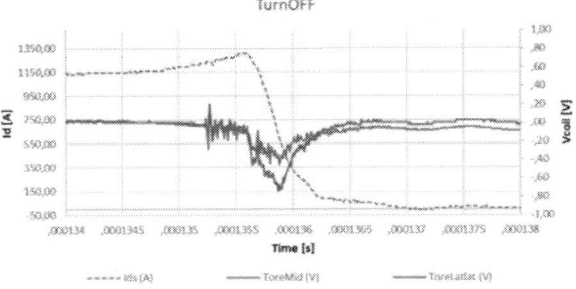

Figure 13: Coupling coil measurement during Turn-off event.

The di/dt during turn-off is -3.2kA/µs, which generates a measured voltage across the middle coil of -0.46V and -0.72V across the lateral coil. In order to compare these results to the voltages measured during a short-circuit phase, we carried out the short-circuit test by boosting the lower switch with a continuous voltage of 20V. This has the consequence of forcing the closure of the lower switch while the upper switch switches to allow the short-circuit current to pass from the HV+ terminal to the HV- terminal. The results of the test are presented in Figure 14.

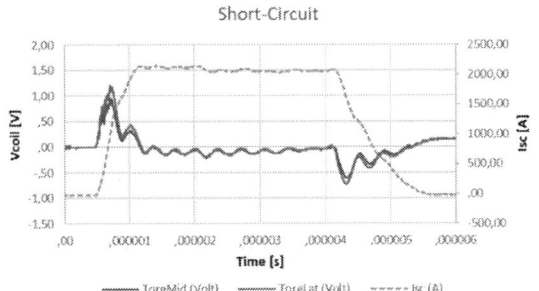

Figure 14: Coupling coil measurement during short-circuit event.

The short-circuit current reaches 2100A with a di/dt of 4.75kA/µs, which places the measurement of the coupling coils in a context comparable to the di/dt measured during the Turn-on switching of the upper switch (4.8kA/µs). At the lateral coupling coil terminals, a voltage of 1.2V is measured, which is similar to that measured during the Turn-on phase. However, at the middle coupling coil terminals, a voltage is measured which is approximately twice that measured during the Turn-on event, namely 0.96V. These results are consistent with those highlighted in the previous section and allow for the consideration of protection against short-circuits using a coupling coil placed at the center of

the HV connectors that would detect short-circuit phases using a simple voltage threshold and reaching very fast reaction time lower than the microsecond can be achieved.

6 Conclusion and perspectives

In this study, we explored the use of a coupling coil to improve short-circuit detection during switching events. Results indicate a promising potential for using coupling coils in power electronics for railway traction and offers avenues for further research and developments.in this field of solution. We compared various coupling coil layout and positioning and have found that small Rogowski coils were particularly useful for measuring power current variation, especially in complex environments with power busbars and power modules. Through Finite Element Method (FEM) simulation, we were able to optimize the sensor position and predict their electromagnetic behavior in complex environments. We also conducted switching tests at different current levels and researched di/dt measurements through High Voltage (HV) main terminals, highlighting the importance of external power busbar geometry on current detection robustness. Our simulation results also identified areas where significant magnetic field level occurs during the short-circuit phase compared to the switching phase. We successfully tested the coupling coil placed on the selected area, this location is placed in the center of HV connectors and can detect short-circuit event with a voltage threshold in a very short time. Some hundreds of nanoseconds can be targeted. This study highlights the potential of coupling coils for sensing applications in railway traction and demonstrates their feasibility for use in larger power electronics application domains.

The work presented in this paper contributes to the ongoing evolution of power electronics, proposing to take benefit of new standards power modules to enhance reliability and system robustness. Improving semiconductor protection, in particular against short circuit events, remains an important requirement. This work is raising some proposals for packaging improvements setting the stage for continued exploration, refinement, and fostering innovation in the realm of coupling coils for improved performance in industries such as railways applications, renewable energy, electric vehicles, and other industrial applications.

References

[R2R16] T. Wiik, "D1.2 New generation power semiconductor - Common specification for traction and market analysis, technology roadmap, and value cost prediction", Roll2Rail project, 2016, R2R-T1.1-D-BTS-030-07.

[RP23] A. Rafiq and S. Pramanick, "Ultrafast Protection of Discrete SiC MOSFETs With PCB Coil-Based Current Sensors," in IEEE Transactions on Power Electronics, vol. 38, no. 2, pp. 1860-1870, Feb. 2023, doi: 10.1109/TPEL.2022.3207594.

[RXMS20] D. Reiff, S. Xie, W. Mei and V. Staudt, "Short-circuit-current protection of SiC power MosFet including characterization by measurement," PCIM Europe digital days 2020; International Exhibition and Conference for Power Electronics, Intelligent Motion, Renewable Energy and Energy Management, Germany, 2020, pp. 1-6.

[YD21] Y. Dumollard « Méthodologie de simulation multiphysique du court-circuit dans les modules de puissance MOSFET SiC composant la chaîne de traction ferroviaire », Thèse en génie électrique, Université de Pau et des Pays de l'Adour, 2021. Français. <tel-03391683>.

[DPAM02] K. Deb, A. Pratap, S. Agarwal and T. Meyarivan, "A fast and elitist multiobjective genetic algorithm: NSGA-II," in IEEE Transactions on Evolutionary Computation, vol. 6, no. 2, pp. 182-197, April 2002, doi: 10.1109/4235.996017.

[MAJ19] M. A. Jamil, A. Alhindi, M. Arif, M. K. Nour, N. S. A. Abubakar and T. F. Aljabri, "Multiobjective Evolutionary Algorithms NSGA-II and NSGA-III for Software Product Lines Testing Optimization", IEEE 6th International Conference on Engineering Technologies and Applied Sciences (ICETAS), pp. 1-5, 2019. Doi: 10.1109/ICETAS48360.2019.9117500.

[CD1859] C. Darwin and L. Kebler, " On the origin of species by means of natural selection, or, The preservation of favoured races in the struggle for life", London: J. Murray, 1859.

[Y18] T. Yanagi et al., "Circuit simulation of a silicon-carbide MOSFET considering the effect of the parasitic elements on circuit boards by using S-parameters," 2018 IEEE Applied Power Electronics Conference and Exposition (APEC), San Antonio, TX, USA, 2018, pp. 2875-2878, doi: 10.1109/APEC.2018.8341425

[YD24] Y. Dumollard,V. Escrouzailles, E. batista, D. Tisne-Grimaud « Exploring coupling coil integration in Power SemiConductor Modules environment for sensing in Railway traction" in CIPS 2024. ISBN: 978-3-8007-6288-0

PCIM Europe 2024, 11– 13 June 2024, Nuremberg   DOI: 10.30420/566262239

Enabling Active Thermal Control via an Adaptive Multi-Voltage Gate Driver

Tianlong B. Albert [1], Lucas Radon[1], Rik W. De Doncker [1]

[1] Institute for Power Electronics and Electrical Drives (ISEA), RWTH Aachen University, Germany

Corresponding author: Tianlong B. Albert, post@isea.rwth-aachen.de
Speaker: Tianlong B. Albert, post@isea.rwth-aachen.de

Abstract

This paper presents a gate driver topology enabling active thermal control of SiC MOSFETs. Active thermal control of SiC MOSFETs reduces device stress, thus increasing their lifetime. The proposed gate driver allows dynamic conduction loss manipulation by modulation of the average on-state resistance. The modulation is achieved utilizing a simple low-cost circuit. The theoretical potential of conduction loss manipulation compared to switching loss manipulation is elaborated. The gate driver is tested and validated during converter operation. The results show the driver's promising loss manipulation capability. Furthermore, the results demonstrate the gate driver's potential for degradation diagnosis and thermal impedance spectroscopy due to sinusoidal loss modulation.

1 Introduction

The transition from fossil fuels towards renewable energy sources in many sectors, e.g. energy and transportation, leads to their electrification. Thus, the demand for power electronics increases. Simultaneously, the requirements for reliability and long lifetime remain high to ensure safe operation and preserve resources. However, power semiconductors and their packaging are the most safety-critical component of power electronic systems because they tend to fail first [1].
These failures occur due to thermal cycles leading to mechanical degradation of the thermal path in power electronic systems [2], [3]. The effects induced by mechanical stress in power semiconductors' packaging include bond wire lift-off [4], degradation of die attach solder [5], gate oxide degradation [6] and the thermal interface material (TIM) [7] respectively. As a consequence, active thermal control techniques have been developed to reduce the thermal stress [8], [9] and therefore the mechanical degradation of power semiconductors and their packaging.
Besides control of the cooling system, the manipulation of device losses is key to enable thermal control. Hence, intelligent gate drivers can be used. There are three main strategies of intelligent gate

driver topologies allowing device loss manipulation. These are the control of the gate current, adaptive adjustment of the gate resistance and control of the gate voltage [10]. Loss manipulation is typically done by one of the these three methods. Although the adaptive adjustment of the gate resistance [11] and manipulation of the gate current [12] provide promising results of device loss manipulation capability, they mainly effect the switching losses.
In this work, a method is proposed which manipulates the conduction losses of SiC MOSFETs. Compared to switching loss manipulation, this method can potentially offer a wider range of loss manipulation capability and low-cost implementation depending on the application and semiconductor choice, see 2.1. Therefore, an adaptive gate driver is developed which controls the gate voltage actively.
This article is organized as follows. After this introduction, Section 2 elaborates on the potential of conduction loss manipulation and the working principle of the developed adaptive gate driver. Next, Section 3 focuses on the experimental validation of the gate driver's loss manipulation capability during converter operation. This article concludes with a short summary of this work and an outlook on potential future research on this topic in Section 4.

2 Working principle of adaptive gate driver

The on-state gate voltage V_{gs} alters the on-resistance $R_{ds,on}$ of SiC MOSFETs and therefore influences conduction losses [13], [14]

$$P_{cond} = V_{on} \cdot I_d \cdot d = I_d^2 \cdot R_{ds,on}(V_{gs}) \cdot d. \quad (1)$$

The potential of conduction loss manipulation compared to switching loss manipulation is elaborated in 2.1 followed by the concept of the adaptive gate driver in 2.2 and 2.3.

2.1 Theoretical potential of conduction loss manipulation

To show the theoretical potential of conduction loss manipulation for MOSFETs, conduction and switching losses are compared to each other in the following.

The turn-on and turn-off switching energy of one switching period can be calculated by integrating the device losses during the switching events [15]

$$E_{on/off} = \int_{t_{on/off}} v_d(t) \cdot i_d(t)\, dt$$
$$\approx \frac{1}{2} \cdot V_d \cdot I_d \cdot t_{on/off}. \quad (2)$$

The resulting switching losses P_{sw} for one switching period can be calculated with

$$P_{sw} = f_{sw} \cdot (E_{on} + E_{off}). \quad (3)$$

$$t_{on+off} := t_{on} + t_{off} \quad (4)$$

The total device losses P_{loss} are the sum of conduction losses P_{cond} and switching losses P_{sw}, see Eq. (5). By inserting Eq. (1) to Eq. (4) into Eq. (5), an expression for an equivalent conduction voltage ν_{cond} and switching voltage ν_{sw} can be derived, see Eq. (6).

$$P_{loss} = P_{cond} + P_{sw} \quad (5)$$
$$= I_d^2 \cdot R_{ds,on}(V_{gs}) \cdot d + f_{sw} \cdot (E_{on} + E_{off})$$
$$= I_d^2 \cdot R_{ds,on}(V_{gs}) \cdot d + \frac{f_{sw}}{2} \cdot V_d \cdot I_d \cdot t_{on+off}$$
$$= I_d \left[\underbrace{I_d \cdot R_{ds,on}(V_{gs}) \cdot d}_{=:\,\nu_{cond}} + \underbrace{\frac{f_{sw}}{2} \cdot V_d \cdot t_{on+off}}_{=:\,\nu_{sw}} \right]$$
$$(6)$$

Comparing both voltages, ν_{cond} and ν_{sw}, there are dependencies of converter operation parameters,

i.e. phase current I_d, DC link voltage V_{DC} — with $V_{DC} = V_d$ — and switching frequency f_{sw}. Changing their values influences converter operation directly which is not always desired. However, both equivalent voltages ν_{cond} and ν_{sw} also depend on one more parameter, the on-resistance $R_{ds,on}$ and the switching times t_{on+off} respectively. Active gate drivers which influence the switching losses by manipulating switching times t_{on+off} need very complex, accurate and fast control circuits which have to be designed carefully [10], [16], [17]. On the other hand, the manipulation of the on-resistance $R_{ds,on}$ can be achieved with a simple gate driver concept presented in the following part.

2.2 Concept of adaptive gate driver

To achieve dynamic conduction loss manipulation, the following concept is proposed. A simplified schematic of the the gate driver is depicted in Fig. 1. The key part are two identical gate driver ICs IXDD614YI from IXYS connected in parallel at the output while being supplied by two different constant voltages, i.e. $V_{cc,1}$ and $V_{cc,2}$ with $V_{cc,1} > V_{cc,2}$. The gate driver ICs have an enable pin (EN) next to the usual PWM input pin (IN). Both gate driver ICs are connected to the same PMW signal with frequency f_{sw}. However, both enable pins are connected to a complementary signal turning on one gate driver IC while turning off the other. This modulation signal is a second PWM signal with a period of T_{mod} with duty cycle k which defines the interval for how long the PWM signal is high. The duty cycle k determines the time share of the period T_{mod} in which the gate driver IC with the higher supply voltage is activated. For the remaining time of T_{mod} the gate driver IC with the lower voltage is active.

The loss modulation is achieved by the switching pattern of this secondary PWM signal. It has two parameters, its constant frequency $f_{mod} = 1/T_{mod}$

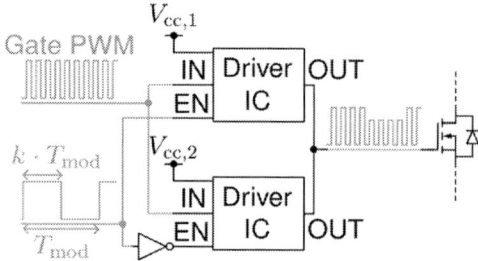

Fig. 1: Simplified schematic diagram of gate driver with $k = 0.5$

and variable duty cycle k. Since thermal sensing above 1 kHz is not necessary due to thermal time constants in power electronic systems [18], [19], the modulation frequency f_{mod} is kept constant at double the frequency with 2 kHz in this study. Hence, the losses are only actively modulated with the duty cycle k. The drivers' resulting switching pattern for $k = 0.5$ is depicted on the right hand side of Fig. 1.

2.3 Selection of gate driver ICs' supply voltages

The values of the two constant supply voltages of the gate driver ICs cannot be changed during converter operation. As a consequence, their selection has to be done with reference to the SiC MOSFETs' on-resistance before operation starts. Hence, the on-resistance $R_{\mathrm{ds,on}}$ of the power module used to validate the gate driver — the Wolfspeed half bridge module CAS480M12HM3 — is analyzed with a curve tracer.

The power module is heated-up to a temperature of 120 °C and then passively cooled-down to 30 °C. At every 10 K step, measurements are taken with a drain current of $I_{\mathrm{d}} = 50\,\mathrm{A}$. Simultaneously, for each temperature step the gate-source voltage V_{gs} is increased stepwise from 6 V to 15 V. The results depicted in Fig. 2 show the on-resistance $R_{\mathrm{ds,on}}$ with respect to the gate-source voltage V_{gs}. This data is used for the choice of the two constant supply voltages. Evidently, the on-resistance $R_{\mathrm{ds,on}}$ shows a negative temperature coefficient for gate-source voltages V_{gs} below 9 V and a positive temperature coefficient for gate-source voltages V_{gs} above 9 V. The higher supply voltage of the gate driver ICs is always chosen as the maximum gate-source voltage V_{gs} of the SiC MOSFETs, i.e. 15 V, to ensure minimal losses.

The lower supply voltage is determined with the $R_{\mathrm{ds,on}}$-characteristic of the SiC MOSFETs. In this work, the lower supply voltage is chosen as 9 V, to ensure a positive temperature coefficient throughout the operating range of the gate driver.

Additionally, the on-resistance $R_{\mathrm{ds,on}}$ at $V_{\mathrm{gs}} = 9\,\mathrm{V}$ is about 30 % higher compared to the $R_{\mathrm{ds,on}}$ at $V_{\mathrm{gs}} = 15\,\mathrm{V}$. This allows for an operating point with higher losses while not leading to extensively high losses. The gate driver introduces the trade of between efficiency and loss manipulation capability. Hence, it is desirable to keep the losses small while loss manipulation is still possible.

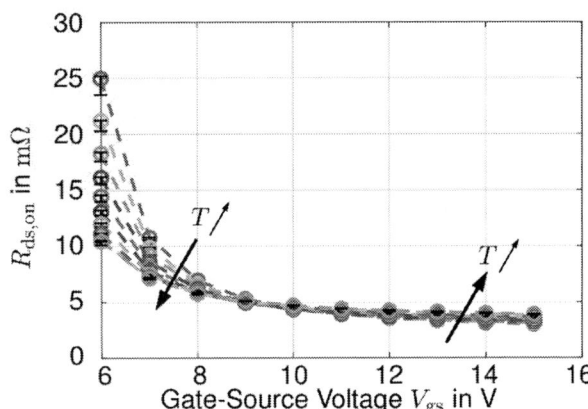

Fig. 2: $R_{\mathrm{ds,on}}$ of high-side of Wolfspeed SiC MOSFET power module CAS480M12HM3 over gate-source voltage V_{gs} and for different device temperatures

3 Experimental validation of gate driver's loss manipulation capability

In this section, the feasibility of the adaptive gate driver is evaluated. Experiments are carried out to analyze the gate driver's loss manipulation capability. The setup and execution of these experiments is explained in 3.1 followed by the experiments in 3.2 and 3.3.

3.1 Setup and execution of experimental validation

The driver is tested in a standard buck converter which is depicted in Fig. 3. The high side S_{high} of a Wolfspeed half bridge module CAS480M12HM3 is used as the main switch of the buck converter. Furthermore, the module's low side S_{low} is used as the freewheeling diode. Thus, the low side MOSFETs are switched off permanently with an identical gate driver. The buck converter is operated at a switching frequency of $f_{\mathrm{sw}} = 100\,\mathrm{kHz}$. The input voltage V_{in} is set to 50 V and a load is set to draw

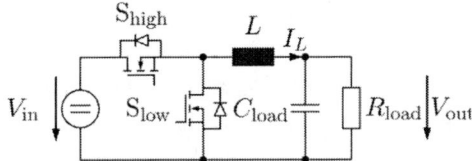

Fig. 3: Schematic diagram of buck converter used for experimental validation

PCIM Europe 2024, 11– 13 June 2024, Nuremberg   DOI: 10.30420/566262239

Fig. 4: Picture of gate driver PCBs connected to SiC half bridge module during lab test

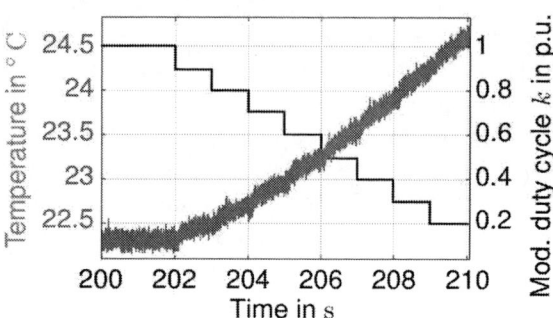

Fig. 5: Temperature measurement results for an experiment with k decreased stepwise from 1 to 0

an average current of $\overline{I}_L = 50\,A$. In this work, the duty cycle of S_{high} is fixed to 0.5 resulting in a output voltage V_{out} of 25 V. Figure 4 shows a picture of the buck converter during the driver validation experiments.

The loss manipulation is validated with a temperature measurement of one of the high side SiC MOSFETs. This is achieved by placing an Opsens fiber optic temperature sensor OTG-PM directly on top of a SiC MOSFET's surface which can be seen on the left hand side in Fig. 4. The only parameter that changes actively during the experiments is the modulation duty cycle k.

For all experiments, the buck converter was initially operated at $k = 1$ to ensure steady state operation regarding the SiC MOSFETs' temperature. After some minutes, the modulation strategies introduced in the next sections were started.

3.2 Stepwise increase of modulation duty cycle k

The gate driver's loss manipulation capability is first validated with this experiment. First, k is kept constant at 1 and then decreased stepwise by 0.1 every second. The resulting device temperature is depicted in Fig. 5. For $k = 1$ only the gate driver IC supplied with 15 V is activated resulting in the lowest losses. Therefore, the device temperature is the lowest measured. With decreasing k, the time the gate driver IC supplied with 9 V is increased with respect to the other one. Thus, the losses increase and lead to a higher temperature.

3.3 Sinusoidal modulation of duty cycle k

In addition to thermal cycle reduction, active loss manipulation enables aging diagnostics and thermal impedance spectroscopy. Degradation diagnosis of power electronic modules can be achieved by

utilizing the thermal response of the system when it is excited with sinusoidal losses [20],[21].

In the second experiment, the duty cycle k is sinusoidally modulated at a frequency of 2 Hz to excite sinusoidal conduction losses and provoke a temperature response. The device temperature in Fig. 6 shows a sinusoidal response with the same frequency.

Figure 7 shows three measurements with k sinusoidally modulated with 0.1 Hz, 1 Hz and 2 Hz. Evidently, the higher the frequency, the smaller the amplitude of the sine wave becomes. Comparing all three measurements, the device losses are modulated with an increasing sine frequency while the thermal impedance of the system is unchanged. This results in a decreased amplitude of the temperature response, since the time for the loss excitation is decreased. Consequently, loss modulation at lower frequencies yields a greater effect regarding thermal control compared to high frequencies. Arguably, this downside is not caused by the gate driving method but much more attributed to physical properties of the system.

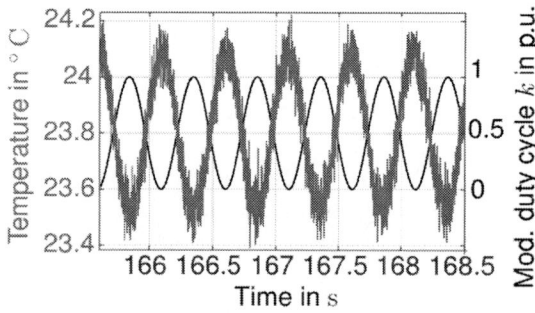

Fig. 6: Temperature measurement results for an experiment with k sinusoidally modulated with 2 Hz

1762

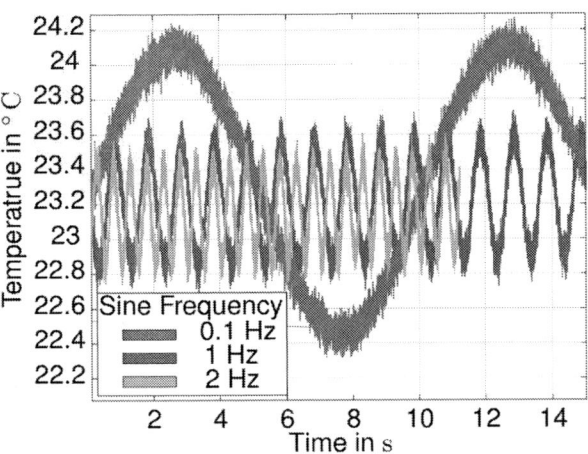

Fig. 7: Temperature measurement results for an experiment with k sinusoidally modulated with 0.1 Hz, 1 Hz and 2 Hz

4 Conclusion

In this work, a method for dynamic loss manipulation for SiC MOSFETs is proposed. A comparison between switching and conduction loss manipulation is made. An adaptive gate driver was build and validated during real converter operation. Evidently, the simple gate driver is capable of manipulating device losses dynamically with little effort regarding auxiliary circuits. In future work, the gate driver can be used in a closed control loop to reduce device temperature cycles and therefore extend its lifetime. Furthermore, the adaptive gate driver can be used in thermal condition monitoring applications enabling thermal impedance spectroscopy.

5 Acknowledgement

This work was funded within the project HiEFFI-CIENT and has received funding from the ECSEL Joint Undertaking (JU) under grant agreement no. 101007281 and the German Federal Ministry of Education and Research (BMBF, Support Code 16MEE0148). The JU receives support from the European Union's Horizon 2020 research and innovation program and Austria, Germany, Slovenia, Netherlands, Belgium, Slovakia, France, Italy, and Turkey.

References

[1] S. Yang, A. Bryant, P. Mawby, D. Xiang, L. Ran, and P. Tavner, "An industry-based survey of reliability in power electronic converters," *IEEE Transactions on Industry Applications*, vol. 47, no. 3, pp. 1441–1451, 2011. DOI: 10.1109/TIA.2011.2124436.

[2] B. Ji, V. Pickert, W. Cao, and B. Zahawi, "In situ diagnostics and prognostics of wire bonding faults in igbt modules for electric vehicle drives," *IEEE Transactions on Power Electronics*, vol. 28, no. 12, pp. 5568–5577, 2013. DOI: 10.1109/TPEL.2013.2251358.

[3] A. Morozumi, K. Yamada, T. Miyasaka, S. Sumi, and Y. Seki, "Reliability of power cycling for igbt power semiconductor modules," *IEEE Transactions on Industry Applications*, vol. 39, no. 3, pp. 665–671, 2003. DOI: 10.1109/TIA.2003.810661.

[4] K. B. Pedersen and K. Pedersen, "Bond wire lift-off in igbt modules due to thermomechanical induced stress," in *2012 3rd IEEE International Symposium on Power Electronics for Distributed Generation Systems (PEDG)*, 2012, pp. 519–526. DOI: 10.1109/PEDG.2012.6254052.

[5] K. Kurabayashi and E. Goodson, "Precision measurement and mapping of die-attach thermal resistance," *IEEE Transactions on Components, Packaging, and Manufacturing Technology: Part A*, vol. 21, no. 3, pp. 506–514, 1998. DOI: 10.1109/95.725215.

[6] W. Lai, M. Chen, L. Ran, S. Xu, N. Jiang, *et al.*, "Experimental investigation on the effects of narrow junction temperature cycles on die-attach solder layer in an igbt module," *IEEE Transactions on Power Electronics*, vol. 32, no. 2, pp. 1431–1441, 2017. DOI: 10.1109/TPEL.2016.2546944.

[7] B. Wunderle, D. May, J. Heilmann, J. Arnold, J. Hirscheider, *et al.*, "A novel concept for accelerated stress testing of thermal greases and in-situ observation of thermal contact degradation," in *2018 17th IEEE Intersociety Conference on Thermal and Thermomechanical Phenomena in Electronic Systems (ITherm)*, 2018, pp. 1071–1080. DOI: 10.1109/ITHERM.2018.8419568.

[8] J. Kuprat, C. H. van der Broeck, M. Andresen, S. Kalker, M. Liserre, and R. W. De Doncker, "Research on active thermal control: Actual status and future trends," *IEEE Journal of Emerging and Selected Topics in Power Electronics*, vol. 9, no. 6, pp. 6494–6506, 2021. DOI: 10.1109/JESTPE.2021.3067782.

[9] D. Murdock, J. Torres, J. Connors, and R. Lorenz, "Active thermal control of power electronic modules," *IEEE Transactions on Industry Applications*, vol. 42, no. 2, pp. 552–558, 2006. DOI: 10.1109/TIA.2005.863905.

[10] J. Henn, C. Lüdecke, M. Laumen, S. Beushausen, S. Kalker, *et al.*, "Intelligent gate drivers for future power converters," *IEEE Transactions on Power Electronics*, vol. 37, no. 3, pp. 3484–3503, 2022. DOI: 10.1109/TPEL.2021.3112337.

[11] C. H. van der Broeck, L. A. Ruppert, R. D. Lorenz, and R. W. De Doncker, "Methodology for active thermal cycle reduction of power electronic modules," *IEEE Transactions on Power Electronics*, vol. 34, no. 8, pp. 8213–8229, 2019. DOI: 10.1109/TPEL.2018.2882184.

[12] J. Henn, "Gate driver integrated closed-loop control for electromagnetic emissions and switching losses of wide bandgap power electronic converters," Published by RWTH Aachen University; Dissertation, Rheinisch-Westfälische Technische Hochschule Aachen, 2023, Dissertation, Rheinisch-Westfälische Technische Hochschule Aachen, Aachen, 2023, 1 Online–Ressource : Illustrationen, Diagramme. DOI: 10.18154/RWTH-2023-07726.

[13] M. Ruff, H. Mitlehner, and R. Helbig, "Sic devices: Physics and numerical simulation," *IEEE Transactions on Electron Devices*, vol. 41, no. 6, pp. 1040–1054, 1994. DOI: 10.1109/16.293319.

[14] D. L. Blackburn and D. W. Berning, "Power mosfet temperature measurements," in *1982 IEEE Power Electronics Specialists conference*, 1982, pp. 400–407. DOI: 10.1109/PESC.1982.7072436.

[15] J. Lutz, H. Schlangenotto, U. Scheuermann, and R. De Doncker, *Semiconductor Power Devices*, 2nd ed. Springer Cham, Feb. 2018, p. 714. DOI: https://doi.org/10.1007/978-3-319-70917-8.

[16] C. Lüdecke, "Compensating asymmetries of parallel-connected SiC MOSFETs using intelligent gate drivers," en, Ph.D. dissertation, RWTH Aachen University, 2022, p. 194. DOI: 10.18154/RWTH-2022-09587.

[17] G. Engelmann, "Reducing device stress and switching losses using active gate drivers and improved switching cell design," en, Ph.D. dissertation, RWTH Aachen University, Aachen, 2018. DOI: 10.18154/RWTH-2018-228973.

[18] M. A. Eleffendi and C. M. Johnson, "Application of kalman filter to estimate junction temperature in igbt power modules," *IEEE Transactions on Power Electronics*, vol. 31, no. 2, pp. 1576–1587, 2016. DOI: 10.1109/TPEL.2015.2418711.

[19] C. H. van der Broeck, R. D. Lorenz, and R. W. De Doncker, "Monitoring 3-d temperature distributions and device losses in power electronic modules," *IEEE Transactions on Power Electronics*, vol. 34, no. 8, pp. 7983–7995, 2019. DOI: 10.1109/TPEL.2018.2882402.

[20] C. H. van der Broeck, T. A. Polom, and R. W. De Doncker, "Degradation diagnosis of power modules based on thermal phase response sensing and artificial neural networks," *IEEE Transactions on Industry Applications*, vol. 60, no. 2, pp. 3438–3448, 2024. DOI: 10.1109/TIA.2023.3334704.

[21] I. Austrup, C. H. van der Broeck, T. B. Albert, S. Kalker, and R. W. de Doncker, "Diagnosing degradation in power modules using phase delay changes of electrical response," in *2022 IEEE 7th Southern Power Electronics Conference (SPEC)*, 2022, pp. 1–6. DOI: 10.1109/SPEC55080.2022.10058410.

PCIM Europe 2024, 11– 13 June 2024, Nuremberg   DOI: 10.30420/566262240

Innovative Gate Drive method TriC3™ for Motor

Hisashi Sugie
ROHM Co., Ltd., Japan

Corresponding author: Hisashi Sugie, Hisashi.Sugie@dsn.rohm.co.jp
Speaker: Hisashi Sugie, Hisashi.Sugie@dsn.rohm.co.jp

Abstract

Innovative Gate Drive method TriC3™ that considers reverse recovery characteristic is devised. At timing when the reverse recovery characteristic has a significant impact to suppress output ringing, the gate current must be reduced.   At other timing, the gate drive current can be increased to suppress heat generation. The result is reduced switching loss with suppressing EMI. Particularly it is useful for PWM motor driving on large current required. We have confirmed the heat suppression of Si MOSFET on actual device. This method shows 51% decreasing of heat effect compared to conventional method with suppressing EMI as well.

1   Introduction

To contribute to energy saving in motor system, we have improved the efficiency of driving MOSFET with suppressing EMI.

The reverse recovery characteristic of the diode is a key factor.

Figure 1 shows the operation of the diode during a forward to reverse bias voltage change. When a bias voltage is applied from forward to reverse, the carriers which are stored in the diode cause a current flow from the cathode to the anode. This is the reverse recovery characteristic.

If the forward current is large or slope of this current is steep, the reverse recovery peak current is also high.

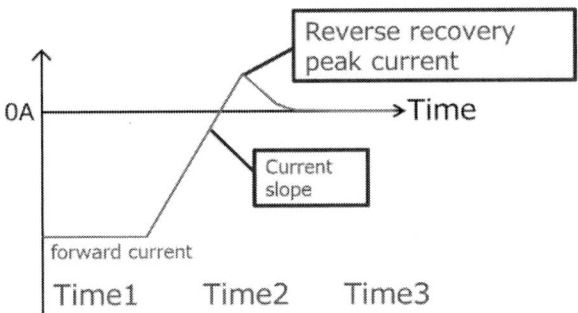

Fig.1(a) Operation when a forward to reverse bias voltage is applied.

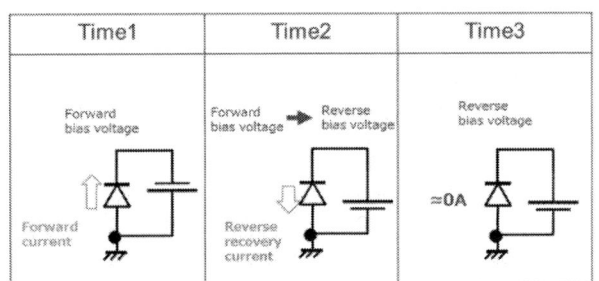

Fig.1(b) Description of Time1~3

Fig.1. Reverse recovery characteristic of the diode

In the motor system, we are doing the Pulse Width Modulation drive during larger motor current, as shown in the Figure 2(a).

Figure 2(b) is the current flowing through the upper and lower MOSFETs at source voltage rising.

At Time1, the Body diode of the lower MOSFET is used.

Time2 is immediately after the lower MOSFET is turned ON from OFF. The influence of reverse recovery is significant.

At Time3, there is a change in OUT voltage, but almost no change in power supply current($\approx$IHO).

The impact of reverse recovery to EMI is significant because there is a PWM switch timing while the motor current is large.

Usually following two topics are well discussed.

1765

- Concern about ringing and high EMI if the reverse recovery peak current is large and change in power supply current is high.
- If the gate current is set uniformly lower to avoid ringing due to reverse recovery effects, power loss of MOSFETs during the transition increases.

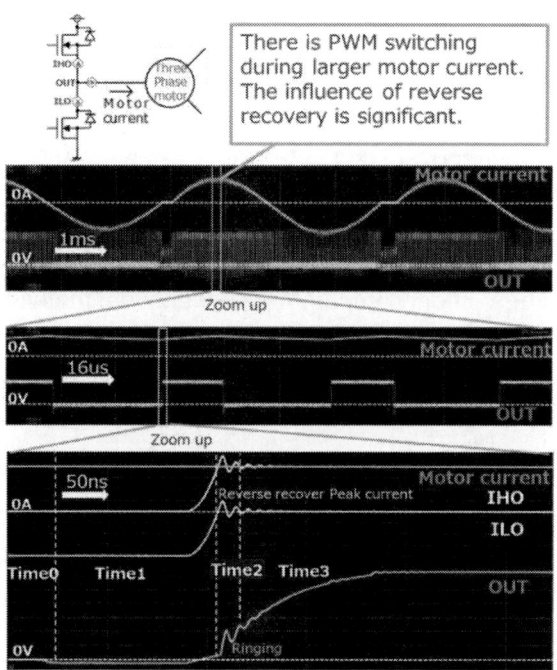

Fig.2(a) Driving three phase motor, one phase Motor waveform.

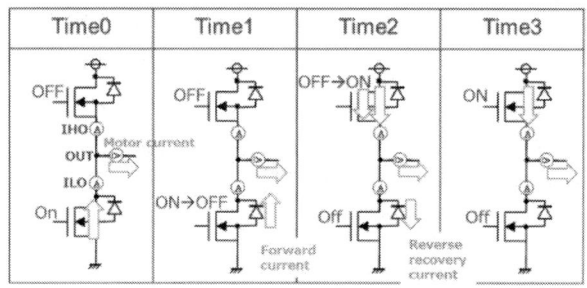

Fig.2(b) Description of Time0~3

Fig.2. PWM switching during motor drive

2 Innovative Gate Drive method TriC3™

2.1 Conventional method

Constant current drive is shown as an example of a conventional system (Figure3(a)). In order to reduce ringing and change in power supply current, the gate current must be reduced uniformly for constant current drive. Therefore, output slew rate is slower, increasing switching loss. The reverse recovery peak current and change in power supply current depends on the magnitude of the gate current at Time2. Especially at Time3, the timing with almost no change in power supply current does not affect the power supply EMI, but the power consumption of the Hi side MOSFETs is high (Figure3(c)).

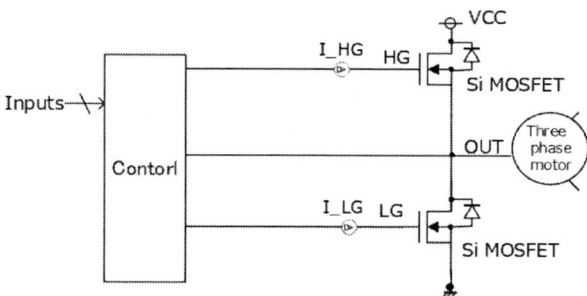

Fig.3(a) Circuit schematic of constant current
Si MOSFET: Ron=1mohm, Gate Charge Total (Qg)=50nC

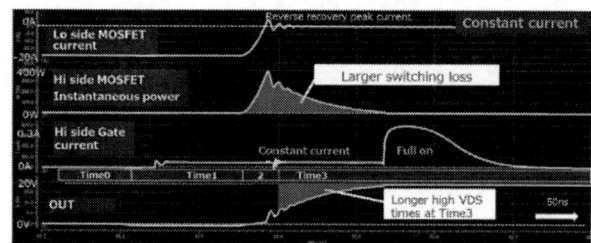

Fig.3(b) Simulation waveform

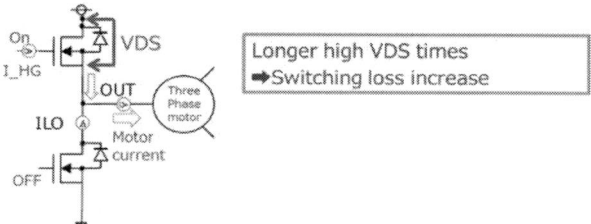

Fig.3(c) Status at Time3

Fig.3. Constant current drive at source voltage rising.

2.2 New method TriC3™

TriC3™ method is able to improve the power consumption of MOSFETs drastically at conventional EMI noise levels. Figure4 is the image.

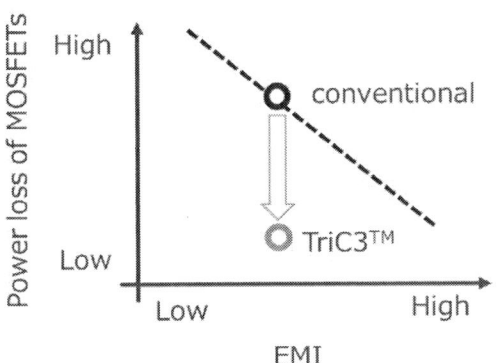

Fig.4 EMI vs. Power loss of MOSFET Image

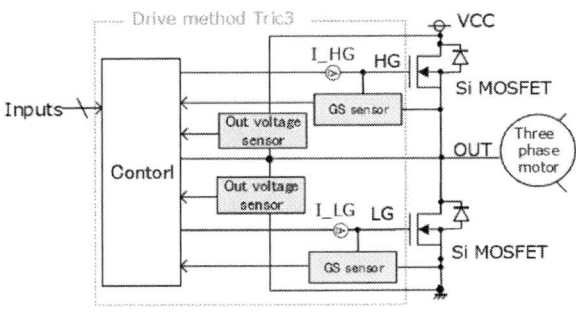

Fig.5(a) Circuit schematic of TriC3™
Si MOSFET: Ron=1mohm, Gate Charge Total (Qg)=50nC

Figure 5(a) shows the circuit configuration of TriC3™. Four sensors are used to actively control the gates. The four sensors are the Gate-Source (GS) sensor of the upper MOSFET, the GS sensor of the lower MOSFET, the lower potential sensor of the OUT voltage and the upper potential sensor of the OUT voltage. These sensors are required to perform at high speed, because this speed of operation has a significant impact to their efficiency.

Usually dead time is set with some margin, and it cause the deteriorate of power consumption. GS sensors are also used in shoot through prevention, so there is no need to consider dead time for the Inputs signals which is generated by controller.

GS sensors reduce dead time and contributes to lower power consumption of the MOSFETs.

The specific gate current switching of TriC3™ is in Figure 5(b). GS sensor of Hi side MOSFET is used from the 1step to the 2step, too. The OUT voltage sensor is used from the 2step to the 3step. Reduce the current in the 2step, where the effect of reverse recovery is particularly significant. In addition to this, reducing the gate current of the 2step also serves to reduce the slew rate of the supply current. This results in low EMI. To reduce power consumption during OUT transitions, increase the current in the 3step.

The comparison between the conventional constant current drive and the TriC3™ in the simulation is shown in Figure 5(c). Since shorter high VDS times at Time3, switching loss decrease (Figure5(d)). Slew rate of Output voltage is high by large gate current at Time3, but little effect on power supply EMI. The effect of TriC3™ is to reduce switching losses with suppressing EMI.

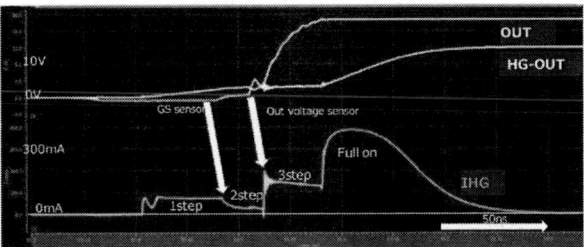

Fig.5(b) The specific gate current switching of TriC3™

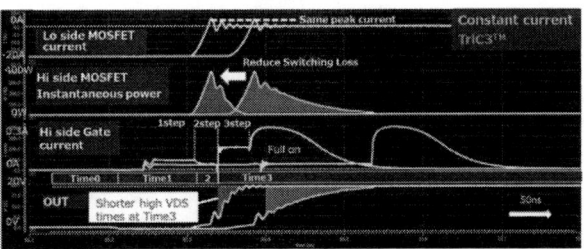

Fig.5(c) Comparison of Conventional Constant current drive vs. TriC3™ in simulation

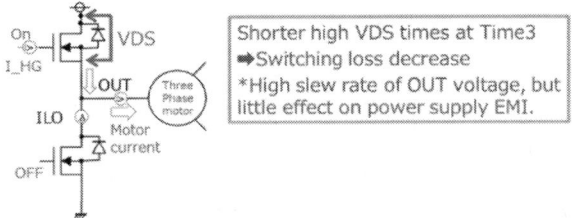

Shorter high VDS times at Time3
➡Switching loss decrease
*High slew rate of OUT voltage, but little effect on power supply EMI.

Fig.5(d) Status at Time3

Fig.5. TriC3™ drive at source voltage rising.

3 Measurement result

Experiments are carried out to compare device implementing TriC3™ with the conventional method. Figure 6(a) shows the measurement conditions, the settings are the same except for gate driver. Conventional one has a dead time setting of 1 us on the Inputs signals, and no dead time for TriC3™. As a result, as shown in Figure 6(b), under the same ringing matched conditions, the TriC3™ reduced the heat generation by -51% compared to the conventional system.

Power loss of Ron element is calculated based on the measured motor current and the measured value of Ron at Vgs=12 V. The actual measured Power loss of MOSFETs minus the calculated Ron is power loss of switching element.

TriC3™ has reduced the number of switching elements.

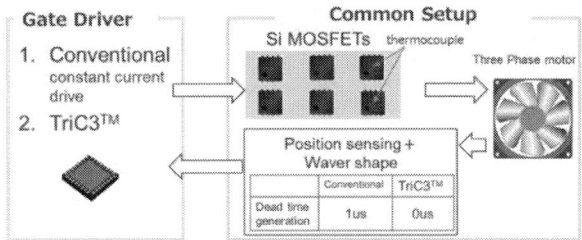

Fig.6(a) Measurement condition
Si MOSFET: Ron=1mohm, Gate Charge Total (Qg)=50nC
PWM frequency: 20kHz

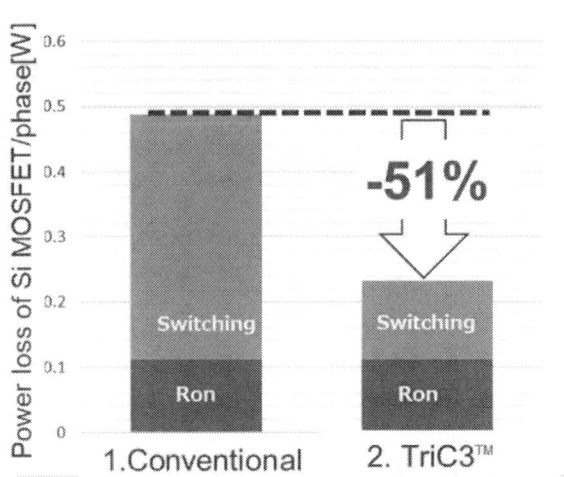

Fig.6(b) Comparison of Conventional (Constant current drive) vs. TriC3™ in Actual device

Fig.6. Measurement result in Actual device

4 TriC3™, an impact to a real application

TriC3™ controls EMI with reducing switching losses. As an application, it is expected to replace MOSFETs with higher Ron ones and still achieve the same power consumption and EMI as conventional (Figure7).

The higher Ron's MOSFET Qg tends to small. So if we can use the higher MOSFET it could be easier to control the gate

This contributes to miniaturization and cost effective solutions.

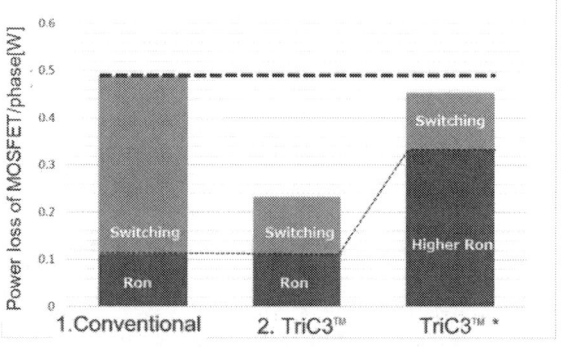

*Assumption of no change in characteristics other than Ron.

Fig.7(a) Adding the case image of higher Ron MOSFETs from Fig.6(b)

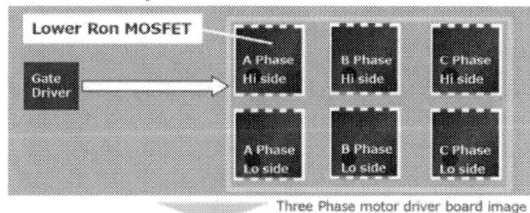

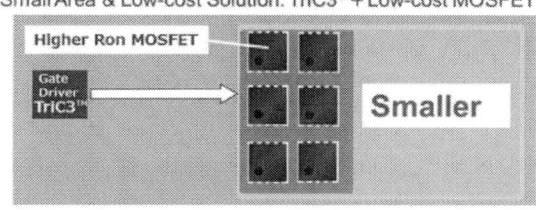

Fig.7(b) Image replaced by a board with a three-phase motor driver.

Fig.7. Image of replacement with higher Ron MOSFETs

5 Conclusion

Innovative Gate Drive method TriC3™ monitors the output status and switches the amount of current flowing through the gate. The effectiveness in reducing switching losses with suppressing EMI has been verified.

Simulation verification of power supply EMI is a future challenge. The correlation between simulation and actual equipment are required accuracy. Once it will be achieved, we could estimate EMI achievement level and power loss without doing try and error by actual device.

PCIM Europe 2024, 11– 13 June 2024, Nuremberg   DOI: 10.30420/566262241

A New Class of Solid State Isolators Enhances the Reliability of Solid State Relays

Wolfgang Frank
Infineon Technologies Germany, Germany

Corresponding author: Wolfgang Frank, Wolfgang.frank@infineon.com
Speaker: Wolfgang Frank, Wolfgang.frank@infineon.com

Abstract

Solid state isolators (SSIs) transmit signal and energy over an isolation barrier for driving the gates of power transistors. This is very convenient as it makes the external output-side voltage supply obsolete. However, the vast majority of SSIs today use optical isolation and only a tiny amount of energy can be transferred. It allows neither the fast turn-on of transistors nor provides added integrated circuits for the protection of power transistors. This paper introduces a new generation of SSIs that can turn power transistors on and off much faster and provide protection against thermal and current overloads. The new SSIs target much higher levels of reliability in solid state relays (SSRs). This paper also explains dynamic Miller clamping, fast turn-on and turn-off, overcurrent and overtemperature protection, and provides measurements to prove the performance of the new SSIs.

1   Introduction

Solid state isolators (SSIs) based on photovoltaic transmission of signal and energy for gate driving are known as the building blocks of solid state relays (SSRs). Various suppliers offer SSI products that are specified with output voltages, V_{OUT}, of up to 10 V and short circuit output currents, I_{SC}, of up to 50 µA [1] – [3]. The turn-on time, therefore, can be very long and last several milliseconds. This means that power transistors need to be over dimensioned with respect to their specified continuous operation when being switched under, for example, capacitive loads. The reverse bias safe operating area (SOA) diagram is, therefore, a relevant diagram for dimensioning power transistors.

Fig. 1 shows the voltage-current (V-I) trajectory of an SSR using the isolator PVI5033 that has a short circuit output current, I_{SC}, of 10 µA. The load current, I_{Load}, is 2 A at a voltage of 320 V with an inductive load of 808 µH. The SOA of the used power transistor IPT60T022S7 [5] is clearly violated as indicated in Fig. 1. In such a case, a power transistor with a lower $R_{ds(on)}$ that offers a larger SOA can be used. However, it would lead to higher costs and limits the application range of SSRs.

Fast turning on and off of the power transistor is, therefore, important for their use in SSRs.

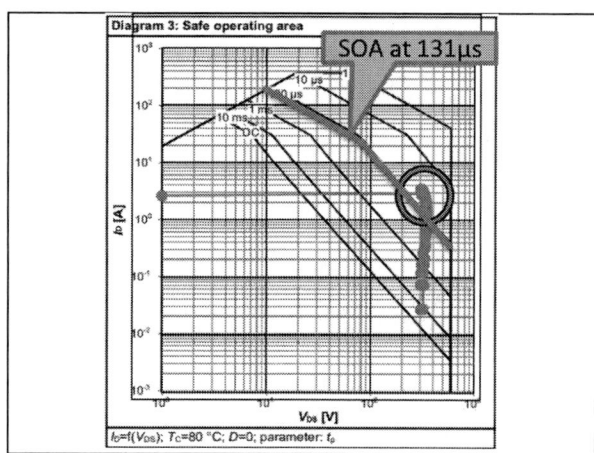

Fig. 1   V-I trajectory of an SSR turn-off using PVI5033 [2] at a load current of 2 A

SSRs integrated with SSIs and switching transistors in the same package are normally not protected. This makes them very sensitive to short circuits or other overcurrent events leading to electrical overstress. If SSRs operate at higher temperatures, unprotected overheating of the switching transistors can reduce their lifetime.

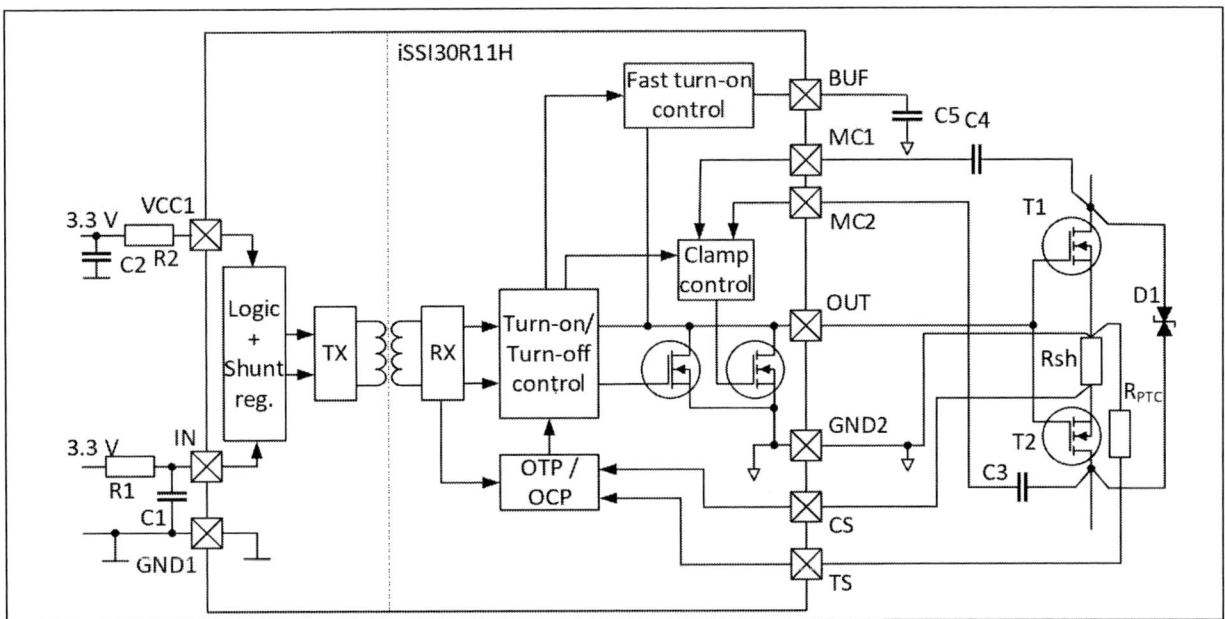

Fig. 2 Application circuit of the proposed solid state isolator with fast turn-on and protection features

It is, therefore, important to monitor protection features with respect to fast overcurrent events in the microseconds range as well as longer lasting overload events. In such cases, the SSR can be turned off to keep it safe.

2 New Class of Solid State Isolators

The proposed isolators from Infineon are based on Infineon's coreless transformer technology and allow for the transmission of much larger amounts of energy, beyond the capability of photovoltaic SSIs. They also provide tailored protection and operating features, such as overcurrent and overtemperature protection, that make it easy to protect power MOSFETs in general and CoolMOS™ S7 in particular. A block diagram of a typical application is shown in Fig. 2.

The input control side of the new solid state isolator (SSI) offers a 3.3 V logic-level compliant control input, IN, to activate and deactivate the energy flow from the input side to the output side. The output side provides protection and gate driving features such as:

- Overcurrent protection (OCP)
- Overtemperature protection (OTP)
- Dynamic Miller clamping (MC1, MC2)
- Fast turn-off

- Fast turn-on

The OCP and OTP functions manage the most sensitive root cause for electrical overstress in SSRs. OCP can handle the load short circuit in microseconds and the OTP manages any kind of overload that might not trigger OCP but leads to overheating of the relay circuit.

2.1 Dynamic Miller clamping

Dynamic Miller clamping helps keep the gates of the external power switches in a safe "off" state when fast dv/dt occurs at the drain terminals of the switches. A capacitor (C3 or C4 shown in Fig. 2) connects the drain of a switch with the input MC1 or MC2 of the iSSI. The dv/dt injects a displacement current into the MCx terminal and activates an integrated MOSFET that pulls the terminal OUT down to GND2. Thus, any kind of detrimental current through the switch's parasitic drain-gate capacitance gets shorted to GND2 and no self-turn-on of the switch is observed. Fig. 3 shows a comparison between the effect of dv/dt events with and without dynamic Miller clamping.

Fig. 3 (a) shows the waveforms without dynamic Miller clamping at a dv/dt of approximately 1.5 V/ns. The yellow waveform shows an offset of the gate voltage during the ramp phase of the power switch's drain voltage (green). This increased gate voltage can be dangerous for logic-level MOSFETs as they can turn on at a gate voltage as low as 0.7 V. The waveforms shown in

MOSFET when the voltage reaches the threshold of 10.4 V. The p-MOSFET connects the stored charge of the external capacitor to the gate, which results in a turn-on gate current of typically 400 mA. The related timing diagram is shown in Fig. 4 (b).

Fig. 5 depicts the test circuit of a capacitive turn-on including a protective turn-off. An SSR in AC-switch configuration connects a buffered voltage source to a load capacitor, C, of 1 μF.

A parallel resistor, R, discharges the capacitor before the turn-on of the SSR. A clamping transient-voltage suppressor (TVS) diode, D1, dissipates the energy of the parasitic stray inductance. The

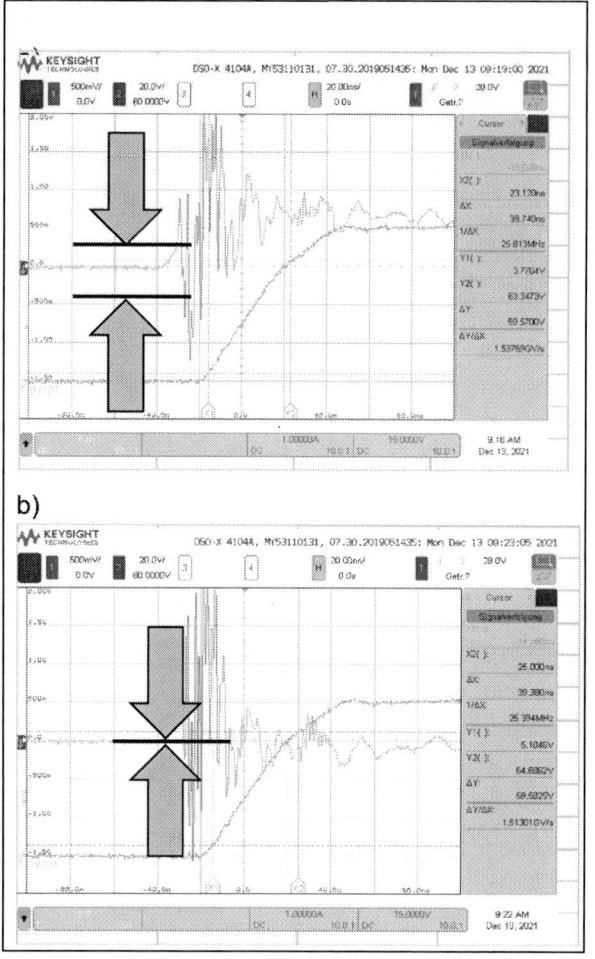

Fig. 3 dv/dt events (a) without and (b) with dynamic Miller clamping (yellow: Ch1 gate voltage 500 mV/div; green: Ch2 drain voltage 20 V/div, 20 ns/div) using a logic level MOSFET, ISZ0803NLS

Fig. 3 (b) are after the dynamic Miller clamp was activated with a coupling capacitor of 18 pF. It can be seen that the gate voltage (yellow) in this case stays at zero during and after the ramp.

2.2 Fast Turn-on and Overcurrent Protection (OCP) with Fast Turn-off

A fast turn-on is essential for switching capacitive loads. A high inrush current is visible if an uncharged capacitive load is activated. Therefore, the transistors of an SSR have to get into the full conduction mode in a very short time, for example within 1 μs. The new SSI incorporates a fast turn-on feature. The transferred energy of the isolator (RX) is parked on an external capacitor C3 as shown in Fig. 4 (a). The fast turn-on control monitors the voltage at this terminal and activates a p-

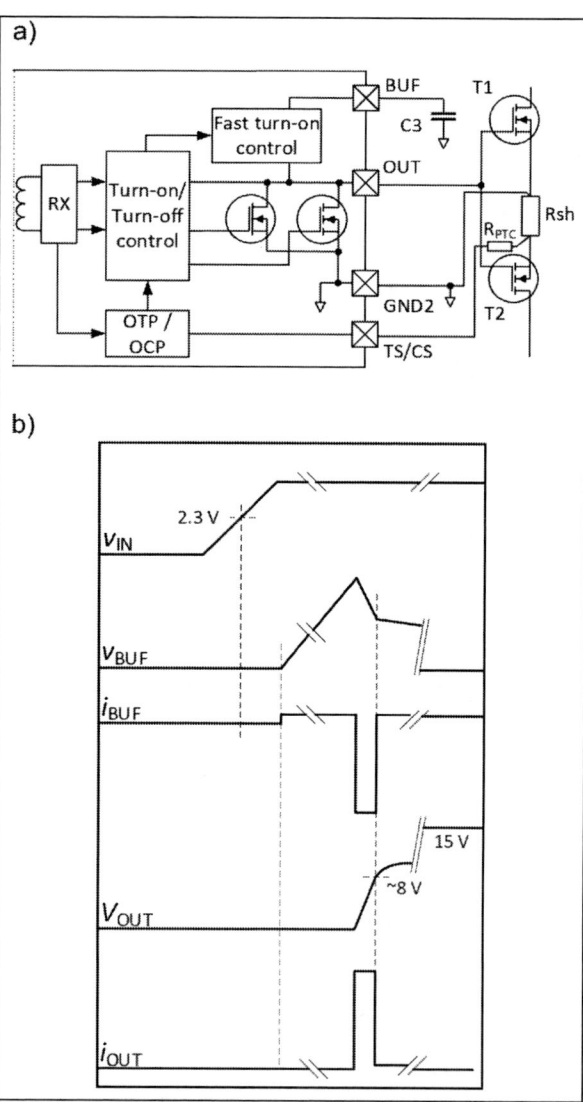

Fig. 4 Fast turn-on control with a) schematic and b) timing diagram

a)

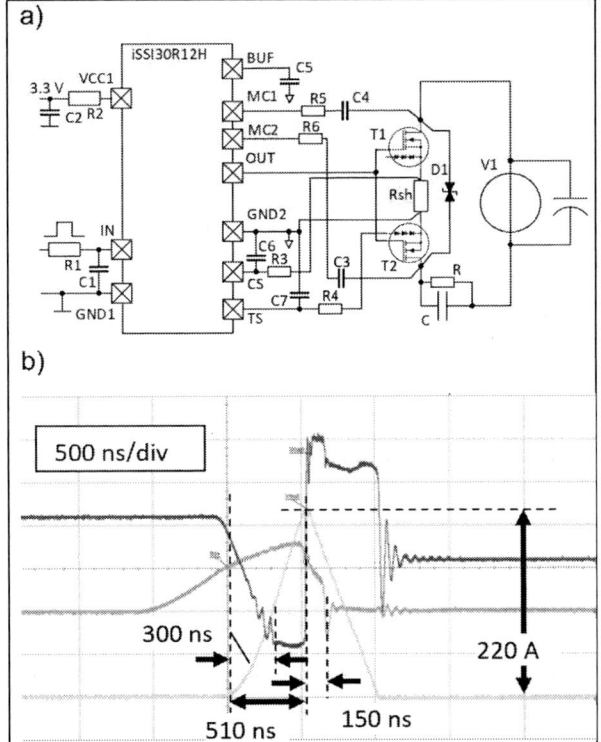

b)

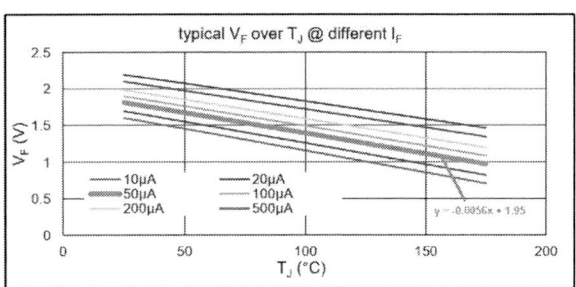

Fig. 7 Forward voltage as a function of the junction temperature in the integrated sensor of IPT60T022S7

threshold, $V_{TS,th}$, of the isolator is 1.095 V and is related to the junction temperature of the switching transistor, $T_J = 155°C$.

The aim of this measurement was to heat the switching transistors rather than switching high voltage under load. Therefore, the overtemperature measurement was conducted by operating an adjustable, 50 Hz transformer without load in a short circuit at a low output voltage, as shown in Fig. 6.

The voltage source V1 represents the transformer. In addition, a hot air fan heated up transistor T2 because the integrated temperature sensor of T2 was connected to the SSI's terminal, TS. The sensor of T1 was left floating.

The measured waveforms shown in b) of

a)

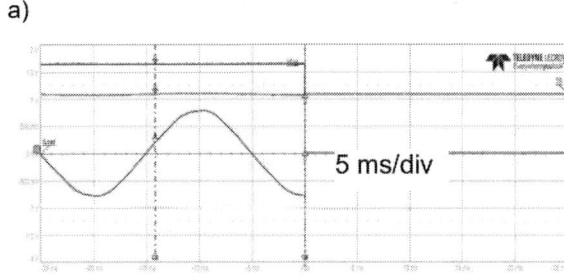

b)

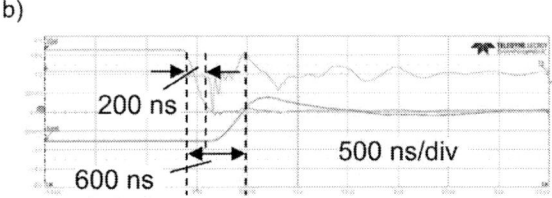

Fig. 5 Setup (a) and waveforms (b) of a capacitive turn-on (yellow: gate voltage 5 V/div, green: load current 50 A/div, blue: drain voltage 100 V/div, 500 ns/div) using CoolMOS™ IPT60T022S7.

uncharged capacitance acts as a short circuit after it is connected to a voltage supply and the inrush current slope is limited by the parasitic stray inductance of the charging path. The drain voltage ramps down to the related on-state level in 300 ns. In this setup, the overcurrent protection turns off the CoolMOS™ by means of a protective turn-off after 510 ns and reaching an amplitude, I_{Load}, of 220 A. Under such critical conditions, a fast turn-on of the power transistors, used in the SSRs, is essential to avoid high instantaneous losses or current filamenting in the active area of the switching transistor [6] – [8]. The gate voltage ramps down from approximately 7.5 V to 2.5 V within 150 ns ensuring that the drain current commutates safely into the TVS diode.

2.3 Overtemperature Protection (OTP)

The test circuit of Fig. 5 (a) shows the switching transistor, IPT60T022S7 (600 V, 22 mΩ), from Infineon [5] that contains a temperature sensor. Its characteristics are shown in Fig. 7. The bold orange line denotes the characteristic obtained with a constant bias current, $I_{TS,bias}$, of 50 µA that matches with the new SSI. The typical trigger

Fig. 6 Overtemperature test waveforms (yellow: v_{gs}, 5 V/div, green: temperature sensor voltage 0.5 V/div, red: load current 10 A/div)
a) Line period at turn-off
b) Zoomed-in turn-off interval

Fig. *6* are the source current i_S (red), the temperature sense signal at terminal TS (green), and the gate source voltage of the switching transistors (yellow). The temperature signal was 1.1 V, which was in the trigger tolerance zone of the over-temperature protection and the fast turn-off was triggered correctly. It can be seen that the gate voltage had ramped down to below 3 V within 150 ns. The source current had commutated to zero in the clamping element, within 600 ns.

3 Conclusion

The new class of SSIs provides excellent protection features and switching capabilities. The overcurrent protection can shut down fast ramping load currents above the protection level within a few 100 nanoseconds. Furthermore, the overtemperature protection is tailored to the combination of the new Infineon SSI and the CoolMOS™ S7 series. The Infineon SSI products make perfect use of the on-chip temperature sensor inside CoolMOS™ providing fast and reliable overtemperature protection.

The fast turn-on feature brings even high-voltage power MOSFETs into full conduction mode in less than 1 µs. This enables the dimensioning of the switching MOSFETs for nominal operation and not for switching transients. Thus, MOSFETs with higher on-state resistance are more often used for specific use cases than photovoltaic SSIs. Thus, a broader range of applications, which had been suffering from poor switching performance until now, can be addressed with the new series of SSIs from Infineon.

References

[1] Toshiba, "TLP3910", datasheet Rev. 1.0, Toshiba Electronic Devices and Storage Corporation, Japan, 2021.

[2] International Rectifier, "PVI5033R", datasheet, International Rectifier, USA, 2015.

[3] Skyworks Solutions, "SI8751", datasheet Rev. A, Skyworks Solutions, USA, 2022.

[4] Littelfuse, "FDA117_R01", datasheet, Littelfuse Inc., 2023.

[5] Infineon Technologies, "IPT60T022S7 600V CoolMOS™ SJ S7 Power Device," datasheet, Infineon Technologies AG, Germany, 2023.

[6] V. D'Alessandro, N. Rinaldi, P. Spirito, "Efficient thermal models of multicellular power devices," *9th THERMal INvestigations of ICs and systems (THERMINIC) Workshop*, 2003.

[7] G. Breglio, N. Rinaldi, P. Spirito, "3D dynamic electro-thermal simulator applied to a new cellular power MOS affected by electro-thermal instability," *6th THERMal INvestigations of ICs and systems (THERMINIC) Workshop*, Budapest, 2000.

[8] V. d'Alessandro, F. Frisina, and N. Rinaldi, "SPICE simulation of electro-thermal effects in new-generation multicellular VDMOS transistors," *2002 23rd International Conference on Microelectronics*, Proceedings (Cat. No.02TH8595), Nis, Yugoslavia, 2002, pp. 201-204 vol.1, doi: 10.1109/MIEL.2002.1003174.

PCIM Europe 2024, 11– 13 June 2024, Nuremberg   DOI: 10.30420/566262242

A Self-Driving 3-Level Active Gate Driver Network to Control the Switching Slew Rate for SiC MOSFETs

Vin Loong Choo[1], Martin Pfost[1]

[1] Chair of Energy Conversion, TU Dortmund University, Germany

Corresponding author: Vin Loong Choo, vinloong.choo@tu-dortmund.de
Speaker: Vin Loong Choo, vinloong.choo@tu-dortmund.de

Abstract

This paper presents a self-driving 3-level active gate driver (3L-AGD) for SiC MOSFETs. It consists of two simple networks that adjust the gate resistance R_G during the switching transient without the need for a µC or FPGA. This results in a less complex system compared to most current solutions. The double-pulse test confirms the operating principle of the self-driving 3L-AGD. In addition, the results with the 3L-AGD show that switching losses can be reduced by 5% to 20% with the same or lower current overshoot compared to a conventional gate driver.

1 Introduction

For a successful transition to a more sustainable and greener society, electric mobility and its infrastructure require systems with high efficiency and power density. To meet these requirements, wide-bandgap devices such as silicon carbide MOSFETs (SiC MOSFETs) are becoming increasingly attractive for high power density and high efficiency systems [1, 2]. SiC MOSFETs have faster switching speeds at high voltages due to lower intrinsic parasitics [3, 4]. However, this results in high dv/dt and di/dt switching transients that lead to increased voltage/current overshoot, higher switching losses and electromagnetic interference (EMI) [1, 4–6]. The conventional method of reducing switching speed is to increase the gate resistance R_G. This reduces overshoot and parasitic ringing, but also results in higher switching losses.

Active gate drivers (AGDs) aim to optimize switching losses and parasitic ringing by adjusting the waveform of the gate-source voltage V_{GS} during the turn-on and turn-off switching transients. Changing the total gate resistance R_G around the Miller plateau has been shown to be an effective way to minimize overshoot and losses [2, 7, 8]. An effective dampening of current overshoot and ringing can be achieved by increasing R_G during the Miller plateau [2,7,8]. Gate driver voltage, gate current, gate-source capacitance C_{GS}, or gate resistance can be used to adjust the switching transient [1]. In addition, current solutions for AGDs can be divided into open-loop and closed-loop control implementations [3].

Open-loop AGDs change the gate voltage, current, or resistance at fixed times during the switching transient. This results in an optimization of the switching transient for an operating point [9–14]. In contrast, closed-loop AGDs use one or more feedback signals to control the gate waveform. The feedback signals are typically the gate-source voltage V_{GS}, drain-source voltage V_{DS}, drain current I_D, or the switching slope of V_{DS} and I_D. When combined with a controller such as an FPGA, the V_{GS} waveform can be shaped into a multi-level waveform depending on the operating conditions. However, the resulting closed-loop AGDs typically result in complex systems that require high-speed components to be used effectively [4, 15].

The 3-level AGD (3L-AGD) presented in this paper is a continuation of the work presented in [16] and [17]. It is a self-driving system that adjusts the gate resistance without feedback signals, control units, digital signals, or fixed timings to control additional active components. In addition, the presented self-driving 3L-AGD is significantly less complex than closed-loop systems and shows reduced switching losses compared to a conventional gate driver.

2 System Overview

The self-driving 3L-AGD network consists of an auxiliary gate driver (AuxGD) network, in parallel with the turn-on gate resistor $R_{G,on}$. The AuxGD dynamically changes the total gate resistance, resulting in an increase in the gate resistance value during the Miller plateau and a decrease in the gate resistance value before and after the Miller plateau. This results in a 3-level gate-source waveform, resulting in an overall faster switching transient with optimized switching losses and reduced overshoot and parasitic ringing.

Fig. 1 shows the double-pulse measurement setup with a simplified view of the 3L-AGD. T1 is a SiC MOSFET, D_1 is a SiC Schottky diode, and L_{Load} is the load inductance of 168 µH. The current I_S is measured with a current viewing resistor shunt from T&M Research Products, V_{GS} is measured with a high-bandwidth low-voltage differential probe, and V_{DS} is measured with a high-bandwidth high-voltage differential probe.

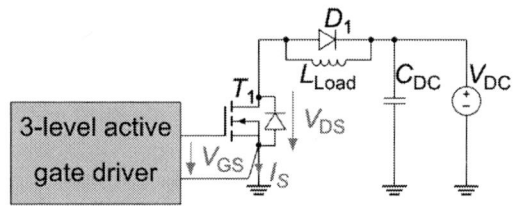

Fig. 1: Schematic of the double-pulse measurement setup and a simplified view of the 3L-AGD.

A more detailed view of the 3L-AGD network, including the gate driver IC with the on/off gate resistors, is shown in Fig. 2. The conventional gate driver (CGD) network consists only of the gate driver IC and the gate resistors $R_{G,on}$ and $R_{G,off}$ shown in Fig. 2. Measurements made with the CGD network are used as a reference to evaluate the performance of the proposed 3L-AGD network. Parallel to the gate resistor $R_{G,on}$ is the AuxGD network, which enables the self-driving gate waveform shaping. The AuxGD network does not affect the gate driver IC and operates independently. The effect of the 3L-AGD network on the turn-off switching transient was shown to be minimal in [17]. Therefore, the turn-off switching is not further investigated in this paper.

The AuxGD network parallel to $R_{G,on}$ is shown in detail in Fig. 3. It consists of two networks, the n-AuxGD with an n-channel enhancement MOS-

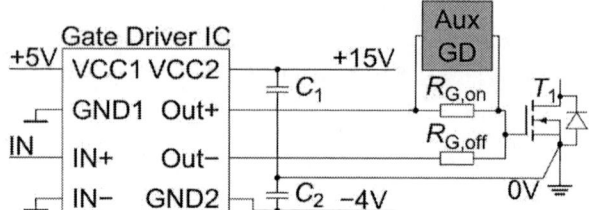

Fig. 2: Schematic of the 3L-AGD network consisting of the gate driver IC, its power supplies and the auxiliary gate driver network AuxGD in parallel to $R_{G,on}$.

FET, the RG-nMOS, parallel to $R_{G,on}$, and the p-AuxGD with a p-channel enhancement MOS-FET, the RG-pMOS, parallel to $R_{G,on}$. Each gate of the RG-nMOS and RG-pMOS is pre-biased and controlled by its own adjustable voltage regulator circuit, see Fig. 3.

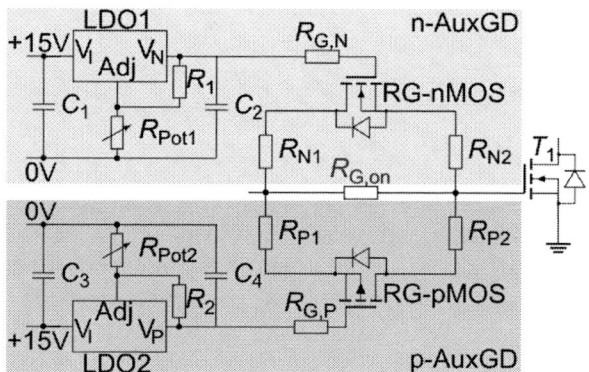

Fig. 3: A detailed view of the auxiliary gate driver network AuxGD, which consists of the two circuits n-AuxGD and p-AuxGD connected in parallel with $R_{G,on}$.

Both adjustable voltage regulator circuits are supplied from the same +15 V and 0 V as the gate driver IC shown in Fig. 2. This results in the bias voltages V_N and V_P, both referenced to the same ground potential of the SiC MOSFET T1, see Fig. 3. The entire system shown in this paper is implemented on the same PCB as the double pulse measurement setup. This allows for a more compact system without the need for an external power supply or software to change the bias voltages V_N and V_P. The AuxGD network shown in Fig. 3 results in a resistor network that can be summarized as

$$R_{Tot} = R_{G,on} \parallel (R_{ON,N} + R_{N1} + R_{N2})$$
$$\parallel (R_{ON,P} + R_{P1} + R_{P2}) . \qquad (1)$$

R_{Tot} is the total turn-on gate resistance applied to T1. The total value of R_{Tot} depends on the

on-resistance of both silicon MOSFETs, $R_{\mathrm{ON,N}}$, $R_{\mathrm{ON,P}}$, and the gate resistor $R_{\mathrm{G,on}}$, see (1).

The value for V_{N} is calculated with

$$V_{\mathrm{N}} = V_{\mathrm{TH,N}} + V_{\mathrm{MIL,MIN}} \tag{2}$$

and for V_{P} with

$$V_{\mathrm{P}} = V_{\mathrm{TH,P}} + V_{\mathrm{MIL,MAX}} \cdot \tag{3}$$

$V_{\mathrm{TH,N}}$ is the threshold voltage of the RG-nMOS, $V_{\mathrm{TH,P}}$ is the threshold voltage of the RG-pMOS, $V_{\mathrm{MIL,MIN}}$ and $V_{\mathrm{MIL,MAX}}$ is the range of the Miller plateau of the SiC MOSFET, all taken from their datasheets.

A picture of the measurement setup with the proposed self-driving 3L-AGD network is shown in Fig. 4.

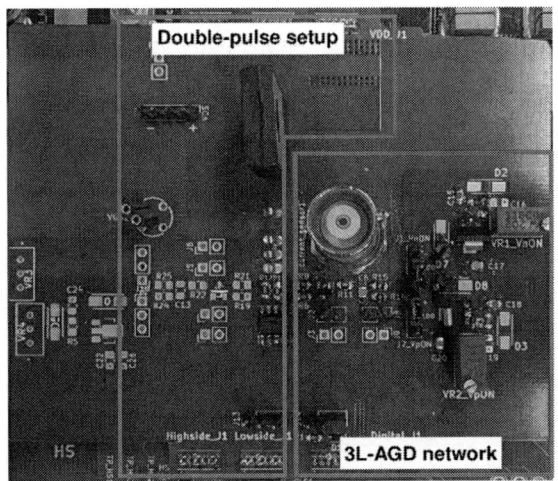

Fig. 4: Image of the double-pulse measurement setup and the investigated 3L-AGD network with its auxiliary gate driver network.

3 Operating Principle

The operating principle of the 3L-AGD, the transition from low gate resistance to high gate resistance before the gate-source voltage reaches the Miller plateau and the transition back to low gate resistance after the Miller plateau, is shown in Fig. 5. Fig. 5 shows the waveform of the gate-source voltage V_{GS} of T1 during an idealized turn-on switching event, without parasitic effects such as ringing or overshoot. The total gate resistance R_{Tot}, according to (1), and its dynamically changing resistance value are also shown in Fig. 5.

The bias voltages V_{N} and V_{P}, see Fig. 3, allow the dynamically changing total gate resistance R_{Tot}

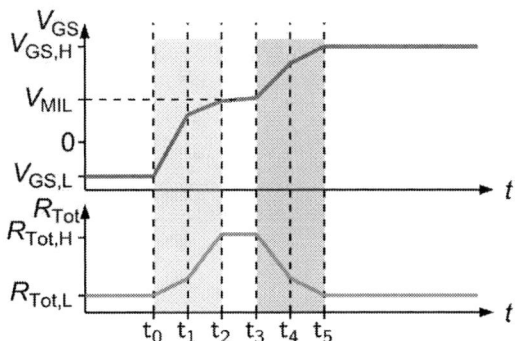

Fig. 5: Idealized waveform of V_{GS} for the self-driving 3L-AGD network and the changing value of R_{Tot} during the turn-on switching transient.

shown in Fig. 5. Due to the parallel connection to $R_{\mathrm{G,on}}$, both RG-nMOS and RG-pMOS source potentials follow the gate source potential V_{GS} of T1, see Fig. 3. Applying the bias voltages V_{N} and V_{P} causes the RG-nMOS to be in the on-state and the RG-pMOS to be in the off-state at the beginning of the turn-on switching event, see Fig. 6.

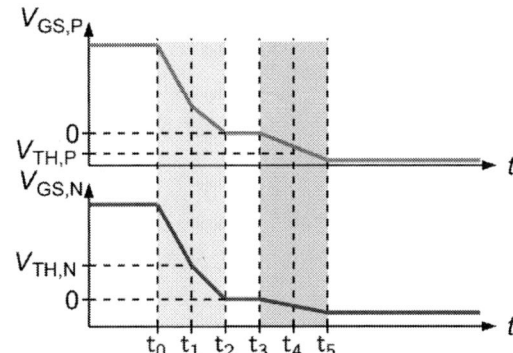

Fig. 6: Idealized waveform of $V_{\mathrm{GS,P}}$ and $V_{\mathrm{GS,N}}$ for the self-driving 3L-AGD network during the turn-on switching transient.

As shown in Fig. 6, the gate source waveforms of RG-nMOS, $V_{\mathrm{GS,N}}$, and of RG-pMOS, $V_{\mathrm{GS,P}}$, are complementary to V_{GS}. Increasing V_{GS} results in decreasing $V_{\mathrm{GS,N}}$ and $V_{\mathrm{GS,P}}$, see Fig. 5 and Fig. 6.

The switching states of RG-nMOS, RG-pMOS and the resulting total values of the gate resistance R_{Tot} during the corresponding time periods can be calculated with (1) and are summarized in Tab. 1.

At the beginning of the turn-on switching transient for $t < t_0$, the switching state of the RG-nMOS is on and off for the RG-pMOS. This results in $R_{\mathrm{Tot}} = R_{\mathrm{Tot,L}}$ and a fast switching transient. From $t_0 - t_2$ the n-AuxGD network transitions R_{Tot} from $R_{\mathrm{Tot,L}}$ to the highest resistance value, reaching $R_{\mathrm{Tot}} = R_{\mathrm{Tot,H}} = R_{\mathrm{G,on}}$ at the beginning of the Miller plateau.

Tab. 1: Overview of switching states of RG-nMOS, RG-pMOS and the total gate value of R_{Tot} during the turn-on switching transient of T1.

Time	RG-nMOS	RG-pMOS	R_{Tot}
$t < t_0$	On	Off	$R_{Tot,L}$
$t < t_2$	On→Off	Off	$R_{Tot,L} \rightarrow R_{Tot,H}$
$t_2 \leq t \leq t_3$	Off	Off	$R_{Tot,H}$
$t \geq t_3$	Off	Off→On	$R_{Tot,H} \rightarrow R_{Tot,L}$

With $R_{Tot} = R_{Tot,H}$ for $t_2 - t_3$ the switching speed is reduced during the Miller plateau, which successfully dampens the overshoot and ringing. During $t_3 - t_5$ the p-AuxGD changes R_{Tot} back to a low resistance value, which increases the switching speed and reduces the switching losses.

This gate-source waveform shaping optimizes the switching transient to reduce switching losses while maintaining the same overshoot and ringing compared to a conventional gate drive strategy with a static value for the gate resistance.

4 Experimental Results

Measurements with the same or different $R_{G,on}$ values for the 3L-AGD and CGD networks are compared to show the effect of the 3L-AGD network. Measurement results of V_{GS}, V_{DS} and I_S for the 3L-AGD and CGD gate driver networks with $R_{G,on} = 10\,\Omega$ are shown in Fig. 7 and Fig. 8.

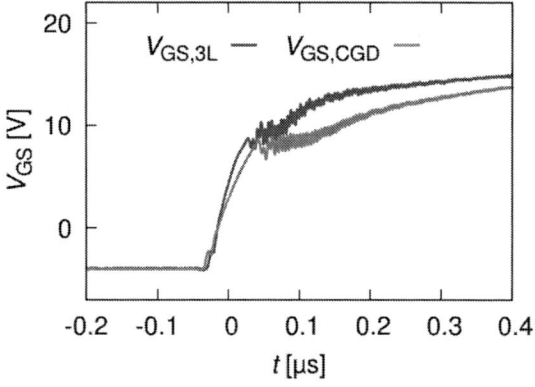

Fig. 7: Measurement of V_{GS} for CGD and 3-level AGD with $V_{DC} = 800\,$V, $I_{Load} = 20\,$A, $V_N = 9.6\,$V, $V_P = 6\,$V, $R_{G,on} = 10\,\Omega$ in the 3L-AGD and CGD network.

The voltages V_N and V_P are set to $V_N = 9.6\,$V and $V_P = 6\,$V so that the RG-nMOS is turned off at the end of t_2 and the RG-pMOS is turned on at the beginning of t_3. Both V_N and V_P are calculated using (2) and (3) to satisfy the operating

conditions shown in Fig. 5 and Fig. 6 and remain at these values for all measurements shown. It can be observed in Fig. 8 that the switching transient of the 3L-AGD is significantly faster, but has a similar peak current $I_{S,PEAK}$ as for the CGD network with $R_{G,on} = 10\,\Omega$. This overall faster switching transient results in a reduction of approximately 18.35% in switching losses, see Fig. 8.

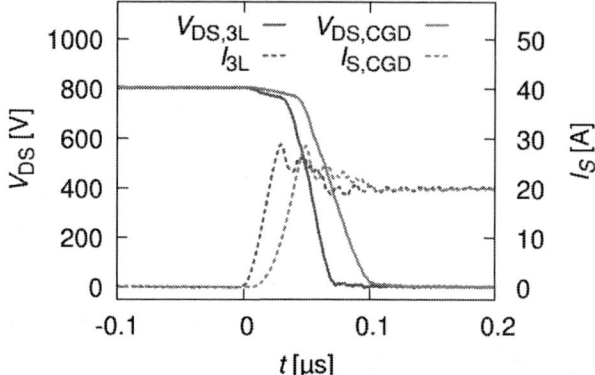

Fig. 8: Measurement of V_{DS} and I_S for CGD and 3-level AGD with $V_{DC} = 800\,$V, $I_{Load} = 20\,$A, $V_N = 9.6\,$V, $V_P = 6\,$V, $R_{G,on} = 10\,\Omega$ in the 3L-AGD and CGD network.

In the measurements shown in Fig. 9 and Fig. 10 the values of $R_{G,on}$ for the 3-level active gate driver and the CGD are chosen for a comparable turn-on switching transient for V_{GS}. This results in $R_{G,on} = 10\,\Omega$ in the 3L-AGD network and $R_{G,on} = 5\,\Omega$ in the CGD network.

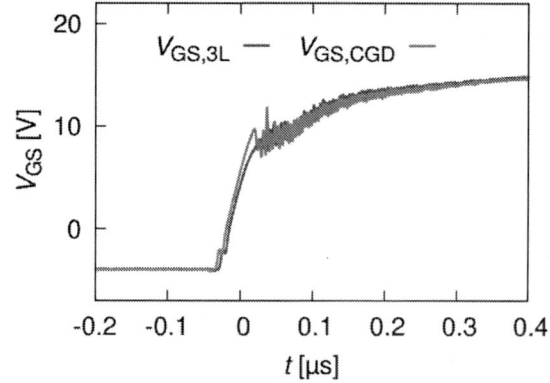

Fig. 9: Measurement of V_{GS} for CGD and 3-level AGD with the same conditions as in Fig. 7, but now with $R_{G,on} = 10\,\Omega$ in the 3L-AGD network and $R_{G,on} = 5\,\Omega$ in the CGD network.

Except for the CGD network around the Miller plateau, which has a faster switching speed, the V_{GS} waveforms are the same for both networks, see Fig. 9. However, the faster switching speed around the Miller plateau of the CGD network results in a significantly higher peak current $I_{S,PEAK}$, see Fig. 10. $I_{S,PEAK}$ for the CGD network is approximately 9.6% higher, but the faster switching speed results in lower switching losses of 14.65% compared to the 3L-AGD network, see Fig. 10.

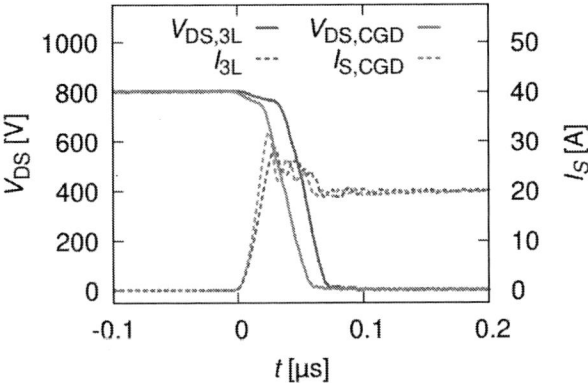

Fig. 10: Measurement of V_{DS} and I_S for CGD and 3-level AGD with the same conditions as in Fig. 8, but now with $R_{G,on} = 10\,\Omega$ in the 3L-AGD network and $R_{G,on} = 5\,\Omega$ in the CGD network.

The $R_{G,on}$ values in Fig. 9 and Fig. 10 were chosen for comparable turn-on transients, but the 3L-AGD network may provide lower switching losses at lower $R_{G,on}$ values. Therefore, measurement with $R_{G,on}$ for both the 3L-AGD and CGD networks is $R_{G,on} = 5\,\Omega$ are shown in Fig. 11 and Fig. 12. Values lower than $R_{G,on} = 5\,\Omega$, e.g. $2.5\,\Omega$, result in a gate-source waveform with a lot of parasitic ringing and are not considered further, see [17].

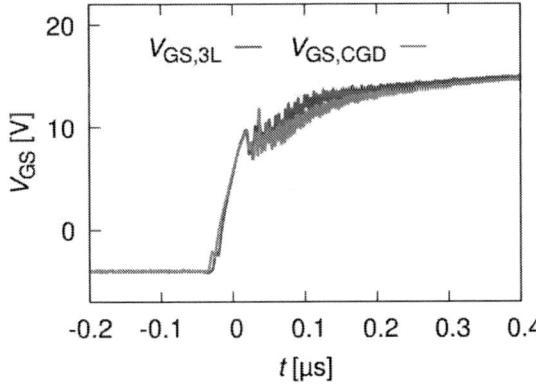

Fig. 11: Measurement of V_{GS} for CGD and 3-level AGD with the same conditions as in Fig. 7, but now with $R_{G,on} = 5\,\Omega$ in the 3L-AGD network and $R_{G,on} = 5\,\Omega$ in the CGD network.

Before and after the Miller plateau, the switching transient of V_{GS} is faster for the 3L-AGD network, see Fig. 11. This shows that for low values of $R_{G,on}$

the 3L-AGD network still shows an improvement in the switching waveform. This improvement can also be seen in Fig. 12, where V_{DS} has a faster transient in the 3L-AGD compared to the measurement with the CGD. As a result, switching losses can be reduced by about 5.8%, while $I_{S,PEAK}$ increases by only 2.2%, see Fig. 12.

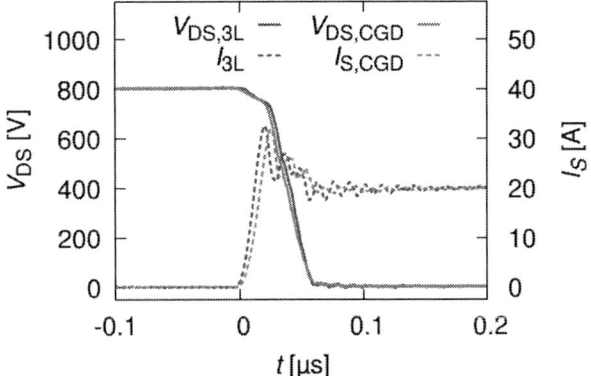

Fig. 12: Measurement of V_{DS} and I_S for CGD and 3-level AGD with the same conditions as in Fig. 8, but now with $R_{G,on} = 5\,\Omega$ in the 3L-AGD network and $R_{G,on} = 5\,\Omega$ in the CGD network.

The faster switching transient of V_{DS} and a comparable $I_{S,PEAK}$ result in an overall reduction in turn-on energy-loss for the 3L-AGD network compared to the CGD network. Measurements show that the 3L-AGD network always results in a decrease in switching losses compared to a CGD measurement with the same or higher $R_{G,on}$. An overview of the measurement results is given in Tab. 2.

Tab. 2: Overview of the measurement results for 3L-AGD / CGD.

$R_{G,on}$ [Ω]	E_{ON} [μJ]	$I_{S,PEAK}$ [A]
10 / 10	600.49 / 735.44	28.89 / 28.60
10 / 5	600.49 / 523.77	28.89 / 31.95
5 / 5	493.34 / 523.77	32.66 / 31.95

5 Conclusion and Future Work

The self-driving 3L-AGD proposed in this work changes the gate resistance during the switching transient of the SiC MOSFET with the presented auxiliary network. This network allows the 3L-AGD to successfully adjust the switching slew rate during the switching transient. In addition, the 3L-AGD network is fully integrated into the PCB design and does not require external control schemes, feedback loops or high-speed components, greatly reducing the complexity of the overall system. Furthermore, switching losses can be reduced due to faster switching transients for the same current overshoot.

Acknowledgment

This project has received funding from the ECSEL Joint Undertaking (JU) under grant agreement No 101007281. The JU receives support from the European Union's Horizon 2020 research and innovation programme and Austria, Germany, Slovenia, Netherlands, Belgium, Slovakia, France, Italy, Turkey.

References

[1] S. Zhao, X. Zhao, Y. Wei, Y. Zhao et al., "A Review of Switching Slew Rate Control for Silicon Carbide Devices Using Active Gate Drivers," *IEEE Journal of Emerging and Selected Topics in Power Electronics*, vol. 9, no. 4, pp. 4096–4114, Aug. 2021.

[2] X. Du, Y. Wei, A. Stratta, L. Du, V. S. Machireddy, and A. Mantooth, "A Four-level Active Gate Driver with Continuously Adjustable Intermediate Gate Voltages," in *2022 IEEE Applied Power Electronics Conference and Exposition (APEC)*, Mar. 2022, pp. 1379–1386.

[3] S. Zhao, A. Dearien, Y. Wu, C. Farnell, A. U. Rashid, F. Luo, and H. A. Mantooth, "Adaptive Multi-Level Active Gate Drivers for SiC Power Devices," *IEEE Transactions on Power Electronics*, vol. 35, no. 2, pp. 1882–1898, Feb. 2020.

[4] R. Li, Z. Hou, T. Liu, M. Elshazly, S. Leung, X. Peng, and W. T. Ng, "Dynamic Gate Drive for SiC Power MOSFETs with Sub-nanosecond Timings," in *2023 IEEE Applied Power Electronics Conference and Exposition (APEC)*, Mar. 2023, pp. 324–330.

[5] A. P. Camacho, V. Sala, H. Ghorbani et al., "A Novel Active Gate Driver for Improving SiC MOSFET Switching Trajectory," *IEEE Transactions on Industrial Electronics*, vol. 64, no. 11, pp. 9032–9042, Nov. 2017.

[6] G. Engelmann, T. Senoner, and R. W. De Doncker, "Experimental investigation on the transient switching behavior of SiC MOSFETs using a stage-wise gate driver," *CPSS Transactions on Power Electronics and Applications*, vol. 3, no. 1, pp. 77–87, Mar. 2018.

[7] W. T. Cui, W. J. Zhang, J. Y. Liang, H. Nishio et al., "A Dynamic Gate Driver IC with Automated Pattern Optimization for SiC Power MOSFETs," in *2022 IEEE 34th International Symposium on Power Semiconductor Devices and ICs (ISPSD)*, May 2022, pp. 33–36.

[8] Y. Yang, Y. Wen, and Y. Gao, "A Novel Active Gate Driver for Improving Switching Performance of High-Power SiC MOSFET Modules," *IEEE Transactions on Power Electronics*, vol. 34, no. 8, pp. 7775–7787, Aug. 2019.

[9] K. Miyazaki, S. Abe, M. Tsukuda et al., "General-Purpose Clocked Gate Driver IC With Programmable 63-Level Drivability to Optimize Overshoot and Energy Loss in Switching by a Simulated Annealing Algorithm," *IEEE Transactions on Industry Applications*, vol. 53, no. 3, pp. 2350–2357, May 2017.

[10] W. J. Zhang, J. Yu, Y. Leng, W. T. Cui et al., "A Segmented Gate Driver for E-mode GaN HEMTs with Simple Driving Strength Pattern Control," in *2020 32nd International Symposium on Power Semiconductor Devices and ICs (ISPSD)*, Sep. 2020, pp. 102–105.

[11] D. Liu, H. C. P. Dymond, S. J. Hollis et al., "Full Custom Design of an Arbitrary Waveform Gate Driver With 10-GHz Waypoint Rates for GaN FETs," *IEEE Transactions on Power Electronics*, vol. 36, no. 7, pp. 8267–8279, Jul. 2021.

[12] Y. Teng, Q. Gao, Q. Zhang, J. Kou, and D. Xu, "A Variable Gate Resistance SiC MOSFET Drive Circuit," in *IECON 2020 The 46th Annual Conference of the IEEE Industrial Electronics Society*, Oct. 2020, pp. 2683–2688.

[13] Y. Wen, Y. Yang, and Y. Gao, "Active Gate Driver for Improving Current Sharing Performance of Paralleled High-Power SiC MOSFET Modules," *IEEE Transactions on Power Electronics*, vol. 36, no. 2, pp. 1491–1505, Feb. 2021.

[14] I. Lee and X. Yao, "Active dv/dt Control with Turn-off Gate Resistance Modulation for Voltage Balancing of Series Connected SiC MOSFETs," in *2021 IEEE Applied Power Electronics Conference and Exposition (APEC)*, Jun. 2021, pp. 1256–1261.

[15] Y. Ling, Z. Zhao, and Y. Zhu, "A Self-Regulating Gate Driver for High-Power IGBTs," *IEEE Transactions on Power Electronics*, vol. 36, no. 3, pp. 3450–3461, Mar. 2021.

[16] V. L. Choo and M. Pfost, "A Variable Gate Resistance SiC MOSFET Driver Network to Mitigate Overshoot and Parasitic Ringing," in *PCIM Europe 2023*, May 2023, pp. 1–7.

[17] ——, "A 3-level Active Gate Driver Network for SiC MOSFETs to Minimize Overshoot and Switching Losses," in *2024 IEEE Applied Power Electronics Conference and Exposition (APEC)*, 2024.

PCIM Europe 2024, 11– 13 June 2024, Nuremberg   DOI: 10.30420/566262244

Analysis of Long-Term Reliability of SiC in Traction Inverter Considering V_{th} Instability

Chi Zhang[1], Riccardo Negri[1], Joachim Härsjö[1]
[1] Volvo Var Group, Sweden

Corresponding author: Chi Zhang, chi.zhang.6@volvocars.com
Speaker: Chi Zhang, chi.zhang.6@volvocars.com

Abstract

Along with the booming of electrical vehicles, wide band gap (WBG) semiconductors, such as silicon carbide (SiC) is receiving more and more attention from both academic and industrial fields. There exist quite some design challenges, such as stray inductance optimization, thermal management enhancement due to smaller die size. Among them, long-term reliability of the gate threshold voltage (V_{th}), especially under AC stress, is of paramount importance. Dynamic gate stress (DGS) test is proposed by AQG324 to give a guideline how to identify the V_{th} drift performance in test stage considering reliability design of traction inverters. Typically, 5% change of R_{DSON} due to DGS test is followed currently as a preliminary acceptance criteria of V_{th} drift test. Nevertheless, the acceptance criteria should be determined based on mission profile of traction inverter. Moreover, how to use the results from DGS in long-term reliability evaluation still require further investigation. In this paper, a design methodology considering long-term reliability is proposed by utilizing the mission profile of electrical vehicles, with which test results from DGS have been taken into account. This not only acts as a way to analyze V_{th} instability impact on SiC lifetime but also can act as a way to justify DGS test results about whether the selected device can satisfy the performance requirement regarding reliability aspects.

1   Introduction

In the past decade, WBG semiconductor brings efficient power converter from vision to reality [1]. Its fast-switching speed, much lower switching loss and higher junction temperature (T_j) capability makes WBG semiconductor become a promising candidate to replace conventional Si-based power semiconductor, such as Insulated Gate Bipolar Transistor (IGBT). SiC Metal-Oxide-Semiconductor Field-effect Transistor (MOSFET), which is one typical WBG semiconductor, is receiving more and more attention from different industries, such as solar energy, automotive application and so on.

However, currently it is still a challenge to fully exploit the advantages of SiC MOSFET [2]. Due to its fast switching-speed, a low stray inductance (L_{stray}) is required regarding the current commutation loop [3]-[5]. A robust, intelligent gate driver design, which can optimize the slew rate, overshoot and switching loss simultaneously, is also beneficial to converter efficiency, EMC and reliability performance [6]-[8]. Moreover, due to the smaller die size of SiC MOSFET, thermal management for both SiC MOSFET package and converter cooling have been emphasized in [9]-[10].

On the other hand, compared with traditional Si based power semiconductor devices, SiC MOSFET requires additional test and qualification to guarantee its reliability performance [11]. Dynamic reverse bias (DRB) is proposed to check the aging due to fast charging of internal structures due to high dv/dt of SiC MOSFET [12]. PC_{min} and PC_{sec} are explained in AQG324 [13] to validate SiC lifetime. Moreover, high temperature forward bias (HTFB) necessity is under discussion.

With increasing numbers of different applications coming to stages brings more concerns regarding the risk of V_{th} drift [14]-[17] with various kinds of gate structures of SiC MOSFET as shown in Fig. 1 [23]. Ephemeral and nonpermanent charge trapping effects results in the V_{th} change [18]. Electrons are captured and emitted when different gate stresses are applied to the gate of SiC MOSFET. Under static stress, it has been found that Vth will drift in a positive way (PBTI) while applying positive gate voltage. With negative gate voltage, Vth drift will show a negative trend (NBTI) shown in Fig. 2. This phenomenon is known as static bias temperature instability (BTI) and it is a permanent drift that can't be recovered. However, static BTI is of low risk and it can be solved by using proper gate voltage [19].

1781

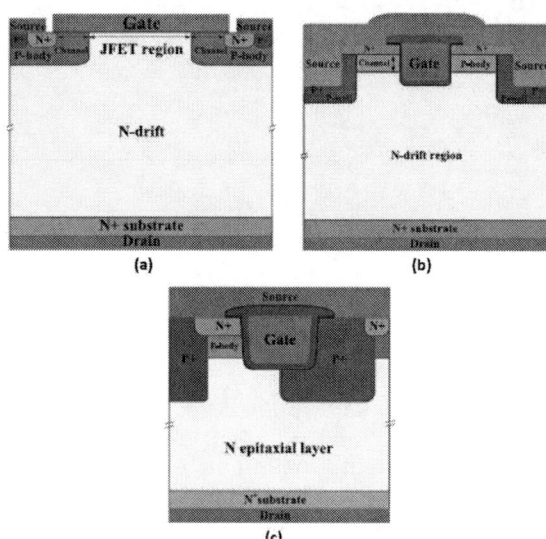

Fig. 1 (a) traditional planar gate structure. (b) DT structure. (c) Asymmetric TG structure. [23]

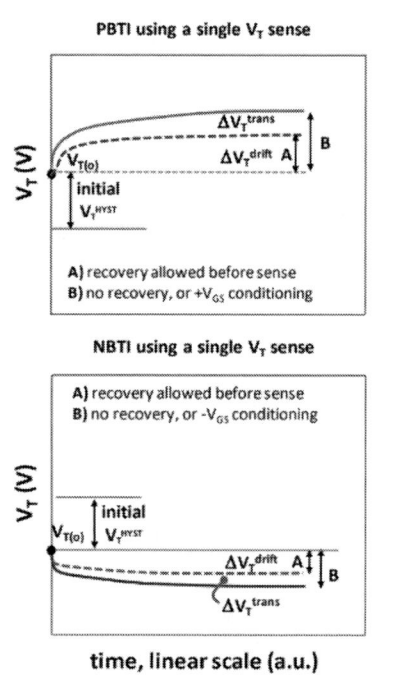

Fig. 2 PBTI and NBTI of SiC MOSFET [22].

In most industrial applications, the gate stress is AC instead of DC. Such kind of time-changing gate stress leads to V_{th} drift that includes both recoverable and unrecoverable components [20], [21]. Conventional high temperature gate bias (HTGB) that is only applying static stress on gate is not sufficient to validate gate instability of SiC MOSFET.

So dynamic gate stress (DGS) test is introduced in both SiC die level and package level qualification. However, it is still unclear that how to define the acceptance criteria for DGS test result since it is highly related with mission profile. Typically, DGS will be performed either >10^11 cycles or with enough cycles that mathematical model can be extracted [22] considering automotive applications. During testing process, gate leakage current of SiC is monitored as a direct indication regarding whether gate of SiC MOSFET fails or not. On the other hand, the V_{th} drift are also recorded during the qualification process. However, how to use the DGS test results to justify whether the selected SiC MOSFET is able to satisfy design or performance requirement still requires further investigation.

In this article, V_{th} drift is considered from perspective of power semiconductor reliability. DGS test results are included in the power semiconductor reliability evaluation cases compared with conventional calculation method. R_{DSON} is acting as a medium to link V_{th} drift test results. Section 2 will describe the proposed evaluation flowchart and evaluation results will be presented in Section 3. Finally, conclusion is discussed in Section 4.

2 Proposed Analysis Flow chart

2.1 DGS and PC_{sec} test impact on V_{TH}

R_{DSON} of SiC devices is one of the critical parameters for both performance evaluation such as Loss and T_j, and reliability evaluation. In PC_{sec} and PC_{min} test proposed in AQG324, R_{DSON} is acting as the main criteria to determine whether the device has failed or not. The channel resistance (R_{ch}) and accumulation region (R_A) are two main parts of R_{DSON}. It has been reported in [24] that V_{th} has direct impact on R_{ch} and R_A, which can be seen from the following two equations,

$$R_{ch} = L_{ch}/[W_{ch}\mu_n C_{ox}(V_{gs} - V_{th})] \qquad (1)$$

$$R_A = L_A/[W_A\mu_{nA}C_{ox}(V_{gs} - V_{th})] \qquad (2)$$

With L_{ch} is the channel length, W_{ch} is the channel width, μ_n is the electron mobility of the inversion layer channel in SiC material, L_A is the accumulation region length, W_A is the accumulation region width, μ_{nA} is the electron mobility of the accumulation layer, V_{gs} is the gate-source voltage and C_{ox} is parasitic capacitance. So an increase of V_{th} leads to a rise of R_{ch} and R_A, which brings a raise of R_{DSON}.

PCIM Europe 2024, 11– 13 June 2024, Nuremberg   DOI: 10.30420/566262244

However, power cycling tests (PC_{sec} and PC_{min}) also leads to R_{DSON} increase. The power cycling capability of today's SiC MOSFET power module is around 50K-100K in case of 100 degC dT_j. This means the tested SiC MOSFET only switches on and off 50,000 – 100,000 times during the tests, which is much less than DGS test, which is typically >10^11 in automotive application. V_{th} drift provides a small contribution to the R_{DSON} increase during power cycling tests. As a conclusion, R_{DSON} drift can be split into two different independent parts, namely V_{th} drift impact and power/thermal cycling impact. In the DGS test, it is recommended to monitor the V_{th} drift and R_{DSON} drift at the same time.

2.2 Proposed analysis process

The proposed reliability evaluation flow process is shown in Fig. 3.

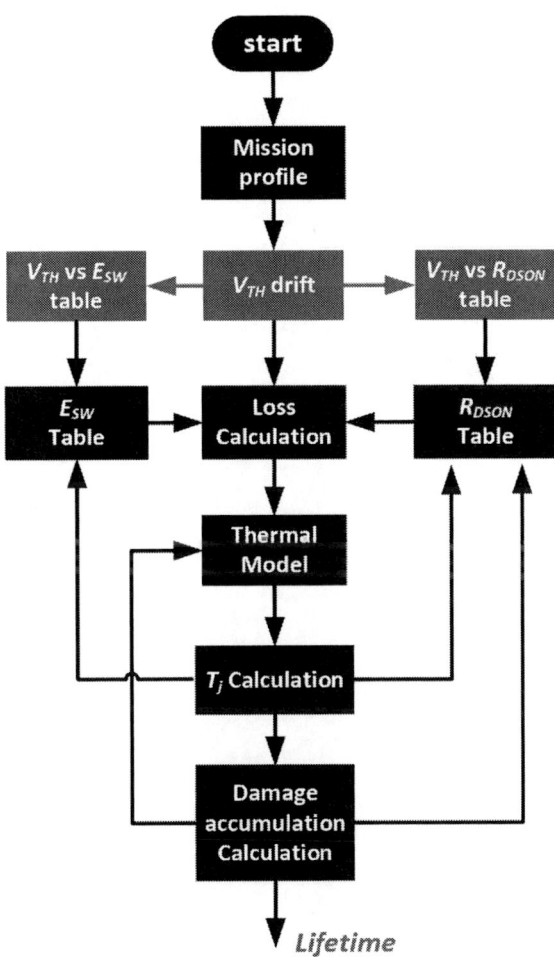

Fig. 3   Proposed design flow chart.

Mission profile is required to start the lifetime calculation. V_{th} drift due to switching on and off based on mission profile is taken into consideration. This will affect not only switching energy (E_{SW}) but also R_{DSON}. The updated E_{sw} and R_{DSON} table will be used to calculate SiC MOSFET loss by considering T_j impact. The T_j is extracted by using thermal model and with the help of rain-flow algorithm, damage status of SiC MOSFET is calculated. Compared with typically reliability/lifetime evaluation, V_{th} impact, as shown by the red part in Fig. 3, is included to take more factors that affect SiC MOSFET lifetime into consideration.

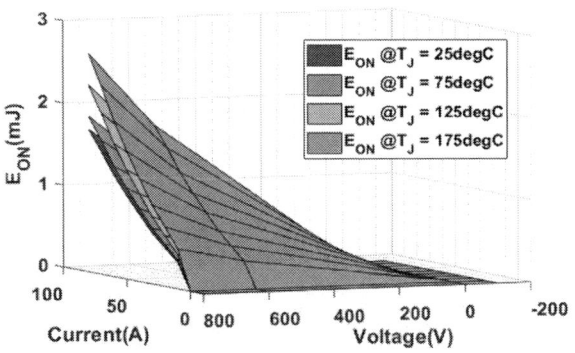

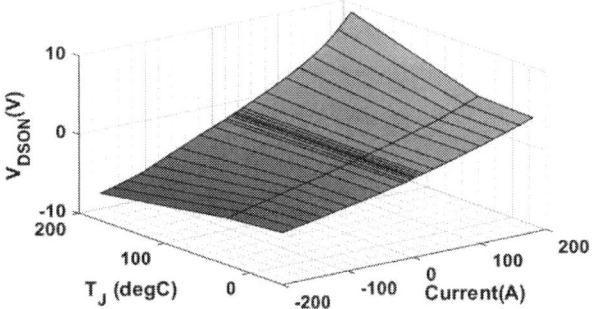

Fig. 4 E_{ON} and V_{DSON} data.

2.2.1   Mathematical model for T_j calculation

In order to achieve efficient calculation considering mass data processing, the loss and T_j calculation is implemented by using python code with mathematical models. E_{SW} and R_{DSON} data shown in Fig. 4 are interpreted to lookup tables in loss/T_j calculation model. V_{th} and T_j impact will be included by using extra math equation to modify the E_{SW} and R_{DSON} data obtained from lookup tables.

As for thermal network, a 5-order foster thermal shown in Fig. 5 is selected to describe the thermal network from junction to ambient (Z_{JA}) and it is implemented by equation (3) – (8),

1783

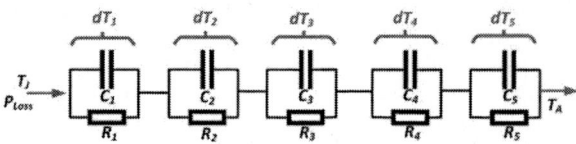

Fig. 5 5-order thermal network for Z_{JA}.

$$dT_1(k) \quad (3)$$
$$= dT_1(k-1) + T_s(dT_1(k-1)/R_1/C_1 + P_{Loss}/C_1)$$

$$dT_2(k) \quad (4)$$
$$= dT_2(k-1) + T_s(dT_2(k-1)/R_2/C_2 + P_{Loss}/C_2)$$

$$dT_3(k) \quad (5)$$
$$= dT_3(k-1) + T_s(dT_3(k-1)/R_3/C_3 + P_{Loss}/C_3)$$

$$dT_4(k) \quad (6)$$
$$= dT_4(k-1) + T_s(dT_4(k-1)/R_4/C_4 + P_{Loss}/C_4)$$

$$dT_5(k) \quad (7)$$
$$= dT_5(k-1) + T_s(dT_5(k-1)/R_5/C_5 + P_{Loss}/C_5)$$

$$T_j(k) = T_A(k) + dT_1(k) + dT_2(k) + dT_3(k) + dT_4(k) \quad (8)$$
$$+ dT_5(k)$$

with T_s is time step for discretized calculation that is equal to switching period, P_{Loss} is calculated loss of SiC MOSFET based on mission profile.

A comparison result between PELCS simulation and the mathematical model used is presented in Fig. 6. Fig. 6(a) shows the T_j swing comparison result. It can be observed that an error that is smaller than 0.5degC has been achieved. Mean T_j estimation result under both transient and static process is shown in Fig. 6(b). Also, a deviation that is lower than 0.5degC has been achieved. As a conclusion, results based on mathematical model shows a good matching with PLECS simulation results in both static and dynamic conditions and it can be used for loss and T_j calculation.

3 Lifetime analysis results

3.1.1 T_j calculation results

Fig. 7 shows the T_j calculation results with and without including V_{th} impact by repeating a mission profile that lasts around 800s. After repeating this mission profile by 500 times, which is around 111 hours operating continuously, the calculated T_j is around 10degC higher than that without considering V_{th} impact. If more mission profiles are taken into account, such as city road, highway, mount type and so on, T_j would have a more pronounced difference with longer operating times.

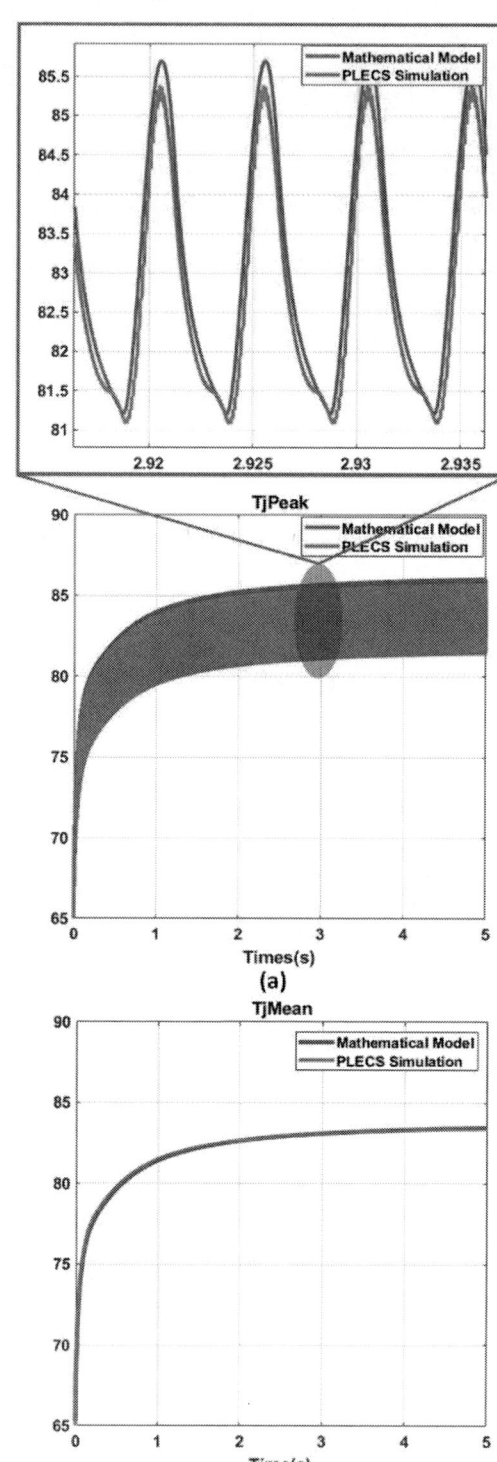

Fig. 6 T_j calculation compare between PLECS and Mathematical model. (a). with T_j swing. (b). w/o T_j swing

PCIM Europe 2024, 11– 13 June 2024, Nuremberg   DOI: 10.30420/566262244

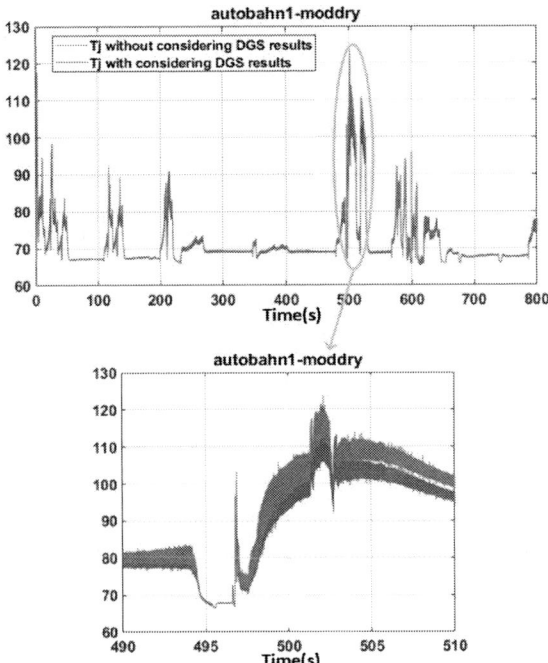

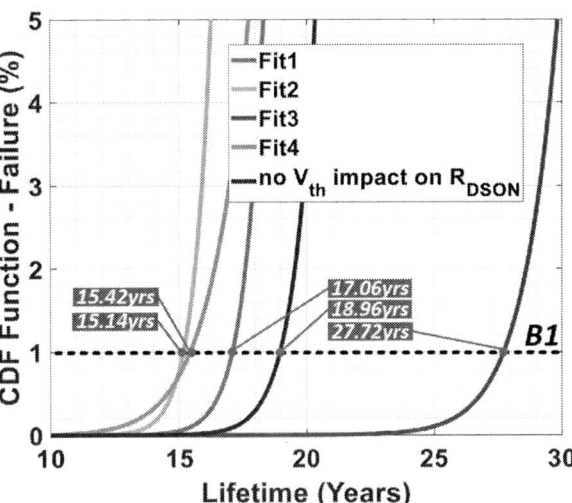

Fig. 9   Lifetime estimation results with Fit 1- 4 with V_{th} drift type 1 (shown in Fig. 10(b)).

Fig. 7 T_j different w/ and w/o V_{th} drift impact after 500 times mission file repeating.

the data distribution bandwidth between Fit2 and Fit3. Fit1 – Fit4 are derived as,

$$R_{DSON\_Fit1}/R_{DSON} = (0.1538V_{th} + 0.6154) \quad (9)$$

$$R_{DSON\_Fit2}/R_{DSON} = (0.1538V_{th} + 0.7308) \quad (10)$$

$$R_{DSON\_Fit3}/R_{DSON} = (0.1538V_{th} + 0.5018) \quad (11)$$

$$R_{DSON\_Fit4}/R_{DSON} = (0.3096V_{th} + 0.1923) \quad (12)$$

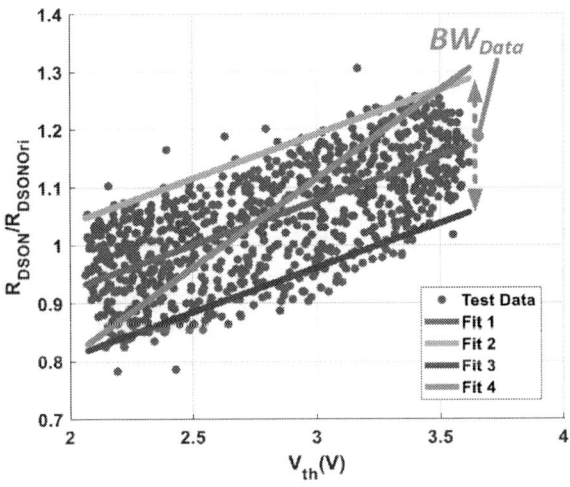

Fig. 8 R_{DSON} test data and data fit options.

3.1.2 Lifetime evaluation

Test data regarding V_{th} impact on R_{DSON} is presented in Fig. 8. It is able to be observed that R_{DSON} is increasing along with V_{th}. In order to verify the impact on reliability/lifetime of SiC MOSFET, linear type curve fit is considered here. Four different fit options, namely Fit1, Fit2, Fit3 and Fit4, have been listed in Fig. 8 and BW_{data} is defined to describe

Lifetime estimation results based on Fit1-4 are shown in Fig. 9 using V_{th} drift type 1 results from DGS test (Fig. 10(a)). The black curve shows the lifetime calculation results without including V_{th} drift impact on R_{DSON}. The estimated lifetime is 18.96 yrs. If V_{th} drift impact on R_{DSON} is included by using Fit 1 type shown by the orange curve in Fig. 9, the lifetime is reduced to 17.06yrs. Fit 2 (yellow curve in Fig. 9) shows the lowest lifetime estimation result, which is 15.14yrs since it utilizes the top boundary of $R_{DSON}/R_{DSONOri}$ in Fig. 8. Moreover, it starts to fail early than Fit 2 but with a relatively slow failure increasing slope when Fit 4 is considered. Nevertheless, Fit 4 shows a higher lifetime estimation result compared with Fit 2 and it is 15.42yrs. As for Fit 3, it results in quite high lifetime estimation result (27.72yrs) since it uses the bottom boundary of $R_{DSON}/R_{DSONOri}$ in Fig. 8. So it can be concluded that Fit2 and Fit 4 lead to the worst cases regarding SiC lifetime estimation result. Either by reducing BW_{data} (shown in Fig. 8) or top boundary of $R_{DSON}/R_{DSONOri}$ will contribute to a higher lifetime of SiC MOSFET.

1785

Fig. 10(b) presents SiC MOSFET lifetime estimation results considering 3 different kinds of V_{th} drift conditions shown in Fig. 10(a). V_{th} drift type 2 shows a smaller drift during DGS test which type 3 shows a much faster V_{th} drift as shown in Fig. 10(a). Fit 2 is considered here since it indicates the lifetime bottle neck as shown in Fig. 9. In case of V_{th} drift type 1, SiC MOSFET lifetime is increased to 16.04yrs as shown by Fit2-V_{th} slower drift (orange curve) in Fig. 10 compared with V_{th} drift type 1 (15.14yrs). However, lifetime of SiC MOSFET will be decreased to 14.96yrs with a fast V_{th} drift as indicated by Fit2-V_{th} faster drift (yellow curve) when V_{th} drift type 3 is used. Considering a 15yrs lifetime requirements, SiC MOSFET with V_{th} drift type 3 offers a less satisfying lifetime results. As a conclusion, a fast V_{th} drift will bring a negative impact on SiC MOSFET lifetime estimation. In the SiC MOSFET selection process, as long as the selected SiC MOSFET shows a slower V_{th} drift compared with V_{th} drift type 1(blue curve in Fig. 10(a)) in DGS test, it will meet the lifetime requirements.

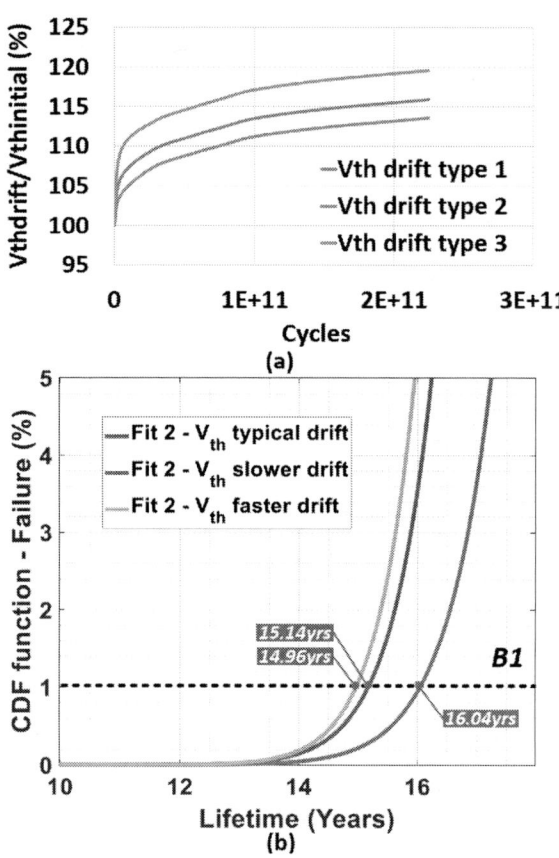

Fig. 10 (a). V_{th} drift type 1,2,3. (b). Lifetime estimation results with V_{th} drift type 1 – 3 with Fit 2.

4 Conclusion

This article introduces an improved reliability evaluation process for SiC MOSFET by including V_{th} drift impact. Based on test data of R_{DSON} and V_{th}, lifetime estimation with different fit options has been performed. Either by reducing the worse R_{DSON} value or data distribution bandwidth (BW_{data}) of R_{DSON} vs V_{th} can contribute to a higher lifetime of SiC MOSFET. On the other hand, investigation regarding SiC MOSFETs with different V_{th} drift results was also included by considering the worst R_{DSON} cases (Fit 2 in Fig. 8). The proposed evaluation process not only can act as an improved way to evaluate SiC MOSFET lifetime but also can derive a more dedicated acceptance criteria for DGS tests.

References

[1] A. I. Emon, Mustafeez-ul-Hassan, A. B. Mirza, J. Kaplun, S. S. Vala and F. Luo, "A Review of High-Speed GaN Power Modules: State of the Art, Challenges, and Solutions," in IEEE Journal of Emerging and Selected Topics in Power Electronics, vol. 11, no. 3, pp. 2707-2729, June 2023.

[2] Z. Guo, H. Li, F. Z. Peng, M. Steurer and K. J. Olejniczak, "Gate Driver Development and Stray Inductance Extraction of 10 kV S iC MOSFET Module for a Switched-Capacit or MMC Application," 2022 IEEE Applied P ower Electronics Conference and Expositio n (APEC), Houston, TX, USA, 2022, pp. 13 87-1393.

[3] Y. Chang et al., "Compact Sandwiched Press-Pack SiC Power Module With Low Stray Inductance and Balanced Thermal Stress," in IEEE Transactions on Power Electronics, vol. 35, no. 3, pp. 2237-2241, March 2020.

[4] S. Hu, M. Wang, Z. Liang and X. He, "A Frequency-Based Stray Parameter Extraction Method Based on Oscillation in SiC MOSFET Dynamics," in IEEE Transactions on Power Electronics, vol. 36, no. 6, pp. 6153-6157, June 2021.

[5] B. Liu et al., "Low-Stray Inductance Optimized Design for Power Circuit of SiC-MOSFET-Based Inverter," in IEEE Access, vol. 8, pp. 20749-20758, 2020.

[6] H. -T. Tang, H. S. -H. Chung and K. J. Chen, "Adaptive Level-Shift Gate Driver With Indirect Gate Oxide Health Monitoring for Suppressing Crosstalk of SiC MOSFETs," in IEEE Transactions on Power Electronics, vol. 38, no. 8, pp. 10196-10212, Aug. 2023.

[7] Q. Yue, H. Peng, Q. Tong and Y. Kang, "A Novel Driving Current Control Approach in Enhanced Current-Source Gate Driver," in IEEE Transactions on Power Electronics, vol. 38, no. 9, pp. 10563-10568, Sept. 2023.

[8] G. L. Rødal and D. Peftitsis, "An Adaptive Current-Source Gate Driver for High-Voltage SiC mosfets," in IEEE Transactions on Power Electronics, vol. 38, no. 2, pp. 1732-1746, Feb. 2023.

[9] G. Tang, T. c. Chai and X. Zhang, "Thermal Optimization and Characterization of SiC-Based High Power Electronics Packages With Advanced Thermal Design," in IEEE Transactions on Components, Packaging and Manufacturing Technology, vol. 9, no. 5, pp. 854-863, May 2019.

[10] X. Sun et al., "Design and Evaluation of a Face-Down Embedded SiC Power Module With Low Parasitic Inductance and Low Thermal Resistance," in IEEE Transactions on Power Electronics, vol. 38, no. 3, pp. 2799-2804, March 2023.

[11] M. Kim et al., "Evaluation of Long-Term Reliability and Overcurrent Capabilities of 15-kV SiC MOSFETs and 20-kV SiC IGBTs During Narrow Current Pulsed Conditions," in IEEE Transactions on Plasma Science, vol. 48, no. 11, pp. 3962-3967, Nov. 2020.

[12] Y. Wang, W. Li, Y. Ding, H. Sun and Y. Yin, "Multi-Physics Coupling Analysis and Optimization Design of SiC MOSFET Power Module Package Insulation," 2022 IEEE Energy Conversion Congress and Exposition (ECCE), Detroit, MI, USA, 2022, pp. 1-5.

[13] ECPE Guideline AQG 324 - Qualification of Power Modules for Use in Power Electronics Converter Units in Motor Vehicles

[14] T. Aichinger, G. Rescher, andG. Pobegen, "Threshold voltage peculiarities and bias temperature instabilities of SiCMOSFETs," Microelectron. Rel.,vol. 80, pp. 68–78, Jan. 2018.

[15] A.I-M. El-Sayed, M. B. Watkins, A. L. Shluger, and V. V. Afanas'ev,"Identification of intrinsic electron trapping sites in bulk amorphous silicafrom AB initio calculations," Microelectron. Eng., vol. 109, pp. 68–71,2013.

[16] W. Shockley andW. T. Read Jr., "Statistics of the recombinations of holes and electrons," Phys. Rev., vol. 87, no. 5, pp. 835–842, 1952.

[17] W. Zhou, X. Zhong, andK. Sheng, "High temperature stability and the performance degradation of SiC MOSFETs," IEEE Trans. Power Electron., vol. 29, no. 5, pp. 2329–2337, May. 2014.

[18] D. B. Habersat and A. J. Lelis, "AC-Stress Degradation and Its Anneal in SiC MOSFETs," in IEEE Transactions on Electron Devices, vol. 69, no. 9, pp. 5068-5073, Sept. 2022.

[19] K. Puschkarsky, T. Grasser, T. Aichinger, W. Gustin, and H. Reisinger, "Review on SiC MOSFETs high-voltage device reliability focusing on threshold voltage instability," IEEE Trans. Electron Devices, vol. 66, no. 11, pp. 4604–4616, Nov. 2019.

[20] "How Infineon controls and assures the reliability of SiC based power semiconductors," Infineon, Neubiberg, Germany, 2020.

[21] H. Jiang, X. Zhong, G. Qiu, L. Tang, X. Qi, and L. Ran, "Dynamic gate stress induced threshold voltage drift of silicon carbide MOSFET,"

[22] JEP 184 – Guideline for evaluating bias temperature instability of Silicon Carbide Metal-Oxide-Semiconductor devices for power electronics conversion.

[23] J. Wei et al., "Review on the Reliability Mechanisms of SiC Power MOSFETs: A Comparison Between Planar-Gate and Trench-Gate Structures," in IEEE Transactions on Power Electronics, vol. 38, no. 7, pp. 8990-9005, July 2023.

[24] E. Ugur, F. Yang, S. Pu, S. Zhao and B. Akin, "Degradation Assessment and Precursor Identification for SiC MOSFETs Under High Temp Cycling," in IEEE Transactions on Industry Applications, vol. 55, no. 3, pp. 2858-2867, May-June 2019.

PCIM Europe 2024, 11– 13 June 2024, Nuremberg   DOI: 10.30420/566262245

Efficient Mapping of On-Demand Drive Load Profiles on Inverter Stress

Zlatko Bosnjic[1], Klaus Krischan[1], Franz Königseder[2], Alexander Loibl[2], Michael Hartmann[1]

[1] Electric Drives and Machines Institute, Graz University of Technology, Graz, Austria
[2] Magna, Austria

Corresponding author: Zlatko Bosnjic, zlatko.bosnjic@tugraz.at
Speaker: Zlatko Bosnjic, zlatko.bosnjic@tugraz.at

Abstract

Mechanical and thermal stresses of EV drive system components strongly depend on how the drive is operated during it's lifetime. In On-Demand drives, expected operation is difficult to quantify as it depends on vehicle design, powertrain design and torque distribution strategy. This paper proposes two methods for generating representative On-Demand drive Load Profiles based on vehicle speed profiles. Additionally, a compact drive system model for efficient simulation of traction inverter thermal stress is derived and explained in detail. The inverter thermal stress is simulated for a measured On-Demand speed/torque profile and is compared to simulation results based on the two proposed methods. The results show good match between measured and generated drive loading profiles.

1   Introduction

Interest in battery electric vehicles (BEVs) has been growing rapidly in recent years [1] as the need for sustainable transportation is becoming more obvious [2]. Benefits of BEVs are quite clear and have been analyzed and discussed in-depth ([1], [3] and [4]). BEVs usually employ only one traction motor (single-motor EVs), however, for applications and driving scenarios where additional performance, stability and reliability is desired, electric vehicles with more than one traction motor (multi-motor EVs) are preferred ([2] and [5]). Dual-motor EVs, with one tracion motor on each vehicle axle are the largest subgroup of multi-motor EVs. In Fig. 1, a schematic depiction of such EV powertrains is shown. Depending on vehicle requirements and system design, the two drive systems in a dual-motor EV can range from being almost equal and working in parallel to substantially different in design where one drive is only used for acceleration and stability. Based on benchmark reports of commercial EVs, the latter is a more common solution. The two drive systems found in dual-motor EVs are designated as:

- Main (primary) drive system, and

- On-Demand (secondary) drive system.

As the On-Demand drive may supply torque only when required for acceleration or stability, the mission profile of such a drive differs from the main drive (and from the drive system in a single-motor EV). Due to this, the lifetime requirements imposed on components of the On-Demand drive system also differ from that of a single-motor EV drive system. The discrepancy of main drive usage and On-Demand drive usage has also been verified by measurements from real-world driving scenarios of a dual-motor EV and is presented in Fig. 2.

Although dual-motor EVs have been in the scope of research for some time, there appears to be a lack of published data regarding the differences in usage and lifetime requirements of the two drive systems. The main drive system in a dual-motor EV is similar in design and usage to a single-motor EV drive system. It is therefore not considered in this work as there is a great deal of publications covering standard single-motor EVs ([6] and [7]). Another topic addressed in this paper is accurate prediction of semiconductor junction temperature profiles. Usually, accurate simulation of junction temperatures relies on FEM and is computationally intensive. Therefore it is not used for system level optimization in EV powertrain design. A computationally efficient electro-thermal model, including

1788

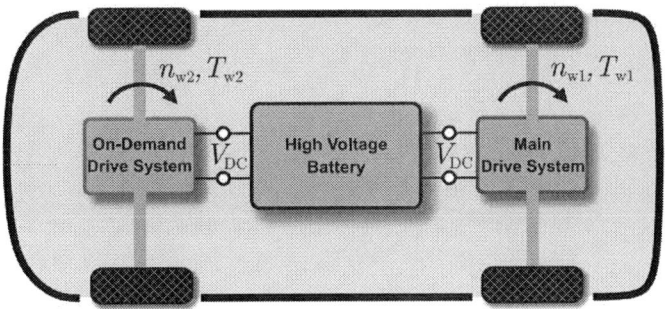

Fig. 1: Schematic of dual-motor EV power-train consisting of main drive system and On-Demand drive system.

thermal coupling between chips, is proposed in this paper as a substitute for FEM.

This paper provides an outline for a more optimal inverter stage design of the On-Demand drive system by addressing the influence of vehicle speed profiles on On-Demand drive inverter thermal stress. The approach presented here can be subdivided into a two step process:

1. Accurate mapping of vehicle speed profiles $v_v(t)$ to On-Demand drive speed $n_2(t)$ and On-Demand drive torque $T_2(t)$ by means of a novel torque mapping method, and

2. Efficient simulation of inverter thermal stress based on drive speed $n_2(t)$ and drive torque $T_2(t)$.

The rest of the paper is structured as follows: **Section 2** provides an in-depth analysis of the derivation of On-Demand drive loading in dual-motor EVs. In **Section 3** the thermal stresses of the On-Demand drive inverter are simulated based on results from **Section 2**. The results of thermal stress simulation for different torque mapping approaches are presented and discussed in **Section 4**. Finally, in **Section 5**, a conclusion and outlook for further work is given.

2 On-Demand Drive Usage

The usage of the On-Demand drive system in dual-motor EVs is not easily defined, especially during the design phase. The usage criteria of the On-Demand drive is a result of a superimposed torque distribution control strategy which determines how the required torque is generated across the two drives in a dual-motor EV. This control strategy is often based on overall system loss minimization ([8], [9]) and requires detailed knowledge on both

the vehicle model and the drive system design(s). Here, the usage of the On-Demand drive is derived from simplified criteria and measurement data without prior information on torque distribution. In Fig. 1, a schematic of a dual-motor EV powertrain is depicted.

For a representative speed/torque profile of the On-Demand drive $(T_2(t), n_2(t))$, the following should be known:

- Representative vehicle speed profile $v_v(t)$ (e.g. WLTC speed profile)

- Longitudinal vehicle model (vehicle mass - m_v, drag area - C_dA, rolling resistance - μ_r, etc.)

- Primary drive specification

- Secondary drive specification

- Torque distribution strategy

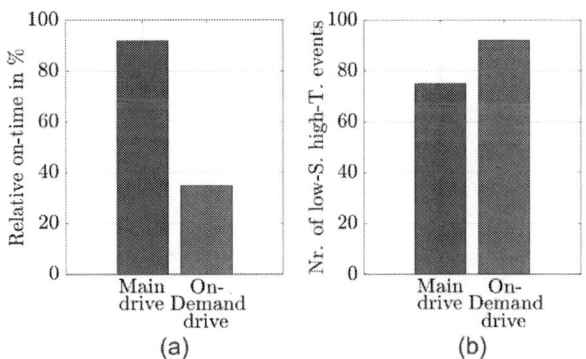

Fig. 2: Primary drive and On-Demand drive usage obtained from measurement data. (a) Relative on-time and (b) Number of low speed - high torque events.

A drive system design based only on standard peak drive torque T_{pk} and peak drive power P_{pk} requirements might lead to oversized drive components.

It is therefore necessary to also address the expected drive system usage as one of the design requirements.

In Fig. 2, a comparison of main drive and On-Demand drive usage is shown. The usage comparison has been extracted from data collected over 10 hours of measurements across different driving scenarios and it can be assumed that the driving scenarios are representative of expected usage for the particular EV. Measurement data indicates that the total on-time of the On-Demand drive is much lower compared to the main drive (by a factor of 2.5 - Fig. 2 (a)). However, the number of "low speed - high torque" events is higher for the On-Demand drive (Fig. 2 (b)). Each operating point of the traction machine for which the machine produces above 80 % of it's rated peak torque at an electrical frequency of 5 Hz or less is considered as "low speed - high torque". These events usually have the largest impact on lifetime degradation [6]. In order to address the topic of On-Demand drive usage and subsequently lifetime degradation of the drive components, representative drive loading profiles should firstly be established. This paper proposes two approaches for generation of representative drive loading profiles:

1. Speed/torque derivation based on vehicle model and simplified On-Demand usage criteria - (**Approach A** in Fig. 3), and

2. Speed/torque derivation based on vehicle model and torque mapping - (**Approach B** in Fig. 3).

For both approaches, a simplified longitudinal vehicle model is assumed. Model dynamics are defined by

$$m_v \dot{v}_v(t) = \frac{T_w(t)}{r_w} - \sum F_i(t) \qquad (1)$$

where m_v is the vehicle mass, r_w is the wheel radius, $v_v(t)$ is the vehicle speed, $T_w(t)$ is the required wheel torque and $\sum F_i$ is the sum of all forces acting on the vehicle in the longitudinal direction. External forces acting on the vehicle can be divided into the aerodynamic drag force, frictional force and gravitational force

$$\sum F_i(t) = \frac{\rho_{air}}{2} C_d A \, v_v^2(t) + m_v g \left(\sin \alpha + \mu_r \cos \alpha \right). \quad (2)$$

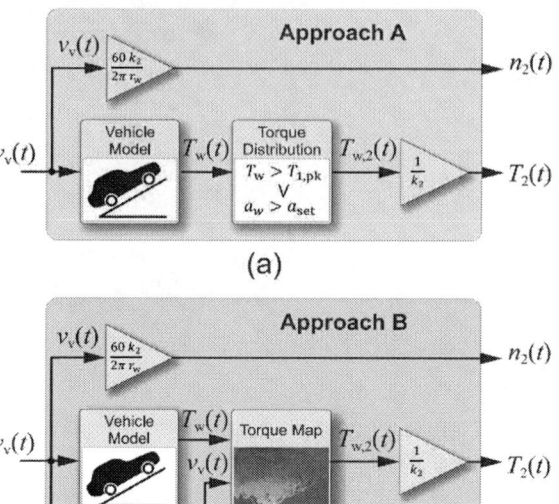

Fig. 3: Generation of On-Demand drive torque profiles based on (a) - vehicle model and control strategy - **Approach A** and (b) vehicle model and On-Demand torque map - **Approach B**.

In (2), ρ_{air} designates the density of air and $\alpha = \alpha(t)$ designates the inclination (slope) profile of the terrain. The required wheel torque is generated by the main drive system and the On-Demand drive system

$$T_w(t) = k_1 T_1(t) + k_2 T_2(t) \qquad (3)$$

where k_1 and k_2 are the gearbox ratios of the main drive and On-Demand drive, respectively. The vehicle dynamics given in (1) provide a value for the required wheel torque, however the distribution of main drive torque T_1 and On-Demand drive torque T_2 is not specified. Employing either **Approach A** or **Approach B**, as presented in Fig. 3, it is possible to obtain this torque distribution for a dual-motor EV ($T_1(t)$ and $T_2(t)$ in (3)).

As an example of drive control strategy for **Approach A**, the On-Demand drive produces torque when the required wheel torque is higher than the peak torque that the main drive can produce

$$T_w(t) > k_1 T_{1,pk}. \qquad (4)$$

As another example, the On-Demand drive is used when the required wheel torque is higher than some defined torque set-point T_{set}

$$T_w(t) > T_{set}. \qquad (5)$$

1790

The mentioned On-Demand drive usage criteria are simplified and are usually not an accurate representation of real-world drive usage. Due to this, a torque mapping approach is presented here which aims to provide better accuracy in estimating secondary drive torque. The derivation of the torque map used in **Approach B**, is shown in Fig. 4.

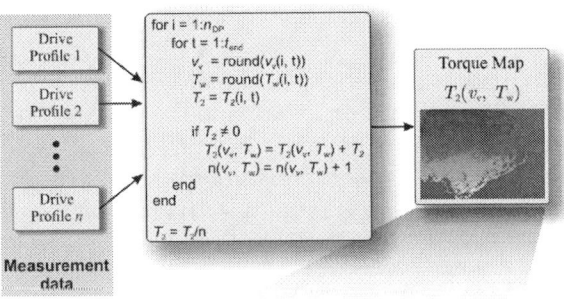

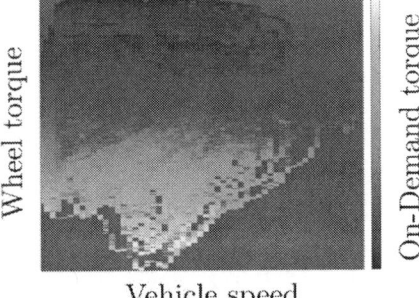

Fig. 4: Generation of On-Demand drive torque profiles based on vehicle model and vehicle measurement data.

This second approach is based on mapping of vehicle speed and vehicle wheel torque to On-Demand drive torque. The mapping has been extracted from vehicle measurement data as shown in Fig. 4 and is based on statistical correlation of the On-Demand drive torque to the vehicle speed and vehicle wheel torque. For each combination of vehicle speed v_v and vehicle wheel torque T_w, the expected value of the On-Demand torque has been calculated by

$$T_2(v_v, T_w) = \frac{\sum T_{2,\text{meas}}(v_{v,\text{meas}} = v_v, T_{w,\text{meas}} = T_w)}{N}. \quad (6)$$

In (6), the On-Demand drive torque map $T_2(v_v, T_w)$ is calculated from measurement data $T_{2,\text{meas}}$, $v_{v,\text{meas}}$ and $T_{w,\text{meas}}$. Measurement data was binned into bins of width 1 m/s and 1 N m respectively. Although this approach inherently introduces "smoothing" into the generated torque profiles, it can serve as

a tool for accurate estimation of On-Demand drive torque. The proposed approach encodes relevant torque distribution data into a single 3D map without the need for *a priori* knowledge of the torque distribution control strategy. Therefore, it is possible to estimate the expected loading of the On-Demand drive without knowing the exact control strategy in a dual-motor EV. This 3D map can ultimately be used as a look-up table for generation of On-Demand drive torque profiles with arbitrary vehicle speed as input.

In Fig. 5, the derived On-Demand torque profiles for **Approach A** and **Approach B** are shown along with measured On-Demand torque for reference. A vehicle speed profile has been isolated from the measurement results in order to compare the validity of the approaches based on torque profile generation from the mentioned approaches. As can be seen, the generated torque profile from **Approach A** does not show good matching to the measured torque as the torque distribution strategy is not known and is only approximated by simple usage criteria. A better match might be achieved if the cut-in criteria is better approximated. **Approach B** shows much better matching with measured results.

For comparison, the first approach offers a lot more flexibility, where arbitrary torque profiles can be generated from vehicle speed profiles and vehicle parameters. This approach allows for freedom of choice where it is possible to iterate through different vehicle segments (e.g. compact vehicle to SUV) and different On-Demand drive usage strategies (e.g. secondary drive activated or continuously in operation in parallel with the main drive) by varying vehicle parameters and and drive usage criteria as shown in (4) and (5). On the other hand, the torque mapping approach limits the flexibility in vehicle choice. As the torque map is derived from measurements in a single vehicle, the mapping is only representative of this vehicle.

3 Drive System Model

In the previous section, two methods for deriving On-Demand drive torque profiles from arbitrary vehicle speed profiles have been presented. The lifetime requirements of drive system components are highly dependent on these profiles as the mechanical and thermal stresses are a direct consequence of drive loading ([10] and [11]). In order to accurately predict thermal stress within a power module,

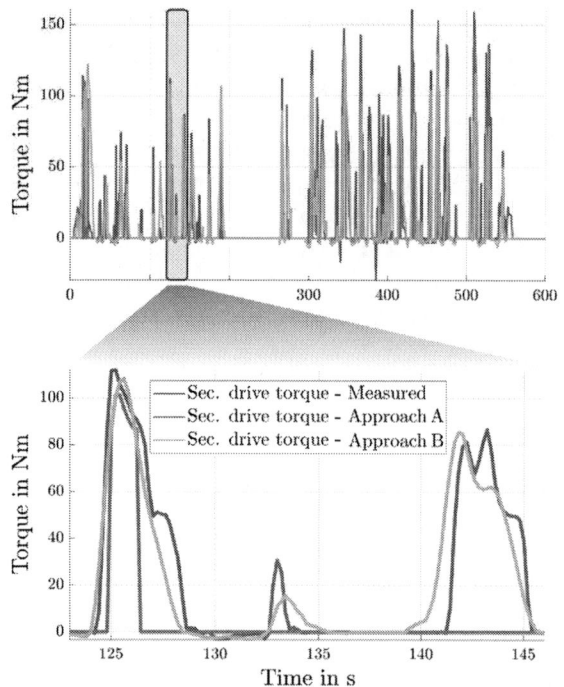

Fig. 5: Comparison of measured On-Demand drive torque with torque derived from cut-in criteria (**Approach A**) and torque derived from torque mapping (**Approach B**).

the expected junction temperature profiles need to be established. This section focuses on efficient simulation of On-Demand drive inverter temperatures for estimation of power module thermal stress. Compared to Finite Element Method simulation models, which are commonly employed for junction temperature estimation, the electro-thermal model presented here is much less computationally intensive, as it is based on a 1D Foster thermal network representation of the thermal system. In the first step, the speed/torque profile of the traction machine needs to be converted into machine voltages and currents. Next, the voltages and currents are to be transferred into power losses within the module and finally into junction temperature profiles of the semiconductors as shown in the block diagram in Fig. 6.

3.1 Traction Machine Model and Speed Control

For the traction machine of the On-Demand drive, a four-pole ($p = 2$) Induction Machine (IM) is assumed. IM voltages $u_m(t)$ and currents $i_m(t)$ are obtained from equivalent circuit model impedance and the machine torque equation

$$\underline{Z} = R_1 + j\,\omega_{el}L_1 + j\,\omega_{el}L_m \| \left(\frac{R_2'}{s} + j\,\omega_{el}L_2' \right) \quad (7)$$

$$T_{el} = \frac{3\,p}{\omega_{el}} \frac{R_2'\,s\,U_m^2}{R_2^2 + (s\,\omega_{el}L_2')^2} \quad (8)$$

where R_1, L_1, R_2', L_2' and L_m are equivalent circuit parameters of the induction machine, p is the number of pole-pairs, $s = \frac{\omega_{el} - p\,\omega_{mech}}{\omega_{el}}$ is the slip and U_m is the magnetizing voltage. The model given in (7) and (8) does not uniquely define the voltage and current wave-forms of the IM for a given mechanical operating point (speed and torque). Therefore, an additional equation is required. The magnitude of the IM voltage as a function of rotational speed is defined as

$$\hat{U}_m = n_2 \frac{\hat{U}_{m,max}}{n_{2,max}}. \quad (9)$$

This is a simplification of speed control of induction machines in which the magnitude of the applied voltage at the stator is proportional to the rotational speed of the machine.

The set of equations (7) - (9) define the electrical operating point of the Induction Machine ($\hat{U}_m$, $\hat{I}_m$, ω_{el} and ϕ) assuming known mechanical operating point (n_2, T_2). Machine voltages and currents are assumed to be sinusoidal

$$\begin{aligned} u_m(t) &= \hat{U}_m \cos(\omega_{el}t) \\ i_m(t) &= \hat{I}_m \cos(\omega_{el}t - \phi). \end{aligned} \quad (10)$$

3.2 Inverter electro-thermal model

With known motor voltages and motor currents, the current stress of inverter power semiconductors can be calculated. Finally, a detailed electro-thermal model of the inverter is implemented for junction temperature simulation. The electro-thermal model is based on a state-space representation of the module junction temperatures. The model is derived from the integral representation of the Foster thermal network [12]

$$T_j(t) - T_C = \int_0^t P(\tau)\dot{Z}_{th}(t - \tau)\,d\tau. \quad (11)$$

and can be extended to include thermal coupling effects of adjacent chips (i-th chip) within a power module, as well as Foster thermal networks of higher order (l_m). The extended model is defined as

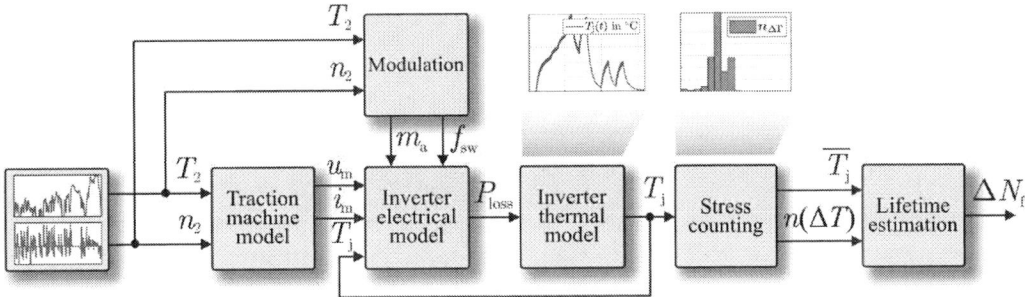

Fig. 6: Schematic overview of On-Demand drive system model used for junction temperature simulation.

$$T_{\mathrm{j}}(t) - T_{\mathrm{C}} = \sum_{i=1}^{n} \sum_{l=1}^{l_{\mathrm{m}}} \frac{R_l}{\tau_l} \int_0^t P_i(\tau) e^{(\tau-t)/\tau_l} \, d\tau \qquad (12)$$

where $T_{\mathrm{j}}(t)$ is the junction temperature of the semiconductor, T_{c} is the coolant temperature, R_l is the thermal resistance of the l-th component in the Foster network of size l_{m} and $P_i(t)$ is the generated power loss within the i-th switch.

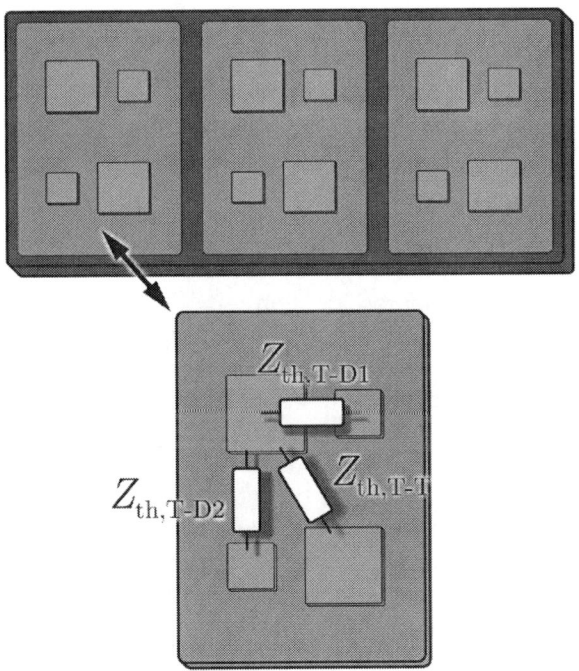

Fig. 7: Schematic of traction inverter power module with indicated thermal impedances.

In Fig. 7, a schematic representation of a six-switch power module is shown with indicated thermal coupling impedances ($Z_{\mathrm{th,T-D1}}$, $Z_{\mathrm{th,T-T}}$ and $Z_{\mathrm{th,T-D2}}$). For simulation, the thermal coupling impedances are modeled with a second order

Foster network, while the junction-coolant thermal impedance $Z_{\mathrm{th,T-C}}$ is represented with a fourth order Foster network

$$Z_{\mathrm{th,T-D1}}(t) = \sum_{l=1}^{2} R_{l,\mathrm{T-D1}} \left(1 - e^{-t/\tau_{l,\mathrm{T-D1}}}\right)$$
$$Z_{\mathrm{th,T-C}}(t) = \sum_{l=1}^{4} R_{l,\mathrm{T-C}} \left(1 - e^{-t/\tau_{l,\mathrm{T-C}}}\right). \qquad (13)$$

The parameters of the Foster networks, as well as network order was determined from transient thermal measurements. Measurements show that there is minimal thermal coupling between adjacent bridge-legs within the power module and therefore only a single bridge-leg is considered although the model in (12) can be extended to include an arbitrary number thermal coupling impedances.

Based on results presented in [13], it is possible to transform the integral representation of the coupled Foster network thermal model into a state-space representation

$$\frac{d\,\Delta\mathbf{T}_i(t)}{dt} = \mathbf{A}_i \Delta\mathbf{T}_i(t) + \mathbf{B}_i \mathbf{P}(t)$$
$$T_{\mathrm{j}}(t) = \mathbf{C}_i \Delta\mathbf{T}_i(t) + T_{\mathrm{C}} \qquad (14)$$

Matrices $\mathbf{A}_i$ and $\mathbf{B}_i$ are derived from the integral representation of Foster thermal networks

$$\mathbf{A}_i = \begin{bmatrix} \mathbf{A_{n1}} & 0 \\ 0 & \mathbf{A_{n2}} \end{bmatrix}, \qquad (15)$$

$$\mathbf{B}_i = \begin{bmatrix} \mathbf{B_{n1}} & 0 \\ 0 & \mathbf{B_{n2}} \end{bmatrix} \qquad (16)$$

where

$$\mathbf{A_{n1}} = \begin{bmatrix} \frac{-1}{\tau_{1,T-C}} & 0 & 0 & 0 \\ & \frac{-1}{\tau_{2,T-C}} & 0 & 0 \\ 0 & 0 & \frac{-1}{\tau_{3,T-C}} & 0 \\ 0 & 0 & 0 & \frac{-1}{\tau_{4,T-C}} \end{bmatrix}, \qquad (17)$$

$$\mathbf{A_{n2}} = \begin{bmatrix} \frac{-1}{\tau_{1,T-D1}} & 0 \\ 0 & \frac{-1}{\tau_{2,T-D1}} \end{bmatrix} \quad (18)$$

and

$$\mathbf{B_{n1}} = \begin{bmatrix} \frac{R_{1,T-C}}{\tau_{1,T-C}} \\ \frac{R_{2,T-C}}{\tau_{2,T-C}} \\ \frac{R_{3,T-C}}{\tau_{3,T-C}} \\ \frac{R_{4,T-C}}{\tau_{4,T-C}} \end{bmatrix}, \mathbf{B_{n2}} = \begin{bmatrix} \frac{R_{1,T-D1}}{\tau_{1,T-D1}} \\ \frac{R_{2,T-D1}}{\tau_{2,T-D1}} \end{bmatrix}. \quad (19)$$

In (15) - (19) system matrices of a simplified state-space model with thermal coupling between the active switch and it's freewheeling diode are shown. This simplified model is outlined here for clarity as the order of the full state-space system is $\left(\sum_{i=1}^{N_i} N_k(i) \times \sum_{i=1}^{N_i} N_k(i)\right)$ where N_k is the number of Foster network elements and N_i is the number of considered thermal couplings. For the case of the model which includes only one thermal coupling (switch to it's diode) for the matrices given in (15) - (19), the order of the system is $6x6$.

The state-space representation of a power module thermal network is easy to implement for simulation and the time-discrete model offers fastest run-time (Fig. 8). The run-time of three model representations has been evaluated for a simplified thermal system - half-bride with full thermal coupling.

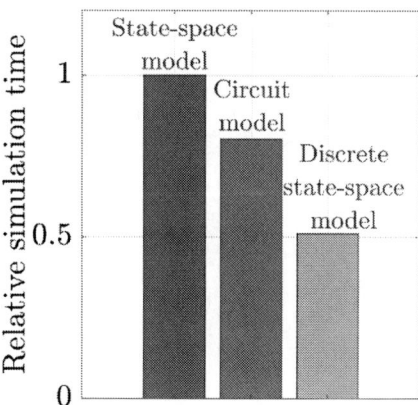

Fig. 8: Comparison of relative run-time for different implementations of the thermal model.

In total, 10 000 electrical periods have been simulated in order to reduce the influence of model initialization on simulation run-time. The model representations have also been compared with different simulation software. Additionally, the accuracy of the different representations has been evaluated against each other and shows a good match.

4 Results and discussion

The simulated drive system is based on a six-pack IGBT power module with a nominal DC-Link voltage of 400 V coupled to a four-pole induction machine (IM). The specification of the drive system is given in Table 1. The model is derived as described in **Section 3** and the speed/torque profile derivation is presented in **Section 2**. The thermal stress of the On-Demand drive inverter is compared for three different speed/torque profiles:

- Measured On-Demand drive machine torque and speed

- Estimated On-Demand drive torque based on simplified torque distribution strategy (**Approach A** in **Section 2**), and

- On-Demand drive torque obtained from torque mapping (**Approach B** in **Section 2**).

Parameter	Value	Unit
Peak torque T_{pk}	165	Nm
Peak power P_{pk}	80	kW
Max. motor speed n_{max}	13500	rpm
System voltage V_{DC}	400	V
Switching frequency f_{sw}	9 .. 11	kHz
Semiconductor type	Si IGBT	-

Tab. 1: On-Demand drive system parameters.

In Fig. 9, simulated junction temperature profiles are shown. The junction temperature profiles are simulated for torque profiles given in Fig. 5. Good match in junction temperatures can be observed between the measured drive cycle and the drive cycle derived using **Approach B** while the results for **Approach A** show deviations in the junction temperature compared to the measured drive cycle. For **Approach A**, the absolute mean error in the simulated junction temperature is 3.4 °C and 2.7 °C for **Approach B** when compared to the simulation results based on the measured speed/torque profile. The simulated temperature profiles are further used to estimate thermal stress of the inverter.

Thermal stress is derived by means of rain-flow counting implemented in MATLAB considering lifetime curves given by the manufacturer of the power module. The thermal stress profiles are depicted in Fig. 10. As can be observed, the stress profiles show very good match between the original torque

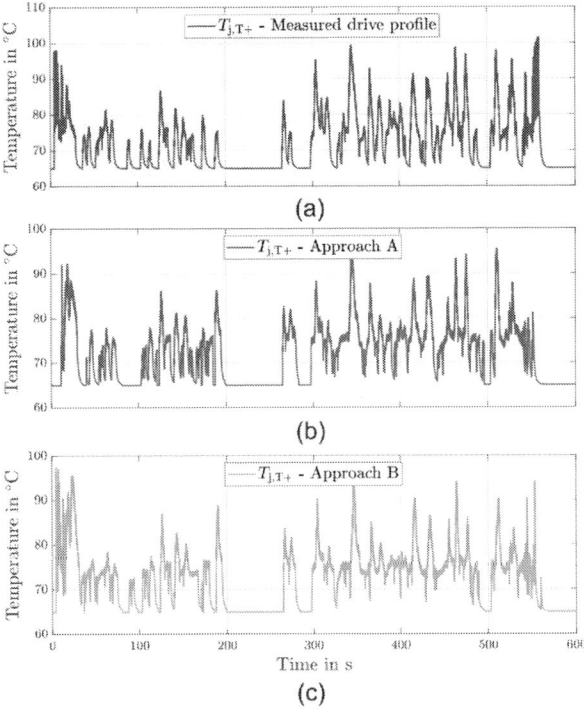

Fig. 9: Comparison of simulated junction temperature profiles for different drive usage approaches. (a) **Measured drive profile**, (b) synthetically derived drive profile based on **Approach A** and (c) derived drive profile based on **Approach B**.

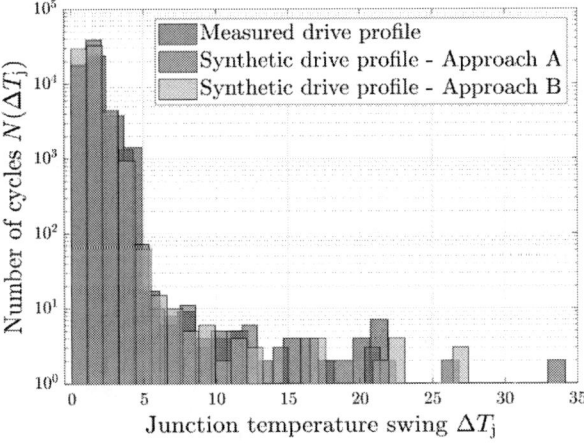

Fig. 10: Comparison of module thermal stress based on junction temperature swing for: **Measured drive profile**, synthetically derived drive profile based on **Approach A** and derived drive profile based on **Approach B**.

profile and the synthetic torque profile generated by means of torque mapping (**Approach B** from **Section 2**), especially for small junction temperature swings. However, for higher junction temperature swings, the synthetic torque profile underestimates the effective thermal stress. This can be attributed to the averaging of the torque map generation approach.

5 Conclusion

This paper presents a procedure for estimating inverter thermal stresses in On-Demand drive systems based on a novel torque mapping approach and efficient electro-thermal simulation. The torque mapping approach shows good agreement with measurement data. Additionally, a detailed system model of the On-Demand drive is presented. The simulation model shows very good accuracy and low simulation time.

Modeling of inverter lifetime degradation based on different torque loading profiles is planned as a next step of the work.

References

[1] M. Yilmaz and P. T. Krein, "Review of battery charger topologies, charging power levels, and infrastructure for plug-in electric and hybrid vehicles," *IEEE Transactions on Power Electronics*, vol. 28, no. 5, pp. 2151–2169, 2013. DOI: 10.1109/ TPEL.2012.2212917.

[2] M. Tong, X. Liu, L. Sun, Z. Xu, M. Cheng, and Q. Zou, "Investigation of an integrated battery charger for EVs based on a dual-motor traction system," in *Proc. of the IEEE Transportation Electrification Conference and Expo, Asia-Pacific (ITEC 2022)*, 2022, pp. 1–5. DOI: 10.1109/ ITECAsia-Pacific56316.2022.9942110.

[3] A. Emadi, Y. J. Lee, and K. Rajashekara, "Power electronics and motor drives in electric, hybrid electric, and plug-in hybrid electric vehicles," *IEEE Transactions on Industrial Electronics*, vol. 55, no. 6, pp. 2237–2245, 2008. DOI: 10.1109/TIE. 2008.922768.

[4] J. A. P. Lopes, F. J. Soares, and P. M. R. Almeida, "Integration of electric vehicles in the electric power system," *Proceedings of the IEEE*, vol. 99, no. 1, pp. 168–183, 2011. DOI: 10.1109/JPROC. 2010.2066250.

[5] O. Nezamuddin, R. Bagwe, and E. Dos Santos, "A multi-motor architecture for electric vehicles," in *Proc. of the IEEE Transportation Electrification Conference and Expo (ITEC 2019)*, 2019, pp. 1–6. DOI: 10.1109/ITEC.2019.8790582.

[6] V. Schwarzer and R. Ghorbani, "Drive cycle generation for design optimization of electric vehicles," *IEEE Transactions on Vehicular Technology*, vol. 62, no. 1, pp. 89–97, 2013. DOI: 10.1109/TVT. 2012.2219889.

[7] M. Shahjalal, T. Shams, S. B. Hossain, M. Rishad Ahmed, M. Ahsan, *et al.*, "Thermal analysis of Si-IGBT based power electronic modules in 50 kW traction inverter application," *e-Prime - Advances in Electrical Engineering, Electronics and Energy*, vol. 3, p. 100 112, 2023. DOI: https://doi.org/10.1016/j.prime.2023.100112.

[8] B. Zhao, N. Xu, H. Chen, K. Guo, and Y. Huang, "Design and experimental evaluations on energy-efficient control for 4WIMD-EVs considering tire slip energy," *IEEE Transactions on Vehicular Technology*, vol. 69, no. 12, pp. 14 631–14 644, 2020. DOI: 10.1109/TVT.2020.3032377.

[9] N. Mutoh, T. Kazama, and K. Takita, "Driving characteristics of an electric vehicle system with independently driven front and rear wheels," *IEEE Transactions on Industrial Electronics*, vol. 53, no. 3, pp. 803–813, Jun. 2006. DOI: 10.1109/tie.2006.874271.

[10] T. Kestler, V. Damec, and M.-M. Bakran, "Differences in dimensioning SiC MOSFETs and Si IG-BTs for traction inverters," in *Proc. of the 20th European Conference on Power Electronics and Applications (EPE 18 ECCE Europe)*, 2018, pp. 1–9.

[11] K. Ma, U.-M. Choi, and F. Blaabjerg, "Prediction and validation of wear-out reliability metrics for power semiconductor devices with mission profiles in motor drive application," *IEEE Transactions on Power Electronics*, vol. 33, no. 11, pp. 9843–9853, 2018. DOI: 10.1109/TPEL.2018.2798585.

[12] M. Ouhab, Z. Khatir, A. Ibrahim, J.-P. Ousten, R. Mitova, and M.-X. Wang, "New analytical model for real-time junction temperature estimation of multichip power module used in a motor drive," *IEEE Transactions on Power Electronics*, vol. 33, no. 6, pp. 5292–5301, 2018. DOI: 10.1109/TPEL.2017.2736534.

[13] J. Ottosson, "Thermal modelling of power modules in a hybrid vehicle application," Ph.D. dissertation, 2013.

PCIM Europe 2024, 11– 13 June 2024, Nuremberg   DOI: 10.30420/566262246

EV Traction Inverter Optimal Design is Dominated by 3-Level ANPC

Delaforge Timothe[1], Moabber Kooros[2]

[1] Bern University of Applied Sciences, Switzerland
[2] Volvocars, Sweden

Corresponding author: Delaforge Timothe, timothe.delaforge@bfh.ch
Speaker: Delaforge Timothe, timothe.delaforge@bfh.ch

Abstract

The paper presents the investigation of 3 levels traction inverter based on wide band gap semiconductors for 800V electric drive. Compared to other converters, the traction inverter operates constantly over an ultra wide range of voltage, current, fundamental frequency and phase shift (speed and torque of the electrical machine). Classical enunciated selection criteria for 3 level topology such as switch count are not pertinent. Price, component stress and thermal management over the driving cycles of the electrical vehicle demonstrate superiority of ANPC.

1   Introduction

Electrical vehicle (EV) main limitations for massive integration into vehicle market are the range autonomy on one charge and the resulting batteries price, and the CO2 footprint of raw materials for battery and electrical machine (EM).

The use of high voltage 800V and soon 1kV [?,?,?] batteries coupled to wide band gap semiconductors (WBGS) [?] and multilevel topologies has proven an optimal technological solution to address both limitations [?].

High voltage batteries limit the metal use, especially copper that represent more than 85kg in a Tesla model S for example. Multilevel topology based on WBGS allow a gain of up to 5% on the whole traction chain (batteries + traction inverter + electrical machine) [?]. Reducing battery need and motor size for same performance and driving range on one charge.

However, often enunciated comparison criteria for 3 level topologies such as simple switch count or basic price comparison of a diode vs an active switch [?] are not the most pertinent for traction inverter.

Traction inverter operates over an ultra wide range of current, voltage, modulation frequency and power factor (speed vs torque) with constant change of operating point depending on the driving cycle especially in city driving. This translate to the switches:

Wide variation of current carrying; Important thermal cycling; Peak stress in case of worst case operation (standstill and active short-circuit).

Considering the operation of 3 levels traction inverter over the use map, the topologies are compared in terms of:

Total required die area for overall current capability; Component stress; Active components and control strategies to improve efficiency.

The Active Neutral Point Clamped (ANPC) proves the optimal one for this application.

2   Multilevel converter modeling

The wide number of operating points for the traction inverter calls for a simplified loss model in order to quickly compare all multilevel topologies and control strategies performances over the driving cycles. In traction application based on WBG semiconductors operating over 20kHz switching frequency, the current ripple is almost canceled and its impact on losses can be ignored without loss of evaluation accuracy.

A simplified model is proposed bellow based on this assumption but also that deadtime impact is ignored and the ratio between switching frequency and EM fundamental frequency, i.e. modulation frequency of the inverter is large enough to ignore the effect of a limited number of pulses within a modulation period.

The model is built from:

- One complete control cycle at the modulation frequency is achieved for a control angle that varies from 0 to 2π.

- The inverter output voltage applied to the EM can be defined as:

$$V_{ph} = \hat{V}_{ph} \cdot sin(\theta) \qquad (1)$$

- Given input DC bus voltage V_{DC} the modulation index can then be defined as:

$$m_{PSPWM} = \frac{\hat{V}_{ph}}{V_{DC}} \cdot sin(\theta) \qquad (2)$$

$$m_{SVPWM} \simeq \frac{\hat{V}_{ph}}{V_{DC}} \cdot sin(\theta) + 0.2 \cdot \frac{\hat{V}_{ph}}{V_{DC}} \cdot sin(3 \cdot \theta) \qquad (3)$$

- The phase current lagging by the phase angle φ delivered to the EM is:

$$i_{ph} = \hat{i}_{ph} \cdot sin(\theta - \varphi) \qquad (4)$$

- Then for each topology, considering the control angles $\alpha 1$ and $\alpha 2$ for each switch, see Fig. 1, the mean and RMS currents across one semiconductor can be define as:

$$i_{MEAN} = \frac{1}{2\pi} \int_{\alpha 1}^{\alpha 2} m_{xPWM} \cdot \hat{i}_{ph} \cdot sin(\theta - \varphi) \, d\theta \qquad (5)$$

$$i_{RMS} = \sqrt{\frac{1}{2\pi} \int_{\alpha 1}^{\alpha 2} m_{xPWM} \cdot \hat{i}_{ph}^2 \cdot sin(\theta - \varphi)^2 \, d\theta} \qquad (6)$$

- The conduction losses for WBG MOSFET are expressed as:

$$P_{cond} = R_{DS_{on}}(T_j) \cdot i_{RMS}^2(\alpha 1, \alpha 2, m_{xPWM}) \qquad (7)$$

- The switching losses are computed using the same approach but with different control angles, as in multilevel topology a semiconductor can conduct without switching:

$$P_{SW} = \frac{f_{sw}}{2\pi} \int_{\alpha 3}^{\alpha 4} (E_{on}(i_{ph}, V_{DS}, T_j)) \, d\theta$$
$$+ \frac{f_{sw}}{2\pi} \int_{\alpha 3}^{\alpha 4} (E_{off}(i_{ph}, V_{DS}, T_j)) \, d\theta \qquad (8)$$
$$+ \frac{f_{sw}}{2\pi} \int_{\alpha 3}^{\alpha 4} (E_{err}(i_{ph}, V_{DS}, T_j)) \, d\theta$$

3 Model angles and modulation index

In order to map the losses for following topologies: classical 2 level, 3 level NPC, 3 level TNPC, 3 level ANPC and 3 level FC using the presented model, the conduction angles, the switching angles and the modulation index are listed for each switch in Fig. 2 to Fig. 8.

4 RMS current distribution and sizing

The EM data and the driving cycles used for the design of a 290kW, 835V DC bus traction inverter are illustrated on Fig. 9.

For each topology, a map representing the share between switches of total inverter RMS current feed to the EM, corresponding to the EM map was drawn, Fig. 10 to Fig. 16. The conclusion from this analysis is that the number of switches in 3L topologies is not the key factor for price and performance comparison. Indeed Flying Capacitor (FC) and T-type NPC (TNPC) have only four switches per phase whereas Active NPC (ANPC) and NPC have six switches, but the total number of semiconductors required is not directly linked to the number of switches.

Looking at the current maps, the criteria for total number of semiconductors in terms of current capability is defined in the operating area with maximum RMS phase current:

- The 2 level must have 1.41 times the total RMS current capability with balanced loss distribution, but with the worst switching performance.

- The TNPC must have 2.45 times the total RMS

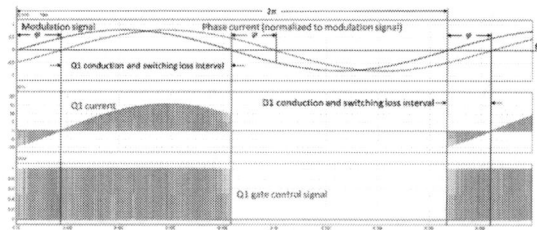

Fig. 1: Electrical waveforms for a semiconductor. Top, the modulation ratio directly in phase with the inverter output voltage and the current that is phase shifted by φ. Middle, the current in the semiconductor, in IGBTs the distinction is made between IGBT and diode conduction angles. Bottom, the duty cycle signal for the semiconductor to find switching angles.

Fig. 2: 2 level inverter angles.

Topology	Device	Modulation	Conduction angles α1	Conduction angles α2	Switching angles α1	Switching angles α2
	Q_1	m_{xPWM}	φ	$\pi+\varphi$	φ	$\pi+\varphi$
	D_1	m_{xPWM}	φ	$\pi+\varphi$	φ	$\pi+\varphi$
	Q_2	$1-m_{xPWM}$ $1+m_{xPWM}$	0 $\pi+\varphi$	φ 2π	0 $\pi+\varphi$	φ 2π
	D_1	$1-m_{xPWM}$ $1+m_{xPWM}$	0 $\pi+\varphi$	φ 2π	0 $\pi+\varphi$	φ 2π

Fig. 3: 3 levels NPC inverter angles.

Topology	Device	Modulation	Conduction angles α1	Conduction angles α2	Switching angles α1	Switching angles α2
	Q_1	m_{xPWM}	φ	π	φ	π
	D_1	m_{xPWM}	0	φ	0	φ
	Q_2	1 $1+m_{xPWM}$	φ π	π $\pi+\varphi$	π	$\pi+\varphi$
	D_2	m_{xPWM}	0	φ	0	φ
	D_5	$1-m_{xPWM}$ $1+m_{xPWM}$	φ π	π $\pi+\varphi$	φ π	π $\pi+\varphi$

Fig. 4: 3 levels TNPC inverter angles.

Topology	Device	Modulation	Conduction angles α1	Conduction angles α2	Switching angles α1	Switching angles α2
	Q_1	m_{xPWM}	φ	π	φ	π
	D_1	m_{xPWM}	0	φ	0	φ
	Q_4	m_{xPWM}	$\pi+\varphi$	2π	$\pi+\varphi$	2π
	D_1	m_{xPWM}	π	$\pi+\varphi$	π	$\pi+\varphi$
	Q_2	$1-m_{xPWM}$ $1+m_{xPWM}$	φ π	π $\pi+\varphi$	φ	π
	D_2	$1-m_{xPWM}$ $1+m_{xPWM}$	0 $\pi+\varphi$	φ 2π	0	φ
	Q_3	$1+m_{xPWM}$ $1-m_{xPWM}$	$\pi+\varphi$ 0	2π φ	$\pi+\varphi$	2π
	D_3	$1+m_{xPWM}$ $1-m_{xPWM}$	π φ	$\pi+\varphi$ $\pi+\varphi$	π	$\pi+\varphi$

Fig. 5: 3 levels ANPC inverter angles. 1st strategy, cell1 : Q1&Q5, cell2 : Q2&Q3, cell3 : Q4&Q6, cell 2 always switches at line frequency, and cells 1 and 3 alternate between line and switching frequency during negative and positive half line cycles respectively.

Topology	Device	Modulation	Conduction angles α1	Conduction angles α2	Switching angles α1	Switching angles α2
	Q_1	m_{xPWM}	φ	π	φ	π
	D_1	m_{xPWM}	0	φ	0	φ
	Q_3	$1-m_{xPWM}$	0	φ	0	φ
	D_5	$1-m_{xPWM}$	φ	π	φ	π
	Q_2	1	φ	π		
	D_2	1	0	φ		

Fig. 6: 3 levels ANPC inverter angles. 1st strategy, cell1 : Q1&Q5, cell2 : Q2&Q3, cell3 : Q4&Q6, cell 2 always switches at the switching frequency, and cells 1 and 3 always at line frequency.

Topology	Device	Modulation	Conduction angles α1	Conduction angles α2	Switching angles α1	Switching angles α2
	Q_1	m_{xPWM}	φ	π		
	D_1	m_{xPWM}	0	φ		
	Q_5	$1+m_{xPWM}$	$\pi+\varphi$	2π		
	D_5	$1+m_{xPWM}$	π	$\pi+\varphi$		
	Q_2	m_{xPWM} $1+m_{xPWM}$	φ π	π $\pi+\varphi$	φ π	π $\pi+\varphi$
	D_2	m_{xPWM} $1+m_{xPWM}$	0 $\pi+\varphi$	φ 2π	0 $\pi+\varphi$	φ 2π

Fig. 7: 3 levels ANPC inverter angles. 1st strategy, cell1 : Q1&Q5, cell2 : Q2&Q3, cell3 : Q4&Q6, all cells switch at the switching frequency.

Topology	Device	Modulation	Conduction angles α1	Conduction angles α2	Switching angles α1	Switching angles α2
	Q_1	m_{xPWM}	φ	π	φ	π
	D_1	m_{xPWM}	0	φ	0	φ
	Q_5	$(1-m_{xPWM})/2$ $(1+m_{xPWM})/2$	0 $\pi+\varphi$	φ 2π	0 $\pi+\varphi$	φ 2π
	D_5	$(1-m_{xPWM})/2$ $(1+m_{xPWM})/2$	φ π	π $\pi+\varphi$	φ π	π $\pi+\varphi$
	Q_2	$(1+m_{xPWM})/2$	φ	$\pi+\varphi$	φ	$\pi+\varphi$
	D_2	$(1+m_{xPWM})/2$ $(1+m_{xPWM})/2$	0 $\pi+\varphi$	φ 2π	0 $\pi+\varphi$	φ 2π

Fig. 8: 3 levels ANPC inverter angles. 1st strategy, cell1 : Q1&Q5&Q3, cell2 : Q2&Q4&Q6, cell 1 and cell 2 alternate between line and switching frequency during negative and positive half line cycles respectively.

Topology	Device	Modulation	Conduction angles α1	Conduction angles α2	Switching angles α1	Switching angles α2
	Q_1	m_{xPWM}	φ	π	φ	π
	D_1	m_{xPWM}	0	φ	0	φ
	Q_5	$(1-m_{xPWM})$ $(1+m_{xPWM})$	0 $\pi+\varphi$	φ 2π	0 $\pi+\varphi$	φ 2π
	D_5	$(1-m_{xPWM})$ $(1+m_{xPWM})$	φ π	π $\pi+\varphi$	φ π	π $\pi+\varphi$
	Q_2	$(1+m_{xPWM})$	φ	$\pi+\varphi$	φ	$\pi+\varphi$
	D_2	$(1+m_{xPWM})$ $(1+m_{xPWM})$	0 $\pi+\varphi$	φ 2π	0 $\pi+\varphi$	φ 2π

PCIM Europe 2024, 11– 13 June 2024, Nuremberg   DOI: 10.30420/566262246

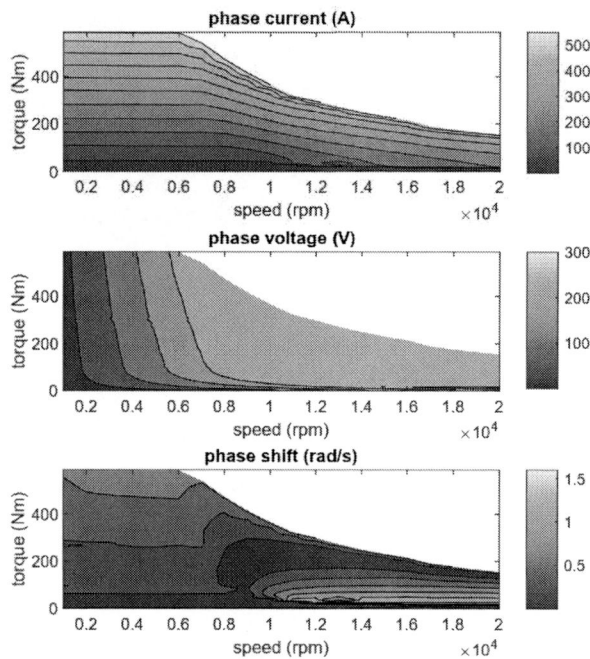

Fig. 9: EM data as input for inverter design.

current capability, with the worst distribution, middle switches experiencing 95% of the total RMS current.

– The ANPC with modulation strategy 4 must have 2.71 times the total RMS current capability, with the best balancing between the switches, from 35% to 55% of the total RMS current per switch.

– The ANPC with modulation strategy 3 must have 3.41 times the total RMS current capability.

– The NPC must have 3.41 times the total RMS current capability but with current share between 50% and 70% per switch.

– The ANPC with modulation strategy 2 must have 3.9 times the total RMS current capability.

– The ANPC with modulation strategy 1 must have 3.91 times the total RMS current capability.

– The FC must have 4.82 times the total RMS current capability, considering flying capacitor for fair comparison with other 3L, but with perfect balancing between switches.

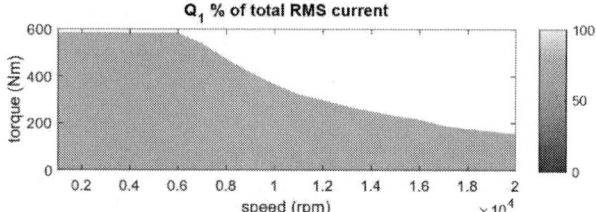

Fig. 10: Percent of total RMS current distribution between switches for 2 level inverter.

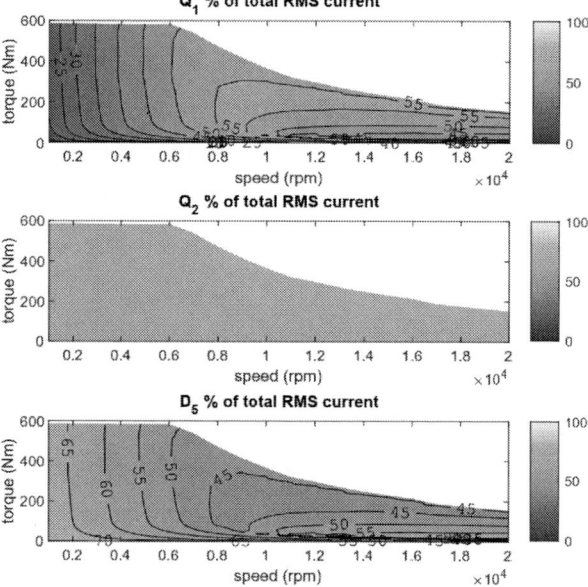

Fig. 11: Percent of total RMS current distribution between switches for 3 level npc inverter.

Those total current capabilities to be installed directly represent the number of required semiconductors and thus the converter price.

5 Loss distribution and thermal stress

The second criteria of selection is the thermal stress of the semiconductors in each topology. While the number of levels being the same the distribution of switching and conduction losses inside the topologies vary. A comparison is made between all the topologies with same total semiconductors' price. The deviation of losses distribution between semiconductors is represented in Fig. 17. In 2 level and flying capacitor, the losses are perfectly balanced between switches. For the other 3 levels topologies, the losses distribution at maximum current and maximum voltage varies a lot. In Fig. 17 the deviation is minimized for the anpc with modulation strategy 4. While TNPC, ANPC 1

1800

PCIM Europe 2024, 11– 13 June 2024, Nuremberg   DOI: 10.30420/566262246

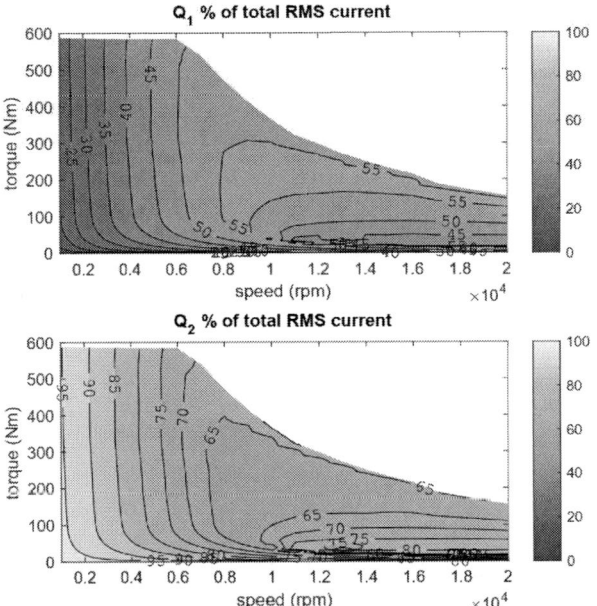

Fig. 12: Percent of total RMS current distribution between switches for 3 level tnpc inverter.

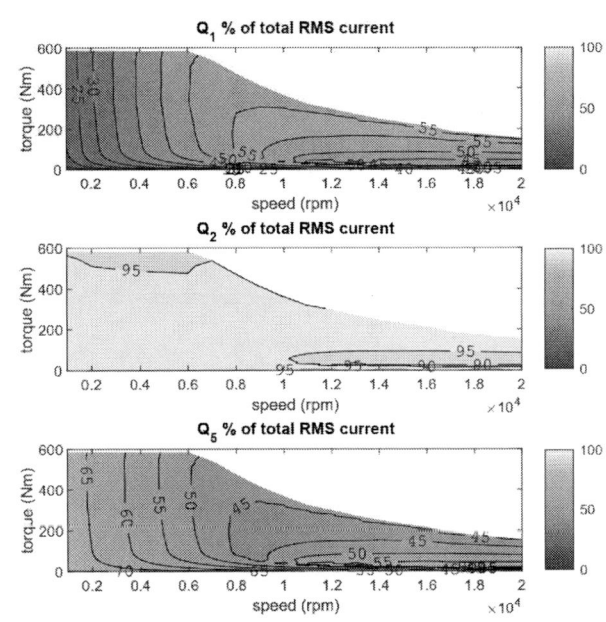

Fig. 14: Percent of total RMS current distribution between switches for 3 level anpc strategy 2 inverter.

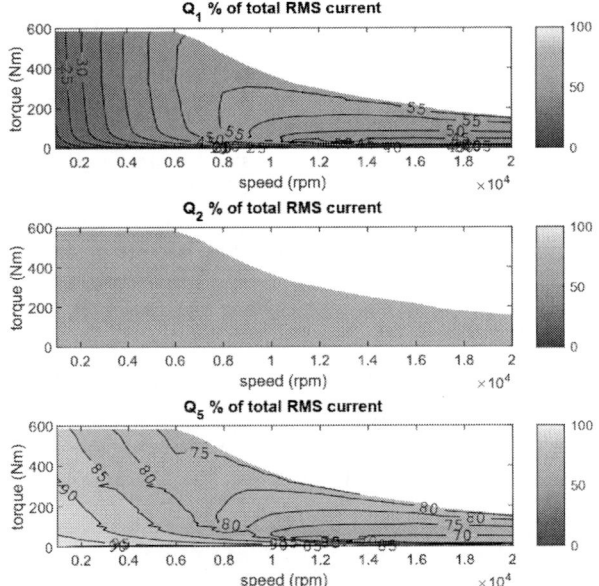

Fig. 13: Percent of total RMS current distribution between switches for 3 level anpc strategy 1 inverter.

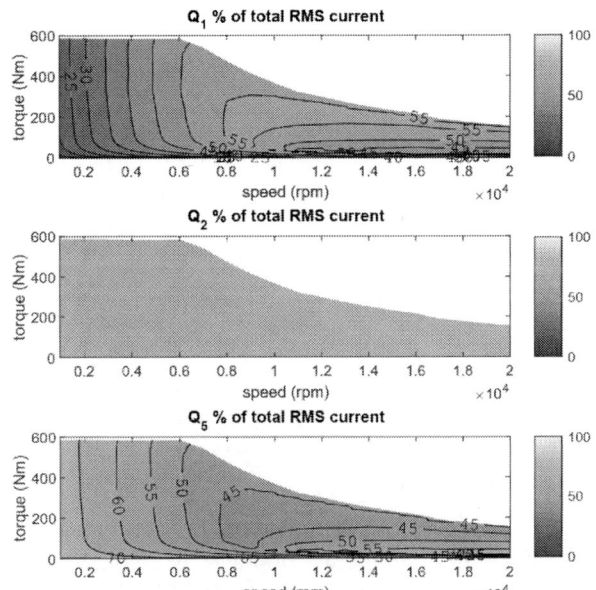

Fig. 15: Percent of total RMS current distribution between switches for 3 level anpc strategy 3 inverter.

1801

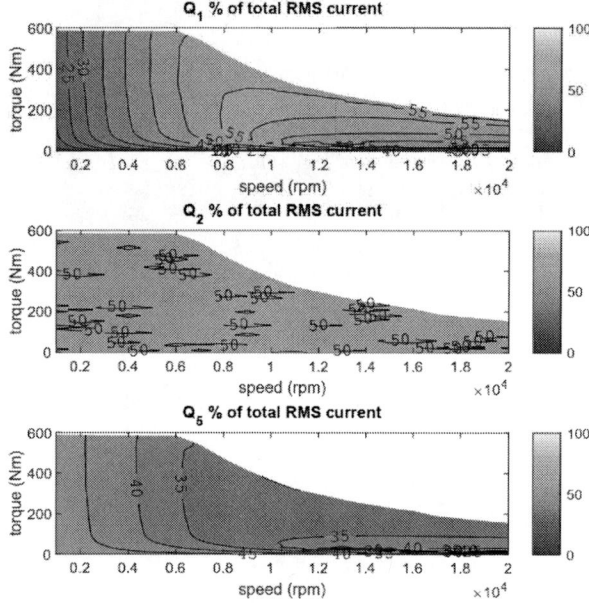

Fig. 16: Percent of total RMS current distribution between switches for 3 level anpc strategy 4 inverter.

and 2 exhibit more than 40% deviation in losses distribution and thus thermal stress. All component being mounted on the same cooling plate, in the end the thermal cycling will be the worst for those topologies while ANPC 4 limits the thermal cycling.

6 Inverter efficiency map

An efficiency comparison is made between all the topologies with same total semiconductors' price. The model presented in first section is used to compute conduction and switching losses for WBG MOSFET. The 1.2kV MOSFET are SiC based while 650V are GaN based. The switching frequency is fixed to 20kHz for optimal EM performances [?].
The resulting efficiency maps are presented in Fig. 18.
As predicted previous sections, the ANPC with the modulation strategy 4 maximizes the efficiency on the whole operating area of the traction inverter.

7 Conclusion

The paper investigates the behavior of semiconductors in 2L and 3L topologies for a 290kW / 835V traction inverter over the whole electrical machine operation area. A simplified model is presented to quickly evaluate performances of an inverter topology over a wide range of operating conditions. The model parameters for 2 level and 3 levels FC, NPC;

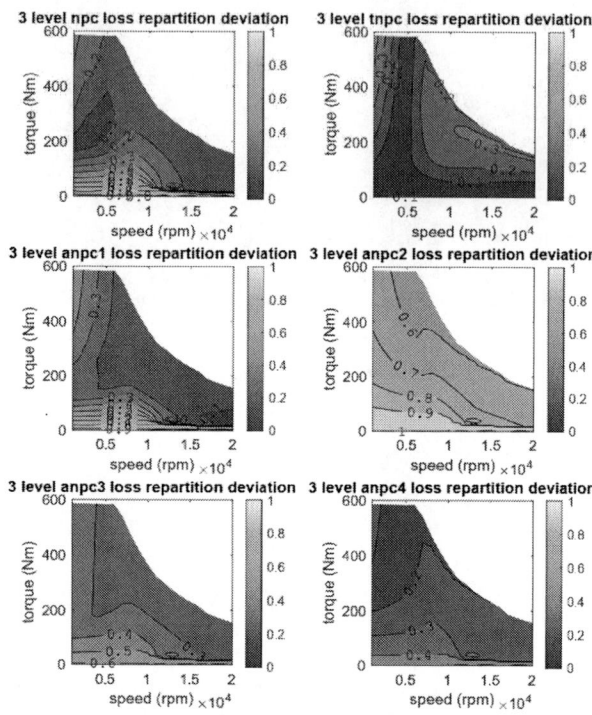

Fig. 17: Representation of losses distribution between semiconductors, 0 is perfect distribution, 1 the worst.

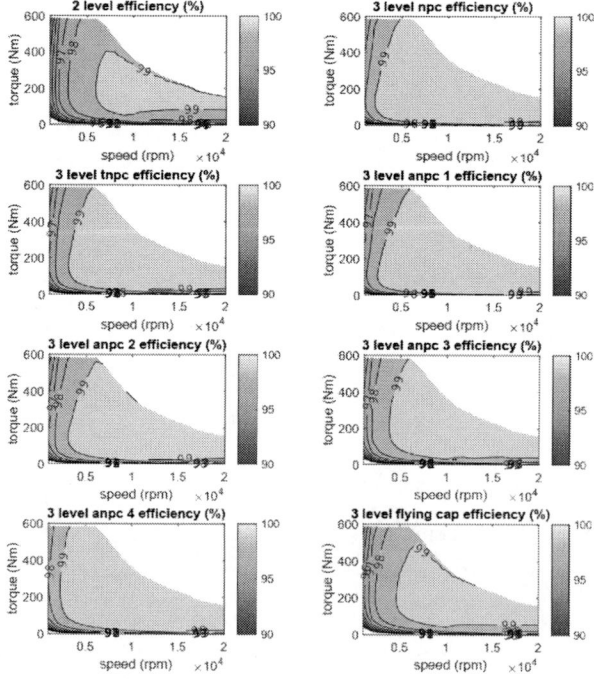

Fig. 18: Representation of investigated converter efficiency for same price of semiconductors.

TNPC and ANPC are given. Using this model, the current distribution (conduction losses), the switching losses distribution and finally the component stress distribution are shown.

This quick analysis highlights the uneven behavior of switches in multilevel topologies and demonstrates that performances and price of a topology is not fixed by the number of switches in the schematic.

The two main criteria used to compare the topologies are the total current capability of semiconductors and the stress distribution between switches in a topology. The total current capability that translates into die area and thus minimum price of semiconductors that must be installed in worst operating conditions shows the superiority of 2 level, 3 level TNPC and 3 level ANPC. However, the second criteria shows that ANPC dominates the other.

In the end, the total performances show that ANPC with the correct modulation strategy improves dramatically traction inverter performances.

Introduction of Power Semiconductor Options for Exciter System of Brushed Electrically Excited Synchronous Motor

Yeriel Bai[1], Kangyoon Lee[1], Bumseung Jin[2]

[1] ONSEMI, South Korea
[2] ONSEMI, U.S.A

Corresponding author: Bumseung Jin, BumSeung.Jin@onsemi.com
Speaker: Yeriel Bai, Yeriel.Bai@onsemi.com

Abstract

This paper introduces the APM (Automotive Power Module) series for the Exciter of EESM (Electrically Excited Synchronous Motor) 1200V motor driver application. EESM is a new EV (electric vehicle) motor solution with higher efficiency compared to the Asynchronous Motor (ASM) due to field-weakening operation. Additionally, EESM does not require permanent magnets (which are made of rare earth materials), making it more environmentally friendly and cost-effective compared to permanent magnet synchronous motors (PMSM). In this paper, 3 kinds of power discrete option will be proposed and provided the power loss comparison through simulation to the best performance option mount into APM for exciter application.

1   Introduction

According to market surveys, permanent magnet synchronous motors account for more than 80% of the automotive motor market. However, the production of permanent magnets requires mining of rare earth elements such as neodymium and dysprosium. To meet the increasing demand for environmental protection and reduce reliance on the development of rare earth elements, research and development of motors that can replace permanent magnets are currently in progress.

The Electrically Excited Synchronous Motor (EESM) [1][2], as shown in Figure 1, is an emerging alternative for EV motor design. The three-phase circuit supplies three-phase alternating current to the motor. The stator coils carry currents that change direction, resulting in changing magnetic fields. As the rotor coil intersects with these changing magnetic field lines, it promotes the rotation of the electric rotor. When the Exciter circuit reaches the steady-state, the inductive load generates DC power and reduces the flux levels when the motor operates at high speed, allowing energy to be stored in the high-voltage battery.

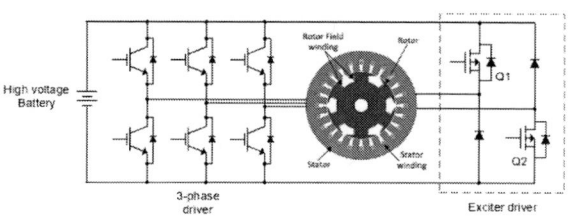

Fig. 1 Electrically Excited Synchronous Motor.

In this paper, we will introduce the Exciter circuit enclosed within the red dotted line of Figure 1. When both Q1 and Q2 are turn-on simultaneously, the high-voltage battery charges the inductive load (motor) of the Exciter. Conversely, when Q1 Q2 are turn-off at the same time, the energy stored in the inductive load charges the high-voltage battery, thereby reducing flux levels during high-speed operation. And current will flow through the freewheeling diode when Q1, Q2 in off-state. This application is to obtain the required load current based on the duty control of Q1 and Q2. The duty of Q1 and Q2 should work between 0.5 and 1, 0.5 < duty < 1. Otherwise, the current of the inductive load could be not reach the steady state to obtain the required load current.

In this paper, 3 options will be proposed for power loss simulation comparison. And the two freewheeling diodes can also be replaced with in Si fast recovery diode (FRD), or SiC technology.

Option 1: SiC MOSFET & Si FRD.
Option 2: SiC MOSFET Full bridge.
Option 3: SiC MOSFET & SiC Schottky Diode.

Then use PSIM for simulation to compare which option has the best thermal performance and lowest power loss than mount into APM for exciter application.

2 Power Loss Simulation

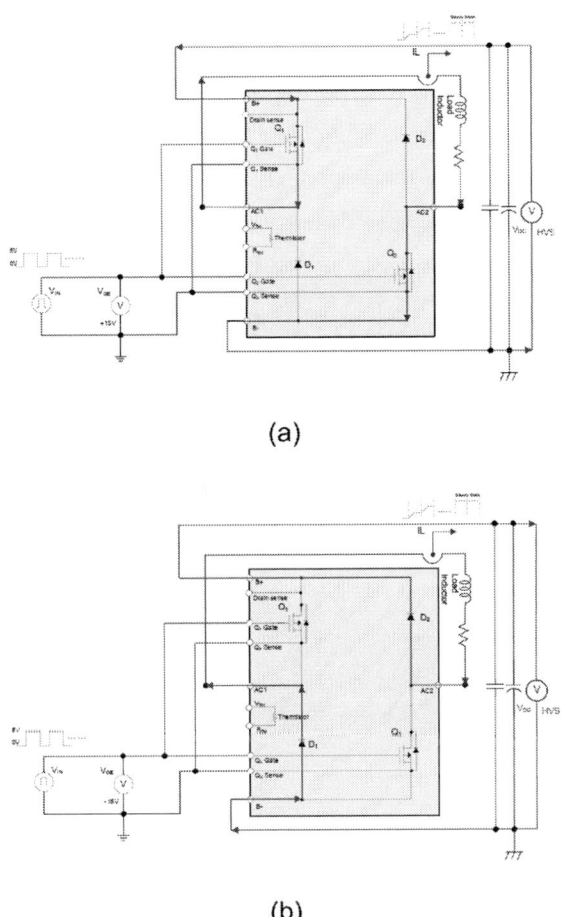

(a)

(b)

Fig. 2 Proposed double pulse switching test circuit for exciter application.

Figure 2 is suggested Double Pulse Tester (DPT) circuit to check the switching losses to find better one among 3 kinds of option.

As mentioned before, the Exciter system operation is when Q1 and Q2 are turned on at the same time to charge the inductive load, the duty of Q1 and

Q2 should work between 0.5 and 1. When Q1 and Q2 are turned off at the same time, current will flow through the freewheeling diode when Q1 Q2 in off-state.

So, the main power loss in Exciter system is switching loss and conduction loss of SiC MOSFET, and the reverse recovery loss and conduction loss of the freewheeling diode. All these 3 options uses SiC MOSFET for switching, so the SiC MOSFET switching loss and conduction loss might be similarities. So, the main difference in power loss might be occurred in freewheeling diode.

First of all, comparison of the reverse recovery loss of the freewheeling diode in 3 options as shown in Fig. 3. It is not difficult to find that SiC diode has the best switching characteristics. Reverse recovery loss: SiC diode < body diode of SiC MOSFET < Si diode.

Secondly, comparison of the forward voltage of the freewheeling diode in 3 options as shown in Fig. 4. SiC diode has the lowest forward voltage. The forward voltage of SiC MOSFET here is simulation under $V_{GS} = 0V$. Forward voltage: SiC diode < Si diode < body diode of SiC MOSFET $_{@Vgs=0V}$.

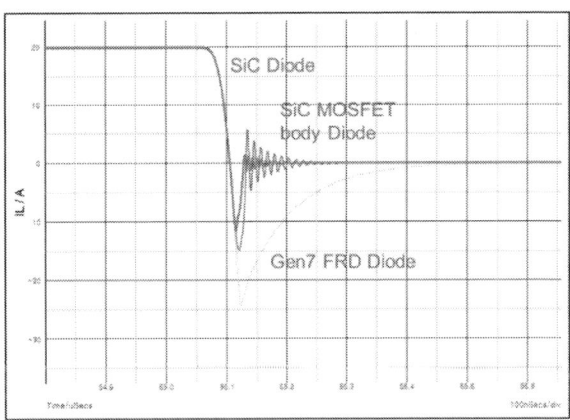

Fig. 3 Comparison of the reverse recovery losses of the diodes in three options.

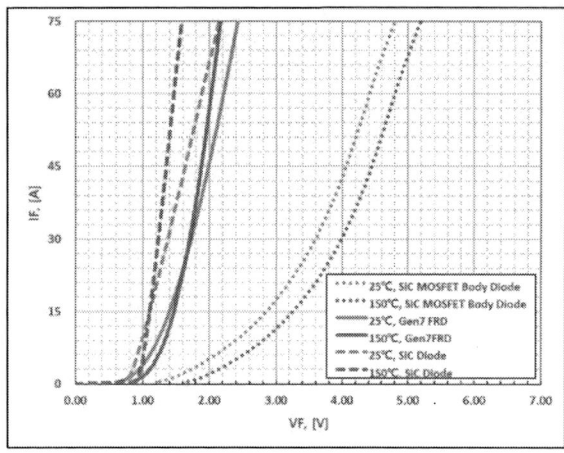

Fig. 4   Comparison of the forward voltage of the diodes in three options.

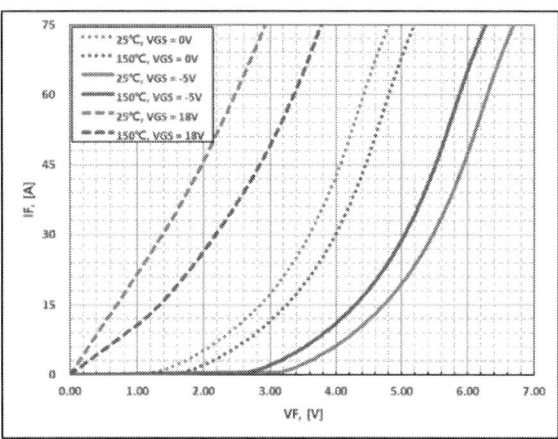

Fig. 5   Comparison of the forward voltage of body diode of SiC MOSFET under different V_{GS}.

Although the body diode of the SiC MOSFET exhibits high V_F value at V_{GS} = 0V, the V_F value decreases when the MOSFET is turned on. As depicted in Figure 5, V_F trends under different V_{GS} values are compared. It is evident that if the MOSFET is turned on during freewheeling instead of conducting directly through the body diode, the conduction loss during this period will be significantly reduced. Although turning on the MOSFET that should be turned off will increase the complexity of the circuit design because gate loops need to be designed for the other two MOSFETs of option 2, this will reduce the system's losses. The simulation results will be discussed in the next chapter.

3   Simulation Result

Among these three options, the DPT simulation results in Figure 3 indicate that the Si FRD has a larger reverse recovery loss compared to the SiC Schottky Diode and the SiC MOSFET body diode. Consequently, option 1 exhibits relatively higher switching losses compared to the other options. Conversely, the body diode of the SiC MOSFET has a higher forward voltage compared to the other options, leading to relatively higher conduction losses for option 2, as illustrated in Figure 4. Therefore, based on these simulation results, we conducted simulations to evaluate the actual performance of these three options within the exciter system. The total power losses in simulation when the exciter circuit reaches the steady state are presented in Table 1.

Condition: VGE=+18/-3V, Fsw=20kHz, 20A, Tf=65deg, duty cycle=60%			External RG [ohm]		
			5	10	30
Total Power Loss [W]	Option 1	Q1, Q2	20.09	20.94	24.46
		D1, D2	80.01	61.33	55.37
	Option 2	Q1, Q2	19.47	20.31	23.72
		Q3, Q4	35.25	34.96	34.66
	Option 3	Q1, Q2	20.57	21.07	24.48
		D1, D2	9.27	9.27	9.31

Table 1   Total power loss of 3 options.

From the results presented in Table 1, it is evident that Q1 and Q2 in all three options utilize the same SiC MOSFET device, resulting in similar losses. However, a significant difference in losses is observed at the freewheeling diode. In option 1, the Si diode incurs higher switching losses, whereas option 2 experiences higher conduction losses due to the body diode of the SiC MOSFET. In comparison to these two options, option 3 demonstrates lower losses in both switching and conduction.

According to equation (1), the junction temperature can be calculated when the fluid temperature is fixed at 65 degrees. The calculation results are shown in Table 2. The thermal resistance values used in the calculations in Table 2 are Rthjf (SiC) = 1.433°C/W for SiC MOSFET and SiC diode, and Rthjf (Si) = 1.669 for the Si Diode.

$$T_j = T_f - R_{thjf} \cdot P_w \qquad (1)$$

Condition: VGE=+18/-3V, Fsw=20kHz, 20A, Tf=65deg, duty cycle=60%			External RG [ohm]		
			5	10	30
Max Tj [deg]	Option 1	Q1, Q2	93.79	95.00	100.04
		D1, D2	200.93	174.30	159.07
	Option 2	Q1, Q2	92.89	94.10	98.99
		Q3, Q4	115.50	115.09	114.67
	Option 3	Q1, Q2	94.47	95.18	100.08
		D1, D2	78.28	78.29	78.34

Table 2 Estimation of the maximum junction temperature.

According to the calculation results in Table 2, it is evident that the junction temperature is very high when using Si diodes in option 1, rendering option 1 unsuitable. Option 3 demonstrates the best thermal performance, and options 2 and 3 will be considered for the project.

However, when option 2 is utilized, the body diode of the MOSFET exhibits a higher value at V_{GS} = 0V, resulting in relatively high conduction losses for option 2, consequently leading to very high junction temperatures for Q3 and Q4. Therefore, Table 3 presents the total power loss and maximum junction temperature of the Exciter system when Q1 and Q2 are turned off and Q3 and Q4 are turned on with V_{GS} = 18/-3V. It can be observed from the results in Table 3 that this application can also achieve better thermal performance results.

Figure 6 depicts the circuit diagram of the Excited Synchronous Motor Automotive Power Module proposed by ONSEMI. And outline overview showed in Figure 7. The product will ultimately be implemented using Option 2. We anticipate good thermal performance based on simulation results.

Condition: VGE=+18/-3V, Fsw=20kHz, 20A, Tf=65deg, duty cycle=60%		External RG [ohm]		
		5	10	30
Total Power Loss [W]	Q1, Q2	20.18	21.04	24.58
	Q3, Q4	7.58	7.58	7.60
Max Tj [deg]	Q1, Q2	93.92	95.14	100.21
	Q3, Q4	75.86	75.87	75.89

Table 3 Total power loss and maximum junction temperature estimation of option 2 turning on the Q3, Q4 when Q1, Q2 off.

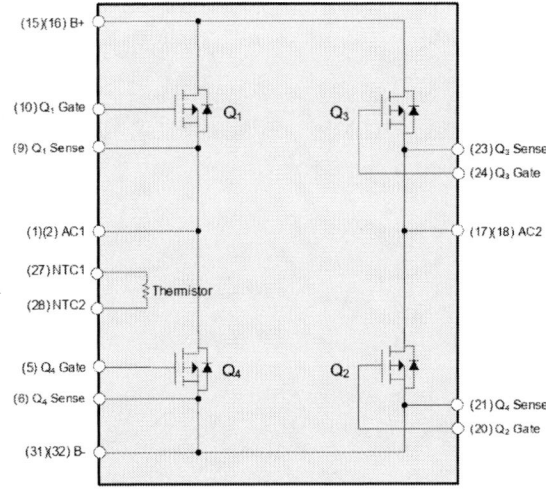

Fig. 6 Circuit diagram of Excited Synchronous Motor Automotive Power Module.

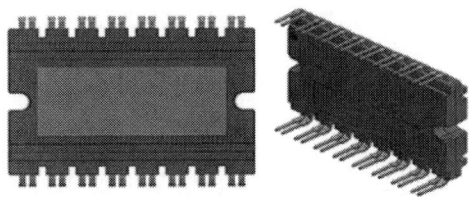

Fig. 7 Outline overview for APM32.

4 Conclusion

In this paper, we utilize PSIM for all simulation results to establish a reference basis for product design through simulation outcomes. The simulations in this article are conducted using discrete methods. During the product design stage, ONSEMI will provide relevant spice models for the simulations. This article was simulated using the M3 series of SiC products, and the relevant spice models can be found on the ONSEMI official website. In further work, the simulations will transition to a modular approach, and after sample assembly, evaluations will be conducted to verify and compare the simulation results.

References

Please follow international scientific citation rules.

[1] J Tang, "Synchronous Machines with High-Frequency Brushless Excitation for Vehicle Applications."

[2] Automotive Application Corner: Magnet-Free Electric Machines and Drives for Electric Vehicles (onsemi.com)

PCIM Europe 2024, 11– 13 June 2024, Nuremberg   DOI: 10.30420/566262248

A Novel High Power Density Three Phase Traction Inverter Architecture for Electric Vehicle (EV) Applications

Yiyang Yan [1], Jiajia Guan[1], Baihan Liu[1], Jianwei Lv[1], Jiaxin Liu[1], Cai Chen[1], Yong Kang[1]

[1] Huazhong University of Science and Technology, China

Corresponding author: Cai Chen, caichen@hust.edu.cn
Speaker: Yiyang Yan, yanyiyang@hust.edu.cn

Abstract

Three-phase traction inverter design needs to realize high power density in EV applications. The inverter architecture design is mainly based on the shape and dimension of power module. In this paper, a novel compact three-phase traction inverter architecture based on double-side-end double-sided bonding power module is proposed and tested. Double-sided direct cooling is realized by soldering the power module on pin-fin baseplate. Low parasitic inductance is also realized by double-side-end busbar design. The proposed three-phase inverter is also established and tested.

1 Introduction

The actual two-level three-phase traction inverter consists of DC-link capacitor, power modules, heat-sink, gate driver boards and DC-link busbars. Inverter architecture design is mainly based on the shape and dimension of power modules. Realizing higher efficiency, higher power density three-phase motor traction inverter design has become one of the developing trends of power electronics due to the space and weight constraints inside the EV. To realize this higher efficiency and higher power density, there are several methods: Utilizing wide bandgap devices and decreasing the thermal resistance of power devices.

In recent years, wide bandgap devices such as silicon carbide (SiC) metal-oxide-semiconductor field effect transistor (MOSFET) has been utilized to replace the silicon insulated gate bipolar transistor (IGBT) in EV applications. This is because novel devices have lower on-state resistance and lower parasitic capacitance, leading to lower conduction loss and switching loss. However, SiC MOSFET is sensitive to the parasitic inductance due to high switching speed. High parasitic inductance will cause voltage spike and EMI issues [1]. The parasitic inductance of devices is contributed by the DC-link busbar and device package. Decreasing parasitic inductance is necessary. Dou-

ble-side-end busbar can decrease parasitic inductance by 40% compared to traditional single-side-end busbar [2]. Power module with double-side-end terminal layout is much more suitable.

Besides, devices with better thermal performance will be able to handle higher power, which will realize higher power density. Compared to traditional power module design, the heat in double-sided cooling power modules can be dissipated from the top and bottom side of the power module. Double-sided cooling power module is suitable for high power density applications. A double-sided cooling power module is proposed in [3], which has double-side-end terminal layout. However, how to design the three-phase architecture based on this power module is also another problem. Traditional architectures are always based on the power module with single-side-end terminal layout. Hence, in this paper, a novel high-power density three-phase traction inverter architecture is proposed.

2 Power Module and Integration Method

In this section, the power module structure is described. Two kinds of water-cooling structures are demonstrated subsequently. Based on the power module structure and the heat-sink. Two kinds of three-phase architectures are proposed. Busbar parameters are also extracted.

2.1 Double-side-end DSC Power Module

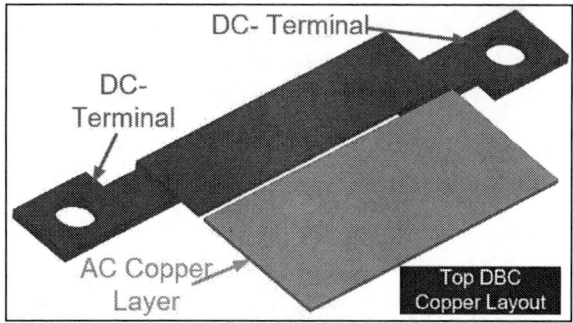

Fig. 1 Top DBC copper layout of the double-side-end DSC power module

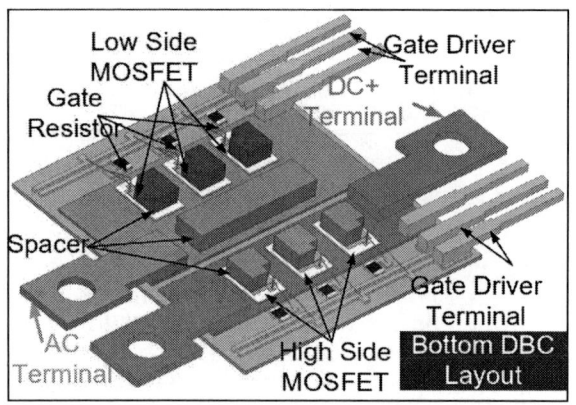

Fig. 1 Bottom DBC layout of the double-side-end DSC power module

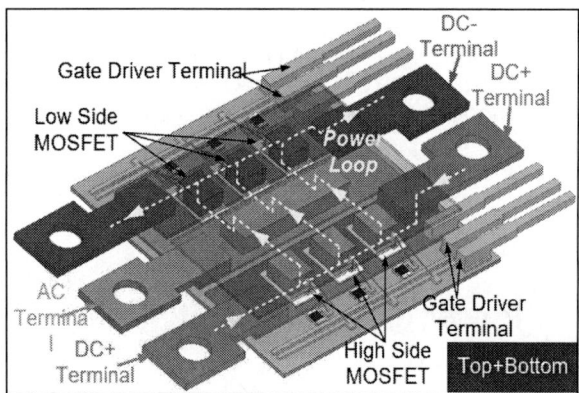

Fig. 3 The double-side-end DSC power module structure with illustration of power loop

The power module structure utilized in this paper is shown in Figure above. The top direct bond copper substrate (DBC) copper layout, bottom DBC layout and the top and bottom DBC layout are shown in Fig. 1, Fig 2, and Fig 3, respectively. All bare dies are soldered on the bottom DBC and spacers are soldered on the bare dies. Bonding aluminum wires with 0.2mm diameter are utilized to realize the driver loop connection between the bare die and DBC copper layers. A long spacer is soldered on the bottom DBC to realize the electrical connection between the top DBC and the bottom DBC.

It can be seen that this power module has two DC+ terminals and two DC- terminals, which realize the double-side-end power loop. The power loop structure of this power module is like the parallel of two power loops with "U" shape shown in Fig. 3, which can decrease the self-inductance of the power loop compared to the structure with a pair of DC terminals. The parasitic inductance of this power module is only 3nH obtained by the simulation result in ANSYS Q3D.

Although this structure realizes low parasitic inductance, the existing three phase integration structure based on double sided cooling power module is only suitable for the power module with single-end terminal layout. For this reason, the novel integration architecture suitable for this double-side-end power module need to be designed and proposed. The busbar design needs to be considered to connect all terminals to realize the double-side-end connection.

2.2 Integration Method

2.2.1 Indirect Water-cooling and Direct Water-cooling

The water-cooling method to realize three-phase integration architecture can be divided into two types, indirect water-cooling and direct water-cooling. The basic structures of these two structures are shown in Fig. 4 and Fig. 5, respectively. Indirect water-cooling method always utilize coldplate and thermal grease to dissipate the loss from power module to the ambient. However, the thermal conductivity of the thermal grease is always ranged from 1W/mK to 10W/mK, which will increase the thermal resistance of the power module.

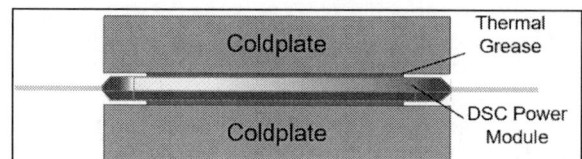

Fig. 4 Indirect water-cooling realization method

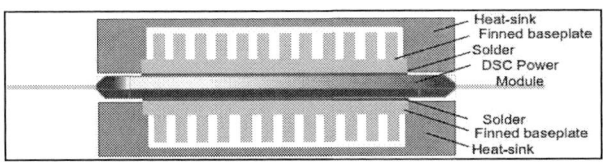

Fig. 5 Structure of direct water-cooling method

Direct water-cooling soldering the power module on the finned baseplate. Thermal grease with low thermal conductivity is replaced by the solder paste with higher thermal conductivity. Rubber ring is also utilized to realize sealing. Some screws will be utilized to provide the clamp force to cause the rubber ring to deform, which can fill the gaps and ensure sealing. The double-sided direct water-cooling method is chosen as the basic heat dissipation structure in this paper. Three DSC power modules are utilized to establish three-phase half-bridge. The thermal simulations are performed to compared these two kinds of water-cooling methods. The temperature of the cooling water is 65℃. Flow rate of the cooling is 10L/min and the loss of each device is 45W. The results are shown in Fig. 6 and Fig. 7, respectively.

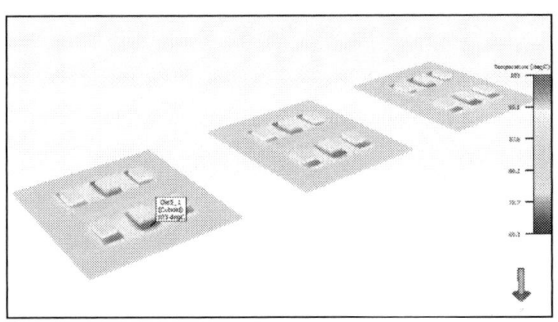

Fig. 6 The maximum temperature is 103℃ with indirect water-cooling structure

Fig. 7 The maximum temperature is 89℃ with direct water-cooling structure

It can be seen that the maximum temperature rises among different SiC MOSFETs with indirect water cooling method is 38℃ and that with direct water cooling method is 24℃, which means that the thermal performance is improved 36.8%.

2.2.2 Busbar and Capacitor Design

To realize the symmetrical connection to the double-side-end terminals. The Busbar and location of the DC-link capacitor needs extra design. Based on the power module and heat-sink structures, the proposed three-phase architecture is shown in Fig. 8 and Fig. 9.

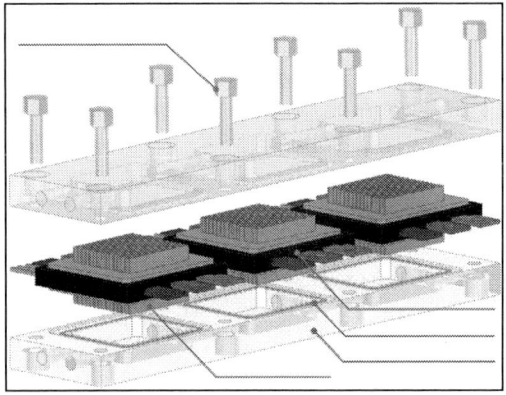

Fig. 8 Direct water-cooling of three DSC power modules

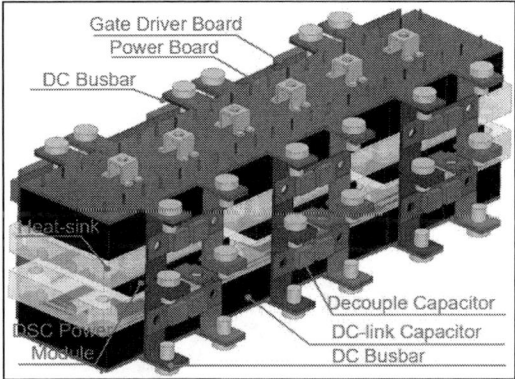

Fig. 9 The integration architecture based on the double-side-end DSC power module.

The DC-link capacitors are placed on the other side of heat-sink. The function of DC-link capacitor is to absorb the current ripple. However, the rms value might exceed the value that capacitors can handle. Hence, this kind of structure will be helpful to heat dissipation of capacitors. The design of the busbar makes the power loop entrance into the power module from the top and bottom of the power module. Each side is double-side-end. Extracting the parasitic inductance containing the busbar and power module in Ansys Q3D and the result is only 9.8nH. Driver boards are soldered with the power module side which places the driver terminals.

3 Assembly and Testing

Assemble the DC-link capacitor with power board, heatsink, driver board and power module into a three-phase inverter system as shown in Fig. 10, Fig. 11, Fig. 12 and Fig. 13. Three-phase reactive power testing is performed. Schematic of three-phase reactive testing is shown in Fig. 14. Three-phase SPWM modulation is utilized for the three-phase system. The outputs of half-bridge power modules are connected to three-phase load inductors. The main frequency component of the output voltage is the fundamental frequency, which results in that the frequency of the load phase current is still the fundamental frequency.

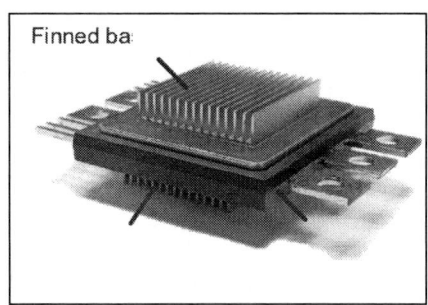

Fig. 10 The DSC power module with finned baseplate

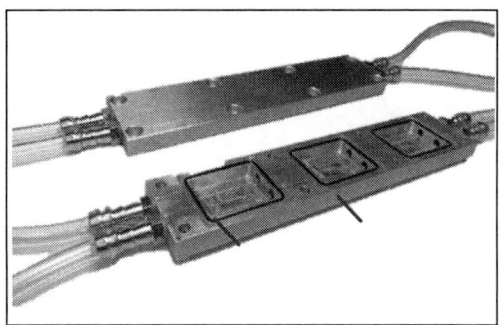

Fig. 11 The heatsink with rubber ring.

The test platform is illustrated in Fig. 15(b). Phase current is measured by current probe with

part number of CP150 with 150A maximum measurement current and 10MHz bandwidth.

Fig. 12 The DC-link capacitor with Power board.

Fig. 13 The assembled three-phase architecture

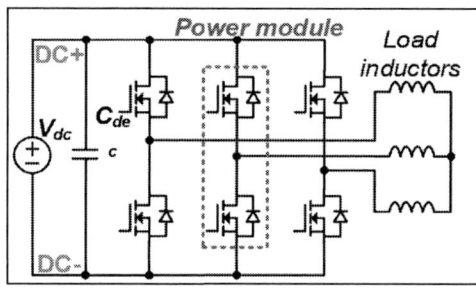

Fig. 14 Schematic of three-phase reactive testing

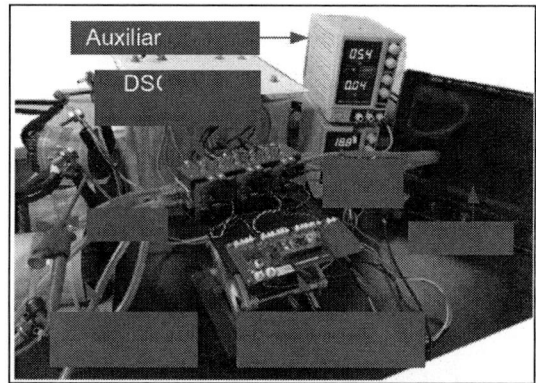

Fig. 15 Testing platform of three phase reactive testing

To measure the output voltage of each power module, an isolated voltage probe is used with

the part number HVD3106A with 1kV maximum measurement voltage and 120MHz bandwidth.
The testing waveform of three--phase system is shown in Fig. 16. DC input voltage is 700V and per phase load current is 100Arms/200Hz. the overshoot voltage of one phase output voltage is much small according to the measurement waveform.

The measured waveform indicates that the proposed three phase architecture can operate stably, which will be helpful to realize the high-power density applications. However, the parasitic inductance of the busbar is still relative large, the busbar need to be optimized and the temperature inside the power module need to be measured to find the operation boundary of this kind of DSC power module under high voltage platfrom.

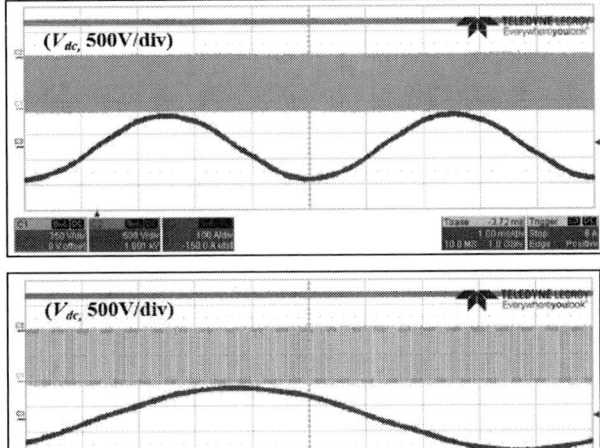

Fig. 16 Waveforms of drain-source voltage and load current

References

[1] J. Noppakunkajorn, D. Han and B. Sarlioglu, "Analysis of High-Speed PCB With SiC Devices by Investigating Turn-Off Overvoltage and Interconnection Inductance Influence," in *IEEE Transactions on Transportation Electrification*, vol. 1, no. 2, pp. 118-125, Aug. 2015.

[2] Z. Yuan *et al.*, "Design and Evaluation of Laminated Busbar for Three-Level T-Type NPC Power Electronics Building Block With Enhanced Dynamic Current Sharing," in *IEEE Journal of Emerging and Selected Topics in Power Electronics*, vol. 8, no. 1, pp. 395-406, March 2020.

[3] X. Liu, Z. Wu, Y. Yan, Y. Kang and C. Chen, "A Novel Double-Sided Cooling Inverter Leg for High Power Density EV Based on Customized SiC Power Module," 2020 IEEE Energy Conversion Congress and Exposition (ECCE), Detroit, MI, USA, 2020, pp. 3151-3154.

PCIM Europe 2024, 11– 13 June 2024, Nuremberg   DOI: 10.30420/566262249

A Modular DC-Link Capacitor Solution for the Main Powertrain Inverter of xEVs

David Olalla[1], Gayatri Kulkarni[2], Fernando Rodriguez[3],

1 TDK Electronics AG, Germany

2 TDK India Private Limited, India

3 TDK Electronics, SAU

Corresponding author and speaker: David Olalla, david.olalla@tdk.com

Abstract

This paper describes a novel approach for a DC-link film capacitor for the main power train of xEVs. With an understanding of the challenges and requirements, the approach targets standardizing the capacitor design for this application. A design concept of the capacitor is validated with simulations and laboratory test data. The paper does not intend to disclose all aspects of product design and process, but it will show why it can be a flexible solution for system engineers by its modularity and usage. These capacitors are connected through an external busbar to the power modules (and these to the motor) and additional circuitry, such as the battery or EMI filters. This standardized approach will simplify the development efforts for circuit designers as well as engineering costs.

1 Introduction: The DC-link capacitor in the main traction inverter of an electric vehicle

The DC-link capacitor acts as an energy buffer between input and output in the power converter. Its core mission is to keep the DC voltage stable, which means within the defined voltage limits by the system (also called smoothing function). While being charged from the input, this capacitor supplies the necessary current to the power semiconductors (and these to the output load with the target voltage-current waveform). This general concept of DC-link capacitor applies to any power converter system and thus end application, from small power adaptors to wind power converters in the megawatt range.

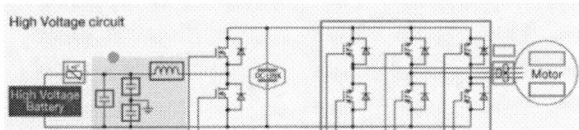

Fig. 1: *Block diagram example of a motor inverter including EMI filter, DC-DC converter, and main traction inverter. The DC-link capacitor is highlighted with a blue hexagon.*

Multiple capacitor technologies can be used as DC-link, and their suitability depends on voltage, power range, application, etc. [1]. Even several capacitor technologies can be used in the same system, one to cope with lower-frequency current har-

monics and another with higher frequencies, responding to the nature and demand of every system design in connection with the properties of every capacitor technology.

For example, the snubber capacitor, which is parallel to the DC-link capacitor, is placed very close to the power semiconductor, targeting a minimum parasitic inductance. This snubber capacitor manages the highest frequency current, reducing voltage overshooting by providing the necessary current when required. So, it targets to compensate for the higher parasitic inductance of the main DC-link capacitor bank. Thus, it is not always needed, as shown in Figure 2. The DC-Link bank of film capacitors supplies the high-frequency harmonics of current with minimum parasitic inductance, so the film capacitors assume the DC-link and snubber roles, the same as the DC-link capacitor in the main traction inverter of xEVs.

Fig. 2: *Demonstrator of a 1500 V solar inverter by Infineon. Easy 3B module: 1200V TRENCHSTOP™ IGBT7 + 1200 V CoolSiC™ MOSFET. TDK Electronics: Hybrid DC-link capacitor bank: High-frequency film capacitor (blue) + aluminum electrolytic (black) [2].*

1814

Capacitance, voltage, current capability, and temperature range are the initial key parameters to identify which capacitor technologies are suitable as DC-link for each system requirement. When multiple technologies are available, system designers must also consider the pros and cons of every technology because they must cope with the non-ideal characteristics of these components:

- Parasitic inductance (ESL) and resistance (ESR).
- Thermal behavior and limits.
- Reliability.
- Complex behaviors with frequency, voltage, temperature, aging, etc., when applicable.
- Mounting.
- On top of all this, size and cost impact the whole system.

As a result, biaxially oriented polypropylene (BOPP) film capacitors are, as of today, the dominant technology solution used for the DC-link of the main traction inverter in xEV systems [3].

Among other well-established technologies for DC-link, polypropylene film capacitors offer (for this application and target power range) the necessary voltage ranges, energy, current density, lifetime, temperature handling, and reliability. In addition, they are solving well the challenges of introducing wide bandgap (WBG) semiconductors, proving that they can handle the increasing frequencies, A per µF, dV/dt, and ESL reduction strategies.

2 The problem: Customizing the DC-link capacitor

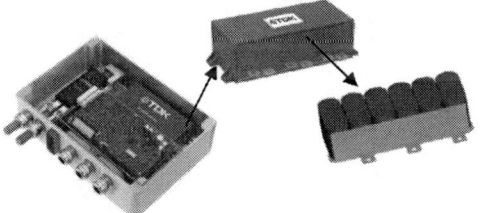

Fig. 3: *Reference design by Mankel-Engineering in collaboration with TDK and Infineon Technologies. 240 V DC to 475 V DC with a peak power of 120 kW at 400 V (DC). The capacitor rating is 650 µF/500 V [4] with an overview of internal construction made with 6 parallel flat winding elements of polypropylene film capacitors.*

The DC-Link capacitor is one of the bulkiest and most customized components in these xEV systems. The terminals are typically divided into three pairs, each for every half-bridge phase, as shown

in Figure 1. The terminals (position, shape, size) must be adapted to every power semiconductor supplier and model to minimize stray inductance. As for DC-input terminals (not shown in the figure), their relative position and size depend on the overall system design, i.e., the power and position of the battery mains. The internal busbar design, winding element number, and arrangement must be adapted to the system requirements (Capacitance, Voltage, IRMS, inductance, cooling, space and geometry, etc.).

The cooling system influences the internal design of the capacitor because it will have effect on the mission operation profile expected for the capacitor: voltage-capacitor temperature—% of lifetime. Vice versa, the internal capacitor design will determine the actual hot spot temperature for every IRMS condition and, therefore, also the cooling system. This applies to everything with continuous changes coming from other subsystems or operation conditions during the design of a new vehicle because the main traction inverter is only a part of it.

The consequence is iterative and long design cycles until system requirements are merged with the final capacitor design. A demand for higher capacitance to manage, for example, higher IRMS, will result in the modification of the whole capacitor with its internal and external construction. Every seemingly minor change in the capacitor design significantly impacts engineering cost and time to market.

For example, Plastic injection tooling for encapsulation, cutting-punching-bending tooling for busbars, related production processes, and all the production equipment for future automation—all these changes must be supported with the required validation testing and evaluation to maintain the required performance, quality, and reliability in a mass production scenario, not only at component level but at whole system level.

The Infineon evaluation kit "EVAL KIT HPD G1 SiC" [5] refers to a TDK capacitor with 300 µF/855 V (B25655P8307K351). That solution would need external busbar adaptation to add DC terminals.

3 The proposal: Design-for-application of a modular capacitor solution

This bulk customized capacitor solution is divided into standardizable capacitor units, as shown in Figure 4.

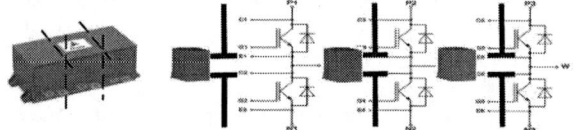

Fig. 4: *The basic concept is to divide the bulk customized capacitor into standardizable units.*

Every capacitor unit cannot provide the full capacitance and power required by the traction inverter. Still, paralleling them through a laminated busbar with a compensated inductance arrangement makes it possible. This solution can be easily scaled to any power level, power semiconductor supplier, model, system geometry, or cooling. DC terminals and other accessories like EMI filters can be incorporated into the busbar.

Critical challenges were observed while designing this application's modular DC-link film capacitor. Firstly, mitigating parasitic inductance requires internal overlapping of busbars, which adds cost to the raw materials and production process. Secondly, based on the application frequency, there is a high chance of a non-homogeneous distribution of RMS currents inside the winding elements, as shown in Figure 5. That can limit current per capacitance for systems employing wide bandgap semiconductors.

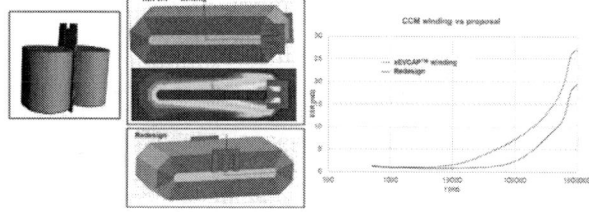

Fig. 5: Left: *Initial Internal structure with 2 winding elements in parallel and overlapped busbar.* **Center:** *Redesign proposals and thermal study.* **Right:** *ESR behavior depends on the terminal position.*

A redesign of the initial concept of Figure 5 (left) is based on a detailed review of present and future inverter power modules in the automotive domain, the cooling methodologies, and their electrical requirements, as shown in Figure 6: Rated power: 150 kW—300 kW; capacitance range: 200 μF to 600 μF. The target spectrum is divided into Size 1 for low and mid power and Size 2 for mid and high power.

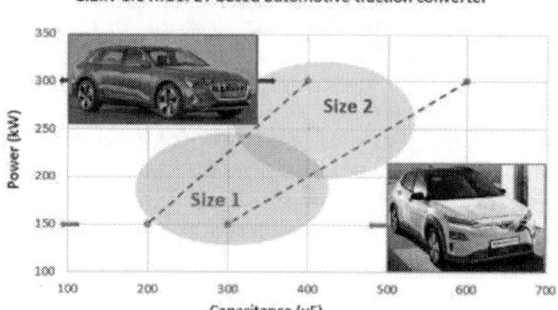

Fig. 6: *Basic system requirements.*

The following table shows the extracted parameters of the DC-link capacitor linked to application parameters that will lead to the optimized design for the application: the ripple current flowing through the capacitors (I_c), the input current from the battery (I_{dc}), the rated voltage of the capacitor (V_R), the power of the system, and the output phase current (I_{ph}; modulation index $0.6 - 0.7$).

Power level	Size 1		Size 2	
	Low	Mid	Mid-High	High
C [μF]	200 – 300	300 – 400	375 – 525	400 – 600
$I_{c\,RMS}$ (max.) [A]	116	155	200	235
I_{dc} (max.) [A]	175	235	295	350
V_R [V]	850	850	850	850
P [kW]	150	200	250	300
$I_{ph.RMS}$ (max.) [A]	200	265	335	400

Table 1: *Design to application (split by Power) of the modular DC-link capacitor approach.*

The DC-link capacitor is divided into individual units, paralleled, and connected through a busbar to the power semiconductors, as shown in Figure 7. All this is done with the support of simulation tools to verify its feasibility for the target specifications, such as the ESL, which is measured at the contact points of the semiconductor with the busbar.

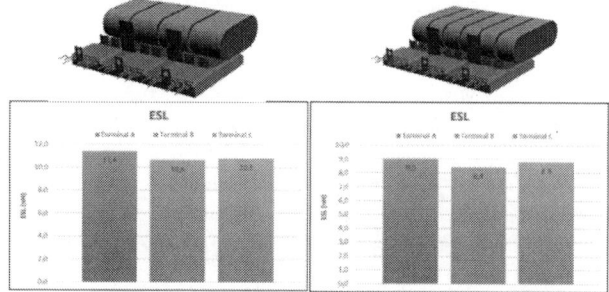

Fig. 7: *Modular concept design for DC-link applications; standardized solution compatible with power semiconductor modules. Left: Size 1 (400 μF); right: Size 2 (500 μF)*

4 The product: xEVCap, design-for-manufacturing

Afterward, a design for manufacturing is developed, targeting standardization in terminal pitches and box sizes, similar to what is available for PCB components. The solution must be flexible and scalable for the target applications. Thus, a limited number of encapsulation box sizes and internal elements are proposed, maximizing capacitance and current handling for every module.

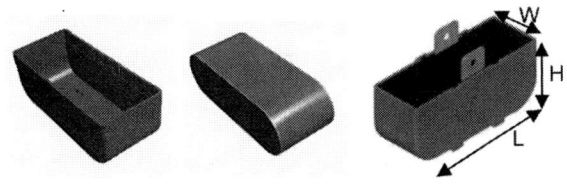

Fig. 8: *Encapsulation box (left), internal winding element (center), final xEVCap (right) for the DC-Link modular approach*

The following table provides the spectrum range and critical electrical and mechanical specifications of xEVCap. Four voltage levels, targeting 400 V (650 V semiconductors) and 800 V systems (1200 V semiconductors), are defined such that, for example, a 920 V rating in the 800 V system provides a longer service life for demanding mission profiles, especially when temperature and voltage are high.

C_R [μF]	$V_{R,DC}$ [V]	I_{C_RMS} (max.) @ 10 kHz [A]	L_s @ 1 MHz [nH]	ESR @ 10kHz [mΩ]	Dimension L x W x H [mm]		
200	475	40*	17	1.13	85	46.5	40.5
270	475	50*	17	0.89	109	46.5	40.5
115	650	60	14	0.51	97.5	35	42.5
130	650	42	17	0.89	85	46.5	40.5
175	650	55	17	0.66	109	46.5	40.5
325	650	53	24	1.16	111.5	77	41.5
75	850	56	14	0.57	97.5	35	42.5
97	850	40	17	1.04	85	46.5	40.5
130	850	50	17	0.78	109	46.5	40.5
242	850	42	24	1.35	111.5	77	41.5
60	920	55	14	0.65	97.5	35	42.5
75	920	35	17	1.18	85	46.5	40.5
100	920	45	17	0.89	109	46.5	40.5
185	920	40	24	1.54	111.5	77	41.5

Table 2: *xEVCap proposed range.*

A narrower width (W) will increase the IRMS per capacitance, while a longer length or height (L or H) will increase the energy density.

A system designer must consider the systems' electrical and mechanical requirements and choose an appropriate capacitor module to match these requirements. Paralleling these units and connecting them through laminated busbars reduces the system ESR and self-inductance (ESL). This solution offers flexibility and scalability to system engineers because capacitance, IRMS, ESL, and power could be adapted and scaled by placing multiple capacitor units in parallel.

The validation testing is done according to the following standards: ZVEI "Basic Qualification of DC-Link Capacitors for Automotive Use" [6] and AEC-Q200 [7] table for film capacitors. Some test results are illustrated in the following figure:

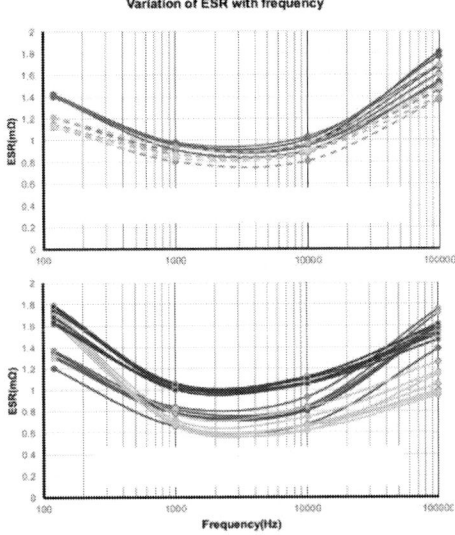

Fig. 9(a): *ESR vs. frequency; laboratory measurement of a device with 270 μF at 500 V (blue), with 115 μF at 700 V (yellow), with 130 μF (brown) and 175 μF both at 700 V (blue)*

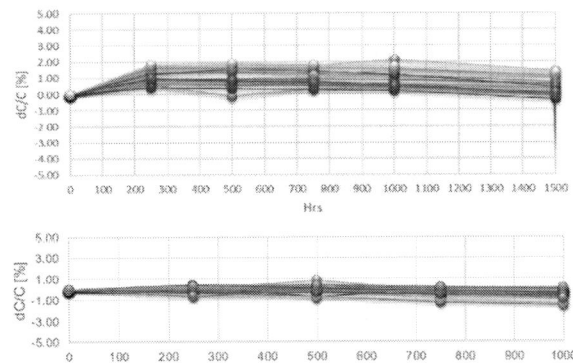

Fig. 9(b): *High-temperature operating lifetime according to AEC-Q200 [7], +105 °C, 1000 hours; 115 μF and 175 μF at 700 V (up), 75 μF at 850 V (bottom).*

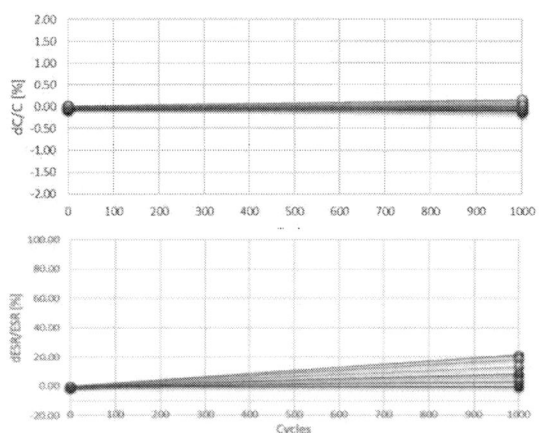

Fig. 9(c): *Thermal shock test according to ZVEI [6] from -40 °C to +105 °C for different sizes of 850 V rating.*

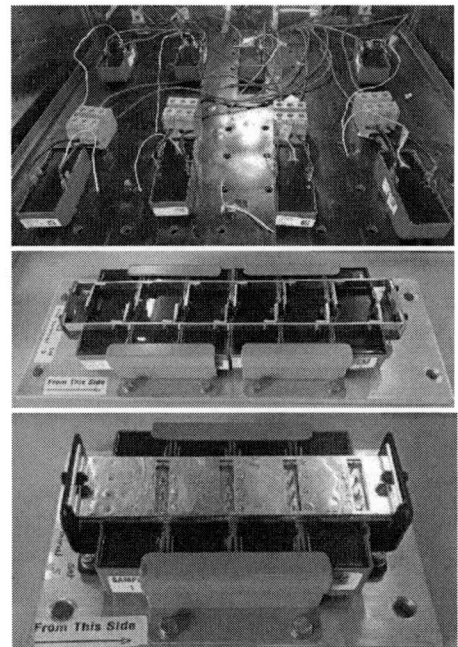

Fig. 9(d): *xEVCaps during testing: High-temperature operating lifetime (top). Test fixtures for AEC-Q200 vibration test (center and bottom)*

5 A real case: System design with xEVCap

The analysis of the inverter system requirements (like in Table 1) is the necessary starting point to scope the feasible modular approach for the DC-Link capacitor solution with the xEVCap range shown previously in Table 2.

C (µF)	Minimum 450
$I_{c\ RMS}$ (max.) [A]	250 to 300
I_{dc} (max.) [A]	250 to 300
V_R [V]	800 (900 for limited time)
P [kW]	250
$I_{ph\ RMS}$ (max.) [A]	400 to 500 (Max 650 for 10 seconds)
ESL [nH]	~7

Table 3: *Analysis of system requirements (collaboration with STMicroelectronics).*

Knowing how much electrical stress is applied in reality is important for voltage rating selection. Will, for example, a voltage level of 900 V occurs continuously or in very short time periods? Will higher temperatures occur for long or short periods or when the voltages applied to the capacitor are higher?

It is then beneficial to have an estimation of the mission profile over established operation cycles and lifetime, for example, driving (8000 h), preconditioning (5000 h), and charging (40,000 h). That will optimize the internal capacitor design through an adequate voltage rating defined by the dielectric thickness required to manage the expected voltages and temperatures applied during the whole lifetime. Higher safety margins will lead to larger and more costly solutions, while insufficient safety margins will require a redesign iteration. These issues can be mitigated by using a standardized modular approach.

To fulfill the requirements in Table 3, two proposals from Table 2 with an 850 V rating are studied, one using 6 components of 75 µF each, expected to comply well with the electrical requirements; the other, with 4 components of 130 µF each, has fewer parts more for cost-size efficiency, but with less performance. These are summarized in Table 4.

C (µF)	75	130
$V_{R,\ DC}$ [V]	850	850
I_{c_RMS} (max.) @ 10 kHz [A]	56	50
L_s @ 1 MHz, [nH]	14	17
ESR @ 10 kHz [mΩ]	0.57	0.78
Dimension L x W x H [mm]	97.5x35x42.5	109x46.5x40.5
Proposals Number of units (C_{total}, I_{RMS})	x6 (450 µF, 336 A)	x4 (520 µF, 200 A)

Table 4: *xEVCap with two proposed options*

Simulation tools are then used to analyze the feasibility of matching the system requirements and calculate the values for ESR and ESL. In the real application, the current in the capacitor (I_c) shall be decomposed into frequency harmonics (see Fig. 10). An evaluation considering only the fundamental switching frequency would underestimate the power losses because the ESR tends to be higher at higher frequencies. As a result, the thermal behavior and current balancing may be underestimated.

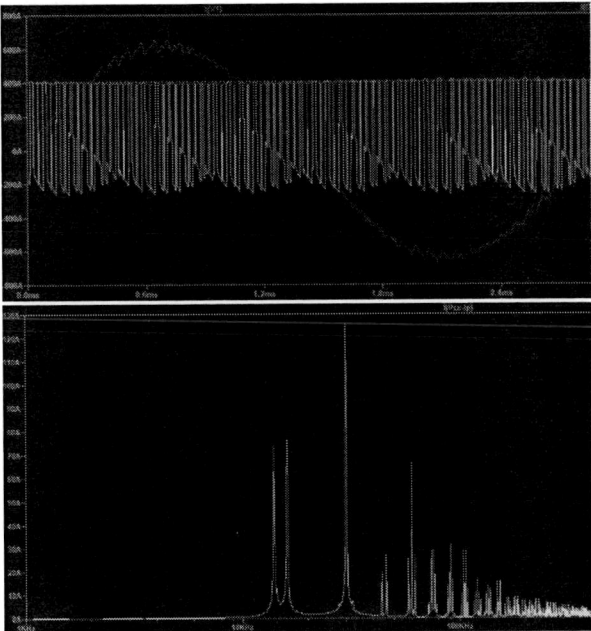

Fig. 10: (Up) *Simulation of inverter waveforms with I_{ph} (blue), I_{dc} (red), and capacitor current I_c (green).* **(Bottom)** *The frequency spectrum of I_c.*

The simulation results (Fig. 11) show very low ESR values (note: scale is different), but the most important characteristic is almost perfect balancing from any of the three pairs of terminals corresponding to every phase. This indicates a balanced current for the studied frequency spectrum. Unbalancing would lead to uncontrolled hot spots in some areas of the capacitors that will limit their lifetime.

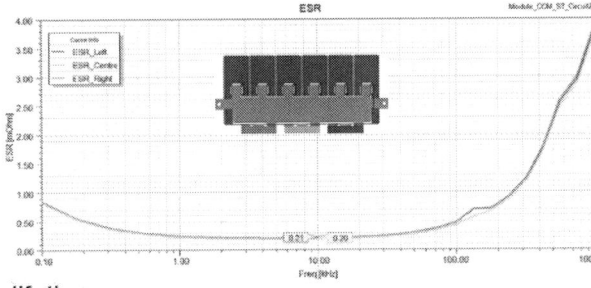

Fig. 11 (a): *ESR vs frequency curve for 6 x 75 µF at 850 V*

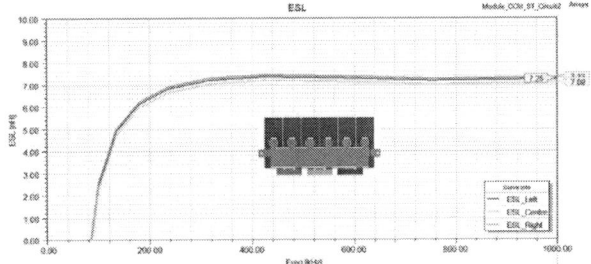

Fig. 11(b): *ESL vs frequency for 6 x 75 µF at 850 V*

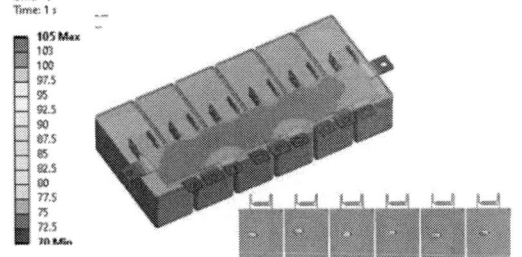

Fig. 11(c): *Thermal simulation with cooling at +70 °C, ambient at +85 °C, and terminals at +105 °C. Realistic $I_{C,RMS}$ spectrums for 300 kW with 300 A, harmonics from 8 kHz to 110 kHz.*

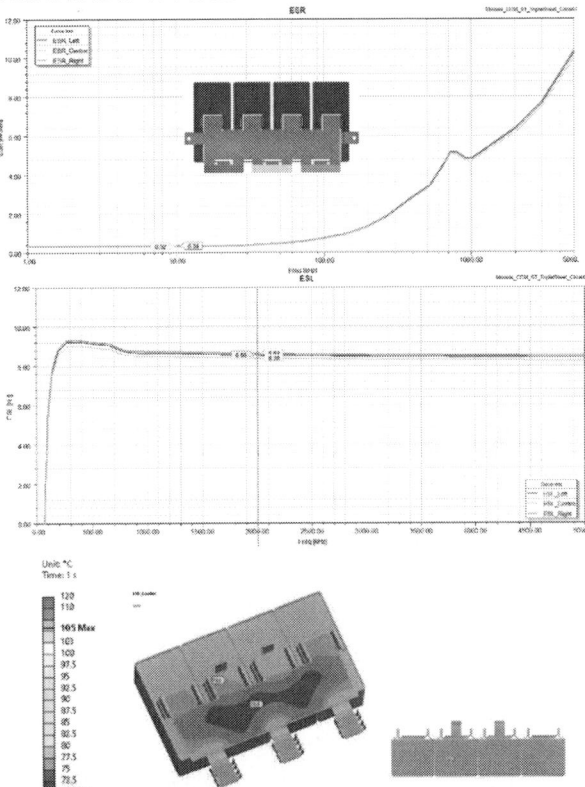

Fig. 11(d): *ESR **(top)** ESL **(center)** vs. frequency curves for 4 x 130 µF at 850 V. Thermal simulation **(bottom)** with cooling at +70 °C, ambient at +85 °C, and terminals at +105 °C. 200 kW with 200 A, harmonics from 20 kHz to 160 kHz.*

The evaluation is done with SiC power semiconductors, proving that the xEVCap is fully compatible with WBG semiconductors in terms of inductance and current density. The DC-link solution with 4 x 130 µF offers 200 kW maximum power while the one with 6 x 75 µF reaches up to 300 kW. This demonstrates its scalability and easy adaption to a wide variety of systems, even those evolving during system design.

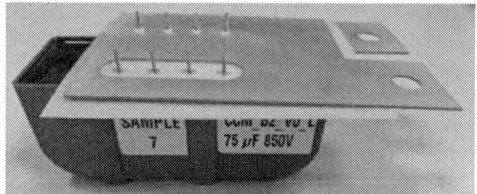

Fig. 13: *Future work-study: Study of joint technologies and terminations.*

References

[1]   Shajjad Chowdhury, Emre Gurpinar "Capacitor Technologies: Characterization, Selection, and Packaging for Next-Generation Power Electronics Applications" IEEE Transactions On Transportation Electrification, Vol. 8, No. 2, June 2022.

[2]   Manuel Gomez, "Innovative film capacitor technologies for wide band-gap semiconductors," EEE PSMA Capacitor Committee Workshop, April 2020.

[3]   Yole Group, "Capacitors for Power Converters 2023", Market and Technology Report 2023

[4]   Wolfgang Rambow, Fabian Beck, Elvis Keli, Katharina Mankel, Mankel-Engineering "Compact, Modular Inverter Manufactured Using Standard Components," Bodo's Power October 2022.

[5]   Infineon Application Note "Quickstart Manual for EVAL KIT HPD G1 SiC," www.infineon.com, 2022-12.

[6]   ZVEI - German Electrical and Electronic Manufacturers' Association "Basic Qualification of DC-Link Capacitors for Automotive Use," March 2020, version 3.0

[7]   AEC Automotive Electronics Council, AEC-Q200- Rev E "Stress Test Qualification for Passive Components," March 2023

Fig. 12: *Collaboration with STMicroelectronics. Double Side Cooling Power Module High Power traction inverter and xEVCap mounted on the laminated busbar. The double-pulse test confirmed the simulated values of 8 nH for the DC-link capacitor solution with 4 x 130 µF at 850 V.*

6   Conclusion and future work

A modular standard capacitor provides a flexible solution that reduces development time and simplifies design efforts while meeting the present performance requirements of DC-link capacitors for inverter power modules in xEVs – also for WBG semiconductors. Modularity ensures lower ESR, homogenous thermal distribution, and flexible adoption of cooling methodologies for the system designer. A future scope includes the use of higher temperature dielectrics with automatic temperature upgrades and higher energy densities, as well as a deeper study of the assembly techniques of capacitors to busbars, power semiconductors, and other circuitry like EMI filters.

PCIM Europe 2024, 11– 13 June 2024, Nuremberg   DOI: 10.30420/566262250

Fault Identification Testing Methods for a Commercial Traction Inverter

Anna Corbitt[1], Chris Farnell[1], Justin Jackson[1], Shailesh Joshi[2], Raymond Viviano[2], Yohei Iwahashi[2], H. Alan Mantooth[1]

[1] University of Arkansas, USA
[2] Toyota Research Institute North America, USA

Corresponding author: Anna Corbitt, amcorbit@uark.edu
Speaker: Anna Corbitt, amcorbit@uark.edu

Abstract

As electric traction drive systems continue to improve in power density and performance, a method for detecting failures and increasing the overall reliability of the system's power control unit is essential. This article proposes a test methodology for fault identification of a commercially available 55-kW traction inverter. The objective of the proposed test method is to accelerate device failure in an environment that replicates realistic driving conditions. This paper details the control strategy, test procedure, and measurement techniques for the prognostic testing of a commercial inverter, while demonstrating the practical need for fault identification testing in the electric vehicle industry.

1   Introduction

The initiative for the electrification of transportation to mitigate fossil fuels has seen an increase in industry support recently. In fact, over the past decade, there has been a significant increase in the production and acceptance of electric vehicles in society. The manufacturing of reliable electric vehicles (EVs) is crucial to the continued improvement and growth of the EV industry.

As EVs gain popularity, a method for detecting faults in the vehicle's main power control unit is crucial for reliability as well as driver safety. If a trustworthy method of fault identification is established, It could be used to predict a failure timeline for the switching devices in a vehicle's traction inverter. The ultimate goal is to utilize fault identification results to alert vehicle owners of potential weaknesses and failures at the power control unit before it stops working. This would drastically improve vehicle safety. Since manufacture lead times on spare parts can be significant, failure prediction using fault identification also has the potential to reduce vehicle service time. If a driver is alerted that the power control unit of the car is reaching the end of its lifespan, the replacement parts could be ordered before device failure occurs eliminating time that the vehicle spends at the service center.

In this paper researchers at the University of Arkansas and Toyota Research Institute of North America collaborated to develop a methodology to test and accelerate the device failure of a 55-kW commercial traction inverter for a failure predictive study. A custom printed circuit board (PCB) was designed by the researchers for the inverter to control the switching devices and measure the device signals for data analysis. This custom PCB was paired with a unified control board that includes a DSP and an FPGA for programming the control algorithm. An OPAL-RT simulation was used to generate data points for realistic driving conditions for this specific traction inverter. Fig.1 shows an overview of the test configuration. Researchers purposely increased the temperature levels of the switching devices to evaluate failure modes and accelerate device failure [1].

The traction inverter was tested on the dynamometer testbed at the University of Arkansas' 6 MVA test facility, the National Center for Reliable Electric Power Transmission (NCREPT). The main objective of this work was to develop a testing method that could be used to evaluate the lifecycle of a traction inverter with extensive data collection and analysis. Specifically, this manuscript details the control design, system model, test procedure, and data acquisition process for the fault identification testing of a commercial traction inverter. The data collected using this fault identification testing method was shared with project collaborators for further diagnostics.

PCIM Europe 2024, 11– 13 June 2024, Nuremberg   DOI: 10.30420/566262250

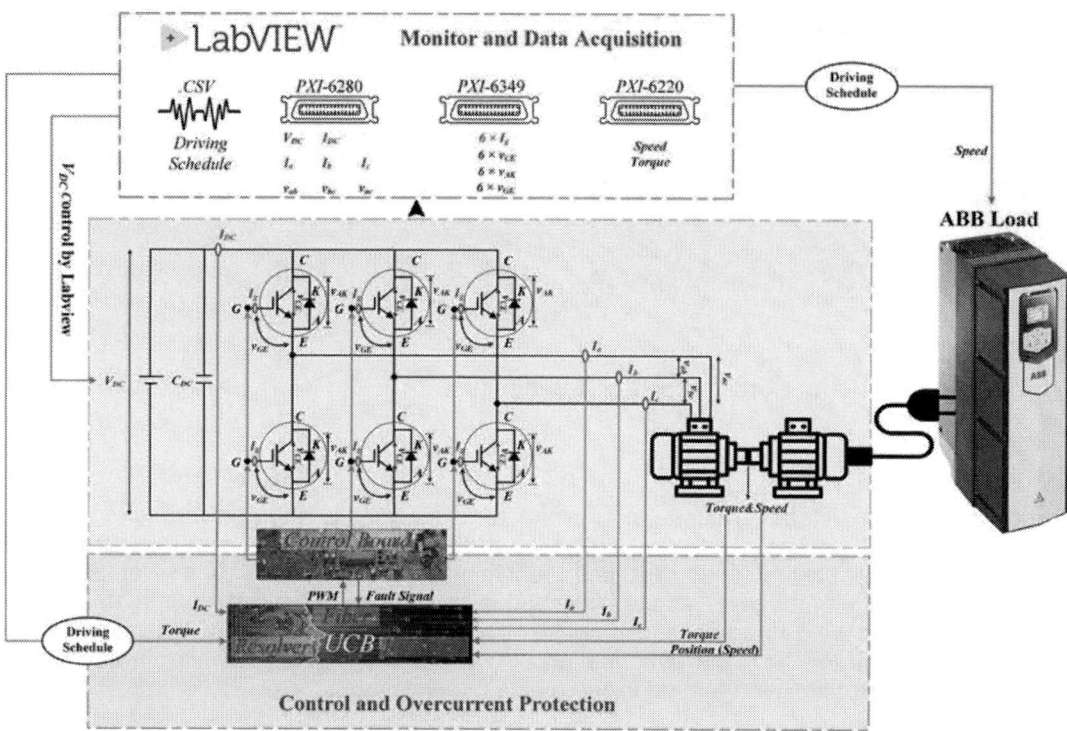

Fig. 1   Overview of the test bench system.

2   Control Design

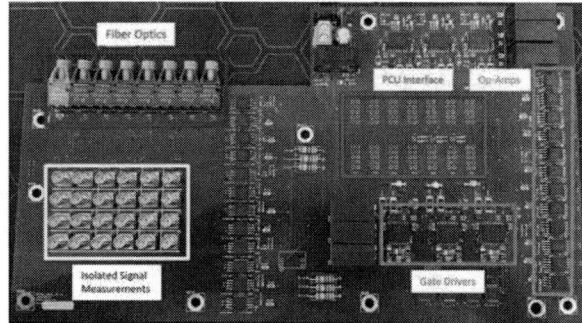

Fig. 2   Custom PCB designed for the 55-kW traction inverter.

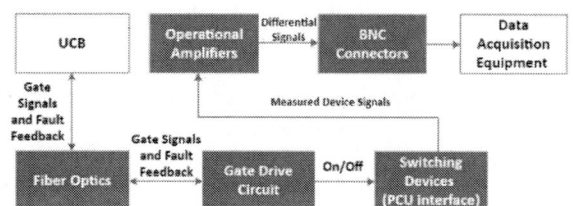

Fig. 3   Block diagram representing the functionality of the custom PCB design.

A printed circuit board (PCB) shown in Fig. 2 was designed custom for this traction inverter. The manufacture's existing control board could not be used for this application, so the development of a custom board was necessary. The PCB was designed to interface directly with the device pins on the traction inverter or power control unit (PCU). The interface connection between the PCB and PCU is shown in Fig. 4. The gate driver circuits on the PCB utilize the UCC21750DW gate driver chip from Texas Instruments that includes features such as fault and reset signaling, DESAT protection, and fault and ready indicators. These features allow for device protection that helps eliminate failures due to operational issues during testing. LEDs were added to the PCB to indicate when a gate driver fault occurs. Fiber optic receivers were used to receive the gate signals from a programmed microcontroller to command the switching sequence. The PCB also has a fiber optic transmitter available to transmit fault and reset signals from the gate drivers back to the microcontroller. Protections were implemented in the controller to safely shut off if a fault occurs. Amplifier circuits were designed for isolated voltage and current sensing directly on the PCB. The differential measurement signals were then routed to BNC

1822

connectors that could be connected to external data acquisition equipment. The data acquisition will be discussed more in Section 4 of this article. This isolated sensing allowed for twenty-four of the thirty device signals shown in Table 1 to be measured directly using the PCB. This was by design to eliminate the need for an excess number of external probes. The gate current waveforms for each device were measured only using external Rogowski coils. Researchers were unable to find an integrated current sensor chip with the desired bandwidth large enough for the switching frequency of the inverter. Fig. 3 shows a block diagram that outlines the functions of the custom control PCB.

Fig. 4 Interface connection between the PCB and PCU.

The custom printed circuit board was designed to work with the University's unified controller, the UCB. The UCB includes Texas Instruments' F28335 DSP and the Lattice MachXO2 FPGA from Lattice semiconductor. It also includes accessory boards for fiber optic communications, voltage and current sensing feedback, and resolver feedback for measuring motor position. The UCB is shown in Fig. 5. The DSP is programmed with a Field Oriented Control (FOC) algorithm that the researchers selected to drive the permanent magnet synchronous machine (PMSM) used for this testing. The controls for this inverter were modified from a previous traction inverter project and based on a Texas instruments application note [2]. The previous project and control algorithms are explained in more detail in [3]. The control algorithm

included additional protective measures such as overvoltage and overcurrent warnings. If the overvoltage or overcurrent reached an unsafe threshold a fault would register, and the control algorithm would command the devices to turn off safely.

Fig. 5 UCB with daughter cards installed: Fiber optic communication (outlined in green), voltage and current feedback (red), and resolver feedback for motor position (blue).

3 System Modelling

An OPAL-RT simulation was used to generate realistic driving profiles for testing the traction inverter [4]. The simulation was created as part of a student's master's thesis and was modified to fit the vehicle conditions for this specific commercial inverter [5]. The OPAL-RT simulation uses a MATLAB Simulink model to represent the vehicle and other variables. Different parameters such as vehicle model characteristics, road conditions, wind, and traction are specified to develop an accurate model. Driving profiles from the U.S. Environmental Protection Agency (EPA) were selected and used to generate the testing points for the speed and torque of the dynamometer [6]. Test points were created for the US06, UDDS, and NYCC dynamometer driving schedules. The driving profiles produced with OPAL-RT were exported to .csv files to allow for more versality and offline testing in labs without an OPAL-RT unit. Different versions of the driving schedules were produced for testing at different speeds, torque values, and power levels. Ultimately, the US06 driving profile was selected as the primary driving schedule for this testing. Fig. 6 shows the US06 driving profile and part of the vehicle model in Simulink. Fig. 7 shows an example of the US06 driving profile running on the experimental test set up with the traction inverter under test driving the electric motor.

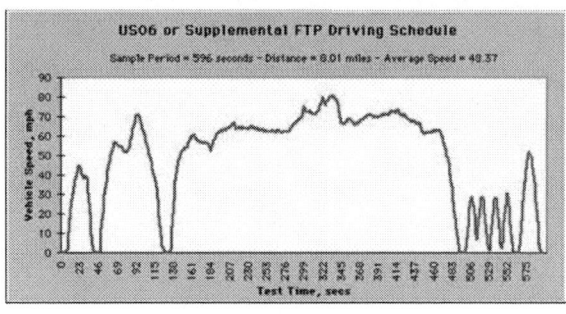

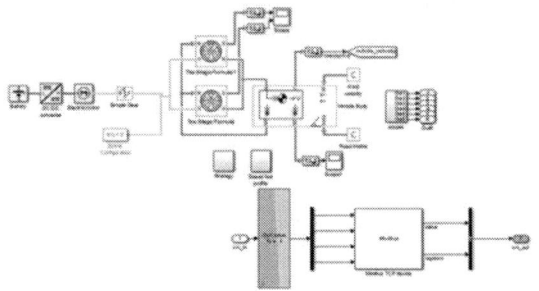

Fig. 6 US06 driving profile and OPAL-RT vehicle model [6].

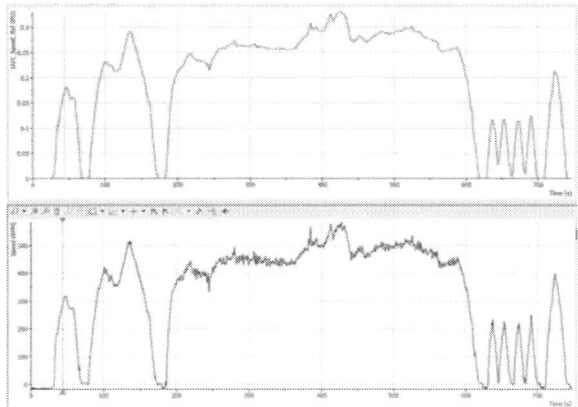

Fig. 7 Experimental test running the US06 Driving schedule at low power with the 55-kW traction inverter. The commanded speed is shown in red while the measured speed is plotted in blue.

4 Dynamometer Testing

Fig. 8 Dynamometer test bed at NCREPT.

The testing was completed using a 100-kW dynamometer test bed at the University of Arkansas NCREPT facility. The test bed is shown in Fig. 8 and includes a 75 HP permanent magnet synchronous machine (PMSM) driven by the unit under test. This is coupled with a torque transducer to a 100 HP induction machine that is driven by a 150 HP commercial motor drive from ABB. For this configuration the traction inverter under test was torque controlled, while the ABB motor drive was operating in speed control mode. This configuration was selected to be compatible with our collaborators' future test set up. The inverter was first tested with an open loop control to verify the custom control board configuration. After the functionality was verified, the closed loop field-oriented control was programmed. The inverter was preliminary tested by commanding constant torque and speed values to validate operation. Once the closed loop feedback control was validated, the PI controller parameters were tuned to accommodate the torque and speed changes of the US06 driving schedule so that the traction inverter could be tested under realistic conditions.

A LabVIEW control interface was used to conduct the dynamometer testing. This custom LabVIEW interface allowed for the inverter, ABB drive, DC power supplies, and National Instruments data collection equipment to be controlled all on one interface. The US06 driving profile was also loaded by using the custom LabVIEW configuration. The speed, torque, device temperature, fault registers, and cooling loop parameters could be monitored on the interface as well. The LabVIEW interface is shown in Fig. 9.

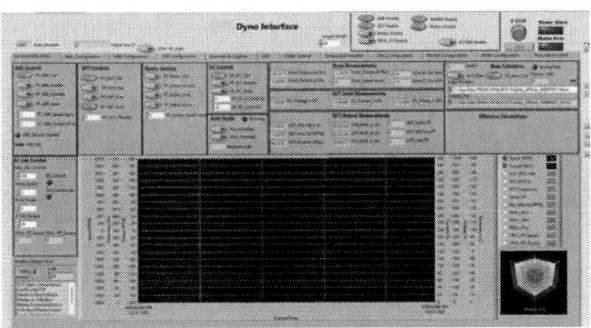

Fig. 9 Custom LabVIEW interface for dynamometer.

Two data acquisition units shown in Fig. 10 were used during this testing. A Yokogawa DL950 ScopeCorder measuring signals at 1 MS/sec per channel was used to collect the thirty inverter device signals at a very high sample rate. This was necessary due to the high switching frequency of the devices. Data from the DL950 was converted to MATLAB format using a custom automated MATLAB script to be shared in the desired format for the collaborators data driven model. NCREPT's CATS-2000 data acquisition unit uses National Instruments PXIe cards. The CATS-2000 was used to collect the device signals at 500 kS/sec per channel to have multiple recordings of the inverter data. The CATS-2000 was also used to collect less dynamic signals, such as speed and torque, at a lower sample rate. Table 1 summarizes the important signals that were collected and the sample rates at which they were measured.

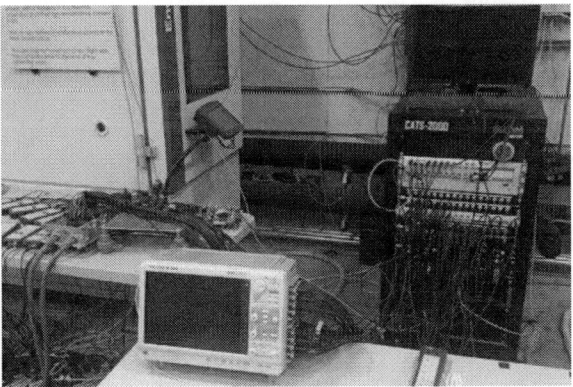

Fig. 10 Traction inverter under test connected to the CATS-2000 and DL950 data acquisition equipment.

Signal Name		Descriptions	Sample Rate
Inverter switching device signals (x6)	V_{CE}	Collector-emitter voltage	1 MS/Sec
	V_{GE}	Gate-emitter voltage	1 MS/Sec
	V_{AK}	Voltage of internal temperature diode	1 MS/Sec
	I_G	Gate current	1 MS/Sec
	I_{CE}	Collector-emitter current	1 MS/Sec
Dynamometer signals	V_{dc}	DC input voltage	5 kS/Sec
	I_{dc}	DC input current	5 kS/Sec
	I_a, I_b, I_c	RMS motor currents	5 kS/Sec
	V_{ab}, V_{bc}, V_{ca}	RMS phase voltages	5 kS/Sec
	Speed	Measured speed	5 kS/Sec
	Torque	Measured torque	5 kS/Sec
Chiller system		Pressure, flow rate, and temperature measured at both the inlet and outlet	250 S/Sec
Control signals		Refence speed and reference torque commands	250 S/Sec

Table 1 Table outlining the important signals measured and their sample rates.

The inverter was tested operating the US06 driving schedule at twenty-five percent of its total speed and seventy-five percent of its total torque. The goal of this testing was to thermally stress the devices until a fault occurred. The switching device temperatures increased during points of large torque swings in the driving profile. The dynamometer torque measurement directly relates to the output current of the inverter, therefore the large torque variations found in the US06 driving schedule make it a suitable method for stressing the devices [7].

Researchers determined elevated temperature testing would be the safest method to stress the devices further. The power level and torque could not be increased due to the limitations of the dynamometer and the safety of personnel and equipment. A 105°C Mokon chiller was used to elevate the cooling loop temperature of the devices and accelerate failure rates while running the driving schedule [8]. The inverter was tested with cooling loop temperatures of 25°C, 35°C, 45°C, 55°C, 65°C, and 75°C. Table 2 shows an example of the maximum and minimum temperatures for each of the six switching devices during the US06 driving cycle with a cooling loop temperature of 75°C.

Switching Device	Minimum Temperature	Maximum Temperature
PC2_U_VAK	71.9°C	119.7°C
PC2_L_VAK	80.2°C	138.8°C
PC3_U_VAK	76.7°C	134.6°C
PC3_L_VAK	72.5°C	116.7°C
PC4_U_VAK	75.5°C	129.2°C
PC4_L_VAK	74.9°C	126.2°C

Table 2 Table showing the inverter device temperatures during the 75°C test where a fault occurred.

After multiple tests, a fault that led to a failure occurred during one of the 75°C tests. The failure happened at approximately the 490-second mark of the US06 driving cycle. The fault occurred in the dc switch module for the power control unit's boost converter. Fig. 11 shows the fault location circled on the inverter schematic. The complete dataset with the fault occurrence and multiple datasets without faults were sent to our collaborators to be used with the data driven model. Their objective is to determine if the collected data can be used to indicate and predict device failure before it occurs. The data will be extensively analyzed to search for anomalies that could help predict a fault occurrence leading up to the point of failure.

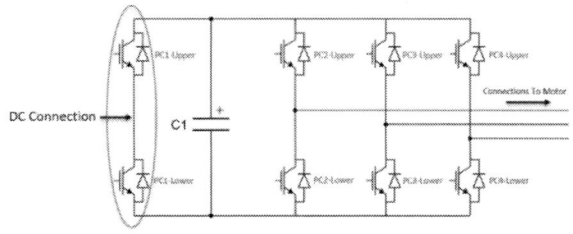

Fig. 11 Inverter topology with the fault region circled in red.

5 Conclusion and Future Work

In this work, researchers successfully developed a test methodology and high-speed data collection process for failure predictive analysis of a commercial traction inverter. Researchers fabricated a custom control PCB and developed a control algorithm to accomplish this testing. The custom PCB included operational amplifiers for on board signal measurements. Custom driving profiles were generated for this specific traction drive using an OPAL-RT unit and vehicle Simulink model. The custom PCB, UCB, control code, LabVIEW interface, data conversion script, and generated driving schedule files were shared so that testing could be continued at our collaborator's lab. The data that has been collected during this testing was also shared with collaborators for evaluation and identification of failure predictive analysis. An article describing the data analysis, diagnostic methods, and fault detection results is currently under development.

The testing methodology described in this article could be further improved by isolating the fault detection to the inverter devices as that was originally the focus of the project. This could be accomplished by bypassing the boost converter module to test only the failure rate of the inverter. Depending on our collaborators test bench configuration, tests could be accomplished at a higher power rating with increased torque and speed to further stress the devices at new parameters. Other driving profiles could be programmed to test the effects of various driving conditions on the longevity of the devices. Future analysis of a more well-rounded dataset could lead to new fault identification results.

Acknowledgements

The authors would like to express gratitude to Hiroshi Ukegawa for his significant contributions to this work. Also thank you to Xia Du for creating Fig.1 used in this paper and for her other work on this project. The authors would like to acknowledge Toyota for funding this project. The views and opinions of the authors expressed herein do not necessarily state or reflect those of Toyota Motor Corporation. Testing for this project was conducted at the National Center for Reliable Electric Power Transmission (NCREPT), the University of Arkansas' high-power 6 MVA test facility.

References

[1] Y. Wu, M. Mahmud, Y. Zhao, and A. Mantooth, "An Optimized Silicon Carbide Based 2×250 kW Dual Inverter for Traction Applications," 35th Annual IEEE Applied Power Electronics Conference & Exposition (APEC), New Orleans, LA, 2020

[2] R. Ramamoorthy, B. Larimore, and M. Bhardwaj "Sensored Field Oriented Control of 3-Phase Permanent Magnet Synchronous Motors Using TMS320F2837x" SPRABZ0A – FEBRUARY 2016 – REVISED MAY 2021

[3] C. Farnell, J. Jackson, A. Corbitt, H. A. Mantooth, "Rapid Prototyping of a SiC-Based PMSM Motor Drive for Aerospace Applications" IEEE Design Methodologies Conference (DMC), Miami, FL, 2023

[4] F. Xu, L. Chen and N. Narayanachar, "Traction inverter evaluation method based on driving cycles for electric and hybrid electric vehicles," 2016 IEEE Energy Conversion Congress and Exposition (ECCE), Milwaukee, WI, USA, 2016, pp. 1-6, doi: 10.1109/ECCE.2016.785553

[5] Daniel Schwartz. Developing a hil-based software platform for testing electric and hybrid vehicle powertrains, May 2018. Graduate Theses and Dissertations Retrieved from https://scholarworks.uark.edu/etd/2735.

[6] United States Environmental Protection Agency. Vehicle and fuel emissions testing, dynamometer drive schedules, epa, 2023. https://www.epa.gov/vehicle-and-fuel-emissions-testing/dynamometer-drive-schedules [Accessed: March 26, 2024].

[7] B. M. Nafis, D. Huitink, A. Iradukunda, "Drive Schedule Impacts to Thermal Design Requirements and the Associated Reliability Implications in Electric Vehicle Traction Drive Inverters," ASME 2018 International Technical Conference and Exhibition on Packaging and Integration of Electronic and Photonic Microsystems, San Francisco, California, USA, 2018.

[8] Hoque, M. J., Günay, A., Stillwell, A., Gurumukhi, Y., Pilawa-Podgurski, R., and Miljkovic, N. (December 8, 2020). "Modular Heat Sinks for Enhanced Thermal Management of Electronics." ASME. J. Electron. Packag. doi: https://doi.org/10.1115/1.4049294

PCIM Europe 2024, 11– 13 June 2024, Nuremberg   DOI: 10.30420/566262251

Short Circuit Robustness for Traction Inverters from an Application Point of View

Karl Oberdieck[1], Semy Ben Khelifa[2], Manuel Horvath[1], Sebastian Strache[1,] Matthias Bösing[3]

[1] Robert Bosch GmbH, Mobility Electronics, Germany
[2] Robert Bosch GmbH, Electrified Motion, Germany
[3] Robert Bosch GmbH, Corporate Research, Germany

Corresponding author: Karl Oberdieck, karl.oberdieck@de.bosch.com
Speaker: Karl Oberdieck, karl.oberdieck@de.bosch.com

Abstract

SiC MOSFETs haven proven a large potential of energy savings, especially in the $V_{DS} = 1200\,V$ voltage class. In automotive drive applications, robustness for short-circuit events is mandatory, which is often primarily defined by a short-circuit withstand time t_{scwt}. During the last decade, t_{scwt} was reduced in each IGBT generation down to $\approx 3\,\mu s$. Nowadays, t_{scwt} for SiC MOSFETs is in the range from 1 to 5 μs, depending on test conditions which must be covered for all operating points. Step by step, we separate a short-circuit event into different phases from turn on until detection and finally turning off the current. While reducing the application-required t_{scwt}, the blanking time and turn-off time become dominant factors, and the gate driver delay time t_{delay} becomes a minor factor. In an example we show, depending on the gate driver IC a reaction time t_{react} of 890 ns to 930 ns under real application conditions and parameter variations can be achieved resulting in a required t_{scwt} between 1.2 μs and 1.3 μs for the applied SiC MOSFETs.

1. Introduction

Saving fossil energy is the driving force for electrified mobility. To further improve efficiency and reduce invested resources, SiC MOSFETs are becoming more and more attractive. Using SiC-MOSFET technology, the losses in an automotive drive train can be reduced between 3-8% compared to conventional Si-IGBTs [1, 2, 11, 13].

Pushing SiC-MOSFET technology forward for optimal utilization inside the traction inverter, the performance will be further improved over the generations. The key performance indicator of static losses is the technology parameter $R_{DSon}A$. A limit to its optimization is set by the trade-off against short-circuit withstand time t_{scwt}.

To ensure safe operation over lifetime, the robustness against temporary overload conditions (short circuit, active short circuit and more), radiation hardness, and driving power demand are essential functional parameters [3, 4, 6, 13]. In this paper, we focus on the application-dependent short-circuit reaction time leading to robustness requirement.

First, we briefly summarize the common test standards. Then, we derive an application driven step-by-step approach for deriving the short circuit load and present an example of this approach.

2. Short Circuit Cases in Applications and Standards

During the design of a specific application, the short circuit cases shall be deducted from a safety analysis of the inverter designer (failure tree analysis, FTA). The defined cases in standards (AQG, IEC, JEDEC) or academic literature are covering single specific events. We first summarize the standards and then discuss it on both AQG 324 events.

Standards common in industry are:

1. ECPE Guideline AQG 324 [5] from 2021 describes two cases of failures for half-bridge power modules: SC 1 (hard switch failure) and SC 2 (failure and load).
2. IEC 60747-8 [7] from 2021 describes a test setup for a single MOSFET.
3. JESD 24-9 [9] reaffirmed in 2002, is similar to IEC 60747-8.
4. JEP-xxx (Proposed in Mar. 2024, Version 0.2) [10] is a draft which extends IEC 60747-8.

Further short circuit standards AEC-Q are today under development for wide bandgap devices.

Source	Name	DUT	DUT Signal	di/dt
AQG 324 [5]	SC type 1 (hard switch failure)	Half bridge	turn-on SC event	high (low L_{CC})
AQG 324 [5]	SC type 2 (failure under load)	Half bridge	turn-on SC event	low (load inductance) v_{DS} rise> 5 µs
IEC 60747-8 [7]	SC safe operating area (SCSOA)	Single device	turn-on SC event	high (low L_{CC}) with $i_{SC,peak}$ in first 25% of pulse
JESD 24-9 [9]	-	Single device	turn-on SC event	high (low L_{CC})
Publication for SiC MOSFETs [8]	SC type 2 (forward conduction)	Single device with auxiliary switch	DUT turned on with 1st quadrant conduction	high (low L_{CC})
Publication for SiC MOSFETs [8]	SC type 3 (body-diode conduction)	Single device with auxiliary switch	DUT turned on with 3rd quadrant conduction	high (low L_{CC})
JEP-xxx [10]	SC type 1 (hard switch failure)	Single device	turn on SC event	high (low L_{CC})
JEP-xxx [10]	SC type 2 (failure under load)	Single device with auxiliary switch	DUT turned on with 1st quadrant conduction	high (low L_{CC})

Table 1: Overview on short-circuit cases.

Common among the literature is the definition of short circuit type 1 (SC1) which usually has a high di/dt due to a low commutation loop inductance L_{CC} as shown in Figure 1. In academic literature, there is also defined further short circuit cases. One example is a short circuit failure with a single semiconductor (as IEC and JEDEC), while the device is already turned on (in IEC and JEDEC the tested device switches on for themselves). This has been named for the different current direction of a semiconductor the forward (type 2) and diode conduction (type 3) [8]. Table 1 lists different short circuit events described in the standards. Three examples on the AQG 324 guideline elaborate limits and ambiguities in today's standards.

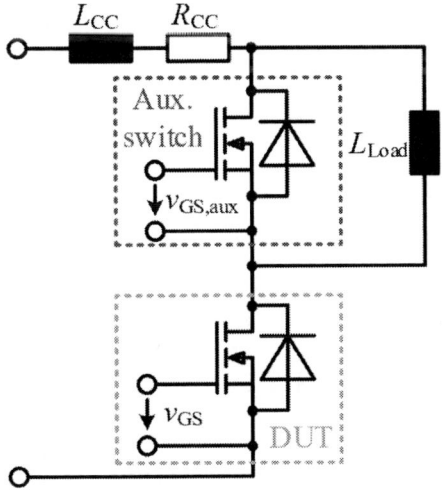

Figure 1: Schematic of half bridge.

First example: AQG 324 SC type 1 is a low inductive (mainly L_{CC}-limited) failure with a high di/dt. It can occur due to two different root causes: either a control error (a half-bridge switch is turning on while the complementary switch is already turned on) or a switch defect (insulation failure, flash-over fault; semiconductor breakdown due to cosmic ray event or gate oxide breakdown). The standard solely defines a test circuit and partially boundary conditions. But they usually lacking specific test conditions (for example DC-link voltage, turn-on speed or starting temperatures).

In addition, also the operation of the auxiliary switch is not clearly defined. At the control error of a power module, high- and low-side switch share the DC-link voltage between each other. The voltage between high- and low-side switch is not shared equally but dependent on turn-on speed and semiconductor parameters.

Figure 2 shows the impact of the auxiliary gate-source voltage $v_{GS,aux}$ on the short-circuit energies. By simply varying this parameter, the dissipated energy during the same pulse length changes from 3.3 J for $v_{GS,aux} = 15$ V to 4.2 J for $v_{GS,aux} = 18$ V.

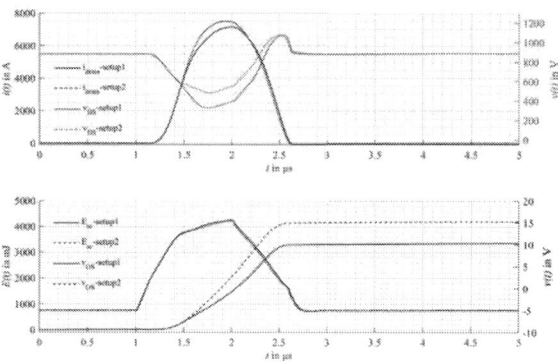

Figure 2: Short circuit type 1 for different test condition.
Setup 1: $v_{GS,aux} = 15$ V of auxiliary switch.
Setup 2: $v_{GS,aux} = 18$ V of auxiliary switch.

In comparison, an insulation failure the full DC-link voltage is either at high- or low-side with unknown failure resistance R_{CC}. The failure resistance can reduce the semiconductor stress. Standards that only testing a single device try to reproduce the conditions on a separate single chip or discrete (TO-247, D2PAK) vehicle, which has other parasitic elements (R_{CC} and L_{CC}).

Second example: The SC type 2 in AQG 324 is defined as a low di/dt with a v_{DS} increase earliest after 5 µs. Due to the low di/dt the induced energy during turn-on time can even be higher compared to other events until the v_{DS} monitoring is detecting and turning off the event. Unfortunately, the SC type 2 definition between different standards and authors are ambiguous. The SC type 2 can also occur as a low-inductive error, while the switch is already conducting [10]. While the definition of SC type 2 in AQG 324 should cover a case, where the insulation failure is not within the power module but at the phase output of the half bridge, the high di/dt SC type 2 definitions cover cases within the power module. Both SC type 2 definitions usually lacking in complete list of required parameters to deduce a pass/fail criteria.

Third example: The start and stop conditions for t_{scwt} itself also differs depending on standard. Most common in the standards is 50%/50% of gate-source voltage as start and stop condition [7, 9, 10]. Another definition is 10%/10% of the short-circuit peak current $i_{SC,peak}$ as start and stop conditions [5, 10].

3. Phases of a short circuit

Section 2 outlines, that different types for traction inverter application of short circuits need to be considered. For typical traction inverter applications, the turn on against an existing short-circuited switch is the most critical one from a load-energy point-of-view. Hence, we use this case, known as short-circuit type 1, for the detailed design discussion of the different phases.

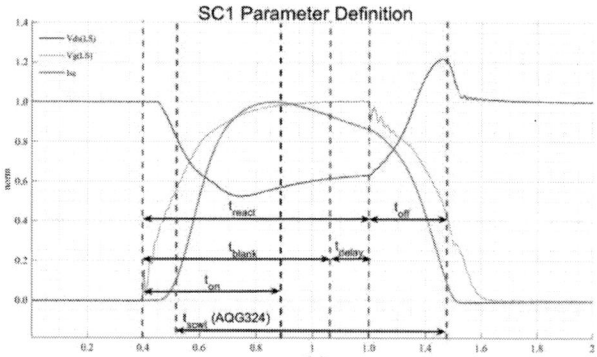

Figure 3: Definition of short circuit phases.

The short-circuit event is separated in different phases, as depicted in Figure 3 [3]. Figure 3 shows, that first the gate-source voltage v_{GS} of the controlled SiC MOSFET rises. Once, the v_{GS} is above the threshold voltage $v_{GS,th}$, a drain-source current starts to conduct and the drain-source voltage v_{DS} drops due to the parasitic inductance. After the blanking time t_{blank} has passed, the short-circuit detection circuit (v_{DS} monitoring) of the gate driver is active an detects the too high v_{DS} voltage. After the internal delay t_{delay}, the turn off process starts (t_{off}).

In the following section, the three parts: turn-on time t_{on}, reaction time t_{react} and turn-off time t_{off} are derived in detail. Finally, as Figure 3 depicts, the t_{scwt} stress as an aggregated time will be discussed.

3.1 Turn-on time t_{on}

During normal operation of the inverter, the turn-on time t_{on} is determined by the maximum allowable turn-on switching speed of the active switch over the whole v_{DC}-, i_{phase}-, and T_j-range of the SiC MOSFETs. In voltage-controlled gate drivers, the turn-on speed is set by the gate resistor $R_{G,on}$; in current-controlled gate drivers by the gate current $i_{G,on}$.

Switching speed limits are given by electro-magnetic compatibility (EMC), common-mode transient immunity (CMTI) of hardware components

(e. g., gate driver integrated circuits (ICs)), and semiconductor safe-operating area (SOA) of the power module (e. g., break down voltage $v_{BR,DSS}$ or gate-voltage pulse limitations). More precisely, this is about dv/dt and overvoltage $v_{DS,max}$ of the body-diode during turn-on transition of the active switch and is strongly influenced by the body-diode characteristic but also commutation inductance L_{CC}.

Once $R_{G.on}$ (or $i_{G,on}$) have been determined, the operation point with the slowest turn-on transition time needs to be derived. In most SiC inverter applications, this occurs in combination with high DC-link voltages, high phase currents and low junction temperatures of the SiC-MOSFETs. Depending on the decision of the inverter designer, special operating points are used for the selection (e. g., maximum DC-link voltage and phase currents up to the overcurrent detection threshold of inverter).

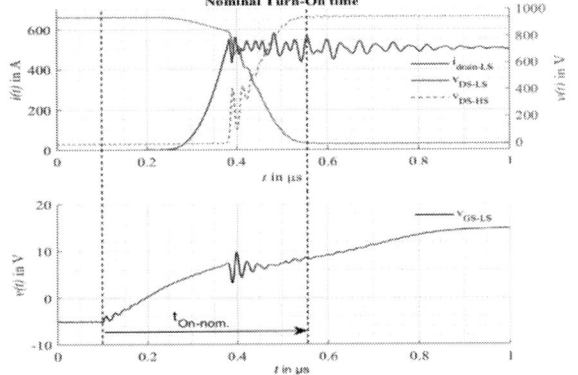

Figure 4: Turn-on time of active turn on sequence. Red-solid line: v_{DS} active switch low-side. Blue-solid-line: drain current. Red-dashed-line: v_{DS} passive switch high-side (body diode).

Figure 4 shows nominal t_{on} for a Bosch inverter application with its inhouse gate-driver IC *EG 120* [12]. A dv/dt limit 100 V/ns and CMTI up to 150 V/ns is used for the example. Turn-on time is defined as the time from the rising edge of v_{GS} until v_{DS} is securely lower than the short-circuit detection threshold (e. g., below 10 V). In the selected example this yield to a time of t_{on} = 430 ns.

Considering semiconductor (gate charge, threshold, and plateau voltage) and gate driver (gate voltage, logic) variations an additional margin of 35 % on t_{on} is added for this example. This results in $t_{on,max}$ = 580 ns.

3.2 Blanking time t_{blank}

The blanking time t_{blank} is the period where the short circuit detection is disabled during the turn-on process to prevent unintended triggering of

SC detection while the voltage is still commutating. Based on $t_{on,max}$ the minimum blanking time is calculated according to:

$$t_{blank,min} > t_{on,max} + t_{CompDelay,max}$$

Eq. 1: Determination of minimum blanking time

$t_{CompDelay,max}$ represents the propagation delay time of the comparator sensing the v_{DS} signal as well as additional delay times. In our example, $t_{CompDelay,max}$ is 130 ns.

The minimum needed blanking time thus becomes 710 ns. Gate drivers typically do not allow arbitrary but only quantized blanking time settings. According to the datasheet of the gate driver IC, next blanking time value which can be configured results in $t_{blank,min}$ = 732 ns. Additionally, the gate driver tolerances of the blanking time with ≈4% needs to be considered. The maximum blanking time includes this tolerance as follows:

$$t_{blank,max} = t_{blank,min} \cdot \rho_{blank}$$

Eq. 2: $t_{blank,max}$ calculation

This results in $t_{blank,max}$ = 762 ns for the maximum blanking time occurring in the application.

3.3 Reaction time t_{react}

The reaction time t_{react} consists of the blanking time t_{blank} and a delay time t_{delay}. The delay time t_{delay} is the sum gate-driver internal propagation delays once the blanking time has elapsed and v_{DS} has exceeded the configurated threshold until beginning of turn-off event. In Eq.3 reaction time is calculated for the maximum needed $t_{react,max}$ reaction time, which includes all tolerances:

$$t_{react,max} = t_{blank,max} + t_{delay,max}$$

Eq. 3: $t_{react,max}$ calculation

Table 2 summarizes all the timings that lead to t_{react} for two different gate driver ICs. First, the current-controlled gate driver IC from (Bosch EG 120) [6] secondly a voltage-controlled gate driver IC.

Gate driver	$t_{on,max}$	$t_{blank,min}$	$t_{blank,max}$	t_{delay}	$t_{react,max}$
1	580 ns	732 ns	762 ns	125 ns	887 ns
2	580 ns	715 ns	807 ns	120 ns	927 ns

Table 2 – Comparison reaction time for two gate driver ICs.

3.4 Turn-off time t_{off}

During the turn-off event, the energy of the commutation loop inductance L_{CC} is dissipated in the semiconductors. Ideally, $v_{DS,max}$ equals the maximum specified voltage $v_{BR,DSS}$ of the device during the complete turn-off duration as shown in Figure 5. This leads to a constant slope of the drain current when neglecting series resistances:

$$t_{off} = \frac{I_{off} \cdot L_{CC}}{v_{DS,max} - V_{DC}}$$

The dissipated energy can be calculated as:

$$E_{off} = v_{DS,max} \cdot \frac{I_{off}}{2} \cdot t_{off} = \frac{1}{2} \cdot \frac{L_{CC} \cdot I_{off}^2 \cdot v_{DS,max}}{v_{DS,max} - V_{DC}}$$

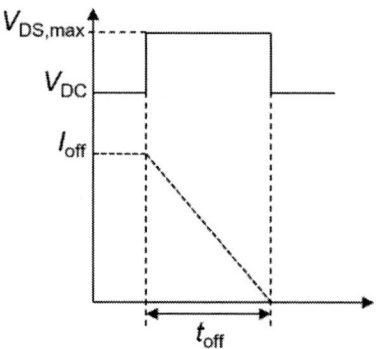

Figure 5: Ideal turn-off waveform

In a measurement, the time from when the gate driver IC starts to turn off the device until the current is below 10 % of its peak value is called turn-off time t_{off}. The turn-off time is adjusted via the turn-off gate resistance $R_{G,off}$ or the gate current $i_{G,off}$. The speed, as shown in the equation above, depends on the commutation loop inductance L_{CC}, the device current at turn-off, and the transfer curve of the SiC MOSFET.

To ensure safe operation of the semiconductor, it should be considered, that the selected turn-off gate resistance $R_{G,off}$ or gate current $i_{G,off}$ shall also keep $v_{DS,max}$ within the limits for the other short circuit cases (for example SC type 2 with low and high di/dt). This is mandatory, because a short-circuit detection, like v_{DS} monitoring, cannot distinguish between the different short-circuit cases.

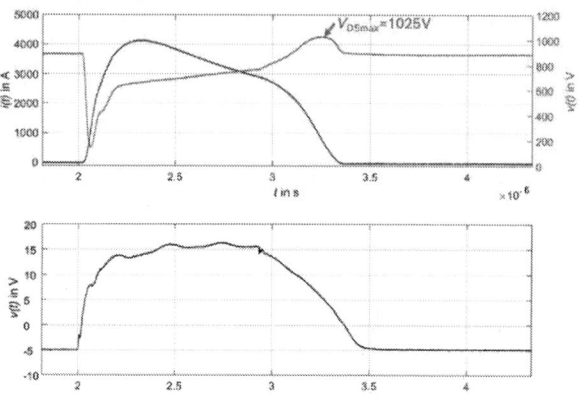

Figure 6: Turn-off measurement of short circuit type 1.

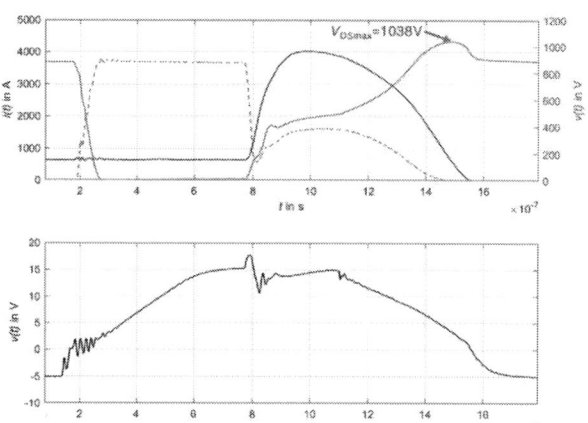

Figure 7: Turn-off measurement of short circuit type 2 with high di/dt.

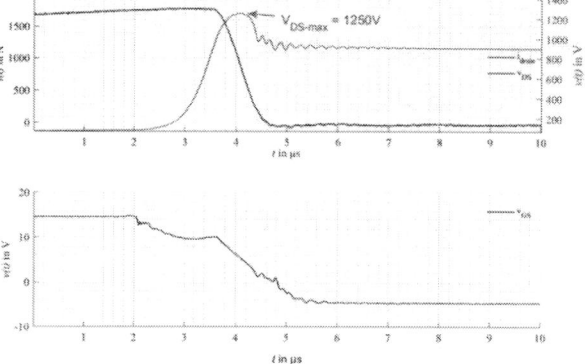

Figure 8: Turn-off measurement of short circuit type 2 with low di/dt.

For this example, the turn-off overvoltage of three different short-circuit types were measured:

1. Short circuit type 1:
   1025 V (Figure 6)

2. Short circuit type 2 with high di/dt:
   1038 V (Figure 7)

3. Short circuit type 2 with low di/dt:
   1250 V (Figure 8)

The measurements show that the SC type 2 with a low di/dt reaches the highest over voltage peak for same DC-link voltage and turn-off gate current compared to the other two types. For our example t_{off} results according $R_{g,off}/i_{g,off}$ to 350 ns.

The real breakdown voltage of SiC MOSFETs is often higher than the rated voltage of the device ($v_{BR,DSS}$ = 1.2 kV). Further investigation can utilize this margin to increase $v_{DS,max}$ to ensure a safe turn-off also in worst-case scenarios.

3.5 Aggregation of required short-circuit withstand time t_{scwt}

The individual time periods of this chapter are now compared with a SC type 1 measurement and explained in the context of a t_{scwt} according to AQG 324. For this measurement, a Bosch power-module generation 6, depicted in Figure 9, utilizing first generation 1200V-SiC MOSFETs [13] is selected.

Figure 9: Bosch power-module generation 6. Source: www.bosch-mobility.com

Figure 10 shows the individual time periods t_{react}, t_{off} and t_{scwt} during a short-circuit type 1 measurement. The dissipated energy E_{SC} in the short-circuit event between device turn on and reaction of the gate driver (t_{react}) is ≈1.8 J. During the turn-off phase t_{off}, the dissipated energy is ≈0.9 J. In the shown example, approximately 2/3 of the total energy is dissipated during the turn-on and reaction (t_{react}) and 1/3 during the turn-off phase (t_{off}). Based on the numbers in section 2, above 85 % of the

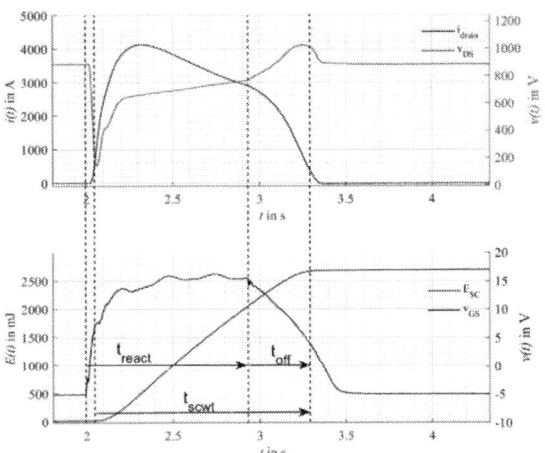

Figure 10: SC type 1 measurement with short-circuit withstand time.

reaction time is due to blanking out the high drain-source voltage during normal turn on. In the blanking time t_{blank} almost 75% comes from the turn-on behavior of the power module equipped with SiC MOSFETs. Therefore, this parameter has the largest influence on short-circuit load.

Depending on the used gate driver IC (see 2), a reaction time t_{react} of 890 ns to 930 ns can be achieved under real application conditions and parameter variations. This results in a required t_{scwt} between 1.2 µs to 1.3 µs for the applied SiC MOSFETs. This value shall be guaranteed by the semiconductor and power module including scattering of peak saturation current due semiconductor parameter variation (gate charge, threshold, and plateau voltage) and gate driver (gate voltage supply).

4. Conclusion

We analyzed the applicability of short circuit events in traction inverters detected with v_{DS} monitoring. During the design of a traction inverter, the most relevant short circuit events depend on the safety concept. Standards for components (power modules or semiconductors) only cover selected safety relevant cases. The standards usually define a short-circuit withstand time t_{scwt} to assess if a product matches its requirements of a certain application. Furthermore, the t_{scwt} definitions in the industry are not unique nor complete.

To design for a safe short-circuit operation, we separated the short circuit event into different phases. During the design process, the influencing factors (gate driver, commutation loop, power module and semiconductor) for different phases are deducted. It is derived which operation point or failures cases shall be considered to determine the minimal required timings e. g. for the $t_{on,max}$ and t_{off}. Based on a Bosch inverter application with application conditions and parameter variations all timing requirements are illustrated resulting in a t_{react} of 890 ns to 930 ns depending on the applied gate driver. This results in a required t_{scwt} between 1.2 µs to 1.3 µs for the applied SiC MOSFETs. This t_{scwt} requirement enables a stronger focus in the $R_{DSon}A./\ t_{scwt}$ trade-off towards a $R_{DSon}A$ performance optimization.

References

[1]   C. T. Banzhaf, M.Grieb, A.Trautmann, A. J. Bauer and L. Frey, „Characterization of Diverse Gate Oxide on 4H-SiC 3D Trench-MOS Structures", in Material Science Forum Vols. 740-742 (2013)

[2]   Klaus Heyers, Stephan Schwaiger, Christian Banzhaf, Michael Grieb, „SiC-Trench-MOSFETs for automotive drive application", in Bauelemente der Leistungselektronik und ihre Anwendungen 2017 - 7. ETG-Fachtagung.

[3]   M. Boesing and D. Schweiker, "Design Aspects in SiC MOSFET based High Performance Automotive and Commercial Vehicle Inverters," PCIM Europe 2023; International Exhibition and Conference for Power Electronics, Intelligent Motion, Renewable Energy and Energy Management, Nuremberg, Germany, 2023, doi: 10.30420/566091063.

[4]   S. Schwaiger, K. Heyers, A. Martinez-Limia, K. Oberdieck and C. Foerster, "Advanced SiC Trench-MOS Technology for Automotive Application," PCIM Europe 2023; International Exhibition and Conference for Power Electronics, Intelligent Motion, Renewable Energy and Energy Management, Nuremberg, Germany, 2023, doi: 10.30420/566091096.

[5]   ECPE, "Qualification of Power Modules for Use in Power Electronics Converter Units in Motor Vehicles – ECPE AQG 324 Automotive Qualification Guideline – Release 03.1/2021

[6]   M. Riefer, J. Winkler, S. Strache and I. Kallfass, "Implementation of Current-Source Gate Driver with Open-Loop Slope Shaping for SiC-MOSFETs," PCIM Europe digital days 2021; International Exhibition and Conference for Power Electronics, Intelligent Motion, Renewable Energy and Energy Management, 2021

[7]   IEC 60747-8 - Semiconductor devices – Discrete devices – Part 8: Field-effect transistors

[8]   X. Liu, X. Li and T. Basler, "Short Circuit Type II and III Behavior of 1.2 kV Power SiC-MOSFETs," 2022 24th European Conference on Power Electronics and Applications (EPE'22 ECCE Europe), Hannover, Germany, 2022

[9]   JEDEC STANDARD – JESD 24-9 - Short Circuit Withstand Time Test Method, AUGUST 1992 (Reaffirmed: OCTOBER 2002)

[10]   JEDEC PUBLICATION (PROPOSED) – JEP-XXX – Guideline for Short Circuit Testing of Wide Bandgap Devices (Version 0.2), Mar. 2024

[11]   W. Jakobi et al., "Benefits of new Cool-SiCTM MOSFET in HybridPACKTM Drive package for electrical drive train applications," CIPS 2018; 10th International Conference on Integrated Power Electronics Systems, Stuttgart, Germany, 2018

[12]   Robert Bosch GmbH, EG120 Gate Driver, https://www.bosch-semiconductors.com/system-ics/powertrain/eg120/

[13]   Robert Bosch GmbH, SiC Technology, https://www.bosch-semiconductors.com/power-semiconductors-and-modules/

PCIM Europe 2024, 11– 13 June 2024, Nuremberg   DOI: 10.30420/566262252

The Impact of The Dead time on The Stability of 1.2kV SiC MOSFET Body Diode Under Hard Switching with synchronous Rectification

Mohammed Amer Karout[1] , Ahmed Topkil[1], Abdul Haleem Malik[1], Craig Fisher[1], Philip Mawby[1], Olayiwola Alatise[1] and Mohamed Taha[1,2]

[1] School of Engineering, University of Warwick, Coventry, CV4 7AL, United Kingdom

[2] Faculty of Engineering, Cario University, Giza, Egypt

Corresponding author: Mohammed Amer Karout, Mohammed-Amer.Karout@Warwick.ac.uk

Speaker: Mohammed Amer Karout, Mohammed-Amer.Karout@Warwick.ac.uk

Abstract

In this paper, the impact of the dead time on the performance and stability of 1.2 kV SiC MOSFET body diodes is characterised considering the variation of three factors (I) Gate turn-off voltage, (II) Junction temperature and (III) gate resistance. When synchronous rectification is used, experimental measurements show that decreasing the dead time reduces the peak reverse recovery current, reverse recovery charge, the turn-on losses and improves switching stability by reducing the dI/dt. In certain scenarios, utilising a dead time of 140 ns can reduce the turn-on losses of SiC MOSFET by 23.3% and enhance the switching stability due to a reduction in the dI/dt by more than 50%. Moreover, turning off the synchronous FET with a negative gate voltage leads to higher peak reverse recovery current, reverse recovery charge and switching losses relative to zero gate turn-off voltage. Utilising the short dead time with either zero gate turn-off voltage or negative gate turn-off voltage will reduce the aforementioned parameters, however, the reduction in the negative gate turn-off voltage case is more than the zero gate turn-off voltage. The impact of the junction temperature and gate resistance variation is extensively analysed. Increasing the former can increase the reduction in the losses and reverse recovery, while increasing the latter has the opposite effect.   Moreover, the experimental results confirm that the threshold dead-time, at which a reduction begins to appear in the peak reverse recovery current, reverse recovery charge, and turn-on losses, depends on both temperature and gate resistance. This threshold time increases with higher temperature and with higher gate resistance and it can exceed 540 ns in some cases.

1   Introduction

Although it is preferable to switch SiC MOSFETs with their body diodes, this is still limited due to its high reverse recovery and the snappiness problem at a fast switching rate and high temperature [1]. In half-bridge configuration, the body diode turn-off characteristics affect the turn-on characteristics of the complementary MOSFET, hence it is desirable to maximize the switching rate to reduce switching losses. However, there are limits as to how fast body diodes can be safely switched without causing excessive ringing and potential failure from parasitic BJT latching [2][3]. Parallel Schottky barrier diode (SBD) has proved its capability to suppress these problems, give better conduction losses during 3rd quadrant operation and

enhance the switching losses [4]. However, utilising a parallel SBD increases the cost, requires more packaging space and introduces more parasites [5]. Synchronous rectification (SR) is well known operation used to turn on the MOSFET during 3rd quadrant operation which allows the current to pass through the channel rather than the body diode. Since the body diode of the SiC-MOSFET has a higher voltage drop compared with body diode of Si-MOSFET, SR has a significant impact on reducing the conduction losses of SiC MOSFET. Moreover, it has been found that SR operation does not only impact the conduction losses but also affects the switching losses due to the body diode operation during the dead time only [6]. Although dead time must be sufficient to prevent shoot-through in the half-bridge setup, it has been observed that decreasing it can effectively reduce switching losses and significantly enhance stability. In [1], It is shown that under asynchronous rectification (ASR), body diode turn-off at high junction temperatures with high dI/dt and turn-off V_{GS} of zero-volt results in electromagnetic instability, excessive EMI and high shoot-through currents. This problem is suppressed either by using a parallel Schottky diode or using negative VGS. In [7], the impact of the dead time on reducing the reverse recovery and the switching losses of low voltage Si-MOSFET is extensively investigated. While in [8], a similar study has been conducted on SiC-MOSFET to address the impact of the dead time on the switching losses. In [4] and [5], it is proved that utilising a parallel Schottky diode can enhance the turn-on losses significantly. While in [9], it has been demonstrated that for 3.3 kV SiC MOSFET, using SR can provide comparable enhancement in the turn-on losses compared to utilising parallel SBD. In [10], a 3.3 kV half-bridge design optimisation for modular multi-level converter is provided. It has been demonstrated that by optimising the dead time the reduction in the switching losses can reach 54% at high current operation. In this study, an extensive investigation is conducted to examine the influence of the dead time on the SiC MOSFET body diode turn-off characteristics, the switching losses of the complementary transistor and the overall switching stability. Furthermore, the study extends to explore the effects of temperature, gate turn-off voltage and gate resistance on the aforementioned characteristics.

2 Experimental Setup

Using a double pulse test rig (DPT) [11] shown in Fig. 1, the switching performance of SiC MOSFET body diodes is investigated under hard switching with synchronous rectification as a function of the dead time range from 1.04 µs down to 100 ns. To observe the impact of the dead time on the switching behaviour under different conditions, three main parameters are considered for this study including junction temperature (T_J), gate turn-off voltage ($V_{gs,off}$) and the gate resistance (R_g). Three different temperatures are used in this research (25 °C, 75 °C and 150 °C) combined with three different gate resistances (2 Ω, 7.6 Ω and 12 Ω). Moreover, zero and negative turn-off gate voltage (0 V, -4 V) are considered in this investigation. Two 1.2 kV SiC MOSFETs from Wolfspeed with datasheet reference C2M0080120D are used as low-side and high-side switch in the DPT. All tests are performed at V_{DS}=800 V, ID= 30 A. To realise the synchronous rectification, GD3160 gate drive from NXP semiconductor   is used due to the

flexibility of changing the dead time accurately. For each specific combination of gate resistance, junction temperature and gate turn-off voltage, a DPT is performed without synchronous rectification (Asynchronous rectification (ASR)). Then six more DPT are performed with synchronous rectification using six different dead time (1.04 µs, 0.54 µs, 0.4 µs, 0.24 µs, 0.14 µs and 0.1 µs). Figure 2 shows a closer view to the double pulse test PCB including the DUT, the measurement probes and the gate drive.

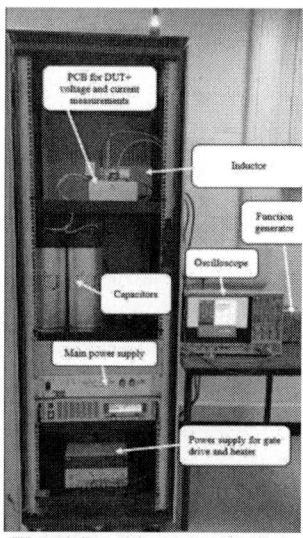

Fig. 1 Double pulse test rig

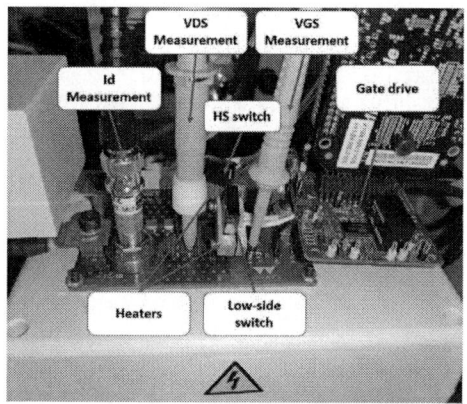

Fig. 2 Double pulse PCB

3 Results Analysis

3.1 Impact of the dead time on the switching performance at specific temperature and gate resistance

Figure 3(a) shows the high-side body diode turn-off current under different dead time with T_J=

150 °C, R_g= 12 Ω and $V_{gs,off}$= 0 V compared with the turn-off current in ASR case. While Fig. 3(b) depicts the same current when utilising negative turn-off gate voltage $V_{gs,off}$= -4 V. It is clear that in both cases the shorter the dead-time, the lower the peak reveres recovery, the shorter reverse recovery time and the lower the reverse recovery charge. This reduction in the reverse recovery charge (Q_{rr}) and the peak reverse recovery current ($I_{rr,max}$) with shorter dead time can be attributed to the shorter operating time of the body diode and lower plasma formation accordingly [8]. Moreover, this reduction in $I_{rr,max}$ and Q_{rr} is also reflected in the drain current of the complementary transistor (low-side switch) during turn-on as shown in Fig. 4(a) and Fig. 4(b). This reflection is well-known as the turn-off characteristics of the diode impact the turn-on transient of the transistor that commutates the current. It is also obvious from Fig. 4(a) and Fig. 4(b) that shorter dead time can reduce the current overshoot (caused by body diode reverse recovery) and enhance the current stability. A similar trend in $I_{rr,max}$ and Q_{rr} has been observed at different gate resistances and junction temperatures. The impact of temperature and the gate resistance will be discussed in the next two sections.

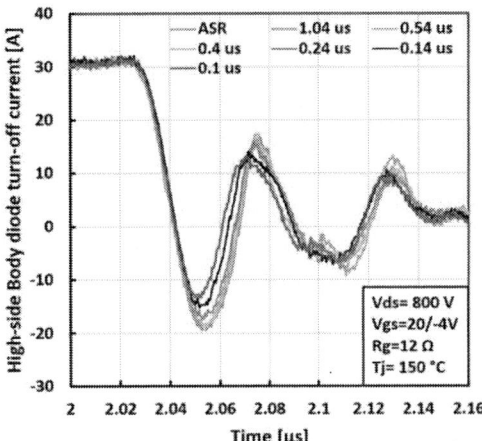

Fig. 3(b) body diode turn-off current under different dead time at 150°C using R_g=12 Ω and $V_{gs,off}$= -4 V

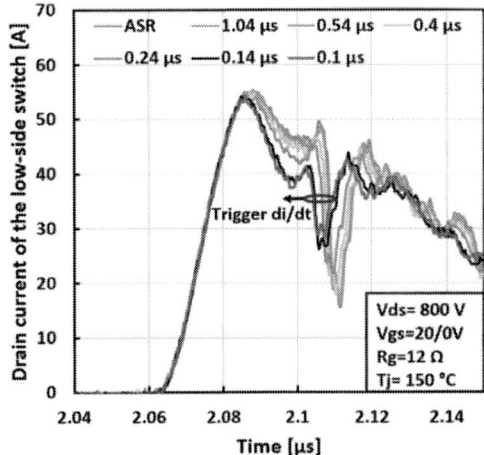

Fig. 4(a) Drain turn-on current under different dead time at 150°C using R_g=12 Ω and $V_{gs,off}$= 0 V

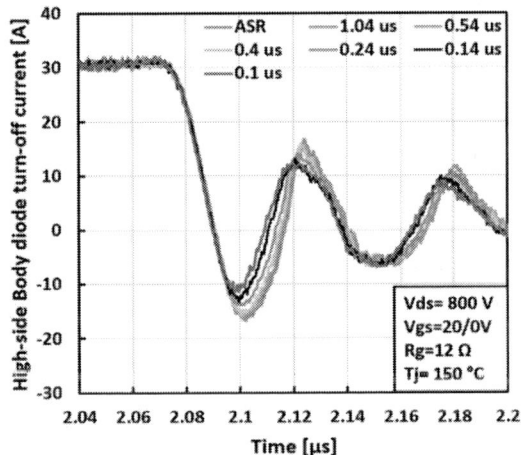

Fig. 3(a) body diode turn-off current under different dead time at 150°C using R_g=12 Ω and $V_{gs,off}$= 0 V

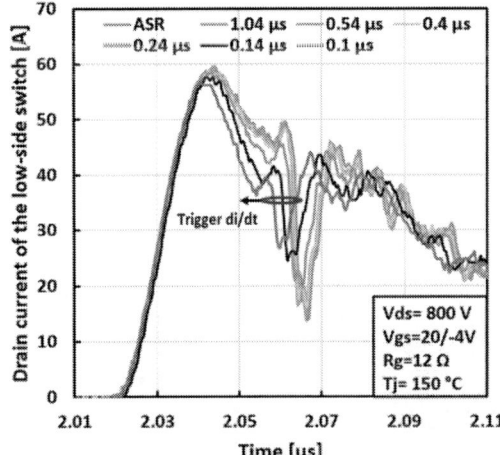

Fig. 4(b) Drain turn-on current under different dead time at 150°C using R_g=12 Ω and $V_{gs,off}$=-4 V

To give a better understanding of the impact of the dead time on the turn-on transient of the low-side transistor, Fig. 5(a) shows the drain-source voltage and the drain current during turn-on for the case of using 100 ns dead time compared with ASR case when utilising $V_{gs,off}$= 0 V. While Fig. 5(b) shows the same transient in the case of using $V_{gs,off}$= -4 V. It is clear that not only the current overshoot reduces when using 100 ns dead time but also faster voltage falling rate (higher dv/dt) which reduces the switching losses consequently.

To quantify the impact of the dead time on the body diode turn-off characteristics and the turn-on characteristics of the complementary transistor, the $I_{rr,max}$ and Q_{rr} are measured and calculated for each dead time as shown in Fig. 6 and Fig. 7. It can be seen that utilising negative gate turn-off voltage cause higher peak reverse recovery current and higher reverse recovery charge compared with using zero gate turn-off voltage (This also can be seen in Fig. 3(a) and Fig. 3(b)). This can be explained as a result of the removal of subthreshold channel conduction which increases the charge density in the PiN diode (body diode) thereby increasing the forward recovery time. Moreover, it is obvious that in both cases the reduction in $I_{rr,max}$ and Q_{rr} only starts at 540 ns and gets lower with shorter dead-time.

While the dead time higher than 540 ns gives almost similar performance to the case of not using synchronous rectification (ASR). This can be attributed to the forward recovery time or plasma formation time of the body diode, which is the time it takes for the injected carriers to build up in the drift region until it gets fully modulated [12]. Therefore, when using dead time lower than 540 ns, the body diode turns on for a period lower than the time needed to accumulate the carriers in the drift region which will reduce the reverse recovery charge that appears during the turn-off. Figure 8 shows the calculated turn-on losses of the low-side transistor as function of different dead-time. It is also clear that the turn-on losses follow the same reduction trend with shorter dead time which increases the overall efficiency consequently. Moreover, as a higher reverse recovery

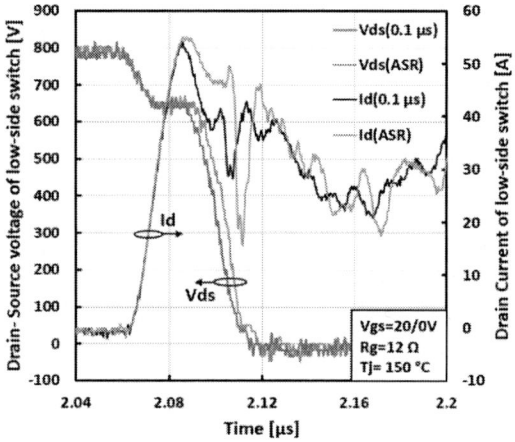

Fig. 5(a) Turn-on transient of the low-side transistor for 100 ns dead time and ASR at 150°C using R_g=12 Ω and $V_{gs,off}$= 0 V

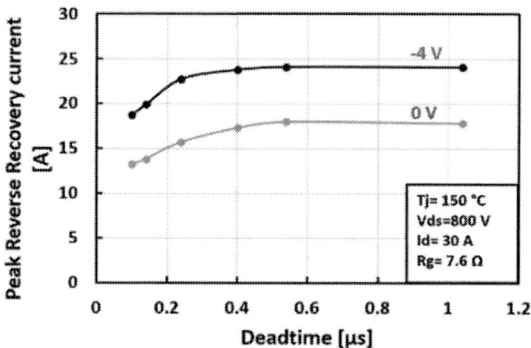

Fig. 6 Peak reverse recovery current under different dead time using $V_{gs,off}$= 0 V and $V_{gs,off}$= -4 V

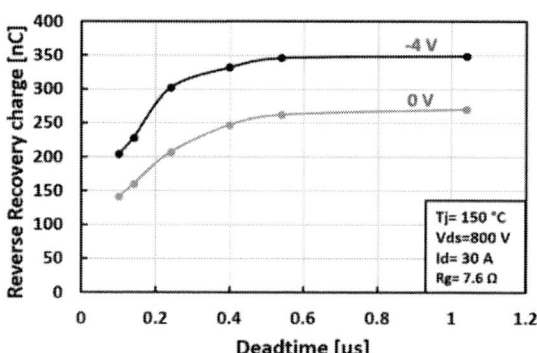

Fig. 7 Reverse recovery charge under different dead time using $V_{gs,off}$= 0 V and $V_{gs,off}$= -4 V

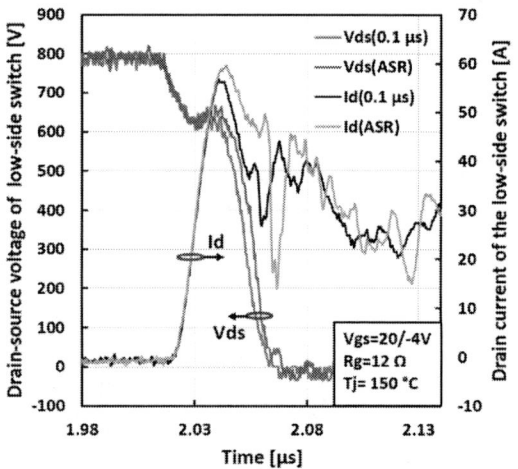

Fig. 5(b) Turn-on transient of the low-side transistor for 100 ns dead time and ASR at 150°C using R_g=12 Ω and $V_{gs,off}$= 0 V

charge obtained when using negative gate turn-off voltage, higher turn-on losses also appear in the complementary transistor. However, using 100 ns dead time can lead to higher reduction with 21 % in the turn-on losses (referenced to the ASR case) in the case of using $V_{gs,off}$= -4 V compared with 19%

reduction (referenced to the ASR case) when using $V_{gs,off}$= 0 V. Figures 9(a) and 9(b) show the drain turn-on current when using 100 ns dead time compared with ASR at $V_{gs,off}$= 0 V and $V_{gs,off}$= -4 V, respectively. It is noted that utilising the short dead time enhances the stability of the switching and reducing the ringing. This enhancement can be attributed to the reduction of the trigger di/dt (see Fig. 4(a) and Fig. 4(b)) that appears in the drain current due to the reverse recovery of the body diode. Figure 10 shows the trigger di/dt as a function of the dead-time. It can be seen that using $V_{gs,off}$= 0 V causes higher di/dt. While using negative voltage can suppress the instability as discussed in [1]. It also can be recognised that shorter dead time reduces the trigger di/dt that can reach more than 50% reduction in case of using 100 ns. Therefore, utilising short dead time can enhance the switching stability of SiC MOSFET and reduce the EMI.

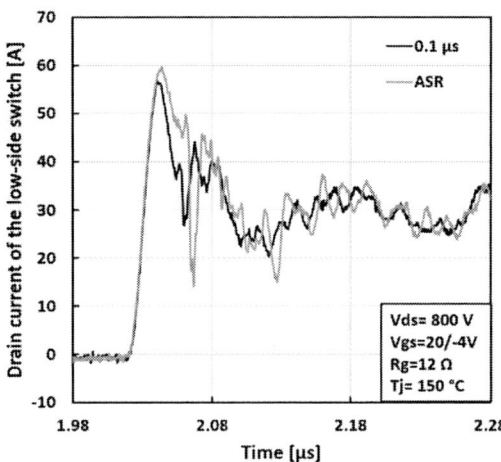

Fig. 9(b) Drain turn-on current using 100 ns dead time compared with ASR at 150°C using R_g=12 Ω and $V_{gs,off}$=-4 V

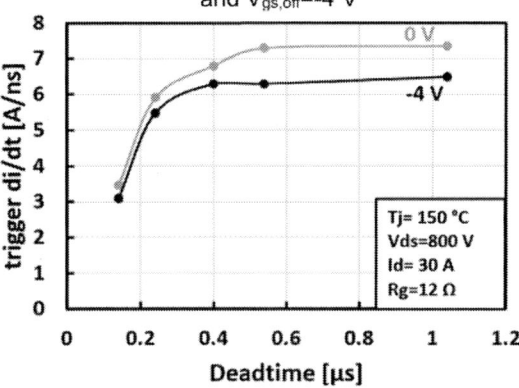

Fig. 10 Trigger di/dt in low-side drain current under different dead time using $V_{gs,off}$= 0 V and $V_{gs,off}$= -4 V

3.2 Impact of the dead time on the switching performance with different junction temperatures

In order to investigate the influence of short dead time with different junction temperatures, 7.6 Ω gate resistance is selected. In addition, the other test conditions are maintained as before (800 V, 30 A) and the junction temperature has changed from 25 °C to 150 °C. Figures 11(a) and 11(b) show the $I_{rr,max}$ and Q_{rr} of the body diode at different dead time using 25°C, 75°C and 150°C combined with utilising zero turn-off gate voltage. While Figures 11(c) and 11(d) show the same results when utilising negative turn-off gate voltage. It can be seen that at 150 °C, the $I_{rr,max}$ and Q_{rr} reduction knee point (threshold dead-time) is 540 ns. While this knee point reduces to 400 ns and 240 ns in the case of 75 °C and 25 °C, respectively. The

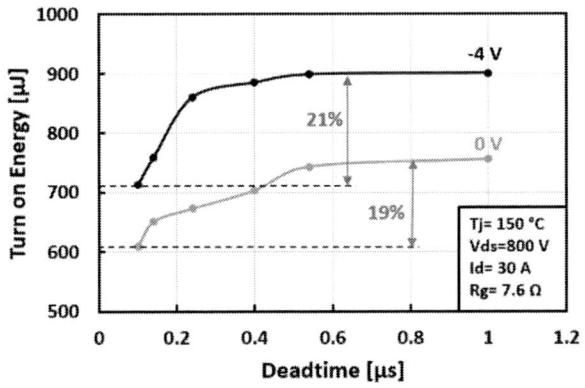

Fig. 8 Turn-on losses of the low-side transistor under different dead time using $V_{gs,off}$= 0 V and $V_{gs,off}$= -4 V

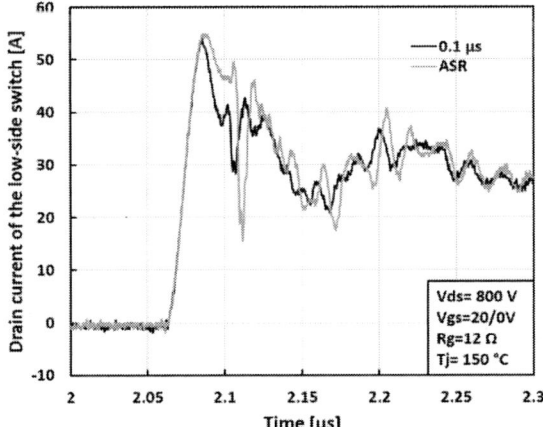

Fig. 9(a) Drain turn-on current using 100 ns dead time compared with ASR at 150°C using R_g=12 Ω and $V_{gs,off}$=0 V

se differences in the reduction knee point for different junction temperatures can be attributed to the increase in the hole forward recovery time at higher temperatures [13][14].

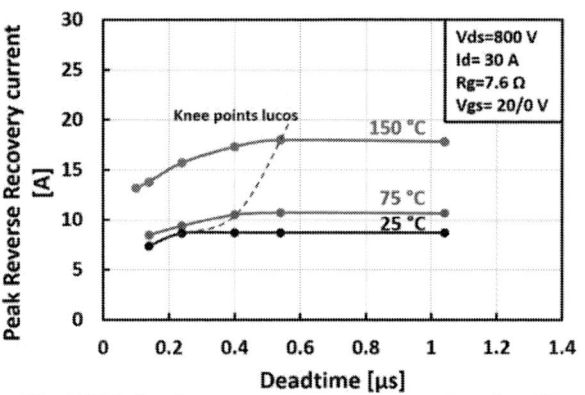

Fig. 11(a) Peak reverse recovery current under different dead time and different junction temperature using $V_{gs,off}= 0$ V

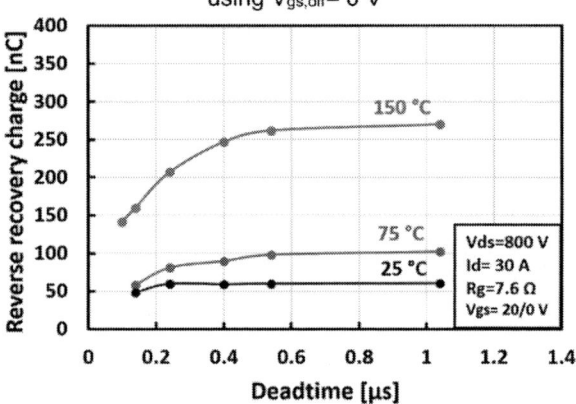

Fig. 11(b) Reverse recovery charge under different dead time and different junction temperature using $V_{gs,off}= 0$ V

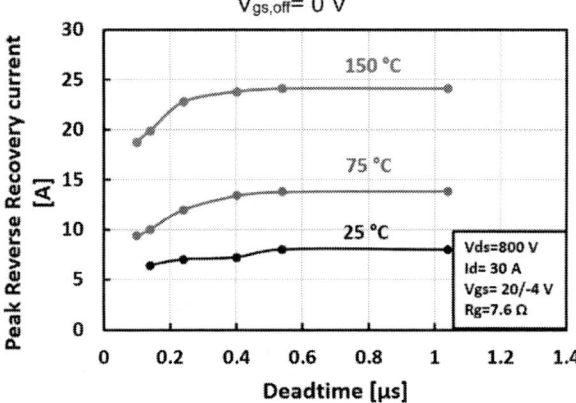

Fig. 11(c) Peak reverse recovery current under different dead time and different junction temperature using $V_{gs,off}= -4$ V

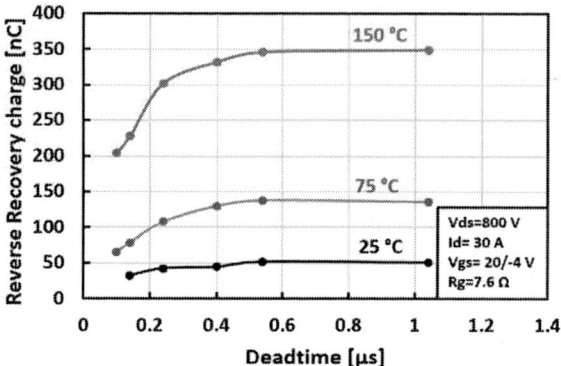

Fig. 11(d) Reverse recovery charge under different dead time and different junction temperature using $V_{gs,off}= -4$ V

Therefore, it is clear that shorter dead time is needed at low temperature to achieve a substantial reduction in the reverse recovery charge. Since 100 ns was creating shoot-through in some cases, 140 ns is used to quantify the reduction in peak reverse recovery current, the reverse recovery charge of the body diode and the turn-on energy of the low side transistor. Figures 12(a) and 12(b) show the reduction in $I_{rr,max}$ and Q_{rr} (referenced to ASR) under different junction temperatures. It is obvious that as the temperature increases, there is a corresponding increase in the reduction of $I_{rr,max}$ and Q_{rr}. The most significant reduction occurs at 150 °C with 4.2 A and 122 nC reduction for the case of using $V_{gs,off}= -4$ V compared with 3.8 A, 109 nC decrease for $V_{gs,off}= 0$ V. Figure 12(c) shows the reduction in the turn-on losses (calculated reference to ASR case) of the low-side switch under different junction temperatures for the case of using $V_{gs,off}= -4$ V and $V_{gs,off}= 0$ V. It can be seen that the reduction in E_{on} follows the same trend and increases at higher temperatures.

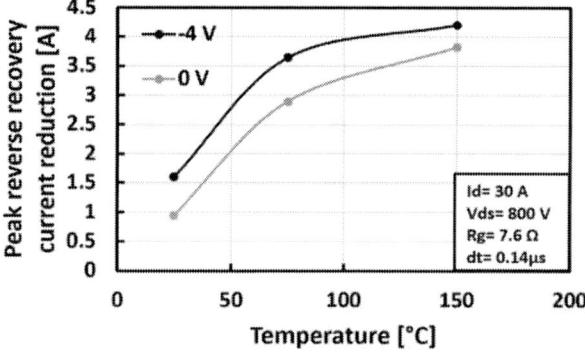

Fig. 12(a) Reduction in the $I_{rr,max}$ under different junction temperature when using 140 ns for $V_{gs,off}= -4$ V and $V_{gs,off}= 0$ V

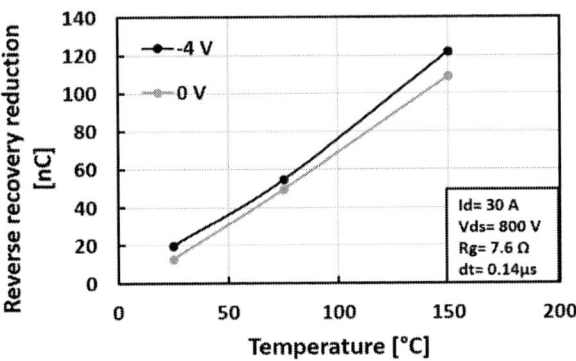

Fig. 12(b) Reduction in the Q_{rr} under different junction temperature when using 140 ns for $V_{gs,off}$= -4 V and $V_{gs,off}$= 0 V

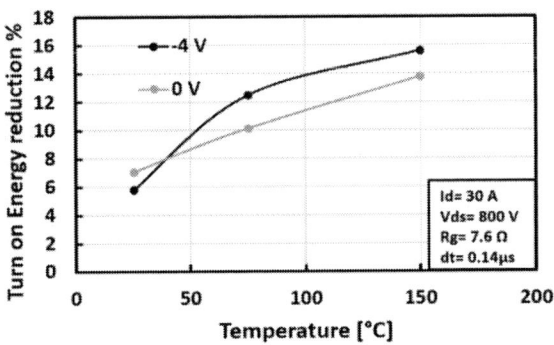

Fig. 12(c) Reduction in the turn-on energy of the low side switch under different junction temperature when using 140 ns for $V_{gs,off}$= -4 V and $V_{gs,off}$= 0 V

This reduction can reach up to 15.63% and 13.8% at 150 °C in the case of using $V_{gs,off}$= -4 V and $V_{gs,off}$= 0 V, respectively. This reduction should not conflict with the reduction calculated in Fig. 8 as the former is calculated at a dead time of 140 ns, while the latter is calculated with a dead time of 100 ns.

3.3 Impact of the dead time on the switching performance with different gate resistance

To investigate the impact of using short dead time at different gate resistances, the junction temperature is maintained at 150 °C as it has the highest impact on the characteristics. Then, the test is repeated at three different gate resistances (2 Ω, 7.6 Ω and 12 Ω). Figures 13(a) and 13(b) show the $I_{rr,max}$ and Q_{rr} under different dead time for R_g=2Ω, 7.6Ω and 12Ω. It is obvious that the reduction knee point is 400 ns for 2 Ω gate resistance. While this knee point increases to 540 ns when using R_g=7.6 Ω a

nd it potentially becomes higher with R_g=12 Ω. However, the knee point at R_g=12 Ω cannot be determined due to the lack of measurement at dead time between 540 ns and 1 µs. This change in the reduction knee points can be attributed to the fact that the hole forward recovery increases with gate resistance increase (slower di/dt). This fact can also be seen in the reverse recovery as shown in Fig. 14. It is clear that reverse recovery time decreases for lower gate resistance.

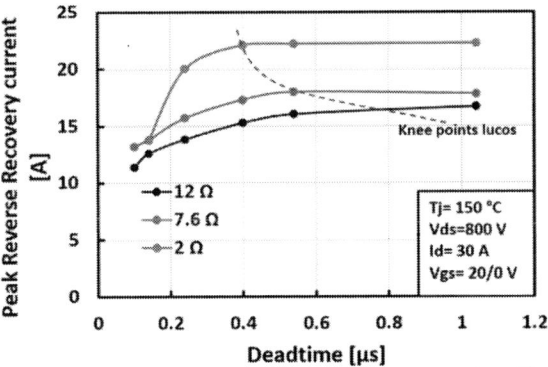

Fig. 13(a) Peak reverse recovery current under different dead time and different gate resistance using $V_{gs,off}$= -4 V

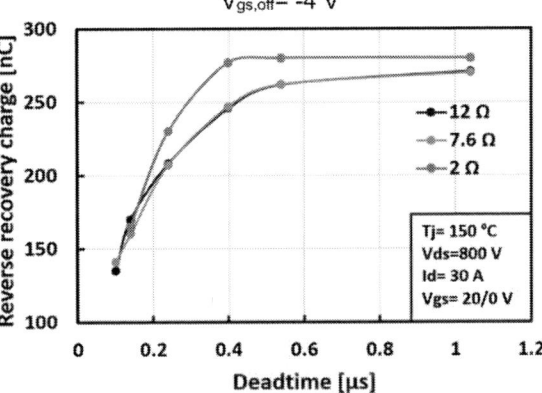

Fig. 13(b) Reverse recovery charge under different dead time and different gate resistance using $V_{gs,off}$= -4 V

Similar to the previous investigation and to quantify the impact of the short dead time at different gate resistances, Fig. 15(a) and Fig. 15(b) show the values of the reduction in the $I_{rr,max}$ and Q_{rr} for different gate resistances, respectively. It is evident that the peak reverse recovery current is significantly decreased at 2 Ω with 6.3 A and 5.8 A reduction when using $V_{gs,off}$= -4 V and $V_{gs,off}$= 0 V, respectively. The reduction in the reverse recovery charge is slightly impacted by decreasing the gate resistance due to the fact that the peak reverse recovery current increases and the reverse recovery time decreases at lower gate resistance (see Fig. 14). Figure 16 illustrates the reduction in the turn-on losses (calculated reference to ASR case) for different gate resistance with using dead time of 140 ns and gate turn-off voltage of zero and -4 V.

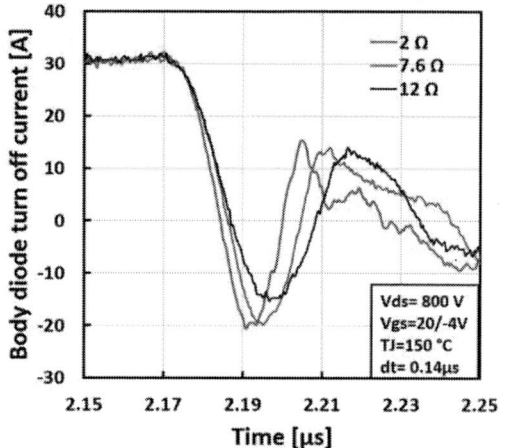

Fig. 14 body diode turn-off current under different gate resistance at 150°C using dt=140 ns and $V_{gs,off}$= -4 V

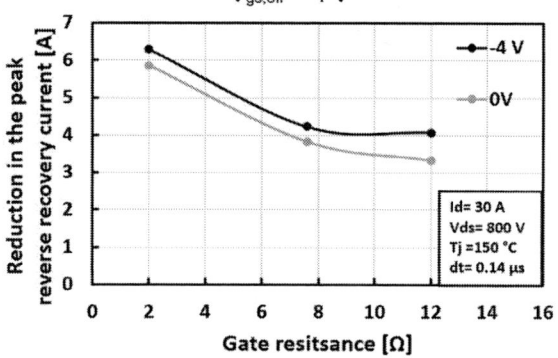

Fig. 15(a) Reduction in the $I_{rr,max}$ under different gate resistance when using 140 ns for $V_{gs,off}$= -4 V and $V_{gs,off}$= 0 V

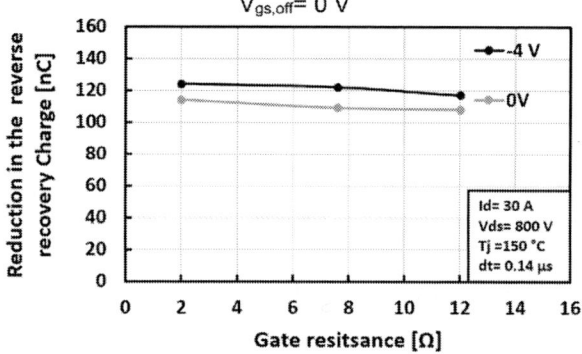

Fig. 15(b) Reduction in the Q_{rr} under different gate resistance when using 140 ns for $V_{gs,off}$= -4 V and $V_{gs,off}$= 0 V

It is clear that the reduction becomes higher at low gate resistance due to the significant reduction in the peak reverse recovery current which contributes to the turn-on losses of the low-side transistor. This reduction can reach 23.3% in the case of using $V_{gs,off}$= -4 V at 2 Ω gate resistance. While it is 15.5% in the case of using zero gate turn-off voltage.

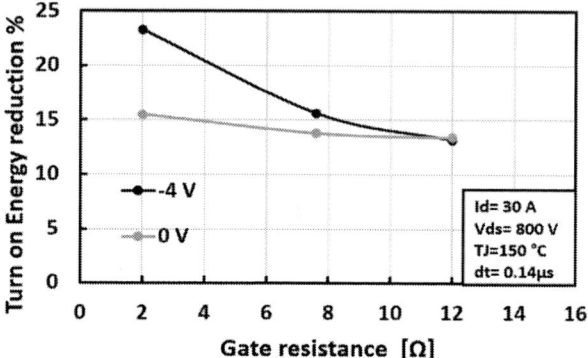

Fig. 16 Reduction in the turn-on energy of the low side switch with different gate resistance when using 140 ns for $V_{gs,off}$= -4 V and $V_{gs,off}$= 0 V

4   Conclusion

The impact of the dead time on the switching performance of 1.2 kV SiC MOSFET is extensively investigated considering the change of the junction temperature, the gate turn-off voltage and the gate resistance. It is proved that decreasing the dead time can reduce the peak reverse recovery current, reverse recovery charge, the turn-on losses and enhance the switching stability. Utilising dead time of 140 ns can achieve a reduction of 23.3% in the turn-on losses and more than 50% in dI/dt. In all cases, the shorter dead time with negative gate turn-off voltage leads to a higher reduction in the switching losses and the reverse recovery. Moreover, the influence of junction temperature variation along with the dead time is also investigated. A higher junction temperature can lead to a higher reduction in the reverse recovery and the switching losses, and it can shift the threshold dead time to a higher value. Furthermore, the impact of changing the gate resistance with the dead time is examined accordingly. Unlike the temperature impact, higher gate resistance can decrease the reduction in the reverse recovery and the switching losses. However, it can shift the threshold dead time to a higher value.

References

[1] M. A. Karout et al., "Characterizing the Stability Limits of 1.2 kV SiC MOSFET Body Diodes Under Hard Switching," *PCIM Europe 2023; International Exhibition and Conference for Power Electronics, Intelligent Motion, Renewable Energy and Energy Management*, Nuremberg, Germany, 2023, pp. 1-7, doi: 10.30420/566091205.

[2] H. Yilmaz, K. Owyang, P. O. Shafer and C. C. Borman, "Optimization of power MOSFET body diode for speed and ruggedness," i n *IEEE Transactions on Industry Applications*, vol. 26, no. 4, pp. 793-797, July-Aug. 1990

[3] S. Jahdi et al., "An Analysis of the Switching Performance and Robustness of Power MOSFETs Body Diodes: A Technology Evaluation," in IEEE Transactions on Power Electronics, vol. 30, no. 5, pp. 2383-2394, May 2015

[4] M. A. Karout, M. Taha, C. A. Fisher, A. Deb, P. Mawby and O. Alatise, "Impact of Diode Characteristics on 1.2 kV SiC MOSFET and Cascode JFET Efficiency: Body Diodes Vs SiC Schottky Barrier Diodes," *2023 IEEE Applied Power Electronics Conference and Exposition (APEC)*, Orlando, FL, USA, 2023, pp. 202-208, doi: 10.1109/APEC43580.2023.10131399.

[5] A. E. Awwad and S. Dieckerhoff, "Operation of planar and trench SiC MOSFETs in a 10kW DC/DC-converter analyzing the impact of the body diode," *2017 IEEE Energy Conversion Congress and Exposition (ECCE)*, 2017, pp. 917-924

[6] P. Sochor, A. Huerner, M. Hell and R. Elpelt, "Understanding the Turn-off Behavior of SiC MOSFET Body Diodes in Fast Switching Applications," *PCIM Europe digital days 2021; International Exhibition and Conference for Power Electronics, Intelligent Motion, Renewable Energy and Energy Management*, Online, 2021, pp. 1-8.

[7] D. Polenov, T. Reiter, R. Baburske, H. Probstle and J. Lutz, "The influence of turn-off dead time on the reverse-recovery behaviour of synchronous rectifiers in automotive DC/DC-converters," *2009 13th European Conference on Power Electronics and Applications*, Barcelona, Spain, 2009, pp. 1-8.

[8] X. Liu, X. Li, C. Herrmann and T. Basler, "The Impact of the Dead time on the Reverse Recovery Behavior of SiC-MOSFET Body Diodes," *2023 35th International Symposium on Power Semiconductor Devices and ICs (ISPSD)*, Hong Kong, 2023, pp. 322-325, doi: 10.1109/ISPSD57135.2023.10147719

[9] A. Pal, N. K. Pilli, C. Klumpner, and M. R. Ahmed, "Improved switching performance of 3.3kv sic mosfets using synchronous rectification in a voltage source inverter," in 2022 IEEE Applied Power Electronics Conference and Exposition (APEC), 2022, pp. 01–06.

[10] Lukas Bergmann, Mark-M. Bakran, "Design and validation of a novel semiconductor area optimised 3300 V SiC half bridge for MMC", IET Power Electronics, 2024.

[11] M. A. Karout, O. Alatise, H. Ayala, C. A. Fisher, P. Mawby and M. Taha, "On the Design Procedure of the Double Pulse Test Rig for WBG devices," *2023 IEEE Conference on Power Electronics and Renewable Energy (CPERE)*, Luxor, Egypt, 2023, pp. 1-8, doi: 10.1109/CPERE56564.2023.10119544.

[12] B. Baliga, Fundamentals of Power Semiconductor Devices. Springer US, 2019. [Online]. Available: https://books.google.co.uk/books?id=UiqrU-WrYZXkC

[13] P. Losee et al., "Soft recovery diodes with snappy behavior," *2015 17th European Conference on Power Electronics and Applications (EPE'15 ECCE-Europe)*, Geneva, Switzerland, 2015, pp. 1-10

[14] S. K. Ghandhi, Semiconductor Power Devices, Physics of Operation and Fabrication Technology, John Wiley & Sons, NY, p. 100, 1977.

PCIM Europe 2024, 11– 13 June 2024, Nuremberg   DOI: 10.30420/566262253

RC-DC Snubber Implementation for Suppression of Diode Voltage Peak and Ringing in a Full SiC Half-Bridge Power Module

Emanuela Alfonzetti[1], Debora Crimi[1], Andrea Cusumano[1]

[1] STMicroelectronics, Italy

Corresponding author: Emanuela Alfonzetti, email: emanuela.alfonzetti@st.com
Speaker: Emanuela Alfonzetti, email: emanuela.alfonzetti@st.com

Abstract

This paper investigates about the benefits of introducing an RC-DC snubber in a full SiC Half-Bridge multi-chip power module from STMicroelectronics. Differently than existing literature on this topic, here the focus is on the reduction of voltage peaks and ringing on the body-diode during turn-on. First, simulations have been performed in Ansys Electronic Desktop tool in order to design the RC network. Then, experimental measurements have been performed and compared with simulations to validate the design.

1 Introduction

To reduce carbon footprint, using more efficient electric technologies has been gaining much attention [1]. Hence, high power density converters are being adopted widely [1],[2]. In this context, one of the main concerns is about the reduction of power losses to achieve better performances. In recent years, the efficiency of power modules has been improved by reducing switching time [3].

SiC devices allow to drastically reduce the switching losses compared to Si IGBTs or Si MOSFETs. Moreover, compared with silicon-based power devices, SiC MOSFETs possess superior electrical and thermal properties, such as lower on-state resistance and higher operating temperature [4]-[6]. At the same time, the very high speed transients produce high voltage overshoots, that can lead to device failure, and high-amplitude voltage ringing which can produce EMI issues at system level.

There are several techniques that can be used to suppress ringing, including reducing parasitic inductance, inserting ferrite beads, use of direct current DC link capacitors, or the use of resistor capacitor (RC) snubbers. The last is a widely used technique that can be expected to provide an effective level of ringing suppression [7]. There are two kinds of RC snubbers that are commonly used: DC snubbers and turn-off snubbers.

DC snubbers can be implemented with an RC series connected in parallel with the half-bridge module. They are typically designed to reduce the equivalent power loop stray inductance (L_{stray})

seen by the half-bridge leg to improve the switching behavior by reducing ringing during turn-off.

A turn-off snubber is an RC series connected in parallel with the device under test (DUT) [8]. Turn-off snubbers are essentially used to limit switching energy losses during turn-off by limiting the voltage rise across the transistors during transients. Anyhow, the turn-off snubber affects the turn-on transient and switching losses in a negative manner, as the current overshoot increases due to the discharge of the snubber capacitor through the snubber resistor during turn-on [8]. Moreover, the implementation of two turn-off snubbers for both the low side (LS) and the high side (HS) switch in a half-bridge power module requires more space compared to a single RC-DC snubber solution. Therefore, in this paper, the effects of an RC-DC snubber on a 5 mΩ, 1.2 kV SiC MOSFET Single-Side-Cooling (SSC) half-bridge power module has been investigated.

The rest of the paper is divided into three sections. Section 2 gives an overview of the literature about the snubber design for half-bridge structures. Section 3 reports a real case study where Section 2 concepts are applied, considering the previously mentioned half-bridge module. A preliminary analysis has been performed in Ansys tool to define the best RC design for the power module under analysis. Then, experimental tests have been done to improve the consistency of the design. The last section summarizes the main conclusions and gives some considerations on future developments.

1844

2 Snubber design theorethical approach

The switching transient of SiC power modules typically show high voltage overshoots both at turn-on and turn-off, due to the combination of high switching speed and parasitic inductance contributions that are linked to the module layout and to the DC-link structure and connection. Although research about the oscillation suppression by RC snubber is relatively complete, the method of solving characteristic equations is cumbersome and hard to correspond the actual physical parameters in the circuit [9].

An half-bridge power module with an implemented RC-DC snubber can be modeled by means of a lumped element circuit as shown in Figure 1. In the proposed schematic, the LS switch is actively controlled to work as device under test (DUT), while the HS switch is kept in off-state and works as free-wheeling diode. L_{bus} is the equivalent parasitic inductance of the DC-link capacitor, while L_M lumps the contribution of the power module itself (bus bars and internal DBC layout).

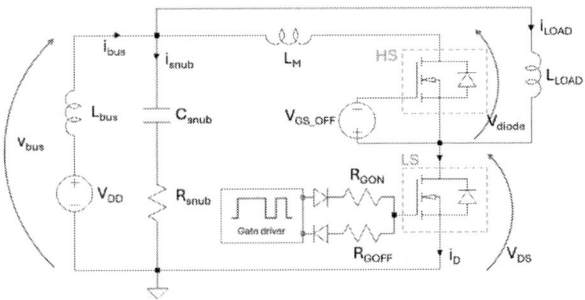

Fig. 1. Electrical scheme for Double Pulse Test setup on a half-bridge power module with implemented RC-DC snubber.

The total stray inductance L_{stray} of the system without the RC-DC snubber is equal to:

$$L_{stray} = L_{bus} + L_M \qquad (1)$$

A properly designed RC-DC snubber reduces the equivalent stray inductance seen by the DUT. For the snubber to work properly, it would be the best if stray inductance of the snubber branch is minimized as much as possible. However it is not always realistic since it might make the heat dissipation worse [10]. Instead, placing the snubber as close as possible to the dies can be a good solution to minimize the inductance of the circuit.

A simple and quick way to design an RC-DC snubber is reported in [7]. Considering a full SiC half-bridge power module, the ringing frequency of the system at turn-off is related to the output capacitance of the active device C_{oss} and the overall power loop stray inductance L_{stray}:

$$f_r = \frac{1}{2\pi\sqrt{L_{stray}C_{oss}}} \qquad (2)$$

The desired damping coefficient for the configuration with the snubber can be expressed as:

$$\zeta = \frac{1}{2R_{snub}}\sqrt{\frac{L_{stray}}{C_{oss}}} \qquad (3)$$

From which the snubber resistor value can be calculated:

$$R_{snub} = \frac{1}{2\zeta}\sqrt{\frac{L_{stray}}{C_{oss}}} \qquad (4)$$

By imposing the damping coefficient in order to obtain the critical damping condtion (i.e. $\zeta = 1$, that is the condition in which the system goes more rapidly to stady state without oscillations), the R_{snub} value is obtained [7].
Then, the value of C_{snub} can be calculated considering that the RC snubber should work as an alternative path for the inductive current during turn-off [7]:

$$C_{snub} = \frac{1}{2\pi f_r R_{snub}} \qquad (5)$$

Other methods are reported in literature: in [8] the authors introduce an analytical technique to design RC snubber components by using the Root Locus Method, analyzing the state-equations of the high-frequency equivalent circuit of a double pulse circuit. The obtained small-signal model describes the drain-source voltage (v_{DS}) oscillations, therefore the authors are mainly focused on reducing the turn-off ringing. In [9], the oscillation time-domain model of RC snubber in the turn-off and turn-on stages is defined, but no experimental evidence of the action of the RC snubber on the voltage peaks reduction at turn-on is given.

As a matter of fact, the effective benefits of introducing an RC snubber are not quite easy to predict at turn-on, since in this stage the ringing has to be observed as oscillations on the diode voltage (v_{diode}). In turn, the v_{diode} behavior is strongly related to the action of the di_D/dt and the di_R/dt on the bus voltage v_{bus}. Therefore, in a full-SiC half-bridge structure, considering the SiC body diode variability during the reverse recovery phase, it is difficult to exactly predict the turn-on v_{diode} overvoltages. In order to better understand the impact of the RC-

DC snubber on the v_{diode}, an example of turn-on commutation is shown in Fig. 2, in which three different stages are highlighted.

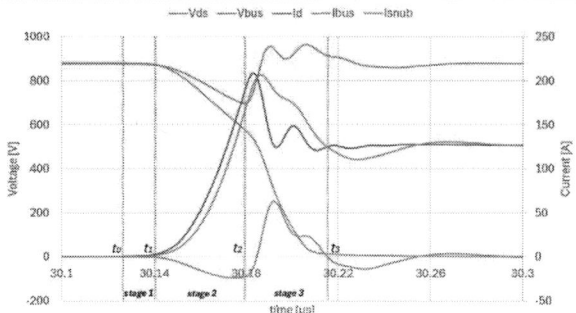

Fig. 2. Simulated switching waveforms during turn-on transient with RC snubber.

Stage 1 [$t_0;t_1$]: the DUT turns on and the rising drain current i_D generates a voltage drop on both v_{bus} and v_{DS} that are proportional to the parasitic inductance on the respective branches:

$$v_{bus}(t) = V_{DD} - L_{bus}\frac{di_D(t)}{dt} \qquad (6)$$

$$v_{DS}(t) = V_{DD} - (L_{bus} + L_M)\frac{di_D(t)}{dt} \qquad (7)$$

In this stage, $i_D = i_{bus}$. The drop on v_{bus} begins to generate a negative current on the RC snubber branch, i_{snub}.

Stage 2 [$t_1;t_2$]: the slope of i_D on the active device is imposed by the driving circuit, and the inductive load can be considered as a constant current source during transients. Therefore, considering that:

$$i_{bus}(t) = i_D(t) + i_{snub}(t) \qquad (8)$$

i_{bus} results lower than i_D.
In this stage, the voltage drop on v_{bus} and v_{DS} are:

$$v_{bus}(t) = V_{DD} - L_{bus}\frac{di_{bus}(t)}{dt} \qquad (9)$$

$$v_{DS}(t) = V_{DD} - L_{bus}\frac{di_{bus}(t)}{dt} - L_M\frac{di_D(t)}{dt} \qquad (10)$$

This explains why the voltage drop on v_{DS} is higher than that on v_{bus}. The voltage variation on v_{bus} makes the current i_{snub} increase in amplitude.

Stage 3 [$t_2;t_3$]: during the reverse recovery phase, the drain current slew rate begins to slow down, and when i_D reaches I_{peak}, v_{bus} begins to rise.

This, in turn, causes also the rise of i_{snub} which returns positive. Therefore, recalling Eq. 8, i_{bus} becomes greater than i_D.

In this stage the impact of i_{snub} on i_{bus} is crucial to explain the benefit in terms of overvoltage reduction. We can say that after the I_{peak}, v_{bus}, being related to the di_{bus}/dt (which is much slower than the di_D/dt on the active device just after the the I_{peak}), presents a smooth profile with very low oscillations and low voltage peaks.

3 RC-DC snubber case study

3.1 Snubber design and electrical simulations

The test vehicle for the case-study treated in this paper is a 5 mΩ, 1.2 kV SiC MOSFET Single-Side-Cooling (SSC) half-bridge module from STMicroelectronics.

The system level simulation has been implemented considering the parasitic contribution matrix of the multi-chip power module under test, the behavioral SPICE model of the SiC dies.

The gate driving circuit, the DC-link capacitor and the inductive load models have been considered too. The equivalent electrical scheme implemented for simulation analysis is reported in Fig.1.

Double Pulse Tests (DPT) have been performed to evaluate the effectiveness of the RC-DC snubber. The simulations have been executed considering as operating conditions V_{DD} = 875 V, I_{LOAD} = 125 A, V_{GS_OFF} = -5 V, V_{GS_ACTIVE} = -5÷18 V. The gate resistance (R_G) setting has been established only by considering the limitations at system level, i.e. the maximum allowed dv/dt that can be reached on the switch without causing issues at system level (EMI, insulation), that in this case is around 30 V/ns. This is speficied since, in some cases, voltage peaks over the nominal breakdown voltage of the device have been reached.

As it is typically true for SiC devices, the worst case in terms of dv/dt and voltage spikes happens to be at high current, high temperature for the diode voltage and at high current, room temperature for the active device voltage.

First, simulations without the RC-DC snubber have been performed (Fig. 3).

The peak of the v_{DS} is sightly lower than the nominal breakdown voltage ($V_{(BR)DSS}$), but it is followed by a high amplitude ringing due to parasitic components inside the power module (Fig. 3a).

The voltage and current ringing is also remarkable during the turning on of the switch. In this commutation, the voltage peak reached by the diode exceeds the $V_{(BR)DSS}$ by nearly 100 V (Fig. 3b).

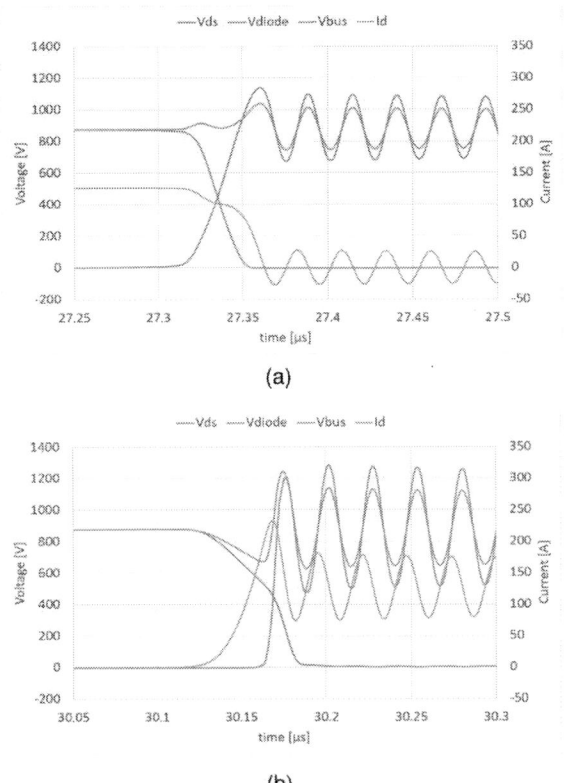

(a)

(b)

Fig. 3. Simulated switching waveforms during turn-off (a) and turn-on (b) without RC snubber.

All these criticisms can be mitigated by the RC-DC snubber insertion. The capacitor and resistance values of the snubber circuit have been chosen to guarantee a tradeoff between the maximum voltage peak and low ringing, without affecting the switching performances. The preliminary design of the snubber has been done following the approach discussed in Section 2. From the device datasheet it is known that C_{oss} = 500 pF (at V_{DS} = 800 V) and L_M = 14 nH, while the DC-link presents a L_{bus} = 16 nH. This implies that L_{stray} = 30 nH. Therefore, recalling Equations (2)-(5):

$$f_r = \frac{1}{2\pi\sqrt{L_{stray}C_{oss}}} \approx 41\ MHz$$

$$R_{snub} = \frac{1}{2\zeta}\sqrt{\frac{L_{stray}}{C_{oss}}} \approx 3.9\ \Omega$$

$$C_{snub} = \frac{1}{2\pi f_r R_{snub}} \approx 1\ nF$$

These snubber parameter values come from an approximated approach, since no parasitic contributions are considered. Therefore, the obtained values of R_{snub} and C_{snub} should be considered as a starting point for a fine-tuning that can be done in simulation. For this reason, some other investigations have been carried out to improve the snubber effectiveness. The snubber is modeled considering the real SPICE model of the capacitor provided by the manufacturer; it includes the ESR, ESL and the capacitance variation related to frequency, temperature and bias voltage.

In the examined case, the best snubber solution is constituted by an equivalent capacitance C_{snub} = 1.5 nF and an equivalent series resistance R_{snub} = 3.3 Ω.

The switching waveforms in Fig. 4 show a noticeable improvement of the power module behavior. Fixing the same switching speed, the turn off transient experiences a reduced peak voltage, and the ringing in current and voltage waveforms almost disappears (Fig. 4a). Moreover, at the turn on transient, the long duration ringing is suppressed as the oscillations are quickly damped (Fig. 4b). Furthermore, the diode voltage peak reduction can be highlighted; it is now in a safer operating condition, since the peak stays around 1 kV.

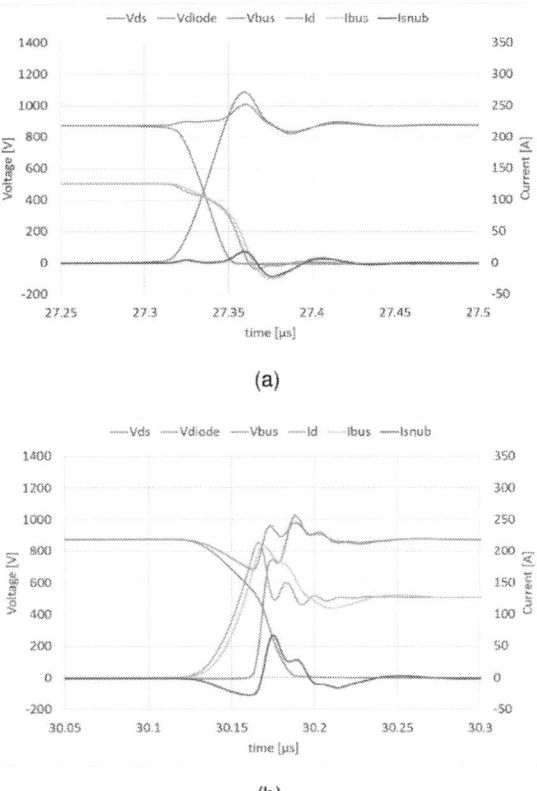

(a)

(b)

Fig. 4. Simulated switching waveforms during turn-off (a) and turn-on (b) with RC snubber.

1847

3.2 Experimental results

Experimental DPTs have been performed to confirm simulation results.

The operating conditions for the tests are the same of the simulation ones; the target switching speed of 30 V/ns has been reached by setting R_{GON} = 2.2 Ω, R_{GOFF} = 1.8 Ω .

As for simulations, a preliminary R_G setting has been done without the insertion of the RC-DC snubber, as shown in Fig. 5.

It can be seen that, in this operating point, without the snubber the voltage on the passive device at turn-on reaches peaks close to 1.3 kV, i.e. 100 V more than the nominal $V_{(BR)DSS}$. As it is well known, these high-amplitude oscillations are triggered by the reverse recovery of the diode: the higher the di_R/dt, the stronger the ringing. The mismatch with simulations is linked to the variability of the reverse recovery behavior.

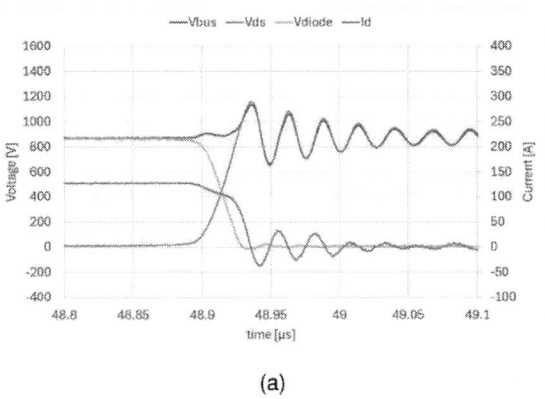

(a)

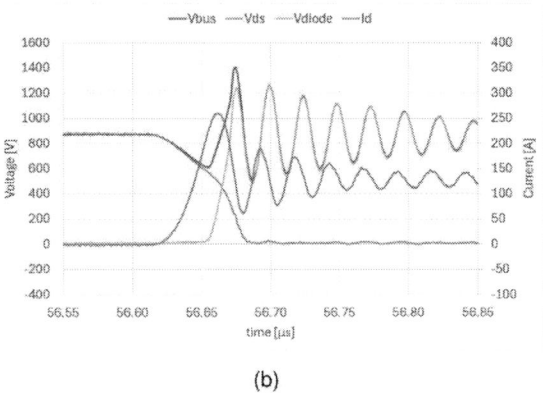

(b)

Fig. 5. Experimental switching waveforms during turn-off (a) and turn-on (b) without RC snubber.

Anyhow, the insertion of the designed RC-DC snubber (R_{snub} = 3.3 Ω, C_{snub} = 1.5 nF), as it can be seen from Fig. 6 and predicted by simulations, drastically reduces both the active switch and the diode ringing and, in particular, it makes the first

diode voltage peak disappear. It is also evident that the RC-DC snubber does not affect the switching times. Both the cases with and without the RC-DC snubber have been obtained by maintaining the same R_G setting.

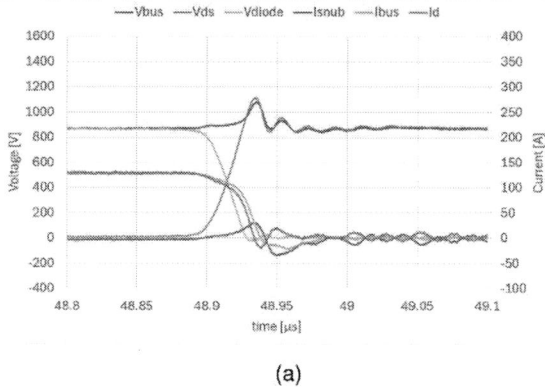

(a)

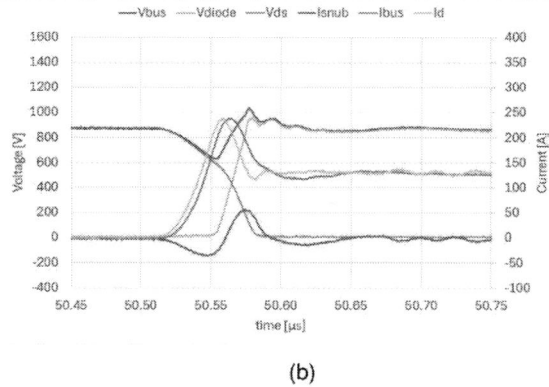

(b)

Fig. 6. Experimental switching waveforms during turn-off (a) and turn-on (b) with RC snubber.

3.3 Switching losses

The insertion of the RC-DC snubber does not affect strongly the switching losses. Experimental DPTs have been performed to measure the turn-on and turn-off switching losses with and without the snubber, maintaining the same R_G setting. For the configuration with the snubber, also the switching losses on the snubber resistor have been measured. The results are reported in Fig. 7.

As a matter of fact, the snubber allows to reduce both the R_{GON} and R_{GOFF}. A different R_G setting has been done in order to reach the same voltage peaks on the two configurations:

- R_{GON} = 2.7 Ω without snubber, R_{GON} = 1.5 Ω with snubber; v_{diode} reaches peak values around 1130 V in both cases (at I_{LOAD} = 125 A, T_j = 175°C).

- R_{GOFF} = 1.8 Ω without snubber, R_{GOFF} = 0 Ω with snubber; v_{DS} reaches peak values around 1150 V in both cases (at I_{LOAD} = 125 A, T_j = RT).

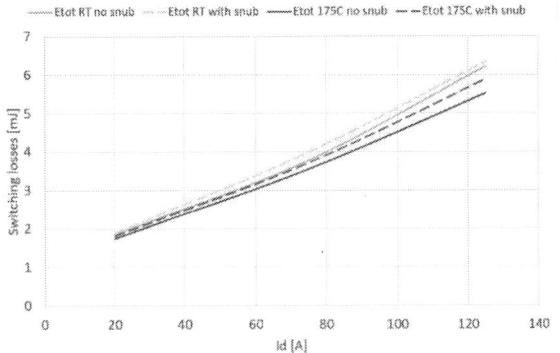

Fig. 7. Experimental total switching losses measured with and without the snubber mainteing the same R_{GON} and R_{GOFF}, at two different temperatures: room temperature (blue) and 175°C (red).

This setting highlights how the snubber implementation can boost the performances of the system. Figure 8 shows a reduction in energy losses around 35% at T_j = RT and around 30% at T_j = 175°C.

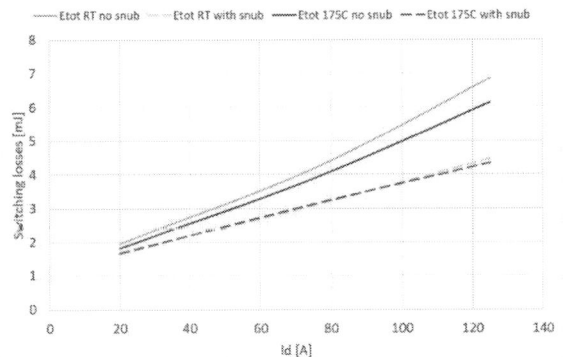

Fig. 8. Experimental total switching losses measured with and without the snubber, at two different temperatures: room temperature (blue) and 175°C (red).

The reduction of the switching losses is lower at high temperature due to the increase in the turn-on energy losses on the snubber resistor, linked to the increase of i_{snub}. In fact, as previously said, the i_{snub} amplitude is linked to the drop of v_{bus}, that is a consequence of i_{bus} and di_{bus}/dt variations. Figure 9 points out a comparison of the v_{bus}, i_{bus} and i_{snub}

profiles at room temperature and high temperature (175°C).

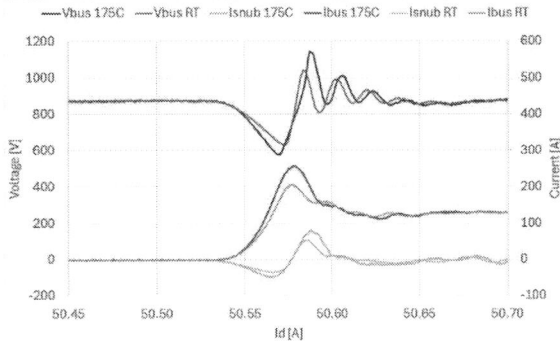

Fig. 9 V_{bus}, i_{bus} and i_{snub} variation: room temperature vs high temperature.

4 Conclusions

In this paper the benefits of introducing an RC-DC snubber on the switching behavior of a full-SiC half-bridge power module from STMicroelectronics have been highlighted by means of simulations and experiments. The impact of the RC-DC snubber on both current and voltage ringing has been analyzed. The reduction of the active switch voltage peak has been proven and a particular focus has been given on the diode peak voltage reduction, strictly related to the turn-on switching losses and the device reliability. It has also been shown that a properly designed RC-DC snubber does not affect the switching times and the switching losses. Actually, the introduction of the snubber allows to rise up the switching speed of both turn-on and turn-off transient without affecting the device reliability in terms of voltage peaks. As a result, a consistent decrease of the switching losses can be achieved.

The experimentally validated SPICE model could be used for future investigations about the snubber potentiality in different working conditions.

5 References

[1] Y. Liu *et al.*, "Switching Devices Comparison and RC Snubber for Ringing Suppression in one Single Leg T-type Converter," *2018 Asian Conference on Energy, Power and Transportation Electrification (ACEPT)*, Singapore, 2018, pp. 1-5.

[2] A. Nawani, C. F. Tong, S. Yin, A. Sakanowa, Y. Liu, et al., "Design and Demonstration of

High Power Inverter for Aircraft Applications", *IEEE Transactions on Industry Applications*, vol. 53, pp. 1168-1176, 2017.

[3] N. Boettcher, H. Afewerki, I. Kallfass and C. Lautensack, "A novel approach for optimal design of monolithic integrated RC snubbers," *2017 International Symposium on Electromagnetic Compatibility - EMC EUROPE*, Angers, France, 2017, pp. 1-6.

[4] L. Zhang, X. Yuan, X.Wu, C. Shi, J. Zhang, and Y. Zhang, "Performance evaluation of high-power SiC MOSFET modules in comparison to Si IGBT modules", *IEEE Transactions on Power Electronics*, vol. 34, no. 2, pp. 1181-1196, Feb. 2019.

[5] D. Jiang, R. Burgos, F. Wang, and D. Boroyevich, "Temperature-Dependent Characteristics of SiC Devices: Performance Evaluation and Loss Calculation*", IEEE Transactions on Power Electronics*, vol. 27, no. 2, pp. 1013-1024, Feb. 2012.

[6] L. Che, Y. Dong and G. Lei, "Design Optimization of RC Snubber Circuit for A SiC Power Module," *2022 23rd International Conference on Electronic Packaging Technology (ICEPT)*, Dalian, China, 2022, pp. 1-6.

[7] B. N. Torsæter, S. Tiwari, R. Lund and O. -M. Midtgård, "Experimental evaluation of switching characteristics, switching losses and snubber design for a full SiC half-bridge power module," *2016 IEEE 7th International Symposium on Power Electronics for Distributed Generation Systems (PEDG)*, Vancouver, BC, Canada, 2016, pp. 1-8.

[8] K. Yatsugi, K. Nomura and Y. Hattori, "Analytical Technique for Designing an RC Snubber Circuit for Ringing Suppression in a Phase-Leg Configuration," in *IEEE Transactions on Power Electronics*, vol. 33, no. 6, pp. 4736-4745, June 2018.

[9] Y. Ding, S. Mao, H. Liu, Q. Tan, S. Yang and P. Liu, "Modeling of SiC MOSFETs Switching Oscillation for Dynamic Optimization with RC Snubber in the Half-bridge Circuit," *2023 IEEE Energy Conversion Congress and Exposition (ECCE)*, Nashville, TN, USA, 2023, pp. 5970-5975.

[10] "Snubber circuit design methods" (Rohm Appl. Note No. 62AN037E, Rev.002, April 2020). Available: https://fscdn.rohm.com/en/products/databook/applinote/discrete/sic/mosfet/sic-mos_snubber_circuit_design_an-e.pdf

PCIM Europe 2024, 11– 13 June 2024, Nuremberg   DOI: 10.30420/566262254

Sub-Five Second Wide-Bandgap Power Device Calorimetric Measurements Utilizing Optical Sensors and Peltier Elements

Ruben Schnitzler ©[1], Tobias Fink ©[1], Kevin Munoz Baron ©[1], Dominik Koch ©[1], Petar Rasic[1], Ingmar Kallfass ©[1]

[1] Institute of Robust Power Semiconductor Systems, University of Stuttgart

Corresponding author: Ruben Schnitzler, ruben.schnitzler@ilh.uni-stuttgart.de
Speaker: Ruben Schnitzler, ruben.schnitzler@ilh.uni-stuttgart.de

Abstract

For an ever-increasing demand of highly efficient highly power-dense system soft-switching converters are gaining more and more attraction. Especially for wide-bandgap power devices, electrical characterization is not sufficiently accurate, while calorimetric measurement methods suffer from high measurement times. To improve the hardware-in-the-loop development, this paper proposes a new measurement method utilizing an optical sensor and Peltier elements to reduce measurement and cooldown times below 5 s each hence enabling time-efficient parameterized characterization of soft-switching losses without the necessity for dedicated calorimetric measurement setups.

1   Introduction

Power converters for electric vehicles and portable chargers must balance high efficiency and compactness. To achieve high efficiency while increasing power density soft-switching topologies are crucial for passively cooled systems, as they limit semiconductor losses at high switching frequencies. Multiobjective optimization, digital twins, and design automation are used to optimize these systems, relying on accurate component models for efficiency and thermal design [1]. While measuring hard-switching losses by electrical methods like the double pulse test is sufficiently accurate, measuring wide bandgap devices soft-switching losses electrically can suffer from poor accuracy due to deskewing errors [2]. Hence soft-switching losses are preferably determined through calorimetric methods, which measure the thermal dissipation and hence do not require deskewing. While the accuracy is within 10 % if all parasitic heat sources such as gate drivers and parasitic loss components such as third quadrant losses and conduction losses are considered the measurement times (t_{meas}) and cooldown times ($t_{cooldown}$) impose a big hurdle for practical implementations [3]. Furthermore, the requirement for dedicated circuitry hinders the implementation of calorimetric mea-

surement. Two general calorimetric measurement methods exist. These are static and transient measurement methods [3]–[12]. For static measurement methods the thermal resistance (R_{th}) is used to determine the occurring losses. This can take between 20 minutes for dedicated water-cooled measurement systems and 2 hours for traditional calorimeters [4], [5]. On the other hand, the transient measurement methods either utilize a large thermal capacitance (C_{th}) resulting in a linear slope of the temperature loss for a power step function or use the thermal impedance (Z_{th}) to calculate the losses[3], [6]–[12]. For large thermal capacitances the cooldown time can increase the total measurement duration significantly and measurement times of 2-10 minutes and $t_{cooldown}$ of 10-60 minutes are reached. This cooldown time can be reduced if a thermo-electric cooling (TEC) is utilized during cooldown phases [9]. In recent publications an automated calorimetric method, with directly measured thermal impedance calibration, t_{meas} of 10 s were reached [3] with $t_{cooldown}$ of 30 s. This paper presents a new approach combining an optical fiber sensor temperature measurement in [12] enabling online measurement and hence thermal impedance curve fitting without modification of the PCB layout. Additionally, a Peltier element-assisted cooling to reduce $t_{cooldown}$ to below 5 seconds each. For this, the thermal accuracy is investigated extensively and different cooldown methods are compared. The paper is structured as follows. In section two, the measurement set-up and thermal layout of the experience are explained in detail. Subsequently, the accuracy and sensitivity of the measurement as well as the cooldown progress is investigated. In section three a comparison of different 1200 V SiC-MOSFETs is done for soft-switching losses in dependence of the current (I_D) and negative gate-source voltage (V_{GS-}). Finally, a conclusion of the work is given.

2 Method

2.1 Thermal layouts

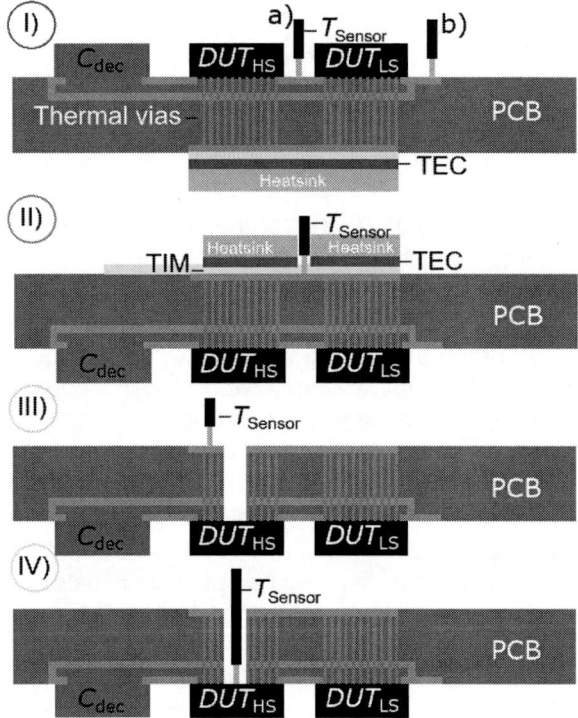

Fig. 1: Cross-section of the different thermal and temperature measurement set-ups for calorimetric investigation.

In this work, an automation optimization of an automated modular calorimetric test bench is conducted. The calorimetric test bench consists of a modular half-bridge with a capacitive divider and a variable inductive load with values between 2-245 µH and can vary drain-source voltage (V_{DS}), I_D, V_{GS}, switching frequency (f_{sw}), and dead-time (t_{dt}) fully automated, which is explained in detail in [12], [13]. For this work, a special focus was set on the thermal layout and thermal measurement sensi-

tivity. Different measurement configurations as well as cooldown strategies were conducted. For all different temperature measurement positions and configurations the temperature is measured with an optical fiber sensor in the vicinity of the device under test (DUT) in such a way, that the thermal responsitivity is maximized. The different DUT board configurations are depicted in Fig. 1.

(I) Close to the case temperature, but not within the main thermal heat-path similar to [12] depicted in Fig. 1 I_a and I_b.

(II) A new PCB was designed and the measurements were conducted within the coupling between both DUTs in the main heat path on the bottom side of the PCB (cf. Fig. 1 II). Here, the cooling is done utilizing thermo-electric cooling (TEC) with a Peltier Element.

(III) In III the temperature is measured directly at the bottom of the PCB on a thermal pad.

(IV) On the same PCB the temperature sensor is placed on the case of the DUT through a hole in the PCB specifically designed for those measurements(cf. Fig. 1 IV).

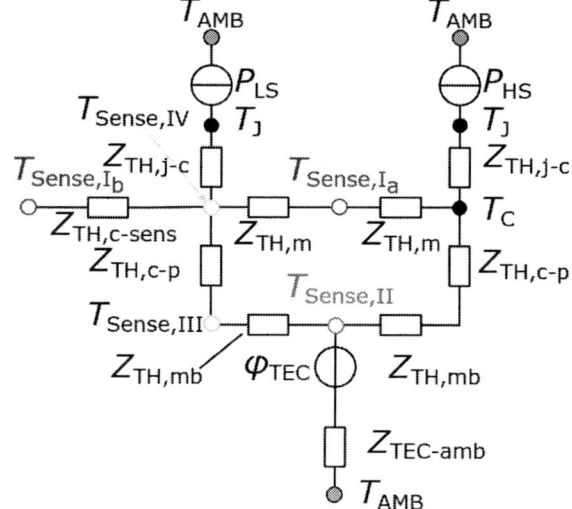

Fig. 2: Different thermal and temperature measurement set-ups for calorimetric investigation.

The thermal equivalent circuit for the different configurations is given in Fig. 2. Within this figure the direct thermal heat paths from the transistors to ambient temperature T_{amb} are neglected. The different sensor positions are shown. It should be mentioned

PCIM Europe 2024, 11– 13 June 2024, Nuremberg   DOI: 10.30420/566262254

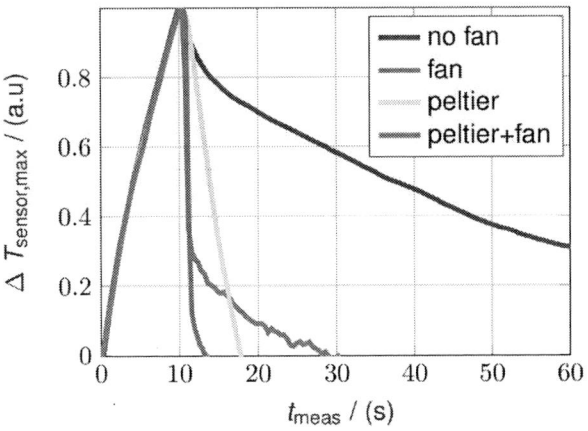

Fig. 3: Comparison of the cooldown between different cooling methods for different cooling strategies.

that the thermal impedance from case to thermal sensor ($Z_{th,c\text{-}sens}$) in I_a and I_b is not within the main thermal heat path and therefore much larger than that from case to PCB bottom ($Z_{th,c\text{-}p}$). It can be concluded that the sensitivity of IV is the highest since it measures directly in the main heat path thereby avoiding additional thermal capacitances.

2.2 Cooldown comparison

To accelerate the measurement procedure of calorimetric measurements two aspects have to be investigated. On the one hand the measurement times and on the other hand the cooldown times. In this section, different cooldown approaches are compared. Consisting of:

1. Air-cooled cooldown

2. Forced convection using a fan

3. Using a Peltier element

4. A combination of Peltier element and cooling fan

As can be seen from Fig. 3 for the same power pulse initially the forced thermal convection of the fan improves the cooldown the most and significantly accelerates the cooldown compared to no forced air cooling. For the last percentages, the Peltier element thermal conduction dominates the cooldown process. If the Peltier element and fan are used simultaneously the temperature follows initially that of the fan and for the last few percentages it follows the Peltier element slope. Reducing $t_{cooldown}$ from more than 20 s for a ventilator-cooled version to 7.5 s for a combination of Peltier element

and ventilator-assisted cooldown. While this approach has the advantage of accelerated cooling it comes with the challenge that a cooldown below room temperature is possible. To avoid a cooldown below room temperature from the Peltier element a detection algorithm is implemented to stop the cooling when the room temperature is almost reached as depicted in Fig. 4. The cooldown process starts with the end of the calorimetric measurement. After this, the Peltier element and fan are turned on while the oscilloscope saves the data. The program defines the room temperature as a threshold to interrupt the cooling progress and subsequently turns off the Peltier element upon reaching the room temperature threshold. Finally, it waits for a variable time to ensure the cooled temperature is not below room temperature. This whole automated cooldown process is dependent on t_{meas} and P_{avg}, but for small and medium power $t_{cooldown}$ is far below t_{meas} typical around 5 s.

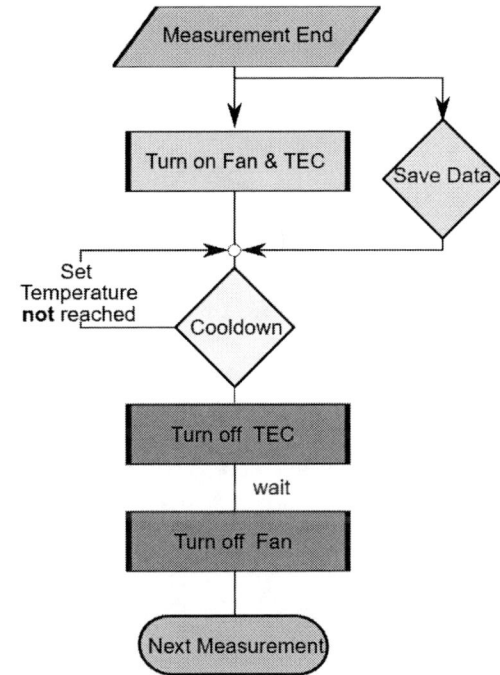

Fig. 4: Comparison of the cooldown between different cooling methods for different cooling strategies.

2.3 Measurement-time reduction

2.3.1 Measurement accuracy

In this section, the time-dependent thermal accuracy is investigated. For this, calibration measurements were conducted at various power levels for several different t_{meas} by utilizing an automated cal-

1853

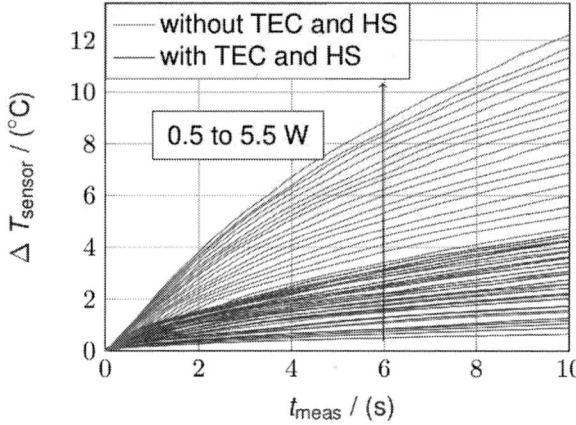

Fig. 5: Comparison of thermal sensitivity of Sensor Position I with (blue) and without heat sink (red).

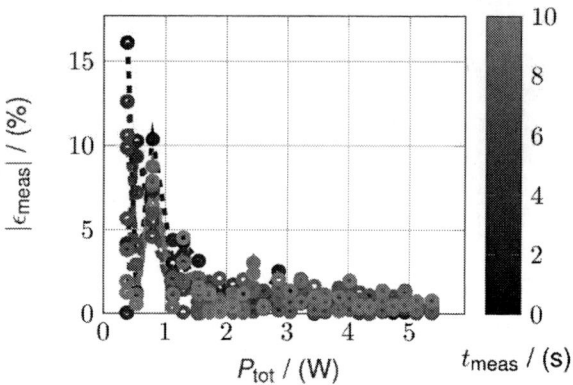

Fig. 6: Thermal accuracy of the temperature model in dependence of t_{meas}.

ibration with a constant power source. The results of these calibrations are shown in Fig. 5 for sensor positions I_a with and without heat sink marked in Fig. 1. It is clear to see that the improved cooldown capabilities of the Peltier elements come with the penalty of having a lower thermal sensitivity due to the additional thermal impedance $Z_{th,mb}$.

Out of these calibrations, a second-order Foster Model is built as described in [12]. The average thermal model is extracted out of the calibration curves and is then compared to the original measurement power (P_{meas}). The normalized difference in extracted loss power (P_{ext}) and P_{meas} is given by equation 1 in which ϵ_{meas} is the thermal measurement error.

$$|\epsilon_{meas}| = \frac{|P_{ext} - P_{meas}|}{P_{meas}} \tag{1}$$

For several different t_{meas}, this thermal accuracy is plotted in Fig. 6. As can be seen for high powers the measurement error of the thermal model is independent of the measurement time below 5 %. For very small measurement times and low loss powers the measurement error increases up to 15 %. This effect is especially pronounced for small measurement times. This can be attributed to a second-order Foster model being unable to fully describe the thermal behavior for very low P_{meas} due to smaller thermal time constants being relevant at those operating points. While increasing the order of the Foster model is possible to alleviate this problem, overfitting has to be avoided to maintain high accuracy at higher power. Hence an increase in measurement sensitivity correlated to lower thermal impedance measurement is required.

2.3.2 Sensitivity analysis

For further reduction of measurement time and an increase in accuracy, the sensitivity of the different sensor positions is investigated. This is depicted in Fig. 7 for sensor positions II-IV. It can be seen that for the same power sensor position IV has a higher sensitivity than III due to the closer measurement without additional thermal capacitance. Position II has the lowest sensitivity for this board layout due to the increased thermal impedance of the heat sink and Peltier element, Compared to sensor position I in Fig. 5 a much lower susceptibility to the degradation of thermal sensitivity is observed if an additional heat sink is placed due to the sensing in the main heat-path.

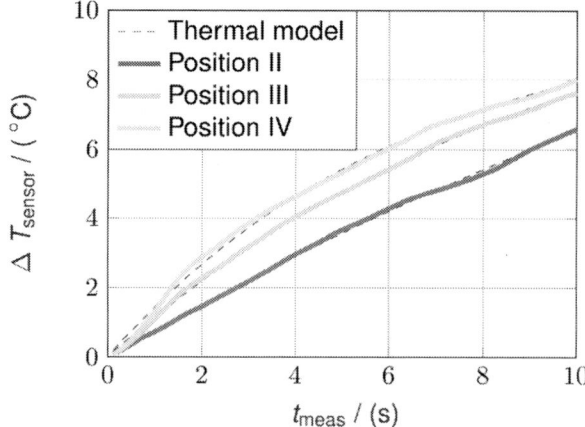

Fig. 7: Thermal sensitivity depending on the of the temperature model in dependence of t_{meas} for P_{ext}=5 W

Although sensor position IV shows the highest sensitivity, an optimum between sensor sensitivity and maximum possible power capability of the set-up has to be found since some sensor positions III and IV can impede the ability for heat sink layout. Additionally, for the measurements on the drain pad at sensor position IV, it has to be mentioned that the repeatability is reduced since the positioning of the optical sensor is quite difficult. Since it has to be ensured that on the one hand, the measurement contact is sufficient and on the other hand not too much pressure is put on the optical sensor to avoid destruction.

3 Calorimetric Measurements

In this section, a comparison between three different SiC-MOSFETs concerning their soft-switching losses is conducted. A list of the used DUTs and their properties are given in table 1.

Tab. 1: Overview of the investigated SiC DUTs

DUT	V_{bd} [V]	R_{dson} [mΩ]	$R_{g,in}$ [Ω]	$\frac{Q_{gd}}{Q_{gs}}$ [a.u.]
A	1200	75	1.3	1.067
B	1200	62	1	1.923
C	1200	75	9	0.667

For each DUT a calibration was done as described in the previous section. For all DUTs working points of V_{GS-} and I_{DS} five linearly spaced different frequencies were measured and the switching energy was calculated out of the slope of the switching losses as described in [3], [14]. An example of this is shown in Fig. 8 for V_{DS}=400 V and I_D= 7 A. For

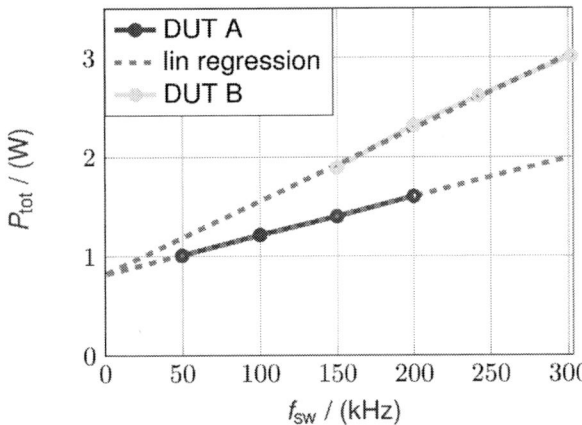

Fig. 8: Frequency dependent P_{tot} for DUT A and B at V_{DS}=400 V and I_D= 7 A.

both devices a linear slope with negligible deviation can be seen, hence enabling an extraction of E_{sw} with high accuracy. It can be seen that although conduction losses are reached for an extrapolation to 0 kHz the slope of the total losses (P_{tot}) and hence E_{sw} can vary greatly, for the different devices.

3.1 Current Dependency

In Fig. 9 the I_D-dependent soft-switching energy for V_{DS} of 400 V, 15 V V_{GS+}, and V_{GS-} of –5 V. The gate-source voltage was chosen as to the vendors' recommended specifications to fully deplete the channel. A clear dependency of the different DUTs with respect to the I_D can be seen. While DUT A and C have quite low losses for small I_D for higher I_D the losses significantly increase up to 4 µJ. This is different in behavior of DUT B behaves quite differently, the soft-switching losses already start with an increased switching energy initially and has subsequently a much slower slope for higher currents. In [15] the hard-switching losses of the same DUTs were analyzed and it was shown that due to the high $\frac{Q_{gd}}{Q_{gs}}$ ratio a much higher current overshoot and switching energies were seen. This might have an impact on the soft-switching losses as well.

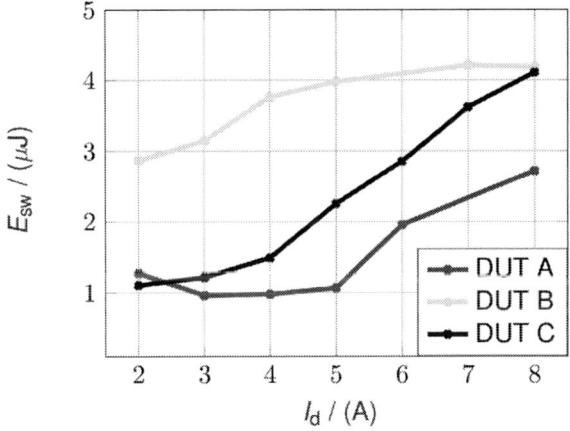

Fig. 9: Current dependent soft-switching energies for various DUTs.

3.2 Gate- Source dependency

In [12] the V_{GS} dependency of a 1200 V was investigated. It was shown that the positive gate-source voltage V_{GS+} has no impact as long as the gate current is sufficiently high. Unlike V_{GS+}, it was observed that V_{GS-} has an impact on the switching energy without any influence on the voltage slope-

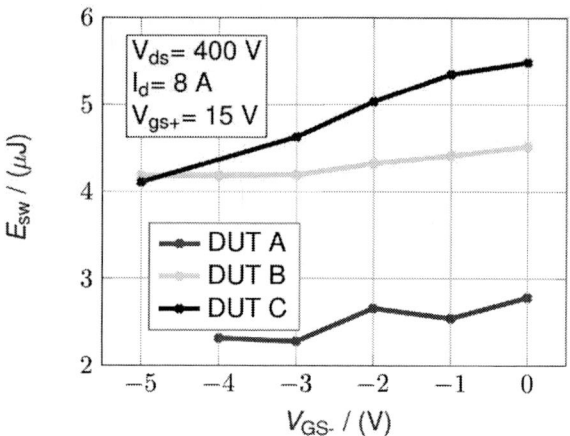

Fig. 10: Gate-source dependent soft-switching energies for various DUTs at V_{DS} =400 V, I_D= 8 A at V_{GS+}=15 V.

which is usually an important parameter. A comparison of soft-switching losses for DUT A-C is done for V_{DS} =400 V, I_D= 8 A at V_{GS+}=15 V in dependence of V_{GS-} and depicted in Fig. 10. It is clear to see that for smaller V_{GS-} the soft-switching losses can be reduced for all DUTs and increase significantly with higher V_{GS-} especially at -2 to 0 V. DUT A and DUT B exhibit a monotonic increase of switching energy with higher gate-source voltage. However, the slope of the V_{GS-}-dependency is different for all DUT technolgy. This increase in V_{GS-} dependency can be explained with a defect-rich channel which is not fully closed at high V_{GS-} [16], [17]. This effect is furthermore dependent on the vendors' technology.

4 Conclusion

This paper presents a new automated calorimetric measurement utilizing optical fiber online measurement and Peltier elements. It is shown that the measurement time can be significantly reduced to less than five seconds while maintaining high thermal accuracy. To further reduce the cooldown time a new automated cooldown process was implemented to utilize Peltier elements combined with a fan for faster cooldown without violating steady state conditions. Furthermore, three different silicon carbide technologies are compared with respect to their soft-switching losses for different V_{GS-}. It is shown that although the voltage slopes remain the same for all DUTs a higher V_{GS-} voltage significantly increases soft-switching losses due to a semi-open channel for all investigated SiC technologies.

References

[1] J Biela, J W Kolar, A Stupar, U Drofenik, and A Muesing, "Towards virtual prototyping and comprehensive multi-objective optimisation in power electronics," 2010.

[2] J. Weimer, D. Koch, and I. Kallfass, "Accuracy study of calorimetric switching loss energy measurements for wide bandgap power transistors," in *2021 23rd European Conference on Power Electronics and Applications (EPE'21 ECCE Europe)*, IEEE, 2021. DOI: 10.23919/ epe21ecceeurope50061.2021.9570481.

[3] J. Weimer, R. Schnitzler, D. Koch, and I. Kallfass, "Thermal impedance calibration for rapid and noninvasive calorimetric soft-switching loss characterization," *IEEE Transactions on Power Electronics*, vol. 38, no. 7, pp. 8472–8485, 2023. DOI: 10.1109/tpel.2023.3267982.

[4] S. Tiwari, J. K. Langelid, O.-M. Midtgård, and T. M. Undeland, "Soft switching loss measurements of a 1.2 kv sic mosfet module by both electrical and calorimetric methods for high frequency applications," in *2017 19th European Conference on Power Electronics and Applications (EPE'17 ECCE Europe)*, IEEE, 2017, P–1.

[5] S. Bolte, L. Keuck, J. K. Afridi, N. Fröhleke, and J. Böcker, "Calorimetric measurements with compensating temperature control," in *2017 IEEE 26th International Symposium on Industrial Electronics (ISIE)*, IEEE, 2017, pp. 636–639.

[6] M. Guacci, J. Azurza Anderson, K. L. Pally, D. Bortis, J. W. Kolar, *et al.*, "Experimental characterization of silicon and gallium nitride 200 v power semiconductors for modular/multi-level converters using advanced measurement techniques," *IEEE Journal of Emerging and Selected Topics in Power Electronics*, vol. 8, no. 3, pp. 2238–2254, 2020. DOI: 10.1109/jestpe.2019.2944268.

[7] A. Jafari, M. S. Nikoo, N. Perera, H. K. Yildirim, F. Karakaya, *et al.*, "Comparison of wide-bandgap technologies for soft-switching losses at high frequencies," *IEEE Transactions on Power Electronics*, vol. 35, no. 12, pp. 12 595–12 600, 2020. DOI: 10.1109/tpel.2020.2990628.

[8] D. Neumayr, M. Guacci, D. Bortis, and J. W. Kolar, "New calorimetric power transistor soft-switching loss measurement based on accurate temperature rise monitoring," in *2017 29th International Symposium on Power Semiconductor Devices and IC's (ISPSD)*, IEEE, 2017. DOI: 10.23919/ ispsd.2017.7988914.

[9] D. Koch, S. Araujo, J. Weimer, and I. Kallfass, "Automated calorimetric measurement with a peltier element for switching loss characterization," in *PCIM Europe digital days 2021; International Exhibition and Conference for Power Electronics, Intelligent Motion, Renewable Energy and Energy Management*, VDE, 2021, pp. 1–8.

[10] D. Nayak, Y. R. Kumar, and S. Pramanick, "A comparative study of loss measurement techniques for sic mosfet based pe converters," in *IECON 2022–48th Annual Conference of the IEEE Industrial Electronics Society*, IEEE, 2022, pp. 1–6.

[11] J. Haarer, M. Eckstein, P. Ziegler, P. Marx, D. Hirning, and J. Roth-Stielow, "A calorimetric and electrical method for measuring loss energies of half-bridges," in *2022 24th European Conference on Power Electronics and Applications (EPE'22 ECCE Europe)*, IEEE, 2022, pp. 1–9.

[12] R. Schnitzler, D. Koch, M. C. Weiser, J. Weimer, and I. Kallfass, "Comparison of gate-source-dependent soft- and hard-switching losses of wide-bandgap semiconductor utilizing electrical and rapid heatsinkless calorimetric measurements," in *IEEE Applied Power Electronics Conference and Exposition.2024*, 2024.

[13] R. Schnitzler, D. Koch, E. D. S. Gomes, and I. Kallfass, "Fully modular, dynamic sic and gan testbench with automated temperature and gate-voltage characterization," in *2023 IEEE Design Methodologies Conference (DMC)*, IEEE, 2023, pp. 1–6.

[14] J. Weimer and I. Kallfass, "Soft-switching losses in gan and sic power transistors based on new calorimetric measurements," in *2019 31st International Symposium on Power Semiconductor Devices and ICs (ISPSD)*, IEEE, 2019. DOI: 10.1109/ispsd.2019.8757650.

[15] R. Schnitzler, E. Gomes, D. Koch, R. Petar, and I. Kallfass, "Multidimensional switching loss analysis of various 1200 v sic - power mosfet technologies," in *IEEE Conference on Integrated Power Electronics Systems.2024*, 2024.

[16] L. Yang, Y. Bai, C. Li, H. Chen, Z. Han, *et al.*, "Analysis of mobility for 4h-sic n/p-channel mosfets up to 300°c," *IEEE Transactions on Electron Devices*, vol. 68, no. 8, pp. 3936–3941, 2021.

[17] A. März, T. Bertelshofer, M. Helsper, and M.-M. Bakran, "Comparison of sic mosfet gate-drive concepts to suppress parasitic turn-on in low inductance power modules," in *2017 19th European Conference on Power Electronics and Applications (EPE'17 ECCE Europe)*, IEEE, 2017, P–1.

PCIM Europe 2024, 11– 13 June 2024, Nuremberg   DOI: 10.30420/566262255

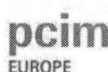

SiC Trench MOSFETs in Avalanche Mode with RC Snubber Circuit

Sebnem Tuncay[1], Guang Zeng[1], Guangye Si[1], Thomas Basler[2]

[1] Infineon Technologies AG, Germany

[2] Professorship of Power Electronics, Chemnitz University of Technology, Germany

Corresponding author: Sebnem Tuncay, sebnem.tuncay@infineon.com
Speaker: Sebnem Tuncay, sebnem.tuncay@infineon.com

Abstract

Existing stray inductances in circuits may force power switches to enter the avalanche condition. Under avalanche condition, a power switch like SiC trench MOSFET can handle certain amount of current and energy before it gets destroyed. Ideas to utilize RC snubber circuits in parallel to the SiC switch or a half bridge are more and more discussed and helpful to reduce the effective stray inductance for switching and support a switching with less ringing. Such a circuit in parallel to SiC trench MOSFET provides current sharing and thereby, the SiC trench MOSFET is not fully stressed in given avalanche condition. This enables higher load currents that the SiC trench MOSFET can survive during turn-off. The use of an RC snubber circuit together with a SiC trench MOSFET may reduce the system total cost and increase the reliability of systems. This paper shows experimental and simulation results of SiC trench MOSFETs in avalanche mode with and without the use of RC snubber circuit.

1   Motivation and test conditions

1.1   Introduction

The avalanche event which is caused by not-free-wheeling circuit inductances like stray inductances is an overload condition for SiC trench MOSFETs which may cause device destruction and damage other elements in circuits. The SiC trench MOSFET enters the avalanche mode, if the overvoltage caused by the stray inductances is higher than the breakdown voltage of the SiC trench MOSFET. That's the point where protection circuits play a role for safer and more reliable operation. Melting fuses and electromechanical relays (EMR) are commonly integrated in protection circuits to control the power distribution and prevent fatal failures within the circuit if the semiconductor is not able to master the failure condition itself. On the other hand, SiC trench MOSFETs can be used in protection circuits [1] [2] [3] [4] [5] instead of solid-state relays or fuses since they have faster reaction time. SiC trench MOSFETs can be controlled and monitored easily, they are more flexible and can be more reliable.

1.2   Test method

A test bench consisting of a SiC trench MOSFET half-bridge that is used for the investigation is shown in Fig. 1 (a). Pulse patterns used for high

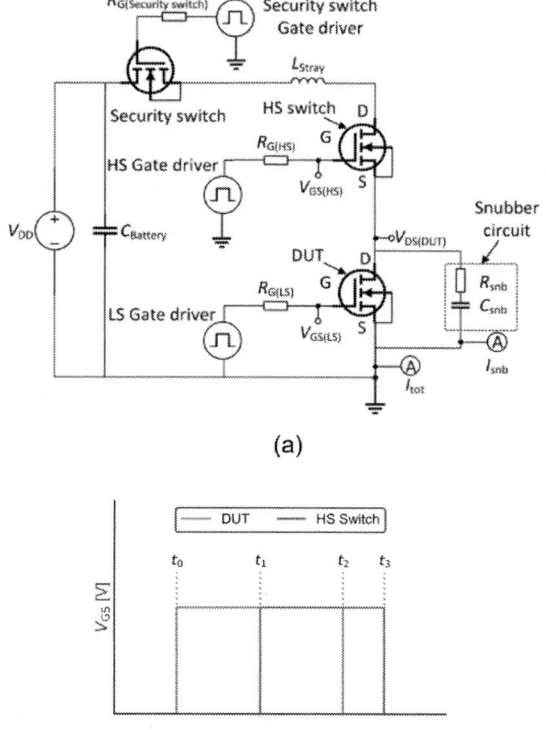

Fig. 1 Schematic of test bench (a) and pulse patterns of V_{GS} of HS switch and DUT for avalanche measurements (b)

side (HS) switch and device under test (DUT) on low side are depicted in Fig. 1 (b).

First, the DUT is turned on to enable the RC snubber circuit to be discharged before the avalanche event to simulate the real application condition when the fault happens. Once the HS switch is turned on, the load current starts increasing with the rate of di/dt which is determined by stray inductance L_{stray} and DC-link voltage V_{DD}. When the desired load current is reached, the DUT is turned off and enter the avalanche mode due to overvoltage caused by stray inductance L_{stray} and the missing free-wheeling diode across it.

Since RC snubber elements share a part of the load current during switching, the DUT is not fully stressed with the stored energy in the inductance and not yet at its destruction limit as without RC snubber circuit elements. Therefore, the DUT can sustain higher load currents before it gets destroyed.

Table 1 Test conditions

Parameters	Device A	Device B
L_{stray} [µH]	9	46
R_G [Ohm]	15	41
T_{start} [K]	298	298
	398	398
V_{DD} [V]	800	800

Table 2 Snubber circuit configurations

Combination	Case			
	I	II	III	IV
R_{snb} [Ohm]	5	10	5	10
C_{snb} [nF]	100	100	300	300

To investigate the impact of RC snubber circuit on avalanche behavior, SiC trench MOSFETs of same technology but having different sizes of diode area (respective p-doped area) with a relation of $A_{diode, Device\ A} = 3.2 * A_{diode,\ Device\ B}$ (in other words Device A is 3.2 times larger than Device B) are chosen as DUT on the low side. Similarly, a SiC trench MOSFET is used as HS switch. The maximum avalanche breakdown voltage of the used DUTs is around 1650 V. DUT and HS switch are placed on a cooling/heating unit. Avalanche measurements are based on incremental current pulses. Test conditions used for DUTs are adjusted considering their different chip sizes and are listed in Table 1.

To show the efficiency of RC snubber circuit four different snubber circuit configurations are defined as in Table 2 and avalanche measurements are compared as with and without RC snubber circuits.

1.3 Current sharing between RC snubber circuit and SiC trench MOSFET

To relieve stress on SiC trench MOSFET in avalanche mode, the load current is shared between snubber circuit and SiC trench MOSFET. The snubber capacitor C_{snb} is charged by the snubber current which flows through the snubber resistance R_{snb} in the snubber circuit. The maximum current flowing through snubber circuit $I_{snb,max}$ can be calculated according to Eq.1

$$I_{snb,max} = \frac{V_{ava,max}}{R_{snb}} \tag{1}$$

where $V_{ava,max}$ is the voltage across the snubber circuit which equals the maximum avalanche breakdown voltage of the SiC trench MOSFET at given condition and R_{snb} is the value of resistance in the snubber circuit.

Snubber current I_{snb} decays according to the time constant of the RC circuit τ_{snb}, which is given by Eq. 2, as

$$\tau_{snb} = R_{snb} * C_{snb} \tag{2}$$

The shared part of the total load current in the snubber circuit is increasing with increasing time constant of the snubber circuit, accordingly.

1.3.1 Device A in avalanche mode

In Fig. 2 waveforms of total load current I_{tot}, current flowing through snubber circuit I_{snb} and current through DUT I_{dut} for Device A are shown using (a) R_{snb} = 5 Ohm with C_{snb}= 300 nF (case III) and (b) R_{snb} = 10 Ohm with C_{snb}= 300 nF (case IV).

For each measurement, the total load current I_{tot} is set to 335 A. As a comparison, current flowing through the DUT without usage of the snubber circuit is also demonstrated. When there is no snubber circuit applied, the total load current I_{tot} equals to the current flowing through DUT I_{dut}. The snubber current reaches its maximum value in each case fast. To achieve this, the snubber circuit is connected with the smallest parasitic to the DUT.

According to Eq. 1, $I_{snb,max}$ is found to be around 137 A and 255 A for R_{snb} = 10 Ohm and R_{snb} = 5 Ohm, respectively. The snubber current decays according to the time constant given in Eq. 2 and reaches to 0 A. When the DUT dissipates the current which is shared, the avalanche event is completed. Followingly, when the voltage through the DUT starts to decrease, the snubber capacitor begins to discharge and the voltage across the DUT reaches finally the DC-link voltage V_{DD} as shown in Fig. 2 (a). Thereby the snubber current passes to the negative current region.

298 K and 398 K, since the DUTs are not yet at their destruction limits. A comparision of snubber circuit configurations shows that the DUT is less stressed at smaller values of snubber resistance keeping the snubber capacitor value constant since the snubber circuit takes more current of the total current.

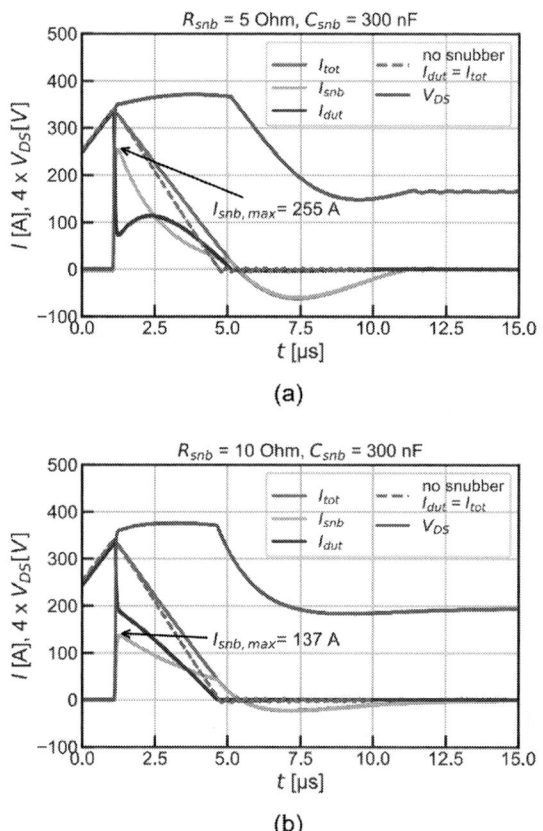

(a)

(b)

Fig. 2 Current waveforms of Device A and total load current with and without RC snubber circuit having values of (a) R_{snb} = 5 Ohm, C_{snb}= 300 nF (case III) (b) R_{snb} = 10 Ohm, C_{snb}= 300 nF (case IV)

Figure 3 shows the total load current and the shared currents between the snubber circuit and Device A during avalanche event. The current values are taken at the beginning of the turn-off event where the snubber current is at its maximum. There is no significant difference between current values at starting temperatures of

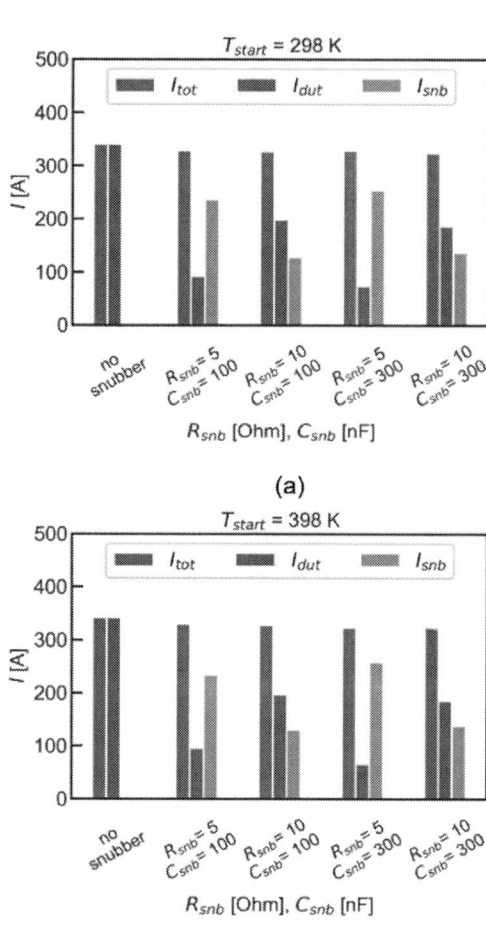

(a)

(b)

Fig. 3 Current values showing total load current and current values shared between snubber circuit and Device A during avalanche event at (a) T_{start} = 298 K and (b) T_{start} = 398 K (Current values are taken at the beginning of the turn-off event)

The power dissipation in Device A in avalanche mode and in snubber circuit having different values of R_{snb} and C_{snb} is shown in Fig. 4 and Fig. 5, respectively. Since the current flows through the resistor, not in the capacitor, the power is dissipated by the resistor. When keeping C_{snb} constant, the maximum energy within Device A in avalanche mode can be shifted by changing R_{snb}. Figure 4 shows that the maximum of power dissipated by

Device A is shifted to a later time point with decreasing R_{snb} at constant C_{snb}.

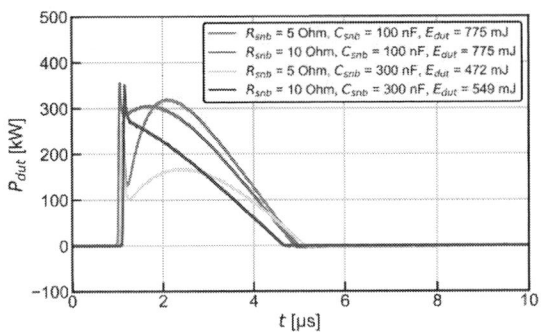

Fig. 4 Power dissipation over time of Device A obtained with different values of R_{snb} and C_{snb}

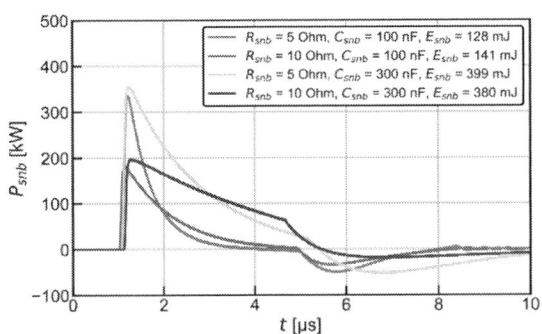

Fig. 5 Power dissipation over time of snubber circuit obtained with different values of R_{snb} and C_{snb}

In applications, the capacitor can be chosen as small as possible to save space within the circuit and to save costs, but energy overtaken by the capacitor will be decreasing accordingly as well. This case can be seen in Fig. 5. At constant R_{snb} = 5 Ohm, the snubber circuit takes 14% and 46% of the total energy with C_{snb} = 100 nF and C_{snb} = 300 nF, respectively. By keeping R_{snb} = 10 Ohm constant, energy values achieved by the snubber circuit are 15% and 41% of the total energy at C_{snb} = 100 nF and 300 nF, respectively. Similarly, energy values achieved by the snubber circuit with C_{snb} = 300 nF at each constant R_{snb} is by factor of 3 higher than with C_{snb} = 100 nF. On the other side, the influence of R_{snb} on the energy is small, therefore, R_{snb} can be used to limit the current in the snubber circuit.

1.3.2 Device B in avalanche mode

As for Device A, the same test conditions (Table 1) and snubber configurations (Table 2) are used for Device B which has a 3.2 times smaller area. Figure 6 shows the waveforms of the total load current I_{tot}, which is set to 73 A and the current flowing through the snubber circuit I_{snb} at R_{snb} = 10 Ohm with C_{snb} = 300 nF and R_{snb} = 5 Ohm with C_{snb} = 300 nF, and the current through the DUT I_{dut} for Device B at a start temperature of 298 K.

At R_{snb} = 10 Ohm with C_{snb} = 300 nF (Fig. 6 (b)), the DUT is not yet in avalanche and the whole current is dissipated by the snubber resistance. By decreasing R_{snb} from 10 Ohm to 5 Ohm at the same snubber capacitance C_{snb} = 300 nF, the DUT enters into avalanche mode and shares a small part of the total load current as shown in Fig. 6 (a).

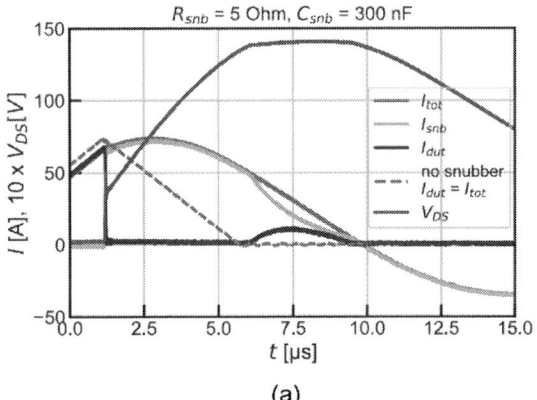

(a)

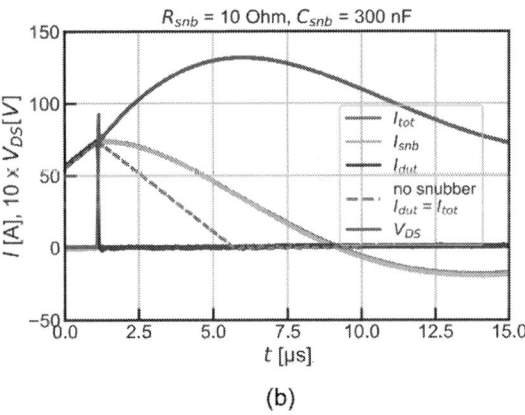

(b)

Fig. 6 Current waveforms of Device B and total load current with and without RC snubber circuit having values of (a) R_{snb} = 5 Ohm, C_{snb} = 300 nF (case III) (b) R_{snb} = 10 Ohm, C_{snb} = 300 nF (case IV)

In Fig. 7 current values shared between the snubber circuit and the DUT are demonstrated at T_{start} = 298 K. The current values are extracted at the time point of the maximum current in the snubber circuit. The decreasing value of the snubber capacitor reduces the efficiency of the snubber circuit; therefore, the DUT takes more current.

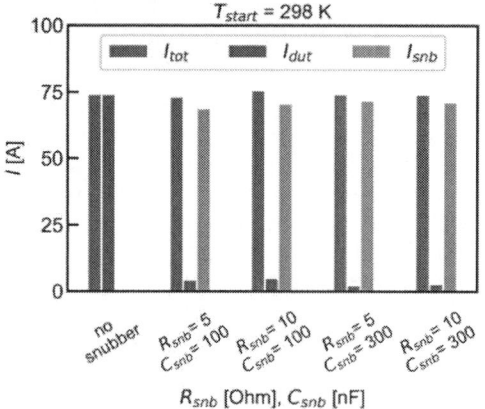

Fig. 7 Current values showing total load current and current values shared between snubber circuit and Device B during avalanche event at T_{start} = 298 K (Current values are taken at the beginning of the turn-off event)

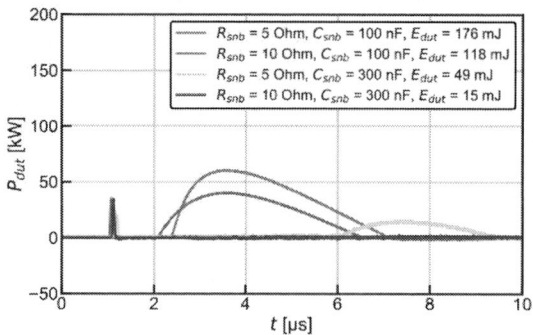

Fig. 8 Power dissipation over time of Device B obtained with different values of R_{snb} and C_{snb}

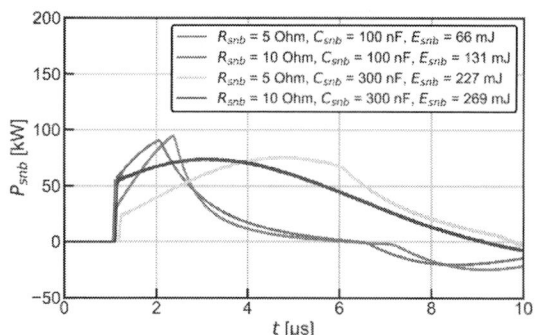

Fig. 9 Power dissipation over time of snubber circuit obtained with different values of R_{snb} and C_{snb}

Using snubber circuits with different values of R_{snb} and C_{snb}, the power dissipation of Device B and the snubber circuit is shown in Fig. 8 and Fig. 9, respectively. The power dissipation in the snubber resistor increases with the value of the snubber capacitance. Decreasing snubber resistance at the same snubber capacitor C_{snb} = 300 nF enables Device B to enter the avalanche mode and leads to the power dissipation of the SiC trench MOSFET.

2 Simulated junction temperature

In Fig. 10, waveforms of V_{DS}, V_{GS}, I_{dut} and I_{snb} of Device A obtained by simulations using Simetrix are demonstrated as a comparison to the measurements. The total load current is 335 A for both simulation and measurement. The results show good correlation. Simulations can be used to predict the behavior of the DUT and the snubber circuit at different test conditions. As can be seen, a part of the current is shared by the RC snubber circuit and relieves the SiC trench MOSFET.

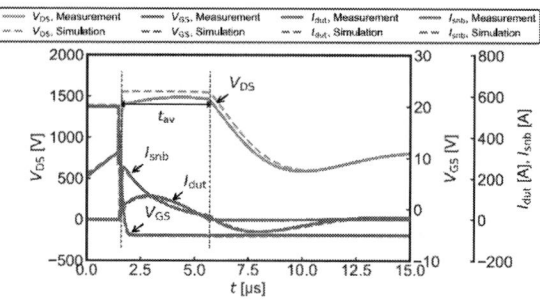

Fig. 10 Waveforms of measurement and simulation results with RC snubber circuit having values of R_{snb} = 5 Ohm and C_{snb} = 300 nF (case III) at total load current I_{tot} = 335 A (V_{DD} = 800 V, L_{stray} = 9 µH and T_{start} = 298 K)

The SiC trench MOSFET device failure occurs mainly due to increasing junction temperatures T_j within the device [6]. 1D - Cauer model from [6] can be used to estimate the maximum junction temperature $T_{j,max}$ of the DUT under given avalanche conditions. Figure 11 shows the Cauer model used for Device A. The SiC drift zone is separated into 5 regions to get a more practical imprint of the losses in this short time interval. The thermal conductivity of SiC was kept constant but defined at the elevated temperature of 650 K.

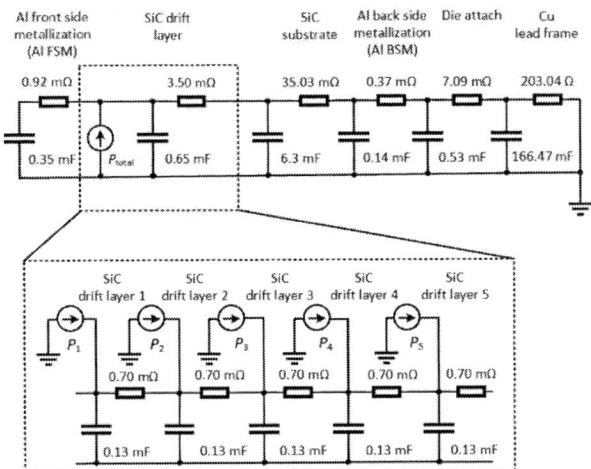

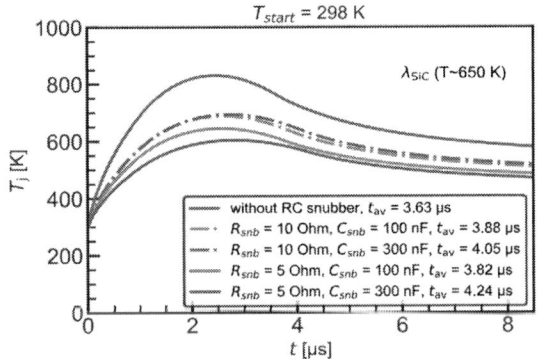

Fig. 11 Cauer thermal network for Device A adapted from [6]

Figure 12 shows simulation results of maximum junction temperatures obtained with and without snubber circuits. Power inputs in the Cauer model are supplied from measurement results at I_{dut}= 335 A and T_{start}= 298 K together with the corresponding avalanche time values.

Fig. 12 Simulated junction temperatures of Device A with and without snubber circuit at T_{start} = 298 K and I_{dut} = 335 A

When snubber elements are not used in the circuit, the total load current is flowing through Device A during avalanche event. Snubber circuits as already seen in Fig. 4 and Fig. 5, share a part of the load current and dissipate energy where Device A has less power dissipation to handle. Consequently, the highest junction temperature is reached without snubber circuits. The lowest increase of the junction temperature is found where the snubber circuit takes the highest current with

R_{snb} = 5 Ohm and achieves its highest energy with C_{snb} = 300 nF.

In [6], simulation results show that the maximum junction temperature $T_{j,max}$ reaches up to 1000 K before the device is destroyed by thermally-driven failures. Thus, by using snubber circuits, higher load currents can be applied during an avalanche event.

3 Conclusion

This paper has described the impact of RC snubber circuits on the SiC trench MOSFET in avalanche mode. Some current reduction for the SiC trench MOSFET has been obtained with the use of RC snubber elements which are in parallel to the SiC trench MOSFET. An increasing value of the snubber capacitor has shown higher efficiency of snubber circuits for SiC trench MOSFET in avalanche. Energy values achieved by SiC trench MOSFET have decreased with smaller snubber resistance. In other words, the ideal case of the snubber circuit has been obtained with smaller R_{snb} and bigger C_{snb} where the snubber circuit takes its highest energy. The avalanche energy, which the SiC trench MOSFET can handle before destruction, is limited but with the use of RC snubber elements the SiC trench MOSFET is relieved at given avalanche condition and can survive higher load currents. The highest maximum junction temperature has been obtained without the use of the snubber circuit, whereas the lowest maximum junction temperature has been reached with the smaller snubber resistance and the bigger snubber capacitance. A precise design of a snubber circuit is necessary to increase avalanche robustness effectively. Additionally, temperature rise of RC snubber elements during an clamping event should also be taken into account.

References

[1] X. She, A. Q. Huang, Ó. Lucía und B. Ozpineci, „Review of Silicon Carbide Power Devices and Their Applications, "IEEE *Transactions on Industrial Electronics,* Bd. 64, Nr. 10, pp. 8193-8205, 2017.

[2] H. Li, R. Yu, Y. Zhong, R. Yao, X. Liao und X. Chen, „Design of 400 V Miniature DC Solid State Circuit Breaker with SiC MOSFET, "Micromachines, Bd. 10 (5), 2019.

[3] D. Marroqui, A. Garrigós und J. M. Blanes, „LVDC SiC MOSFET Analog Electronic Fuse with Self-Adjusting Tripping Time Depending

on Overcurrent Condition, "IEEE *Transactions on Industrial Electronics,* Bd. 69, pp. 8472-8480, 2022.

[4] H. Qin, Y. Zhang, Y. Dong, K. Xu, C. Zhao und S. Wang, „Research on Solid State Circuit Breaker Based on SiC MOSFET with Soft Switch off Method, "in *PCIM Europe*, Nuremberg, 2018.

[5] Y. Zhang und Y. C. Liang, „Over-current protection scheme for SiC power MOSFET DC circuit breaker, "in *2014 IEEE Energy Conversion Congress and Exposition (ECCE)*, Pittsburgh, 2014.

[6] S. Tuncay, G. Zeng, G. d. F. de Falco und T. Basler, „Avalanche ruggedness and failure mode of SiC trench MOSFETs, "Microelectronics *Reliability,* 2023.

PCIM Europe 2024, 11– 13 June 2024, Nuremberg   DOI: 10.30420/566262256

High-Frequency Oscillations in SiC MOSFET Power Modules During Turn-on Switching Transient – Analysis Based on Simulations and Mitigation Methods

Rajani Kumar Thirukoluri[1], Roveendra Paul[2]

[1] onsemi, Germany
[2] onsemi, USA

Corresponding author: Rajani Kumar Thirukoluri, rajani.thirukoluri@onsemi.com
Speaker: Rajani Kumar Thirukoluri, rajani.thirukoluri@onsemi.com

Abstract

The continuously increasing trend of design and development of SiC based automotive traction inverter power modules exhibiting features such as, more compactness, high robustness & reliability, high power densities and able to accommodate newer generation die technologies always comes with several design challenges. To meet load current requirements, it is general practice to parallel multiple SiC dies per switch in a half-bridge configuration. The switching transients (during t_{on}, t_{off}) in these modules are associated with high di/dt and dv/dt, and together with module layout RLC parasitics produces high frequency and high amplitude gate oscillations in the range between 300 MHz and 450 MHz. This potentially causes EMI and reliability issues at system level. This paper tears down the oscillations phenomenon during predominantly turn-on switching transients, due to phase shifted reverse recovery of parallel SiC body diodes and its dependency on junction temperature, external R_{gonext} and SiC MOSFET technologies. Several potential mitigation methods are proposed, such as optimizing module layout and increasing internal R_G (embedded inside the die), which are later demonstrated and validated in a half-bridge SiC power module with 8 devices in parallel. This entire analysis and validation are carried out using Ansys Q3D, SIMetrix simulation tools and the results are accordingly presented.

1   Introduction

Using SiC power modules in traction inverter application are increasingly gaining momentum due to several advantages that they present in terms of size, efficiency, and reliability. Due to their higher switching speeds caused by inherent lower device parasitic capacitances, paralleling of multiple SiC dies in modules can lead to high frequency and high amplitude oscillations between gate-source terminals especially during turn-on switching event with an inductive load. These gate-source oscillations, if not addressed, will degrade the thin gate oxide layer that is under higher electric field stress, potentially leading to long-term failure mechanisms that affect reliability. It could also cause parasitic turn-on of the complimentary switch, leading to a half-bridge shoot through. It also presents both conductive and radiation EMI related issues at the inverter system level and overall leads to compromised control of the gate-driver signals.

To demonstrate gate oscillations, we considered a prototype half-bridge 1200 V SiC power module, assembled with 8 x 25 mm² dies in parallel per switch. The module package and its internal structure along with interconnect method and unique node names assigned to each respective chip are shown in **Fig. 1**. The parasitics of the physical module are extracted using Ansys Q3D Extractor software, and the netlist exported to SPICE simulation tool SIMetrix. The SPICE model of the SiC chip itself was formulated and imported by SIMetrix. The compact SPICE model of the power module was formulated, and external components added to the circuit to complete the schematic to run traditional double pulse testing (DPT) as shown in **Fig. 2** [1]. The complete electrical equivalent circuit with total parasitic RLC lumped components of the power module is not presented in this paper rather a simplified segment of the circuit consisting of only lumped parasitic stray self-inductances in the commutation loop identified between DC+ node and the virtual equipotential node (equal potential

1865

node for all devices) in HS domain of half-bridge is shown in **Fig. 3**, which will be referenced in the upcoming sections. The gate-source waveforms of power module exhibit the high-frequency oscillations during second pulse turn-on switching event as depicted from **Fig. 4** under the given conditions: V_{DC}= 850 V, I_{LOAD}= 1100 A, R_{gonext}=8 Ω, $R_{goffext}$=2 Ω, T_J=25°C, DUT=LS, Internal R_{gint} of die=5 Ω.

From the figures **Fig. 4**, **Fig. 5**, **Fig. 6** & **Fig. 7**, we can see high frequency and high amplitude oscillations appearing on module external gate-source pins of both high-side (HS) and low-side (LS) switches during the second pulse turn-on event of the LS switch, which is here device under test (DUT) at the above mentioned conditions. The SiC MOSFET devices used in this power module have very low parasitic capacitances for enhancing switching speeds and improving power module efficiency, but at the trade-off of causing higher sensitivity to oscillations. The LS switching waveforms shown in **Fig. 8** depicts the high-frequency oscillations which are also appearing on module external drain-source pins.

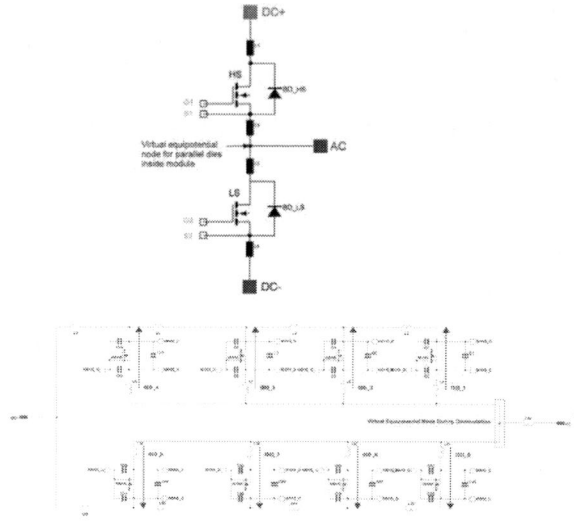

Fig. 3   Electrical circuit schematic of 8 die SiC PM

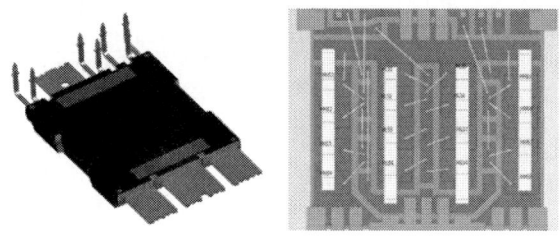

Fig. 1   1200 V, 1.6 mΩ Half-bridge (B2) Module

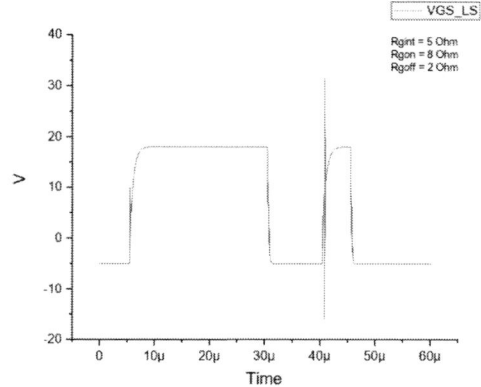

Fig. 4   LS switch gate-source voltage

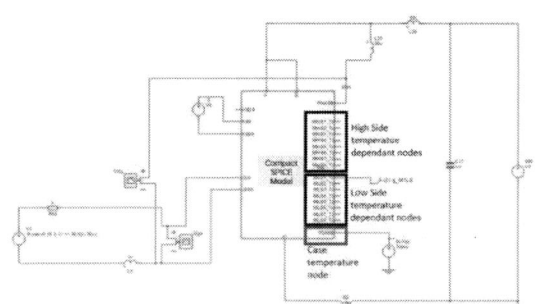

Fig. 2   Compact SPICE model with auxiliary components for LS switching simulation

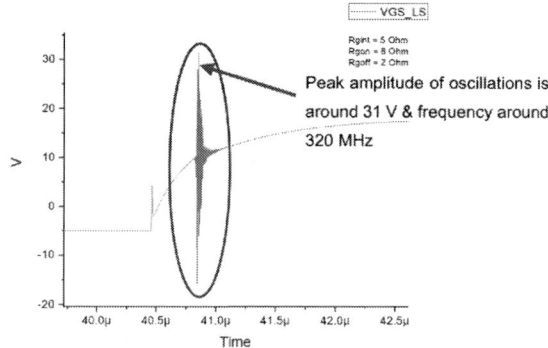

Fig. 5   LS switch gate-source voltage (magnified view)

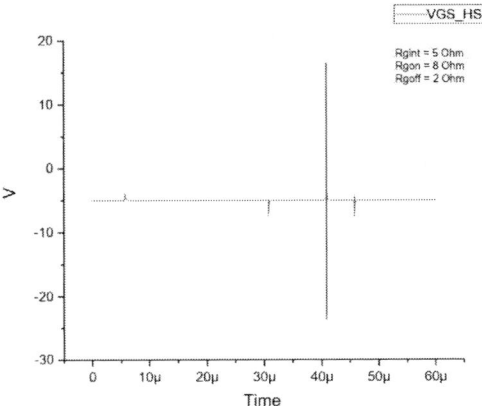

Fig. 6 HS switch gate-source voltage

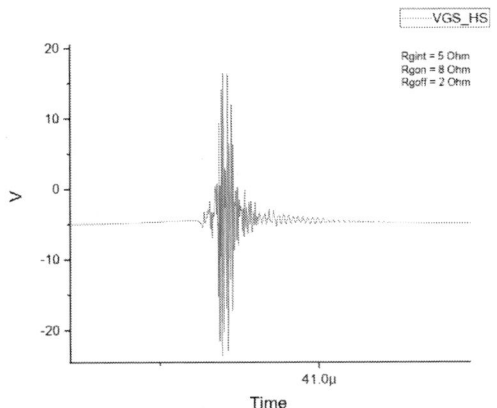

Fig. 7 HS switch gate-source voltage (magnified view)

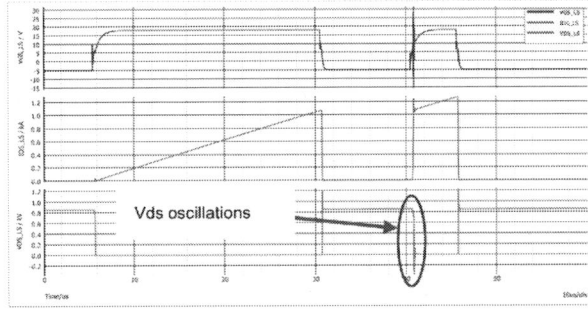

Fig. 8 LS switch V_{GS}, I_{DS} and V_{DS} Waveforms

The frequency of the LS gate-source voltage oscillations was measured at approximately 320 MHz with a peak amplitude of 31 V.

2 Turn-on Oscillations

Before the initiation of the second pulse turn-on switching event on LS switch, the load inductor current is freewheeling over the HS SiC devices body diode. The total steady load current is shared among the 8 parallel SiC body diodes in a way that the device with highest forward voltage drop V_F conducts maximum portion of the total current and similarly, the device with lowest V_F conducts minimum portion of total current. In this simulation, all the devices are operating at their nominal parametric values, implying the V_F of all devices are equal and hence, steady state current sharing is balanced. Soon after the second pulse turn-on switching event is initiated for the LS switch, the drain current rapidly increases at di/dt defined by external R_{gonext}, internal R_{gint}, load current, DC-link voltage and junction temperature of the device. The total drain current is shared among the parallel SiC devices inside the module as shown in **Fig. 9**. As LS devices start conducting drain currents, simultaneously body diode currents of HS devices start decreasing from their steady state currents. In actual scenario, the current sharing among LS devices during turn-on transient is not balanced and hence, results in a different turn-on di/dt at individual chip level. The root cause of this current imbalance is not in the scope of this paper, hence not covered. For the sake of simplification, we consider the currents I_{DS1} to I_{DS8} of LS chips are perfectly balanced with a common di/dt. This in turn ensures common di/dt of the complementary HS body diode currents at the start of commutation. As time elapses, the HS body diode currents turn asymmetrical and attains an un-cohesive rate of decrease towards their zero. This is due to the dynamically changing V_F of these devices due to the imbalances in inductive voltage drops within their individual voltage loops, measured between DC+ and the virtual equipotential node as shown in **Fig. 3**. The device with the lowest body diode forward voltage drop V_F during commutation will transition faster into reverse recovery as compared to the device with a highest value of V_F. It is evident from the curves in **Fig. 9**, the currents I_{SD4}, I_{SD8} will almost simultaneously recover minority charge carriers in their P+& N- regions first by reaching their reverse recovery peak current faster. This is due to the lower internal geometric position of chips in positions 1 and 8. This phenomenon leads to a time lag reverse recovery phase for the HS parallel devices and always fulfils the V_F relation among them which is given by following equations and are derived from **Fig. 3**.

$$V_{SD4} = V_{L2} + V_{L6} \tag{1}$$

$$V_{SD3} = V_{L2} + V_{L1} + V_{L7} \tag{2}$$

$$V_{SD2} = V_{L2} + V_{L1} + V_{L4} + V_{L8} \tag{3}$$

$$V_{SD1} = V_{L2} + V_{L1} + V_{L4} + V_{L5} + V_{L9} \tag{4}$$

$$V_{SD5} = V_{L85} + V_{L50} + V_{L44} + V_{L32} + V_{31} \tag{5}$$

$$V_{SD6} = V_{L85} + V_{L50} + V_{L44} + V_{L37} + \tag{6}$$

$$V_{SD7} = V_{L85} + V_{L50} + V_{L36} \tag{7}$$

$$V_{SD8} = V_{L85} + V_{L30} \tag{8}$$

As we initially assumed all body diode currents at the very beginning of commutation have common di/dt, also by assuming and assigning one common value L_σ for all the parasitic inductances we can then further simplify the equations Eq. (1) until Eq. (8) as shown below.

$$V_{SD4} = 5 \cdot L_\sigma \cdot \frac{di}{dt} \tag{9}$$

$$V_{SD3} = 8 \cdot L_\sigma \cdot \frac{di}{dt} \tag{10}$$

$$V_{SD2} = 10 \cdot L_\sigma \cdot \frac{di}{dt} \tag{11}$$

$$V_{SD1} = 11 \cdot L_\sigma \cdot \frac{di}{dt} \tag{12}$$

$$V_{SD8} = 5 \cdot L_\sigma \cdot \frac{di}{dt} \tag{13}$$

$$V_{SD7} = 8 \cdot L_\sigma \cdot \frac{di}{dt} \tag{14}$$

$$V_{SD6} = 10 \cdot L_\sigma \cdot \frac{di}{dt} \tag{15}$$

$$V_{SD5} = 11 \cdot L_\sigma \cdot \frac{di}{dt} \tag{16}$$

$$V_{SD4} < V_{SD3} < V_{SD2} < V_{SD1} \tag{17}$$

$$V_{SD8} < V_{SD7} < V_{SD6} < V_{SD5}$$

From Eq. (17), we can deduce which body diode enters the reverse recovery phase chronologically depending on the V_F relation. Once the currents I_{SD4}, I_{SD8} reaches their reverse recovery peak at time instant "T1", they tend to decrease transiently with the defined snappiness of body diode until time instant "T2". During this transition from "T1" till "T2" period, there is a sudden increase in I_{SD1}, I_{SD2}, I_{SD3}, I_{SD5}, I_{SD6}, I_{SD7} currents to compensate for the total reverse recovery current, as I_{SD3} and I_{SD7} are the next in line following Eq. (17) they will take a bigger portion of current during this dynamic current transient, and as a result induce a higher slope and reaches their reverse recovery peak at "T2" as shown in **Fig. 9**. As this phenomenon progresses further, the total reverse recovery current is dynamically shared or oscillates among the parallel HS body diodes with a defined frequency. The oscillations seen in the body diode currents will be coupled on to the drain-source currents of LS parallel devices, and hence indirectly leads to V_{DS} oscillations and V_{GS} oscillations via miller capacitance coupling with the source side stray inductance induced voltage drop coupling. Until all die level V_{GS} oscillations are fully dampened, the individual gate currents I_G will continue circulating among parallel SiC devices on the LS. Hence, these oscillations are identified as inter-bench oscillations within the power module.

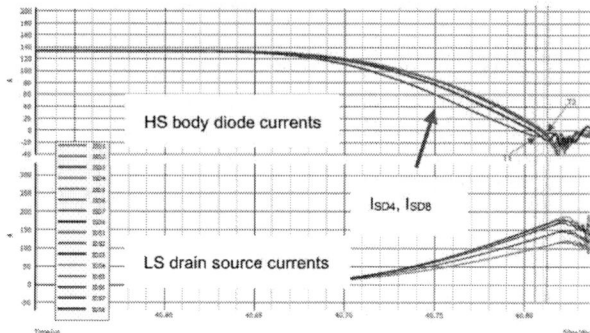

Fig. 9   LS: I_{DS} currents and HS: I_{SD} currents during commutation

2.1   Influence of External R_{gonext}

By increasing the external R_{gonext} placed on gate driver board increases the amplitude of these high-frequency gate oscillations and provides weak damping, but conversely decreases the amplitude of low-frequency gate oscillations which are usually due to interaction of L_σ *$C_{bodydiode}$ once the body diode currents reach to 0 A.

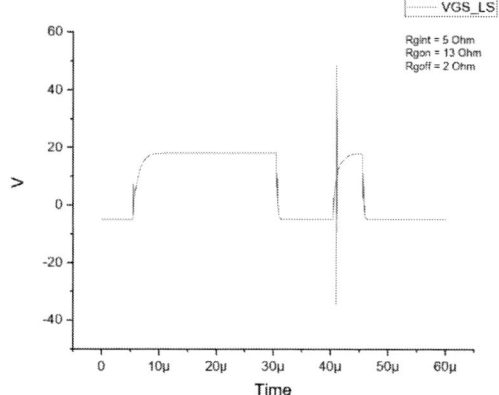

Fig. 10   LS switch gate-source voltage

Under the same test conditions V_{DC}= 850 V, I_{LOAD}= 1100 A, R_{gonext}=13 Ω, $R_{goffext}$= 2 Ω, T_J=25°C, DUT=LS, Internal R_{gint} of die=5 Ω, an increased R_{gint} from 8 Ω to 13 Ω can alter the peak amplitude of gate oscillations. It is measured to be 48 V with a frequency of 429 MHz as seen in **Fig. 10** and **Fig. 11**. Increasing external R_{gonext} counter effects oscillations following the theory established in previous sections. These oscillations are purely inter-bench oscillations but not the external gate loop LC oscillations.

PCIM Europe 2024, 11– 13 June 2024, Nuremberg   DOI: 10.30420/566262256

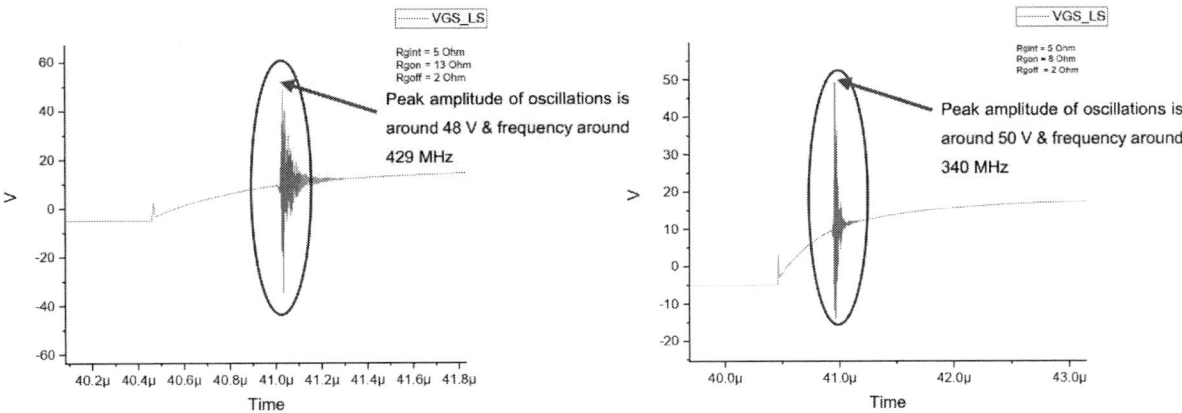

Fig. 11   LS switch gate-source voltage (magnified view)

Fig. 13   LS switch gate-source voltage (magnified view)

2.2   Influence of Junction Temperature

With higher junction temperature of 175°C compared to 25°C, the body diode snappiness behavior worsens and results in higher di/dt that are associated with reverse recovery current, and hence leads to higher amplitude gate oscillations. Under the same test conditions V_{DC}= 850 V, I_{LOAD}= 1100 A, R_{gonext}= 8 Ω, $R_{goffext}$= 2 Ω, T_J=175°C, DUT=LS, Internal R_{gint} of die=5 Ω, the peak amplitude of gate oscillations is measured at 50 V with a frequency of 340 MHz as seen in **Fig. 12** and **Fig. 13**.

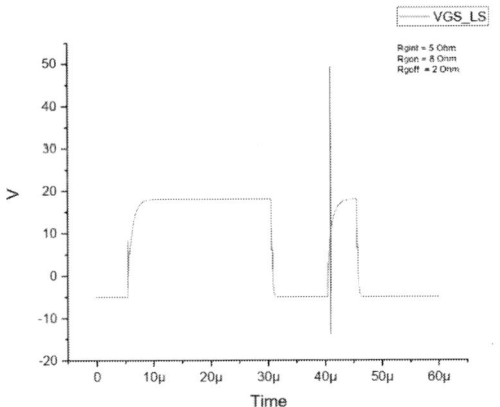

Fig. 12   LS switch gate-source voltage

3   Mitigation and Validation

After establishing the motivation in mitigating the oscillations issue in section (1) and considering the well understood phenomenon of these oscillations in section (2), via theoretical analysis and simulations, this section describes potential mitigation measures, verified with supporting simulation analysis.

3.1   Internal R_{gint}

Internal R_{gint} is a series poly resistor embedded inside the SiC die, usually beneath the gate metallization pad, and is in series with the gate current path. This resistance is part of total R_{gontot} (R_{gonext} + R_{gint}) during turn-on switching events, and $R_{gofftot}$ ($R_{goffext}$ + R_{gint}) during turn-off switching events. As we understand high-frequency gate oscillations captured between module level gate & source pins, there are circulating gate currents which are flowing in and out of the parallel SiC devices in both HS and LS domains inside the half-bridge module. These high frequency circulating currents can be effectively dampened by the internal R_{gint}. SPICE based simulation tools can greatly assist in predicting the right value of R_{gint} for any application. Although this being an iterative approach, it is nonetheless straightforward and simple.

The 8-die 1200 V SiC prototype sample is tested under the same conditions of V_{DC}= 850 V, I_{LOAD}= 1100 A, R_{gonext}=8 Ω, $R_{goffext}$= 2 Ω, T_J=25°C, DUT=LS. Internal gate resistor R_{gint} of the chip is increased from 5 Ω to 15 Ω. From simulations shown in **Fig. 14** and **Fig. 15**, the peak amplitude of

1869

gate oscillations is measured at 13 V with a frequency of approximately 300 MHz.

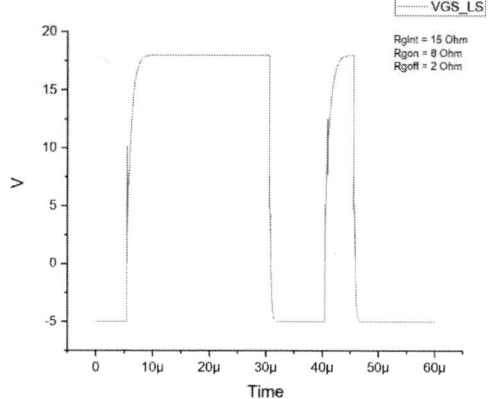

Fig. 14 LS switch gate-source voltage

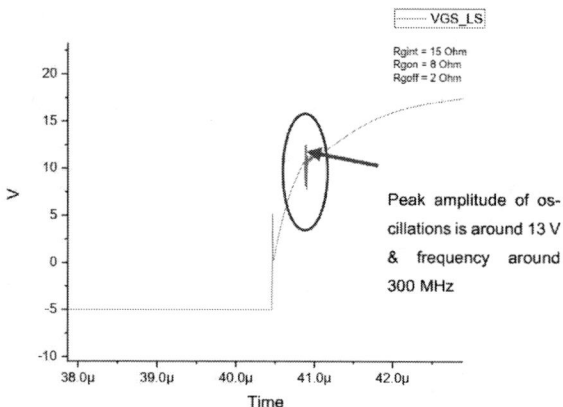

Fig. 15 LS switch gate-source voltage (magnified view)

3.2 Module Layout Optimization

The other method of addressing these oscillations is to alter the internal module layout to enable paralleled SiC body diodes forward voltage drops V_F to be symmetrical, not only during steady state conduction, but especially at the inception of the commutation event to reduce the time-lag gap between these devices. This ensures symmetrical reverse recovery transients, ideally at identical time instances. This approach can theoretically significantly improve the dynamic or transient reverse recovery currents alternating between multiple devices in parallel and leads to lower amplitude

and adequately dampened high-frequency oscillations.

In the presented SiC power module sample prototype, minor internal layout changes were implemented, and the equivalent electrical circuit is presented in **Fig. 16**. In **Fig. 3**, the parasitic inductances L2 and L85 are relocated at nodes between L4, L3 and L10, L44 as seen in **Fig. 16**. The new location of these stray inductances will induce almost equal parasitic voltage drop across the voltage loops of individual SiC devices V_{SD4}, V_{SD1}, V_{SD8}, V_{SD5} as a unified group. V_{SD3}, V_{SD2}, V_{SD7}, V_{SD6} is another distinct group. The HS domain voltage drop is measured respectively between DC+ and virtual equipotential node as depicted from the set of voltage equations shown below.

$$V_{SD4} = V_{L2} + V_{L4} + V_{L1} + V_{L6} \tag{18}$$

$$V_{SD3} = V_{L2} + V_{L4} + V_{L7} \tag{19}$$

$$V_{SD2} = V_{L2} + V_{L3} + L_8 \tag{20}$$

$$V_{SD1} = V_{L2} + V_{L3} + V_{L5} + V_{L9} \tag{21}$$

$$V_{SD5} = V_{L85} + V_{L44} + V_{L32} + V_{L31} \tag{22}$$

$$V_{SD6} = V_{L85} + V_{L44} + V_{L37} \tag{23}$$

$$V_{SD7} = V_{L85} + V_{L10} + V_{L36} \tag{24}$$

$$V_{SD8} = V_{L85} + V_{L10} + V_{L50} + V_{L30} \tag{25}$$

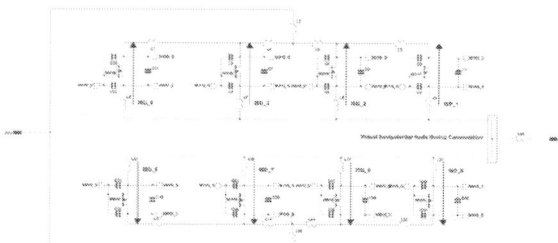

Fig. 16 Electrical circuit schematic of new layout

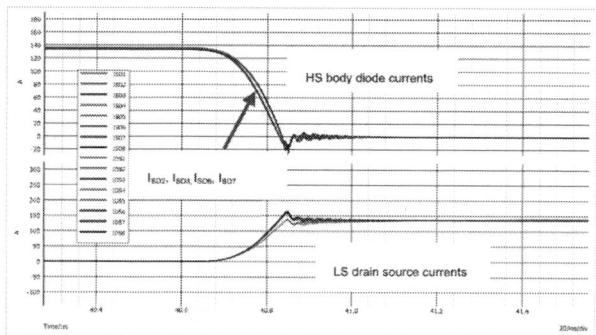

Fig. 17 LS: I_{DS} currents and HS: I_{SD} currents during commutation (optimized layout)

From the simulated optimized layout power module waveforms under similar conditions of V_{DC}= 850 V, I_{LOAD}= 1100 A, R_{gonext}=8 Ω, $R_{goffext}$= 2 Ω, T_J=25°C, DUT=LS, Internal R_{gint} of die=5 Ω, the peak amplitude of gate oscillations is measured at 10 V with a frequency of around 300 MHz as shown in **Fig. 18**.

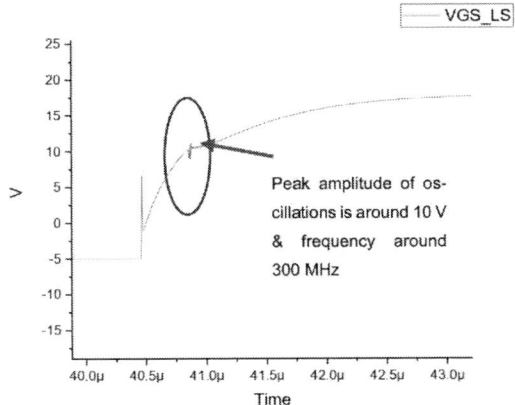

Fig. 18 LS switch gate-source voltage (magnified view)

4 Conclusion

The high-frequency gate turn-on oscillations in SiC MOSFET power modules for automotive traction inverter applications is formulated and the underlying phenomenon is established based on theory and simulation results with an actual power module prototype. The imbalance in the SiC device body diode forward voltage drops V_F of the complimentary switch results in asymmetrical transient switching currents among parallel devices during the commutation period. This further couples with module parasitics to induce circulating gate currents among parallel SiC MOSFETs, leading to high frequency oscillations on V_{GS}, V_{DS} and I_{DS}. Optimizing the module layout or using internal chip level R_{gint} were successfully implemented and validated as potential mitigation solutions.

References

[1] Roveendra Paul, Leon Zhang, Rajani Thirukoluri and James Victory, "A Modelling Approach Towards High Power Module Design," 2023 PCIM.

PCIM Europe 2024, 11– 13 June 2024, Nuremberg   DOI: 10.30420/566262257

A Dynamic Current Balancing Method Using Full-Coupled Inductors in Paralleled Gate Branches

Jianwei Lv[1] , Jiaxin Liu[1], Yiyang Yan[1], Zexiang Zheng[1], Baihan Liu[1], Cai Chen[1] , Yong Kang[1]
[1] State Key Laboratory of Advanced Electromagnetic Technology, Huazhong University of Science and Technology, China

Corresponding author: Cai Chen, caichen@hust.edu.cn
Speaker: Jianwei Lv, jianweil@hust.edu.cn

Abstract

Dynamic currents between paralleled SiC MOSFETs have to be balanced to extend the device's lifetime. Inserting driving-loop full-coupled inductors is a promising method with low volumes, but cannot reduce the current difference caused by unbalanced parasitic power-driving mutual inductances. To address the problem, this paper presents a method with full-coupled inductors in the paralleled gate branches, which can simultaneously reduce the unbalanced dynamic currents under unbalanced parasitic power inductances and power-driving mutual inductances. The experimental results show a 90.3 % turn-on current difference reduction compared to the non-optimized circuit, and a 71.5 % reduction compared to the existing method.

1   Introduction

Recently, SiC devices have been widely used in many applications [1]. Limited by the manufacturability, the maximum current rating of SiC MOSFETs is lower than 120 A [2]. So, paralleled SiC MOSFETs are usually used to meet the current capacity in high power asplications. However, unbalanced parasitic self- and mutual inductances can cause unbalanced dynamic currents among the paralleled chips, which can cause unbalanced switching losses and junction temperatures, degrading device reliability and lifetime [3]. It is essential to balance the dynamic current. The current balancing method can be divided into three directions: symmetrical layout design, passive methods using inserted passive components, and active methods using active auxiliary circuits. Passive methods don't require complex symmetrical circuit layout design, do not change the original circuit layout and terminal location, and do not require additional active devices and complex auxiliary circuits compared with the active methods. [4] and [5] inserted full-coupled inductors into the power branches to achieve balanced dynamic currents. However, the inductors have to bear large power currents and have large volumes. Moreover, the linkage inductance of the inductors can increase the loop parasitic inductance, inducing larger turn-off overvoltage. [6] added full-coupled inductors into the paralleled driving branches, which only

bare small driving currents and have small volumes. However, this method can only reduce the imbalanced currents caused by the unbalanced self-inductances, and cannot reduce that caused by the unbalanced mutual inductances. To address this problem, this paper presents a new passive current balancing method adding full-coupled inductors into paralleled gate branches, which can simultaneously reduce the current imbalance caused by the unbalanced self- and mutual inductances.

2   Causes of dynamic current imbalances

Figure 1(a) shows a typical DPT(double pulse test) circuit with two paralleled SiC MSFETs and parasitic inductances. There are parasitic mutual inductances between the power branches(L_{d1}, L_{d2}, L_{s1}, L_{s2}) and the driving branches(L_{gin1}, L_{gin2}, L_{k1}, L_{k2}). During the current communicating period, Q1 and Q2 can be considered as current sources i_1 and i_2 controlled by v_{gs1} and v_{gs2}:

$$\begin{cases} i_1 = g_m \ (v_{gs1} - V_{th}) \\ i_2 = g_m \ (v_{gs2} - V_{th}) \end{cases}, \quad (1)$$

where g_m and V_{th} are the MOSFET transconductance and threshold voltage, which should be symmetrical between the two paralleled MOSFETs.

1872

PCIM Europe 2024, 11– 13 June 2024, Nuremberg DOI: 10.30420/566262257

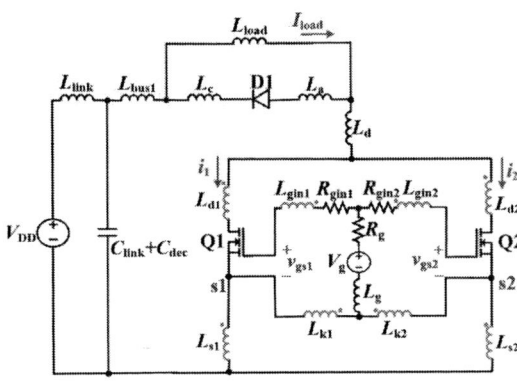

(a) Circuit model.

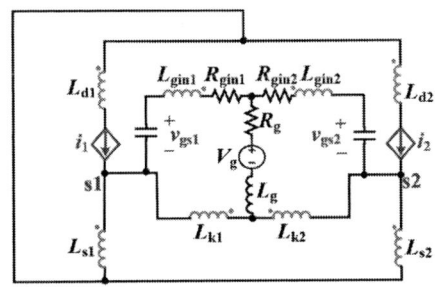

(b) Equivalent circuit model.

Fig. 1 DPT circuit with two paralleled SiC MOSFETs.

The MOSFET gate-source electrodes can be equivalent to input capacitance $C_{iss}=C_{gd}+C_{gs}$. Based on the analysis, and ignoring the power impedances outside the paralleled branches, the circuit can be equivalent into Fig. 1(b). From Eq. (1), the unbalanced current can be written as

$$\Delta i = g_m \, \Delta v_{gs}. \tag{2}$$

So, the dynamic current imbalance is caused by unbalanced v_{gs}. It can be concluded from Fig. 1(b) that there are two effects of the power currents influencing the driving circuit and inducing Δv_{gs}:

1) The unbalanced parasitic power inductances can induce unbalanced voltage potential v_{s1s2} between s1 and s2, which is equal to $v_{s1s2}=(L_{s1}-L_{s2}+M_{d1s1}+M_{d2s1}-M_{d1s2}-M_{d2s2})di_d/dt$ when supposing $di_1/dt=di_2/dt$. Then v_{s1s2} causes unbalanced v_{gs}, as shown in Fig. 2(a), where the parasitic mutual inductances between the gate branches and the drive-source branches are ignored, and the parameters in the driving loop are supposed to be symmetrical [2]. The induced Δv_{gs-1} is expressed as

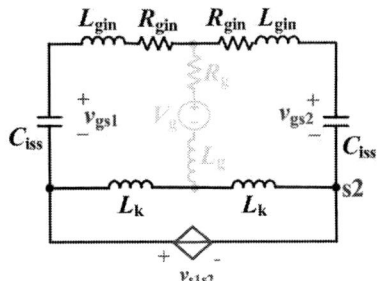

(a) Under influence of v_{s1s2}.

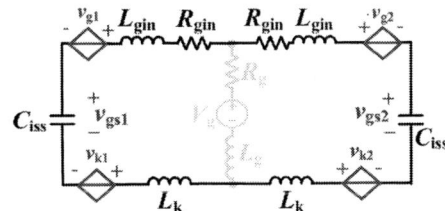

(b) Under influence of v_g and v_k.

Fig. 2 Driving circuit models.

$$\Delta V_{gs-1} = -2\Delta I_{g-1}\frac{1}{sC_{iss}} = -\frac{V_{s1s2}}{Z_g}\frac{1}{sC_{iss}} \tag{3}$$

Where $Z_g=R_{gin}+sL_{gin}+1/sC_{iss}$ is the gate impedance.

2) The unbalanced mutual inductance between the power branches and the driving branches can induce unbalanced induced voltages, causing Δv_{gs}, as shown in Fig. 2(b). v_{g1}, v_{g2}, v_{k1}, and v_{k2} are the induced voltages, and can be expressed as Eq. (4) when supposing $di_1/dt=di_2/dt$.

$$\begin{cases} v_{g1}=\left(M_{d1g1}+M_{s1g1}+M_{d2g1}+M_{s2g1}\right)di_d/dt \\ v_{g2}=\left(M_{d1g2}+M_{s1g2}+M_{d2g2}+M_{s2g2}\right)di_d/dt \\ v_{k1}=\left(M_{d1k1}+M_{s1k1}+M_{d2k1}+M_{s2k1}\right)di_d/dt \\ v_{k2}=\left(M_{d1k2}+M_{s1k2}+M_{d2k2}+M_{s2k2}\right)di_d/dt \end{cases} \tag{4}$$

The induced Δv_{gs-2} can be expressed as

$$\Delta V_{gs-2} = -2\Delta I_{g-2}\frac{1}{sC_{iss}} = -\frac{\Delta V_M}{Z_g+Z_k}\frac{1}{sC_{iss}}, \tag{5}$$

where $Z_k=sL_k$ is the drive-source impedance. ΔV_M is the unbalanced induced voltage and is caused by the unbalanced power-driving mutual inductances ΔM, and can be expressed by

$$\Delta V_M = V_{g1}-V_{g2}-(V_{k1}-V_{k2})=s\Delta MI_d, \tag{6}$$

where ΔM is expressed as

1873

PCIM Europe 2024, 11– 13 June 2024, Nuremberg   DOI: 10.30420/566262257

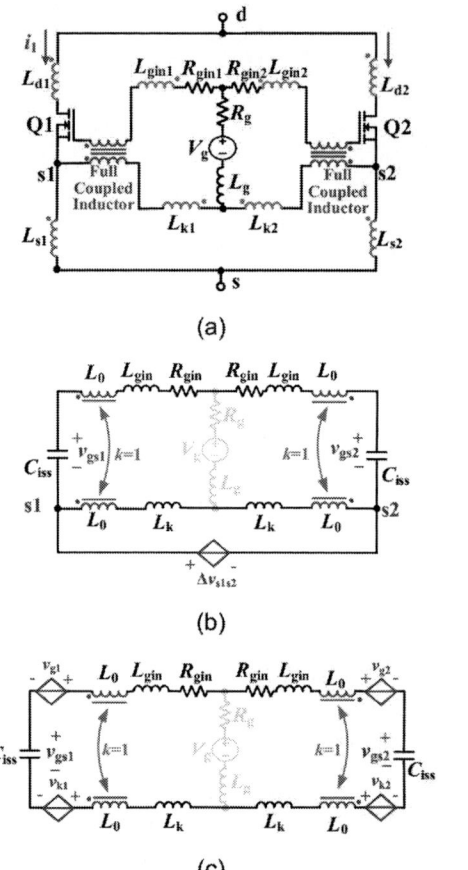

(a)

(b)

(c)

Fig. 3   Method in [6] with full-coupled inductors in the paralleled driving loops.

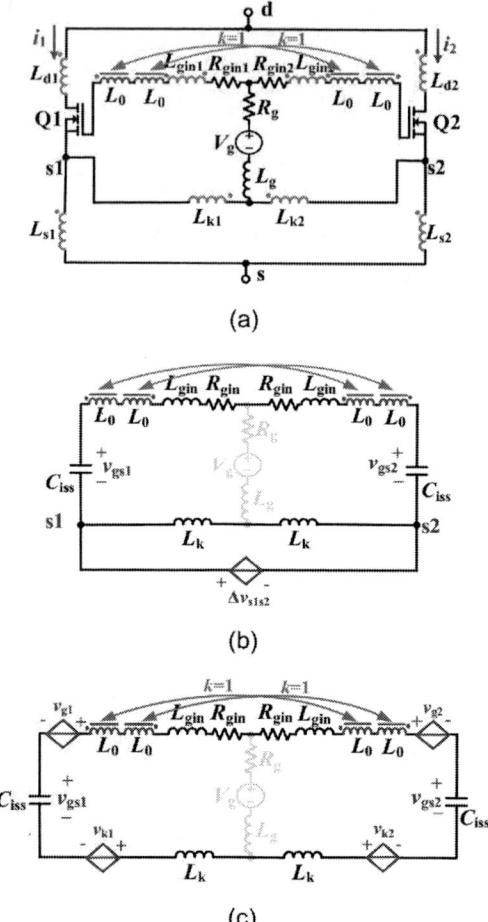

(a)

(b)

(c)

Fig. 4   Proposed method with two full-coupled inductors in the paralleled gate branches.

$$\Delta M = ((M_{d1g1} + M_{s1g1} + M_{d2g1} + M_{s2g1}) - (M_{d1g2} + M_{s1g2} + M_{d2g2} + M_{s2g2})) - ((M_{d1k1} + M_{s1k1} + M_{d2k1} + M_{s2k1}) - (M_{d1k2} + M_{s1k2} + M_{d2k2} + M_{s2k2})). \quad (7)$$

So, the unbalanced power-source inductances and power-driving mutual inductances can cause unbalanced v_{gs} and then cause unbalanced dynamic currents.

3   Dynamic current balancing method

The passive method presented by [6] is shown in Fig. 3(a), where full coupled inductors are added into each paralleled driving loop. The driving circuit under difference effects are shown in Fig. 3(b) and Fig. 3(c). supposing sL_0 is much larger than Z_g, $\Delta v_{gs\text{-}1}$ and $\Delta v_{gs\text{-}2}$ can be expressed by

$$\begin{cases} \Delta v_{gs\text{-}1} = -2\Delta I_{g\text{-}1} \dfrac{1}{sC_{iss}} = -\dfrac{\Delta V_{s1s2}}{sL_0(Z_g/Z_k + 1)} \dfrac{1}{sC_{iss}} \\[4mm] \Delta v_{gs\text{-}2} = -2\Delta I_{g\text{-}2} \dfrac{1}{sC_{iss}} = -\dfrac{\Delta V_M}{Z_g + Z_k} \dfrac{1}{sC_{iss}} \end{cases}, \quad (8)$$

where L_0 is the self-inductance of the full coupled inductors. The $\Delta v_{gs\text{-}1}$ induced by the unbalanced power source inductances can be reduced by increasing L_0, but L_0 has no influence on $\Delta v_{gs\text{-}2}$. It can be explained by Fig. 3(b) that increasing L_0 can increase the impedance under v_{s1s2}, reducing $\Delta v_{gs\text{-}1}$. However, the full-coupled inductor has no influence on the impedance under the induced voltages, so $\Delta v_{gs\text{-}2}$ cannot be reduced.

To simultaneously reduce $\Delta v_{gs\text{-}1}$ and $\Delta v_{gs\text{-}2}$, this paper presents a new passive method as shown in Fig. 4(a). The two full-coupled inductors are moved into the paralleled gate branches. The driving circuit under difference effects are shown in Fig. 4(b) and Fig. 4(c). supposing sL_0 is much

1874

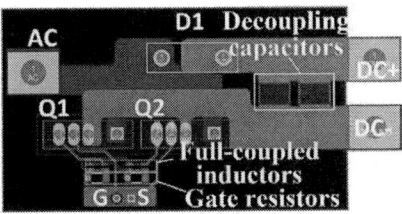

(a) Referenced method in [6].

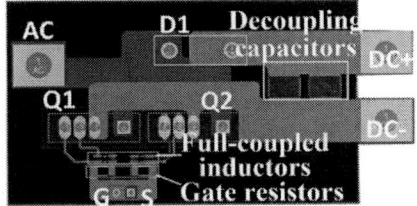

(b) Proposed method.

Fig. 5 Experimental PCB.

larger than Z_g and Z_k, $\Delta v_{gs\text{-}1}$ and $\Delta v_{gs\text{-}2}$ can be expressed by

$$
\begin{cases}
\Delta v_{gs\text{-}1} = -2\Delta I_{g\text{-}1}\dfrac{1}{sC_{iss}} = -\dfrac{\Delta V_{s1s2}}{s\cdot 4L_0}\dfrac{1}{sC_{iss}} \\[2mm]
\Delta v_{gs\text{-}2} = -2\Delta I_{g\text{-}2}\dfrac{1}{sC_{iss}} = -\dfrac{\Delta V_M}{s\cdot 4L_0}\dfrac{1}{sC_{iss}}
\end{cases}
\tag{9}
$$

By increasing L_0, $\Delta v_{gs\text{-}1}$ induced by unbalanced power-source inductances and $\Delta v_{gs\text{-}2}$ induced by unbalanced power-driving mutual inductances can be largely decreased. It can be explained by Fig. 4(a) and (b) that by inserting full-coupled inductors into the gate branches, the circuit impedances under v_{s1s2} and the unbalanced induced voltages can be largely increased, reducing Δv_{gs}. The average gate current is not influenced by the inserted inductors because of the magnetic couplings.

4 Simulation and Experiment Verification

Figure 5 shows the experimental unbalanced PCB layouts with the referenced and proposed method, where two Wolfspeed C3M060065K SiC MOSFETs and one GeneSiC GD60MPS06H SiC diode are used as Q1, Q2, and D1. By Q3D simulation, the circuit parasitic parameters are obtained and the dynamic currents are then simulated. Figure 6 shows the simulated relative maximum dynamic current differences. When there is no coupled inductor (L_0=0), there are large turn-on and turn-off current differences under the unbalanced circuit layouts. By using the referenced method,

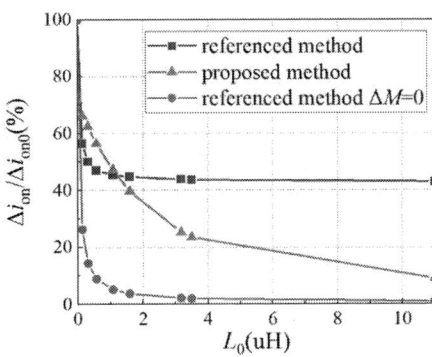

(a) Turn- on results.

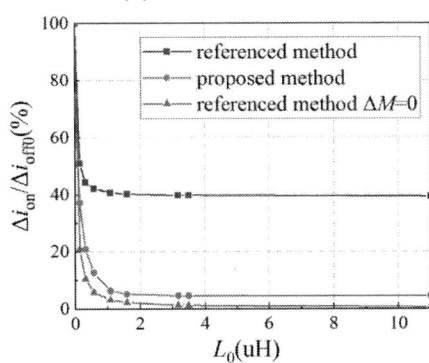

(b) Turn-off results.

Fig. 6 Simulated relative current differences under different L_0.

the unbalanced currents induced by -5.1 nH ΔL_s can be reduced, but there are still 43.1 % and 39.3 % unbalanced turn-on and turn-off currents caused by -0.72 nH ΔM when Ls_0=11 uH. If eliminating ΔM, the dynamic current can be well balanced as the green curve. When using the proposed method, the turn-on and turn-off current difference can be reduced by 91 % and 95.5 % under 5.1 nH ΔL_s and 0.5 nH ΔM.

The circuits in Fig. 5 are fabricated, where commercial common-mode chokes are used as the full-coupled inductors. The test setup is shown in Fig. 7. The dynamic currents of Q1 and Q2 are tested by two 120 A/30 MHz Rogowski coils. The tested current waveforms under 10 uH L_0 under 400 V bus voltage and 72 A load current are shown in Fig. 8(a). The tested turn-on and turn-off maximum dynamic current differences of the two circuits are depicted in Fig. 8(b) and Fig. 8(c). It is proved that under ΔL_s and ΔM, the proposed method can achieve more balanced dynamic currents. When L_0=3.5 uH, the turn-on and turn-off current differences can be reduced by 90.3 % and 89.9 % compared to the circuit without coupled in-

Fig. 7 Tested setup.

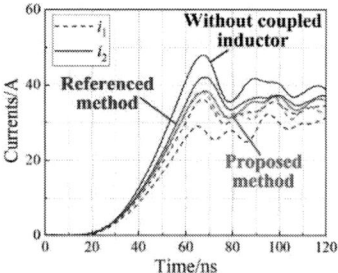

(a) Turn-on waveforms under 3.5 uH L_0.

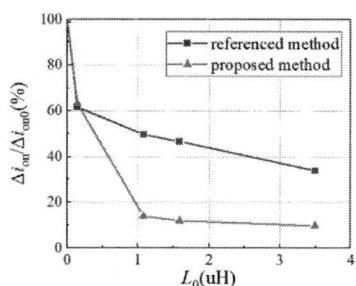

(b) Turn-on current relative differences.

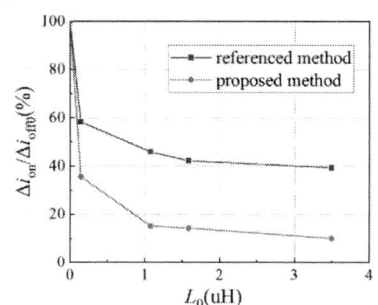

(c) Turn-off current relative differences.

Fig. 8 Tested results under 400 V bus voltage and 72 A I_{load} under different L_0.

ductors (L_0=0 nH), and by 71.5 % and 74 % compared to the existing referenced method. Results under more L_0 values and switching losses results will be presented in the final paper.

5 Summary

This paper presents a passive dynamic current balancing method to simultaneously reduce the current imbalances caused by unbalanced parasitic power inductances and power-driving mutual inductances. By inserting full-coupled inductors into gate branches, the circuit impedance under the two effects can be simultaneously increased. Then the imbalance of v_{gs} and i_d can be reduced. Simulation and experimental verifications are conducted. The unbalanced currents are well reduced under unbalanced power-source inductances and unbalanced power-driving mutual inductances. The experimental results show that the turn-on and turn-off current differences can be reduced by 90.3 % and 89.9 % compared to the circuit without coupled inductors, and by 71.5 % and 74 % compared to the referenced method. Detailed analysis and experimental results will be presented in the final paper.

References

[1] J. Millán, P. Godignon, X. Perpiñà, A. Pérez-Tomás and J. Rebollo, "A Survey of Wide Bandgap Power Semiconductor Devices," IEEE Transactions on Power Electronics, vol. 29, no. 5, pp. 2155-2163, May 2014.

[2] J. Lv, C. Chen, B. Liu, Y. Yan and Y. Kang, "A Dynamic Current Balancing Method for Paralleled SiC MOSFETs Using Monolithic Si-RC Snubber Based on a Dynamic Current Sharing Model," IEEE Transactions on Power Electronics, vol. 37, no. 11, pp. 13368-13384, Nov. 2022.

[3] Y. Ge, Z. Wang, Y. Yang, C. Qian, G. Xin and X. Shi, "Layout-Dominated Dynamic Current Balancing Analysis of Multichip SiC Power Modules Based on Coupled Parasitic Network Model," IEEE Transactions on Power Electronics, vol. 38, no. 2, pp. 2240-2251, Feb. 2023.

[4] Z. Zeng, X. Zhang and Z. Zhang, "Imbalance Current Analysis and Its Suppression Methodology for Parallel SiC MOSFETs with Aid of a Differential Mode Choke," IEEE Transactions on Industrial Electronics, vol. 67, no. 2, pp. 1508-1519, Feb. 2020.

[5] Y. Mao, Z. Miao, C. -M. Wang and K. D. T. Ngo, "Balancing of Peak Currents Between Paralleled SiC MOSFETs by Drive-Source Resistors and Coupled Power-Source Inductors,"

IEEE Transactions on Industrial Electronics, vol. 64, no. 10, pp. 8334-8343, Oct. 2017.

[6] J. Liu and Z. Zheng, "Switching Current Imbalance Mitigation for Paralleled SiC MOSFETs Using Common-mode Choke in Gate Loop," in 2020 IEEE Energy Conversion Congress and Exposition (ECCE), Detroit, 2020, pp. 705-710.

PCIM Europe 2024, 11– 13 June 2024, Nuremberg   DOI: 10.30420/566262258

Quantitative Performance Comparison of Large-Format SiC MOSFET and Si IGBT Modules

Arthur Boutry ⊚[1], Sergio Jimenez[1], Andrew Lemmon[1], Calvin Flack[1]

[1] The University of Alabama, USA

Corresponding author: Arthur Boutry, ajboutry@ua.edu
Speaker: Arthur Boutry, ajboutry@ua.edu

Abstract

In this article, the most recent 1.7 kV SiC MOSFET and Si IGBT large-format modules (high current ratings, 900 to 1000 A) are compared, focusing on their dynamic characteristics. First, the losses and overvoltage characteristics are analyzed using the same Double-Pulse Test platform and the same external gate resistance. The limitations of this study are discussed, and a second comparison is proposed. This second comparison involves normalizing the total gate resistance of each module, including the internal gate resistance. This method leads to narrower differences between modules, although significant variation is still observed.

1   Introduction

SiC (Silicon Carbide) MOSFET devices and modules are becoming widely implemented in power converters, especially at 1200 V and below, due to the potential for dramatic reductions in switching losses. Indeed, at these voltage levels, SiC MOSFET technology is extremely mature, and initial barriers to commercial adoption have been overcome. Recent developments have led to the industrial development of medium-voltage modules (from 1700 V to 3300 V) [1], [2]. While technical challenges remain in this voltage class, this category of applications is considered by many to represent the next frontier for the adoption of SiC technology. This article compares recent large-format power modules with a 1.7 kV voltage rating and current ratings from 900 A to 1000 A. The modules considered in this study all correspond to the industry-standard 100 mm × 140 mm footprint that is shown in Fig. 1. These modules were sourced from three manufacturers and are designated Modules A, B, C (SiC MOSFET) and Modules M, N (Si IGBT) for confidentiality reasons. Some modules are pre-production samples; thus, their electrical and thermal performance may differ from the final production parts. However, this study still provides valuable insight into the current landscape of large-format Si IGBT and SiC MOSFET power modules. This study aims to provide insight into the mod-

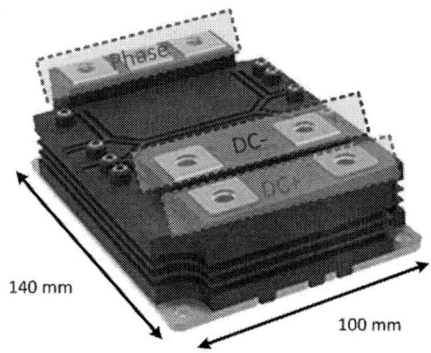

Fig. 1: Industry-standard large-format module footprint.

ules' dynamic characteristics and discuss quantitative performance comparison methods. Indeed, dynamic characteristics are the key to unlocking converter efficiency and design improvements and, therefore, will serve as the focus of this study.

While existing literature often compares a single SiC MOSFET module with a single Si IGBT module, the present study offers a broader perspective by evaluating three SiC MOSFET and two Si IGBT modules having the same package, similar to the one presented in Fig. 1. Moreover, this study focuses on large-format modules with high current ratings (900 or 1000 A). This is another distinction from previous studies involving 1.7 kV modules. For example, the 1.7 kV modules evaluated in [3], [7], [8] have current ratings lower than 450 A. A list of existing articles comparing SiC MOSFET and Si

1878

Article	Voltage Rating (V)	Current Rating (A)	Number of Modules	Year of Publication
[3]	1700	450	1 MOSFET, 1 IGBT	2017
[4]	1700	300	1 MOSFET, 1 IGBT	2015
[5]	1200	300	1 MOSFET, 1 IGBT	2016
[6]	1200	100	1 MOSFET, 1 IGBT	2013
[7]	1700	300	1 MOSFET, 1 IGBT	2020
[8]	1700	325	1 MOSFET, 1 IGBT	2019
[2]	3300	450-600	1 MOSFET, 1 IGBT	2023
This work	1700	900-1000	3 MOSFET, 2 IGBT	2024

Tab. 1: Non-exhaustive but representative listing of publications comparing SiC MOSFET and Si IGBT modules.

IGBT power modules is presented in Table 1.

As far as methodology is concerned, the module comparisons presented in this article are approached in two ways. The first, which is the most classic, is performed using the same external gate resistance (on the gate driver) for both turn-on and turn-off. This approach permits a comparison of multiple modules with the rest of the system remaining constant. However, this comparison might be somewhat unrealistic and biased for multiple reasons discussed later in this article. Due to these limitations, a second comparison is performed, which involves normalizing the total gate resistance for each module, including the contribution from the internal gate resistance (which may be made up of physical resistors within the module as well as the intrinsic gate resistance of the semiconductors themselves).

The contributions of this work are provided by the following elements: the number of modules evaluated, which provides a broader perspective compared to previous studies; the ratings and size of the modules evaluated (1700 V/900-1000 A with a 100 mm × 140 mm footprint); and, the use of multiple methodologies for normalizing the gate resistance within these comparisons.

2 Methodology

2.1 Experimental setup

To evaluate the modules' dynamic characteristics considered herein, a double-pulse test setup (DPT) was used, which is presented in Fig. 2. The same setup was used in a similar study with large-format 3.3 kV modules [2]. This setup includes a capacitor bank of 210 µF, a load inductor of 0.25 mH, and features a total stray inductance of only 12 nH. The V_{DS} voltage measurement is performed using a differential probe (Tektronik THDP0100 6 kV); the

V_{GS} voltage measurement is performed using an IsoVu probe (Tektronik TIVH05); and, the I_D current measurement is performed with a shunt resistor of 2.5 mΩ (T&M Research W-2-0025-4FC). A test matrix describing all the operation points of the DPT experiments conducted as part of this study is presented in Table 2. The Si IGBT modules were only evaluated up to a temperature of 125 °C, representing their maximum rated junction temperature. The SiC MOSFET modules were evaluated up to 175 °C. The current and voltage values were selected to reflect typical working voltages and currents for the modules' ratings. Each module's internal gate resistance (R_{Gin} in Fig. 2) was experimentally evaluated using an E4990A impedance analyzer. The physical resistor used in the gate-drive circuit (called "external gate resistance" in this paper) is displayed as R_{Gext} in Fig. 2. The total gate resistance, which is the sum of R_{Gin} and R_{Gext}, is designated R_{Gtot} in this paper.

	SiC MOSFET	Si IGBT
Voltage (V)	800, 1000, 1200	
Current (A)	250, 500, 750, 1000	
Temp. (°C)	25, 100, 125, 175	25, 100, 125

Tab. 2: Test Parameters (Voltage, Current and Temperature)

2.2 Measurement Methodology

Several figures of merit (FOMs) are utilized for this study. First, switching loss values are quantified using the method developed in [9], [10], which involves applying a Gaussian curve fit to the power waveform. This method has been validated against the standard loss measurements from the JEDEC, IEC, and manufacturers' methods in [10]. Total switching losses (E_{tot}) is the sum of swiching losses at turn-on and turn-off. This method is also ad-

PCIM Europe 2024, 11– 13 June 2024, Nuremberg   DOI: 10.30420/566262258

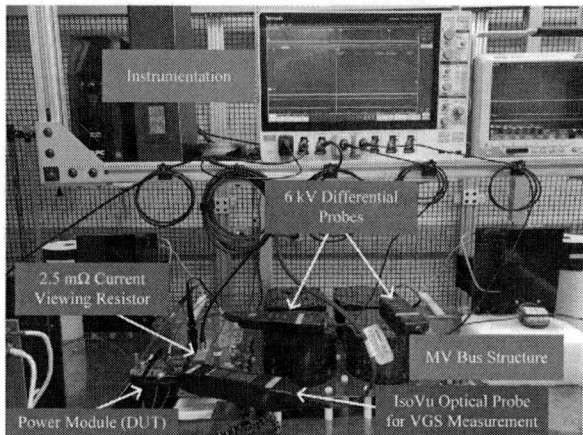

(a) Double-pulse test setup.

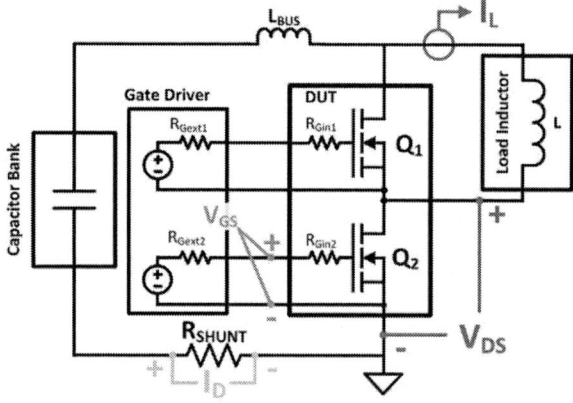

(b) Test setup schematic.

Fig. 2: Double-pulse test setup used in this study.

vantageous for processing large quantities of DPT data in an automated fashion. Second, the overvoltage value is calculated as the difference between $max(V_{ds})$ and V_{bus} (voltage of the capacitor bank) during the turn-off switching transition. Finally, the di/dt and dv/dt values at turn-on or turn-off are calculated using 10% and 90% thresholds for the waveform steady-state values in question.

2.3 Experimental Plan

This study is divided into two main parts. For the first part, each module's external gate resistance is normalized without consideration for the internal gate resistance (R_{Gin}). This first part compares all five modules discussed in this article (i.e. SiC MOSFET and Si IGBT).

For the second part, each module's total gate resistance is normalized by adjusting the external

gate resistance (R_{Gext}) to account for the differences in internal gate resistance (R_{Gin}). The goal of this study is to use a fixed value of the total gate resistance of each module (R_{Gtot}) to extract and compare each module's semiconductor device performance. This study is also more realistic than the first in another way. Power electronic converter designers generally adapt the control circuitry, including R_{Gext}, to the module selected for the design while controlling for other criteria (e.g., maximum dv/dt). The Si IGBT modules are excluded from this second study in order to streamline the analysis. For this second part, the test matrix was also reduced to accelerate the testing procedure. Only temperatures of 25 and 125 °C were explored in this study.

3 Study One

3.1 Measurements and Results

Fig. 3 compares the modules' performance when R_{Gext} is 2 Ω. With regard to switching losses, the SiC MOSFET modules (solid lines) exhibit lower losses than the Si IGBT modules (dashed lines) as expected. Under the most demanding conditions (1.2 kV/1 kA/125 °C), the losses are lower than 0.8 J for the SiC modules, with the lowest losses obtained for Module A (0.21 J). However, this module exhibits the biggest overvoltage (340 V) under the same conditions. These two observations can be explained by observing that the commutation speed of Module A is higher than that of the other modules. The voltage slew rate of Module A is almost three times faster than that of Module B (second fastest dv/dt). At 1 kA, the current slew rate of Module A is almost 3 A/ns higher than Module M (second fastest di/dt). Only the di/dt and dv/dt at turn-on are displayed in Fig. 3, but the same phenomenon is observed at turn-off. The findings are also similar across different temperatures, voltage, and current values. However, these additional results were omitted from this article for clarity and conciseness.

3.2 Limitations

A variety of factors can influence the power module switching performance. One of the most obvious factors is the gate resistance value. The internal gate resistance value, R_{Gin}, is essential to consider; its influence is cumulative with the external gate resistance, R_{Gext}. R_{Gext} was held to 2 Ω in this section. Each SiC MOSFET module internal

1880

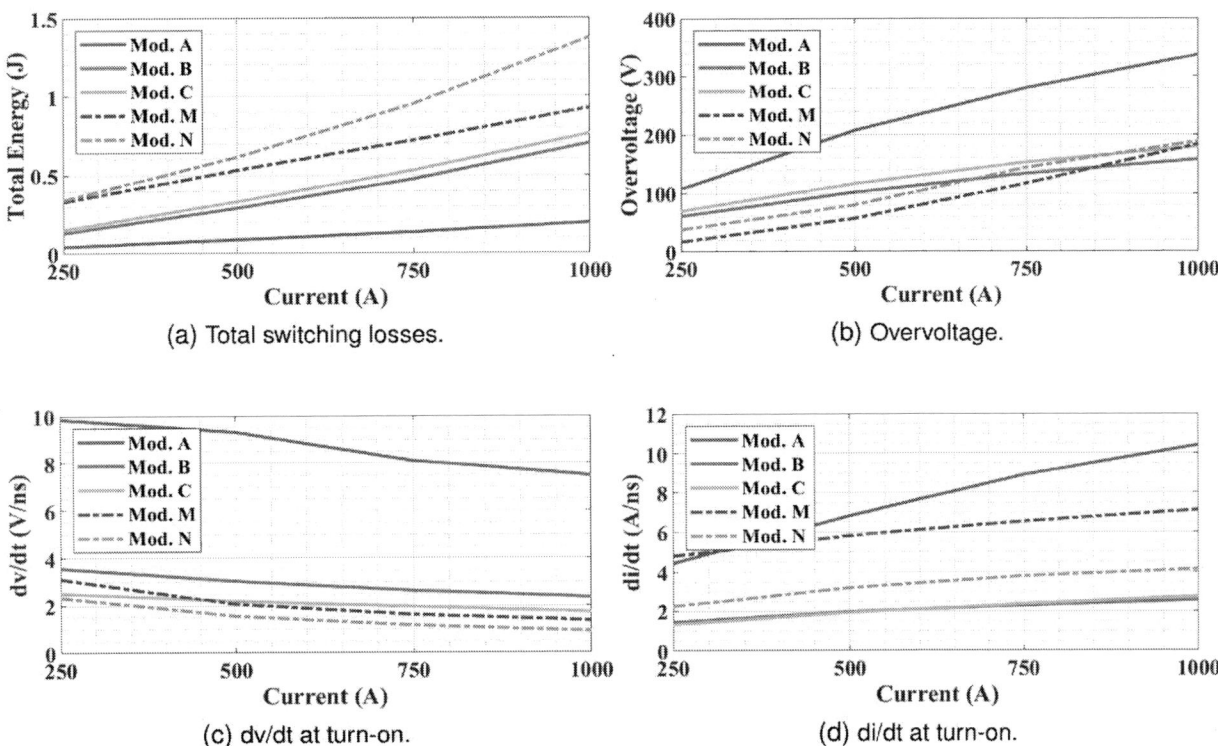

Fig. 3: All modules dynamic measurements, with V = 1.2 kV, $R_{Gext} = 2\,\Omega$ and T = 125 °C.

gate resistance R_{Gin} considered in this study was evaluated experimentally. The evaluation results are displayed in Table 3. These results demonstrate that the R_{Gin} of Module A is $2\,\Omega$ lower than Module B or Module C. Therefore, one natural question is whether the reduced switching losses observed for Module A in this study are solely the result of this difference in R_{Gin}. To quantitatively answer this question, a second experimental study was performed.

Module	Internal Gate Resistance
Mod. A	0.5
Mod. B	3
Mod. C	2.5

Tab. 3: SiC MOSFET module internal gate resistance R_{Gin} (Ω).

4 Study Two

4.1 Direct Comparison

For the second study, only the SiC MOSFET modules were compared. The configurations selected

Module	Config. I	Config. II	Config. III
Mod. A	2.5	3.9	4.5
Mod. B	4.5	3.9	4.5
Mod. C	5	4	4.6

Tab. 4: Total gate resistance for each module in the different configurations, R_{Gtot} (Ω).

for this comparison represent an effort to normalize the total gate resistance value (R_{Gtot}) for each module. Three configurations (I, II, and III) were selected for this study, which are summarized in Table 4. Configuration I is the same configuration used in the first study, and this configuration serves as a point of reference for comparing the results of the two studies. The results for the second study are displayed in Fig. 4. In this study, Module A still exhibits the smallest losses in both configurations, although the difference between Module A and the other modules is reduced compared to Configuration I (see Fig. 3). The edge rate plots of Fig. 4(e)-(h) confirm that the switching speed of Module A is higher than that of Modules B and C. This result indicates that total gate resistance R_{Gtot} is only one of many design parameters influencing commutation speed. Other factors likely influencing this behavior include the semiconductors' transconductance and non-linear capacitance profiles. However, the plots of Fig. 4(c)-(d) demonstrate that the overvoltage difference between modules is significantly different than that observed in Configuration I. For this study, only minor differences in overvoltage are observed (around 30 V difference in the most extreme case). This result suggests that overvoltage may be more strongly influenced by the value of R_{Gtot} than either switching losses or edge rates.

4.2 Introduction of the relative distance to average

A FOM is presented in this section to deliver a clear and concise conclusion for the studies in this paper. This FOM is referred to herein as the "Relative Distance to the Average" (RDA). The goal of this FOM is to present data from different operating conditions, in a unified way, without having to represent every sample point, as done previously in Fig. 3 and Fig. 4. The RDA of a given variable is simply the value of that variable divided by the average of this variable among all the modules at the same operating condition. This FOM can be applied to any of the metrics considered, including switching losses, overvoltage, and edge rates. Equation 1 provides an example of applying RDA to the total switching losses E_{tot}, for Module i, at a given (V_{DS}, I_D and temperature), where n_{mod} is the number of modules being compared.

$$RDA(E_{tot}) = \frac{E_{tot}(Mod_i)}{\frac{\sum_i E_{tot}(Mod_i)}{n_{mod}}} \quad (1)$$

An example may help to clarify the intended use of RDA. If the $RDA(E_{tot})$ for Module A is 0.6 for a given operating point, this would mean that the switching losses for Module A are 40% lower than the average switching losses for Modules A, B, and C considered collectively at this operating point. This simple calculation permits the compact representation of a large quantity of results from different configurations on the same plot. Such a volume of data is effectively displayed using box plots. Box plots are particularly useful for clearly presenting the median value and the variability and spread around it. This approach also permits leveling the importance of any particular operating point, as the same importance is given to all considered conditions. This stands in contrast to a direct comparison, in which disproportionate importance is often given to high voltage and high current operation. However, this method can only be used as a relative comparison. In this context, the use of RDA permits the investigation of the following questions:

- What is the overall relative performance of each module, broadly considered across all operating conditions?

- What are the evolutions of the spread for each measured quantity across the considered configurations?

- What measured quantities are most and least affected by the configuration changes considered herein?

4.3 Analysis through relative distance to average

Figure 5 presents the RDA FOM distribution for the total switching energy E_{tot} and overvoltage for the three configurations discussed in the previous section. For Configurations I and II, only the measurements at 25 °C are included; while for Configuration III, only the 125 °C measurements are included. In an extension of this study, additional measurements, configurations, and temperatures will be considered. However, this preliminary comparison already provides considerable insight into the questions raised in the last subsection.

First, the switching loss trend for the three SiC modules is observed to be independent of the configuration. Irrespective of the methodology used for normalizing the gate resistance, Module A exhibits lower losses than Module B, and Module B exhibits lower losses than Module C. However, the difference between Module A and Modules B and C is narrower for Configurations II and III than for Configuration I (around 30% for Config. III, around 40% for Config. II, and around 60% for Config. I). This indicates that comparing SiC modules at the same R_{Gtot} levels the switching loss differences between the modules, but only to some extent. Another interesting result is that the distribution of the total switching energy RDA for a given module is narrow. This indicates that the modules always perform the same way relative to one another in terms of losses, and this trend is observable among numerous operating conditions and gate-resistor configurations.

Second, for both the total energy RDA and the overvoltage RDA, the distribution changes from

PCIM Europe 2024, 11– 13 June 2024, Nuremberg   DOI: 10.30420/566262258

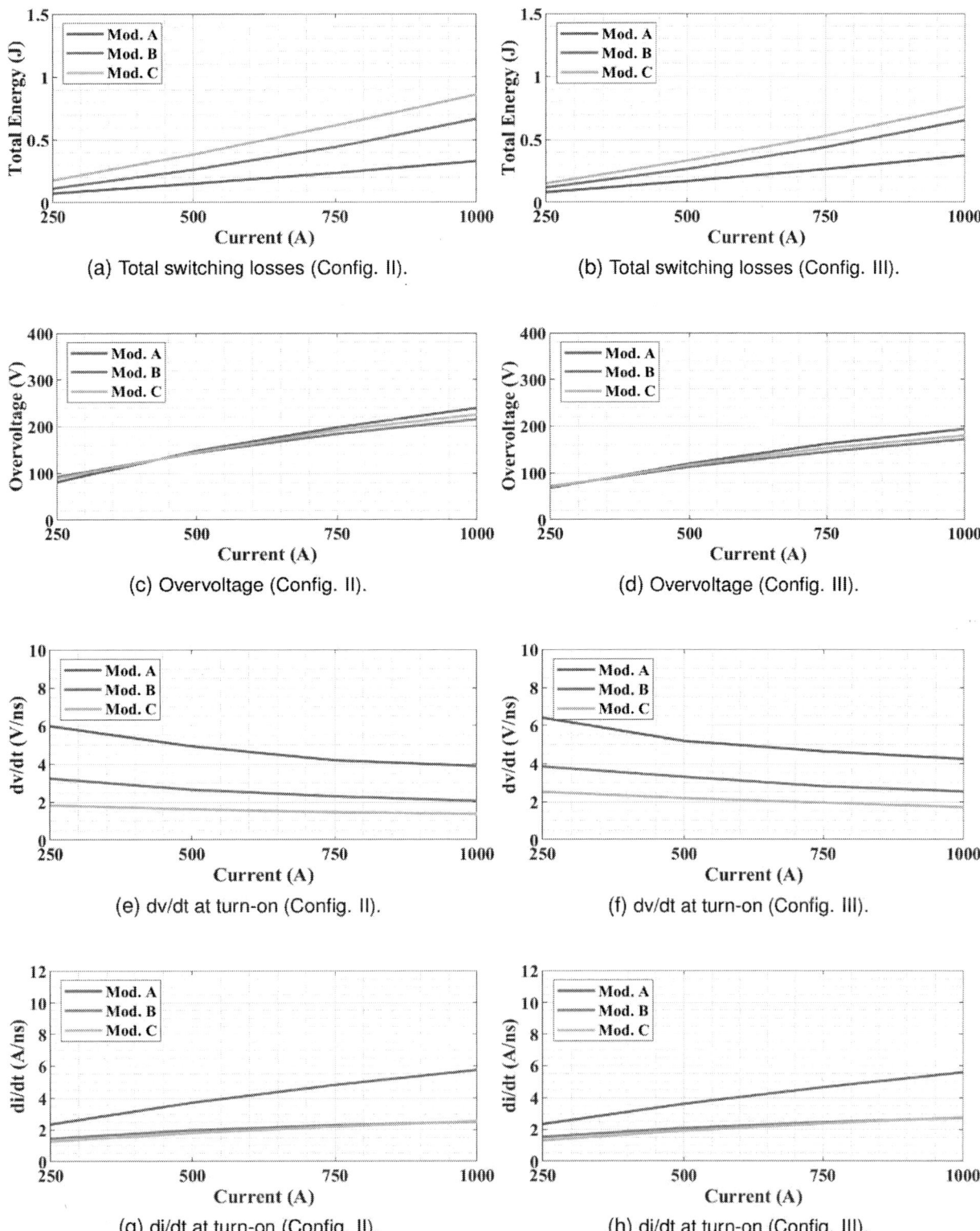

(a) Total switching losses (Config. II).

(b) Total switching losses (Config. III).

(c) Overvoltage (Config. II).

(d) Overvoltage (Config. III).

(e) dv/dt at turn-on (Config. II).

(f) dv/dt at turn-on (Config. III).

(g) di/dt at turn-on (Config. II).

(h) di/dt at turn-on (Config. III).

Fig. 4: SiC MOSFET modules' dynamic characteristics measurements when similar total gate resistance values are used. Left Column: Configuration II (25 °C), Right Column: Configuration III (125 °C). V_{DS} for all points is 1200 V. Scales are the same as Fig. 3 to facilitate the comparison with Configuration I.

1883

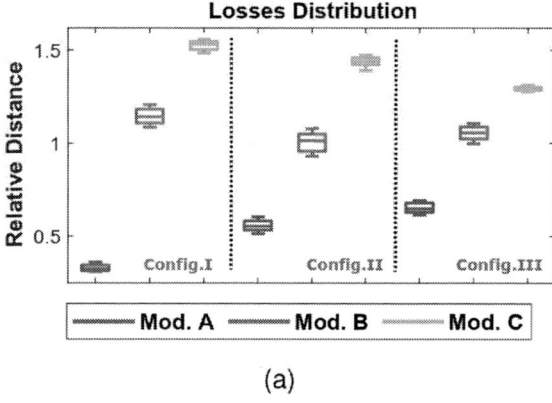

(a)

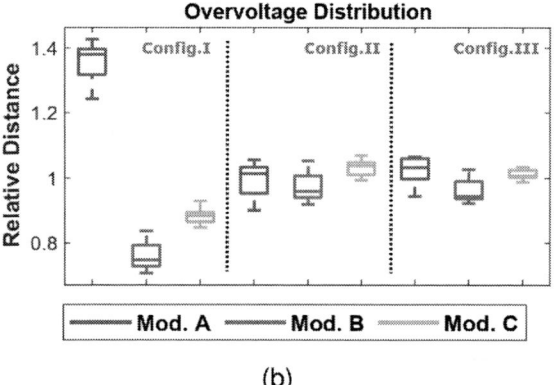

(b)

Fig. 5: Distribution of the Relative Distance Average (RDA) for different measurements in multiple configurations.

the normalized-R_{Gext} configuration (Config. I) to normalized-R_{Gtot} configurations (Config. II and Config. III). These differences illustrate the value of using this method for comparing modules' dynamic characteristics. For example, the difference in losses shrinks for the normalized-R_{Gtot} configurations relative to the normalized-R_{Gext} configuration. However, the same hierarchy is observed clearly for all three cases. On the other hand, the overvoltage values are very similar for the normalized-R_{Gtot} configurations, while a clear hierarchy is apparent for the normalized-R_{Gext} configuration. These results demonstrate that the measured quantities evolve differently when the configuration changes from one with the same external gate resistance value to configurations with similar total gate resistance.

5 Conclusion

This article describes an experimental comparison of the latest generation large-format 1.7 kV SiC MOSFET and Si IGBT power modules. The stud-

ies included in this article provide insight into the expected performance of the latest generation of these power modules in terms of switching losses, overvoltage, and switching edge rates. In addition, these studies evaluate the impact of normalizing the gate resistance in two different ways when performing comparisons between modules. This evaluation demonstrates that a portion of the observed performance difference can be accounted for by normalizing the total gate resistance (R_{Gtot}). However, this approach does not fully explain the observed differences in dynamic performance among the considered modules. The residual differences are believed to be associated with the relative performance of the semiconductors themselves. Additional research is needed to explore advanced methods for comparing the performance of emerging large-format power modules and to reveal the underlying trends that are present in such comparisons.

References

[1] Y. Zhang, R. Wu, F. Iannuzzo, and H. Wang, "Aging investigation of the latest standard dual power modules using improved interconnect technologies by power cycling test," *Microelectronics Reliability*, vol. 138, p. 114 740, Aug. 2022. DOI: 10.1016/j.microrel.2022.114740.

[2] C. Flack, A. Lemmon, J. Helton, S. Jimenez, B. Deboi, and C. New, "Evaluation of next generation MV SiC power modules," in *PCIM Europe 2023; International Exhibition and Conference for Power Electronics, Intelligent Motion, Renewable Energy and Energy Management*, 2023, pp. 1–7. DOI: 10.30420/566091116.

[3] S. Acharya, X. She, R. Datta, M. H. Todorovic, and G. Mandrusiak, "Comparison of 1.7kV, 450A SiC-MOSFET and Si-IGBT based modular three phase power block," in *2017 IEEE Energy Conversion Congress and Exposition (ECCE)*, 2017, pp. 5119–5125. DOI: 10.1109/ECCE.2017.8096862.

[4] J. Rąbkowski and T. Płatek, "Comparison of the power losses in 1700V Si IGBT and SiC MOSFET modules including reverse conduction," in *2015 17th European Conference on Power Electronics and Applications (EPE'15 ECCE-Europe)*, 2015, pp. 1–10. DOI: 10.1109/EPE.2015.7309444.

[5] A. Albanna, A. Malburg, M. Anwar, A. Guta, and N. Tiwari, "Performance comparison and device analysis between Si IGBT and SiC MOSFET," in *2016 IEEE Transportation Electrification Conference and Expo (ITEC)*, 2016, pp. 1–6. DOI: 10.1109/ITEC.2016.7520242.

[6] G. Wang, F. Wang, G. Magai, Y. Lei, A. Huang, and M. Das, "Performance comparison of 1200V 100A SiC MOSFET and 1200V 100A silicon IGBT," in *2013 IEEE Energy Conversion Congress and Exposition*, 2013, pp. 3230–3234. DOI: 10. 1109/ECCE.2013.6647124.

[7] M. J. Rogers, E. R. Motto, and M. Steiner, "Performance comparison of state-of-the-art 300A/1700V Si IGBT and SiC MOSFET Power Modules," *IEEE Power Electronics Magazine*, vol. 7, no. 3, pp. 44–51, 2020. DOI: 10.1109/ MPEL.2020.3011776.

[8] L. Zhang, X. Yuan, X. Wu, C. Shi, J. Zhang, and Y. Zhang, "Performance evaluation of high-power SiC MOSFET Modules in comparison to Si IGBT Modules," *IEEE Transactions on Power Electronics*, vol. 34, no. 2, pp. 1181–1196, 2019. DOI: 10.1109/TPEL.2018.2834345.

[9] B. Bryant, A. Lemmon, B. DeBoi, and C. New, "Improved methodology for estimating switching losses of wide band-gap semiconductors using gaussian curve fitting," in *2022 IEEE Applied Power Electronics Conference and Exposition (APEC)*, 2022, pp. 915–922. DOI: 10.1109/ APEC43599.2022.9773708.

[10] B. Bryant, A. Lemmon, B. DeBoi, and C. New, "Improved methodology for estimating switching losses of wide band-gap semiconductors using gaussian curve fitting," in *2022 IEEE Applied Power Electronics Conference and Exposition (APEC)*, 2022, pp. 915–922. DOI: 10.1109/ APEC43599.2022.9773708.

PCIM Europe 2024, 11– 13 June 2024, Nuremberg   DOI: 10.30420/566262259

Solder Preform Technology for Improved Thermomechanical Performance in Molded Power Module Package-Attach

Joseph Hertline[1], Ryan Mayberry[1], Andreas Karch[3], Aaron Hutzler[4]

[1] Indium Corporation, USA
[2] Indium Corporation, USA
[3] Indium Corporation, Germany
[4] BondPulse GmbH, Germany

Corresponding author: Joseph Hertline, jhertline@indium.com
Speaker: Joseph Hertline, jhertline@indium.com

Abstract

For power modules, the attach to the heat sink plays a major role in overall performance. Soldering is a cost effective and reliable solution, and SnSb5 is most commonly used in this application. Nevertheless, state-of-the-art SnSb5 alloys can cause mold delamination of the module due to the high peak temperature required. Therefore, a low-temperature Pb-free alloy alternative offering similar reliability is discussed. This SAC-In alloy technology reduces peak temperatures, removing the risk of delamination with molded package power modules. This Bi-free alloy provides favorable mechanical properties to achieve reliability performance that is comparable to SnSb5 for thermal shock testing -40°/+125°C.

1   Introduction

To this point, a critical trade-off between cost, complexity, and performance has been required when designing in a thermal management/mechanical attach solution for power modules. Thermal management and long-term reliability are two critical aspects in nearly all areas of electronic packaging. With the drive for reduced size, weight, and power consumption enabled by SiC or GaN technologies, system designs are focused on consolidating to a reduced number of power module packages [1, 2, 3]. This is especially the case for water cooled systems such as automotive and e-mobility applications with a maximum fluid temperature of 80°C - 90°C. Consequently, these individual modules must achieve higher performance (voltage, power density) to achieve system objectives. One key factor in this design is the thermal interface material (TIM) used to reduce the bottleneck in heat dissipation between the power module itself and the heat sink, see Figure 1.

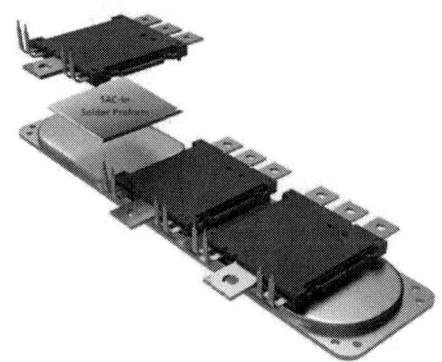

Figure 1: Molded-Package Power Module and Cooler with Thermal Interface Material

1.1   Traditional Thermal Interface Materials

For existing power electronics and inverter platforms, indirect cooling is a commonly-used thermal management strategy. In this case, a thermal interface material is applied between the module and cold plate, and the assembly is mechanically fastened in a clamping fashion. Traditional solutions such as organic, silicone, and/or carbon based (TIM's) are widely used in these designs. However, these traditional TIM's do not always meet the performance criteria due to their low thermal conductivity, particularly in the z-axis for carbon materials [4, 5, 7]. Refer to Figure 2 for a thermal conductivity comparison of various TIM's. Further, these traditional TIM's lack long term stability in the harsh conditions encountered in e-mobility power electronics applications, impacting the life of the system [6].

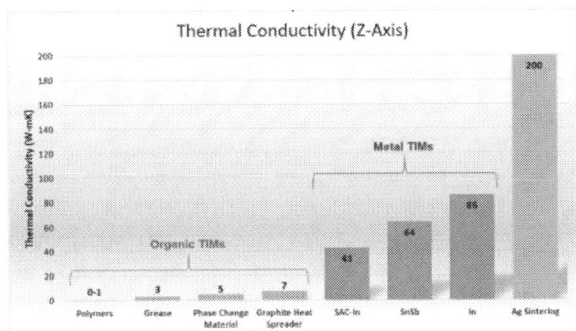

Figure 2: Thermal Interface Material Comparison

1.2 Soldering as a Thermal Interface Material

Alternatively, a soldering process can be used to serve as a thermomechanical interface to attach a SiC molded-package power module to a cooler (Figure 3). In this case, the solder serves as a means for direct cooling of the package through the heatsink.

Figure 3: Cross-Section Example of Molded Package Power Module and Cooler

Most mold epoxies employed in these SiC module packages have a glass transition temperature (Tg) around 200°-220°C [2]. High temperature materials are under development [3] but are either not yet ready for mass production or will lead to increased costs. If the Tg is exceeded during the soldering process, delamination and increased thermo-mechanical stresses occur between the molding and the die/substrate [4,5]. Current tin-based solders such as SAC305 or SnSb require soldering peak temperatures of ~240°C or higher, far exceeding this critical Tg. Therefore, a novel solder alloy technology with a melting range from 190°C to 205°C has been invented. This low temperature solder alloy addresses critical factors including solderability to various surface finishes, high reliability mechanical properties, high thermal conductivity, and avoidance of low melting phases that may compromise the performance of this alloy at elevated operating temperatures.

1.3 Alloy Technology Motivation

The primary novelty of this low temperature solder for package-attach applications is the specific temperature range it can be processed while still maintaining mechanical integrity and thermal performance compared to state-of-the-art SnSb5 alloys and at reduced cost to a sintering alternative. Across various industries, the primary solution for a downstream step soldering process that can be reflowed below 210 - 220°C has been a bismuth-tin based alloy, which has significant drawbacks. Particularly, the brittle nature of bismuth-based alloys lead to poor mechanical strength and reliability [8, 9]. The peak temperature range of 205 - 215°C for this alloy is critical due to its intended primary application of molded power electronics. For these types of packages, epoxy molding compounds are the most common materials for high volume manufacturing due to their glass transition temperature (Tg) of 210 - 220°C [10-12]. This is important since the coefficient of thermal expansion (CTE, α) has a different value depending on whether the temperature is below or above the Tg. Using this information, we can calculate a theoretical situation based on a commercially available EMC with Tg of 210°C. The discrepancy between heating these packages to 210°C (proposed alloy) versus 250°C (standard Pb-free alloy) is drastic regarding warpage and therefore overall material stress.

2. Experimental Methods

The novel SAC-In solder alloy was assembled, reflowed, and tested under thermal shock conditions established in alignment with AQG324 for -40 to +125 °C operating range. Assembly materials were chosen to represent relevant material sets for package-attach application including direct-bonded copper (module base) and Electroless Nickel-plated Copper (cold plate), and then soldered through a formic acid/vacuum reflow process.

The other alloys tested were SnAgIn which is a common low temperature/indium containing alloy, SAC305 which is the industry standard Pb-free alloy, and SnSb which is a popular higher temp. alloy used in various applications including power electronics die and lead frame-attach. The full test matrix in Table 1. The assembly materials are outlined in Table 2, and Figure 4 below shows an image of a cold plate and DBC test sample.

Alloy	Solidus (°C)	Liquidus (°C)
SnAgIn (77.2Sn/20In/2.8Ag)	175	187
SAC-In	190	205
SAC305 (96.5Sn/3Ag/0.5Cu)	217	221
SnSb (95Sn/5Sb)	237	240

Table 1: Test matrix including the alloys tested and their respective solidus and liquidus temperatures

2.1 Materials

Cold Plate	
Plate LxW	50mm x 35mm
Plate Thickness	2mm
Base Metal	Copper
Plating Finish	Electroless Ni, Med. Phosphorus (5-7%)
Plating Thickness	0.5 µm
Direct Bonded Copper (DBC)	
Cu Pad LxW	24.25mm x 24.25mm
Cu Pad Thickness	0.30mm
Core LxW	25.4mm x 25.4mm
Core Thickness	0.30mm
Core Material	Al2O3 (alumina)

Table 2: Full material details for both cold plate and DBC substrates used in this experiment

Figure 5: Cold Plate and DBC Test Sample

2.2 Solder Reflow

The solder reflow process, in order to physically solder the DBC to the cold plate, was carried out using a PiNK VADU100 Vacuum/Formic Acid batch reflow system. The exact reflow profile used for the SAC-In alloy which defines the temperature at the solder location measured by Type K thermocouples, the chamber pressure and the atmosphere at each stage. Not explicitly labelled is the double gas exchange (vacuum to N2) at both the beginning and end of the profile. Figure 4 provides full details of this reflow profile including temperature measurement, chamber pressure and atmosphere conditions (HCOOH = Formic Acid).

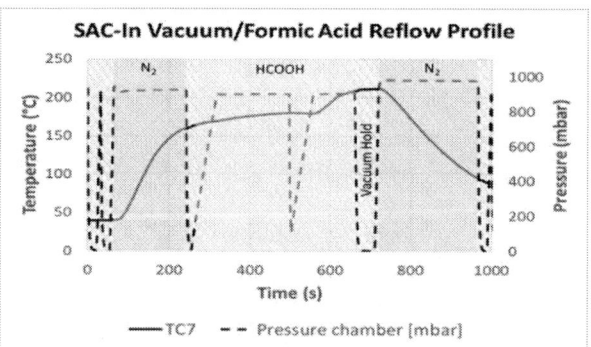

Figure 4: Vacuum/Formic Acid Reflow Profile

2.3 Acoustic Microscopy

The acoustic microscopy imaging was carried out using a Sonix Echo VS Scanning Acoustic Microscope (SAM). For inspection of these assemblies, a 75 MHz transducer was used. Samples were measured before and after thermal shock.

2.4 Thermal Shock Testing (TST)

Thermal shock testing was performed using an ESPEC liquid-to-liquid thermal shock chamber with a target of 1000 cycles over the -40 to +125 °C operating range in accordance with AQG324 guidelines.

2.5 Cross-Section Imaging

After TST, each alloy was cross sectioned through the solder joint and potted in epoxy. From here, the samples were grinded and polished in order to get a clear image of the solder bond between cold plate and DBC. These were imaged using a high magnification optical microscope.

3. Results

As shown by SAM imaging in Figure 6, SAC-In, SAC-305, and SnSb exhibit no degradation after TST. SnAgIn shows clear evidence of delamination and increased voiding after TST.

Further failure analysis was performed with cross-section imaging of each alloy after thermal shock testing. The results in Figure 7 demonstrate correlation with acoustic microscope voiding analysis, indicating robust performance for SAC-In compared to SAC-305 and SnSb. Significant cracking and degradation with the traditional low melting point SnAgIn alloy is also noteworthy.

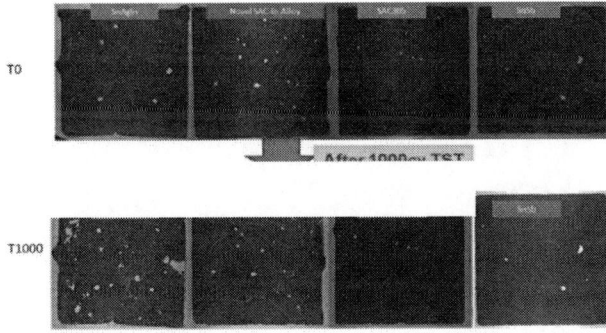

Figure 6: SAM images of each alloy tested before and after 1000 cycles of -40 to 125°C thermal shock

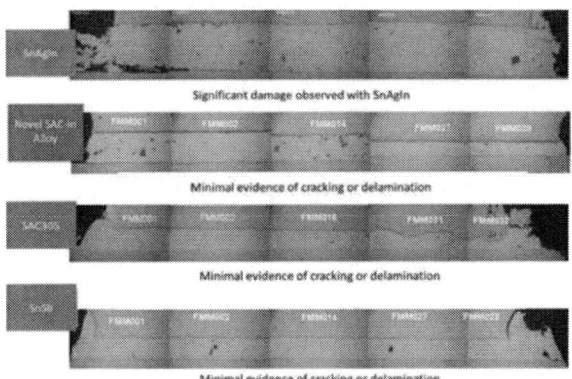

Figure 7: High magnification optical images of each alloy tested after 1000 cycles of -40 to 125°C thermal shock

In summary, the SAC-In alloy achieved reliability performance comparable to SnSb and SAC305 soldering alloys over the -40 to +125 °C operating range while maintaining a lower processing temperature. When compared to another low temperature alloy, SnAgIn, the SAC-In solution achieved far more robust reliability performance.

4. Conclusions and Future Work

The above data demonstrates the feasibility of this SAC-In alloy technology to achieve reduced processing temperatures while sustaining acceptable reliability performance in the -40 to +125 °C operating temperature range. When used as a thermomechanical interface to attach molded package power modules to a cooler, SAC-In sol-der preforms can provide superior thermal performance compared to traditional TIM's to satisfy the requirements in Automotive and e-mobility applications. Finally, a reduced peak reflow temperature prevents delamination within the package, which is a common failure mode with SnSb in this application. This alloy technology can enable manufacturers to leverage proven solder perform reflow techniques and equipment, while minimizing yield loss from delamination, reducing the cost of ownership. Future work is planned to evaluate the soldering performance across additional assembly metallization sets and with alternative reflow environments including flux-assisted formic acid and conventional reflow processes.

In addition, characterization in a representative power cycling use case is also planned as part of the ongoing re-search into this solder alloy technology.

Acknowledgements

The authors would like to recognize the following key contributors to this work:

James McCoy, Indium Corporation
Dr. Hongwen Zhang, Indium Corporation
Dr. Francis Mutuku, Indium Corporation
Kyle Aserian, Indium Corporation
Andreas Karch, Indium Corporation

References

[1] G. Moreno, S. Narumanchi, X. Feng, P. Anschel, S. Myers, P. Keller, "Electric-Drive Vehicle Power Electronics Thermal Management: Current Status, Challenges, and Future Directions," ASME Journal of Electronics Packaging, vol. 10.1115, Jan. 2021

[2] Haksun Lee, Vanessa Smet, and Rao Tummala. "A Review of SiC Power Module Packaging Technologies: Challenges, Advances, and Emerging Issues," IEEE JOURNAL OF EMERGING AND SELECTED TOPICS IN POWER ELECTRONICS, vol 10.1109, 2019

[3] R. Yawger, "PSMA Power Technology Roadmap [PSMA Corner]," IEEE Power Electronics Magazine, vol. 9, no. 2, pp. 10-12, Jun. 2022

[4] H. Wang, D. Ihms, S. Brandenburg, J. Salvador, "Thermal Conductivity of Thermal Interface Materials Evaluated By a Transient Plane Source Method," Journal of Electronic Materials, vol. 48, iss. 7, Sep. 2019

[5] D. Blazej, Nov. 2003, "Thermal Interface Materials," https://www.electronics-cooling.com/2003/11/thermal-interface-materials/, accessed Jul. 2023

[6] C. Johnstone, Jul. 2020, "Thermal Interface Materials: Why TIMs suffer at High Power Density," https://jetcool.com/post/thermal-interface-materials-why-tims-suffer-at-high-power-density/, accessed Jul. 2023

[7] https://www.sciencedirect.com/science/article/pii/S0264127521004901

[8] S. Chantaramanee, P. Sungkhaphaitoon, "Influence of bismuth on microstructure, thermal properties, mechanical performance, and interfacial behavior of SAC305-xBi/Cu solder joints," Transactions of Nonferrous Metals Society of China, vol. 31, iss. 5, pp. 1397-1410, May 2021

[9] M. El Amine Belhadi, S. Hamasha, A. Alahmer, "Effect of Bi content and aging on solder joint shear properties considering strain rate," Microelectronics Reliability, vol. 146, 115020, Jul. 2023

[10] Ohmsemi, "Resin Encapsulation Combined with Insulated Metal Baseplate for Improving Power Module Reliability," https://www.ohmsemi.com/Blog/resin-encapsulation-combined-with-insulated-metal-baseplate-for-improving-power-module-reliability.html, accessed Jul. 2023

[11] A. Teverovsky, "The Significance of Glass Transition Temperature of Molding Compounds for Screening and Reliability Qualification of COTS PEMs," NASA, Jan. 2003

[12] Shin-etsu Chemical Co, Ltd., https://www.shinetsu-encap-mat.jp/e/product/03e.html, accessed Jul. 2023

PCIM Europe 2024, 11– 13 June 2024, Nuremberg   DOI: 10.30420/566262261

Effect of Flip-Chip Die-Attach on the Thermal Behavior of Power GaAs Diodes

Felix Steiner[1], Jens Kowalsky[2], VolkerDudek[2], Thomas Blank[1]

[1] Karlsruhe Institute of Technology, Germany
[2] 3-5 Power Electronics GmbH, Germany

Corresponding author: Felix Steiner, felix.steiner@kit.edu
Speaker: Felix Steiner, felix.steiner@kit.edu

Abstract

Gallium arsenide (GaAs) has a thermal conductivity three times lower than silicon (Si). This makes thermal management of GaAs power devices an important aspect when working with such devices. Flip-chipping can shorten the thermal path and lower the thermal resistance. The thermal behavior GaAs pin-diodes rated for 1200 V and 20 A was evaluated in finite element method (FEM) simulations and experimentally. The experimental measurements show that a standard non-flipped device reaches a junction temperature (T_j) of up to 117°C while the flipped variant heats up to 104°C under the same conditions. The simulation results show the same tendency and are in good accordance with the non-flipped devices, jet yield a much lower temperature for the flipped devices as observed in reality.

1   Introduction

Due to their fast switching and high current-carrying capabilities, GaAs pin diodes could become a cost-effective alternative to SiC diodes for high-frequency power electronics applications.[1]
A remaining challenge is the thermal management of GaAs diodes. The thermal conductivity of GaAs is three times lower than that of Si. The thermal conductivity of Si is around 139.4 W/m*K at room temperature,[2] while GaAs is around 50 W/m*K at room temperature and decreases further with higher temperatures.[3] The standard approach would be to shorten the thermal resistance of the die by reducing the die thickness, where the higher critical field strength of GaAs would allow diodes of the same voltage class to be roughly 20 % thinner than a Si alternative.[1] However, the low fracture toughness of GaAs would make producing and handling thinned GaAs diodes challenging.
Shortening the thermal path from the junction to the heatsink can not only be achieved by reducing the substrate thickness but also by mounting the diode flipped. In this way, the heat generated mainly at the diode junction does not need to travel through the thick substrate, making its thickness mostly irrelevant to the thermal performance. For the device in question flip chip mounting could theoretically reduce the thermal resistance of the semiconductor to ~12 % compared to traditional mounting.

In this study, we want to evaluate the effectiveness of a flip-chip assembly for an improved thermal performance of GaAs pin-power-diodes. FEM simulations and an experimental setup will be used to evaluate the "normal" and flip-chip configurations.

2   Simulation

A FEM simulation is carried out in COMSOL Multiphysics. The simulation domain represents the experimental setup and comprises a cutout of the cooler, the thermal interface material, TO-247 lead frame, solder, diode and wire bonds.
The water cooling is modeled as a heat flux resulting from a forced convection flow of water through the cooler. The water temperature is set to 35°C. A further cooling effect is implemented by a natural convection heat flux affecting all surfaces open to air.
The diode is modeled in five domains: Metallization (top and bottom), passivation, substrate and junction. The simulation domain is shown in Fig 1 and 2.
The heating of the diode is approximated by joule heating. A current is imprinted from lead frame leg to lead frame leg, passing through the diode and bond wires. The diode junction materials electrical conductivity is adjusted so the voltage between the lead frame terminals matches the voltage measured in real life, so that the dissipated power is equal in simulation and experiment. The values are fitted to the standard orientation device at 21 A

heating current. No values are changed when the die is flipped.

For the temperature measurement of the die three values are used. These are: The die top-side temperature, an outside temperature as seen by the thermal camera. It is used for comparison with the thermal imaging results. And the maximum and average junction temperature. The average junction temperature is used for comparison with the virtual junction temperature acquired from the experimental setup.

3 Experimental Setup

3.1 Device under Test

GaAs pin power diodes rated for 1200 V and 20 A were used for this evaluation. The dies possess a silver/gold metallization on anode and cathode, which allows soldering on both sides. The test vehicles are built by soldering a GaAs diode die to a TO-247 lead frame with either their cathode or anode facing the substrate. In the following, these variants are called "flipped" (cathode down) and "standard" (anode down). The top side connection is then established by wire bonding, using four 300 μm Al wires. The device under test is shown in fig 3. The lead frame is not molded to enable an analysis using infrared thermography. For this purpose, black paint is applied to the chip topside to achieve a high emissivity. The device is then mounted on a water-cooled aluminum cooler with thermal interface material applied between the two surfaces. The water temperature is held constant at 35 °C.

3.2 Heating

Heating of the die is achieved by the application of a heating current supplied by an electronic load. The load (EA-EL 9750-40 B) applies a rectangular current pulse with rise and fall times of ~ 50 μs. The pulse amplitude is chosen to be 21 A, where

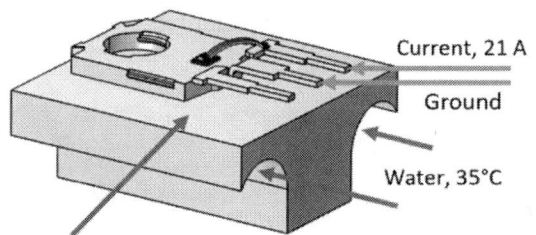

Fig. 1 Simulation domain as implemented in COMSOL Multiphysics.

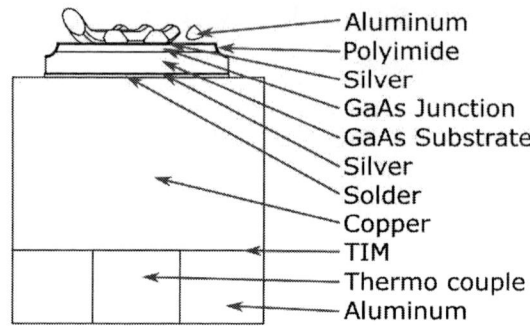

Fig. 2 Simulation domain cross section and material designation

the thermal camera reads 145 °C on the standard device.

3.3 Measurement

A thermocouple sensor is attached to the lead frame back-side through a 1 mm hole in the water cooler.

The infrared camera's emissivity setting is then adjusted so the temperature of the diode is measured to the temperature of the water cooler in unpowered steady state.

The temperature measurement using the thermal camera cannot yield precise temperatures inside the die, due to the temperature gradient from the paint topside to the diode junction. It is also limited to 30 frames per second which is too slow to record the cool-down phase after the heating current is shut off. To measure the diode junction temperature more accurately, the diode forward voltage will be used as temperature sensitive electrical parameter (TSEP).

For this, a small current is constantly forced through the device under test, and the resulting forward voltage is measured. To equate the diode forward voltage to the junction temperature, a cal-

Fig. 3 DUT experimental implementation. 20 A GaAs die in flipped orientation, soldered onto a TO-247 lead frame.

ibration measurement is needed. For this the voltage must be measured at different die temperatures while the measurement current is applied. For this, the devices are placed in an oven where they are heated up to 30, 70, 110 and 150°C. At each step the temperature is left to equalize for 30 minutes. Then measurement currents of 1, 5, 10, 15 and 20 mA are applied, and the forward voltage is measured. For this case, a 10 mA measurement current is chosen, as the self-heating effect is minimal while smaller currents produce a non-linear response at higher temperatures. A schematic of the experimental setup is shown in fig. 4.

As this measurement is only valid when only the 10 mA measurement current flows, only the cooldown phase can be measured. It is important, that the measurement is started quickly after the heating current is turned off, to get an accurate temperature reading. However, due to the slow turn-off of the current source and semiconductor related factors such as the diode capacitance, some delay is required for the measurement to produce realistic results. From that point in time the temperature at the turn-off point can be extrapolated. The extrapolation is done by plotting the temperature over the square root of the time for the first five milliseconds after turn-off. As soon as the temperature begins to drop linearly over the square root of the time, the measurements are valid. The linear temperature can then be extrapolated to get the junction temperature at the time of switch-off.

Due to an inhomogeneous distribution of the current and temperature inside the junction, this value, representing the entire junction, does not supply the maximum junction temperature but an average value, called virtual junction temperature.

4 Results

4.1 Simulation results

A heating current pulse with 10 s on and 10 s off-time is used for comparability with the experiment. At 21 A, the maximum temperature in the diode junction reaches 150.5°C with the normal orientation and 89.1°C when the die is flipped. The die-top-side temperature, as would be seen by the thermal camera, reaches 144.3°C and 90°C. The average junction temperatures, as would be measured by TSEP, reach 121.1°C and 84.6°C. The result is visualized in fig. 5.

4.2 DUT setup

Three devices are manufactured per die orientation variant. The diodes are soldered using SAC preforms, flux and a stencil around the die too

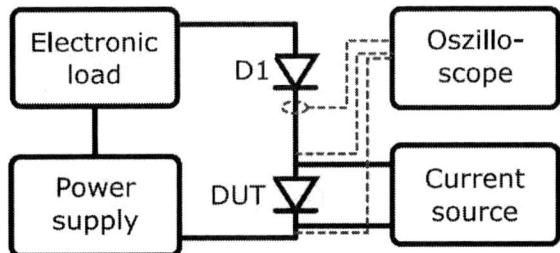

Fig. 4 Schematic of the experimental measurement setup with heating current supply on the left and measurement on the right side. D1 is required to force the measurement current from the current source through the DUT only.

keep the die position stable for optimal comparability with the simulation. The soldering process involves a vacuum section during the first half of time above the liquidus temperature of the solder to minimize voids. The joints are analyzed by scanning acoustic microscopy (SAM). The SAM results in fig. 6 show a low level of voiding with no voids large enough to significantly skew the assumptions made for the simulations.

4.3 TSEP calibration

All six devices were calibrated as described above. Figure 7 shows the voltage to temperature curves obtained during the measurement. In most applications only a few devices are measured, and then average values are used to calculate the slope of

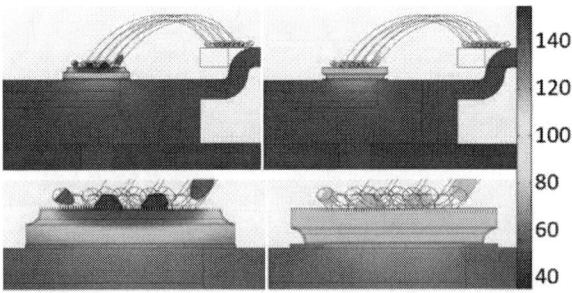

Fig. 5 Simulation result: cross-section through a normal and a flipped device at the end of a ten-second-long current pulse.

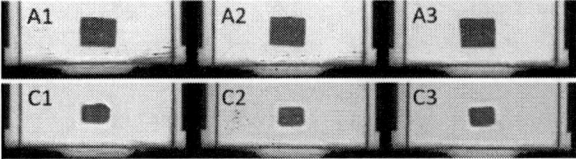

Fig. 6 Scanning acoustic microscopy images of the die bonds. A: normal orientation, B: flipped orientation. The clear contrast and dark soldering areas indicate a good soldering result.

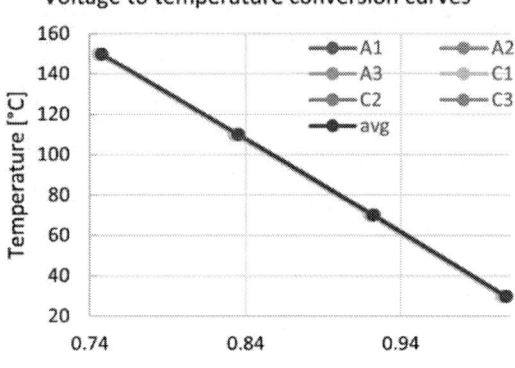

Fig. 7 Forward voltage to temperature conversion curves obtained from calibration at 10 mA and 30, 70, 110 and 150°C.

the conversion curve. An offset correction measurement for each tested die at any temperature is then deemed good enough for thermal characterization. In this case, we use the individual curves measured for the respective device for the best accuracy. Using an averaged method yields a measurement error of more than one Kelvin for some devices.

4.4 Experimental results

The forward voltage signal recorded during cooldown is transformed to a temperature signal using the calibration data acquired in the previous chapter. The temperature is then plotted over the square root of the time. An exponential decay, as is to be expected from the cool-down event, will be represented by a straight line in this setting. Also, any part of the signal, that does not exhibit the characteristics of an exponential decay, is not a valid signal, but an artifact generated by the power source or the diode itself. The extrapolation of an exemplary device is shown in fig. 8. The blue line shows the virtual junction temperature of the die according to the measurement. It is clearly visible, that the first, rising part of the signal is not related to the die temperature. The first straight-line part of the curve is accepted as valid measurement. The virtual junction temperature is then extrapolated to the turn-off time using a linear fit. The extrapolated line is shown in orange. For the shown

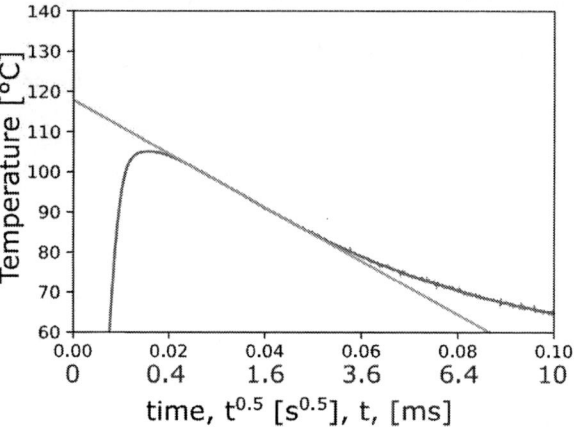

Fig. 8 Temperature measurement by TSEP (blue). And Extrapolation to T_{jmax} by linear curve fit in transformed time frame.

device, a maximum virtual junction temperature of 117.6°C is calculated.

This step is repeated for every device.

The thermography measurement yields measurements of 145°C for the normal device and 119°C for the flipped device.

5 Discussion

The acquired measurements are collected in table 1.

Considering, that the thermal camera does not measure the junction temperature, but the die topside. And that the TSEP measurement corresponds to an average junction temperature, the results for the normal devices fit very well. The die top surface is simulated accurately, and the average junction temperature is also close to the TSEP-measured value.

For the flipped devices, the results do not fit as well. The TSEP-measured values deviate from the simulated result by 15 to 20 K. While the thermal camera shows temperatures, 30 K higher than simulated.

A possible issue with the TSEP measurement is the slow turn-off of the electronic load, forcing a long dead-time and an equally long extrapolation distance, which is can easily introduce large errors.

Table 1 Temperature measurements for normal and flipped devices from different measurements methods.

	Simulation			Experimental			
	T_{jmax}	T_{javg}	$T_{surface}$	IR-cam	Die 1	Die 2	Die 3
normal	150.5°C	121.1°C	144.3°C	145°C	117.6°C	116.1°C	113.6°C
flipped	89.1°C	84.6°C	90.1°C	119°C	104.6°C	101.1°C	104.5°C

The simulation results show that the wire-bonds directly on the die reach higher temperatures than are present in the junction, due to the current flowing through them. This could in part explain the large discrepancy between the measurements of the flipped devices, but should also affect the normal devices, where no significant self-heating of the wires occurs.

When comparing thermal resistance junction-to-water, the simulation predicts a reduction to 47 % by flip-chipping. The experiment yields a reduction down to 82 % at best.

For a better understanding of the differences between simulation model, experimental measurement by TSEP and thermal imaging, a better understanding of the thermal path needs to be acquired. For this, an optimized test bench is required to be able to measure the earlier part of the cooldown more precisely. And thus be able to extract thermal resistances and capacitances of the die and die-attach layers.

6 Conclusion

The effect of the die orientation on the thermal performance of GaAs diodes for power electronics applications was evaluated in a FEM simulation and experimentally. Experimental measurements using a thermal camera and forward voltage measurements yield good results showing that the flipped diode runs 10-15 K cooler than without flip-chipping. The simulation fits the measurements for the normal device very well, but the flipped device is predicted to perform a lot better which leads to a large discrepancy between model and reality. For further analysis the simulation as well as the experimental setup need to be improved upon.

References

[1] J. Kowalsky *et al.*, 'GaAs pin Diodes as Possible Freewheeling Diodes', May 2013.

[2] E. Yamasue, M. Susa, H. Fukuyama, and K. Nagata, 'Thermal conductivities of silicon and germanium in solid and liquid states measured by non-stationary hot wire method with silica coated probe', *Journal of Crystal Growth*, vol. 234, no. 1, pp. 121–131, Jan. 2002, doi: 10.1016/S0022-0248(01)01673-6.

[3] A. Amith, I. Kudman, and E. F. Steigmeier, 'Electron and Phonon Scattering in GaAs at High Temperatures', *Phys. Rev.*, vol. 138, no. 4A, pp. A1270–A1276, May 1965, doi: 10.1103/PhysRev.138.A1270.

Influences of Solder Delamination on the Thermal Performance in Automotive Traction Module

Hansol Seo[1], Changsun Yun[1], Nick Bridwell[2], Nasyriq Khaliddi Zainudin[3]

[1] onsemi, Republic of Korea
[2] onsemi, the United States
[3] onsemi, Malaysia

Corresponding author: Hansol Seo, hansol.seo@onsemi.com
Speaker: Hansol Seo, hansol.seo@onsemi.com

Abstract

Estimating the die temperature (T_{vj}) during the operation is essential to determine the operating limit and predict the lifetime. Thermal resistance (R_{th}) and Thermal impedance (Z_{th}) are commonly used to estimate the die temperature at a static and dynamic level. These thermal properties are determined by the thermal stack up, geometry and material properties. The solder is widely used to attach different materials because it provides good thermal resistance and reliable connection in the module stack up. This paper shows the thermal influences of the solder delamination between the substrate and heatsink in the silicon carbide (SiC) power module. To evaluate the impact of solder delamination, the thermal resistance (R_{th}) was measured at the switch level and individual die temperatures were carefully assessed using an Infrared (IR) camera with a black-painted module.

1 Introduction

The automotive traction module comprises multiple dies to handle a high output current in horizontal geometries and vertical stacks that can withstand various environments and dissipate the heat from the die. There are several methods to dissipate the heat from the substrate to the heatsink in the Automotive traction module, but solder is widely used for the six-pack power module that is pre-assembled heatsink by the semiconductor manufacturer.

The solder has a better thermal conductivity than the thermal interface material (TIM) and provides a reliable connection between two different materials. However, solder is known to present certain issues, such as void and delamination [1] [2] which may arise due to factors such as an improper reflow process or surface contamination. When the solder voids that exceed a certain level exist in the module, it can be easily screen out by X-ray inspection method to provides a good product to the end customer. However, if delamination occurs within a solder joint, detecting it using X-ray methods can be challenging due to its thin nature.

Therefore, the Scanned Acoustic Microscopy (SAM) inspection method effectively examines the solder delamination because it can access the target layer without destroying the module [3].

This paper aims to assess the thermal performance of the onsemi NVXR17S90M2SPC [4], which integrates 8 of EliteSiC™ Power MOSFET dies in a switch, specifically focusing on solder delamination. To effectively evaluate thermal performance, samples with solder delamination are identified using SAM inspection, and thermal evaluation is conducted using two distinct methods. First, thermal resistance measurement is a widely accepted approach for obtaining the thermal characteristics of the power module. Second, in order to precisely examine the geographical temperature distribution, die temperature is measured using an infrared camera with the black painted module during inverter operation through the Active Load Emulator (ALE) test system. [5]

2 Experimental Analysis

2.1 Device Under Test (DUT)

The onsemi NXR17S90M2SPC module was utilized to assess the thermal effects of solder delamination. This module comprises 900V SiC

MOSFETs in a 6-pack configuration designed for Electric Vehicle (EV) Traction inverter applications and each switch consists of eight parallel dies, allowing for easy evaluation of its geographical thermal performance.

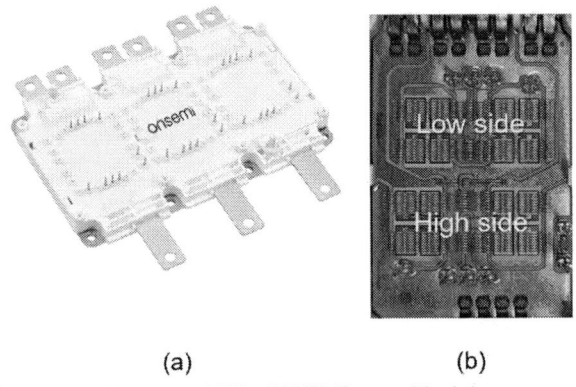

(a)   (b)

Fig. 1 (a) onsemi EliteSiC™ Power Module NVXR17S90M2SPC (b) die configuration of single phase

2.2 Solder Void and Delamination

The solder has good thermal conductivity and the reflow process is widely used in the manufacturing site. Figure 2 represents the usage of the solder in the SiC power module. The SiC MOSFET, the dies are sintered to the Active Metal Braze (AMB) substrate and it attach to the heatsink through the solder for an efficiency cooling.

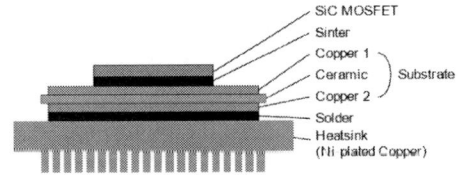

Fig. 2 Material stacks up of NVXR17S90M2SPC

It is worth noting that solder voids or delamination can occur because of an unoptimized reflow process or a contaminated surface. Solder voids are a widely recognized issue in electronic assemblies. These voids lead to an increased geographical thermal resistance [3] and contribute to accelerated thermal degradation. Specifically, the growth of solder cracks under thermal shock testing exacerbates the problem. The solder void between the substrate and heatsink can be discovered by X-ray inspection. In Fig. 3(a), the red circle highlights the solder void area as observed in the X-ray image,

while Fig. 3(b) presents the cross-sectional image of the solder void.

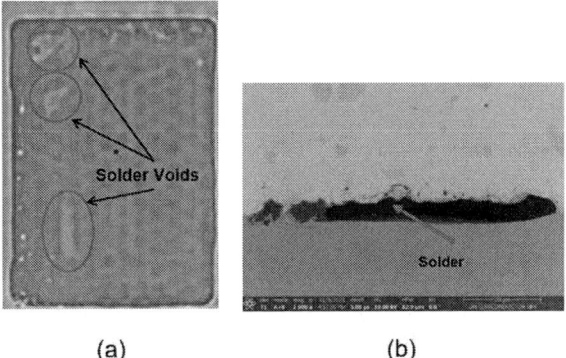

(a)   (b)

Fig. 3 Solder void. (a) X-ray image (b) Cross-sectional image

Solder delamination, characterized by the partial separation of layers between the substrate and heatsink, poses a critical challenge in electronic assemblies. Traditional X-ray inspection methods fail to discover it due to the thin surface of the affected area (Fig. 4 (a)). However, an alternative approach—the Scanning Acoustic Microscopy (SAM) C-scan—proves effective in capturing the delaminated layer without causing any damage.

In Fig. 4 (b), the brighter region stands out against other locations, clearly indicating the presence of delamination between the solder and heatsink. To further verify this, high-power microscopy (range from 250x to 1000x) is required to capture a thin surface area. Figure 5 shows cross sectional image of the red dashed area extracted from the SAM image Fig. 4(b)

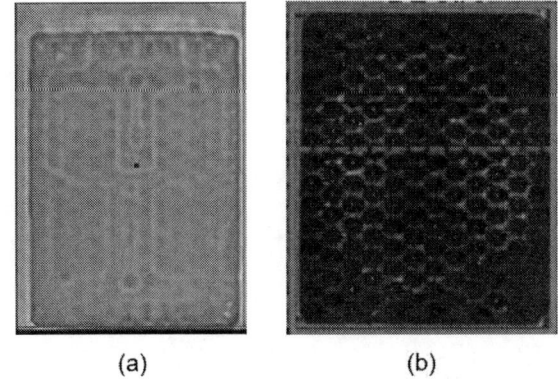

(a)   (b)

Fig. 4 Solder delamination sample (a) SAM image (b) X-ray image

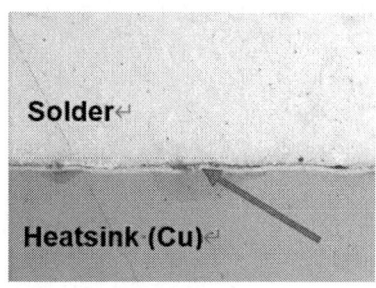

Fig. 5 High power microscope image

2.3 Thermal Resistance Measurement

It is crucial to measure the thermal resistance between the junction to fluid ($R_{th(j-f)}$) for the NVXR17S90M2SPC due to its integrated heatsink .Under the specified thermal resistance measurement conditions, T_f = 65°C, EGW 50:50, Flow rate = 10L/min, and heating current = 200 A from source to drain.(Fig. 6).

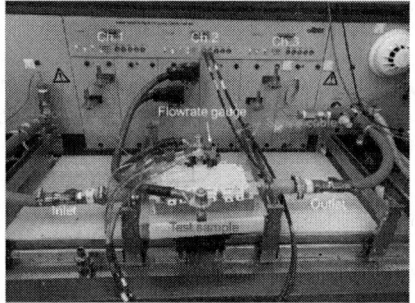

Fig. 6 Thermal resistance measurement setup

In order to comprehensively assess the impact of solder delamination on thermal resistance, the status of solder joints was examined using Scanning Acoustic Microscopy (SAM) images and categorized into two groups: normal and delaminated samples.

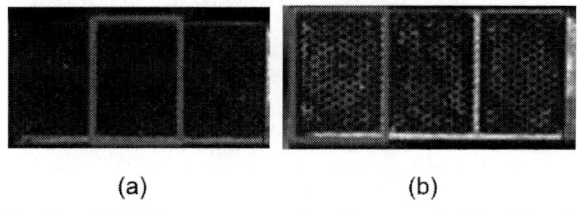

(a)   (b)

Fig. 7 SAM image example (a) normal (b) delaminated sample

The thermal resistance ($R_{th(j-f)}$) of the normal samples indicates an average of 0.0965 K/W and a maximum of 0.100 K/W (Table 1), consistent with the NVXR17S90M2SPC datasheet value.

Normal Sample	High Side	Low Side
Sample 1	0.095 K/W	0.100 K/W
Sample 2	0.094 K/W	0.097 K/W

Table 1 $R_{th(j-f)}$ measurement result of the normal group

To conduct a thorough evaluation of the delamination's impact, $R_{th(j-f)}$ was measured using samples twice the size of the standard samples. (Table 2)

Delaminated Sample	High Side	Low Side
Sample 3	0.098 K/W	0.107 K/W
Sample 4	0.094 K/W	0.100 K/W
Sample 5	0.096 K/W	0.102 K/W
Sample 6	0.096 K/W	0.101 K/W

Table 2 $R_{th(j-f)}$ measurement result of the delaminated group

The thermal analysis of the delaminated sample indicates an average thermal resistance of 0.099 K/W and a maximum of 0.107 K/W. While the low side of sample 3 exhibits a 7% higher $R_{th(j-f)}$ than the normal sample. Notably, the most of measurement results for the delaminated samples close with those of the normal sample, making it challenging to differentiate between the two based solely on the $R_{th(j-f)}$ measurement."

2.4 Inverter Testing

The measurement of thermal resistance conclusively does not provide a meaningful result to assess the solder delamination. To comprehensively evaluate the influence of the solder delamination for each die at the system level, which involves the utilization of an inverter test system and a special module that black-painted module without gel for ease of capture the temperature through Infrared (IR) camera.

The inverter test system is comprised of the Active Load Emulator (ALE) and battery simulator. The ALE simulates the various electric motors and load conditions without a physical load connection as a dynamometer, offering increased flexibility for testing the inverter system. And battery simulator is designed to simulate the high voltage battery of the Electric Vehicle (EV) without using a battery pack.

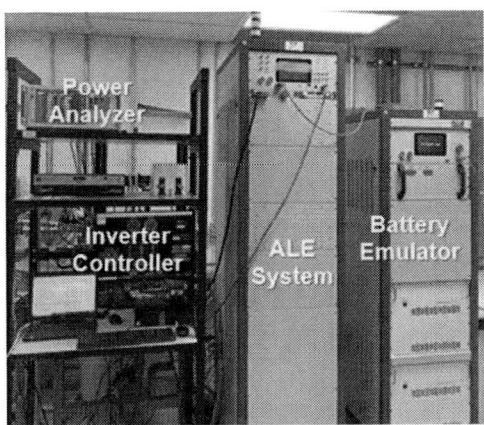

Fig. 8 ALE System and battery emulator

Motor Model	PMSM 150 kW
Battery Voltage (V)	450
Switching Frequency (kHz)	10
Load Current (Arms)	100-400
Output Frequency (Hz)	33
Coolant Type	EGW 50/50
Coolant Temperature (°C)	65
Flow Rate (l/min)	10

Table 3 Inverter operating condition for the experiment

To achieve precise measurements of individual die temperatures, an IR camera is positioned above the black-painted module. This configuration enables real-time monitoring of individual die temperatures during inverter operation. (Fig. 9)

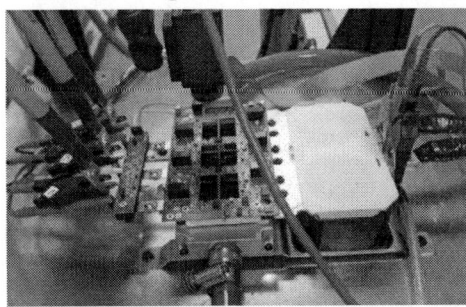

Fig. 9 DUT and IR camera configuration for the temperature monitoring

Identifying two Device Under Tests (DUTs) based on the solder delamination in the SAM scan. Initially, select the DUT without solder delamination. Operate the inverter under the following conditions and monitor all sixteen die's temperatures in a single phase.

Fig. 10 Die numbering for the temperature monitoring

The normal sample, identified using SAM, exhibits a narrow temperature distribution across its dies. Based on the IR camera data recorded during inverter operation (as shown Fig. 11), the average maximum junction temperature ($T_{vj(max)}$) among the 16 monitored dies was 95.5°C. The highest recorded die temperature reached 97.1°C. Notably, the temperature gap between the minimum and maximum dies was 2.9°C, observed when the inverter output current was set to 400A rms.

(a)   (b)

Fig. 11 (a) Normal sample SAM image, (b) IR camera image

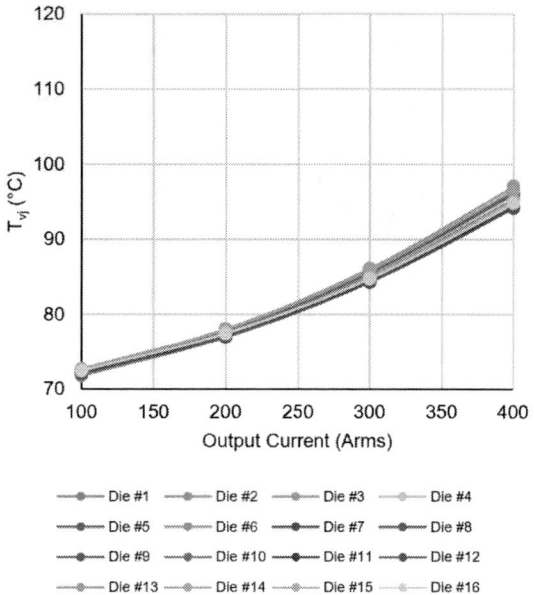

Fig. 12 Normal sample's die temperature distribution

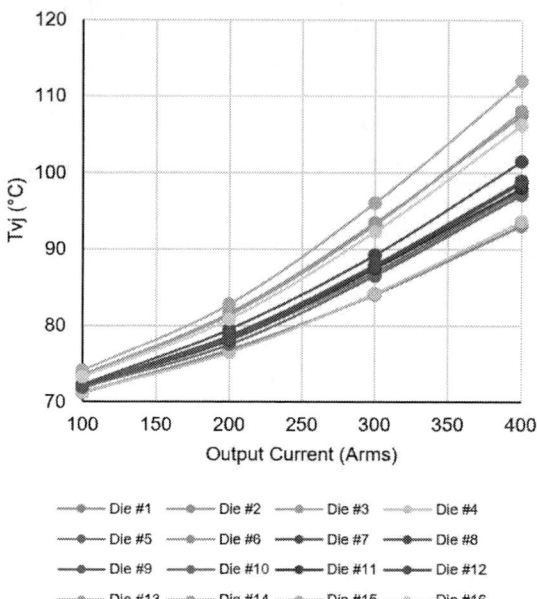

Fig. 14 Delaminated sample's die temperature distribution

Focused on a delaminated sample identified using the SAM image. Notably, this sample showed a significantly broader temperature distribution across its dies compared to the normal sample based on the recorded IR camera data during the inverter operation . Among the 16 dies monitored, the average maximum junction temperature ($T_{vj(max)}$) was 100°C, with the highest recorded die temperature reaching 112°C. The temperature gap between the minimum and maximum dies was 19°C, observed when the invert-er output current was set to 400 Arms.

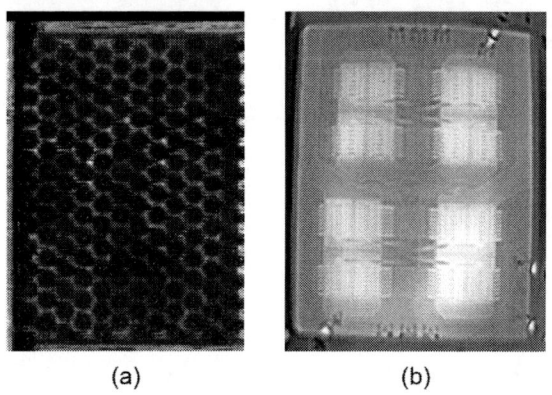

| (a) | (b) |

Fig. 13 (a) Delaminated sample SAM image, (b) IR camera image

2.5 Gap analysis

An inverter simulator was employed to evaluate the thermal resistance gap between the dies in the study. Because of the several dies are parallelly connected inside the module, it is not easy to measure each die's thermal resistance with a conventional thermal resistance measurement method without destruction or special sample.

1) Add the loss model of the power module NVXR17S90M2SPC to the inverter simulation tool. The loss model consists of conduction loss, switching loss, and thermal impedance. Additionally, it has a current range from 0 to 1000 A at 3 different temperatures and voltage ranges for precise results.

2) Extract the output parameters from the power analyzer, such as DC-link voltage, power factor, phase current, and modulation index. Then, input the other inverter operating conditions, such as switching frequency and coolant conditions, into the inverter simulation tool.

3) Fine tune the inverter operating parameters in the simulation tool to fit the simulated $T_{vj(max)}$ curve is aligned with the normal sample's maximum temperature curve.

4) Adjust the $R_{th(j-f)}$ value in the simulation tool until the simulated $T_{vj(max)}$ curve meets the delaminated sample's maximum temperature curve.

Comparing the maximum temperature between the normal and delaminated samples, the inverter test result shows a difference of 14.9°C, while the simulation, adjusted 1.45 times to the $R_{th(j-f)}$, indicates a difference of 14.7°C. (Fig. 15)

The measured $R_{th(j-f)}$ gap between normal and delaminated samples is 45% by inverter testing while a matched or maximum 7% gap was shown in the thermal resistance measurement result.

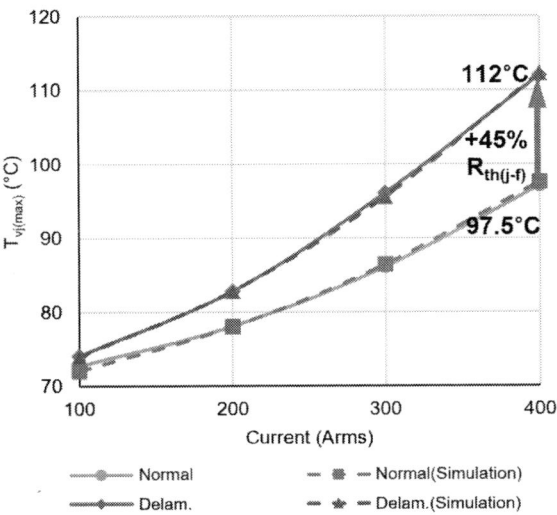

Fig. 15 Delaminated sample's die temperature distribution

3 Conclusion

The maximum junction temperature ($T_{vj(max)}$) decides the maximum output power of the traction inverter system, therefore it is important that secure the die temperature below the maximum operating junction temperature.

It is crucial to address solder delamination in the power module. The delaminated module can experience thermal runaway, even within the normal operating range, due to the high thermal resistance of a single die. This issue can lead to a decrease in the lifetime and reliability of the module, as higher thermal resistance results in a wider temperature range. Furthermore, the delaminated area can expand when it encounters solder cracks caused by Coefficient of Thermal Expansion (CTE) mismatch.

In this paper, the proposed inverter level assessment method of the solder delamination able to estimate the risk correctly. While conventional thermal resistance measurement can be applied for a single die device, it cannot accurately represent the geographical temperature distribution in a multiple die case. This is because it can only provide average data among the dies and cannot access the temperature of individual die directly.

4 Conclusion

[1] J. Fang, M. Wang and X. Chen, "The Influence Of Soldering Voids In Power Devices," 2021 22nd International Conference on Electronic Packaging Technology (ICEPT), Xiamen, China, 2021, pp. 1-4, doi: 10.1109/ICEPT52650.2021.9568145.

[2] Baylakoglu, I., Hedin, E. (2011). PCB Delamination. In: Grossmann, G., Zardini, C. (eds) The ELFNET Book on Failure Mechanisms, Testing Methods, and Quality Issues of Lead-Free Solder Interconnects. Springer, London. https://doi.org/10.1007/978-0-85729-236-0_12

[3] K. Mugunan, L. C. Ying and C. C. Fei, "Verification of Delamination Observed in SAM Transmission Mode (Thru-Scan) Using Reflection Mode (C-SAM Bottom Scan)," 2018 IEEE 38th International Electronics Manufacturing Technology Conference (IEMT), Melaka, Malaysia, 2018, pp. 1-6, doi: 10.1109/IEMT.2018.8511783.

[4] https://www.onsemi.com/products/discrete-power-modules/power-modules/silicon-carbide-sic-modules/nvxr17s90m2spc

[5] U. Vangaveti, A. Sesha and D. Cho, "Measurement and Validation of Junction Temperatures of Dies in Automotive Traction Modules at Inverter Level," PCIM Europe 2023; International Exhibition and Conference for Power Electronics, Intelligent Motion, Renewable Energy and Energy Management, Nuremberg, Germany, 2023, pp. 1-7, doi: 10.30420/566091348.

PCIM Europe 2024, 11–13 June 2024, Nuremberg DOI: 10.30420/566262264

Development of a Passive Capillary-Pumped Cooling System for High-Performance Electronics

Boris Schilder[1], Justin Fey[1], Josip Bosnjak[1]

[1] Frankfurt University of Applied Sciences, Germany

Corresponding author: Boris Schilder   b.schilder@fb2.fra-uas.de
Speaker: Justin Fey   justin.fey@fb2.fra-uas.de

Abstract

An innovative passive capillary-pumped cooling system for high-performance electronics is presented. Similar to heat pipes, evaporation and condensation of a working fluid is employed for heat transfer. In order to avoid low operating pressure as well as complex evacuation and degassing of the system, a low saturation temperature heat transfer fluid is employed. Compared to a heat pipe, the heat can be transported over longer distances using flexible hoses as the fluid lines. Employing 3M Novec 649 as the heat transfer fluid, the passive cooling system is capable of dissipating heat fluxes of 7 W/cm². The performance of the passive system in cooling a central processing unit (CPU) is compared to a conventional air based active cooling system.

1 Introduction

The performance of electronic components, especially in the semiconductor industry, is constantly increasing, while the installation space is becoming smaller. Current heat fluxes of 50 to 1.000 W/cm² make high-performance cooling systems necessary [1], [2]. However, the requirements for high efficiency, low maintenance and low-noise operation often lead to conflicting objectives.

One method for dissipating such high heat fluxes is the heat pipe, which has been undergoing development and improvement for decades and is nowadays widely used in commercial products such as personal computers. In order to increase the distance between heat source and heat sink, two-phase circuits with separated vapor and liquid phase have been developed.

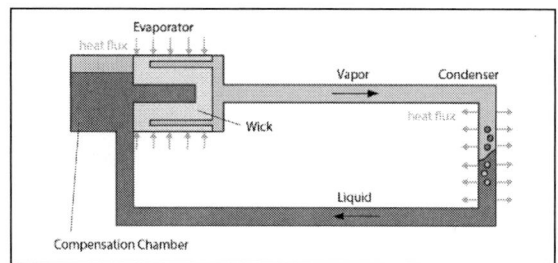

Fig. 1   Operating principle of a loop heat pipe [3].

In this so called capillary pumped loops (CPLs) or loop heat pipes (LHPs) the capillary effect is no longer utilised in the liquid line of the device, but is replaced by a separate wick in the evaporator section. In the loop heat pipe, a compensation chamber connected to the wick ensures that the wick is constantly wetted with liquid, as shown in Fig. 1. In the capillary pumped loop, a reservoir with a supply line is arranged in front of the wick, as illustrated in Fig. 2 [4].

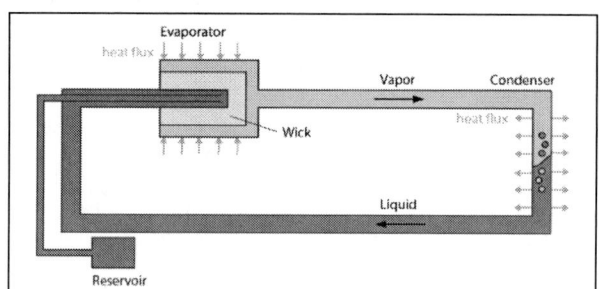

Fig. 2   Operating principle of a capillary pumped loop [3].

In these two-phase circuits, hydrocarbons or water are often used as the heat transfer fluid. Due to high boiling points, it is necessary to reduce the system to a low pressure level, which in turn requires complex filling and evacuation of the system. The passive capillary-pumped cooling system presented in this publication uses synthetic fluids with a low boiling point and can therefore be operated at moderate pressures close to atmospheric pressure without complex evacuation, using flexible hose lines instead of metal pipes. In addition, the system has a second circuit for removing non-condensable gases (NGCs). Figure 3 shows the operating principle of the new cooling system.

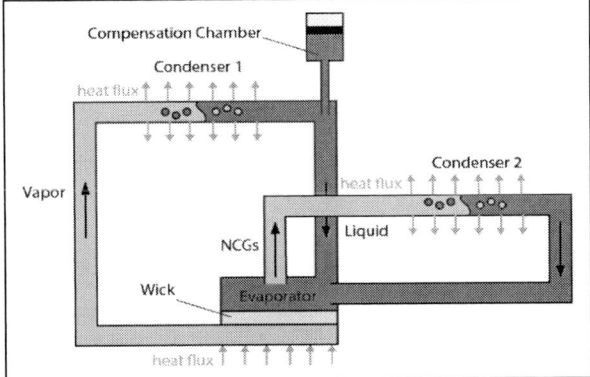

Fig. 3   Operating principle of the innovative cooling system

2   Description of the Cooling System

The cooling system consists of an evaporator, two condensers for convective heat dissipation, an compensation chamber and flexible hoses for the fluid transport. The cooling concept is employing capillary forces for transporting the heat transfer fluid. Neither pumps nor fans are employed, so that the system requires no external energy supply. The general concept of the cooling system has been already described in previous publications [5], [6]. However, the system described in this work is employing improved components based on new manufacturing methods and a new heat transfer fluid.

The evaporator is manufactured using 3D printing, which eliminates the need for assembly and a sealing concept and therefore allows for an extremely compact design. The evaporator contains two fluid chambers that are separated by a capillary structure. The research currently focuses on the quality of the porous structure inside the evaporator and not on the thermal material properties.

Since 3D printing employing stainless steel is well developed and the industry has experience in printing delicate porous structures, stainless steel is the chosen material for the evaporator.

Figure 4 shows a 3D rendering of the evaporator. Excluding the connectors, the evaporator has a length of 37.5 mm, a width of 32.5 mm and a height of 14.5 mm. Compared to the active air cooling system supplied with the processor, consisting of a heat sink and a fan, the installation dimensions of the evaporator are significantly smaller. The air cooling system has a diameter of 94 mm with a total height of 44.9 mm. Therefore, the evaporator requires approximate 90% less space compared to the standard air cooling system.

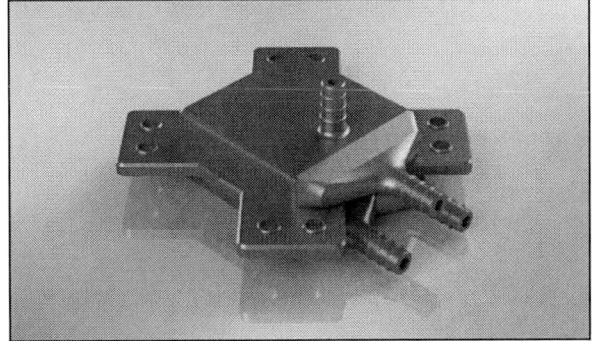

Fig. 4   Rendering of the 3D printed stainless steel evaporator

Two radiators from Alphacool (NexXxoS XT45 Full Copper X-Flow 240 mm) are used as the condensers. These radiators can be equipped with fans in order to allow for forced convective cooling. However, since our target is to develop a passively cooled system, no fans are mounted on the condensers. As the compensation chamber a Zilmet ZILFLEX Hydro Plus Inox reservoir with a volume of 0.5 liters is used.

The employed heat transfer fluid is 3M Novec 649 with a boiling point of 49°C at a pressure of one atmosphere. As mentioned above, the cooling system can be employed for all kinds of high-performance electronics e.g. power electronics, processors and LEDs. As an example, the cooling systems has been characterized in a test bench for computer processors at Frankfurt University of Applied Sciences. Within the setup, the temperatures are measured at different positions of the system using seven thermocouples. In addition, the processor temperature and its performance are measured using the Intel Power Gadget software. The processor offers a maximum heat flow output of 65 W, which is dissipated over an area of

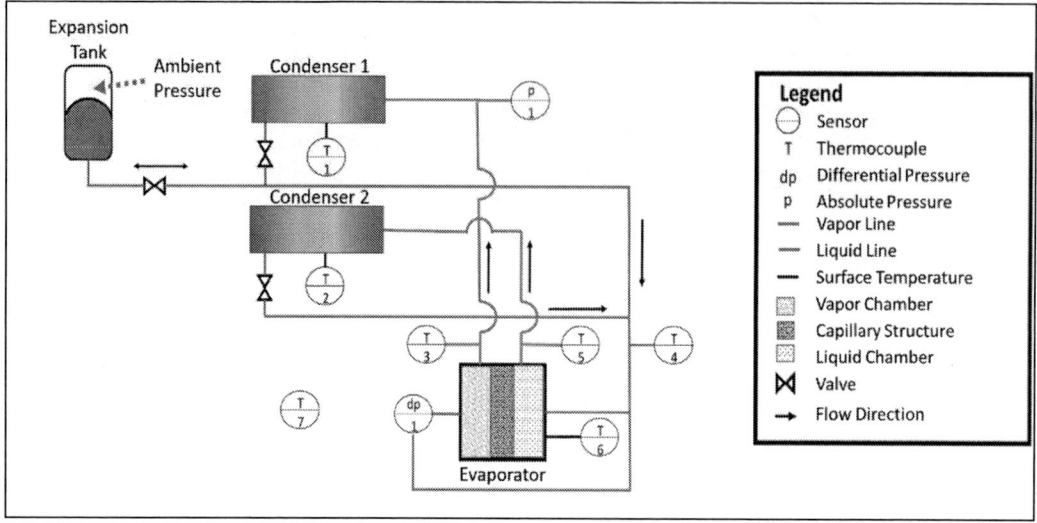

Fig. 5 Simplified flow chart of the test bench

28.3 mm x 32 mm. The absolute pressure of the heat transfer fluid and the differential pressure between inlet and outlet of the evaporator are monitored. The simplified flow chart of the setup is shown in Fig. 5.

The heat flow coming from the processor causes the evaporation of the heat transfer fluid within the evaporator. The phase change causes a pressure increase in the evaporator due to the much larger specific volume of the vapor compared to the liquid. This pressure increase is equalised by the menisci of the liquid inside the porous structure, acting as a barrier for the vapor. Therefore, the vaporized fluid is leaving the evaporator via the vapor line towards condenser 1 where the fluid is condensed by convective heat transfer to the ambient air. The liquid heat transfer fluid is then transported back into the evaporator. This backflow is additionally supported by the hydrostatic pressure of the liquid column in the transport line from condenser 1 to the evaporator since condenser 1 is installed 35 cm above the evaporator.

3 Measurements and Results

3.1 Temperatures and Performance

With the help of a rendering engine, the processor is subjected to different average heat flow levels on the test bench. The tests are carried out at 13, 26, 42, 54 and 64 W processor heat flow and five repetitions were performed for each condition. The following diagram shows the evaluation of five tests with a processor heat flow of 54 W using the cooling system with the fluid 3M Novec 649. The mean values of the temperatures at the various measuring points and the processor performance as a function of time are shown in Fig. 6.

The tests were carried out at an ambient temperature of 20°C. Once the rendering program is switched on after four minutes, the processor temperature as well as the evaporator outlet temperature of the first circuit rises. Two minutes later an increase of the condenser temperature can be observed. Stationary operation of the system is reached after around one hour.

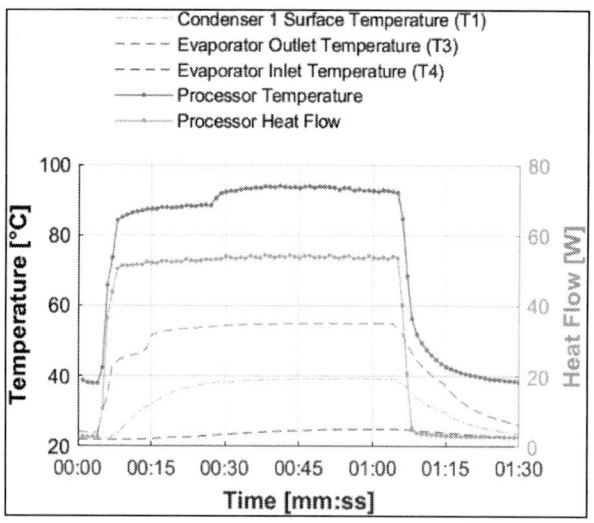

Fig. 6 Average temperatures of five repetitions at a heat flow of 54 W

With a heat flow of 54 W, the absolute pressure of the system in steady state operation is 1.3 bar, as a consequence the boiling temperature is slightly increased compared to a pressure of one atmosphere. The outlet temperature at the evaporator is 54°C, and, thanks to the transparent fluid lines, it is observed that vaporized heat transfer fluid leaves the evaporator. The outside temperature of condenser 1 is below 40°C and therefore well below the boiling temperature.

Purely liquid heat transfer fluid is leaving the condenser. With 25°C, the temperature of the evaporator inlet is only slightly higher than the ambient temperature. The heat transfer performance of the condenser is therefore sufficient to condense the fluid completely and furthermore cool the liquid down to a temperature well below saturation temperature. A considerable cool down is required in order to avoid an unwanted evaporation of the fluid already in front of the capillary structure of the evaporator. Even at maximum processor performance, the processor's limit temperature of 100°C is not reached. Thanks to the compensation chamber, low operating pressures in between one and two bars absolute have been observed within all test runs.

Figure 7 shows evaporator outlet temperatures (T3) and system absolute pressures (p1) at different heat flows.

For comparison, the vapor pressure curve of the heat transfer fluid Novec 649 is plotted additionally. At a low heat flow of 13 W, the temperature of the fluid leaving the evaporator is considerably lower than the saturation temperature while at higher heat flows the outlet temperature is close to saturation temperature. It is assumed that a mixture of liquid and vaporized heat transfer fluid is leaving the evaporator at low heat flows causing the low temperature measured right after the evaporator. At larger heat flows of 26 W and above, only vapor is leaving the evaporator, resulting in temperatures close to saturation temperature.

3.2 Comparison with Intel Air-Cooling

Tests were also carried out with Intel's thermal solution shipped with the processor. This consists of a conventional aluminum heat sink that is ventilated by a fan (Intel Thermal solution E97379-003). Further information can be found in the data sheet of this fan heat sink assembly [8].

The corresponding CPU temperatures are displayed in Fig. 8. The novel cooling system is able to remove the heat from the processor even at maximum heat flow while keeping the chip temperature below the operation limit of 100°C. However, the measured CPU temperatures are 15-20°C higher for the novel cooling system compared to active air cooling, indicating that further improvement of the system is required.

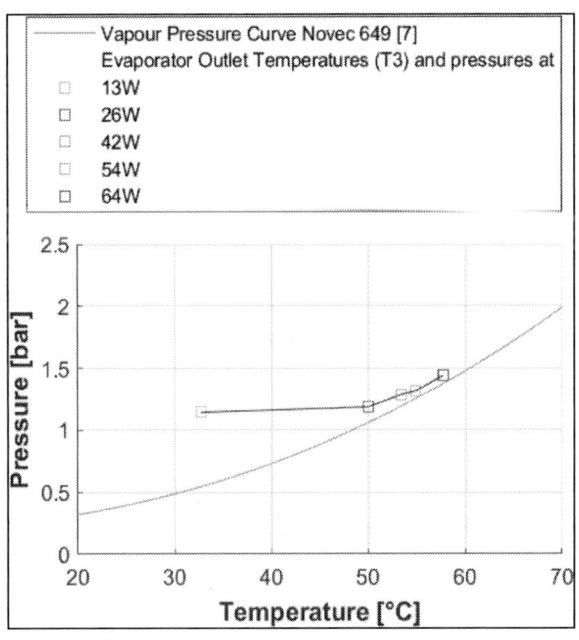

Fig. 7   Evaporator outlet temperatures (T3) and system absolute pressures (p1) at different heat flows

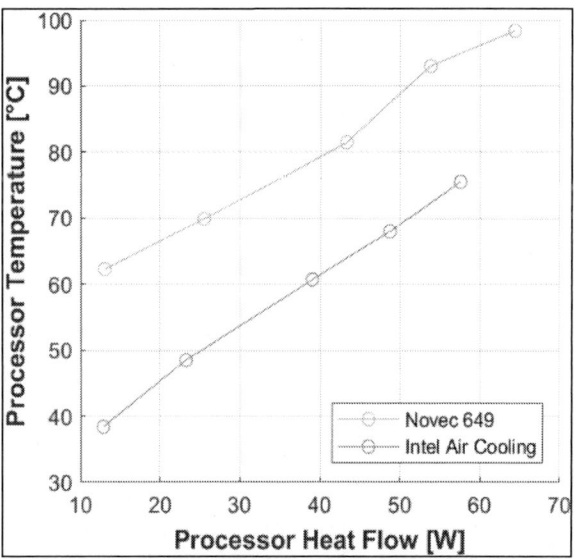

Fig. 8   Comparison of the processor temperatures of the two cooling systems at different performance levels

4 Conclusion

A novel passive cooling system has been introduced. The test results shown that the passive system is able to cool high-performance electronics with a heat flux of 7 W/cm². The required installation space at the heat source is significantly smaller compared to conventional air cooling systems.

The system will be further improved and investigated by testing optimized components, such as an evaporator with a smaller thermal resistance. To achieve this, the thickness of the baseplate could be reduced and a material with a higher thermal conductivity such as aluminium or copper can be used. In addition, the capillary structure in the evaporator can be analysed in detail and improved in order to increase the capillary pressure and reduce the flow pressure losses. Other manufacturing processes for the capillary structure, such as sintering technology, should also be considered. Higher mass flows can increase the performance of the overall system.

Additionally, it will be investigated which low-boiling fluids are suitable for the system and how they affect the cooling performance.

References

[1] C. W. Chan, E. Siqueiros, J. Ling-Chin, M. Royapoor, und A. P. Roskilly, „Heat utilisation technologies: A critical review of heat pipes", Renew. Sustain. Energy Rev., Bd. 50, S. 615–627, Okt. 2015, doi: 10.1016/j.rser.2015.05.028.

[2] X. Chen, H. Ye, X. Fan, T. Ren, und G. Zhang, „A review of small heat pipes for electronics", Appl. Therm. Eng., Bd. 96, S. 1–17, März 2016, doi: 10.1016/j.applthermaleng.2015.11.048.

[3] HAMDAN, MOHAMMAD OMAR. Loop heat pipe (Lhp) modeling and development by utilizing coherent porous silicion (Cps) wicks. Diss. University of Cincinnati, 2003.

[4] D. Butler, J. Ku, und T. Swanson, „Loop heat pipes and capillary pumped loops-an applications perspective", in AIP Conference Proceedings, Albuquerque, New Mexico (USA): AIP, 2002, S. 49–56. doi: 10.1063/1.1449707.

[5] Schilder, Boris (2011). Entwicklung eines kapillar gepumpten Wärmeübertragersystems für einen Mikroenergiewandler. Technische Universität. Dissertation im Fachbereich Maschinenbau.

[6] Schilder, Boris and Stephan, Peter (2011). Design and Operation of a Novel Capillary Pumped Two-Loop System for Cooling of Electronic Devices. Heat Transfer Engineering Volume 33, 2012 - Issue 1 Pages 12-20 | Published online: 06 Sep 2011.

[7] G. Kasapis, S. Yang, Z. Falgout, und M. Linne, „A study of Novec 649TM fluid jets injected into sub-, trans-, and supercritical thermodynamic conditions using planar laser induced fluorescence and elastic light scattering diagnostics", Phys. Fluids, Bd. 34, Nr. 10, S. 102106, Okt. 2022, doi: 10.1063/5.0106473.

[8] Specifications and Datasheets of Intel Thermal Solution. https://www.intel.com/content/dam/support/us/en/documents/processors/E97379-Foxconn-Datasheet.pdf. Accessed on 09 Apr 2024.

PCIM Europe 2024, 11– 13 June 2024, Nuremberg   DOI: 10.30420/566262266

Advanced cooling of power electronics with copper cold sprayed aluminium heatsinks & busbars.

Reeti Singh[1], Ján Kondás[1], Max Meinicke[1], Leonhard Holzgaßner[1], Markus Brotsack[1]

[1] Impact-Innovations GmbH, Bürgermeister-Steinberger-Ring 1, D-84431 Haun/Rattenkirchen, Germany

Corresponding author: Dr. Reeti Singh, rs@impact-innovations.com
Speaker: Michael Dasch, md@impact-innovations.com

Keywords: cold spray; copper; electrical applications; heat sinks; busbars

Abstract

Cold spray (CS) technology is a well-established technique for metallic deposits in various industries. CS process is known to deposit the powder particles in a solid state far below the melting point of the materials; as a result, common problems associated with temperatures, such as high-temperature oxidation, thermal stresses, and phase transformation, can be avoided. Moreover, cold spray offers short production times, unlimited component size capability, and flexibility for localized deposition. In recent years, CS technology has been extensively used for electrical applications, e.g., cold-sprayed copper for heat sinks, busbars, heat exchangers, and refrigeration units. The present work demonstrates the properties of cold-sprayed copper and its utilization for hybrid heat sink and busbar applications. The results illustrate that the properties of cold-sprayed Cu in the as-sprayed state are comparable with bulk-Cu with 98% IACS electrical conductivity and thermal conductivity of 368 W/mK. Perfectly gas-tight Cu-deposits with a He-leakage rate smaller than 1×10^{-7} mbar-l/s have been produced.

1   Hybrid Heat Sink

Electronic devices, e.g., in telecommunications and high power systems, generate heat during regular operation that must be dissipated to avoid junction temperatures exceeding tolerable limits, as this can lead to performance inhibition and deterioration of reliability. It has been shown that every 10 K reduction in the junction temperature will increase the device's life and performance. Thus, maintaining the junction temperature below the maximum allowable limit is a primary issue.

The most common way to cool devices has been air/liquid cooling using a heat sink. Conventionally, copper and aluminum heat sinks are combined with such cooling systems. Copper is always a preferred choice for heat sinks due to its cooling capacity superior to aluminum; however, copper's weight and cost limit the size, especially for large electronics systems. Due to lower thermal conductivity, aluminum heat sinks do not spread the heat quickly enough; thus, a large surface area or taller fins are required, which is not a plausible option in many cases. Moreover, a problem arises if a heat sink is substantially more

significant than the integrated circuit devices it resides on. If the electronic device generates heat faster than the heat sink spreads, portions of the heat sink far away from the device do not contribute much to heat dissipation. In other words, if the base is a poor heat spreader, much of its surface area is wasted. Furthermore, to connect the aluminum heat sink with electronic devices, a thermal interface material is generally used because soldering of aluminum with direct bond copper of the electronic device is difficult. Typically, this material has a very low thermal conductivity, affecting the overall aluminum heat sink's performance. A hybrid heat sink, combining the thermal benefits of copper with lightweight aluminum, presents an exciting alternative to overcome the issues associated with conventionally available copper and aluminum heat sinks. In such a concept, the portion of the heat sink that comes in contact with the electronic device is made of copper, while the other part is made of cheaper and lighter aluminum. A cold-sprayed copper layer was deposited on a base plate of a commercially available extruded aluminum heat sink (as shown in Figure 2). The thickness of such a copper layer can be adjusted to the electronic devices' design and operational temperature.

To demonstrate the performance of hybrid-heat sinks, Impact Innovations conducted experiments to compare the performance of identically structured copper, aluminum, and hybrid-heat sinks. The experiment was performed three times, each time with a different heat-sink design.

Thermal impedance and thermal resistance were measured. The thermal impedance of heat sinks was evaluated by running power cycles at specific load currents, heating the device until reaching the thermal equilibrium. Then, the load current was switched off, and the voltage drop was recorded. When an aluminum heat sink was tested, a maximum temperature of 438 K was registered.

Fig. 1. Cold sprayed hybrid-heat sinks.

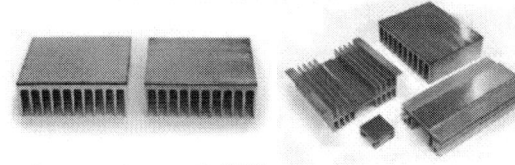

Fig. 2. Cold sprayed hybrid-heat sinks.

This value corresponds to a thermal resistance of 0.7 K/W. For the copper heat sink, the maximum temperature was just 348 K, and the corresponding thermal resistance was 0.33 K/W. Testing the hybrid-heat sink, the maximum temperature was slightly higher at 349 K, and the thermal resistance was 0.36 K/W.

These results show that the copper and hybrid heat sinks have almost identical thermal results and substantially outperformed the aluminum heat sink, thus showing the importance of quick heat spreading along the base. At the same time, the hybrid heat sink weighed less than the copper heat sink.

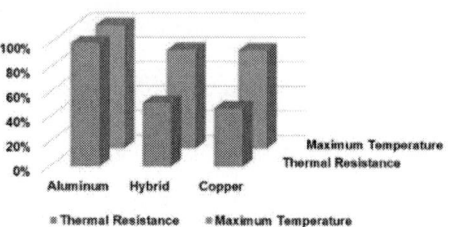

Fig.2. Thermal resistance and maximum temperature obtained at the device using aluminum, hybrid, and copper heat sinks.

Indeed, hybrid heat sinks manufactured by cold spraying have higher production costs than commercially available aluminum heat sinks; however, adding a layer of copper to an aluminum heat sink decreases its thermal resistance by 48%, as shown in Figure 3. This has a direct effect on the production costs since the semiconductor area can be reduced by 94%. Besides, the deposition efficiency and deposition rates of copper powder by the cold spray process are 95% (including overspray) and 10 kg/h, respectively, indicating the potential of the CS process to realize a cost-effective large-scale industrial production.

2 Busbars

Busbars are sophisticated technology that simplifies, reduces costs, and increases the flexibility of intricate power distribution. A busbar is vital for transmitting significant current levels between functions inside the assembly in power-intensive electrical applications. Copper and aluminum are the two most common conductors used in electrical equipment, including busbar trunking systems. Traditionally, copper is the conductor of choice for busbar trunking systems; however, in recent years, aluminum conductors have become more prevalent in the global busbar trunking market, offering specific advantages over copper. When compared by volume, copper outperforms aluminum regarding electrical ratings—boasting a lower electrical resistance, lower power loss, lower voltage drop, and higher ampacity. All of which contribute to the electrical efficiency of the busbar trunking system. However, when compared by weight, aluminum is more electrically efficient. Again, this can be attributed to aluminum having a density 70% lower than copper, making it the perfect choice where busbar sizing is a non-issue.

Impact Innovations' cold spray system presents an exciting alternative, where a copper busbar trunking system can be replaced by aluminum, combining the thermal benefits of copper with lightweight aluminum. Impact Innovations' cold spray system with a special powder injecting assembly can deposit flat copper tracks on aluminum busbar profiles without masking.

Cu-tracks were produced on an Al-alloy plate without masking, demonstrating the sharp edges of the coating tracks, as shown in Figure 1a, with the following characteristics:

- Adjustable coating thickness with narrow tolerance (e.g., ±50 µm)
- Al plate length 580 mm
- No masking used
- Deposition efficiency >98%
- Porosity <0,5%
- Uniform coating thickness
- Deposition rate up to 10 kg/h
- Sharp edges of the spray spot

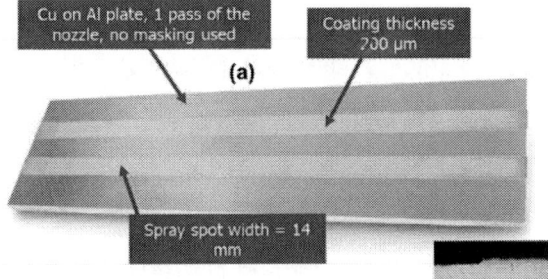

Fig. 2 (a) Cold Spray deposited Cu contact tracks on Aluminum

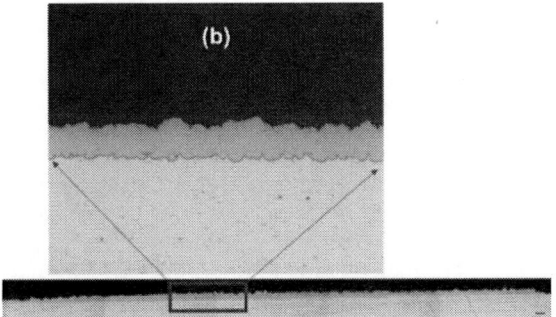

Fig. 1 (b) A uniform and dense Cu coating on Al plate.

Reference

[1] T. Stoltenhoff, H. Kreye, H.J. Richter, An ng Analysis of the cold spray process and its on coatings, J. Therm. Spray Technol. 11(2002) 542-550.

[2] H.H. Manko, Solders and Brazing, McGraw-Hill, New York, 2001.

[3] Busbars: copper vs aluminum, https://www.datacenterdynamics.com/en/opinions/busbars-copper-versus-aluminum/.

[3] Heat sink manufacturing technology, https://www.qats.com/.

PCIM Europe 2024, 11– 13 June 2024, Nuremberg   DOI: 10.30420/566262267

Cold Plate Design for Cooling LV100 Silicon Carbide Power Module Packaging

Wahid Cherief[1], Mariya Pektova[1], Jean-François de Palma[1].

[1] Cooling Division, Mersen EP, France.

Corresponding author: wahid.cherief@mersen.com

Abstract

An experimental and numerical investigation on forced convective heat transfers in cold plate for cooling LV100 power modules packaging is presented. An original design is used to build the cold plate. This study is conducted for a flow regime varying from 1500 < Reynolds < 7000. Thermal results show that the use of this efficient design allows an increase in heat exchanges of around 60% compared to a basic geometry. In addition, this optimized design shows a good homogeneous temperature distribution between power module chips. The maximum deviation between chips is lower than 4%.

1   Introduction

The current and future reduction in the availability of non-renewable energy resources has led to significant developments in the field of energy optimization. In this context, cooling techniques have seen significant development and interest at the industrial level in recent years. Indeed, in the field of power electronics, the dissipated heat causes an increase in the junction temperature of the electronic components. Thus, excessive heating degrades the performance of the component, reduces its lifetime, and can cause his failure. In this case, the installation of a cooling system is therefore essential and helps to increase the lifetime of these components.

Nevertheless, link to the latest technological development, the heat flux densities to be evacuated in power converters are becoming increasingly significant and can, for example, reach several hundred Watts per square centimeter. Faced with such power densities, conventional cooling systems based on liquid cooling (water, oil, glycol) present some limits. In addition, the cooling of the power modules is link to the respect of the manufacture maximum junction temperature and the homogeneity of the temperature within the chips of the same module and between modules. An uneven temperature distribution caused by the cooling system affects the electrical properties of the static components, which affects the overall quality of the electrical conversion especially if they are used in parallel.

For this reason, several developments are being carried out with the aim of intensifying heat exchange in these systems. From a global point of view, these investigations concern either the exploitation of latent heat of phase change, or the use of sensible heat. The latter is often used and finds several industrial applications thanks to its ease of use and integration. Thus, several heat exchange enhancement technologies have been developed for single phase heat transfer with liquids.

These devices aim to optimize convective thermal resistance. This resistance can be optimized by increasing the exchange surface and/or the heat transfer coefficient. These two parameters are intimately linked. The coefficient h will depend on the geometry of the surface exchange, the flow regime, and the thermo-physical properties of the fluid. In the same way, the modification of the surface exchange is associated by an increase in the coefficient h but also by an increase in the friction factor.

Mersen cooling division uses several geometric configurations to intensify the heat transfer in cold plate. In fact, from literature point of view, offset strip fins (OSF) and wavy technologies show better thermal performance. The wavy geometry was studied by many authors since than more twenty years ago. Gong et al [1] indicated an increase of thermal performance by 55% compared to straight channels. Aneesh et al [2] studied different wavy geometry. They found that trapezoidal channel highlight very high thermal performances, but with higher pressure drops. In addition, manufacturing this kind of geometry could be very difficult for industrial point of view. Numerical simulation between smooth wavy and straight channels was carried out by [3]. The added value of this study was the use of pin fin in the bulk of the wavy and straight channels. This leads to significant improvement of heat transfer coefficient by 2.3 times and an increase of pressure drop of 13.6 times.

Nevertheless, it can be noticed that many studies were conducted on the fluid flow distribution in a cooling system. Temperature uniformity in power module can be ensured by the geometry intensification and the fluid distribution strategy. Xia et al [4], shows that the I-type configuration allows a better distribution of the fluid compared to the Z-type configuration. In the same way, Thonon et al [5] shows that the U-type distribution allows better flow balancing, and a lower pressure drops at the same total flow rate than the Z-type configuration. In addition, the header shapes of the cold plate (distribution chambers) play significant role in the uniformity of the flow [4]. However, on a practical level the use of an I or Z type distribution configuration involves an increase in the volume of the system due to the hydraulic connections.
Thus, depending on the applications, and to optimize power converter integration, the fluid supplies are generally arranged on the same side. This implies a U-shaped distribution [6].

The choice of a wavy geometry as heat intensification technology will results from a compromise between thermal and hydraulic performance, industrial implementation, mass, volume, and the application type. Studies conducted on wavy geometries propose an integration in the form of straight grooves. Thus, the main purpose of this study is to study a new cooling architecture (single-phase liquid) based on wavy spiral geometry. This geometry will be applied to LV100 power modules packaging. This solution will aim to obtain homogeneous cooling between the chips and between modules on the cold plate.

2 Methodology

2.1 Cold plate manufacturing

Mersen cold plates are assembled by vacuum brazing (aluminum material). This well-controlled process makes it possible to obtain high-end plates in terms of reliability and durability. In fact, the average lifetime of these plates is more than 20 years. This method of assembling cold plates by brazing allows to easily integrate intensification of heat exchanges geometries and consequently obtain efficient cold plates. In this sense, other technical assembly of cold plates such as FSW (Friction stir welding) could show some limits and

can be used for simple plates (without intensification geometry like offset strip fins).

Furthermore, Mersen cold plates with indirect cooling obtained by vacuum brazing, show excellent resistance to thermal cycling issue thanks to the cover plate which allows a better heat spreading. Likewise, the leaks risk is non-existent thanks to the monolithic composition of the cold plate and the grain-to-grain connection. Finally, the risks linked to corrosion are considerably reduced and in the same way the maintenance of this type of plate is almost non-existent.

These few features enable to Mersen to offer reliable indirect-cooled cold plates for the power electronics market. In addition, the ever-increasing power density dissipated by semiconductors, and their miniaturization, call for more elaborate cooling solutions that can be adapted to the variety of power module packaging available on the market. Figure 1 shows the LV100WP cold plate studied in this paper, and the associated LV100 power modules.

Fig. 1. Cold plate LV100WP fitted with LV100 power module.

2.2 LV100 power module

LV100 packaging is increasingly used in power conversion, particularly with the advent of silicon carbide transistors.
Mersen is demonstrating added value in its participation as a strong player in the growth of

silicon carbide technologies. In this way, Mersen's cooling division is one of the partners in the European collaborative project AdvanSiC. The aim of this project is to develop silicon carbide technologies, and in this case for Mersen better thermal management of these components.

Figure 2 shows an example of the power module considered in this study. It contains 32 Mosfet transistor chips mounted on a DBC providing 3300 V isolation. The assembly is connected to a copper baseplate. The study of this power module is carried out in this paper by numerical simulation. Unlike the cold plate, which was used for experimental characterization.

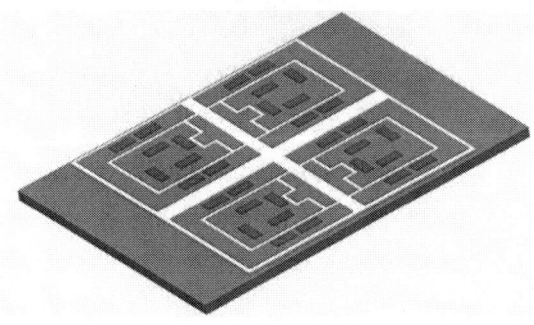

Fig. 2. LV100 power module and chips example position.

2.3 Characterization tools

Mersen cold plates are characterized on thermal-hydraulic benches to determine their hydraulic and thermal behavior. An example of a thermal-hydraulic loop is shown in Figure 3. This closed loop system consists of a tank, a centr

ifugal pump, a heat exchanger, two flow meter : Bamo (accuracy ± 1%) and Gems sensors F T-110 (accuracy ± 3%), two pressure transducers: AEP (accuracy ± 0,5%), test section and two K-type thermocouples used for the measurement of fluid temperature between inlet and outlet of the device under tests. The accuracy of the temperature measurements is evaluated to be 1°C.

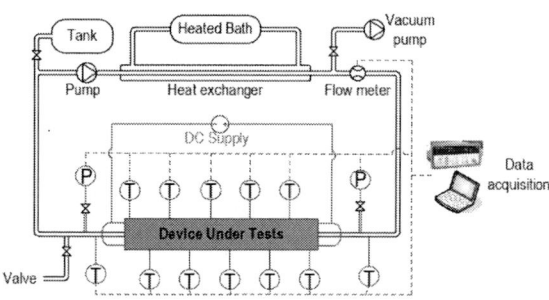

Fig. 3. Experimental setup: thermal hydraulic loop.

In addition to the thermal-hydraulic loop, Mersen cooling division has a test laboratory for mechanical, microstructural analysis, burst and dynamic pressure cycling tests. These tools enable us to better control our cooling products and meet our commitments to our customers.

2.4 Numerical simulation

Numerical simulations are also carried out during the development process of the cold plate. For this study, the three-dimensional numerical simulations were solved using EFD Lab and Ansys software based on the finite volume method. The momentum equation for the coolant, the inlet volume flow rate is specified, while the outlet pressure is set to 0 Pa. We assume that there is no slip at the wall. For energy equation, the inlet temperature is given. The k-epsilon model is used to simulate flow characteristics for turbulent flow conditions.

The heat transfer between plates and ambient, viscous dissipation, gravity and body forces are not taken into consideration. Finally, the numerical model will be compared to experimental results and discussed.

The fluid used in this study consists of a mixture of water ethylene glycol (EG50%). This type of fluid is widely used in cooling of power electronics. Table 1 shows the thermo-physical properties of the coolant as a function of temperature.

EG50%	ρ kg.m⁻³	Cp J.kg⁻¹.K⁻¹	λ W.m⁻¹.K⁻¹	μ Pa.s
T = 50°C	1042	3456	0.418	0.00174
T = 40°C	1048	3409	0.418	0.00214

Table 1. Thermo-physicals properties of EG50%

3 Hydraulic behavior

3.1 Pressure drops

The pressure drops measurement across the device under tests for the cold plate LV100WP with different volume flowrate at isothermal condition (T = 50°C) are carried out. Figure 4. highlights that the wavy spiral geometry causes turbulence in the fluid with small flow rate (Reynolds number). This is an interesting result regarding thermal stress. In fact, with small flow rate, turbulence could start and heat flux from chips is more evacuated. This highlights the possibility of working with low flow rates to evacuate high power densities.

Otherwise, the experimental results of figure 4 are also compared with the simulation results obtained by numerical simulation. This comparison shows an excellent fit between the numerical and experimental results. A variation of less than 6% is observed.

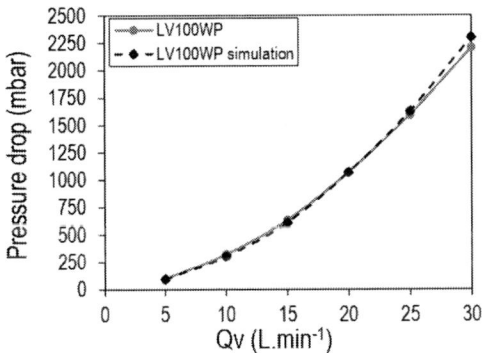

Fig. 4. Pressure drops comparison between experimental and numerical results of LV100WP at T = 50°C.

3.2 Flow rate distribution

The fluid distribution in the LV100WP plate is a very important parameter for electronics cooling. This will improve the current sharing and loss distribution of the chips and enhance the lifetime of the power module. Thus, to check the fluid distribution in the LV100WP, we used numerical model. Figure 5 gives the variation of the flow rate in each part of the cold plate and between power modules. These results show a very good distribution of the fluid in the plate, the maximum variation in flow rate between each part is less than 2%.

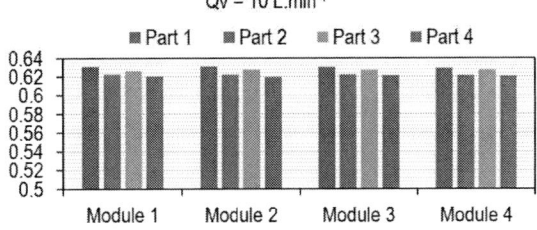

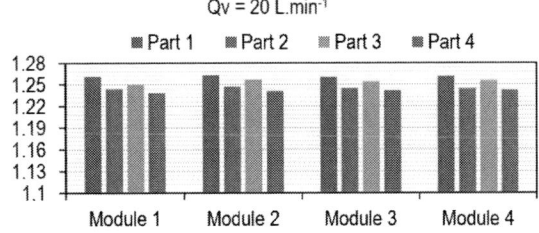

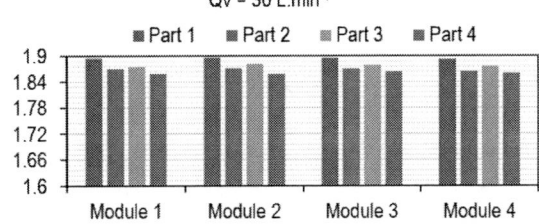

Fig. 5. Flow rate distribution in each part of the plate and between each module for 3 different flow rates: Qv = 10 L.min⁻¹, Qv = 20 L.min⁻¹, Qv = 30 L.min⁻¹. The maximum variation in flow rate between each spiral is less than 2%.

4 Cold plate thermal performances

4.1 Geometry cooling add value

The added value of the wavy spiral is studied by comparing it to a smooth spiral (without wavy). The comparison is made at iso dimension, material, flow rate and power dissipation. Figure 6 shows the variation of the average heat transfer coefficient versus the pumping power for the two studied geometry: spiral wavy and spiral smooth plate. This comparison shows that at iso-volume of plate, an improvement in heat exchange exceeds 60% in favor of the wavy geometry. Moreover, for an identical h = 10000 W.m⁻².K⁻¹, the pumping power of the wavy spiral is equal to 1.25 W against 3.25 W for the smooth spiral. It is therefore necessary to increase the pumping power by around 300% for the smooth spiral in order to obtain the cooling performance of the wavy spiral. This highlights the interest of using a wavy geometry for cooling in power electronics.

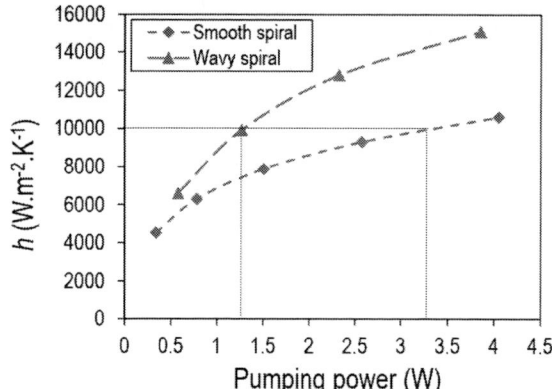

Fig. 6. Comparison of geometry wavy cooling with smooth geometry as a function of pumping power.

4.2 Cold plate thermal resistance

Figure 7 shows comparison between thermal resistance for each module at T_inlet = 40°C. A maximum deviation of 7% of thermal resistances between under each module is observed. It will be noticed that for high value of flow rate, the effect of intensification of heat exchanges induced by the wavy geometry fades. In fact, in this case, the flow controls most of the heat transfer, making thus the increase of heat transfer through the use of wavy geometry less efficient. The wavy geometry shows more interesting characteristics with small flow rate.

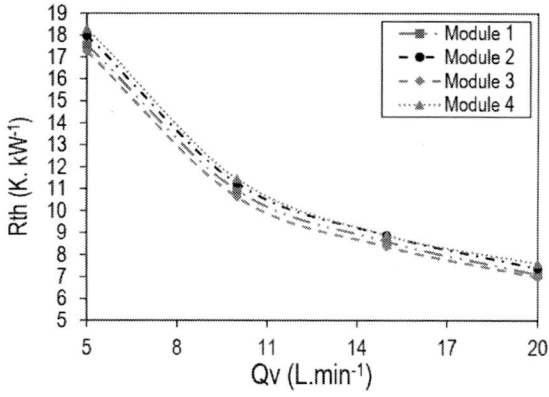

Fig. 7. Comparison between thermal resistance of LV100WP under each module. T_inlet = 40°C.

4.3 Cold plate temperature uniformity

Temperature distribution on the cold plate will be studied numerically on two hydraulically connected spirals. The flow rates considered are Qv = 15 L.min⁻¹ and Qv = 20 L.min⁻¹. The power dissipated in each spiral is similar.

Figure 8 and figure 9 show the mapping temperature at the upper face of the cold plate. In both cases the temperature is homogeneous over the entire face of the plate in contact with the electronic component. A maximum variation of 0.5°C is observed. On the periphery of the plate, we observe a slight variation in temperature due to the fin effect. This area is generally used for cooling additional components such as capacitors and/or resistors.

Fig. 8. Temperature distribution on the cover of cold plate for Qv = 15 L.min⁻¹.

Fig. 9. Temperature mapping on the cover of cold plate for Qv = 20 L.min⁻¹.

5 LV100 power module thermal performances

As previously mentioned, the study of the LV100 power module is carried out using numerical simulation. In this context, and in order to simplify the study, we're going to use a reduced model that optimizes simulation time.

Figure 10 shows the reduced model considered. This approach is motivated by the parallel distribution of the fluid within the LV100WP plate. By imposing the right boundary conditions on the reduced model, the expected results would be similar to those obtained in the case of the LV100WP plate simulation.

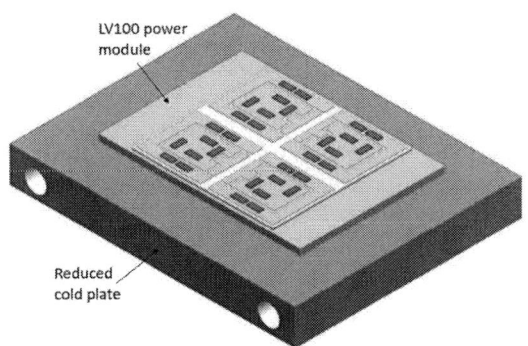

Fig. 10. Reduced studied model: only one power module is considered.

Figure 11 and Figure 12 show the temperature distribution in the LV100 module and in the cold plate for Qv = 13.6 L.min⁻¹ et Qv = 16 L.min⁻¹. The power dissipated in the module is 3000 W. We observed that the maximum chip temperature reached in the case where Qv = 13.6 L.min⁻¹ is equal to 147.5°C and that for Qv = 16 L.min⁻¹, Tmax = 143.08°C.

Furthermore, the temperature variation between the different chips is less than 3% for Qv = 16 L.min⁻¹. For Qv = 13.6 L.min⁻¹, the temperature variation between chips is equal to 3.72%. These results confirm the uniformity of the temperature distribution in the module. They also validate the hydraulic approach observed previously.

Finally, we carried out another analysis (not presented in this paper) on other heat exchange intensification geometry. The result shows temperature variations of over 12% between chips. This clearly demonstrates the benefits of using spiral wavy and its contribution to increasing power module lifetime.

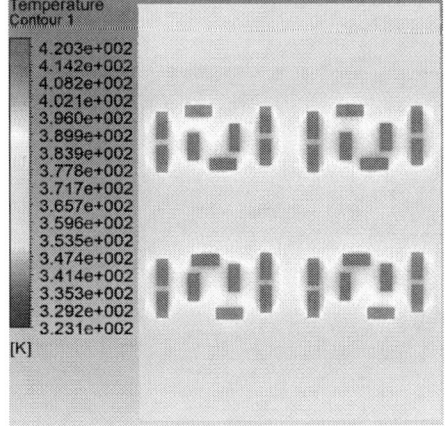

Fig. 11. Temperature distribution in the LV100 module and the reduced cold plate model. Qv = 13.6 L.min⁻¹. Power losses = 3 000 W.

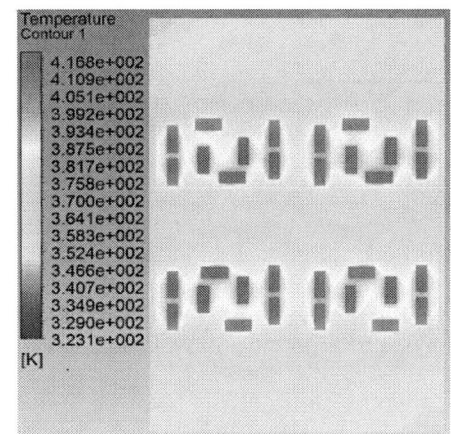

Fig. 12. Chip temperature in the LV100 module for Qv = 16 L.min⁻¹. The power losses in the power module = 3 000 W.

6   Conclusion

Wavy geometry allows a 60% improvement of convective heat transfers comparing to the smooth geometry. The arrangement of the spiral geometry ensures good temperature distribution. A maximum deviation of temperature between spiral is less than 0.5°C. In addition, the spirals arrangement allows better management of hydraulic stresses.

Likewise, this flexibility provides better cooling management of the various configurations according to chips positioning. In fact, the studied LV100 power module shows small variation of temperature between chips. The maximum deviation is less than 4%.

In this study, the arrangement of spirals was optimized according to the module chip layout. Different power module chip layout will require a custom cooling layout to get full benefit of the wavy spiral design. Synergies between power module manufacturers and cooling pattern design is more than ever require. With the new generation of SiC device, we can no longer consider the module base plate like a uniform heat generation source. High power density areas will be present. Only a custom cooling design pattern matching the chip layout will provide the full benefit of the cooling plate device.

The perspective of this study consists of measuring the performance of the cold plate with a real LV100 power module. The junction temperature homogeneity of the chips will be measured using a classical thermo-sensitive electrical parameter: the forward voltage under a low current.

References

[1] L. Gong, K. Kota, W.Q. Tao, Y. Joshi, "Parametric numerical study of flow and heat transfer in microchannels with wavy walls", J Heat Trans-T ASME 133 (5) (2011) 746–758.

[2] A.M. Aneesh, A. Sharma, A. Srivastava, P. Chaudhury, "Effects of wavy channel configurations on thermal-hydraulic characteristics of Printed Circuit Heat Exchanger (PCHE)", Int. J. Heat Mass Tran. 118 (2018) 304–315.

[3] Z. Chamanroy, M. Khoshvaght-Aliabadi, "Analysis of straight and wavy miniature heat sinks equipped with straight and wavy pin-fins, International Journal of Thermal Sciences", Volume 146, 2019, 106071, ISSN 1290-0729.

[4] D.D. Ma, G.D. Xia, J. Wang, Y.C. Yang, Y.T. Jia, L.X. Zong, "An experimental study on hydrothermal performance of microchannel heat sinks with 4-ports and offset zigzag channels", Energy Convers. Manag. 152 (June) (2017) 157–165.

[5] B. Thonon, P. Mercier, "Les échangeurs à plaques: dix ans de recherche au GRETh: Partie 2. Dimensionnement et mauvaise distribution", Revue Générale de Thermique, Volume 35, Issue 416, 1996, Pages 561-568, ISSN0035-3159.

[6] Z. Zhang, X. Wang, Y. Yan, "A review of the state-of-the-art in electronic cooling", e-Prime - Advances Electrical Engineering, Electronics and Energy, Volume 1, 2021,100009, ISSN 2772-6711.

PCIM Europe 2024, 11– 13 June 2024, Nuremberg   DOI: 10.30420/566262268

An Improved Double-layer Spacer in Double-sided Cooling Power Module

Linhao Ren[1], Zexiang Zheng[2], Jiaxin Liu[3], Heng Zhang[4], Cai Chen[5], Yong Kang[6]

[1] State Key Laboratory of Advanced Electromagnetic Technology, Huazhong University of Science and Technology, China

Corresponding author: Cai Chen, caichen@hust.edu.cn
Speaker: Linhao Ren, ringo@hust.edu.cn

Abstract

This paper focuses on enhancing the thermal performance and reliability of double-sided cooling (DSC) power modules by introducing an improved double-layer spacer. The proposed spacer aims to minimize thermal resistance and improve reliability. The lower layer of the spacer facilitates the connection of bonding wires, while the upper layer maximizes heat dissipation by effectively utilizing the surface area of the top DBC (direct bonding copper), which is typically underutilized. Through simulation verification and experiments, the study demonstrates that implementing the improved spacer results in a significant 24% reduction in thermal resistance and longer lifetime compared to conventional power modules. Additionally, this paper explores the impact of changing the spacer material from copper to Molybdenum copper (MoCu). This material substitution shifts the solder layer in which solder failure is most likely to occur, with the new layer being predominantly influenced by the size of the spacer.

1   Introduction

Power modules play a crucial role in power electronic systems. Over the past few decades, double-sided cooling (DSC) methods have garnered increasing attention as a means to enhance the thermal performance of power modules [1]. DSC packaging can make full use of the heat dissipation potential on both sides of the chip, resulting in a notable reduction in junction-to-case resistance (R_{thj-c}). Furthermore, it contributes to an 80% reduction in parasitic inductance, facilitated by the elimination of bonding wires [2].

One of the unique components of this structure is its metal spacers, which are responsible for both electrical and thermal conductions via the connecting embedded power semiconductor dies and their top-side substrates. These spacers, as reported, commonly exhibit a cuboid shape, ensuring a large solder-contact area with the power dies [3]. While this large contact area is beneficial in thermal performance, it increases thermomechanical stress and accelerate the deterioration of die-bonded solders, thereby reducing the lifetime of modules.

As shown in Fig.1, the DSC module has a complex multi-layer structure, composed of diverse materials having different coefficients of thermal expansion (CTEs). CTE mismatch immediately causes

thermomechanical stress at critical locations, particularly on the solder layers interfacing with the spacers and chips.

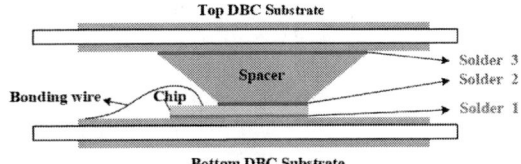

Fig. 1 Cross-section of the power module with the pyramidal spacer

In the few reported studies which aim to reduce the stress from spacers, there are two major approaches: material selection and shape modifications [3] Due to the CTE mismatch, modules with Cu spacer is not competitive, although it has great thermal conductivity. Therefore, Molybdenum copper (MoCu) spacers have been proposed to extend the lifetime of modules [4] Another approach is to reduce the size and contact area of spacers. However, as mentioned in [5] the modification also increases the thermal resistance, highlighting the need for balancing the reliability and thermal resistance which is a challenge given their contradictory design requirements.

As shown in Fig. 1, a pyramidal spacer is used, to electrically connect the bottom DBC substrate to

1917

the top DBC substrate (junction temperature reduces by ~ 6% as compared to 90-degree connection spacers) [6] The size of spacer can be altered by adjusting its tilt angle.

Through temperature cycling FEM simulation and experiments, it can be found that with an increase in the size of the Cu spacer, there is a corresponding decrease in the thermal resistance of the module. However, this reduction in thermal resistance comes at the expense of the solder layer's lifetime adjacent to the spacer, which is diminished. After changing the material to Mo85Cu15, the solder layer most prone to failure shifts to the layer beneath the chip. Notably, the reliability of this solder layer is largely independent of the material of the spacer.

What's more, in conventional multi-chip module layouts, the distance between parallel chips is typically set at 8-10mm to minimize thermal coupling [7]. Consequently, the top DBC without chips layout presents untapped potential for utilization in double-sided cooling modules. Additionally, the shape, size, and placement of the spacer within DSC power modules significantly impact their thermal characteristics and reliability [3].

Based on this, this paper proposes an improved double-layer spacer to further reduce thermal resistance and enhance reliability. The shape of the spacer can be adjusted to accommodate the module layout without impacting its thermal performance. Furthermore, the larger-sized spacers tend to accelerate the fatigue of solders, highlighting the need to balance thermal resistance improvements with potential negative effects on solder fatigue. This paper also explores the relationship between the size, material of the spacer and the lifetime of DSC module, respectively.

2 Double-sided Cooling Module with spacer

Fig.2 illustrates the model of a double-sided power module. The type of SiC MOSFET chip is S4661 from ROHM with a size of 5 mm ×5 mm ×0.15mm. The power module has a Cu/Cu15Mo85 spacer

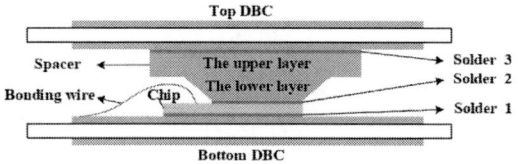

Fig. 2 Cross-section of the power module with the improved spacer

and three solder layers which bond the bottom DBC to the power die (solder_1), the power die to the spacer (solder_2), and the spacer to the top DBC (solder_3).

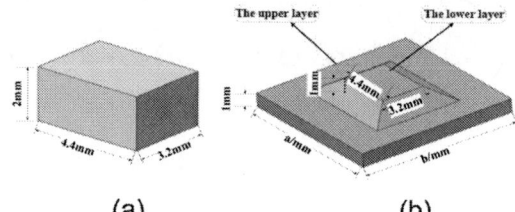

(a)   (b)

Fig. 3 (a) Conventional spacer (b) Improved spacer

Fig.3 shows two spacers. The cross-sectional area of the first spacer is equal to the source pad area of the chip, and the second spacer consists of two layers. The lower layer uses a pyramid-shaped connection block to facilitate the connection of bonding wires while providing maximum heat dissipation path as much as possible. Since the voltage between gate and source is very small, the insulation spacing here is absolutely sufficient. The upper layer then makes full use of the area of the Top DBC, as mentioned above. Thermal performance and reliability of both spacers are analyzed and compared below.

Fig. 4 the tilt angle (α) of the lower layer of the improved spacer

At the same time, as the tilt angle (α) increasing, the size of the lower layer proportionally expands, as shown in Fig. 4. In other words, by changing (α), the relationship between the size of spacer within DSC power module and its thermal characteristics and reliability can be studied.

3 Thermal Performance

To compare the thermal performance of the both spacers quantitatively, in this section, some thermal FEM simulations are performed and results are also discussed in this section. The thickness of each kind of spacer is set to 2mm normally used to keep comparisons fair. Steady-State Thermal component of ANSYS Workbench software is used. The environment temperature is $22°C$; The heat transfer coefficient of bottom and top heatsink is $3000W/(m^2 \cdot K)$ to simulate the water-cooled conditions; The loss of each bare die is 60W.

As shown in Fig. 5, it is observed that the node temperatures of the top and bottom cases, along with the heat flux form junction to the cases in the DSC module are different, which allowing for the calculation of R_{thj-c} (junction-to-case thermal resistance) from the junction to both the top and bottom cases, according to Eq. (1). Alternatively, it is possible to calculate R_{thj-a} (the thermal resistance from the junction to the ambience) directly.

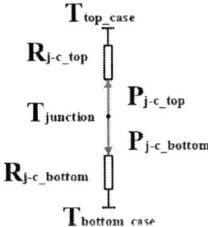

Fig. 5 Diagram of the thermal resistance of DSC modules

$$R_{thj-a} = \frac{T_j - T_a}{P} \quad (1)$$

Where R_{thj-a} is the thermal resistance from junction to the ambience; T_j, T_a is the temperature of junction and ambience respectively;P is the power loss of the chip.

In order to analyze the influence of lower layer on thermal resistance, the length and width of upper lower are set to 12mm (its area is $144mm^2$). Tilt angle increase from 0 deg to 60 deg; Only the top source pad with dimension of 3.2mm×4.4mm is soldered in simulations. Copper and Copper-Molybdenum alloy is chosen as the spacer material.

Given the correlation between longitudinal heat transfer and the heat dissipation area, an objective comparison of the thermal resistance among spacers with varying tilt angles can be achieved by utilizing the heat dissipation area at the top of the lower layer, as shown in Fig.6.

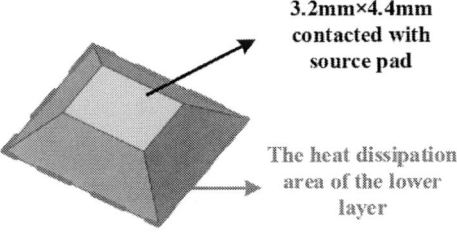

Fig. 6 the lower layer

The R_{thj-a} comparison results among different materials and sizes are shown in Fig. 7.

And then, to analyze the upper layer, its area remains unchanged at $144mm^2$, and only the ratio of length (a) and width (b) is changed, with the 45 deg tilt angle (lower layer).

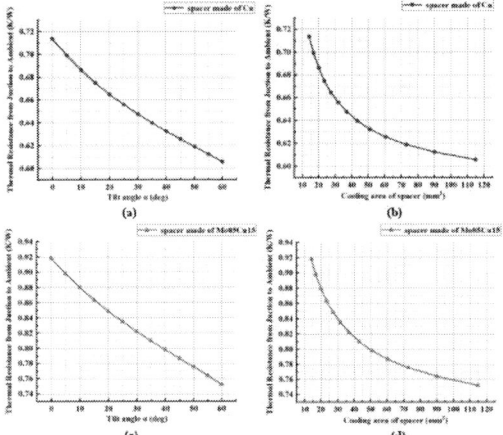

Fig. 7 R_{thj-a} of the power module under different tilt angles and cooling area

If the thermal resistance of the conventional structure spacer made of Cu15Mo85 is normalized to 1, the thermal resistance comparison results among different sizes and structures are shown in Fig. 8.

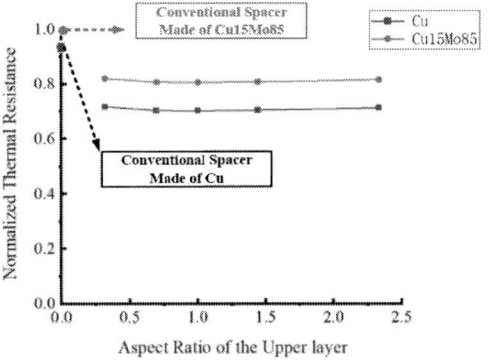

Fig. 8 Thermal Resistance from junction to ambience

It is not difficult to see that when the area of the upper layer is constant, the thermal resistance of the improved spacer under different aspect ratios of the upper layer is almost unchanged, which means that the size of the improved spacer can be adjusted according to the actual chip layout. Moreover, the change of spacer balances the heat flow and temperature distribution of the top and bottom DBC, which means better mechanical and thermal properties.

4   Reliability

The main factor of module failure is attributed to the solder layer. As shown in Fig. 1, there are totally 3 solder layers in DSC module. The blue solder layers (2 and 3) are adjacent to the spacer, while the red solder layer 1 corresponds to the layer adjacent to the chip but without direct contact with the spacer. To assess solder reliability, a temperature cycling FEM simulation is performed. The obtained range of equivalent plastic strain ($\Delta\varepsilon_p$) from the last temperature cycle is used to evaluate solder lifetime [8]. Most of the methods for predicting low-cycle fatigue lifetime using plastic strain are based on the Coffin-Manson equation [9].

$$N_f = c_1 \Delta\varepsilon_p^{-c_2} \qquad (2)$$

Where N_f is the number of cycles to failure, c_1 and c_2 are constants related to material properties. In general, the value of $\Delta\varepsilon_p$ is bigger, the lifetime of solder will become shorter.

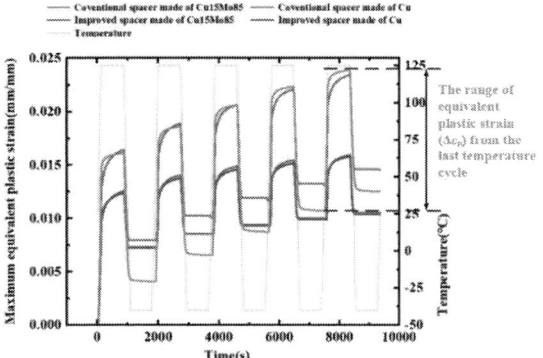

Fig. 9 Maximum equivalent plastic strain of solder_1

It is easy to see from Table I that when the module uses copper as the spacer material, solder_2 is the most possible to fatigue due to the mismatch of CTE. However, after replacing the material of spacer to Cu15Mo85, the solder most likely to fatigue becomes solder_1. Moreover, the change of the material of the spacer primarily affects solder 2 and solder 3, while the solder 1 is mainly affected by the shape of the spacer. Counter-intuitively, the larger the size of the spacer, the longer the lifetime of solder 1, which means the improved spacer result in longer lifetime.

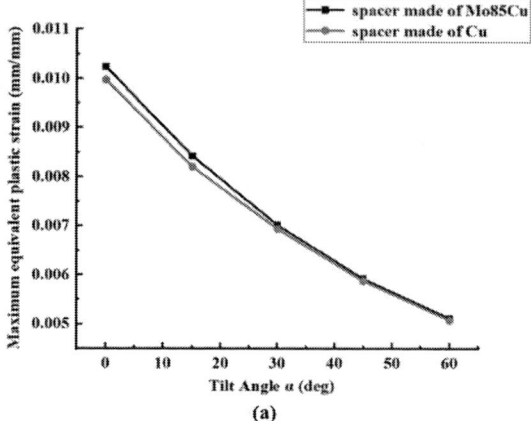

(a)

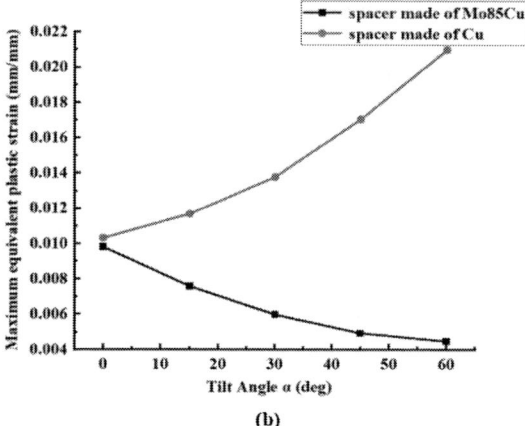

(b)

Fig. 10 (a) Maximum equivalent plastic strain of solder_1 at different tilt angle α
(b) Maximum equivalent plastic strain of solder_2 at different tilt angle α

5 Experimental Results

The experimental platform shown in Fig. 11 was built to accurately measure the thermal resistance of the two different spacer shapes made of copper.

Table I. the range of equivalent plastic strain under different shapes and materials

The range of equivalent plastic strain from the last temperature cycle (mm/mm)	Conventional spacer made of Cu	Improved spacer made of Cu	Conventional spacer made of Cu15Mo85	Improved spacer made of Cu15Mo85
Solder 1	9.973×10^{-3}	5.8739×10^{-3}	$\mathbf{1.0237 \times 10^{-2}}$	$\mathbf{5.923 \times 10^{-3}}$
Solder 2	$\mathbf{1.0328 \times 10^{-2}}$	$\mathbf{1.7056 \times 10^{-2}}$	9.821×10^{-3}	4.962×10^{-3}
Solder 3	5.948×10^{-3}	1.3138×10^{-2}	5.753×10^{-3}	5.446×10^{-3}

The experiment adopted the transient double interface method. The separation point of the curves of the two working conditions is the thermal resistance from junction to the case. It can be clearly seen that the improved spacer reduces the junction-to-case thermal resistance of the module by about 24%.

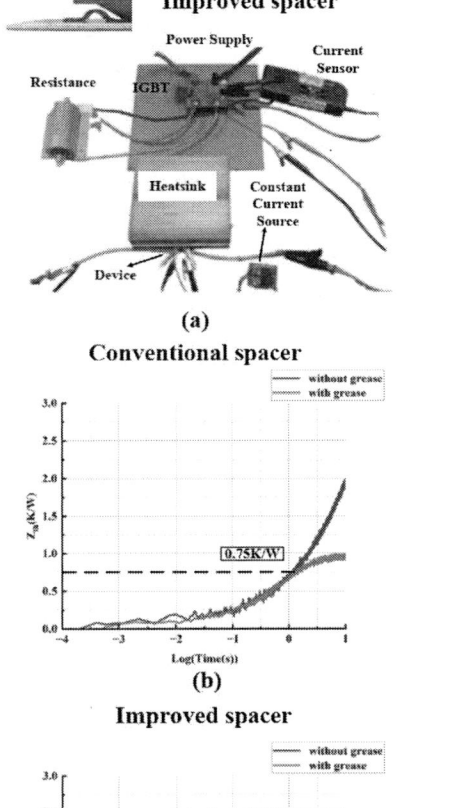

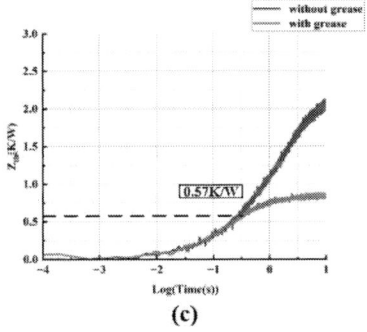

Fig. 11 (a) Thermal resistance test platform
(b) R_{th} of conventional spacer
(c) R_{th} of improved spacer

6 Conclusion

This paper proposes an improved spacer to enhance the thermal performance and reliability of DSC modules. In terms of thermal management, the improved spacer makes full use of the top DBC area and the 45-degree copper block maximize the use of heat dissipation, ensuring the connection of bonding wires. Through steady thermal simulation, it can be found that there is a marginal impact on reducing thermal resistance with the increase in spacer size. Moreover, the size of the upper layer can be adjusted without affecting the thermal resistance.

What's more, this paper explores the relationship between the size of the spacer and the reliability of the module. Through temperature cycling FEM simulation and experiment, employing Mo85Cu15 as the material for the spacer proves effective in diminishing the thermal stress experienced by the solders on both sides of the spacer. Nevertheless, the lifetime of the solder layer beneath the chip is primarily influenced by the size of the spacer. Counterintuitively, the larger the size of the spacer, the longer the lifetime of this solder layer.

References

[1] M. Liu, A. Coppola, M. Alvi and M. Anwar, "Comprehensive Review and State of Development of Double-Sided Cooled Package Technology for Automotive Power Modules," in IEEE Open Journal of Power Electronics, vol. 3, pp. 271-289, 2022.

[2] S. Seal and H. A. Mantooth, "High performance silicon carbide power packaging-past trends, present practices, and future directions," Energies, vol. 10, no. 3, pp. 1-30, 2017.

[3] J. Jeon et al., "Finite Element and Experimental Analysis of Spacer Designs for Reducing the Thermomechanical Stress in Double-Sided Cooling Power Modules," in IEEE Journal of Emerging and Selected Topics in Power Electronics, vol. 9, no. 4, pp. 3883-3891, Aug. 2021.

[4] M. Wang et al., "Reliability Improvement of a Double-Sided IGBT Module by Lowering Stress Gradient Using Molybdenum Buffers," in IEEE Journal of Emerging and Selected Topics in Power Electronics, vol. 7, no. 3, pp. 1637-1648, Sept. 2019.

[5] X. Cao, G. -Q. Lu and K. D. T. Ngo, "Planar Power Module with Low Thermal Impedance and Low Thermomechanical Stress," in IEEE Transactions on Components, Packaging and Manufacturing Technology, vol. 2, no. 8, pp. 1247-1259, Aug. 2012.

[6] A.B. Lostetter, F. Barlow, A. Elshabini, "An overview to integrated power module design for high power electronics packaging," in Microelectronics Reliability, Volume 40, Issue 3, 2000, pp. 365-379.

[7] N. Kim et al., "A Direct Bond Fabrication Process for Compact GaN Intelligent Power Modules on Liquid Coolers for EV Applications," 2021 33rd International Symposium on Power Semiconductor Devices and ICs (ISPSD), Nagoya, Japan, 2021, pp. 131-134.

[8] P. Ning, J. Liu, D. Wang, Y. Zhang and Y. Li, "Assessing the Fatigue Life of SiC Power Modules in Different Package Structures," in IEEE Access, vol. 9, pp. 12074-12082, 2021.

[9] G. B. Thomas, J. Bressers, and D. Raynor, "Low cycle fatigue and life prediction methods," in High Temperature Alloys for Gas Turbines, Liège Belgium, 1982, pp.291-317.

PCIM Europe 2024, 11– 13 June 2024, Nuremberg

DOI: 10.30420/566262269

Power Cycling of 1.7kV Multi-Chip Power Modules – SiC MOSFETs vs Silicon IGBTs

Nick Baker, Vishesh Vikas, Andrew Lemmon, Chase Fortin, Justin Conzola, Arthur Boutry, John Austin

University of Alabama, USA

Corresponding author: Nick Baker, nbaker2@ua.edu
Speaker: Nick Baker, nbaker2@ua.edu

Abstract

In this paper we compare the lifetime and thermal resistance of Silicon Carbide MOSFET and Silicon IGBT multi-chip power modules. Five module variants are presented, all rated for 1.7 kV and 1000 A. Module lifetime is compared using active power cycling with temperature swings of between 80°C to 100°C, a 2-second on-time, and 6-second off-time. We observed large variation in performance between module variants, as well as between samples of the same module variant. The paper describes the test procedure and steps taken to achieve a fair comparison between each module. Results are presented anonymously.

1   Introduction

SiC MOSFETs provide lower switching losses and higher current density than Si IGBTs. This can enable substantial reductions in size and weight of power electronic systems. Nevertheless, power semiconductor modules are not sized solely based on their nominal power density. Instead, they are sized according to the expected lifetime of the power electronic system.

The thermo-mechanical reliability of SiC MOSFETs is reported to be lower than Si IGBTs, with SiC MOSFET power cycling lifetime less than 30% of identically packaged Si IGBTs [1]. This is generally explained by SiC's Young's Modulus being 3× higher than Si, which induces higher stress into the packaging materials [2]. The coefficient-of-thermal expansion of SiC is also slightly higher, at 4.3 ppm/K vs. 3 ppm/K for Si.

The thermo-mechanical reliability of a power module can have an impact on the power density of the system since the module (and system) may need to be derated to ensure the required lifetime.

SiC MOSFET and Si IGBT modules are both commercially available at 1.7 kV and up to 1000 A. However, high current SiC MOSFET module availability is limited, and thermo-mechanical reliability experiments are inherently expensive and time consuming. For example, power cycling at low temperature swings (below 40°C) can take over 200 days to complete [3].

In this paper, we report the power cycling lifetime of commercial 1.7 kV SiC MOSFET and Si IGBT power modules with datasheet continuous current ratings of between 900 A to 1000 A. The studied half-bridge modules are in "XHP" style housing with a baseplate size of 100 mm x 140 mm [4]. Since the modules are sourced from multiple manufacturers, the internal construction of each module will vary. The paper presents results from the five module variants listed in Table 1. To maintain anonymity, they are described as "Module A" or "Module B", and the precise current rating is not provided.

2   Power Cycling

2.1   Power Cycling Conditions

Power cycling was performed using a Siemens Mentor Graphics PowerTester 2400A. A Julabo FLW20006 provides coolant to the heatsinks. The experimental setup is shown in Figure 1.

Module gate voltages ranged from 15 V to 20 V and were selected based datasheet recommendations. To maintain anonymity, these voltages are not reported alongside the module variants. Table 2 displays the power cycling test conditions. All

Module	Technology	Voltage Rating (V)	Current Rating (A)	Power Cycling Current (%)	ΔT$_J$ Min (°C)	ΔT$_J$ Max (°C)	# of SW
A	SiC MOSFET	1700	900 – 1000	112%	81	96	6
B	SiC MOSFET	1700	900 – 1000	70%	76	85	4
C	SiC MOSFET	1700	900 – 1000	82%	91	99	2
D	Si IGBT	1700	900 – 1000	88%	80	90	2
E	Si IGBT	1700	900 – 1000	89%	90	90	1

Table 1. List of Power Modules.

Figure 1. Power Cycler and Chiller.

Parameter	Value
T$_{ON}$	2 seconds
T$_{OFF}$	6 seconds
Coolant Temperature	15°C
Failure Criteria	5% V$_{ON}$ Increase
Target ΔT$_J$	80°C - 100°C
Current	Module Specific
Gate Voltage	Module Specific
Cycling Strategy	Constant Current

Table 2. Power Cycling Test Conditions.

modules in this study are half-bridge configurations; however, the upper and lower switch positions were studied individually. Due to module sourcing limitations, in some cases it was only possible to source one sample. Therefore, a maximum of two switch positions was available for study.

Each switch position was cycled with a target ΔT$_J$ between 80°C to 100°C. Power cycling was performed in constant current mode, and the required current to achieve the desired temperature swing was set on the first switch cycled. This current was subsequently used for all remaining switch positions of the same module type.

A first observation is that the modules required different currents as a percentage of the datasheet

value to achieve the desired temperature swing. This is listed in Table 1. For example, Module A required 112% of the datasheet DC current rating to induce a temperature swing of 81°C to 96°C. Module B on the other hand, required 70% its rated current.

We also observed large variations in ΔT$_J$ within module variants despite each switch being subject to the same cycling current. Since each module was placed in the same heatsink position, we believe this is due to the manufacturing variability of electrical and thermal parameters of the module. For Modules A, B, C, and D – where at least 2 switch positions were available – we observed a minimum 9°C difference in ΔT$_J$.

2.2 Calibration

The V$_{SD}$(T) (or V$_{CE}$(T) for IGBTs) method was used to estimate the junction temperature during power cycling. To estimate the junction temperature at the end of the 2-second on-time, we used an extrapolation vs. the square-root-of-time using datapoints between 1000 µs and 3000 µs after turn-off.

For selecting the SiC MOSFET turn-off voltage for the V$_{SD}$(T) measurement, we reduced the gate voltage in 0.5 V steps until the MOSFET body-diode voltage showed no change [5, 6]. This voltage ranged between 5 V to 8 V for the three SiC MOSFET modules. The IGBTs sensing current

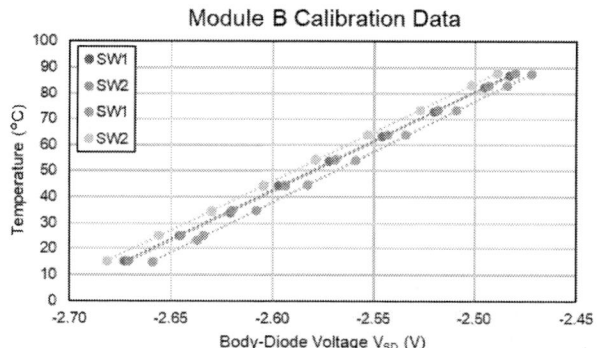

Figure 2. V$_{SD}$(T) Calibration for Module B.

was chosen to be $1/1000^{th}$ of the rated current (900 mA – 1000 mA). For SiC MOSFETs, since the sensing current conducts through the body-diode, the sensing current was $1/2000^{th}$ of the rated current to reflect the lower current carrying capacity of the modules in body-diode mode.

Each switch position was calibrated individually on the voltage measurement channel used for power cycling. Figure 2 shows calibration data from Module B. Four switch positions in total are shown. From this data we can see a spread of approximately 20mV. Given the sensitivity of the body-diode voltage is approximately 2.7mV/°C, this could represent an error of 7.5°C on the estimated temperature during power cycling if only one module was calibrated.

Calibration was performed from 15°C up to the 90°C limit of the FLW20006. The chip temperature was assumed to be equal to a thermocouple placed module baseplate rather than the coolant fluid temperature [6].

2.3 Choice of Thermal Interface Material (TIM)

In preparation for this study, the original chosen TIM was a thermal paste. The thermal paste was applied using a stencil. However, pump out of this paste was observed during initial tests. This is shown in Figure 3. Pump out is observed only on one side of the module in Figure 3 due to only a single switch position being operated.

As a result, we switched to a graphite pad with a thickness of 200µm. Graphite pads have been shown to be more robust to aging (and do not experience pump out – however they may experience stress relaxation and changes in thickness during the test) [7]. In addition, the cleanliness of a Graphite TIM was an advantage in this series of experiments.

3 Results

3.1 General Comments

The target ΔT_J was limited to 80°C - 100°C to maintain temperature measurement accuracy since calibration was possible only up to 90°C.

In total we were able to complete power cycling experiments for all three SiC modules (A, B, and C). All modules failed with a 5% increase in V_{ON}. Due to the unexpectedly long lifetime of some modules, we were unable to complete power cycling experiments for the IGBT modules (D and E).

Figure 3. Observed pump-out from one side of a 100 mm x 140 mm power module during power cycling. The chosen TIM for this experiment was subsequently changed to a graphite.

However, Module D is currently cycling and the total number of cycles at the time of writing this paper (approximately 471,000) is reported in Figure 6. For Module E, only thermal resistance data is available at present.

3.2 Thermal Resistance (R_{TH})

The junction-to-ambient R_{TH} for Modules C, D, and E, is reported in Figure 4. In preliminary investigations for this research, we performed R_{TH} measurements under different operating conditions. For Modules D and E (IGBT) we used two gate voltages (15 V and 11 V). For Module C (SiC), we operated the module in MOSFET mode and body-diode mode (i.e. the heating current conducts through the MOSFET body-diode). The measurements were all performed using a 200 µm Graphite TIM, and a 360-second heating and cooling period.

We observed a clear dependency of R_{TH} on the power dissipation within the module. In addition, the operating condition had an impact on the measured R_{TH}. For the SiC MOSFET, operating the module in body-diode mode resulted in an increased R_{TH}, and this increased with the power dissipated. For the IGBTs, operating with a lower gate voltage resulted in a higher R_{TH}. For Module D, the R_{TH} decreased with increasing power.

PCIM Europe 2024, 11– 13 June 2024, Nuremberg   DOI: 10.30420/566262269

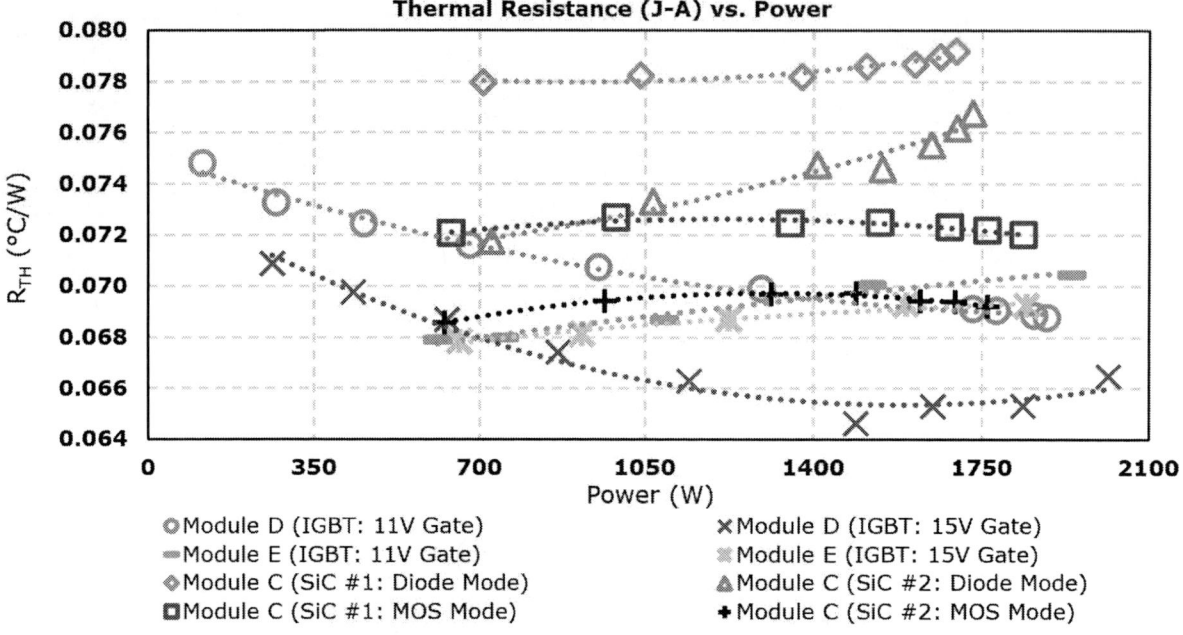

Figure 4. Thermal Resistance vs. Power Level for Modules C, D, and E.

Dependency of power semiconductor thermal resistance on power level and temperature has been observed before [8] and may be explained by the temperature dependence of the thermal conductivity of the semiconductor and packaging materials. However, past literature on the topic is limited and the mechanism behind the trends seen in Figure 4 will be the subject of future research.

A first hypothesis is that the different operating conditions and power dissipation alter the chip temperature gradient within the module. Since the $V_{SD}(T)$ method sensing current predominantly flows through the hotter cells in the semiconductor chip, its correlation to the mean chip temperature is likely dependent on the temperature gradient within the module [9].

Figure 5 displays R_{TH} data calculated by averaging the first 500 cycles of the power cycling test. This is calculated by dividing the ΔT_J by the power dissipated during the 2-second on-time during power cycling. The thermal resistances ranged from a minimum of 0.041 K/W to 0.049 K/W.

It is known that SiC MOSFET chip sizes are in general smaller than Si IGBTs [10]. We used a Manncorp MX1 X-Ray machine to estimate the chip area in each switch position and then used this estimation to normalize the R_{TH} to the total chip area. The results are displayed in Figure 6.

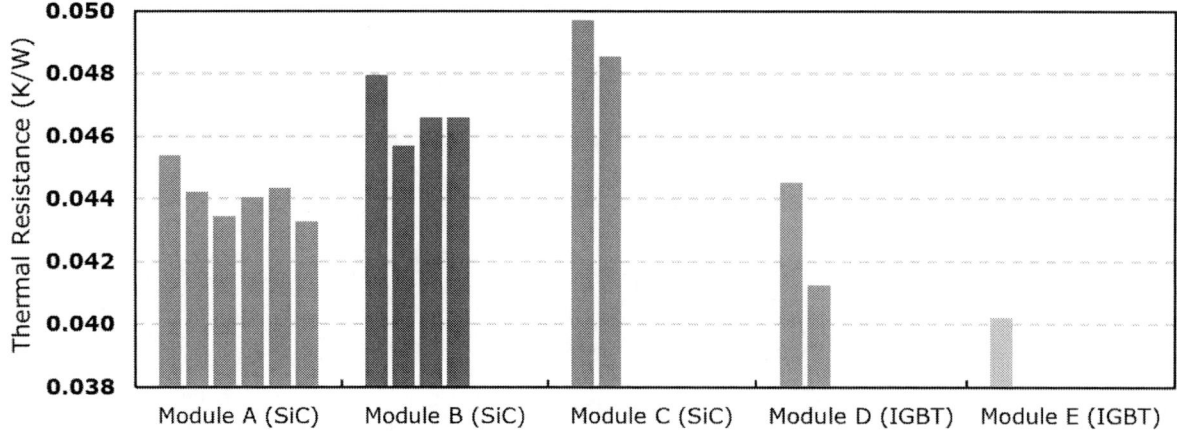

Figure 5. Absolute R_{TH} calculated from the first 500 cycles during power cycling for each Module A – E.

1926

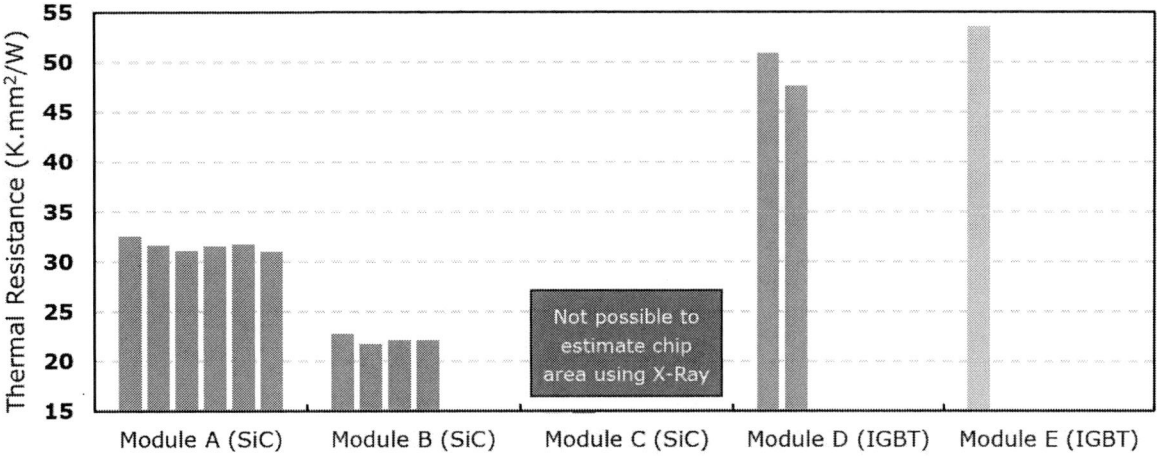

Figure 5. R_{TH} normalized for chip area for each Module A – E.

For Module C, it was not possible to estimate the chip area through X-Ray; it will be estimated during post-failure analysis. However, from Figure 6 it can be observed that the IGBT modules have the highest thermal resistance when normalizing for the chip area (despite having lower absolute R_{TH} as depicted in Figure 5).

3.3 Power Cycling

Figure 6 displays the power cycling lifetimes of Modules A to D. The ΔT_J value for each module is calculated using the average of the first 500 cycles of the power cycling test.

For Modules A and B, where at least four switch positions have been cycled, a best fit equation using a power function with the ΔT_J as the sole parameter is plotted. The ΔT_J coefficient for Module A is -2.089, while for Module B it is -3.989. In previous literature, including models such as CIPS-08,

the ΔT_J coefficient is quoted to be typically between 3.5 and 5.0 [11, 12, 13]. Module A therefore appears to be an anomaly. Nevertheless, we have not included several other parameters such as on-time, current per bond-foot, mean temperature, etc.

Module C is the best performing of the SiC modules. For the IGBT (Module D), the test is still ongoing at a ΔT_J of 81.2°C and 90.7°C. After 471,000 cycles, the V_{ON} has increased 1.9% and 2.4% respectively. Using this data we have extrapolated linearly to a 5% increase and predicted that the modules will fail after 1,250,000 and 990,000 cycles. This is approximately 46x the lifetime as Module B, which would be expected to survive 21400 cycles at 90.7°C from the data in Figure 6.

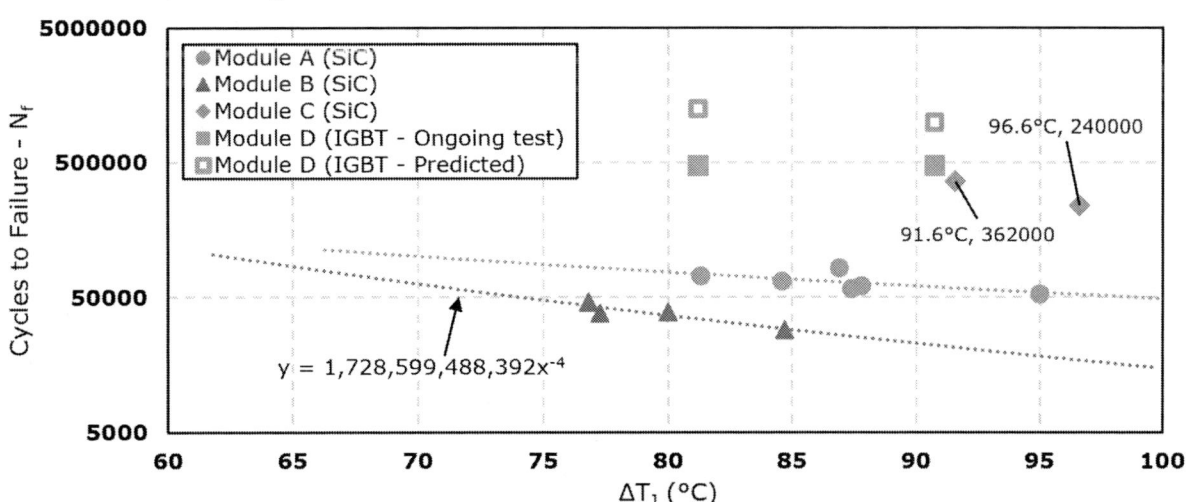

Figure 6. Power Cycling Lifetimes for Modules A – D. Includes a predicted lifetime for Module D from on-going tests.

PCIM Europe 2024, 11– 13 June 2024, Nuremberg   DOI: 10.30420/566262269

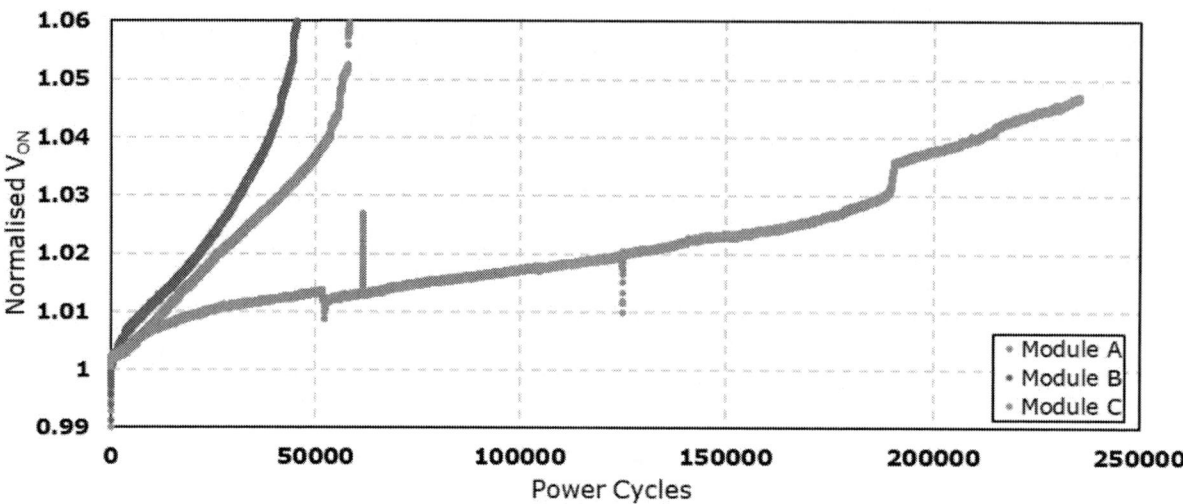

Figure 7. Evolution of V_{ON} (normalized) for one sample of Modules A, B and C.

3.4 Module lifetime and derating – A simplified example

Using Module B's lifetime extrapolation and a ΔT_J coefficient of -3.989, Module A would require a ΔT_J of approximately 35°C to survive the 990,000 cycles of Module D at 90.7°C.

Given Module B's average absolute R_{TH} of 0.04671 K/W, derating would have to be applied such that if both Module B and Module D were applied in the same application, the system would require 2.27 modules of Module B for each one of Module D to achieve the same lifetime. This significantly decreases the power density and adds further weight and volume with the required additional heatsink space.

3.5 Runaway Degradation at End-of-Life

Further analysis of the completed lifetime tests on the SiC modules shows that Modules A and B both experience runaway degradation at the end of their lifetime. Module C on the other hand does not show this. We would therefore expect that Module C is capable of operating beyond the standard 5% increase in V_{ON}. As a result, the module effective lifetime may extend beyond the data shown in Figure 6. Figure 7 displays the normalized evolution of V_{ON} for one sample of Module A, B, and C.

4 Conclusion

This paper compares thermal and reliability aspects of 1.7 kV, 900 A – 1000 A power modules manufactured with SiC MOSFETs and Si IGBTs.

We observed dependency of module thermal resistances on power level, as well as operating mode, for both SiC MOSFET and Si IGBT variants. The absolute thermal resistance of each module is similar, with Si IGBTs resistances slightly lower. However, when normalized for overall chip area, Si IGBT modules have larger thermal resistances. Power cycling lifetime is assessed, and the lowest performing SiC MOSFET module has a lifetime 98% less than that of the highest performing Si IGBT at a ΔT_J of approximately 90°C.

References

[1] F. Hoffmann, N. Kaminski and S. Schmitt, "Comparison of the Power Cycling Performance of Silicon and Silicon Carbide Power Devices in a Baseplate Less Module Package at Different Temperature Swings," 2021 33rd International Symposium on Power Semiconductor Devices and ICs (ISPSD), Nagoya, Japan, 2021, pp. 175-178.

[2] J. Lutz, "Packaging and Reliability of Power Modules," CIPS 2014; 8th International Conference on Integrated Power Electronics Systems, Nuremberg, Germany, 2014, pp. 1-8.

[3] Hernes M, D'Arco S, Antonopoulos A, Peftitsis D. Failure analysis and lifetime assessment of IGBT power modules at low temperature stress cycles. IET Power Electron. 2021; 14: 1271–1283.

[4] W. Brekel, W. Rusche, A. Hoehn and W. Buecker, "XHPTM 2 – The low-inductive, multi-package housing for the next generation of high-power applications," PCIM Europe digital days 2020; International Exhibition and Con-

ference for Power Electronics, Intelligent Motion, Renewable Energy and Energy Management, Germany, 2020, pp. 1-7.

[5] Y. Zhang et al., "Figures-of-Merit Study for Thermal Transient Measurement of SiC MOSFETs," in IEEE Transactions on Power Electronics

[6] N. Baker, A. Lemmon, F. Iannuzzo, S. Bęczkowski, J. Austin and L. Ostrander, "Temperature Monitoring of Multi-Chip SiC MOSFET Modules: On-Chip RTDs vs. VSD(T)," 2023 25th European Conference on Power Electronics and Applications (EPE'23 ECCE Europe), Aalborg, Denmark, 2023

[7] J.P. Ousten, Z. Khatir, "Investigations of thermal interfaces aging under thermal cycling conditions for power electronics applications," Microelectronics Reliability, Volume 51, Issues 9–11, 2011, Pages 1830-1835,.

[8] J. C. J. Paasschens, S. Harmsma, and R. van der Toorn, "Dependence of Thermal Resistance on Ambient and Actual Temperature," Philips Research Laboratories.

[9] Scheuermann U., Schmidt R. Investigations on the VCE(T)-Method to Determine the Junction Temperature by Using the Chip Itself as Sensor; Proceedings of the PCIM Europe 2009; Nuremberg, Germany. 12–14 May 2009

[10] Amol Deshpande, Riya Paul, Asif Imran Emon, Zhao Yuan, Hongwu Peng, Fang Luo, Si-IGBT and SiC-MOSFET hybrid switch-based 1.7 kV half-bridge power module, Power Electronic Devices and Components, Volume 3, 2022.

[11] U. Scheuermann and R. Schmidt, "A New Lifetime Model for Advanced Power Modules with Sintered Chips and Optimized Al Wire Bonds," 2013.

[12] Bayerer, R., et al.: Model for Power Cycling lifetime of IGBT Modules - various factors influencing lifetime. In: 5th International Conference on Integrated Power Electronics Systems, Nuremberg, Germany, pp. 1–6 (2008).

[13] F. Hoffmann, S. Schmitt and N. Kaminski, "Lifetime Modeling of SiC MOSFET Power Modules During Power Cycling Tests at Low Temperature Swings," 2023 35th International Symposium on Power Semiconductor Devices and ICs (ISPSD), Hong Kong, 2023.

PCIM Europe 2024, 11– 13 June 2024, Nuremberg   DOI: 10.30420/566262270

Power Cycling Capability of Discrete SiC MOSFET Devices with Different Designs

Luhong Xie[1], Erping Deng[2], Yushan Zhao[3,4], Ying Zhang[1], Dianjie Gu[1], Hao Liu[1], Yongzhang Huang[1]

[1] State Key Laboratory of Alternate Electrical Power System with Renewable Energy Sources (North China Electric Power University), China

[2] State Key Laboratory of High Efficiency and High Quality Electric Energy Conversion (Hefei University of Technology), China

[3] NARI Group Corporation (State Grid Electric Power Research Institute), Nanjing, China

[4] Nanjing NARI Semiconductor Co., Ltd, Nanjing, China

Corresponding author: Erping Deng, erping.deng@hfut.edu.cn
Speaker: Luhong Xie, xieluhong07@163.com

Abstract

The purpose of this paper is to evaluate the power cycling capability of the commercial discrete SiC MOSFET devices as there are many differences in the designs. Six groups of PCT(Power Cycling Test) with the 1200V discrete SiC MOSFETs from six different manufacturers are carried out under the same test conditions. The power cycling lifetime data show that the chip area and the rated current have a significant influence on the lifetime, while that of other design parameters is small. As the chip area and the rated current decide the power density, the conclusion is obtained that the power density is the key factor that impacts the power cycling capability of discrete SiC MOSFET devices.

1 Introduction

SiC MOSFET devices reach widespread attention due to their high power density, high thermal conductivity and low power losses [1]. As different designs and production technologies are applied by various manufacturers, the varying levels of package reliability of commercial discrete SiC MOSFET devices are presented. Because for the discrete devices, the bond wire failure has supplanted solder layer degradation as the primary failure mechanism [2]. However, the reliability of the bond wire is affected by many design factors, such as the bond wire diameter, the joint length, the chip area, and the rated current [3]. Power cycling capability is normally used to quantitatively describe the reliability of the device since the power cycling test is the most important aging test to evaluate the package reliability of devices.

In order to evaluate the power cycling capability of the commercial discrete SiC MOSFET devices, six types of commercial discrete SiC MOSFET from different manufacturers are selected in this paper as research objects. By comparing the power cycling test results of these six SiC MOSFET devices with different designs, the key factors affecting the power cycling capability of discrete SiC MOSFET devices are analyzed.

2 Power Cycling Test

2.1 Test setup

The test circuit diagram and the control strategy are shown in Fig. 1(a) and Fig. 1(b), respectively. There are three phases, and the load current I_{load} will flow through each phase controlled by switches S1, S2 and S3. In order to avoid the oscillation of the output load current, the overlap time with 10μs is set between the adjacent phases. As $V_{SD}(T)$ method is utilized to measure the junction temperature, gate driver V_{GS} is applied to provide an adjustable voltage of a positive voltage during t_{on} and a negative voltage during t_{off}. In order to prevent the DUT (Devices Under Test) from switching the load current, a delay time t_d of 50μs is set before the load current is switched on and after the load current is switched off.

Note that the measurement current I_m should be small enough to create negligible power losses, normally $I_m \approx 0.001 \cdot I_L$ is advised [4]. A measurement current with 100 mA is chosen in this paper. Considering the inevitable carrier recombination process, the maximum junction temperature T_{jmax} is measured with a delay time t_{delay} of 200μs after the load current is switched off [5]. And the minimum junction temperature T_{jmin} is measured before the load current of the next cycle is switched on.

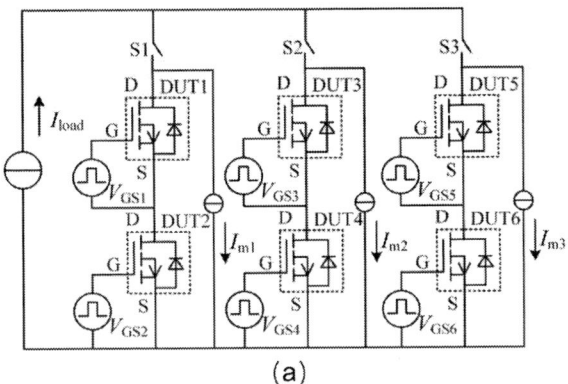

(a)

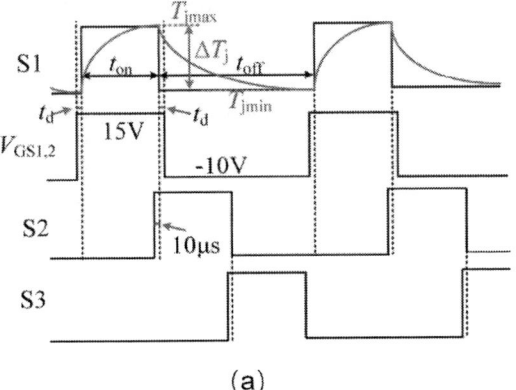

(a)

Fig. 1 The power cycling test circuit. (a) The test circuit diagram. (b) The schematic of the control strategy.

2.2 Devices under Test

The SiC MOSFET devices with TO247 package from six manufacturers are investigated in this paper, labeled from A to F. The rated voltage of the devices is 1200 V. Considering that the geometry parameters of the chip and the bond wire influence the lifetime, the geometry parameters included the chip area, the number of the bond wire, the diameter and the joint length of the bond wire are obtained by removing the epoxy mold of each SiC MOSFET devices. It should be noted that the chip thickness is obtained from the scanning electron microscope (SEM) pictures after cutting the devices from the longitudinal section. And the details are shown in Tab.1.

Type	A	B	C	D	E	F
Rated current in 25°C/	26	29	37	36	24	19

(A)						
Chip area/(mm²)	4.62	8.76	11.24	13.07	6.04	3.41
Chip thickness/(µm)	110	192	356	237	351	146
The number of bond wire	2	2	4	2	1	2
The diameter of bond wire/(mm)	300	580	300	480	400	350
The joint length of bond wire/(mm)	0.65	1.22	0.63	1.37	1.05	0.75

Table 1 Details of the six types of SiC MOSFET devices from six manufacturers

3 Results and Investigation

Six groups of PCTs with different DUTs are carried out. And 6 devices are tested at the same in each group considering the individual differences. The junction temperature ΔT_j and the maximum junction temperature T_{jmax} are the most critical factors affecting the power cycling capability [6]. Therefore the DUTs in each group are tested under the same temperature conditions of ΔT_j =90 K, T_{jmax}=150 ℃. Table 2 shows the test settings and the measured temperature conditions of each PCT. Noting that the presented temperature is the mean value of the six DUTs.

Type	I_{load}/(A)	P/(W)	ΔT_j/(K)	T_{jmax}/(°C)
A	18.8	73.4	94.3	156.4
B	19.5	81.3	89.2	153.9
C	26.5	92.3	88.7	157.5
D	26.8	84.8	89.0	153.4
E	17.7	67.0	89.5	153.3
F	16.0	65.8	82.7	150.1
The other test settings: t_{on}=1 s, t_{off}=2 s, t_{delay}=200µs				

Table 2 PCT settings and measured temperature conditions of each group.

3.1 Experimental results

During the test, the voltage drop at load current V_{DS} and junction-to-heatsink thermal resistance R_{thjs} are monitored and recorded after each cycle. Using one DUT in each group as an example, Figure 2 presents the record parameters of six DUTs from six groups. For a better comparison of the power cycling capability of the DUTs, the lifetimes are all normalized to the maximum value between the presented six DUTs.

It is clear that all DUTs present the increment of V_{DS} and reach the failure criteria of 105% V_{DS} of its initial value. The thermal resistance R_{thjs} decreases slowly during the test due to the setting of thermal interface material(TIM) used for electric insulation between DUT and the heatsink. Since the increase of V_{DS} represents the bond wire failure while the increase of R_{thjs} represents the solder layer degradation, it can be deduced that all DUTs fail with bond wire fatigue but no aging in the chip solder layer. The pictures of the scanning acoustic microscope (SAM) can further verify this conclusion, as shown in Fig.2. The left picture is the picture before PCT while the right is that after PCT. Compared with the picture before PCT, it can be seen that the black point representing a healthy contact state of the bond wire joint of devices A, E and F has been turned white after PCT. But the joints of devices B, C and D are still partially bonded.

In addition, note that there are steps on the V_{DS} curves. In a power module with bond wire failure, the steps are normally considered caused by the bond wire lift-off [7], however in discrete devices, cracks is the main failure mechanism because the lift-off is limited by the epoxy mold. Figure 3 presents the SEM pictures of the bond wire of the 6 types of DUTs. It is clear to see that cracks occur on the heel region or the joint interface between the bond wire and the aluminum metal layer.

Therefore, it can be concluded that the steps on the V_{DS} curves are due to the crack propagation.

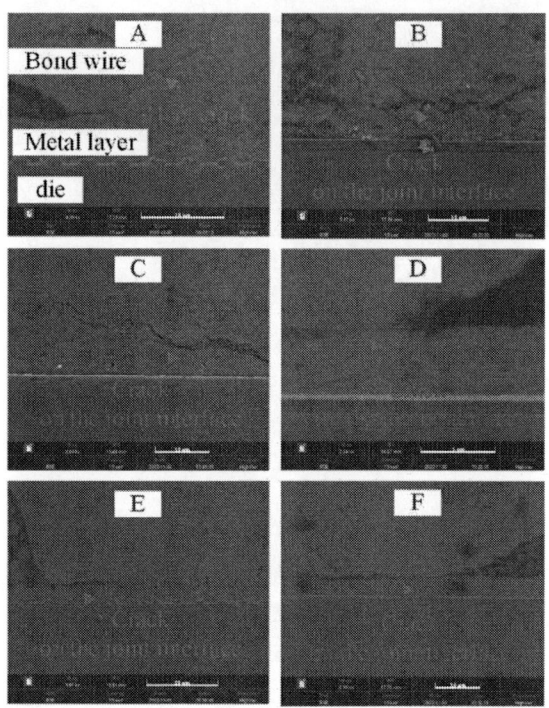

Fig. 3 The SEM pictures of the bond wire of the DUTs in each group.

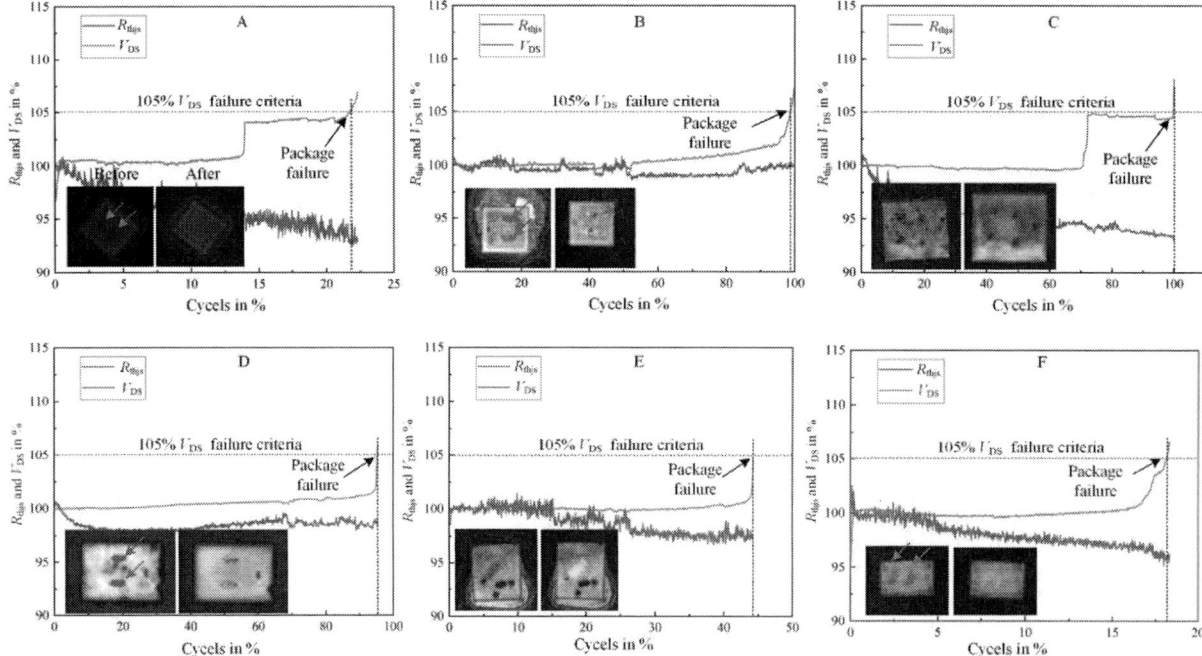

Fig. 2 The change trends of R_{thjs} and V_{DS} of six SiC MOSFET devices from six manufacturers during the PCT and their SAM pictures before and after PCT.

Since this paper aims to evaluate the power cycling capability of the SiC MOSFET devices with different designs, the power cycling lifetime should be compared fairly. Therefore, the power cycling lifetime of the DUTs should be normalized to the same test condition ΔT_j=90 K, T_{jmax}=150 °C. Although the lifetime model for discrete Si IGBT devices is established by Zeng [8], significant differences are presented as SiC MOSFET devices are used here. Therefore, LESIT lifetime model [9] with the parameters A=1.34e10, α=-2.7478, E_a=9.89e-20J, k_b=1.38e-23J/K, as shown in formula (1) is used in this paper.

$$N_f = A \cdot \Delta T^\alpha \cdot e^{\frac{E_a}{k_b \cdot T_m}} \quad (1)$$

After normalization, the Weibull distribution of the power cycling lifetime of the six types of discrete SiC MOSFET devices are presented in Fig.4. The shape parameter and the scale parameter, which represents the dispersion and the median value respectively, are also listed in Fig.4. The shape

parameters of each PCTs are all in the range of 1 to 4, indicating that the DUTs in each type devices have roughly the same level of dispersion. The median value of the Weibull distribution can effectively represent the lifetime of the devices. Note that the lifetime of the DUTs are all normalized to the maximum value between the DUTs.

Comparing the median value of the Weibull distribution, We can conclude that $N_{fC}>N_{fD}>N_{fB}>N_{fE}> N_{fA} >N_{fF}$. That means the power cycling capability of the type C SiC MOSFET is the largest, which has 4 bond wires with diameter of 300mm and joint length of 0.63mm, followed by the type D and type B SiC NOSFET devices, which have 2 bond wires with a larger diameter. Therefore, it can be concluded that the discrete SiC MOSFET devices with four thinner bond wires have a higher power cycling capability than those with two thicker bond wires. This conclusion is the same as that presented in [10].

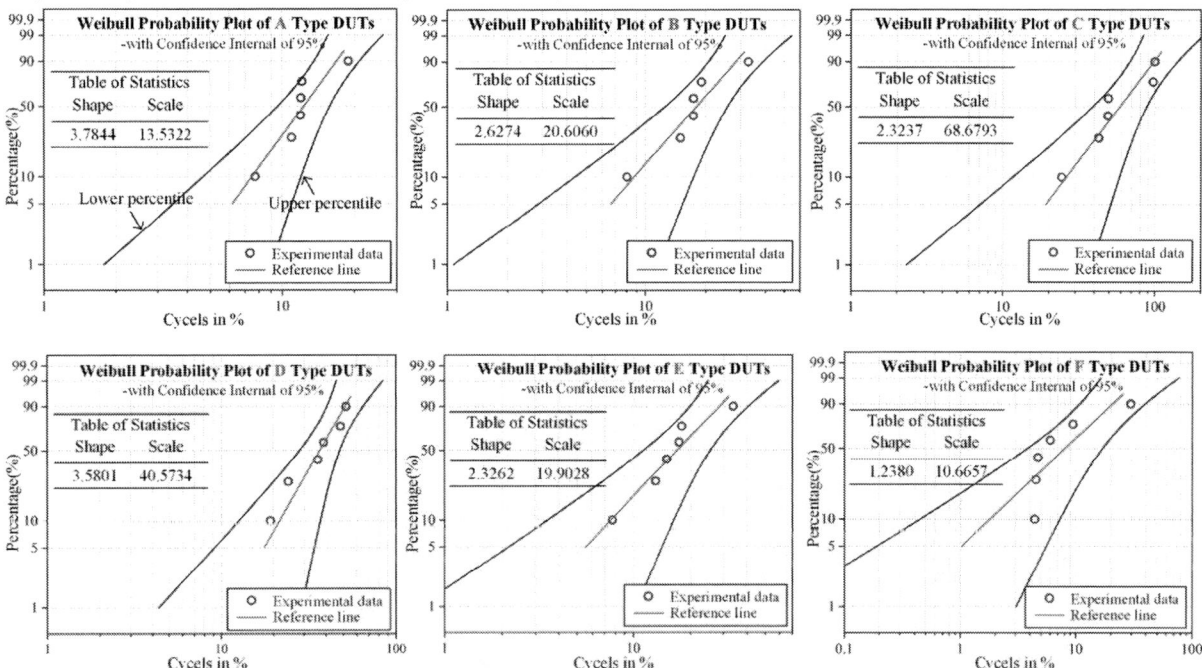

Fig. 4 The Weibull distribution of the power cycling lifetime of the discrete SiC MOSFET devices from six manufacturers.

3.2 Investigation of the influencing factors

In order to further investigate the relationship between the power cycling capability and design parameters, the power cycling lifetime and design parameters are plotted by logarithm, as shown in Fig.5. The power cycling lifetime is the median value of the Weibull distribution of each PCT. The

design parameters mainly contain the rated current, the number of bond wires, the chip area, the chip thickness, the bond wire diameter, and the joint length.

It is clear to see that an obvious linear relationship is presented in Fig5(a) and Fig.5(c). In contrast, a large spread of the data is presented in Fig.5(d) to Fig.5(f). That means the rated current and the chip area are the main factors that affect the power cycling capability. The spread of data indicates the impact of other factors. Consequently, a roughly linear increase trend is presented in Fig.(d) to Fig.(f) due to the effect of the rated current and the chip area. Therefore, it can be concluded that the effect of the chip thickness, the bond wire diameter, and the joint length is smaller than that of the rated current and the chip area.

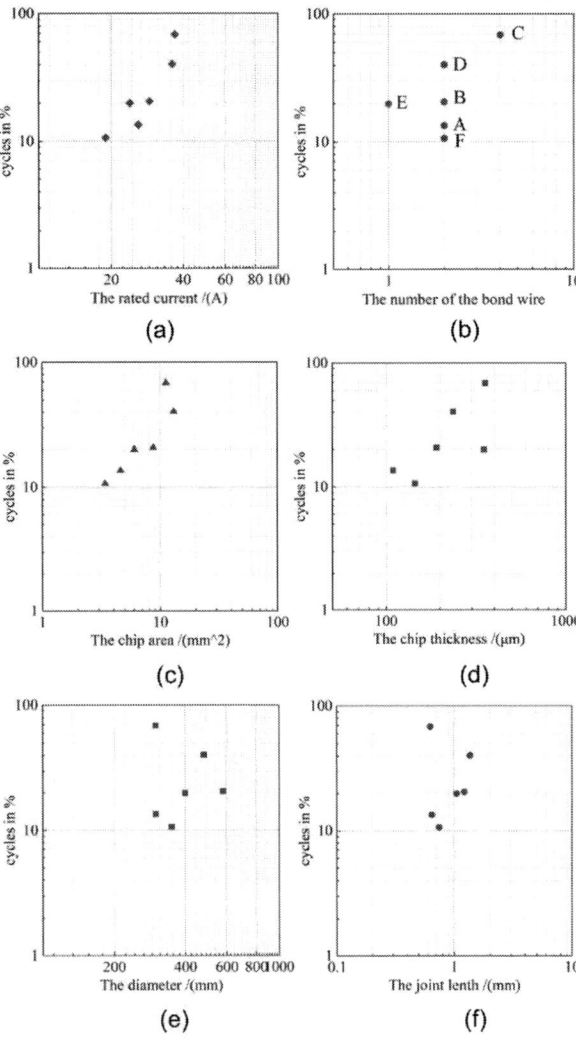

Fig. 5 The power cycling lifetime data plotted over different design parameters. (a) The rated current. (b) The number of the bond wire. (c) The chip area. (d) The chip thickness. (e)The diameter of the bond wire. (f) The joint length of the bond wire.

As for the number of the bond wire, the device with more bond wires has a higher power cycling lifetime. That is because the more bond wires there are, the lower the load current of the bond wire for the same rated current, which means the self-heat of the bond wire is less. So the type C device with four bond wires has the highest power cycling capability. If the number of bond wire is the same, the power cycling capability of the devices increases with the chip area increase, as shown in Fig.5(b). The type D, B, A and F devices all have two bond wires, but the chip area of type D devices is the largest, followed by type B, A and F devices. This sequence is exactly the same with the power cycling lifetime between type D, B, A and F devices.

In conclusion, compared with the other design parameters, the rated current and the chip area are the most important factors that affect the power cycling capability of the discrete SiC MOSFET devices. It is worth noting that the rated current and the chip area decide the current density, which can be considered as the power density that is more normally used to describe the characteristics of semiconductor devices. Therefore, the power density under the PCT is used in this paper.

Figure 6 presents the relationship between the power cycling lifetime and the power density. The blue line is the fitted curve based on the six points which represent the power cycling lifetime of six SiC MOSFET devices. The power cycling lifetime of type C devices is strongly influenced by the number of bond wires, leading to a significant deviation between point C and the blue line. Therefore, the relationship between the lifetime and the power density is obtained by fitting the experimental results except the power cycling lifetime of type C devices, represented by the orange line. The corresponding fitted formula is also presented in Fig. 6 with orange words. And the fitting accuracy of 0.96 is reached. That means the fitted formula can effectively describe the relationship between the power cycling lifetime and power density.

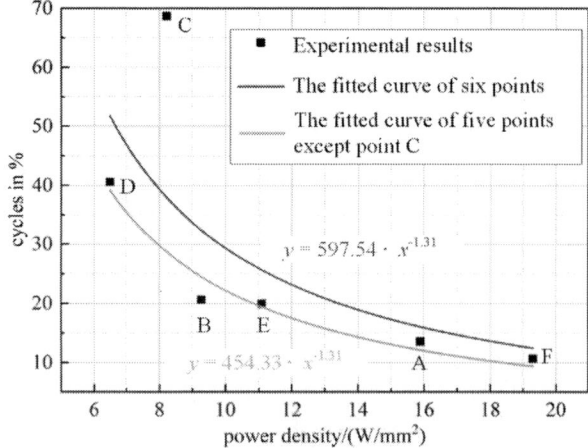

Fig. 6 The power cycling lifetime data plotted over power density.

4 Conclusion

This paper uses six types of commercial discrete SiC MOSFET devices with different design parameters to evaluate the power cycling capability. The key factor influencing the power cycling capability is investigated by analyzing and comparing the power cycling lifetime under different design parameters, mainly containing the rated current, the number of bond wires, the chip area, the chip thickness, the bond wire diameter and the joint length. The following shows the results.

(a) Increasing the number of bond wires can effectively improve the power cycling capability due to the decreased current per bond wire.

(b) Not taking into account the effect of the number of bond wires, the power density is the main factor influencing the power cycling capability of the discrete SiC MOSFET devices. The power cycling capability shows a negative power exponential relationship with power density, indicating that the higher the power density, the lower the capability.

Therefore, the balance between power cycling capability and high power density needs to be considered when designing discrete SiC MOSFET devices.

Acknowledgment

This work was supported by the Science and Technology Project of State Grid under Grant 5500-202440127A-1-1-ZN.

References

[1] J. A. Cooper, M. R. Melloch, R. Singh, A. Agarwal and J. W. Palmour, "Status and prospects for SiC power MOSFETs," *in IEEE Transactions on Electron Devices*, vol. 49, no. 4, pp. 658-664, April 2002.

[2] P. Heimler, N. Thönelt, J. Lutz and T. Basler, "Impact of Bond Wire Configuration on the Power Cycling Capability of Discrete SiC-MOSFET Devices," *2022 24th European Conference on Power Electronics and Applications (EPE'22 ECCE Europe)*, Hanover, Germany, 2022, pp. 1-9.

[3] L. Xie, E. Deng, S. Yang, Y. Zhang, Y. Z hong, Y.H. Wang, Y.Z. Huang, " State-of-t he-Art of the Bond Wire Failure Mechanis m and Power Cycling Lifetime in Power E lectronics," *Microelectronics Reliability*, vol. 147, 2023.

[4] J.Lutz, H.Schlangenotto, U.Scheuermann, et al. "Semiconductor Power Devices." Ge rmany: Springer International Publishing A G, 2018, 506.

[5] C. Herold, J .Franke, R. Bhojani , A. Schl eicher, J. Lutz. "Requirements in power cy cling for precise lifetime estimation. " *Micr oelectronics Reliability*, vol.58, pp:82-89, 2 016.

[6] R. Bayerer, T. Herrmann, J. Lutz. "Model for Power Cycling lifetime of IGBT Modules - various factors influencing lifetime," *5th International Conference on Integrated Power Electronics Systems*. Nuremberg, Germany, pp. 1-6, 2008.

[7] E.P Deng, L.H Xie, L.X Wu, H.Y. Cao, M.Y Pan, Y. Zhang, and et al, "Load Current Injection Modes Affected Power Cycling Lifetime and Failure Mechanism of IGBTs," *in IEEE Journal of Emerging and Selected Topics in Power Electronics*, vol. 11, no. 3, pp. 3525-3534, June 2023.

[8] G. Zeng, L. Borucki, O. Wenzel, O. Schilli ng and J. Lutz, "First Results of Develop ment of a Lifetime Model for Transfer Mol ded Discrete Power Devices," *PCIM Europ e 2018; International Exhibition and Confe rence for Power Electronics, Intelligent Mo tion, Renewable Energy and Energy Mana gement*, Nuremberg, Germany, 2018, pp. 1-8.

[9] F. Hoffmann and N. Kaminski, "Impact of Device Design on the Power Cycling Cap ability of Discrete SiC MOSFETs at Differ ent Temperature Swings," *2020 32nd Inter national Symposium on Power Semicondu ctor Devices and ICs (ISPSD)*, 2020, pp. 533-536.

[10] P. Heimler, N. Thönelt, J. Lutz and T. Basler, "Impact of Bond Wire Configuration on the Power Cycling Capability of Discrete SiC-MOSFET Devices," *2022 24th European Conference on Power Electronics and Applications (EPE'22 ECCE Europe)*, Hanover, Germany, 2022, pp. 1-9.

PCIM Europe 2024, 11– 13 June 2024, Nuremberg   DOI: 10.30420/566262271

Model-based Parameter Tuning of Semiconductor Devices in DC Power Cycling Test

Yichi Zhang ⊚, Yi Zhang ⊚, Bo Yao ⊚, Huai Wang ⊚

AAU Energy, Aalborg University, Aalborg, Denmark

Corresponding author: Yi Zhang, yiz@energy.aau.dk
Speaker: Yi Zhang, yiz@energy.aau.dk

Abstract

Power cycling test serves as a widely employed accelerated test for semiconductor package reliability analysis, and has received considerable research attention in this area in the last few decades. Before the test is run formally, how to set test parameters to efficiently achieve the desired thermal stress is a crucial yet underreported aspect. The several attempts may be feasible but invariably result in a time-consuming process. To address this issue, this paper presents an approach based on the established thermal model, which quantifies the required thermal stresses, junction temperature fluctuation (ΔT_j), and maximum junction temperature (T_{jmax}). Moreover, the adjustable parameters affecting two key thermal indicators are revealed. Then a flowchart for the parameter tuning is proposed. In addition, the model also can provide the thermal stress range of the tested module within the constraints of these adjustable parameters, which offers a valuable reference for implementing various testing conditions. Finally, the effectiveness of the method is validated through the experimental case study.

1   Introduction

DC power cycling test is widely recognized as an effective method for assessing the reliability of the packaging of semiconductor devices [1]. Over the past two decades, extensive power cycling tests have been conducted. However, the existing research has primarily focused on the post-result analysis, such as performance validation [2], lifetime modeling [3, 4], failure mechanism analysis [5], and so on [6], instead of the test itself. How to efficiently achieve the desired test condition, a key point, has not received more attention although power cycling tests appear to be straightforward.

The existing standards lack comprehensive frameworks for the precise tuning of parameters in power cycling tests. The test-related standards and guidelines, IEC 60749-34 and AQG324, only offer general directives on test equipment, procedures, and conditions [7]. Moreover, the diverse modules and cooling systems used in practice necessitate adjustments to operating parameters based on specific circumstances. Therefore, the absence of a detailed tuning guide represents a significant gap in current research.

Parameter tuning poses additional challenges, notably the time-intensive nature of the process. Thermal stress, the key metric in testing, depends on both electrical (power loss) and thermal parameters (thermal impedance). While adjusting electrical parameters is relatively straightforward, optimizing these parameters without guidance can lead to inefficient testing, especially under various conditions and with high t_{on} values. Moreover, the adjustment process of thermal impedance often involves disassembling the module, applying thermal grease, and reassembling, which is particularly burdensome for high-power modules. Additionally, the heterogeneity of devices and the non-uniform heat dissipation in heat sink systems can cause deviations from expected thermal stresses [8]. Testing multiple samples in series may exacerbate these discrepancies. One solution to minimize the overall deviations is to connect devices with similar thermal stresses in series. This necessitates the precise characterization of thermal stress in the device.

Despite these aforementioned challenges being infrequently discussed, they are pivotal in the power cycling test and are expected to be addressed in this study. Initially, a thermal model is provided, which is capable of precisely characterizing the thermal stress experienced by the device. Subsequently, the parameters that are adjustable during testing are determined and this study introduces a streamlined flowchart for effective parameter tuning. Moreover, by defining the constraints of the adjustable parameters, it elucidates the range of thermal stresses the device can withstand under a given test system, thereby offering valuable guidance for the specifications of the expected test conditions.

The rest of this paper is organized as follows. In section II, a brief introduction about the power cycling test and the details of the thermal model are presented. Section III provides the flowchart for the parameter tuning while all possible scenarios in the test are considered. In section IV, the effectiveness of the proposed method is experimentally validated. Concluding remarks are drawn at last.

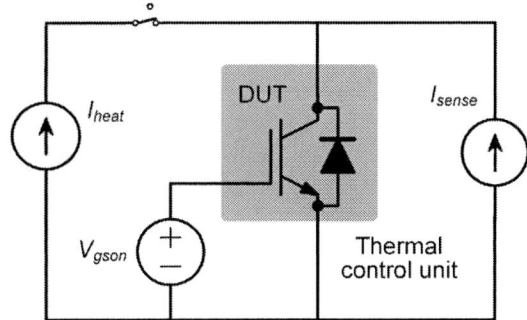

Fig. 1: DC power cycling test circuit.

2 DC Power Cycling Test and Thermal Estimation Model

Figure 1 provides the test circuit of the DC power cycling test for insulate-gate bipolar transistor (IGBT) modules. It can be seen that the device under test (DTU) is still on during the whole test, and the heating and cooling are achieved utilizing the on-off control of the heating current (I_{heat}). The junction temperature information is derived based on V_{ce} (T) under thermal sensing current (I_{sense}) [9]. For the test design, the maximum junction temperature (T_{jmax}) and temperature fluctuation (ΔT_{j}) are typically two key target values.

During the testing phase, the collector current passing through the device can be represented as a series of square wave signals, as depicted in Fig. 2. This is attributed to the fact that the device incurs only conduction losses, allowing these losses to be similarly approximated as square waves. Furthermore, it is customary for the two measured values during steady state to align closely with the predetermined target value.

As shown in Fig. 2, the device's junction temperature incrementally increases until it transiently achieves a steady state. ΔT_{j} can be quantitatively depicted as illustrated in Eq. (1), which is primarily influenced by the power losses and thermal impedance. The on-state voltage (V_{ce}) is governed by the heating current and the turn-on gate voltage (V_{gson}). It should be noted that the variation of V_{gson} is only permissible within the saturation range of the power semiconductor, which aligns

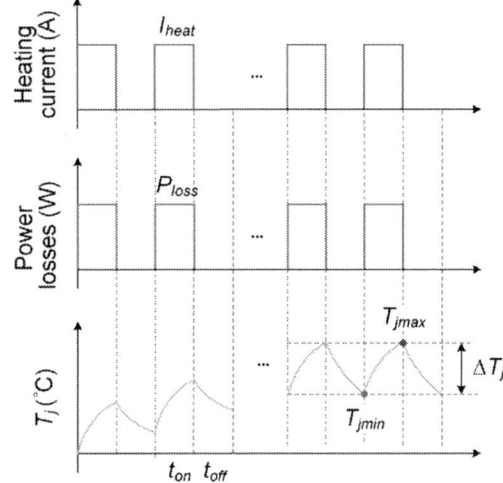

Fig. 2: Schematic diagram of the electrical and thermal transition from the transient stage to the steady-state stage.

with the requirements in AQG 324 [7]. Regarding the maximum junction temperature, through the application of iterative modeling theory as proposed in our study [10], it can be quantitatively expressed as demonstrated in Eq. (2) with the assumption that t_{on} and t_{off} are equal. Therefore, it can be concluded that I_{heat}, V_{gson}, Z_{th}, t_{on}, and T_{coolant} are the adjustable parameters in power cycling test. Absolutely, it is important to note that when these parameters are determined and cannot be modified during the testing process.

$$\begin{aligned} \Delta T_{\text{j}} &= P_{\text{loss}} \cdot Z_{\text{th}}(t_{\text{on}}) \\ &= I_{\text{heat}} \cdot V_{\text{ce}}(I_{\text{heat}}, V_{\text{gson}}) \cdot Z_{\text{th}}(t_{\text{on}}) \end{aligned} \tag{1}$$

$$T_{\text{jmax}} = \sum_{j=1}^{m} \left[\frac{P_{\text{loss}} R_{\text{thj}} \left(e^{-\frac{t_{\text{on}}}{2\tau_{\text{thj}}}} - e^{-\frac{t_{\text{on}}}{\tau_{\text{thj}}}} \right)}{1 - e^{-\frac{t_{\text{on}}}{\tau_{\text{thj}}}}} \right] + \underbrace{\qquad}_{\Delta T_j} + T_{\text{coolant}} \qquad (2)$$

where Z_{th} represents the thermal impedance from the device junction to the coolant, which is expressed in terms of the Foster thermal network. R_{thj}, and τ_{thj} indicate the thermal resistance and thermal time constant of the j-th RC lump, respectively, m is the order of the RC lumps.

3 Parameter Tuning Strategy

Upon identifying the adjustable parameters, the subsequent step involves determining the sequence for their adjustment. As previously mentioned, modifying different parameters entails varying degrees of effort. Consequently, two practical and empirical principles are used to guide the subsequent tuning process:

- prioritize adjusting ΔT_j since T_{jmax} can be controlled by changing the temperature of the coolant.

- when adjusting ΔT_j, prioritize adjusting P_{loss} over Z_{th}, because the adjustment of the electrical parameters is easier compared to the thermal impedance.

During the tuning process, both ΔT_j and T_{jmax} need to reach the expected values and t_{on} can be slightly adjusted. It is important to emphasize that significant alterations in the t_{on} will result in modifications to the failure mechanism. It is advisable that the adjustment of t_{on} adhere to the guidelines provided in AQG324. Specifically, adjustments to less than 5 seconds are predominantly associated with the failure of chip-near interconnections. Moreover, adjustments exceeding 15 seconds are likely to induce failure not only in chip-near but also in chip-remote interconnections.

Figure 3 presents a flowchart to provide the details. Firstly, when t_{on} is smaller than the thermal time constant of the IGBT module, ΔT_j only can be altered by adjusting P_{loss}, namely V_{gson} and I_{heat}. In addition, these variables typically have a constrained range. V_{gson} should ensure that the device operates in the saturation region and avoids the linear region for the IGBT module. The maximum

heating current also should be chosen carefully, as excessive current may lead to different failure mechanisms [11]. Secondly, when t_{on} exceeds the thermal time constant of the IGBT module, it signifies another possibility of adjusting ΔT_j by changing the thermal impedance. Various effective approaches, such as incorporating aluminum or copper blocks or modifying the thermal interface material (TIM) between the module and the heat sink, can be employed [11]. In cases where none of the options meet the test requirements, the adjustment of t_{on} can be the final choice.

Furthermore, it is pertinent to acknowledge that tests do not often focus on a singular test condition, for example, the establishment of the lifetime model. Therefore, it is logical to establish the range of thermal stresses that the device is capable of withstanding within the constraints of the adjustable parameters. When the thermal stress requirement is under the deduced range, the operating parameters can be directly ascertained using the provided equations. In instances where these requirements are not met, modifications are implemented following the aforementioned adjustment guidance. Consequently, the proposed method substantially reduces the effort required for parameter tuning, thereby efficiently identifying the requisite parameters.

4 Experimental Validation

To validate the effectiveness of the proposed method, a case study is conducted. Figure 4 shows the test platform, and the test is carried out using the power tester (PWT) and the desired temperature and cooling capability can be achieved through the thermal control unit. The module under test is a commercial product, with a rated current of 50 A and a rated voltage of 1.2 kV. The expected test conditions are ΔT_j =100 K, T_{jmax} =150 °C, and t_{on}=3 s.

The thermal time constant identified from the datasheet for the tested module is 0.356 s, markedly lower than the required t_{on}, indicating the presence of multiple parameters available for adjusting junction temperature fluctuations. The initial step involves obtaining the output characteristic curves of the device from the datasheet, as illustrated in Fig. 5. It is specified that during testing, the adjustable range for the heating current should not fall below 0.85 times nor exceed 1.25 times the rated current (I_{rated}). It is observed that at a gate

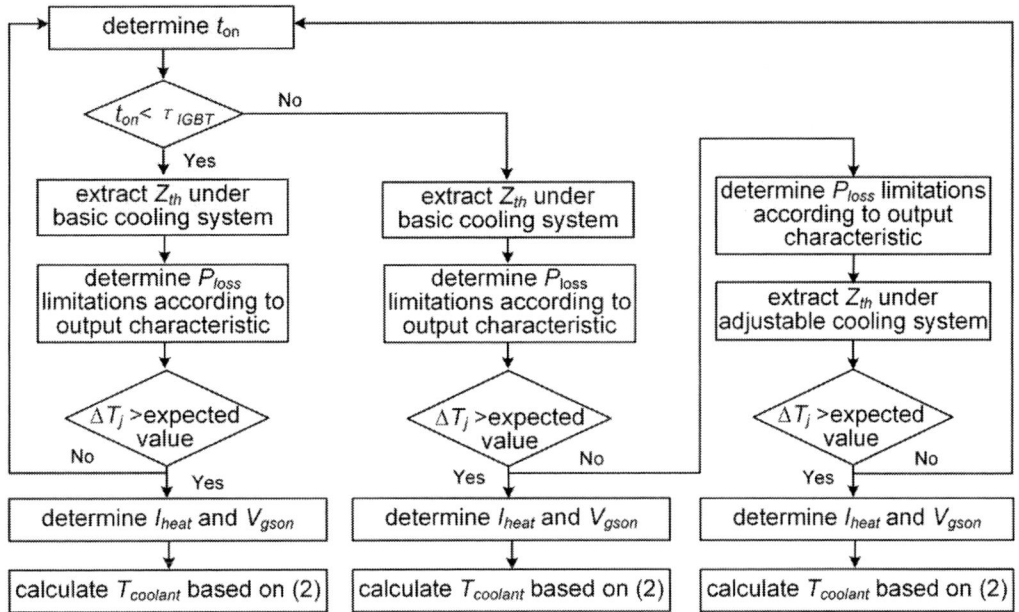

Fig. 3: Flowchart for the parameter tuning in the DC power cycling test.

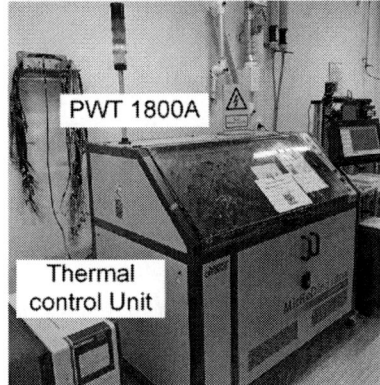

Fig. 4: Power cycling test platform.

voltage of 10V, the device swiftly transitions into the linear region, which deviates from the requirement. Therefore, a gate voltage range of 12-20V is deemed preferable.

Moreover, thermal impedance serves as another adjustable parameter. This research demonstrates that modifications in thermal impedance can be facilitated through the alteration of the thermal interface material (TIM). Figure 6 displays the thermal impedance curves with three scenarios, revealing the significant impact of TIM material on thermal impedance.

The range of P_{loss} is deduced by extracting the minimum gate voltage and maximum heating current, as well as the maximum gate voltage and minimum

heating current from Fig. 5. Subsequently, the thermal impedance value at 3 s is ascertained from Fig. 6. This enables the calculation of the junction temperature fluctuation range the device can withstand under various t_{on} and TIMs, depicted in Fig. 7. The diagram demonstrates that at 3 s, the device can withstand a junction temperature fluctuation range of 60-150 °C. Given the target of 100 °C falls within this range, it can be met irrespective of the TIM material chosen. Furthermore, it is observed that the range of thermal stresses remains unaffected by the TIM when the t_{on} value is less than the thermal time constant of the module. Additionally, the fluctuation in junction temperature is relatively small under specified gate voltage and heating current conditions. Therefore, to induce more significant junction temperature fluctuations, the introduction of switching losses is commonly required in millisecond power cycling tests if the excessive heating current needs to be avoided.

If the power cycling test is performed using TIM3 and V_{gson} is set to 15 V, the test current can be calculated as 54.2 A. Additionally, the parameters of the Foster thermal network can be obtained by fitting the thermal impedance curve from Fig. 6, then the coolant temperature can be calculated as 39.3 °C using Eq. (2). With these parameters set, the experiment is implemented, resulting in an actual ΔT_{j} of 103.3 K and T_{jmax} of 150.4 °C, which closely aligns with the expected values.

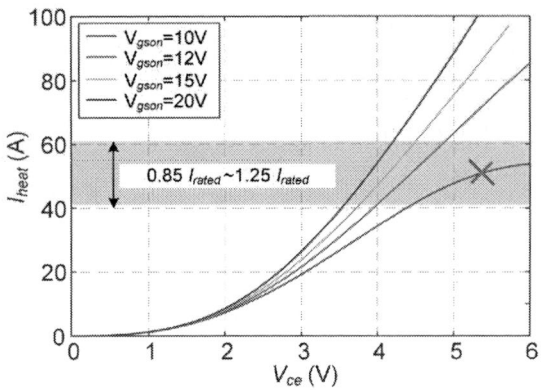

Fig. 5: Output characteristic of the tested IGBT module.

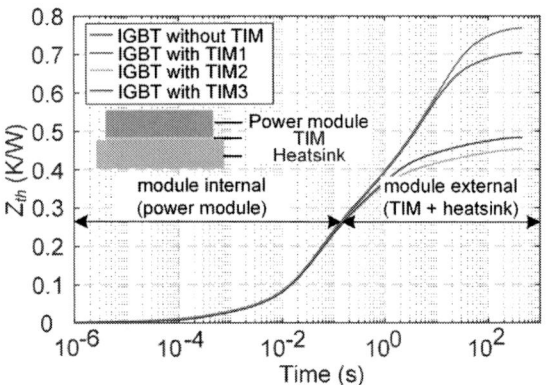

Fig. 6: Transient thermal impedance with different TIMs.

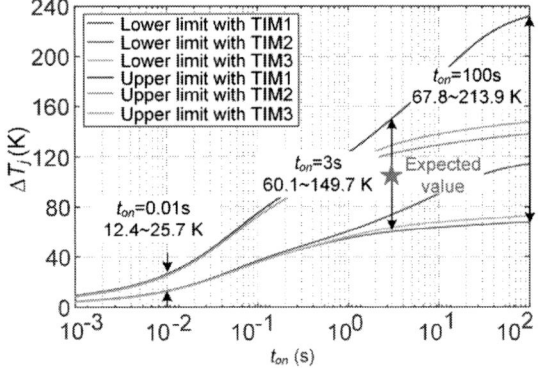

Fig. 7: Thermal stress range under a different t_{on}.

5 Conclusion

This paper proposes a guideline to effectively illustrate the parameter tuning process before the DC power cycling test is run formally. Firstly, based on the established thermal estimation model, it is determined that I_{heat}, V_{gson}, t_{on}, Z_{th} and $T_{coolant}$ are the adjustable parameters. Then a systematic

and specific instruction for setting test parameters, along with a corresponding flowchart is presented. In addition, the thermal stress range can be obtained, which not only ensures the parameter selection for the current test condition but also serves as a valuable reference for others. All dependent information is only the output characteristics of the tested module provided in the data sheet and the thermal impedance of the cooling system. The consistency of theoretical expectations (ΔT_j =100 K, T_{jmax} =150 °C) and actual values (ΔT_j =103.3 K, T_{jmax} =150.4 °C) in an experimental case validates the effectiveness of the proposed method.

References

[1] C. Durand, M. Klingler, D. Coutellier, and H. Naceu, "Power cycling reliability of power module: A survey," *IEEE Trans. Device Mater. Rel.*, vol. 16, no. 1, pp. 80–97, Mar. 2016. DOI: 10.1109/PEDS.1997.618742.

[2] Y. Zhang, R. Wu, F. Iannuzzo, and H. Wang, "Aging investigation of the latest standard dual power modules using improved interconnect technologies by power cycling test," *Microelectronics Reli.*, vol. 138, pp. 114740, 2022. DOI: 10.1016/j.microrel.2022.114740.

[3] M. Held, P. Jacob, G. Nicoletti, P. Scacco, and M. .-. Poech, "Fast power cycling test of IGBT modules in traction application," in *Proceedings of Second International Conference on Power Electronics and Drive Systems*, vol. 1, pp. 425–430, May. 1997. DOI: 10.1109/PEDS.1997.618742.

[4] R. Bayerer, T. Herrmann, T. Licht, J. Lutz, and M. Feller, "Model for power cycling lifetime of IGBT modules-various factors influencing lifetime," in *Proc. 5th Int. Conf. Integr. Power Syst.*, Nuremberg, Germany, 2008, pp. 1–6.

[5] H. Oh, B. Han, P. McCluskey, C. Han and B. D. Youn, "Physics-of-failure, condition monitoring, and prognostics of insulated gate bipolar transistor modules: A review," *IEEE Trans. Power Electron.*, vol. 30, no. 5, pp. 2413-2426, May 2015. DOI: 10.1109/TPEL.2014.2346485.

[6] S. Zhao, S. Chen, F. Yang, E. Ugur, B. Akin and H. Wang, "A composite failure Precursor for condition monitoring and remaining useful life prediction of discrete power

devices," *IEEE Trans. Ind. Informatics*, vol. 17, no. 1, pp. 688-698, Jan. 2021. DOI: 10.1109/TII.2020.2991454.

[7] ECPE, "AQG 324 — qualification of power modules for use in power electronics converter units in motor vehicles," ECPE, 2019.

[8] M. Junghaenel, R. Schmidt, J. Strobel, and U. Scheuermann, "Investigation on isolated failure mechanisms in active power cycle testing," in *Proc. PCIM Eur.*, Nuremberg, Germany, 2015, pp. 1–8.

[9] U. Scheuermann, R. Schmidt, "Investigations on the VCE(T) method to determine the junction temperature by using the chip itself as sensor," in *Proc. PCIM Eur.*, Nuremberg, Germany, 2009, pp. 802–807.

[10] Y. Zhang, X. Ge, Y. Zhang, D. Xie, B. Yao, and H. Wang, "A novel three pulse equivalent power loss profile for simplified thermal estimation," *IEEE J. Emerg. Sel. Topics Power Electron.*, vol. 9, no. 6, pp. 6875–6885, Dec. 2021. DOI: 10.1109/JESTPE.2021.3070994.

[11] Z. Ying, Y. Zhao, E. Deng, M. Pan, L. Xie, and Y. Huang, "Power cycling method for power modules with low thermal resistance," *IEEE J. of Emerg. and Sel. Topics in Power Electron.*, vol. 11, no. 1, pp. 1121-1131, Feb. 2023. DOI: 10.1109/JESTPE.2022.3212107.

PCIM Europe 2024, 11– 13 June 2024, Nuremberg   DOI: 10.30420/566262272

Influence of Transfer Molding on the Reliability of DCM SiC Power Modules

Jacek Rudzki[1], Henning Ströbel-Maier[1], Martin Becker[1], Patrick Heimler[2], Dong Xie[2], Mohamed Ala-luss[2], Thomas Basler[2], Anu Mathew[2][3], Sven Rzepka[3]

[1] Semikron Danfoss, Germany

[2] TU Chemnitz, Germany

[3] Fraunhofer ENAS, Micro Materials Center, Germany

Corresponding author: Jacek Rudzki, jacek.rudzki@semikron-danfoss.com

Abstract

Nowadays, SiC components are increasingly used in the design of power modules. The packaging of such semiconductors is challenging due to the higher Young's modulus of SiC devices. The resulting forces are affecting the reliability of such modules which is usually lower than modules with Si technology. In the past, the introduction of Danfoss BondBuffer® technology (DBB) and sinter layers have significantly increased the reliability of Si-based modules. The combination of a SiC device with DBB within molding compound in a DCM module has shown a full potential in its lifetime increment. For this work, active power cycle test on unmolded and molded SiC power modules have been performed and analyzed via simulation. The results from both experiment and simulation showed that the lifetime of molded SiC module is significantly increased compared to reference unmolded test samples.

1   Introduction

Packaging technologies for power modules are a field of constant development. Due to high thermo-mechanical stress on packaging and interconnection technologies, lifetime of power modules is still a concern. State of the art power modules with a frame-based package, soldered semiconductors and aluminum wire bonds are still commonly used even though the lifetime demands of the interconnections are increasing due to new semiconductor materials and higher demands in automotive applications [1,2]. Although innovations like silver sintering and Danfoss BondBuffer® combined with copper wire bonding have significantly improved the robustness, silicon carbide semiconductors are still challenging. The higher Young's modulus of the semiconductor material induces higher mechanical stress on the semiconductor and its interfaces during temperature cycles. Therefore, the reliability of SiC modules is usually lower compared to modules with Si technology, which in higher lifetime demanding applications is a severe issue. A possible and best solution would be to de-crease the thermo-mechanical stress levels on interconnections and semiconductors which then would lead to increased lifetime. Such a support structure can be realized in the shape of a mold casing.

This article describes in detail the physical influence of a solid mold housing on the stress distribution at the different interconnection layers and thus the impact on the lifetime. It is widely known that the molding compound increases the reliability of Si modules by a factor of up to 4 for the standard Si base package [3]. However, it was a question of how much the molding compound increases the lifetime of SiC modules. In this article, the question is answered based on the existing power cycling data and thermo-mechanical simulations.

2   Test structure

For this investigation, the DCM platform was used to perform power cycling test (PC_{sec}) and conduct the respective thermo-mechanical FEM analysis. The selected semiconductor was a 750V SiC MOSFET which was utilized in a single die setup

1942

and investigated in PCT (power cycling test). The structure used is depicted in Fig. 1.

The top side contacts are consisting of Danfoss BondBuffer® [4,5], which is sintered on the top-side Al/Ag metallization layer of the SiC MOSFET, and Cu wire bonds (400um diameter). The semiconductor is sintered onto the Si3N4 AMB substrate with an Ag sinter layer. The Si3N4 substrate is soldered to the Cu-ShowerPower® cooler.

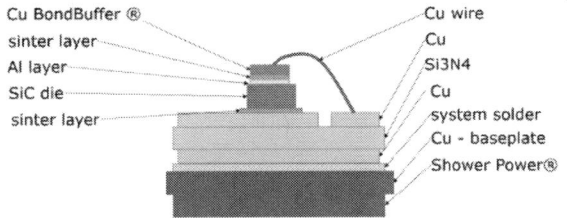

Fig. 1 Structure for power cycling test

Two different variants of test modules were assembled. One variant was encapsulated in mold resin while the other one was not covered. The unmolded variant was not additionally protected with a silicone encapsulation. In the previous work it was shown that the influence of the silicone gel on the power cycling results for Si modules with BondBuffer® is neglectable. The power cycling results are around 10% lower for silicone encapsulated modules [6]. The molded DCM module used for power cycling is shown in Fig. 2

Fig. 2 Test sample based on DCM platform for power cycling test.

3 Power Cycling Test

3.1 Test conditions

PCTs have been performed on unmolded and molded power modules to estimate their respective lifetimes. These two technologies were tested with approximately the same test conditions, shown in Table 1 and a total of six test specimens

used for each test series. For this purpose, a turn-on time of 1 s and a turn-off time of 2 s were selected. The target was a maximum junction temperature T_{jmax} of 175°C with a temperature swing ΔT_j of 110 K and a load current I_L of about 115 A. T_{jmax} was measured after a delay time of 100 µs (PCT 1) or 200 µs (PCT 2) after the load current was switched off. A sufficiently negative gate voltage of - 10 V was selected to completely close the n-channel to avoid temperature measurement errors [7,8,9]. The gate voltage during the turning on period should be carefully concerned when discussing the power cycling capability [21].

Table 1 The PCT conditions for unmolded and molded power modules.

Parameters	PCT 1 (unmolded)	PCT 2 (molded)
I_L [A]	115.4	115.6
ΔT_j [K]	110	
$P_{v,hot}$ [W]	220	240
$V_{GS,on}$ [V]	12.5-20V	16-20V
$T_{vj,min}$ [°C]	65	
$T_{vj,max}$ [°C]	175	
t_{on} [s]	1	
t_{off} [s]	2	
$V_{GS,off}$ [V]	-10	

3.2 PCT results

Fig. 3(a) illustrates exemplarily the progression of R_{thjhs} of one device of each PCT group during the power cycling test. The failure criterion of an increase of 20 % is not reached. However, a slight R_{thjhs}-increase is recognizable at the end of the test which is related to cooler contamination.

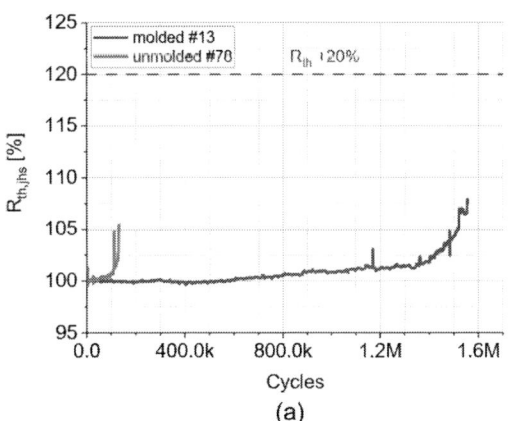

(a)

PCIM Europe 2024, 11– 13 June 2024, Nuremberg   DOI: 10.30420/566262272

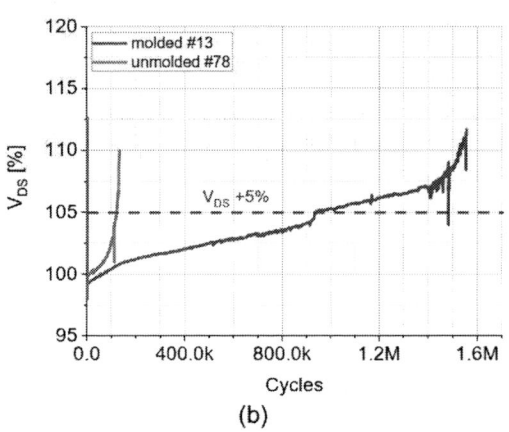

(b)

Fig. 3 PCT results (Raw Data) until the end of lifetime (EOL) for both groups: (a) R_{thjhs} in % and (b) V_{DS} in % of unmolded DUT #78 and molded DUT #13

The development of forward voltage V_{DS} is also shown for one device of each PCT group in Fig. 3(b). A V_{DS} increase of 5 % is used as a failure criterion for all modules and all devices failed with this failure criterion according to the AQG324 standard [22]. In the future, this end of lifetime (EOL) limitation must be re-discussed for SiC chips in more advanced modules.

The corresponding lifetimes with and without of fset correction are shown in the diagram in Fig. 4. The results show that the lifetime of molded devices is significantly higher (up to ten times) compared to unmolded test samples. The offse t correction was done with the square-root-t m ethod according to the following formula and th e material parameters are used at Tjmax [10, 1 1]:

$$\Delta T(t) = \frac{P}{A} \cdot \frac{2}{\sqrt{\pi c \rho \lambda}} \cdot \sqrt{t}$$

where P is power loss in W, A is the chip active area in mm2, c is the specific heat in J/(kg*K), ρ is the density in kg/m3, λ is the thermal conductivity in W/(m*K) and t is the time in s. The reason for the difference in the offset is that the delay time and the power loss were slightly different. For PCT 1 the offset is about 6 K and for PCT2 approx. 9 K.

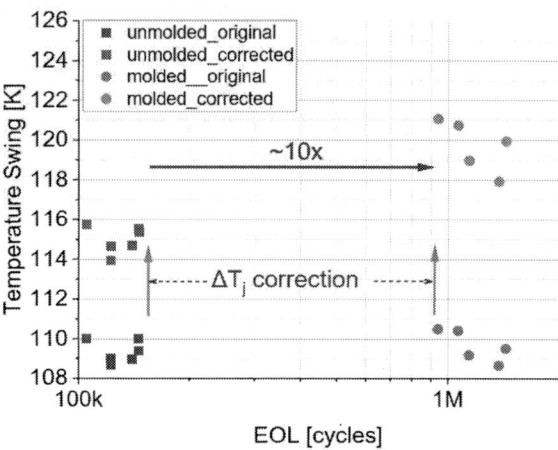

Fig. 4 EOL comparison with and without offset correction for unmolded and molded devices under a temperature swing of about ΔT=110K. PCT-EOL: PCT 1: 100k-150k and PCT 2: 940k-1.45M. Note: 1 molded DUT didn't reach a hard EOL and the test was stopped after 2M cycles.

3.3 Statical parameters comparison

The JEDEC standard JEP 184 [12, 13, 14] is adopted to measure the threshold voltages ($V_{GS,th}$) for unmolded and molded devices, in which the measurement sequence is described in Fig. 5. The preconditioning time is used for minimizing recoverable fast charge-trapping effects, and 100ms is recommended for SiC to generate the proper gate stress. $V_{GS,th}^{up}$ is extracted under $V_{GS}=V_{DS}$ and a drain current in mA range after a negative gate stress V_{GS}=-10V, while $V_{GS,th}^{down}$ is extracted after a positive gate stress V_{GS}=10 V. Between the measurements, the delay time with V_{GS}=0 V, i.e., $t_{float,up}$ and t_{short}, is required. The measurement current (I_D=10 mA) flows via the sense terminal to obtain measured values of $V_{GS,th}^{up}$ and $V_{GS,th}^{down}$ after an acquisition delay of 1 µs. A Keysight SMU B2901A was used as a measuring instrument for this investigation.

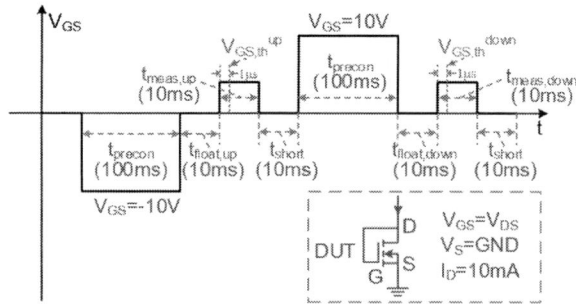

Fig. 5 $V_{GS,th}$ measurement sequence with the preconditioning time.

1944

Based on such $V_{GS,th}$ measurement sequence, $V_{GS,th}{}^{up}$ and $V_{GS,th}{}^{down}$ are obtained for two types of devices under pre- and post-PCT conditions. The results in Fig. 6 have illustrated that the threshold voltage has a positive bias temperature instability (PBTI), causing a slight increase in the range of 0.1 V for $V_{GS,th}{}^{up}$ and $V_{GS,th}{}^{down}$. Compared to $V_{GS,th}$ shifts in unmolded DUTs, the larger $V_{GS,th}$ drift is observed in molded DUTs due to much longer cycles/run-time for PCT2. $V_{GS,off}$ of -10 V can make the channel closed completely to result in the negligible effect of $V_{GS,th}$ drift on the temperature determination during the PCT [14, 15]. Due to a damaged gate after PCT, molded DUT #14 and #15 cannot be effectively measured.

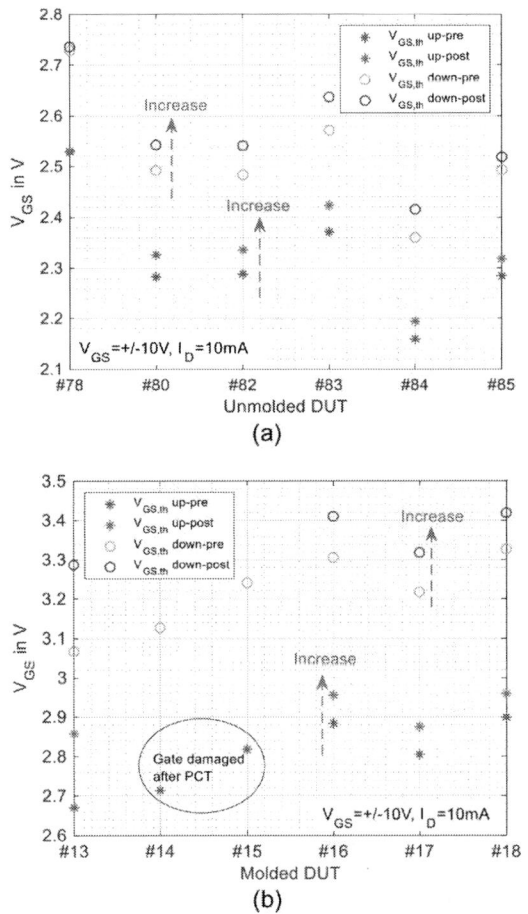

Fig. 6 Measured values of $V_{GS,th}{}^{up}$ and $V_{GS,th}{}^{down}$ under pre- and post-PCT conditions: (a) unmolded DUT and (b) molded DUT.

Furthermore, the $R_{DS,on}$ is measured by using the IWATSU CS-5300 under V_{GS}=18 V and I_D=80 A for the two types of housing under pre- and post-PCT

conditions, shown in Fig .7. Using the pre-PCT results as a reference, $R_{DS,on}$ values for unmolded and molded DUTs have increased positively, which is directly related to the crack through the aluminum layer (see chapter "Failure Analysis") and the PBTI from the threshold voltage [8, 14].

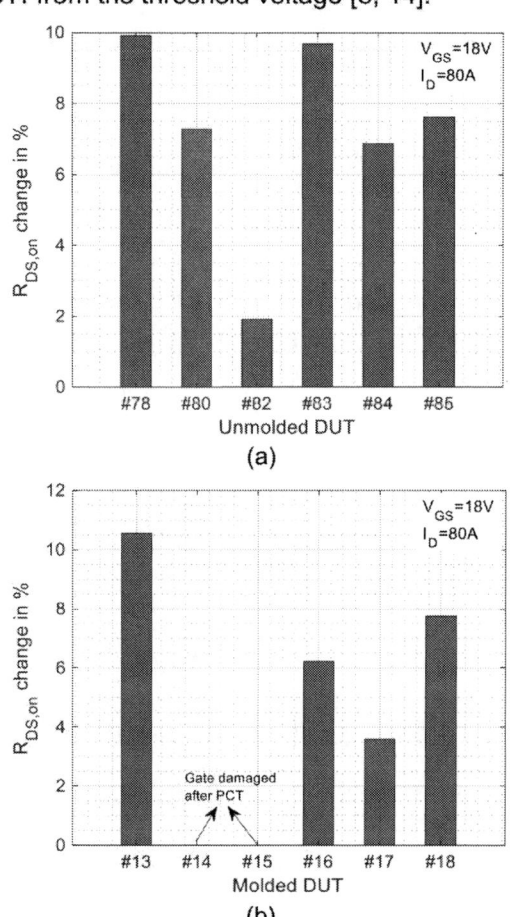

Fig. 7 Change ratio of $R_{DS,on}$ under the post-PCT condition: (a) unmolded DUT and (b) molded DUT.

4 Failure Analysis

4.1 Analysis of unmolded devices

After the power cycling test, the tested modules were thoroughly analysed. Previous work has shown that the absence of e.g. silicone encapsulation has no significant influence on long-term reliability. Microsections were made to analyse the degradation of the unmolded modules which failed after approx. 100k cycles due to 5% V_{DS} increase. It can be concluded that the cause of failure is related to top-side contacting. The metallographic section reveals that the weak point is the Al metallisation of the SiC chip shown in Fig. 8 as already stated in [5]. Other areas, such as the Ag sinter layer below the chip and the system soldering were not affected, see Fig. 9 and Fig. 10.

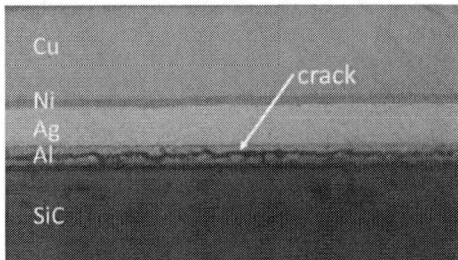

Fig. 8 Crack propagation in the aluminum layer on the upper side of the SiC device for unmolded module

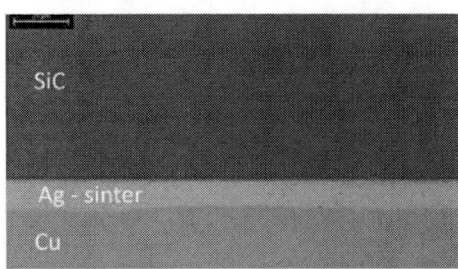

Fig. 9 Sinter layer bottom side of the SiC device for unmolded module.

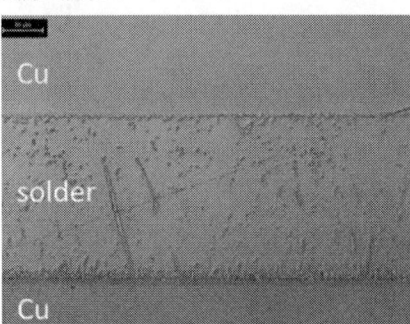

Fig. 10 Solder layer between AMB and cooler for unmolded module.

4.2 Analysis of molded devices

Molded modules were analysed after power cycling test the same way. It could be seen that the cause of V_{DS} increase was the degradation of aluminium layer in the upper side of the SiC components, as was already the case with unmolded modules in Fig. 11. The crack spreads from the edge towards the centre of the bond buffer. The sintering layer after power cycling test was not affected, Fig.12.

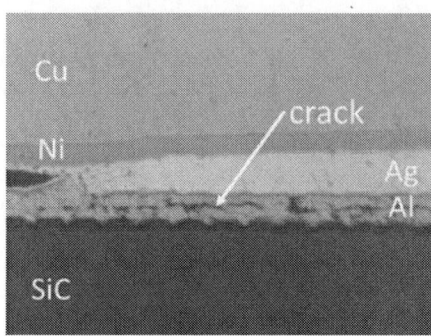

Fig. 11 Crack propagation in the aluminum layer after power cycling test for molded modules.

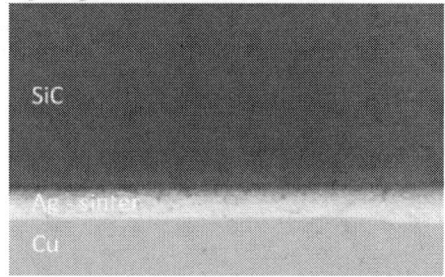

Fig. 12 Bottom side sinter layer of the SiC device for molded module after power cycling test.

5 Simulation Results

5.1 Finite element analysis

Finite element analysis (FEA) has been performed to investigate the thermo-mechanical behavior of the power module based on power cycle test. Numerical simulations are performed by using finite element (FE) software ABAQUS release R2023. Coupled electro-thermo-mechanical simulation is conducted on power modules with and without mold compound (MC). The FE models are constructed for both power modules with and without mold compound, which is shown in Fig. 13 respectively. Quarter symmetry of the power module is considered for better computational efficiency. The material properties are assigned to each layer of power module according to its material stack.

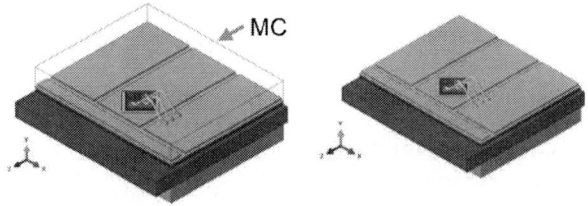

(a) Module with mold compound (MC)

(b) Module without mold compound

Fig. 13 FE model for numerical simulation.

5.2 Electro-thermal simulation

Initially, electro-thermal simulation has been carried out to obtain the temperature distribution of power module during the PCT. The simulation is performed according to the test condition which is used for power cycle test by experiment. The test conditions are mentioned in Table 1. During the electro-thermal simulation, the materials in the FE model is assigned with electrical and thermal material properties. The electrical properties of the SiC dies are estimated from the datasheet and are given as the material parameters for the numerical simulation. Load current is given as the input parameter and corresponding temperature distributions in the power module are observed from the simulation shown in Fig. 14. It has been noticed that the higher temperature is at the center of the chip with a delta temperature of 110 K represented in Fig. 14b. The delta temperature is calculated as the difference at end of t_{on} and t_{off}. The voltage drop of the power module at the end of t_{on} time is observed to be ~1.75 V from the simulation. The temperature obtained from the electro-thermal simulation (Fig. 14a) is applied as input temperature profile for thermo-mechanical analysis.

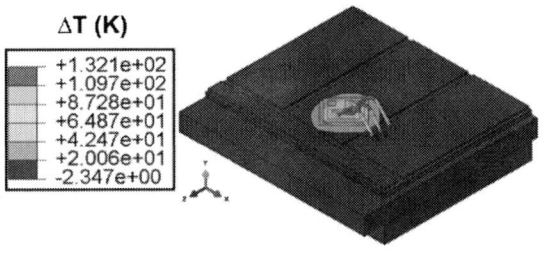

(a) Delta temperature distribution on power module

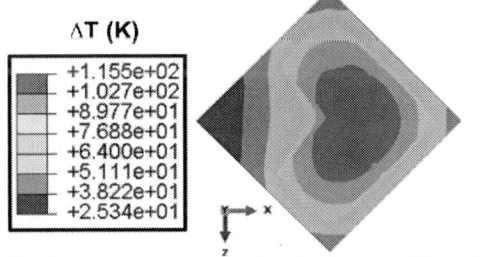

(b) Delta temperature distribution on SiC chip

Fig. 14 a) Delta temperature distribution on whole power module and b) Delta temperature distribution on chip.

It has been noticed that the power module without mold compound has higher temperatures (i.e, ~ 10 K) than the power module with mold compound which is represented in Fig. 15. Therefore, the power module without mold compound could experience higher stress distribution and strain accumulation at the active region of the power module. The mold compound helps to protect the active region from environmental contamination, moisture, vibrations etc. It also helps to transfer heat from active region [16]. Therefore, the power module without mold is expected to have more reliability issues than the module with mold compound. A detailed thermo-mechanical analysis is also performed to investigate the mechanical behavior of the power modules with respect to the temperature obtained from electro-thermal simulation.

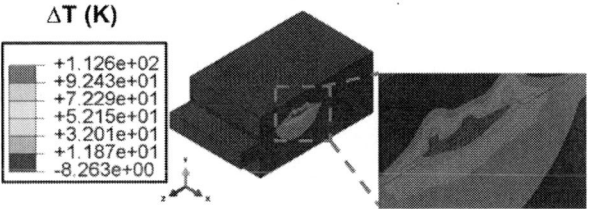

(a) Power module with mold compound

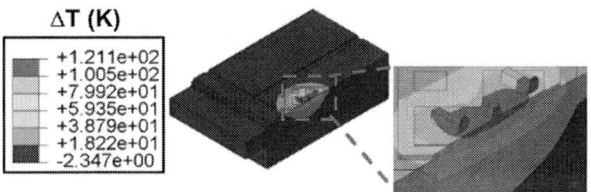

(b) Power module without mold compound

Fig. 15 Delta temperature distribution in power modules. Higher temperatures are observed for the module without mold compound at the active region of the die (zoomed image).

5.3 Thermo-mechanical analysis on power module

Thermo-mechanical analysis has been performed on power modules with and without mold compound. The temperature obtained from the electro-thermal simulation is applied to the power modules for the mechanical analysis. During the mechanical analysis, stress free temperature of 210°C is considered and it is cooled down to room temperature in 1 hour before the PCT temperature profile from the electro-thermal simulation is applied, which is graphically represented in Fig. 16. Thermo-mechanical material properties are assigned to the material stack. Some of the material data used for the numerical simulation are from different literatures [17], [18].

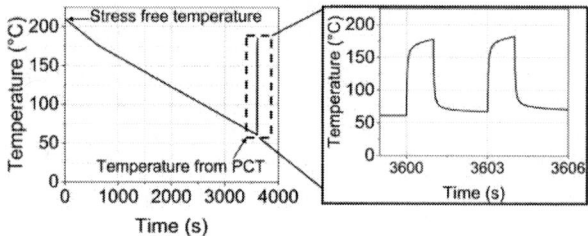

Fig. 16 Temperature profile for mechanical analysis

In the experimental PCT, failures of the power modules are observed at the top metallization layer (Aluminum) of the die. Therefore, from the numerical simulation, the top side metallization layer of the chip is focused for further evaluation. Out of plane stresses are observed at the metallization layer exhibited in
Fig. **17**. A peel tensile stress distribution at the edges of the metallization layer is observed in both modules.

Out of plane stress (MPa)

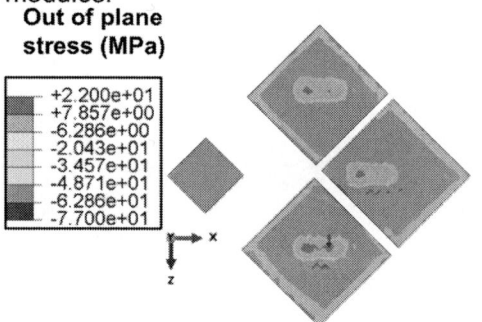

(a) Out of plane stress in power module with mold compound

Out of plane stress (MPa)

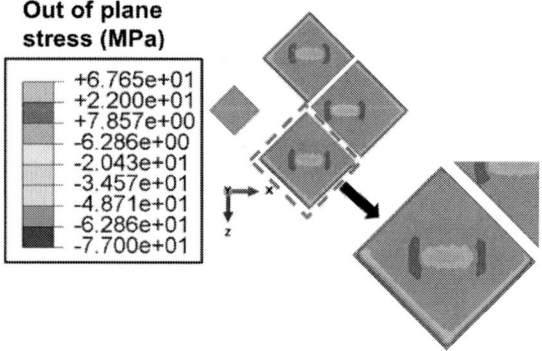

(b) Out of plane stress in power module without mold compound

Fig. 17 Comparison of out of plane stresses of top side metallization layer in power modules.

It has been noticed that modules without mold compound have a higher peel stress at the edges of the metallization layer than with mold compound. The scale of the color bar in
Fig. **17** is scaled to 22 MPa and it gives an inference that the peel stresses in unmolded modules is three times larger than in molded modules.

In-plane tensile stress in the metallization layer leads to the development of a peel stress component and it exhibits singularity issues towards the edges of the layer. In-plane tensile stress acts as an opening for the crack (Mode I – tensile stress normal to the plane of crack) which is critical for the failures. In molded power modules, mold compound compresses the peel tensile stress which is occurring in the metallization layer and leads to stress reduction in the module which enhances the lifetime. So, modules without mold compound tend to earlier failure of the system as the metallization on the module has higher peel stress that leads to the opening for the crack.
Moreover, plastic deformation on the metallization is also examined during numerical analysis. Cyclic equivalent plastic strain (PEEQ) is investigated, which will be used for estimating the lifetime of the power modules based on Coffin-Manson law, where N_f is the number of mean cycles to failure, $\Delta\varepsilon_{pl}$ is the cyclic equivalent plastic strain, C_1 and C_2 are the Coffin-Manson coefficients [19], [20].

$$N_f = C_1 \cdot (\Delta\varepsilon_{pl})^{-C_2} \qquad (1)$$

Higher cyclic equivalent plastic strain accumulation is observed on modules without mold compound at the metallization layer which is depicted in
Fig. **18**. The color bar is scaled to 0.3% of cyclic equivalent plastic strain. If the maximum cyclic equivalent plastic strain is considered, then the strain accumulation in the power module without mold compound is four times greater than module with mold compound.

PEEQ

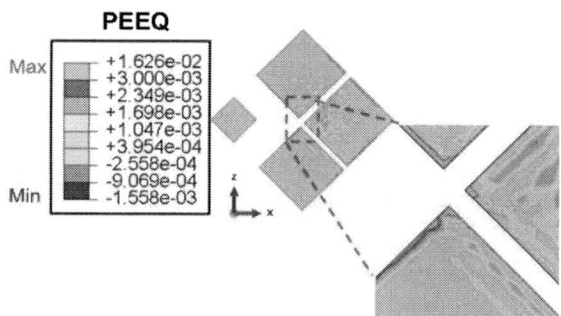

(a) PEEQ in power module with mold compound

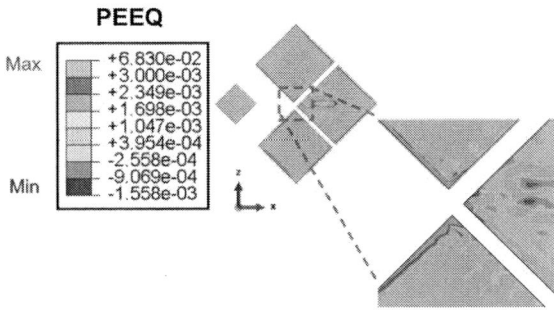

(b) PEEQ in power module without mold compound

Fig. 18 Cyclic equivalent plastic strain in metallization layer of power modules. Grey colour at the corners of the metallization layer denotes cyclic equivalent plastic strain greater than 0.3%.

The fracture type is in the transition region between brittle or ductile fracture. Therefore, local area for cyclic equivalent plastic strain averaging is considered for the cyclic equivalent plastic strain evaluation. The average cyclic equivalent plastic strain in the metallization layer of the power module without mold compound is 2.3 times larger than for the molded version, which is shown in Fig. 19.

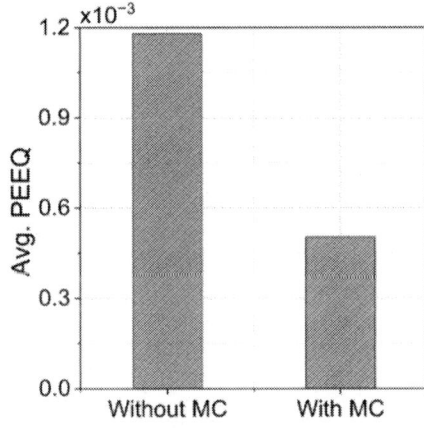

Fig. 19 Average cyclic equivalent plastic strain in metallization layer of power modules.

6 Summary

This article reveals a strong impact of the mold compound on the lifetime of power modules with isolating substrate and BondBuffer®. The power cycling robustness can be improved by a factor of approx. 10. The aluminum layer of SiC semiconductors was observed to be a vulnerable point.

Based on FEM simulations, it could be observed that the mold compound relieves the aluminum thermo-mechanically. Based on this data, it can be concluded that the use of the mold compound enables the use of aluminum based SiC components especially for applications with high reliability requirements.

7 Acknowledgement

This work is supported by the H2020 - KDT JU program of the European Union under the grant of the TRANSFORM project 'Trusted European SiC Value Chain for a greener Economy' (KDT Grant No. 101007237).

8 Reference

[1] Peter Beckedahl, Jacek Rudzki, Matthias Spang; Trends in Power Module Packaging and Impact of Wide Bandgap Semiconductors; Bauelemente der Leistungselektronik und ihre Anwendung; Bad Nauheim 2023

[2] A. Wintrich, U. Nicolai, W. Tursky, T. Reimann; Application Manual Power Semiconductors; Semikron International GmbH *2nd* release 2015, ISBN 978-3-938843-83-3

[3] Guang Zeng; Some aspects in lifetime prediction of power semiconductor devices; PhD Thesis; Chemnitz 2019

[4] J. Rudzki, F. Osterwald, M. Becker, R. Eisele; Novel Cu-bond contacts on sintered metal buffer for power module with extended capabilities; PCIM 2012

[5] A. Hinrich, et.al.: Failure Mechanisms of Sintered Die Top Systems under Power Cycling Tests; PCIM 2019

[6] A. Schiffmacher; A. Bashiti; D. Strahringer; J. Wilde; C. Kempiak; A. Lindemann; J. Rudzki; H. Stroebel-Maier; New Lifetime Model for Advanced Power Semiconductor Interconnects; ECTC 2022

[7] C. Herold, J. Sun, P. Seidel, L. Tinschert and J. Lutz, "Power cycling methods for SiC MOSFETs," 2017 29th International Symposium on Power Semiconductor Devices and IC's (ISPSD), Sapporo, Japan, 2017, pp. 367-370, doi: 10.23919/ISPSD.2017.7988994.

[8] Hoffmann, Felix & Kaminski, Nando. (2019). Evaluation of the V SD-method for temperature estimation during power cycling of SiC-MOSFETs. IET Power Electronics. 12. 10.1049/iet-pel.2018.6369.

[9] P. Heimler, M. Alaluss, C. Schwabe, X. Liu, J. Lutz and T. Basler, "Online Threshold Voltage

Monitoring at SiC Power Devices during Power Cycling Test and Possible Consequences," 2023 25th European Conference on Power Electronics and Applications (EPE'23 ECCE Europe), Aalborg, Denmark, 2023, pp. 1-10, doi: 10.23919/EPE23EC-CEEurope58414.2023.10264579.

[10] H. P. Felsl, "Silizium- und SiC Leistungsdioden unter besonderer Berücksichtigung von elektrisch-thermischen Kopplungseffekten und nichtlinearer Dynamik," PhD thesis, Faculty ET/IT,, TU Chemnitz, Chemnitz, Germany, November, 2009 . [Online]. Available: https://monarch.qucosa.de/api/qucosa%3A19253/attachment/ATT-0/

[11] D. L. Blackburn and F. F. Oettinger, "Transient thermal response measurements of power transistors," IEEE Trans. Ind. Electron. Control Instrum., vol. IECI-22, no. 2, pp. 134—141, May 1975.

[12] JEDEC, "Guideline for evaluating bias temperature instability of silicon carbide metal-oxide-semiconductor devices for power electronic conversion," in JEDEC JEP 184, 2021.

[13] R. Boldyrjew-Mast, C. Baumler, F. B. Wenisch-Kober, X. Liu, and T. Basler, "Impact of degradation mechanisms in gate stress tests on the hard-switching behavior of 1.2 kV SiC power MOSFETs," in 2022 IEEE 34th International Symposium on Power Semiconductor Devices and ICs (ISPSD), Vancouver, BC, Canada: IEEE, May 2022, pp. 229–232.

[14] C. Schwabe, X. Liu, T. N. Wassermann, P. Salmen, and T. Basler, "SiC MOSFET threshold voltage stability during power cycling testing and the impact on the result interpretation," in 2023 IEEE International Reliability Physics Symposium (IRPS), Monte Mar. 2023.

[15] P. Salmen and P. Friedrichs, "Qualifying a Silicon Carbide Power Module: Reliability Testing Beyond the Standards of Silicon Devices," in International Conference on Integrated Power Electronics Systems, Berlin, 2022

[16] Kokatev, A. N., et al. "Monitoring of properties of epoxy molding compounds used in electronics for protection and hermetic sealing of microcircuits." IOP Conference Series: Materials Science and Engineering. Vol. 665. No. 1. IOP Publishing, 2019.

[17] Dudek, R., et al. "Combined experimental-and FE-studies on sinter-Ag behaviour and effects on IGBT-module reliability." 2014 15th International Conference on Thermal, Mechanical and Mulit-Physics Simulation and Experiments in Microelectronics and Microsystems (EuroSimE). IEEE, 2014.

[18] Sauveplane, Jean-Baptiste, Emmanuel Scheid, and A. Deram. "On the accurate determination of the thermomechanical properties of micro-scale material: Application to AlSi1% chip metallization of a power semiconductor device." *Microelectronics Reliability* 49.5 (2009): 499-505.

[19] Roellig, Mike, et al. "Fatigue analysis of miniaturized lead-free solder contacts based on a novel test concept." Microelectronics Reliability 47.2-3 (2007): 187-195.

[20] A. Mathew et al.; Lifetime modelling of sintered silver interconnected power devices by FEM and experiment; 2021 22nd International Conference on Thermal, Mechanical and Multi-Physics Simulation and Experiments in Microelectronics and Microsystems (EuroSimE); St. Julian, Malta, 2021; pp. 1-9, doi: 10.1109/EuroSimE52062.2021.9410877.

[21] Patrick Heimler, Christian Schwabe, Nick Thönelt, Sören Gesell, Josef Lutz, Thomas Basler, "Influence of the Gate Voltage During On-Time on the Power Cycling Capability of SiC MOSFETs", Proceedings PCIM2024.

[22] "ECPE Guideline AQG 324: Qualification of Power Modules for Use in Power Electronics Converters Units in Motor Vehicles," Tech. Rep., May 2021, pp. 1–71.

PCIM Europe 2024, 11– 13 June 2024, Nuremberg   DOI: 10.30420/566262273

Damp Heat Behavior of High Heat Capacitors for Applications in Electric Vehicles

Adel Bastawros[1], Tetsuya Motohashi[2], Koichi Nakashima[2], Takamune Sugawara[2], and Hisao Katsuta[2]
[1] SABIC, USA
[2] SABIC, Japan

Corresponding author and speaker: Adel Bastawros, Adel.Bastawros@SABIC.com

Abstract

Aging behavior of new generation capacitors intended for use at high temperatures in AC-DC inverter applications have been evaluated under damp heat conditions (85°C and 85% relative humidity (RH)) and applied voltage over 1000 hours of aging.
Two capacitor designs were tested, using 3µm and 5µm ELCRES™ HTV150A high-heat films, respectively. Test voltages were 0, 300 and 500 volts for the capacitors made with 3µm film and 0, 500 and 800 volts for the capacitors made with 5µm film. Capacitance change (ΔC), dissipation losses (tan δ) and insulation resistance (IR) were tracked as indicators of aging performance.

For both designs, ΔC, tan δ and IR showed stable response with little or no change over the 1000-hour test duration. Tan δ and IR were independent of the applied voltage level. The change in capacitance (ΔC) showed a slight increase (gain) of 1~2% with applied voltage.

Meeting industry requirements for damp heat performance on the component level increases the confidence in employing the high heat capacitors in incumbent and new AC-DC inverter designs for electric vehicles.

1 Introduction

Aging performance of power electronics systems, such as AC-DC inverter modules in electric vehicle applications, is key for reliable extended operation of the entire system. Such performance is typically evaluated by means of accelerated reliability testing of the whole system. Although the architecture of the entire system with all its assembled components is what determines the system performance; reliability testing on the individual component level could provide early indications of impact on the whole system behavior when a new component is used in an existing or new module design. Capacitors are an example of passive components in inverter modules that receive considerable reliability testing on the component level.
Adoption of SiC technology in AC-DC inverter modules in electric vehicles has signified the need for capacitors that can operate at higher temperatures reaching 150°C. Operating at higher temperatures beyond the limits of incumbent film materials, such as BOPP (biaxially oriented polypropylene), has been an ongoing industry and academic challenge [1,2,3].

SABIC has been developing new materials for use as ultra-thin film in dielectric applications with operating temperatures upwards of 150°C [4]. A recently introduced film grade, ELCRES™ HTV150A, was engineered to provide a new generation of ultra-thin dielectric films and offer a different solution to meet the need for high temperature capacitor films [5].

The base resin in HTV150A film is an amorphous engineering thermoplastic resin with high glass transition temperature (T_g) of 205°C and relatively stable loss factor (D_f) at temperatures up to 150°C, avoiding the issues seen in some crystalline resins where the loss factor can change significantly at higher temperatures, or at temperatures near the glass transition temperature of the material. The polymer's aromatic carbon/aliphatic carbon ratio and net polarizability are balanced to deliver a higher dielectric constant than conventional BOPP while maintaining adequate self-clearing for many high voltage applications [6].

In the current work, high heat capacitors made with HTV150A film are tested against damp heat aging requirements for the AC-DC inverter modules. Meeting industry requirements on the component level increases the confidence in employing the capacitors in existing AC-DC inverter designs that already meet the requirements. The damp heat performance demonstrated here also supports implementation of the high-heat capacitors in new demanding module designs (e.g., SiC based modules) that would benefit from using high-heat components.

2 Experimental

In the current work, damp heat aging behavior of new generation high-heat capacitors is evaluated. Damp heat testing is performed according to AEC-Q200 (REV D) standard published by JEITA (Japan Electrics and Information Technology Industries Association). In this test, powered capacitors are aged in a humidity-controlled chamber at 85°C temperature and 85% relative humidity for 1000 hours.

2.1 Test capacitors

Metallized HTV150A films of 3μ and 5μm thicknesses were used to build two groups of capacitors for testing. A segmented metallization pattern was applied to rolls of film at Machine Technologies Co., Ltd. Film width was 30mm with resistivity of 20 Ω/sq on an aluminum body and 5 Ω/sq on a zinc heavy edge, Fig.1.

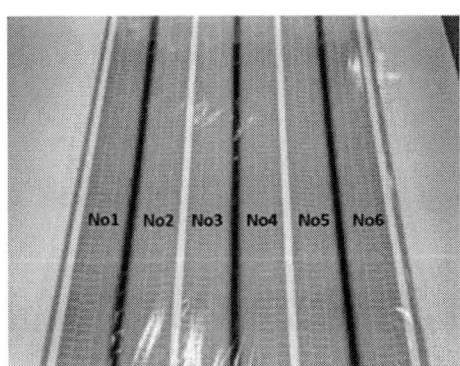

Fig. 1 Segmented metallization on HTV150A film

Pairs of film were wound into round capacitors that were later flattened to an oval geometry for a better shape factor. Zinc thermal spray was applied to both ends of the flattened elements, followed by terminal soldering for electrical connectivity. The elements were placed in a PPS casing and potted with a high temperature thermoset epoxy resin to provide the test capacitors, Fig. 2.

Design capacitance was 5μF for 5μm film capacitors and 10μF for 3μm film capacitors.

2.2 Aging Test

Damp heat exposure was done on powered test capacitors in a controlled environmental chamber (ESPEC, PSL-2J) set to 85°C and 85% relative humidity, Fig. 3. A group of 6 identical capacitors was used for each test condition.

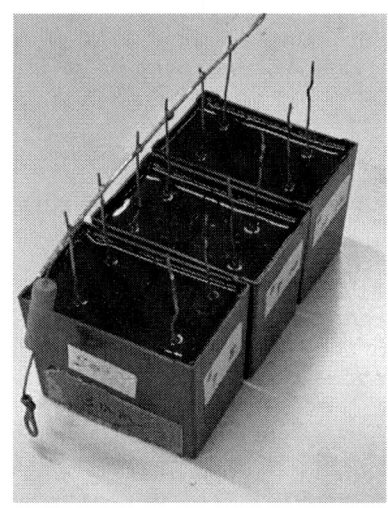

Fig. 2 Test capacitor made with HTV150A film

Fig. 3 Controlled environmental test chamber

Throughout the exposure DC voltage was applied by external power supplies; TEXIO, PA600 and Matsusada HJPM series. Three voltage levels were applied to each group during aging: 0, 300 and 500 volts for capacitors with 3µm film, and 0, 500 and 800 volts for the capacitors with 5µm film. The maximum applied voltage is consistent with target operating voltages for the two dominant segments of AC-DC inverters for EV applications namely the 500 and 800 volt segments.

Capacitance change (ΔC), dissipation losses (tan δ) and insulation resistance (IR) were tracked throughout the test as indicators of aging performance. Capacitance change (ΔC), dissipation losses (tan δ) were measured with an LCR meter (Keysight Technology, E4980A) at 1kHz frequency. IR was measured with an IR meter (HIOKI, SM-8220). For IR measurements the capacitors were briefly removed from the chamber, stabilized and returned to the chamber after measuring.

3 Capacitor Aging Behavior

Aging test results are depicted in Fig. 4 for the 5µm film capacitors and in Fig. 5 for the 3µ film capacitors.
Fig. 4(a) shows stable capacitance over 1000 hours of aging. The graphs are the averages for 6 test capacitors. Change in capacitance ΔC/C (%) shown in Fig. 4(b) increased slightly over time with highest increase of about 2% at 800V. Capacitance gain is typically caused by additional tightening of the winding element when thermal stress is applied. However, it is not clear why higher voltage would induce additional tightening. Nevertheless, the 1~2% of change remains a harmless gain.
Fig. 4(c) shows stable average dielectric losses, tan δ, of only 0.0015. The losses were the same at the 0, 500 and 800V applied voltages.
Similarly, insulation resistance IR remained stable and unaffected by applied voltage.

Capacitance for 3µm film capacitors remained stable over 1000 hours of damp heat aging, Fig. 5(a). A harmless gain of ~2% in capacitance at the highest voltage level (500V) was also present here, Fig. 5(b). Dielectric losses (tan δ) and insulation resistance (IR) remained stable as shown in Figs. 5(c) and 5(d), respectively.

The demonstrated stable capacitor performance for both gauges meets damp heat test requirements per AEC-Q200 standard.

4 Conclusions

High-heat capacitors made with 3µm and 5µm HTV150A film pass 1000-hour aging test at 85°C and 85%RH. Capacitance change, dielectric losses and insulation resistance had little or no change over the duration of the test when 0, 300 and 500 volts were applied to the 3µm film capacitors and when 0, 500 and 800 volts were applied to the 5µm film capacitors.

Meeting industry requirements on the component level increases the confidence in employing the high-heat capacitors in existing AC-DC inverter designs that already meet the requirements. The damp heat performance demonstrated here supports implementation of the high-heat capacitors in new demanding module designs using SiC technology.

5 Acknowledgements

The authors are indebted to SABIC's Global Technology Teams for valuable insights on capacitor testing and film characterization. Special thanks to Machine Technologies for assistance with metallization and building of test capacitors.

6 References

[1] L. Caliari, P. Bettacchi, E. Boni, D. Montanari, A. Gamberini, L. Barbieri, and F. Bergamaschi, "KEMET film capacitors for high temperature, high voltage and high current". Capacitor and Resistor Technology Symposium (CARTS) International Proc. ECA (Electronics Components, Assemblies & Materials Association), pp. 1-15, 2013.

[2] I. W. Clelland and R. A. Price, "Polymer Film Capacitors", APEC 2011, Special Session 1.3.4, pp. 2-4, 2011.

[3] D. Tan, L. Zhang, Q. Chen, et al, "High-Temperature Capacitor Polymer Films". Journal of Electronic Materials, Vol. 43, 4569–4575, 2014.

[4] N. Pfeiffenberger, F. Milandou, and M. Niemeyer, "High Temperature Polyetherimide Film Development," IEEE Trans. Dielectrics and Electrical Insulation Vol. 25, No. 1, pp 120-126, 2018.

[5] A. Bastawros, A. Pingitore, J. Mahood, M. Niemeyer, F. Yu and T. Sugawara, "New Generation Capacitor Films for 150°C High Voltage

AC-DC Inverter Applications," PCIM Europe 2022, pp. 1517-1523, 2022.

[6] A. Bastawros, A. Pingitore, C. Grabowski, P. Flanagan, and M. Buratto, "High Temperature Capacitor Films with Reduced Dissipation Losses for High Voltage AC-DC Inverters," proc. PCIM Europe 2023, pp. 1314-1317, 2023.

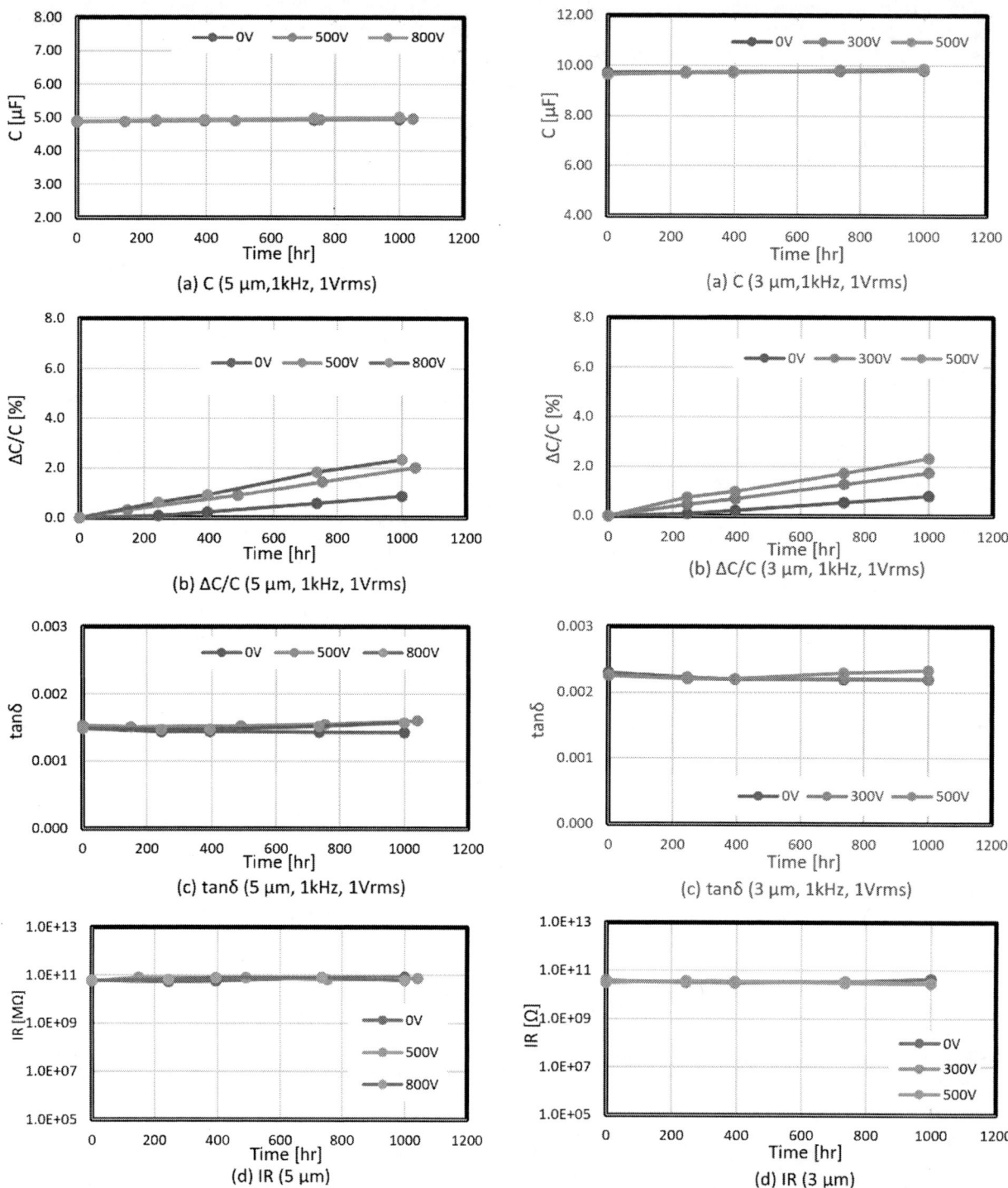

(a) C (5 µm,1kHz, 1Vrms)

(b) ΔC/C (5 µm, 1kHz, 1Vrms)

(c) tanδ (5 µm, 1kHz, 1Vrms)

(d) IR (5 µm)

Fig. 4 Aging Test for 5µm film capacitors

(a) C (3 µm,1kHz, 1Vrms)

(b) ΔC/C (3 µm, 1kHz, 1Vrms)

(c) tanδ (3 µm, 1kHz, 1Vrms)

(d) IR (3 µm)

Fig. 5 Aging Test for 3µm film capacitors

PCIM Europe 2024, 11– 13 June 2024, Nuremberg   DOI: 10.30420/566262274

Influence of the Gate Voltage during On-Time on the Power Cycling Capability of SiC MOSFETs

Patrick Heimler[1], Christian Schwabe[1], Nick Thönelt[1], Sören Gesell[1], Josef Lutz[1], Thomas Basler[1]

[1]Chemnitz University of Technology, Germany

Corresponding author: Patrick Heimler, patrick.heimler@etit.tu-chemnitz.de
Speaker: Patrick Heimler, patrick.heimler@etit.tu-chemnitz.de

Abstract

In this work, the power cycling capability of SiC MOSFETs in the TO-247 package from three different manufacturers (chip structure: planar, single trench and double trench) was investigated at different positive gate voltages during on-time ($V_{GS,on}$ = + 8 V … + 20 V). It was found that the TCP (temperature compensation point) of the $R_{DS(ON)}$ in a gate voltage range of 10 V to14 V is depending on the chip technology. The test results show that the power cycling capability decreases strongly when using gate voltages below the TCP. Therefore, the choice of the gate voltage should be considered for the test conditions. The cause of end-of-life failure is an increase in forward voltage by 5 %, resulting from distinct bond wire degradation.

1   Introduction

Typically, in power cycling tests (PCT), several DUTs (e.g. up to 5 DUTs per branch in 3 parallel branches) are tested simultaneously. The setting of the final test parameters, such as equal maximum junction temperature or temperature swing, is often done by matching the voltage drop in the individual phases and DUTs with the help of a gate voltage variation. This is standard for IGBTs - but what impact have these variations of the positive gate voltage on the cycle stability and number to failure of SiC MOSFETs?

2   Test Specimens

Test specimens from three different manufacturers (A, B and C) with the same blocking voltage of 1.2 kV in TO-247 package (see Figure 1(a)) have been selected for this investigation. The technology of the chips is different. This should clarify the impact of the device structure including the most common concepts (see Figure 1(b to d)) on the power cycling capability using different gate voltages. According to the data sheet, A has an $R_{DS(ON)}$ of 75 mΩ, B of 80 mΩ and C of 60 mΩ, respectively. This results in a continuous current (T_C = 100 °C) of 23 A for A, 22 A for B and 26 A for C. An overview of the continuous current at a temperature of T_C = 100 °C and the associated current density in the bond wire is given in Table 1. For the calculation of the current density, the bond wire configuration can be found in Table 2.

Table 1: Overview of the continuous load current and the resulting current density in the bond wire at a temperature of T_C = 100 °C for the respective manufacturer. Note: *assumed bond wire diameter

Manufacturer	Continuous current [A]	Current density [A/mm²]
A	23	233.4
B	22	155.6*
C	26	117.7

PCIM Europe 2024, 11– 13 June 2024, Nuremberg   DOI: 10.30420/566262274

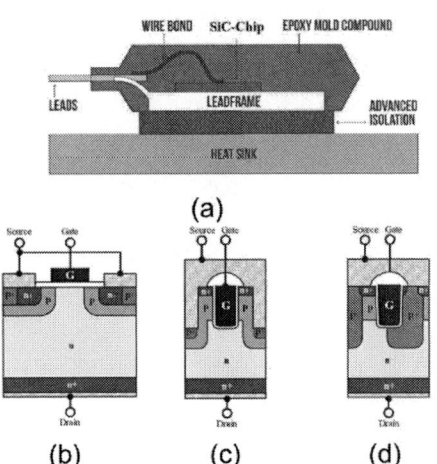

(a)

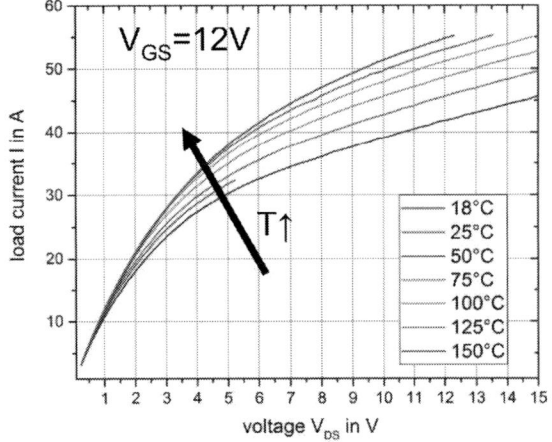

(b)   (c)   (d)

Figure 1: (a) schematic cross section of the TO-247 [1]; (b) planar structure [2]; (c) double trench structure [3]; (d) trench structure with asymmetric channel [4]. Figures are taken from [1, 4]

3 Static characteristic - A question of the temperature coefficient of the $R_{DS(ON)}$

Initially, forward measurements were carried out with the TEKTRONIX 371B from Sony to determine the TCP (temperature compensation point) of the $R_{DS(ON)}$. The temperatures were varied from 18 °C to 150 °C and the gate voltages from 12 V to 18 V. The measurements were carried out in an isolated chamber. The waiting time was taken long enough to reach a thermal equilibrium in the device. Figure 2 shows these forward measurements of one DUT from manufacturer C and it can be seen that a negative temperature coefficient occurs at a gate voltage of 12 V and a positive temperature coefficient at a gate voltage of 18 V.

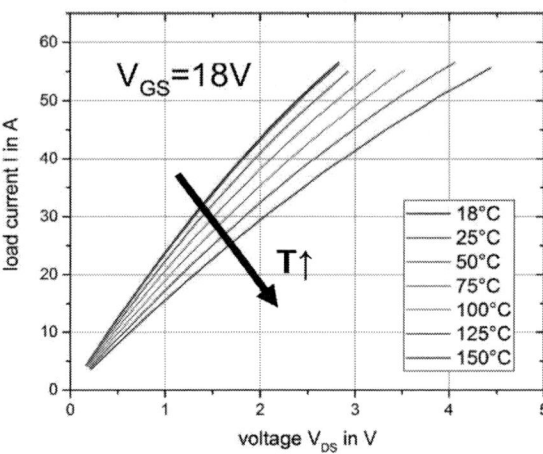

(b) positive temperature coefficent (PTC)

Figure 2: Forward measurements at different gate voltages and temperatures of one test specimen of test series C.

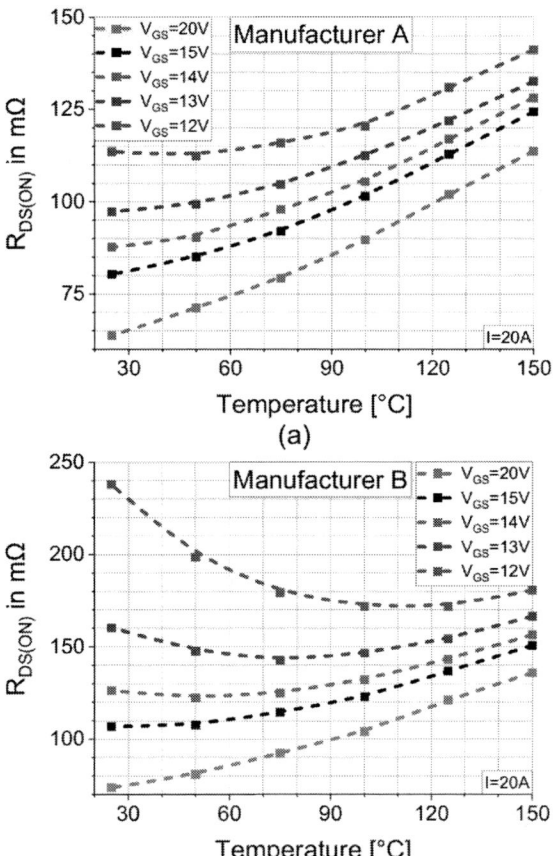

Figure 3: Dependency of $R_{DS(ON)}$ at a current of 20 A on the temperature at different gate voltage for one specimen of each test series A and B.

The dependency of the $R_{DS(ON)}$ on different temperatures at various gate voltages for one device of

(a) negative temperature coefficent (NTC)

1956

each manufacturer A and B is shown in Figure 3. It can be seen that the temperature coefficient of the $R_{DS(ON)}$ is significantly more negative at the same gate voltage for the trench structures B and C (see e.g. $V_{GS} = 12$ V curve). C is not shown in Figure 3, because the device shows a similar behavior like B. The change in resistance as a function of temperature and various gate voltages strongly depends on the gate geometry and channel design (see equation 3).

The following equation describes the composition of the $R_{DS(ON)}$:

$$R_{DS(ON)} = R_{S*} + R_{n+} + R_{CH} + R_a + R_{JFET} + R_{epi} + R_{Sub} \tag{1}$$

with R_{S*}: package, R_{n+}: n-source layer, R_{CH}: n-channel, R_a/R_{JFET}: accu. layer + JFET, R_{epi}: n⁻-layer, R_{Sub}: substrate.

With channel resistance

$$R_{CH} = \frac{1}{\varkappa * \cdot (V_{GS} - V_{th})} \tag{2}$$

and channel conductivity parameter $\varkappa$

$$\varkappa = \frac{W \cdot \mu_n \cdot C_{OX}}{L} \tag{3}$$

including W: total width of the channel, L: channel length, C_{ox}: capacitor through the gate oxide layer and gate electrode, μ_n: mobility.

Figure 4(a) illustrates the $R_{DS(ON)}$ of one DUT of each device concept (A, B and C) at two different temperatures (25 °C and 150 °C) in a gate voltage range from 10 V to 20 V to determine the TCP more precisely. At the intersection of these two curves, the TCP can be found. For higher V_{GS}, the R_{DSON} has a positive temperature coefficient (PTC) and for lower V_{GS} a negative temperature coefficient (NTC) behavior, respectively. Further, it was observed that the TCP also varies due to device scattering. It was found that the TCP of the $R_{DS(ON)}$ is around 10 V to 14 V depending on the chip technology and therefore on the manufacturer. As a comparison to the SiC-MOSFET, Figure 4(b) illustrates the dependency of the forward voltage V_{CE}

on the temperature and gate voltage of a commercially available Si-IGBT with a rated current of 15 A. In contrast to the SiC-MOSFET, the Si-IGBT shows only a PTC behavior in a gate voltage range of 11 V to 20 V.

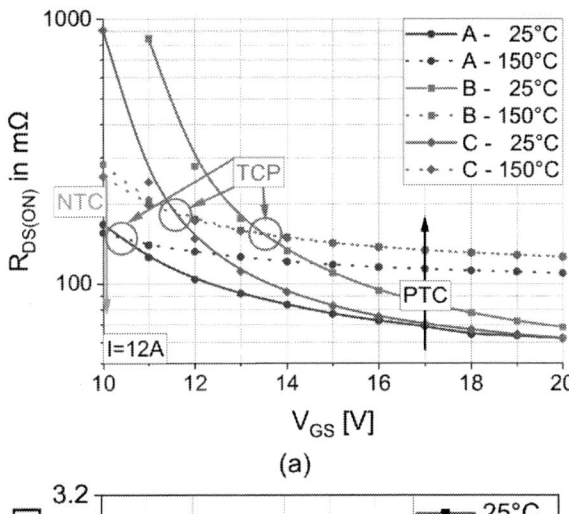

(a)

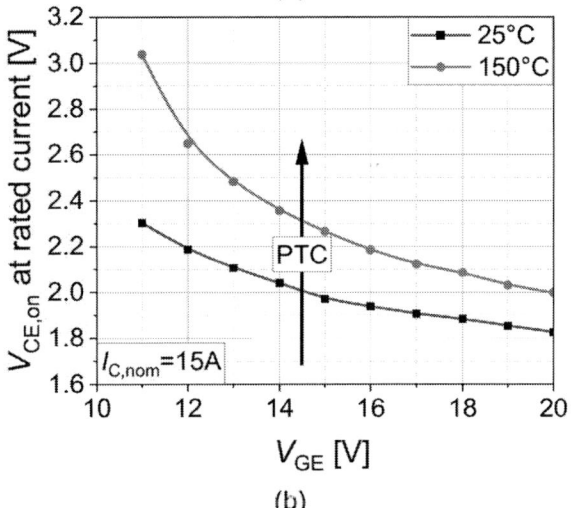

(b)

Figure 4: (a) Determination of the TCP of the $R_{DS(ON)}$ of the three different manufacturers; (b) Forward voltage $V_{CE,on}$ as a dependency of the temperature and gate voltage of a commercially available IGBT with a rated current of 15 A.

4 Power Cycling Capability

The test specimens were mounted on heat sink adapter plates using commercially available clips and the used thermal interface material was a silicone foil (Kerafol 86/60), which is also electrically insulating. The first power cycling test was performed with a t_{on} duration of 2 s and a t_{off} time of 4 s. The maximum junction temperature was around 150 °C and a temperature swing of approx.

90 K was realized for each test series at different positive gate voltages. A sufficiently negative gate voltage to completely close the n-channel of the DUTs of $V_{GS,off}$ = - 8 V for A, $V_{GS,off}$ = - 10 V for B and $V_{GS,off}$ = - 8 V for C has been applied to avoid measurement errors in the temperature determination [5, 6, 7, 8]. Since the positive gate voltage has been changed over a wide range ($V_{GS,on}$ = 8 ... 20 V), the load current also had to be adjusted to achieve the desired test conditions. This results in a load current of 20 A to 22 A for A, 16 A to 23 A for B and 13 A to 23 A for C (see Figure 5(a)) according to the respective test conditions. This means that the current per bond foot as well as the current density in the bond wire is reduced as the gate voltage decreases. An overview of the final test conditions without delay time correction can be found in Table 2.

Table 2: Overview of the test conditions for classic t_{on} – time experiments.

	A	B	C
T_{jmax} [°C]	~150°C		
ΔT_j [K]	~90K		
I_{load} [A]	Variable		
bond config.	3*200µm (double stitch)	2*300µm (assumed)	2*375µm
t_{on} [s]	2		
t_{off} [s]	4		
t_{delay} [µs]	200		80/160
$V_{GS(on)}$ [V]	Variable		
$V_{GS(off)}$ [V]	-8	-10	-8

Figure 5(b) shows the raw data of the power cycling test. Due to this measurement delay, the test data must be corrected in temperature swing and maximum junction temperature. This is done with the square-root-t-method according to the following equation and the parameter properties for the calculation have been applied at T_{jmax} [10, 11]:

$$\Delta T(t)= \frac{P}{A} \cdot \frac{2}{\sqrt{\pi\rho\lambda c}} \cdot \sqrt{t_{delay}}$$

where P is power loss in W, A the chip active area in mm^2, c the specific heat in J/(kg*K), ρ the density in kg/m^3, λ the thermal conductivity in W/(m*K) and t_{delay} the time in s.

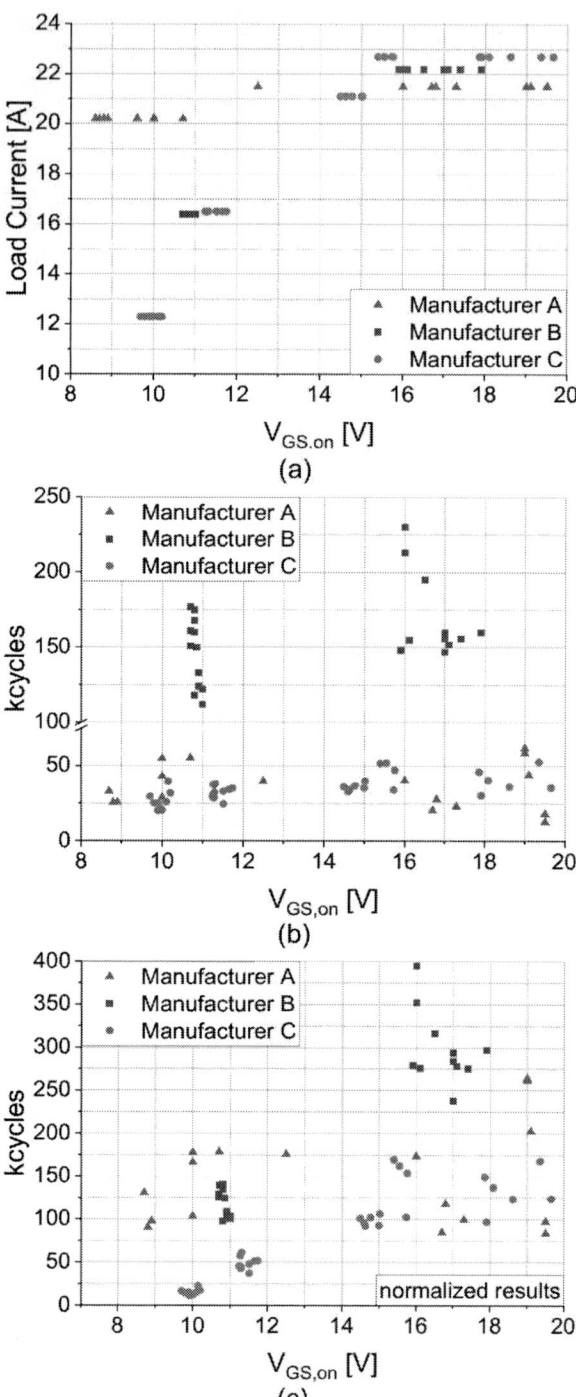

Figure 5: a) Overview of the required load current per gate voltage; (b) Raw data of the power cycling test results (c) Normalized power cycling test results, normalization based on [9] adapted to the respective group (T_{jmean} = 105 °C, ΔT_j = 100 K and I_{load} = 12A ... 17 A).

The test results were normalized to the same temperature swing of 100 K and an average temperature of 105 °C with the model from [9]. The current was normalized to the respective application-near current (with Iso-foil) of 12 A for device A, 17 A for B and 13 A for C. These results are shown in Figure 5(c). The lifetime decreases with lower gate voltages, while the current per bond foot also reduces. Normally, a strong current dependency of the lifetime according to [9] is recognized in TO housing for IGBTs and usually the lifetime should increase with lower current. Therefore, an additional influencing parameter on the lifetime is existing in this case. The hypothesis is that the dominance of the NTC of the R_{DSON} is the cause of the reduction in lifetime. However, to investigate the influence of the gate voltage $V_{GS,on}$ also at smaller t_{on} times, a reduction was done from 2 s to 40 ms. To ensure that the test specimens, which were also mounted on an adapter plate with standard clips, always reach the same T_{jmin} during the short switch-off phase and no thermal runaway occurs, the heat conducting foil was replaced for this test to the improved ICT-BFG20-A type. For this purpose, a maximum temperature of 150 °C with a temperature range of 80 K was targeted. In these tests, the 5 % increase in V_{DS} is also the cause of failure. The same switch-off gate voltages $V_{GS,off}$ were selected for the temperature determination as in the previous tests. An overview of the test conditions can be found in Table 3.

Table 3: Overview of the test conditions for short t_{on} – time.

	B	C
T_{jmax} [°C]	~150°C	
ΔT_j [K]	~80K	
I_{load} [A]	Variable	
t_{on} [ms]	40	
t_{off} [ms]	80	
t_{delay} [µs]	200	
$V_{GS(on)}$ [V]	Variable	
$V_{GS(off)}$ [V]	-10	-8

Figure 6 illustrates the progression of $V_{DS,hot}$ and the associated power loss $P_{V,hot}$ during the power cycling test using both the sense and the load contact for one test specimen of test series B. This is possible because these test specimens are 4-pin devices, which have an additional sense source pin. This sense source pin is directly connected with a small bond to the chip surface. This allows to determine the losses of the chip itself and those including the load bonding wires.

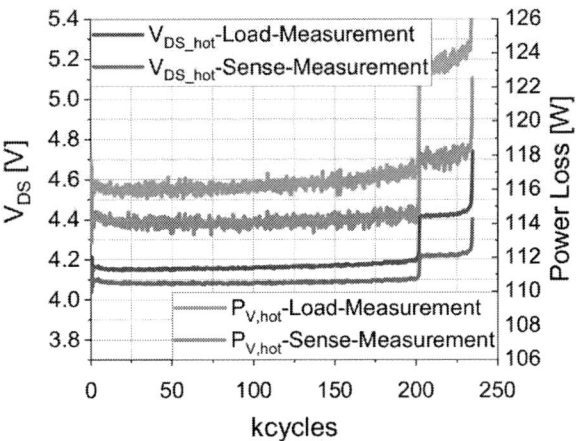

B-#51-$V_{GS,on}$ = 18.3 V – PTC

Figure 6: Progression of the forward voltage VDS, during the power cycling test and the associated power loss $P_{V,hot}$ - measured at the load and sense terminals of one DUT of manufacturer B (#51-$V_{GS,on}$ = 18.3 V).

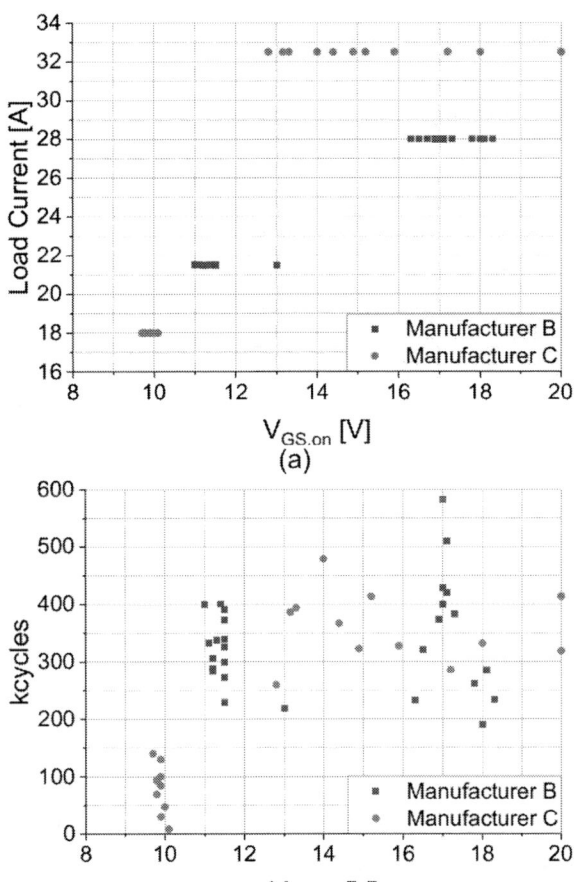

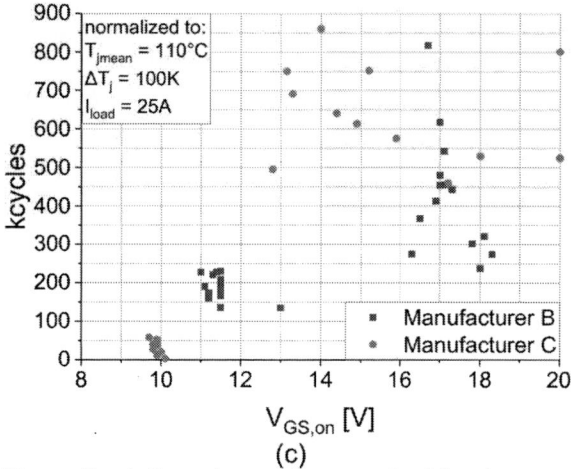

Figure 7: a) Overview of the required load current per gate voltage for the t_{on} = 40 ms test; (b) Raw data of the power cycling test results (c) Power cycling test results for both tests with focus on the gate voltage, cycles are normalized based on [9] (T_{jmean} = 110 °C, ΔT_j = 100 K and I_{load} = 25 A).

This means that approximately 114 W of the total power dissipation of about 116 W is generated in the chip. Therefore, about 2 % of the total power loss is generated by the bond wire.

Figure 7(a) shows the load current required per gate voltage $V_{GS,on}$ to achieve the test conditions. It ranges from 21 A to 28 A for manufacturer B and from 18 A to 33 A for C accordingly. The raw data of the power cycling test is presented in Figure 7(b). As before, a high spread in lifetime for discrete devices visible. Before the normalization is carried out, the temperature offset must be calculated. According to this measurement delay of T_{jmax} (200 μs for both test series B and C) an offset of up to 25 K can occur when using the square-root-t-method in this short t_{on}-case, which should be considered for future power cycling tests to avoid an impermissible temperature range according to data sheet. When calculating the temperature offset using the power loss measured via the sense and load contact, differences in the range of 1 K for the test series B should be considered in an error analysis when determining the temperature. The basis for the normalization is also in this case the lifetime model for discrete devices [9]. The normalization was done to the same temperature swing of 100 K and an average temperature of 110 °C and a load current of 25 A. These results are shown in Figure 7(c). The strong influence of the gate voltage is also recognizable with shorter switch-on times. For a more accurate comparison between the different t_{on}-times, the mean values from the lifetimes in the same gate voltage range, on the one hand from 9...11/12 V (NTC) and on

the other hand from 15...20 V (PTC), were used. These results are shown in Table 4. However, when interpreting the results, it is important to be careful, as the lifetime and R_{DSON} = f(T,V_{GS}) behavior depends strongly on the gate voltage in the test (see Figure 4(a)). An abrupt changeover to the NTC of the $R_{DS(ON)}$ can be detected at this point which strongly influences the power-cycling results. This effect appears much stronger for SiC-MOSFETs than for Si-IGBTs (see Figure 4 (a) for SiC-MOSFETs form different manufacturers and (b) for one Si-IGBT).

Table 4: Mean value of the lifetime at different t_{on} times and gate voltage range for manufacturer B and C. Note: normalized data is used. Note: colormap: B – black and C – red.

t_{on}-time	mean lifetime @9...12 V – B @9...11 V – C	mean lifetime @15...20 V
2 s	121,000	299,000
40 ms	187,000	429,000
2 s	15,500	135,000
40 ms	33,500	607,000

Further, it can be recognized in the millisecond as well as in the second t_{on}-range of the power cycling test that the scattering in lifetime in the NTC-regime is lower than with PTC-mode. This can be explained by the fact that a partial current crowding around the bond wire occurs in the NTC-mode as the resistance is reduced by increasing the temperature. Whereas with the PTC-mode, the resistance increases as the temperature rises and therefore the current flow is not distributed to a specific region like in the NTC-mode [12].

This normalization was carried out according to the following equation with the specific parameters for discrete Si IGBT devices from [9]:

$$N_{f, normalized\,(\Delta T_j, T_{jm}, I_b)} =$$

$$N_{f, measured} \cdot \frac{\Delta T_{j, desired}^{\alpha}}{\Delta T_{j, measured}^{\alpha}}$$

$$\cdot \frac{e^{\frac{E_A}{k \cdot T_{jm, desired}}}}{e^{\frac{E_A}{k \cdot T_{jm, measured}}}} \cdot \frac{I_{b, desired}^{\gamma}}{I_{b, measured}^{\gamma}}$$

whereas E_A is an activation energy, k the Boltzmann constant, the Coffin-Manson exponent α and the exponent γ for the load current.

The reason for using this model is that the failure mode (degradation of the bond wire) is the same and the model was developed with test specimens in TO-247 package.

5 Infrared-Measurements

To investigate the influence of the gate voltage during the t_{on}-phase on the temperature distribution at the metallization/bond foot, measurements were carried out with the infrared camera. Therefore, the gate voltage was varied in a range from 5 V to 20 V and the maximum junction temperature T_{jmax}, which is approximately 88 °C in this case, was measured after a measuring delay of 100 µs after switching off the load current. The load current was between 3 A and 20 A. An overview of the settings can be found in Table 5.

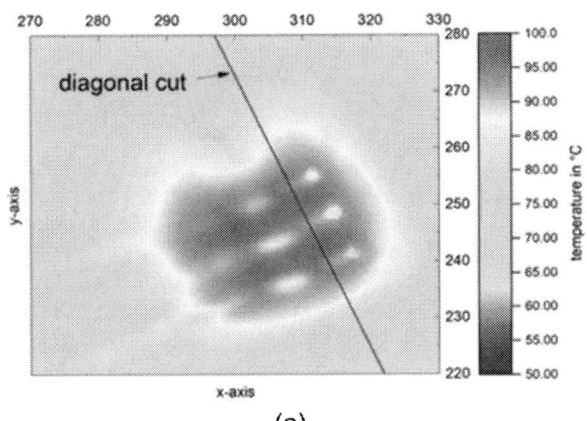

(a)

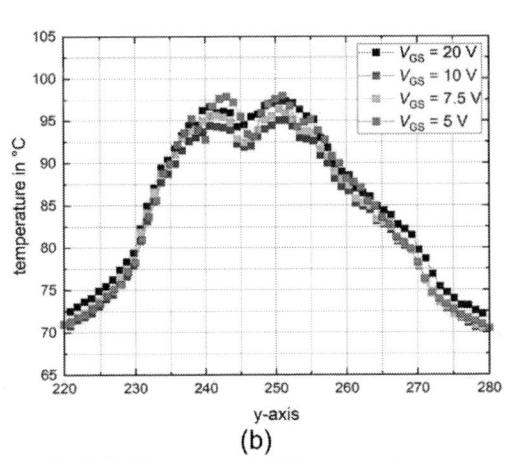

(b)

Figure 8: (a) IR-image (b) Temperature curve at different gate voltages extracted from the diagonal cut in (a) — Note: No lock-in thermography was available.

Figure 8(a) shows the temperature profile with focus on the chip surface of an infrared measurement of a test specimen from test series A. The results of the IR-measurements are illustrated in Figure 8(b) and it can be seen that there is no significant difference in the temperature profile for the diagonal cut - in other words no indication for strong current crowding is found. Unfortunately, just an A-device was available for this investigation, remembering that device A showed the smallest V_{GS} influence and the NTC starts only at lower V_{GS}. In future research, IR-measurements will be performed for different heating scenarios (times and currents) and gate voltages for manufacturer B and C.

Table 5: Overview of the test conditions for the IR-measurements.

V_{GS} [V]	T_{jmax} [°C]	t_{md} [µs]	I_L [A]
20	88.5		20
10	86.9	100	15.7
7.5	87.8		11.2
5	89.3		3

6 Failure Analysis

Before and after the power cycling test, scanning acoustic microscopy (SAM) images (see Figure 9) were carried out to determine the cause of the failure. No solder degradation is visible for all chip technologies. Bond wire degradation is noticeable as growing white spots for test series B and C, but this area was not focusable for A.

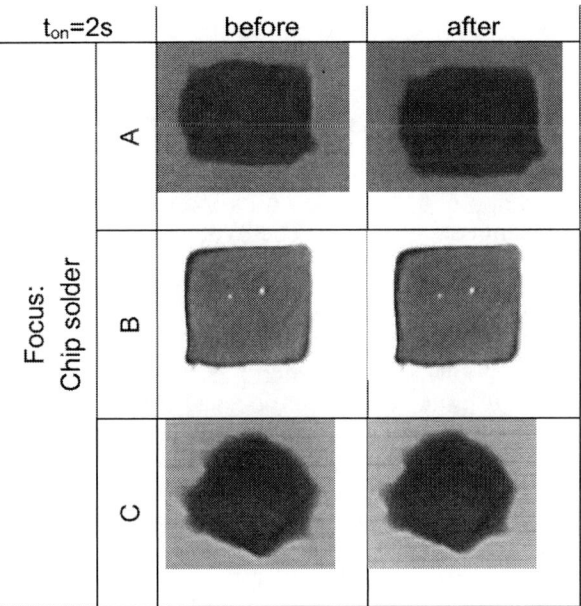

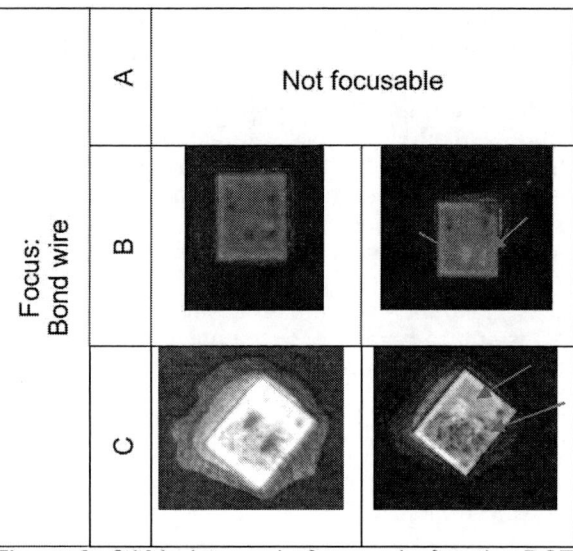

Figure 9: SAM pictures before and after the PCT with a t_{on}-time of 2 s.

7 Conclusion

In this study, the power cycling capability of SiC MOSFETs at different gate voltages during on-time has been investigated. Operation points above and below the TCP of the $R_{DS(ON)}$, which is in the range of a gate voltage of 10 V to 14 V depending on the chip technology, were tested. All DUTs failed with an increase in forward voltage by V_{DS} + 5 %, which correlates to bond wire fatigue which has been confirmed by the failure analysis. The percentage of the total losses caused by the bond wire at short switch-on times was also investigated. This was possible because the test specimens had an additional sense-source pin. The additional power loss amounts to around 2% as an example at the selected test conditions for manufacturer B. As a main outcome, the power cycling capability decreases strongly at low gate voltages. This occurs in the millisecond as well as in the second range of the t_{on}-time and appears much more significant as for IGBTs. The hypothesis is that the pronounced negative temperature coefficient of the $R_{DS(ON)}$ leads to partial current crowding during the heating phase and reduces the lifetime of the bond wire/bond wire interface to power metal. Consequently, the operation of SiC MOSFETs with low gate voltage, especially, if the TCP of $R_{DS(ON)}$ is at high V_{GS} values, should be avoided during power-cycling testing and in the application. Future research will attempt to prove this using lock-in thermography and IR measurements on the other technologies.

8 Reference

[1] Jeffrey Jenkins, in CHARGED ELECTRIC VEHICLES MAGAZINE, „Here's why Tesla transitioned to a semi-custom power module design in Model 3 inverter", URL: https://chargedevs.com/features/heres-why-tesla-transitioned-to-a-semi-custom-power-module-design-in-model-3-inverter/ (Access: 20.07.2023), 3.3.2022

[2] A. Agarwal and S.H. Ryu.: "Status of SiC Power Devices and Manufacturing Issues", Proc. CSMANTECH Conference, pp. 215-218, Vancouver, 2006

[3] T. Nakamura, Y. Nakano, M. Aketa, R. Nakamura, S. Mitani, H. Sakairi and Y. Yokotsuji.: "High Performance Trench SiC Devices with Ultra-low Ron", Proc. IEDM, pp.599-601, Washington, 2011

[4] R. Siemieniec et al., "A SiC Trench MOSFET concept offering improved channel mobility and high reliability," 2017 19th European Conference on Power Electronics and Applications (EPE'17 ECCE Europe), Warsaw, Poland, 2017, pp. P.1-P.13, doi: 10.23919/EPE17ECCEurope.2017.8098928.

[5] R. Schmidt, R. Werner, J. Casady, B. Hull and A. Barkley, "Power Cycle Testing of Sintered SiC-MOSFETs," PCIM Europe 2017; International Exhibition and Conference for Power Electronics, Intelligent Motion, Renewable Energy and Energy Management, Nuremberg, Germany, 2017, pp. 1-8.

[6] C. Herold, J. Sun, P. Seidel, L. Tinschert and J. Lutz, "Power cycling methods for SiC MOSFETs," 2017 29th International Symposium on Power Semiconductor Devices and IC's (ISPSD), Sapporo, Japan, 2017, pp. 367-370, doi: 10.23919/ISPSD.2017.7988994.

[7] F. Hoffmann and N. Kaminski, "Evaluation of the VSD-method for temperature estimation during power cycling of SiC-MOSFETs," IET Power Electronics, 2019, 12. Jg., Nr. 15, S. 3903-3909

[8] F. Hoffmann and N. Kaminski, "Investigation on the Accuracy of the VSD-Method for Different SiC MOSFET Designs Considering Different Measurement Parameters," 2021 IEEE 8th Workshop on Wide Bandgap Power Devices and Applications (WiPDA), Redondo Beach, CA, USA, 2021, pp. 18-23, doi: 10.1109/WiPDA49284.2021.9645142

[9] G. Zeng, L. Borucki, O. Wenzel, O. Schilling and J. Lutz, "First Results of Development of a Lifetime Model for Transfer Molded Discrete Power Devices," PCIM Europe 2018; International Exhibition and Conference for Power Electronics, Intelligent Motion, Renewable Energy and Energy Management, Nuremberg, Germany, 2018, pp. 1-8.

[10] H. P. Felsl, "Silizium- und SiC Leistungsdioden unter besonderer Berücksichtigung von elektrisch-thermischen Kopplungseffekten und nichtlinearer Dynamik," Ph.D Dissertation, Faculty of Elec. and Info. Engineering, TU Chemnitz, Chemnitz, Germany, November, 2009. [Online]. Available: https://monarch.qucosa.de/api/qucosa%3A19253/attachment/ATT-0/

[11] D. L. Blackburn and F. F. Oettinger, "Transient thermal response measurements of power transistors," IEEE Trans. Ind. Electron. Control Instrum., vol. IECI-22, no. 2, pp. 134—141, May 1975

[12] J. Abuogo, C. Schwabe, J. Lutz, T. Basler, "Influence of Current Density on Power Cycling Test of Low Voltage MOSFETs in DC Body-Diode Mode and Switching MOSFET-Mode, "proceedings CIPS2024

PCIM Europe 2024, 11– 13 June 2024, Nuremberg   DOI: 10.30420/566262275

Investigation of the Temperature Measurement via $V_{SD}(T)$-Method applied to Paralleled SiC MOSFET Chips during Power Cycling

Kevin Ladentin[1], Andreas Lindemann[1], Carsten Kempiak[1], David Strahringer[2]

[1] Otto-von-Guericke-University Magdeburg, Germany
[2] University of Freiburg, Germany

Abstract

Accurate junction temperature measurements are essential to obtain meaningful power cycling results. The purpose of this paper is to investigate the application of the $V_{SD}(T)$-method to parallel-connected SiC MOSFET chips during power cycling. This shall answer the question how the temperature which is determined by measuring the junction temperature via the reverse voltage across parallel connected chips — as is common practise when modules with internally paralleled SiC MOSFETs are power cycled — correlates with the individual and potentially different chip temperatures. For this purpose, the junction temperatures are determined separately as well as considering the parallel connection like a single chip.

1 Introduction

SiC MOSFETs are increasingly becoming industry standard. Therefore it is necessary to specify their long-term reliability [1]. This will typically include DC power cycling whereby package-related failure mechanisms can be determined and the lifetime can be estimated [2]. For an accurate interpretation of such power cycling results, a precise temperature measurement is necessary [3]. For this purpose, the $V_{SD}(T)$-method is applied to SiC MOSFETs, which utilises the voltage drop across the body diode when a small measurement current flows through it, while the channel is fully closed [4],[5],[6]. For high current applications MOSFET chips need to be paralleled; this can either be achieved by externally connecting components or by a parallel connection within a component such as a module. Ideally the current distribution between the chips and the power dissipation would be equal, but differences are to be expected [7],[8],[9] e. g. because of tolerances, asymmetrical layout or different cooling conditions. A resulting inhomogeneous temperature distribution can only be measured with the $V_{SD}(T)$-method when the single chips are accessible, while a parallel connection in a module would just yield a single temperature value. As the measurement current may distribute inequally between the chips while the reverse voltage is the same, this is not conclusive to deter-

mine the potentially different junction temperatures. This work aims at investigating this interrelationship more in detail to achieve meaningful temperature measurements from parallel connected SiC MOSFETs.

2 Consideration of the principle

As exemplarily illustrated in fig. 1 power cycling is effectuated by heating the MOSFET chips T1 and T2 — being turned on with $V_{GS0} > V_{GS,th}$ and parallel connected through the closed switches S1 and S2 — with a high forced current i_L. In the subsequent cooling phase, $i_L = 0$ and $V_{GS0} < 0$ in order to fully close the channel. The MOSFETs thus are turned off and only a small measurement current originating from the current sources i_M will flow solely through the body diodes. When S1 and S2 remain closed, $i_{M1} + i_{M2} = -(i_{D1} + i_{D2})$ will apply which corresponds to the case when a single read-out is taken from the parallel connection. When in contrast S1 and S2 are open, $i_{M1} = -i_{D1}$ and $i_{M2} = -i_{D2}$ apply which permits to obtain the individual chip temperatures through V_{DS1} and V_{DS2}.
This relation is derived from the reverse conduction characteristics as exemplarily shown in fig. 2 for $30\,°C$ and $150\,°C$ which permits to derive the calibration curve in fig. 3 for a low measurement current. As an extreme example to illustrate the investigated problem it is assumed that one chip would be heated to $150\,°C$ and the other only to $30\,°C$. When measuring the chips separately as explained

1964

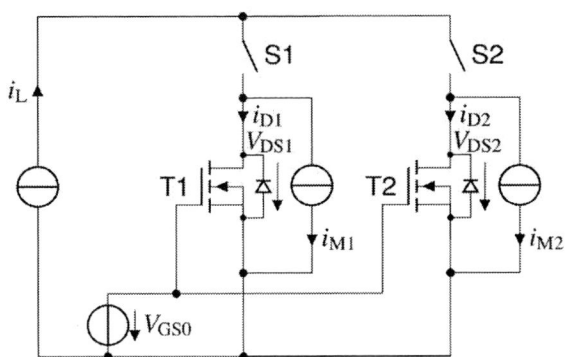

Fig. 1: Principle circuit with paralleled chips

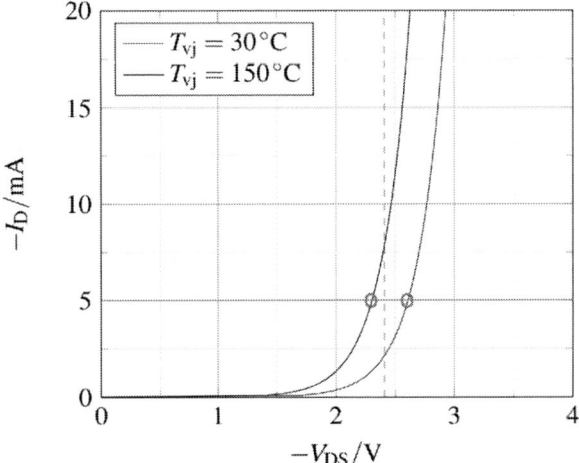

Fig. 2: Simulated exemplary body diode characteristics

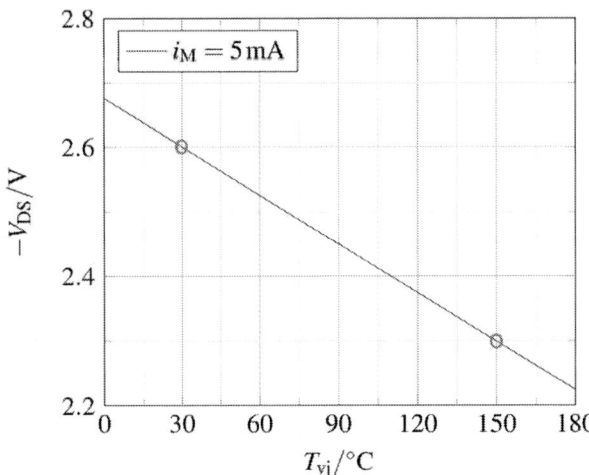

Fig. 3: Calibration curve for a single chip; $V_{GS,off} = -10\,V$

above, these temperatures would be determined through the $-V_{DS}$ readout at an individual current $-I_D = 5\,mA$. When however measuring the parallel connection, the operating points as marked with the vertical line in fig. 2 would be met with an unsymmetric distribution of measurement current at equal reverse voltage $-V_{DS}$, yielding a single temperature readout of $105\,°C$. This would reflect a temperature slightly above the area related average and corresponds to the findings with respect to the $V_{CE}(T)$ method in [10] and [11]. As a result, the hottest chip is overrepresented and a hotter temperature compared to the area related mean value can be read.

Consequently this kind of temperature measurement can only lead to meaningful results if the actual temperature differences between the chips are small enough to have no significant impact on lifetime. In the considered example this would not be the case because, in a power cycling test, the hotter chip would fail much earlier than expected from the common temperature readout.

3 Simulative study

The following considerations are extended to three parallel devices whose body diodes are depicted in fig. 4. The current distribution between those as well as the obtained voltage and temperature readings have been exemplarily simulated.

A LTSpice model of the conduction characteristics of the diodes was created for this purpose. The knee voltage V_{F0} where current flow begins is an important model parameter. For slightly higher voltages the curve is bent; this area can be scaled using a parameter "ε" in a dedicated diode model. At higher voltage and current levels the characteristic curve becomes almost linear, representing the device's differential resistance r_0.

To simulate the temperature-dependent behaviour of the parallel connection at a low measurement current parameter sets which are typically used to describe device behaviour around the nominal current are not useful because such operating points are situated in the linear part of the curve while the temperature measurement is taken close to the knee voltage in the bent part. The models have been suitably adapted to approximate this part of the characteristic curve using the aforementioned parameter ε, leading to the values given in table 1 where the following is summarised:

D_2 is at reference temperature, leading to the in-

PCIM Europe 2024, 11– 13 June 2024, Nuremberg   DOI: 10.30420/566262275

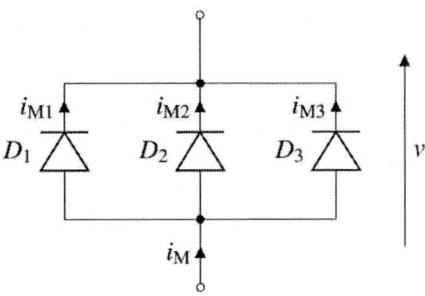

Fig. 4: circuit diagramme

Tab. 1: Comparison of currents, voltages and temperature readouts between single chip and paralleled measurement of three diodes with three different temperatures

Diode	D_1	D_2	D_3
r_0	$3.63\,\text{m}\Omega$	$3.63\,\text{m}\Omega$	$3.63\,\text{m}\Omega$
V_{F0}	$2.59\,\text{V}$	$2.6\,\text{V}$	$2.61\,\text{V}$
ε	21.842	21.842	21.842
single voltages at $i_\text{M} = 5\,\text{mA}$	$2.618\,\text{V}$	$2.628\,\text{V}$	$2.638\,\text{V}$
ΔT	$4\,\text{K}$	$0\,\text{K}$	$-4\,\text{K}$
single currents at $i_\text{M} = 15\,\text{mA}$	$8.6\,\text{mA}$	$4.6\,\text{mA}$	$1.8\,\text{mA}$
total voltage at $i_\text{M} = 15\,\text{mA}$		$2.6269\,\text{V}$	
ΔT		$0.11\,\text{K}$	

dicated V_{F0} and r_0. D_1 is $4\,\text{K}$ hotter; V_{F0}'s typical temperature sensitivity of $-2.5\,\frac{\text{mV}}{\text{K}}$ leads to a $10\,\text{mV}$ lower value while r_0 is assumed to remain constant. Correspondingly D_3 is $4\,\text{K}$ colder leading to $10\,\text{mV}$ more V_{F0}. These values could be determined measuring each diode individually.

When in contrast measuring the parallel connection, the current distribution will be significantly unequal, leading to a voltage drop across the parallel connection which would deliver a temperature readout which is slightly higher than the average junction temperature and thus lower than the hottest diode's which confirms the results from section 2.

4 Experimental validation

4.1 Power cycling test setup

A sixpack module with SiC MOSFETs has been chosen as exemplary device under test because it permits to measure the three phaselegs in common as well as separately. This applies to the heating current as well as to the temperature using the $V_{\text{SD}}(T)$-method as explained with reference to

Tab. 2: Power cycling parameters; ΔT_{vj} was set according to the measured temperature of the parallel connection.

t_{on}	t_{off}	i_L	$T_{\text{vj,max}}$	ΔT_{vj}	$V_{\text{GS,on}}$	$V_{\text{GS,off}}$
$3\,\text{s}$	$6\,\text{s}$	$57.7\,\text{A}$	$120\,^\circ\text{C}$	$80\,\text{K}$	$15\,\text{V}$	$-10\,\text{V}$

fig. 5: The test strategy was chosen such that the "upper" chips of the three phaselegs are always paralleled by auxiliary switches S_{12} and S_{23} when heating current $i_\text{L} \gg 0$ flows. In the subsequent cooling phase with $i_\text{L} = 0$, the individual chip temperatures are measured in each odd cycle, opening the switches S_{12} and S_{23}. In each even cycle the switches S_{12} and S_{23} in contrast remain closed in the cooling phase and the temperature of the parallel connected chips can be measured. For this configuration, each single chip and the module with active parallel connection and three times the measurement current $i_{\text{M1}} + i_{\text{M2}} + i_{\text{M3}}$ was calibrated, resulting in four different calibration curves for the DUT. The switches S_{12} and S_{23} are $1\,\text{m}\Omega$ MOSFETs. These were selected as they do not cause any significant asymmetry between the phaselegs. The DUT was cycled according to the parameters indicated in tab. 2.

4.2 Results

Fig. 6 depicts the individual heating currents as measured in the course of an initial power cycling test. They are obviously not equal and even diverge during the test run which underlines the importance of the investigation. The heating currents and the drain source voltage are measured simultaneously in each cycle. This allows to localise faults via the on-resistance $R_{\text{DS,on}}$ even in the parallel connection. Fig. 7 shows the maximum junction temperature over the entire lifespan. There is quite a large deviation between the individual chips, which can be expected considering the current distribution according to fig. 6. This means that the three MOSFETs within the module are cycled with different maximum junction temperatures $T_{\text{vj,max}}$ and therefore also different temperature swing ΔT_{vj}. This in turn leads to different lifetimes: It is to be expected that the chip with the highest temperature swing will fail first [12].

Comparing the individual junction temperatures with the temperature readout for the parallel connection according to fig. 8 confirms the findings in chapters 2 and 3: The temperature of the hottest chip is initially underestimated, here by approxi-

1966

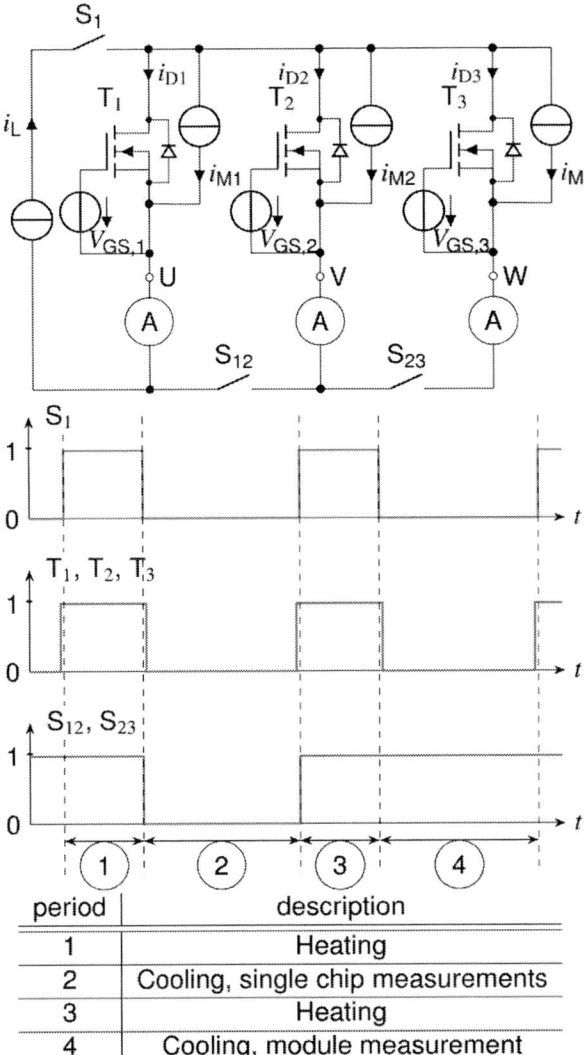

Fig. 5: Used power cycling setup with corresponding control signals

period	description
1	Heating
2	Cooling, single chip measurements
3	Heating
4	Cooling, module measurement

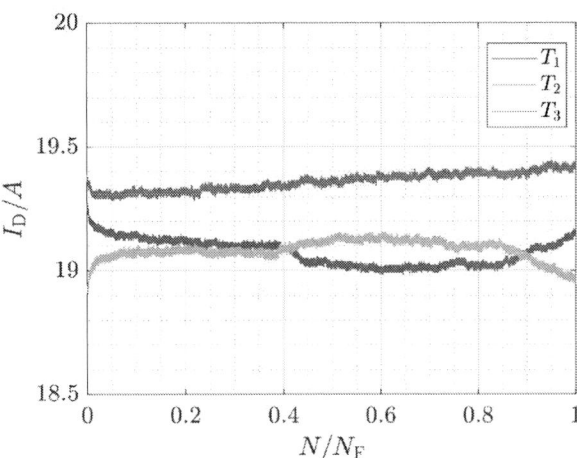

Fig. 6: Measured heating current distribution during power cycling

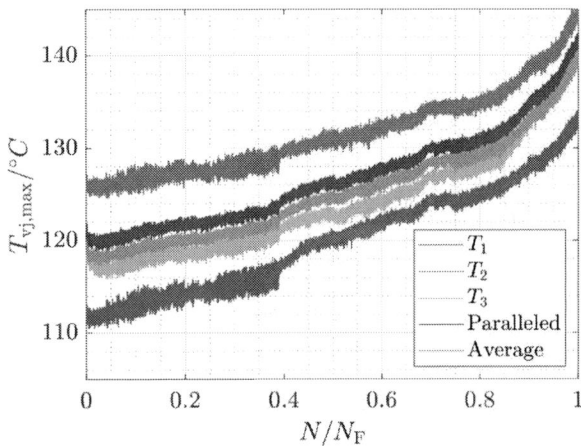

Fig. 7: Maximum junction temperature comparison

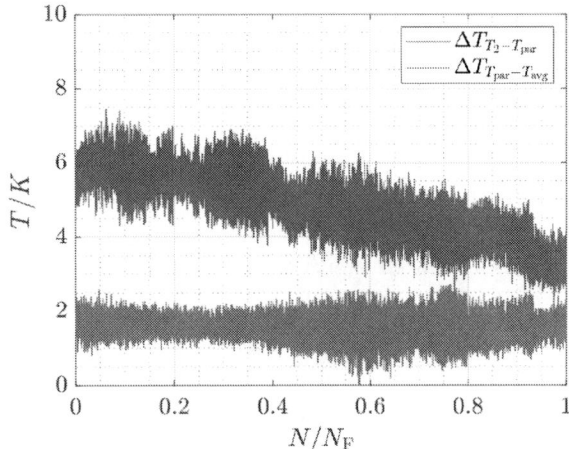

Fig. 8: Relevant temperature deviations

mately 6 K. Additionally the temperature measurement of the parallel connection yields a value which is higher than the junction temperature averaged across the three chips.

5 Conclusion

The application of the $V_{SD}(T)$-method to determine the junction temperatures of paralleled SiC MOSFETs during power cycling has been investigated, taking into account possible asymmetries. It has been shown that the considered asymmetries will lead to a readout of the temperature of the parallel connection which is slightly higher than the average value of the individual junction temperatures, but lower than the junction temperature of the hottest chip. Such a misleading temperature readout will consequently make devices achieve a lower number of power cycles than expected from the technology.

To minimise this effect, designs should obviously be as symmetric as possible regarding the electrical and thermal paths as well as the chip properties. When individual parallel connected chips are still accessible, the methods demonstrated in this paper can be used to determine the performance and parameters individually as well as for the parallel connection. This will allow to characterise the degree and effect of asymmetries and to localise certain failures.

Acknowledgement

This research project was carried out in the framework of the industrial collective research programme (IGF no. 22187 BG). It was supported by the Federal Ministry for Economic Affairs and Climate Action based on a decision taken by the German Bundestag.

The authors also gratefully acknowledge the contribution of Infineon Technologies AG for providing materials, services and consulting.

References

[1] N. Baker, S. Munk-Nielsen and S. Bęczkowski, "Test setup for long term reliability investigation of Silicon Carbide MOSFETs", 2013 15th European Conference on Power Electronics and Applications (EPE), Lille, France, 2013

[2] C. Kempiak and A. Lindemann, "A Method for the Measurement of the Threshold-Voltage Shift of SiC MOSFETs During Power Cycling Tests", in IEEE Transactions on Power Electronics, vol. 36, no. 6, June 2021

[3] J. Chen, E. Deng, Z. Zhao, Y. Wu and Y. Huang, "Power Cycling Capability Comparison of Si and SiC MOSFETs under Different Conduction Modes," 2020 32nd International Symposium on Power Semiconductor Devices and ICs (ISPSD), Vienna, Austria, 2020

[4] Hoffmann, F. and Kaminski, N.: "Evaluation of the VSD-method for temperature estimation during power cycling of SiC-MOSFETs", Oct. 2019, IET Power Electronics

[5] C. Kempiak, A. Lindemann, S. Idaka and E. Thal, "Comparative study of determining

junction temperature of SiC MOSFETs during power cycling tests by a Tj sensor and the VSD(T)-method", CIPS 2020; 11th International Conference on Integrated Power Electronics Systems, Berlin, Germany, 2020

[6] C. Herold, J. Sun, P. Seidel, L. Tinschert and J. Lutz, "Power cycling methods for SiC MOSFETs," 2017 29th International Symposium on Power Semiconductor Devices and IC's (ISPSD), Sapporo, Japan, 2017

[7] H. Li et al., "Influences of Device and Circuit Mismatches on Paralleling Silicon Carbide MOSFETs", in IEEE Transactions on Power Electronics, vol. 31, no. 1, Jan. 2016

[8] Y. He, X. Wang, J. Zhang, S. Shao, H. Li and C. Luo, "Analysis on Static Current Sharing of N-Paralleled Silicon Carbide MOSFETs," 2021 IEEE Energy Conversion Congress and Exposition (ECCE), Vancouver, BC, Canada, 2021

[9] N. Baker, A. Lemmon, F. Iannuzzo, S. Bęczkowski, J. Austin and L. Ostrander, "Temperature Monitoring of Multi-Chip SiC MOSFET Modules: On-Chip RTDs vs. VSD(T)," 2023 25th European Conference on Power Electronics and Applications (EPE'23 ECCE Europe), Aalborg, Denmark, 2023

[10] R. Schmidt and U. Scheuermann, "Using the chip as a temperature sensor — The influence of steep lateral temperature gradients on the Vce(T)-measurement," 2009 13th European Conference on Power Electronics and Applications, Barcelona, Spain, 2009

[11] F. Nehr and U. Scheuermann, "Consequences of Temperature Imbalance for the Interpretation of Virtual Junction Temperature Provided by the VCE(T)-Method," PCIM Europe 2022; International Exhibition and Conference for Power Electronics, Intelligent Motion, Renewable Energy and Energy Management, Nuremberg, Germany, 2022

[12] F. Hoffmann, S. Schmitt and N. Kaminski, "Lifetime Modeling of SiC MOSFET Power Modules During Power Cycling Tests at Low Temperature Swings," 2023 35th International Symposium on Power Semiconductor Devices and ICs (ISPSD), Hong Kong, 2023

PCIM Europe 2024, 11– 13 June 2024, Nuremberg   DOI: 10.30420/566262276

Approaches of Tsep Measurements for Power Semiconductors

Philipp Hauenschild[1], Regine Mallwitz[1]

[1] Technische Universität Braunschweig, Germany

Corresponding author: Philipp Hauenschild, p.hauenschild@tu-braunschweig.de
Speaker: Regine Mallwitz, r.mallwitz@tu-braunschweig.de

Abstract

This paper discusses different approaches to measuring temperature-sensitive electrical parameters (TSEPs). The construction of a modular research platform for testing different TSEP measurement circuits is described and, using the example of the TSEP forward voltage, offline as well as online measurement results are presented.

1   Introduction

Prediction of the expected lifetime of power semiconductors is usually based on the LESIT [1] or CIPS-08-model [2]. The models infer the expected lifetime in the application from accelerated aging, known as power cycling [3]. As the electrical and thermal properties are determined under fixed ambient and test conditions, even the smallest parameter changes can be detected and measured. Failures of power semiconductors in accelerated aging are usually indicated by an increase in junction temperature and thus in forward voltage, as well as in transient thermal impedance (Zth) [4]. Furthermore, accelerated aging reveals that power semiconductors of the same model and production batch achieve different lifetimes [4]. Thus, it is of interest for reliability critical applications to predict an imminent failure in order to take preventive measures if necessary. The power semiconductors can be checked when the device is dismantled. It is advantageous to monitor the semiconductors using a suitable measuring method during operation. This is referred to as online condition monitoring.

2   Known Methods

For online condition monitoring, certain parameters must be continuously recorded and evaluated. Before a power semiconductor fails, this can usually be seen in the Zth or the forward voltage [4]. Some electrical parameters of the power semiconductor provide information on the junction temperature. These are referred to as temperature-sensitive electrical parameters TESPs. The following seven reviewed briefly methods are known for recording TESPs:

2.1   Temperature sensing

Temperature sensors are designed as diodes or NTC sensors. They require space on the substrate and specially adapted power semiconductors [5].

2.2   Gate plateau voltage sensing

The gate plateau voltage depends on the junction temperature, the forward current and the transconduction parameter. It can be used to determine the junction temperature in mV per K [6].

2.3   Gate plateau time sensing

The gate plateau time depends on the junction temperature, the forward current and the forward voltage. It can be used to determine the junction temperature in ns per K [7].

2.4   Delay time sensing

The on-delay-time depends on the temperature and the reverse voltage and can be used as measured values in ns per K. The switch-off delay time can also serve as a measured value in ns per K and is dependent on the temperature, the reverse voltage and additionally on the forward current [8].

2.5   Current slope of turn-on

The current slope of turn-on can serve as a measured value in A/(µs*K) and is dependent on the temperature, the reverse voltage and additionally on the forward current [8].

2.6   Gate resistor

Measurement of the internal gate resistance with an RCL meter integrated in the driver circuit. The measurement provides a voltage in mV per K which corresponds to a junction temperature dependent resistance [9].

2.7 Forward voltage

The forward voltage depends on the temperature and the forward current. The measured value is mV per K. A small measuring current of up to 100 mA is used in offline measurements. For online temperature detection, the forward current of the application must be used [10, 11].

3 Test setup

The test setup for investigating TESP measurements during operation is shown in Fig 1. and consists of a Lauda ECO RE 1050G refrigerated circulator for temperature control of the power semiconductor from 5 °C to 175 °C. Voltage sources and sinks, as well as the demonstrator and, as a reference, a parameter analyzer.

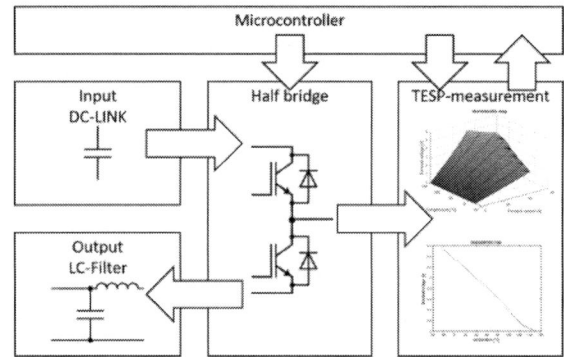

Fig. 1: Laboratory measuring setup

3.1 Modular TSEP research platform

A modular research platform (Fig. 2) was designed for testing and validating various TSEP measurement circuits for different power semiconductors.

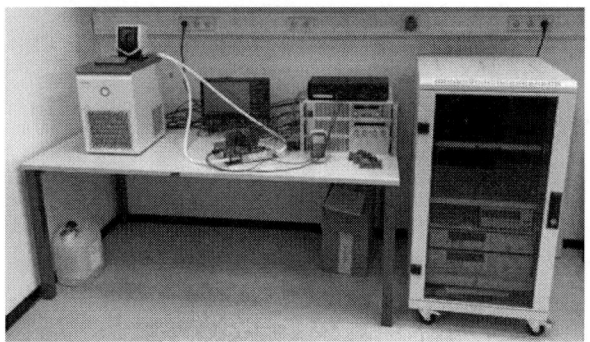

Fig. 2: Modular Research Platform Structure

The demonstrator (Fig. 3) consists of 3 PCBs: the logic PCB with DC-LINK and output filter, the DUT PCB with the half bridge and driver circuit and the measurement PCB for the TESP measurement.

The half-bridge can be equipped with different power semiconductors and drivers. The half-bridge can be connected to the demonstrator via PCB. The TSEP measurement is also interchangeable as a PCB module and pluggable onto the research platform as well as the half-bridge PCB. The control and data acquisition are realized by a microcontroller.

The power semiconductors are also arranged in a close-meshed arrangement on a heat sink. The heat sink, DUT PCB and measurement PCB are shown in Fig. 3 in an exploded view for clarity.

Fig. 3: Constructed demonstrator in exploded view

Temperature measurement is carried out with sensors under the semiconductors in the heat sink.

3.2 Used Method

For the investigations in this publication, the TESP measurement was carried out using the forward voltage discussed in Section 2.7. The measuring circuit for recording the forward voltage in the demonstrator is designed with a current source and two diodes in the same package (Fig. 4). The forward voltage of the diodes is therefore assumed to be identical. The voltage drop across the IGBT thus corresponds to the voltage of the current source added with both diode voltages. This means that the measurement voltage V_{ADC} is equal to the voltage drop V_{CE} across the IGBT.

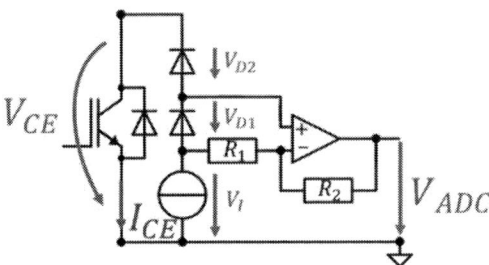

Fig.4: Forward voltage measurement [10]

$$V_{CE} = V_{D1} + V_{D2} + V_I \text{ with } V_{D1} = V_{D2}$$
$$V_{CE} = 2 * V_{D1} + V_I$$

$$V_{ADC} = k * (V_{D1} + V_I) - V_I$$
$$k = 1 + R_1/R_2 \text{ with } R_1 = R_2$$
$$k = 2$$

$$V_{ADC} = V_{CE} = 2 * V_{D1} + V_I$$

To validate the measurement data of the demonstrator a characteristic map of the forward voltage over temperature and current was recorded with the parameter analyzer as a reference (Fig. 5). For this purpose, the gate was permanently switched on at 15 V. Furthermore, the conduction characteristics were recorded with individual current pulses for the power semiconductors at different temperatures. For these measurements with the parameter analyzer, the DUT PCB was disconnected from the DC link, output filter, measurement PCB and gate driver. The voltage across the power semiconductors was increased in 20 mV steps up to a forward current of 50 A.

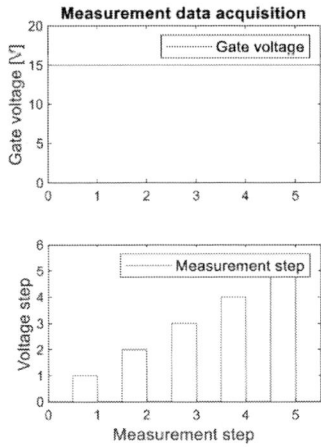

Fig. 5: Parameter analyzer measuring point

The individual measured value was recorded as a pulse to minimize self-heating of the semiconductor. Each period is 300 ms long and each pulse is 500 µs long. This corresponds to a duty cycle of 0.1667 % (Fig. 6).

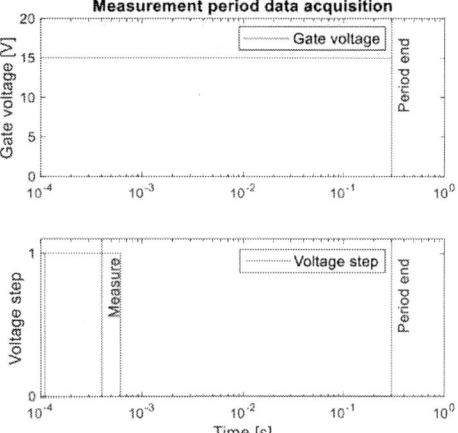

Fig. 6: Parameter analyzer measuring pulse

For these measurements with the demonstrator, the DUT PCB was connected to the DC link, output filter, measurement PCB and gate driver. The measurement is carried out for the forward voltage in the conducting state by injecting a measuring current of 20 mA. The phase current is also measured using a shunt and assigned to the upper and lower switches of the half bridge. The resistive load is between 0.7 and 2.5 Ω at the output of the demonstrator. The switching frequency is 15.625 kHz and the duty cycle is increased from 0 to 80 % in 0.5 % increments every second. For each duty cycle, 100 periods are initially not measured for settling. A further 100 periods are used for averaging (Fig. 7).

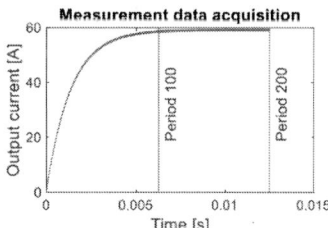

Fig.7: Demonstrator measuring

The junction temperature is calculated using the transient thermal resistance, which depends on the switching frequency and the duty cycle, as well as the power loss in the semiconductor.

$$T_J = T_{Case} + R_{th,j-c} * P_v$$
$$P_v = V_{CE} * I_{CE}$$

The duty cycle does not change during measurements with the analyzer parameter, so $R_{th,j-c}$ must be regarded as constant. Due to the low duty cycle, the $R_{th,j-c}$ can be assumed to be the same as for a single pulse.

In the demonstrator, the forward current is set by modulating the duty cycle. The transient thermal resistance is thus determined by the duty cycle. The relationship is linear and is taken from the data sheet.

4 Measurement

Characteristic curves were recorded for four power semiconductors in TO-247-3 housings of class 600/650 V and 47 A using the measurement methods described in Chapter 3 for the on-state characteristics. An IGBT, a MOSFET, a SIC-MOSFET and a GaN power semiconductor. First, the principle and the correction are illustrated using the IGBT.

The junction temperature depends on the case temperature and the self-heating. The self-heating depends on the thermal resistance and the power dissipation of the power semiconductor. As the junction temperature is only measured indirectly, all TESP measured values must be corrected on this basis. For this purpose, the transient thermal impedance must be taken from the data sheet of the respective power semiconductor.

$$T_J = T_{Case} + R_{th,j-c} * P_v$$
$$P_v = V_{CE} * I_{CE}$$

For the parameter analyzer (PA), the measured values were recorded with a period length of 500 µs and a duty cycle of less than 1 %. The transient thermal impedance can therefore be assumed to be a single pulse. The data sheet states a value of $R_{th,j-c} = 0.135$ K/W.

Furthermore, interpolation between the measuring points is carried out to obtain a uniform step size. The transfer characteristic of the IGBT is shown in Fig. 8.

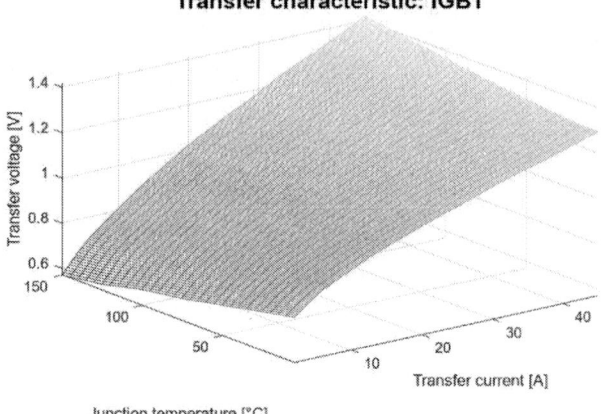

Fig.8: Transfer characteristic IGBT (PA)

The measured values of the demonstrator are corrected using the same method with the addition that the transient thermal impedance changes linearly to the duty cycle due to the duty cycle variation. The characteristic curve is taken from the data sheet.

$$R_{th,j-c} = 0.5041 \, K/W * Dutycycle + 0.0315 \, K/W$$

The transfer characteristic of the IGBT with the demonstrator is shown in Fig. 9.

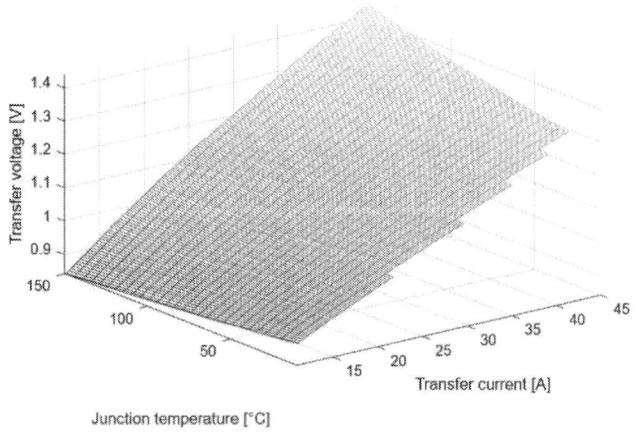

Fig.9: Transfer characteristic IGBT (Demonstrator)

To estimate the measurement deviation, the difference between the parameter analyzer and demonstrator is shown in absolute (Fig. 10) and percentage (Fig.11) terms at each point.

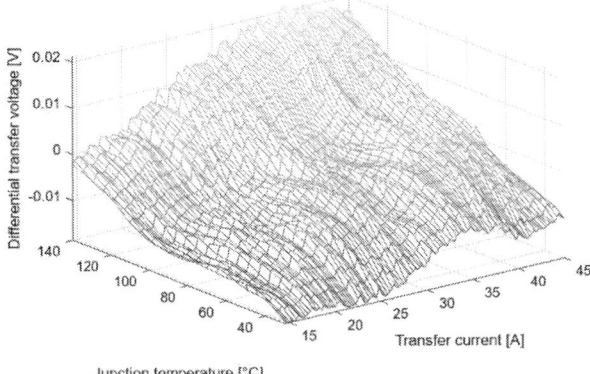

Fig.10: Deviation as difference (PA – Demonstrator)

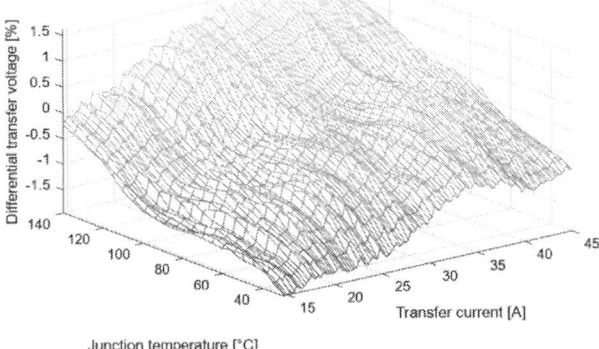

Fig.11: Deviation percentage 100*(PA/Demonstrator)

The separate diode in the IGBT is calibrated with the following characteristic curve.

$$R_{th,j-c} = 0.6889\,K/W * Dutycycle + 0.0667\,K/W$$

As well as a single shot for the parameter analyzer with $R_{th,j-c} = 0.234$ K/W.

The measured values for the power semiconductors used are corrected for $R_{th,j-c}$ according to Tab. 1:

Tab. 1: Rth correction values

	parameter analyzer	Demonstrator
IGBT	0.135	0.5041x+0.0315
IGBT DIODE	0.234	0.6889x+0.0667
MOSFET	0.02	0.2023x+0.0027
SIC	0.316	0.9065x+0.0674
GAN	0.165	0.7327x+0.0586

The x stands for the duty cycle from 0 to 100 % and is specified between 0 and 1.

Figs. 12 to 14 show the behavior of the diode installed in the IGBT.

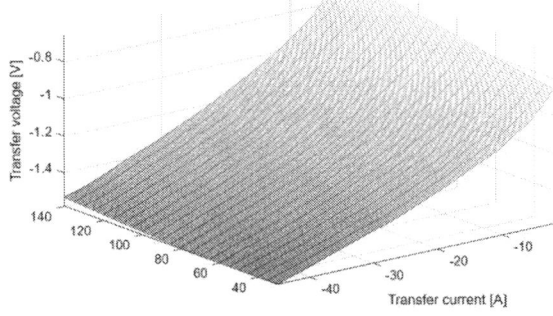

Fig.12: Transfer characteristic IGBT Diode (PA)

In contrast to the IGBT, the diode has an offset of approx. 70 mV between the measuring systems. The deviation is +-20 mV (Fig. 13). In percentage terms, this is reflected in an offset of 5 % and a deviation of +- 0.5 % (Fig. 14).

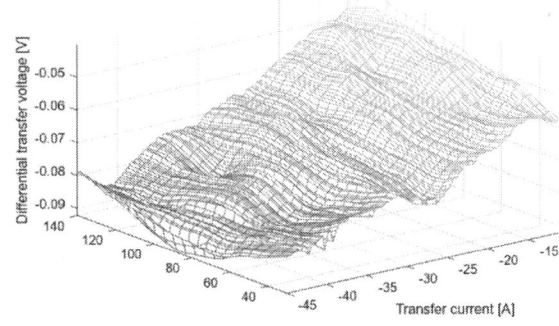

Fig.13: Difference IGBT Diode

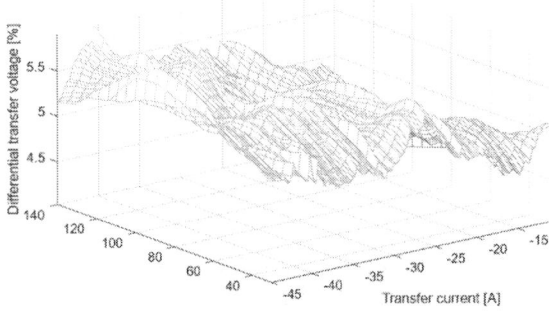

Fig.14: Percentage IGBT Diode

In the following, the figures for each component are shown in a series starting from the left with the transfer characteristic recorded with the parameter analyzer, as well as the difference in the middle figure and the percentage deviation in the right-hand figure. For clarity of the captions, the labels are abbreviated as follows: transfer characteristic (TC) and parameter analyzer (PA). Furthermore, due to the small deviation, only the transfer characteristic map of the parameter analyzer is shown; the map of the demonstrator has been omitted. Figures 15 to 41 provide an overview of the measurement method with other power semiconductors and are summarized in Chapter 5.

Figs. 15 to 17 show the behavior of the MOSFET. The range in which the MOSFET is considered was adjusted to up to 30 A and 100 °C.

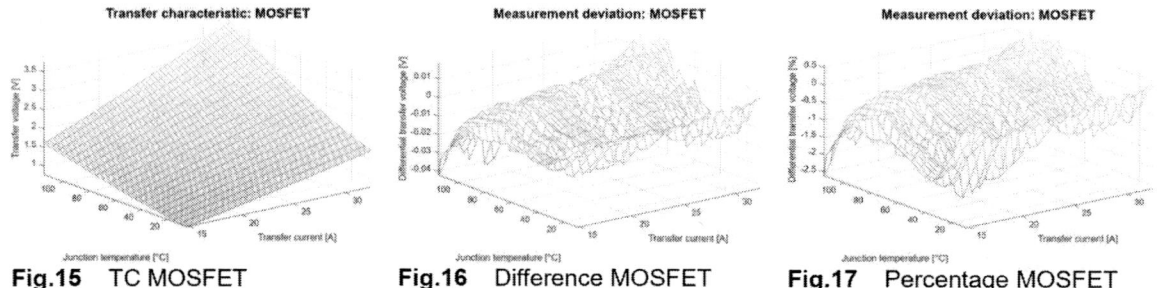

Fig.15 TC MOSFET   **Fig.16** Difference MOSFET   **Fig.17** Percentage MOSFET

Fig. 18 to 20 show the diode behavior of the MOSFET at 0 V gate voltage.

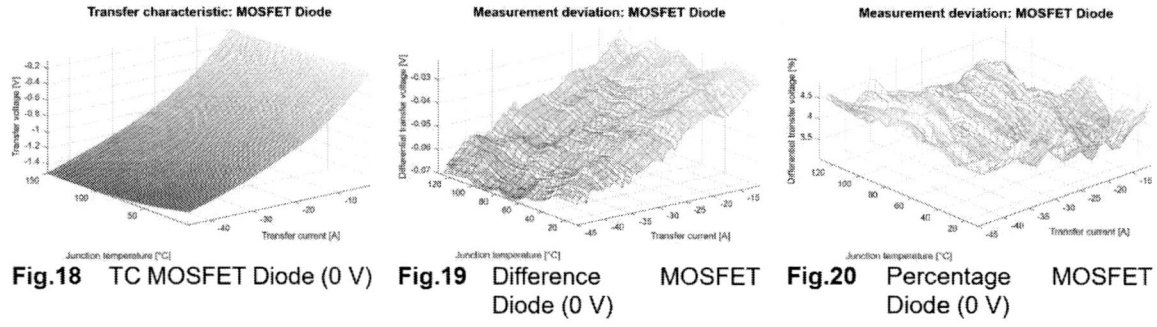

Fig.18 TC MOSFET Diode (0 V)   **Fig.19** Difference   MOSFET   **Fig.20** Percentage   MOSFET
Diode (0 V)   Diode (0 V)

Fig. 21 to 23 show the diode behavior of the MOSFET at 15 V gate voltage.

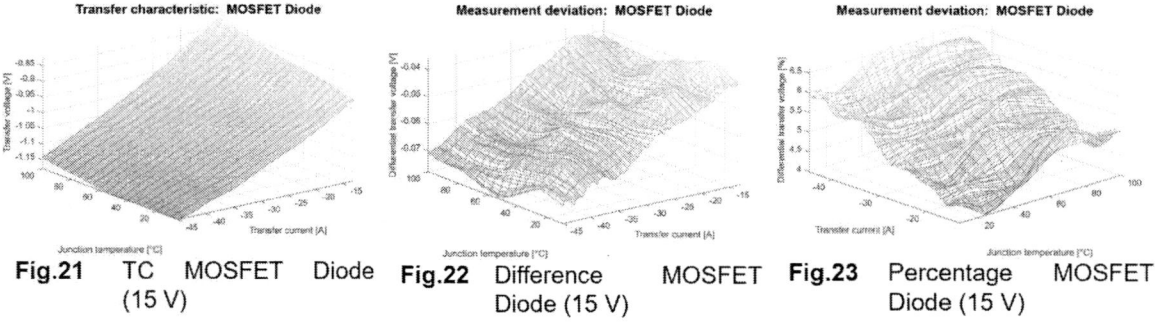

Fig.21 TC   MOSFET   Diode   **Fig.22** Difference   MOSFET   **Fig.23** Percentage   MOSFET
(15 V)   Diode (15 V)   Diode (15 V)

Figure 24 to 26 show the behavior of the SiC-MOSFET.

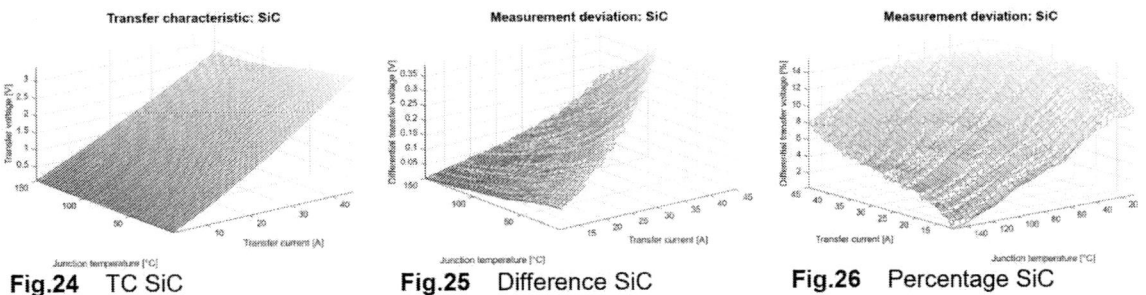

Fig.24 TC SiC

Fig.25 Difference SiC

Fig.26 Percentage SiC

Fig. 27 to 29 show the diode behavior of the SiC-MOSFET at 0 V gate voltage.

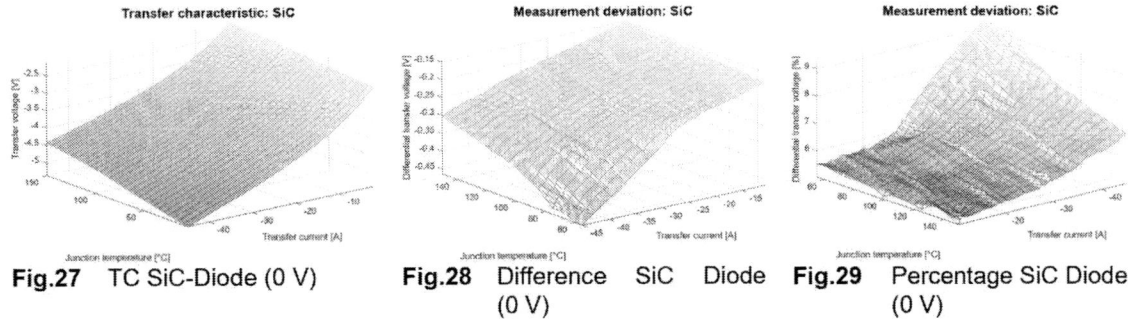

Fig.27 TC SiC-Diode (0 V)

Fig.28 Difference SiC Diode (0 V)

Fig.29 Percentage SiC Diode (0 V)

Fig. 30 to 32 show the diode behavior of the SiC-MOSFET at 15 V gate voltage.

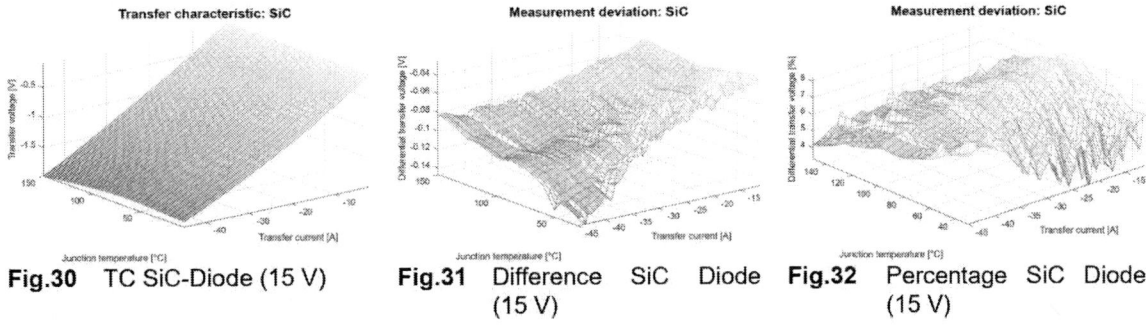

Fig.30 TC SiC-Diode (15 V)

Fig.31 Difference SiC Diode (15 V)

Fig.32 Percentage SiC Diode (15 V)

Fig. 33 to 35 show the behavior of the GaN FET.

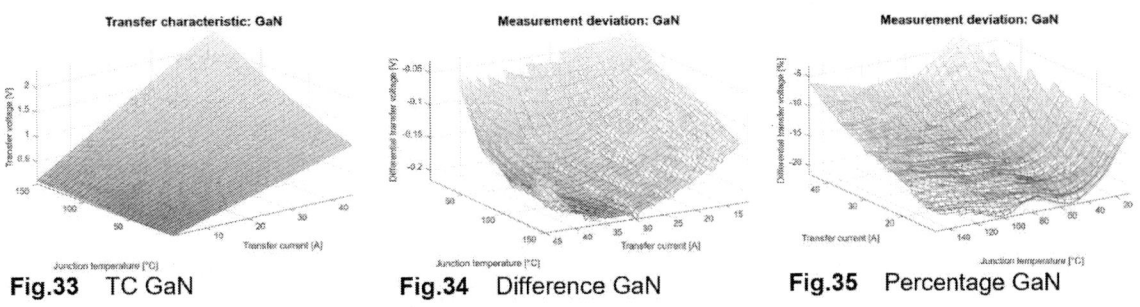

Fig.33 TC GaN

Fig.34 Difference GaN

Fig.35 Percentage GaN

Fig. 36 to 38 show the diode behavior of the GaN-FET at 0 V gate voltage.

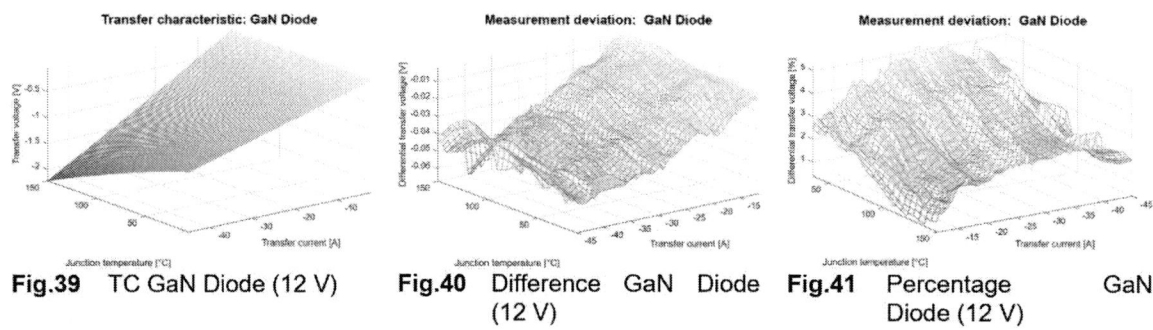

Fig.36 TC GaN Diode (0 V)

Fig.37 Difference GaN Diode (0 V)

Fig.38 Percentage GaN Diode (0 V)

Fig. 39 to 41 show the diode behavior of the GaN-FET at 12 V gate voltage.

Fig.39 TC GaN Diode (12 V)

Fig.40 Difference GaN Diode (12 V)

Fig.41 Percentage GaN Diode (12 V)

5   Evaluation

With both measuring systems, it is possible to create a map for the forward characteristic. With the IGBT, these deviate from the parameter analyzer with the demonstrator in the range shown of -1.5 % to 1.5 % or a few tens of millivolts. The diode installed in the IGBT shows an offset shift.

The characteristic map of the MOSFET shows a similar behavior. If the MOSFET is now operated as a diode, the deviations with and without a switched gate are almost identical. The percentage deviations are significantly greater when the MOSFET is actively switched on, as the voltage drop across the component is lower.

The differences with the SiC MOSFET are much more pronounced with a deviation of 2 % to 14 %.

For the GaN FET, the deviation between the measurement systems is also very pronounced. In diode mode, the deviation for SiC and GaN is not as pronounced, but still much more noticeable than for the IGBT.

6   Conclusion

The demonstrator makes it clear that the junction temperature can be deduced from the forward voltage and the forward current with a known temperature in the thermal path using simple means in switching operation. This can be done for different types of power semiconductors. The evaluation shows small to large deviations for different semiconductors. On the one hand, this can be explained by the uncertainty of reading the transient thermal resistance from the data sheets. Furthermore, self-heating cannot be neglected when averaging due to the low thermal time constants in the semiconductor. For reliable verification of the measuring circuit, the thermal path must be measured and the measured values corrected with real measured values. Furthermore, the switching frequency and the duty cycle for creating the flow characteristics should be fixed and the change in forward current should be made via the DC link voltage or the load.

References

[1]   M. Held, P. Jacob, G. Nicoletti, P. Scacco und M.-H. Poech, „Fast power cycling test of IGBT modules in traction application" in *Second International Conference on Power Electronics and Drive Systems*, Singapore, 1997, S. 425–430, doi: 10.1109/PEDS.1997.618742.

[2] Reinhold Bayerer, Tobias Herrmann, Dr. Thomas Licht, Prof. Dr. Josef Lutz, Marco Feller, *Proceedings / CIPS 2008, 5th International Conference on Integrated Power Electronics Systems: March, 11 - 13, 2008, Nuremberg, Germany.* Model for Power Cycling lifetime of IGBT Modules – various factors influencing lifetime. Berlin: VDE-Verl., 2008.

[3] A. Wintrich, U. Nicolai, W. Tursky und T. Reimann, Hg., *Applikationshandbuch Leistungshalbleiter,* 2. Aufl. Ilmenau: ISLE Verlag, 2015.

[4] Florian Lippold, Philipp Hauenschild, Regine Mallwitz, *2022 24th European Conference on Power Electronics and Applications (EPE'22 ECCE Europe): Comparison of Power Cycling Results of discrete GaN Cascodes for Automotive Power Electronics with high Temperature Swings.* Piscataway, NJ: IEEE, 2022. [Online]. Verfügbar unter: https://ieeexplore.ieee.org/servlet/opac?punumber=9907022

[5] Eric R. Motto, John F. Donlon, Hg., *2012 Twenty-Seventh Annual IEEE Applied Power Electronics Conference and Exposition (APEC).* IGBT Module with User Accessible On-Chip Current and Temperature Sensors. IEEE: IEEE, 2012.

[6] Christoph H. van der Broeck, Alexander Gospodinov and Rik W. De Doncker, Hg., *2017 IEEE Energy Conversion Congress and Exposition (ECCE).* IGBT Junction Temperature Estimation via Gate Voltage Plateau Sensing. IEEE: IEEE, 2017.

[7] V. Sundaramoorthy, E. Bianda, R. Bloch, I. Nistor, G. Knapp, A. Heinemann, Hg., *2013 15th European Conference on Power Electronics and Applications (EPE).* Online estimation of IGBT junction temperature (Tj) using gate-emitter voltage (Vge) at turn-off. IEEE: IEEE, 2013.

[8] A. M. Harald Kuhn, *2009 13th European Conference on Power Electronics and Applications: EPE 2009 ; Barcelona, Spain, 8 - 10 September 2009.* On-line Junction Temperature Measurement of IGBTs based on Temperature Sensitive Electrical Parameters. Piscataway, NJ: IEEE, 2009.

[9] M. Denk, „In-Situ-Zustandsüberwachung von IGBT-Leistungshalbleitern mittels Echtzeit-Sperrschichttemperaturmessung". Dissertation.

[10] S. Beczkowski, P. Ghimre, A. R. de Vega, S. Munk-Nielsen, B. Rannestad und P. Thogersen, „Online Vce measurement method for wear-out monitoring of high power IGBT modules" in *2013 15th European Conference on Power Electronics and Applications (EPE)*, Lille, 2013, S. 1–7, doi: 10.1109/EPE.2013.6634390.

[11] A. M. Aliyu, S. Chowdhury und A. Castellazzi, „In-situ health monitoring of power converter modules for preventive maintenance and improved availability" in *2015 17th European Conference on Power Electronics and Applications (EPE'15 ECCE-Europe)*, Geneva, 2015, S. 1–10, doi: 10.1109/EPE.2015.7309348.

PCIM Europe 2024, 11– 13 June 2024, Nuremberg   DOI: 10.30420/566262277

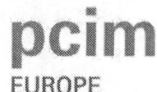

Realtime Junction Temperature Estimation in SiC Power Modules Based on Multiple TSEP Acquisition

Kevin Muñoz Barón [1], Sarthak Swaroop Dash [1], Dominik Koch [1], Kanuj Sharma [1], Ingmar Kallfass [1]

[1] University of Stuttgart, Germany

Corresponding author: Kevin Muñoz Barón, munoz-baron@ilh.uni-stuttgart.de
Speaker: Kevin Muñoz Barón, munoz-baron@ilh.uni-stuttgart.de

Abstract

Accurate junction temperature estimation is crucial for optimizing SiC power module performance and reliability. Temperature-sensitive electrical parameters (TSEPs) can be used to implicitly determine the junction temperature. In this work, a novel acquisition concept based on four TSEPs is proposed, using the intrinsic parasitic inductance between the power source and Kelvin source terminals to acquire the virtual threshold voltage and the virtual time delay during the turn-on and turn-off transitions. Experimental results reveal maximum errors in temperature estimation of ±10 K, offering a promising non-invasive real-time junction temperature estimation solution for SiC power modules.

1 Introduction

In the field of power electronics, ensuring reliability and performance of switches is of utmost importance, especially in comparatively modern semiconductor technologies utilized in high-voltage applications such as SiC MOSFETs. These devices, often considered the critical link in power electronic systems [1], demand attention to ensure robust operation. Traditionally, efforts have been directed towards lifetime estimation and enhancement through accelerating power cycling tests and health monitoring strategies in application. These approaches rely heavily on temperature estimation approaches such as measuring temperature-sensitive electrical parameters (TSEPs).

For SiC MOSFETs, characterized by low on-state resistance $R_{\mathrm{DS,on}}$ and the potential for high switching frequency f_{SW}, understanding their reliability is critical for effective deployment in various applications. Despite the growing interest in SiC technology, there remains a scarcity of reliability data concerning parameter degradation throughout its operational lifespan. In this context, online health monitoring emerges as a crucial tool for designing reliable power electronic converters employing SiC-based power devices.

Among the parameters influencing the reliability of the power semiconductor, junction temperature T_{J} stands out as a primary factor. The maximum junction temperature and its rate of change induces thermo-mechanical stress, causing a gradual degradation in performance over time. This phenomenon is observed in transistors, inclusive of their mounting and packaging technologies, whether employed in discrete switches or integrated power modules. Conventionally, its estimation relies on either direct and indirect methods [1]. Direct methods involve placing dedicated temperature sensors near the chip, which entails modifications to the device design. Conversely, indirect estimation through TSEPs offers a more economical alternative, facilitated by specialized acquisition circuits integrated into existing gate driver boards. This non-invasive approach eliminates the need for device or package modifications, thus minimizing costs and complexity on the system level.

However, acquiring TSEPs under practical operating conditions poses significant challenges. Conventional measurement techniques, effective for offline assessments, fall short when applied in dynamic application environments [2]. Addressing these challenges necessitates the development of isolated measurement setups capable of transmitting parameter values across isolation barriers, ensuring accurate monitoring and control in real-time applications.

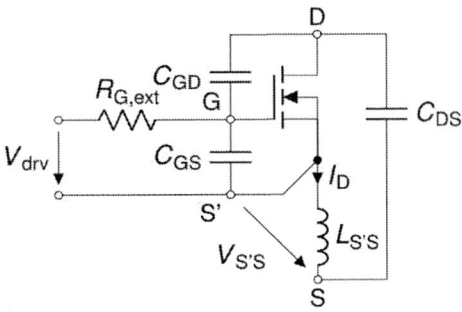

Fig. 1: Equivalent circuit of a power MOSFET with the intrinsic capacitances and parasitic source inductance drawn.

1.1 TSEPs in Power MOSFETs

Various TSEPs have been observed for SiC MOS-FETs [2] each exhibiting distinct sensitivities, linearities, and complexities of implementation. The most common TSEPs are listed in table 1 with their dependencies on other parameters. To be able to successfully implement a temperature measurement, it is advantageous to have as little dependencies on other electrical parameters as possible, to simplify the measurement complexity. At the same time, however, the feasibility of the implementation of a certain TSEP greatly influences its suitability in the application. While the on-state resistance may not exhibit notable linearity with respect to the junction temperature for SiC, it remains an important indicator for detecting bond-wire failure [5], [6]. Conversely, the peak gate current, influenced by the resistances within the gate loop, directly reflects the degradation in the gate path [8]–[10]. The virtual threshold voltage V_{vth} presents itself as a suitable candidate for junction temperature estimation due to its high linearity and sensitivity [13]. At the same time, however, one of the main effects that can influence the operation of a SiC MOSFET is the bias temperature instability of the threshold voltage [17]. Measurement of the threshold voltage in operation proves to be difficult, as an interruption of normal operation is necessary. Therefore, a virtual threshold voltage has to be defined, which offers similar properties, but can be measured relatively easily by using the intrinsic parasitic inductance of the device [14], [18].

2 Parameter Extraction Using the Intrinsic Parasitic Inductance

Devices with a Kelvin source (S') possess an intrinsic parasitic inductance between the Kelvin source

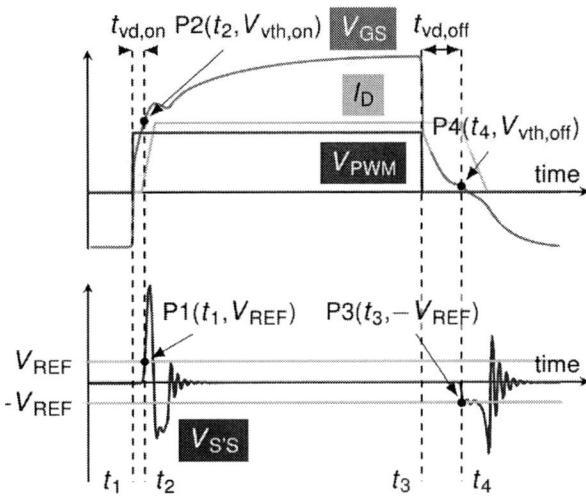

Fig. 2: Relevant signals for one switching cycle of the transistor in a double pulse test at 600 V/52 A, using 3.9 Ω as external gate resistance.

and the power source as shown in Fig. 1. Four TSEPs can be acquired by leveraging the intrinsic parasitic inductance: The virtual threshold-on voltage $V_{\text{vth,on}}$ and virtual turn-on delay $t_{\text{vd,on}}$, the virtual threshold-off voltage $V_{\text{vth,off}}$ and the virtual turn-off delay $t_{\text{vd,off}}$. The voltage drop over the parasitic inductance can be expressed as per Eq. (1), given that the resistance over the parasitic path $R_{\text{S'S}}$ is negligible.

$$V_{\text{S'S}} = L_{\text{S'S}} \cdot \frac{\text{d}I_{\text{D}}}{\text{d}t} \tag{1}$$

where I_{D} is the drain current flowing through the device.

Figure 2 shows the voltage drop over the parasitic inductance during a complete switching cycle together with the gate-source voltage, the drain current and PWM signal. To mathematically derive these TSEPs, equations must be solved within different time intervals during the turn-on and turn-off process. The simplified gate-source voltage can be seen as an exponential charging waveform of a capacitor and as an example for the turn-on can be written as Eqs. (2) and (3).

$$V_{\text{GS}} = V_{\text{GS,off}} + \left(V_{\text{GS,on}} - V_{\text{GS,off}}\right) \cdot \left(1 - e^{-\frac{t}{\tau}}\right) \tag{2}$$

$$\tau = \left(R_{\text{G,int}}\left(T_{\text{J}}\right) + R_{\text{G,ext}}\right) \cdot \left(C_{\text{GS}} + C_{\text{GD}}\right) \tag{3}$$

where $V_{\text{GS,on}}$ and $V_{\text{GS,off}}$ are set by the gate driver, $R_{\text{G,int}}$ is the internal gate resistance, C_{GS} is the gate-source capacitance and C_{GD} is the gate-drain

Tab. 1: Common Temperature-sensitive electrical parameters and their dependencies on other parameters. (PN = forward voltage of a pn junction; DYN = Dynamic switching behavior)

Category	Parameter	Symbol	Dependencies [3]	Reference
PN	On-state resistance	$R_{DS,on}$	T_J, I_D	[4]–[6]
PN	Forward voltage	V_F	T_J, I_D	[7]
$R_{G,int}$	Peak gate current	$I_{G,Peak}$	T_J	[8]
$R_{G,int}$	High frequency measurement signal	$R_{G,int,HF}$	T_J, $R_{G,ext}$	[9], [10]
DYN	Rate of change of the load current	di_L/dt	T_J, V_{DC}, $R_{G,ext}$	[11]
DYN	Threshold voltage	V_{th}	T_J, V_{GS}	[3], [12]–[14]
DYN	Turn-on/-off delay	$t_{d,on}$, $t_{d,off}$	T_J, I_D, V_{DC}, $R_{G,ext}$	[15], [16]

capacitance. By solving Eq. (2) for the time, the turn-on delay can be extracted:

$$t_{d,on} = -\tau \cdot \ln\left(\frac{V_{GS,on} - V_{th}(T_J)}{V_{GS,on} - V_{GS,off}}\right) \quad (4)$$

Here, the threshold voltage is defined as the gate-source voltage at which point the device current starts flowing. Unlike this definition, the virtual threshold-on voltage that is acquired within the frame of this work, an additional time delay has to be added, as the measurement is based on the rate of change of the load current as described in Eq. (1). Once the device enters the saturation region, current starts to flow and voltage drop over the parasitic inductance can be measured. In this region the current can be described by Eq. (5) according to the Shichman-Hodges Model [19].

$$I_D = \frac{1}{2} \cdot K_n(T_J) \cdot (V_{GS}(t) - V_{th}(T_J))^2 \quad (5)$$

$$K_1 = \mu_c \cdot \frac{W}{L} \quad (6)$$

where K_1 is a MOSFET geometry parameter, μ_c is the carrier mobility, W is the width of the transistor, C is the specific capacitance of the gate and L is the length of the transistor. It increases with an increase in temperature for a SiC MOSFET [20]. The additional time delay until reaching a specific reference voltage is described by the second term in Eq. (7) and has to be added to Eq. (4) to receive

the total virtual turn-on delay.

$$t_{vd,on} = t_{d,on} - \tau \cdot \ln\left(\frac{1}{2} + V_a\right) \quad (7)$$

$$V_a = \sqrt{\frac{1}{4} - \frac{\tau \cdot V_{REF}}{L_{S'S}K_n(T_J) \cdot \left(V_{GS,on} - V_{th}(T_J)\right)^2}} \quad (8)$$

where V_{REF} is the reference voltage of the voltage drop over the parasitic inductance.
Similarly for the turn-off process, equations can be extracted:

$$t_{d,off} = \tau \cdot \ln\left(\frac{V_{GS,on}}{V_p(T_J)}\right) + \tau \cdot \left(\frac{V_{DC}}{V_p(T_J)}\right) \quad (9)$$

$$t_{vd,off} = t_{d,off}$$
$$+ \ln\left(\frac{L_{S'S} \cdot K_1(T_J) \cdot (V_p(T_J) - V_{th}(T_J))^2}{\tau \cdot V_{REF}}\right)^{\frac{\tau}{2}} \quad (10)$$

where V_{DC} is the DC-link voltage and V_p is plateau voltage. The virtual turn-on delay has a negative temperature coefficient due to the fact that it depends on the MOSFET geometry parameter and the threshold voltage which have negative temperature coefficients. The nature of the temperature coefficient of the virtual turn-off delay cannot be commented directly by looking at Eq. (10) because of the cumulative effect of multiple parameters.

2.1 Simulation of the temperature dependency

Simulation using LTspice is conducted to verify whether these TSEPs exhibit linear or non-linear behavior. The simulation utilizes a 1200 V, 134 A power module (BSM120D12P2C005). The DC link

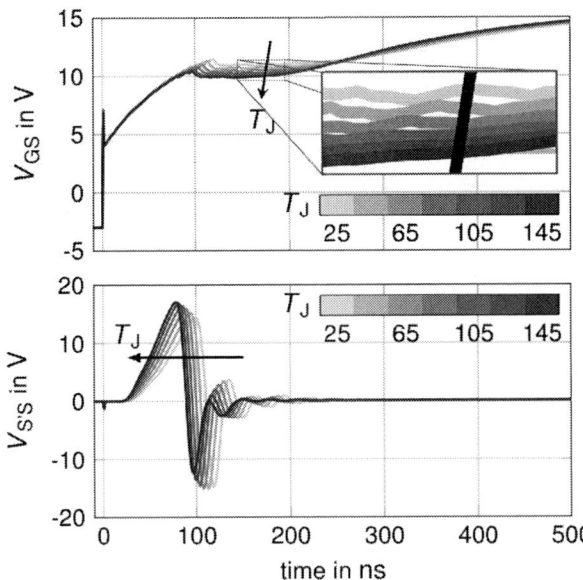

Fig. 3: Simulated results of the turn on at 400 V and 80 A showing variation with respect to junction temperature.

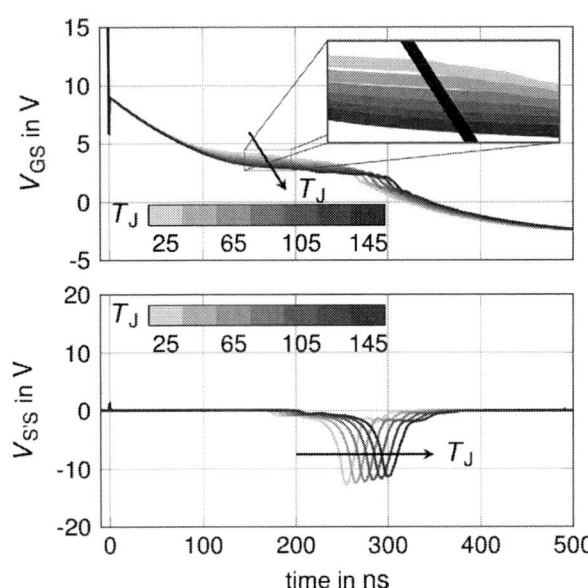

Fig. 4: Simulated results of the turn off at 400 V and 80 A showing variation with respect to junction temperature.

voltage is maintained at a constant 400 V, while a steady load current of 80 A passed through the device during operation. The temperature is changed from 25 °C to 145 °C in 20 K steps. The plots of V_{GS} and $V_{S'S}$ are depicted in Figs. 3 and 4. Observing the plot of $V_{S'S}$ during turn-on in Fig. 3, it is evident that the slope of $V_{S'S}$ increases with rising temperature, leading to a decrease in $t_{vd,on}$ when a V_{REF} of 14 V is utilized. Adhering to the acquisition method detailed earlier, $V_{vth,on}$ also decreases with increasing temperature. Conversely, during turn-off, V_p decreases with rising temperature, prolonging the $t_{vd,off}$. Simulation results illustrate that $t_{vd,off}$ exhibits a positive temperature coefficient when V_{REF} is −14 V , implying an overall positive cumulative impact of temperature on K_1, V_{th} and V_p. For this scenario, the variations of TSEPs with respect to change in T_J are represented in Fig. 5. The sensitivity of $t_{vd,on}$, $V_{vth,on}$, $t_{vd,off}$ and $V_{vth,off}$ is −174 ps/K, −10.8 mV/K, 291 ps/K and −8.64 mV/K respectively.

3 Implementation

The circuit for acquiring virtual threshold-on voltage and virtual turn-on delay is adapted from [12] and [16], and it is modified to acquire the turn-off TSEPs: virtual threshold-off voltage and virtual turn-off delay. To achieve this with just one time-to-digital converter (TDC), logic gates are added to create a pulse generation circuit enabling time interval ex-

traction using lap mode. During operation, the input signal is compared to a reference voltage, and subsequent pulses are generated to calculate virtual turn-on delay and virtual turn-off delay.

Using only a single reference voltage for measurement, the phase of the voltage drop over the parasitic inductance $V_{S'S}$ is altered by 180° using an amplifier circuit to produce $-V_{S'S}$. The choice of the reference voltage depends on the device and system itself (±1.25 V in this work): It must be chosen to be high enough so that a false trigger is prevented and lower than the peaks of the voltage drop over the parasitic inductance waveform.

The PWM signal, utilized for activating the gate driver, acts as the first trigger for the TDC at $t = t_1$ of Fig. 2. The subsequent three pulses are generated under specific conditions: When the voltage drop over the parasitic inductance equals the reference voltage ($t = t_2$), when the falling edge of the PWM signal occurs ($t = t_3$), and when the voltage drop over the parasitic inductance equals the negative reference voltage ($t = t_4$).

Figure 6 presents the block diagram of the implemented circuit. When the voltage drop over the parasitic inductance equals the reference voltage, a D-type flip-flop (D-F/F) is triggered, latching this event. Simultaneously, a differential amplifier feeds the gate-source voltage to a sample and hold cir-

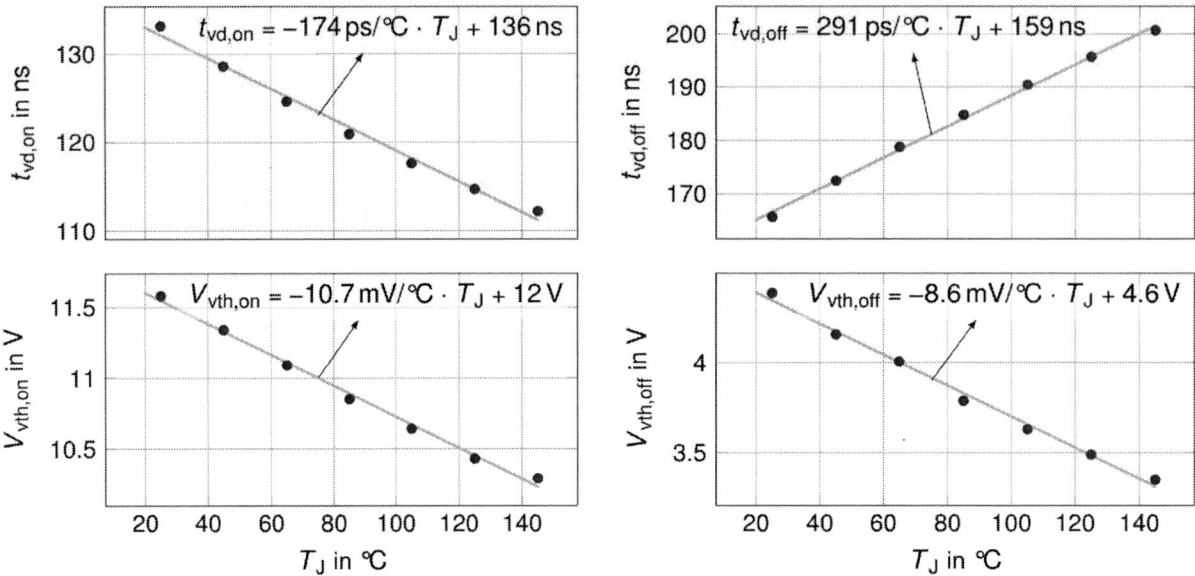

Fig. 5: Simulated results of TSEPs showing variation with respect to junction temperature.

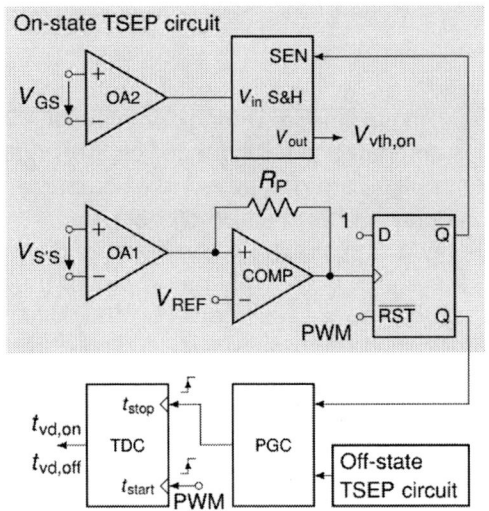

Fig. 6: Block diagram of the connection of the acquisition circuit for the TSEPs.

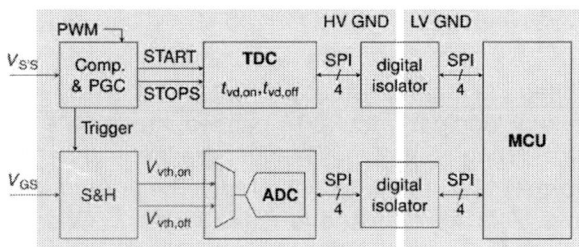

Fig. 7: Block diagram of the connection of the acquisition circuit to the MCU

cuit triggered by the D-F/F output. The moment the output goes high, the voltage at the sample and hold circuit is held, resulting in the acquisition of the virtual threshold-on voltage. Additionally, at this moment, the D-F/F initiates the pulse generation circuit to generate a pulse for the TDC to save the time. The virtual turn-on delay is then measured as the time interval between the rising edge of the PWM and this stop pulse. A similar acquisition concept is applied using a negative reference voltage to obtain the turn-off TSEPs.

To enable readout of the circuit described in Fig. 6, the signals have to be transmitted from the HV GND of the device to a LV GND, where they can be used in a microcontroller unit (MCU) or FPGA. The realized connection in shown in the block diagram of Fig. 7. For all of the parameters a digitization on the HV GND is aimed for due to the higher robustness and speed of a digital isolation interface. The virtual time delay t_{vd} is digitized directly in the TDC7200 TDC and can be transmitted through an SPI interface. For the virtual threshold voltage, a AD7389 16 bit analog-to-digital converter is used as an interface. Both of these ICs are read through independent SPI interfaces by an MCU.

4 Experimental Verification

To verify the proposed acquisition circuit in operation, an H-bridge inverter using the same power modules as the simulation is built. The setup is

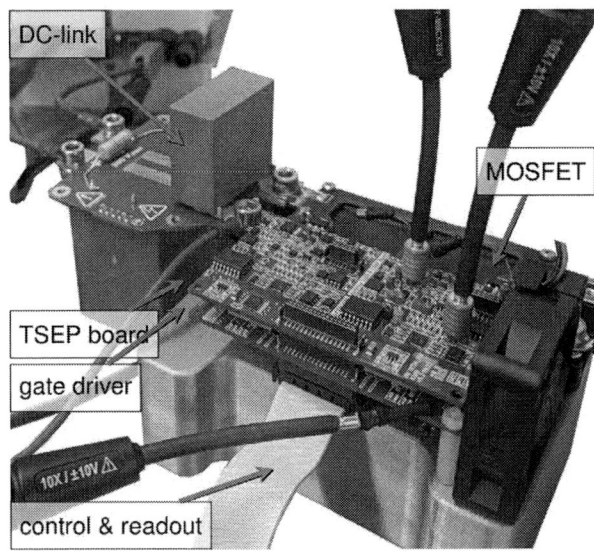

Fig. 8: Image of the double pulse test setup showing the TSEP board connected directly to the gate driver of a 1.2 kV / 120 A SiC half bridge module.

Tab. 2: System parameters for the setup shown in Fig. 8. Device parameters are specified for 25 °C and 18 V.

Parameter	Symbol	Value
DC-link voltage	V_{DC}	400 V
Load current	I_L	80 A
Total DC-link capacitance	C_{DC}	752 µF
Load inductance	L_{load}	100 µH
Switching frequency	f_{SW}	40 kHz
Nominal on-state resistance	$R_{DS,on}$	17 mΩ
External gate resistance	$R_{G,ext}$	4.7 Ω

operated at a constant DC-link voltage of 400 V with short current regulated burst at an amplitude of 80 A, whereby the junction temperature is varied. The external gate resistance is set to 4.7 Ω. An image of one of the half-bridge modules with the acquisition circuit connected to the gate driver is depicted in Fig. 8 with the system parameters summarized in table 2.

The combined waveform of a turn on and off is shown in Fig. 9. As soon as the voltage drop over the parasitic inductance waveform reaches the set trigger values at +1.25 V for the turn-on or −1.25 V for the turn-off, a trigger pulse is generated that is used both by the quasi-threshold voltage circuitry to hold the V_{GS} value and as a STOP signal for the TDC. This way, using the generated signal, a measurement of all four parameters is acquired within

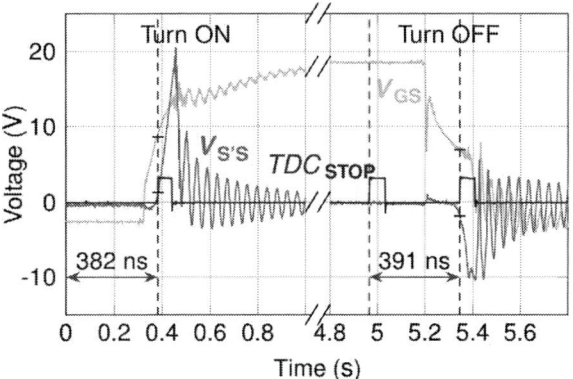

Fig. 9: Turn ON & OFF waveform at 400 V / 100 A, 25 °C. The $V_{S'S}$ waveform reaches the trigger voltage of ±1.25 V at V_{GS} = 8.6 V and 6.9 V respectively. The trigger for the delay calculation is generated at the same time.

Tab. 3: Sensitivity comparison between simulation and measurement of the temperature-sensitive electrical parameters.

TSEP	Simulation	Measurement	Difference
$V_{vth,on}$	−10.7 mV/K	−11.5 mV/K	−6.7%
$V_{vth,off}$	−8.6 mV/K	−14.3 mV/K	−65%
$t_{vd,on}$	−174 ps/K	−153 ps/K	12%
$t_{vd,off}$	291 ps/K	370 ps/K	−27%

one PWM cycle. Attention was given to ensure that the threshold voltage levels do not fall within the Miller Plateau, as dependencies on additional parameters are anticipated.

To use the acquired values as TSEPs, it is essential to calibrate the circuits. The calibration curves for the four TSEPs used for the calculation are shown in Fig. 10.

The temperature sensitivities of all four TSEPs are summarized in table 3. While the overall temperature sensitivity of the measured values agrees with the simulation, a clear deviation is apparent. There are multiple reasons for the unconclusive temperature sensitivity: The device model used in the simulative approach was never optimized to properly reflect the TSEP values. For an optimized simulation, only limited parasitics are applied in the simulation, especially in the gate loop, leading to diverging gate-source voltage waveforms. Furthermore, the choice of reference voltage has a further influence on the sensitivity, as the change in the waveforms of Figs. 3 and 4 with respect to the junction temperature is not consistent. Each active

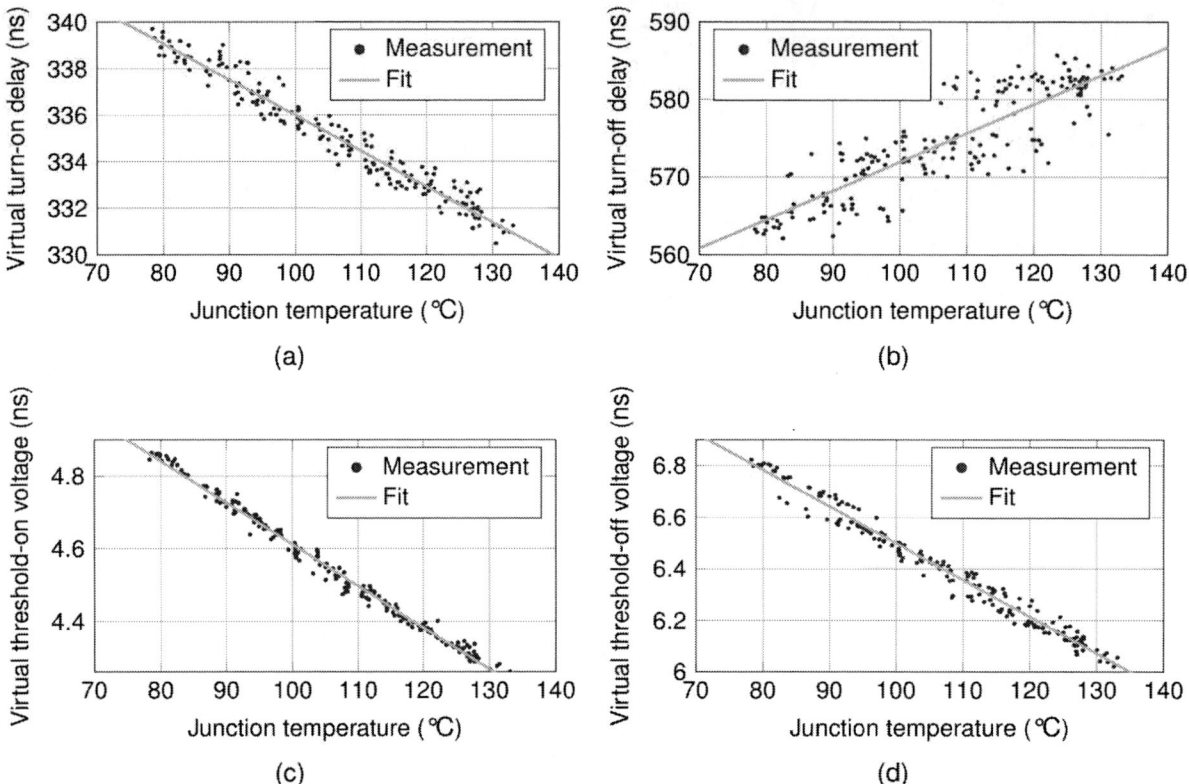

Fig. 10: Measured results of the TSEPs showing their dependency on the junction temperature at 400 V and 80 A.

component in the system as shown in Fig. 7 introduces a slight propagation delay within its signal chain, leading to a certain skew of the gate-source voltage with respect to the voltage drop over the parasitic inductance, causing waveforms that are not properly aligned anymore. In addition, in the simulation the virtual turn-on delay and the virtual turn-off delay are calculated starting from the point the gate-source voltage changes it value at the beginning of the switching. In the implementation, however, the PWM signal is used as a starting reference to improve the precision of the measurement due to its sharper edges when compared to the gate-source voltage. However, all of these differences do not cause any issues with the usage of the parameters as TSEPs.

To analyze the propagation delay in the implementation of the circuit, it is tested with a sine wave as input and the expected value of the TDC is calculated to be 370 ns for the virtual turn-on delay and 155 ns for the virtual turn-off delay. The maximum error from the TDC for the virtual turn-on delay is determined to be 474 ps, resulting in a temperature error of 7.28 K. Similarly, for the virtual turn-off

delay, a maximum error of 1451 ps is found, resulting in a temperature error of 7.39 K. The median error in the virtual turn-on delay and the virtual turn-off delay is found to be 84 ps (1.22 K) and 323 ps (1.67 K) respectively. To verify the acquisition of the MCU, the waveforms are also analyzed with an oscilloscope. There is a small additional error introduced by the limited accuracy of the TDC component used on the board. However, the total deviation from the linearly fitted calibration curve reaches less than 10 K in all cases, save for the virtual turn-off delay. Here, the inaccuracy is much higher than for the other parameters, indicating a systematic error in its acquisition. The implemented measurement mode of the TDC7200 TDC only offers limited accuracy for the time frames used in this work. Therefore, to use this TSEP effectively, post-processing in the form of a moving averaging filter or similar is necessary.

5 Conclusion

The proposed acquisition circuit successfully generates three stop pulses per cycle, enabling the acquisition of both virtual turn-on delay and virtual turn-off delay. Considering the median values, the temperature error remains within the range of 10 K which can be further improved by averaging multiple values. The magnitude of the measured time error depends on the sensitivities of virtual turn-on delay and virtual turn-off delay, with higher sensitivities resulting in reduced errors. This innovative concept holds significant promise for real-time applications, facilitating the simultaneous measurement of all four TSEPs. By obtaining multiple parameters concurrently, the overall error is significantly reduced since it is determined by the root of the sum of squared individual errors. Incorporating this approach in temperature monitoring and characterization of power MOSFETs promises improved performance and reliability across various applications.

References

[1] D. Blackburn, "Temperature measurements of semiconductor devices - a review," in *Twentieth Annual IEEE Semiconductor Thermal Measurement and Management Symposium (IEEE Cat. No.04CH37545)*, Mar. 2004, pp. 70–80. DOI: 10.1109/STHERM.2004.1291304.

[2] F. Yang, E. Ugur, and B. Akin, "Evaluation of Aging's Effect on Temperature-Sensitive Electrical Parameters in SiC mosfets," *IEEE Transactions on Power Electronics*, vol. 35, no. 6, pp. 6315–6331, Jun. 2020. DOI: 10.1109/TPEL.2019.2950311.

[3] K. Sharma, S. Kamm, K. Muñoz Barón, and I. Kallfass, "Characterization of Online Junction Temperature of the SiC power MOSFET by Combination of Four TSEPs using Neural Network," in *2022 24th European Conference on Power Electronics and Applications (EPE'22 ECCE Europe)*, Sep. 2022, pp. 1–8.

[4] F. Stella, G. Pellegrino, E. Armando, and D. Daprà, "Online Junction Temperature Estimation of SiC Power MOSFETs Through On-State Voltage Mapping," *IEEE Transactions on Industry Applications*, vol. 54, no. 4, pp. 3453–3462, Jul. 2018. DOI: 10.1109/TIA.2018.2812710.

[5] U.-M. Choi, F. Blaabjerg, and S. Jørgensen, "Power Cycling Test Methods for Reliability Assessment of Power Device Modules in Respect to Temperature Stress," *IEEE Transactions on Power Electronics*, vol. 33, no. 3, pp. 2531–2551, Mar. 2018. DOI: 10.1109/TPEL.2017.2690500.

[6] E. Ugur, C. Xu, F. Yang, S. Pu, and B. Akin, "A New Complete Condition Monitoring Method for SiC Power MOSFETs," *IEEE Transactions on Industrial Electronics*, vol. 68, no. 2, pp. 1654–1664, Feb. 2021. DOI: 10.1109/TIE.2020.2970668.

[7] K. Muñoz Barón, K. Sharma, and I. Kallfass, "Virtual Junction Temperature Estimation during Dynamic Power Cycling Tests," in *2023 11th International Conference on Power Electronics and ECCE Asia (ICPE 2023 - ECCE Asia)*, May 2023, pp. 3193–3199. DOI: 10.23919/ICPE2023-ECCEAsia54778.2023.10213642.

[8] N. Baker, S. Munk-Nielsen, F. Iannuzzo, and M. Liserre, "IGBT Junction Temperature Measurement via Peak Gate Current," *IEEE Transactions on Power Electronics*, vol. 31, no. 5, pp. 3784–3793, May 2016. DOI: 10.1109/TPEL.2015.2464714.

[9] M. Denk and M.-M. Bakran, "Junction Temperature Measurement during Inverter Operation using a TJ-IGBT-Driver," in *Proceedings of PCIM Europe 2015; International Exhibition and Conference for Power Electronics, Intelligent Motion, Renewable Energy and Energy Management*, May 2015, pp. 1–8.

[10] J. Ruthardt, D. Hirning, K. Sharma, M. Nitzsche, P. Ziegler, *et al.*, "Investigations on Online Junction Temperature Measurement for SiC-MOSFETs Using the Gate-Signal Injection Method," in *2021 IEEE Energy Conversion Congress and Exposition (ECCE)*, Oct. 2021, pp. 5354–5359. DOI: 10.1109/ECCE47101.2021.9595166.

[11] J. O. Gonzalez, O. Alatise, J. Hu, L. Ran, and P. Mawby, "Temperature sensitive electrical parameters for condition monitoring in SiC power MOSFETs," in *8th IET International Conference on Power Electronics, Machines and Drives (PEMD 2016)*, Apr. 2016, pp. 1–6. DOI: 10.1049/cp.2016.0267.

[12] A. Griffo, J. Wang, K. Colombage, and T. Kamel, "Real-Time Measurement of Temperature Sensitive Electrical Parameters in SiC Power MOSFETs," *IEEE Transactions on Industrial Electronics*, vol. 65, no. 3, pp. 2663–2671, Mar. 2018. DOI: 10.1109/TIE.2017.2739687.

[13] K. Muñoz Barón, K. Sharma, M. Nitzsche, P. Ziegler, D. Koch, and I. Kallfass, "Characterization of Threshold Voltage for Application-Oriented Power Cycling Conditions for Wide-Bandgap Power Devices," in *PCIM Europe Digital Days 2020; International Exhibition and Conference for Power Electronics, Intelligent Motion, Renewable Energy and Energy Management*, Jul. 2020, pp. 1–7.

[14] H. Yu, X. Jiang, J. Chen, J. Wang, and Z. J. Shen, "A Novel Real-Time Junction Temperature Monitoring Circuit for SiC MOSFET," in *2020 IEEE Applied Power Electronics Conference and Exposition (APEC)*, Mar. 2020, pp. 2605–2609. DOI: 10.1109/APEC39645.2020.9124486.

[15] Z. Zhang, J. Dyer, X. Wu, F. Wang, D. Costinett, *et al.*, "Online Junction Temperature Monitoring Using Intelligent Gate Drive for SiC Power Devices," *IEEE Transactions on Power Electronics*, vol. 34, no. 8, pp. 7922–7932, Aug. 2019. DOI: 10.1109/TPEL.2018.2879511.

[16] K. Sharma, K. Muñoz Barón, J. Ruthardt, and I. Kallfass, "Online Junction Temperature Monitoring of Wide Bandgap Power Transistors using Quasi Turn-on Delay as TSEP," in *2021 IEEE 8th Workshop on Wide Bandgap Power Devices and Applications (WiPDA)*, Nov. 2021, pp. 129–134. DOI: 10.1109/WiPDA49284.2021.9645132.

[17] D. Peters, T. Aichinger, T. Basler, G. Rescher, K. Puschkarsky, and H. Reisinger, "Investigation of threshold voltage stability of SiC MOSFETs," in *Proceedings of the 30th International Symposium on Power Semiconductor Devices and ICs*, Piscataway, NJ: IEEE, 2018, pp. 40–43. DOI: 10.1109/ISPSD.2018.8393597.

[18] K. Sharma, D. Dayanand, K. Muñoz Barón, J. Ruthardt, F. Münzenmayer, *et al.*, "A Robust Approach for Characterization of Junction Temperature of SiC Power Devices via Quasi-Threshold Voltage as Temperature Sensitive Electrical Parameter," in *2020 IEEE Applied Power Electronics Conference and Exposition (APEC)*, Mar. 2020, pp. 1532–1536. DOI: 10.1109/APEC39645.2020.9124609.

[19] H. Shichman and D. Hodges, "Modeling and simulation of insulated-gate field-effect transistor switching circuits," *IEEE Journal of Solid-State Circuits*, vol. 3, no. 3, pp. 285–289, Sep. 1968. DOI: 10.1109/JSSC.1968.1049902.

[20] S. Kalker, C. H. van der Broeck, and R. W. De Doncker, "Online Junction-Temperature Sensing of SiC MOSFETs with Minimal Calibration Effort," in *PCIM Europe Digital Days 2020; International Exhibition and Conference for Power Electronics, Intelligent Motion, Renewable Energy and Energy Management*, Jul. 2020, pp. 1–7.

PCIM Europe 2024, 11– 13 June 2024, Nuremberg   DOI: 10.30420/566262278

Enhanced Current Measurement Approach for Non-Isolated 6.5 kV Silicon Carbide MOSFETs

Xinyuan Du[1], Ahmed H. Ismail[1], Zhuxuan Ma[1], Yue Zhao[1]

[1] University of Arkansas, USA

Corresponding author: Xinyuan Du, xd006@uark.edu
Speaker: Yue Zhao, yuezhao@uark.edu

Abstract

In this paper, comprehensive dynamic characterizations were conducted for the latest 6.5 kV silicon carbide (SiC) MOSFETs from room temperature to 175 °C. Since the 6.5 kV device sample has non-isolated baseplate, it is challenging to achieve accurate switching characterizations. A custom clamped inductive load (CIL) test platform is developed and optimized for the medium voltage (MV) SiC devices to achieve the high-bandwidth noise-immunity measurement. The results of the reverse recovery current measured by Rogowski coil, current probe, and current viewing resistor (CVR) are compared and analyzed. Body diode performance under different junction temperature is evaluated.

1   Introduction

The medium voltage (MV) silicon carbide (SiC) MOSFETs, e.g., 3.3 kV, 6.5 kV or higher, are gaining increasing interest in the MV and high power applications such as solid-state transformers, renewable and grid-tied applications. With the development of the MV SiC MOSFETs technology, the body diode performance is essential to evaluate the device performance. Double pulse test (DPT) is widely used for the dynamic characterization of power devices. Different from the low voltage (LV) DPT, the high voltage level (> 1000 V) and high dv/dt have a significant impact on the measurement of MV DPT.

This paper discusses the reverse recovery performance of the latest 6.5 kV SiC MOSFETs up to 175 °C. Different from the low voltage (LV) SiC devices, the reverse recovery of SiC MOSFET body diode cannot be neglected due to a large capacitive charge stored on C_{oss}. Meanwhile, due to high dv/dt and voltage level, the MV double pulse test (DPT) is challenging compared with the LV DPT and needs a high-bandwidth noise-immunity measurement [1]. In this paper, the results of three different current measurement probes, which use Rogowski coil, current probe, and Current Viewing Resistor (CVR), are compared with each other. The reverse recovery charge, i.e., Q_{rr}, at different junction temperature is analyzed.

2   Current measurement Methods

2.1   Setup

The custom clamped inductive load (CIL) setup is shown in Fig. 1 (a). A 0-4 kV dc power supply is adopted to establish the dc bus with voltage up to the rated 3.6 kV. To evaluate the body diode performance at different junction temperatures, the hotplate is used to regulate the case temperature. The baseplate of the SiC MOSFETs is drain terminal. Both SiC MOSFETs are attached to the hotplate through two half inch thick aluminum nitride (AlN) ceramic sheets for electrical insulation and low parasitic capacitance to the ground. A 2.5 mH MV air core inductor with single layer winding is built to minimize the parasitic capacitance of the inductor. A 10 Ω external gate resistance for turn on and 8 Ω for turn off are selected for the devices. The 6 kV differential voltage probe and the IsoVu Tektronix TIVP05L are used to measure the drain-to-source voltage V_{ds} and the gate-to-source voltage V_{gs}, respectively. Three current measurement methods are implemented and compared in the tests, which are Rogowski coil CWT Mini50HF/03, the current probe Tektronix TCP0030A and current viewing resistor (CVR) T&M SSMA-414-10 with IsoVu Tektronix TIVP05L.

Two 6.5 kV SiC MOSFETs are shown in Fig. 1 (b). The baseplate is the drain pad of the

1987

MOSFET. Terminal 1 and 2 are the gate and source terminal of the MOSFET. These two terminals are mounted on the DPT board and the drain pad is connected to the board by copper foil.

(a)

(b)

Fig. 1   (a) Custom CIL setup with device heating features; (b) devices under test.

2.2 Current Measurement Comparison

The device under test (DUT) is the upper MOSFETs of the DPT setup. The drain current of DUT is measured by three different methods which are CVR + IsoVu, current probe and Rogowski coil. Fig. 2 shows the double pulse test results of the 6.5 kV SiC MOSFETs under the dc link voltage V_{dc} of 800 V, 1800 V and 3600 V. In Fig. 2 (a), the currents measured by three different probes are consistent with each other. The oscillation on the diode current I_{diode} is caused by the relatively larger

C_{oss} under LV. In Fig. 2 (b), the diode current I_{diode_Rog} which is measured by the rogowski coil shows a smaller peak value of the reverse recovery when the dc link voltage of 1800 V is applied. Since the Rogowski coil is easy to be coupled with noise, the oscillation of the measured current is more obvious than others. When the dc link voltage rises to 3.6 kV, besides the oscillation and the peak reverse recovery current mismatching, an offset of the measured current can be observed in Fig. 2 (c).

It can be observed that the current measured by the Rogowski coil becomes more and more distorted with the increasing of V_{dc}. A detailed figure of the reverse recovery is shown in Fig. 2 (d). When the inductor current flows through the body diode, the current measured by Rogowski coil is around 2 A higher than those measured by CVR and current probe. The peak reverse recovery current measured by Rogowski coil is around 2.4 A lower than the others. Owing to the high bandwidth and good noise immunity, the results of the CVR and current probe are consistent from 800 V to 3600 V.

The performance of these three current measurement methods are summarized in Table 1. For CVR with IsoVu, it has high band width, low noise sensitivity but high cost. The Rogowski coil is vulnerable under MV due to high noise sensitivity.

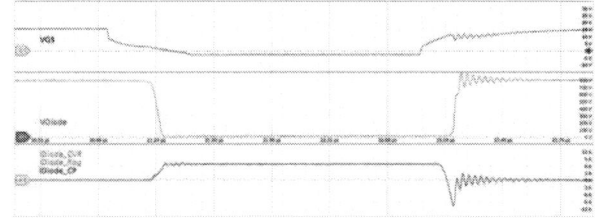

(a)

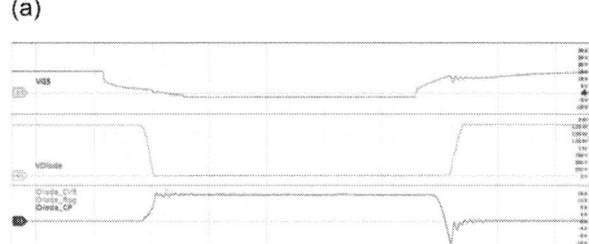

(b)

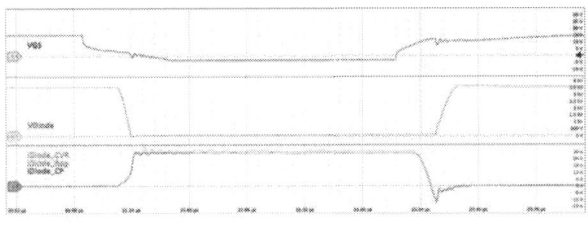

(c)

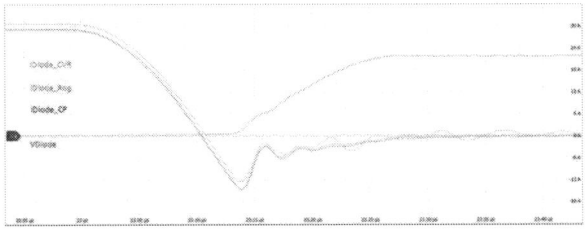

(d)

Fig. 2 Experimental waveforms using different probes to measure the body diode current. (a) Vdc = 800V. (b) Vdc = 1800V. (c) Vdc = 3600V. (d) reverse recovery at 3600V.

	CVR + IsoVu	Current probe	Rogowski coil
Band width	High	medium	Low
Noise sensitivity	Low	Low	High
Connection difficulty	Moderate	Moderate	Easy
Cost	expensive	cheap	cheap

Table 1 Summary of the current measurement methods

3 DPT Results

3.1 Results

To evaluate the body diode performance at different junction temperatures, the DPT is conducted under 25 °C, 125 °C, 150 °C and 175 °C. Fig. 3 shows the reverse recovery current measured by Rogowski coil and current probe. Due to low bandwidth and noise sensitivity, the current waveforms measured by Rogowski coil are distorted so that the calculation of the reverse recovery charge Q_{rr} based on this data is not accurate. However, the current waveforms measured by current probe shows a strong correlation with the junction temperature, which is consistent with the theoretical analysis. It can be found that the reverse recovery of the body diode worsens with the rise of the junction temperature.

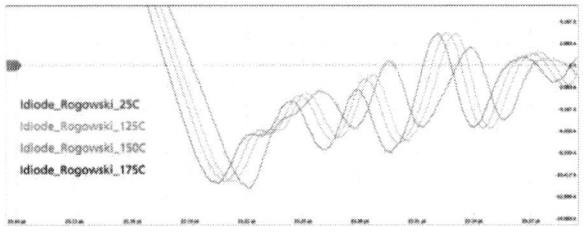

(a)

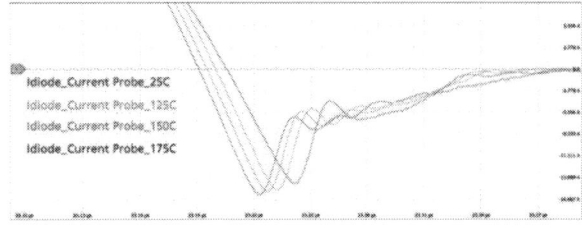

(b)

Fig. 3 Experimental waveforms using different probe to measure the body diode current: (a) Rogowski coil and (b) current probe.

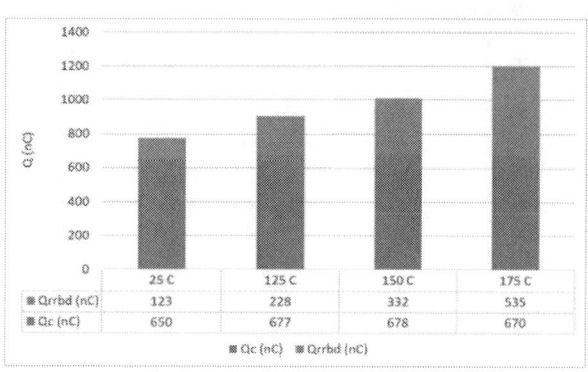

Fig. 4 Measured Q_{rr} of the SiC MOSFETs under different junction temperatures

Fig. 4 shows the Q_{rr} of the body diode at different junction temperature. The measured Q_{rr} consists of the charge stored on C_{oss} and the actual body diode reverse recovery charge Q_{rrbd}. The capacitive charge Q_c, which is the charge stored on the C_{oss}, can be calculated by the first turn-on instant of the MOSFETs. As temperature increases, Q_{rrbd} rises from 165 nC to 336 nC. But Q_c does not change obviously since C_{oss} and the capacitance between the device's baseplate and PCB change slightly with temperature.

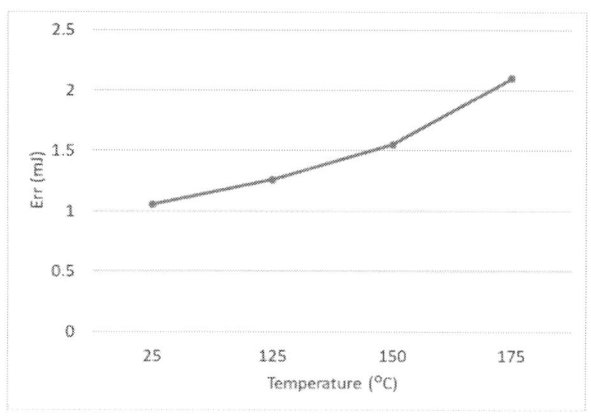

Fig. 5 Reverse recovery energy E_{rr} of the SiC MOSFETs under different junction temperatures

In Fig. 5, the results of the reverse recovery energy at different junction temperatures are shown. Compared with the reverse recovery energy at room temperature, the E_{rr} is doubled at 175 °C

4 Conclusion

This paper proposed a high-bandwidth noise-immunity measurement for reverse recovery of the latest 6.5 kV SiC MOSFETs up to 175 °C. A custom CIL setup is built to reduce the parasitic inductance while accommodating the clearance and creepage requirements for MV SiC device testing. The reverse recovery current measured by Rogowski coil, current probe, and CVR are compared with each other. The current probe and CVR can be a good candidate to achieve high-bandwidth noise immunity measurement. Using the current probe, the relationship of Q_{rr} and E_{rr} with different junction temperatures is analyzed.

References

[1] H. Li, Z. Gao, R. Chen and F. Wang, "Improved Double Pulse Test for Accurate Dynamic Characterization of Medium Voltage SiC Devices," *IEEE Transactions on Power Electronics*, vol. 38, no. 2, pp. 1779-1790, Feb. 2023.

[2] C. Qian, Z. Wang, D. Zhou, Y. Ge, Y. Zhou, X. Yan, G. Xin and X. Shi, "Investigation of Reverse Recovery Phenomenon for SiC MOSFETs in High-Temperature Applications," *IEEE Transactions on Power Electronics*, vol. 38, no. 11, pp. 14375-14387, Nov. 2023.

[3] D. P. Nayak, R. K. Yakala, M. Kumar and S. K. Pramanick, "Temperature-Dependent Reverse Recovery Characterization of SiC MOSFETs Body Diode for Switching Loss Estimation in a Half-Bridge," *IEEE Transactions on Power Electronics*, vol. 37, no. 5, pp. 5574-5582, May 2022.

PCIM Europe 2024, 11– 13 June 2024, Nuremberg   DOI: 10.30420/566262279

New 2 kV SiC-MOS Technology for Application Fields in the Industrial Landscape

Carlos D. Fuentes [1], Igor Kasko [1], Keigo Minode [1], Ryo Yoshida[2], Seiya Nakazawa[2], Tomonori Hoki[2], Andreas Thamm[1], Christian Felgemacher [1]

[1] ROHM Semiconductor GmbH, Germany
[2] ROHM Co., Ltd., Japan

Corresponding author: Carlos D. Fuentes, carlos.fuentes@de.rohmeurope.com
Speaker: Igor Kasko, igor.kasko@de.rohmeurope.com

Abstract

A new 2 kV SiC-MOSFET based on ROHMs 4th generation trench cell has been developed to support the growing requirements of renewable energy applications driven by the photovoltaic sector. The longevity of solar panels, the search for higher efficiencies and the range of conditions that outdoor-mounted converters must withstand impose tough requirements on converter manufacturers and module makers alike, which in turn require semiconductor devices up to the task. Results show that this device presents excellent reliability and humidity robustness characteristics while portraying the low loss characteristics proper of ROHMs 4th gen trench technology, enabling simpler two-level topologies with 1500 V DC-Links.

1 Introduction

Renewable energies have been and are continuously at the forefront of key solutions to address current energy demands while tackling the environmental issues of our time. Solar Photovoltaic (PV), as one of the leading components of the renewable mix in regards of yearly new installed capacity [1], has seen substantial growth in the last few years due to its constant reductions in cost, improvements in efficiency and ever more competitive leveraged costs of electricity when compared to traditional fossil fuel alternatives in several parts of the world. These advantages have been enabled by advancing solar panel technology, the introduction of new converter topologies and the development of new semiconductor technologies over the years. Among these trends in utility scale, $1500\,V$ DC voltage has quickly grown in popularity, as it reduces installation costs, improves plant flexibility (string-wise, curtailment-wise, and temperature-wise), and reduces losses by the same amount of power [2]. For the inverter to work with this higher DC-Link voltage, solutions have mainly relied on multilevel topologies based on Si-IGBTs and have been improved over the years by hybridizing these approaches using SiC devices or going for full-SiC

solutions to improve overall efficiency and reduce system costs. To address the target of using $1500\,V$ DC-Links, multilevel topologies have been preferred by manufacturers. More simple topologies must rely on serialization of semiconductors [3] (which is possible, but at the cost of additional complexity) due to the lack of available voltage classes between $1700\,V$ and $3300\,V$. This while at the same time not providing the advantages multilevel converters bring to the table in power quality, filter requirements, etc.

Technological improvements are not always straightforward to implement, as every step is met with ever increasing reliability requirements. Solar inverters are desired to fulfill 25 years of lifetime operation [4] with corresponding mission profiles to match solar panel longevity in alternating temperature and humidity conditions. These reliability goals are among the highest in industry, arguably higher than automotive applications, and hence careful consideration of several factors such as, for example, cosmic radiation robustness, gate oxide reliability and humidity robustness is necessary.

The 2 kV - 2.3 kV voltage class has already been released to the market in the forms of Si-IGBT and SiC-MOSFETs to cater to the needs of the industrial sector [5, 6]. To further support this trend ROHM introduces its new 2 kV device voltage class.

In this paper, a 2 kV SiC-MOSFET reference device is presented. The corresponding demonstration device has been developed with an integrated gate resistance to address in-module die parallelization requirements for high-power inverters. Although in principle designed with PV applications in mind, the usage of this device into alternative industrial applications such as, for example, wind turbines, auxiliary power supplies and energy storage, among others, would also be possible.

ROHM 2 kV technology has been developed based on 4th generation ROHM SiC-MOSFET trench technology [7], extending operation voltage range and optimizing the MOSFET cell structure and edge termination to operate in this voltage class. First, the main characteristics of this new device are presented, with a focus on key static parameters and dynamic behavior. Subsequently, a system-level performance evaluation is performed. Afterwards, key reliability tests results indicating the suitability of this device to fulfill the demanding requirements of PV applications are introduced. Finally, analysis of cosmic radiation robustness under a given mission profile is presented.

2 Device Characteristics

The device that is presented here is based on ROHMs 4th generation SiC-MOSFETs. This particular demonstrator device has been optimized for parallelization targeting module usage. Therefore, its switching speed has been limited by design, presenting an equivalent internal gate resistance higher than $10\,\Omega$. Devices designed to operate as single switches can switch significantly faster than the results that are presented in this work. In this section selected static characteristics of the device are presented, to then focus on dynamic behavior characterization and results. Finally, a system level performance assessment is presented by means of simulating a three-phase two-level voltage source inverter based on the device with a PLECS model created from the obtained results.

2.1 Static characteristics

The presented static values in this paper were obtained by measurement on technology samples and are therefore considered as reference values only. These results can be found in Tab. 1. Interesting remarks of this device are its high threshold voltage, low C_{GD}/C_{GS} ratio (reduced PTO risk), low gate charge, and blocking voltage capability rated up to $-40^\circ C$ ambient temperature.

Param.	Value	
$R_{DS(on)}$	$22\,m\Omega$	@$V_{GS} = 18\,V$, $I_D = 33\,A$
$V_{BDSS,min}$	$2000\,V$ for $T_{amb} >= -40^\circ C$	
$T_{j,max}$	$175 - 200^\circ C$	
$V_{GS(th)}$	$4\,V$	@ $I_D = 40\,mA$
V_{SD}	$3.8\,V$	@ $I_D = 33\,A$
C_{GS}	$6900\,pF$	$\mid V_{DS} = 1500\,V$
C_{DS}	$111\,pF$	@$\mid V_{GS} = 0\,V$
C_{GD}	$6\,pF$	$\mid f = 100\,kHz$
Q_g	$240\,nC$	

Tab. 1: Static characteristics of the 2 kV sample device ($T_{vj} = 25^\circ C$ unless otherwise specified).

2.2 Dynamic characteristics

To dynamically characterize this device, a double pulse test experiment has been carried out with two devices in half bridge configuration using a load inductor parallel to the device that is operated as free-wheeling diode as described in [8]. The device was packaged in an experimental package, similar to a TO-247-4L, and was characterized using the Double Pulse Test (DPT) test bench presented in Fig. 1. The test bench features an 8 kV voltage source, three $250\,\mu H$ air-core inductors, temperature control by means of a hotplate, a DC-Link featuring electrolytic and film capacitors adding to a total capacitance of $1.87\,mF$ and an optical connection to enable remote operation. Pulses are sent to the pulse generator from a computer. A summary of the used probes can be found in Tab. 2. In all DPT measurements only the low side DUT was investigated, while the switched device was alternated between high side MOSFET and low side MOSFET. When the low side MOSFET was switched, the load inductor was connected between midpoint and DC+, and then the low side MOSFET was characterized. When the high side MOSFET was switched, the inductor was connected between midpoint and DC-, and then the low side body diode was characterized. Dynamic characterization was performed with DC-Link voltages ranging from $700\,V$ to $1500\,V$ DC and junction temperatures up to $150^\circ C$. Selected measurement points (as summarized in Tab. 3) are presented in this paper.

2.3 Dynamic characterization results

The corresponding dynamic characterization results presented here were measured through DPT events as per the parameters described in Tab. 3. In Fig. 2 the turn-on event of the SiC-MOSFET

PCIM Europe 2024, 11– 13 June 2024, Nuremberg   DOI: 10.30420/566262279

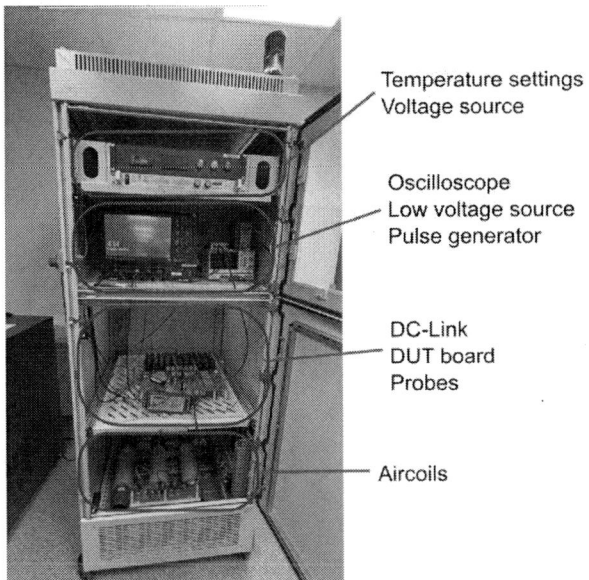

Fig. 1: Description of the high voltage double pulse test (DPT) test bench.

Param.	Probe	Comments
$V_{DS,LS}$	PHVS 662-6	$B_W = 400\,MHz$
$V_{GS,LS}$	HVFO0103	Optical probe $60\,MHz$
$I_{DS,LS}$	$100\,m\Omega$ coax. Shunt	$B_W = 2\,GHz$
$V_{DS,HS}$	TT-SI 9110	Diff. Probe $100\,MHz$

Tab. 2: Probes employed for the characterization of the SiC-MOSFET.

at several currents ranging from 10 to 160 A can be observed. Due to its optimization for parallel connection, the device features a slower transition than standard 4G devices, which is desired for parallel connection of several dies in module and produces less excitation for oscillations. The di/dt and dv/dt while during the transition at operation point $V_{DS} = 1500\,V$, $I_{DS} = 160\,A$, and $25°C$ are $3\,A/ns$ and $28\,V/ns$ respectively.

In Fig. 3, the diode turn-off event (corresponding to

Parameters	Value
I_{DS}	10 to 160 A
V_{DS}	1500 V
V_{GS}	$-2\,V$, 18 V
T_{vj}	$25°C$ and $150°C$

Tab. 3: Parameters and measurement range of presented DPT Experiments results.

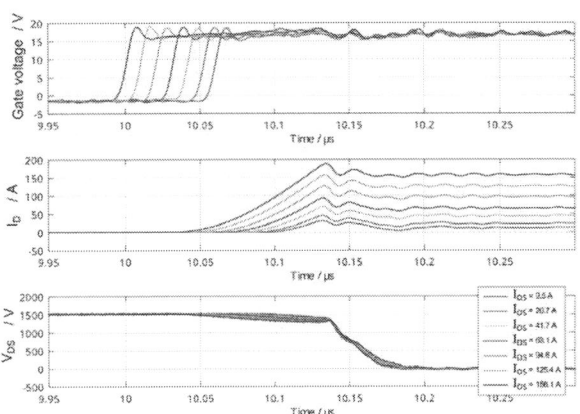

Fig. 2: MOSFET Turn-on switching waveforms at $V_{DC} = 1500\,V$ and $T_{vj} = 25°C$. From top to bottom: V_{GS} voltage, I_{DS} current and V_{DS} voltage. $V_{GS(off)} = -2\,V$, $V_{GS(on)} = 18\,V$.

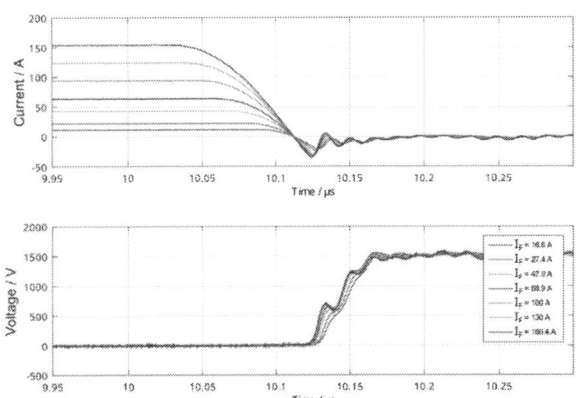

Fig. 3: Body diode Turn-off switching waveforms at $V_{DC} = 1500\,V$ and $T_{vj} = 25°C$. From top to bottom: I_F current and V_{DS} voltage. $V_{GS(off)} = -2\,V$, $V_{GS(on)} = 18\,V$.

MOSFET turn-on event) can be observed. Here a slightly higher ringing on the waveforms can be observed when comparing with MOSFET waveforms. These ringing oscillations ($78\,MHz$) originate from the circuit stray inductance resonating with the output capacitance of the blocking device in the loop. Overall, transitions here are fairly clean with almost zero reverse recovery effects as it is characteristic for SiC technology.

In Fig. 4 the turn-off event of the SiC-MOSFET can be observed. There it can be observed that turn-off events are very smooth, with almost no ringing observed across full current range. The di/dt and dv/dt during these transitions at $V_{DS} = 1500\,V$, $I_{DS} = 160\,A$ and $25°C$ were $5\,A/ns$ and $37\,V/ns$ respectively.

1993

PCIM Europe 2024, 11– 13 June 2024, Nuremberg   DOI: 10.30420/566262279

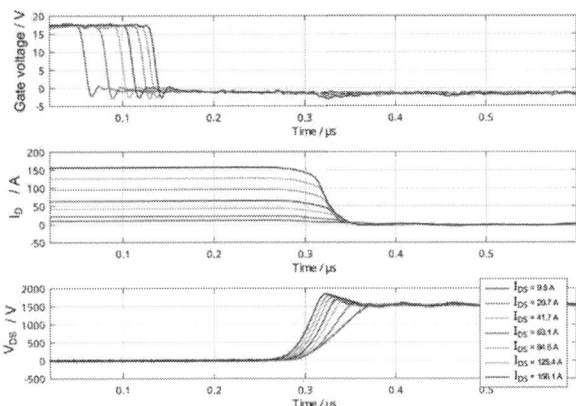

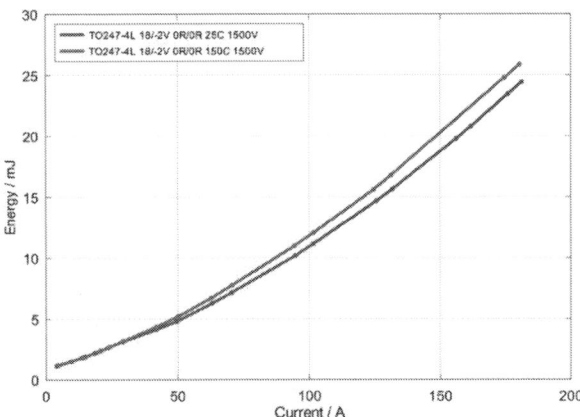

Fig. 4: Turn-off switching waveforms at $V_{DC} = 1500\,V$ and $T_{vj} = 25°C$. From top to bottom: I_{DS} Current and V_{DS} Voltage. $V_{GS(off)} = -2\,V$, $V_{GS(on)} = 18\,V$.

Fig. 5: Total switching losses (Turn-on, Turn-off and reverse recovery losses) at $V_{DC} = 1500\,V$ and T_{vj} = 25°C and 150°C. $V_{GS(off)} = -2\,V$, $V_{GS(on)} = 18\,V$.

Finally, a summary of the added total switching loss (turn-on, turn-off and reverse recovery loss) can be observed in Fig. 5. There, overall switching losses at the maximum characterized current are close to $25\,mJ$ total, and the temperature dependency of the losses, although slightly inclined to increase with temperature, shows little dependency on it. Furthermore, when considering that the maximum rating of the device should be close to half the measured currents in DPT test events, the effect of temperature takes an even less relevant role. The reason for this behavior rests in the fact that in ROHM 4G SiC MOSFETs, turn-on energy losses decrease with temperature, while turn-off losses and reverse recovery losses increase with temperature, balancing overall loss behavior versus temperature.

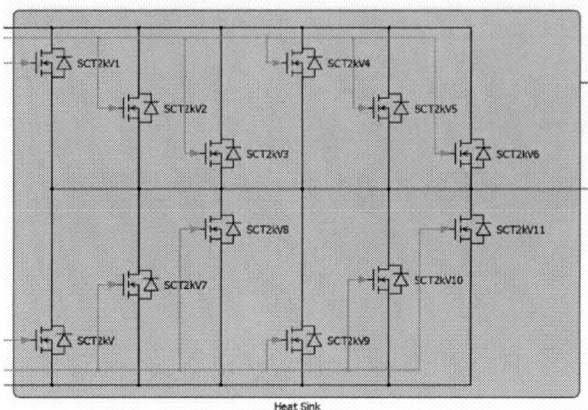

Fig. 6: Model of a single module representing one leg of the simulated inverter.

2.4 Inverter performance simulation

With these results and the measured static characteristics of the device, a PLECS model has been developed with the goal to simulate the device performance in converter operation. To that end, the efficiency of a three-phase two-level voltage source inverter (3Ph-2L-VSI) over several currents has been calculated. The converter operates three half-bridge modules with six devices per switch (twelve per module). Only one of the three modules has been simulated for simplicity, and it can be found in Fig. 6.

To create the simulation some assumptions have been necessary. The simulation model assumes a fixed modulation index of $m = 1$, and the efficiency is modeled while assuming $cos(\phi) = 1$ in inverter operation (power to grid) as it presents the worst case scenario (less utilization of active rectification). A case temperature of $120°C$ has been assumed for the purposes of this simulation, and the R_{th} and C_{th} values of the cauer model for the case-to-junction temperature have been obtained by assuming a thermal stack based on a standard module set-up. Current distribution in the module is assumed perfectly symmetrical as well. A summary of the simulation parameters is presented in Tab. 4, and results can be found in Fig. 7.

As it can be observed in Fig. 7, a converter based on these devices under the defined constrains would present high efficiency values across simulated power range, above 99 % (only considering semiconductor losses). Moreover, the switching speed can still be improved, as the equivalent internal gate resistance in the current sample is over

1994

Param.	value
Topology	3Ph-2L-VSI
DC-Link voltage	$1500\,V$
Switching frequency	$20\,kHz$
Converter peak power	$472.5\,kW$
$V_{GS,op}$	$[-2\,V\,,\,18\,V]$
Dead time	$500\,ns$
$Cos(\phi)$	1, inverter operation
Peak current per MOS	$70\,A$

Tab. 4: Simulation parameters considered in the corresponding PLECS simulation.

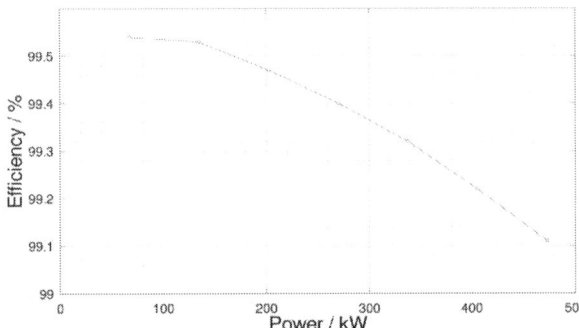

Fig. 7: Efficiency of a three-phase, two-level standard voltage source inverter using three modules per leg based on twelve 2 kV devices each. Results are according to the parameters described in Tab. 4. Only semiconductor losses are considered in this calculation.

10 Ω. At the highest tested current, the average T_{vj} in the devices was $143.8°C$.

3 Reliability tests

Considering that the presented 2 kV devices are to target industrial applications, and particularly among them, photovoltaic applications, the long-term reliability requirements including particularities related to outdoor conditions needed to be examined. To demonstrate the aptitude of the device to comply with these reliability requirements, reliability tests have been carried out. As reference, AQG-324 will be referred upon, as it is used as reference by industry regarding reliability in general. However, this being a standard of automotive nature, tests have been extended to consider the extended longevity that industrial applications require. Related to the demands of the application: HTGB+/-, HTRB and HV-H3TRB have been performed and additionally, an analysis of cosmic radi-

ation robustness under an example of a plausible mission profile is also presented.

The summary of mentioned reliability tests can be found in Tab. 5. There, all parameters that measure beyond reference are presented in the column "extended parameters".

3.1 HTGB

The objective of High Temperature Gate Bias HTGB+/- is twofold. To evaluate if the device threshold voltage drifts due to gate stress and to estimate the oxide lifetime at operation voltage. In this case, both HTGB+/- tests were performed using devices packaged in discrete 3-pin packages for test purposes. Results show no failure after 1100 hrs. and all relevant parameters rested within test specification requirements during and after test.

3.2 HTRB

High Temperature Reverse Bias is a test dedicated to determine the weakpoints in the chip passivation and edge termination over time. In this case, the test has been extended to apply $T_{vj} = 200°C$ for an additional 100 hrs., negative gate bias has been applied to stress the gate oxide further, and nominal blocking voltage has been used for the total duration of the test. Devices were tested while packaged in standard industrial modules with silicon gel and 30 devices were tests in this manner. Results show no failure after 1100 hrs. and all relevant parameters rested within test specification requirements during and after test.

3.3 HV-H3TRB

High Voltage-High Humidity High Temperature Reverse Bias (HV-H3TRB) is a test dedicated to detect weak points in modules due to humidity degradation while under voltage. At the chip level, it also stresses the passivation layer and the edge termination of the chips, while under the influence of humidity that has permeated the medium in which the devices are.

HV-H3TRB has been tested on the device while packaged in standard industrial modules with silicon gel. Over 90 chips were tested within the full available time thermal chambers were available for the purpose, reaching over 1900 hrs. Results show no failure after testing time and all relevant parameters rested within test specification requirements during and after test.

Test	Standard Parameters	Extended Parameters	Result
HTGB+	$V_{GS} = V_{GS,on,max}$, $T = T_{j,max}$, $t = 1000\,hrs$	$+T = 200°C$, $t = 100\,hrs$	Passed
HTGB-	$V_{GS} = V_{GS,off,min}$, $T = T_{j,max}$, $t = 1000\,hrs$	$+T = 200°C$, $t = 100\,hrs$ $V_{GS} < V_{GS,off,min}$	Passed
HTRB	$V_{DS} = 0.8\,V_{DS,max}$, $V_{GS} = 0\,V$, $T = T_{j,max}$, $t = 1000\,hrs$	$+T = 200°C$, $t = 100\,hrs$ $V_{GS} = V_{GS,off,min}$, $V_{DS} = V_{DS,max}$	Passed
HV-H3TRB	$V_{DS} = 0.8\,V_{DS,max}$, $RH = 85\%$, $T = 85°C$, $t = 1000\,hrs$	$t = 1900\,hrs$	Passed

Tab. 5: Preliminary results of reliability tests and associated parameters. Standard parameters are extracted from automotive design guideline AQG-324 requirements, and extended parameters were set to consider the longevity that industrial applications require.

3.4 Cosmic radiation robustness

Single event burnouts occur due to neutrons hitting semiconductor atoms in the device drift region while it is in blocking state. These neutrons, which are part secondary radiation from cosmic rays, if carrying enough energy have an associated probability of hitting an atom in the crystal lattice and create a plasma channel within the semiconductor that leads to device failure. This failure mechanism is relevant for power electronic applications in general, but its particularly important for applications targeting high longevity, as they will inevitably observe longer exposure to neutrons while under voltage. The probability that this failure mechanism occurs depends on neutron flux (altitude, geomagnetic conditions, solar activity) and semiconductor parameters. The most relevant among them being the blocking voltage the device experiences while exposed to neutrons. This failure mode, after reaching a determined voltage from which the failure has enough energy to occur, increases its chances of occurring in an exponential manner as voltage increases [9]. To consider this effect, accelerated tests in particle accelerators have been performed to estimate the failure rate for a given voltage with a defined level of confidence, to then use mission profiles and necessary semiconductor area per converter to estimate the corresponding failure rate the converter would present due to this effect.

ROHM has consistently developed devices that fare well against cosmic radiation robustness, and this experience has also been applied to the 2 kV device. To demonstrate this, two efforts are presented: First, a comparison against competitors using normalized nominal voltages is shown as reference. Second, an evaluation for a virtual inverter with a given plausible mission profile of a

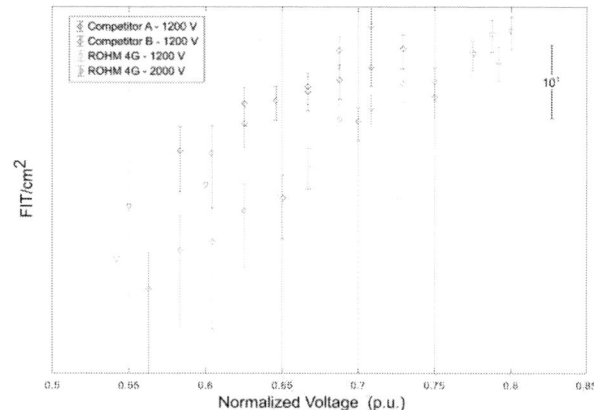

Fig. 8: Qualitative log-scale reference of failure rates for ROHM 4G 2 kV, ROHM 4G 1200 V and competitor devices normalized by nominal blocking voltage. One order of magnitude is shown as reference at the top right section of the figure.

PV inverter has been performed.

The comparison can be found in Fig. 8. There it can be observed that ROHM 4G devices fare well against cosmic radiation when compared to competitor devices in the same voltage class. ROHM 4G 1200 V devices can present FIT rates of around one order of magnitude lower for the same tested voltage than some of the competitors presented in this analysis. The 2 kV device have been designed under the guidelines of 4G technology, and when normalized by nominal breakdown voltage it can be observed that it matches the robustness of ROHMs 1200 V devices at its own voltage class.

Finally, a converter design and a yearly mission profile are assumed to estimate its failure rate due to this effect. In this case, the converter is assumed to be a standard three-phase two-level inverter, targeting $3.6\,m\Omega @ T_{vj} = 25°C$. In other words, three half-bridge modules are used with 6 devices in parallel on each switch. This yields 36 semiconductors in operation. During nights the DC-Link is assumed discharged. Therefore, half of the year is considered to occur in this manner. The other 4380 hrs., equivalent to the other half of the year, are distributed as in the mission profile shown in Fig. 9.

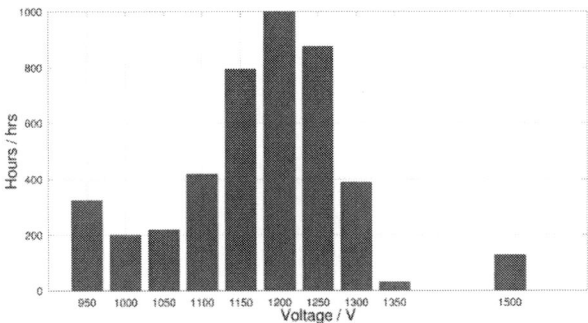

Fig. 9: Proposed mission profile during 12 daylight hours of a PV application with a $V = 1500\,V$ DC-Link.

The calculation of the failure rate λ is described by the expression:

$$\lambda = S \cdot \frac{\Sigma_i^n \left(expfit(V_i) \cdot Hours(V_i) \right)}{\Sigma Hours} \qquad (1)$$

where S is the amount of semiconductors blocking voltage at any given time, expfit is an exponential fit of the upper confidence intervals (confidence = 90%) evaluated at V_i, and hours are the given hours that the converter stays at the given voltage V_i. Using the failure rate results obtained for the 2 kV device, the result yields

$$\lambda = 1.67\,FIT \qquad (2)$$

for the whole converter at sea level, due to this effect alone. This is a very low FIT value. To put it in context, in 25 years of operation as per mission profile in Fig. 9, the converter should have an associated failure rate of 359 PPM due to cosmic radiation effects at sea level.

4 Conclusion

In this paper a new 2 kV trench SiC-MOSFET device prototype based in ROHMs fourth generation technology is presented. The device has been designed with PV applications in mind, as it is a representative industrial application with high, long-term reliability requirements. To demonstrate the operation and suitability of the device for the application, device characterization has been performed

and theoretical performance has been evaluated through simulations using models based on the measured data. Regarding reliability requirements: extended HTGB, HTRB and HV-H3TRB have been performed and passed successfully. Finally, analysis of cosmic radiation robustness under a defined mission profile have been performed for the device, showcasing high cosmic radiation robustness characteristics and low failure rates under the proposed mission profile. Overall, the presented results endorse the suitability of the 2 kV prototype for this and similarly demanding industrial applications requiring this voltage class.

References

[1] R. Secretariat, "REN21. renewables 2023 global status report collection, renewables in energy supply," Tech. Rep., 2023.

[2] E. Serban, M. Ordonez, and C. Pondiche, "DC-Bus voltage range extension in 1500 V photovoltaic inverters," *IEEE Journal of Emerging and Selected Topics in Power Electronics*, vol. 3, no. 4, pp. 901–917, 2015.

[3] C. Noeding, C. Felgemacher, B. Dombert, and P. Zacharias, "Advantages of IGBT series connection in 1.500 V PV inverters," in *Proceedings of PCIM Europe 2015; International Exhibition and Conference for Power Electronics, Intelligent Motion, Renewable Energy and Energy Management*, 2015, pp. 1–8.

[4] Z. Tang, Y. Yang, and F. Blaabjerg, "Power electronics: The enabling technology for renewable energy integration," *CSEE Journal of Power and Energy Systems*, vol. 8, no. 1, pp. 39–52, 2022.

[5] T. Radke, E. Wiesner, K. Masuda, M. Miyazawa, and S. Miyahara, "2000 V class LV100 IGBT module enabling higher power density and design simplification in renewable 1500 V inverter systems," in *PCIM Europe 2022; International Exhibition and Conference for Power Electronics, Intelligent Motion, Renewable Energy and Energy Management*, 2022, pp. 1–6.

[6] F. Schiffer, "Infineon expands CoolSiC™ portfolio, 2 kV voltage class to enable simple, high-power density solutions for 1500 VDC applications," Accessed 30.10.2023, May 2022. [Online]. Available: https://www.infineon.com/

cms/en/about-infineon/press/market-news/2022/INFIPC202205-079.html

[7] ROHM Co. Ltd., "ROHM Co., Ltd., application benefits of using 4th generation SiC MOSFETs," Accessed 03.03.2024, 2022. [Online]. Available: https://fscdn.rohm.com/en/products/databook/applinote/discrete/sic/mosfet/4g_sic_mos_app_benefits_an-e.pdf

[8] International Electrotechnical Commission, "IEC 60747-8: Semiconductor devices - Discrete devices - Part 8: Field effect transistors,," IEC, May 2021.

[9] C. Felgemacher, S. Vasconcelos Araujo, C. Noeding, and P. Zacharias, "Benefits of increased cosmic radiation robustness of sic semiconductors in large power-converters," in *PCIM Europe 2016; International Exhibition and Conference for Power Electronics, Intelligent Motion, Renewable Energy and Energy Management*, 2016, pp. 1–8.

PCIM Europe 2024, 11– 13 June 2024, Nuremberg　　　DOI: 10.30420/566262280

High Temperature Experimental Characterizations of C_{OSS} of 3.3 kV SiC MOSFET for Medium Voltage PV Applications

Paul Schmidt[1,2], Van-Sang Nguyen[1], Stephane Catellani[1]

[1] Univ. Grenoble Alpes, CEA, Liten, Campus INES, 73375 Le Bourget du Lac, France
[2] Univ. Grenoble Alpes, CNRS, Grenoble INP*, G2Elab, 38000 Grenoble, France

Corresponding author:　Paul Schmidt, paul.schmidt@cea.fr
Speaker:　　　　　　　Paul Schmidt, paul.schmidt@cea.fr

Abstract

The rapid development of Wide Band Gap semi-conductors (WBG) in medium voltage area (MV) contributes to the enhancement and the integration of power electronics on the electrical power grid. Indeed, high frequency systems become crucial for any MVDC implementation. Then, the losses mechanism of the entire system requires an accurate investigation in order to design high performant equipment. Focusing on the power switches, semi-conductors include switching losses and conduction losses, which can be determined by conventional methods. In Metal-Oxide Semiconductor Field-Effect Transistors (MOSFETs), switching losses rely directly on the output capacitance (C_{OSS}), which is the sum of the Drain-Source capacitance (C_{DS}) and the Drain-Gate capacitance (C_{DG}). Therefore, at every switching cycle the output capacitance is discharged conducting to additional power dissipation. Moreover, for soft switching configuration, the output capacitance can be a key element in the design of ZVS or ZCS topologies, as in DC-DC Phase-Shifted Full Bridge converter. Thus, parasitic elements, including C_{OSS} capacitance, are usually identified using static characterization methods.

In this article, the extracted measurements from Silicon Carbide (SiC) MOSFETs are revised by two dynamic characterization methods. Besides, high temperature static measurement have been carried out using an automatic industrial equipment the Keysight B1506A, coupled with a ThermoStream. Concerning experimental measurements, this paper presents one direct practical technic and an applied resonant method in order to identify the output capacitance of medium voltage SiC transistors. This work employs discreet 3.3 kV/11 A (400 mΩ) SiC MOSFETs in a TO-247-4-lead package while the measurements take into account the parasitic elements of the circuit.

1　Introduction

Today's AC power grid implements together with the DC power grid for interoperability. As part of the distribution grids, medium voltage DC systems are expected to facilitate a growing number of distributed renewable energy sources. This requires power electronics interfaces to increase largely, along with a growing number of DC loads. Besides, technological breakthroughs continue to emerge, expending semi-conductors limits and attaining the medium voltage environment (above 1.5 kV). MVDC systems are investigated to inject renewable energies sources directly on the medium voltage collector by using the medium voltage transistors [1-3]. For photovoltaic (PV) applications, the limitations of the switching speed requires proper recognition when using transistors in medium voltage environment, particularly in hard switching topologies. It becomes necessary to identify the output capacitance C_{OSS}, along with other parasitic elements participating in these limitations. This work visualizes the sensitivity of the output capacitance value under high temperature in the range of medium voltage through three methods:

- First method is a static measurement of C_{OSS}, using the Keysight B1506A from 25°C to 150°C up to 3 kV.
- Second characterization method proposes a dynamic approach. In a half bridge configuration, the high side transistor triggers a voltage step of 3 kV, applied simultaneously to the low side transistor. During the voltage pulse, the Drain-Source current (I_{DS}) and the Drain-Source voltage (V_{DS}) are measured, deducing the output capacitance knowing the charging current and the dV/dt of each given value of V_{DS} [4-5].
- Final measurement characterizes equally C_{OSS} in a dynamic way, establishing a RLC resonant circuit with C_{OSS} and identify the oscillations frequency along with the output capacitance by knowing the inductor value.

---

* Institute of Engineering Univ. Grenoble Alpes

PCIM Europe 2024, 11– 13 June 2024, Nuremberg   DOI: 10.30420/566262280

The device under test (DUT) is a discreet 3.3 kV/11 A (400 mΩ) SiC MOSFET in a TO-247-4-lead package, the MSC400SMA330B4 from Microchip. In this package, the device offers an additional Kelvin-Source terminal reducing the parasitic line inductor at the Source terminal. At the beginning, a third experiment has been considered and implemented. Based on a RC circuit, the output capacitance replaced the charging capacitor within a RC circuit and the time constant was identified. Inspired by the resonant RLC method, the results conclude to irrelevant capacitance calculations, since during that charging period the C_{oss} value changes over time, as it is illustrated in [6].

2 Static characterization of C_{oss} at high temperature

The considered device to undertake static characterization is the B1506A from Keysight. The possible applied conditions by the analyzer allow a wide operating range to characterize power devices. However, the actual 3.3 kV SiC MOSFET's datasheet, gives the output capacitance C_{oss} up to 2.3 kV. The B1506A setup gives a maximum measured voltage of 3 kV. Thus, the obtained data by the analyzer complete the manufacturer's datasheet. In addition, a ThermoStream is attached to the B1506A for the high temperature static characterization, as it is demonstrated in Fig. 1. In this work, temperature starts from 25°C, and reaches 150°C.

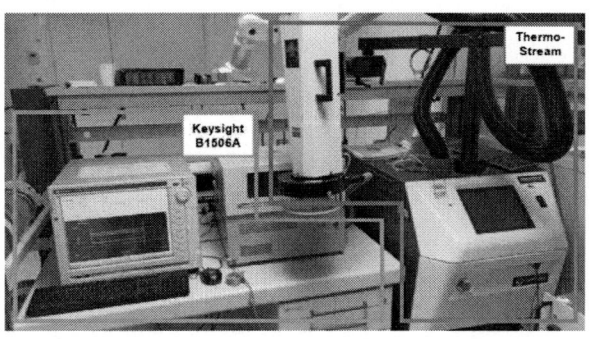

Fig. 1   High temperature static characterization setup.

Figure 2 shows the static characterization results of the DUT. These present the output capacitance C_{oss} function of the applied Drain-Source voltage, with different range of temperature, and compared to the data from the documentation.

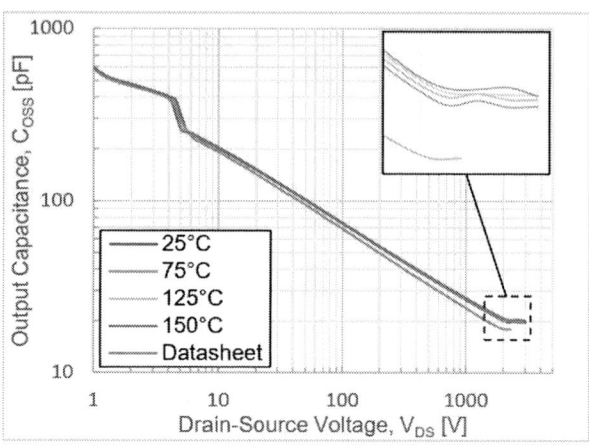

Fig. 2 C_{oss} measurement with Keysight B1506A from 25°C to 150°C.

Measurements have been conducted with a unique device; these static results are the average of at least three repeated experiments. The experimental results concur with the datasheet's curve. Although temperature appears to have a minor impact in the calculation of C_{oss} [7]. Bellow 10 V, all data plummet, this phenomena is brought by the increase of the depletion width when V_{DS} rises [8-9]. Then, a slight offset starts to appear when increasing the applied voltage, conducting to split the datasheet's curve and experimental results. Finally, the curves near 2 kV illustrate a plateau on both data; the explanation behind this event is presumably due to the increase of the leakage current, which is V_{DS} dependent in the MOSFET channel, and then perturbs the C_{oss} measurements from 2 kV. The value of the leakage current I_{DSS} is provided by the datasheet at V_{GS}=0 V and 3.3 kV at temperature 25°C and 125°C.

The following section focuses on the dynamic characterizations of this unique DUT, through two different experimental methods. Thereby, tests and results are analyzed besides datasheet of the DUT.

3 Dynamic characterizations of C_{oss}

As explained, the following methods reflect two experimental measurements of the output capacitance in the medium voltage environment.

3.1 1st method: charging current under fast medium voltage step

This first implemented method, as proposed in [4] and [5], is based on the Eq. (1), establishing the current, the capacitance and the temporal derivative of the applied voltage.

$$i = C \cdot \frac{dV}{dt} \qquad (1)$$

From Eq. (1), considering C as the output capacitance C_{OSS}, when a voltage step is exerted at the Drain and Source terminals, it becomes feasible to calculate the capacitance by knowing the Drain-Source current and the applied voltage's variations over time, as demonstrates in Eq. (2).

$$C_{OSS} = i_D \cdot \frac{dt}{dV_{DS}} \qquad (2)$$

Considering the following circuit diagram for this experiment, illustrated in Fig. 3. The experimental setup consists of two SiC MOSFETs, T_1 and T_2 in Half Bridge configuration, with a paralleled resistor R_{disch} (of 500 kΩ), in order to discharge C_{OSS}. The source is a DC power supply V_{BUS} which can deliver up to 3 kV, referenced to the earth potential. Furthermore, it is associated in parallel with a decoupling capacitor C_{bulk}. The system's principals lies in T_2, which remains blocked and serves as DUT. Thus, T_1 is the only transistor driven through an isolated gate driver, delivering V_{ctrl} and is immune to 3 kV pulses.

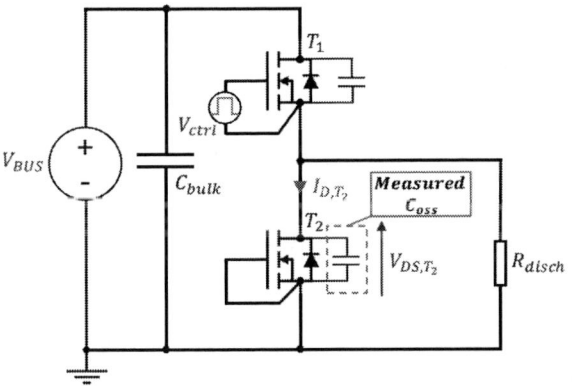

Fig. 3   Circuit diagram for the first dynamic characterization method of C_{OSS}.

Figure 4 shows the experimental setup including two SiC MOSFETs, T_1 as the high side switch (HS) and T_2 as the low side switch (LS). The gate driver is identical both HS and LS transistors using fibers optic cables.

Concerning the measurements equipment, the voltage probe is a 6 kV 100 MHz Lecroy HVD3605A, which is used to determine $V_{DS,T2}$. Likewise, i_D is measured with a current transformer (CT) made out of a 10 mm diameter toroid ferrite (3E25) with 10 winding turns; placed directly at T_2's Drain terminal with a proper insulation. Due to potential current fluctuations, the gain of the toroidal ferrite has been characterized as a function of the current for greater accuracy. Finally, a 30 A 120 MHz HYOKI CT6711 probe measures the Drain current, i_D.

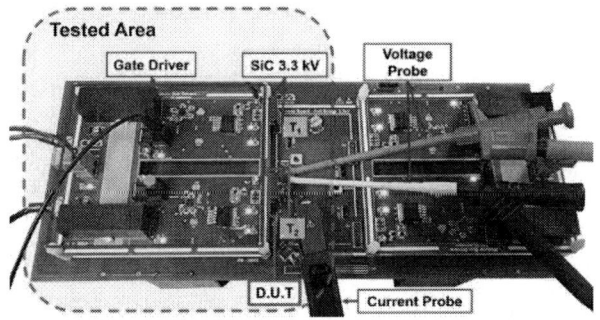

Fig. 4   First method setup for C_{OSS} dynamic characterization.

Initially, the power supply is set to 3 kV and C_{OSS} is discharged, leading to V_{DS} equal to 0V. From the moment the HS transistor T_1 starts to conduct, a rising voltage can be observed between the Drain and Source terminals of T_2, as it is indicated in Fig. 5.

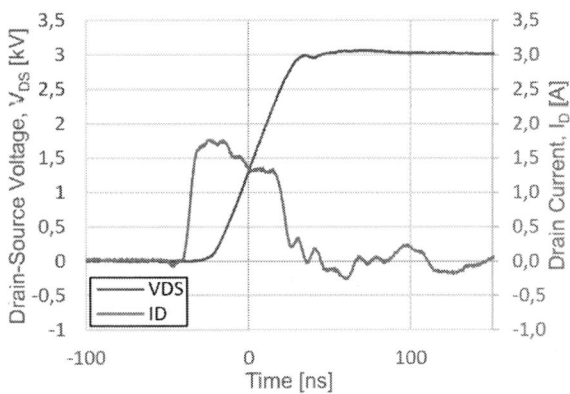

Fig. 5   Drain-Source voltage and drain current measurement from T_2 at applied middle voltage slew rate.

T_1's rising time (10% ~ 90%) has been measured to 40.4 ns. A gate-driving resistor of 22 Ω is selected especially for that experiment, in order to slow down the applied step and have greater precision. Prior to the rising V_{DS} of T_2, the transistor's Drain current started to grow, reaching 1.82 A (see Fig. 5). Then, it begins to glide down, whilst V_{DS} continues to increase, then this current finally collapses around 0 A once the voltage reaches 3 kV.

The explained process is the result of the first experimental measurement method of C_{OSS}. Thus, Fig. 5 presents the charging current flowing to the output capacitor C_{OSS}, composed of both C_{GD} and C_{DS}. However, it is crucial to calibrate and adjust both current and voltage probes. In fact, 20 ns time difference separate i_D and V_{DS} curves. In order to adjust them properly, the moment the voltage slew rate returns to zero (0 V.s^{-1}) attaining 3 kV, then it is assumed that the current is null similarly, see in Eq. (1).

Subsequently, it becomes feasible to calculate the output capacitance C_{OSS} in function of the applied V_{DS}, with a single voltage step. As it is presented in the Fig. 6, the red dots correspond to the single measured value of C_{OSS}, by employing Eq. (2); and the blue line is provided by the transistors datasheet. The obtained results meet accurately the data. However, various calculated capacitances drift away from the referenced values, mostly due to the probes uncertainty of measurement; it appears especially at low voltage, bellow 300 V. Then, the measurements precision become thinner until it enters into the medium voltage range.

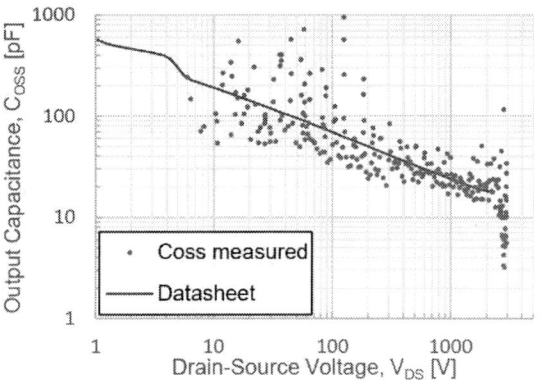

Fig. 6   Results of the first method dynamic characterization of C_{OSS}.

From 2.4 kV, the measurements become instable due to the low slew rate value, simultaneously with the rapid fall of the current. In order to increase the precision of the calculated capacitance values with the slew rate method, Fig.7 recommends a measurable zone for this direct method with a high precision where the dynamic voltage is between 25% and 75% of the DC applied voltage. Thereby, it corresponds to most accurate area, where it prevents from the diminution of the slew rate and the abrupt current falling area.

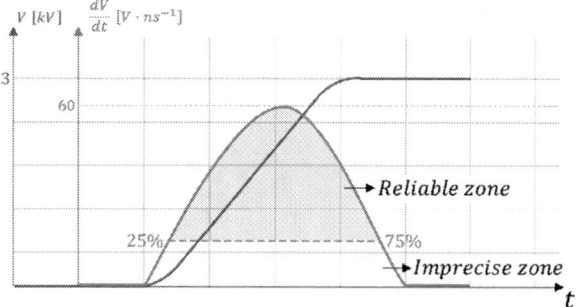

Fig. 7   Reliable zone and imprecise zone for the direct method.

This primary approach has the benefit to be the simplest and most direct method to obtain the output capacitance C_{OSS}. Nevertheless, this method should be applied only in a predefined voltage zone of the measurement. Additionally, both current and voltage must be measured and time adjusted.

In order to increase the precision, the next part presents another dynamic characterization method, consisting of a resonant circuit.

3.2   2nd method: resonant RLC series circuit

In this section, the dynamic characterization of the SiC MOSFETs output capacitance is implemented through a RLC series circuit. The RLC series circuit is a common combination of passives components, where components enter into resonance. The resonance phenomenon produced by the inductor capacitor circuit (LC) has a resonant frequency described in Eq. (3).

$$f_0 = \frac{1}{2\pi\sqrt{LC}} \tag{3}$$

The aiming of this experiment is to substitute the formula's capacitor with the SiC MOSFET's output capacitance. Under a high slew rate voltage, the

C_{OSS} capacitance resonates with a full-custom inductor, generating a voltage oscillation, where the resonant frequency is determined. In other terms, by using the Eq. (4), it is possible to deduce the output capacitance of the SiC MOSFET, with the identified resonant frequency and knowing the inductor's value.

$$C_{OSS} = \frac{1}{L \cdot (2\pi f_0)^2} \qquad (4)$$

Nevertheless, in medium voltage range, high voltage peak of the oscillations can be hazardous for the experiment setup. Hence, an additional resistor is introduced in the resonant circuit known as damping resistor. This resistor grows the decay in order to decrease the voltage peak and attenuates the oscillations. Figure 8 illustrates the circuit diagram for the final experiment.

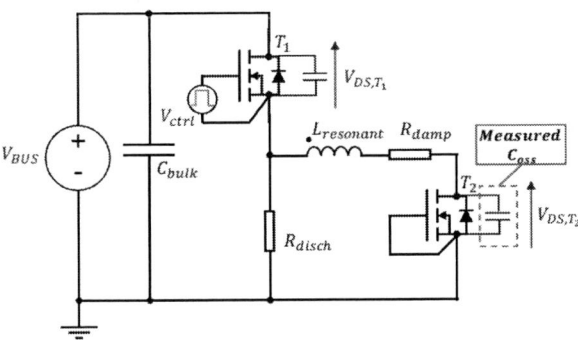

Fig. 8   Circuit diagram for the second dynamic characterization method of C_{OSS}.

Succeeding the first dynamic characterization, the experimental setup has been modified in order to add the resonant circuit. The driven transistor T_1 is identical to the previous dynamic characterization, and the second transistor T_2 remains as the DUT. However, T_2 has been moved to the second inverter leg, where in the first leg a discharge resistance of 90 kΩ (R_{disch}) took its place. In the middle of the two switching nodes of the inverter legs, there are an inductor of 500 nH ($L_{resonant}$) in series with the damping resistor of 2.4 Ω (R_{damp}). In the other hand, the power source stays a ground referring DC power supply of 3 kV, with a C_{bulk} capacitance in parallel (see Fig. 9).In order to prevent saturation, the inductor is an air core manufactured with an ETD plastic former, consequently having nearly a constant value in the range of 20 Hz to 120 MHz. Thus, this inductor has been statically characterized as a function of the frequency, with an impedance analyzer, the Keysight E4990A. The process of the experiment

consists of a fast voltage pulse at a set voltage level by the DC power supply, applied to the Drain-Source terminals of transistor T_2. This become feasible acknowledged by the driven transistor T_1, connected to the isolated gate driver. When the fast voltage step is engaged at $V_{DS,T2}$, oscillations would be observed instantly. Thereby, resonant frequency can be deduced along with C_{OSS} value.

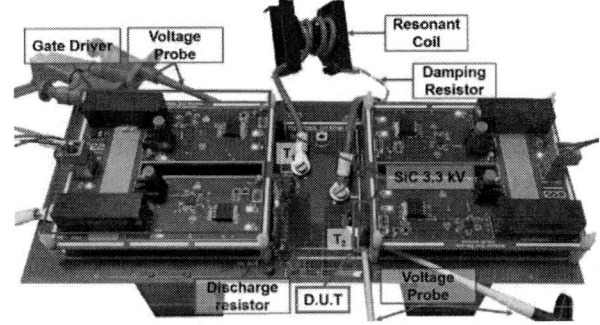

Fig. 9   Second method setup for C_{OSS} dynamic characterization.

All the elements are depicted on Fig. 9, with the same test bench as the last experiment. Similar to the inductor, it is likely to analyze the PCB's capacitance for the right resonant frequency, in order to obtain accurate C_{OSS} measurements. Hence, this value requires to be subtracted to the total output capacitance calculated. Likewise, the input impedance of the voltage probe, reaching 5 pF, would be considered and subtracted. Concerning the measurement setup, both transistors Drain-Source voltage are measured. Normally, only T_2's Drain-Source voltage has to be estimated. Nevertheless, to verify the shape of the applied step, it was decided to include T_1 switching curve. To obtain the most theoretical step voltage, the gate driver resistor is reduced to a value of 3 Ω, conducting to an acceleration of the step. Figure 10 is an example of what it can be observed at a 2.8 kV applied step. Despite the damping resistor, the peak voltage of $V_{DS,T2}$ reaches 4.032 kV. In the future work, a clamp circuit with several high voltage diodes in series should be considered to prevent this over-voltage. Moreover the $V_{DS,T1}$'s falling time has been measured to 18.1 ns. For this method, 2.8 kV was the maximum conducted experiment, in order to not risk damages to the equipment. Thereby, the resonant frequency is determined. In the case of 2.8 kV, it happens to be approximately 29.07 MHz. At that exact frequency, the inductor value is 492.1 nH, and the PCB's capacitance equals to 29.6 pF.

Applying the Eq. (4), the capacitance obtained is therefore 26.3 pF.

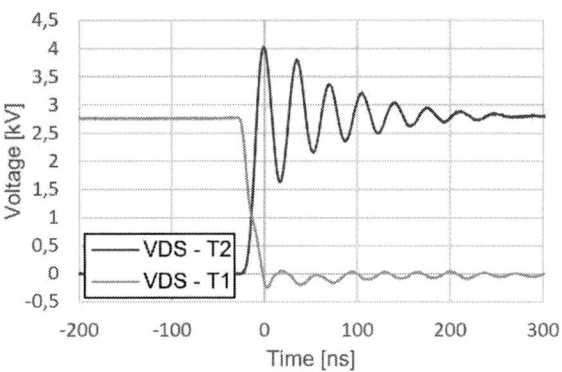

Fig. 10 Drain-Source voltages from transistors T_1 and T_2 (DUT) at 2.8 kV.

By a similar process, the tests start at low voltage, deducing the C_{OSS} capacitance while increasing gradually the voltage level. The red measured curve demonstrates in Fig. 11 gather various C_{OSS} measurements. Likened with the reference values from the datasheet, it appears that the results concur clearly with the component's documentation, leading equally to a plateau.

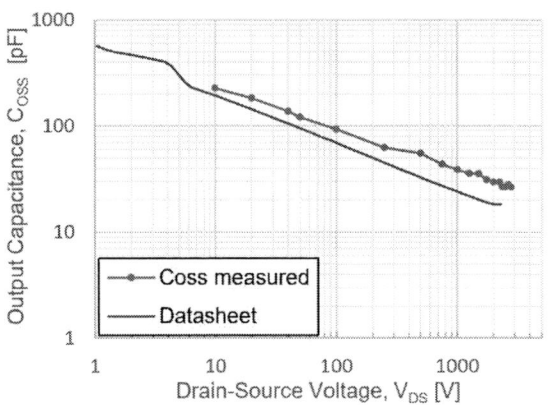

Fig. 11 Results of the second method dynamic characterization of C_{OSS}.

However, it illustrates a slight offset between the two data. It suggests that additional parasitic capacitance is present, or some elements have been neglected. For instance, input impedance of the voltage, which is frequency dependant, remains unknown. Equally, further investigations suggest that the damping resistor decreases the oscillations frequency, resulting in an overestimation of the calculated C_{OSS}. Thus, to confirm the damping resistor's influence, extra experiments have been conducted in absence of the serial resistor. These results conclude to a greater concordance with the manufacturer's measurements and so higher precision. Nevertheless, in order to preserve equipment from irreversible damages, it is still recommended to include the damping resistor and guarantee an attenuation of the resonance.

The convenience of the resonant method mainly lies in the simplicity of the measurement setup, measuring only the DUT's Drain-Source voltage. Furthermore, this experiment demonstrates a high precision of C_{OSS} measurement and applicable to a specific voltage level. Concerning the method's limitations, the resonance might cause serious damages to the equipment and the DUT itself, depending on the design of the resonant circuit. Although the DUT's Drain-Source voltage is the sole realized measurement, elements like the chosen inductor and the PCB's capacitance have to be determined over a certain frequency range. Equally, in order to obtain C_{OSS} measurements over a wide voltage range, this method appears more time-consuming than the first direct technic.

In this section, the dynamic characterizations of C_{OSS}, are implemented by two experimental methods at the ambient temperature. Nevertheless, high temperature characterization of C_{OSS} is more complicated to implement for dynamic approaches. Environment installed by medium voltage can be hostile, in particular for electrical insulation. This implies to take additional requirements in terms of electrical clearances, insulation materials and safety procedures. In the future, this would be the dynamic characterization under wide temperature range of C_{OSS} SiC MOSFET using an infrared beam heating exclusively the DUT [10].

4 Conclusion & Perspectives

This work proposed a static and two experimental characterizations of the output capacitance C_{OSS} of a medium voltage SiC transistor. Concerning the static characterization, the measurements were undertaken with the Keysight B1506A from 25°C to 150°C up to 3 kV. The results presents a minor influence of the surrounding temperature to the DUT, adding a slight positive offset compare to the device's datasheet, where the manufacturer's curve stops at 2.3 kV. First experimental measurements express a fast and simple method, giving a good precision considering a single voltage pulse applied to the DUT. This work proposed an accurate measurable zone around the dV/dt peak. The resonant method proposes an excellent accuracy at any desired voltage. However, for a large voltage interval, this method is not considered as the most timesaving; along with the presence of a resonance peak, that might cause disastrous damages to the whole experiment setup.

Moreover, the high temperature static characterizations delivers the reference results to the two dynamic methods; proposing efficient dynamic strategies to manufacturers and users within the power electronics design phases. In a near future, an adiabatic calorimetric transient method is considered to measure C_{OSS} losses in the medium voltage range [11].

Acknowledgment

This work was supported by the French National Program "Programme d'Investissements d'Avenir – INES.2S" under Grant ANR-10-IEED-0014-01.

This work was possible thanks to Microchip Technology who provided the SiC devices for testing.

References

[1] L.-A. Gomez, L. G. Alves Rodrigues, G. Gateau, and S. Sanchez, 'Characterization of 3.3 kV Discrete SiC MOSFETs in Synchronous Rectification Mode for PV Current Source Inverter Applications', in *PCIM Europe 2022; International Exhibition and Conference for Power Electronics, Intelligent Motion, Renewable Energy and Energy Management*, May 2022, pp. 1–10. doi: 10.30420/565822037.

[2] M. N. Ngo *et al.*, 'Implementation and Characterization of a 200-kW Full-SiC Isolated DC/DC Converter for Future Medium Voltage PV Plants', in *PCIM Europe 2023; International Exhibition and Conference for Power Electronics, Intelligent Motion, Renewable Energy and Energy Management*, May 2023, pp. 1–9. doi: 10.30420/566091092.

[3] V. Kremer, A. Lacarnoy, and T. A. Meynard, 'New Multi-Level Multiplexed Power Converter Topology for Medium-Voltage Power Drives', in *PCIM Europe digital days 2021; International Exhibition and Conference for Power Electronics, Intelligent*

Motion, Renewable Energy and Energy Management, May 2021, pp. 1–8.

[4] J. Rąbkowski, M. Zdanowski, R. Kopacz, F. Gonzalez-Hernando, I. Villar, and U. Larrañaga, 'From the Measurement of COSS–VDS Characteristic to the Estimation of the Channel Current in Medium Voltage SiC MOSFET Power Modules', *IEEE Trans. Instrum. Meas.*, vol. 72, pp. 1–10, 2023, doi: 10.1109/TIM.2023.3291788.

[5] J. Rabkowski, F. Gonzalez-Hernando, M. Zdanowski, I. Villar, and U. Larrañaga, 'Measurement of Coss-V characteristic of the 1.7kV/900A SiC power module and estimation of the channel current', in *2022 24th European Conference on Power Electronics and Applications (EPE'22 ECCE Europe)*, Sep. 2022, pp. 1–9.

[6] 'A More Realistic Characterization of Power MOSFET Output Capacitance Coss'.

[7] 'Application note en 20230209 AKX00063.pdf'.

[8] N. Mohan, T. M. Undeland, and W. P. Robbins, *Power electronics: converters, applications, and design*, 2nd ed. New York: Wiley, 1995.

[9] 'Microsemi-14692-mosfet-tutorial.pdf'.

[10] N. Van Sang, A. Bier, R. Escoffier, S. Catellani, J. Martin *et al*, 'A High Precision Dynamic Characterization Bench with a Current Collapse Measurement Circuit for GaN HEMT Operating at 175°C', in *PCIM Europe digital days 2021; International Exhibition and Conference for Power Electronics, Intelligent Motion, Renewable Energy and Energy Management*, May 2021, pp. 1–8.

[11] M. Guacci *et al.*, 'On the Origin of the Coss-Losses in Soft-Switching GaN-on-Si Power HEMTs', *IEEE J. Emerg. Sel. Top. Power Electron.*, vol. 7, no. 2, pp. 679–694, Jun. 2019, doi: 10.1109/JESTPE.2018.2885442.

Impact of Gate Control on the Switching Performance of 3.3kV SBD-Embedded SiC-MOSFET

Junya Sakai[1], Daniel He[1], Shota Yamamoto[2], Ryo Tsuda[2], Kenji Hatori[2], Nils Soltau[1]

[1] Mitsubishi Electric Europe B.V., Germany
[2] Mitsubishi Electric Corporation, Japan

Corresponding author: Junya Sakai, Junya.Sakai@meg.mee.com
Speaker: Junya Sakai, Junya.Sakai@meg.mee.com

Abstract

This paper presents the voltage gradient dv/dt when switching a 3.3 kV Schottky barrier diode (SBD) embedded silicon carbide (SiC) MOSFET module rated for 800 A. Different measurements are performed to identify the influences of temperature, dc-link voltage, stray inductance, gate drive conditions and drain current value on dv/dt. Finally, the controllability of voltage transients and the impact on switching losses are analyzed.

1 Introduction

Silicon carbide (SiC) power semiconductors are key devices for carbon neutrality. Since several years, SiC MOSFET modules have become more common in many applications due to their significant benefits in performance compared to conventional silicon (Si) IGBT devices. In 2013 Mitsubishi Electric has already introduced its first commercial Full SiC device rated for 1200 V/1200 A. Since then Mitsubishi Electric has expanded its SiC module lineup also for higher voltage classes like 3.3 kV [1]. Conventional SiC MOSFET modules utilize MOSFET chips combined with antiparallel SBD chips in order to avoid the bipolar degradation. In 2023 Mitsubishi Electric released an SBD-embedded MOSFET module in the 3.3 kV class as depicted in Fig. 1 [2-3]. Using the new SBD-embedded MOSFET structure, the power density of the power module has been increased while ensuring a high reliability by maintaining the SBD and thus, avoiding the risk of bipolar degradation [4].

The switching speed of SiC increases compared to Si and with it the voltage gradient dv/dt during commutation [5]. Since years system engineers fear the impact of rising dv/dt on insulation materials. Still today, different research activities are ongoing [6-8]. One recent research work has found no significant impact of dv/dt on the partial discharge inception voltage of motor-winding insulation. However, if the insulation system is not free of partial discharge, insulation lifetime will decrease with increasing dv/dt [9]. Moreover, local overvoltage due to voltage reflections or unequal voltage distribution among windings become more pronounced with increasing dv/dt.

This paper demonstrates the dv/dt characteristics of the new SBD-embedded SiC MOSFET. Subsequently, the impact of different parameters such as temperature, dc-link voltage, stray inductance etc. are examined and the controllability by different gate resistances on the voltage transients is analyzed. Finally, the trade-off characteristics for the switching losses under limited voltage transient operation is investigated.

2 Definition of Reference Point and dv/dt

The device under test (DUT) is an SBD embedded SiC MOSFET module rated for 3.3 kV/800 A in the standardized LV100 package with the type name "FMF800DC-66BEW" as depicted in Fig. 1.

Fig. 1: Device under test (DUT): SBD embedded SiC MOSFET module rated for 3.3 kV/800 A in LV100 package: (FMF800DC-66BEW)

At first, the turn-off and turn-on waveforms of the DUT at the N-Side are measured at nominal conditions and room temperature. Therefore, the dc-link voltage is set to V_{DD} = 1800 V, the switching current is I_D = 800 A and the channel temperature is T_{ch} = 25 °C. The gate resistances are selected to be $R_{G(off)}$ = 1.5 Ω and $R_{G(on)}$ = 1.5 Ω during turn-off and turn-on respectively which is the minimum recommended gate resistance value as given by the datasheet. The gate is driven by V_{GS} = +17 V/-7 V. The dc-link stray inductance is about $L_s \approx$ 40 nH. The measurement results of these nominal conditions are used as a reference point in the following sections.

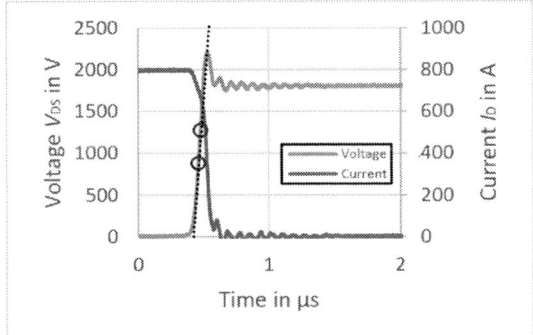

Fig. 2: Turn-off waveform at nominal conditions: V_{DD} = 1800 V, I_D = 800 A, $R_{G(on)}$ = 1.5 Ω, $R_{G(off)}$ = 1.5 Ω and T_{ch} = 25 °C

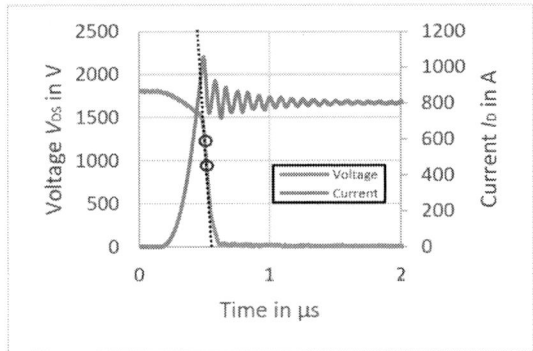

Fig. 3: Turn-on waveform at nominal conditions: V_{DD} = 1800 V, I_D = 800 A, $R_{G(on)}$ = 1.5 Ω, $R_{G(off)}$ = 1.5 Ω and T_{ch} = 25 °C

The switching waveforms at the defined nominal conditions during turn-off and turn-on are given in Fig. 2 and Fig. 3 respectively.

A common approach to determine the dv/dt is by using two points, where the drain-source voltage V_{DS} equals 50 % and 70 % of the applied dc-link voltage ($V_{DD,50}$ and $V_{DD,70}$). In Fig. 2 and 3, small circles at the V_{DS} curve are marking these reference points and an easy estimation of the dv/dt is possible using the slope between the points $V_{DS} = V_{DD,50}$ and $V_{DS} = V_{DD,70}$. By using this 50/70

dv/dt approach, the dv/dt values during turn-off and turn-on at nominal conditions and room temperature can be determined respectively based on Fig. 2 and 3:

$$\left.\frac{\mathrm{d}v}{\mathrm{d}t}\right|_{off} \approx \left.\frac{\Delta v}{\Delta t}\right|_{off} = 24.30 \ \frac{\mathrm{kV}}{\mu s}$$

$$\left.\frac{\mathrm{d}v}{\mathrm{d}t}\right|_{on} \approx \left.\frac{\Delta v}{\Delta t}\right|_{on} = -21.62 \ \frac{\mathrm{kV}}{\mu s}$$

3 Measurement Results of dv/dt in dependency on various application parameters

In the following section, the dv/dt characteristics of the DUT for the turn-off and turn-on switching event will be analyzed under different application conditions. Based on the measurement results, the impact of the different application parameters on the dv/dt and thus, the operation condition with the highest dv/dt can be determined.

3.1 Drain Current Dependency on dv/dt

The impact by increasing the drain current I_D on the dv/dt value has been characterized for nominal conditions (V_{DD} = 1800 V, T_{ch} = 25°C, $R_{G(on)}$ = 1.5 Ω, $R_{G(off)}$ = 1.5 Ω). The result is shown in Fig. 4 for the turn-off and Fig. 5 for the turn-on switching event respectively.

Based on the measurement results, the dv/dt for turn-off and turn-on is increasing with increasing drain current. Therefore, the voltage slope during the switching events, especially during the turn-off become more steep with higher drain current. It can be noted that the dv/dt during turn-on increases up to a drain current of approximately 300 A. Afterwards, it stays almost constant at a level of around 24 kV/µs independent on current. Nevertheless, the highest dv/dt for both turn-off and turn-on switching event within an application can be expected at the highest drain current operation.

PCIM Europe 2024, 11– 13 June 2024, Nuremberg   DOI: 10.30420/566262281

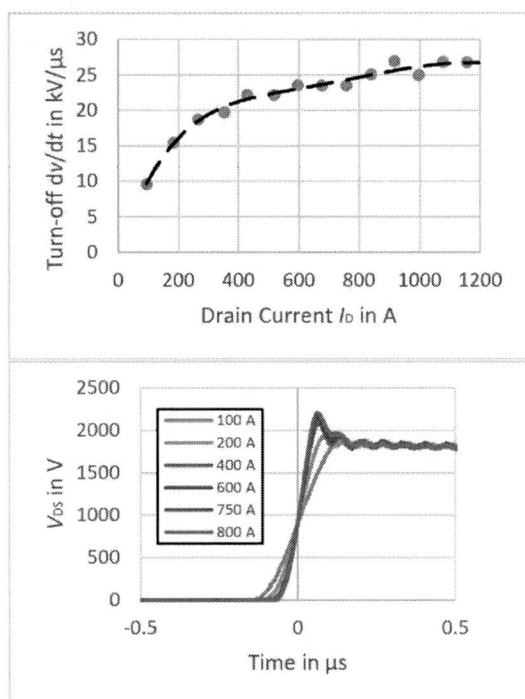

Fig. 4: Turn-off I_D dependency at nominal conditions: V_{DD} = 1800 V, $R_{G(on)}$ = 1.5 Ω, $R_{G(off)}$ = 1.5 Ω and T_{ch} = 25 °C

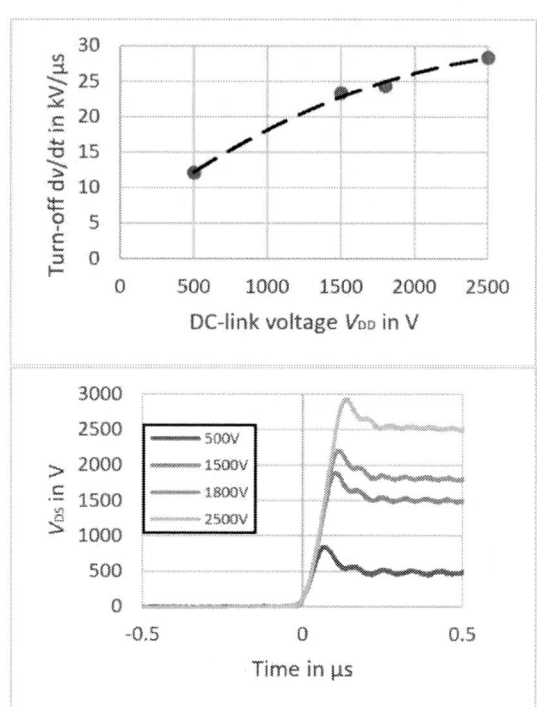

Fig. 6: Turn-off V_{DD} dependency at nominal conditions: I_D = 800 A, $R_{G(on)}$ = 1.5 Ω, $R_{G(off)}$ = 1.5 Ω and T_{ch} = 25 °C

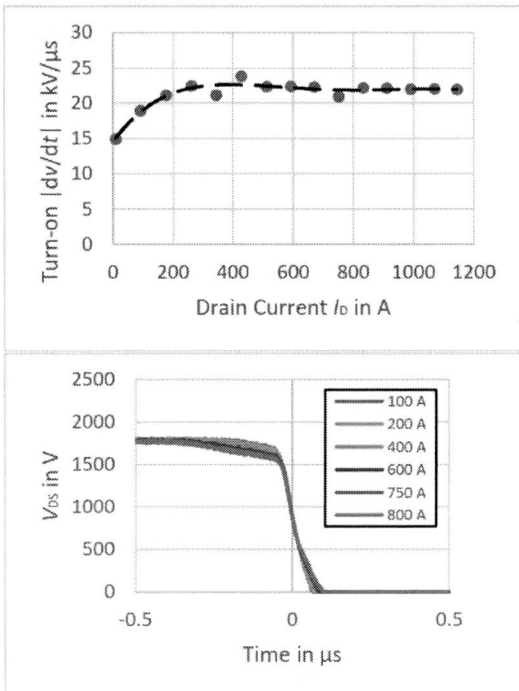

Fig. 5: Turn-on I_D dependency at nominal conditions: V_{DD} = 1800 V, $R_{G(on)}$ = 1.5 Ω, $R_{G(off)}$ = 1.5 Ω and T_{ch} = 25 °C

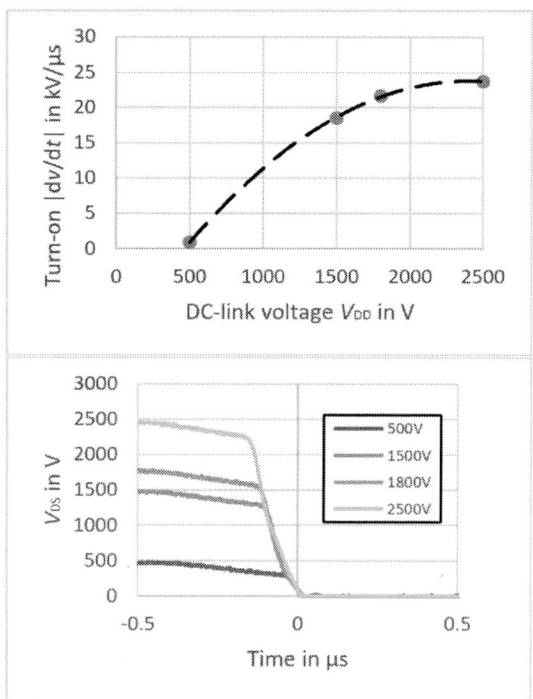

Fig. 7: Turn-on V_{DD} dependency at nominal conditions: I_D = 800 A, $R_{G(on)}$ = 1.5 Ω, $R_{G(off)}$ = 1.5 Ω and T_{ch} = 25 °C

2008

3.2 DC-Link Voltage Dependency on dv/dt

The DC-link voltage V_{DD} has been varied up to a maximum voltage of 2500 V as recommended for 3.3 kV devices in order to investigate the quantitative impact on the dv/dt characteristics of the DUT. The measurement results at nominal conditions and room temperature (V_{DD} = 1800 V, I_D = 800 A, $R_{G(on)}$ = 1.5 Ω, $R_{G(off)}$ = 1.5 Ω and T_{ch} = 25 °C) are given in Fig. 6 and Fig. 7 for the turn-off and turn-on switching event respectively.

It is confirmed that the dv/dt for turn-off and turn-on is increasing with higher dc-link voltage. Based on the measurements at nominal conditions, the dv/dt at turn-off increases from 12.1 kV/μs to 28.4 kV/μs and at turn-on from 0.88 kV/μs up to 23.7 kV/μs showing a strong dependency by the DC-link voltage. The highest dv/dt can be expected with the highest V_{DD} in the corresponding application accordingly.

Since the electric field strength in the drift layer within the chip increases depending on V_{DD}, this nearly linear dependency is consistent with the conventional understanding.

3.3 Temperature Dependency on dv/dt

The operating temperature will usually vary between room temperature up to the maximum allowed operation temperature which is 175°C for the FMF800DC-66BEW depending on the overall system design. Therefore, the temperature dependency during the turn-off and turn-on switching event has been characterized in Fig. 8 and 9.

Fig. 8 shows a decreasing dv/dt during turn-off for increasing channel temperature. Subsequently, the highest dv/dt for turn-off is expected at low channel temperatures. On the contrary, the highest dv/dt for the turn-on switching event can be seen at high temperature up to 175°C. The overall dependency is almost linear in both cases and shows a lower impact considering absolute values compared to the dc-link voltage and drain current dependency.

3.4 Stray Inductance Dependency on dv/dt

Depending on the system design, the overall stray inductance L_s can vary between different applications. The impact of the stray inductance value on the dv/dt has been analyzed in Fig. 10 and 11 for the turn-off and turn-on switching event respectively under nominal conditions (V_{DD} = 1800 V, I_D = 800 A, $R_{G(on)}$ = 1.5 Ω, $R_{G(off)}$ = 1.5 Ω and T_{ch} = 25 °C). Therefore, the stray inductance L_s has been varied between 40 nH, 70 nH and 95 nH.

As Fig. 10 shows, the dv/dt for turn-off is almost independent of the overall stray inductance value. However, the design should still aim for low stray inductance, as the surge voltage will increase with higher L_s. Fig. 11 also shows the turn-on dv/dt having a weak or almost non-existent dependency on the stray inductance L_s as well. Subsequently, it can be concluded that the L_s has almost no or only a weak impact on the dv/dt which may also be within the measurement tolerances.

3.5 Summary Table for the dv/dt dependencies of the DUT

The different operation parameter influences on the dv/dt have been summarized in Table 1. Based on the measurement results from the previous section, the highest dv/dt operation conditions can be easily determined.

	Turn-Off	Turn-On
I_D	Strong(↑ 177%)	Medium(↑ 47%)
(100A→1200A)	(9.6 kV/μs→26.6 kV/μs)	(14.9 kV/μs→21.9 kV/μs)
V_{DD}	Strong(↑ 135%)	Strong(↑ 2593%)
(500V→2500V)	(12.1 kV/μs→28.4 kV/μs)	(0.88 kV/μs→23.7 kV/μs)
T_{ch}	Weak(↓ 12%)	Medium(↑ 37%)
(25°C→175°C)	(24.3 kV/μs→21.4 kV/μs)	(21.6 kV/μs→29.6 kV/μs)
L_s	Weak(↑ 3%)	Weak(↓ 15%)
(40nH→95nH)	(24.3 kV/μs→25 kV/μs)	(21.6 kV/μs→18.3 kV/μs)

Table 1: Summary of |dv/dt| dependencies

The highest dv/dt for turn-off is expected at room temperature during high drain current and high dc-link voltage operation. The highest dv/dt during turn-on will occur during high temperature, high drain current and high dc-link voltage operation.

4 Analysis of Gate Resistance dependency on dv/dt and switching energy

A possibility to reduce the dv/dt known from Si IGBT modules is to increase the gate resistances $R_{G(on)}$ and $R_{G(off)}$. Due to an increase in gate resistance, the overall switching speed will be reduced resulting in a lower dv/dt.

For the analysis of the gate resistance impact on dv/dt, the gate resistance has been varied from 1.5 Ω up to 20 times the minimum recommended R_G value as defined by the datasheet of the DUT. The result on the dv/dt at continuous operation conditions is shown in Fig. 12 and 13. The temperature has been set to T_{ch} = 25°C for turn-off and T_{ch} = 175°C for the turn-on switching event as the highest dv/dt is expected under these temperature conditions based on the previous analysis.

PCIM Europe 2024, 11– 13 June 2024, Nuremberg   DOI: 10.30420/566262281

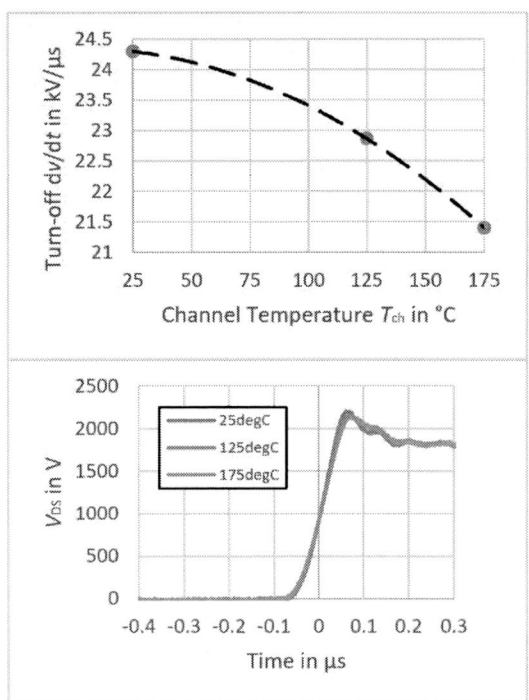

Fig. 8: Turn-off T_{ch} dependency at nominal conditions: V_{DD} = 1800 V, I_D = 800 A, $R_{G(on)}$ = 1.5 Ω and $R_{G(off)}$ = 1.5 Ω

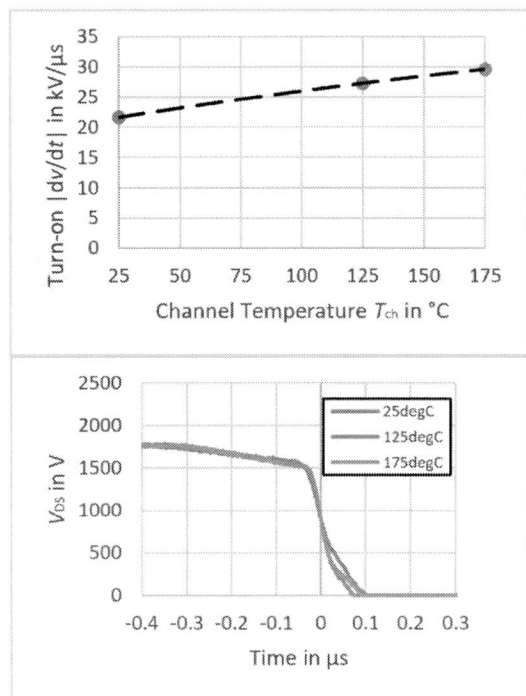

Fig. 9: Turn-on T_{ch} dependency at nominal conditions: V_{DD} = 1800 V, I_D = 800 A, $R_{G(on)}$ = 1.5 Ω and $R_{G(off)}$ = 1.5 Ω

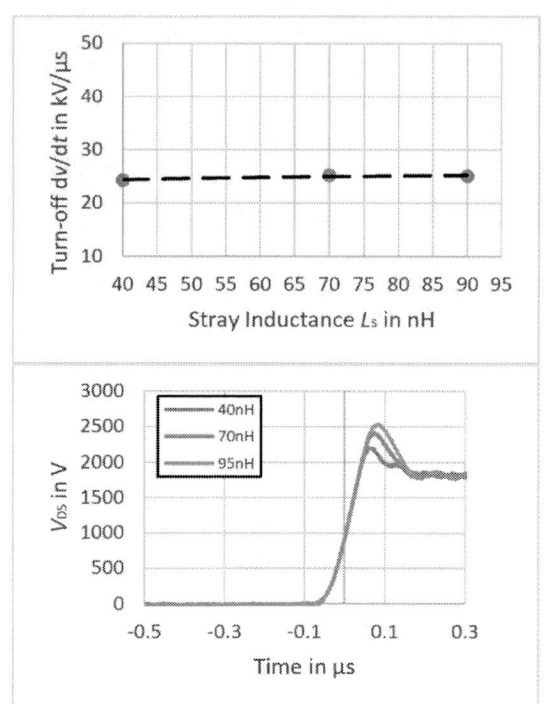

Fig. 10: Turn-off L_s dependency at nominal conditions: V_{DD} = 1800 V, I_D = 800 A, $R_{G(on)}$ = 1.5 Ω, $R_{G(off)}$ = 1.5 Ω and T_{ch} = 25 °C

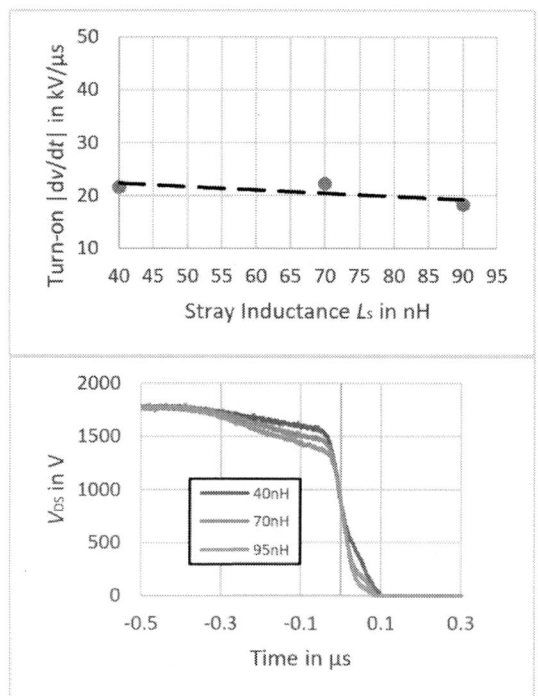

Fig. 11: Turn-on L_s dependency at nominal conditions: V_{DD} = 1800 V, I_D = 800 A, $R_{G(on)}$ = 1.5 Ω, $R_{G(off)}$ = 1.5 Ω and T_{ch} = 25 °C

2010

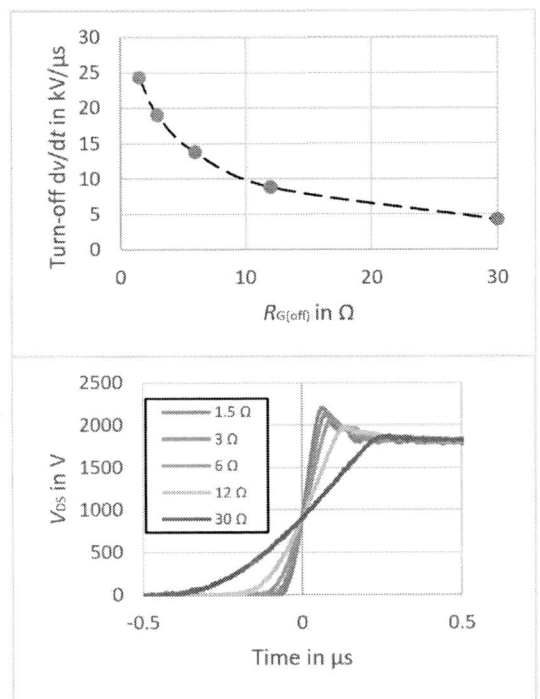

Fig. 12: $R_{G(off)}$ dependency at nominal conditions: V_{DD} = 1800 V, I_D = 800 A and T_{ch} = 25 °C

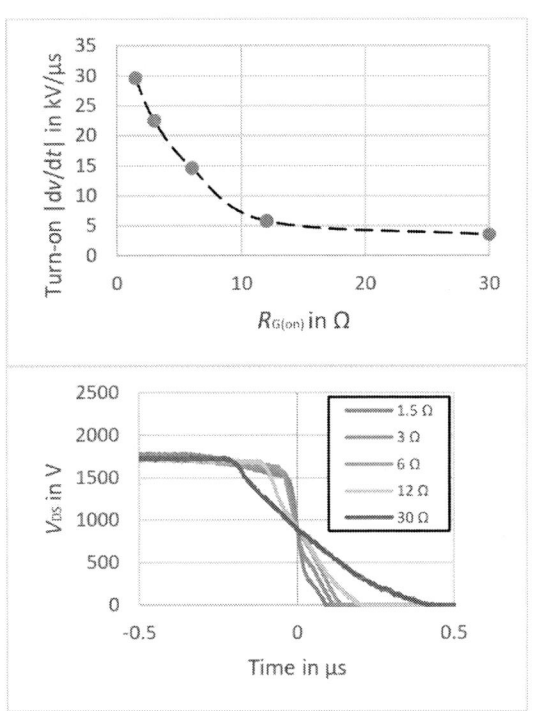

Fig. 13: $R_{G(on)}$ dependency at nominal conditions: V_{DD} = 1800 V, I_D = 800 A and T_{ch} = 175 °C

The measurement results verify the possibility to control the dv/dt of the FMF800DC-66BEW by varying the gate resistance. The initial increase in the gate resistance reduces the switching speed and thus, the dv/dt at turn-off as well as turn-on event. However, by increasing the gate resistance and lowering the switching speed, the energy losses during the switching event increases due to the longer overlap between drain current I_D and drain-source voltage V_{DS} during the switching event. Based on the previous results, the trade-off curve between dv/dt and switching energy can be derived as shown in Fig. 14.

If for example the system designer requires a certain maximum dv/dt at turn-on and turn-off of 10 kV/µs for nominal operation, the Fig. 12 and 13 can be used to determine the required $R_{G(on)}$ and $R_{G(off)}$. In this case, the values can be determined to $R_{G(on)}$ = 8 Ω and $R_{G(off)}$ = 10 Ω. Although the dv/dt can be even further lowered by increasing the R_G value, the switching energy and thus, the capability to dissipate the resulting losses need to be confirmed. In this example the switching energy can be determined to be E_{on} = 0.45 J and E_{off} = 0.3 J. Therefore, the system designer needs to ensure that this switching energy is able to be dissipated by the thermal system design.

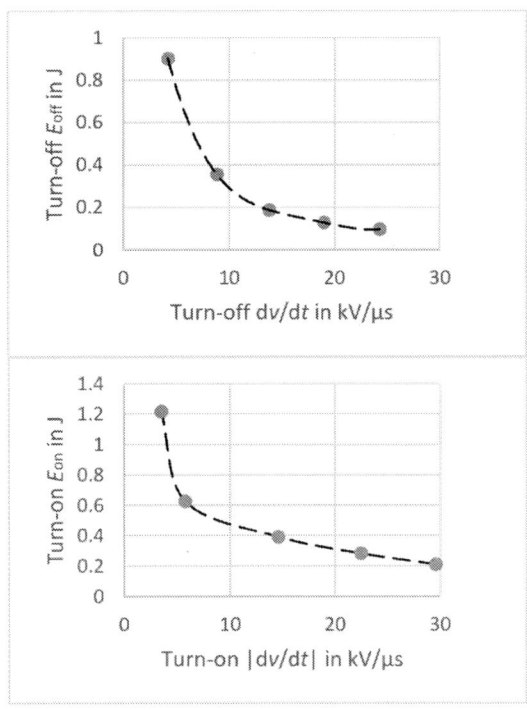

Fig. 14: dv/dt vs. switching energy trade-off curve at nominal conditions for turn-off at T_{ch} = 25°C and turn-on at T_{ch} = 175°C

5 dv/dt Value of SiC Power Semiconductor Products

Having talked about the flexibility of dv/dt, it is likely that it will continue to increase as future power semiconductor products continue to approach ideal switching behaviour.

The switching speed of SiC MOSFET chips, including dv/dt, exceeds that of Si by a significant margin. This disparity arises from the remarkably higher breakdown electric field inherent to SiC material, approximately tenfold greater than that of Si, enabling the fabrication of thinner device layers. Moreover, with a saturation drift velocity exceeding twice that of Si, the resulting increase in carrier diffusion velocity is believed to be instrumental in reducing the propagation time of the switching state[8]. These properties are also the reason why the performance of power semiconductors can be dramatically improved by using SiC materials. As long as we continue to aim to improve the performance of power semiconductors, dv/dt will continue to rise.

6 Conclusion

This paper introduced the new 3.3 kV/ 800 A SBD-embedded SiC-MOSFET module and analyzed the module regarding its dv/dt characteristics. Although the SiC MOSFET modules allow lower switching losses compared to their Si IGBT module counterpart, higher voltage transients are expected as a result of the higher switching speed. This paper indicates the different impacts on dv/dt by various influence factors and demonstrates the flexible adjustment of dv/dt via the gate resistor. The trade-off curve between switching energy and dv/dt has been determined through measurements allowing system designers for an tuning for their individual application. As an example, an external resistance of 30 Ω reduces the turn-off dv/dt from about 24.4 kV/µs to 4.2 kV/µs while resulting in a switching energy increase at turn-off from 0.09 J to 0.90 J at nominal conditions. Based on the requirements of certain applications, this paper allows the reader to easily determine the required compromise between switching performance and dv/dt.

References

[1] T. Negishi, R. Tsuda, K. Ota, S. Iura and H. Yamaguchi, "3.3 kV All-SiC Power Module for Traction System Use," in *PCIM Europe 2017; International Exhibition and Conference for Power Electronics, Intelligent Motion, Renewable Energy and Energy Management*, Nuremberg, Germany, 2017.

[2] Mitsubishi Electric Corporation "Mitsubishi Electric to Ship Samples of SBD-embedded SiC-MOSFET Module" News Release, May 2023.

[3] Mitsubishi Electric Corporation "Mitsubishi Electric Develops SBD-embedded SiC-MOSFET with New Structure for Power Modules" News Release, June 2023.

[4] N. Soltau, E. Wiesner. "Switching Performance of 750A/3300V Dual SiC-Modules", *Bodo's Power Systems*, Feb. 2019.

[5] N. Soltau et al, "Impact of Gate Control on the Switching Performance of a 750A/3300V Dual SiC-Module" *Proceedings of EPE'18 ECCE Europe*, pp. 1-7

[6] R. Möller, J. Vocke, A. Mühlbeier and R. Puffer, "Development of a Test Bench for the Investigation of the Breakdown Voltage of Insulation Materials at Medium Frequency Rectangular Voltages," in *IEEE Transactions on Power Electronics*, vol. 36, no. 3, pp. 2621-2628, March 2021.

[7] A. Claudi, G. Braun, R. Klengel, S. Klengel, P. Zacharias and X. Yu, "Aging of Insulation Materials under Repetitive Impulse Voltage Stress with high dv/dt," *CIPS 2022; 12th International Conference on Integrated Power Electronics Systems*, Berlin, Germany, 2022, pp. 1-6..

[8] M. Fuerst and M. -M. Bakran, "Design of an Easy-to-Use Standalone PD Measurement System for Pulsed Voltage and High dV/dt Setups," *PCIM Europe 2023; International Exhibition and Conference for Power Electronics, Intelligent Motion, Renewable Energy and Energy Management*, Nuremberg, Germany, 2023, pp. 1-9

[9] M. Fuerst and M.-M. Bakran, "Is the Motor Insulation System Really Affected by High Voltage Transients?," *Proc. of the 13th International Conference on Integrated Power Electronics Systems (CIPS)*, Duesseldorf, Germany, 2024, pp. 699-706

[10] Y. Sadafumi, "Aspects and Prospects of the Researches on the Development of SiC Devices" *Hyomen-kagaku* Vol.21, Japan, 2000

PCIM Europe 2024, 11– 13 June 2024, Nuremberg   DOI: 10.30420/566262282

Comparative Assessment of Overloadability Potential of 3.3 kV Si-IGBTs and SiC-MOSFET Power Modules

Muhammad Nawaz[1], Bochen Liu[1], Virgiliu Botan[2] and Tobias Keller[2]

[1] Hitachi Energy Research (HER), Forskargränd 7, SE-721 78 Västerås, Sweden
[2] Hitachi Energy, Fabrikstrasse 3, CH-5600 Lenzburg, Switzerland

Corresponding author: Muhammad Nawaz,   Muhammad.nawaz@hitachienergy.com
Speaker: Muhammad Nawaz,   Muhammad.Nawaz@hitachienergy.com

Abstract

We present overload capability assessment of 3.3 kV (450 A) Si-IGBT and 3.3 kV (500 A) SiC-MOSFET based power modules for potential STATCOM application using chain-link MMC topology. Static tests were first performed for both Si and SiC power modules up to 250 °C. Double pulse dynamic tests have been performed at a supply voltage of 2.0 kV and up to 175 °C. Compared to Si, SiC MOSFETs present 7 – 10 x lower losses under overload condition and with higher overload factor and overload duration under nominal operating condition (i.e., V_{DC} = 2000 V, f_{sw} = 200 Hz).

1   Introduction

One critical aspect associated to the long-term reliability of power converters is the unexpected appearance of voltage or current transient during converter operation. Here, the power converter may exceed the current/voltage rating of the active power device for limited time, thus crosses the safe operating area (i.e., SOA) boundary. This phenomenon of so-called converter "overloadability" has recently been getting much attention in the scientific literature, not only because it is considered as the main precursor that limits the life-time of the converter but also it poses a great threat to premature converter failure leading to a possible explosion under extremely stressed environment. From a system perspective, an overload capability is generally considered as a desirable requirement of power converters for multi-gigawatt power systems [1].

Various overload protection methods are discussed for LLC resonant converters for MVDC applications [2]. A control-algorithm for achieving the overloading capability for MMC VSC is proposed in [3] where a temperature dependent current controller is added to the current control block of the converter plus a sub-module temperature regulator as an additional item to the sub-module capacitor voltage balancing controller. A novel fault-tolerant topology of T-Type NPC inverter is introduced with increased thermal overload capability in [4]. For traction applications, IGBTs are stressed

under overload condition to study the clamped inductive turn-off failures induced in the power modules [5]. Analysis shows that the mismatches in the electrothermal properties of the IGBT device during transient operation are the potential pre-cursors that can lead to uneven power dissipation, and hence significantly enhance the risk of failure and reduce the lifetime of the semiconductor power module. Similarly, a novel thermal management algorithm scheme for improved lifetime and overload capabilities for traction converters is introduced in [6]. With improved overload capability potential in mind, experimental performance comparison of several cooling solutions for power semiconductor devices to increase their thermal performance and reliability during surge current events is proposed using hybrid cooling approach [7]. Some authors [8, 9] have recently proposed phase change material to be integrated into the package design for "press-pack IGBTs" as a viable option to improve the power surge handling capability. More recently, a well-structured review of potential base technology enablers (i.e., metallic phase change materials, Peltier materials, graphene, diamond) to achieve overcurrent capability of a grid-connected voltage source inverters is presented in [10]. An overview of the limiting factors for the overcurrent levels and their duration for SiC power modules in power electronics converters has been addressed. Device failure due to transient overstress induced by press-pack packaged IGBTs, a short-circuit capability optimization

PCIM Europe 2024, 11– 13 June 2024, Nuremberg   DOI: 10.30420/566262282

method that improves the active edge heat dissipation by using silver conductive adhesive (SCA) is proposed in [11].

This work deals with the comparative assessment of overloadability potential of 3.3 kV (450 A) Si-IGBT [12] and 3.3 kV (500 A) SiC-MOSFET [13] power modules for STATCOM applications. Both power modules offer ultra-low stray inductance L_s of 10 nH and are well-suited for paralleling of several power modules as per requirement for the increased power rating of converter. The power modules are in the half-bridge (HB) phase-leg configuration with an $R_{th,j-c}$ of 31 (54) K/kW for Si-IGBT (Si-diode) and 32.9 K/kW for SiC-MOSFET per switch position. The maximum junction temperature limit for Si and SiC power module is set to 150 °C according to the datasheet. The overall LinPak module configuration and its switching behavior have earlier been reported in [14, 15]. Figure. 1 shows the physical footprint of Si/SiC IGBT/MOSFET LinPak power module.

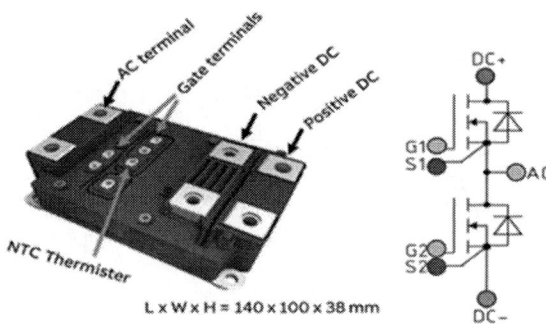

Fig. 1 Physical footprint and electrical circuit of half bridge phase leg of Si-IGBT or SiC-MOSFET module.

2 Module characterization

2.1 Static measurements

Both Si-IGBT and SiC-MOSFET LinPak modules have first been characterized under extreme conditions (up to 250 °C) using B1505A curve tracer (from Keysight/Agilent).

Figure 2 shows the typical transistor characteristics (i.e., I_{CE} versus V_{CE} at V_{GE} of 15 V). An ON-resistance of diode of 1.74 mΩ was obtained at 25 °C with a temperature coefficient of resistance of 0.0082 mΩ/°C. An extracted ON-resistance of 2.92 mΩ at 25 °C with a temperature coefficient of ON-resistance of 0.0141 mΩ/°C is obtained (i.e., R_{ON} increases with temperature) for Si-IGBT module within the linear range. Similarly, an extracted threshold voltage defined at 1.0 A was 6.4 V at 25 °C with a temperature coefficient of V_{TH} shift of -0.0121 V/°C (i.e., V_{TH} decreases with temperature).

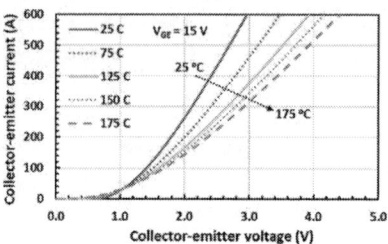

(a)

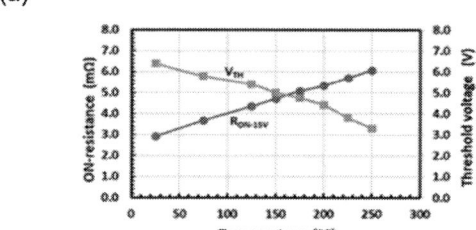

(b)

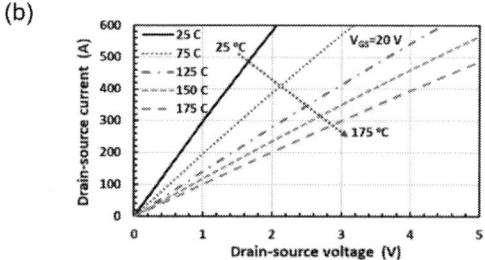

(c)

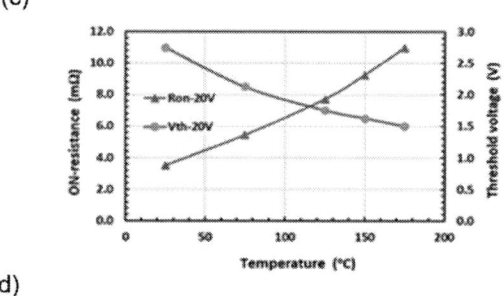

(d)

Fig. 2 $I_{CE} - V_{CE}$ (a), R_{ON}/V_{TH} vs temperature (b) of Si-IGBT LinPak and $I_{DS} - V_{GS}$ (c) and R_{ON}/V_{TH} vs temperature (d) of SiC-MOSFET LinPak module.

Static characterization of 3.3 kV, 500 A SiC-MOSFET modules are performed up to 175 °C. An extracted ON-resistance of 3.52 mΩ for V_{GS} of 20 V at 25 °C with a temperature coefficient of ON-resistance of 0.0425 mΩ/°C is obtained (i.e., R_{ON} increases with temperature) for SiC-MOSFET module when measured up to 175 °C. Similarly, an extracted threshold voltage defined at 0.1 A was 2.75 V at 25 °C with a temperature coefficient of V_{TH} shift of -0.009 V/°C (i.e., V_{TH} decreases with temperature). An ON-resistance of the body diode of 3.376, 2.945 and 2.6 mΩ at 25, 125, and 175 °C (defined at V_{GS} = -5.0 V) respectively, has been

2014

obtained for SiC-MOSFETs. Compared to SiC-MOSFET, a significant higher sub-threshold leakage current and off-state leakage current (not shown here) is observed with increasing temperature beyond 150 °C for Si-IGBT modules. Note that the extracted ON-resistances of both LinPaks also include the resistance of the measurement cables and connectors/terminals.

2.2 Dynamic characterization

The dynamic characterization setup for overloadability is based on a symmetrical busbar, designed originally for paralleling investigation of 4 similar power modules, along with 8 units of 2.0 kV rated DC capacitors (C_{DC} = 1.16 mF). A load inductance of 50 µH has been used in the measurements. A total stray inductance in the half-bridge (HB) phase-leg commutation process stays around 22 nH in our test setup. Note that, lower switch was used as DUT for all dynamic tests for both Si and SiC power modules, while the top switch has been kept in the off-state. Experimental test setup is shown in Figure 3.

A typical double pulse transient at 2.0 kV and 500 A is shown in Figure 4 for 3.3 kV (450 A) Si-IGBT and 3.3 kV (500 A) SiC-MOSFET power module for 25 °C. An extracted turn-on, turn-off, and E_{rec} losses of Si-IGBTs (SiC-MOSFETs) were 955.5 (134.9), 446.5 (93.5) and 383.8 (7.0) mJ, respectively, at 2.0 kV and 25 °C. Note that the SiC-MOSFETs offer a dv/dt (di/dt) of 14.6 (9.6 A/ns) V/ns during turn-off phase, and dv/dt (di/dt) of 14.6 (3.75 A/ns) V/ns during turn-on transition. Similarly, Si-IGBT module presents a dv/dt (di/dt) of 8.9 (3.9 A/ns) V/ns during turn-off transition and dv/dt (di/dt) of 3.08 (5.5 A/ns) V/ns during turn-on transition. Overall, SiC-MOSFET modules present 7 – 10 x lower switching losses compared to Si-IGBT modules under similar test conditions and this is consistent to our previous measurements [15].

Fig. 3 Experimental test setup for double pulse tests.

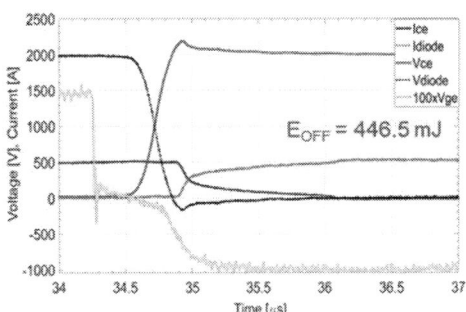

(a)

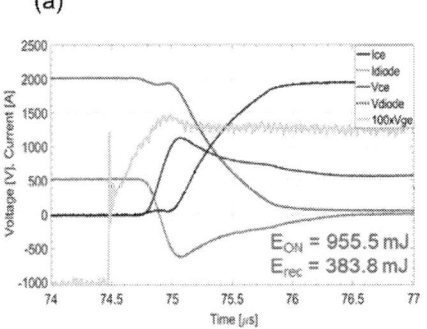

(b)

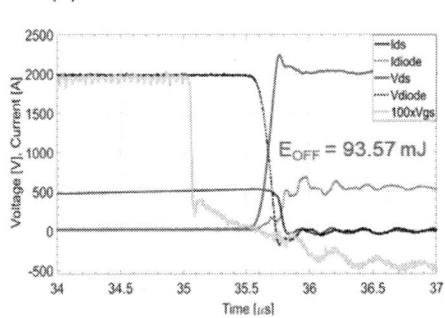

(c)

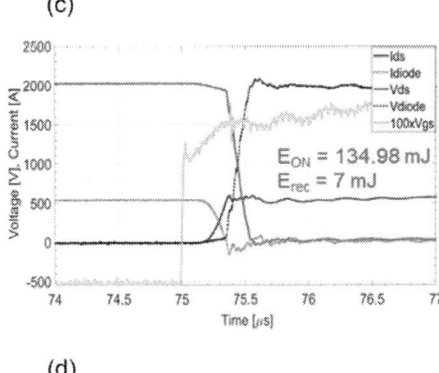

(d)

Fig. 4 Turn-off (a, c) and turn-on (b, d) transient at 2.0 kV, 500 A for Si-IGBT (a, b) and SiC-MOSFETs (c, d) modules at 25 °C for a supply voltage of 2.0 kV.

The SOA tracing during turn-on and turn-off phase of dynamic tests is presented in Figure 5 for Si-

IGBT (@ 25, 150 °C) and SiC-MOSFETs (@ 25, 150,175 °C) with increasing load current. Note that the turn-off process is safe and complete (i.e., black lines) for Si-IGBTs. The turn-on of the IGBT tracing clearly shows the short-circuit initiation (i.e., blue lines > 1100 A). Here, the V_{CE} is gradually increased (i.e., minimum V_{CE} point is not reaching to 0 V after turn-on, thus the de-saturation mode) with the increase of the load current commutation during second turn-on transient beyond 1100 A. This second but incomplete turn-on process (i.e., V_{CE} should simply reach to 0 V after turn-on) is further strengthened at higher temperatures. Note that it is generally difficult to perform dynamic tests for Si-IGBT modules beyond 150 °C due to the increased leakage current in the drift layer of the device that leads to significant voltage drop across the DC capacitor during energization phase of the test. For a fixed temperature and supply voltage V_{DS}, a safe turn-off and turn-on has been observed under all test conditions up to 1.5 kA, 2.5 kV and up to 175 °C for SiC-MOSFET modules. Contrary to Si-IGBT modules, de-saturation behavior was not visible at all in these tests for SiC-MOSFETs, thus producing a good confidence for overload condition. The Si-IGBT modules may be weakened under overload as a result of desaturation when switched beyond 2.2 kV (@ >1100 A) at 150 °C. Our tests at various temperatures and V/I levels lead to a safe overloadability boundary of Si-IGBT modules up to a maximum supply voltage of 2.2 kV and load current limited to a maximum of 900 A at 150 °C, which is consistent to the Si-IGBT LinPak datasheet [12]. Compared to SiC-MOSFETs, extracted energy losses of Si-IGBT modules under overload condition are 7 – 10x higher at 125 °C as illustrated in Figure 6. Turn-on losses of Si-IGBTs increase rapidly at higher current level due to de-saturation during turn-on process. Some oscillations due to circuit parasitics and fast switching have been noticed at high temperatures for SiC-MOSFETs. Overall, switching losses remain approximately constant with variation of temperature for SiC-MOSFET modules.

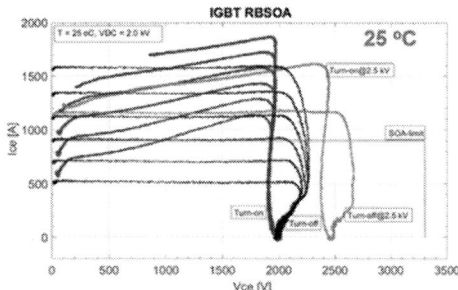

(a)

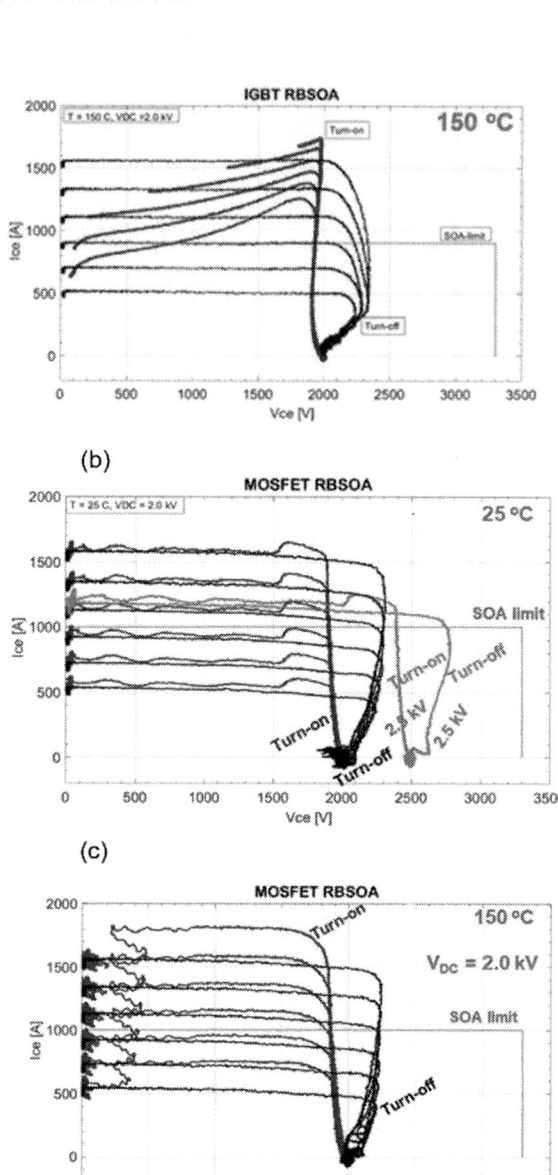

(b)

(c)

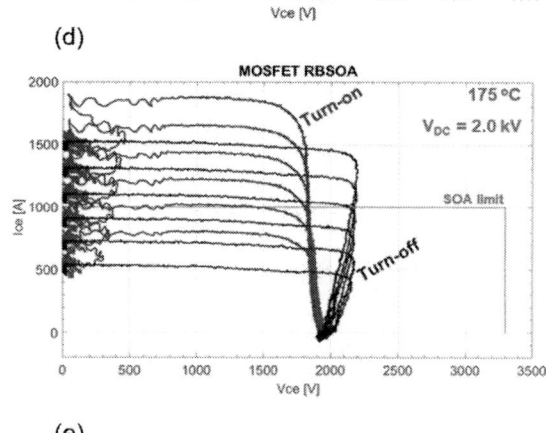

(d)

(e)

Fig. 5 SOA tracing during turn-on and turn-off transient of Si-IGBT (a, b) and SiC-MOSFET modules (c, d, e).

PCIM Europe 2024, 11– 13 June 2024, Nuremberg   DOI: 10.30420/566262282

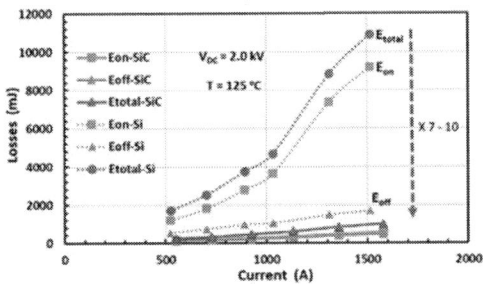

Fig. 6 Switching losses of Si-IGBT and SiC-MOSFETs under overload condition at 125 °C and 2.0 kV.

3 Overloadability model

With experimental base data, overloadability assessment has been carried out for Delta-connected chain-link MMC with full-bridge cell structure as illustrated in Figure 7. A cell DC voltage of 2.0 kV is used here for grid connected line voltage of 15 kV and targeted power range of 10 – 80 MVA. For a given line voltage level, a set of maximum 4 modules have been placed in parallel to meet the required power level starting from single power module at 10 MVA. The targeted arm switching frequency of converter was set to 8.8 kHz with a device switching frequency of 200 Hz. Modulation index was set to 0.85 and using grid frequency of 50 Hz.

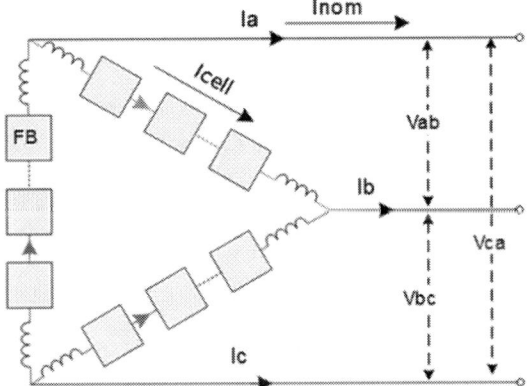

Fig. 7 Delta-connected chain-link MMC topology for overloadability investigation.

The thermal impedance curves of Si-IGBT and SiC MOSFET power modules is shown in Figure 8. The detailed conduction and switching loss data and thermal impedance parameters are given in the datasheet of the respective power modules [12, 13]. Based on the experimental findings, the thermal limit for Si-IGBT and SiC-MOSFET modules was set to 125 ($T_{Jmax-Si}$) and 175 °C ($T_{Jmax-SiC}$)

respectively, meaning that the module temperature should not exceed under overload condition beyond these values. The case temperature during simulation was kept constant to 80 °C. The overload margin (%) is defined as

For Si module: $(I_{@125°C} - I_{rating})/I_{rating}$

For SiC module: $(I_{@175°C} - I_{rating})/I_{rating}$

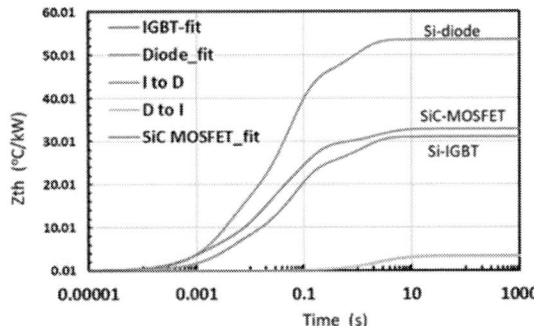

Fig. 8 Thermal impedance Z_{th} curves for 3.3 kV Si-IGBT and 3.3 kV SiC-MOSFET power modules.

Note that the rating current I_{rating} is 450 A for Si-IGBT module and 500 A for SiC-MOSFET module. The definitions of $I_{@125°C}$ and $I_{@175°C}$ represent the overload device current at which the maximum junction temperature of 125°C and 175°C is reached for Si and SiC devices, respectively, where the converter operation is permitted for very limited short-duration. The overload margin ($I_{@125°C}$ - I_{rating})/I_{rating} as a function of device switching frequency is illustrated in Figure 9 for a maximum junction temperature of 125 °C for Si-IGBT module and 175 °C for SiC-MOSFET module at a DC supply voltage of 2000 V and case temperature of 80 °C. A noticeable higher overload margin is obtained for SiC-MOSFET than that of Si-IGBT. Overload margin difference of Si vs SiC gradually increases with the increase of switching frequency.

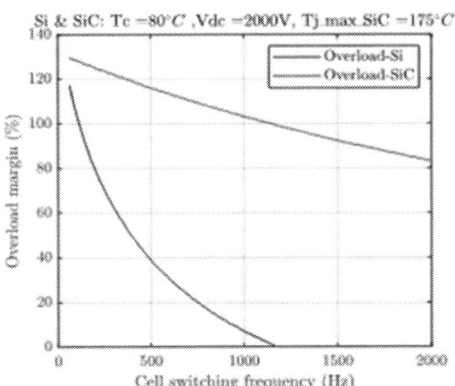

Fig. 9 Overload margin comparison for Si-IGBT and SiC-MOSFET LinPak power modules at 2.0 kV.

PCIM Europe 2024, 11– 13 June 2024, Nuremberg   DOI: 10.30420/566262282

An overload duration of a converter is predicted from the appropriate information of Z_{th} of a device during converter operation. For a converter that is operating at a switching frequency of 6 kHz and a grid voltage of 15 kV, the resultant switching frequency for individual device is around 136 Hz. For this case, considering the converter is under full-power situation (the device operates at rating current, i.e. 450 A for Si, 500 A for SiC), the thermal transition for Si and SiC devices are simulated in Figure 10. Starting from the same case temperature of 80°C under the given cell voltage of 2000 V, the final steady junction temperature T_j is 94 °C and 93 °C for Si and SiC devices, respectively meaning that the converter may operate indefinitely **"allowed operation"** under rated condition as long as T_j is less than $T_{j\text{-max}}$ of Si and SiC power modules.

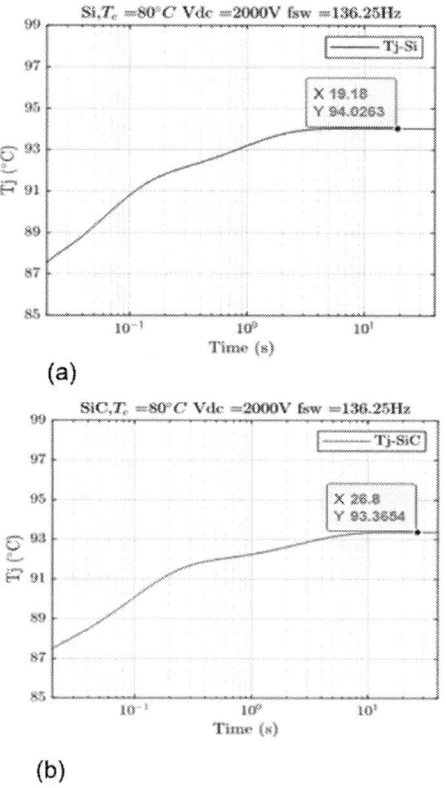

(a)

(b)

Fig. 10 Thermal transition of Si (a) and SiC (b) devices operating at rated current, i.e. 450 A and 500 A, respectively, corresponding to a converter working at 6 kHz switching frequency and 15 kV grid voltage.

Thermal transition from normal steady-state at rated current to another steady-state with higher current than rated current is illustrated in Figure 11. Assuming that the device is initially operating normally (say from -20 to 0 s in Fig. 11) at rated current (i.e., 450 A for Si-IGBT and 500 A for SiC-

MOSFET module). When the overload happens, the thermal transition from normal steady-state to another steady-state with double rated current is also shown in Figure 11 (i.e., say from 0 to 20 s).

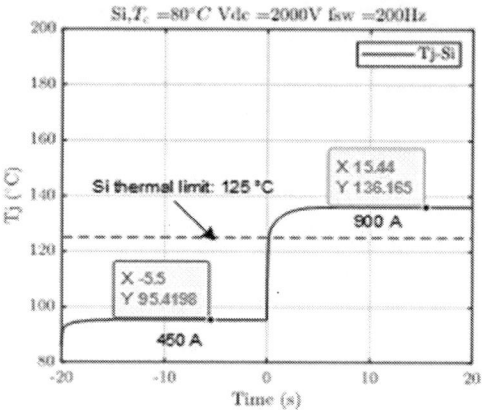

(a)

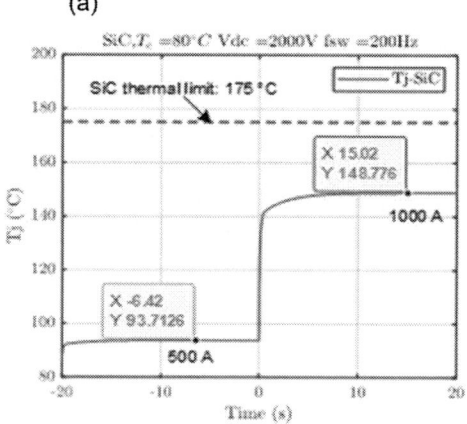

(b)

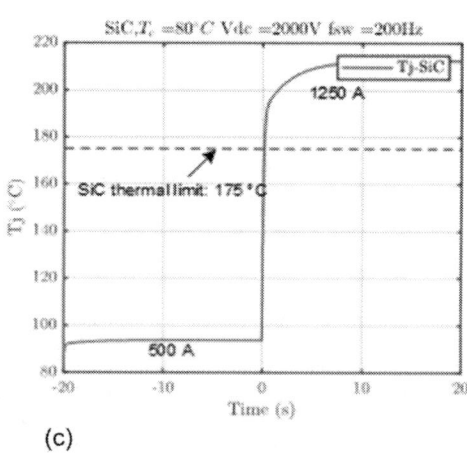

(c)

Fig. 11 Simulated thermal transition of Si (a) and SiC (b, c) devices operating from rated current (i.e., -20 – 0 s) to overload operation (i.e., 0 - 20 s), i.e., overload current equal to 2x I_{rated} in (a, b), and 2.5x I_{rated} in (c).

2018

If the safe thermal limit is set constant (i.e., 125 °C for Si and 175 °C for SiC), it can be seen from Figure 11 that the overload duration is 0.18 s for Si, and infinite for SiC as the maximum junction temperature of SiC is 148 °C (< 175 °C) under this certain operating condition (i.e. f_{sw} = 200 Hz, overload current = 1000 A, case temperature = 80 °C, cell DC voltage = 2000 V). If the overload current is increased to 1250 A for SiC while keeping the other parameters constant, the SiC junction temperature will exceed the thermal limit as shown in Figure 11(c), and similarly, the overload duration can be extracted as 0.16 s for this operating condition.

Taking the converter operation using 15 kV grid voltage at 8.8 kHz arm switching frequency, the overload duration is plotted as a function of overload factor (i.e., device operating current under overload/device rated current) in Figure 12. It can be seen that the overload duration is decreasing with the overload factor as expected. For the same overload duration of 5 s, the maximum overload current SiC can operate at is 2.262 (overload factor) x 500 (rated current) = 1131 A, which is clearly higher than that of Si at 1.794 x 450 = 807 A. For another overload duration of 2 s, the maximum overload currents for SiC and Si devices are 2.301 x 500 = 1151 A and 1.823 x 450 = 820 A, respectively. Note that the dashed part of curves in Figure 12(a) represent infinite overload duration, meaning that the thermal limit (125 °C for Si, 175 °C for SiC) will never be reached when the overload factor is lower than certain values. In this case, the junction temperatures under these lower overload factors are plotted in Figure 12(b). It indicates that the SiC-MOSFET and Si-IGBT based converters can operate permanently (only in theory) if the overload current is lower than 2.246 x 500 = 1123 A and 1.792 x 450 = 806 A, respectively. Despite this, it is not recommended to run the converter at overload condition for a long duration in practice (say up to 20s), considering that the cooling system is usually designed for rated operation (perhaps some margin is covered), and the simulated junction temperature here is the average value without considering the junction temperature fluctuation happening in one switching period.

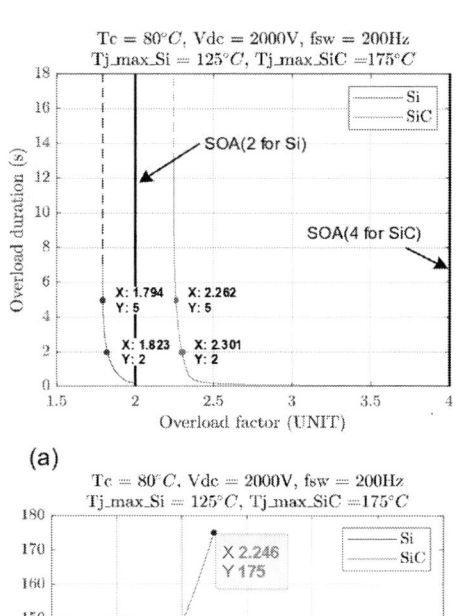

(a)

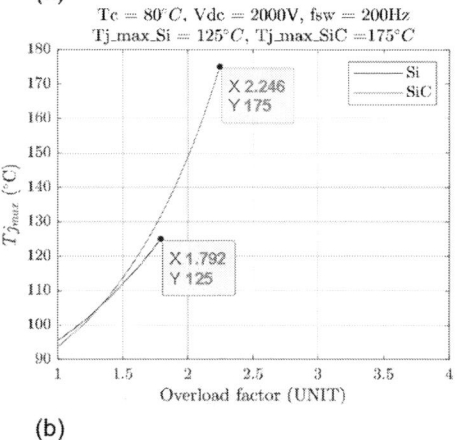

(b)

Fig. 12 Overload duration as a function of overload factor for Si and SiC modules. Also shown is the T_{jmax} with the increase of overload factor for Si and SiC modules.

Converter overload factor versus normal power level is shown in Figure 13 for a grid line voltage of 15 kV. Since paralleled devices are commonly employed in the converter for a given power level and each device does not always operate at rated current. The steep jump-down in each polyline happens when one more device is paralleled to fulfill the power requirement. A current derating factor of 5% is also considered for device paralleling, and as a result, the maximum current for each parallel device does not reach the device rating current (450 A for Si, 500 A for SiC). The converter overload factor for an overload duration of 5 s (Fig. 13a) and 250 ms (Fig. 13b) are displayed for a grid line voltage of 15 kV at arm switching frequency of 6.0 and 10 kHz. It can be clearly seen that the SiC based D-MMC has clearly larger overload factor than that of Si.

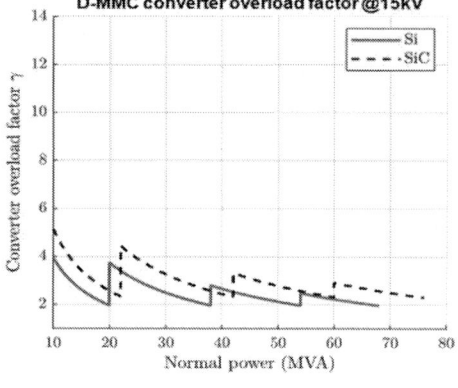

(a)

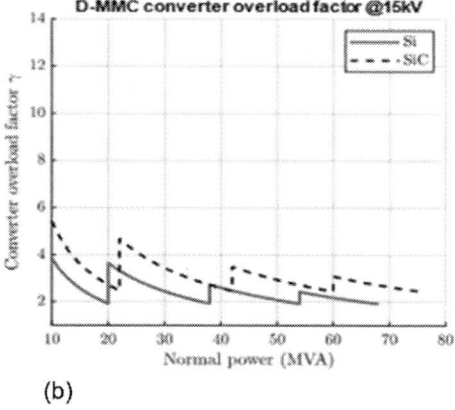

(b)

Fig. 13 Converter overload factor versus power level for overload duration of 5 s, 6 kHz (a) and 250 ms, 10 kHz (b). The steep jumps indicate that more power modules in parallel are required to meet the converter power requirement.

4 Conclusions

Static and dynamic measurement have first been performed for 3.3 kV (450A) Si-IGBT and 3.3 kV (500A) SiC-MOSFET LinPak power modules at various temperatures under overload test conditions. Higher subthreshold leakage current and off-state leakage current is observed beyond 150 °C for Si-IGBT module. Compared to Si-IGBT modules, SiC devices show a stable transient behavior with safe turn-on/turn-off up to 2.5 kV and 175 °C with measured load current over 1.5 kA. SiC-MOSFET presents overall $7 - 10x$ lower switching losses under overload condition compared to equivalent Si-IGBT modules.

For a studied chain-link MMC topology under nominal operating condition (i.e., V_{DC} = 2000 V, f_{sw} = 200 Hz), SiC power modules can operate at higher overload current (i.e., 1131 A, overload factor 2.26) than Si-IGBT modules (i.e., 807 A, overload factor 1.80) for the same overload duration of 5 s. A longer overload duration (say up to 20 s) is even expected as long as overload factor is below certain value (i.e., 1.80 and 2.25 for Si and SiC modules).

Note that, overloadability potential is strictly limited by the T_{jmax} limit set by any power module. Besides extra overload margin, SiC in general provides additional performance advantages such as higher allowed switching frequency and better temperature ruggedness (i.e., potential for higher T_j of SiC device package in future). In certain operation range, considering the need for overload margin, SiC version converter might need less power modules in parallel than Si counterpart (e.g. SiC can offer the power rating with 3 in parallel while Si need 4 in parallel), or in other words, SiC enables wider power range (including both normal power and overload power) compared to Si under the same number of paralleled devices.

5 References

[1] X. Aidong, W. Xiaochen, H. Chao, J. Xiaoming and Li Peng, "Study on Overload Capability and Its Application of HVDC Transmission System in China Southern Power Grid," IEEE Power Engineering Society Conference and Exposition in Africa - PowerAfrica, Johannesburg, South Africa, pp. 1-4, 2007.

[2] H. Bishnoi, S. Alvarez, G. Ortiz and F. Canales, "Comparison of overload protection methods for LLC resonant converters in MVDC application", ECCE-2018, Portland, OR, USA, pp. 3579-3586, 2018.

[3] N. B. Kadandani, M. Dahidah and S. Ethni, "On extending the overloading capability of modular multilevel converter", IEEE PEDS 2019, Toulouse, France, 9 – 12 July 2019.

[4] J. He, N. Weise, L. Wei, N. A.O. Demerdash, "A fault-tolerant topology of T-Type NPC Inverter with increased thermal overload capability", APEC, pp. 1065 – 1070, 2016.

[5] X. Perpiñà, J-F. Serviere, J. Urresti-Ibañez, I.Cortés, X. Jordà, S. Hidalgo, J. Rebollo, and M. Mermet-Guyennet, "Analysis of clamped inductive turnoff failure in railway traction IGBT power modules under overload conditions", IEEE Trans On Industrial Electronics, Vol. 58, No. 7, pp. 2706-2714, July 2011.

[6] D. Kaczorowski, B. Michalak and A. Mertens, "A novel thermal management algorithm for improved lifetime and overload capabilities of traction converters," 17th European Conference on Power Electronics and Applications (EPE'15 ECCE-Europe), Geneva, Switzerland, pp. 1 – 10, 2015.

[7] R. Rodrigues, T. Jiang and D. Das, "Comparison of Cooling Solutions to Improve Overload Capability of Power Semiconductor Devices," IEEE Energy Conversion Congress and Exposition (ECCE), Portland, OR, USA, pp. 5074-5080, 2018.

[8] G. Hao et al., "Study on Improving the Short-Time Overcurrent Capability of Press- pack IGBTs Using Phase Change Materials", 2020 IEEE 9th International Power Electronics and Motion Control Conference (IPEMC2020-ECCE Asia), Nanjing, China, pp. 2173-2177, 2020.

[9] H. Ren et al., "A phase change material integrated press pack power module with enhanced overcurrent capability for grid support—a study on FRD," in IEEE Transactions on Industry Applications, Vol. 57, no. 4, pp. 3956-3968, July-Aug. 2021.

[10] S. Bhadoria, F. Dijkhuizen, R. Raj, X. Wang, Q. Xu, E. Matioli, K. Kostov, H-P Nee, "Enablers for overcurrent capability of silicon-carbide-based power converters: an overview", IEEE Trans. On Power Electronics, Vol. 38, No. 3, pp. 3569-3589, 2023.

[11] Y. Yu, H. Li, R. Yao, F. Iannuzzo, Z. Zhu and X. Chen, "Short-Circuit capability optimization of press-pack IGBT by improving active edge heat dissipation", IEEE Trans on Power Electronics, Vol. 38, No. 5, pp. 6143-6156, May 2023.

[12] Si-IGBT LinPak power module, Data Sheet, "5SNG 0450X330300", Doc. No. 5SYA 1458-03 May 20

[13] SiC MOSFET LinPak power module, Data Sheet, "5SFG 0500X330100".

[14] R.Schnell, S.Hartmann, D.Trüssel, F.Fischer, A.Baschnagel, M.Rahimo, "LinPak, a new low inductive phase-leg IGBT module with easy parallelling for high power density converter Design", PCIM-2015, 19-21 May 2015, Germany.

[15] S. Kicin, R. Burkart, J-Y. Loisy, F. Canales, M. Nawaz, G.t Stampf, P. Morin and T. Keller, "Ultra-fast switching 3.3kV SiC high-power module", PCIM 2020, pp. 1-8, Germany.

PCIM Europe 2024, 11– 13 June 2024, Nuremberg   DOI: 10.30420/566262283

Improved Reliability of a 2200 V SiC MOSFET Module with an Epoxy-Encapsulated Insulated Metal Substrate

Hiroshi Kono[1], Shun Takeda[1], Eitaro Miyake[1], Tomohiro Iguchi[2], Teruyuki Ohashi[2], Georges Tchouangue[3] and Kazuya Kodani[4].

1 Toshiba Electronic Devices & Storage Corporation, 2 Toshiba corporation, 3 Toshiba Electronics Europe GmbH, 4 Toshiba Infrastructure Systems & Solutions Corporation.

Corresponding author: Hiroshi Kono, hiroshi.kono@toshiba.co.jp

Abstract

The impact of combining epoxy-potting encapsulation with an insulated metal substrate (IMS) on the performance and reliability of SiC MOSFET modules was investigated. Static and dynamic characteristics, thermal resistance, and power cycle tolerance were measured, and a high-temperature bias test and a high-temperature humidity test were carried out. An IMS module with epoxy-potting encapsulation was compared with a conventional ceramic insulated substrate with silicone-gel encapsulation. The IMS module was found to have a higher thermal cycling tolerance, which allowed for a more flexible copper pattern layout. The optimized copper pattern layout enabled reduced conduction loss in the IMS module. In addition, the IMS module exhibited improved power cycling tolerance compared with the conventional ceramic insulated substrate. This improved performance and reliability are expected to contribute to the realization of higher-density power units.

1 Introduction

Silicon carbide power devices offer lower power loss and higher power density compared with silicon due to their superior material properties. Our group has previously developed 3.3 kV class SiC metal-oxide-semiconductor field effect transistors (MOSFETs) and 2.2 kV class SiC MOSFETs modules and showed that these modules make it possible to reduce the volume of the cooling system compared with silicon insulated gate bipolar transistors (Si-IGBTs) [1,2]. As discussed in these studies, thermal management is a key factor in further reducing the volume and weight of power conversion units. For example, power units with higher power densities are subject to higher thermal cycling stress, which reduces the life of the power unit. Therefore, it is important to reduce losses, and thermal resistance and to improve thermal cycling life in order to increase the power density of the unit.

Traditionally, ceramic substrates have been used for power modules owing to their excellent insulation and thermal properties. With the recent development of resin insulation materials, insulated metal substrates are also being considered as promising candidates for power module insulation materials. Insulating resin has a linear expansion coefficient similar to that of copper, and thus maintains thermal cycling reliability even as substrate size increases. Epoxy-potting encapsulations can be used for insulating resin substrates and can improve power cycle tolerance compared with conventional silicone-gel encapsulations. However, since the thermal conductivity of insulating resin is lower than that of ceramic substrates, thus increasing the thermal resistance of the device, the overall thermal design is an important factor for improving reliability.

In this study, we fabricated modules with reduced thermal resistance by optimizing the chip layout and investigated the impact on characteristics and reliability. Modules were fabricated using a 2.2 kV Schottky barrier diode (SBD)-embedded SiC-MOSFET chip [2], and 2.2 kV modules using ceramic substrates and standard gel-encapsulated packages [2] were used for reference.

2 SBD-embedded SiC MOSFET device

Degradation of on-resistance caused by bipolar operation has been a problem for SiC MOSFETs [3,4]. SBD-embedded MOSFETs are an essential solution to this problem because they are able to

2022

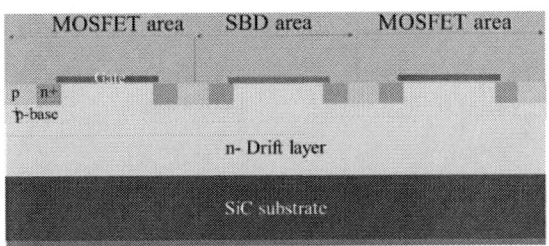

Fig. 1: Schematic cross-section of the fabricated SBD-embedded SiC MOSFET.

Fig. 3: 2200 V, 250A, 2in1 all SiC IMS module with an epoxy-potting encapsulation.

suppress the bipolar operation of the body diode of a MOSFET [5,6]. Figure 1 shows a schematic cross-section of the fabricated 2.2 kV class SBD-embedded SiC MOSFET that is based on a planar gate structure.

Figure 2 shows the reverse current characteristics of the developed MOSFETs when V_g = −5 V and the temperature is 150°C. The SBD-embedded SiC MOSFET does not start bipolar operation until the current exceeds 450 A, even at 150°C.

3   Insulated metal substrate module

In order to reduce the size of a power converter, it is important to reduce the power consumption and package size of the power module. As mentioned above, IMS can achieve sufficient reliability even if

with increased substrate size and copper pattern thickness which help to achieve a low resistance package. In this study, to take advantage of these features and overcome the low thermal conductivity, the thermal resistance of the module was reduced by optimizing the chip layout. This was accomplished by suppressing the thermal interference between the chips. Figure 3 shows the fabricated 2.2 kV, 250 A, 2-in-1 all SiC IMS module with an epoxy-potting encapsulation.

Figure 4 shows a comparison of the thermal resistances estimated from the structure function between the IMS module and the conventional ceramic module. The thermal resistance of the IMS module was 16% lower than that of the conventional package.

Figure 5 shows a comparison of the power cycling test between the IMS module and the ceramic module. The input power was set such that T_{vjmax}, ΔT, and t_{on} were 150°C, 80°C, and 2 s, respectively. The vertical axis shows the rate of change of V_{sd}. Results are normalized with respect to the lifetime of the conventional ceramic module, which was defined as the point at which the rate of change exceeds 1.05. The lifetime of the IMS module was six times longer than that of the

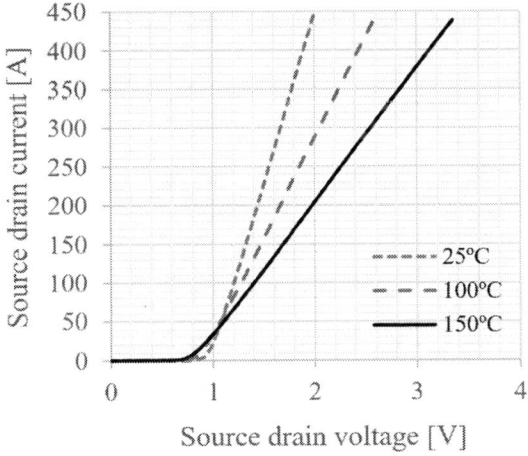

Fig. 2: I_{sd}-V_{sd} characteristics of the SBD-embedded SiC MOSFET when V_g = −5 V at 150 °C.

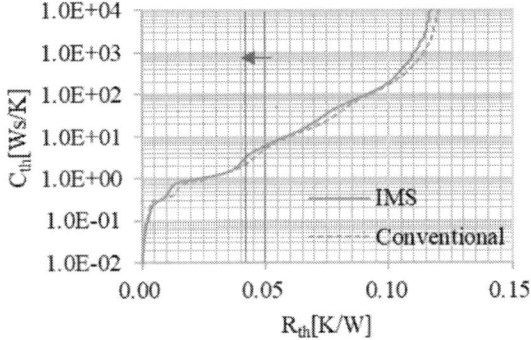

Fig. 4: Comparison of structure functions between the conventional and IMS modules.

PCIM Europe 2024, 11– 13 June 2024, Nuremberg   DOI: 10.30420/566262283

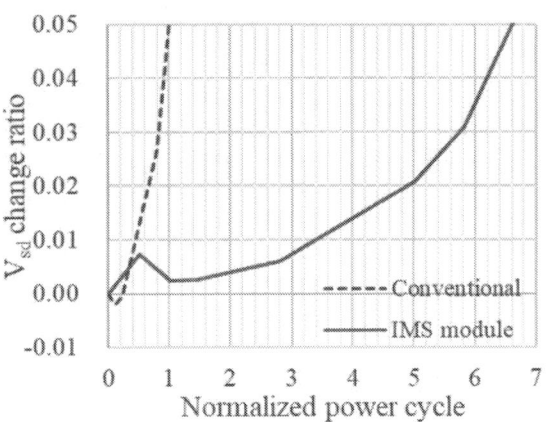

Fig. 5: Comparison of power cycling tolerance between the conventional and IMS modules. Tests were caried out with T_{jmax} = 150°C, d_{tj} = 80°C, and t_{on} = 2 s.

ceramic module. These experimental results demonstrate that the developed module had higher performance and reliability than did the conventional modules.

Figures 6 and 7 show scanning electron microscope (SEM) images of the bonding wires at the end of the power cycling test. Whereas the ceramic module wire separated at the interface between the bonding wire and the surface metal, the IMS module wire cracked vertically. It is thought that the epoxy-encapsulation prevented wire lift-off and improved the lifetime.

The thermal resistance was calculated from the power cycling test results as shown in Fig. 8. This result shows that the developed IMS module had lower thermal resistance than did the conventional

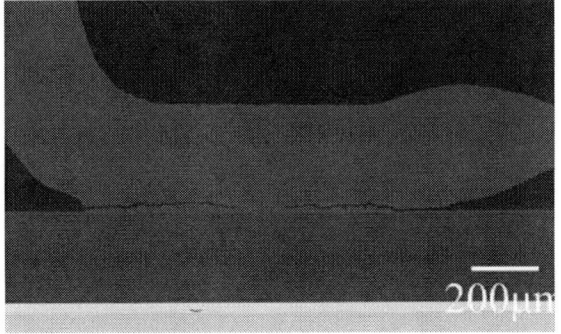

Fig. 6: SEM image of the bonding wire at the end of the power cycling test for the ceramic module.

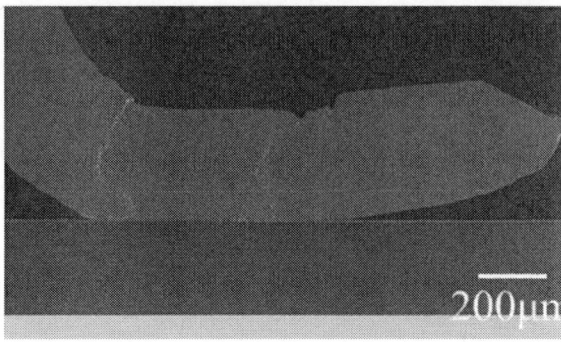

Fig. 7: SEM image of the bonding wire at the end of the power cycling test for the IMS module.

module, which matches the results evaluated by the structure function.

In addition to thermal reliability, we also investigated electrical reliability. Although there

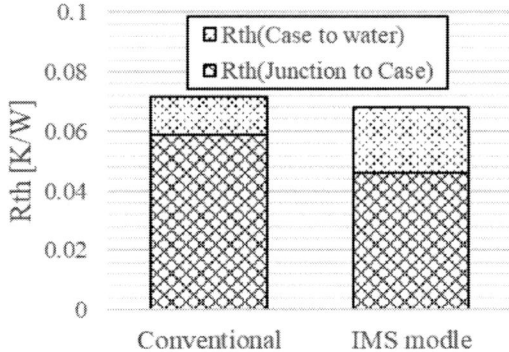

Fig. 8: Comparison of thermal resistivity evaluated by a power cycling test between the conventional and IMS modules.

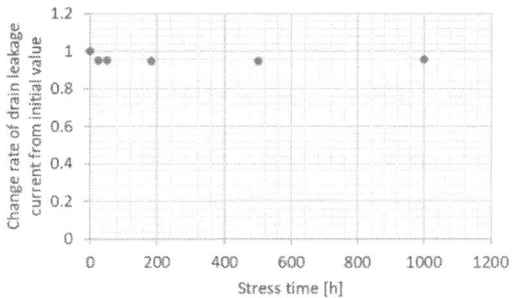

Fig. 9: Time dependence of leakage current evaluated by a high-temperature drain vias stress test (V_d = 2200 V, V_g = −10 V, 150°C).

was a concern that the moisture resistance of the resin-insulated substrate would be lower than that of the ceramic substrate, no failures occurred during a 1000 h high-temperature humidity test conducted under conditions of 85°C, 85% humidity, and 1760 V. In addition, as shown in Fig. 9, no degradation was observed in the high temperature drain bias stress test where V_d = 2200 V, V_g = −10 V, and the temperature was 150°C. These results indicate that the developed epoxy-potting encapsulated IMS had low thermal resistance and high reliability.

4 Comparison of device characteristics

Figures 10 and 11 show comparisons of the I_d-V_d characteristics and on-resistance between the conventional and IMS modules. The resistances were measured at the main terminals. The copper pattern can be made thicker on the IMS, thereby the package loss was reduced.

Figures 12 shows the turn-on and turn-off switching waveforms of the conventional module and IMS module. Figures 13 shows the comparison of the switching loss. No obvious difference in switching losses was observed.

The power consumption of inverters using the conventional Si-IGBT module and the developed SiC module were calculated and compared. In the inverter loss calculation, we assumed the case of Si-IGBTs are in a T-type three-level inverter and SiC-MOSFETs in a two-level inverter [1, 7].

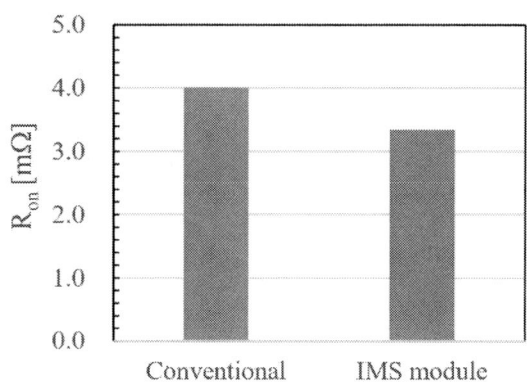

Fig. 11: Comparison of on-resistance between the conventional and IMS modules.

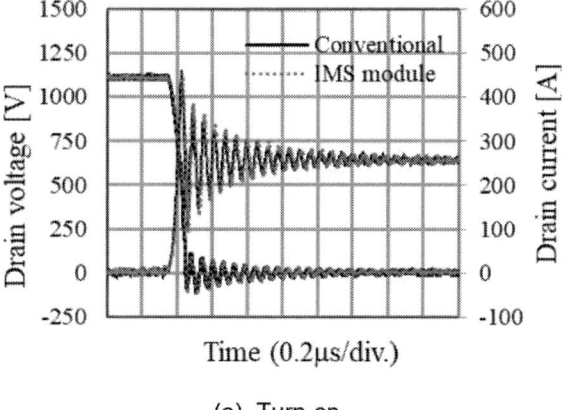

(a) Turn-on

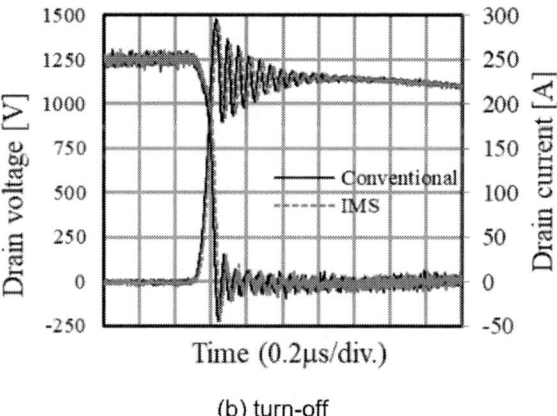

(b) turn-off

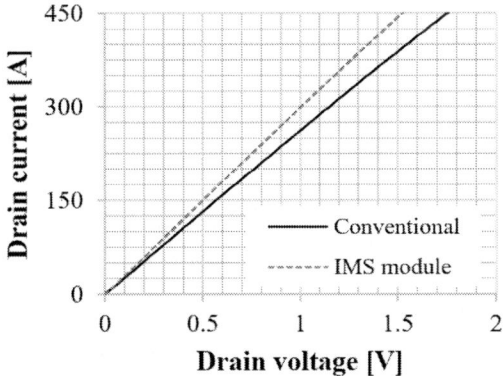

Fig. 10: I_d-V_d characteristics of the conventional and IMS modules when V_g = 20 V.

Fig. 12: Comparison of the switching waveforms between the IMS module and the conventional module at 150°C. Test condition: V_{ds}=1100V, I_d = 250 A, V_{gs} = −6 /+20 V, R_{gon} = 0.5 Ω, R_{goff} =3 Ω.

PCIM Europe 2024, 11– 13 June 2024, Nuremberg   DOI: 10.30420/566262283

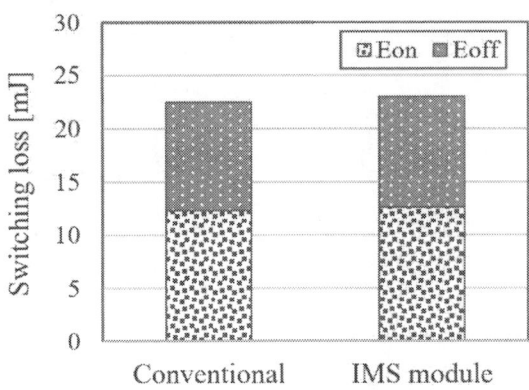

Fig. 13: Comparison of switching losses between conventional and IMS modules.

Figure 14 shows the calculation results for the power dissipation of an inverter arm. The calculation was carried out under conditions of V_{ds} = 1200 V, I_d = 200 A, cosϕ = 1, f_c = 10 kHz, and T_j = 150°C for SiC MOSFET. For Si-IGBTs, the calculation was carried out under the conditions of V_{ce} =600 V, Id = 200 A, cos = 1, fc = 5 kHz, and Tj = 150°C. The inverter loss was 32% lower in the SiC MOSFET module compared with the Si-IGBT module.

We also estimated the impact of the developed SiC MOSFET module on reducing cooler volume. For this, we used the cooling system performance index (CSPI) [W/K L] [8], which is calculated as

$$\mathrm{CSPI} = \frac{1}{\{R_{th(s-a)}V_{CS}\}}$$

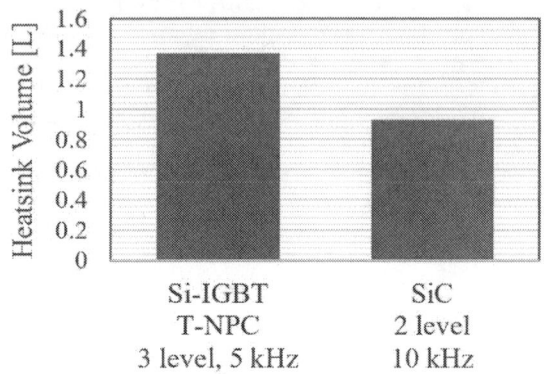

Fig. 15: Heatsink volume of the inverter with Si IGBT module and the developed SiC MOSFET module when CSPI is assumed to be 5.

where $R_{th(s-a)}$ is the thermal resistance between the heatsink and ambient and V_{cs} is the volume of the heatsink. We assumed a common cooling system and a CSPI value of 5. The operating conditions of the inverter were the same as in Fig. 14. The calculated cooling system volumes at T_j = 150°C and T_a = 40°C are shown in Fig. 16. These results show that the volume of the cooling system for the developed SiC MOSFET module was 27% lower than that of the Si IGBT module.

In high-load applications, the operating conditions ΔT_j and T_{jmax} can be suppressed to ensure the power cycle life of the module. Assuming that lowering ΔT_j by 20°C improves the power cycle endurance by a factor of about 3, the developed module allows ΔT_j to be increased by 20°C. The improved power cycle tolerance allows a reduction

Fig. 14: Comparison of an inverter arm power dissipation between the Si IGBT modules and the developed SiC MOSFET modules.

	Conventional	IMS module
Current [A]	100	133
Number of modules	4	3
Loss per module [W]	66.1	92.5
Total loss [W]	264.2	277.6
ΔT_j [°C]	50	69.6

Table 1   Comparison of temperature rise when the number of modules in parallel is reduced.

2026

in the number of modules connected in parallel, whereas conventional modules need to be arranged in parallel to ensure reliability by reducing the current per module. Table 1 shows this comparison. The temperature rose by 50°C with 4 modules in parallel and an output current of 100 A and by 69.6°C when the number of modules in parallel was 3 and the current was increased to 133 A.

5 Conclusion

We have developed an IMS module SiC MOSFET that has lower conduction and thermal resistance and six-times higher power cycle life compared with conventional ceramic modules. The developed SiC MOSFET module exhibited a 32% reduction in power dissipation compared with Si-IGBTs. The impact of the developed module on volume reduction of the cooling system was also evaluated. The results showed that the volume of the heat sink can be reduced by 27%. It was also demonstrated that improvement in power cycle endurance contributed to reducing the number of modules in parallel. Therefore, the combination of third-generation SiC MOSFETs and IMS modules can contribute to reducing the volume of power units.

6 References:

[1] H. Kono, T. Iguchi, T. Hirakawa, H. Irifune, T. Kawano, M. Furukawa, K. Sano, M. Yamaguchi, H. Suzuki, and G. Tchouangue, "3.3 kV All SiC MOSFET Module with Schottky Barrier Diode Embedded SiC MOSFET", Proceedings of PCIM Europe 2021, pp. 914-919, (2021).

[2] T. Ogata, H. Kono, H. Irifune, S. Fujii, T. Tanaka and G. Tchouangue, "A Novel 2200 V Schottky Barrier Diode-Embedded SiC MOSFET Module", Proceedings of PCIM Europe 2023, pp. 1-6 (2023).

[3] A. Agarwal, H. Fatima, S. Haney, and S. H. Ryu, I, "A New Degradation Mechanism in High-Voltage SiC Power MOSFETs", IEEE Electron Device Lett. 28, 587, (2007).

[4] K. Konishi, S. Yamamoto, S. Nakata, Y. Nakamura, Y. Nakanishi, T. Tanaka, Y. Mitani, N. Tomita, Y. Toyoda, and S. Yamakawa, "Stacking fault expansion from basal plane dislocations converted into threading edge dislocations in 4H-SiC epilayers under high current stress", Journal of Applied Physics, vol. 114, p. 014504, (2013).

[5] K. Kawahara, S. Hino, K. Sadamatsu, S. Tomohisa, and S. Yamakawa, "Impact of embedding Schottky barrier diodes into 3.3kV and 6.5kV SiC MOSFETs", Materials Science Forum, vol. 924, p. 727, (2018).

[6] M. Furukawa, H. Kono, K. Sano, M. Yamaguchi, H. Suzuki, T. Misao and G. Tchouangue, "Improved reliability of 1.2kV SIC MOSFET by preventing the intrinsic body diode operation", Proceedings of PCIM Europe 2020, pp.1-5, (2020).

[7] I. Staudt, "3L NPC & TNPC Topology. Semikron Application Note: An11001 SEMIKRON" in, 2015.

[8] U. Drofenik, G. Laimer and J. W. Kolar, "Theoretical Converter Power Density Limits for Forced Convection Cooling", Proceedings of PCIM2005, pp. 608-619, (2005).

* Company names, product names, and service names may be trademarks of their respective companies.

PCIM Europe 2024, 11– 13 June 2024, Nuremberg   DOI: 10.30420/566262284

Paralleling 3.3-kV/800-A rated SiC-MOSFET Modules: An Optimization Method

Hiroyuki Irifune[1], Shinichi Hiroshige[1], Hiroshi Matsuyama[1], Tsuguhiro Tanaka[1], Hiroshi Kono[1]
and Georges Tchouangue[2]

[1] Toshiba Electronic Devices & Storage Corporation, Japan
[2] Toshiba Electronics Europe GmbH, Germany

Corresponding author: Hiroyuki Irifune, hiroyuki.irifune@toshiba.co.jp
              Shinichi Hiroshige, shinichi1.hiroshige@glb.toshiba.co.jp
Speaker: Hiroyuki Irifune, hiroyuki.irifune@toshiba.co.jp

Abstract

When power semiconductor modules are connected in parallel, the switching characteristics and current imbalance of each device need to be aligned. This paper focuses on the mutual inductance between the gate driver and the main circuit and the stray inductance of the main circuit when a silicon carbide (SiC) metal–oxide–semiconductor field-effect transistor (MOSFET) module with a rating of 3.3 kV/800 A is operated in parallel, to examine the effect on the switching operation and describe the optimum method. First, the effect of mutual inductance between the gate wiring of devices in parallel and the main circuit was evaluated by performing actual measurements, and it was shown that gate voltage fluctuations could be suppressed by bringing the gate wiring of each device closer together. Next, when an external capacitor C_{gs} was inserted as a gate-noise countermeasure, it was shown that reducing the loop between the main circuit and the capacitor could decrease the mutual inductance and lessen the difference in switching characteristics. Finally, the difference in stray inductance of the main circuit of each device was evaluated. Differences in stray inductance of the main circuit cause current imbalance, resulting in differences in the loss of each device. As a result, there is a difference in the thermal fatigue life of each device, and the expected life of the system is shorter than anticipated. It is therefore important to match stray inductances of the main circuits within the range where turn-off surges are allowed.

1   Introduction

In inverters with capacities exceeding several hundred kilowatts, such as those used in the rail, industrial, and energy sectors, it is common practice to connect them in parallel when it is difficult for a single power semiconductor module to meet the capacity of the equipment [1] [2]. When power semiconductor modules are connected in parallel, the current imbalance needs to be reduced, and the switching characteristics need to be matched in order to ensure uniform electrical and thermal stress and to prevent malfunctions due to resonance between modules [3] [4]. This section describes the general issues to be considered in parallel drives using metal-oxide-semiconductor-field effect transistors (MOSFETs) in terms of devices, gate drivers, and main circuits.

First, it is important to match the characteristics of the devices constituting the parallel circuit as closely as possible. In particular, deviations in the MOSFET threshold voltage V_{th} lead to resonance and other problems due to timing deviations during switching, while on-resistance generates current imbalance [5].

Next, there are two methods for connecting gate drivers: one in which each device connected in parallel is driven by an individual gate driver, and the other in which a common gate driver is used. When using individual gate drivers, the gate-drive circuits for each device are separated, thereby reducing the risk of malfunctions due to parasitic vibrations. However, the circuit is more complex, which necessitates a larger number of components because a gate driver is installed for each device. In the common method, only one driver is needed, but resonance between the gate wiring inductance and the input capacitance may occur, causing the MOSFETs to malfunction during switching. Therefore, as a means of preventing malfunctions, gate resistors and gate-source capacitances are connected to each device in order to suppress resonance.

Finally, with regard to the main circuit, the stray inductance L_s between each module and the direct-current link capacitor leads to a current imbalance, which causes differences in the losses of each device because the current is diverted by the ratio of L_s [6]. The current imbalance due to stray inductance in the main circuit has been discussed when implementing parallel drive of SiC MOS modules with high speed operation and high density [5] [6]. On the other hand, the effect of mutual inductance between the gate driver and the main circuit has not been discussed. This paper describes a method for optimizing parallel drives by clarifying the influence of mutual inductance between the gate driver and the main circuit and the stray inductance of the main circuit when parallel drive is performed using the 3.3-kV/800-A all-SiC MOSFET module in the iXPLV package developed by the company.

2 Construction of parallel-driven evaluation systems

2.1 Devices in use and evaluation circuits

Figs. 1 and 2 show the 3.3-kV/800-A all-SiC module in the iXPLV package developed by the company, which features low stray inductance and high-power density [7]. The basis of the evaluation circuit configuration in this paper is shown in Fig. 3. A common gate-drive system was used. V_{DD} represents the main voltage supply, C represents DC link capacitor, and L_m represents inductive loads. Gate resistors R_{g1-4} and external gate-to-source capacitances C_{gs1-4} were inserted at each gate. L_{s1-4} represents the stray inductance of the main circuit and $M_{1,2}$ represents the mutual inductance of the stray inductance L_s of the main circuit and the external capacitance C_{gs}.

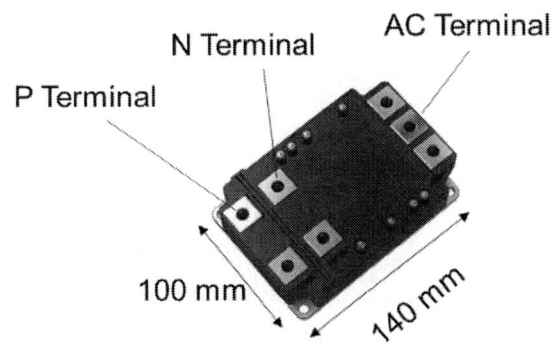

Fig. 1 3.3-kV/800-A, 2-in-1 all-SiC module (iXPLV).

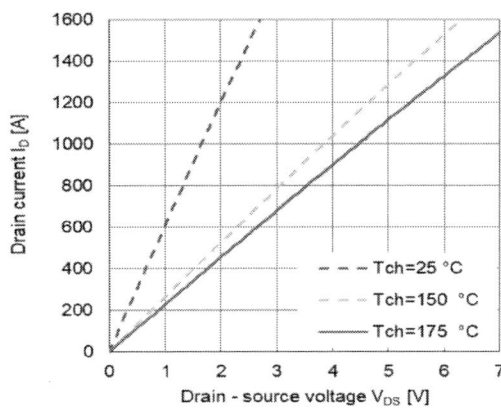

Fig. 2 Forward I_D-V_{DS} characteristics in the on-state (V_{gs} = +20 V) of the 3.3-kV/800-A, 2-in-1 all-SiC MOSFET module (iXPLV).

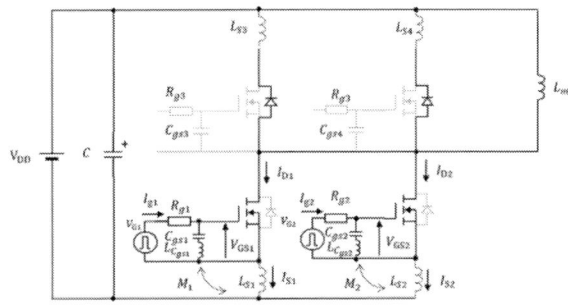

Fig. 3 Parallel drive evaluation circuit.

2.2 Influence of mutual inductance between gate wiring and main circuit

When devices to be driven in parallel are driven by a common gate driver, attention must be paid to the arrangement of the gate drive wiring. The effect of mutual inductance between the gate wiring and the main circuit was investigated using the evaluation circuit shown in Fig. 3. The measurement conditions were V_{DD}=1920 V, I_d=2300 A, $R_{g(on/off)}$=10 Ω/10 Ω, T_{ch}=175 °C, L_m=100 µH, $V_{GG(on/off)}$=+20 V/−6 V, C=1500 µF, C_{gs}=100 nF, and L_s=250 nH. The switching waveforms are shown in Fig. 4(b) when the gate wiring of each device is separated, as shown in Fig. 4(a). Large variations in V_{gs}, I_d, and V_{ds} were observed. This is because the area of the low-impedance loop due to the source line is larger, and when the magnetic flux due to the main circuit current penetrates this loop, current flows in the loop. As a result, the gate voltage fluctuates due to the resistance and inductance components, leading to device breakdown. However, when the gate wiring is placed close together, as shown in Fig. 5(a), the V_{gs} fluctuations subside and improve, as shown in Fig. 5(b).

This is attributed to the gate wirings being close to each other and having a small loop area, which reduces the mutual inductance of the main circuits and thus reduces the gate voltage fluctuations due to magnetic flux. It is therefore important to keep the gate wirings of devices that are driven concurrently as close together as possible.

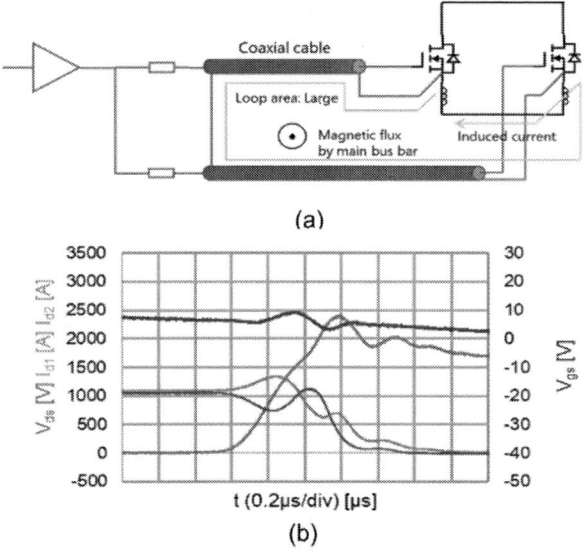

(a)

(b)

Fig. 4 Circuit concept diagram with gate wiring separated (a) and turn-off switching waveforms (b).

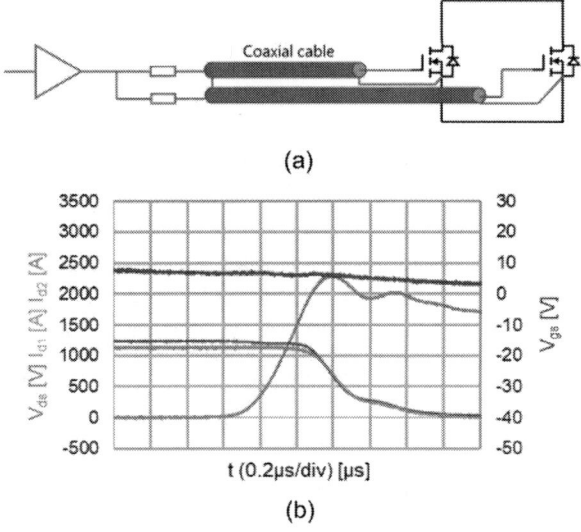

(a)

(b)

Fig. 5 Circuit concept with gate wiring in close proximity (a) and turn-off switching waveforms (b).

2.3 Influence of mutual inductance between the external gate-source capacitor and main circuit

In devices with fast switching capability, such as SiC MOSFETs, a capacitor may be connected between the gate and source in order to reduce gate noise. However, it is important to pay attention to the mutual induction between the inductance of the capacitor connection loop and the stray inductance of the main circuit. Devices such as SiC-MOSFETs that can operate at high speeds generate a back electromotive force due to mutual induction which causes the gate voltage to fluctuate. Equation (1) shows the gate voltage Vgs of the equivalent circuit in Fig 3.

$$V_{gs} = \frac{1}{C_{gs}} \int I_{c_{gs}} \, dt + \left(L_{cgs} \frac{dI_{c_{gs}}}{dt} + M \frac{dI_d}{dt} \right) \quad (1)$$

Where: $i_{c_{gs}}$ is current of C_{gs}.

From equation (1), as dI/dt increases due to high-speed operation, the effect of mutual inductance increases. This can result in unintended false turn-on during off state. To check the effect of switching due to mutual inductance with the external capacitor C_{gs}, simulations were performed using the LT Spice Simulator. Simulations were performed with the evaluation circuit simulating Fig. 3 and the measurement conditions V_{DD} = 1800 V, I_d = 800 A, $R_{g(off)}$ = 3.9 Ω, T_{ch} = 175 °C , $V_{GG(on/off)}$ = +20 V/−6 V, L_s = 70 nH, C_{gs} = 100 nF and L_{cgs} = 15 nH with a single drive. Fig. 6 shows the relationship between dI/dt when the mutual inductance M is varied. As the mutual inductance M shifts towards the positive side, dI/dt tends to become lower and switching loss increases. Meanwhile, dI/dt increases as the shift towards the negative side, and at M = −3 nH, the electromotive force due to mutual induction becomes significantly peaks sharply, as shown in Fig. 7.

Simulations were performed in parallel drives in order to investigate the effect of the difference between the mutual inductance M and the device threshold voltage V_{th} in parallel drives. The measurement were performed with two devices driven in parallel at I_D = 1600 A. Other conditions were the same as those for the single drive. If there is no difference in V_{th} and the mutual inductance is M = −3 nH, then the waveform will have two devices falsely turning-on simultaneously, as shown in Fig. 8. When there is a difference in V_{th}, as shown in Fig. 9, the switching timing shifts and the waveform repeats alternating falsely turning-on. This state is an overcurrent condition, given that the current should be shared by two devices, but is borne by one, which can lead to device breakdown. Therefore, in parallel drives, the influence of the mutual inductance of the main circuit and C_{gs} is more pronounced, and the mounting method of the external C_{gs}, which determines the mutual inductance, must be fully considered.

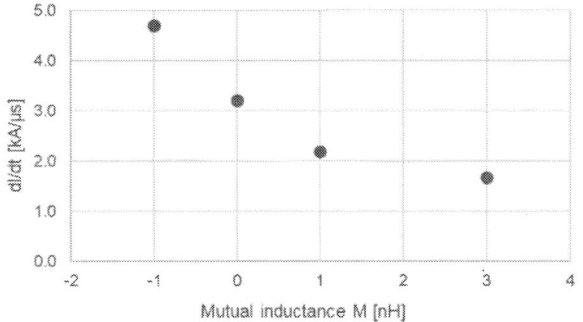

Fig. 6 Relationship between mutual inductance M and dI/dt.

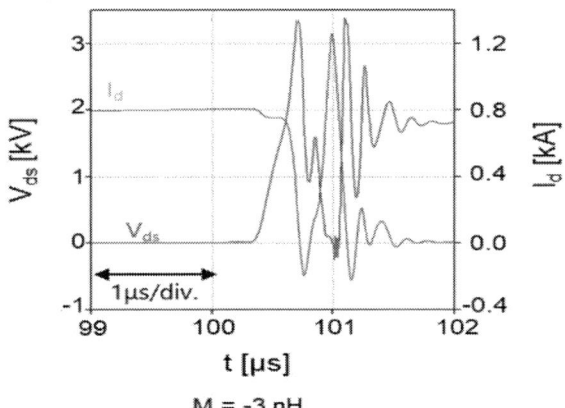

M = -3 nH

Fig. 7 Turn-off switching waveform at mutual inductance M = -3 nH in single operation.

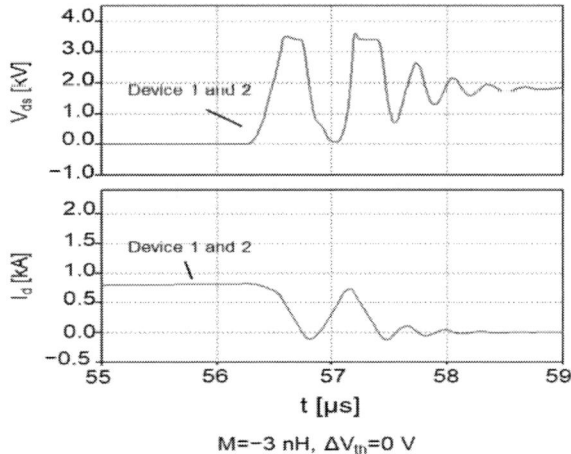

M=−3 nH, ΔV_{th}=0 V

Fig. 8 Turn-off switching waveform at a threshold voltage difference of 0 V and mutual inductance M = −3 nH in parallel operation.

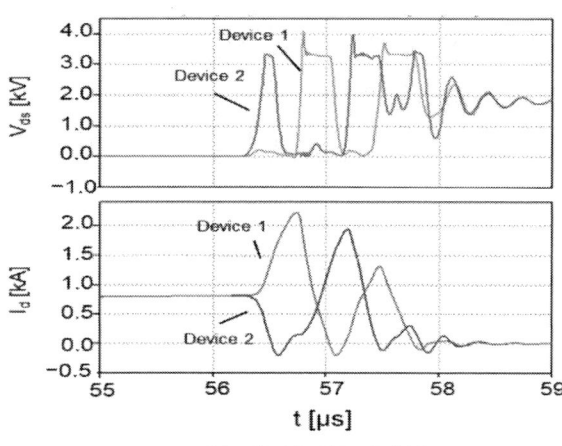

M=−3 nH, ΔV_{th}=0.5 V

Fig. 9 Turn-off switching waveform at a threshold voltage difference of 0.5 V and mutual inductance M = −3 nH in parallel operation.

2.4 Influence of stray inductance in the main circuit

As discussed above, stray inductance in the main circuit causes current imbalance, which may lead to reliability degradation due to variations in device temperature. Attempting to align the inductance of the main circuit can increase the total inductance. The effect of increased inductance on switching was confirmed by simulation.

Fig. 10 shows the waveform results of the simulation of inductive load switching with a main circuit inductance of L_s = 70 nH and 140 nH under identical conditions. The measurement conditions were V_{DD}=1800 V, I_d=800 A, $R_{g(on)}$/$Rg_{(off)}$=2.2 Ω/3.9 Ω, T_{ch}=175°C, L_m=100 µH, $V_{GG(on/off)}$=+20 V/−6 V and C_{gs1-4}=100 nF. At turn-on, when L_s was large, the voltage dropped significantly due to di/dt caused by the rise in current.

Consequently, turn-on loss E_{on} became small, as shown in Fig. 11(a). In contrast, during turn-off, the energy stored in L_s generated a turn-off surge, which was proportional to L_s. As L_s increased, the turn-off loss E_{off} also increased, as shown in Fig. 11(b).

Fig. 12 shows the calculated loss for each device, assuming a two-level inverter, and confirms that they decreased as L_s increased. In contrast, the turn-off surge increased as L_s increased. Therefore, it is important to design equal stray inductance, even if it increases the total inductance, in order to align the imbalanced current within the range where turn-off surge is acceptable.

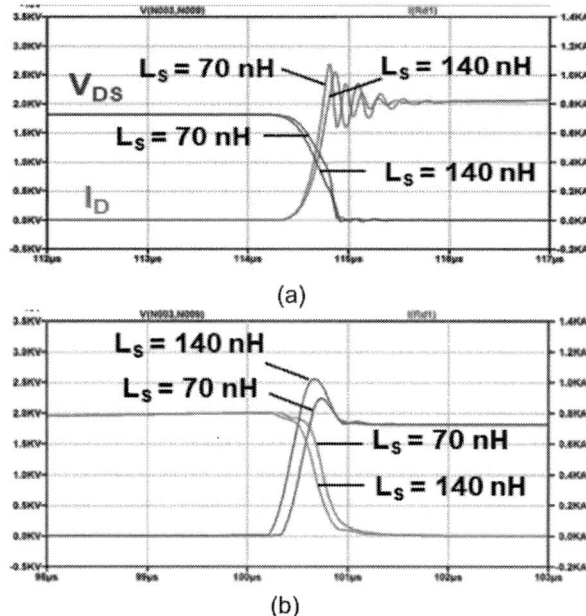

(a)

(b)

Fig. 10 (a) Turn-on and (b) turn-off waveform of the Inductive load switching simulation.

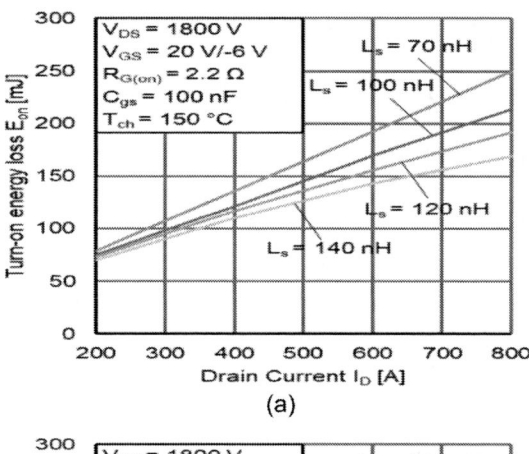

(a)

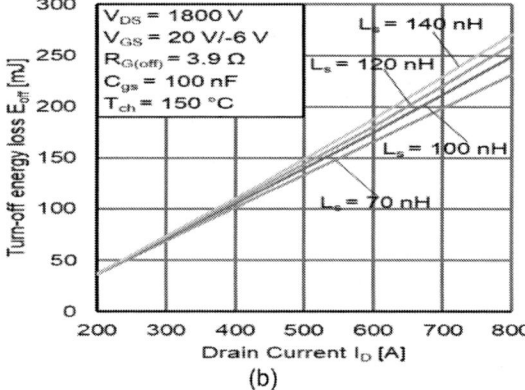

(b)

Fig. 11 Dependence of switching losses on stray inductance of busbar.

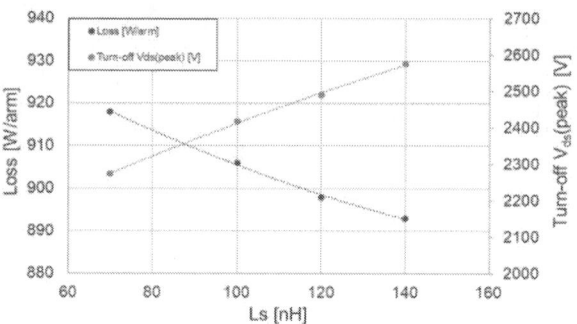

Loss calculation conditions: Two-level sinusoidal inverter, S_{out} = 1800 kVA, V_{DD} = 1800 V, V_{out} (line-to-line) = 1100 V, p.f. = 0.8, f_{sw} = 3 kHz, T_f = 100 °C

Fig. 12 Stray inductance dependence of loss and Turn-off V_{DS} (peak).

3 Evaluation of parallel drives on actual equipment

To demonstrate the influence of the mutual induction M between the external capacitance C_{gs} and the main circuit and the influence of the difference in stray inductance L_s of the main circuit, the circuit shown in Fig. 3 was constructed and evaluated in an actual machine. The gate wiring of each device connected to the gate drive circuit is as equidistant and as close as possible to the gate wiring of each device described in section 2.2.

As shown in Table 1, the devices to be driven in parallel are selected with the threshold voltage V_{th} and the on-voltage aligned. The evaluation circuit and device were used for switching evaluation by double-pulse testing under two conditions, as shown in Table 2.

The measurement conditions were V_{DD}=1800 V, I_d=1600 A, T_{ch}=175 °C, $R_{g(on)}/R_{g(off)}$=1.5 Ω/3.6 Ω, $V_{GG(on/off)}$=+20 V/−6 V, L_m=100 μH and $C_{gs1–4}$ =100 nF. To check the effect of the external capacitors C_{gs}, lead-type capacitors were used in Condition 1 and chip-type capacitors were used in Condition 2.

The busbars connected to each device were designed differently in Conditions 1 and 2 in order to vary the stray inductance of the main circuit. Each busbar is made of Cu and is 1.0 mm thick. The structure of the P- and N-terminal busbars is shown in Fig. 12. The distance d between the busbars of the P and N terminals of Condition 2 is narrower than that in Condition 1, which has the effect of reducing the total stray inductance of the main circuit by causing the magnetic fluxes at the P and N terminals to cancel each other out.

The busbars for the AC terminals are shown in Fig. 13. As shown in Fig. 13(a), Condition 1 was designed with a simple busbar of equal length, while Condition 2 was designed so that the AC terminals are of equal length from the load-connection point. This structure reduces the difference in stray inductance for each element in Condition 1 compared with Condition 2.

	V_{th} [V] T_{ch} = 175°C V_{ds} = 10 V I_d = 0.8 A	$V_{ds(on)}$ [V] T_{ch} = 175°C V_{gs} = 20 V I_d = 800 A
SiC MOSFET module 1	3.75	3.73
SiC MOSFET module 2	3.77	3.74
Δ	0.02	0.01

Table 1 Characteristics of devices to be driven in parallel.

	Condition 1	Condition 2
	Ls [nH]	Ls [nH]
SiC MOSFET module 1	120	97
SiC MOSFET module 2	151	105
Total Ls	67	50
ΔLs	31	8
External C_{gs} type	Lead	Chip

Table 2 Stray inductance and external gate-source capacitor types for Conditions 1 and 2.

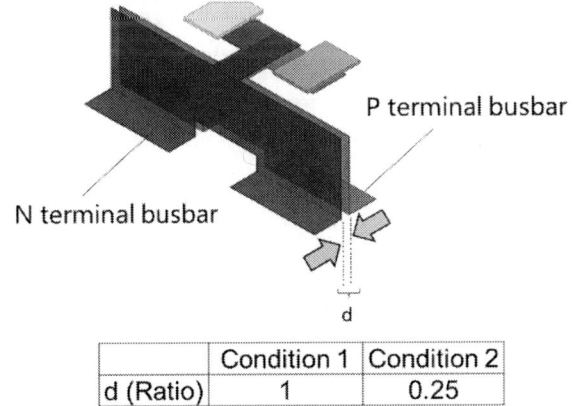

	Condition 1	Condition 2
d (Ratio)	1	0.25

Fig. 12 Design of P and N terminal busbar for conditons1 and 2

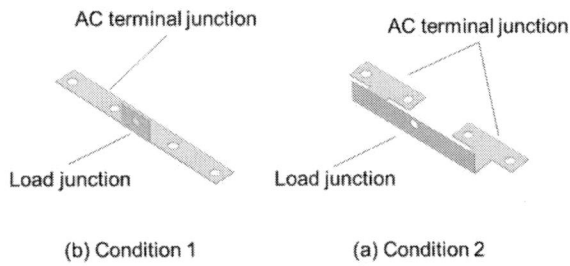

(b) Condition 1   (a) Condition 2

Fig. 13 Design of AC terminal busbars for Conditions 1 and 2.

Fig. 14 show the turn-off switching waveform for Condition 1, in which the difference in the steady-state on-state current imbalance at turn-off is 8%. This is thought to be due to the large difference in the main circuit Ls and the large steady-state unbalance. In turn-off, the difference in dI/dt is thought to be due to the large mutual inductance of the lead type C_{gs}.

Fig. 15 show the turn-off switching waveform of Condition 2, in which the current imbalance in the steady-state on-state of the two devices in parallel from the turn-on waveform is small, at about 1%. This is considered to be due to the small difference in the main circuit Ls, which results in a small steady-state unbalance. In addition, no difference in dI/dt can be seen between the devices. This is considered to be because the external C_{gs} is a chip type and the loop formed by the main circuit and C_{gs} is small and the mutual inductance with the main circuit is also small, so there is no difference in dI/dt. The slight shift during switching is considered to be due to a slight difference in the V_{th} of the device's high temperature.

Based on the measured switching waveforms, the losses and the average channel temperature difference were calculated assuming a two-level inverter. The losses are calculated per inverter arm and assume no load fluctuations. The results for the losses are shown in Fig. 16 and the results for the average channel temperature are presented in Table 3. For Condition 1, the temperature difference between the two devices is large at 9.6°C due to the large loss difference. For Conditon2, due to the small in difference loss, the average channel temperature difference between the two devices is small at 1.6°C. A large temperature difference between each device affects the life of the device [4] [8] [9]. Particularly in applications with large load fluctuations, such as railway applications, there are failure modes such as bonding wires being subjected to stress and breaking as the device temperature rises and falls. It is important to adjust the current imbalance by aligning the stray inductances of each device because differences in device life may result in a shorter-than-expected lifetime for the entire system.

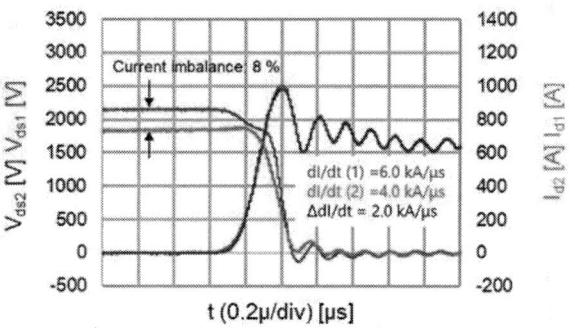

Fig. 14 Turn-off switching waveforms for Condition 1.

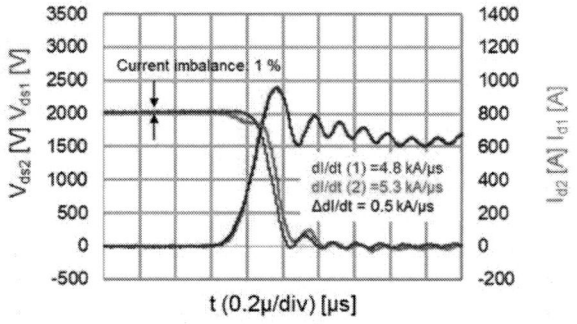

Fig. 15 Turn-off switching waveforms for Condition 2.

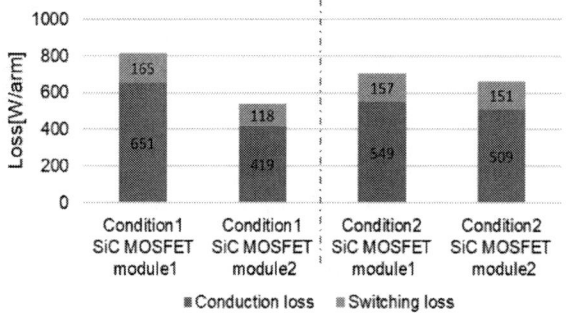

Loss calculation conditions: two-level sinusoidal inverter, S_{out} = 1800 kVA, V_{DD} = 1800 V, V_{out} (line-to-line) = 1100 V, p.f. = 0.8, f_{sw} = 3 kHz, T_f = 100°C

Fig. 16 Two-level inverter loss simulation results based on switching waveform measurements for Conditions 1 and 2.

	Condition 1		Condition 2	
	SiC MOSFET module 1	SiC MOSFET module 2	SiC MOSFET module 1	SiC MOSFET module 2
T_{ch} [°C]	128.2	118.6	124.4	122.8
ΔT_{ch} [°C]	9.6		1.6	

Table 3 Average channel temperatures for each device and the difference between them when two-level inverters are used.

4 Conclusion

When implementing parallel drives using SiC MOS modules, which can be densified and operated at high speed, the effects of mutual inductance in the gate driver and main circuit and the effects of current imbalance due to stray inductance in the main circuit were identified and the optimum method was described.

The influence of mutual inductance between the gate driver and the main circuit lies in the low impedance loop of the gate wiring, and it is important to make the gate wirings close to each other to reduce the low impedance loop.

When an external capacitor C_{gs} is inserted as a gate noise countermeasure, the mutual inductance between the main circuit and C_{gs} can be reduced by making the loop between the main circuit and C_{gs} smaller. As a result, there is no difference in dl/dt and the timing is less likely to shift.

The difference in stray inductance of each device in the main circuit causes a current imbalance, resulting in differences in the losses of each device. This results in a difference in the thermal fatigue life of each device, which shortens the expected life of the system. It is therefore important to match stray inductances of the main circuits within the range where turn-off surges are allowed. The above results demonstrate the most effective method for driving SiC MOSFET modules in parallel.

References

[1] J. Fabre, P.Ladoux, "Characterization and Implementation of Resonant Isolated DC/DC Converters for Future MVdc Railway Electrification Systems", Transactions on Transportation Electrification, 7(2) , (2021), pp. 854–869.

[2] J. Fabre, P.Ladoux, "Parallel Connection of 1200-V/100-A SiC-MOSFET Half-Bridge Modules", IEEE Transactions on Industry Applications, 52(2) , (2015), pp. 1669–1676.

[3] H. Li., S. Munk-Nielsen, X. Wang, R. Maheshwari, S.Beczkowski. et al. "Influences of Device and Circuit Mismatches on Paralleling Silicon Carbide MOSFETs," IEEE Trans.Power Electron., vol. 31, no. 1, (2016), pp. 621–634.

[4] K. Mainali, R. Wang, J. Sabate, S. Klopman, "Current Sharing and Overvoltage Issues of Paralleled SiC MOSFET Modules". 2019 IEEE Energy Conversion Congress and Exposition, ECCE 2019, (2019), pp. 2413–2418.

[5] H. Li., S. Zhao, X. Wang, L. Ding, H. Mantooth, "Current Sharing and Overvoltage Issues of Paralleled SiC MOSFET Modules". 2019 IEEE Energy Conversion Congress and Exposition, ECCE 2019, (2019), pp. 2413–2418.

[6] A. Jauregi, D. Garrido, I. Baraia-Etxaburu, A. Garcia-Bediaga, A. Rujas, "Static Current Unbalance of Paralleled SiC MOSFET Modules in the Final Layout", 2020 IEEE Vehicle Power and Propulsion Conference, (2020), VPPC 2020 – Proceedings.

[7] H. Kono, T. Iguchi, T. Hirakawa, H. Irifune, T. Kawano, M. Furukawa, K. Sano, M. Yamaguchi, H. Suzuki, and G. Tchouangue, "3.3 kV All SiC MOSFET Module with Schottky Barrier Diode Embedded SiC MOSFET", Proceedings of PCIM Europe 2021, (2021), pp. 914-919.

[8] T. Methfessel, F. Sauerland, K. Mainka, O. Schilling, "Enhanced lifetime and power-cycling modelling for PrimePACK™ .XT power modules", Proceedings of PCIM Europe 2020, (2020), pp. 583–590

[9] M. Held, P. Jacob, G. Nicoletti, P. Scacco, M.-H. Poech, "Fast Power Cycling Test for IGBT Modules in Traction Application", Proceedings of the International Conference on Power Electronics and Drive Systems, 1, (1997), pp. 425–430

---

* Company names, product names, and service names may be trademarks of their respective companies.

PCIM Europe 2024, 11– 13 June 2024, Nuremberg   DOI: 10.30420/566262285

Performance Assessment of 10 kV SiC MOSFET and PiN Diode in 3L-NPC Converter Topology

Lucas B. Spejo[1], Lars Knoll[2], Renato A. Minamisawa[1]

[1] Institute of Electric Power Systems, University of Applied Sciences and Arts Northwestern Switzerland, Switzerland

[2] Formerly with ABB Switzerland Corporate Research, Switzerland

Corresponding author: Lucas B. Spejo, lucas.spejo@gmail.com
Speaker: Renato A. Minamisawa, renato.minamisawa@fhnw.ch

Abstract

Medium voltage SiC devices present great potential to be implemented in high-power converters to improve system performance and reduce the number of switches required in multi-level topologies. This work systematically evaluates the performance of 10 kV SiC MOSFETs and PiN diodes in a 3-level NPC topology. The static and dynamic characteristics have been experimentally characterized, and electro-thermal converter simulations were performed to evaluate maximum converter power and system efficiency for different switching frequencies. The results show that the converter can operate at high switching frequencies (up to 10 kHz) with an efficiency higher than 99 %. Furthermore, a significant degree of freedom to choose higher switching frequencies designs (500 up to 2000 Hz) without heavily compromising converter efficiency (~ 99.5 % nominal efficiency) and output nominal power (~ 15 % downrating of nominal power with a four times increase in the switching frequency) is demonstrated.

1   Introduction

Commercial Si IGBT and diode devices present limited voltage ratings up to 6.5 kV. On the other hand, SiC technology enables the design of higher voltage rating devices with promising performance in MV drives and grid-tie converters [1-3]. Such devices can simplify multi-level topologies by reducing the number of required switches, thus increasing system reliability and reducing system complexity. Furthermore, these devices enhance performance at higher switching frequencies, improving power converter density by reducing filter requirements and improving output voltage harmonic distortion. The 3-level neutral point clamped (3L-NPC) converter is a typical topology used in commercial MV drives. For converters with DC link voltage up to 12 kV, where a series connection of two 6.5 kV Si IGBTS is required [4], implementing 10 kV devices can reduce the number of switches by half.

In this work, we evaluate the performance of state-of-the-art 10 kV SiC MOSFET and PiN diodes in a 3L-NPC topology. The static and dynamic charac-

terization of the devices has been performed under identical conditions, followed by electro-thermal system simulations to evaluate the maximum system power (according to thermal criteria) and efficiency for different frequencies. The analysis was performed at the chip level, with the SiC MOSFET and PiN diode presenting the same active area.

2   Methodology

2.1   Power Devices Characterization

2.1.1   Static Characterization

The experimental characterizations were performed on the 10 kV SiC MOSFET and PiN diode chips, presenting the same active area of 0.4 cm x 0.4 cm. The devices have been soldered and wire-bonded on a copper base plate with an insulator-based package, as demonstrated in Fig. 1. The characterization was performed with a Keysight B1505A Power Device Analyzer. The temperature was controlled with a hotplate and K-type thermocouple placed on the drain and cathode terminals.

Curves at 25 and 125°C were obtained for both devices.

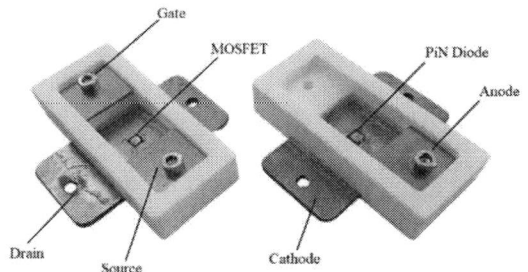

Fig. 1 10 kV MOSFET and PiN diode in a home-made packaging.

2.1.2 Dynamic Characterization

The dynamic characterization was performed in a double pulse tester with an estimated main loop stray inductance (L_σ) of around 200 nH and gate voltages (V_G) of + 15 V / -10 V. The gate resistance value is 50 Ω to prevent increased ringing during switching. The switching curves and losses have been characterized at 25°C, current range between 3 up to 8 A and DC-link voltage of 5 kV.

2.1.3 Simulation

The electro-thermal converter simulations were performed for a 3-phase 3L NPC converter with one phase leg represented in Fig. 2A. The NPC diodes (D_{npc}) are the 10 kV SiC PiN Diodes, and the other diodes (D_{out} and D_{in}) are the SiC MOSFET body diode. An electro-thermal model was defined in software PLECS with the thermal impedance of the chips thermally modelled as individual chips soldered on typical substrates used in power modules, as shown in Fig. 2B. A multi-layer Cauer-type thermal network was considered with an assumed heat spreading of 45° [6]. Table 1 shows each layer's calculated RC lumped values considering the chip's and substrate's dimensions and materials properties [6]. A total thermal resistance from junction to case of 2.1 K/W was estimated. The simulation model was performed assuming a constant heat sink temperature of 80°C, typical of MV drives systems [7]. The model considered the experimental switching losses data at 25°C equal to the switching losses at 125°C as an approximation because SiC MOSFET devices present an almost constant temperature behavior regarding switching losses [7,8]. The experimental static curves at 25 and 125°C were implemented in the model and interpolated to obtain the device static data at distinct junction temperatures. Considering a nominal output voltage of 7.2 kV, typical

of MV drives [4], a DC link of 12 kV has been considered to obtain such output voltage with a 2% margin in a Sinusoidal PWM (SPWM). The converter details are shown in Table 2. A linear escalation of the switching energies [9] to obtain the switching losses at 6 kV has been performed, and a SPWM modulation technique has been employed.

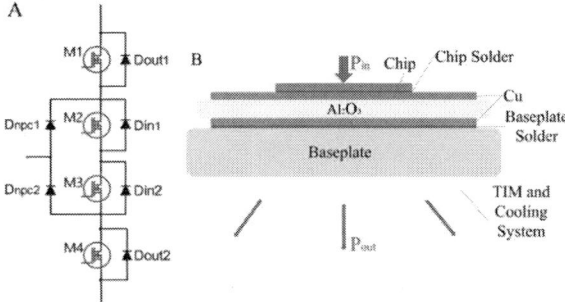

Fig. 2 **A.** Phase-leg of a 3L-NPC topology, **B.** Substrate schematic.

Table 1: Thermal impedance of the chip-baseplate

Layers	Thickness (mm)	Thermal resistance (K/W)	Thermal capacitance (J/K)
Chip	0.45	8.03E-02	1.73E-02
Chip solder	0.05	4.00E-02	2.02E-03
Copper	0.25	3.58E-02	1.59E-02
Al$_2$O$_3$	1	1.83E+00	9.85E-02
Copper	0.25	1.42E-02	4.02E-02
Base solder	0.1	2.54E-02	1.27E-02
Baseplate	3	7.47E-02	1.10E+00

Table 2: Converter specifications

DC link voltage	12 kV
Output frequency	50 Hz
Output Voltage	7200 V
Heat sink temperature	80°C
Max junction temperature	150°C
Switching frequency	500 – 10000 Hz

3 Results and Discussion

3.1 Static Characterization

Figures 3 and 4 show the static curves of the MOSFET, PiN and body diode, respectively. The SiC MOSFET presented 0.7 Ω and 1.7 Ω on-state resistance at 25 and 125°C, respectively. This on-state resistance provides a chip with an estimated current rating within the range of 3 up to 8 A. The PiN diode presented a knee voltage reduction that improves its current conduction at high temperatures similar to lower voltage rating SiC PiN diodes [5]. The SiC PiN diode is advantageous at higher currents (> 5 A), as it presents lower on-state voltages than the body diode and, consequently, lower on-state losses in this regime. For current ranges lower than around 2 A, the body diode presents a lower ohmic resistance due to the SiC PiN diode's high built-in voltage, which reduces its current capability in the low current regime. The diode with lower on-state losses in the current range of 2 up to 5 A strongly depends on the device's junction temperature, with the PiN diode presenting better performance at higher temperatures.

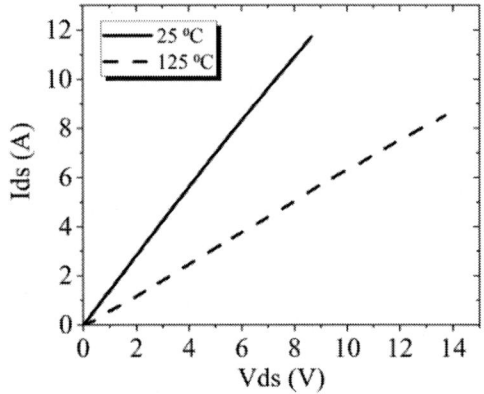

Fig. 3 Forward curves of the SiC MOSFET at gate voltage of 15 V and junction temperatures of 25°C and 125°C.

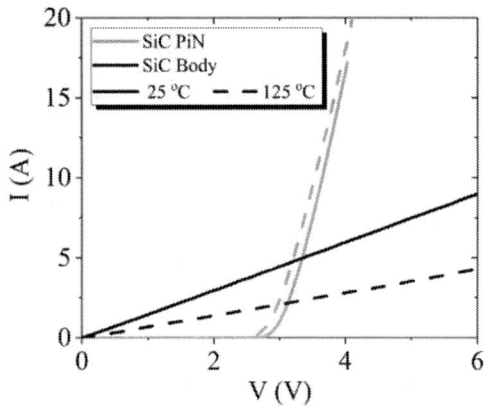

Fig. 4 PiN diode and body diode (V_G=15 V) forward curves at junction temperatures of 25°C and 125°C.

3.2 Dynamic Characterization

Figure 5 shows the turn-on curve of the MOSFET at a switched current of about 13 A at 6 kV. The presence of a reverse recovery current is characteristic of the PiN diode technology and is caused by slow carrier recombination during diode turn-off. A systematic characterization of the switching losses for different switched currents (3 up to 8 A) at room temperature and voltage of 5 kV is shown in Fig. 6 for the MOSFET and PiN diode. The turn-on losses are the dominant losses in SiC MOSFETs, around four times higher than the turn-off losses. The PiN diode presents the smallest losses of around 2 mJ for the range investigated.

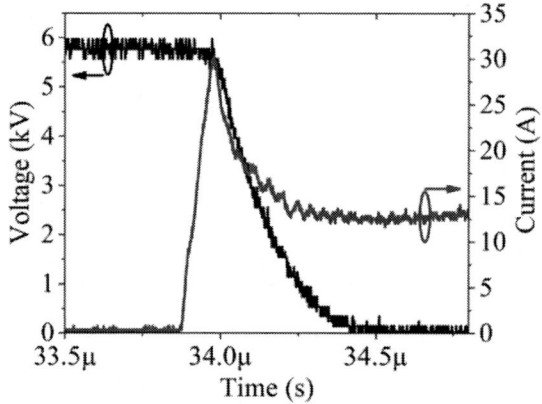

Fig. 5 Turn-on curve of the SiC MOSFET at 6 kV, 25 °C, R_G: 50 Ω.

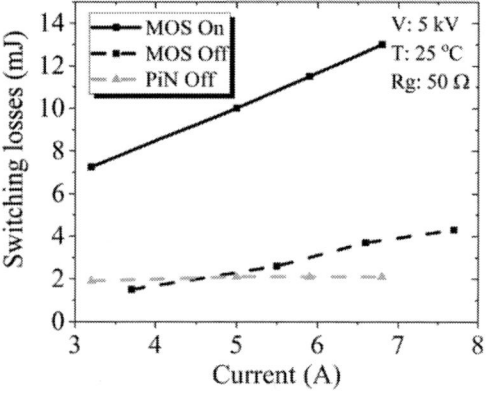

Fig. 6 Switching losses of the MOSFET and PiN diode at 5 kV, 25 °C, R_G: 50 Ω.

3.3 3L-NPC Converter Simulation

Since the devices in this topology present unequal temperature distribution, we have considered the most critical operating condition limiting the converter's maximum power rating. Such a condition occurs when one device achieves the maximum allowed junction temperature (150 °C) at the maximum output current allowed. This condition is reached at a maximum modulation factor (m = 1) and load power factor of 1, as demonstrated by [5]. Figure 7 shows the converter losses distribution of at 2 kHz switching frequency and output current of 5 A. The inner switches do not present switching losses in this operating condition, as demonstrated in [5]. Figure 8 shows the case study to evaluate the maximum switching frequency possible for such a converter sweeping the output current. These devices can switch up to around 10 kHz at an output current of 2 A and 500 Hz at 6 A. The efficiency for such frequencies and current range is higher than 99 %, reaching a maximum value of 99.47 % at 6 A output current, presenting significant potential to extend switching frequency operation and improve the efficiency compared to Si IGBT based MV drives as well as to reduce the number of switches. Interestingly, the investigated SiC converter can operate at 500 Hz switching frequency at 6 A nominal load current, a typical switching frequency used in MV drives with 3.3 and 6.5 kV Si IGBT devices. The SiC technology allows the designer to increase the switching frequency by 4 times (from 500 up to 2000 Hz) by reducing the output nominal current by around 15 % (6 down to 5 A). Such characteristic allows the designers to choose the optimum switching frequency without heavily compromising the converter's nominal power. This characteristic can be used for designs with higher switching frequencies, reducing output filter sizes and improving voltage quality. The converter presents excellent nominal efficiency for designs with higher output nominal currents, reaching values up to ~99.5%.

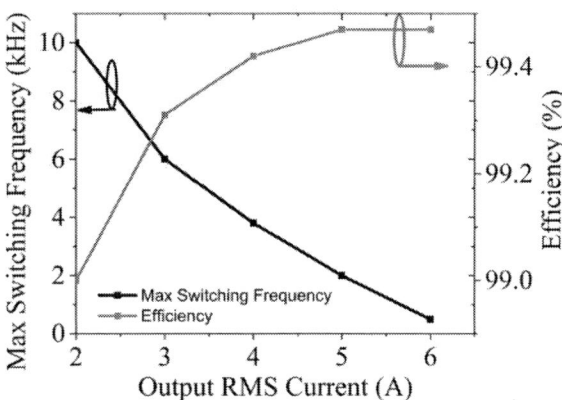

Fig. 8 Maximum switching frequency allowed for distinct nominal output currents (thermal criteria of 150 °C) and converter efficiency.

4 Conclusion

The performance of 10 kV SiC MOSFETs and PiN diodes in a 3-level NPC topology has been evaluated based on experimental device characteristics and system simulations. The converter can operate at a maximum switching frequency of 10 kHz at a 2 A output current and 500 Hz at a 6 A output current. The efficiency for such limit cases is higher than 99 %, reaching a maximum of 99.47 % in the range investigated. The SiC converter also provides a significant degree of freedom for the designers to choose higher switching frequencies (500 up to 2000 Hz) without heavily compromising converter efficiency (~ 99.5 %) and output nominal power (downrating of ~ 15 % by increasing four times the switching frequency). In a nutshell, the investigated 10 kV SiC devices presented great potential to improve switching frequency and efficiency in MV applications as well as reduce the number of switches required compared to lower voltage class Si devices, improving system reliability and reducing complexity.

References

[1] S. Mocevic, J. Yu, B. Fan, K. Sun, Y. Xu et al., "Design of a 10 kV SiC MOSFET-based high-density, high-efficiency, modular medium-voltage power converter," IEnergy, vol. 1, no. 1, pp. 100–113, Mar. 2022, DOI: 10.23919/IEN.2022.0001

[2] S. Ji, X. Huang, L. Zhang, J. Palmer, W. Giewont et al., "Medium Voltage (13.8 kV) Transformer-less Grid-Connected DC/AC Converter

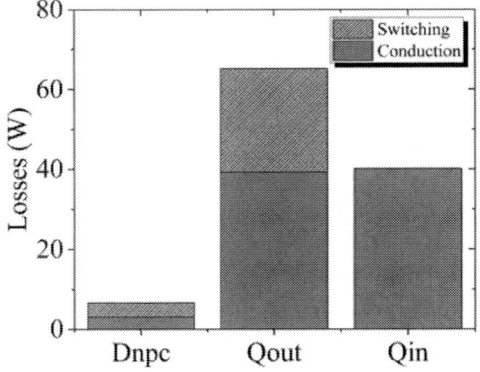

Fig. 7 Converter losses for each group of components (Output current: 5 Arms, fs: 2 kHz).

Design and Demonstration Using 10 kV SiC MOSFETs," IEEE Energy Convers. Congr. Expo., 2019, DOI: 10.1109/ECCE.2019.8912882

[3] V. Pala, E. V. Brunt, L. Cheng, M. O'Loughlin, J. Richmond et al., "10 kV and 15 kV silicon carbide power MOSFETs for next-generation energy conversion and transmission systems," IEEE Energy Convers. Congr. Expo., 2014, DOI: 10.1109/ECCE.2014.6953428

[4] C. Dietrich, S. Gediga, M. Hiller, R. Sommer, and H. Tischmacher, "A new 7.2kV medium voltage 3-Level-NPC inverter using 6,5kV-IGBTs ," European Conf. on Power Electron. and Appl., 2007, DOI: 10.1109/EPE.2007.4417307

[5] F. Filsecker, R. Alvarez, and S. Bernet, "The Investigation of a 6.5-kV, 1-kA SiC Diode Module for Medium Voltage Converters," IEEE Trans. Power Electron., vol. 29, no. 5, pp. 2272-2280, May 2014, DOI: 10.1109/TPEL.2013.2278190

[6] M. Xu, K. Ma, Q. Zhong, and M. Liserre, "Frequency-Domain Thermal Modeling of Power Modules Based on Heat Flow Spectrum Analysis," IEEE Trans. Power Electron., vol. 38, no. 2, pp. 2446-2455, Feb 2023, DOI: 10.1109/TPEL.2022.3210505

[7] A. Marzoughi, J. Wang, R. Burgos, and D. Boroyevich, "Characterization and Evaluation of the State-of-the-Art 3.3-kV 400-A SiC MOSFETs," IEEE Trans. Ind. Electron., vol. 64, no. 10, pp. 8247-8257, Oct. 2017, DOI: 10.1109/TIE.2017.2694380

[8] S. Ji, S. Zheng, F. Wang, and L. M. Tolbert, "Temperature-Dependent Characterization, Modeling, and Switching Speed-Limitation Analysis of Third-Generation 10-kV SiC MOSFET," IEEE Trans. Power Electron., vol. 33, no. 5, pp. 4317-4327, May 2018, DOI: 10.1109/TPEL.2017.2723601

[9] J. Biela, M. Schweizer, S. Waffler, and J. W. Kolar, "SiC versus Si-Evaluation of Potentials for Performance Improvement of Inverter and DC-DC Converter Systems by SiC Power Semiconductors," IEEE Trans. Ind. Electron., vol. 58, no. 7, pp. 2872-2882, July 2011, DOI: 10.1109/TIE.2010.2072896

AUTHOR INDEX

Abbas, Khizra 764
Ackermann, Martin 1336
Aiello, Giuseppe 1217
Akbari, Saeed 2094
Akturk, Akin 739
Alauzet, Louis 2811
Albert, Tianlong 1759
Alfonso, Irene Maria Torres 2503
Alfonzetti, Emanuela 1844
Allioua, Abdelmoumin 2128
Ammar, Ahmed 1087
Appleby, Matthew 3276
Arai, Nobuhide 298
Araujo, Lucas 1673
Arnaudov, Dimitar 2268
Askan, Kenan 1545
Aspalter, Paul 2258
Augustin, Tim 3086
Aunon, Fernando 1467
Ausseresse, Pierrick 1082
Austrup, Isabel 2956
Babaki, Amir 1227
Bagheribavaryani, Mohammadreza 1418
Baharizadeh, Mehdi 378
Bai, Yeriel 1804
Baker, Nick 1923
Bándy, Kristóf 403, 2566
Barcelos, Renan Pillon 264
Barón, Kevin Muñoz 1978
Barth, Henry 2838
Basso, Christophe 3096
Bastawros, Adel 440, 1951
Batista, Emmanuel 2394
Baudais, Briac 3187
Behrendt, Stefan 361
Beiranvand, Hamzeh 1105
Beyerle, Raphael 958
Bhatia, Tamanna 1259
Bicer, Ekin Alp 40
Bimmel, Luc 3206
Blechinger, Christoph 1717
Block, Marius 2217
Bockholt, Yannick 3334
Böhning, Lukas 2208
Boldyrjew-Mast, Roman 723
Bosnjic, Zlatko 1788
Boutry, Arthur 1878
Bouzerd, Souhila 581

Branas, Christian 2286
Brandl, Anja Katerina 1613
Breidenstein, Daniel 1634
Bürger, Matthias 863
Cairnie, Mark 599
Calmels, Alain 3305
Cammarata, Federica 1289
Campos, Adriana 2663
Cannone, Marco 502
Capobianco, Thomas Anthony 1168
Çay, Yunus 3247
Cepin, Simon 1051
Chaisakdanugull, Chanuch 3067
Chatroux, Daniel 2278
Chatterjee, Bhaskar 774
Chen, Mengxing 424
Cherief, Wahid 1910
Cho, Wonjin Dylan 1046
Choo, Vin Loong 1775
Chorfi, Ilias 2175
Cinik, Sadik 2453
Colak, Baris 490
Colomer, Pau 456
Conilh, Christophe 2227
Corbitt, Anna 135, 1123, 1821
Croston, Jose Andres Aguilar 3150
Curbow, Austin 1475
Cusumano, Andrea 1627
Czerwenka, Philipp 1139, 3034
Daire, Baptiste 3110
Dasch, Michael 1907
Davoodi, Hossein 1013
Debbadi, Karthik 2963
Deboy, Gerald 15
Dedew, Mohamed Lemine 34
Delaforge, Timothé 1797
Denk, Marco 1192
Despesse, Ghislain 797
Diz, Sergio De Lopez 411
Do, Nguyen Nghia 1428
Dresel, Lars 2737
Du, Xinyuan 1987
Duijsen, Peter Van 1658, 2248, 2657, 3213
Dumollard, Yannick 1751
Dupont, Max 93
Dusmez, Serkan 383, 2334, 3060
Eichler, Felix 3020
Eyama, Takaaki 56

Fabian, Benjamin	190
Fenske, Florian	3390
Fey, Justin	1902
Fleck, Soenke	338
Förster, Nikolas	3237
Fotteler, Oleg	3328
Fräger, Lukas	926
Frank, Michael	754
Frank, Wolfgang	1770
Frei, Steffen	2478, 3007
Fuchs-Gade, Jannik	2632
Fuhrmann, Jan	1315
Gackowski, Bartosz	1504
Gandluru, Veera Bharath Chandra Reddy	2167
Gavin, Serge	1101
Gebhard, Thomas	1128
Gebhardt, Mathias	2769
Gellman, Ziv	608
Gendrin, Martin	909
Ghanbari, Alireza Ramezan	3175
Ghosh, Priyanka	1523
Gick, Sebastian	1264
Gioda, Alexis	3400
Girgin, Mehmet Oguz	3353
Giuffrida, Simone	248
Giuffrida, Vittorio	1065
Gleissner, Michael	2803
Goff, Gregoire Le	3160
Gomez, Antonio Miguel Munoz	625
Gottardo, Davide	2461
Gragger, Johannes	2104
Graham, Robert	1410
Groon, Fabian	3380
Groos, Gerhard	986
Guan, Jiajia	2591, 3395
Gudala, Bhavana	2524
Guiot, Eric	1604
Gunes, Ekrem R.	3221
Gupta, Gaurav	534
Gürlek, Yavuz	745
Haake, Daniel	2538
Haas, Tobias	2326, 3017
Haehre, Karsten	214
Haensel, Stefan	230
Hanf, Michael	351, 571
Harmand, Thomas	2138
Hasegawa, Kazunori	3002
Hauenschild, Philipp	1969
Hegarty, Timothy	1092
Hegde, Niranjan	1374
Heimler, Patrick	1955
Hellinger, Rolf	1
Hepp, Maximilian	3045
Herrera, Adolfo	1057
Herrmann, Clemens	731
Hertline, Joseph	1886
Herzog, Fabian	3136
Hirao, Takashi	1007
Hironaka, Yoichi	699
Hoffmann, Lennart	3264
Horat, Andreas	480
Hornbuckle, Malachi	2724
Hosseinzadehlish, Mana	1402, 1610
Hu, Jhih-Cheng	791
Huber, Jonas	254
Huerner, Andreas	681
Huselstein, Jean-Jacques	2547
Husev, Oleksandr	893
Igartuburu, Daniel San Laureano	2303
Imai, Ayano	180
Ippisch, Matthias	2638
Irifune, Hiroyuki	2028
Jahn, Simon	883
Jamal, Adeel	1346
Jappe, Tiago	2843
Jegal, Junhyeok	1590
Jha, Kunal	2930
Jia, Minli	2730
Jo, David	1732
Jones, Jeremy	1031
Kaiser, Jeremias	1538
Kampert, Erik	2342
Kanatzar, Paul	1361
Kangjia, He	62
Karout, Mohammed Amer	1835
Kasko, Igor	1991
Kato, Koji	1368
Kaufmann-Bühler, Marius	2400
Kawabata, Junya	2049
Keilmann, Robert	2972
Kempitiya, Asantha	497
Klever, Severin	1561
Knappstein, Lukas	1745
Knecht, Martin	3142
Koch, Jan-Niklas	2240
Koczy, Dawid	1651
Kohlhepp, Benedikt	2316
Koi, Kenichi	67
Kono, Hiroshi	2022
Kopischke, Ruben	2796
Körner, Patrick	615
Kragl, Robert	1385
Kreppel, Thomas	2416
Krigar, Tim	174

Kroics, Kaspars	510
Kugener, Jeff	3315, 3318
Kurukuru, Varaha Satya Bharath	875
Kuzmanoska, Sara	2745
Ladentin, Kevin	1964
Lambert, Adrien	1574
Langfermann, Sascha	1516
Lavery, Melanie	1485
Lee, Chih Hui	1152
Lee, Jongmu	1712
Lee, Kihyun	1724, 1737
Lemaitre, Damien	2596
Lenz, Travis	1352
Lenzen, Patrick	903
Leung, Wing Tai	74
Liao, Xinyuan	322
Lim, Alex	2937
Lindner, Lars	2370
Lippold, Florian	2981
Liu, Baihan	1072
Liu, Iris	1222
Liu, Yusi	3181
Lottis, Christian	3347, 3358
Lotz, Marc René	1457
Lu, Juncheng	19, 837
Lucia, Oscar	2448, 2513
Lutzen, Hauke	976
Lv, Jianwei	1872
Ma, Kwokwai	2778
Machtinger, Katharina	2119
Madloch, Sonja	369
Maheshwari, Ramkrishan	3118
Mai, Annette	284
Maier, Jannik	1642
Mandrioli, Riccardo	2576
Mannen, Tomoyuki	831
Mari, Jorge	843
Marie, Alexandre	2819
Martano, Emanuele	2874
Martínez, Alfonso	2359
Masuda, Akiyoshi	1018
Mauromicale, Giuseppe	2751
Mazzer, Simone	2162, 2532
McRae, Tim	2190, 2364, 3042
Medina-Garcia, Alfredo	1207
Meligy, Ahmed	1495
Menzel, Steffen	933
Merrouche, Abdennour	1113
Minamisawa, Renato Amaral	2036
Mirkovic, Nikola	2488
Mo, Xianghao	2386
Mochizuki, Yo	870

Mönch, Stefan	167
Mueller, Lukas	2425
Mühlfeld, Christian	3340
Muralikrishna, Ajay Krishna Voppu	2886
Nachete, Idriss	2408
Nakako, Hideo	49
Nawaz, Muhammad	2013
Nehmer, Dominik	24
Neira, Sebastian	2700
Neuner, Matthias	3296
Nikiforidis, Ioannis	916, 2718
Nkembi, Armel Asongu	2469
Oberdieck, Karl	1828
O'Keeffe, Rosemary	1249
Olalla, David	1814
Ong, Shu Ee	2942
Orlando, Stefano	1434, 2673
Otori, Daichi	518
Otte, Raphael	2234
Ouhab, Merouane	589, 2948
Owzareck, Michael	3371
Palma, Marco	1568
Panchal, Pranav	315
Paradkar, Sachin Shridhar	2627
Patterson, Andrew	1330
Paul, Indrajit	1133
Peng, Hujun	665
Petzold, Tom	1158
Pham, Thanh-Toan	2786
Philippe, Antoine	1441, 1449
Phung, Thanh Hai	1555, 2612
Piccioni, Andrea	2680
Piepenbrock, Till	463
Poller, Tilo	2914
Porpora, Francesco	222
Pouresmaeil, Mobina	419
Prince, Aswathy M.	2850
Rabay, Battist	2082
Radix, Bryan	2112
Radomsky, Lukas	3286
Randerath, Joschka	3256
Raßmann, Rando	2350
Rauh, Michael	690
Rebenklau, Lars	2088
Reddy, Niranjan Suravarapu	2273
Rehlaender, Philipp	2686
Reimann, René	2377
Reiner, Richard	557
Reißenweber, Lukas	447, 525
Reitz, Niclas	1393
Ren, Linhao	1917
Ren, Xufu	803

Rendek, Karol	2831
Reymond-Laruina, Frédéric	103
Rezaeizadeh, Amin	1686
Ribarich, Tom	1254
Ribeiro, Kelly	2294
Rillo, Oriol Subirats	290
Ringelmann, Tim	2708
Rodrigues, Luis Alves	635
Rodriguez, Manuel Escudero	812
Rodruigez, Manuel Escudero	3077
Rosensaft, Boris	2620
Rudzki, Jacek	1942
Ruoff, Dominik	1277
Ruppert, Lukas	1703
Sakai, Junya	2006
Salomez, Florentin	2431
Samura, Koki	2764
Sankari, Rasched	197
Sawada, Takashi	161
Schindler, Stefanie	1147
Schindler, Tobias	140
Schmidhuber, Michael	2438
Schmidt, Matthias	397
Schmidt, Paul	1999
Schmitz, Laurids	2995
Schnell, Raffael	855
Schnitzler, Ruben	1851
Schulte, Felix	3364
Schulz, Martin	1077
Schwab, Stefan	343
Schwarz, Niklas	1200
Scuto, Alfio	2921
Seber, Elizabeth	888
Sekar, Ajith Kumar	2041
Sen, Gokhan	784
Seo, Hansol	1896
Sheikhan, Alireza	549, 2606
Shi, Sanbao	1212, 3029
Sifoune, Sarah	390
Singer, Mehyeddine	2309
Solomakha, Oleksandr	1174
Somarin, Hasan Mousavi	2494
Sos, Carlos Costas	205
Sousa, Gean	2557
Srikrishna, N. H	3269
Steenbock, Liska	3169
Steiner, Felix	1891
Stone, David A.	949
Subotic, Stefan	274
Sugie, Hisashi	1765
Sun, Qing	2791
Suzuki, Keita	2053

Syed, Hadiuzzaman	2758
Talits, Kevin	997
Tan, John Emmanuel	150
Tanikawa, Kohei	1272
Tarmoom, Ehab	942
Tekir, Bünyamin	672
Tengvall, Sebastian	1380
Thamm, Merlin	1532
Thekemuriyil, Tanya	645, 2986
Thirukoluri, Rajani Kumar	1865
Thomas, Mark	564
Thönnessen, André	1584
Tigira, Sandu	3130
To, Pham Ha Trieu	707
Tobler, Stefan	472
Tokorozuki, Takeshi	849
Torrisi, Marco	822
Tranchero, Maurizio	1322
Troudi, Rami	1185
Tuncay, Sebnem	1858
Uemura, Hirofumi	433
Ueno, Masaki	2909
Ugur, Abdulkerim	3229
Uhlemann, Andre	1283
Urbaneck, Daniel	2152
Varadarajan, Kamal	1598
Vemulapati, Umamaheswara Reddy	1025
Vinciguerra, Vincenzo	1039
Vobecky, Jan	1002
Vogelsberger, Markus	1307
Vogt, Michael	2866
Vuletic, Radovan	305
Walter, Michael	1297
Wang, Hamlin	2185
Wang, Hao	2693
Wang, Lei	1242
Wang, Lisheng	2060
Wang, Qilei	113
Wang, Rui	84
Wang, Yushi	966
Watanabe, Hiroki	3125
Weckbrodt, Julien	121
Wei, Frank	2146
Wei, Suhang	2074
Weihe, Sven	330
Wen, Jin	2182, 3103
Wessel, Wilfried	655
Wietschel, Martin	7
Wille, Christopher	128
Winkler, Paul	1511
Xie, Dong	2880
Xie, Luhong	1930

Yadav, Sachin..2583
Yan, Xingda 1680
Yan, Yiyang..1809
Ye, Yijun 2826
Ye, Zhong 1621, 2646
Yoshida, Satoshi..543
Yoshioka, Kentaro..2067
Yu, Renze 717
Yu, Sean 1180, 2518
Yu, Sheng-Yang 2200
Zeng, Chenhang 1693
Zhang, Chi 1781
Zhang, Hongpeng 238
Zhang, Huaiyuan...2901
Zhang, Yi 1936
Zhao, Yue..3197
Zheng, Zexiang 2860
Zhu, Shiwu..2893
Zipperstein, David 2651
Zipprich, Robert..1664
Zocher, Markus 1233